W9-BUL-464

Base (fundamental) SI units

Physical quantity	SI unit*	Symbol
Length	meter	m
Mass	kilogram	kg
Time	second	s
Electric current	ampere	A
Temperature	kelvin	K
Amount of substance	mole	mol

Derived SI units

Physical quantity	SI unit	Symbol for Si unit	Unit in terms of base units	Unit in terms of other SI units
Velocity (speed)	. . . .	. . . .	m/s	
Acceleration	. . . .	. . . .	m/s^2	N/kg
Force	newton	N	$kg \cdot m/s^2$	J/m
Pressure	pascal	Pa	$kg/(m \cdot s^2)$	N/m^2
Energy (work; quantity of heat)	joule	J	$kg \cdot m^2/s^2$	$N \cdot m$
Power	watt	W	$kg \cdot m^2/s^3$	J/s
Momentum	. . . .	. . . .	$kg \cdot m/s$	. . . .
Angle	radian	rad	m/m (dimensionless)	. . . .
Frequency	hertz	Hz	s^{-1}	. . . .
Electric charge	coulomb	C	$A \cdot s$	$V \cdot F$
Voltage (emf)	volt	V	$kg \cdot m^2/(A \cdot s^3)$	W/A; C/F
Electric field strength	. . . .	. . . .	$kg \cdot m/(A \cdot s^3)$	V/m; N/C
Electric resistance	ohm	Ω	$kg \cdot m^2/(A^2 \cdot s^3)$	V/A
Capacitance	farad	F	$A^2 \cdot s^4/(kg \cdot m^2)$	C/V
Inductance	henry	H	$kg \cdot m^2/(A^2 \cdot s^2)$	Wb/A; $V \cdot s$/A
Magnetic flux	weber	Wb	$kg \cdot m^2/(A \cdot s^2)$	$V \cdot s$
Magnetic field strength (magnetic flux density)	tesla	T	$kg/(A \cdot s^2)$	Wb/m^2; $N/(A \cdot m)$

*Definitions of the base units are given at appropriate places in the text.

INTRODUCTORY
College
Physics

As an additional learning tool, McGraw-Hill also publishes a study guide to supplement your understanding of this textbook. Here is the information your bookstore manager will need to order it for you: 44040-9 STUDY GUIDE TO ACCOMPANY INTRODUCTORY COLLEGE PHYSICS

INTRODUCTORY
College
Physics

Joseph F. Mulligan
University of Maryland Baltimore County

McGraw-Hill Book Company

New York   St. Louis   San Francisco   Auckland   Bogotá
Hamburg   Johannesburg   London   Madrid   Mexico   Montreal
New Delhi   Panama   Paris   São Paulo   Singapore   Sydney   Tokyo   Toronto

Cover

The cover shows parts of the mural *La Fée Electricité* (The Magic of Electricity) painted by the French artist Raoul Dufy (1877-1953) in 1936-1937 for the Electrical Pavilion at the 1937 International Exposition in Paris. This 200-ft by 35-ft mural, one of the largest ever made, is now in the Museum of Modern Art in Paris. It includes the figures of Marie Curie, Lorentz, Kelvin, Helmholtz, Maxwell, Faraday, Henry, Ampère, Joule, Franklin, Galileo, and many other physicists whose work is discussed in this text. (*Courtesy of the Museum of Modern Art in Paris and Giraudon/Art Resource, New York.*)

Introductory College Physics

1 2 3 4 5 6 7 8 9 0 V N H V N H 8 9 8 7 6 5

ISBN 0-07-044036-0

This book was set in Times Roman by Black Dot, Inc.
The editors were Stephen Zlotnick, Kathleen M. Civetta, and David A. Damstra;
the designer was Joseph Gillians;
the production supervisor was Joe Campanella.
The drawings were done by J & R Services, Inc.
Von Hoffmann Press, Inc., was printer and binder.

Library of Congress Cataloging in Publication Data

Mulligan, Joseph F. (Joseph Francis), date
   Introductory college physics.

   Includes bibliographies and index.
   1. Physics. I. Title.
QC21.2.M85 1985   530   84-14369
ISBN 0-07-044036-0

*To the memory of
my father, Joseph Lawrence Mulligan
and my mother, Mary Collins Mulligan*

Contents

Chapter 3   Forces and Newton's Laws of Motion   50

Chapter 4   Equilibrium   78

Chapter 5   Motion in Two Dimensions   105

Chapter 18

Magnetism 498

Chapter 19

Electromagnetic Induction 534

Chapter 20

Alternating Current Circuits 564

List of Tables

Preface

The scientist does not study nature because it is useful; he studies it because he delights in it, and he delights in it because it is beautiful. If nature were not beautiful, it would not be worth knowing, and if nature were not worth knowing, life would not be worth living.

Henri Poincaré (1854–1912)

The above quotation from the great French mathematician Poincaré reveals the purpose of this textbook: to help students understand—and find delight in understanding—the physical universe about them. This text presents the physics that medical doctors, dentists, architects, and other well-informed citizens must know in order to understand the world in which we live. Ours is a technological world, and to comprehend many of its problems and become alive to its prospects a basic knowledge of physics is required. Without this knowledge the Jeffersonian ideal of an informed citizenry as the foundation of a free society is unattainable.

The material covered in this text is that of the standard two-semester, noncalculus, introductory college physics course. The text emphasizes the basic facts and laws of mechanics, thermodynamics, electromagnetic theory, the theory of relativity, and quantum theory, together with the conservation laws and other integrating principles. No previous physics courses are required, and the only mathematical prerequisites are high-school algebra, geometry, and some trigonometry.

A unique feature of the book is its attention to the men and women who have made significant contributions to the development of physics, in an attempt to show that physics is a vital, developing subject created by the intuition, hard work, and sacrifice of physicists like Isaac Newton, Albert Einstein, and Marie Curie, and still being advanced by similarly dedicated scientists today. This text also includes many applications of physics to health, medicine, energy, technology, and everyday life. An understanding of the electron microscope, solar cells, and the use of lasers and optical fibers in medical diagnosis and treatment points up the relevance of physics in contemporary life. This is, however, a physics text and health-related applications which include more biology or medicine than physics have been avoided.

Order of the Text

This textbook attempts to present physics in a clear, logical fashion that stresses the development of physical ideas and the relationships among the various fields of physics. The order followed is in great part the traditional one of mechanics, sound, heat, electricity, magnetism, light, and modern physics. One unusual feature is Chapter 9—on atoms, molecules, and gases—which introduces at an early stage the atomic picture of matter to make more meaningful the discussion, in subsequent chapters, of liquids, solids, and their behavior. Chapter 21 presents Maxwell's equations in a simplified form without using calculus. This chapter also serves as a review and summary of the facts and laws of electricity and magnetism contained in Chapters 16 to 20. Once Maxwell's equations and electromagnetic waves have been introduced, Chapter 22 moves directly into wave optics. Geometric optics is then presented in Chapter 23 as an approximation that is valid when the wavelength of light is short compared with the size of objects interacting with light.

The flow of the material culminates in the Bohr theory of the hydrogen atom in Chapter 27. The last three chapters, which deal with nuclear, solid-state, and elementary-particle physics, are not essential to the logical development of the course, but are crucial for awakening the student to the importance of physics in our contemporary world, and to its ongoing nature. In the words of microbiologist and noted science writer Lewis Thomas, "We have a wilderness of mysteries to make our way through in the centuries ahead."

The text includes more than enough material for the standard course in physics, including all topics required for MCAT and other tests. There are no sections marked as optional, and no physics material relegated to appendixes, because teachers are the best judges of what is important for their students. Some more mathematical topics such as Gauss' law and Ampère's circuital rule (which are needed for the discussion of Maxwell's equations), and the mathematical treatment of wave motion in Sections 11.7 to 11.10, can be omitted in courses in which a more qualitative approach to these topics seems desirable. Much of the last three chapters can be assigned as reading material if necessary. Finally, where feasible, applications have been placed in special sections, often at the end of chapters. Depending on class interest, sections such as 15.8, 18.10, 20.8, 21.9, 21.10, 22.8, 23.9, 24.7, 24.8, 28.7 to 28.9, 29.6, and 29.7 can be omitted or assigned as outside reading without loss to the continuity of the course.

The appendixes include a brief review of important mathematical concepts, some useful ideas about graphs, and hints on the solution of problems, including a discussion of order-of-magnitude calculations. Students having difficulty with mathematical derivations in the text or with problem solving should refer to these appendixes. Students may find additional help in a *Study Guide* prepared by Professor David A. Jerde, which uses a question-and-answer approach to lead students through the solution of physics problems similar to those in this text.

Other Features of the Textbook

Experimental Approach No equations or laws are presented without mathematical, logical, or experimental justification. Where our knowledge is still limited or uncertain, that fact is simply stated. Order-of-magnitude calculations are used throughout the text, as is dimensional analysis.

Units   SI (Système International) units are used almost exclusively in the text. At the beginning some British engineering units are introduced to convey a feeling for the size of the SI units, and to provide practice in converting units. Other units are occasionally introduced for informational purposes.

Worked Examples   Over 300 worked examples appear within the chapters to provide models for solving the problems at the ends of the chapters.

Questions and Problems   Each chapter contains at least 10 discussion questions and about 50 problems divided into four categories: multiple-choice, easy, standard, and difficult problems. No assignment of problems to chapter sections has been made, because a most important element in solving physics problems is the ability to recognize what previously acquired knowledge is important for a problem's solution. Assigning problems to individual text sections merely helps students locate an equation that "works" in the designated section, whether the student understands *why* it works or not. However, in the *Instructor's Manual* problems are listed by chapter section for the convenience of the instructors in assigning homework.

Illustrations   The book contains over 100 photographs and about 900 line drawings. These are intended to be an integral part of the student's learning experience and are referred to at appropriate places in the text.

Summaries   Each chapter ends with a brief summary, which includes all important definitions, principles, and equations in the chapter.

Tables   Over 60 tables of useful experimental data are provided at pertinent places throughout the book. A list of these tables follows the table of contents. Tables frequently needed, e.g., those containing conversion factors and the physical constants, are placed inside the front and back covers.

Biographies   The text contains 22 one-page biographies of famous physicists, chosen for the contributions they have made to the development of the physics in the chapter containing the biography. There is clearly some arbitrariness in the choice of only 22 out of the thousands of physicists who have made substantial contributions to the field. Briefer accounts of the contributions of other important physicists, including many still active in research, are given in the captions for the photographs scattered throughout the book. The historical data for the biographies have been taken from the books listed at the end of the chapters, and have been checked against the accounts in the *Dictionary of Scientific Biography*.

Suggestions for Further Reading   The list of additional readings at the end of each chapter consists of relatively popular books and articles that can be read profitably by students taking their first physics course. Emphasis is on biographical materials about the men and women who created and are creating physics, and on popular articles of the kind written for *Scientific American* by experts in the field on their current or past researches. Not included are a vast array of books and articles consulted by the author and probably of interest to instructors in the course, but too advanced for students in an introductory course such as this one.

Answers to Odd-Numbered Problems   The answers to odd-numbered problems are provided with some reluctance since, as Alfred Bork has recently pointed out, the student is inclined to appeal to the authority of the textbook's answer to decide whether he or she has done the problem correctly—an approach which should be anathema in a science like physics. The aim of a course such as this is to build sufficient problem-solving skills and confidence in the student that, if a discrepancy appears between the student's answer and that given in the book, the student will question the book's correctness. Despite the cogency of this argument, students' strongly expressed desires for answers to assigned problems has led the author to provide them for the odd-numbered problems. Instructors who agree with Bork's views on this subject can assign mostly even-numbered problems.

The Dutch physicist Paul Ehrenfest (1880–1933) once wrote that he found great joy first in making something clear to himself, and then in making it clear to others so that they could share his joy. As students work through this book, I hope that they may share the joy I found in writing it.

Acknowledgments

I would like to thank the many people who contributed in various ways to making this a better book. Among those to whom I am grateful are the physics professors who read parts of this book in manuscript form and made helpful comments on how it could be improved. These include the following physicists: Angelo Armenti, Jr., Villanova University; Stanley Bashkin, University of Arizona; Robert C. Bearse, University of Kansas-Lawrence; Rhoda Berenson, Nassau Community College; Bennet B. Brabson, Indiana University; John J. Brennan, University of Central Florida; Alex F. Burr, New Mexico State University; Paul R. Byerly, Jr., University of Nebraska; George Carr, University of Lowell; George W. Coyne, Valencia Community College; Don Chodrow, James Madison University; Richard Dalven, University of California, Berkeley; Dewey I. Dykstra, Jr., Boise State University; James B. Gerhart, University of Washington; Edward Graff, Lake Michigan College; H. Kimball Hansen, Brigham Young University; Paul L. Lee, California State University-Northridge; David F. Measday, University of British Columbia; Gregor M. Novak, Purdue University, Indianapolis; Robert F. O'Connell, Louisiana State University; William E. Vehse, West Virginia University; and Alfons Weber, National Bureau of Standards. Also, from my own institution, the University of Maryland Baltimore County, I would like to thank Professors Alfonso Campolattaro, Ivan Kramer, Robert Rasera, Morton Rubin, and Ellen Yorke.

In particular I would like to thank Professor David A. Jerde of St. Cloud State University and Dr. Terrence C. Dymski of UMBC, who read the entire manuscript with great care and made many constructive suggestions for its improvement. Professor Jerde has also prepared the *Study Guide* to accompany the text. Any errors which remain in the text are, of course, fully my responsibility, and I welcome corrections and comments to improve future editions.

I am also especially grateful to Janet Gethmann, who typed many drafts of the manuscript, to Peggy King, Paul Ciotta, and Tom Beck of UMBC; and to the editorial staff at the McGraw-Hill Book Company, especially to Kathleen Civetta, David Damstra, Joe Gillians, and Nancy Warren, for their assistance in bringing the manuscript into this final book form. Finally, I would like to thank my wife, Eleanor, for much good advice, for help with proofreading and preparing the index, and for her constant encouragement and understanding.

Joseph F. Mulligan

B ut occasionally one gets a man like Enrico Fermi, the Italian genius who rose to fame in 1927 as a theoretician and then surprised us all by the breathtaking results of his experiments with neutrons and finally by engineering the first nuclear reactor. On December the second, 1942, he started the first self-sustaining nuclear chain reaction initiated by man and thus became the Prometheus of the Atomic Age.*

Otto R. Frisch (1904–1979)

Chapter 1

Physics as a Science

During college you will study many different subjects—literature, history, psychology, mathematics, music, biology, art, computer science, physics, philosophy. Each of these disciplines has its own subject matter and its own unique approach to reality, and each will make an important contribution to your knowledge of the world in which you live. In this chapter we will try to define the scope and approach of one such discipline—physics. This is only an introduction, however; the nature of physics probably will only be clear to you after you have completed this course.

1.1 The Scope of Physics

Physics is often defined as the science of matter and energy, and of the relations between them. As a science, physics is rooted and grounded in observation and experiment, and the validity of its theories and laws must stand the test of continuing comparison with the quantitative results of observation and experiment.

The domain of physics includes *matter* in all its forms—solids, liquids, gases, plasmas, molecules, atoms, and the particles out of which atoms are made. It also includes *energy* in all its forms—mechanical energy, electromagnetic energy, nuclear energy—and the manifestations of these basic kinds of energy in the form of heat, sound, light, gravitation, and chemical energy. This is what we mean when we say that physics is the science of matter and energy. Still, such a definition does not clearly distinguish physics from other disciplines, since chemistry is also a science which deals with molecules and atoms, and electrical engineering is concerned with the production, transmission, and control of electromagnetic energy. How, then, can we arrive at a clearer picture of the scope of physics?

First, physics is the oldest and the most basic of the sciences, since it attempts to study and explain all the different kinds of particles and radiations which make up the universe. Over the course of history other sciences

---

*Credits for miscellaneous chapter quotes appear in the Additional Readings sections at the ends of chapters.

branched off from physics to become sciences in their own right. Nonetheless, the findings of physics are still often used to elucidate the findings in other branches of sciences.

Physics and Other Sciences

Physics and the Biological Sciences   The biologist studies matter which possesses the peculiar property we call life. The physicist has no interest in living matter as such, even though physics is concerned a great deal with the particles from which living matter is constructed. Physicists have made important contributions to biology and medicine by uncovering the physical principles underlying the biological sciences, and by developing sensitive instruments of great utility in biology and medicine, such as the electron microscope and the CAT scanner (see Fig. 1.1). Some eminent biologists, such as F. H. C. Crick, one of the two developers of the Watson-Crick double-helix theory of DNA (see Fig. 1.2), and Rosalyn Yalow (Fig. 1.3),

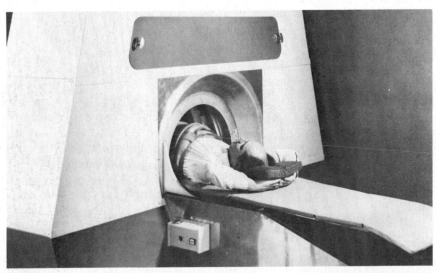

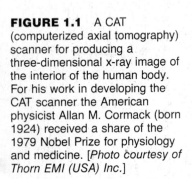

FIGURE 1.1   A CAT (computerized axial tomography) scanner for producing a three-dimensional x-ray image of the interior of the human body. For his work in developing the CAT scanner the American physicist Allan M. Cormack (born 1924) received a share of the 1979 Nobel Prize for physiology and medicine. [*Photo courtesy of Thorn EMI (USA) Inc.*]

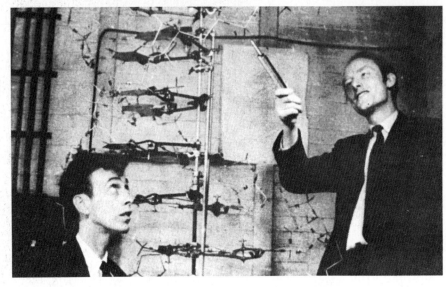

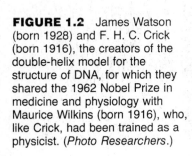

FIGURE 1.2   James Watson (born 1928) and F. H. C. Crick (born 1916), the creators of the double-helix model for the structure of DNA, for which they shared the 1962 Nobel Prize in medicine and physiology with Maurice Wilkins (born 1916), who, like Crick, had been trained as a physicist. (*Photo Researchers.*)

FIGURE 1.3 Rosalyn Yalow (born 1921), who shared the 1977 Nobel Prize for physiology and medicine for her part in creating the technique of radioimmuno-assay, which uses radioactive tracers to locate antibodies and other biologically active substances that are present in the human body in quantities so minute that they are detectible in no other way. (*Photo courtesy of AIP Niels Bohr Library, Meggers Gallery of Nobel Laureates.*)

winner of the Nobel Prize for medicine, were trained as physicists. Also, medical doctors like Hermann von Helmholtz (see biography, Chap. 12), Julius Robert Mayer (see biography, Chap. 13), and Thomas Young (see biography, Chap. 22) made significant contributions to physics by research done while still engaged in their medical practices.

Physics and Chemistry   The closest discipline to physics in interest and approach is chemistry. The main difference between these two sciences is that chemistry deals with matter at the molecular level and with molecular interactions, whereas physics is more concerned with the atoms which make up the molecules, with the protons, neutrons, and electrons out of which atoms are constructed, and with the macroscopic properties of matter. In recent years much of physics has separated further from chemistry by devoting its attention chiefly to reactions which occur at very high energies and which produce new particles that play no role in chemical reactions. That physics and chemistry are very closely related remains clear from the fact that hybrid disciplines like physical chemistry and chemical physics exist and flourish.

Physics and Mathematics   Mathematics as a discipline is more concerned with the proper ordering of mathematical concepts and constructs than with physical reality. Physics takes many of the results of mathematics and uses them to better describe reality, but is more concerned with the application of mathematical ideas than with the ideas themselves. Theoretical physicists use advanced mathematics continually in their work, but they use it as a tool to understand the physical universe and not as a road to further mathematical discoveries. In some cases (as with Einstein's general theory of relativity) new branches of mathematics have been created to fill a need in physics research.

Physics and Engineering   Engineering stands somewhat in the same relationship to physics as does physics to mathematics. Just as physics uses mathematics to elucidate the physical universe, so engineering applies the laws and discoveries of physics to develop practical devices like automobiles, computers, electric generators, nuclear reactors, bridges, tunnels, and space shuttles. All modern engineering is rooted in the laws of physics, but physicists are interested in discovering these laws, not in applying them. Since technology is engineering applied to large-scale productive processes, physics has essentially the same relationship to technology as it does to engineering. Physicists uncover data and develop physical theories and laws which technologists then apply to society's needs.

The distinctions made above may help somewhat to clarify the scope of physics compared with other sciences. These distinctions will never be perfectly sharp, however, and hence the scope of physics will never be perfectly clear. There will always be mathematical physicists, chemical physicists, bioengineering physicists, applied physicists, and space physicists to render the dividing line between physics and other disciplines fuzzy and uncertain. But despite these ambiguities and uncertainties, our definition of physics as the science of matter and energy, and of the relations between them, is still a useful one that adequately describes the contents of this book.

1.2 The Approach of the Physicist

Physics is one of the so-called hard sciences, not because it is difficult, but because it is based on hard, quantitative data and makes predictions in quantitative (i.e., numerical) form. These predictions can then be checked against measurements made in the laboratory. (The so-called soft sciences include disciplines like psychology and sociology in which human behavior plays a crucial role, and in which the precision that is characteristic of physics does not play a major role.) The physicist is able to extract quantitative data from nature only by focusing on the purely physical and quantitative aspects of a problem to the exclusion of other considerations. This fundamental, in some ways oversimplified, approach does not take into account the interaction of physical processes with the economy, political events, or the quality of human life, but concentrates on matter and energy and their interrelationships. This is both the secret of the great success physics has enjoyed and the reason its applicability to real-life problems involving human beings and other living creatures is limited.

Once a reasonable amount of experimental data has been collected, say, in a physics research laboratory of the type shown in Fig. 1.4, physicists try to develop a theory to correlate and explain this data. In so doing they rely on

FIGURE 1.4 Laser research. The two long rectangular boxes from which intense light is being emitted are gas lasers used for research at the Lawrence Livermore Laboratory in California. (*Photo courtesy of U.S. Department of Energy.*)

FIGURE 1.5 Two famous physicists, Albert Einstein (1879–1955) and J. Robert Oppenheimer (1904–1967), at work on a theoretical problem in physics. (*International Communication Agency; courtesy of AIP Niels Bohr Library.*)

the known laws of physics and their own creative imaginations. At times their explanation of the available data may involve nothing more than an application of a simple physical law such as Newton's second law of motion, which relates forces to accelerations; in other cases a break with existing physical theories may be required, as was the case with Einstein's theory of relativity (see Fig. 1.5).

The test of the success of a physical theory is twofold: first, it must fit the available quantitative data which have been collected; and second, it must be fruitful in making predictions about phenomena which have never been observed or measured. If these predictions are confirmed in the real world, the theory is retained; if not, it is modified or replaced by a better theory which does agree with the available data.

Physics has been extremely successful over the years in explaining matter and energy at ever-increasing levels of difficulty and sophistication. In all cases, however, its secret of success has been the same—the ability to abstract from a complicated situation in the real world the crucial physical elements which determine what is going on. In doing this, much of the situation may be passed over as of no consequence for physics. For example, a physicist may see a ball falling from the window of a building (Fig. 1.6). Who dropped the ball, its color, and the material out of which it is made are usually of little importance in analyzing the physics of the falling ball. From a physical point of view, the only things of importance are the mass of the ball (although it will turn out that even this is not really important), the height from which the ball falls, the force of gravity pulling the ball down, and the resistance of the air to the ball. Since the force of air resistance is often very small compared with the force of gravity, physicists normally neglect air resistance in calculating a first approximation to the speed with which the ball falls. Then at a later time they can, if necessary, modify the theory to include the effects of air resistance. It is this ability to abstract the "guts" of a physical problem from the nonessentials which is the key to the progress of physics. We will see many examples of this technique in the pages that follow. Solving the problems at

FIGURE 1.6 A ball being pushed from the window of a building and falling freely to the ground.

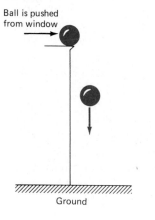

Ball is pushed from window

Ground

the ends of the chapters should help you develop a similar technique—a technique, by the way, which may be of considerable assistance to you in other facets of life.

Physics is still far from complete. There are many things we do not know about the elementary particles, about the behavior of matter at very high and very low temperatures, about plasma physics and astrophysics, etc. All over the world physicists are hard at work in their offices and laboratories, trying to solve very difficult problems in these and other fields. It will be their ability to penetrate to the heart of a complicated physical situation, and then to express the guts of the situation in the form of an equation, that will be the key to their success in solving these difficult problems.

1.3 Physical Theories and Laws

Books have been written on how physics arrives at definitive descriptions of reality like those found in Newton's laws of motion or Maxwell's equations for the electromagnetic field. It would probably be more honest to say that individual physicists often arrive at important conclusions in very different ways, and hence any attempt to systematize their procedures runs the risk of oversimplification. Still it may be worthwhile to attempt some kind of broad outline of how physics progresses from experimental data to hypothesis to theory to physical law.

As we have seen, the first step in developing a valid physical understanding of some aspect of reality is to collect as much good, unbiased data as possible, preferably data gathered by different physicists, perhaps in different parts of the world. Physicists then start providing and rejecting possible explanations of the data before them. Such tentative explanations, assumed for the sake of argument, are called *hypotheses*. These hypotheses usually include intellectual *models* constructed in an attempt to make sense of the available data. Finally one hypothesis seems to fit the data well and is advanced to the stage of a *physical theory*, i.e., a theoretical explanation, perhaps in mathematical terms, which correlates and makes understandable the data already collected.

Once a theory has been developed, more experiments are needed to test it, perhaps by varying the circumstances of the original experiments or extending their range. Also, if the theory is a good one, it should make predictions about other aspects of reality which have not been measured or understood. If these predictions are verified in reality, the theory is well on its way to acceptance by the physics community.

Theories which agree so well with experimental observations that they seem to reflect the constant behavior of nature under a variety of conditions are called *physical laws*. Newton's law of universal gravitation, which deals with the attractive force exerted by every object in the universe on every other object, is an example of such a law. Newton did not create this law; he found it in the world about him and, by the strength of his great intellect, was able to express it in the form of a simple equation. The fact that this law explained not merely the behavior of apples falling from trees, but also the motion of the planets around the sun, provided the convincing evidence needed to convert a theory into a universally accepted law of physics. There are not too many physical laws of this kind, but those which do exist include some of the greatest achievements of the human mind.*

Laws in physics are very different from the laws of a country or moral and religious laws. The latter are *prescriptive* laws; they tell us how we are expected to behave. Physical laws, on the other hand, are *descriptive*; they summarize how nature operates. Moral laws are frequently broken; physical laws can never be broken or they would cease to be physical laws. To paraphrase Galileo, "The Bible tells us how to go to heaven, not how the heavens go."

1.4 Theory and Experiment in Physics

As we have seen, physics is marked by the constant interplay between theory and experiment. Experimental physicists carry out observations and perform controlled experiments to collect the physical data which any theory must explain. Theoretical physicists seek a theory, preferably in the form of an equation, to explain these data (Fig. 1.7). If they are successful, their theories make new predictions which send the experimentalists back into their laboratories to confirm these additional predictions. If these predictions are not completely verified by experiment, the theory may have to be modified somewhat to fit the experimental data. Further predictions of the modified theory are then again checked in the physics laboratories of the world. Physics thus progresses by the constant interplay of theory and experiment. Without one, the other would be of limited value.

---

*It should be noted that sometimes in physics universally accepted laws or principles of physics are loosely referred to as "theories." Good examples are the theory of special relativity and the quantum theory of specific heats.

FIGURE 1.7 Three famous theoretical physicists, Victor F. Weisskopf (born 1908), Maria Goeppert-Mayer (1906–1972; Nobel Prize in physics, 1963), and Max Born (1882–1970; Nobel Prize in physics, 1954) relax on their bicycles in Germany in 1930. (*Photo courtesy of Niels Bohr Library, AIP.*)

Enrico Fermi (1901–1954)*

FIGURE 1.8 Enrico Fermi (*Courtesy of AIP Niels Bohr Library*).

Enrico Fermi was born in Rome in 1901. He obtained his Ph.D. degree in 1922 from the University of Pisa, the city in which Galileo had been born in 1564. His dissertation was on x-rays, and he had already published several research papers before he received his degree. He then did postdoctoral research at the University of Göttingen in Germany under Max Born and at Leiden in Holland under Paul Ehrenfest.

After returning to Italy Fermi taught for 2 years at the University of Florence, where he soon established a reputation by developing what are now known as the Fermi-Dirac statistics. In 1926, at the age of only 25, he was appointed to a full professorship in physics at the University of Rome. He gathered a group of outstanding young faculty members and students around him and immediately began to make a name for Rome in the fields of nuclear physics and quantum mechanics. In 1933 his theory of beta decay caused a great stir in world physics circles. He won the 1938 Nobel Prize in physics for his experimental research on neutron absorption by heavy elements, in the course of which he discovered 40 new artificial radioactive isotopes and observed the crucial importance of water and paraffin in slowing down neutrons. This was one of the key elements in the first working nuclear reactor.

Because Fermi's wife, Laura, was Jewish, their life in Italy became increasingly difficult as the power of Mussolini and Hitler grew during the 1930s. Hence when Enrico and Laura Fermi left for Sweden for the Nobel Prize ceremony in 1939, they resolved never to return to Italy. Fermi immigrated to the United States and taught at Columbia University in New York City from 1939 to 1942 and at the University of Chicago from 1942 to 1954.

On December 2, 1942, about 1 year after the bombing of Pearl Harbor and the entry of the United States into World War II, Fermi directed the first self-sustaining chain reaction in uranium at Stagg Field in Chicago. Arthur Compton telephoned the good news of this achievement to James Conant at Harvard by the coded message: "The Italian naviga-tor has landed in the New World." That evening Laura Fermi gave a dinner party for some of her husband's colleagues and their spouses. As the guests arrived each physicist warmly congratulated Fermi on his great accomplishment. Laura was mystified by this but was unable to find out until years later why congratulations were in order. Such was the secrecy surrounding the atomic weapons project, to which Fermi continued to make significant contributions both during and after World War II.

Fermi had great skill in carrying out what physicists call *order-of-magnitude calculations* to test the approximate correctness of a physical theory or the likely success of a physics experiment.† With a small slide rule and the information he had stored in his head he could solve a problem more accurately and faster than lesser physicists could with large computers and libraries filled with books.

Fermi was a warm and friendly person who loved physics and the great outdoors. He was universally admired and loved by his many students and colleagues. The clarity of his ideas made him a great teacher; his probing imagination and his ability to get to the heart of any problem made him a great research physicist. In the words of Emilio Segrè, Fermi's pupil and colleague from Rome, "Fermi gave science his utmost, and with him disappeared the last individual of our times to reach the highest summits in both theory and experiment and to dominate all of physics."

*For the 23 brief biographies scattered throughout this book outstanding physicists have been chosen whose work has become part of the accepted physics taught in courses like the one for which this book is intended. In the heading of each biography are given the physicist's full name and the dates of birth and death. Biographies are not included for living physicists, but the last few chapters include photographs and accomplishments of some living Nobel Laureates.
†For order-of-magnitude calculations see Appendix 3C.

The abilities and personal qualities of theoretical and experimental physicists are often quite different. Theorists must have high mathematical ability, be very imaginative and creative in analyzing experimental data and mathematical equations, and have a broad knowledge of as many fields of physics as possible, for analogies are important in developing physical theories. Experimentalists, on the other hand, must be creative in deciding the best way to measure an unknown physical quantity, possess considerable manual dexterity, be aware of the potentialities and limitations of physics equipment, be gifted in designing experimental apparatus, be patient and persevering in collecting data, and be completely honest in publishing the results of experiments for the use of other physicists, whether or not these results agree with accepted theories or previous experimental results.

For this reason few physicists have been equally good at theory and experiment. Those who have include Galileo Galilei, Isaac Newton, and Enrico Fermi. Fermi (see Fig. 1.8 and accompanying biography) had an ability to turn from theory to experiment and then back again to theory in a manner almost unique among twentieth-century physicists.

Of course, for a student starting out in physics both theory and experiment are equally important. It is only after students majoring in physics have finished their undergraduate studies that they must decide between experimental and theoretical physics as a career path.

1.5 Physical Measurements and Experimental Error

When experimental physicists carry out measurements in a physics laboratory, they express their results in a form which includes three essential elements: a number giving the magnitude of the quantity measured, a unit (e.g., meters or seconds) in terms of which the property is being measured, and an estimated error in the measured value. For example, a measurement of the diameter d of a coin with a meterstick might lead to a value $d = 2.41$ centimeters (cm), but the measurement of the same diameter with a precision measuring instrument called a micrometer caliper (Fig. 1.9) might yield a value $d = 2.406$ cm. Here the same property is measured in the two cases, but the larger number of digits given in the second measurement indicates that it was judged more accurate than the first by the person making the measurement.

FIGURE 1.9 A micrometer caliper. The instrument is designed so that the movable rod moves 1 millimeter (mm) in a horizontal direction for each complete turn of the screw. A dial reads fractions of a turn to the nearest 1/100 of a turn. Hence, the micrometer can measure distances to 0.01 mm, or 0.001 cm. (*Photo courtesy of Sargent-Welch Scientific Co.*)

Significant Figures

The digits given in reporting the results of an experiment are called *significant figures*. For example, 2.41 cm has three significant figures, whereas 2.406 cm has four significant figures. The greater the number of significant figures, the more accurate the experiment is presumed to be. It is misleading for an experimental physicist (or for the student) to include more significant figures in an experimental result or the result of a calculation than is justified by the apparatus used and the conditions under which the experiment was performed.

Zeros written at the right end of numbers are assumed to be significant figures, since they are included to indicate that the measurement was carried out to this last place, and that this last place was determined to be zero. For example, the charge on an electron is 1.60×10^{-19} coulombs (C), where the last zero is a significant figure. Zeros before the number are not significant, since they merely locate the decimal point and say nothing about the accuracy of the number given. Thus 0.00164 is the same as 1.64×10^{-3} in powers-of-10 notation,* and in both cases the number has only three significant figures. Because the use of powers-of-10 notation makes clear the number of significant figures, it will always be used in examples and problems in this book.

When the diameter of a coin is written as 2.41 cm, it is generally accepted that there may be an uncertainty of one or two units in the last digit, because of errors in the measuring instrument or difficulties in estimating fractions of the smallest division on the meterstick scale. Hence the actual value of the diameter might be anywhere between 2.40 and 2.42 cm. Under these circumstances it would be completely misleading to state the measured diameter as 2.410 cm, for this would indicate that the experimenter believed the diameter to be between 2.409 and 2.411 cm, when all that was really certain was that the value was between 2.40 and 2.42, an uncertainty range 10 times larger than would be indicated by the value 2.410 cm.

In adding or subtracting measured values of physical quantities care must be taken to avoid carrying insignificant figures over into the result. For example, suppose a long brass rod has a square handle at one end, as in Fig. 1.10. The length of the handle is measured to be 3.019 cm with a micrometer, while the length of the rest of the rod is measured to be 36.6 cm with a meterstick. The correct length of the full rod is then 39.6 cm, not 39.619 cm, since the last two digits in 3.019 are added to nonsignificant (actually completely unknown) digits and hence the result has nonsignificant digits in those two places.

Similarly, in multiplication and division the number of significant figures in the result can never be more than the least number of significant figures in the quantities multiplied or divided. For example, the area of a rectangle of length 4.61 cm and width 3.5 cm is 16 cm, not 16.135, 16.14, or 16.1 cm. There are only two significant figures in 3.5 and hence there can only be two significant figures in the product.

The final result of a multiplication or division should have only as many digits as the quantity with the least number of significant digits used in the calculation.

FIGURE 1.10   A long brass rod with a square handle at one end.

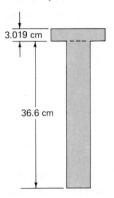

3.019 cm

36.6 cm

Rule for Handling Significant Figures

---

*Students not familiar with powers-of-10 notation should consult Appendix 3B at this time. Problems involving powers-of-10 notation are included at the end of this chapter.

In the examples and problems given in this book, the data will usually be given only to two or three significant figures. Hence all answers are expected to include only this number of significant figures.

Units

The second important element in the result of a physics experiment is the unit in terms of which the result is given. For example, you may be accustomed to measuring the length of a house in feet and inches; a physicist is more likely to measure the same distance in meters and centimeters. The length of the house is the same in the two cases, but both the numerical value and the units differ when the distance is measured in meters instead of feet. Thus lengths of 100 feet (ft) and 30.5 meters (m) are both valid measurements for the length of a house, and are completely equivalent.

In stating the answers to problems in this text, proper units must be given along with the numerical answers, or else the answer is *wrong*, since in physics a number without its proper unit is meaningless. The units to be used in this book are the metric SI units, which will be discussed in detail in the next chapter.

Experimental Errors

Finally, any physical measurement should specify a *probable experimental error* for the measured quantity. This indicates the limits within which the experimenter is confident that the actual value of the measured quantity is likely to fall. For example, the statement that the length of a house is 30.5 ± 0.3 m means that it is considered highly probable that the actual length is between 30.2 and 30.8 m. The number of significant figures used gives some measure of the experimenter's estimate of the error in the result, but explicitly stating the error is a more complete way of indicating the accuracy of the measurement.

How is this experimental error obtained? A full discussion of this topic is normally given in laboratory courses in physics, but it may be useful to give a brief outline of the process here.

Statistical Errors

The number reported for a measured physical property is usually an average of a large number of measurements made to balance out any slight uncertainties introduced by inadequacies in the apparatus used or the observing powers of the person doing the experiment. For example, two students trying to measure the length of a metal block to 0.1 mm with a meterstick will get slightly different results. The *statistical error* is a measure of the scatter of the individual measured values about the average value. If all these values cluster close together, the statistical error is small; if they scatter widely about the mean value, then the statistical error is large. The statistical error thus measures the consistency of the data collected in the experiment. It is obtained by calculating a special kind of average of the individual deviations of each measurement from the best average value of the quantity measured. In some cases the only probable experimental error attached to a final experimental result is this statistical error.

Systematic Errors

On the other hand, it is possible to have very consistent data and hence a very small statistical error and still obtain a measured value that is far from the actual value of the quantity being measured. (This presupposes that we know the actual value from other experiments. In original physics research this is, of

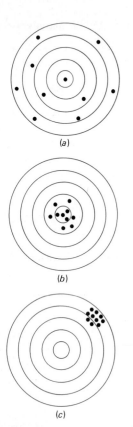

FIGURE 1.11 Comparison of results of a physics experiment with shooting at a target: (a) Low-precision experiment: Scatter of different measurements of the same quantity is great. (b) High-precision, high-accuracy experiment: Scatter is small, and all measurements are close to actual value of quantity being measured. (c) High-precision, low-accuracy experiment: Scatter is small, but a systematic error of some sort in the apparatus or the design of the experiment leads to a result far from the actual value.

course, not the case, since the property under investigation has presumably never been measured accurately before.) Such a discrepancy between the measured and the actual value is due either to some conceptual error in the design of the experiment, to failure to control some variable such as temperature during the experiment, or to some defect in the apparatus used. In the latter case the error is referred to as an *instrumental error*. Errors of this sort which lead to differences between the average value of the measured quantity and its actual value, even when the statistical error for the experiment is small, are referred to as *systematic errors*.

To illustrate the difference between statistical and systematic errors, let us consider the analogy of a gun with a defective gunsight. A good sharpshooter using this defective gun might fire a series of shots clustered together in a very small area of the target, but this area might be very far removed from the bull's-eye at which the sharpshooter was aiming, as in Fig. 1.11c. When the scatter in the measurements is small, the statistical error is small, and the experiment is said to be *precise*; when the measured results agree closely with the actual value, the systematic error is small, and the experiment is said to be *accurate*. Hence the results of a physics experiment can be very precise, and still terribly inaccurate (in other words, wrong).

To allow for this possibility experimental physicists usually attempt to reduce as much as possible all conceivable experimental errors, and then to estimate the maximum possible errors that could be introduced into the result by the instruments used and the conditions under which the experiment is performed. The statistical error is then combined with these estimated systematic errors into one number which gives the physicist's best estimate of the accuracy of the measurement. This is what is usually added on to any measured quantity as the probable experimental error of the experiment. For example, in 1958 the speed of light was measured by K. D. Froome with a microwave interferometer. His result was $c = 299{,}792.5 \pm 0.1$ kilometers per second (km/s), where the error includes an allowance for all sources of error that Froome thought possible in his measurement. The smaller the stated error, the more accurate the experiment is presumed to be. It is extremely important for any experimental physicist to estimate the experimental error, so that other physicists can judge the accuracy of the experiment, and theorists can see if the measured value agrees with theoretical predictions within the estimated error of the experiment.

No physical quantity is ever known with perfect exactness. All physical measurements are approximations, but improvements in equipment and technique make these approximations better and better as time goes on, and they finally converge toward the actual value of the property being measured. This is shown in Table 1.1 for measurements of the speed of light. The accuracy of any measurement depends on the apparatus used, the technique of the experimenter, and his or her skill in analyzing the possible errors inherent in any physics experiment. Even outstanding physicists have been known to underestimate seriously the stated errors in their experimental results. Thus Robert A. Millikan's value for the charge on the electron, obtained in his famous oil-drop experiment, turned out to be in error because the value he used for the viscosity of air was inappropriate to the conditions of his experiment.

TABLE 1.1   Results of Some Important Measurements of the Speed of Light

Date	Experimenter	Method	Velocity of light (in a vacuum), km/s	Experimental error, km/s
1862	Foucault	Rotating mirror	298,000	±500
1925	Michelson	Rotating mirror	299,796	±4
1950	Essen	Cavity resonator	299,792.5	±3
1950	Bergstrand	Geodimeter	299,793.1	±0.4
1958	Froome	Microwave interferometer	299,792.5	±0.1
1972	Bay, Luther, and white	Wavelength and frequency measurements of 0.633-μm He–Ne laser line	299,792.462	±0.018
1973	Evenson et al.	Wavelength and frequency measurements of 3.39-μm He–Ne laser line	299,792.4574	±0.0011
1974	Blaney et al.	Wavelength and frequency measurements of 9.32-μm CO_2 laser line	299.792.4590	±0.0008

Example 1.1

How many significant figures are there in the following results for quantities measured in the laboratory?
(a) 2.997924×10^8 m/s
(b) 3.0120 s
(c) 0.00124 m
(d) 100 s

SOLUTION

(a) Seven significant figures.
(b) Five significant figures. The final zero is significant.
(c) Three significant figures. The initial zeros are not significant. This result would be better written as 1.24×10^{-3} m, or 1.24 mm. (See Appendix 3B for a review of powers-of-10 notation.)

(d) Three significant figures. The final zeros are significant. To make sure that this is what the experimenter intended, it would be clearer if the result were given as 1.00×10^2 s.

Example 1.2

(a) Use Table E (inside back cover) to find the mass difference in kilograms (kg) between a proton and a neutron.

(b) If seven significant figures are used for the proton and neutron masses, how many significant figures does the mass difference have?

SOLUTION

(a) From Table E we have:

$m_p = 1.672649 \times 10^{-27}$ kg $m_n = 1.674954 \times 10^{-27}$ kg

The mass difference is then:

$\Delta m = (1.674954 - 1.672649) \times 10^{-27}$ kg

$$= \boxed{0.002305 \times 10^{-27} \text{ kg}}$$

(b) The mass difference has only four significant figures and is better written as 2.305×10^{-30} kg. This example illustrates the fact that significant figures can be quickly lost in calculations involving differences between almost-equal numbers.

Example 1.3

If a proton and an electron combine to form an atom of hydrogen, what is the expected mass of a hydrogen atom?

SOLUTION

From Table E we have:

$m_p = 1.672649 \times 10^{-27}$ kg

$m_e = 9.10953 \times 10^{-31}$ kg $= 0.000910953 \times 10^{-27}$ kg

Then the sum is:

$(1.672649 + 0.000910953) \times 10^{-27}$ kg

$$= \boxed{1.673560 \times 10^{-27} \text{ kg}}$$

Notice that in the addition m_e can be rounded off to 0.000911×10^{-27} kg, since the last three digits in m_e would be added to nonsignificant (actually unknown) digits in m_p.

Example 1.4

(a) Write 0.00000027 in powers-of-10 notation.
(b) Calculate $(2.56 \times 10^3) + (7.92 \times 10^4)$.
(c) Calculate $(2.56 \times 10^3) \times (7.92 \times 10^4)$.

(d) Calculate $\dfrac{2.56 \times 10^3}{7.92 \times 10^4}$.

SOLUTION

(a) $\boxed{2.7 \times 10^{-7}}$

(b) $(2.56 \times 10^3) + (7.92 \times 10^4) = (2.56 + 79.2) \times 10^3$

$$= \boxed{81.8 \times 10^3}$$

Notice here that on adding 2.56 and 79.2 the second decimal place no longer is significant, since 79.2 has only one significant figure to the right of the decimal point. Hence we round off 2.56 to 2.6 and add it to 79.2 to obtain 81.8.

(c) $(2.56 \times 10^3) \times (7.92 \times 10^4) = 20.3 \times 10^7$

$$= \boxed{2.03 \times 10^8}$$

Here we multiply the two numbers, rounding off to three significant figures, and add the powers of 10 to obtain 10^7.

(d) $\dfrac{2.56 \times 10^3}{7.92 \times 10^4} = 0.323 \times 10^{-1} = \boxed{3.23 \times 10^{-2}}$

Here we divide the numbers, rounding off again to three significant figures, and subtract the powers of 10 to obtain $10^{3-4} = 10^{-1}$.

Summary: Important Definitions and Equations

Physics: The science of matter and energy, and of the relations between them.

Hypothesis: A tentative explanation, assumed for the sake of argument, to correlate data obtained from observation and experiment.

Physical theory: A theoretical explanation, perhaps in mathematical terms and based on a physical model, which correlates and makes understandable data obtained from observation and experiment.

Physical law: A physical theory which agrees so well with experimental observations that it seems to reflect the constant behavior of nature.

Physical measurements:

Significant figures: A measure of the accuracy with which some measured quantity is known. A small number of significant figures indicates low accuracy; a large number of significant figures, high accuracy.

Rules for handling significant figures: The final result of a multiplication or division should have only as many digits as the quantity with the least number of significant digits used in the calculation.

Probable experimental error: A range of values specified for the results of a physics experiment, within which the experimenter is confident the actual value is likely to fall. Probable experimental error includes both statistical errors and systematic errors.

Statistical error: A measure of the scatter of the individual measured values about the average measured value of a physical quantity. Measurements with low statistical errors are said to be precise.

Systematic error: A consistent discrepancy between the actual and the measured values of a physical quantity, due either to instrumental errors or failures in the design or execution of the experiment. Measurements with low systematic errors are said to be accurate.

Questions

1 How does physics differ from philosophy? Which would you say leads to more certain results? Why?

2 How does physics differ from a science like astronomy? Is there much overlap between these two sciences?

3 What would you see as the main difference between a natural science like physics and a social science like political science?

4 Which would you expect to be greater, the contributions made by biology to physics or the contributions made by physics to biology? Why?

5 Physics has led to the development of a large number of precision instruments of great usefulness to chemists, biologists, and medical scientists, such as electron microscopes, nuclear magnetic resonance (NMR) spectrometers, infrared spectrometers, and x-ray diffraction units.

(*a*) Is it necessary for a biologist or chemist to understand all the physics behind the working of such instruments in order to use the instruments effectively in research?

(*b*) Is it helpful if the biologist or chemist understands at least the basic physical principles behind the working of such instruments?

6 Why is biology in at least one respect a much more difficult science than physics?

7 François de La Rochefoucauld (1613–1680) once wrote: "One of the tragedies of life is the murder of a beautiful theory by a brutal gang of facts." Comment on this statement as it applies to the progress of the science of physics.

8 In a speech delivered on receiving the 1923 Nobel Prize in physics the American physicist Robert A. Millikan made the following comment:

The fact that Science walks forward on two feet, namely theory and experiment, is nowhere better illustrated than in the two fields for slight contributions to which you have done me the great honor of awarding me the Nobel Prize in Physics for the year 1923. Sometimes it is one foot which is put forward first, sometimes the other, but continuous progress is only made by the use of both—by theorizing and then testing, or by finding new relations in the process of experimenting and then bringing the theoretical foot up and pushing it on beyond, and so on in unending alternations.

Discuss how well Millikan's statement describes the progress of physics.

9 Do you think that physics will ever be "complete" in the sense that there will no longer be any need for theoretical and experimental research in physics?

10 (*a*) Distinguish carefully between *precision* and *accuracy* in a physics experiment.

(*b*) Can an experiment be precise without being accurate?

(*c*) Can an experiment be accurate without being precise?

11. One of Enrico Fermi's favorite questions on Ph.D. oral examinations in physics was "How many piano tuners are there in New York City?" How would you go about providing an order-of-magnitude answer to this question? (Looking up the number listed in the New York City phone book is against the rules of the game!)

Problems

A 1 The number 0.03210 has the following number of significant figures:
(*a*) 6 (*b*) 5 (*c*) 4 (*d*) 3
(*e*) None of the above

A 2 The measured value of the length of a rug is 3.60 m. This is usually taken to indicate that the length of the rug is between:
(*a*) 3.599 and 3.601 m (*b*) 3.59 and 3.61 m
(*c*) 3.5 and 3.7 m (*d*) 3 and 4 m
(*e*) None of the above

A 3 The number of significant figures in pi (π), the ratio of the circumference to the diameter of a circle, is:
(*a*) 3 (*b*) 5 (*c*) 2 (*d*) 4
(*e*) As many as you want to calculate

A 4 A metal block of height 3.75 cm is resting on a metal shim whose thickness is known accurately to be 0.1250 cm. The height of the combination of shim and metal block is:
(*a*) 3.875 cm (*b*) 3.8750 cm (*c*) 3.9 cm
(*d*) 3.88 cm (*e*) None of the above

A 5 The electric power P (in watts, W) in an electric circuit is the product of the current I (in amperes, A) and the voltage V (in volts, V): $P = VI$. If the voltage is 115 V and the current is 2.6 A, the electric power is:
(a) 299 W   (b) 2.99×10^2 W   (c) 29.9×10^2 W
(d) 30×10^2 W   (e) 3.0×10^2 W

A 6 The length of an athletic field is 106.51 m and its width is 36.32 m. The surface area (in square meters, m^2) of the field, in powers-of-10 notation, is:
(a) 3.86844×10^3 m^2 (b) 3.868×10^2 m^2
(c) 3.868×10^4 m^2 (d) 3.87×10^3 m^2
(e) 3.868×10^3 m^2

A 7 The sum of 3.89×10^3 and 2.56×10^5 is:
(a) 2.60×10^5 (b) 2.5989×10^5 (c) 2.60×10^3
(d) 2.5989×10^3 (e) 6.45×10^8

A 8 The product of 6.915×10^2 and 2.16×10^{-3} is:
(a) 1.4936   (b) 1.49×10^{-1} (c) 1.4936×10^{-1}
(d) 4.9   (e) 1.49

A 9 The result of the calculation
$$\left(\frac{3.98 \times 10^5}{6.31 \times 10^{-3}}\right)(5.6 \times 10^{-2})$$
is:
(a) 3.5322×10^6 (b) 3.53   (c) 3.53×10^4
(d) 3.5×10^6 (e) 3.5×10^{-6}

A10 The result of the calculation
$$\left(\frac{2.460 \times 10^3}{3.981 \times 10^{-6}}\right)\left(\frac{6.4196 \times 10^2}{2.468 \times 10^{-3}}\right)$$
is
(a) 1.6073×10^{13} (b) 1.6073×10^{14}
(c) 16.07×10^{13} (d) 1.607×10^{-14}
(e) 16.07×10^{-13}

Additional Readings

Conant, James B.: *On Understanding Science: An Historical Approach*, Yale University Press, New Haven, Conn., 1951. This book emphasizes the importance of the history of science to our understanding of culture and civilization.

Fermi, Laura: *Atoms in the Family*, University of Chicago Press, Chicago, 1954. A delightful, perceptive book by Enrico Fermi's wife, Laura.

Feynman, Richard: *The Character of Physical Law*, MIT Press, Cambridge, Mass., 1965. This stimulating book is well worth reading in its entirety, but the first three chapters are particularly relevant to this course.

Morrison, Philip, Phylis Morrison, and the Office of Charles and Ray Eames: *Powers of Ten*, Scientific American Library, Freeman, San Francisco, 1982. A fascinating trip through the universe in steps of powers of 10.

Segrè, Emilio: *Enrico Fermi, Physicist*, University of Chicago Press, Chicago, 1970. An excellent biography by a friend and scientific colleague of Enrico Fermi.

The above references cover interesting historical material or contain more detailed discussions of specialized topics than is possible in an introductory textbook. There are a large number of other books to which the student may wish to turn for clarification of difficult topics or for additional information. Listed here are only a few of those that the author knows from experience will be helpful.

Noncalculus Physics Textbooks

Atkins, Kenneth R.: *Physics*, 2d ed., Wiley, New York, 1970.

Bueche, Frederick: *Principles of Physics*, 4th ed., McGraw-Hill, New York, 1982.

Cooper, Leon N.: *An Introduction to the Meaning and Structure of Physics*, Harper & Row, New York, 1968.

Giancoli, Douglas C.: *Physics: Principles with Applications*, Prentice-Hall, Englewood Cliffs, N.J., 1980.

Holton, Gerald: *Introduction to Concepts and Theories in Physical Science*, 2d ed., revised and with new material by Stephen G. Brush, Addison-Wesley, Reading, Mass., 1973.

Miller, Franklin, Jr.: *College Physics*, 5th ed., Harcourt Brace Jovanovich, New York, 1982.

Sears, Francis W., Mark W. Zemansky, and Hugh D. Young: *College Physics*, 4th ed., Addison-Wesley, Reading, Mass., 1974.

Smith, Alpheus W., and John N. Cooper: *Elements of Physics*, 9th ed., McGraw-Hill, New York, 1979.

Calculus-Based Physics Textbooks

Halliday, David, and Robert Resnick: *Physics*, 3d ed., Wiley, New York, 1978.

Sears, Francis W., Mark W. Zemansky, and Hugh D. Young: *University Physics*, 5th ed., Addison-Wesley, Reading, Mass., 1976.

More Advanced Introductory Textbook

Feynman, Richard P., Robert B. Leighton, and Matthew Sands: *The Feynman Lectures on Physics*, Addison-Wesley, Reading, Mass., 1963, 3 vols.

Historical Aspects of Physics

Asimov, Isaac: *Asimov's Biographical Encyclopedia of Science and Technology*, 2d rev. ed., Doubleday, Garden City, N.Y., 1982. This contains the lives and achievements of 1510 outstanding scientists. It is a gold mine of information and is written in Asimov's lively, interesting style.

Boorse, Henry A., and Lloyd Motz (eds.): *The World of the Atom*, Basic Books, New York, 1966, 2 vols. These volumes discuss the lives and accomplishments of scientists who had a connection with the development of atomic physics in the broadest sense, and include excerpts from important papers written by these scientists.

Dampier, W. C.: *A History of Science and its Relations with Philosophy and Religion*, 4th ed., reprinted with a postscript by I. Bernard Cohen, Cambridge University Press, New York, 1966. A classic work on the history of science, with considerable emphasis on physics.

Frisch, Otto R.: *What Little I Remember*, Cambridge University Press, New York, 1979. The quotation at the beginning of this chapter is taken from page 22 of Frisch's book (with permission).

Magie, William F.: *Source Book in Physics*, Harvard University Press, Cambridge, Mass., 1963. An excellent source for brief biographical sketches and excerpts from the original writings of major characters in the history of physics.

Shamos, Morris H.: *Great Experiments in Physics*, Holt, Rinehart and Winston, New York, 1959. Contains accounts from original papers of some of the most famous experiments in the history of physics, together with brief commentaries and biographical material.

Throughout the rest of this course we will not take the time to refer explicitly to any of the above books. It is presumed that it will always be helpful to the student to obtain a different perspective on the material presented in this text by consulting one of them.

*P*hilosophy is written in this grand book—I mean the universe—which stands continually open to our gaze, but it cannot be understood unless one first learns to comprehend the language and interpret the characters in which it is written. It is written in the language of mathematics . . . without which it is humanly impossible to understand a single word of it.*

Galileo Galilei (1564–1642)

Chapter 2
Rectilinear Motion and Its Laws

Herbert Butterfield, the great British historian, in his *Origins of Modern Science* (1965), aptly indicates the importance of this chapter on motion: "Of all the intellectual hurdles which the human mind has been faced with and has overcome in the last fifteen hundred years, the one which seems to me to have been the most amazing in character, and the most stupendous in the scope of its consequences, is the one relating to the problem of motion."

From the time of Aristotle (384–322 B.C.) to that of Galileo very learned people believed and taught some very wrong ideas about the motion of inanimate objects. Such objects were thought to be intrinsically heavy or light, and to fall or rise with a speed proportional to their heaviness or lightness, since they always sought their "natural places" in the universe. It was only when Galileo and Newton combined careful observation and experiment with mathematical analysis that a correct view of the motion of objects emerged. In this chapter we will discuss the motion of objects without considering the causes of that motion. This is *kinematics*, one area of the broader field of physics called *mechanics*. The other area of mechanics, called *dynamics*, takes up the *causes* of motion and will be discussed in the next chapter.

In this chapter we will limit our discussion to rectilinear motion, i.e., to motion in a straight line. Examples include the motion of an automobile along a straight superhighway, and the motion of an electron down the straight 2-mile-long evacuated tube of the Stanford linear accelerator at Palo Alto, California. The ideas to be developed here are as important for our understanding of the world of automobiles and airplanes as they are for comprehending the world of the atom. They form the foundation of the imposing structure we call physics.

2.1 The Fundamental Quantities of Mechanics

The motion of objects can be discussed in terms of three fundamental physical quantities—length, time, and mass. Other mechanical quantities such as velocity, acceleration, and force can be easily expressed in terms of these three fundamental quantities.

---

*From Butterfield (see Additional Readings).

Length

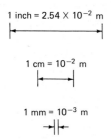

Distance traveled by light in a vacuum in 1 s = 299,792,458 m

1 m

Distance traveled by light in 1/299,792,458 s

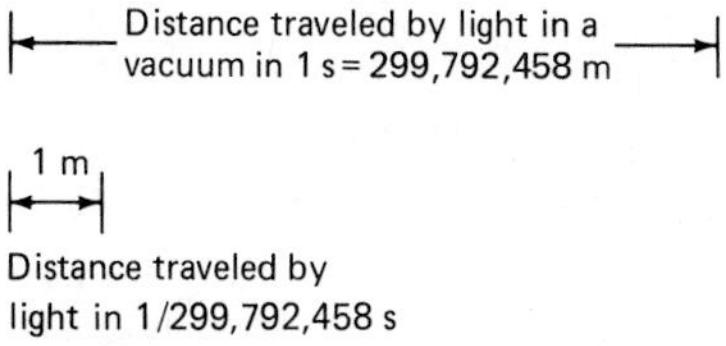

FIGURE 2.1 The definition of the meter in terms of the speed of light. Since in 1 s light travels 299,792,458 m, the meter is now defined as the distance traveled by light in 1/299,792,458 s.

1 inch = 2.54 × 10^{-2} m

1 cm = 10^{-2} m

1 mm = 10^{-3} m

FIGURE 2.2 Actual size of some common units of length.

Time

Definition

We all have some feeling for what we mean by the *length* of an object. It is the distance from one end of the object to the other, as measured with a ruler or tape measure. The ruler or tape measure is divided by the manufacturer into equal divisions such as inches or centimeters, and the length of the object is measured by counting the number of these divisions between one end of the object and the other. Even though the basic idea of length is the same for all objects, the units in terms of which length is measured are completely arbitrary. For example, we can measure the length of a pendulum in meters, centimeters, feet, or inches, and the pendulum's length is still the same even though its numerical value may differ greatly depending on the units chosen.

In this book we will use exclusively metric SI (for the French *Système International*) units. In the SI system lengths are measured in meters (m). All our metric rulers and tape measures are ultimately checked against some standard meter at a laboratory like the National Bureau of Standards (NBS) at Gaithersburg, Maryland. A new definition of the meter was introduced by the General Conference on Weights and Measures meeting in Paris in November 1983. Since light travels at a speed of 299,792,458 m/s (see Table 1.1), *the meter is now officially defined as the distance traveled by light through space in 1/299,792,458 s* (see Fig. 2.1). The nine figures given here are all significant, and indicate that the meter is now believed known to better than 1 part in a billion (10^9). This definition of the meter relates it to the unit of time, the second, which is known to better than 1 part in 10 trillion.

The British engineering units of length, which are still used in the United States, are defined in terms of the standard meter. Thus 1 inch (in) is exactly equal to 0.0254 m, and 1 yard (yd) is 0.91440 m. The actual sizes of some length units are shown in Fig. 2.2. For those not yet familiar with the metric system it may be worthwhile to indicate that a millimeter (10^{-3} m) is about the thickness of the wire in a large paper clip, a centimeter (10^{-2} m) is about the size of the nail on the little finger, and a kilometer (10^3 m) is the length of 11 football fields.

St. Augustine once wrote that we all think we know what time is until we attempt to define it. Fortunately for us, it is much easier to measure the passage of time than it is to define its nature precisely. Time can be measured in terms of the swings of a clock's pendulum or the vibrations of the quartz crystal in a watch. The *period*, or time for one complete vibration, usually designated by the symbol *T*, is then the fundamental unit in terms of which unknown times can be measured. For years the time for the earth to make one complete revolution on its axis, which determines the length of a day, was used as the time standard. The fundamental unit of time, the second, was then defined as equal to $1/(24 \times 60 \times 60) = 1/86,400$ of a mean solar day.

In 1967 a new definition of the second was adopted by the Thirteenth General Conference on Weights and Measures. *The second is now defined as the duration of 9,192,631,770 periods of a particular radiation emitted by the cesium atom.* An oscillator is tuned to this frequency* in a device called an

---

*By *frequency* we mean the number of vibrations per unit time. Frequencies are measured in *hertz* (Hz), or cycles per second. The frequency of a vibration is the reciprocal of its period, or $f = 1/T$. We will discuss this more thoroughly in Chap. 11.

FIGURE 2.3 A commercial cesium atomic clock occupying its own seat on an airplane. The standard atomic clock for timekeeping in the United States is located at the National Bureau of Standards in Boulder, Colorado. Time is measured in terms of a particular vibrational frequency of the cesium atom. The atomic clock shown in this photograph was flown around the world on a jetliner to test Einstein's theory of relativity. (*Photo courtesy of U.S. National Bureau of Standards.*)

atomic clock. This atomic clock (Fig. 2.3) is believed to be accurate to one part in 10^{13}, or to about 3 s in 1 million years. Because of the extreme accuracy of this standard, times and frequencies are the quantities in physics which can be most accurately measured. Often experimental physicists try to arrange their experiments so that the ratio of two masses, or of two velocities, can be reduced to the ratio of two frequencies, for then the accuracy of the experiment can be enormously improved. To aid in such comparisons the NBS radio station WWV in Boulder, Colorado, continuously broadcasts accurate frequencies of 2.5, 5, 10, 15, 20, and 25 megahertz (MHz), or 10^6 cycles per second, stabilized to 1 part in 100 billion (10^{11}) by comparison with a cesium atomic clock.

Mass

The third fundamental quantity of mechanics is *mass.* Mass is a measure of the amount of matter present in an object. For example, a large rock has more mass than a small rock of the same kind. We know from experience that it is much more difficult to move a large rock by pushing on it than it is to move a small rock. Physicists say that the large rock has more *inertia* than the small rock, where by inertia they mean the resistance the rock offers to a change in its condition of rest or motion. *Mass is therefore a measure of inertia.*

The standard mass is that of a platinum-iridium cylinder, which is preserved at Sèvres, near Paris, and is exactly equal (by definition) to one kilogram (1 kg), the SI unit of mass. Also, 1 kg = 10^3 grams (g). To get some feeling for the units here, it is worth noting that a jelly bean has a mass very close to 1 g, while a quart of milk has a mass a little less than a kilogram (actually 975 g).

Secondary standards of mass exist at NBS and at other standards laboratories throughout the world (see Fig. 2.4). Unknown masses can then be compared with these secondary standards to a few parts in 10^8 by using sensitive equal-arm balances similar to the one shown in Fig. 2.5.

FIGURE 2.4 A secondary mass standard maintained at the National Bureau of Standards in Gaithersburg, Maryland. (*Photo courtesy of U.S. National Bureau of Standards.*)

FIGURE 2.5 An equal-arm balance similar to those used for the comparison of masses. (*Photo courtesy of Sargent-Welch Scientific Co.*)

2.2 Frames of Reference

In the remainder of this chapter we will be discussing the position and the speed of various moving objects. To do this scientifically we must first answer the questions "Position with respect to what?" and "Speed with respect to what?" For example, at Eastport, Maine, on the Bay of Fundy, the mean difference between high and low tide is 5.52 m (18.1 ft). Hence a flag flying from a house on the dock is apparently 5.52 m higher at low tide than it is at high tide, if its position is measured with respect to the water level. Such a changing frame of reference is not very useful. Before we can indicate the position of an object like the flag with any accuracy, we must first specify an unchanging frame of reference (such as the dock) with respect to which the flag's position can be specified.

Similarly, if a person walks down the aisle toward the front of an Amtrak train at 2 m/s with respect to the train, and the train is moving at 25 m/s with respect to the ground, the person's speed with respect to the ground is 27 m/s. Hence to state the person's speed without specifying what reference frame that speed is being measured against can be quite misleading.

When we discuss motion on the surface of the earth, it is convenient to take the earth's surface as our frame of reference. For the motion of the planets, the fixed stars are a good choice, whereas for the motion of the electrons in an atom, the nucleus of the atom is preferred. The choice of a frame of reference is an arbitrary one, but in all cases it is necessary to specify what reference frame is being used since all motion is specified relative to that frame.

Coordinate Systems

In physics a frame of reference is usually pictured in terms of a *coordinate system*, consisting of three mutually perpendicular axes, called the *X*, *Y*, and *Z axes*, relative to which positions in space can be specified. These three axes

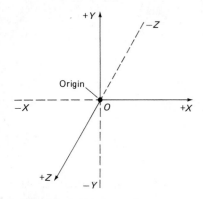

FIGURE 2.6 A rectangular coordinate system.

intersect at the *origin* (O) of the coordinate sytem, as shown in Fig. 2.6.* If we want to describe the position of a ball moving along the positive X axis, we simply mark a dot on the X axis, indicating its position at some instant of time. As time goes on, this point moves to greater and greater positive values of x along the X axis. If the object is moving to the left instead of the right, then its position will be marked by larger and larger negative values of x as time goes on. Rectilinear motion of an object will therefore be described by a straight line in this coordinate system. This straight line could coincide with the X axis, the Y axis, or the Z axis, or could be at some angle with respect to these axes, and we would still have rectilinear motion. For such motion it is always simpler to make the direction of motion of the object coincide with one of the three coordinate axes, as can always be done by choosing the coordinate system properly.

2.3 Vectors and Scalars; Velocity and Speed

If we have a sack of flour with a mass of 10 kg, that mass is in no way dependent on where the flour is, whether it is at rest in some particular reference frame or in motion in some particular direction. The mass is what we call a *scalar* quantity; i.e., it is a physical property of the flour which can be completely described by one number (10 kg). Another example of a scalar quantity is energy, which again can be completely specified by just one number [say, 10 joules (J)].

On the other hand, a quantity such as velocity is quite different. If a train is moving at 25 m/s, and a woman on the train wants to go from Baltimore to New York, it obviously makes a big difference to her whether the train is moving in the direction of Washington, D.C., or of New York. Here both the *speed* and the *direction* are vitally important. Quantities like velocity, which have both a magnitude (25 m/s) and a direction (from Baltimore to New York), are called *vectors*. For clarity let us reiterate the difference between these two important quantities:

Definitions

Scalar: A quantity for which only the magnitude is important, and direction is unimportant or meaningless.

Vector: A quantity for which not merely the magnitude but also the direction is important and must be specified.

Quantities such as mass, time, energy, and temperature are scalars; velocity, acceleration, force, and momentum are vectors.

Displacement Vectors

The simplest example of the use of vectors is to specify the position of one object with respect to some other object, or to the origin of a coordinate system. For example, suppose we are asked to describe the motion of a car that starts at the origin of a coordinate system and moves a distance of 2 km in a northeasterly direction. We can do this as in Fig. 2.7 by selecting a scale

---

*The coordinate system shown in Fig. 2.6 is a *right-handed* coordinate system, in which a right-handed screw (an ordinary carpenter's screw), if turned from $+X$ into $+Y$, would move in the direction of $+Z$. This is the only kind of coordinate system we will use in this book.

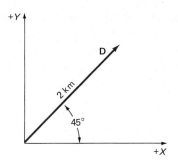

FIGURE 2.7 A displacement vector, where the displacement is measured with respect to the origin of the coordinate system.

(say, 1 cm on the paper corresponds to a distance of 1 km), and drawing an arrow of length 2 cm (for 2 km) at an angle of 45° with respect to the *X* axis. This arrow is called a *displacement vector*. Its length indicates the net distance traveled, and it points in the direction in which the car traveled. (We will normally use boldface type, as in **v**, to describe vector quantities.)

It should be pointed out that the displacement is not the same as the total distance traveled, even in one dimension. Thus if a student bikes from one building to another on campus, and then turns around and returns to the starting point, the total distance traveled might be 400 m, but the displacement is zero, since the student has returned to where he or she started.

Velocity and Speed

Similarly, a velocity is characterized by two quantities, the magnitude of the velocity, which we call the *speed*, and its *direction*. We can therefore describe a velocity **v** by an arrow pointing in the correct direction and with a length corresponding to the speed, where we can use some convenient scale to fix this length. Thus a train going at 100 kilometers per hour (km/h), that is, about 60 mi/h, from St. Louis to Madison, Wisconsin, might be indicated by the vector **v** in Fig. 2.8, whereas the same train going from St. Louis to Kansas City, Missouri, at the same speed might be illustrated by the vector **v′** in the same figure.

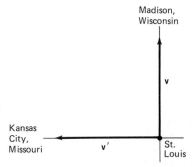

FIGURE 2.8 Two velocity vectors of equal length, indicating that the speeds are the same but the directions different.

Addition of Velocities

Scalar quantities can be added by using simple arithmetic. Thus two sacks of flour, each with a mass of 10 kg, have a combined mass of 20 kg. The same is not true, however, of vector quantities. If a passenger is walking down the aisle of a train at a rate of 2 meters per second (m/s), and the train is moving at 10 m/s with respect to the ground, the passenger's velocity with respect to the ground is either 8 or 12 m/s, depending on whether the passenger is moving toward the back or the front of the train. This is illustrated in Fig. 2.9,

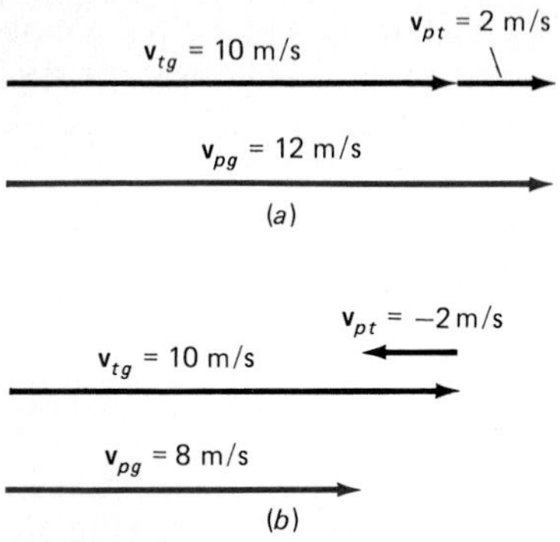

$v_{pt} = 2$ m/s

$v_{tg} = 10$ m/s

$v_{pg} = 12$ m/s

(a)

$v_{pt} = -2$ m/s

$v_{tg} = 10$ m/s

$v_{pg} = 8$ m/s

(b)

FIGURE 2.9   Addition of two velocities in one dimension: (a) Passenger on train is walking in same direction in which train is moving. (b) Passenger is walking in opposite direction to that of train's motion. Note the difference in the resultant velocity of the passenger with respect to the ground.

where the velocity of the passenger with respect to the ground is $\mathbf{v}_{pg}$, the velocity of the train with respect to the ground is $\mathbf{v}_{tg}$, and the velocity of the passenger with respect to the train is $\mathbf{v}_{pt}$.

The negative of a vector $\mathbf{A}$ is the vector $-\mathbf{A}$, a vector with the same magnitude as $\mathbf{A}$ but in the opposite direction. For example, if a train is moving forward with a velocity $\mathbf{v}$ with respect to the ground, and a passenger is moving with a velocity $-\mathbf{v}$ with respect to the train, then the velocity of the passenger with respect to the ground is given by the equation

$$\mathbf{v}_{pg} = \mathbf{v}_{pt} + \mathbf{v}_{tg}$$

Now in this case $\mathbf{v}_{pt} = -\mathbf{v}$ and $\mathbf{v}_{tg} = +\mathbf{v}$. Hence

$$v_{pg} = -\mathbf{v} + \mathbf{v} = 0$$

Hence the passenger is at rest with respect to the ground, even though he or she is walking down the aisle of a moving train.

Two velocities can combined to obtain a resultant velocity by *vector addition*. For motion in one dimension the velocities are either in the same or in opposite directions, and the resultant velocity can be easily obtained by simple addition or subtraction of the two speeds*; for velocities in more than one dimension, where one velocity can have any possible direction with respect to the other, vector methods developed in Chap. 4 must be used.

Since velocity is a vector, a change in a velocity can be a change either in its magnitude (the speed), in its direction, or in both. This idea will become especially important when we discuss acceleration.

2.4 Average and Instantaneous Speed

The speedometer on a car tells us how many kilometers we would cover in an hour if we continued to travel for the whole hour at the indicated rate. For example, if a car moves at a constant speed of 88 km/h (or 55 mi/h) in a straight line, in 1 h it will travel 88 km. In the SI system a speed is commonly expressed in meters per second, where, for example,

$$88 \text{ km/h} = \frac{88 \times 10^3 \text{ m}}{3600 \text{ s}} = 25 \text{ m/s}$$

Hence a speed of 55 mi/h is approximately equal to 25 m/s.

If in a time interval $\Delta t = t_2 - t_1$ a car travels a straight-line distance $\Delta s = s_2 - s_1$ at a *constant speed*, this speed is:

Constant Speed

$$v_{\text{const}} = \frac{\Delta s}{\Delta t} = \frac{s_2 - s_1}{t_2 - t_1} \tag{2.1}$$

Here the symbol Δ is the Greek capital *delta* and the above symbols are read as "delta s" and "delta t"; Δs is the change in s from its original position, and Δt is the change in t during the time interval involved.

---

*The introduction of vectors in this chapter may seem out of place since for one-dimensional motion it is possible to get along without vectors merely by using plus and minus signs to indicate direction (as we do in much of this and the following chapter). Despite this fact, vectors are introduced here to clarify at the earliest possible time the vector nature of velocities, accelerations, and forces, without complicating the discussion by introducing rules for the addition of vectors in more than one dimension.

For the case of constant speed, a plot of distance traveled (plotted along the vertical axis) against time (plotted along the horizontal axis) leads to the graph shown in Fig. 2.10. The graph is seen to be a straight line. The slope of the straight line is the ratio of the displacement of the object $\Delta s = s_2 - s_1$ to the corresponding time interval $\Delta t = t_2 - t_1$. Hence

$$\text{Slope of graph} = \frac{s_2 - s_1}{t_2 - t_1} = \frac{\Delta s}{\Delta t}$$

The slope of the straight line is therefore equal to $\Delta s/\Delta t$ and hence is also equal to v_{const}.

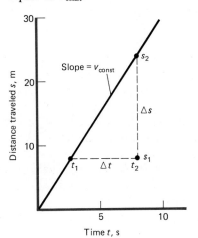

FIGURE 2.10 One-dimensional motion with constant speed.

Average Speed

In practical situations the speed of one-dimensional motion is seldom constant but changes with time. For example, the reading on a car's speedometer is usually changing from one instant to another, as the driver speeds up or slows down to adjust to the flow of traffic and the presence of stoplights and stopsigns. When the speed is changing in time, it becomes important to distinguish between the average speed and the instantaneous speed for the motion. We define the *average speed* as follows:

Average Speed

$$v_{\text{av}} = \bar{v} = \frac{\text{total distance covered}}{\text{total time elapsed}} = \frac{\Delta s}{\Delta t} \tag{2.2}$$

where Δs and Δt are as previously defined.

In this case the total distance traveled is the reading of the car's odometer, which reflects the total distance covered whether it is forward or backward, or whether the driver has turned around in the middle of the trip and returned to the starting point.

On the other hand, the *average velocity* for the same trip would be defined as follows:

Average Velocity

$$\mathbf{v}_{\text{av}} = \bar{\mathbf{v}} = \frac{\text{total displacement}}{\text{total time elapsed}} = \frac{\Delta \mathbf{s}}{\Delta t}$$

where $\Delta \mathbf{s}$ is here the vector displacement from the point where the car started to where it finally came to a stop.

In Fig. 2.11 the position of a car as a function of time is graphed. Since the graph is not a straight line, the speed is changing with time. In this case

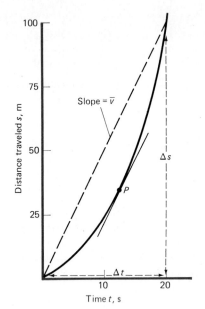

FIGURE 2.11   Motion with changing speed. The instantaneous speed is equal to the average speed only at point *P*.

the speed is increasing, since the slope of the graph constantly increases. The average speed of the car during the 20 s shown is, however,

$$\bar{v} = \frac{\Delta s}{\Delta t} = \frac{100 \text{ m} - 0}{20 \text{ s} - 0} = 5 \text{ m/s}$$

This is the slope of the straight line drawn on the graph.

In this situation the car's speedometer would gradually increase from zero to some maximum value much larger than 5 m/s during the first 20 s of the trip. The value of the speed at any instant of time, or the instantaneous speed *v*, would be constantly increasing, and would be equal to the average speed of 5 m/s for only a single instant of the 20-s interval. This is at point *P* where the slope of the black line is equal to the slope of the dashed line representing $\bar{v}$.

Instantaneous Speed

Instantaneous Speed

The *instantaneous speed* is obtained by evaluating the average speed over a very, very short time interval Δt near the time *t* at which the speed is desired. Then we can write:

$$v_{\text{instant}} = \lim_{\Delta t \to 0} \frac{\Delta s}{\Delta t} \tag{2.3}$$

Let us see what this means. In Fig. 2.12 we show part of the same graph as in Fig. 2.11. Suppose we want to know the instantaneous speed at time *t* = 5 s. We calculate the average slope of the graph between the time 5 s and a series of later times, each time shortening the time interval Δt involved. We find the results given in Table 2.1. We see from the table that as the time interval gets smaller, i.e., as Δt approaches zero ($\Delta t \to 0$), the average velocities get smaller and smaller and also get closer and closer together in value. In the limit where Δt becomes infinitesimally small, the chords drawn to show the average slope approach the tangent to the curve at the point *t* = 5 s. This limiting value of the slope is the instantaneous speed at the time *t* = 5 s.

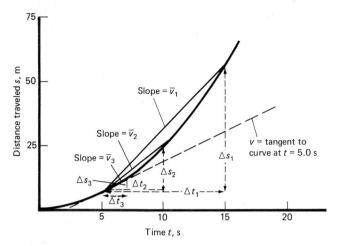

FIGURE 2.12 Definition of instantaneous speed. It is the limiting value of the average speed as the time interval over which the average speed is evaluated gets smaller and smaller.

TABLE 2.1 Average Values of the Speed for Time Intervals Near $t = 5$ s in Fig. 2.12.

Time interval Δt, s	Distance covered Δs, m	Average speed $\bar{v} = \dfrac{\Delta s}{\Delta t}$, m/s
15.0−5.0	56.3−6.3	$\bar{v}_1 = 5.0$
10.0−5.0	25.0−6.3	$\bar{v}_2 = 3.7$
7.0−5.0	12.5−6.3	$\bar{v}_3 = 3.1$
6.0−5.0	9.0−6.3	$\bar{v}_4 = 2.7$
limit $\Delta t \rightarrow 0$		$v = 2.3$

The tangent to a position-versus-time graph at any point gives the slope of the graph at that point and hence the instantaneous speed v at the time corresponding to that point.

On the graph the tangent is indicated by the line of slope v. Its value is

$$v = \frac{25 \text{ m} - 0}{13 \text{ s} - 2 \text{ s}} = 2.3 \text{ m/s}$$

The limit therefore of the series of values obtained for the average speed for different time intervals: $\bar{v}_1 = 5.0$ m/s, $\bar{v}_2 = 3.7$ m/s, $\bar{v}_3 = 3.1$ m/s, $\bar{v}_4 = 2.7$ m/s, and so on, is $v = 2.3$ m/s, and this is the instantaneous speed of the car at time $t = 5$ s.

Of course, if the speed is constant, then the position-versus-time graph is a straight line, and so $v = \bar{v} = v_{\text{const}}$; that is, the average and instantaneous values of the speed are the same and are equal to the constant speed.

Distance Traveled

If the car moves with average speed $\bar{v}$ in the time interval from $t = 0$ to $t = t$, then

$$\bar{v} = \frac{\Delta s}{\Delta t} = \frac{s - s_0}{t - 0} = \frac{s - s_0}{t}$$

where s_0 is the position of the car with respect to the origin at time $t = 0$. On solving this equation for s, the car's position at some given time t, we have

$$\boxed{s = s_0 + \bar{v}t} \tag{2.4}$$

This is the equation of a straight line with slope v and intercept s_0. Such a straight line is shown in Fig. 2.13. The quantity s_0 is called the *initial position* of the car at time $t = 0$. In many cases this position can be made zero by the proper choice of the coordinate system in which the object moves.

The same equation applies for a constant speed. Equation (2.4) cannot be used for motion in which the speed is changing in time, however, unless it is possible to find the average speed $\bar{v}$ from the instantaneous speeds given. In the next section we will see situations in which this can be done.

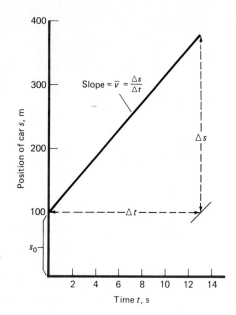

FIGURE 2.13 Position of a car moving with average speed $\bar{v}$ as a function of time. The car starts from a point $s_0 = 100$ m.

Example 2.1

In 1982 Franz Weber set the downhill speed skiing record at Velocity Peak in the San Juan Mountains of Colorado, traveling at an incredible speed of 56.428 m/s (126.24 mi/h). How long did it take Weber to cover the measured 100-m stretch over which his average speed was measured?

SOLUTION

We have for his average speed

$$\bar{v} = \frac{\Delta s}{\Delta t}$$

and so

$$\Delta t = \frac{\Delta s}{\bar{v}} = \frac{100 \text{ m}}{56.428 \text{ m/s}} = \boxed{1.77 \text{ s}}$$

When you consider that it takes a very fast sprinter about 10 s to run 100 m, you can see how fast this skier was traveling.

2.5 Acceleration

When a driver presses down on the gas pedal of a car, the car accelerates; i.e., its speed changes from zero to a large value in a short time. In this case of one-dimensional motion, by *acceleration* we mean the rate at which the car's speed changes with time. Just as the slope of a position-versus-time graph yields the speed v, so too the slope of a speed-versus-time graph yields the acceleration a. For example, Fig. 2.14 shows a plot of the speed of a car against time for the first 10 s of its motion. The graph in this case is a straight line; its slope is therefore constant and hence the acceleration is also constant. The magnitude of this constant acceleration is therefore given by the equation

Constant Acceleration

$$a_{\text{const}} = \frac{\Delta v}{\Delta t} = \frac{v_2 - v_1}{t_2 - t_1} \qquad (2.5)$$

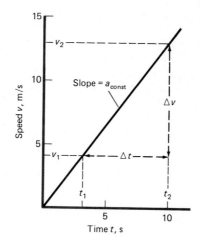

FIGURE 2.14 Graph of the speed of a car against time for the first 10 s of its motion.

Average Acceleration

If the acceleration is not constant, we must distinguish between the average acceleration and the instantaneous acceleration in the same way we distinguished between average and instantaneous speed. If the acceleration of a car is changing, say, because the driver is changing the pressure on the gas pedal, then the acceleration changes from instant to instant. The average acceleration can be found by taking the average change in the speed of the car during some chosen time interval. For example, Fig. 2.15 shows the speed v of a car as a function of the time t. The speed does not increase linearly with time, and hence the acceleration is not constant. The *average acceleration* of the car during the time interval $\Delta t = t_2 - t_1$ is simply

Average Acceleration

$$\bar{a} = a_{av} = \frac{\Delta v}{\Delta t} = \frac{v_2 - v_1}{t_2 - t_1} \qquad (2.6)$$

For example, if in the time interval between 5 and 10 s the speed of the car increased from 3.8 to 12 m/s, its average acceleration would be

$$\bar{a} = \frac{\Delta v}{\Delta t} = \frac{12.0 \text{ m/s} - 3.8 \text{ m/s}}{10 \text{ s} - 5 \text{ s}} = \frac{1.6 \text{ m/s}}{\text{s}} = 1.6 \text{ m/s}^2$$

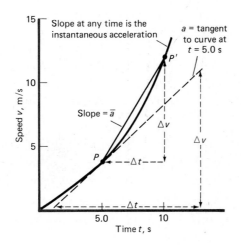

FIGURE 2.15 Average and instantaneous accelerations. The average acceleration in the interval between 5 and 10 s is the slope of the straight line $\bar{a}$. The instantaneous acceleration a is the tangent to the curve of speed against time at the point of interest.

The units for acceleration $\dfrac{m/s}{s}$ = m/s^2 (read either as "meters per second per second" or as "meters per second squared") can be easily understood. Since a speed is a rate of change of distance with time, units for time must occur in the denominator of the expression for a speed. But the acceleration is the rate at which the speed is changing in time, and so time occurs again in the denominator. Hence the SI unit for acceleration is (m/s)/s, or m/s^2.

Instantaneous Acceleration

The *instantaneous acceleration* at any instant during the 10-s interval is different from the average acceleration. It is found (as in the previous section) by taking the slope of the curve at the instant of time at which the acceleration is desired. This slope is given by:

Instantaneous Acceleration

$$a = a_{\text{instant}} = \lim_{\Delta t \to 0} \frac{\Delta v}{\Delta t} \qquad (2.7)$$

As we have seen in the preceding section, the slope is equal to the tangent to the curve at the point of interest. For example, in Fig. 2.15 the value of the tangent at P is

$$\frac{\Delta v}{\Delta t} = \frac{11 \text{ m/s} - 0 \text{ m/s}}{13 \text{ s} - 1 \text{ s}}$$

and so the acceleration is 0.92 m/s^2.

If the acceleration is constant, then $a = \bar{a} = a_{const}$.

Vector Nature of Acceleration

Thus far we have limited ourselves to motion in one dimension and have concentrated on the magnitude of the velocity, i.e., the speed, and ignored its direction. Even in one dimension, however, velocity is a vector, and so too is acceleration. Thus, if a car is moving in the positive X direction at 5 m/s, the driver can either speed up or slow down by using the gas pedal and the brakes. In the first case the vector acceleration **a** is in the direction of the vector velocity **v**, i.e., along the positive X axis; in the second case, i.e., that of braking the car, **a** is directed along the negative X axis. Hence in the first case the velocity is increased; in the second it is decreased.

For example, if the average acceleration is 2 m/s^2 and this acceleration lasts for 2 s, then from Eq. (2.6) the change in the speed is

$$\Delta v = \bar{a}\,\Delta t = \left(2\frac{m}{s^2}\right)(2 \text{ s}) = 4 \text{ m/s}$$

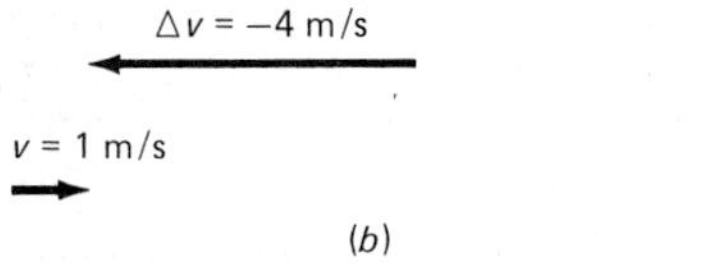

Suppose the original velocity was 5 m/s in the direction of the positive X axis. Then, if the acceleration is in the same direction as the velocity, the change in velocity $\Delta \mathbf{v}$ is in the same direction, and we have for the final velocity

$$\mathbf{v}' = \mathbf{v} + \Delta \mathbf{v} = 5 \text{ m/s} + 4 \text{ m/s} = 9 \text{ m/s}$$

Thus the final velocity is 9 m/s along the positive X axis.

On the other hand, if the car is slowed down by applying the brakes, and the (now negative) acceleration is again 2 m/s^2, we have for the final velocity

$$\mathbf{v}'' = \mathbf{v} - \Delta \mathbf{v} = 5 \text{ m/s} - 4 \text{ m/s} = 1 \text{ m/s}$$

$v_0 = 5$ m/s $\Delta v = \bar{a}\,\Delta t = 4$ m/s

$v = 9$ m/s

(a)

$v_0 = 5$ m/s

$\Delta v = -4$ m/s

$v = 1$ m/s

(b)

FIGURE 2.16 (a) Positive acceleration (in the direction of the velocity). (b) Negative acceleration (deceleration, in the direction opposite to that of the velocity).

This is illustrated in Fig. 2.16. Braking a car in this manner is known as *deceleration*.

Since acceleration is a vector, as this simple example shows and as will become clearer in subsequent chapters, a more general and accurate definition of acceleration should allow for the possibility of change in both the magnitude and the direction of the velocity. Hence:

Definition

Acceleration is the rate of change of velocity (either in its magnitude or its direction or both) with time.

Example 2.2

Figure 2.17 shows the position of a locomotive which moves along a perfectly straight railroad track at a speed that varies with time. Describe the motion of the locomotive, particularly emphasizing what it is doing at points A to G. Be as specific as possible about the motion.

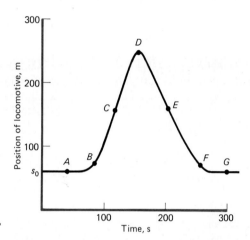

FIGURE 2.17

SOLUTION

At point A: The locomotive is at rest at a position on the track 60 m from the origin of the coordinate system. Hence $s_0 = 60$ m. The locomotive's position remains fixed as time changes.

B: The locomotive is accelerating and picking up speed in a forward direction. Its velocity is going from 0 to its value at point C.

C: The locomotive is moving at a constant velocity in the forward direction. The magnitude of this velocity is approximately $(200 - 100 \text{ m})/(130 - 100 \text{ s}) = 3.3$ m/s. Since the train went from rest to 3.3 m/s in about 40 s, as read from the graph, the average acceleration of the locomotive in the vicinity of B is about

$$\frac{3.3 \text{ m/s} - 0}{(100 - 60) \text{ s}} = 8.3 \times 10^{-2} \text{ m/s}^2$$

D: Here the locomotive stops at a position about 250 m from the origin and begins to move backward toward its original position.

E: The velocity is now negative, and is equal to $(120 - 220 \text{ m})/(225 - 173 \text{ s}) = -1.9$ m/s. Hence the locomotive moves at a slower speed in reverse than it did in a forward direction.

F: The locomotive is slowing down, i.e., decelerating.

G: The locomotive stops when it has returned to its original position on the track, i.e., when its position is 60 m from the origin.

Hence the overall motion of the locomotive described by the graph is that it starts from rest, reaches a constant forward speed of about 3.3 m/s, then slows down, stops, reverses its direction, reaches a constant reverse speed of about 1.9 m/s, decelerates, and finally stops at the same place where it started.

Example 2.3

A car is moving along at a constant speed of 20 m/s. The driver presses down on the accelerator and maintains a constant acceleration of 2.0 m/s² for 5.0 s. At the end of this time the driver hears a police siren and is worried that the police are after him for speeding.

(a) What was the car's final speed?
(b) Was the driver exceeding the speed limit of 55 mi/h?

SOLUTION

(a) The increase in speed of the car during the 5-s interval was

$$\Delta v = a\, \Delta t = \left(2.0\, \frac{\text{m}}{\text{s}^2}\right)(5.0\text{ s}) = 10\text{ m/s}$$

Since the original speed was 20 m/s, the car's final speed was

$$v = 20\text{ m/s} + \Delta v = 20\text{ m/s} + 10\text{ m/s} = \boxed{30\text{ m/s}}$$

(b) $55\text{ mi/h} = \left(55\, \frac{\cancel{\text{mi}}}{\cancel{\text{h}}}\right)\left(1.61\, \frac{\text{km}}{\cancel{\text{mi}}}\right)\left(\frac{1\, \cancel{\text{h}}}{3600\text{ s}}\right) = 25\text{ m/s}$

Hence at a speed of 30 m/s the driver was indeed breaking the speed limit. *Note:* To avoid repeatedly having to convert speeds in miles per hour to meters per second, it is much simpler to note that:

$$1\, \frac{\text{mi}}{\text{h}} = \left(1\, \frac{\cancel{\text{mi}}}{\cancel{\text{h}}}\right)\left(1.61\, \frac{\text{km}}{\cancel{\text{mi}}}\right)\left(\frac{1\, \cancel{\text{h}}}{3600\text{ s}}\right) = 0.447\text{ m/s}$$

and so $1\text{ m/s} = 2.24\text{ mi/h}$

Here are some useful conversion factors:

$$1\text{ mi/h} = 0.447\text{ m/s} = 1.61\text{ km/h}$$

$$1\text{ m/s} = 2.24\text{ mi/h} = 3.61\text{ km/h}$$

2.6 Uniformly Accelerated Motion in One Dimension

We now want to apply these ideas of velocity and acceleration to the motion of an object in a straight line, say, a train moving along a straight stretch of track. We assume that we are dealing with *uniformly accelerated motion*, which is defined as follows:

Definition

Uniformly accelerated motion: Motion in a straight line with an acceleration whose magnitude remains constant.

Since the magnitude of the acceleration is constant, the average and instantaneous accelerations are both equal to this constant acceleration. We want to find the position of the train and its speed at any given instant of time.*

If the acceleration is constant, the speed of the train changes linearly with time (i.e., in direct proportion to the time) from some initial value v_0 (which might be zero) to a final value v. Hence the average speed during the interval in which the acceleration is uniform is, according to the definition of an arithmetic average,

$$\boxed{\bar{v} = \frac{v_0 + v}{2}} \tag{2.8}$$

Let us now try to find the position of the train at time t. We cannot assume in general that the train starts from the origin of our coordinate system, and hence its position s will not necessarily be zero when t is zero. Instead we assume that at time $t = 0$, the train has an initial position s_0 and an initial speed v_0. It is then accelerated during the time t to a new speed v.

The final position s can be found from Eq. (2.4),

---

*Note that even though the motion is confined to one dimension by the track, the direction of motion is still important since the train can move either forward or backward on the track. For simplicity we will not use vector notation in this section but simply indicate the direction of the velocity and acceleration by + and − signs, where needed.

$$s = s_0 + \bar{v}t$$

where the average speed v is given by Eq. (2.8), and so

$$s = s_0 + \left(\frac{v_0 + v}{2}\right)t \tag{2.9}$$

Now the constant acceleration is given by Eq. (2.5), which in this case becomes

$$a = \frac{\Delta v}{\Delta t} = \frac{v - v_0}{t - 0}$$

and so
$$\boxed{v = v_0 + at} \tag{2.10}$$

On substituting this value of v in Eq. (2.9), we have

$$s = s_0 + \left(\frac{v_0 + v_0 + at}{2}\right)t$$

or
$$\boxed{s = s_0 + v_0t + \frac{at^2}{2}} \tag{2.11}$$

This is our perfectly general equation for the position of the train (or any other object) moving with uniform acceleration a, and with initial position s_0 and initial speed v_0 at time $t = 0$.

We have now developed most of the equations governing uniformly accelerated motion. Sometimes it is also important to have an equation connecting v and s in which the time does not explicitly appear.

To find this, we replace t in Eq. (2.4) by its value from Eq. (2.10), $t = (v - v_0)/a$, and obtain

$$s = s_0 + \bar{v}\left(\frac{v - v_0}{a}\right)$$

But, from Eq. (2.8), $\bar{v} = (v_0 + v)/2$. Hence

$$s - s_0 = \left(\frac{v + v_0}{2}\right)\left(\frac{v - v_0}{a}\right) = \frac{v^2 - v_0^2}{2a} \qquad \text{or}$$

$$\boxed{v^2 = v_0^2 + 2a(s - s_0)} \tag{2.12}$$

We have now found five equations governing uniformly accelerated motion in one dimension. They are

Equations for Uniformly Accelerated Motion (a = constant)

$s = s_0 + \bar{v}t$	(2.4)	$v = v_0 + at$	(2.10)
$s = s_0 + v_0t + \dfrac{at^2}{2}$	(2.11)	$\bar{v} = \dfrac{v_0 + v}{2}$	(2.8)
$v^2 = v_0^2 + 2a(s - s_0)$	(2.12)		

These five equations enable us to find the values of the position s and the speed v of an object moving in one dimension as a function of time t, if we know the constant acceleration a and the initial values of s_0 and v_0.

Note that Eq. (2.10) does not contain s, (2.4) does not contain a, (2.11) does not contain v, and (2.12) does not contain t. Hence it is always possible to select that one equation which relates a desired unknown quantity to quantities given in the statement of the problem.

Figure 2.18 shows how in uniformly accelerated motion the graph of position versus time departs from the straight line which would result if the speed were constant. It also shows the graph of velocity and acceleration versus time for uniformly accelerated motion.

FIGURE 2.18 Uniformly accelerated motion: (a) Position as a function of time. (b) Velocity as a function of time. (c) Acceleration as a function of time.

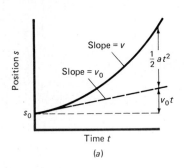
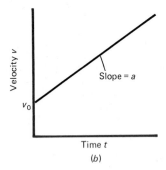
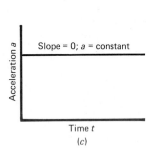

(a) (b) (c)

Nonuniformly Accelerated Motion

The above equations clearly do not apply when the magnitude of the acceleration is not constant in time. In such cases it is still possible to find the speed at the end of a time interval Δt, if the average acceleration is known over that time interval. For example, from Eq. (2.6) we have

$$\bar{a} = \frac{\Delta v}{\Delta t} = \frac{v - v_0}{t - 0}$$

so that $\quad v = v_0 + \bar{a}t \qquad\qquad (2.13)$

from which the final speed can be obtained.

In what follows we will assume that the acceleration is uniform unless the contrary is explicitly stated.

Figure 2.19 shows how the velocity depends on time for three different situations: no acceleration, uniform acceleration, and an acceleration increasing with time.

FIGURE 2.19 Velocity versus time for three different accelerations: (a) $a = 0$; (b) $a =$ constant (uniform acceleration); (c) nonuniform acceleration; a increases with time.

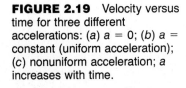
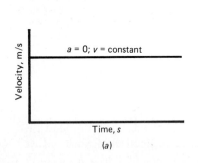
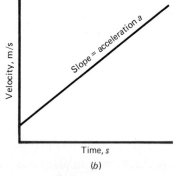
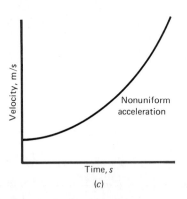

(a) (b) (c)

Example 2.4

The Chrysler G24 sports car is advertised as being able to accelerate from rest to 60 mi/h in 8.2 s.
(a) What is the car's acceleration (assuming that it is uniform) in meters per second squared?
(b) How far must the car travel before reaching 60 mi/h?

SOLUTION

(a) To obtain the acceleration in SI units, the final speed must first be converted to SI units. We have

$$v = (60 \text{ mi/h}) \left(0.447 \frac{\text{m/s}}{\text{mi/h}} \right) = 27 \text{ m/s}$$

The most useful equation for finding the acceleration will be one containing only one unknown a, and the three known quantities $v_0 \, (= 0)$, $v \, (= 27 \text{ m/s})$, and $t \, (= 8.2 \text{ s})$. This is $v = v_0 + at$ or

$$a = \frac{v - v_0}{t} = \frac{27 \text{ m/s} - 0}{8.2 \text{ s}} = \boxed{3.3 \text{ m/s}^2}$$

(b) The equation needed is one containing $s - s_0$, the distance traveled, and the three known quantities v_0, v, and a. This is

$$v^2 = v_0^2 + 2a(s - s_0)$$

Hence

$$s - s_0 = \frac{v^2 - v_0^2}{2a} = \frac{(27 \text{ m/s})^2 - 0}{2(3.3 \text{ m/s}^2)} = \boxed{110 \text{ m}}$$

The car will therefore reach a speed of 60 mi/h in a distance of a little over 100 m, about the length of an American football field.

Example 2.5

A car is moving along at a constant speed of 20 m/s. The driver presses down on the accelerator and maintains a constant acceleration of 2.0 m/s² for 5.0 s. How far does the car travel in the 5.0-s interval during which it is accelerated? Do this problem in as many different ways as possible to check your results.

SOLUTION

Here we have $v_0 = 20$ m/s, $t = 5.0$ s, $a = 2.0$ m/s², and $s - s_0 = ?$ Since $v = v_0 + at$, we find immediately that

$$v = 20\frac{\text{m}}{\text{s}} + \left(2.0\frac{\text{m}}{\text{s}^2} \right) (5.0 \text{ s}) = 30 \text{ m/s}$$

1 From Eq. (2.11) we have

$$s - s_0 = v_0 t + \frac{at^2}{2}$$

$$= \left(20\frac{\text{m}}{\text{s}} \right) (5.0 \text{ s}) + \left(2.0\frac{\text{m}}{\text{s}^2} \right) \frac{(5.0 \text{ s})^2}{2}$$

$$= 100 \text{ m} + 25 \text{ m} = \boxed{125 \text{ m}}$$

2 Another approach is to use Eq. (2.12). Then

$$s - s_0 = \frac{v^2 - v_0^2}{2a}$$

$$= \frac{(30 \text{ m/s})^2 - (20 \text{ m/s})^2}{2(2.0 \text{ m/s}^2)}$$

$$= \frac{500 \text{ m}^2/\text{s}^2}{4 \text{ m/s}^2} = \boxed{125 \text{ m}}$$

3 A third way is to calculate first the average speed for the 5.0-s interval, using Eq. (2.8):

$$\bar{v} = \frac{v_0 + v}{2} = \frac{20 \text{ m/s} + 30 \text{ m/s}}{2} = 25 \text{ m/s}$$

The distance traveled is then, from Eq. (2.4),

$$s - s_0 = vt = \left(25\frac{\text{m}}{\text{s}} \right) (5.0 \text{ s}) = \boxed{125 \text{ m}}$$

These different approaches to the solution of the same problem are introduced to indicate that there is more than one correct way to do most physics problems. One of the best approaches to problem solving is, after arriving at an answer by one method, to try an entirely different method and see if the same answer is obtained This approach will gradually build a mastery of problem-solving techniques, and instill confidence in your ability to solve complicated problems not merely in physics but in other fields as well.

Example 2.6

A particular make of small airplane must reach a ground speed of 50 m/s before it can become airborne. This same plane can accelerate at 10 m/s². What is the minimum length of a runway which will allow the plane to take off without accident, if it maintains this acceleration over its entire takeoff run?

SOLUTION

Here $v_0 = 0$, $v = 50$ m/s, and $a = 10$ m/s². We want to find $s - s_0$. The most useful equation for this uniformly accelerated motion is Eq. (2.12):

$$v^2 = v_0^2 + 2a(s - s_0)$$

from which

$$s - s_0 = \frac{v^2 - v_0^2}{2a} = \frac{(50 \text{ m/s})^2 - 0}{2(10 \text{ m/s}^2)} = \boxed{125 \text{ m}}$$

2.7 Free Fall and the Acceleration Due to Gravity

Clear ideas about uniformly accelerated motion originated with the great Italian scientist Galileo Galilei (see Fig. 2.20 and accompanying biography). Galileo applied his ideas to the free fall of objects on the surface of the earth and was able to show that, in the absence of air resistance, all objects should fall with the *same* acceleration. This means that the velocities and the distances traveled by all falling objects should increase in the same way at any particular place on the earth's surface. This was in direct contradiction to the teaching of Aristotle, who held that the velocity of freely falling objects should depend on their weights.

The constant acceleration with which all objects fall at the surface of the earth (when we neglect air resistance, as we do in what follows) is called the *acceleration due to gravity*, and is designated by the symbol g. Why g is a constant for every object at a given point on the earth's surface and why its value varies slightly with latitude and elevation will become clear when we discuss the nature of the gravitational force in Chap. 3. On the earth's surface g is approximately 9.80 m/s², as determined by a variety of laboratory experiments.

The fall of an object under gravity, with negligible air resistance, can be completely explained by the equations of the preceding section. Thus suppose a camera falls out of a Goodyear blimp at some height h above the earth, so that the initial downward velocity is $v_0 = 0$ and $a = g$. Here the best choice of a coordinate system is one with an origin at the height of the blimp, and all displacements, velocities, and accelerations taken as positive in a downward direction. (If you have any problem with this, consider motion along the positive X axis, for which all displacements, velocities, and accelerations are positive to the right. Then simply rotate the coordinate system clockwise through 90°.) In such a coordinate system Eq. (2.11) becomes

$$s = s_0 + v_0 t + \frac{gt^2}{2}$$

Here s_0 is assumed to be zero because it is at the origin of the coordinate system, and v_0 is also zero. Hence

$$s = \frac{gt^2}{2} \quad \text{or} \quad t = \left(\frac{2s}{g}\right)^{1/2}$$

Galileo Galilei (1564–1642)

FIGURE 2.20 Galileo Galilei.

Galileo Galilei was born in Pisa, Italy, on February 15, 1564. His father, Vincenzio Galilei, was an able mathematician and musician. Galileo studied Latin, Greek, and logic as a boy, and in 1581 entered the University of Pisa to study medicine. He found medicine little to his liking, however, and turned to mathematics. In 1583 he was forced to leave the university without a degree since his parents could no longer afford his tuition. This ended his formal education.

In 1589 a wealthy patron, the Marchese del Monte, recognized Galileo's mathematical abilities and had him appointed a lecturer in mathematics at the University of Pisa. Here he initiated experiments on the laws of motion of bodies, which led to results so contradictory to those of Aristotle that they aroused the strong antagonism of his Aristotelian colleagues at the university. He defended his ideas with such determination and argued with such vehemence that he earned the nickname of "the Wrangler." He remained at Pisa until 1592, when he was appointed to teach mathematics at the University of Padua. There he remained for 18 years (1592–1610), in what was to prove the happiest and most productive period of his life.

In 1609 Galileo heard of a Dutch lens grinder who had developed an optical instrument which made distant objects seem close at hand. He immediately set to work to construct one himself, using his knowledge of the refraction of light, and in a few days succeeded in combining two glass lenses in a tube to produce sights beyond his wildest dreams. He turned his "telescope" to the skies and discovered all sorts of wonderful things: the rough surface of the moon, the satellites of Jupiter, the phases of Venus, sunspots, the huge number of stars in the Milky Way. The publication of these results in his *Sidereus Nuncius* (1610) soon made him famous all over Europe.

Galileo's observations made him more and more certain that the Copernican theory, which hypothesized that the earth moves around the sun, was correct. In 1616 he was admonished by Cardinal Bellarmine not to uphold or teach Copernicanism as a "true" picture of the universe, although it was permissible as a mathematical hypothesis. In 1623 Galileo's old friend Cardinal Maffeo Barberini was elected Pope Urban VIII. This encouraged Galileo to believe that his ideas would eventually prevail, and in 1632 he published his major treatise on the Copernican system, the *Dialogue Concerning the Two Chief World Systems*.

After the *Dialogue* appeared, Galileo's world collapsed around him. Pope Urban VIII was furious, perhaps because advisers told him that Simplicio (the "simpleton") in the dialogue was really intended to be the Pope. Galileo was summoned to Rome and in April 1633 was made to stand trial before the Holy Office. In June it was decided that Galileo had rendered himself "vehemently suspect of heresy" because of his defense of Copernicanism. Galileo was forced to abjure publicly all beliefs and writings upholding the Copernican system, and was sentenced to imprisonment for the rest of his life. This was later commuted to permanent house detention at the home of his friend Ascanio Piccolomini, the Archbishop of Siena.

Later Galileo received permission to move to his villa at Arcetri near Florence. There he resumed his scientific work on mechanics, and in 1636 published what is probably his best work, *Two New Sciences*, which summarized most of his work on the physics of motion, his single most important contribution to physics. After completing this book his eyesight began to fail; in 1638 he went completely blind, and he died at Arcetri at age 78 on January 8, 1642, the same year in which Isaac Newton was born.

More important than any of Galileo's discoveries in physics was his approach to the study of nature, an approach which combined mathematics with experiment in a manner so in tune with physicists today that he is rightly considered the first modern physicist.

FIGURE 2.21 Acceleration, velocity, and position of a falling object as a function of time.

FIGURE 2.22 Flash photograph of a falling billiard ball. The ball is illuminated by rapid light flashes at equal time intervals of $\frac{1}{30}$ s as it falls freely. The position scale is in centimeters. The increased distances traveled in equal time intervals show that the motion is accelerated. Measurements on the photograph show that the acceleration is constant and equal to the acceleration due to gravity g. (*Courtesy of Educational Development Center, Newton, Mass.*)

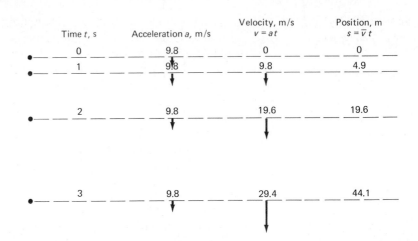

Time t, s	Acceleration a, m/s	Velocity, m/s $v = at$	Position, m $s = \bar{v}\,t$
0	9.8	0	0
1	9.8	9.8	4.9
2	9.8	19.6	19.6
3	9.8	29.4	44.1

This equation enables us to find the time t it takes the camera to reach the earth, which is a distance s below the blimp. When the camera hits the earth, its speed is given by Eq. (2.10) as

$$v = v_0 + at \qquad \text{or} \qquad v = 0 + gt$$

Hence, on putting in the above value of t, the speed of the camera at the earth's surface is found to be

$$v = +gt = g\left(\frac{2s}{g}\right)^{1/2} = +\sqrt{2gs}$$

The positive sign has been taken on the square root, since the motion is directed downward toward the earth. Because s is numerically equal to the initial height h, this can also be written as

$$v = \sqrt{2gh} \tag{2.14}$$

It may be helpful to diagram here, as is done in Fig. 2.21, the acceleration, velocity, and position of the camera as a function of time. As soon as the camera falls, i.e., as soon as the force holding it up is removed, its acceleration changes instantaneously from 0 to 9.8 m/s² and remains at that value until it strikes the earth. Hence its velocity increases by 9.8 m/s in a downward direction for each second of fall, leading to the velocities shown in the diagram. The positions at the ends of the 1-s intervals can be obtained by finding the average velocity $\bar{v} = (v_0 + v)/2$ during that interval, and multiplying by the time, to find the distance $s = \bar{v}t$ traveled during that interval. The resulting positions are also shown on the diagram. Positions measured experimentally agree well, within the limits of experimental error, with those predicted by the theory presented here, indicating the essential correctness of the theory. Figure 2.22 shows an actual flash photograph of a falling ball.

Since Eq. (2.10) does not include any property peculiar to the falling objects, all objects should fall with the same velocity at any point on the earth's surface, in agreement with Galileo's predictions. It is interesting to note that Galileo came to this conclusion long before Newton's time and hence long before anything was known about the gravitational attraction between objects.

Objects with Initial Upward Velocities

Consider a situation in which a baseball catcher throws a baseball high in the air and catches it when it returns to earth. The vertical motion of the ball in this case is again an example of uniformly accelerated motion under gravity. Since the vector nature of velocities and accelerations is particularly important in this situation where the ball moves both up and down, it is essential that we be clear about the coordinate system used and the sign convention adopted.

The best choice of a coordinate system in this case is one with origin at the height of the catcher's arm, so that the original position of the baseball can be assumed to be $s_0 = 0$ at that height. We choose a sign convention in which velocities and accelerations are positive when directed upward and negative when directed downward.

Then, using the equations for uniformly accelerated motion, we have for the position of the ball at any instant of time:

$$s = s_0 + v_0 t + \frac{at^2}{2} = 0 + v_0 t - \frac{gt^2}{2}$$

$$\text{or} \qquad s = v_0 t - \frac{gt^2}{2} \tag{2.15}$$

where the acceleration is negative because it is directed downward. Similarly,

$$v = v_0 + at = v_0 - gt \tag{2.16}$$

These two equations enable us to find s and v at any instant of time. If s should turn out to be negative, it means that the ball has fallen below the level of the catcher's arm. If v comes out negative, it means that the velocity is directed downward.

For example, to find the time at which the ball stops at the top of its flight, we can make use of the fact that $v = 0$ at this point. (Note that the acceleration a is *not* zero but equal to g at this point.) Hence

$$v = v_0 - gt = 0$$

$$\text{or} \qquad t = \frac{v_0}{g}$$

The height to which the ball rises before stopping is then

$$s = v_0 t - \frac{gt^2}{2} = v_0 \left(\frac{v_0}{g}\right) - \frac{g}{2}\left(\frac{v_0}{g}\right)^2 = \frac{v_0^2}{2g}$$

The ball then falls back to the earth from this height. This is illustrated in Fig. 2.23.

For the descent the ball falls to the ground from an initial height $h = v_0^2/2g$, where v_0 is the speed with which the ball was originally thrown upward. The ball's speed on reaching the ground is given by Eq. (2.14):

$$v = \sqrt{2gh} = \left(2g\frac{v_0^2}{2g}\right)^{1/2} = (v_0^2)^{1/2} = \pm v_0$$

The ball therefore returns to the catcher's mitt with the same speed with which it left the catcher's hand. We shall see later that the two speeds must be the same if mechanical energy is to be conserved in the process. The negative sign of the square root must be chosen in this case, since the ball is falling

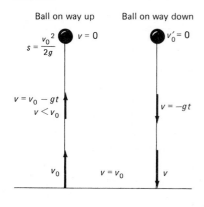

FIGURE 2.23 The motion of a ball thrown upward by a baseball catcher.

Ball on way up   Ball on way down

$s = \frac{v_0^2}{2g}$ $v = 0$ $v_0' = 0$

$v = v_0 - gt$
$v < v_0$ $v = -gt$

v_0 $v = v_0$ v

down, and the velocity is therefore negative, by assumption. The positive sign indicates that the initial velocity was $+v_0$ at this same height.

The same result can be obtained more easily from Eq. (2.12),

$$v^2 = v_0^2 + 2a(s - s_0) = v_0^2 - 2g(s - s_0)$$

When the ball returns to the ground, its position s is equal to its initial position s_0. Hence $s - s_0 = 0$ and $v^2 = v_0^2$, and so $v = \pm v_0$, as found above. Again the negative sign is the proper choice in this case.

In handling problems of this sort the sign convention used is extremely important. We are free to call downward displacements, velocities, and accelerations either positive or negative. The choice can be made in the way most convenient for solving a particular problem. But, once made, this choice must be maintained consistently throughout the working of the problem.

Example 2.7

A very good baseball pitcher can throw a baseball at speeds over 90 mi/h.
(a) If such a pitcher threw a ball at this speed straight up in the air, how high would it go, if we assume that we can neglect air resistance?
(b) With what speed would it return to the ground?

SOLUTION

(a) Here we have $v_0 = +90$ mi/h $= +40$ m/s, $a = -g = -9.8$ m/s², $v = 0$; $s - s_0 = h = ?$. Note that we have taken the upward direction as positive. The most useful equation in this case is Eq. (2.12):

$$v^2 = v_0^2 + 2a(s - s_0)$$

or $h = s - s_0 = \dfrac{v^2 - v_0^2}{2a} = \dfrac{0 - (40 \text{ m/s})^2}{2(-9.8 \text{ m/s}^2)} = \boxed{82 \text{ m}}$

This is the height to which the ball rises before it stops and falls back to earth.

(b) We have already seen that the ball will return to the ground with exactly the same speed with which it was thrown upward, in the absence of air resistance. Hence

$$v = \boxed{40 \text{ m/s}}$$

Example 2.8

With what speed must a rock be thrown upward if it is to return to the thrower in 5.0 s?

SOLUTION

Here we have $t = 5.0$ s, $a = -g = -9.8$ m/s², $s - s_0 = 0$; $v_0 = ?$. The equation containing all the known quantities and the one unknown v_0 is

$$s = s_0 + v_0 t + \frac{at^2}{2} \quad \text{or} \quad s - s_0 = 0 = v_0 t + \frac{at^2}{2}$$

from which

$$v_0 = \frac{-at}{2} = -\left(-9.8 \, \frac{\text{m}}{\text{s}^2}\right)\left(\frac{5.0 \text{ s}}{2}\right) = \boxed{25 \text{ m/s}}$$

(Try solving the same problem by finding the time for the rock to reach its highest point, and then the time for the rock to fall from this point to the ground, and adding the results.)

Example 2.9

A person throws a stone straight up in the air with a speed of 15 m/s.
(a) How long will it take for the stone to reach a point 10 m above the ground on its way down?
(b) How fast will the stone be moving at this point?

SOLUTION

Here we have $v_0 = +15$ m/s, $s - s_0 = 10$ m, $a = -9.8$ m/s^2; $t = ?$. Again the upward direction is taken as positive.

In this case trial and error shows that it is simpler to solve part (b) first, since part (a) leads to a quadratic equation which may take some time to solve.

(b) Using Eq. (2.12), which contains all the known quantities and the one unknown v, we have

$$v^2 = v_0^2 + 2a(s - s_0)$$

$$= \left(15\frac{m}{s}\right)^2 + 2\left(-9.8\frac{m}{s^2}\right)(10 \text{ m})$$

$$= 225\frac{m^2}{s^2} - 196\frac{m^2}{s^2} = 29\frac{m^2}{s^2}$$

and so $v = \pm5.4$ m/s

The plus sign applies to the motion in the upward direction when the stone is at the height of 10 m. Since we are interested in the downward motion at the same point, the correct answer is

$$v = \boxed{-5.4 \text{ m/s}}$$

(a) Now it is simple to solve this part of the problem. We have

$$v = v_0 + at \quad \text{ or}$$

$$t = \frac{v - v_0}{a} = \frac{-5.4 \text{ m/s} - 15 \text{ m/s}}{-9.8 \text{m/s}^2} = \boxed{2.1 \text{ s}}$$

2.8 Useful Techniques in Solving Problems

There are a great variety of practical problems that can be solved using the basic ideas and equations of this chapter, as illustrated in the preceding examples. In solving physics problems like these as well as a great variety of other problems, the following techniques are particularly useful.*

The Conversion of Units

The units in which a physical quantity is measured are completely arbitrary. We could, for example, decide to measure the length of a fence in "pickets," where a picket is the arbitrarily chosen length of one of the upright posts of the fence. In principle we are free to choose any convenient system of units so long as we use this system consistently.

In this book we use SI units, with distances in meters, masses in kilograms, and time in seconds. For this reason the SI system is sometimes called the mks (meter-kilogram-second) system as opposed to the older metric cgs (centimeter-gram-second) system. The derived units of mechanics can then all be expressed in combinations of these three basic units.

Sometimes it is necessary to convert nonmetric units such as feet and miles into SI units, as has been done earlier in this chapter. Take, for example, a distance of 50 mi. According to Table D (inside back cover), 1 mi is equal to 1.61 km, or 1.61×10^3 m. Hence

$$50 \text{ mi} = (50 \text{ mi}) \left(\frac{1.61 \times 10^3 \text{ m}}{1 \text{ mi}}\right) = 8.1 \times 10^4 \text{ m}$$

Note how in converting from miles to meters we have multiplied by a factor $(1.61 \times 10^3 \text{ m})/(1 \text{ mi})$ which is equal to 1, since numerator and denominator are equal. Units are treated like any other algebraic factors in multiplication and division. We can cancel similar units in the numerator and denominator and obtain our answer in the desired units (meters), as above. In this section lines will be drawn through the units which cancel, in the same way that equal numerical factors in the numerator and denominator of fractions can be

---

*See also Appendix 3: Solving Physics Problems.

canceled. In the rest of the book the units will be written at every step of a problem for clarity, but it is assumed that you can cancel the units for yourself without having the cancellation explicitly indicated.

Using this technique and a table of conversion factors such as Table D, converting units becomes a straightforward process. For example, suppose we want to convert a speed of 55 mi/h into meters per second. We have

$$55 \text{ mi/h} = \left(55 \frac{\text{mi}}{\text{h}}\right)\left(\frac{1.61 \times 10^3 \text{ m}}{1 \text{ mi}}\right)\left(\frac{1 \text{ h}}{3600 \text{ s}}\right) = 2.46 \times 10^1 \frac{\text{m}}{\text{s}} = 25 \frac{\text{m}}{\text{s}}$$

We can be sure that this answer is correct if the original units cancel out as they should and the answer is expressed in the desired units. Sometimes confusion can arise about whether a conversion factor should be used as a multiplier or a divisor. If the units are explicitly written out and the units do not cancel as they should, then the conversion factor is being used improperly.

The convenience of the SI system is that all commonly used units of measurement are equal to powers of 10 times the basic units such as the meter or kilogram. Commonly used prefixes for powers of 10 in the SI system are given in Table A1 (inside front cover). These prefixes are used not merely for lengths but for masses, energies, and all other physical quantities.

Dimensions

The physical length of a fence remains the same no matter what units are used to measure this length. The property of the fence described by the word *length* we call a *dimension*. When we are talking about this physical property in the abstract without assigning a numerical value to it, we may put square brackets around the symbol and write, for example, $[L]$ to mean the dimension of length. Clearly the width of an object is physically the same kind of a quantity as length, and so is its height. Hence any area has the dimensions $[L] \times [L] = [L^2]$, and all volumes have the dimensions $[L^3]$.

In addition to length, the other two fundamental physical quantities in mechanics are time $[T]$ and mass $[M]$. Length, mass, and time are very different kinds of physical quantities. They have different dimensions, and hence different units; lengths cannot be measured in seconds, nor masses in meters.

The other derived quantities of mechanics—velocity, acceleration, force, energy, power—are all combinations of the three fundamental dimensions of mass, length, and time. For example, velocity is dimensionally the ratio of a distance to a time, and hence has the dimensions $[L]/[T]$. Similarly acceleration has the dimensions $[L]/[T^2]$, since it is the rate of change of velocity with time.

Numerical constants such as π, which is the ratio of one length (the circumference of a circle) to another length (the diameter), have no dimensions and are referred to as *dimensionless constants*.

Dimensional Analysis

The usefulness of introducing dimensions becomes immediately clear when we consider how they may be used to check the validity of calculations or equations, and even to predict the functional dependence of complicated physical properties on simpler properties of a system. The technique used to accomplish these goals is called *dimensional analysis*.

Suppose you are given a very involved problem to solve in which you are asked to find the velocity of an object at the end of a certain period of time during which specified forces act on it. You set up the problem, solve it, and conclude that the velocity is 64 m/s² in some specified direction. Now this is certainly wrong, since the answer has the dimension $[L]/[T^2]$, which, as we have seen, is proper to an acceleration, not to a velocity. Hence the answer is dimensionally incorrect. There is no point in checking your arithmetic to see if the numerical answer is right, since the dimensions indicate that some more serious error has been committed in solving the problem. *In doing all problems in this book check your answers to see if they are at least dimensionally correct.* If they are not, you should redo the problem from the beginning.

A second use of dimensional analysis is to check whether an equation being used is dimensionally correct. In this case all terms on both sides of the equation must have the same dimensions. Otherwise they cannot be added, just as apples and oranges cannot be added. For example, suppose that during a physics examination, you remember an equation $a = g + 2vt$ and want to test it for correctness. Here a and g are accelerations and v is a velocity. You write a dimensional equation corresponding to the equation above by replacing each physical quantity by its proper dimension, ignoring all constants, since they have no dimensions. You obtain

$$\frac{[L]}{[T^2]} = \frac{[L]}{[T^2]} + \frac{[L]}{[T]}[T]$$

or

$$\frac{[L]}{[T^2]} = \frac{[L]}{[T^2]} + [L]$$

This equation is dimensionally incorrect and cannot be used, since the third term has different dimensions from the first two. A dimensional check such as this will only tell you when an equation is wrong; it will never tell you that it is right, since it can never yield the numerical factors in the equation.

Another, more advanced, use of dimensional analysis is to predict the functional dependence of important physical properties of a system on measurable properties of the system. For example, dimensional analysis can show that the dependence of the period of a pendulum on its length l and on the acceleration due to gravity g is $T \propto \sqrt{l/g}$, as we will see in Sec. 11.3. But it will never lead to $T = 2\pi\sqrt{l/g}$, which is the fully correct formula. For this, a more complete application of the laws of mechanics or an experimental measurement is needed.

Example 2.10

Use the conversion factors given in Table D to convert:
(a) 20 horsepower (hp) into watts (W)
(b) 170 kilocalories (kcal) into joules (J)
(c) 20 gallons (gal) into cubic meters (m³)

SOLUTION

The fact that we have not yet discussed any of these units should present no problem here. The conversion of units is a routine process which can be carried out even if the units being converted are completely new.

(a) From Table D, 1 hp = 746 W. Hence

$$20 \text{ hp} = (20 \text{ hp})\left(\frac{746 \text{ W}}{1 \text{ hp}}\right) = \boxed{1.5 \times 10^4 \text{ W}}$$

(b) In the same way,

$$170 \text{ kcal} = (170 \text{ kcal})\left(\frac{4184 \text{ J}}{1 \text{ kcal}}\right) = \boxed{7.11 \times 10^5 \text{ J}}$$

(c) $20 \text{ gal} = (20 \text{ gal})\left(\frac{3.785 \times 10^{-3} \text{ m}^3}{\text{gal}}\right) = \boxed{7.6 \times 10^{-2} \text{ m}^3}$

Note that we have consistently limited the number of significant figures in the answers to the number of significant figures in the least well known factor entering into these computations.

Example 2.11

Prove that the following equation for a simple pendulum is dimensionally correct:

$$T = 2\pi \sqrt{\frac{l}{g}}$$

where l is the length of the pendulum, g the acceleration due to gravity, and T the time for one complete vibration of the pendulum.

SOLUTION

The left side of this equation has the dimension of time $[T]$. The right side has the dimensions

$$\left[\frac{[L]}{[L]/[T^2]}\right]^{1/2} = \frac{[L^{1/2}]}{[L^{1/2}]}[T] = [T]$$

Hence both sides have the same dimensions, and the equation is dimensionally correct.

Summary: Important Definitions and Equations

Frame of reference: A set of coordinate axes in terms of which the position of objects in space may be specified.

Scalar: A physical quantity which can be completely specified merely by giving its magnitude.

Vector: A physical quantity which requires both a magnitude and a direction for its complete specification.

Uniformly accelerated motion: Motion in a straight line with an acceleration whose magnitude remains constant. Equations governing uniformly accelerated motion:

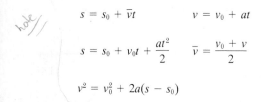

$$s = s_0 + \bar{v}t \qquad\qquad v = v_0 + at$$

$$s = s_0 + v_0 t + \frac{at^2}{2} \qquad \bar{v} = \frac{v_0 + v}{2}$$

$$v^2 = v_0^2 + 2a(s - s_0)$$

Dimension: A basic physical property of an object (such as length or mass) in the abstract, independent of any units or numerical value being assigned to it.

Dimensional analysis: A technique for checking the validity of a solution to a problem, a unit, or an equation for correctness and consistency by dealing only with dimensional factors and neglecting numerical values.

Units: Standard quantities like the second, the kilogram, and the meter in terms of which physical properties of objects can be measured and assigned numerical values.

SI units (Système International d' Unités): The internationally adopted system of metric units used for scientific purposes. The basic SI units in mechanics are the meter, the kilogram, and the second. Table 2.2 shows how the SI units are used in kinematics. A complete list of SI units may be found in Table B (inside front cover).

TABLE 2.2 SI Units Used in Kinematics

Physical quantity	Dimension	SI unit
Fundamental		
Length	$[L]$	meter (m)
Mass	$[M]$	kilogram (kg)
Time	$[T]$	second (s)
Derived:		
Velocity: velocity is a vector, with both magnitude (the speed) and direction.	$[L]/[T]$	meters per second (m/s)
$\mathbf{v}_{av} = \bar{\mathbf{v}} = \dfrac{\text{total displacement}}{\text{total time elapsed}} = \dfrac{\Delta \mathbf{s}}{\Delta t}$		
$v_{av} = \bar{v} = \dfrac{\text{total distance covered}}{\text{total time elapsed}} = \dfrac{\Delta s}{\Delta t}$		
$v_{instant} = v = \lim\limits_{\Delta t \to 0} \dfrac{\Delta s}{\Delta t}$		
Acceleration: acceleration is a vector, with both magnitude and direction. It is the rate of change of velocity (either in its magnitude or its direction) with time.	$[L]/[T^2]$	meters per second squared (m/s²)
$a_{av} = \bar{a} = \dfrac{\Delta v}{\Delta t}$		
$a_{instant} = \lim\limits_{\Delta t \to 0} \dfrac{\Delta v}{\Delta t}$		

Questions

1 Is it possible to measure masses to the same accuracy as lengths and times? Why?

2 One automobile increases its speed from 5.0 to 10 m/s. A second automobile increases its speed from 20 to 25 m/s in the same time. Is the acceleration the same in these two cases?

3 Must the velocity of a moving object always be in the same direction as its acceleration? Give some examples to confirm your answer.

4 Can the velocity of an object be zero when its acceleration is not zero? If so, give an example to illustrate your point.

5 A car driving along a straight highway slows down as it enters a small town. It stops for the one traffic light in the town, then gradually picks up speed, and finally resumes its original highway speed when it leaves the town. Draw graphs, one above the other, of the car's position, its velocity, and its acceleration, as a function of time.

6 (*a*) Draw a velocity-versus-time graph for a ball thrown straight up in the air and caught on its return to the ground.

(*b*) Draw an acceleration-versus-time graph for the same motion of the ball.

7 If you have a displacement-versus-time graph for a moving particle, is it possible to use this graph to obtain a velocity-versus-time graph, and then to use this velocity-versus-time graph to obtain a graph of the acceleration as a function of time? How would you proceed to do this?

8 Is it possible for an object to have a downward acceleration greater than 9.8 m/s²? If so, what would be required to accomplish this?

9 A person throws a tennis ball straight up in the air and catches it again when it comes down. What is the average velocity of the ball during the up-and-down trip? What is its average speed for the trip?

10 What is the basic difference between the dimension of a physical quantity and the units in which this quantity is measured?

11 Is it possible for an equation to be dimensionally correct and still be wrong? If so, indicate a number of ways in which this might happen.

Problems

In this and subsequent chapters the end-of-chapter problems are broken down into four kinds, indicated by the letters **A**, **B**, **C**, and **D**. These four kinds are the following:

A *Answers given* (indicated by A1, A2, etc.): Multiple-choice questions, mostly with numerical answers, including some order-of-magnitude calculations (see Appendix 3C).

B *Basic problems* (indicated by B1, B2, etc.): Problems involving little more than substitution in the formulas developed in the chapter.

C *Common, or standard, problems* (indicated by C1, C2, etc.): Problems involving the ability to reason, combine equations and ideas, and apply important ideas developed in the chapter, but without excessive difficulty.

D *Difficult problems* (indicated by D1, D2, etc.): More sophisticated problems which require somewhat more involved reasoning or deeper understanding of the physics involved. These problems can be done using only the contents of the book up to the chapter involved, together with algebra and trigonometry, but they are not easy. Some are derivations requiring considerable use of algebra; others require the relation of material in the chapter under study to earlier chapters in the book.

Before tackling these problems you are urged to read over Some General Hints on Problem Solving in Appendix 3A.

A 1 The order-of-magnitude of the thickness of a penny is:
(*a*) 10^{-4} m   (*b*) 10^{-3} m   (*c*) 10^{-2} m
(*d*) 10^{-1} m   (*e*) 10^{-5} m

A 2 The number of seconds in a year is, as to order of magnitude,

(*a*) 10^7 (*b*) 10^5 (*c*) 10^3 (*d*) 10^4 (*e*) 10^6

A 3 The time it takes for light, traveling at a speed of 3.0×10^8 m/s, to cover a distance of 300 m is:
(*a*) 10^6 s   (*b*) 10^{-6} s   (*c*) 10^{11} s
(*d*) 10^{-11} s   (*e*) None of the above

A 4 A speed of 40 mi/h is equivalent to:
(*a*) 34 m/s   (*b*) 10 m/s   (*c*) 340 m/s
(*d*) 18 m/s   (*e*) 25 m/s

A 5 A car accelerates from rest to 60 mi/h in 6.0 s. Its acceleration in SI units is:
(*a*) 4.5 m/s   (*b*) 4.5 m·s²   (*c*) 27.0 m/s²
(*d*) 9.0 m/s²   (*e*) 4.5 m/s²

A 6 A sailboat starts from rest and accelerates uniformly to a speed of 10 knots (a knot is a speed of 1 nautical mile per hour, or 6067 ft/h) in 5 min. Its average speed during the 5.0-min period was:
(*a*) 2.6 m/s   (*b*) 5.2 m/s   (*c*) 1.8×10^3 m/s
(*d*) 0.26 m/s   (*e*) 0.52 m/s

A 7 A car traveling with an initial speed v_0 is stopped by the driver applying its brakes. The deceleration during the time t is constant. The distance covered by the car during the time t is:
(*a*) $v_0 t$ (*b*) $v_0 t^2/2$ (*c*) $3v_0 t/2$ (*d*) $v_0 t/2$
(*e*) $v_0/2t$

A 8 A car starts from rest and accelerates uniformly for 10 s. The distance traveled in the 10 s is greater than the distance traveled in the first second by a factor of:
(*a*) 5   (*b*) 50   (*c*) 25   (*d*) 10   (*e*) 100

A 9 A sailboat is moving at a speed of 3.0 m/s. A strong wind blows up and accelerates the boat with a constant acceleration of 0.20 m/s² for 10 s. The final speed of the sailboat at the end of the 10-s period is:

(a) 30 m/s   (b) 5.0 m/s   (c) 10 m/s
(d) 5.0 m/s²   (e) 10 m/s²

A10 A stone is dropped from a cliff 200 m high. When the stone is 100 m above the ground, its downward acceleration is:

(a) 0   (b) 9.8 m/s²   (c) 19.6 m/s²
(d) 100 m/s²   (e) 9.8 m/s

A11 A flowerpot falls from a third-story window at a height of 10 m above the ground. When it passes a first-story window 2 m above the ground, its speed is:

(a) 9.8 m/s   (b) 8.0 m/s   (c) 13 m/s
(d) 13 m/s²   (e) 1.6 × 10² m/s

B 1 Express the following distances in meters:
(a) The length of a 100-yd dash
(b) The length of a marathon run (26 mi, 385 yd)
(c) The distance from Seattle, Washington, to Miami, Florida (3273 mi).

B 2 In 1978 Bill Rodgers won the 26.2-mi Boston Marathon in a time of 2 h, 10 min.
(a) What was his average speed in meters per second?
(b) What was his average time for 1 km?

B 3 A quarter-miler runs one lap around a 400-m track in 65 s.
(a) What was the runner's average speed?
(b) What was her average velocity?

B 4 The stoplights on the north-south avenues in New York City are designed to keep traffic moving at 30 mi/h (13.4 m/s). The average length of a block between traffic lights is about 80 m. What must be the time delay between green lights on successive blocks to keep the traffic moving continuously?

B 5 It is believed that the universe began with a big bang about 20 billion years ago. Since light travels faster than anything else in the universe, the farthest that the cosmic debris produced in the big bang could have spread by now must be less than the distance light could have traveled in 2.0 × 10¹⁰ years. If the speed of light is 3.0 × 10⁸ m/s, what is the maximum possible radius of the universe at the present?

B 6 Good baseball pitchers can throw a baseball with speeds of more than 90 mi/h. (On August 20, 1974, Nolan Ryan threw a pitch which was timed at 100.9 mi/h, the fastest ever recorded.) If the distance from the pitcher's mound to home plate is about 60 ft:
(a) How long does it take the ball moving at 90 mi/h to make the trip from pitcher to catcher?
(b) If the batter has to make contact with the ball during the 2 ft or so of its path when it is approximately over home plate in order to hit the ball into fair territory, how much time does the batter really have to make contact with the ball? (This may give you some evidence for the opinion of sportswriters that the single most difficult feat in all sports is to hit a baseball cleanly with a bat, especially when the ball is thrown by a good pitcher.)

B 7 A window washer drops a brush from a scaffold on a tall office building.
(a) What is the speed of the falling brush at the end of 3.0 s?
(b) How far has it fallen in that time?

B 8 A car starts from rest and accelerates for 5.0 s with an acceleration of 2.0 m/s². How far does it travel during these 5.0 s?

B 9 A car is moving at a constant speed of 15 m/s when the driver presses down on the gas pedal and accelerates for 10 s with an acceleration of 1.0 m/s². What is the average speed of the car during the 10-s interval?

B10 A ball is moving upward at a speed of 8.0 m/s. At the end of 3.0 s, what will be its velocity (both magnitude and direction), if we neglect any air resistance?

B11 A wrench falls out of the gondola of a balloon which is 500 m above the ocean. What will its speed be just before it hits the water, if we neglect air resistance?

B12 A girl throws a frisbee by accelerating it, while it is held in her hand, through a distance of 1.3 m before releasing it. If the frisbee leaves the girl's hand at a speed of 3.0 m/s, what acceleration was imparted to the frisbee during the throwing motion?

B13 A reasonable value for the deceleration of a skidding car (with brakes locked) on a dry pavement is about 7.0 m/s². How long will it take for a car going 25 m/s to come to rest, once the brakes are engaged?

B14 The elevators in the John Hancock building in Chicago move 900 ft in 30 s.
(a) What is the speed in meters per second?
(b) What is this speed in miles per hour?

B15 During a baseball game being televised from Japan it was announced that the pitcher for the Yokohama Whales threw a pitch at a speed of 138 km/h. What was this speed in miles per hour? in meters per second?

B16 Measure the length of the line below in meters, centimeters, millimeters, kilometers, feet, and inches.

←——————————————————————→

B17 For the falling ball shown in Fig. 2.22 calculate for the $\frac{1}{30}$-s time intervals shown the following quantities: the displacement of the ball during each successive $\frac{1}{30}$ s, the average speed during the time interval, the change in the average speed from one time interval to the next. Use these data to obtain the best possible average value for the acceleration due to gravity g. What is your value for g?

C 1 The Pepsi Challenge 10,000-m national championship road race was held in New York City in July 1982. The winner in the men's division was Rod Dixon of New Zealand with a time of 28 min, 12.3 s. The women's division was won by Ellen Hart of Boulder, Colorado, with a time of 32 min, 59.7 s. What is the difference in meters per second between the speeds of Dixon and Hart?

C 2 A train starts from rest with a uniform acceleration of 0.20 m/s².
(a) How fast is the train moving after 1.0 min?
(b) At what time during the first minute was its instantaneous speed equal to its average speed?

C 3 A racing car is moving along at a constant speed of 50 m/s. To pass the lead car the driver presses down on the accelerator and maintains a constant acceleration of 3.0 m/s² for 5.0 s.
(a) At the end of the 5.0 s, what is the car's speed?
(b) During the 5.0 s, how much ground does the car cover?

C 4   One racing car is 1 lap behind the lead car with 50 laps to go in a race. If the speed of the lead car is 60 m/s, what must be the average speed of the second car to catch the lead car before the end of the race, if 1 lap = 1.5 km?

C 5   A skier starts from rest, accelerates uniformly, and slides 10 m down a slope in 5.0 s.
(a)   What is the skier's acceleration?
(b)   At what time after starting will the skier acquire a speed of 20 m/s?

C 6   A train is moving at a speed of 80 km/h when the engineer applies the brakes and the train decelerates at a rate of 2.0 m/s². How long does it take for the train to travel the next 100 m after the brakes are applied?

C 7   A Washington, D.C., subway train leaves a metro station with a constant acceleration of 2.5 m/s². Find (a) its speed at the end of 5.0 s; (b) its average speed during the 5.0-s interval; (c) the distance it travels in the 5.0 s.

C 8   An express train traveling at a speed of 15 m/s passes through a subway station without stopping. It constantly accelerates between this station and the next station, and its average velocity for the trip between the two stations is 20 m/s. How fast is the train going when it passes the second station?

C 9   In a TV set the electrons which produce the TV picture are accelerated by being subjected to a very large electric force as they pass through a small region of space. If such space is 1.5 cm in length, and the electrons enter with a speed of 1.0×10^5 m/s and leave with a speed of 1.5×10^8 m/s:
(a)   What is their acceleration over this 1.5 cm length?
(b)   How long does the acceleration take?

C10   Figure 2.24 is the graph of the position of a train on a straight track as a function of time. Calculate approximate values of the speed of the train at points A, B, C, and D.

C11   Figure 2.25 is the graph of the speed of a train on a straight track as a function of time. Calculate approximate values of the acceleration of the train at points A, B, C, and D.

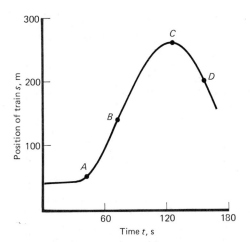

FIGURE 2.24   Diagram for Prob. C10.

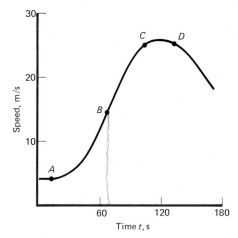

FIGURE 2.25   Diagram for Prob. C11.

C12   A train is moving along a straight track from Washington, D.C., to Baltimore. When it is 1.0 km from the Washington station, it is moving with a speed of 10 m/s. It then accelerates uniformly at a rate of 0.10 m/s² for 5.0 min.
(a)   What is the train's final speed?
(b)   How far is it from the Washington station at the end of the 5.0-min time interval?

C13   A car initially traveling at a constant speed accelerates at the rate of 1.0 m/s² for 15 s. If the car travels 200 m during this 15-s interval, what was the speed of the car when it started to accelerate?

C14   A racing car can accelerate from rest to 45 m/s in 6.0 s.
(a)   If we assume that the acceleration is constant, what is the magnitude of this acceleration?
(b)   How far must the racing car travel before reaching 45 m/s?

C15   An airport has runways only 100 m long. A small plane must reach a ground speed of 40 m/s before it can become airborne. What average acceleration must the plane's engines provide if it is to take off safely from this airport?

C16   A Boeing 747 jet can accelerate at 4 m/s². If it requires 40 s to acquire the speed necessary to become airborne,
(a)   What is its takeoff speed?
(b)   How long must the airport runway be?

C17   An engineer is driving a locomotive along a railroad track and sees a car stuck on the track at a railroad crossing in front of the train. When the engineer first sees the car, the locomotive is 200 m from the crossing and moving at a speed of 30 m/s. If the engineer's reaction time is 0.60 s, how rapidly will the locomotive have to decelerate to avoid an accident?

C18   Check by your own calculations the values of the acceleration, velocity, and position of a falling object shown in Fig. 2.21.

C19   A person drops a stone from the top of a building, and it takes 5.1 s for the stone to reach the ground. How tall is the building?

C20 A baseball player throws a ball straight up in the air with an initial speed of 40 m/s.

(*a*) How high does the ball go?

(*b*) What is its speed just before it hits the ground on its way down?

(*c*) How long was it in the air?

C21 Show, by dimensional analysis, that the following equations are dimensionally correct:

(*a*) $s = s_0 + v_0 t + \dfrac{a\,t^2}{2}$

(*b*) $v^2 = v_0^2 + 2a(s - s_0)$

C22 The equation for the distance d_S required for a car to stop after the driver is alerted to danger ahead is:

$$d_S = v_0 t_R - \frac{v_0^2}{2a}$$

where v_0 is the initial speed of the car, t_R is the reaction time of the driver, and a is the acceleration. Prove that this equation is dimensionally correct.

C23 A baseball is thrown horizontally at a large gong 40 m away. The gong is heard to sound 4.0 s later. The speed of sound in air is about 330 m/s. With what speed was the ball thrown?

C24 The Russian juggler Sergei Ignatov can juggle 11 hoops at one time. To do this he must throw the hoops high into the air to allow himself time to catch and throw aloft each of the 11 hoops in turn. Show that doubling the height reached by a hoop from 5.0 to 10 m only increases the time required for a hoop to leave the juggler's hand and return again by about 40 percent.

D 1 Two girls, Mary and Jane, are racing each other on foot. Jane starts 10 m behind Mary, but runs at an average speed of 8.0 m/s, whereas Mary runs at a speed of only 7.0 m/s.

(*a*) How far will Jane have to run before she catches up with Mary?

(*b*) How long will it take Jane to catch Mary?

D 2 A police car is at rest beside a road, looking for speeders. A car passes, traveling at a constant speed of 29 m/s. The police car starts up 4.0 s later, and accelerates from rest with an acceleration of 1.8 m/s². If the police car maintains this acceleration until it reaches the speeding car:

(*a*) How far would the police car have to travel before catching the speeding car?

(*b*) How fast would the police car be traveling when it caught the car?

D 3 A train moving at 20 m/s decelerates uniformly to a speed of 12 m/s in a time of 4.0 s.

(*a*) Determine the acceleration of the train.

(*b*) Determine the distance it moves during the third second of the 4.0-s interval.

D 4 Prove that the equation for the stopping distance of a car is

$$d_S = v_0 t_R - \frac{v_0^2}{2a}$$

where v_0 is the initial speed of the car, t_R is the reaction time of the driver, and a is the acceleration provided by the car's brakes.

D 5 A rock is dropped from a cliff on the rockbound coast of Maine and is heard to hit the water 2.0 s later. If the speed of sound is 330 m/s, how high is the cliff?

D 6 A person leans out a window of an apartment house and throws a stone straight up in the air at a speed of 4.0 m/s. The stone rises, stops, and falls to the courtyard below the window, which is 20 m above the ground.

(*a*) What is the speed of the stone when it reaches the ground?

(*b*) How much time elapses from the moment the person throws the stone until it hits the ground.

D 7 A camera falls from a Goodyear blimp which is 500 m above the ground, and rising at a speed of 10 m/s. Find (neglecting air resistance):

(*a*) The maximum height reached by the camera with respect to the ground.

(*b*) Its position and velocity 5.0 s after it leaves the blimp.

(*c*) The time it takes to reach the ground.

(*d*) The downward speed with which it strikes the ground.

D 8 A stone is dropped from a bridge at a point 50 m above a river, and 1 s later another stone is thrown downward after the first one. Both stones hit the water at exactly the same time. What was the initial speed of the second stone?

D 9 The acceleration due to gravity on the moon is about 1/6 its value on earth, i.e., about $g/6$. If a baseball reaches a height of 50 m when thrown upward by someone on the earth, what height would it reach when thrown in the same way on the surface of the moon?

D10 A ball is thrown straight up with an initial speed v_0 from a point s_0 m above the ground. Show that the time required for the ball to reach the ground is:

$$t = \frac{v_0}{g}\left[1 + \left(1 + \frac{2s_0 g}{v_0^2}\right)^{1/2}\right]$$

Additional Readings

Astin, Allen V.: "Standards of Measurement," *Scientific American*, vol. 218, no. 6, June 1968, pp. 50–62. A very informative discussion of the standards used for length, mass, time, and temperature.

Butterfield, Herbert: *The Origins of Modern Science*, rev. ed., Free Press, New York, 1965. This stimulating book is the source of the quotation used at the beginning of this chapter.

Cohen, I. Bernard: *The Birth of a New Physics*, Doubleday Anchor, Garden City, N.Y., 1960. An account of the development of physics from Copernicus to Newton by a well-known historian of science.

De Santillana, Georgio: *The Crime of Galileo*, University of Chicago Press, Chicago, 1955. An interesting discussion of Galileo's problems with the Roman Inquisition. Santillana draws some intriguing parallels between the Galileo case and the case of J. Robert Oppenheimer.

Geymonat, Ludovico: *Galileo Galilei*, McGraw-Hill, New York, 1965. A very good brief biography of Galileo.

Gingerich, Owen: "The Galileo Affair," *Scientific American*, vol. 247, no. 2, August 1982, pp. 132–143. A balanced account of Galileo's confrontation with the church.

Jespersen, James, and Jane Fitz-Randolph: *From Sundials to Atomic Clocks: Understanding Time and Frequency*, U.S. Government Printing Office, Washington, D.C., 1977. Published by the National Bureau of Standards, a good popular account of frequency, time, and clocks.

Lyons, Harold: "Atomic Clocks," *Scientific American*, vol. 196, no. 2, February 1957, pp. 71–82. An account of the development of atomic clocks by one of the pioneers in the field.

McCloskey, Michael: "Intuitive Physics," *Scientific American*, vol. 248, no. 4, April 1983, pp. 122–130. Tests show that the misconceptions of people, including physics students, about motion neatly parallel the beliefs about motion in the three centuries *before* Newton.

McMahon, Thomas A., and John Tyler Bonner: *On Size and Life*, Scientific American Library, W. H. Freeman and Co., New York, 1983. This fascinating book (especially for premedical students) contains a good discussion of dimensional analysis in chapter 3.

Schwartz, Clifford: *Prelude to Physics*, John Wiley and Sons, New York, 1983. A more extended treatment of many of the topics treated in this chapter and in the appendices to the present text, with emphasis on getting the student past the most common stumbling blocks in the introductory physics course. Particularly good on graphic techniques.

Nature to him [Newton] was an open book, whose letters he could read without effort. In one person he combined the experimenter, the theorist, the mechanic and, not least, the artist in expression.

Albert Einstein (1879–1955)

Chapter 3

Forces and Newton's Laws of Motion

We have all probably wondered at one time or another why certain physical events occur. Why does the earth continue to move in its orbit about the sun instead of going off into the darkness of space? Why do magnetic compasses in automobiles always point north? Why do electrons in our TV sets strike the fluorescent screen with sufficient speeds to produce a picture on the screen? In every case the answer is that *forces* act which control the motion of the objects in question. A *gravitational force* keeps the earth in its orbit around the sun. *Electromagnetic forces* align the compass needle in a north-south direction and accelerate the electrons in the TV tube. Other kinds of forces, the strong and weak *nuclear forces*, explain the stability (and the occasional instability) of atomic nuclei.

In this chapter we begin a discussion of such forces which will continue throughout the rest of the book. We will also consider the motion produced by these forces, and in particular Newton's laws which describe this motion. In so doing, we will again limit ourselves to motion in one dimension. Once the basic ideas of forces and motion are clear, it will be easier to apply them to two- and three-dimensional motion in the chapters ahead.

3.1 The Nature of Force

FIGURE 3.1 A man pushing a shopping cart in a supermarket. He exerts a force on the cart to set it into motion.

Suppose a man wants to move a shopping cart, initially at rest, down an aisle in a supermarket at a speed of 1 m/s, as in Fig. 3.1. What does he do? He pushes or pulls the cart and thus accelerates it from zero speed to a speed of 1 m/s. In more technical terms we say that he has exerted a *force* on the cart and changed its condition of motion. Such a force satisfies the following definition*:

---

*A more complete definition will be given after we introduce the idea of *momentum* in Chap. 7.

Definition

A force is that which, when applied to an object, tends to produce an acceleration of that object.

Vector Nature of Force

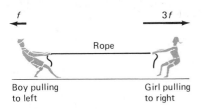

FIGURE 3.2 Tug-of-war between a boy and a girl. The net force acting on the rope determines its resultant motion.

Force is a *vector*, since the direction in which the force is applied is clearly very important in determining the acceleration which results. The force applied by a golf club to a golf ball must be in the direction in which it is desired to accelerate the golf ball. We will designate forces by the symbol **F**. If we are only interested in the magnitude of the force, we will write simply *F*.

Since forces are vectors, they must be added as vectors. For example, if a boy and a girl are engaged in a tug-of-war with a rope, and the girl exerts a force of magnitude $3f$ on the rope to the right, and the boy exerts a force of f to the left, as shown in Fig. 3.2, then the net force on the rope is:

$$\mathbf{F}_{net} = \mathbf{F}_1 + \mathbf{F}_2$$

and the magnitude of this force is

$$F_{net} = 3f - f = 2f$$

This force is in the same direction as $\mathbf{F}_1$, i.e., toward the right. In what follows, we will use the symbol $\mathbf{F}_{net}$ for the resultant of all the forces acting on any body of interest, or the *net force* acting.

Forces can be measured with spring balances of the sort shown in Fig. 3.3. Pulling on the end of the spring elongates it to an extent which is in direct proportion to the force applied. Hence the scale on the spring balance can be marked to indicate forces directly. In the SI system, the unit of force is the newton (N), which will be defined in Sec. 3.3. In the British engineering system, the unit of force is the pound, where 1 lb = 4.45 N.

FIGURE 3.3 A spring balance used to measure forces. (Courtesy of Sargent-Welch Scientific Co.)

3.2 Newton's First Law of Motion

The contributions to the study of motion made by many physicists, but especially those made by Galileo and Sir Isaac Newton (see Fig. 3.4 and accompanying biography), were combined by Newton in his *Principia* of 1687 into three generalizations now called Newton's laws of motion.* The first of these is the following:

Newton's First Law of Motion

An object at rest will remain at rest and an object in motion will continue to move in a straight line at constant speed forever unless some net external force acts to change this motion.

(Rest and motion are here measured with respect to some frame of reference such as the earth's surface.)

The second part of the above statement may seem to contradict our experience that a shopping cart will soon come to a stop unless we keep pushing it. The reason is that frictional forces (i.e., the rubbing together of rough spots on the hub of the wheel and the wheel supports) act to slow the cart down. If we lived in a world without friction or air resistance, however, the cart would keep moving with constant velocity forever once it was given its initial motion.

The general validity of Newton's first law can be seen if we examine situations in which we gradually reduce the frictional forces opposing the motion. For example, if we push a metal puck over a smooth metal surface, it will soon come to a stop after being released, because the rubbing of the uneven surfaces produces a retarding, "frictional" force. If we pour water over the surface, the friction will be somewhat reduced and the puck will travel farther after being released. If we freeze the water on the surface, the ice will slow down the puck even less. Finally, if we float the puck on a layer of air between it and the table, as is done in air hockey games at arcades, a puck set in motion will travel a very long distance on the table before coming to rest. In the limit, i.e., in the ideal case of no friction and no air resistance, we can therefore expect that Newton's first law will be satisified; the puck will indeed continue to move in a straight line at a constant speed forever.

Before Galileo's time it was believed that a force was necessary to keep an object moving with constant velocity. The great contribution of Galileo and Newton was to show that no net force is necessary to do this; forces are only necessary to produce *changes* in velocities, i.e., to produce *accelerations*, not to keep an object moving with constant speed in a straight line. For example, a baseball hit by a batter would continue to move in a straight line forever if there were no air resistance to slow it down and no force of gravity to pull it toward the earth. No force is required to keep it moving in a straight line; that force was applied once and for all when the batter imparted an acceleration to the ball with the bat. What keeps the baseball moving is its inertia, not any other force applied to it.

---

*Although the name *Newton's laws of motion* is commonly used, a better title might be *Newton's generalizations about motion*. They describe how nature behaves under certain conditions. Although these conditions are most often met, there are situations (very high velocities, very great distances) where Newton's laws are inadequate and a more exact description of motion requires Einstein's theory of relativity.

Sir Isaac Newton (1642–1727)

FIGURE 3.4 Sir Isaac Newton. [Courtesy of American Institute of Physics (AIP) Niels Bohr Library.]

Isaac Newton, considered by many the greatest scientist who ever lived, was born in 1642 at Woolsthorpe, Lincolnshire, England. He was a delicate child who lived a somewhat solitary life, preferring to construct a variety of mechanical models and devices rather than play games with his schoolmates.

In 1661 he entered Trinity College of Cambridge University, but was forced to interrupt his studies when the plague broke out in London. This was severe enough to close down the university. Newton therefore spent the years 1665 to 1666 at his home in Woolsthorpe, devoting all his energies to private study in mathematics and physics. This was the period which saw some of his greatest discoveries in mechanics, optics, and the calculus, although it was not until much later that he completed and published his findings in these fields. As an old man Newton reflected as follows on the rough-

ly 18 months in which he accomplished so much: "For in those days I was in the prime of my age for invention, and minded mathematics and philosophy more than at any time since."

Two years after Newton returned to Cambridge, his professor, Isaac Barrow, became convinced that Newton already knew more mathematics than he did, and resigned the Lucasian Professorship of Mathematics in Newton's favor. Newton assumed this prestigious position in 1669 at the age of 26. During the following years he lectured at the university, although few students attended his classes, since he was a poor lecturer who often seemed to be talking to himself rather than to a class. He made his own lens-grinding equipment, ground lenses and mirrors, and built the first successful reflecting telescope. For this work he was elected a Fellow of the Royal Society at the age of 30.

In 1687 Newton published his *Principia*. This was probably the greatest single event in the history of science. The *Principia* combined Newton's three laws of motion and his law of universal gravitation into a system capable of explaining a great variety of phenomena both on earth and in the heavens. Here was the great synthesis of mechanics that scientists had hoped for.

Newton's election in 1689 as a member of Parliament from Cambridge University began the more public side of his life. He moved to London, served inconspicuously in Parliament (an associate claimed that

the only speech he ever made was to ask the usher to close the window), suffered a slight nervous breakdown, and in 1695 was appointed warden and later master of the Mint. In these posts he discharged his duties as custodian of Britain's currency in an efficient and dedicated fashion.

In 1703 he was elected president of the Royal Society of London, and held that post until his death in 1727. His best scientific days were now behind him, but in 1704 he published the *Opticks*, an account of some of his ingenious work on the physics of the spectrum, interference, color vision, and the rainbow. Increasingly, however, his thoughts were directed to biblical and theological studies and to historical chronology rather than to natural science.

Newton never married, and at one time protested bitterly to John Locke, the famous philosopher, about Locke's efforts to "embroil him with women." Newton was absent-minded and inconsistent in his behavior, sometimes open and generous, at other times irritable and spiteful. He was extremely diffident about publishing his important ideas, but vicious in attacking those who claimed priority in discoveries he believed to be rightly his. The most famous incident along these lines was his controversy with Leibnitz over the invention of the calculus.

Newton died in 1727 at the age of 84, and was buried with great pomp and ceremony in Westminister Abbey, London.

Newton's first law is sometimes referred to as the *law of inertia*, where by *inertia* we mean the property of a body which makes it difficult to change its velocity, or accelerate it. As we saw earlier, mass is the measure of an object's inertia. Any frame of reference in which Newton's first law is valid is called an *Inertial Frame of Reference*.

3.3 Newton's Second Law of Motion

Newton's first law tells us that, if no external force acts, an object at rest will remain at rest and an object with a constant velocity **v** will retain that velocity indefinitely. If a net external force acts on an object, then the implication is that it will change the velocity of the object. But a change in velocity is an *acceleration*. Hence *the effect of a net external force is to produce an acceleration.*

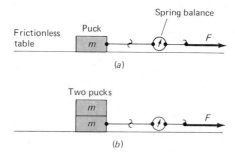

FIGURE 3.5 Measuring the relationship between the force applied and the acceleration produced in a mass on a frictionless table. (a) If *F* is doubled, *a* is doubled. (b) If *m* is doubled, *a* is cut in half, for the same applied force.

Suppose we have a metal puck of mass m on a frictionless air table and apply known forces in newtons to the puck by pulling the puck with an accurately calibrated spring balance, as in Fig. 3.5. We first apply a force of magnitude F_1 and find that the accleration is a_1 in the direction of the force. We then double the force to $F_2 = 2F_1$ and find that the resultant acceleration is $a_2 = 2a_1$. If we do this for a large number of forces, we find always that the acceleration is directly proportional to the net force applied and is in the same direction. Hence we can write:

$$\mathbf{a} \propto \mathbf{F}$$

If now we choose a constant force **F** but put two identical pucks on top of each other so as to double the mass, we find that any force **F** produces an acceleration that is exactly half of what it was for one puck. If we use three pucks and the same force, the acceleration is found to be one-third of what it was originally. Hence we can conclude that the acceleration a is inversely proportional to the mass accelerated, for the same applied force, or

$$a \propto \frac{1}{m}$$

If we combine these two results, we have:

$$\mathbf{a} \propto \frac{\mathbf{F}}{m} \qquad \text{or} \qquad \mathbf{F} = km\mathbf{a}$$

with k a constant of proportionality still to be determined.* We now define the SI unit of force, the *newton* (N), as that force that gives a mass of 1 kg an acceleration of 1 m/s². Or, using the above equation,

---

*On this change from a direct proportion to an equality, see the section on direct proportions in Appendix 2A.

$$1 \text{ N} = k(1 \text{ kg})(1 \text{ m/s}^2)$$

Hence the constant k must be unity in the above expression, and we have:

Newton's Second Law

$$\boxed{\mathbf{F} = m\mathbf{a}} \tag{3.1}$$

This is Newton's second law of motion, which states that the acceleration of a body, produced by a force, is directly proportional to the force and in the same direction and is inversely proportional to the mass of the object.

Newton's second law of motion is the most important equation in mechanics and one of the most important equations in all physics. In applying it to particular problems, it is important to isolate one particular object or system of objects, and then to consider all the external forces acting on this object or system; we will call the vector sum of the forces acting $\mathbf{F}_{\text{net}}$. Newton's second law applies equally well to an automobile and to each passenger in that automobile.

Newton's second law, in the form $\mathbf{F}_{\text{net}} = m\mathbf{a}$, can be broken up into three equations along three mutually perpendicular axes in space, yielding

$$(F_{\text{net}})_x = ma_x \qquad (F_{\text{net}})_y = ma_y \qquad (F_{\text{net}})_z = ma_z \tag{3.2}$$

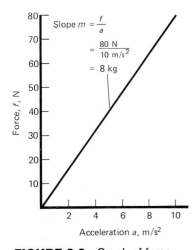

A useful way to look at this result is to say that, in one dimension, if we graph F_{net} as a function of a for constant m, we obtain a straight line of slope m, as in Fig. 3.6, since $F_{\text{net}} = ma$ is the equation of a straight line of slope m. The larger m is, the steeper the slope, and hence the larger the force needed to produce a given acceleration. In other words, the greater an object's *inertia*, the more difficult it is to accelerate it.

There are a great variety of different kinds of forces in physics in addition to the pushes and pulls produced by human or animal muscles. Some, such as frictional forces, act by making immediate contact with the object to be accelerated. Others, such as gravitational and electromagnetic forces, act without contact by producing *fields* of force which fill all space. These forces all differ in their intrinsic natures, but have the same dimensions and are therefore all the same kind of physical quantity. They produce results qualitatively the same (i.e., accelerations), although quantitatively much different. For example, the gravitational force between two electrons is about 42 orders of magnitude smaller (that is, smaller by a factor of 10^{42}) than the electric force between the same two electrons. Hence the acceleration of the electrons produced by these two very different forces would be 10^{42} times less in the gravitational case than in the electrical case, since the mass is the same in the two cases. Equation (3.1) is still perfectly applicable in the two cases, however, despite this huge difference in the strength of the two forces and in the resulting accelerations.

FIGURE 3.6 Graph of force applied to an object of mass m against the acceleration produced by that force. In this case the slope of the straight line shows that $m = 8.0$ kg.

Relationship between First and Second Laws

In learning Newton's laws of motion it is sometimes useful to look at the relationship between Newton's first two laws. The second law, $\mathbf{F}_{\text{net}} = m\mathbf{a}$, states that, if $\mathbf{F}_{\text{net}} = 0$, then $\mathbf{a}$ is necessarily zero for any value of m. If $\mathbf{a}$ is zero, the velocity of a given object does not change either in direction or in magnitude. Hence an object at rest will remain at rest. An object with initial velocity will continue to move in a straight line, for any deviation from a straight line would correspond to a change in direction; it will also keep the same speed it had originally, since a change in speed would represent a change in the magnitude of the velocity. This then is Newton's first law.

Example 3.1

(a) What force is required to give a 2000-kg automobile an acceleration of 2.0 m/s²?
(b) How would this force change if the mass of the car were reduced to 1000 kg?

(c) How far has the car traveled after 10 s, if it starts from rest and the acceleration remains constant over the 10-s period?

SOLUTION

(a) $F = ma = (2000 \text{ kg})(2.0 \text{ m/s}^2) = 4.0 \times 10^3 \text{ kg·m/s}^2$

$$= \boxed{4.0 \times 10^3 \text{ N}}$$

since 1 N is equal to 1 kg · m/s².

(b) If the mass is reduced to 1000 kg, the force required is also cut in half and becomes

$$\boxed{2.0 \times 10^3 \text{ N}}$$

Since the force required to accelerate a car determines in great part the energy which must be provided by the gasoline the car uses (especially for city driving), reducing

the mass of the car by one-half is equivalent to reducing the car's fuel consumption by one-half.

(c) From Eq. (2.11) for uniformly accelerated motion, we have

$$s = s_0 + v_0 t + \tfrac{1}{2} a t^2$$

Hence, with $v_0 = 0$,

$$s - s_0 = \tfrac{1}{2} a t_2 = \tfrac{1}{2} \left(2.0 \, \frac{\text{m}}{\text{s}^2} \right) (10 \text{ s})^2$$

$$= \boxed{100 \text{ m}}$$

3.4 Newton's Third Law of Motion

Another important generalization arrived at by Newton was the following:

Newton's Third Law of Motion

Whenever an object A exerts a force on a second object B, object B exerts a force back on A; these two forces are equal in magnitude but opposite in direction.

In this law the force A exerted on B is sometimes called the *action*, and the force exerted by B on A is called the *reaction*. For example, the sun exerts a force on the earth according to Newton's law of universal gravitation (see Sec. 3.5). This can be called the action force. The earth, in turn, exerts an equal and opposite force back on the sun. This is the reaction force.

For this reason Newton's third law is often referred to as the *action-reaction law*, which states that to every action there is an equal and opposite reaction. This is a less satisfactory statement of the third law, since it gives the impression that there is one preferred force which must be called the action, with the other force then called the reaction. Actually either force can be chosen to be the action, and the other is then the reaction.

Note that these two forces, the action and the reaction, *are not exerted on the same object*. For example, a boy pushing a box across a rough floor exerts a force on the box which is equal and opposite to the force exerted by the box on the boy's hands, according to Newton's third law (see Fig. 3.7). This is true whether the box moves or remains at rest. What is important for the motion

FIGURE 3.7 Newton's third law applied to a boy pushing a box over a rough floor.

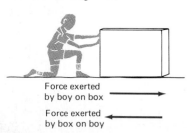

Force exerted by boy on box ⟶

Force exerted by box on boy ⟵

of the box is the size of the force the boy exerts to move the box forward compared to any forces acting on the box in the opposite direction. If the force the boy exerts is larger than any resisting forces, the box will accelerate. If the box is at rest and the force exerted is not sufficient to overcome the opposing forces, then the box will remain at rest. In both cases, however, the force exerted *on the boy's hand* by the box is *always* equal and opposite to the force the boy exerts *on the box*, according to Newton's third law. Figure 3.8 gives some other important examples of Newton's third law.

An interesting example of Newton's third law is an early form of the steam engine, called Hero's engine, shown in Fig. 3.9. This engine was developed in Greece in the first century. When the water boils in the spherical container, the steam jets emerging from the two sides drive the container in rotational motion at high speed. This is because the force on the steam driving it out of the container is accompanied by an equal and opposite reaction force exerted back on the container by the emerging steam, according to Newton's third law. It is this reaction force which drives Hero's engine. A similar kind of reaction force drives rockets and jet planes. Here the emerging gases at the tail of the rocket or plane produce reaction forces in the opposite direction which push the rocket or plane forward.

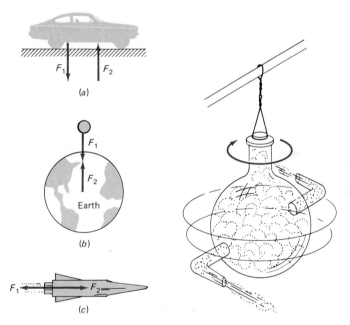

FIGURE 3.8 Some other examples of Newton's third law of motion: Action equals reaction. (*a*) Car pushes down on road; road pushes up on car. (*b*) Earth pulls ball down; ball pulls earth up. (*c*) Rocket pushes gas back; gas pushes rocket forward.

FIGURE 3.9 Hero's engine. The reaction forces from the steam jets on the spherical container set it into rapid rotation.

3.5 Newton's Law of Universal Gravitation

The most familiar force on our planet is the force of gravity, which binds all objects to the earth. Sir Isaac Newton first explained the nature of this gravitational force. The story goes that Newton was sitting under an apple tree, saw an apple falling to the ground, and was struck not by the apple, but by the insight that the force pulling the apple to the ground must be the same kind of force that keeps the moon moving in its orbit about the earth. Whether or not this story is factual is unclear, but there is no doubt that it is accurate in spirit. Newton certainly was the first physicist to write an equation which explained both the fall of an apple and the motion of the moon, and he was able to use this insight to calculate the period of the moon's motion around the earth.

Newton's great discovery is incorporated in his famous law of universal gravitation:

Law of Universal Gravitation

Every mass in the universe attracts every other mass by a force which is directly proportional to the product of the two masses and inversely proportional to the square of the distance between them.

The magnitude of the gravitational force F_G is thus given by

$$F_G \propto \frac{m_1 m_2}{r^2}$$

or, since any proportionality can be converted into an equation by inserting a constant of proportionality,

$$F_G = G \frac{m_1 m_2}{r^2} \tag{3.3}$$

In this equation the direction of F_G is along the line between the masses m_1 and m_2, and r is the distance between m_1 and m_2, as shown in Fig. 3.10. Body 1 pulls body 2 toward it, and body 2 pulls body 1 toward it, in turn. If the objects are not point masses but large spherical masses such as the planets, then r is the distance between their centers and the force is directed along the line connecting their centers, as Newton was able to show after considerable intellectual effort.

For example, the gravitational force between the earth and the sun is directed along the line connecting the center of the earth with the center of the sun, and its magnitude is inversely proportional to the square of the distance between their two centers. The force exerted by the sun on the earth is exactly the same in magnitude as the force exerted by the earth on the sun, as can be seen from Eq. (3.3), but is in the opposite direction. Hence these two forces are a good example of an action-reaction pair, in the sense of Newton's third law of motion.

FIGURE 3.10 The gravitational attraction between two masses m_1 and m_2.

The Gravitational Constant

FIGURE 3.11 A gravitational torsion balance.

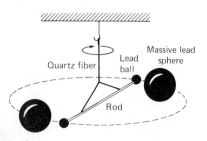

In Newton's equation for universal gravitation, the constant G must be determined experimentally. This was first done in 1798 in a famous experiment performed by Henry Cavendish (1731–1810), after whom the Cavendish Physics Laboratory at Cambridge University in England is named. In this experiment two small lead balls are placed at the ends of a thin rod which is suspended at its center by a very thin quartz fiber hung from a support. This allows the system to rotate about the fiber as an axis (see Fig. 3.11). Two large lead balls are then brought up at opposite sides of the small balls. It is found that the large balls attract the small balls and twist the suspended system about its axis. By attaching a small mirror to the fiber and bouncing a light beam off the mirror, the motion of the system can be studied. The twist of the quartz fiber can be obtained from the deflection of the light beam, and the force needed to produce this twist found by producing the same twist with a known force. In this way the force F_G can be determined.

Since this experiment yields a value for F_G, and since the masses m_1 and m_2 and the distance r can be easily measured, we can solve Eq. (3.3) for G. The experimental value obtained is $G = 6.67 \times 10^{-11}$ m³/(kg·s²). Note that these must be the units of G if Eq. (3.1) is to be dimensionally correct. The units of F_G are newtons, or kg·m/s². Hence from Eq. (3.3) the units of G must be

$$\frac{\text{kg·m}}{\text{s}^2} \cdot \frac{\text{m}^2}{\text{kg}^2} = \frac{\text{m}^3}{\text{kg·s}^2}$$

G, called the *universal gravitational constant*, is one of the least accurately known constants of nature. The gravitational force is so small between most objects on earth that it is difficult to measure it with any real accuracy. In astronomical systems the gravitational force is much greater, but in this case it becomes difficult to measure accurately the masses and the distance between them.

This gravitational force holds our complex universe together: it keeps the moon in orbit around the earth; it holds the planets in orbit around the sun; it anchors all of us to the earth so that we don't float off into space; and it explains the presence of globular star clusters or galaxies.

Newton's great contribution was the recognition of the truly *universal* nature of the gravitational force. The constant G has the same value at every place in the universe. Newton's discovery of the law of universal gravitation is sometimes called the greatest single discovery in the history of science, although future historians of science may give the edge to the unraveling of the genetic code, or to some other discovery yet to be made.

Example 3.2

Calculate the gravitational force exerted by the sun on the earth, using the data given in Table A2 (inside front cover).

SOLUTION

From Newton's law of universal gravitation we have for the magnitude of F_G:

$$F_G = G \frac{m_S m_E}{r^2}$$

$$= \left(6.67 \times 10^{-11} \frac{\text{m}^3}{\text{kg·s}^2}\right) \frac{(1.99 \times 10^{30} \text{ kg})(5.98 \times 10^{24} \text{ kg})}{(1.49 \times 10^{11} \text{ m})^2}$$

where m_S and m_E are the masses of the sun and the earth,

respectively, and r is the distance between their centers. Hence

$$F_G = 3.58 \times 10^{22} \text{ kg·m/s}^2 = \boxed{3.58 \times 10^{22} \text{ N}}$$

It is this enormous force of 3.58×10^{22} N which keeps the earth moving in its elliptical orbit about the sun. If it ceased to exist, the earth would go flying off in a straight line tangent to its original orbit.

3.6 Mass and Weight

Probably the most familiar example of Newton's law of gravitation is the attractive force exerted by the earth on objects at its surface. If m_E is the mass of the earth and m the mass of a book on a shelf in the library, then the earth exerts a gravitational force on the book of magnitude

$$F_G = \frac{Gm_Em}{r^2} \qquad (3.4)$$

where r is the distance from the center of the earth* to the center of the book.

As long as the shelf supports the book, the downward force F_G is balanced out by the upward force of the shelf on the book. If, however, the book is pushed off the shelf, an unbalanced force of magnitude F_G produces an acceleration, according to Newton's second law, given by:

$$a = \frac{F_G}{m} \qquad (3.5)$$

The book then falls with this acceleration.

Careful experiments have shown that *this acceleration is the same for all objects, no matter what their mass, at any given point on the earth, as long as air resistance is neglected.* Even two objects as different as a feather and a coin will fall with the same acceleration in a glass tube from which the air has been evacuated.

How do we explain this surprisingly consistent behavior? Suppose we replace the one book by two books of different masses m_1 and m_2. They are attracted by the earth with forces of magnitudes given by Eq. (3.4).

$$F_{G1} = \frac{Gm_Em_1}{r^2} \qquad \text{and} \qquad F_{G2} = \frac{Gm_Em_2}{r^2} \qquad (3.6)$$

If the two books are pushed off the same shelf, they will fall with accelerations a_1 and a_2, where, from Eq. (3.5), the magnitudes of the accelerations are:

$$a_1 = \frac{F_{G1}}{m_1} \qquad \text{and} \qquad a_2 = \frac{F_{G2}}{m_2}$$

On substituting the values of F_{G1} and F_{G2} from Eq. (3.6),

$$a_1 = \frac{Gm_Em_1}{m_1r^2} \qquad \text{and} \qquad a_2 = \frac{Gm_Em_2}{m_2r^2}$$

Then on canceling the masses m_1 and m_2,[†]

$$a_1 = \frac{Gm_E}{r^2} \qquad \text{and} \qquad a_2 = \frac{Gm_E}{r^2} \qquad (3.7)$$

and so $a_1 = a_2$.

This is true for all objects at the same place on the earth's surface. We give the acceleration due to gravity the symbol g to indicate its universal nature. Here

$$g = \frac{Gm_E}{r^2} \qquad (3.8)$$

g is approximately equal to 9.80 m/s², and so the velocity of a falling body on the earth's surface increases by approximately 9.8 m/s for each second of fall.

---

*For purposes of calculating gravitational forces, Newton was able to show that the mass of a sphere can be considered as concentrated at its geometric center. The argument used is similar to that used for spherical charges in Chap. 16.

†There is a subtle assumption here that the mass m_1 appearing in Newton's second law is the same mass m_1 appearing in the gravitational force equation. This may seem like a reasonable assumption, but it was only after Einstein's theory of general relativity was developed that it was possible to justify it.

The value of g differs slightly with location on the earth's surface and with altitude, since these factors change the value of the distance r in Eq. (3.7). The value of g is a maximum at sea level, and decreases with increasing altitude. Its value varies from about 9.832 m/s² at the north and south poles to about 9.780 m/s² at the equator, indicating that the earth is not a perfect sphere but bulges out at the equator.

Weight—A Gravitational Force

As we just saw, all objects at the same place on the earth's surface fall with the same acceleration g. According to Newton's second law, this acceleration must be produced by a force which in this case is the *weight* of the object.

Definition

Weight: The downward gravitational force exerted on an object.

Hence in this case Newton's second law, $\mathbf{F}_G = m\mathbf{a}$, becomes

$$\boxed{\mathbf{w} = m\mathbf{g}} \tag{3.9}$$

where both $\mathbf{w}$ and $\mathbf{g}$ are directed downward toward the center of the earth.

Hence we see the relationship between the mass of an object and its weight. On earth, to obtain the weight of an object in newtons from its mass in kilograms we must multiply by $g = 9.80$ m/s². Since on the moon g is about one-sixth as large as it is on the earth, the weight of a 200-lb astronaut on the moon is only about 33 lb, even though the astronaut's mass is exactly the same on the earth as it is on the moon. Near the earth's surface the earth is pulling down at 9.8 N/kg, while on the moon's surface the pull is only 9.8/6 = 1.6 N/kg, because of the difference in the masses and the radii of the earth and the moon, as can be seen from Eq. (3.3).

This shows us the essential difference between mass and weight:

Mass and Weight

Weight is a force (a gravitational force), whereas mass is a measure of inertia.

Mass is therefore an "invariant quantity"; that is, it does not vary from place to place,* whereas weight changes as g changes from one place to another.

If, however, two objects are weighed on a spring balance at the same place, their masses will be in the same proportion as their weights. Since $w_1 = m_1g$ and $w_2 = m_2g$, we have $w_1/w_2 = m_1/m_2$. Hence spring balances can be used to compare masses as well as weights.

Example 3.3

(a) What is the force exerted on a freely falling ball, which has a mass of 2.0 kg, by the gravitational attraction of the earth?

(b) What reaction force is exerted on the earth by the ball?

(c) What acceleration of the earth would this force produce?

SOLUTION

(a) $w = mg = (2.0 \text{ kg})(9.8 \text{ m/s}^2)$

$= 1.96 \times 10^1 \text{ kg·m/s}^2 = \boxed{19.6 \text{ N}}$

(b) The ball, by Newton's third law, exerts an equal and opposite force of 19.6 N on the earth, pulling the earth toward the ball!

---

*This statement is true in classical, nonrelativistic physics. It will require modification for high speeds to fit the special theory of relativity, to be discussed in Chap. 25.

(c) Since (from Table A2) the mass of the earth is 5.98×10^{24} kg, the acceleration of the earth produced is, using $F = ma$,

$$a = \frac{F}{m} = \frac{19.6 \text{ N}}{5.98 \times 10^{24} \text{ kg}} = 3.2 \times 10^{-24} \frac{\text{kg·m/s}^2}{\text{kg}}$$

$$= \boxed{3.2 \times 10^{-24} \text{ m/s}^2}$$

This infinitesimal acceleration is too small to be observable.

Example 3.4

In the British engineering system of units the pound is a measure of the weight of an object on the surface of the earth, and hence is a *force* unit, not a mass unit.
(a) If a woman astronaut weighs 125 lb on the surface of the earth, what is her mass on the surface of the earth?

(b) If she were on the surface of the moon, where the acceleration due to gravity would be only one-sixth that on earth, what would be her mass? Her weight?

SOLUTION

(a) Newton's second law, $F = ma$, applied to the force of gravity, becomes as we have seen

$$w = mg$$

Here

$$w = 125 \text{ lb} = (125 \text{ lb})\left(\frac{4.45 \text{ N}}{1 \text{ lb}}\right) \qquad \text{(from Table D, inside back cover)}$$

$$= 556 \text{ N}$$

and so

$$m = \frac{w}{g} = \frac{556 \text{ N}}{9.8 \text{ m/s}^2} = \frac{556 \text{ N}}{9.8 \text{ N/kg}} = \boxed{57 \text{ kg}}$$

(Note that, since $1 \text{ N} = 1 \text{ kg·m/s}^2$, $1 \text{ m/s}^2 = 1 \text{ N/kg}$.) Hence the astronaut's mass on the surface of the earth is 57 kg.

(b) Since the mass is an invariant property, it does not change on the surface of the moon or anywhere else. Hence the astronaut's mass on the surface of the moon is again

$\boxed{57 \text{ kg}}$. Her weight, however, is different. Since g on the

surface of the moon is only one-sixth its value on earth, we have

$$w_M = mg_M = (57 \text{ kg})\left(\frac{9.8 \text{ m}}{6 \text{ s}^2}\right) = \boxed{93 \text{ N}}$$

and her weight has been reduced by a factor of 6. In pounds her weight on the surface of the moon would be

$$w_M = (93 \text{ N})\left(\frac{1.0 \text{ lb}}{4.45 \text{ N}}\right) = \boxed{21 \text{ lb}}$$

Example 3.5

An Atwood's machine (shown in Fig. 3.12) is sometimes used in physics laboratories to explore the behavior of falling objects and to measure the acceleration due to gravity g. It consists of two masses tied together by a light cord which is hung over a pulley. For simplicity we neglect the mass of the pulley and the cord, and assume that the pulley is frictionless. If the masses are $m_2 = 20$ kg and $m_1 = 10$ kg, what is the acceleration of the masses?

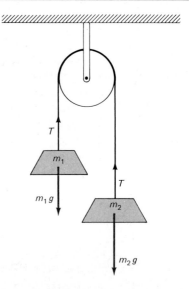

FIGURE 3.12 Diagram for Example 3.5: An Atwood's machine.

SOLUTION

This is a good example of a situation in which we have to isolate particular parts of the system and apply Newton's second law to each part individually. In this case the two parts are the two masses m_1 and m_2. Since the larger mass (m_2) will move down and the smaller mass (m_1) up, as is clear from the diagram, we choose motion in those directions as positive. Hence motion downward is positive for m_2, and motion upward is positive for m_1. The application of Newton's second law then yields:

For mass 2: $\qquad (F_{\text{net}})_2 = m_2 g - T = m_2 a$ $\qquad$ [1]

For mass 1: $\qquad (F_{\text{net}})_1 = T - m_1 g = m_1 a$ $\qquad$ [2]

Now the acceleration of the two masses must be the same, since they are tied together by the cord. Also the tension T must be the same throughout the cord since the cord is assumed to have no mass, and so Newton's second law for the cord is

$$F_{\text{net}} = T_2 - T_1 = ma = 0 \qquad \text{or} \qquad T_2 = T_1 = T$$

Hence we can add the two equations above and obtain:

$$m_2 g - m_1 g = (m_1 + m_2)a \qquad [3]$$

Here m_1, m_2, and g are known, and a is the unknown to be determined. We have

$$a = \frac{(m_2 - m_1)g}{m_1 + m_2}$$

$$= \left(\frac{20 \text{ kg} - 10 \text{ kg}}{20 \text{ kg} + 10 \text{ kg}}\right)(9.8 \text{ m/s}^2) = \frac{10}{30}(9.8 \text{ m/s}^2)$$

$$= \boxed{3.3 \text{ m/s}^2}$$

A check on this answer is provided by the fact that, from Eq. [1] above,

$$T = m_2 (g - a) = (20 \text{ kg})(9.80 - 3.27 \text{ m/s}^2) = 131 \text{ N}$$

and also from Eq. [2],

$$T = m_1(g + a) = (10 \text{ kg})(9.80 + 3.27 \text{ m/s}^2) = 131 \text{ N}$$

These two values for T agree, and hence our assumption that T is the same throughout the cord is confirmed. Note that the tension has a value 131 N, which is between the weight of mass 2 (196 N) and that of mass 1 (98 N).

The tension is the same throughout the cord only if the cord can be considered to have zero mass and if the pulley bearing has no friction. If the cord has a sizable mass, a net unbalanced force must be applied to accelerate it, and hence the tension at the two ends of the cord must be different. Also, the pulley must be considered without mass, for the presence of any mass would require a net difference in the forces on the two sides of the pulley to accelerate the pulley. Some of these complications will be considered in subsequent chapters.

The secret of physics' success in describing reality is precisely its ability to omit initially complications like the masses of the cord and the pulley in order to solve highly idealized problems such as this one. Once the physical principles are grasped, the omitted complications can be reintroduced at a later stage and the complete problem solved.

3.7 The Force of Friction

Before discussing some important applications of Newton's three laws of motion, it is important that we understand more fully some of the forces which oppose the motion of objects. These we call *resistive* or *dissipative* *forces*. Examples are the force of friction, air resistance, and the viscosity of liquids and gases. The effect of all these forces is to resist the motion of objects and in so doing to dissipate energy, i.e., to convert mechanical energy into some other kind of energy, such as heat, which is usually less useful.

Friction is a good example of a resistive force. A frictional force may be defined as follows:

Definition

Force of friction: A force, due to the unevenness of two surfaces in contact, which opposes the relative motion of one object with respect to the other.

The force on each of the objects is opposite to the direction of its motion relative to the other object. Hence the forces on the two objects make up an action-reaction pair. Figure 3.13 shows a trash can being pushed from left to right over the surface of a driveway. In this case, the frictional force of the driveway on the trash can acts towards the left to oppose the motion of the trash can to the right; the trash can similarly tries to "pull" the driveway to the right.

Frictional forces may also act when there is no relative motion. A horizontal push given a stalled automobile may not be enough to move it because the frictional force exerted by the road on the tires of the car just balances out the push given the car. Hence the sum of the forces on the car is zero and the car remains at rest, according to Newton's first law. There is, of course, some maximum limit to the frictional force possible, which depends on the surfaces of the tires and the road, and on the weight of the car. When this maximum frictional force is exceeded, the car will be accelerated and start to move, according to Newton's second law.

It is found experimentally that the force of friction depends on the nature of the two surfaces, e.g., whether they are rough or smooth, and also on how intimately they come in contact. The great artist and engineer Leonardo da Vinci first found that the frictional force depends on the normal force between the two surfaces, i.e., the force exerted by one on the other perpendicular to their surface of contact. In the example of the trash can in Fig. 3.13, this would be the weight of the trash can pressing down on the driveway and is designated as $F_N\,(= mg)$ in the figure. We can write in general that

$$F_{\text{friction}} \propto F_N$$

or $\qquad F_{\text{friction}} = \mu F_N$ $\hfill$ (3.10)

where F_N is the normal force and μ is the *coefficient of friction*, which depends on the nature of the two surfaces in contact.

It is important to distinguish between static and sliding friction:

Definitions

Static friction: The frictional force exerted by one surface on another when there is no relative motion of the two surfaces.

Sliding (kinetic) friction: The frictional force exerted by one surface on another when one surface slides over the other.

Experimentally it is found that sliding friction is usually less than the maximum value of the static friction, since it takes a smaller force to keep an object moving with constant velocity against a frictional force than it does to start the object moving from rest. Also, as noted above, the actual static frictional force exerted on an object always just balances out the applied force until the point is reached at which the object starts to move. Hence the static frictional force can assume any value from zero up to the maximum value at which sliding begins. Once sliding starts, the sliding frictional force does not usually depend on the speed of the sliding object. We can summarize these facts in the form of two equations:

$$F_{\text{sliding friction}} = F_k = \mu_k F_N \hfill (3.11)$$

$$F_{\text{static friction}} = F_s \leq \mu_s F_N \hfill (3.12)$$

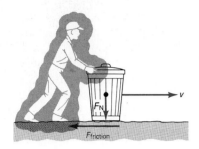

FIGURE 3.13   A man pushing a trash can over a driveway. The frictional force between the driveway and the trash can opposes the force exerted by the man.

where μ_k is the coefficient of sliding (kinetic) friction and $\mu_s F_N$ is the maximum value of the force of static friction. μ_k is almost always less than μ_s, since the surface irregularities are usually less important once the object is moving. As a consequence, the frictional forces can be graphed very approximately as in Fig. 3.14. Table 3.1 contains coefficients of friction for various substances in contact. Note that the coefficient of friction is a pure number, since it is the ratio of two forces. The very low value of μ for Teflon indicates why Teflon is often used as a lubricant in place of oil or grease, and why Teflon pots and pans have nonstick surfaces.

FIGURE 3.14 Schematic diagram for frictional forces. The applied force is gradually increased until the object starts to slide, at which point the force of static friction is $\mu_s F_N$. The force required to keep it moving at a constant speed then falls to some constant value $\mu_k F_N$ determined by the coefficient of sliding friction μ_k.

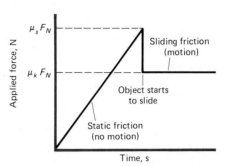

TABLE 3.1 Coefficients of Static and Sliding Friction

Material	Coefficient of static friction μ_s	Coefficient of sliding friction μ_k
Dry, unlubricated surfaces:		
Hard steel on hard steel	0.78	0.42
Mild steel on lead	0.95	0.95
Aluminum on aluminum	1.05	1.4
Glass on glass	0.94	0.4
Oak on oak (parallel to grain)	0.62	0.48
Oak on oak (perpendicular to grain)	0.54	0.32
Teflon on Teflon	0.04	0.04
Teflon on steel	0.04	0.04
Rubber tire on concrete (low speed)	0.9	0.7
Rubber tire on concrete (high speed)	0.6	0.35
Racing tires on concrete		~1.5
Wet surfaces:		
Rubber tire on wet concrete (low speed)	0.7	0.5
Steel on ice (ice skates)	0.02	0.01
Waxed skis on wet snow	0.14	0.10

Although friction has many disadvantages and produces heat and wear, it also has many important uses. It makes the wheels of a locomotive grip the track and push the train forward. It enables us to walk by keeping our shoes from slipping on the sidewalk. Walking on ice is difficult because the coefficient of friction is too low—0.01 as opposed to 0.6 for leather on concrete. For the same reason it is hard to walk on dry sand.

Example 3.6

A steel puck slides over a steel surface. The coefficient of sliding friction between the two steel surfaces is 0.42.
(a) If the mass of the puck is 2.0 kg, what is the force of friction retarding the motion?

(b) If the initial velocity of the puck is 3 m/s, how long will it take for the puck to be brought to rest by the frictional force?

SOLUTION

(a) From Eq. (3.4), the force of sliding friction is

$$F_k = \mu_k F_N$$

Here the normal force is the weight of the puck, which is

$$F_N = w = mg = (2.0 \text{ kg})(9.8 \text{ m/s}^2) = 19.6 \text{ N}$$

Then

$$F_k = \mu_k F_N = 0.42 \times 19.6 \text{ N} = \boxed{8.2 \text{ N}}$$

(b) The effect of this retarding force is to produce an acceleration (negative) given by Newton's second law as

$$a = \frac{F_k}{m} = \frac{-8.2 \text{ N}}{2.0 \text{ kg}} = -4.1 \text{ m/s}^2$$

where the minus sign indicates that this force opposes the motion. Using Eq. (2.10), we see that the time required for the puck to be brought to rest is:

$$t = \frac{v - v_0}{a} = \frac{0 - 3 \text{ m/s}}{-4.1 \text{ m/s}^2} = \boxed{0.73 \text{ s}}$$

Hence the force of friction is so high that the puck will come to rest in less than 1 s, unless more energy is provided from outside to keep it going or the force of friction is reduced by lubricating the surfaces.

3.8 Other Resistive Forces

The resistance of the air to the flight of a golf ball, and of water to the passage of a boat, are similar. In both cases the fluid substance, the air or water, has to be pushed out of the way by the moving object. This sets up a resisting force which depends in a complicated way on the size and shape of the moving object and on the fluid friction, or *viscosity*, of the fluid.

Definition

Viscosity: The resistance of a fluid to flow caused by the internal friction of the fluid's molecules moving against each other.

For example, molasses is much more viscous than water; i.e., it has a higher viscosity. Air, on the other hand, has a very low viscosity but still resists the motion of objects through it, as you learn from experience when you put your hand out the window of a moving car. It is found experimentally that this resistive force at low speeds is, to a good approximation, directly proportional to the velocity of the moving object, but in the opposite direction. We can therefore write

$$F_{\text{vis}} = -Kv \tag{3.13}$$

where K is a constant which varies with the size and shape of the moving object and with the nature of the fluid.

Freely Falling Bodies and Terminal Velocities

When air resistance is considered, we find that the velocity of a freely falling object does not increase without limit. Rather it approaches a limiting or *terminal velocity* which depends on the nature of the falling object and the resisting medium. Once that velocity is reached, no further acceleration occurs and the object continues to move at its terminal speed.

To see this, let us write Newton's second law of motion for an object falling in a viscous medium which produces a resistive force $-Kv$. Then $F_{\text{net}} = ma$ becomes

$$F_{\text{net}} = mg - Kv = ma \tag{3.14}$$

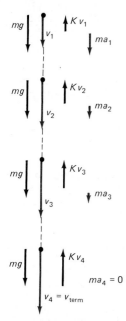

FIGURE 3.15 Fall of a ball under gravity in a viscous medium. The acceleration is gradually reduced because the resistive force increases with the speed of the falling object. Finally, when $v_4 = v_{\text{term}}$, the resistive force is exactly equal to the force of gravity on the object, and the resulting acceleration is zero.

since the net force acting here is the sum of the downward force of the object's weight (mg) and the upward force due to the resistance of the air ($-Kv$). (Here, since the motion is downward, we have, for simplicity, taken all velocities and accelerations as positive when directed downward.) What happens then is that v continually increases, and so Kv also increases and approaches the value mg, as shown in Fig. 3.15. When the resistive force $-Kv$ finally becomes equal to the gravitational force mg, we have

$$Kv = mg \qquad \text{and} \qquad F_{\text{net}} = mg - Kv = 0$$

$$\text{and so} \qquad ma = 0 \qquad \text{or} \qquad a = 0$$

Hence there is no further acceleration, and the object continues to move with a terminal velocity obtained from the condition that

$$mg - Kv = 0$$

$$\text{or} \qquad v_{\text{term}} = \frac{mg}{K} \tag{3.15}$$

We see therefore that the terminal velocity is proportional to the weight of the object and inversely proportional to the viscous resistance of the medium. In this sense Aristotle's belief that the velocity of a falling object was proportional to its weight had some real validity.

It turns out experimentally, in accordance with Eq. (3.15), that for objects of the same shape and density falling in the same resistive material, the heavier the object, the greater its terminal velocity. For example, a raindrop 2 mm in diameter will fall with a terminal velocity in air of about 4.5 m/s, whereas a hailstone of the same size has a somewhat lower terminal velocity of 4 m/s, because ice has a lower density than water. For the same reason large raindrops fall more rapidly than small ones. Sky divers falling head-first in air, without parachutes, reach terminal velocities of about 55 m/s (125 mi/h), whereas divers with open parachutes have terminal velocities of about 11 m/s (25 mi/h).

On the other hand, the terminal velocity of red blood cells in water is only about 10^{-6} m/s because of the buoyancy and viscosity of the water. In this case the terminal velocity is referred to as the *sedimentation velocity*, that is, the velocity with which substances like blood cells settle to the bottom of a container of water or other liquid. In the case of blood cells we would have to wait 10^5 s, or 28 h, for the cells to settle to the bottom of a 10-cm-high beaker of water. This sedimentation rate can be enormously increased with centrifuges, as we will see in Sec. 5.5.

Example 3.7

A raindrop of mass 4.2×10^{-6} kg falls freely in air, with a terminal velocity of 4.5 m/s.
(a) What is the value of the constant K which determines the resistive force in this case?
(b) If the raindrop is replaced by an iron pellet of exactly the same size and shape as the raindrop, but of mass 3.3×10^{-5} kg, what will the terminal velocity of the iron pellet be?

(c) How long does it take the iron pellet to reach this terminal velocity?
(d) For how long would the iron pellet have to fall in the absence of air resistance to acquire a terminal velocity equal to its terminal velocity in air?

SOLUTION

(a) Since the terminal velocity is always in the downward direction, we will concentrate on its magnitude (i.e., the speed). We have

$$v_{\text{term}} = \frac{mg}{K}$$

and so

$$K = \frac{mg}{v_{\text{term}}} = \frac{(4.2 \times 10^{-6} \text{ kg})(9.8 \text{ m/s}^2)}{4.5 \text{ m/s}}$$

$$= 9.1 \times 10^{-6} \frac{\text{kg·m}}{\text{s}^2} \cdot \frac{\text{s}}{\text{m}} = \boxed{9.1 \times 10^{-6} \text{ kg/s}}$$

(b) In the case of the iron pellet, K remains the same, since the pellet has the same size and shape as the raindrop and is falling through the same fluid. Hence the terminal speed is:

$$v_{\text{term}} = \frac{mg}{K} = \frac{(3.3 \times 10^{-5} \text{ kg})(9.8 \text{ m/s}^2)}{9.1 \times 10^{-6} \text{ kg/s}} = \boxed{36 \text{ m/s}}$$

(c) With air resistance, the acceleration of the pellet varies linearly from an initial value of 9.8 m/s^2 to zero when the terminal speed is reached. Hence the average acceleration is

$$\bar{a} = \frac{a_0 + a}{2} = \frac{9.8 \text{ m/s}^2 + 0}{2} = 4.9 \text{ m/s}^2$$

Then the final speed is given by the expression $v = v_0 + \bar{a}t$, and so

$$t = \frac{v - v_0}{a} = \frac{36 \text{ m/s} - 0}{4.9 \text{ m/s}^2} = \boxed{7.4 \text{ s}}$$

(d) In the absence of air resistance the acceleration is the same, 9.8 m/s^2, for the whole trip. Hence we have motion with uniform acceleration under gravity, and so

$$v = v_0 + at = v_0 + gt$$

or

$$t = \frac{v - v_0}{g} = \frac{36 \text{ m/s} - 0}{9.8 \text{ m/s}^2} = \boxed{3.7 \text{ s}}$$

Hence in this case it takes the iron pellet twice as long to reach its terminal speed in the presence of air resistance as it would take to reach the same speed in the absence of air resistance.

3.9 Using Newton's Laws of Motion

Because of the importance of Newton's laws of motion in so many fields of physics, we now focus our attention on how these laws may be used to solve a great variety of practical problems in physics. (For some general ideas about solving physics problems, see Appendix 3.)

In applying Newton's laws, we must first isolate one particular object in whose motion we are interested. We then consider all the forces acting *on that object*.

Free-Body Diagrams

In dealing with force vectors, it is often useful to eliminate the extraneous detail in a diagram of the physical situation, and simply draw all the force vectors necessary to describe the situation from the point of view of physics. Such diagrams are called *free-body diagrams*. Thus if a boy tries to lift a book off his desk by exerting a force **F** upward, and the weight of the book is **w** (directed downward) (Fig. 3.16a), then in a free-body diagram we omit the book and the desk and merely show the two vector forces acting on the book, as in Fig. 3.16b. This preserves the essentials of the physical situation. We have abstracted these essentials from the full picture, which would include the shape and color of the book, the kind of wood in the table, etc., since these have no effect on what happens in this particular case. The ability to abstract the physical essentials of a situation from the surrounding complexities, as is done here, is the secret of the successful physicist or physics student.

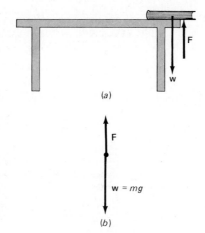

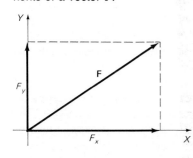

FIGURE 3.16 (a) A book resting on the edge of a desk. (b) The corresponding free-body diagram for the forces acting *on the book*.

Components of Forces

The forces, which are vectors, can then be broken up into components along axes chosen to make the solving of the problem as easy as possible. The same is true for the accelerations and resulting velocities produced by the forces. For example, if we are considering motion in a horizontal direction, it is most convenient to break up Newton's second law into two equations, one applying to the horizontal (X) direction, and the other to the vertical (Y) direction (because the force of gravity acts in this direction). Then

$$\mathbf{F}_{net} = m\mathbf{a}$$

becomes $(F_{net})_x = ma_x$ and $(F_{net})_y = ma_y$

These two equations, together with the equations for uniformly accelerated motion in one dimension, will enable us to solve a great variety of problems. The components of the forces and accelerations can be found graphically, as illustrated in Example 3.11 below, or by the principles of trigonometry discussed in Appendix 1C. We find that

FIGURE 3.17 The components of a vector **F**.

$$(F_{net})_x = F_{net} \cos \theta \qquad \text{and} \qquad (F_{net})_y = F_{net} \sin \theta$$

and similarly,

$$a_x = a \cos \theta \qquad \text{and} \qquad a_y = a \sin \theta$$

where θ is the angle between the force $\mathbf{F}_{net}$ (or the acceleration $\mathbf{a}$) and the X axis, as shown in Fig. 3.17.

If the motion is not along the horizontal but down an inclined plane, then the best choice of components is one along the plane and the other normal to the plane. The proper choice of a coordinate system for a particular problem can often simplify its solution greatly.

Example 3.8

A trunk of mass 20 kg rests on a floor. What constant horizontal force is required to give it a velocity of 10 m/s in 20 s, if the coefficient of sliding friction between trunk and floor is 0.65?

SOLUTION

First draw a clear diagram of the situation, and then draw a free-body diagram, identifying all the forces. Here, in addition to the applied force f and the frictional force F_k, which are in the X direction, we have two vertical forces, the weight w of the trunk pulled down by the earth, and the reaction force F_N of the floor pushing upward on the trunk, as in Fig. 3.18.

Since all the motion in this case is in the X direction, a_y is equal to zero, and our two basic equations become:

$$(F_{net})_x = ma_x \qquad (F_{net})_y = ma_y = 0$$

or $\quad f - F_k = ma_x \qquad F_N - w = 0$

From the second equation

$$F_N = w = mg = (20 \text{ kg})(9.8 \text{ m/s}^2) = 196 \text{ N}$$

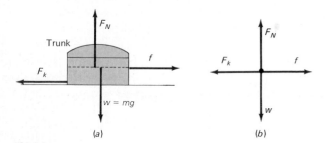

Trunk

(a)

(b)

FIGURE 3.18 Diagram for Example 3.8: (a) A trunk sliding on a floor with friction. (b) Free-body diagram of the same physical situation.

Hence the normal force is equal to the weight of the trunk, 196 N. Then

$$F_k = \mu_k F_N = (0.65)(196 \text{ N}) = 127 \text{ N}$$

For the X component we then have, from above,

$$f = F_k + ma_x$$

where

$$a_x = \frac{\Delta v_x}{\Delta t} = \frac{10 \text{ m/s} - 0}{20 \text{ s}} = 0.50 \text{ m/s}^2$$

Hence

$$f = 127 \text{ N} + (20 \text{ kg}) (0.50 \text{ m/s}^2) = \boxed{137 \text{ N}}$$

This is the force which must be applied to overcome friction and produce the required acceleration.

In general, in solving problems like this it is a good idea to make a list of the known and unknown quantities. This list often provides a clue to the equations needed to solve the problem. In this case m is given, a_x can be obtained from the given velocities and times, and F_k and f are unknowns. Since there are two unknowns we need two equations to determine them. The first equation $F_k = \mu_k F_N$, together with the fact that F_N is numerically equal to the weight w, yields the value of F_k. Newton's second law then becomes $(F_{net})_x = f - F_k = ma_x$, and this equation can then be solved for the applied force f.

Example 3.9

An elevator and its contents add up to a mass of 1000 kg. The elevator is initially moving downward at 4.0 m/s, and is brought to rest in a distance of 30 m by a constant force applied by the support cable of the elevator. Find the upward force which must be applied by the support cable (see Fig. 3.19).

FIGURE 3.19 Diagram for Example 3.9: (a) An elevator being brought to rest. (b) Free-body diagram of the same situation.

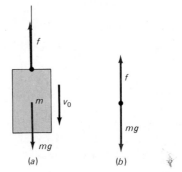

(a)

(b)

SOLUTION

Here there are no horizontal forces because the elevator does not touch the walls of the elevator shaft, and so Newton's second law need only be applied to the vertical forces. We have

$$(F_{net})_y = ma_y$$

or $\quad f - w = ma_y$

where f is the applied force. We choose a coordinate system in which the upward direction is positive. Hence f and a_y are positive because they are directed upward, and w is negative because it is directed downward. Let us make a list of the knowns and unknowns to be used in solving this problem. In all cases the proper sign must be attached to each quantity.

Knowns　　　　　　　　*Unknowns*
$m = 1000$ kg　　　　　$a_y = ?$
$v_0 = -4.0$ m/s; $v = 0$　$f = ?$
$y_0 = 30$ m; $y = 0$

Since there are two unknowns in this case, we need two equations to find the two unknowns.

The first step in the solution is to find a_y, since once a_y is known, f can be found from Newton's second law applied to the vertical motion. To find a_y we need an equation containing positions, velocities, and acceleration, but not time. The only such equation is Eq. (2.12):

$$v^2 = v_0^2 + 2a_y(y - y_0)$$

from which we get

$$a_y = \frac{v^2 - v_0^2}{2(y - y_0)} = \frac{0 - (-4.0 \text{ m/s})^2}{2(0 - 30 \text{ m})}$$

$$= \frac{-16 \text{ m}}{-60 \text{ s}^2} = 0.27 \text{ m/s}^2$$

where this acceleration is positive and hence in the upward direction.

To find f we use the equation above, $f - w = ma$, or

$$f = mg + ma_y = m(g + a_y)$$

$$= (1000 \text{ kg})(9.80 + 0.27) \text{ m/s}^2 = 1.01 \times 10^4 \text{ kg} \cdot \text{m/s}^2$$

$$\boxed{= 1.01 \times 10^4 \text{ N}}$$

This upward force is therefore greater than the weight by 0.27×10^3 N, and this net unbalanced force produces the upward acceleration, which causes the elevator to stop.

Example 3.10

A block of mass $m_2 = 2.0$ kg moves on a level, frictionless surface. It is connected by a light, flexible, nonstretching cord passing over a light frictionless pulley to a falling block of mass $m_1 = 1.0$ kg, as in Fig. 3.20.

(a) What is the acceleration of the system?
(b) What is the tension in the cord between the two blocks?

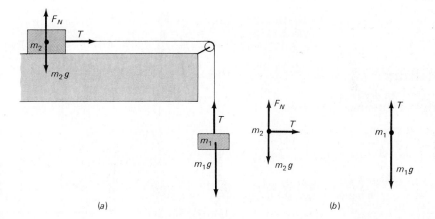

FIGURE 3.20 Diagram for Example 3.10: (a) Two masses at the end of a string. (b) Free-body diagrams.

(a)

(b)

SOLUTION

This may seem like a two-dimensional problem, but it really is not since each block moves along a straight line, although their motions are connected because they are tied together by the cord.

(a) Applying Newton's second law to m_1, we have

$$(F_{\text{net}})_y = w_1 - T = m_1 a$$

where m_1 and w_1 are known, and T and a are to be found. For m_2 we have

$$(F_{\text{net}})_x = T = m_2 a$$

where m_2 is known, but T and a are again unknown. The tension in the cord must be the same throughout, since the cord is assumed to have no mass and therefore no force is required to accelerate it. We can therefore eliminate T between the two equations above by adding

$$w_1 - T = m_1 a \quad \text{and} \quad T = m_2 a$$

to obtain

$$w_1 = m_1g = (m_1 + m_2)a$$

or

$$a = \frac{m_1g}{m_1 + m_2} = \left(\frac{1\ \text{kg}}{2\ \text{kg} + 1\ \text{kg}}\right)(9.80\ \text{m/s}^2) = \boxed{3.27\ \text{m/s}^2}$$

(b) Since $T = m_2a$, we have

$$T = (2\ \text{kg})\ (3.27\ \text{m/s}^2) = \boxed{6.54\ \text{N}}$$

This checks out well, since 6.54 N produces the same acceleration in the 2-kg mass as does $9.80 - 6.54\ \text{N} = 3.27\ \text{N}$ in the l-kg mass. Note that here the 1-kg mass acted on by gravity is able to move the 2-kg mass along the surface of the table, even though the 1-kg mass is smaller. This should not surprise us, since there is assumed to be no frictional or other opposing force in the horizontal direction. Hence even the smallest tension in the string (say, 0.0001 N) would still be enough to accelerate the 2-kg block along the table.

If there were friction between the block and the horizontal surface, our equation for the motion of m_2 would have to be modified to

$$(F_{net})_x = T - F_k = m_2a$$

Example 3.11

A child pulls a sled of mass 50 kg by exerting a force **F** of 60 N on a rope tied to the sled at an angle of 35° with respect to the ground, as shown in Fig. 3.21. (Assume that frictional forces are very small.)
(a) What is the acceleration of the sled?
(b) What is the normal force f_N of the ground on the runners of the sled?

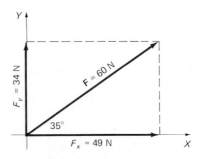

(a)

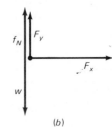

(b)

FIGURE 3.21 Diagram for Example 3.11: (a) A child pulling a sled. (b) Free-body diagram.

SOLUTION

Here we need components of a force **F** along the X and Y axes.* These components are:

$$F_x = F \cos 35° = (60\ \text{N})(0.82) = 49\ \text{N}$$

$$F_y = F \sin 35° = (60\ \text{N})(0.57) = 34\ \text{N}$$

as shown in Fig. 3-22.

Then, applying Newton's second law of motion,

$$(F_{net})_x = F \cos 35° = ma_x$$

$$(F_{net})_y = f_N + F \sin 35° - w = 0$$

Hence

(a) $a_x = \dfrac{F \cos 35°}{m} = \dfrac{49\ \text{N}}{50\ \text{kg}} = \boxed{0.98\ \text{m/s}^2}$

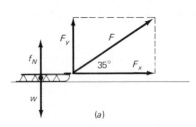

FIGURE 3.22 Components of a vector found either by graphing or by the use of trigonometry.

---

*Students not familiar with trigonometry should consult Appendix 1C at this point.

(b) $f_N = w - F \sin 35° = mg - F \sin 35°$

$\quad\quad = (50 \text{ kg})(9.8 \text{ m/s}^2) - 34 \text{ N} = 490 \text{ N} - 34 \text{ N}$

$\quad\quad = \boxed{456 \text{ N}}$

In this case the pulling force both accelerates the sled in the X direction and reduces the force in the Y direction exerted on the ground by the sled's runners. Then, since action equals reaction, the reaction force f_N is also reduced. This reduction in the normal force can be important in a problem where there is friction between the sled and the ground, since the decrease in the normal force f_N means a corresponding reduction in the frictional force, for the force of sliding friction is $f_k = \mu_k f_N$, according to Eq. (3.11). For this reason it is easier to pull a wheelbarrow than to push one, for pulling reduces the normal force whereas pushing increases the normal force and hence the force of friction.

Summary: Important Definitions and Equations

Force: That which, when applied to an object, tends to produce the acceleration of that object. Force is a vector. One newton (N) is the magnitude of that force which will give a one-kilogram mass an acceleration of one meter per second squared.

Force of gravity:

Newton's law of universal gravitation:

$$F_G = \frac{G m_1 m_2}{r^2}$$

Universal gravitational constant:

$$G = 6.67 \times 10^{-11} \text{ m}^3/(\text{kg} \cdot \text{s}^2)$$

For earth's gravitation: $g = \dfrac{F_G}{m} = \dfrac{G m_E}{r^2}$

The acceleration due to gravity, g, is the same for all objects at any point on the earth's surface, if air resistance is neglected.

Special forces:

Weight: The downward gravitational force exerted on an object as a consequence of Newton's law of universal gravitation. The magnitude of this force is $w = mg$. Weight is a force (a gravitational force); mass is a measure of inertia.

Force of friction: A force, due to the unevenness of two surfaces in contact, which tends to make the objects stick together and opposes the relative motion of one object with respect to the other.

Static friction: The frictional force exerted by one surface on another when there is no relative motion of the two surfaces.

$$F_s \leq \mu_s F_N$$

where μ_s is the coefficient of static friction and F_N is the normal force between the two surfaces.

Sliding (kinetic) friction: The frictional force exerted by one surface on another when one surface is in relative motion with respect to the other.

$$F_k = \mu_k F_N$$

where μ_k is the coefficient of sliding friction.

Air resistance: The resistance of a mass of air to the passage of an object through it.

Viscosity: The resistance of a fluid to flow, caused by the internal friction of the fluid molecules moving against each other.

$$F_{\text{vis}} = K v$$

(which is an approximation valid for small objects moving at slow speeds).

Terminal velocity: The limiting downward speed reached by objects falling in a resistive medium when the resistance to the motion becomes equal to the force of gravity on the object.

$$v_{\text{term}} = \frac{mg}{K}$$

Newton's laws of motion:

1 An object at rest will remain at rest and an object in motion will continue to move in a straight line at constant speed forever unless some net external force acts to change this motion.

2 A net force $\mathbf{F}_{\text{net}}$ acting on a mass m produces an acceleration $\mathbf{a}$ in the direction of the force, where

$$\mathbf{F}_{\text{net}} = m\mathbf{a}$$

3 Whenever an object A exerts a force on a second object B, object B exerts a force back on A; these two forces are equal in magnitude, but opposite in direction, and act on different objects.

Inertial frame of reference: Any reference frame in which Newton's laws of motion are valid.

Questions

1 (*a*) If a rock is buried deep beneath the surface of the earth, is the force of gravity on the rock greater or less than it would be at the surface of the earth?

(*b*) What would be the force of gravity on the rock at the exact center of the earth?

2 Suppose a shaft could be drilled through the earth from the north to the south pole. A ball might then be dropped down the shaft at the north pole.

(*a*) Describe the motion of the ball as a function of time.

(*b*) What would be the ball's final position and velocity?

3 Newton's third law tells us that if a horse pulls on a cart with a force of 100 N, then the cart pulls back on the horse with a force of 100 N. If this is so, why do the cart and horse move forward?

4 Why is it necessary to push harder on the pedals of a bicycle when first starting out than when the bike is moving at a constant speed?

5 A man holding a bouquet of flowers is riding in an elevator when the cable breaks. In his panic he drops the flowers. Will the flowers stay near his hand? Hit the floor of the elevator? Hit the roof of the elevator? Give reasons for your answer.

6 Automobile riders often suffer whiplash when their car is suddenly struck from the rear by another car. In whiplash the head is thrown violently backward. Explain why this happens, using Newton's laws of motion.

7 A brick hangs by a thin cord from a ceiling and another thin cord dangles from the bottom of the brick.

(*a*) If the bottom cord is suddenly jerked, what will happen?

(*b*) If a slow, steady pull is applied to the bottom cord, what will happen? Why?

8 A frictionless puck is given a push along a very long, flat plane which makes contact with the earth's surface at one point. Describe the motion of the puck along the plane.

9 A trick that courageous party-goers sometimes perform is to pull the tablecloth out from under the china and glassware. If the trickster is lucky enough to pull this off, which of Newton's laws provides the secret to his or her success? Explain your answer.

10 Discuss the difference you would expect to notice in the force required to lift a person off the surface of the moon, and the force required to knock the person down with a football block, again on the surface of the moon.

11 A ball is thrown vertically up into the air. If air resistance is taken into account, will the time during which the ball rises be equal to the time during which the ball falls? If not, will the time of rise be longer or shorter than the time of fall? Why?

12 A glider slides down a metal track which is frictionless except for one spot near its center where it has been roughened up to slow down the glider. At the bottom of the track is a coiled bumper spring, as shown in Fig. 3.23. The glider hits the spring, compresses it, is pushed by the spring back up the track, and then falls back to the bumper again.

(*a*) Draw a graph of the position of the glider along the track (call this *x*) versus time.

(*b*) Do the same for the velocity of the glider as a function of time.

(*c*) Do the same for the acceleration of the glider as a function of time.

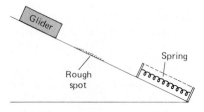

FIGURE 3.23 Diagram for Question 12.

Problems

A 1 The acceleration due to gravity *g* on the surface of the sun is larger than on the surface of the earth by a factor of about:
(*a*) 10^2 (*b*) 10^3 (*c*) 10^4 (*d*) 10 (*e*) 10^5

A 2 On the moon the acceleration due to gravity is about one-sixth its value on earth. The mass on the moon of an astronaut whose mass on earth is 100 kg is:
(*a*) 16.7 kg (*b*) 980 kg (*c*) 163 kg
(*d*) 100 N (*e*) 100 kg

A 3 The acceleration of the earth produced by the gravitational attraction of a rock whose weight is 20 N on the surface of the earth is:
(*a*) 9.8 m/s^2 (*b*) 3.3×10^{-2} m/s^2 (*c*) 196 m/s^2
(*d*) 0.49 m/s^2 (*e*) None of the above

A 4 A rock of mass 50 kg is resting on a concrete driveway. The force exerted by the driveway on the rock is:
(*a*) 0 (*b*) 50 N (*c*) 490 N (*d*) 5.1 N
(*e*) None of the above

A 5 A 2000-kg car is moving along a road with a constant speed of 20 m/s. The net force exerted on the car in the direction of its motion is:
(*a*) 4.0×10^4 N (*b*) 100 N (*c*) 2.0×10^4 N
(*d*) 1.96×10^4 N (*e*) None of the above

A 6 The force required to accelerate a 200-kg car from rest to a speed of 30 m/s in 10 s is:
(*a*) 600 N (*b*) 6000 N (*c*) 6.0×10^5 N
(*d*) 6.0×10^4 N (*e*) 667 N

A 7 The weight of a woman is 550 N. Her mass is:
(*a*) 550 kg (*b*) 5.4 × 10³ kg (*c*) 5.4 × 10² kg
(*d*) 56 lb (*e*) 56 kg

A 8 A man has a mass of 80 kg. His weight in newtons is:
(*a*) 80 N (*b*) 250 N (*c*) 780 N (*d*) 32 N
(*e*) 9.8 N

A 9 A man pushes a 2000-kg car by exerting a force of 50 N. Neglecting frictional forces, the car's acceleration is:
(*a*) 9.8 m/s² (*b*) 40 m/s² (*c*) $\frac{1}{40}$ m/s²
(*d*) 0.025 m/s (*e*) 32 m/s²

A10 A sky diver has a mass of 100 kg. She jumps from a plane and at a certain point in her descent the air resistance is 250 N. Her downward acceleration at that point in the descent is:
(*a*) 9.8 m/s² (*b*) 7.3 m/s² (*c*) 0 (*d*) 9.8 m/s
(*e*) 4.9 m/s²

B 1 Calculate the gravitational attraction of the sun for the moon when the moon is at its closest position to the sun.

B 2 A trunk of mass 50 kg rests on a floor. What constant horizontal force is required to give it a velocity of 20 m/s in 10 s, if we neglect air resistance and friction?

B 3 A force is applied to a 2.0-kg mass and produces an acceleration of 4.0 m/s². The same force applied to a 10-kg mass will produce what acceleration?

B 4 A tow truck is towing a car with a short rope. If the rope has negligible mass, the tow truck is accelerating at 1.0 m/s², and the mass of the car is 1500 kg, what is the tension in the rope?

B 5 What is the mass of a 250-lb football player?

B 6 (*a*) An elevator is accelerating upward with an acceleration of 1.2 m/s². What is the upward force exerted on the elevator passengers by the floor of the elevator?
(*b*) If the same elevator accelerates downward with the same acceleration as in (*a*), what is the upward force exerted by the elevator floor on the passengers?

B 7 A disk of mass 3.0 kg is supported by a jet of air on a metal table. The disk is connected by a light, flexible cord passing over a light, frictionless pulley to a 4.0-kg mass which is free to fall. If the 4.0-kg mass is falling, what is the acceleration of the system?

B 8 A boy is pushing on a heavy table which weighs 500 N with a force of 100 N. The table refuses to budge. (*a*) What is the force of friction acting on the table? (*b*) What is the coefficient of static friction at the instant that the applied force is 100 N?

B 9 A box of 5.0-kg mass is being pulled by a horizontal rope along a floor. If the coefficient of sliding friction between box and floor is 0.62, what is the frictional retarding force on the box?

B10 The force of air resistance on a raindrop is 5.0 × 10⁻⁵ N when it falls with a terminal velocity of 5.5 m/s. What is the mass of the raindrop?

C 1 The earth has a bulge at the equator which makes its radius at the equator larger than its radius at the poles by about 21 km. What is the difference in the value of *g* at the north pole and at the equator? (Hint: Use the power series expansion discussed in Appendix 1D)

C 2 How far above the earth's surface would an earth satellite have to be for the force of gravity on it (due to the earth) to be reduced to one-tenth its value on the earth?

C 3 A force of 2.0 × 10³ N is applied to a 1500-kg car being towed by a tow truck. If this force is applied for 5.0 s and the car starts from rest:
(*a*) What is the acceleration of the car?
(*b*) What is its average speed during these 5 s?

C 4 In a test of the safety of automobiles a car is deliberately crashed into a brick wall. If the 2000-kg car is traveling at 40 m/s when it hits the wall, and its speed is reduced to zero in 0.01 s:
(*a*) What force does the wall exert on the car?
(*b*) What force does the car exert on the wall?
(*c*) How far would the car travel in the 0.01 s, if the wall were not completely rigid?

C 5 A rocket-propelled sled on a level track is sometimes used to investigate the physiological effects of large accelerations on human beings. One such sled can reach a speed of 1500 km/h in 2.0 s; starting from rest.
(*a*) If the acceleration is constant, how much larger is this acceleration than the acceleration due to gravity *g*? (This is usually expressed by saying that the person being tested was subjected to so many *g*'s.)
(*b*) How far will the sled travel in these first 2.0 s?

C 6 A 90-kg astronaut lies on a couch in a rocket accelerating upward with an acceleration of 9.8 m/s². What force does the astronaut exert on the couch?

C 7 A golfer exerts an average force of 50 N for 0.10 s on a golf ball weighing 0.40 N.
(*a*) Find the average acceleration of the golf ball.
(*b*) Find the speed of the ball at the end of the 0.10-s interval.

C 8 An average person jumping straight up in the air by pushing off from the floor with his or her legs can exert a force on the floor roughly equal to twice the person's weight. If this force is applied over a distance of about 0.50 m (from a crouching to a stretching position):
(*a*) With what speed does the person leave the ground?
(*b*) How high will the person be able to jump?

C 9 In the same situation outlined in Prob. C8, suppose the person were not on the earth but on the moon. If the muscle forces of the person were the same in the two cases, about how high could the person jump on the moon?

C10 A light cord passes over a light, frictionless pulley. It has a 5.0-kg mass at one end and a 10-kg mass at the other. Find (*a*) the acceleration of the system; (*b*) the tension in the cord.

C11 A tow truck is pulling a disabled car by using a rope which will only stand a tension of 1200 N. If the car has a mass of 1000 kg, what is the largest acceleration possible for the car?

C12 A steel shuffleboard puck slides over a concrete surface, where the coefficient of sliding friction between the two surfaces is 0.40. If the mass of the puck is 0.30 kg, how fast must the puck be pushed by a player if it is to come to rest in the scoring area, which is 15 m away?

C13 A 1000-kg car is coasting along a level road at 25 m/s.

(*a*) How large a retarding force is required to stop it in a distance of 60 m, if the force is constant over the whole 60 m?

(*b*) If the brakes stop the wheels completely and the car skids the entire 60 m, what is the minimum coefficient of sliding friction between tires and road which will make this possible?

C14 A 50-kg crate is pushed along a level floor. The coefficient of static friction is 0.45; the coefficient of kinetic friction is 0.40.

(*a*) What horizontal force will start the crate in motion?

(*b*) If the crate is pushed horizontally with a force of 400 N, how far does it move in 3.0 s?

C15 A bookcase of mass 10 kg is being moved from one room to another by sliding the bookcase along the floor on a piece of cardboard which acts to reduce the friction that would otherwise result between wood and wood. If the coefficient of sliding friction between cardboard and floor is 0.35, what constant horizontal force is required to give the bookcase a speed of 0.50 m/s in 5.0 s?

C16 Sky divers with open parachutes are often said to hit the ground with terminal velocities about equal to the velocity reached by a person who jumps out of a second-story window.

(*a*) Approximately what is this velocity?

(*b*) What is the value of the drag coefficient K in this case?

C17 A raindrop of mass 6.0×10^{-6} kg falls in air and reaches a terminal speed of 5.5 m/s.

(*a*) How long does it take the raindrop to reach this terminal speed?

(*b*) If there were no air resistance, how long would it take the raindrop to reach the same speed?

C18 An elevator starts from rest with a constant upward acceleration and moves 1.0 m in the first 4.0 s. A passenger in the elevator is holding a 10-kg bundle at the end of a vertical cord. What is the tension in the cord as the elevator accelerates?

C19 A 100-kg mass hangs by a cord from the ceiling of an elevator. If we neglect the weight of the cord, what is the tension in the cord when:

(*a*) The elevator is accelerating upward at 1.0 m/s²?

(*b*) The elevator is accelerating downward at 1.0 m/s²?

(*c*) The elevator is accelerating downward at 9.8 m/s²?

C20 A person is standing on a spring bathroom scale on the floor of an elevator which is being accelerated. If the person's mass is 90 kg and the elevator accelerates at 2.5 m/s², what does the scale read for the person's weight (*a*) when the elevator is accelerating upwards; (*b*) when the elevator is accelerating downward?

C21 A spring balance is hung from the roof of an elevator cab, and a 10-kg mass is hung from the bottom of the spring balance.

(*a*) If the balance is calibrated in newtons, what does the balance read for the weight of the mass when the elevator is stopped at a floor?

(*b*) What does the balance read when the elevator is accelerating upward at 1.5 m/s²?

(*c*) What does the balance read when the elevator is accelerating downward at 1.5 m/s²?

D 1 The distance between the earth and the moon is 3.84×10^8 m. At what point between the earth and the moon is the net gravitational force on any object equal to zero?

D 2 Prove that if Newton's second law of motion is valid in one inertial reference frame, it is also valid in any reference frame moving with constant velocity with respect to the first.

D 3 Two blocks A and B are connected by a heavy uniform rope of mass 2.0 kg, as shown in Fig. 3.24. The masses of A and B are 10 and 5.0 kg, respectively. If an upward force of 200 N is applied to the top block:

(*a*) What is the acceleration of the system?

(*b*) What is the tension at point P_1 on the heavy rope?

(*c*) What is the tension at point P_2 on the heavy rope?

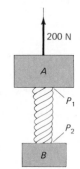

FIGURE 3.24 Diagram for Prob. D3.

D 4 A U-Haul trailer is being pulled by a heavy chain attached to a car. The mass of the trailer is 100 kg and the mass of the chain is 5.0 kg. If the car pulls on the chain with a force of 50 N:

(*a*) What is the acceleration of the trailer?

(*b*) What is the force applied to the trailer by the chain?

(*c*) Why is the force different than the force applied by the car?

D 5 Two blocks A and C are attached to a third block B resting on a level surface, as in Fig. 3.25. There is friction between block B and the surface with a coefficient of sliding friction $\mu_k = 0.20$. If block A has a mass of 2.0 kg and block C a mass of 10.0 kg, and block B moves to the right with an acceleration of 3.0 m/s²:

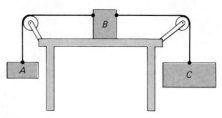

FIGURE 3.25 Diagram for Prob. D5.

(*a*) What is the magnitude of the mass *B*?
(*b*) What is the tension in each cord?

D 6 A 1500-kg car is traveling at 25 m/s when the driver stops accelerating. If it takes 5.0 s for the car to slow down to 20 m/s, how large is the force slowing down the car?

D 7 A person tries to move a 2000-kg car by attaching a rope to its front bumper and pulling up on the rope at an angle of 30° with respect to the horizontal. The coefficient of static friction between the car's tires and the road is 0.80. How great a force would the person have to exert to move the car?

D 8 In the state of California, San Francisco and Mount Hamilton have about the same latitude and longitude, but Mount Hamilton is 1282 m above sea level whereas San Francisco is only 114 m above sea level. If the acceleration due to gravity in San Francisco is 9.7997 m/s^2, what would you expect its value to be at the top of Mount Hamilton?

Additional Readings

Andrade, E. N. da C.: *Sir Isaac Newton*, Doubleday Anchor, Garden City, N.Y., 1958. A good brief biography of Newton.

Asimov, Isaac: *Understanding Physics: Motion, Sound and Heat,* New American Library, New York, 1969. A popular, accurate and interesting presentation of mechanics.

Cohen, I. Bernard: "Newton's Discovery of Gravity," *Scientific American*, vol. 244, no. 3, March 1981, pp. 166–173. A careful presentation of the historical facts about Newton's discovery of the law of universal gravitation.

Gamow, George: *Gravity*, Doubleday Anchor, Garden City, N.Y., 1962. A popular account of both classical and modern theories of the gravitational field.

Heiskanen, Weikko A.: "The Earth's Gravity," *Scientific American*, vol. 193, no. 3, September 1955, pp. 164–174. An interesting account of why *g* varies from place to place on the earth's surface.

Palmer, Frederic: "Friction," *Scientific American*, vol. 184, no. 2, February 1951, pp. 54–58. An article pointing out how difficult it is to explain one of our most common physical phenomena.

Westfall, Richard A.: *Never at Rest*, Cambridge University Press, New York, 1981. This is the latest and most scholarly biography of Newton.

From being the preoccupation of a few curious spirits science has grown to be a universal study, on the fruits of which peace among people and the prosperity of nations depend, but the great principles enunciated by Newton and their orderly development by him remain as the foundation of the discipline and as a shiny example of the exalted power of the human mind.*

E. N. da C. Andrade

Chapter 4
Equilibrium

The preceding chapter might lead us to think that forces always lead to motion. Nothing could be further from the truth. Any net unbalanced force acting on an object accelerates that object. But a large number of forces may act on an object and it may still remain at rest, as long as the vector forces applied add to zero.

This balancing out of forces, a physical condition usually referred to as equilibrium, is the main theme of this chapter. *Equilibrium* is that condition in which two or more forces acting on an object produce no acceleration because they balance each other out.

Some examples of equilibrium are of great practical importance. A gigantic ocean-going oil tanker at anchor in a harbor is at rest, but it experiences both the very large gravitational pull of the earth and the equally large buoyant force of the water. Since these two forces are equal and opposite, the oil tanker remains at rest. Similarly, the arches in a cathedral are subjected to forces in a variety of directions. But the architects have so planned the cathedral that all the forces on the arches cancel each other out. If this were not the case, the arches would soon shift position and the cathedral would collapse.

In this chapter we will first discuss ways to add vectors in more than one dimension to obtain their resultant, and we will then apply these ideas to force vectors and to equilibrium situations.

4.1 Graphical Addition of Vectors

A number of techniques can be used to find the resultant of two or more vectors, whether they are displacements, velocities, accelerations, or forces. A simple and reliable technique is a graphical one. It requires only a ruler, a protractor, and a sharp pencil.

Suppose we are told that a cyclist pedaled 40 mi to the east and then 40 mi to the northeast with respect to the origin of a coordinate system. What is the cyclist's final position? These two displacement vectors are shown in Fig.

*See Additional Readings for source of quote.

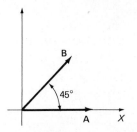

FIGURE 4.1 Two displacement vectors **A** and **B** drawn from the same origin.

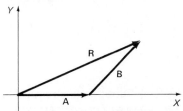

FIGURE 4.2 Triangle method for adding the two displacement vectors **A** and **B** in Fig. 4.1.

FIGURE 4.3 Graphical method of adding five vectors **A**, **B**, **C**, **D**, and **E** to obtain the resultant vector **R**, using the tail-to-tip method.

Subtraction of Vectors

FIGURE 4.4 The subtraction of two vectors. **R** is the difference vector **A** − **B**, obtained by reversing the direction of **B** and then adding the resulting vector to **A**.

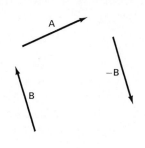

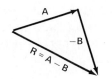

4.1, drawn from the same origin, using a convenient scale of 1 cm = 20 mi. Then each vector is 2 cm in length, one along the X axis, the other at an angle of 45° with respect to the X axis, as laid off with a protractor. The resultant of these two vectors can be obtained by drawing them with the tail of **B** starting at the tip of **A**, as in Fig. 4.2, and then joining the tail of **A** to the tip of **B**. Then all we need do is complete the triangle to find the magnitude and direction of the resultant, which can be measured with a ruler and protractor. In this case **R** is 74 mi at an angle of 23° with respect to the Y axis.

Such an approach, usually called the *triangle method*, conforms well with the idea of one displacement following the other in succession, as is the case in this example.

If we have more than two vectors, and draw them one after the other, tail to tip, then the resultant is the vector obtained by joining the tail of the first vector to the tip of the last vector. Figure 4.3 shows this for a situation where five vectors are added. The resultant of the five vectors **A**, **B**, **C**, **D**, and **E** is then the vector **R**, constructed by connecting the tail of **A** to the tip of **E**. This is sometimes referred to as the *tip-to-tail method* of adding vectors. Probably tail-to-tip method would be a more accurate description.

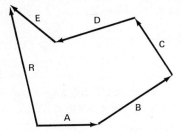

The above discussion shows us how to add two vectors **A** and **B** to get the resultant **R**; i.e.,

$$\mathbf{A} + \mathbf{B} = \mathbf{R} \tag{4.1}$$

where the boldface type indicates that we are adding *vectors*.

What about the subtraction of two vectors, say **A** − **B**? Since

$$\mathbf{A} - \mathbf{B} = \mathbf{A} + (-\mathbf{B}) \tag{4.2}$$

all we need do to subtract **B** from **A** is to reverse the direction of **B** and then add this vector graphically to **A**. An example is given in Fig. 4.4, where two vectors **A** and **B** are shown, and we are asked to find their difference **A** − **B**. The diagram shows **B** and −**B**, and shows how by adding **A** and −**B** together we can get the resultant **R** = **A** − **B**.

If a parallelogram is constructed with vectors **A** and **B** as its sides, Fig. 4.5 shows that the long diagonal of the parallelogram is the sum **A** + **B**. The short diagonal is then **A** − **B**, since it is clear from the figure that **B** + (**A** − **B**) = **A**. It can be seen from the figure that this result for **A** − **B** is the same in both magnitude and direction as was obtained in Fig. 4.4, using the triangle of forces.

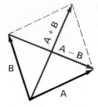

FIGURE 4.5 Parallelogram method: The long diagonal is the sum **A** + **B**; the short diagonal is the difference **A** − **B**.

Example 4.1

A surveyor is attempting to find the width of the Hudson River at a point north of New York City. She sets up her transit at a point A on the east bank of the river and sights directly across the river to a flag on the west bank at point B. She then moves up the river 500 m to point C and again sights on the same flag. She finds that this line of sight makes an angle of 70° with respect to the line from A to C. What is the width AB of the river?

SOLUTION

This is the method surveyors call *triangulation*. We plot the 500 m on graph paper on some convenient scale, as in Fig. 4.6. We also lay down a perpendicular to the line AC at point A, and a line making an angle of 70° with the line AC at point C. These two lines will intersect at the flag used as a marker at point B. Once the graph is complete, the distances can be read directly off the graph and the width of the river determined. In this case it turns out to be 1400 m, as can be seen from Fig. 4.6.

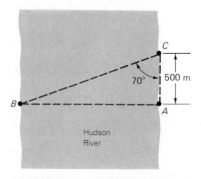

FIGURE 4.6 Diagram for Example 4.1.

Example 4.2

We want to combine two velocity vectors **A** and **B**. Vector **A** has a magnitude of 10 m/s and is directed along the positive X axis. Vector **B** has a magnitude of 20 m/s and is directed at an angle of 60° above the X axis, as shown in Fig. 4-7. Find the magnitude of **A** − **B**.

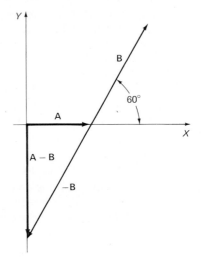

FIGURE 4.7 Diagram for Example 4.2.

SOLUTION

We first draw the two vectors **A** and **B** to scale, as in Fig. 4.7. To subtract **B** from **A**, we first find −**B** and add this to **A**. On the graph we find −**B** by reversing the direction of **B** while keeping its magnitude the same. The resultant is found from the graph to have a magnitude of 17 m/s and to be directed at an angle of −90° with respect to the positive X axis; i.e., **A** − **B** is along the negative Y axis.

4.2 Resolution of Vectors into Rectangular Components

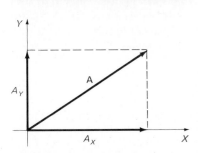

FIGURE 4.8 Rectangular resolution of a vector **A** into components A_X and A_Y.

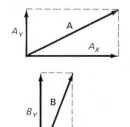

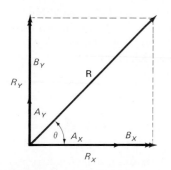

FIGURE 4.9 Adding two vectors **A** and **B** by using rectangular resolution. Here $R^2 = R_X^2 + R_Y^2$, and $\tan\theta = R_Y/R_X$.

Another approach to the addition of vectors in two dimensions is to break up each vector into its rectangular components along the X and Y axes, and then combine all the X components to get the X component of the resultant and all the Y components to get the Y component of the resultant vector **R**. Thus suppose we have a vector **A** in the XY plane, as shown in Fig. 4.8. It is always possible to break up this vector into two components, one along the X axis and the other along the Y axis. We call these components A_X and A_Y. These components are found by dropping perpendiculars from the tip of the vector **A** to the X axis to find A_X, and to the Y axis to find A_Y. It should be clear that, using our graphical rules for combining vectors, the vector sum of A_X and A_Y is the vector **A**, as Fig. 4.8 shows. Hence **A** can always be replaced by its two component vectors A_X and A_Y.

Similarly if we have a second vector **B**, we can resolve it into its rectangular components B_X and B_Y. To add **A** and **B** we can then add A_X and B_X to get R_X, and A_Y and B_Y to get R_Y, as in Fig. 4.9. We now have the two rectangular components R_X and R_Y of the vector **R**. Hence, by using the pythagorean theorem from geometry, we have for the magnitude of **R**:

$$R^2 = R_X^2 + R_Y^2$$

or
$$R = \sqrt{R_X^2 + R_Y^2} \tag{4.3}$$

where $R_X = A_X + B_X$ and $R_Y = A_Y + B_Y$

The direction of **R** (i.e., the angle θ) can then be obtained from the graph in which R_X and R_Y are combined to get **R**.

The advantage of rectangular resolution is that any number of vectors can be combined easily by this technique. For example, suppose we have a series of vectors $\mathbf{A}_1, \mathbf{A}_2, \mathbf{A}_3, \ldots$, or, in general $\mathbf{A}_i$. Here the subscript i can stand for any of the vectors by letting $i = 1, 2, 3, \ldots$. Then we can find the X and Y components of all the original vectors. We find

$$R_X = A_{1X} + A_{2X} + \cdots \qquad R_Y = A_{1Y} + A_{2Y} + \cdots$$

In more compact form this can be written

$$R_X = \sum_{i=1}^{N} (A_i)_X \qquad R_Y = \sum_{i=1}^{N} (A_i)_Y \tag{4.4}$$

This is called *summation notation*, and means that we sum over the index i, which runs through the values $1, 2, \ldots, N$, where N is the number of vectors being added. Once R_X and R_Y are known, the magnitude of **R** can be obtained from Eq. (4.3).

4.3 Trigonometric Methods for Vectors

The X and Y components of any vector can also be found from trigonometry. For example, in Fig. 4.10 the vector **A**, which makes an angle θ with the X axis, is seen to have rectangular components:

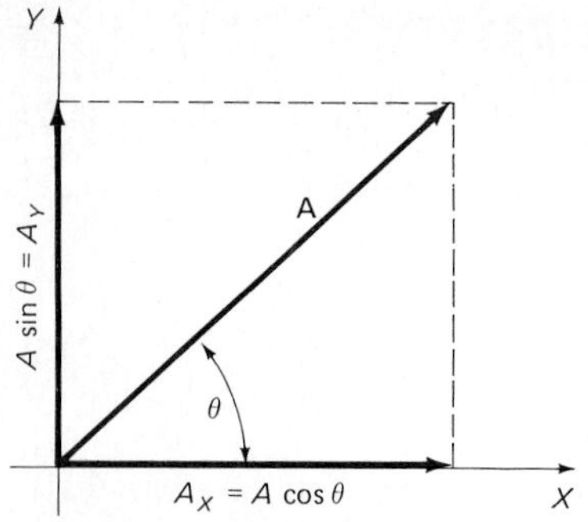

FIGURE 4.10 Use of trigonometry to obtain the rectangular components A_X and A_Y of a vector **A**.

$$A_X = A \cos \theta \qquad A_Y = A \sin \theta$$

Here the sine of an angle is the ratio of the side opposite the angle θ to the hypotenuse of the right triangle. Hence $\sin \theta = A_Y/A$, or $A_Y = A \sin \theta$. Similarly, the cosine of θ is the ratio of the side adjacent to the angle θ to the hypotenuse, so that $\cos \theta = A_X/A$, or $A_X = A \cos \theta$. Hence the resultant **R** of the vectors $\mathbf{A}_1$, $\mathbf{A}_2$, $\mathbf{A}_3$, etc., can be found from

$$R_X = A_1 \cos \theta_1 + A_2 \cos \theta_2 + A_3 \cos \theta_3 + \cdots$$

and $\quad R_Y = A_1 \sin \theta_1 + A_2 \sin \theta_2 + A_3 \sin \theta_3 + \cdots$

or $\quad R_X = \sum_{i=1}^{N} A_i \cos \theta_i \qquad$ and $\qquad R_Y = \sum_{i=1}^{N} A_i \sin \theta_i$

Using these results we can find the magnitude of **R** from

$$R = \sqrt{R_X^2 + R_Y^2} \tag{4.5}$$

The direction of **R** can then be found from the fact that

$$\tan \theta_R = \frac{R_Y}{R_X} \tag{4.6}$$

Once the magnitude and direction of **R** have been found, **R** is completely determined.

It should be pointed out that the set of rectangular coordinates chosen in combining vectors is completely arbitrary. The resultant has a magnitude and direction determined completely by the magnitudes and directions of the individual vectors being added. The use of a set of rectangular axes to combine these vectors is a mathematical device which plays no role in the final result. The individual vectors have the same magnitude no matter what axes are chosen for the resolution, since

$$\sqrt{A_X^2 + A_Y^2} = \sqrt{A^2 \sin^2 \theta + A^2 \cos^2 \theta} = A\sqrt{\sin^2 \theta + \cos^2 \theta} = A$$

and this is true for all angles, since $\sin^2 \theta + \cos^2 \theta = 1$ for all values of θ (see Appendix 1C). The direction of each vector will be different, of course, with respect to the rectangular axes chosen. But as long as the same set of axes is used for all the vectors, the direction of the vectors *with respect to one another* does not change, and hence the direction of the resultant relative to the individual vectors also does not change.

For this reason the axes X and Y used for the rectangular resolution should be chosen to make the calculation as simple as possible. For example, one vector can be chosen to be along the X axis and so have no Y component. For motion on an inclined plane, the direction along the inclined plane can be taken as the X axis, and the axis normal to the plane as the Y axis.

It is also possible to use trigonometry to calculate the magnitude and direction of the resultant of two vectors directly without using rectangular components. This involves using the law of sines and the law of cosines given in Appendix 1C.

Example 4.3

An airplane travels 120 km on a straight course, making an angle of 25° east of due north. It then changes direction and flies 180 km due east. How far has it traveled from the airport where it started its flight? In what average direction has it traveled?

SOLUTION

Call the two vectors **A** and **B**. Then, since the angle of **A** with respect to the X axis is 65°, we have the components

$A_X = (120 \cos 65°)$ km $\qquad A_Y = (120 \sin 65°)$ km
$B_X = 180 \cos 0° = 180$ km $\qquad B_Y = 180 \sin 0° = 0$

Hence

$R_X = A_X + B_X = 120(0.423)$ km $+ 180$ km $= 231$ km
$R_Y = A_Y + B_Y = 120(0.906)$ km $= 109$ km

Then

$$R = \sqrt{R_X^2 + R_Y^2} = \sqrt{(231)^2 + (109)^2} = \boxed{255 \text{ km}}$$

$$\tan \theta = \frac{R_Y}{R_X} = \frac{109}{231} = 0.472$$

$$\theta = \boxed{25°}$$

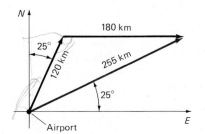

FIGURE 4.11 Diagram for Example 4.3.

Hence the plane has traveled 255 km in an average direction 25° north of east, as shown in Fig. 4.11.

4.4 Force Vectors; Forces of Tension and Compression

Since forces have both magnitude and direction, they are vectors. A force can therefore be represented by an arrow pointing in the direction of the force and of a length representing the magnitude of the force on some suitable scale. Thus a force of 5 N in the positive X direction could, in a particular case, be represented by a vector pointing to the right and of length 5 cm, where we have arbitrarily chosen a scale according to which 1 cm is equivalent to 1 N.

The resultant of two vector forces can then be obtained by graphical analysis or by trigonometry, using the techniques discussed above. Thus forces of 4 N in the positive X direction, and of 3 N in the positive Y direction, have a resultant of 5 N at an angle of 37° with respect to the positive X direction, as Fig. 4.12 shows.

For more complicated problems the best approach is usually to use rectangular resolution to combine the force vectors.

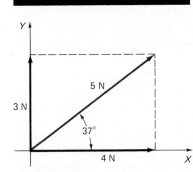

FIGURE 4.12 The composition of two vector forces to find the resultant force.

Forces of Tension and Compression

Definitions

Two important kinds of forces which occur frequently in nature are tensile forces and compressive forces (see Fig. 4.13).

Tensile force (tension): A force which acts to stretch a substance like a metal rod, a spring, or a cord.

Compressive force (compression): A force which acts to reduce the length of a substance like a metal rod or a spring by forces exerted at its two ends.

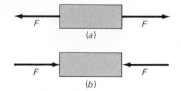

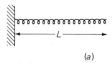

FIGURE 4.13 (*a*) Tensile forces acting on a metal rod. (*b*) Compressive forces acting on a metal rod.

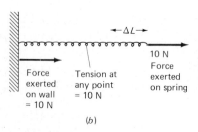

FIGURE 4.14 (*a*) A spring attached to a wall and at its equilibrium length *L*. (*b*) The forces which arise when the spring's length is increased by Δ*L*.

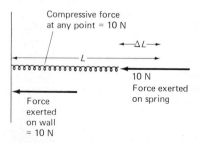

FIGURE 4.15 The compression of a spring through a distance Δ*L*.

Such forces may be understood most easily by considering a metal spring, as in Fig. 4.14. If this spring is attached to a wall and is at its normal length (*L*), and we then pull it out and thus increase its length by an amount Δ*L*, we are setting up in the spring a tension due to the forces between molecules tending to restore the spring to its original length. This force is of the same magnitude at every point along the spring. If we exert a force of 10 N on the spring, then this force is transmitted along the spring, and it in turn exerts an equal force of 10 N on the wall. We say that the *tension* is the same (10 N) throughout the spring.

Similar tensile forces can be set up in a piece of rope or string which is stretched by weights. In all cases in which the rope or cord is in equilibrium, the forces at the two ends are equal in magnitude and opposite in direction, and the tension in the rope is equal to the force applied at either end. This is not true, however, if the rope has a finite mass and is being accelerated. For a mass *m* to be accelerated, a net unbalanced force must act. Hence in this nonequilibrium case the tensions at the two ends of the rope must be *different*.

Forces of compression are the opposite of forces of tension. Thus we can compress a spring by pushing its coils together, as in Fig. 4.15. The spring reacts by exerting a force which tends to increase its length back to its original value. Here the force exerted on the spring at its free end is transmitted undiminished to the wall. Again in this case the compressive force is the same at any point along the length of the spring, and equal to the force applied at the free end.

Clearly a metal rod can sustain a compressional force. Cords and ropes cannot sustain compressional forces since they will simply buckle under such forces. They can, however, sustain forces of tension if the stretching force is not too large, and can therefore be used, together with systems of pulleys or wheels, to change the directions of applied forces without changing their magnitudes.

An interesting application of these ideas is provided by *concrete*, a frequently used building material. Concrete is quite strong under compression, but is easily broken when subjected to tension (stretching). For this reason it is necessary to reinforce concrete when it is used for beams in buildings. The reinforcement is provided by iron rods which are embedded in the concrete before it hardens. An even better solution is to prestress the concrete, that is, to keep the iron rods under tension when the concrete is being poured. When the concrete dries, the tension is removed from the rods, which tend to revert to their original length and thus subject the concrete to compressive forces. The amount of compression thus built into the prestressed concrete is calculated to be greater than the tensile forces to which a beam might be subjected in a building. In this way the concrete is never subjected to net tensile forces, and therefore will not crack when in place in the building.

4.5 Reaction Forces

Newton's third law of motion becomes of great importance in dealing with equilibrium problems, since many of the forces which must be taken into

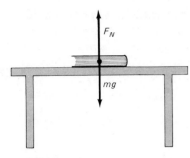

FIGURE 4.16 A book resting on a table.

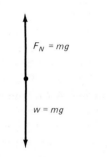

FIGURE 4.17 Free-body diagram for the situation shown in Fig. 4.16.

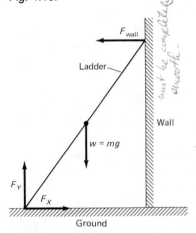

FIGURE 4.18 Forces on a ladder in an equilibrium position against a wall.

account in discussing the equilibrium of an object turn out to be reaction forces.

For example, consider a book resting on a table, as in Fig. 4.16. The book is pulled toward the center of the earth by the force of gravity, which is equal to the weight of the book w. Since the book is not accelerating, the table must be pushing up with a force F_N to balance the weight of the book. The reaction to this force exerted by the table on the book is a force of the same magnitude exerted by the book on the table. Here, however, we are interested only in the forces acting *on* the book. These are the gravitational force pulling it down and the force F_N exerted by the table to hold it up. Since these forces are equal and opposite, the book is in equilibrium and remains at rest. This is shown in the free-body diagram of Fig. 4.17.

Another example is provided by a ladder up against a wall, where the ladder is at rest with respect to both the ground and the wall. For the ladder to remain at rest, all the forces on the ladder must cancel out as they did with the book in our first example.

In this case, however, there are forces in both the horizontal (X) direction and the vertical (Y) direction. Figure 4.18 shows all the forces acting on the ladder. First we have the weight of the ladder, which is a force directed downward. This force must be balanced by the Y component of the force exerted by the ground on the base of the ladder, so that this Y component (F_Y) and w are equal in magnitude but in opposite directions. This assumption that F_Y and w are equal is valid only if we assume that the wall is completely smooth, so that there is no frictional force parallel to the wall keeping the ladder from moving up or down the wall. Then the push F_{wall} of the wall is horizontal and has no component parallel to the wall.

On the other hand, the ground cannot be smooth, for if it were, the ladder would merely slide until it assumed a horizontal position parallel to the ground, because of the force of the wall pushing out on the ladder. The friction between the ground and the ladder keeps the ladder from slipping. The base of the ladder "tries" to move along the ground, and therefore sets up a horizontal force on the ground directed away from the wall. The reaction force to this is the X component (F_X) of the force exerted by the ground on the ladder, a force which is directed toward the wall. For equilibrium the sum of the horizontal forces acting on the ladder must be zero, irrespective of where these forces are applied. Hence F_X is equal to F_{wall}, the outward push exerted on the ladder by the wall. This force F_{wall} is a reaction force to that exerted on the wall by the ladder. Since no net unbalanced forces act, the ladder is not accelerated but is in equilibrium.

4.6 Translational Equilibrium

In this section we will consider the equilibrium of objects that are not changing their positions in space, i.e., objects that do not undergo translational motion. If an object is at rest, it is clearly not being accelerated, and hence, from Newton's second law, it can have no net unbalanced force acting on it. The ladder resting against the wall in the preceding section is a good illustration here. If we confine our discussion to two dimensions, then we must have, for *translational equilibrium*,

$$(F_{\text{net}})_X = \sum_{i=1}^{N} (F_i)_X = 0 \qquad (F_{\text{net}})_Y = \sum_{i=1}^{N} (F_i)_Y = 0 \qquad (4.7)$$

no matter where these forces are applied. These two equations are sometimes called the *first condition of equilibrium.*

First Condition of Equilibrium

For translational equilibrium the components in any given direction of all the forces acting on an object must add to zero.

(The second condition of equilibrium applies to rotational equilibrium and will be discussed in Sec. 4.8.)

Equations (4.7) tell us that, if an object is at rest, then the sum of the X components of all the forces acting on it must add to zero when taken with their proper signs. The same is true for the Y components of all the forces acting. Hence, if in any problem the magnitude or direction of any one of the forces is unknown, it can be found from Eq. (4.7). It is surprising how many real-life problems can be solved using these basic ideas, as we will see from the examples which follow.

In solving problems on objects in translational equilibrium, it is useful to adopt the following straightforward procedure to arrive at a solution:

1 Draw a rough sketch of the objects involved in the problem.

2 Select one object as the one in equilibrium. Draw a free-body diagram for *all* the forces acting *on* this object, including all reaction forces. Label each force with its own letter or symbol.

3 Select a set of convenient rectangular coordinates for the problem, and resolve all the forces acting on the object into X and Y components.

4 Apply Eqs. (4.7) to the X and Y components of the forces acting. These equations will provide two independent relationships which can then be solved for any two unknown quantities in the problem. These may be the magnitudes or the directions of the forces.

5 Repeat this procedure, as needed, for all objects of interest in the problem. In so doing, you may have to use results obtained from earlier applications of the equilibrium conditions.

Examples 4.4 and 4.5 illustrate this procedure.

Example 4.4

A patient in a hospital has a leg in traction. A pulley arrangement of the type shown in Fig. 4.19a applies traction to the leg. A 25-kg mass acted on by gravity produces the tension in the rope that applies the traction. If the angles are as shown in the diagram, what is the horizontal force applied to the leg? Assume that the pulleys are all frictionless.

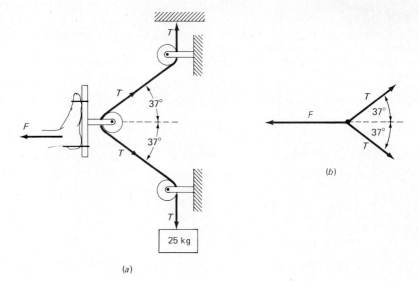

FIGURE 4.19 (a) Diagram for Example 4.4. (b) Free-body diagram for same situation.

(a)

SOLUTION

Let us first isolate the 25-kg mass and consider the forces on it. For equilibrium we must have

$$\sum F_Y = 0 \qquad T - mg = 0$$

$$T = mg = (25 \text{ kg}) (9.8 \text{ m/s}^2) = 245 \text{ N}$$

If the pulleys are frictionless, this tension remains the same throughout the rope, and the rope pulls down on the ceiling with a force of 245 N. The *reaction force* of the ceiling on the cord is what ultimately supports the 25-kg mass. The tension in the rope is the same at every point since the rope is not accelerated.

Now let us isolate the patient's leg, and consider the forces acting on it. These are the pull of the body to the left (*F*) and the pull of the two ropes (above and below the central pulley) to the right, as shown in Fig. 4.19*b*.

$$\sum F_X = 0 \qquad -F + T \cos 37° + T \cos 37° = 0$$

$$F = 2T \cos 37°$$

$$= 2(245 \text{ N})(0.80) = \boxed{392 \text{ N}}$$

Hence the traction on the patient's leg consists of the pull of the ropes with a force of 392 N to the right.

Example 4.5

A painting of mass 10 kg hangs on a wall, as illustrated in Fig. 4.20. The supporting wires are attached to the frame of the painting by eyelets, and then pass over a hook in the wall. The supporting wires make angles of 60° with the vertical.
(a) What is the tension in the supporting wires?
(b) What is the vertical force on each eyelet?
(c) What is the horizontal force on each eyelet?

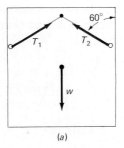

(a)

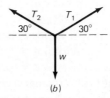

FIGURE 4.20 Diagram for Example 4.5: (a) Forces on a painting hanging on a wall. (b) Free-body diagram of the forces on the painting.

(b)

SOLUTION

(a) Let us isolate the painting and consider it as the object in equilibrium. Then the horizontal forces are the horizontal components of the tensions in the wires, or $T \cos 30°$. For equilibrium

$$\sum F_X = 0 \qquad T_1 \cos 30° - T_2 \cos 30° = 0$$

or $\qquad T_1 = T_2$

as we would expect. [Here we are using the abbreviated notation $\sum F_X$ for $\sum_{i=1}^{N} (F_i)_X$, since it is clear what is intended.]

The vertical forces on the painting are the vertical components of the tensions in the two wires directed upward, and the weight of the painting directed downward, as in the free-body diagram of Fig. 4.20*b*. For equilibrium

$$\sum F_Y = 0 \qquad T_1 \cos 60° + T_2 \cos 60° - w = 0$$

Setting $T_1 = T_2 = T$, $2T \cos 60° - mg = 0$, and

$$T = \frac{mg}{2 \cos 60°} = \frac{(10 \text{ kg})(9.8 \text{ m/s}^2)}{2(0.5)} = \boxed{98 \text{ N}}$$

Hence the tension is 98 N in each supporting wire.

(b) The vertical force on each eyelet is then

$$T_Y = T \cos 60°$$

$$= (98 \text{ N}) \left(\tfrac{1}{2}\right) = \boxed{49 \text{ N}}$$

(c) The horizontal force on each eyelet is

$$T_X = T \sin 60°$$

$$= (98 \text{ N})(0.866) = \boxed{85 \text{ N}}$$

Note that in this case the horizontal forces on the eyelets are greater than the vertical forces (half the weight of the painting) which each eyelet must support. The smaller the angle the wires make with the horizontal, the larger the horizontal forces on the eyelets. Thus, suppose the angle with the horizontal is only 10°. Then

$$T = \frac{mg}{2 \cos 80°} = \frac{(10 \text{ kg})(9.8 \text{ m/s}^2)}{2(0.17)} = 288 \text{ N} \qquad \text{and}$$

$$T_X = (288)(\sin 80°) = (288 \text{ N})(0.98) = 284 \text{ N}$$

In this case the eyelets must be able to sustain forces *3 times as great* as the weight of the painting itself, because the near-horizontal direction of the supporting wires requires that the tension in the wires be very great to produce vertical components sufficient to support the weight of the painting. This large tension and the small angle made by the wires with respect to the horizontal then make the horizontal forces very large. This must always be considered when hanging paintings or mirrors on walls, since the horizontal forces set up can become great enough to pull eyelets out of the frame or even to break the frame, if the supporting wire is too close to being horizontal.

4.7 Forces that Produce Rotation: Torques

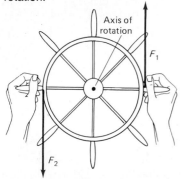

FIGURE 4.21 The forces exerted by the captain's hands on a ship's wheel. The result is a rotation of the wheel about its axis of rotation.

For translational equilibrium it is of no importance at what point on the body the forces are applied. If we consider the possibility that the object may rotate, however, then the point where the forces are applied becomes of great importance. Consider, for example, the wheel with which a ship's captain controls the motion of the ship. This wheel is mounted securely in such a way that it cannot be moved from one position in space to another without moving the whole ship along with it. The captain, however, can easily move the steering wheel circularly about the shaft on which it is mounted by using both hands to move, for example, the right side of the wheel up and the left side down, as in Fig. 4.21. This causes the wheel to move, but the motion is a rotation about the axis of the wheel, not a translation in space.

The reason for this is that the two forces are applied *at different points*. If the two forces F_1 and F_2 are equal in magnitude and opposite in direction, the first condition of equilibrium tells us that no translational motion can result. But the effect of these two equal and opposite forces in this case is to produce a counterclockwise rotation about the axis of the wheel, as Fig. 4.21 shows. We now want to consider what additional requirement must be satisfied by two forces to make both translation and rotation impossible. To do this we first need some way to measure the effectiveness of a given force in producing rotation about an axis.

Moment of a Force

The rotational effect produced on a body by a force of given magnitude and direction depends not only on the size of the applied force but also on its line of action with respect to the axis about which rotation is desired. Thus a force of 20 N may be enough to open a door easily if applied at the edge of the door away from the hinges, but will not be very effective in opening the door if applied very near to the hinges. The rotational effectiveness of any given force depends therefore both on the magnitude of the force and on the distance of its line of action from the axis of rotation.

We make these ideas quantitative by introducing the idea of the *moment of a force*, or *torque*.

Definition

Moment of a force (or torque): The product of the magnitude of the force (F) and of its lever arm (l)

$$\tau = Fl$$

(4.8)

where l is the perpendicular distance from the line of action of the force to the axis of rotation.

We use the symbol τ (Greek *tau*, the equivalent of t in English) for the torque, or moment of a force.

By convention we call torques producing counterclockwise rotations positive ($+$) and those producing clockwise rotations negative ($-$), although the opposite convention would work equally well.

The lever arm, being the perpendicular distance from the line of action of the force to the axis of rotation, is the *shortest distance* from the axis of rotation to the line of action of the force. Notice that the line of action of the force extends indefinitely in both directions along the force vector. Hence in Fig. 4.22 the lever arm for the force **F** about an axis at right angles to the paper at A is shown, as is that for a different axis of rotation B. It can be seen that both lever arms make right angles with the line of action of the force vector at the place where they meet. In this case the torque about A is negative, while the torque about B is positive.

The units for torque, or the moment of a force, are the units of force times distance, and hence in the SI system they are newton-meters (N·m). In Chap. 6 we will clarify the distinction between the moment of a force and work and energy, which have the same dimensions and units as torque has, but which are very different kinds of physical quantities.

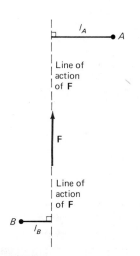

FIGURE 4.22 Lever arms of the force **F** about two different axes of rotation, one at A, the other at B. The corresponding lever arms are l_A and l_B. About point A the torque is clockwise; about B it is counterclockwise.

Couples

A special kind of torque consists of two equal and opposite forces applied at equal distances from an axis of rotation. Such a pair of torques we call a *couple*, and a good example of such a couple might be the forces applied to turn a ship's wheel in a counterclockwise direction, as discussed above. In this case the two forces are assumed to be equal, and are applied at distance $l/2$ from the axis of the wheel, where l is the diameter of the wheel.

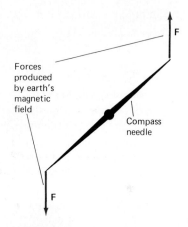

FIGURE 4.23 The couple exerted by the earth's magnetic field on a compass needle.

Forces produced by earth's magnetic field

Compass needle

Such a couple has no effect in producing translation, since $F + (-F) = 0$. The only effect of the couple is to produce rotation. The sum of the moments of the two forces, i.e., of the two torques, is

$$\tau = \tau_1 + \tau_2$$

$$= \frac{Fl}{2} + \frac{Fl}{2} = Fl \qquad (4.9)$$

Hence the torque in this case is the magnitude of either force (since they are equal) times the perpendicular distance between the lines of action of the two forces. This turns out to be true in general for couples. If the steering wheel is rotated clockwise, the couple τ is negative because of the convention we have adopted.

Some examples of couples are the forces exerted by a wrench on the two sides of a bolt, or by a magnetic field on the two ends of a compass needle, as shown in Fig. 4.23.

Example 4.6

A bus driver makes a bus turn a corner by exerting forces of 50 N with each hand on the steering wheel of the bus, as in Fig. 4.24. The wheel has a radius of 25 cm. What is the couple exerted by the driver on the wheel?

SOLUTION

The torque produced by a couple is the magnitude of either force times the perpendicular distance between the lines of action of the two forces. This distance is in this case the diameter of the wheel, which is

$$l = 2R = 50 \text{ cm} = 0.50 \text{ m}$$

Hence we have

$$\tau = Fl = (50 \text{ N})(0.50 \text{ m}) = \boxed{25 \text{ N·m}}$$

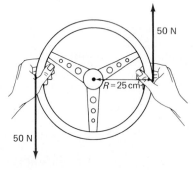

FIGURE 4.24 Diagram for Example 4.6.

Example 4.7

A flat piece of tile in the shape of an equilateral triangle with sides 20 cm long is subjected to the three forces shown in Fig. 4.25, where $\mathbf{F}_1 = 17.3$ N, and $\mathbf{F}_2 = \mathbf{F}_3 = 10$ N.
(a) What is the resultant force on the piece of tile?
(b) What is the resultant torque on the tile, with respect to an axis through the center of the tile at right angles to its plane surface?

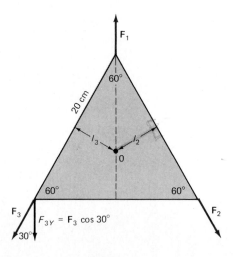

FIGURE 4.25 Diagram for Example 4.7.

SOLUTION

(a) By symmetry the horizontal components of forces $\mathbf{F}_2$ and $\mathbf{F}_3$ are equal and opposite, and hence $\sum F_X = 0$. The vertical components of the three forces are then

$$\sum F_Y = 17.3\ \text{N} - 0.866(10\ \text{N}) - 0.866\,(10\ \text{N})$$

$$= 17.3\ \text{N} - 17.\,3\ \text{N} = 0$$

and the tile is in *translational equilibrium*.

(b) For the torques we have

$$\sum \tau_0 = F_1 l_1 - F_2 l_2 + F_3 l_3$$

$$= (17.3\ \text{N})(0) - (10\ \text{N})(l_2) + (10\ \text{N})(l_3)$$

Since $l_2 = l_3$, $\sum \tau_0 = 0$, and the torques all cancel. In such a case we say that the tile is also in *rotational equilibrium*. Note that the lever arm of $\mathbf{F}_1$ is zero, since its line of action passes through the axis of rotation.

4.8 Rotational Equilibrium

The requirement that a body be in rotational equilibrium under the action of a number of forces means that the net effect of the forces is to produce no rotation (as in Example 4.7), no matter what axis is chosen about which to take moments of the forces. For this to be true, the clockwise moments must cancel out the counterclockwise moments exactly, so that we can write

$$\sum \tau = 0 \quad \text{(about any arbitrary axis)} \tag{4.10}$$

This is the *second condition of equilibrium*, the condition which must be satisfied if no rotational motion is to result.

Second Condition of Equilibrium

For rotational equilibrium the sum of the clockwise torques must be equal to the sum of the counterclockwise torques.

This condition is completely general and applies to all axes of rotation. We must, of course, use the *same* axis for calculating the moments of each force, and we must be sure to include *all* the forces which act on the body under consideration, including all reaction forces.

Since we are free to choose the axis about which moments are taken in problems of rotational equilibrium, we can often eliminate unknown forces from the torque equation by choosing an axis of rotation about which these forces have zero lever arms. Sometimes the problem cannot be solved unless it is simplified in this way.

For an object free to rotate about any axis, the second condition of equilibrium puts very strict limitations on the magnitudes and lines of action of the forces which can act to produce rotational equilibrium. For two forces acting, the only way that the torque will be zero for all axes is for the two forces to be equal in magnitude and opposite in direction, and to have the same line of action. For an object such as a wheel where the axis is fixed, it becomes much simpler to satisfy the second condition of equilibrium. In all cases, however, the sum of the moments of all forces acting must equal zero.

In solving problems involving torques the following procedures are useful in arriving at a solution:

1 Draw a rough sketch of the objects involved in the problem.

2 Use the procedure described in Sec. 4.6 to find relationships between the forces which act in the problem.

3 Select a convenient axis of rotation. (This need *not* be an axis about which the system could really rotate.) Choose an axis through which the lines of action of as many unknown forces as possible pass, to avoid having to calculate the torques of these forces.

4 Apply Eq. (4.10) to all torques acting about the chosen axis of rotation, being careful to assign the correct signs to clockwise and counterclockwise torques. This should provide enough equations, when used in conjunction with Eq. (4.7), to solve for all the unknowns.

5 Choose a different axis of rotation and solve the problem of the torques about this new axis as a check on your answer.

Examples 4.8 and 4.9 illustrate these procedures.

Example 4.8

A seesaw of the type shown in Fig. 4.26 (for which the weight of the seesaw plank can be neglected, either because it is very light or because its center of mass is at the axis of rotation) has a child of weight 200 N on the right side and the child's mother of weight 600 N on the left side.
(a) If the child is 3.0 m from the fulcrum, i.e., the support of the seesaw, how far from the fulcrum must the mother sit to balance out the torques acting?
(b) Solve the same problem by taking moments about point B, which is 1 m to the right of A.

SOLUTION

(a) Here the easiest way to solve the problem is to take moments about the fulcrum at A. We have

$$\sum\tau_A = (600 \text{ N})(x) - (200 \text{ N})(3.0 \text{ m}) = 0$$

from which

$$x = \left(\frac{200 \text{ N}}{600 \text{ N}}\right)(3.0 \text{ m}) = \boxed{1.0 \text{ m}}$$

Hence the mother must sit 1.0 m from the fulcrum to balance out the torque set up by the weight of the child, who is 3.0 m from the fulcrum.
(b) If we now decide to take moments about B, there is an additional force present. This is the reaction of the fulcrum to the forces acting due to the weight of the mother and child. Applying the first condition of equilibrium to forces in the vertical direction, we have

$$\sum F_Y = 600 \text{ N} + 200 \text{ N} - R = 0$$

or $R = 800 \text{ N}$

Taking moments about B, we then have

$$\sum\tau_B = (600 \text{ N})x' - (800 \text{ N})(1.0 \text{ m}) - (200 \text{ N})(2.0 \text{ m}) = 0$$

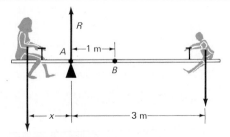

FIGURE 4.26 A seesaw with a child on one end and its mother on the other side.

or $$x' = \left(\frac{800 \text{ N} + 400 \text{ N}}{600 \text{ N}}\right)(1 \text{ m}) = 2.0 \text{ m}$$

and the mother should sit 2.0 m from B, which is the same position as found in part (*a*).

It is recommended that you solve the same problem by taking moments about point C at the right end of the seesaw, where the child is sitting.

Example 4.9

A girl of mass 80 kg stands on the end of a diving board 4.0 m long. The diving board is supported by two pedestals, one 3.0 m from the end and the other 4.0 m from the same end, as in Fig. 4.27.

Calculate the magnitude and direction of the forces exerted on the board by the two pedestals. Assume that the weight of the diving board is negligible compared with the other forces acting.

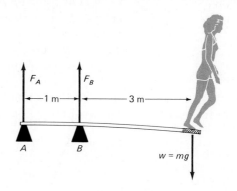

FIGURE 4.27   Diagram for Example 4.9.

SOLUTION

From the first condition of equilibrium

$$\sum F_Y = 0$$

$$F_A + F_B - w = 0$$

$$F_A + F_B = w = mg$$

$$= (80 \text{ kg})(9.8 \text{ m/s}^2) = 7.8 \times 10^2 \text{ N}$$

Here we have assumed that $\mathbf{F}_A$ and $\mathbf{F}_B$ are both directed upward, which may turn out not to be true.

Since the board is also in rotational equilibrium, we must apply our second condition of equilibrium. In this case we will take torques about B to eliminate one unknown force from the problem. Then the force at A must be directed downward if the moments about B are to cancel. We must still keep our original assumption about the direction of F_A and F_B, however, if we are to be consistent in solving the problem. Still assuming that F_A is directed upward, and taking clockwise torques as negative, we have

$$\sum \tau_B = 0$$

$$-F_A(1.0 \text{ m}) - w(3.0 \text{ m}) = 0$$

$$F_A = -3.0w = -3.0(7.8 \times 10^2 \text{ N})$$

$$= \boxed{-2.3 \times 10^3 \text{ N}}$$

Hence F_A has a magnitude of 2.3×10^3 N and is directed *downward*, contrary to our original assumption.

Then, since

$$F_A + F_B = 7.8 \times 10^2 \text{ N}$$

$$F_B = 7.8 \times 10^2 \text{ N} + 2.3 \times 10^3 \text{ N} = \boxed{3.1 \times 10^3 \text{ N}}$$

Hence the upward thrust of pedestal B is just sufficient to balance the downward forces exerted on the board by pedestal A (2.3×10^3 N) and by the girl's weight (7.8×10^2 N). If a heavier diver uses the board, the upward thrust at B increases to satisfy the condition of translational equilibrium, and the downward pull at A increases to satisfy the condition of rotational equilibrium.

Note that it is possible to make any arbitrary assumption about the direction of unknown forces in solving problems. As long as the problem is solved correctly, the assumption may merely result in the unknown force having a negative sign—indicating that the actual direction of the force is opposite to its originally assumed direction. We must, of course, stick to our original assumption consistently throughout the problem.

4.9 Center of Gravity

Normally if a single force is applied to one point on an object, the result is some mixture of translation and rotation, as when a frisbee is thrown. There is one point in any object, however, at which any applied force produces a pure translation, i.e., translation without rotation. This point is the *center of gravity*.

Definition

Center of gravity: That point of an object about which all gravitational torques cancel.

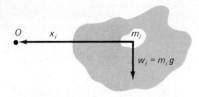

FIGURE 4.28 Defining the center of gravity of an irregularly shaped object. It is the point at which all the mass would have to be located to produce the actual gravitational torque observed about an axis through O. Each bit of mass m_i in the object exerts a torque $\tau_i = w_i x_i = m_i g x_i$ about O.

Let us try to find the center of gravity of an irregularly shaped flat object of mass M, as in Fig. 4.28. (The third dimension may be introduced in the same fashion as is done here for the other two dimensions, if so desired.) Suppose we choose an axis of rotation outside the object, in this case an axis at right angles to the plane of the paper at point O in the diagram. A bit of mass m_1 in the body has a weight $w_1 = m_1 g$, and this weight produces a torque about O of magnitude $\tau_1 = w_1 x_1$. The same is true for all other bits of mass. The total torque on the body about an axis at O is then

$$\tau_O = w_1 x_1 + w_2 x_2 + \cdots = \sum_i w_i x_i = \sum_i m_i g x_i \tag{4.11}$$

while the total weight of the object is

$$W = \sum_i w_i = \sum_i m_i g = g \sum_i m_i = Mg \tag{4.12}$$

We now ask at what point the whole weight would have to act to produce the same torque about the axis that is produced by all the bits of mass acting together. In other words what is the mean distance $\bar{x}$ from the axis of rotation in the expression for the torque $\tau_O = W\bar{x} = Mg\bar{x}$? On equating the two values of τ_O from this expression and from Eq. (4.11), we have

$$\tau_O = Mg\bar{x} = \sum_i m_i g x_i$$

$$\text{or } \bar{x} = \sum \frac{m_i x_i}{M} \tag{4.13}$$

This is the value of x for which the correct gravitational torque would be obtained if all the mass were concentrated there. Hence it is also the X coordinate of the center of gravity.

To find the Y coordinate of the center of gravity we rotate the object through 90° and repeat the process. We obtain

$$\bar{y} = \sum_i \frac{m_i y_i}{M} \tag{4.14}$$

by analogy with Eq. (4.12) for the X coordinate.

The coordinates $\bar{x}$, $\bar{y}$ (and $\bar{z}$, if needed) then determine the location of the center of gravity of the object. The gravitational pull of the earth on any object then acts as if the weight of the object were all concentrated at the center of gravity, i.e., at $\bar{x}$, $\bar{y}$, $\bar{z}$. This gravitational pull thus acts to produce translational motion, but no rotational motion, about an axis through the center of gravity O.

A related concept is the *center of mass*, which is that point at which the whole *mass* of the object may be considered to be concentrated, when Newton's laws are applied to the motion of the object. For an object of uniform density, the center of mass is at the center of symmetry of the object. For example, the center of mass of a sphere, cube, circular disk or rectangular plate is at the exact geometric center of each, while the center of mass for a cylinder is on the axis of the cylinder halfway between the two ends. For odd-shaped objects such as a doughnut or boomerang, however, the center of mass need not even be within the object.

FIGURE 4.29 Finding by experiment the center of gravity of an irregularly shaped object.

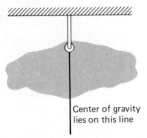

Center of gravity lies on this line

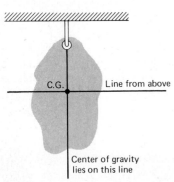

C.G.   Line from above

Center of gravity lies on this line

If the gravitational field in which an object is placed is the same over the whole of the object, then its center of gravity and its center of mass are at exactly the same place. This might not be true, for example, for a large space platform whose most distant point might be some kilometers farther away from the center of the earth than its nearest point. In this case the center of gravity of the platform would be closer to the earth than the center of mass of the platform.

The above analysis leads to a simple technique for finding the center of gravity of an irregularly shaped flat object. If we suspend the object from a hole near one edge, as in Fig. 4.29, the force of gravity will pull down on the center of gravity (C.G.), and it will assume a position directly below the hole. A straight line drawn from the hole and directed straight down toward the center of the earth must then pass through the center of gravity. If the object is then rotated through some angle (say, 90°) and the process repeated, the two lines will intersect at the center of gravity of the object. Hence the center of gravity can be found experimentally in this simple fashion. As a check, the object can be hung from any other point, and it will be found that a vertical line through that point will also pass through the center of gravity.

Stable and Unstable Equilibrium

FIGURE 4.30 (*a*) A tall block of metal resting on a table. (*b*) Stable equilibrium for the block; the two forces acting produce a torque restoring the block to its rest position. (*c*) Unstable equilibrium for the block; the two forces produce a torque which tends to overturn the block.

If a tall, narrow block of metal is at rest on a table, as in Fig. 4.30*a*, it will remain there since all forces and torques acting cancel out and leave it in equilibrium. If the block is tipped up on one edge, however, two different results are possible: the block, when released, will return of its own accord to its original position, in which case we say it was in *stable equilibrium*; or it will continue to move of its own accord in the direction in which it was tilted, in which case we say it was in *unstable equilibrium*. In this case the difference between these two conditions depends on the amount of the initial tilt.

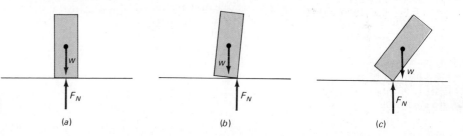

(a)   (b)   (c)

FIGURE 4.31 A ball at the end of a string; the ball is always in stable equilibrium, no matter what its displacement angle θ from its equilibrium position.

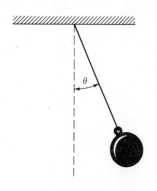

In general, the lower the center of gravity of an object, the more stable it is. For example, a ball hung from the ceiling at the end of a string, as in Fig. 4.31, will always return to its equilibrium position, no matter how far it is displaced from it. Hence it is always in stable equilibrium. The same is not true for the block of metal in Fig. 4.30*a*. If the block is tilted slightly about one edge, Fig. 4.30*b* shows that the torques due to the gravitational force on the center of gravity and the push of the table on the edge still in contact with the table will restore the object to its equilibrium position. If the block is tilted even farther, however, the situation in Fig. 4.30*c* results. The two torques then act to topple the block over, and it falls to the table.

It can be seen that in Fig. 4.30*b* the center of gravity of the block is raised, whereas in Fig. 4.30*c*, after an initial increase in the height of the center of gravity, it is lowered greatly as the block falls over. Because of the gravitational pull of the earth, objects will always try to seek a position where their center of gravity is as low as possible. If other forces and torques prevent a lowering of the center of gravity by the object's falling over, the object is in stable equilibrium; if they do not and the object can move to a position with a lower center of gravity, it will do so. It is then in unstable equilibrium. We will see in Chap. 6 that this behavior is closely related to what happens to the gravitational potential energy of the object.

Example 4.10

(a) Find the center of mass of the piece of flat brass stock shown in Fig. 4.32.
(b) Find the center of gravity of the same piece of brass.

SOLUTION

(a) If we draw a vertical line which divides both of the pieces A and B exactly in half, it is clear that the center of mass must lie on this line, since there is just as much mass to the right of the line as there is to the left. The point on this line which is the center of mass must also have just as much mass above it as it does below. Since the two pieces have the same density and the same thickness, the masses are in the same ratio as the two areas. The ratio of area B to area A is 3/9. The center of mass is that point for which the area above it will be exactly equal to that below it. By trial and error this point is found to be 1 cm below the intersection of A and B, since the area above this point is (3 cm)(1 cm) + (1 cm)(3 cm) = 6 cm², and the area below this point is (3 cm)(2 cm) = 6 cm². Hence the center of mass is on the vertical line drawn to bisect the area of A and B, and 2 cm up from the bottom of piece A, as shown in the figure.
(b) For a small piece of metal like this, the acceleration due to gravity g does not differ to any measurable extent over the object. Hence, since $w = mg$, the center of weight of the object is the same as the center of mass. The point at which all the weight of the object appears to be concentrated is exactly what we mean by the center of gravity.

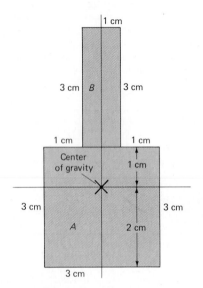

FIGURE 4.32 Diagram for Example 4.10.

Example 4.11

Consider a ladder 10 m long and of mass 50 kg resting against a smooth vertical wall and making an angle of 53° with the ground, as in Fig. 4.33. Find the magnitude of the horizontal force exerted by the wall (f_w) and the magnitude and direction of the force exerted by the ground on the ladder. Assume that the ladder has its center of gravity halfway up the ladder.

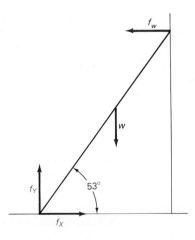

FIGURE 4.33 Diagram for Example 4.11.

SOLUTION

The situation is shown in Fig. 4.33. For equilibrium

$$\sum F_Y = 0 \qquad f_Y - w = 0$$

$$f_Y = w = mg = (50 \text{ kg})(9.8 \text{ m/s}^2) = 490 \text{ N}$$

$$\sum F_X = 0 \qquad f_X - f_w = 0$$

$$f_X = f_w$$

This is all the information we can get from the first condition of equilibrium, but it is not enough, since we do not know either f_X or f_w. We can, however, solve the problem by using the second condition of equilibrium. Suppose we take torques about the point where the ladder rests on the ground. (At this point the lever arms of the forces f_Y and f_X are zero.) From trigonometry (or the use of a scale diagram, ruler, and protractor), we find that this point is 6.0 m from the wall and the top of the ladder is 8.0 m from the ground, since cos 53° = 0.60 and sin 53° = 0.80. There is a clockwise torque produced by the weight of the ladder acting at its center of gravity, and a counterclockwise torque produced by the push of the wall on the ladder (a reaction force). From the second condition of equilibrium

$$\sum \tau_0 = 0$$

$$f_w(8.0 \text{ m}) - w(3.0 \text{ m}) = 0$$

$$f_w = \tfrac{3}{8} w = \tfrac{3}{8} (490 \text{ N}) = \boxed{184 \text{ N}}$$

and $\qquad f_X = f_w = 184 \text{ N}$

The force applied by the ground to the ladder therefore has components of 490 N in the vertical direction and 184 N in the horizontal direction. This leads to a resultant of $\boxed{523 \text{ N}}$ at an angle of 69° with respect to the ground. Note that this force is *not parallel to the ladder*, as shown in Fig. 4.34.

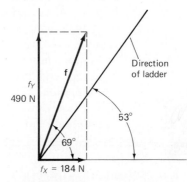

FIGURE 4.34   The force exerted by the ground on the ladder in Example 4.11. Note that the force exerted by the ground is *not* parallel to the ladder.

4.10 Equilibrium of Moving Objects

Thus far in this chapter we have been discussing the conditions under which objects at rest remain at rest and do not undergo translational or rotational motion as a result of forces acting on them. This is because the applied forces cancel out and so no translational motion results, and similarly the torques cancel and so no rotation results. These are examples of equilibrium situations.

It is important to realize, however, that equilibrium can and does exist even *when an object is moving*, provided that motion is of a special kind. Suppose you are in a car moving along a straight road at a constant 25 m/s. Since the car moves with constant speed in a straight line, it must be either because no forces act on it, or because any forces which do act cancel each other out and translational equilibrium results. In practice, retarding forces like friction and air resistance always exist. The function of the gasoline fed to the motor of the car is then just to produce forces which balance out these retarding forces and keep the car moving at a constant speed. These forces are produced by the reaction of the road to the backward push of the wheels as the car's motor turns them. Hence the accelerating and decelerating forces are in equilibrium, and the car keeps moving at constant speed, according to Newton's first law of motion.

This very important point should be kept in mind in the problems on motion in subsequent chapters. *The first condition of equilibrium does not mean that an object is at rest.* It can be either at rest or in motion with constant speed in a straight line. In both cases, either no forces are acting, or the forces which do act cancel each other out and equilibrium results.

In dealing with the second condition of equilibrium, the fact that all torques acting cancel out can mean either that there is no rotational motion, or that there is no change in the existing rotational motion of the object. An example of the latter situation would be the freely spinning front wheel of a bicycle (off the ground), which is moving at a high rotational speed due to someone's hand spinning the rim of the wheel. In this case, if the wheel rotates at a constant speed, then the retarding torques due to friction and air resistance must just be balanced out by the torque applied by the person's hand. For equilibrium these two torques must be equal and opposite. If they are, the wheel will continue to rotate at constant speed and will neither speed up nor slow down. We will see more applications of Newton's laws of motion to rotational motion in Chap. 8.

Example 4.12

A 400-kg wooden crate is being pushed by a tractor at a constant speed of 5.0 m/s along a concrete pier. The tractor exerts a constant force of 1.6×10^3 N on the crate. What is the coefficient of sliding friction between the wood and the concrete?

SOLUTION

Newton's second law applied to this problem yields

$$\sum F_X = f - \mu_k F_N = ma_X = 0$$

Here f is the force applied by the tractor, and the acceleration is zero since the crate moves at constant speed. F_N is the normal force exerted on the ground by the crate, which in this case is equal to the weight of the crate. Hence we have

$$f - \mu_k F_N = 0$$

$$\mu_k = \frac{f}{F_N} = \frac{1.6 \times 10^3 \text{ N}}{(400 \text{ kg})(9.8 \text{ m/s}^2)} = \boxed{0.41}$$

Note that the speed of 5.0 m/s is never used in solving the problem. The only important thing is that this speed is constant.

Another good example of translational equilibrium for a moving object is an object falling at its terminal velocity in a retarding medium, as discussed already in Sec. 3.8.

Example 4.13

A sailboat is tacking into the wind at 45° with respect to the wind direction.
(a) If the resultant force exerted by the wind on the sail is 1000 N, and the boom which contains the sail makes an angle of 25° with respect to the keel of the boat, find the force exerted on the boat in its forward direction.

(b) If the boat continues at a constant speed along its original direction, what other forces must be acting on it and how are they provided?

SOLUTION

The situation is diagrammed in Fig. 4.35, with the wind coming from the east and the sailboat moving in a north-easterly direction. When the sail is set properly, the wind blowing across the canvas exerts a force on the sail normal to its surface, as shown. The two components of this force, one F_X along the direction of motion and the other F_Y at right angles to it, are shown.

(a) If the force is 1000 N, then its components are:

$$F_X = F \sin 25° = (1000 \text{ N})(0.42) = 420 \text{ N}$$

$$F_Y = F \cos 25° = (1000 \text{ N})(0.91) = 906 \text{ N}$$

Hence the force driving the boat forward is 420 N. It should be noted that a sailboat can build up greater speeds tacking into the wind in this way, for the faster the boat moves, the greater the force exerted on the sail by the wind. If the sailboat is moving away from the wind on the other hand, the force of the wind decreases as the sailboat builds up speed.

(b) If the sailboat moves through the water at a constant speed, then the resistance of the water (and of the air) to the motion of the boat must be equal to 420 N in the opposite direction to the boat's progress. Otherwise the sailboat would be accelerating.

Similarly the keel of the boat and its side must experience a resistive force of 906 N to balance out F_Y and keep the boat from moving sideways. Since the force of the wind acts near the middle of the sail, whereas the water resistance acts on the hull of the boat, there is also a couple set up that tends to overturn the sailboat.

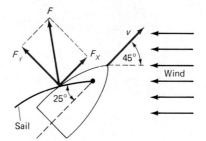

FIGURE 4.35 Diagram for Example 4.13.

Summary: Important Definitions and Equations

Vectors: Quantities which have both magnitude and direction.

Subtraction of vectors:

$$\mathbf{A} - \mathbf{B} = \mathbf{A} + (-\mathbf{B})$$

Rectangular resolution of vectors:

$$R_X = \sum_{i=1}^{N} (A_i)_X = \sum_{i=1}^{N} A_i \cos \theta_i$$

$$R_Y = \sum_{i=1}^{N} (A_i)_Y = \sum_{i=1}^{N} A_i \sin \theta_i$$

$$R = \sqrt{R_X^2 + R_Y^2} \qquad \tan \theta_R = \frac{R_Y}{R_X}$$

Forces (vector quantities):

Compression and tension: Forces which tend to compress or to stretch an object by applying equal and opposite forces at the two ends of the object.

Reaction forces: Forces which arise as a consequence of Newton's third law of motion.

Moment of a force (or torque):

$$\tau = Fl$$

where l is the lever arm, the perpendicular distance from the line of action of the force $\mathbf{F}$ to the axis of rotation. Counterclockwise rotations are commonly taken as positive, clockwise rotations as negative.

Couple: Two equal and opposite forces $\mathbf{F}$ applied at equal distances ($l/2$) from an axis of rotation. For a couple $\tau = Fl$.

Equilibrium: A physical situation in which forces act but no acceleration is produced.

Translational equilibrium: No translational acceleration occurs.

Condition: $\sum F_X = 0 \qquad \sum F_Y = 0$ (first condition of equilibrium)

Rotational equilibrium: No rotational acceleration occurs.

Condition: $\sum \tau = 0$ for any arbitrary axis (second condition of equilibrium)

Center of gravity (center of mass): That point in an object about which all gravitational torques cancel.

Coordinates of center of gravity:

$$\bar{x} = \sum_i \frac{m_i x_i}{M} \qquad \bar{y} = \sum_i \frac{m_i y_i}{M} \qquad \bar{z} = \sum_i \frac{m_i z_i}{M}$$

Questions

1 Why is it always possible to replace a vector $\mathbf{A}$ in two dimensions by its rectangular components A_X and A_Y?

2 A baseball is thrown straight up in the air by a baseball player and momentarily comes to rest at its highest point. Is the baseball in equilibrium at this point?

3 Give an example of a body that is in motion but that is also in equilibrium.

4 If a spring is attached to a wall and a person pulls on the end of the spring with a force of 50 N, this force is transmitted undiminished all the way to the wall, and the wall pulls back on the spring with a force of 50 N. Since the spring is being subjected to a force of 50 N at both ends, why is the tension in the spring not 100 N instead of 50 N?

5 Give an example of a physical situation involving weights tied together by cords passing over pulleys, in which the tension in a cord is not the same throughout the cord.

6 The components in any given direction of all the forces acting on a car are known to be zero.

(*a*) What does this tell us about the car?

(*b*) Is it necessary that the car be at rest in this case? Why?

(*c*) What can you say definitely about the motion (or lack thereof) of the car?

7 (*a*) If two equal and opposite forces are applied to an object, under what conditions will the object be in equilibrium?

(*b*) Under what conditions will the object not be in equilibrium? What will be the nature of the resulting motion?

8 Give an example in which the neglect of a reaction force would completely invalidate the solution of an equilibrium problem.

9 (*a*) Is it possible for an object to be in translational equilibrium and not in rotational equilibrium? Give examples to support your answer.

(*b*) Is it possible for an object to be in rotational equilibrium and not in translational equilibrium? Give examples to support your answer.

10 (*a*) Distinguish clearly between the center of mass of an object and its center of gravity.

(*b*) Under what conditions are these two the same?

(*c*) Give an example of a situation in which an object would have its center of gravity at a different place than its center of mass.

11 (*a*) Where is the center of mass of a doughnut?

(*b*) Where is its center of gravity?

(*c*) How can the force of gravity act at a point where there is no mass?

12 Does the experimental method suggested in Sec. 4.9 for finding the center of gravity of an object require that the object be homogeneous throughout? Why?

13 Why do you tend to lean backward when carrying a heavy load of groceries in your arms, and forward when you are wearing a backpack?

Problems

A 1 Two vectors represent forces of magnitude 7.0 N and 5.0 N. One vector force which cannot result from the vector addition of these two vectors is:
(*a*) 2.0 N (*b*) 12.0 N (*c*) 7.0 N (*d*) 1.0 N
(*e*) 4.0 N

A 2 A vector velocity **A** is 6.0 m/s along the positive *X* axis. A second velocity **B** is −5.0 m/s along the same axis; i.e., it is directed along the negative *X* axis. The vector **B** − **A** is then:
(*a*) 11.0 m/s (*b*) 1.0 m/s (*c*) −1.0 m/s
(*d*) −11.0 m/s (*e*) None of the above

A 3 An airplane flies 400 km due east, makes a right-angle turn, and then flies 300 km due north. The plane's straight-line distance from its starting point is then:
(*a*) 700 km (*b*) 100 km (*c*) 350 km
(*d*) 500 km (*e*) None of the above

A 4 A seesaw has two sides of equal length at either side of a fulcrum (the support of the seesaw). A person of weight 800 N sits halfway from the fulcrum on one side. To balance the person's weight, sandbags piled at the end of the seesaw on the other side of the fulcrum must have a weight of:
(*a*) 800 N (*b*) 1600 N (*c*) 400 N
(*d*) 2400 N (*e*) 200 N

A 5 An automobile has a steering wheel of radius 15 cm. The driver exerts a force of 400 N with each hand on opposite sides of the wheel to turn the car. The couple produced is:
(*a*) 600 N·m (*b*) 1200 N·m (*c*) 12 N·m
(*d*) 6 N·m (*e*) 25 N·m

A 6 The 0.80-m horizontal rod *AB* in Fig. 4.36 is pivoted at *A* and supported by a rope that makes an angle of 40° with the line *AB*. The lever arm of the force exerted by the rope on the rod is:
(*a*) 0 (*b*) 0.80 m (*c*) 1.25 m (*d*) 0.61 m
(*e*) 0.51 m

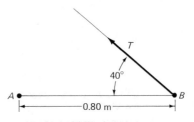

FIGURE 4.36 Diagram for Prob. A6.

A 7 A 1000-kg car is parked on a 30° incline. The force of friction keeping the car from sliding down the incline is:
(*a*) 1000 N (*b*) 500 N (*c*) 4900 N
(*d*) 9800 N (*e*) 8500 N

A 8 A block of weight 400 N resting on a table is pulled at constant speed by a force of 200 N directed at an angle of 30° with respect to the horizontal. The normal force F_N exerted by the table on the block is:
(*a*) 200 N (*b*) 400 N (*c*) 500 N
(*d*) 100 N (*e*) 300 N

A 9 In Prob. A8 the coefficient of sliding friction between the table and the block is:
(*a*) 0 (*b*) 0.35 (*c*) 0.43 (*d*) 0.58
(*e*) 0.26

A10 A uniform beam weighs 200 N and supports a 600-N weight three-fourths of the way down its length. The beam is supported at its two ends by pillars on which its rests. The force exerted by the support nearest to the added weight is:
(*a*) 800 N (*b*) 550 N (*c*) 400 N
(*d*) 250 N (*e*) 600 N

B 1 Two horizontal forces are acting on a small puck on a frictionless table. One is a force of 1.0 N along the positive X axis. The second is a force of 1.5 N at an angle of 45° with respect to the negative X axis.

(*a*) What is the magnitude and direction of the resultant force?

(*b*) If the puck's mass is 0.25 kg, what is the magnitude and direction of its acceleration?

B 2 A force of 50 N is directed at an angle of 83° with respect to the positive X axis. What are the X and Y components of this force?

B 3 A rectangular building lot has a length of 30 m fronting on a street. A surveyor finds that in sighting diagonally across the property the line of sight makes an angle of 55° with the front edge of the property. What is the area of the building lot?

B 4 A sharpshooter is shooting at a target. The gun fires bullets with speeds of 750 m/s. A wind with speed 25 m/s is blowing at right angles to the line connecting the sharpshooter and the target. At what angle must the sharpshooter shoot with respect to this line to hit the target?

B 5 A wrench exerts a force of 50 N on each side of a 4.0-cm-square bolt head. What is the couple produced by the wrench?

B 6 Find the center of gravity of the flat metal right-angle rule (or square) shown in Fig. 4.37.

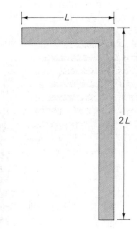

FIGURE 4.37 Diagram for Prob. B6.

B 7 A 4.0-m-long ladder rests up against the wall of a building, making an angle of 60° with the ground. If the ladder has a weight of 300 N, what moment does the ladder's weight produce about an axis at the point of contact of the ladder with the ground?

B 8 A person is trying to lift a rock with a mass of 100 kg by using a wooden beam as a lever. He or she supports the beam on a fulcrum placed so that the distance from the fulcrum to the rock is one-fifth the distance from the fulcrum to the point at which he or she pushes down on the beam. What force must the person exert to lift the rock?

B 9 Calculate the torque about the base support of a diving board exerted by a 100-kg person 2.5 m from this support.

B10 A lighting fixture of mass 100 kg is hung in the corner of a room by suspending it from the ceiling and the adjacent wall by two ropes, as shown in Fig. 4.38.

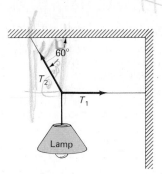

FIGURE 4.38 Diagram for Prob. B10.

(*a*) Find the tension T_1 in the rope attached to the wall.

(*b*) Find the tension T_2 in the rope attached to the ceiling.

C 1 A girl can throw a baseball horizontally with a speed of 30 m/s. She throws the ball out of a convertible which is moving north at 25 m/s. If the girl throws the ball out of the right side of the car at right angles to the direction of motion of the car, what will be the actual speed and direction of the ball with respect to the ground?

C 2 An airplane, whose ground speed in still air is 200 m/s, is flying with its nose pointed due north. There is a crosswind of 30 m/s in an easterly direction.

(*a*) What is the actual ground speed of the plane?

(*b*) In what direction does it actually travel?

C 3 An acrobat hangs from the middle of a tightly stretched horizontal wire so that the angle between the wire and the horizontal is 6.0°.

(*a*) Draw a free-body diagram for this physical situation.

(*b*) If the acrobat's mass is 90 kg, what is the tension in the wire?

C 4 An assembly of five identical springs is mounted on a frame as shown in Fig. 4.39. The tension in the middle spring A is 20 N. What is the tension in springs B, C, D, and E?

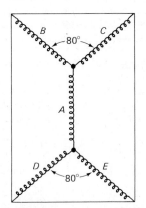

FIGURE 4.39 Diagram for Prob. C4.

C 5 A traffic light of mass 12 kg is hanging from two wires, as shown in Fig. 4.40.

(*a*) Draw a free-body diagram for this physical situation.

(*b*) Compute the tension in the two cables *A* and *B*.

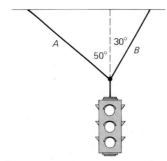

FIGURE 4.40
Diagram for Prob.
C5.

C 6 A 30° wedge is driven into a log, as in Fig. 4.41. If a force of 200 N is applied straight down on the top of the wedge by a blow from a mallet, what are the forces felt by the log at the two points of contact?

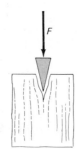

FIGURE 4.41 Diagram for
Prob. C6.

C 7 A man is trying to move a car stuck in the mud by tying a rope between a tree and the bumper of the car, making the rope as taut as possible, and then pushing sideways on the rope at its center with all his strength. If, when the man is pushing, the angle made by the rope with its original direction is 5.0°, and the man is pushing with a force of 500 N, what force is being exerted on the car?

C 8 A 50-kg plank, which is 2.0 m long and of uniform thickness, rests on two knife-edges attached to two balances, as in Fig. 4.42. What are the readings of the two balances on which the knife-edges rest?

0.30 m   0.70 m

FIGURE 4.42 Diagram for
Prob. C8.

C 9 A 500-kg rock is suspended, as shown in Fig. 4.43 from a wooden boom *AB* which has a mass of 100 kg and a length of 2.0 m. Calculate the tension in the supporting cable *C*, and the force exerted on the boom at *A*.

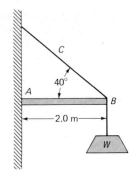

FIGURE 4.43 Diagram for
Prob. C9.

C10 A woman pushes a lawnmower at a steady speed. The lawnmower weighs 25 kg and the woman exerts a 100-N force along a line making an angle of 45° below the horizontal. Find the magnitude of the vertical force the ground exerts on the lawnmower.

C11 A ladder 8.0 m long leans against a vertical frictionless wall and makes an angle of 53° with the ground, which is assumed to be rough. The mass of the ladder is 40 kg, with its center of gravity at its center. Find:

(*a*) The magnitude of the horizontal force exerted by the wall on the ladder.

(*b*) The magnitude and direction of the force exerted by the ground on the ladder.

C12 Repeat Prob. C11 for a situation in which a 100-kg person is standing on the ladder at a point three-fourths of the way from the bottom of the ladder.

C13 A person of mass 80 kg stands at the end of a diving board, as shown in Fig. 4.44. The diving board is of uniform thickness and has a mass of 2.0 kg. Find:

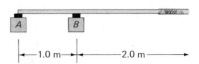

├─1.0 m─┤├──────2.0 m──────┤

FIGURE 4.44 Diagram for
Prob. C13.

(*a*) The magnitude and direction of the force exerted by support *A*.

(*b*) The magnitude and direction of the force exerted by support *B*.

C14 Two flat wooden boards, each of uniform construction and 2.0 m long, and each with a mass of 2.0 kg, are stood on end 1.0 m apart and tilted until they touch at the top. Find:

(*a*) The horizontal force exerted by one board on the other at the top of the assembly, assuming that at that point there are no vertical forces acting.

(*b*) The magnitude of the coefficient of static friction needed to keep the bottom ends of the boards from sliding.

C15 Three girls who weigh 500, 600, and 700 N, respectively, are to be seated on a plank of negligible weight in such a way that the 500-N girl is between the other two girls, the heavier girls are 4.0 m apart, and the center of mass of the system is located at the position of the 500-N girl. What are the distances of the two girls from the first girl?

C16 A uniform meterstick 1.0 m long has a weight of 4.0 N. It is loaded with two pieces of heavy metal, with a 2.0-N piece being attached at the 25-cm line, and a 3.0-N piece at the 85-cm line. Where is the center of gravity of the loaded meterstick?

C17 A uniform boom of weight 400 N is supported as shown in Fig. 4.45. The boom supports a weight W of 1600 N from its top. Find:

(*a*) The tension in the tie rope A.
(*b*) The force exerted on the boom by the pivot at B.

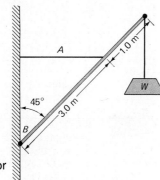

FIGURE 4.45 Diagram for Prob. C17.

C18 A uniform door of weight 500 N is held up by two hinges A and B in Fig. 4.46. The weight of the door is shared equally between the two hinges. The width of the door is 75 cm, and the distance between the two hinges is 120 cm. Find the magnitude and direction of the forces exerted on the door at the hinges A and B.

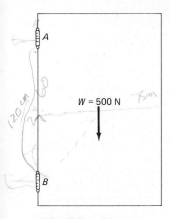

FIGURE 4.46
Diagram for Prob. C18.

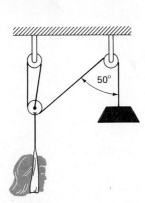

FIGURE 4.47 Diagram for Prob. C19.

C19 Calculate the traction force exerted on the neck of the woman in Fig. 4.47.

C20 Figure 4.48 shows a human arm extended. The center of mass of the arm is located about two-fifths of the way from shoulder to hand. The arm is held up by two other forces. One is the tension T in the deltoid muscle, which acts

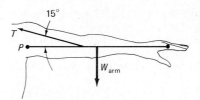

FIGURE 4.48 Diagram for Prob. C20.

at a 15° angle at a point about one-fifth of the way from shoulder to hand. The other force is exerted by the pivot P where the upper arm (humerus) joins the shoulder bone (scapula).

(*a*) If the outstretched hand is used to support a dumbbell of mass 5.0 kg, calculate the tension T in the deltoid muscle required to hold up both the arm and the dumbbell. The mass of the arm can be taken to be about 4.0 kg.

(*b*) How does this tension compare with the weight of the arm?

D 1 Two ropes support a load of 100 kg. The two ropes are perpendicular to each other, and one rope has twice the tension of the other. Find:

(*a*) The tension in each rope.
(*b*) The angle each rope makes with the vertical.

D 2 A block-and-tackle system for lifting heavy weights is shown in Fig. 4.49. If the mass to be lifted is 195 kg and the movable pulley has a mass of 5.0 kg, what force F must be exerted on the rope to lift the weight? (Assume that the tension is the same throughout the rope.)

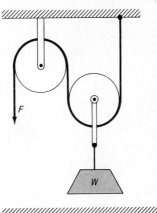

FIGURE 4.49
Diagram for Prob. D2.

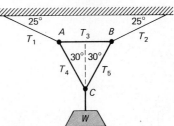

FIGURE 4.50 Diagram for Prob. D3.

D 3 Find the tensions T_1, T_2, T_3, T_4, and T_5 in the ropes shown in Fig. 4.50, if these ropes support a weight of 500 N at the bottom.

D 4 Figure 4.51 shows a system of pulleys and weights hung from a ceiling. If $W_3 = 400$ N, what are the values of W_1, W_2, T_2, and T_3?

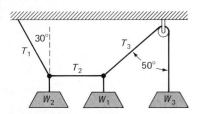

FIGURE 4.51 Diagram for Prob. D4.

D 5 A uniform ladder of mass 20 kg and length L leans against a smooth wall so that it makes an angle of 53° with respect to the ground. A person of mass 80 kg starts to climb up the ladder to a point two-thirds of the way to the top of the ladder. How large must be the coefficient of static friction between the floor and the ladder if the ladder is not to slip?

D 6 An eccentric drive wheel is made of uniform brass of the same thickness throughout. Its radius is 10.0 cm. It has a circular hole cut in it, of radius 2.0 cm, for the drive shaft. The center of this hole is 2.0 cm from the center of the wheel. What is the location of the center of gravity of the drive wheel?

D 7 Show analytically that in applying the second condition of equilibrium the choice of an axis of rotation is completely arbitrary, and can therefore be made to simplify the solution of the problem as much as possible.

D 8 Three forces acting on a particle and keeping it in a state of both translational and rotational equilibrium must be both coplanar and concurrent (i.e., their lines of action must pass through a common point). Show that the vectors representing three such forces form a closed triangle when added together. Also show that the magnitude of any one of the forces divided by the sine of the angle between the lines of action of the other two is the same for all three forces.

D 9 Show that the torque produced by a couple is independent of the axis of rotation about which the couple is taken.

Additional Readings

Andrade, E. N. da C.: "Isaac Newton," in *Newton Tercentenary Celebrations,* Cambridge University Press, New York, 1947. A brief life of the most famous of all physicists, and the source of the quote at the beginning of this chapter (with permission of the Cambridge University Press). This essay is also reprinted in James R. Newman (ed.): *The World of Mathematics,* vol. 1, Simon and Schuster, New York, 1956.

Hobbie, Russell K.: *Intermediate Physics for Medicine and Biology,* Wiley, New York, 1978. The first chapter of this advanced text contains some interesting examples of the forces acting on various part of the human body.

Mark, Robert: "Structural Analysis of Gothic Cathedrals," *Scientific American,* vol. 227, no. 5, November 1972, pp. 90–99. An interesting comparison of the cathedrals at Bourges and Chartres, using optical stress analysis of model structures.

Metcalf, Harold J.: *Topics in Classical Biophysics,* Prentice-Hall, Englewood Cliffs, N.J., 1980. The first chapter of this brief book discusses biomechanics, including equilibrium problems.

Steinman, David B.: "Bridges," *Scientific American,* vol. 191, no. 5, November 1954, pp. 60–71. An account of some famous bridge structures, including the collapse of the Tacoma Narrows Bridge.

*T*he inner logic of Galilean mechanics was so strong
that Newton was able to take the great step of applying it
to the motion of the stars.

Max Born (1882–1970)

Chapter 5

Motion in Two Dimensions

Thus far we have confined our discussion of mechanics to motion in one
dimension. Although this has enabled us to introduce the basic ideas and
laws of mechanics, very few objects in the universe move in only one
dimension. Motion in a plane, i.e., motion on a flat, two-dimensional surface,
is, however, quite common in our experience. If we neglect the slight
curvature of the earth, ice skaters, sailors, and restaurant workers all move in
two dimensions. So too do basketballs, golf balls (if not hooked or sliced!),
and various other kinds of projectiles whose initial straight-line motion is bent
into a plane by the force of gravity.

In this chapter we will discuss a number of interesting kinds of planar
motion, including the motion of carousels, centrifuges, projectiles, and the
most important of all two-dimensional motions—that of the planets, including
our own earth, around the sun.

5.1 Motion in a Plane

In discussing motion in a plane we will use the ideas already introduced in our
discussion of one-dimensional motion. Motion in a plane is merely putting
together two one-dimensional motions to describe the two-dimensional
motion of an object.

Velocity

The definitions of velocity and acceleration for planar motion are very similar
to those for rectilinear motion. Thus, suppose a particle is moving along a
curved path from point A to point B in Fig. 5.1. The direction of the velocity $\mathbf{v}$
is constantly changing between A and B, but we can define the average value
of the velocity between A and B as

FIGURE 5.1 Motion of a particle along a curved path in a
plane.

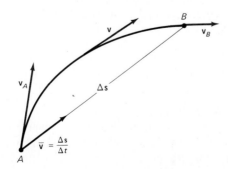

Average Velocity

$$\bar{\mathbf{v}} = \frac{\Delta \mathbf{s}}{\Delta t} \tag{5.1}$$

where $\Delta \mathbf{s}$ is the displacement vector connecting A and B. The instantaneous velocity $\mathbf{v}$ at point A is then, by analogy with the one-dimensional case,

Instantaneous Velocity

$$\mathbf{v} = \lim_{\Delta t \to 0} \frac{\Delta \mathbf{s}}{\Delta t} \tag{5.2}$$

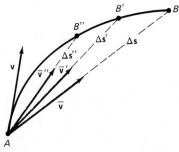

FIGURE 5.2 Definition of instantaneous velocity $\mathbf{v}$ at point A. The average velocity between points A and B is $\bar{\mathbf{v}} = \Delta \mathbf{s}/\Delta t$; between A and B' it is $\bar{\mathbf{v}}' = \Delta \mathbf{s}'/\Delta t'$; between points A and B'' it is $\bar{\mathbf{v}}'' = \Delta \mathbf{s}''/\Delta t''$. As Δt approaches 0, the direction of the chord AB changes from that of $\bar{\mathbf{v}}$ to that of $\bar{\mathbf{v}}'$ to that of $\bar{\mathbf{v}}''$. In the limit the chord has the same direction as the tangent to the curve at A. This is the direction of the instantaneous velocity $\mathbf{v}$ at A; its magnitude is given by Eq. (5.2).

Figure 5.2 shows that $\mathbf{v}$ can be represented by a vector tangent to the curve at A, whereas $\bar{\mathbf{v}}$ is a vector parallel to $\Delta \mathbf{s}$, and hence represents the average direction of the velocity along the curve from A to B. As $\Delta t \to 0$, the chord AB becomes shorter and shorter and finally becomes identical with the tangent to the curve at A. This agrees with the treatment of instantaneous velocity in Chap. 2, where we saw that the instantaneous velocity is the slope of the displacement-versus-time curve at the instant of time the velocity is desired.

Since $\bar{\mathbf{v}}$ and $\mathbf{v}$ are vectors whose directions are constantly changing as the particle moves in the plane, it is useful to find their components along fixed perpendicular axes in the plane by projecting these velocity vectors on the X and Y axes, as in Fig. 5.3. As usual, we have $\bar{v}_x = \bar{v} \cos \theta$ and $\bar{v}_y = \bar{v} \sin \theta$, and also $v_x = v \cos \theta$ and $v_y = v \sin \theta$, where θ is the angle made by $\bar{\mathbf{v}}$ or $\mathbf{v}$ with the X axis. We then find that

Components of Velocity

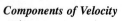
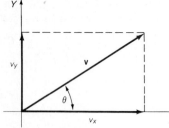

$$\bar{v}_x = \frac{\Delta x}{\Delta t} \qquad \bar{v}_y = \frac{\Delta y}{\Delta t} \tag{5.3}$$

$$v_x = \lim_{\Delta t \to 0} \frac{\Delta x}{\Delta t} \qquad v_y = \lim_{\Delta t \to 0} \frac{\Delta y}{\Delta t} \tag{5.4}$$

If the components v_x and v_y are known for a vector in two dimensions, then the vector is completely determined, for its magnitude is $v = \sqrt{v_x^2 + v_y^2}$, and its direction makes an angle θ with the X axis, where $\tan \theta = v_y/v_x$, as in Fig. 5.3.

FIGURE 5.3 Components of a velocity vector in two dimensions.

5.2 Motion on an Inclined Plane

Motion on an inclined plane is the type of motion Galileo used to deduce the laws of uniformly accelerated motion. The motion is actually confined to one direction along the plane, but since forces act at angles to the plane, we have deferred our discussion of the topic until now. We will start with the idealized case of no friction, and then consider how the motion changes when friction is taken into account.

Motion without Friction

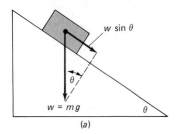

(a)

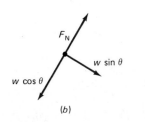

(b)

FIGURE 5.4 (a) A block sliding down a frictionless plane of inclination angle θ. (b) Free-body diagram of forces acting on the block.

Consider a block of mass m sliding down a frictionless plane, as in Fig. 5.4a. The angle of inclination of the plane is θ. Since it is assumed that there is no friction, the only forces acting are the force of gravity, $w = mg$, and the normal force of the plane pushing up on the block. The force of gravity can be broken down into two components, one normal to the inclined plane and the other parallel to the plane. The component normal to the plane is balanced out by the upward push of the plane on the block, as shown in Fig. 5.4b. The component parallel to the plane produces an acceleration, according to Newton's second law, of magnitude $a = F/m$, where F is the component of the force down the inclined plane. This is equal to $w \sin \theta$, where θ is the angle of the inclined plane. Hence

$$F = w \sin \theta = mg \sin \theta$$

$$a = \frac{F}{m} = \frac{mg \sin \theta}{m} = g \sin \theta \tag{5.5}$$

or $a = g \sin \theta$

Note that in this case the acceleration does not depend on the mass of the block.

Once we know the acceleration along the inclined plane, we can find the velocity and position of the block along the plane. Its speed down the plane at any instant of time, starting from rest, is:

$$v = v_0 + at = at = (g \sin \theta)t \tag{5.6}$$

and its position down the plane is:

$$s = s_0 + v_0 t + \tfrac{1}{2} at^2 = s_0 + \tfrac{1}{2}(g \sin \theta)t^2 \tag{5.7}$$

From these equations all useful information about the motion of the block can be obtained.

Example 5.1

A 10-kg block on a frictionless plane inclined at an angle of 30° is attached to a light cord passing over a pulley and then attached to a 5.0-kg block hanging at the end of the cord.

(a) Describe the motion of this system, including the acceleration of the 10-kg block.
(b) Repeat for a 10-kg block at the end of the cord.

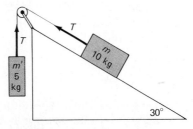

FIGURE 5.5 Diagram for Example 5.1.

SOLUTION

The physical situation is shown in Fig. 5.5. The forces on the 5.0-kg block are $F' = m'g$ directed downward and the tension T in the cord pulling upward in the opposite direction. For the 10-kg block there is a downward force along the plane $F = mg \sin \theta$, as shown above, and an upward force along the plane equal to T, the tension in the cord.

(a) Free-body diagrams for the two blocks are shown in Fig. 5.6. For the motion of the block of mass m', we have, assuming motion downward as positive,

$$F_{net} = m'g - T = m'a$$

For the block on the plane we assume upward motion as positive, since the whole system can only move in one direction, and when m' moves down, m must move up. Hence

$$F_{net} = T - mg \sin \theta = ma$$

Now the tensions are the same, since the cord is assumed to have no mass. Hence adding these two equations, we have

$$m'g - mg \sin \theta = m'a + ma$$

or $$a = \frac{m' - m \sin \theta}{m' + m} g$$

If m' is 5.0 kg and m is 10 kg, we have

$$a = \frac{5 - 10 \sin 30°}{5 + 10} g = \frac{5 - 10(\frac{1}{2})}{15} g = \boxed{0}$$

In this case there is no acceleration since the two forces on each of the blocks just cancel out. Hence the system will either remain at rest or, if in motion with a speed v, will continue its motion with the same speed v.

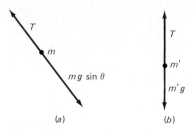

FIGURE 5.6 Free-body diagrams for Example 5.1: (a) Forces on mass m; (b) forces on mass m'.

(b) Here $m' = m = 10$ kg, and

$$a = \frac{m' - m \sin 30°}{m' + m} g = \frac{10 - 5.0}{10 + 10} g$$

$$= \frac{g}{4} = \boxed{2.5 \text{ m/s}^2}$$

Since a is positive, the block m' moves down and the block m is pulled up the plane with this acceleration.

Motion with Friction

In the case of motion with friction, the component of the weight normal to the plane, $F_N = w \cos \theta$, leads to an additional retarding force on the block of magnitude $f_{FR} = \mu_k F_N$. Hence for a single block sliding down a plane, we have, from Newton's second law,

$$F - \mu_k F_N = ma$$

or $$mg \sin \theta - \mu_k mg \cos \theta = ma$$

and so $$a = g \sin \theta - \mu_k g \cos \theta$$

Suppose we have a situation where the block is observed to slide down a plane of variable angle θ at a constant speed. Then we have $a = 0$, or

$$g \sin \theta = \mu_k g \cos \theta = 0$$

and so $$\mu_k = \frac{g \sin \theta}{g \cos \theta} = \tan \theta$$

In a simple laboratory experiment the angle of an inclined plane can be varied until the block moves down the plane at constant speed. Then the coefficient of sliding friction can be obtained from $\mu_k = \tan \theta$. This is a convenient, though not highly accurate, way of measuring μ_k.

Example 5.2

A 5.0-kg block on an inclined plane of inclination angle 40° is attached to a light cord passing over a pulley and then attached to a 10-kg block hanging freely at the end of the cord. Find the acceleration of the 5.0-kg block along the plane. The coefficient of sliding friction between the block and the surface of the plane is 0.35.

SOLUTION

The physical situation is similar to that in Example 5.1 except for the presence of friction. Free-body diagrams for the two blocks are shown in Fig. 5.7. For the block of mass m', we have

$$m'g - T = m'a \qquad [1]$$

where motion down has been taken as positive for m'.

For the block on the plane we have

$$T - mg \sin \theta - f_{FR} = ma$$

or

$$T - mg \sin \theta - \mu_k F_N = ma$$

$$T - mg \sin \theta - \mu_k mg \cos \theta = ma \qquad [2]$$

On adding Eqs. [1] and [2] to eliminate T, we obtain

$$m'g - mg \sin \theta - \mu_k mg \cos \theta = (m' + m)a$$

and so

$$a = \frac{m' - m \sin \theta - \mu_k m \cos \theta}{m' + m} g$$

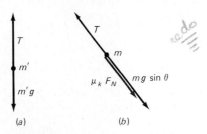

FIGURE 5.7 Free-body diagrams for Example 5.2: (a) Forces on mass m'; (b) forces on mass m.

Putting in numerical values we have:

$$a = \left[\frac{10 - 5.0 \sin 40° - 0.35(5.0) \cos 40°}{10 + 5} \right] (9.8 \text{ m/s}^2)$$

$$= \boxed{3.5 \text{ m/s}^2}$$

5.3 Projectile Motion

FIGURE 5.8 Path of a bullet fired from a gun parallel to the ground.

We live at a time when the motion of projectiles is of great interest and practical importance. The task of NASA engineers and scientists at launching sites like Cape Canaveral is to predict the path which will be followed by the projectiles they launch. This path we call the *trajectory* of the projectile. We have already learned enough physics in this course to calculate the trajectories of some simple projectiles.

As a first example, consider a bullet fired from a gun parallel to the ground, as in Fig. 5.8. Assume that the muzzle velocity of the gun is v_{0x}, that the bullet is fired at a height h above the ground, and that the effect of air resistance is negligible. Then we can break up the motion into two components, one in the X direction, where no forces act, and another in the Y direction, where the force of gravity acts. *Since a force in the Y direction can have no effect on motion in the X direction, the two components of motion are independent of each other*, as shown in Fig. 5.9. The actual motion is then the superposition or combination of the X and Y components of the motion.

If we apply Newton's second law to the two components of the motion, we have

$$\sum F_x = ma_x = 0 \qquad \sum F_y = ma_y = -w = -mg \qquad (5.9)$$

FIGURE 5.9 A flash photo-graph of two golf balls, one falling from rest and the other projected horizontally. The interval between light flashes is $\frac{1}{30}$ s, and the hori-zontal lines are 15 cm apart. It can be seen that the horizontal motion of the right-hand ball has no effect on its vertical motion, since its vertical position is every-where identical with that of the ball falling from rest. (Photo cour-tesy of Educational Development Center, Newton, Mass.)

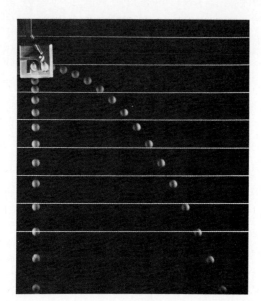

$\vec{V}_{iy} = 0$

$\vec{V}_{ix} = given$

$\vec{V}_x = constant$
   $\quad no\ acceleration$

$\therefore\ x = V_x t .$

$y = y_o + \vec{V}_{yt} + \frac{1}{2} g t^2$
   ↓
   h

where the force of gravity has been given a negative sign because it is directed downward. Hence, since there is no acceleration in the X direction, and the acceleration in the Y direction is $-g$, we have

$$v_x = \text{constant} \qquad a_y = -g$$

Hence *we have a combination of horizontal motion with constant velocity and vertical motion with constant acceleration.*

The distance traveled in the X direction in time t is

$$x = v_{0x}\, t$$

where v_{0x} is the initial velocity of the bullet. For the Y motion, the time it takes for the bullet to fall a distance h to the earth can be obtained from Eq. (2.11),

$$y = h - \tfrac{1}{2}gt^2 = 0$$

or

$$t = \sqrt{\frac{2h}{g}} \tag{5.10}$$

Notice that this is the same result found in Sec. 2.7 for an object falling from the height h. Hence the bullet travels a distance in the X direction of

$$x = v_{0x}\, t = v_{0x}\sqrt{\frac{2h}{g}}$$

before it hits the ground.

Since $y = h - \tfrac{1}{2}gt^2$, we have

$$y = h - \tfrac{1}{2}g\left(\frac{x}{v_{0x}}\right)^2 = h - \frac{\tfrac{1}{2}g}{v_{0x}^2}x^2 \tag{5.11}$$

An equation like this in which one variable y depends on the square of another variable x defines a parabola, as shown in Fig. 5.8.

Example 5.3

A bullet is fired from a rifle at a speed of 1000 m/s. The gun is held parallel to the surface of the earth at a height of 1.5 m.
(a) How long will it take for the bullet to return to the ground?

(b) How far will the bullet travel before it hits the ground, if we neglect the curvature of the earth?

SOLUTION

(a) The time it takes for the bullet to reach the ground is, from Eq. (5.10), simply $t = \sqrt{2h/g}$, the time it would take for the bullet to fall this vertical distance from rest. Hence

$$t = \left[\frac{2(1.5 \text{ m})}{9.8}\right]^{1/2} = \boxed{0.55 \text{ s}}$$

(b) In this time the horizontal distance traveled is

$$x = v_{0x}t = \left(1000 \frac{\text{m}}{\text{s}}\right)(0.55 \text{ s}) = \boxed{550 \text{ m}}$$

Projectile Fired at an Angle

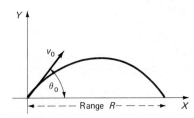

FIGURE 5.10 Path of a projectile fired at an angle θ_0 with respect to the horizontal.

Another interesting case is that of a projectile fired at an angle θ_0 with respect to the horizontal, as in Fig. 5.10. How far does it travel, and how does its maximum range depend on the initial angle θ_0?

We set up a coordinate system in the XY plane with the origin at the point where the projectile begins its flight, i.e., just outside the muzzle of the gun. Then $a_x = 0$ and $a_y = -g$, and, at any instant of time.

$$v_x = v_0 \cos \theta_0 = \text{constant}$$

$$v_y = v_0 \sin \theta_0 - gt \qquad v_2 = v_1 + \vec{a}t$$

The magnitude of the velocity at time t is therefore

$$v = \sqrt{v_x^2 + v_y^2}$$

where v_x is a constant but v_y obviously depends on t. The direction of the shell at time t is determined by

$$\tan \theta = \frac{v_y}{v_x}$$

where θ is the angle v makes with the horizontal. Also, since v_y is constantly decreasing, θ goes from its original value θ_0 to zero at the top of the trajectory, and then becomes progressively more negative as time goes on.

The position of the projectile at time t can be found from Eq. (2.11):

$$x = v_{0x}t = v_0 \cos \theta_0\, t \tag{5.12}$$

$$x = x_0 + vt + \tfrac{1}{2}at^2$$

$$y = v_{0y}t + \tfrac{1}{2}a_y t^2 = v_0 \sin \theta_0\, t - \tfrac{1}{2}gt^2 \tag{5.13}$$

To find the trajectory of the projectile, we must eliminate t between these two equations. From Eq. (5.12) we have

$$t = \frac{x}{v_0 \cos \theta_0}$$

Substituting this value into Eq. (5.13), we obtain

$$y = (v_0 \sin \theta_0)\left(\frac{x}{v_0 \cos \theta_0}\right) - \tfrac{1}{2}g\left(\frac{x}{v_0 \cos \theta_0}\right)^2$$

or

$$y = \tan \theta_0\, x - \frac{1}{2}\frac{g}{v_0^2 \cos^2 \theta_0}x^2 \tag{5.14}$$

Since v_0 and θ_0 are constants, this takes the form

$$y = Ax - Bx^2$$

where A and B are constants. This is again the equation of a parabola. Hence the path of a projectile fired at an angle with respect to the ground and acted on only by the force of gravity is a parabola, as shown in Fig. 5.10.

Range of the Projectile
Definition

The projectile rises to some maximum height and then returns to earth. *The range of a projectile is the distance R along the X axis from the point of firing to the point where it returns to earth.* At the point where the projectile returns to earth, $y = 0$, and so, from Eq. (5.14),

$$y = 0 = \tan \theta_0 \, x - \frac{1}{2} \frac{g}{v_0^2 \cos^2 \theta_0} x^2 \qquad \text{or}$$

$$x = \frac{2v_0^2 \cos^2 \theta_0}{g} \tan \theta_0 = \frac{2v_0^2 \cos^2 \theta_0}{g} \frac{\sin \theta_0}{\cos \theta_0}$$

$$= \frac{2v_0^2}{g} \sin \theta_0 \cos \theta_0$$

Now the value of x at which $y = 0$ is precisely the range R. Hence

$$R = \frac{2v_0^2}{g} \sin \theta_0 \cos \theta_0$$

but, from trigonometry (see Appendix 1C), $2 \sin \theta_0 \cos \theta_0 = \sin 2\theta_0$, and so

$$R = \frac{v_0^2}{g} \sin 2\theta_0 \tag{5.15}$$

This is the general expression for the range of the projectile, which, in the absence of other forces like air resistance and wind, depends only on the initial velocity v_0 and the angle of projection θ_0.

What is the value of θ_0 for which the range R is a maximum? It is the value of R when $\sin 2\theta_0$ takes on its maximum value of 1. This happens when $2\theta_0 = 90°$, or $\theta_0 = 45°$. Hence the range is a maximum when θ_0 is equal to $45°$, and its value is

Maximum Range

$$R_{\text{max}} = \frac{v_0^2}{g} \tag{5.16}$$

For angles above and below $45°$, the range falls off, as shown in Fig. 5.11. Note that the range in Eqs. (5.15) and (5.16) refers to a projectile fired from one elevation and returning to the *same* elevation.

The vertical motion of the shell in this case is identical with that of a ball thrown up in the air with a velocity $v_0 \sin \theta_0$. This must be the case, since if we looked at the problem from an inertial frame of reference moving to the right with a velocity $v_0 \cos \theta_0$, it would appear that the shell was shot straight up in the air, stopped at its highest point, and then fell back to earth again.

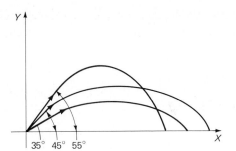

FIGURE 5.11 Range of projectiles fired at different angles with respect to the horizontal.

Example 5.4

Expert sharpshooters often use a technique which makes it almost certain that they will hit their targets in shooting competitions. If a target is thrown straight up in the air, the contestant will wait until it reaches the top of its trajectory before firing directly at the target. The target therefore starts to fall at the exact instant the contestant pulls the trigger.

Will the sharpshooter always hit the target? Does this depend on the speed with which the bullet is shot out of the gun?

SOLUTION

The situation is as diagramed in Fig. 5.12. The target is a height y_0 above the ground and a horizontal distance x_0 along the ground from the gun. The bullet is shot with an initial velocity v_0 at an angle θ_0 with respect to the ground, and is aimed directly at the target when it is stopped at the top of its trajectory.

Here $y_0 = x_0 \tan \theta_0$. Since the bullet is fired at time $t = 0$, the height of the target above the ground at time t, as it falls under gravity, is

$$y_t = y_0 - \tfrac{1}{2}gt^2 = x_0 \tan \theta_0 - \tfrac{1}{2}gt^2$$

The height of the bullet at a horizontal distance x from the gun is, from our discussion in the previous section,

$$y_b = v_0 \sin \theta_0\, t - \tfrac{1}{2}gt^2$$

or, since $t = \dfrac{x}{v_0 \cos \theta_0}$

$$y_b = v_0 \sin \theta_0 \dfrac{x}{v_0 \cos \theta_0} - \tfrac{1}{2}gt^2$$

Hence, when $x = x_0$, $y_b = x_0 \tan \theta_0 - \tfrac{1}{2}gt^2$, and so $y_b = y_t$. The bullet will therefore *always* strike the target, no matter what the value of v_0.

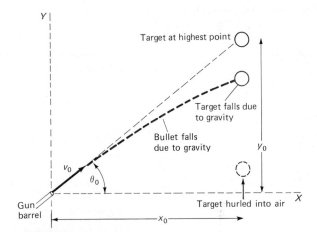

FIGURE 5.12 Diagram for Example 5.4, illustrating how expert sharpshooters make sure they hit their targets.

5.4 Motion in a Circle

In this section we will introduce enough of the basic ideas related to circular motion to enable us to discuss the motion of planets and earth satellites. A more complete explanation of circular motion will be presented in Chap. 8.

As a simple example of circular motion, consider one of the stationary horses on a carousel, or merry-go-round, as in Fig. 5.13. As the carousel rotates, the horse travels around the circumference through a linear distance $2\pi R$, where R is the radius of the circle in which the horse moves, or the distance from the center of the carousel to the horse. If it takes T s for the carousel, moving at constant speed, to make one complete revolution, then the linear speed of the horse is

$$v = \frac{\Delta s}{\Delta t} = \frac{2\pi R}{T} \tag{5.17}$$

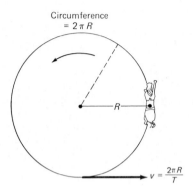

Circumference $= 2\pi R$

$v = \frac{2\pi R}{T}$

FIGURE 5.13 Motion of a horse on a carousel.

Definition

The period T is the time for one complete revolution. Related to the concept of the period in circular motion is the idea of *frequency*. For example, if the carousel takes 10 s for one complete revolution when it is up to constant speed, then its frequency is $\frac{1}{10}$ of a revolution per second, or $\frac{1}{10}$ rev/s. *The frequency is the number of revolutions per second.* Frequency is related to the period by the equation

Definition

$$f = \frac{1}{T} \tag{5.18}$$

Frequencies are usually measured in *cycles per second*, which in the SI system are called *hertz* (Hz).

Radian Measure of Angles

FIGURE 5.14 Definition of a radian: If the arc s ($= AB$) is equal to the radius R, then the angle θ is equal to one radian.

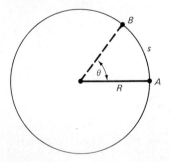

As the horse moves in a circle, a line connecting the horse to the center of the carousel sweeps out an angle θ, as in Fig. 5.14. We could measure the rate at which the angle is swept out, or the *angular velocity*, in degrees per second or in revolutions per second, but it is more useful to introduce the *radian* as our angle measure and use radians per second as the unit for angular velocity.

Consider the circle in Fig. 5.14. If an object moves a distance s along the arc of the circle from A to B, then the angle swept out is θ. We measure θ in radians (rad), where *one radian is the angle subtended by an arc equal to the radius of the circle R.* That is, if $s = R$, then θ is equal to 1 rad. The radian is a pure number, without units or dimensions, since it is the ratio of one length (the arc) to another length (the radius).

The value of θ in radians for any given arc s is then

$$\theta = \frac{s}{R} \tag{5.19}$$

Thus, if $s = 3$ m and $R = 2$ m, $\theta = (3\text{ m})/(2\text{ m}) = 1.5$ rad. Since for a full circle $s = 2\pi R$, the number of radians in a full circle is

$$\theta = \frac{2\pi R}{R} = 2\pi \text{ rad}$$

This means that, since there are 360° in a circle,

$$2\pi \text{ rad} = 360°$$

or $\boxed{1 \text{ rad} = \dfrac{360°}{2\pi} = 57.3°}$

Angular Velocity

By analogy with the linear case we can also define the average angular velocity $(\bar{\omega})$ and the instantaneous angular velocity (ω) as follows:

Angular Velocity

$$\bar{\omega} = \frac{\Delta\theta}{\Delta t} \qquad \omega = \lim_{\Delta t \to 0} \frac{\Delta\theta}{\Delta t} \tag{5.20}$$

where $\Delta\theta$ is the angular displacement (in radians) in time Δt. If the angular velocity is constant, then the average and instantaneous velocities are the same. The units of angular velocity are radians per second (rad/s) if θ is in radians.

The relationship between angular and linear velocities for circular motion is easily found, since, from Eq. (5.19),

$$\Delta s = R \, \Delta\theta$$

and so $\qquad R\omega = \lim\limits_{\Delta t \to 0} \dfrac{R\Delta\theta}{\Delta t} = \lim\limits_{\Delta t \to 0} \dfrac{\Delta s}{\Delta t}$

But, from Eq. (2.3), the right-hand side is equal to the linear velocity v. Hence

$$v = R\omega \tag{5.21}$$

which parallels our previous equation

$$s = R\theta \tag{5.19}$$

Equation (5.21) is valid whether we are dealing with constant, average, or instantaneous velocities.

Another way to measure the angular velocity for motion with constant speed is to notice that in a time equal to the period T, the rotating object sweeps out an angle of 2π rad. Hence the constant angular velocity is

$$\omega = \frac{\Delta\theta}{\Delta t} = \frac{2\pi}{T} \tag{5.22}$$

Now $f = 1/T$, and so

$$\omega = \frac{2\pi}{T} = 2\pi f$$

or $\boxed{\omega = 2\pi f}$ $\qquad\qquad$ (5.23)

Equation (5.23) states that, since in each complete revolution 2π rad is swept out, the rate at which the angle is swept out (in rad/s) is 2π times the number of revolutions per second.

Note that both the radian and the cycle are pure numbers without dimensions. They therefore play no role in the dimensional analysis of equations.

Example 5.5

(a) What is the period of the hour hand on a clock?
(b) What is the frequency of the motion of the minute hand?

(c) What is the angular velocity of the minute hand?

SOLUTION

(a) It takes the hour hand on a clock 12 h to make one complete revolution. Hence its period T is 12 h, or

$$12 \times 3600 \text{ s} = \boxed{4.32 \times 10^4 \text{ s}}$$

(b) Since the minute hand makes one complete revolution every hour, the frequency of the minute hand is

$$f = \frac{1 \text{ rev}}{3600 \text{ s}} = \boxed{2.78 \times 10^{-4} \text{ rev/s}}$$

(c) The angular velocity is

$$\omega = \frac{\Delta\theta}{\Delta t} = 2\pi f$$

For the minute hand,

$$\omega = \left(2\pi \frac{\text{rad}}{\text{rev}}\right)\left(2.78 \times 10^{-4} \frac{\text{rev}}{\text{s}}\right) = \boxed{1.75 \times 10^{-3} \text{ rad/s}}$$

The "radian" is included in the answer even though the answer is equally correct without it.

Example 5.6

A car in one of the rides at an amusement park is rotating in a circle of radius 5.0 m, and makes one complete revolution every 3 s.
(a) What is the angular velocity of the car in radians per second?

(b) What is the angular velocity of the car in degrees per second?
(c) What is its linear speed?

SOLUTION

(a) Since the car makes one complete revolution every 3 s and each revolution corresponds to an angle of 2π rad, the angular velocity is

$$\omega = \left(\frac{1 \text{ rev}}{3 \text{ s}}\right)\left(2\pi \frac{\text{rad}}{\text{rev}}\right) = \tfrac{2}{3}\pi \text{ rad/s}$$

$$= \boxed{2.1 \text{ rad/s}}$$

(b) $\omega = \left(2.1 \dfrac{\text{rad}}{\text{s}}\right)\left(\dfrac{57.3°}{1 \text{ rad}}\right) = \boxed{120°/\text{s}}$

which is just the $\tfrac{1}{3}$ rev/s stated in the problem.
(c) The car's linear speed can be found from Eq. (5.21):

$$v = R\omega = (5.0 \text{ m})(2.1 \text{ rad/s}) = \boxed{11 \text{ m/s}}$$

Angular Acceleration

In this section we confine ourselves to changes in the *magnitude* of the angular velocity ω. Then we can define the average and instantaneous angular accelerations, by analogy with the average and instantaneous accelerations for rectilinear motion, as follows:

Angular Acceleration $\qquad \bar{\alpha} = \dfrac{\Delta\omega}{\Delta t} \qquad \alpha = \lim_{\Delta t \to 0} \dfrac{\Delta\omega}{\Delta t}$ (5.24)

In most cases considered in this book the angular acceleration will be constant, and there will be no difference between the above quantities. Angular accelerations are measured in radians per second squared (rad/s^2) since they are the ratio of an angular velocity (rad/s) to a time (s).

Since $\quad \alpha = \lim\limits_{\Delta t \to 0} \dfrac{\Delta \omega}{\Delta t}$

$$R\alpha = \lim\limits_{\Delta t \to 0} \dfrac{R\,\Delta\omega}{\Delta t} = \lim\limits_{\Delta t \to 0} \dfrac{\Delta v}{\Delta t} = a$$

Hence $\quad a = R\alpha$ $\hfill$ (5.25)

Therefore angular displacements, velocities, and accelerations for circular motion (expressed in radian measure) are all connected to the corresponding linear quantities by parallel expressions involving the radius R. That is,

Angular and Linear Quantities $\quad\quad s = R\theta$ $\hfill$ (5.19)

$\quad\quad\quad\quad\quad\quad\quad\quad\quad\quad\quad\quad\quad\quad v = R\omega$ $\hfill$ (5.21)

$\quad\quad\quad\quad\quad\quad\quad\quad\quad\quad\quad\quad\quad\quad a = R\alpha$ $\hfill$ (5.25)

These simple relationships are often of great importance in solving practical problems, as we shall see in the sections ahead.

Example 5.7

A rotating gyroscope increases its speed from 0 to 25 rad/s in 10 s.
(a) What is its angular acceleration during this 10-s interval, if we assume that this acceleration is constant?

(b) What is the linear acceleration of a point on the rim of the gyroscope 0.25 m from the axis of rotation?

SOLUTION

(a) From Eq. (5.24),

$$\alpha = \frac{\Delta\omega}{\Delta t} = \frac{25 \text{ rad/s} - 0}{10 \text{ s}} = \boxed{2.5 \text{ rad/s}^2}$$

(b) From Eq. (5.25),

$$a = R\alpha = (0.25 \text{ m})(2.5 \text{ rad/s}^2) = \boxed{0.63 \text{ m/s}^2}$$

5.5 Centripetal Acceleration and Centripetal Force

In the preceding section we discussed angular displacements, velocities, and accelerations, and the corresponding linear quantities which relate to motion along the arc of a circle. But what produces motion in a circle in the first place? Consider, for example, a model airplane flying in a horizontal circle at the end of a string. If no forces act on the plane, it would simply fly off in a straight line tangent to the circle, as Newton's first law requires. If the plane moves in a circle, some force must be provided by the string to produce this circular motion. We now consider the nature of this force, its magnitude and direction, and the acceleration it produces.

In Fig. 5.15 we show the airplane moving *with constant speed* in a horizontal circle at the end of a string. Here the magnitude of the linear velocity is constant, but its direction is continually changing. Hence *the*

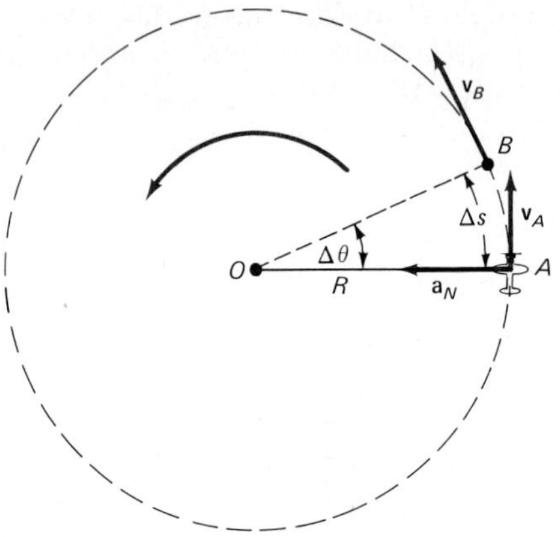

FIGURE 5.15 A model air-
plane flying in a horizontal circle
at the end of a string.

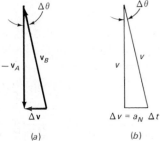

FIGURE 5.16 (a) Vector diagram showing the direction of the
change in velocity $\Delta\mathbf{v}$ of the airplane in Fig. 5.15. (b) Diagram
used to obtain the magnitude of
the centripetal acceleration a_N.

motion is accelerated. The acceleration of the airplane is, as always, equal to
the change in velocity divided by the change in time. This is

$$\mathbf{a} = \frac{\Delta\mathbf{v}}{\Delta t} = \frac{\mathbf{v}_B - \mathbf{v}_A}{\Delta t} \tag{5.26}$$

where $\mathbf{v}_A$ and $\mathbf{v}_B$ are the velocities at points A and B in Fig. 5.15. Now, since
the magnitude of the velocity does not change, $\mathbf{a}$ can have no component
parallel to $\mathbf{v}$; $\mathbf{a}$ must therefore be directed at right angles to $\mathbf{v}$ at every point on
the orbit, and is therefore directed toward the center of the circle. This can be
seen from the vector diagram in Fig. 5.16a. The direction of $\Delta\mathbf{v}$, and hence
the direction of $\mathbf{a}$, is toward the center of the circle. Such an acceleration is
called a *centripetal acceleration*, where *centripetal* means "center-seeking"
from its Latin equivalent and aptly describes this acceleration.

Since the only acceleration acting in this case is the centripetal acceleration, we will use for it the symbol a_N (N for *normal*) and try to find its
magnitude; we already know its direction. From Eq. (5.26) we then have

$$a_N = \frac{\Delta v}{\Delta t}$$

or $\quad \Delta v = a_N\,\Delta t \tag{5.27}$

Now $\mathbf{v}_A$ and $\mathbf{v}_B$ are both perpendicular to the radii at points A and B, and
therefore the angle between $\mathbf{v}_A$ and $\mathbf{v}_B$ is the same as between OB and OA,
that is, $\Delta\theta$. Hence, as can be seen from Fig. 5.16b,

$$\Delta\theta = \frac{\Delta v}{v}$$

where v is the constant magnitude of the velocity. But

$$\Delta\theta = \frac{\Delta s}{R} = \frac{v\Delta t}{R}$$

and so $\quad \dfrac{\Delta v}{v} = \dfrac{v\Delta t}{R}$

From Eq. (5.27) we have $\Delta v = a_N \, \Delta t$, and so

$$\frac{a_N \, \Delta t}{v} = \frac{v \Delta t}{R}$$

or $$\boxed{a_N = \frac{v^2}{R}}$$ (5.28)

This is the magnitude of the centripetal acceleration.

Centripetal Acceleration

The centripetal acceleration for circular motion is directed toward the center of the circle and is of magnitude v^2/R.

The dimensions of v^2/R are

$$\frac{[L/T]^2}{[L]} = \frac{[L]}{[T^2]}$$

which are the correct dimensions for an acceleration. Since $v = R\omega = R(2\pi f)$, we can write

$$a_N = \frac{R^2(2\pi f)^2}{R} = 4\pi^2 f^2 R$$ (5.29)

where f is the frequency, or number of revolutions per second.

In addition to the centripetal acceleration, which is normal to the velocity, it is also possible that the *speed* of a car may change as the car moves along. This change in the magnitude of the velocity can then be defined as:

Tangential Acceleration

$$a_T = \lim_{\Delta t \to 0} \frac{\Delta v}{\Delta t}$$

where Δv is the change in the *magnitude* of the velocity **v**. Then the *tangential* component of the acceleration measures a change in the *magnitude* of the velocity **v**, whereas the *normal* or centripetal acceleration a_N arises from a change in the *direction* of the velocity.

The resultant acceleration is the vector combination of these two possible kinds of changes in the velocity with time. Since a_N is perpendicular to a_T, we have for the magnitude of the total acceleration:

$$a = (a_T^2 + a_N^2)^{1/2}$$ (5.30)

The direction of the resultant acceleration can be seen from Fig. 5.17 to be given by:

$$\tan \theta = \frac{a_N}{a_T}$$ (5.31)

θ is then the angle made by the resultant acceleration vector **a** with a_T and hence with the direction of motion at the instant **a** is calculated. Note that, if a centripetal acceleration exists, the direction of the resultant acceleration is *not* that of the velocity.

FIGURE 5.17 Acceleration for circular motion: The resultant acceleration is the vector sum of the tangential acceleration a_T (due to a change in speed) and the centripetal acceleration a_N (due to a change in direction).

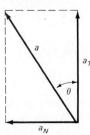

Example 5.8

A racing car is going around a circular track. At a certain instant of time it is changing speed at a rate of 8.0 m/s², and its centripetal acceleration is 10 m/s².
(a) What is the magnitude of the resultant acceleration at that instant?

(b) What is the direction of this acceleration with respect to the direction of motion of the car at that instant?

SOLUTION

(a) Since the components of the acceleration are at right angles to each other, we have for the magnitude of the acceleration

$$a = \sqrt{a_T^2 + a_N^2}$$

or $\quad a = \left[\left(8.0 \frac{m}{s^2} \right)^2 + \left(10 \frac{m}{s^2} \right)^2 \right]^{1/2} = \boxed{13 \text{ m/s}^2}$

(b) The component a_T is in the same direction as the velocity of the car at the instant a_T is being calculated. This is parallel to the curve describing the car's motion. The

angle θ made by the acceleration **a** with respect to a_T is then given by Eq. (5.31) as

$$\tan \theta = \frac{a_N}{a_T} = \frac{10}{8} = 1.25$$

and so $\quad \theta = \boxed{51°}$

Hence the acceleration of the car is directed at an angle of 51° with respect to the direction of the car's motion at the instant when the acceleration is calculated.

Centripetal Force

If an acceleration is produced in circular motion, there must be a corresponding net force to cause this acceleration. Since, from Newton's second law, any force is the product of a mass and an acceleration, here we have

Centripetal Force

$$\boxed{F_N = \frac{mv^2}{R}}$$

(5.32)

This is the *centripetal force*, which is also directed toward the center of the circle. For the airplane moving in a horizontal circle, m is the mass of the plane.

A centripetal force produces a change in the direction of the linear velocity of an object, without any change in its magnitude. The equation for the centripetal force, however, tells us nothing about how this force is provided physically. It merely says that if an object is in circular motion, some force of magnitude mv^2/R must be acting to keep it moving in a circle if its linear velocity is v. For a toy airplane at the end of a string, this force is provided by the string, whereas for the planets the gravitational attraction of the sun provides the necessary force.

Note that, since $v = R\omega$, we can write the expression for centripetal force also as

$$F_N = \frac{mv^2}{R} = \frac{m(R\omega)^2}{R} = mR\omega^2 = 4\pi^2 f^2 mR$$

(5.33)

where ω is the angular speed of the object in an orbit of radius R, and $\omega = 2\pi f$.

Banking of Curves

A car going around a curve requires a centripetal force to keep it moving in a circle. This force can be provided by the friction between the road and the car's tires, even on a flat road, but, to reduce the chance of skidding, it is

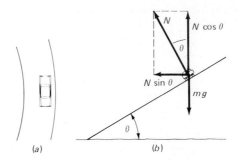

FIGURE 5.18 (a) Top view of a car going around a curve. (b) End-on view of a car going around a curve.

better to bank the curve. Let us see if we can find the angle θ which the banked curve should make with the flat road surface to reduce the chances of skidding to a minimum.

Figure 5.18a shows a top view of the situation. Figure 5.18b provides an end-on view which shows the banking angle θ.

To simplify the problem, we neglect friction in this case. If the curve is banked at an angle θ, the road exerts a normal force of magnitude N perpendicular to its surface. This force has two components. $N \cos \theta$ is vertical and just balances out the weight of the car, mg. $N \sin \theta$ is horizontal and is directed inward along the radius of the circle of which the curve is a part. This provides the centripetal force to keep the car moving smoothly around the curve. Hence we have

$$N \cos \theta = mg$$

$$N \sin \theta = \frac{mv^2}{R}$$

On dividing the second equation by the first, we have

$$\tan \theta = \frac{v^2}{Rg} \tag{5.34}$$

Hence the angle of banking, θ, is determined by the sharpness of the curve (the smaller R is, the larger the amount of banking needed) and by the speed of the car. For a given radius R, the tangent of the banking angle varies as v^2. Hence it is impossible to bank a curve for all speeds. Therefore curves are banked for the average speed at that point on the road. It is clear that at speeds in excess of the speed for which the curve was banked, there is a greater chance that skidding and accidents will occur.

Example 5.9

A ball at the end of a string is rotating in a horizontal circle of radius 1.5 m. The mass of the ball is 0.25 kg, and the string makes an angle of 10° with the horizontal.
(a) Draw a free-body diagram showing *all the forces* acting on the ball.

(b) What is the rotational velocity of the ball?
(c) How many revolutions does the ball make per second?

SOLUTION

(a) The free-body diagram is shown in Fig. 5.19. The direction of the string cannot be perfectly horizontal, since there must be a vertical component of the tension in the string sufficient to balance out the weight of the ball.

(b) We write Newton's second law of motion for the vertical and horizontal components, where the tension in the string provides the centripetal force needed to keep the ball moving in a circle.

M = 0.25 kg

$$T \sin \theta - mg = 0$$

$$T \cos \theta = F_N = \frac{mv^2}{R} = mR\omega^2$$

Hence

$$T = \frac{mg}{\sin 10°} = \frac{(0.25 \text{ kg})(9.8 \text{ m/s}^2)}{0.174} = 14 \text{ N}$$

and $\quad \omega^2 = \dfrac{T \cos \theta}{mR} = \dfrac{(14 \text{ N})(\cos 10°)}{(0.25 \text{ kg})(1.5 \text{ m})} = \dfrac{36.8}{\text{s}^2}$

$$\omega = \boxed{6.1 \text{ rad/s}}$$

(c) $\quad f = \dfrac{\omega}{2\pi} = \dfrac{6.1 \text{ rad/s}}{2\pi \text{ rad/rev}} = \boxed{0.97 \text{ rev/s}}$

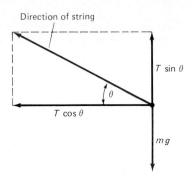

FIGURE 5.19 Free-body diagram for Example 5.9.

Hence the ball must make about 1 rev/s. The faster the ball goes, the greater the tension in the string, and hence the smaller the angle made by the string with the horizontal, since $\sin \theta = mg/T$.

Centrifuges and Ultracentrifuges

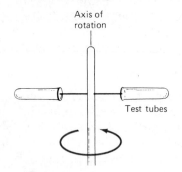

FIGURE 5.20 A side view of a rotating centrifuge.

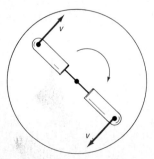

FIGURE 5.21 Top view of a rotating centrifuge.

The centrifuge is a device frequently used in biological, chemical, and medical laboratories to separate substances of different densities. It provides an interesting example of centripetal acceleration.

One type of centrifuge, the so-called swinging bucket rotor, consists of a rotating cylinder (the rotor) to which holders for test tubes are attached. The test-tube holders hang vertically when the rotor is stopped, but swing out to horizontal positions when the rotor gets up to speed, as shown in Fig. 5.20.

It is often important in medical laboratories to take a sample of whole blood and determine the ratio of the volume of red blood cells to total blood volume. The blood cells are slightly more dense than the rest of the liquid and hence would slowly fall through the liquid, under the force of gravity, and collect at the bottom of the test tube. This process might normally take days, or weeks. With a centrifuge the same process can be carried out in minutes or even seconds, since the centrifuge can increase the effective value of g, the acceleration due to gravity, by a factor of thousands.

When a blood sample in the test tube is whirling around in the centrifuge at a very high speed, as in Fig. 5.21, a red blood cell behaves somewhat like a passenger going around a sharp turn in a car. The passenger feels as if he or she is being pushed against the side door, even though actually the body merely tends to continue motion in a straight line, according to Newton's first law. The car turns so that the door gets in the passenger's way and pulls the body around the corner with the car. Similarly, the red blood cells in the centrifuge test tube want to keep moving in a straight line, but the whirling test tube gets in the way. The more dense blood cells therefore fall quickly through the less dense liquid to the bottom of the tube, where the glass wall of the tube provides the centripetal force to keep the blood cells moving in a circular path. The force which must be provided by the test tube wall is mv^2/R, and this provides a centripetal acceleration of $v^2/R = R\omega^2$. This is very much larger than the acceleration due to gravity for the blood cells, and this acceleration would normally determine the rate at which the cells would fall through the liquid. The effective acceleration in this case is:

$$g_{eff} = R\omega^2$$

For a blood cell 10 cm from the center of the rotor, rotating at 100 rev/s this is:

$$g_{eff} = (0.10 \text{ m}) \left[(100)\left(2\pi \frac{\text{rad}}{\text{s}} \right) \right]^2 \simeq 40,000 \text{ m/s}^2 \simeq 4000g$$

It is because of this greatly increased effective value of g that the red blood cells can be separated from the solution so quickly in a centrifuge.

Ordinary centrifuges rotate at speeds up to 100 rev/s and thus produce effective g values in the thousands.

To separate out the lipoproteins from the heavier proteins and albumin in a blood sample, an ordinary centrifuge is inadequate and ultracentrifuges become necessary.

Ultracentrifuges have rotating parts supported by air jets or magnetic fields to reduce frictional heating, and are constructed to withstand the forces produced by speeds of as much as 3000 rev/s. They can produce effective g values above 10^6.

A technique frequently used with ultracentrifuges is to use a sucrose or a saline solution in which the density increases from top to bottom of the test tube. Substances immersed in such a solution and spun in an ultracentrifuge then sort themselves out to match the density of the solution along the axis of the tube, since then the buoyant force of the solution is just able to provide the centripetal force needed to keep molecules of the right density moving in a circle. Hence the more dense molecules are found at the end of the centrifuge tube, and the density decreases inward along the test-tube axis. Such techniques are now used routinely in research laboratories in molecular biology and biochemistry.

Physiological Effects of Increasing g

Large centrifuges are often used to study the effect of large values of g on the human body. In these devices men or women are rotated rapidly in a circle in carriages at the end of a rigid arm. Such research is particularly important in the space program (see Fig. 5.22), since astronauts are subjected to large g's when they are blasting off from the earth or making their return landing on earth.

FIGURE 5.22 Large centrifuge used by NASA to study the effect of large g values on the human body at the Manned Spacecraft Center, Houston, Texas. The centrifuge has a 50-ft-long arm at the end of which a gondola can hold three astronauts. The rapidly rotating arm creates g values similar to those that astronauts experience during lift-off and reentry into the earth's atmosphere. (Photo courtesy of NASA.)

Example 5.10

An ultracentrifuge rotates at 180,000 rev/min. Red blood cells are suspended in the liquid of the rotating tube. They are driven outward and come to rest against the glass end of the test tube at a distance of 10 cm from the axis of rotation.

(a) What acceleration must be imparted to the blood cells by the wall of the tube to keep them moving in a circle?
(b) In terms of the acceleration due to gravity g, how large is this acceleration?

SOLUTION

(a) The required centripetal acceleration is

$$a_N = \frac{v^2}{R} = \frac{(R\omega)^2}{R} = R\omega^2$$

and, since $\omega = 180,000$ rev/min $= 3000$ cycles/s,

$$\omega = (3000)(2\pi \text{ rad/s}) = 1.88 \times 10^4 \text{ rad/s}$$

$$a_N = (0.10 \text{ m})(1.88 \times 10^4 \text{ s}^{-1})^2 = \boxed{3.5 \times 10^7 \text{ m/s}^2}$$

(b) Since $g = 9.80 \text{ m/s}^2$,

$$a_N = (3.5 \times 10^7 \text{ m/s}^2)\left(\frac{g}{9.80 \text{ m/s}^2}\right) = \boxed{3.6 \times 10^6 \, g}$$

Hence a force of over a million times the force of gravity is required in this case to keep the blood cell moving in a circle.

Example 5.11 (Motion in a vertical circle)

A ball attached to a cord of length $R = 30$ cm is being rotated in a vertical circle.
(a) What is the minimum speed the ball must have at the top of the circle if the cord is not to become slack and the ball begin to fall?

(b) If the ball is rotating at this same speed at the bottom of the circle, what is the tension in the cord?

SOLUTION

A diagram of the situation is shown in Fig. 5.23. The forces on the ball at any point are its weight $w = mg$ acting vertically downward, and the tension T in the cord directed toward the center of the circle. The sum of these two forces must provide the centripetal acceleration to keep the ball moving in a circle.

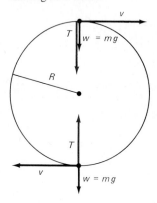

FIGURE 5.23   Motion of a ball rotating in a vertical circle at the end of a string.

(a) At the top of the circle, both w and T are directed downward, as shown in the figure. Hence we have

$$T + mg = \frac{mv^2}{R}$$

or $\quad T = m\left(\frac{v^2}{R} - g\right)$

For the ball to continue moving in a circle, there must be some tension in the cord; otherwise the ball would leave its circular path. The limiting case is when T is zero and v is equal to its critical value v_c. Then we have

$$T = m\left(\frac{v_c^2}{R} - g\right) = 0 \quad \text{or} \quad \frac{v_c^2}{R} - g = 0$$

and so $\quad v_c = \sqrt{gR}$

Hence for a circle of radius 30 cm, we have

$$v_c = [(9.8 \text{ m/s}^2)(0.30 \text{ m})]^{1/2} = \boxed{1.7 \text{ m/s}}$$

(b) At the bottom of the circle w is directed downward and T upward. Hence we have

$$T - mg = \frac{mv^2}{R}$$

or $\quad T = m\left(\frac{v^2}{R} + g\right)$

If $v = v_c = \sqrt{gR}$, we have

$$T = m\left(\frac{gR}{R} + g\right) = m(2g) = 2w$$

In this case the tension in the cord at the bottom is twice the weight of the ball.

Hence, if the speed of the ball could be kept constant at $v = v_c$, the tension in the cord would go from $2w$ at the bottom of the circle to w halfway up the circle and then to zero at the top. In practice, of course, the ball's speed will increase on the way down and decrease on the way up.

5.6 Planetary Motion: Kepler's Laws

Although Galileo believed that the earth and the other planets rotated about the sun, he was never able to provide completely convincing proof for this Copernican model of planetary motion. It was left for Newton to provide this proof at a later date.

The experimental evidence on which Newton's theory was based was in great part due to the Danish astronomer Tycho Brahe (1546–1601), who had collected astronomical data on the motion of the planets good to about 1 minute of arc (1') or $\frac{1}{60}°$. This precision was all the more remarkable since Brahe's observations were made with the naked eye without the aid of telescopes, which had not yet been invented. Brahe's secret was the use of very large sighting devices mounted on strong, rigid support systems, and of highly precise scales for his position and angle measurements.

The German astronomer Johannes Kepler (1571–1630) took Brahe's data and came up with a detailed description of the motion of the planets about the sun. The word *description* is purposely used here because Kepler's three laws merely *described* the behavior of the planets. We will here just state Kepler's three laws, as a preliminary to presenting Newton's explanation of them.

Kepler's Laws for Planetary Motion

1 *Law of elliptical paths*: The orbit of each planet about the sun is a planar ellipse* with the sun at one focus (see Fig. 5.24).

---

*An ellipse is a curve such that the sum of the distances from two points, the two foci, is a constant for any point on the curve. As the foci get closer and closer together compared with the average radius of the curve, an ellipse approaches a circle. One way to construct an ellipse is to stick two pins a few inches apart in a piece of paper, throw a small loop of string over the two pins and then trace out a complete curve with a pencil inserted inside the loop with the string kept taut by the pencil point.

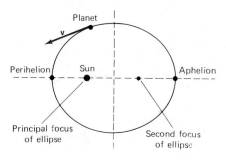

FIGURE 5.24 Elliptical path of a planet around the sun.

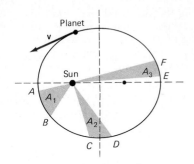

FIGURE 5.25 Kepler's second law, the law of equal areas. The planet sweeps out equal areas in equal times. If area A_1 is equal to A_2 and to A_3, then the times for the planet to go from A to B, from C to D, and from E to F are all equal. Hence the planet travels faster in going from A to B than from E to F.

2 *Law of equal areas*: Each planet moves so that an imaginary line drawn from the sun to the planet sweeps out equal areas in equal times (see Fig. 5.25). Hence the planets move faster when they are closer to the sun.

3 *Harmonic law*: The ratio of the squares of the periods of any two planets revolving about the sun is equal to the ratio of the cubes of their average distances from the sun; i.e.,

$$\frac{T_2^2}{T_1^2} = \frac{R_2^3}{R_1^3} \tag{5.35}$$

These three general principles made sense of the motion of the planets for the first time, but Kepler was not able to derive these principles from the basic laws of mechanics. It was left to Isaac Newton to accomplish this great feat.

5.7 Newton's Explanation of Kepler's Laws

Sir Isaac Newton had, in his three laws of motion and the law of universal gravitation, all that was needed to explain the motion of the planets and to derive Kepler's laws. Here we will only present Newton's proof of Kepler's third law.

In deriving Kepler's third law we introduce the following approximation: Since the elongation of the ellipses is quite small,* we will use a simplified model which assumes that the planets move in circles of radius R about the sun, where R is each planet's average distance from the sun in its elliptical orbit. Some form of physical force is needed to provide the centripetal force required for circular motion. Newton's great discovery was that this force was the same kind of force that causes an apple to fall from a tree to the ground, i.e., a gravitational force.

Let us assume that the gravitational force of the sun on a planet of mass m provides the centripetal force to keep the planet moving in a circle. We then have, using Eqs. (3.4) and (5.32),

$$F_G = \frac{Gm_S m_P}{R^2} = \frac{m_P v^2}{R}$$

Hence, canceling like terms, we have

$$v^2 = \frac{Gm_S}{R} \tag{5.36}$$

or $\quad v^2 R = Gm_S$

or, since $v = 2\pi R/T$,

---

*The earth is about 4.8×10^9 m closer to the sun in January than in July, but the average earth-sun distance is about 1.5×10^{11} m. Hence the change in the earth-sun distance is only about 3 percent.

$$v^2 R = \frac{4\pi^2 R^3}{T^2} = Gm_S$$

or $\qquad \dfrac{R^3}{T^2} = \dfrac{Gm_S}{4\pi^2}$ (5.37)

Now the right side of this equation is independent of the properties of any particular planet, and is thus a constant for all the planets. Hence, for two different planets we have

$$\frac{R_1^3}{T_1^2} = \frac{Gm_S}{4\pi^2} = \frac{R_2^3}{T_2^2}$$

or $\qquad \dfrac{T_2^2}{T_1^2} = \dfrac{R_2^3}{R_1^3}$

which is Eq. (5.35), Kepler's third law.

Before Newton's time many people believed that the laws governing the motion of the heavens were different from those describing motion on earth. Hence Newton's demonstration that his laws worked equally well on earth and in the heavens was one of the great triumphs of the human mind.

5.8 Earth Satellites

FIGURE 5.26 Newton's suggestion of the possibility of an earth satellite. He pointed out that if an object were fired parallel to the surface of the earth, as its initial speed increased it would travel greater and greater distances before returning to earth. In the diagram, as the speeds increase, the object travels first to point A, then to B, then to C, and so on. If a high enough speed $\mathbf{v}_0$ could be reached, the object would circle the earth forever. This speed $\mathbf{v}_0$ is about 8.0 km/s.

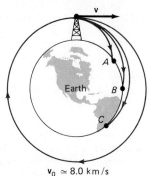

$\mathbf{v}_0 \simeq 8.0$ km/s

The moon was, of course, the first earth satellite. Newton predicted the possibility of satellites orbiting the earth like small planets (see Fig. 5.26), but seventeenth-century technology was not up to putting satellites into permanent orbits about the earth.

Why does a satellite in orbit not fall back toward the earth? The fact is that it does! It is constantly accelerated toward the earth by the gravitational force of the earth, with the acceleration proper to its distance from the center of the earth. This gravitational force provides the centripetal force to keep the satellite moving in a circle or ellipse about the earth. The secret is to get the satellite up to a speed v parallel to the surface of the earth which corresponds to a centripetal force just equal to the gravitational attraction of the earth for the satellite at that height.

Consider a case in which a satellite is in orbit very close to the surface of the earth. The Russian satellite Vostok 6, which was the first satellite to carry a woman into space, had an average height above the earth of about 193 km. Since the radius of the earth is 6360 km, this makes the radius of Vostok's orbit 6.55×10^6 m. Then, equating the centripetal force to the gravitational attraction of the earth for the satellite, we have

$$\frac{mv^2}{R} = \frac{Gmm_E}{R^2} = w = mg$$

since the gravitational force on the satellite is precisely what we mean by its weight. Hence

$$\frac{v^2}{R} = g = \frac{Gm_E}{R^2}$$

where g is the acceleration due to gravity at a height of 193 km above the earth's surface and is, with $R = 6.55 \times 10^6$ m,

$$g = \frac{6.67 \times 10^{-11}\,\text{m}^3}{\text{kg}\cdot\text{s}^2}\,\frac{5.98 \times 10^{24}\,\text{kg}}{(6.55 \times 10^6\,\text{m})^2} = 9.30\,\text{m/s}^2$$

Then

$$v = \sqrt{gR} = \sqrt{(9.30\,\text{m/s}^2)(6.55 \times 10^6\,\text{m})}$$

$$= 7.80 \times 10^3\,\text{m/s} = 7.80\,\text{km/s}$$

Hence a speed of about 5 mi/s was needed to keep Vostok 6 in orbit at a height of 193 km above the surface of the earth.

The period of the satellite in such an orbit was, then,

$$T = \frac{s}{v} = \frac{2\pi R}{v}$$

$$= \frac{2\pi(6.55 \times 10^6\,\text{m})}{7.80 \times 10^3\,\text{m/s}} = 5.27 \times 10^3\,\text{s} = 88\,\text{min}$$

This result agrees well with the observed fact that Vostok 6 completed 1 rev around the earth every 88.34 min.

Weightlessness

We have probably all seen television shots of astronauts floating around inside their spaceship, in which the force of gravity does not seem to be acting. The reason is that a spaceship in orbit about the earth is being accelerated toward the earth with an acceleration g. For an astronaut this is similar to being in an elevator falling freely through space. For this reason we first consider this simpler one-dimensional case.

Forces on a Passenger in an Accelerated Elevator

Suppose an elevator passenger is standing on a spring scale resting on the floor of an elevator, as in Fig. 5.27. What weight does the scale read for the passenger when the elevator undergoes accelerated motion?

We first consider the case where the elevator accelerates upward with a constant acceleration a, e.g., the case in which the elevator moves upward with increasing speed. Then Newton's third law tells us that the downward force exerted by the passenger on the scale (the passenger's apparent weight) is equal to the upward force f exerted by the scale on the passenger.

We are interested in all the forces acting on the passenger and in the passenger's acceleration. From Newton's second law we have

$$\sum F_y = f - W = ma$$

where f is the upward push of the scale on the passenger, W is the passenger's true weight, i.e., the force of gravity acting on his or her mass, and a is the acceleration of the passenger, which is the same as the acceleration of the elevator (f here is the apparent weight of the passenger as read from the scale). Hence we have

$$f = W + ma$$

or the apparent weight is *larger* than the passenger's true weight by an amount ma. If the passenger has a mass of 100 kg, his or her true weight is $mg = 980$ N. Suppose the elevator's upward acceleration is 2.0 m/s². Then

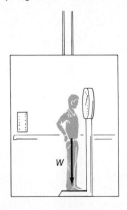

FIGURE 5.27 A man standing on a spring scale in an elevator.

$$f = 980 \text{ N} + (100 \text{ kg})(2.0 \text{ m/s}^2) = 1180 \text{ N}$$

Hence the passenger's apparent weight is *larger* than the actual weight by 200 N.

If, on the other hand, the elevator's acceleration is *downward* at 2.0 m/s², then we have

$$f = W - ma = 980 \text{ N} - (100 \text{ kg})(2.0 \text{ m/s}^2)$$

$$= 980 \text{ N} - 200 \text{ N} = 780 \text{ N}$$

Hence the apparent weight of the passenger is less than the true weight by 200 N. If the elevator cable breaks so that the car falls at a downward acceleration equal to g, we have:

$$f = W + ma = mg - mg = 0$$

Hence the passenger appears to be *weightless*, since there is no upward push from the scale on the passenger's feet. This is true of all objects in any freely falling system like this elevator.

An earth satellite is merely a freely falling spaceship; it falls toward the earth with an acceleration g just as does a freely falling elevator. Hence the astronauts in such a spaceship experience no upward pushes from the floors or rest pads in their chambers. As a result, they don't feel the compression of their vertebrae that is normally produced by a gravitational force, and conclude that they are in a weightless state. This is true whether they are falling freely toward the earth, the moon, or some other planet or star. As long as the spaceship is falling freely in space, i.e., as long as its rocket engines are not operating, everything within the spaceship appears to be weightless.

Since the speed required for the spaceship to remain in a given orbit is independent of the mass of the spaceship, all the contents of the spaceship, including the astronauts, are moving in the same orbit and experience the same condition of weightlessness. Everything therefore appears to float freely in the cabin, as seen in Fig. 5.28.

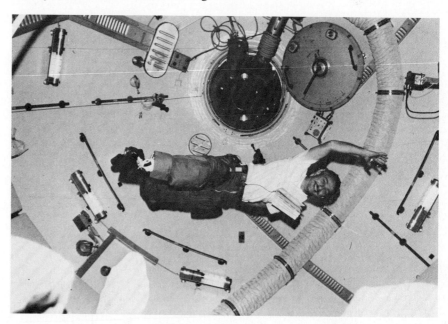

FIGURE 5.28 Astronaut Owen K. Garriott experiencing weightlessness in an earth satellite during the Skylab mission in 1973. (Photo courtesy of NASA.)

Physiological Effects of Weightlessness

Weightlessness gives astronauts a lightsome feeling, since no effort is needed to hold up their arms when they are in free-fall. Prolonged exposure to weightlessness can, however, have bad effects. Red blood cell counts are decreased; bones can become brittle; and muscles can lose their tone. This is because the organs of the body no longer touch and press on each other in the usual way. At present, much research is being done by NASA on this subject in connection with the U.S. space program.

Example 5.12

A communications satellite is used to relay radio and TV signals from one side of the earth to the other. Such satellites keep a fixed position above one spot on the earth's surface and move eastward with the earth as it rotates on its axis. These are sometimes referred to as *stationary satellites*, or *synchronous satellites*.

(a) What must be the distance of such a satellite from the center of the earth?
(b) How fast must it travel in its orbit?
(c) How far above the earth's surface must the satellite be?

SOLUTION

(a) The satellite, like the earth, makes one complete revolution each day, or every 24 h. Since the force needed to keep it in orbit is provided by the gravitational attraction of the earth, the equation governing its motion is the same as that governing the motion of the earth around the sun. Hence we have, from Eq. (5.37), with the mass of the sun replaced by the mass of the earth,

$$R^3 = \frac{Gm_E T^2}{4\pi^2}$$

where $T = 24$ h $= 8.64 \times 10^4$ s. Hence

$$R^3 = \frac{\left(6.67 \times 10^{-11} \frac{\text{N·m}^2}{\text{kg}^2}\right)(5.98 \times 10^{24}\text{ kg})(8.64 \times 10^4\text{ s})^2}{4(3.14)^2}$$

$$= 7.55 \times 10^{22}\text{ m}^3$$

or $R = \boxed{4.23 \times 10^7\text{ m}}$

(b) The speed of the satellite must be

$$v = \frac{2\pi R}{T} = \frac{2\pi(4.23 \times 10^7\text{ m})}{8.64 \times 10^4\text{ s}} = \boxed{3.08 \times 10^3\text{ m/s}}$$

(c) Since the radius of the earth is 6.37×10^6 m, the height of the satellite above the earth's surface must be

$$4.23 \times 10^7\text{ m} - 0.64 \times 10^7\text{ m} = 3.59 \times 10^7\text{ m}$$

In miles this is

$$(3.59 \times 10^7\text{ m})\left(\frac{1\text{ mi}}{1.61 \times 10^3\text{ m}}\right) = \boxed{2.23 \times 10^4\text{ mi}}$$

Hence synchronous communications satellites circle the earth at a height about 22,300 mi above the earth's surface.

Summary: Important Equations and Definitions

Motion in a plane:

$$\bar{\mathbf{v}} = \frac{\Delta s}{\Delta t} \qquad v = \lim_{\Delta t \to 0} \frac{\Delta s}{\Delta t}$$

$$\bar{\mathbf{a}} = \frac{\Delta \mathbf{v}}{\Delta t} \qquad \mathbf{a} = \lim_{\Delta t \to 0} \frac{\Delta \mathbf{v}}{\Delta t}$$

For motion in a plane the acceleration a has a component a_T parallel to the velocity which specifies the change in the magnitude of the velocity, and a component a_N perpendicular to the velocity which specifies the change in the direction of the velocity.

$$a = \sqrt{a_T^2 + a_N^2}$$

Motion on an inclined plane:
    $a = g \sin \theta$, where θ is the angle of the plane.
Projectile motion:
    A combination of horizontal motion with constant velocity and vertical motion with constant acceleration.
Circular motion (motion in a circle of radius R):
    Period (T): The time for one complete revolution
    Frequency ($f = 1/T$): The number of revolutions per unit time (s)

Linear velocity: $v = \dfrac{2\pi R}{T}$

Angular velocity: $\bar{\omega} = \dfrac{\Delta\theta}{\Delta t}$ $\omega = \lim\limits_{\Delta t \to 0} \dfrac{\Delta\theta}{\Delta t}$

where $\Delta\theta$ is the angular displacement (in radians) in time Δt.

Radian measure:

1 rad $= 57.3° =$ the angle subtended by an arc equal to the radius of the circle.

For constant angular velocity, $\omega = 2\pi f$.

Angular acceleration:

$\bar{\alpha} = \dfrac{\Delta\omega}{\Delta t}$ $\alpha = \lim\limits_{\Delta t \to 0} \dfrac{\Delta\omega}{\Delta t}$

Relationship between linear and angular quantities:

$s = R\theta$ $v = R\omega$ $a = R\alpha$

Centripetal acceleration and centripetal force:

$a_N = \dfrac{v^2}{R}$ $F_N = \dfrac{mv^2}{R} = mR\omega^2$

Kepler's three laws of planetary motion:

1. The orbit of each planet about the sun is an ellipse with the sun at one focus.
2. Each planet moves so that an imaginary line drawn between the sun and the planet sweeps out equal areas in equal times.
3. The ratio of the squares of the periods of any two planets revolving about the sun is equal to the ratio of the cubes of their average distances from the sun; i.e., $(T_2/T_1)^2 = (R_2/R_1)^3$.

Earth satellite:

A small mass which is constantly being accelerated by the gravitational force of the earth with the acceleration proper to the satellite's distance from the center of the earth.

Weightlessness:

The condition of apparent loss of all weight which exists in an elevator or earth satellite falling freely in space.

Apparent weight $f =$ real weight $W - ma$. If $a = g$, $f = 0$.

Questions

1 A sailor climbs to the top of a ship's mast to repair a sail while the ship is moving at a fast speed. At the top of the mast a screwdriver falls out of the sailor's pocket. Where will the screwdriver hit the deck of the ship: behind the bottom of the mast, in front of the bottom of the mast, or exactly at the foot of the mast? Why?

2 A block of smooth wood is pushed up an inclined plane with an initial speed v_0 along the plane. There is friction between the block and the plane. The block comes to rest and then slides back down the plane to the bottom.

(a) Draw a free-body diagram showing all the forces acting on the block.

(b) Draw graphs of (1) the acceleration of the block as a function of time; (2) the velocity of the block as a function of time; (3) the position of the block as a function of time. For all graphs use distances as measured along the inclined plane.

3 Suggest a way to measure the coefficient of *static friction* between a block and an inclined plane of variable inclination.

4 Why is the radian a pure number, i.e., a number without dimensions and without units?

5 Show from the formula for the magnitude of the centripetal acceleration ($a_N = v^2/R$) that the centripetal acceleration must be normal to the direction of motion.

6 Suggest why a centrifuge can be used to separate liquids of different densities.

7 Show that the method described in the footnote to Sec. 5.6 for constructing an ellipse produces a curve which satisfies the definition of an ellipse.

8 When an elevator is moving downward with decreasing speed, what is the direction of its acceleration? As the elevator moves in this way, does a spring scale at the bottom of the elevator read a greater or a smaller apparent weight for a person standing on the scale?

9 If the earth is constantly falling toward the sun, why does it not gradually move closer to the sun? (If it did, the increased heating of the earth's surface by solar radiation would soon destroy all life on earth.)

10 Although the distinction between mass and weight may seem trivial on the surface of the earth, it can be a matter of life and death in space. Suppose a space station is being built in orbit around the earth. Astronauts building the station are moving huge girders into position in the space station.

(a) Are these girders weightless?

(b) If they are weightless, are they any easier to move around in space than they would be on earth? Why?

(c) Could an astronaut be crushed to death between two huge girders moving toward each other in space, even if they were weightless?

Problems

A 1 A block is sliding upward along a frictionless plane, where the angle of the plane with the horizontal is 40°. Its downward acceleration along the plane is:
(*a*) 9.8 m/s² (*b*) 32 m/s² (*c*) 6.3 m/s²
(*d*) 7.5 m/s ² (*e*) None of the above

A 2 A block slides down an inclined plane at a constant speed of 0.25 m/s. The angle of inclination of the plane is 20°. The coefficient of sliding friction between the block and the plane is:
(*a*) 0.25 (*b*) 0.36 (*c*) 0.36 N
(*d*) 0.94 (*e*) None of the above

A 3 A golfer drives a ball at an angle of exactly 45° with respect to the horizontal. The ball first hits the ground 210 m from the tee. The ball must have left the tee with an initial speed of about:
(*a*) 2.1 × 10³ m/s (*b*) 14 m/s (*c*) 9.8 m/s
(*d*) 45 m/s (*e*) 21 m/s

A 4 A horse fixed on a carousel is moving with a horizontal linear speed of 5.0 m/s. The horse is 7.0 m from the center of the carousel. The period of the horse's motion is:
(*a*) 1.4 s (*b*) 0.11 s (*c*) 8.8 s (*d*) 0.71 s
(*e*) 4.4 s

A 5 The angular velocity of the carousel horse in Prob. A4 is, in degrees per second:
(*a*) 0.71°/s (*b*) 41°/s (*c*) 57°/s (*d*) 35°/s
(*e*) 80°/s

A 6 The centripetal acceleration of a fly sitting on a phonograph record at a distance of 20.0 cm from the center of the record, while the record is rotating at 33.3 rev/min, is:
(*a*) 2.4 m/s² (*b*) 2.4 m/s (*c*) 0.60 m/s²
(*d*) 1.5 m/s² (*e*) 1.5 m/s

A 7 An ultracentrifuge is rotating at 100,000 rev/min. The effective *g* value experienced by a piece of sand in a test tube at a distance of 12 cm from the center of the rotating system is:
(*a*) 1.6 × 10⁵ *g* (*b*) 3.3 × 10⁴ *g* (*c*) 2.1 × 10⁵ *g*
(*d*) 1.3 × 10⁷ *g* (*e*) 1.3 × 10⁶ *g*

A 8 For an earth satellite to remain in an orbit at a distance of 10,000 km from the center of the earth, it would have to travel at a speed of:
(*a*) 6.3 km/s (*b*) 6.3 × 10⁴ km/s (*c*) 6.3 m/s
(*d*) 6.3 × 10³ km/s (*e*) None of the above

A 9 For the earth satellite in Prob. A8, the period of the satellite would be:
(*a*) 1.0 × 10⁴ s (*b*) 6.4 × 10³ s (*c*) 2.0 × 10³ s
(*d*) 3.2 × 10³ s (*e*) 1.6 × 10⁻⁴ s

A10 A woman with a normal weight of 500 N stands on a spring scale in an elevator which is accelerating upward with an acceleration of 3.0 m/s². The apparent weight of the woman, as read from the spring scale, is:
(*a*) 500 N (*b*) 350 N (*c*) 650 N
(*d*) 0 (*e*) None of the above

B 1 At a certain instant of time an automobile is going around a curve. Its acceleration in its direction of motion is 2.0 m/s² and the acceleration required to enable it to maneuver the curve is 4.0 m/s².
(*a*) What is the magnitude of the resultant acceleration at that instant?
(*b*) What is the direction of this acceleration with respect to the direction of motion of the car at that instant?

B 2 A bullet is fired from a rifle at a speed of 500 m/s. The rifle is held level to the ground and 1.5 m off the surface of the ground. How long will it take for the bullet to hit the earth at the end of its trajectory?

B 3 What is the range of a tennis ball hit at an angle of 5.0° above the horizontal at a speed of 30 m/s? (Neglect spin and other complicating effects such as air resistance.)

B 4 A ball at the end of a string 0.75 m in length rotates in a horizontal circle. It makes 2 complete revolutions per second.
(*a*) What is the period of the ball's motion?
(*b*) What is the frequency of the motion?
(*c*) What is the angular velocity of the ball?

B 5 A stereo turntable accelerates from rest to its final rotational speed of 33.3 rev/m in 3.4 s.
(*a*) What is the angular acceleration of the turntable during this 3.4-s interval, if we assume that the acceleration is constant during this interval?
(*b*) What is the linear acceleration of a point on the turntable 14.0 cm from its center?

B 6 A car is going around a curve of radius 100 m at a speed of 20 m/s. At what angle should this curve be banked to reduce to a minimum the possibility of skidding?

B 7 A grinding wheel is rotating at 1000 rev/min. If the radius of the wheel is 10 cm, find the centripetal force on a 1.0-g particle at the rim of the wheel.

B 8 The moon is rotating in its orbit around the earth at a distance of 3.84 × 10⁸ m from the earth. What is its linear velocity? (*Hint:* Use Table A2 inside front cover).

B 9 A man who normally weighs 900 N stands on a spring scale in an elevator which is falling with an acceleration of 4.9 m/s². What does the scale read for the man's weight?

B10 A synchronous communications satellite is in a circular orbit at a distance of about 22,300 mi above the earth's surface.
(*a*) What is the velocity of the satellite?
(*b*) What is the component of its acceleration parallel to its direction of motion?
(*c*) What is the component of its acceleration normal to its direction of motion?
(*d*) If its mass is 100 kg, to what force is it subjected?

C 1 An archer is trying to hit a bull's-eye 1.5 m off the ground and 20 m from the archer. If the arrow is fired from a point also 1.5 m above the ground and at a speed of 40 m/s, in what direction must the arrow be pointed to hit the bull's-eye?

$v_i = 500 m/s$

15 m

C 2 A plane drops a hamper of medical supplies from a height of 3.2×10^3 m during a practice run over the ocean.

(*a*) If we neglect the effects of air resistance, how long does it take the hamper to reach the water?

(*b*) How fast is the hamper moving downward at the instant it strikes the water?

(*c*) If the plane's horizontal velocity was 200 mi/h at the instant the hamper was dropped, what is the overall velocity of the hamper at the instant it strikes the ground (both speed and direction)?

C 3 A cart travels up a frictionless inclined plane making an angle of 30° with the horizontal. If its initial velocity up the plane is 3.0 m/s, how far will it travel up the plane before it stops?

C 4 A 2000-kg car is moving up a 15° incline at 30 km/h when the gears are shifted into neutral. Neglecting friction:

(*a*) How far does the car move uphill before it stops?

(*b*) How much time will elapse before the car returns to the location where the gear shift occurred?

(*c*) What will be the speed of the car when it returns to that location?

C 5 A 10-kg block on an inclined plane of angle 25° is attached to a light cord passing over a pulley and then attached to a 6.0-kg block hanging freely at the end of the cord, as in Fig. 5.29.

(*a*) Draw free-body diagrams for this system.

(*b*) What is the acceleration of the system, if we neglect friction?

(*c*) How far will the block move in the first 2 s, starting from rest?

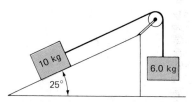

FIGURE 5.29 Diagram for Prob. C5.

C 6 A skier starts down a 25° slope, on which the coefficient of sliding friction between skis and snow is 0.10.

(*a*) What is the skier's acceleration down the slope?

(*b*) What speed does the skier reach at the end of 4.0 s?

C 7 A gun fires a shell at a speed of 75 m/s at an angle of 53° with respect to the horizontal.

(*a*) Find the position and velocity of the shell when $t = 2.0$ s.

(*b*) Find the time at which the shell reaches its highest point and the elevation at that point.

(*c*) Find the range of the shell.

(*d*) Find the velocity when it returns to earth, neglecting air resistance.

C 8 A bullet is fired from a gun at a speed of 400 m/s in a direction parallel to the ground and from an original height of 1.5 m.

(*a*) Find the X and Y coordinates of its position at a time 0.1 s after it was fired.

(*b*) Repeat this calculation for 0.2 s, 0.3 s, 0.4 s, . . . , until the bullet strikes the ground.

(*c*) Graph the position of the bullet as a function of time.

C 9 A car accelerates from 0 to 25 m/s in 10 s. A tire on the car has a radius of 0.40 m.

(*a*) Calculate the angular acceleration of the tire.

(*b*) Plot the angular velocity, the linear velocity of the outer edge of the tire, and the centripetal acceleration of the outer edge as a function of time.

C10 Compare the ranges of golf balls hit at angles of 35°, 40°, 45°, 50°, and 55°, if the initial velocities are the same in all cases.

C11 A pulley of radius 4.0 cm has a belt around it and is attached to a motor revolving at 30 rev/s. The motor then slows down uniformly to 20 rev/s in 2.0 s. Calculate (*a*) the angular acceleration of the motor, (*b*) the number of revolutions it makes in the 2.0 s, (*c*) the length of belt passing over the pulley in that time.

C12 A turntable is accelerating from rest with a constant angular acceleration of 0.10 rad/s².

(*a*) At the end of 4.0 s, what is the angular velocity of the turntable?

(*b*) What is the linear acceleration of a point on the edge of a record 0.15 m from the center of the turntable?

(*c*) What is the linear velocity of the same point 4.0 s after the record begins to accelerate?

(*d*) What is the centripetal acceleration of this same point on the edge of the record, also at the end of 4.0 s?

(*e*) What is the resultant acceleration of this same point 4.0 s after the turntable starts to accelerate?

C13 (*a*) Calculate the centripetal acceleration of a racing car traveling on a circular race course of radius 500 m at a speed such that it makes 1 complete revolution of the track each minute.

(*b*) If the mass of the car is 1500 kg, what is the centripetal force on the car?

(*c*) At what angle would the track have to be banked to prevent skidding at such speeds as that of this car?

C14 A highway designer is planning a curve with a radius of 50 m. How much must the roadway be banked if cars traveling at 25 m/s are to negotiate the curve without any help from friction?

C15 (*a*) Calculate the centripetal force needed to keep your body from leaving the surface of the earth because of the earth's rotation on its axis.

(*b*) Compare this force with the gravitational attraction of the earth for your body.

(*c*) Why are these two forces not equal?

C16 (*a*) Calculate the centripetal force needed to keep your body moving with the earth in orbit about the sun.

(*b*) Show that this force is exactly equal to the gravitational attraction of the sun for your body.

C17 A ball is fastened to one end of a string 0.20 m long and the other end is held fixed at point O, so that the string makes an angle of 30° with the vertical, as in Fig. 5.30. The ball rotates in a horizontal circle with the string kept at this 30° angle. Find the speed of the ball in its circular path required to preserve this 30° angle.

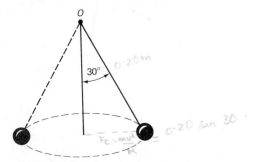

FIGURE 5.30 Diagram for Prob. C17.

C18 A car in a ride at an amusement park does a loop-the-loop in a vertical circle of radius 4.0 m. What is the minimum speed which will keep the car moving in a vertical circle at the top of the loop without the car falling off the tracks?

C19 A bucket of water is to be swung in a vertical circle without a drop of water being spilled. If the circle has a radius of 1.0 m, what must be the speed of the upside-down bucket at the top of the arc in order for the water to stay in the bucket?

C20 An apple on the surface of the earth falls 4.9 m in its first second of fall from rest. The moon is about 60 times farther away from the earth's center than is the apple. How far does the moon fall toward the earth each second?

C21 With what speed would a moon satellite have to rotate in a circular orbit 2.0×10^6 m from the center of the moon?

D 1 A 10-kg block rests on an inclined plane of inclination angle 35°. A light cord attached to the block passes over a pulley and is attached to a hanging block of mass 20.0 kg at the other end, as in Fig. 5.31. The coefficient of sliding friction between the block and the surface of the plane is 0.45.

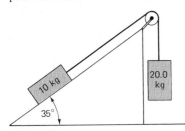

FIGURE 5.31 Diagram for Prob. D1.

(*a*) Find the acceleration of the 10-kg block up the plane.

(*b*) If the block starts from rest, how far does it travel in the first 2 s?

D 2 In a circus act a performer is being shot out of a cannon whose muzzle is at an angle of 30° with the horizontal, as in Fig. 5.32. The circus performer is supposed to land in a net which is 20 m away and at an elevation 1.5 m higher than the gun muzzle. With what minimum speed v must the performer leave the gun in order to land safely in the net?

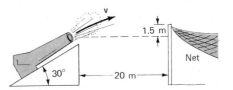

FIGURE 5.32 Diagram for Prob. D2.

D 3 A 1500-kg car travels around the curve of radius 60 m banked at an angle of 10°. The car is traveling at a speed of 25 m/s.

(*a*) Will the banking be sufficient to keep the car on the road?

(*b*) If not, what additional frictional force will be required?

(*c*) What must be the coefficient of friction between the car and the road to keep the car from skidding?

D 4 A 1.0-kg metal sphere hangs at the end of a string 5.0 m long to form a pendulum, as shown in Fig. 5.33. If the sphere is pulled to one side so that it makes an angle of 40° with the vertical and is then released:

(*a*) Find the tension in the cord just after the ball is released.

(*b*) Find the tension in the cord when the ball passes through the bottom of its swing.

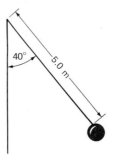

FIGURE 5.33 Diagram for Prob. D4.

D 5 A boy stands at the edge of a merry-go-round 5.0 m from its center. He holds in his hand a string with a ball on its end. When the merry-go-round is up to its full speed of 1 rev every 5 s, the boy notices that the string now makes an angle θ with the vertical.

(*a*)   Explain the cause of this change in position of the pendulum bob.

(*b*)   What is the angle θ in this case?

D 6   A car on a country road in Virginia goes over an old-fashioned humpbacked bridge, which follows the arc of a circle of radius 30 m. If the car has a weight of 15,000 N and is traveling at a speed of 15 m/s:

(*a*)   What is the force exerted by the car on the road at the highest point of the bridge?

(*b*)   At what speed will the car lose contact with the road?

D 7   A miniature ultracentrifuge is driven by air jets at 9.0×10^4 rev/min and produces an effective g value of $1.5 \times 10^5\ g$.

(*a*)   What is the radius of the rotor on the centrifuge?

(*b*)   What is the angular acceleration of the rotor if it comes up to full speed in 60 s?

D 8   An unbanked curve has a radius of 100 m. What is the maximum speed at which a car can make this turn if the coefficient of static friction between tires and road is 0.81?

D 9   (*a*)   A person is standing on a spring scale at the equator when the earth suddenly stops rotating! If the person survived the resulting catastrophe, by what percentage would his or her apparent weight increase?

(*b*)   Answer the same question for a person at the south pole.

D10   A binary star system consists of two stars, with centers separated by 1.2×10^{11} m, rotating about a point midway between them. The stars make 1 complete revolution every 13.1 years.

(*a*)   What keeps the two stars from colliding with each other?

(*b*)   What is the gravitational force between the two stars?

(*c*)   What are the (identical) masses of the two stars?

D11   How large would the earth's period of rotation on its axis have to be to make objects weightless on the equator?

Additional Readings

Beams, Jesse W.: "Ultrahigh-Speed Rotation," *Scientific American*, vol. 204, no. 4, April 1961, pp. 134–147. A fascinating account of work by Beams on high-speed rotation and the ultracentrifuge.

Casper, B. M., and R. J. Noer: *Revolutions in Physics*, Norton, New York, 1972. Contains a good discussion of the Copernican revolution.

Cohen, I. Bernard: "Isaac Newton," in *Lives in Science*, a *Scientific American* book, Simon and Schuster, New York, 1957, pp. 21–30. A brief account of Newton's life and contributions to science.

————:*The Birth of a New Physics*, Doubleday Anchor, Garden City, N.Y., 1960. The contributions of Kepler, Galileo, and Newton to our knowledge of the solar system.

Gray, G.W.: "The Ultracentrifuge," *Scientific American*, vol. 184, no. 6, June 1951, pp. 42–52. An account of the development of the ultracentrifuge and its use in accelerating sedimentation.

Koestler, Arthur: *The Sleepwalkers: A History of Man's Changing View of the Universe*, Macmillan, New York, 1959. A fascinating account of Copernicus, Kepler, and Galileo by a gifted writer.

Wood, Elizabeth A.: *Science for the Airplane Passenger*, Ballantine, New York, 1968. The chapters on frames of reference and quantitative measurements present some interesting material relevant to this chapter.

Energy will remain in some sense the lord and giver of life, a reality transcending our mathematical descriptions. Its nature lies at the heart of the mystery of our existence as animate beings in an inanimate universe.*

Freeman Dyson (1924–   )

Chapter 6

Work and Energy

The past 20 years have witnessed electric power blackouts, long gasoline lines at service stations, insufficient oil and natural gas to heat some homes in winter—all signs of what is sometimes called an energy crisis. Even if *crisis* should prove too strong a word here, it is clear that the energy demands of the world's people will have great social, political, and perhaps military repercussions in the years ahead. To understand the problems associated with energy use and the solutions that are being proposed for them, it is important that we first understand what the word *energy* really means. In this chapter we begin clarifying this concept with a discussion of mechanical energy.

6.1 The Physical Meaning of Work

If we work too hard, we are likely to say that we are tired or that our energy is low. Personal experience therefore indicates some kind of relationship between work and energy. In physics *work* has a more restricted meaning than in everyday language.

Definition

Work: The magnitude of the force applied to an object times the distance the object moves in the direction of the force.

FIGURE 6.1 Work done in moving a book across a desk. One joule of work is done when a force of 1 N moves an object a distance of 1 m in the direction of the force.

For example, if we apply a force of magnitude F to push a book along the top of a desk, and the book moves a distance s in the direction of this force, as in Fig. 6.1, then the work done is:

$$W = Fs$$

If the force is measured in newtons and the distance in meters, the work is in joules (J), where

$$1 \text{ J} = 1 \text{ N} \times 1 \text{ m} = 1 \text{ N·m}$$

In the British engineering system the unit of work is the foot-pound (ft·lb), which is the work done by a force of one pound moving an object through a

*See Additional Readings for source of quote.

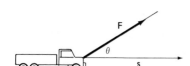

FIGURE 6.2 Pulling a toy truck with a cord which makes an angle θ with respect to the direction of the truck's motion: $W = \mathbf{F} \cdot \mathbf{s} = Fs \cos \theta$.

Work Done by a Force

distance of one foot in the direction of the force. One joule is equal to 0.737 ft·lb.

Sometimes, as, for example, if a toy truck is being pulled along the ground with a cord, the force exerted is at an angle θ with respect to the direction of motion, as in Fig. 6.2. Then the work done is the magnitude of the force in the direction in which the truck moves times the distance the truck moves. Since F_x is the component of $\mathbf{F}$ in the direction of $\mathbf{s}$, and $F_x = F \cos \theta$, this work is $W = (F \cos \theta)s$. Since $\mathbf{F}$ and the displacement $\mathbf{s}$ are in fact both vectors, this is sometimes written in vector notation as

$$W = \mathbf{F} \cdot \mathbf{s} = Fs \cos \theta \qquad\qquad (6.1)$$

where the dot product $\mathbf{F} \cdot \mathbf{s}$ means the magnitude of $\mathbf{F}$ times the magnitude of $\mathbf{s}$ times the cosine of the angle between them. The dot product of two vectors is a scalar quantity. The work W is a scalar that has no direction but only a magnitude, and can be specified completely by assigning to it a single number in some convenient system of units.

Note that $W = \mathbf{F}\cdot\mathbf{s}$ can be written as $F_{\parallel}s = (F \cos \theta)s$, where $F_{\parallel}$ is the component of F parallel to s, or as $Fs_{\parallel} = Fs \cos \theta$, where $s_{\parallel}$ is the component of s parallel to F. Hence it makes no difference, in calculating W, whether we take the component of $\mathbf{s}$ in the direction of $\mathbf{F}$ and multiply by F, or the component of $\mathbf{F}$ in the direction of $\mathbf{s}$ and multiply by s. The result is still the same.

The definition of work leads to the surprising conclusion that, in a technical sense, no work need be done *on a book* to hold it up for a long period of time, or even to move it horizontally at a fixed height above the ground. In the first case $s = 0$, and so no work is done, whereas in the second case the force which must be exerted to hold the book up is at right angles to the direction of motion, $\cos \theta = \cos 90° = 0$, and so $W = 0$. Clearly this physical definition of work is at variance with our biological experience. We all feel we are doing work if we hold a sack of potatoes in our arms for a few minutes. The physical definition has the great advantage, however, of being precise and of providing a formula which will yield the same value for the work done in any particular physical situation, no matter who measures it.

To lift the book, of course, requires the exertion of a force just sufficient to overcome the force of gravity. We assume that this force is only infinitesimally greater than the force of gravity, so that no acceleration is imparted to the book. If the book is lifted up a distance s by applying the force F, then the work done is

$$W = \mathbf{F} \cdot \mathbf{s} = Fs \cos \theta = Fs$$

since $\cos \theta = \cos 0° = 1$. Here the force F is numerically equal to the weight of the book $w = mg$, and hence the work which must be done to lift a book of mass m through a distance h is simply $W = mgh$.

In this case we have calculated the work done on the book by someone who lifts the book in opposition to the force of gravity. The amount of work done can be calculated, however, for each separate force in the system. For example, if we desire the work done on the book by the force of gravity when the book is lifted, we have $\theta = 180°$, since the force of gravity acts *down* but the book is lifted *up*. Hence cos $\theta = $ cos $180° = -1$, and the work is $\mathcal{W} = Fs$ cos $\theta = -Fs$. The work done by the force of gravity on the book is therefore *negative*. If the book were allowed to fall freely under the force of gravity, then the work done by the force of gravity would be positive and equal to Fs, since cos θ would be $+1$.

Example 6.1

A 100-kg man pulls his 30-kg daughter on a sled through the snow by pulling the sled along by a rope which makes an angle of 40° with respect to the ground.
(a) If the man exerts a force of 50 N and the sled moves a distance of 100 m, how much work does the man do?

(b) In order to do the same amount of work by pulling horizontally on the sled, the man would have to exert what force on the sled?

SOLUTION

(a) Here

$$\mathcal{W} = \mathbf{F} \cdot \mathbf{s} = Fs \cos 40°$$

$$= (50 \text{ N})(100 \text{ m})(0.77) = 3.9 \times 10^3 \text{ N·m}$$

$$= \boxed{3.9 \times 10^3 \text{ J}}$$

Note that the masses of the man and his daughter are irrelevant here. They are given merely to give you

practice in separating the "sheep" (the needed data) from the "goats" (the useless data) in solving physics problems.
(b) Here $\mathcal{W} = Fs \cos 0° = Fs$, and so

$$F = \frac{\mathcal{W}}{s} = \frac{3.9 \times 10^3 \text{ J}}{100 \text{ m}} = \boxed{39 \text{ N}}$$

6.2 Work Converted into Kinetic Energy

Energy is frequently defined as the ability to do work. This indicates that energy can be converted into useful work, as, for example, when the electric energy in the storage batteries of a car is used to start the car's engine. The reverse is also true: Work can be converted into stored energy. We can, for example, do work to pump the water from a lake to a higher reservoir, as is done at the Ludington pumped-water storage plant on Lake Michigan. That stored water then has added energy because of its height and is used to drive electric generators and produce electric energy needed in periods of peak demand. Hence work and energy are directly convertible one into the other. They are the same kind of physical quantity and therefore have the same dimensions. In the SI system of units they also have the same unit, the joule (J), where one joule is equal to one newton-meter, as we have seen.

In the sections which follow we will present a number of important examples of the interconversion of work and energy.

Kinetic Energy

Kinetic energy is energy of motion. For example, a moving rocket has kinetic energy. How did the rocket obtain this energy? The rocket engines did work to impart motion to the rocket, and thus gave it kinetic energy. We now want to investigate the quantitative relationship between the work done and the kinetic energy produced. In doing this we will confine ourselves to *translational motion*, where the object actually moves from one position in space to another.

Let us consider a small hockey puck moving slowly with a velocity $\mathbf{v}_0$ on a frictionless surface like those in the air-hockey games found at penny arcades. We can push the puck along the surface by exerting a constant force $\mathbf{F}$ on it for a short period of time, and give it a displacement $\mathbf{s}$. The work done is then $\mathcal{W} = Fs$, since $\mathbf{F}$ and $\mathbf{s}$ are in the same direction. Because there is nothing to resist the applied force $\mathbf{F}$, it must accelerate the puck according to Newton's second law of motion. The constant force $\mathbf{F}$ produces a constant acceleration $\mathbf{a}$, and the velocity increases from its initial value $\mathbf{v}_0$ to some final value $\mathbf{v}$. Hence the work done is converted into energy of motion, or kinetic energy, as in Fig. 6.3.

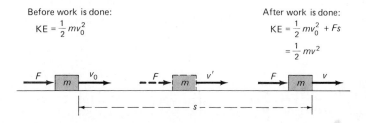

Before work is done:
$$KE = \frac{1}{2} m v_0^2$$

After work is done:
$$KE = \frac{1}{2} m v_0^2 + Fs$$
$$= \frac{1}{2} m v^2$$

FIGURE 6.3 Work converted into kinetic energy. If no forces oppose the motion, all the work done is converted into kinetic energy.

From Eq. (2.12) for uniformly accelerated motion in a straight line, the final velocity is related to the initial velocity by the equation

$$v^2 = v_0^2 + 2a(s - s_0)$$

In our case $s - s_0$ is equal to the displacement s, since we have started the motion at $s_0 = 0$. Hence

$$a = \frac{v^2 - v_0^2}{2s}$$

Now the work done in accelerating the puck is

$$\mathcal{W} = Fs = mas = m \frac{(v^2 - v_0^2)\, s}{2s}$$

or

$$\mathcal{W} = \tfrac{1}{2}mv^2 - \tfrac{1}{2}mv_0^2 \tag{6.2}$$

The effect of the work done is therefore to change the energy possessed by the puck by virtue of its motion from its initial value ($\tfrac{1}{2}mv_0^2$) to its final value ($\tfrac{1}{2}mv^2$). We call this quantity the *kinetic energy* of the moving puck, and the effect of the work in this case has therefore been to increase the kinetic energy of the system.

Definition

Kinetic energy: The energy of motion of an object, equal to one-half the product of its mass and the square of its velocity.

$$KE = \tfrac{1}{2}mv^2 \tag{6.3}$$

Note that, even though $\mathbf{v}$ is a vector, v^2 is not, since it is the dot product $\mathbf{v} \cdot \mathbf{v}$, which is a scalar. Hence, as expected, kinetic energy is a scalar quantity and can be specified by a single number which gives its magnitude.

For Eq. (6.2) to be correct, the dimensions must be the same on both sides. This can easily be shown using dimensional analysis. Work has the dimensions

$$[W] = [F][L] = [M][A][L] = [M]\frac{[L]}{[T^2]}[L] = [M]\frac{[L^2]}{[T^2]}$$

whereas kinetic energy has the dimensions

$$[KE] = [M][V^2] = [M]\frac{[L^2]}{[T^2]}$$

Since these are the same dimensionally, Eq. (6.2) is dimensionally correct. As a matter of fact, dimensional analysis alone would have sufficed to obtain the functional dependence of the kinetic energy on m and v, but the factor $\frac{1}{2}$ could not have been obtained without further mathematical analysis of the kind used above, or an experiment to determine the factor $\frac{1}{2}$ directly.

Variable Forces

Although we have proved Eq. (6.2) only for the simplest kind of force and motion, experimentally it turns out to be true in complete generality. For example, it is valid for variable forces, but a proof, which requires calculus, is beyond the scope of this book. If a force is variable, but it is possible to obtain its average value $\overline{\mathbf{F}}$ during some time interval, then the work done during that time interval is simply

$$\mathcal{W} = \overline{\mathbf{F}} \cdot \mathbf{s} = (\overline{F \cos \theta})s \qquad (6.3a)$$

It is also possible to find the work done by a variable force graphically, as in Fig. 6.4. If we plot the force in the direction of motion, i.e., $F \cos \theta$, as a function of the distance s, then the area under the curve between the points s_0 and s is equal to the work done over the distance $d = s - s_0$. For example, if the force increases linearly between s_0 and s, as in the figure, then the area under the curve is simply the average value of $F \cos \theta$ times the distance d.

FIGURE 6.4 Work done by a variable force. The work done by a variable force in moving an object from s_0 to s is simply $(F \cos \theta)d$, where $d = s - s_0$. This is the area under the curve when the force $F \cos \theta$ is plotted as a function of distance s. $\overline{F \cos \theta}$ is the average value of $F \cos \theta$ over the distance $d = s - s_0$.

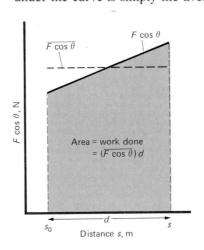

Example 6.2

A 60-kg woman pushes a 20-kg suitcase on wheels a distance of 10 m by exerting a constant force of 2.0 N on it in the direction of motion, starting from rest.

(a) How much work does she do?
(b) If there were no friction acting on the suitcase, what would its final velocity be at the end of the 10-m distance?

SOLUTION

(a) From Eq. (6.1),

$$\mathcal{W} = \mathbf{F} \cdot \mathbf{s} = Fs = (2.0 \text{ N})(10 \text{ m}) = \boxed{20 \text{ J}}$$

(b) If there were no friction, all the work would go into the kinetic energy of the suitcase. Hence, from Eq. (6.2),

$$\mathcal{W} = \tfrac{1}{2}mv^2 - \tfrac{1}{2}mv_0^2$$

or $20 \text{ J} = \tfrac{1}{2}mv^2 - 0 = \tfrac{1}{2}(20 \text{ kg})v^2$

Hence

$$v^2 = 2.0 \text{ J/kg} = 2.0 \text{ N·m/kg}$$

$$= \left(2.0 \ \frac{\text{kg·m}}{\text{s}^2}\right)\left(1\frac{\text{m}}{\text{kg}}\right) = 2.0 \text{ m}^2/\text{s}^2$$

and $v = \boxed{1.4 \text{ m/s}}$

6.3 Work Converted into Gravitational Potential Energy

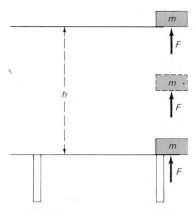

FIGURE 6.5 Work converted into gravitational potential energy. Here the work done is stored as gravitational potential energy in the puck because of its position above the table. Work = mgh = PE$_{\text{grav}}$.

Definition

Just as the work done on the puck in the previous section imparted energy of motion (kinetic energy) to the puck, it is possible under different circumstances for that work to be converted into another kind of energy, which depends on the position of the puck rather than on its velocity.

Suppose that rather than push the puck along the table, we lift it slowly from rest and place it on a shelf directly above the table, as in Fig. 6.5. In this case we presume that no acceleration is imparted to the puck since it is moved at a constant and very slow velocity. Hence the puck is in equilibrium throughout its motion, and there can be no unbalanced force acting on it. This means that the upward force **F** exerted on the puck must be almost exactly equal to the weight of the puck **w** = $m\mathbf{g}$ pushing downward. The work done, therefore, in lifting the puck a distance h above the table against the force of gravity is

$$\mathcal{W} = \mathbf{F} \cdot \mathbf{s} = mgh \tag{6.4}$$

This is a very different situation from the first one we considered. Now, although the puck has no velocity and hence no kinetic energy, it possesses stored energy by virtue of its raised position above the table. We could, for example, push the puck off the shelf and use it, in conjunction with the force of gravity, to crack open a peanut on the table. Hence useful work can be obtained from the system by allowing the puck to fall. We call this stored energy *potential energy* because it has the *potential* for doing work.

Gravitational potential energy: The energy possessed by an object because of its position in a gravitational field.

$$\boxed{\text{PE}_{\text{grav}} = mgh} \tag{6.5}$$

where h is the height above some position chosen as the zero of potential energy, and g varies slightly with altitude and position on the earth's surface. Potential energy is also measured in joules, because it is the product of a force (mg) and a distance (h), and 1 N·m = 1 J.

There is a certain arbitrariness about Eq. (6.5). It says that the potential energy is zero when h is zero. But from what place do we measure the height h—from the ground outside, the floor of the room, or the top of the table?

The answer is that it makes no difference, so long as we consistently use the same height as the zero for potential energy throughout any problem. All we are ever able to measure, or use practically, are *differences* in potential energy, and these are always the same, no matter what zero of potential energy we choose in solving a problem. This is illustrated in Fig. 6.6.

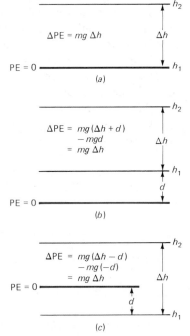

$\Delta PE = mg\, \Delta h$ Δh h_2

$PE = 0$ h_1

(a)

$$\Delta PE = mg\,(\Delta h + d)$$
$$- mgd$$
$$= mg\, \Delta h$$ Δh h_2

h_1

d

$PE = 0$

(b)

$$\Delta PE = mg\,(\Delta h - d)$$
$$- mg\,(-d)$$
$$= mg\, \Delta h$$ Δh h_2

$PE = 0$

d

h_1

(c)

FIGURE 6.6 Arbitrary nature of the zero of potential energy. Whether the zero for the potential energy is taken at h_1, as in (a), below h_1, as in (b), or above h_1, as in (c), the difference in potential energy between h_1 and h_2 is exactly the same, $mg\, \Delta h$, as shown in the calculation of ΔPE in each case. Hence potential energy *differences* are independent of the zero chosen for the potential energy.

A good example of the use of gravitational potential energy is a pendulum clock. The hands of such a clock are usually turned by a falling weight, as in Fig. 6.7. Once a week the owner of the clock lifts the weights by doing work on them. During the week the weights fall slowly, do work, and, in so doing, give up their stored potential energy to turn the clock's hands; i.e., gravitational PE is converted into the KE of the moving hands.

FIGURE 6.7 A pendulum clock, showing the weights whose gravitational potential energy is converted into the energy of motion of the clock's hands. (Photo courtesy of the Smithsonian Institution.)

Potential Energy

$PE = mgh = Fs = W$

Hydrologic Cycle

One of the most important uses of gravitational potential energy is in the production of electricity by hydroelectric power plants. Water is raised by the action of the sun to higher elevations through what is called the *hydrologic cycle* (see Fig. 6.8). The sun beats down on the oceans, rivers, and lakes of the world and raises the water temperature. This gives some water molecules sufficient energy to evaporate as water vapor and form clouds in the atmosphere. When these clouds are blown by the winds (also produced by energy from the sun) over high mountains, they are driven upward, cool off, and release their moisture in the form of snow or rain. Some of this water can then be trapped in mountain reservoirs and used to generate electricity by means of hydroelectric turbines. The water has a potential energy *mgh*, or 9.80 J/kg per meter of height above the point at which the stored energy in the water is to be used (we call this height the *head* of the water). By opening the gates of the reservoir, this stored energy can be converted into kinetic energy of the water and used to drive turbines and generate electricity.

The great advantage of the use of the hydrologic cycle to produce electricity is, of course, that the process is renewable and no money need be spent on fuel. For this reason hydroelectric power is the cheapest form of electric power available at the present time.

FIGURE 6.8 The hydrologic cycle. In this cycle energy from the sun goes through a series of transformations and is finally converted into gravitational potential energy, which can then be transformed into electric energy by hydroelectric turbines.

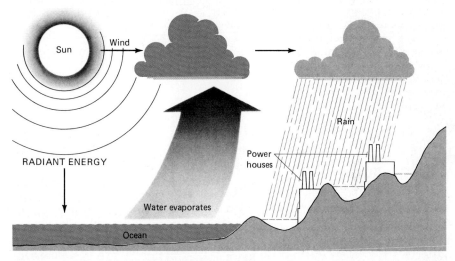

The Conservative Nature of the Gravitational Force

The gravitational force is the prime example of what physicists call *conservative forces*, which differ greatly from the resistive (frictional) forces considered in Secs. 3.7 and 3.8. A force is said to be *conservative* if all the work done against the force is stored as useful energy in the system, as either kinetic energy or potential energy. This energy can then at some later time be reclaimed and used to do an amount of work exactly equal to the original work done against the force. The work done is therefore completely recoverable, if the force is conservative. If, for example, we do a certain amount of work on an object in lifting it slowly from a height h_1 to a height h_2, then when the object falls back from h_2 to h_1, it will deliver exactly the same amount of work initially put into the system. The process is completely reversible, with no loss of mechanical energy occurring.

If some mechanical energy is lost in the form of heat when work is done against a force, then the force is *nonconservative*. A good example is the force of friction. A potential energy can only be defined for a conservative force.

Hence our introduction of a gravitational potential energy has already assumed that the gravitational force is a conservative force. This implies that the total mechanical energy of the system is constant and does not change in time.

Example 6.3

Water is stored in a reservoir 1.0 by 15 km in surface dimensions and 200 m deep. Water flows out of the reservoir through a gate at the bottom which is 100 m above the electric turbine that is being driven by the falling water.

How much electric energy can be generated by such a system if 100 percent of the energy stored in the water can be converted into electric energy?

SOLUTION

The amount of water stored in the reservoir is

$$V = (10^3 \text{ m})(15 \times 10^3 \text{ m})(200 \text{ m}) = 3.0 \times 10^9 \text{ m}^3$$

The density of water is 10^3 kg/m³; i.e., 1 m³ has a mass of 10^3 kg. Hence the mass of the water stored in the reservoir is

$$m = (3.0 \times 10^9 \text{ m}^3)\left(10^3 \frac{\text{kg}}{\text{m}^3}\right) = 3.0 \times 10^{12} \text{ kg}$$

From Eq. (6.5), the gravitational potential energy is mgh, and here the average height h of the water above the turbine is 200 m, so that we have

$$\text{PE}_{\text{grav}} = mgh = (3.0 \times 10^{12} \text{ kg})(9.8 \text{ m/s}^2)(200 \text{ m})$$

$$= \boxed{5.9 \times 10^{15} \text{ J}}$$

Hence at 100 percent efficiency 5.9×10^{15} J of electric energy could be produced by the use of this stored water. (Realistic efficiencies for hydroelectric plants are about 85 to 90 percent.)

It is interesting to observe that, since 1 J = 2.78×10^{-7} kilowatthours (kWh) and electric energy costs about \$0.08 per kilowatthour, the value of this amount of electric energy would be about

$$(5.9 \times 10^{15} \text{ J})\left(2.78 \times 10^{-7} \frac{\text{kWh}}{\text{J}}\right)\left(\frac{\$0.08}{1 \text{ kWh}}\right) = \$1.3 \times 10^8$$

Hence the energy stored in the water would be worth over \$100 million!

6.4 Work Converted into Elastic Potential Energy

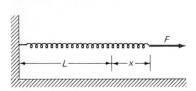

FIGURE 6.9 Work converted into elastic potential energy. When a spring is stretched a distance x by doing work on it, the elastic potential energy stored in the spring is $\text{PE}_{\text{elast}} = \frac{1}{2}kx^2$.

There are other kinds of mechanical potential energy in addition to the gravitational energy discussed in the preceding section. For example, consider a metal spring resting on a table and attached to a wall, as in Fig. 6.9. At its normal length L the spring is in equilibrium, and no net unbalanced force acts on it. Suppose someone seizes the end of the spring and stretches it through a distance x so that its new length is $L + x$. It is found by experiment that, so long as x is not too great, the spring behaves elastically; i.e., its increase in length is proportional to the stretching force, or

$$F \propto x$$

and so $$F = kx \tag{6.6}$$

where k is called the *force constant* of the spring. (Springs which behave in this manner are said to obey *Hooke's law*.) If $F = kx$ is the force the person applies, by Newton's third law of motion the spring exerts an equal and opposite force $(-kx)$ back on the person's hand. If he or she removes the applied force, F becomes equal to zero, and, therefore, by Eq. (6.6), $x = 0$, and the spring returns to its original length L. This is what we mean when we say that the spring is *elastic*.

When the spring is stretched to a length $L + x$, it has some stored energy by virtue of its increased length. We could, for example, attach the spring to a door and use this stored energy to open the door. This energy we call *elastic potential energy*.

Definition   *Elastic potential energy: The energy stored in a spring or other elastic body by virtue of its increase or decrease in length.*

Since the force applied to stretch the spring varied from 0 when the length of the spring was L to kx when the length of the spring was $L + x$, the average force the person applied was

$$\overline{F} = \frac{0 + kx}{2} = \tfrac{1}{2}kx$$

The work done in stretching the spring is, then,

$$W = \overline{F} \cdot d = \tfrac{1}{2}kx \cdot x = \tfrac{1}{2}kx^2$$

Hence $W = \tfrac{1}{2}kx^2$ (6.7)

This work has been converted into stored potential energy in the spring, where

$$\boxed{PE_{elast} = \tfrac{1}{2}kx^2}$$ (6.8)

Here x is the increase (or decrease) in length of the spring. Since $k = F/x$, the units of PE_{elast} are clearly those of $(F/x)x^2$ or Fx. These are newton-meters, or joules, as they must be for any energy.

An example of the use of elastic potential energy is provided by a trampoline, in which the kinetic energy of the jumper is constantly being converted into stored elastic potential energy in the springs of the trampoline, and then immediately changed back into kinetic energy again. This enables the jumper to reverse his or her direction without losing the mechanical energy acquired initially through the use of body muscles.

Note that both gravitational potential energy and elastic potential energy depend on displacements from some position where the potential energy is assumed to be zero, as is clear from Eqs. (6.4) and (6.8).

Example 6.4

(a) A spring is hung from a ceiling and a 500-g mass is hung from it. This increases the length of the spring by 20 cm. What is the force constant k for the spring?
(b) If the spring is then attached to a wall and stretched along a table so that the 500-g mass is displaced 40 cm from its equilibrium position, what is the stored potential energy in the spring?

SOLUTION

(a) Since $F = kx$, and here $F = mg = (0.50 \text{ kg})(9.8 \text{ m/s}^2)$, we have

$$k = \frac{F}{x} = \frac{(0.50 \text{ kg})(9.8 \text{ m/s}^2)}{0.20 \text{ m}}$$

$$= 24.5 \text{ kg/s}^2 = \boxed{24.5 \text{ N/m}}$$

(b) From Eq. (6.8),

$$PE_{elast} = \tfrac{1}{2}kx^2 = \tfrac{1}{2}\left(24.5\frac{\text{N}}{\text{m}}\right)(0.40 \text{ m})^2$$

$$= 1.96 \text{ N}\cdot\text{m} = \boxed{1.96 \text{ J}}$$

6.5 Work Converted into Thermal Energy

Our next example is similar to the one discussed in Sec. 6.2, that of a small air-hockey puck moving on a table. Here, however, we turn off the air jets, and a frictional force acts to prevent the puck from sliding over the table, as in Fig. 6.10. If we now exert on the puck a constant force F which is just sufficient to overcome the force of sliding friction, we will be able to move the puck through a distance d, but without giving it any measurable velocity. Because the speed of the puck is the same at the end as it was at the beginning, its acceleration is zero, and so, from Newton's second law, the sum of all the forces acting on the puck must also be zero. Since we have exerted a force **F** on the puck, and the total force is zero, there must have been an equal and opposite force **F'** exerted on the puck by some other source, so that **F** + **F'** = 0. **F'** must be the frictional force of the table rubbing on the bottom surface of the puck.

FIGURE 6.10 Work converted into thermal energy. If the frictional force **F'** always remains approximately equal in magnitude to the applied force **F**, no acceleration occurs and all the work done is converted into thermal energy (heat) in the puck and the table. The KE is the same at the end as at the beginning, i.e., zero.

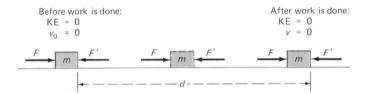

In this case we did an amount of work equal to $W = Fd$. Where did this work go? Certainly not into kinetic energy, since the puck was at rest at the end; not into gravitational potential energy, for the height of the puck was the same at the end as at the beginning; and not into elastic potential energy, for the puck and table were in no way deformed or changed in shape. If we observe this simple experiment more closely, the only significant difference we find between it and the previous cases is that the puck and the table have warmed up a little; i.e., thermal energy (or heat) has been generated while the work was being done. A reasonable explanation would then seem to be that the work done went into thermal energy in the puck and the table. The frictional rubbing of the two surfaces raised their temperatures, in the same way that a matchhead can be ignited by rubbing it against a rough surface.

In a sense it would be correct to say that the work done went into kinetic and potential energies in this case also, if we mean by this the internal energy of the molecules in the two surface layers which are rubbing together. The rubbing sets the molecules into more agitated motion. The motion of the molecules is random and disorganized, however, and in this case not very useful for practical purposes, since it is difficult to convert the disorganized motion of the molecules in the surface layers into the organized form required for useful work.

Nonconservative forces such as the force of friction are sometimes called *dissipative forces*, because mechanical energy is not conserved when such forces act, but rather is dissipated or lost. For this reason problems involving dissipative forces are much more difficult to solve since it is impossible to know the total mechanical energy of the system, for this total energy is constantly decreasing in time.

Thermal Energy and Work

Although the explanation given above for the conversion of work into thermal energy may seem obvious to us, it was not obvious to physicists much

Benjamin Thompson, Count Rumford (1753–1814)

FIGURE 6.11 Benjamin Thompson, Count Rumford. (Photo courtesy of the Granger Collection.)

Although born into a simple American family of farmers in Woburn, Massachusetts, Benjamin Thompson spent most of his life in Great Britain, France, and Germany.

Thompson's education as a boy was largely self-acquired, and he soon gained a reputation as a bright lad interested in books and scientific instruments. In 1772 he was employed as a teacher in Concord, New Hampshire, and while there had the good fortune to marry, at age 19, a wealthy widow 11 years his senior. His newly acquired wealth enabled him to live the life of a country gentleman, managing his wife's estate and helping John Wentworth, the British governor of New Hampshire, with some agricultural experiments. To reward Thompson, the governor commissioned him a major in the New Hampshire militia in 1773. There Thompson acted as an informant for the governor, arousing the rancor of his fellow colonists to such an extent that he was almost tarred and feathered. He prudently left for Boston, leaving his wife and baby daughter behind—and never returned.

In Boston Thompson worked for a while for General Gage, the highest British officer in the Massachusetts Bay Colony. When the American Revolution broke out in 1776, his loyalty was naturally suspect, and he was forced to flee to England.

In London Thompson parlayed his knowledge of the American scene into a comfortable job with the British Foreign Office. Here he returned to his first love, scientific research, and did some important work on the chemical and physical properties of gunpowder.

After the peace of 1783 Thompson left for the continent, and became aide-de-camp and military adviser to Karl Theodor, the ruler of Bavaria. While in Bavaria, Thompson made his most significant contribution to science by his experiments on the boring of cannon and his subsequent realization that the caloric theory of heat was completely inadequate to explain his data. As a consequence of this work, he became one of the most respected men in Bavaria, and in 1792 was raised to the rank of Count of the Holy Roman Empire, ever afterward to be known as Count Rumford.

In 1795 Rumford returned to Great Britain and devoted his energies to starting the Royal Institution of Great Britain, dedicated to the application of science to the common purposes of life. After getting the Royal Institution started, he went to Paris, where he met and in 1805 married Madame Lavoisier, the widow of the great chemist Antoine Lavoisier. These two strong personalities had a very stormy marriage, however, which ended in divorce after only 2 years.

The last years of Count Rumford's life in Paris were devoted to the scientific development of chimneys, stoves, lamps, heating systems, and even the first drip coffeepot. His interest in the basic scientific principles behind devices such as these made Rumford the world's first great applied physicist.

He died suddenly in August 1814, leaving his entire estate to Harvard College to endow a professorship in his name.

Count Rumford's scientific accomplishments merit him a far greater reputation than he has today. However, because he had such severe limitations as a human being, his contemporaries were perhaps unwilling to give him full credit for his great scientific accomplishments.

---

before the end of the eighteenth century. Up to that time mechanics and heat had been treated as two very different disciplines that no one had been able to tie together. Heat was supposed to be material fluid called "caloric," which was added to a body to make it warmer and subtracted from the body to make it colder. In 1778 Benjamin Thompson, while doing some experiments on the boring of cannon in Bavaria (see Fig. 6.11 and the accompanying biography) was led to the idea that the heat produced in the process could not be a material substance, since the rubbing seemed to produce an inexhaustible supply of heat. He concluded that thermal energy (or heat) must be in some way related to *motion*. These ideas were refined by other physicists at a later date to give us the understanding of heat and thermal energy we have today (see Chap. 14).

The examples discussed above show us that work (in the technical sense in which it is used in physics) always produces some form of energy. Sometimes this energy is useful kinetic or potential energy. Sometimes it is stored in a less useful fashion in the molecules of the system as increased thermal energy. In all cases, however, the work done is not lost; it is merely converted into one of many different forms of energy.

6.6 The Extended Work-Energy Theorem

Let us consider a mechanical system which is not necessarily a conservative system. We will denote by $\mathcal{W}$ any work done on the system by an outside agent (say, a person who gives a shove to a hockey puck); by PE the total potential energy of the system, where $PE = PE_{grav} + PE_{elast}$; by KE the kinetic energy of the system; and by ΔQ the amount of mechanical energy converted into heat.

If all the work done on the system is converted into kinetic energy (see Sec. 6.2), then we can state the following *work-energy theorem*:

Work-Energy Theorem

If all the work done in a given system is converted into energy of motion, then the change in the kinetic energy of the system is equal to the work done.

In the form of an equation, this is

$$\boxed{\mathcal{W} = KE_f - KE_i} \qquad (6.9)$$

where KE_i is the initial kinetic energy of the system before the work is done, and KE_f is the final kinetic energy after the force doing the work has ceased to act.

Let us now extend this theorem to include mechanical potential energy, both gravitational and elastic, and also the energy lost in the form of heat, which we designate by ΔQ. Our extended work-energy theorem can then be written as

$$\mathcal{W} = KE_f - KE_i + PE_f - PE_i + \Delta Q$$

Extended Work-Energy Theorem   or $$\boxed{\mathcal{W} - \Delta Q = \Delta KE + \Delta PE} \qquad (6.10)$$

This equation states that the work done minus any energy converted into heat is equal to the total increase in mechanical energy of the system. From a practical point of view, this is still not very useful, since ΔQ is not easy to measure experimentally.

A simpler case is one in which $\Delta Q = 0$; i.e., no mechanical energy is lost as heat, and hence all the work done on the system goes into potential and kinetic energy. In that case Eq. (6.10) becomes

$$\mathcal{W} = \Delta KE + \Delta PE \qquad (6.11)$$

If the work done is known and the change in potential energy is given, the final kinetic energy of the system can then be obtained from its initial value.

Example 6.5

A woman rides a bicycle up a hill which is, at its highest point, 10 m above the level road from which she starts as in Fig. 6.12. The woman and the bike together have a mass of 100 kg. If she does 10^4 J of work in climbing the hill and her initial speed is 4.0 m/s, what is her speed at the top of the hill (neglecting friction and all other dissipative forces)?

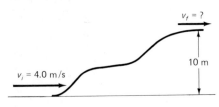

FIGURE 6.12 Diagram for Example 6.5.

SOLUTION

From Eq. (6.11), the work done by the woman is

$$\mathcal{W} = \Delta KE + \Delta PE = KE_f - KE_i + PE_f - PE_i$$

If we take the flat road as our zero of potential energy, $PE_i = 0$, and $PE_f = mgh = (100 \text{ kg})(9.8 \text{ m/s}^2)(10 \text{ m}) = 9.8 \times 10^3$ J. Also the initial kinetic energy is $KE_i = \frac{1}{2}mv_i^2 = \frac{1}{2}(100$ kg)$(4.0 \text{ m/s})^2 = 8.0 \times 10^2$ J. Hence we have

$$10^4 \text{ J} = KE_f - (8.0 \times 10^2 \text{ J}) + (9.8 \times 10^3 \text{ J}) - 0 \quad \text{or}$$

$$KE_f = [10^4 - (9.8 \times 10^3) + (8.0 \times 10^2)] \text{ J} = 10^3 \text{ J}$$

and so

$$v_f = \sqrt{\frac{2KE_f}{m}} = \sqrt{\frac{2 \times 10^3 \text{ J}}{100 \text{ kg}}} = \boxed{4.5 \text{ m/s}}$$

Hence the woman has done enough work to both climb the hill and to increase her bike's speed from 4.0 to 4.5 m/s.

6.7 Conservation of Mechanical Energy

Suppose, now, we have a conservative system on which no external work is done. Then

$$\mathcal{W} = \Delta KE + \Delta PE = 0$$

or

$$PE_f - PE_i = -(KE_f - KE_i)$$

and so

$$PE_f + KE_f = PE_i + KE_i$$

or

$$\boxed{\text{Total mechanical energy} = PE + KE = \text{constant}} \qquad (6.12)$$

This is the *principle of conservation of mechanical energy*:

Conservation of Mechanical Energy   *The sum of the potential and kinetic energies of a given system remains constant for any isolated (no external forces acting), conservative ($\Delta Q = 0$) system.*

Falling Objects   As an illustration of the usefulness of the principle of conservation of mechanical energy, consider the case of a falling ball, as in Fig. 6.13. Suppose that initially the ball is resting on the ground, and that a person does work on it to lift it to the ledge of a third-story window at a height h above the ground. We presuppose that there is no friction and no air resistance. Hence $\Delta Q = 0$. If now the work done on the ball in lifting it slowly from a height $h_i = 0$ to a height $h_f = h$, is $\mathcal{W}$, we have, from Eq. (6.11),

$$\mathcal{W} = \Delta KE + \Delta PE$$

But there has been no change in the kinetic energy of the ball since it is at rest at both the beginning and end of the lifting process. Hence $\Delta KE = 0$, and so

$$\mathscr{W} = \Delta PE = mg\,(h_f - h_i) = mgh$$

The work done is therefore converted into an increase in the potential energy by an amount mgh.

If now we push the ball off the ledge, how rapidly will it be falling just before it hits the ground? Of course, we could find this result by using the laws of uniformly accelerated motion discussed in Secs. 2.6 and 2.7. It turns out to be easier, however, to apply the principle of conservation of mechanical energy, which yields

$$PE_i + KE_i = PE_f + KE_f$$

or $\qquad mgh + 0 = 0 + \tfrac{1}{2}mv^2$

from which $\qquad v = \sqrt{2gh}$

This is the speed of the ball just before it hits the ground, and is the same result found in Sec. 2.7.*

Of course, when the ball collides with the ground, some of its kinetic energy is converted into thermal energy (and sound), and mechanical energy is no longer conserved. Hence the ball bounces up to a height lower than the original ledge. If we consider all forms of energy in the process, however, it still remains true that the total energy of the system is conserved.

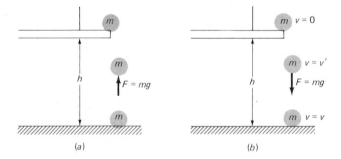

FIGURE 6.13 Energy conservation for the motion of a ball: (a) Ball acquires $PE_{grav} = mgh$ by being lifted to a height h. (b) Ball falls off ledge and PE_{grav} is converted into KE. Since $mgh = \tfrac{1}{2}mv^2$, $v = \sqrt{2gh}$.

Roller Coaster

Another good example of the conservation of mechanical energy is provided by a roller coaster of the kind shown in Fig. 6.14. In discussing its motion we simplify the situation by neglecting any friction in its movement. The roller coaster car is hauled to the top of the first hill, which is the highest point on the track, as in Fig. 6.15. This gives the car and its riders a potential energy equal to mgH, where m is the total mass of car and riders, and H is the height of the first hill above the ground. The car is then released, and its potential energy is gradually converted into kinetic energy as the car descends. At the bottom of the first hill, its kinetic energy is large enough to carry the car to the top of the next hill, which is lower than the first, while in climbing the hill the car's kinetic energy gradually decreases and its potential energy simultaneously increases. The motion of the car therefore becomes a constant conversion from PE to KE and back again, as shown in the figure.

The kinetic energy (and hence the speed) of the car at any point on the track can be obtained from the conservation-of-energy principle:

---

*Note that the "system" here includes both the ball and the earth so that no work is done on the system by an external force. Hence the sum of the kinetic and potential energies is constant.

FIGURE 6.14   A modern roller coaster: the Loch Ness Monster at The Old Country, Busch Gardens, Williamsburg, Virginia. (Photo courtesy of The Old Country, Busch Gardens.)

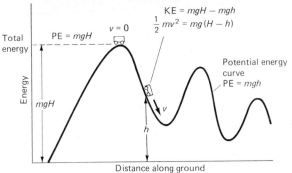

FIGURE 6.15   Energy conservation in a roller coaster.

$$\text{Total energy} = \text{PE} + \text{KE} = \text{constant}$$

$$\text{or} \quad mgh + \tfrac{1}{2}mv^2 = mgH \tag{6.13}$$

where mgH is the initial energy put into the system by hauling the car to the top of the first hill and v is the speed of the car at a height h. For any height h, the velocity can then be found from this equation, since all the other quantities are known. Note that the plot of the height of the roller coaster track as a function of distance along the ground is actually a graph of the potential energy of the system, because the potential energy of the car depends on its height above the ground. Since there actually is friction, energy is lost in each conversion from potential to kinetic energy and back, and eventually the car will not have sufficient energy to climb one of the hills and will be trapped in one of the valleys unless additional energy is supplied to the system. This energy can be supplied by using an electric motor on the ground to lift the car to the top of another hill and start the process over again.

Example 6.6

A book of mass 1.0 kg rests on a window ledge 10 m above the ground. It is then pushed off the ledge and falls to the ground below.
(a) With what speed is it moving down just before it strikes the ground?

(b) What is the book's downward speed when it is just 5.0 m off the ground?

SOLUTION

(a) From Eq. (6.12) we have

$$PE + KE = \text{constant} = mgh$$

$$= (1.0 \text{ kg})(9.8 \text{ m/s}^2)(10 \text{ m}) = 98 \text{ J}$$

When the block hits the ground, its PE is zero, for h is zero. Hence its entire energy is kinetic and equal to 98 J. We therefore have

$$KE = \tfrac{1}{2}mv^2 = 98 \text{ J} \quad \text{or}$$

$$v = \sqrt{\frac{2KE}{m}} = \sqrt{\frac{2(98 \text{ J})}{1.0 \text{ kg}}} = \sqrt{196 \text{ m}^2/\text{s}^2} = \boxed{14 \text{ m/s}}$$

(b) At any point along the path of the book as it falls, we have

$$PE + KE = PE_i + KE_i = 98 \text{ J} + 0 = 98 \text{ J}$$

At any height h' above the ground we therefore have

$$mgh' + \tfrac{1}{2}mv^2 = 98 \text{ J}$$

For $h' = 5.0$ m, we have

$$(1.0 \text{ kg})(9.8 \text{ m/s}^2)(5.0 \text{ m}) + \tfrac{1}{2}(1.0 \text{ kg})v^2 = 98 \text{ J}$$

$$\text{or} \quad 49 \text{ J} + \tfrac{1}{2}v^2 \text{ kg} = 98 \text{ J}$$

$$\text{Hence} \quad v^2 = 2(49 \text{ J/kg}) = 98 \text{ m}^2/\text{s}^2$$

$$\text{or} \quad v = \boxed{9.9 \text{ m/s}}$$

This problem shows clearly that energy considerations are very useful in solving mechanical problems. To impress this on your mind, try to solve the same problem by using Newton's second law and the laws of uniformly accelerated motion.

Example 6.7

A roller coaster car has a speed of 25 m/s at the bottom of the first hill. Will it be able to make it to the top of the second hill, which is 40 m high, if we neglect all energy losses in the system?

SOLUTION

From Eq. (6.12) we have

$$PE_i + KE_i = PE_f + KE_f$$

Let us take the zero of potential energy at the bottom of the first hill, so that $PE_i = 0$ and $KE_i = \tfrac{1}{2}m(25 \text{ m/s})^2$. The height to which the roller coaster car can climb is then h, where PE_f is mgh and KE_f is zero, since the car will stop climbing when it has converted all its initial kinetic energy into potential energy.

Hence we have

$$\tfrac{1}{2}m(25 \text{ m/s})^2 = mgh$$

$$\text{or} \quad h = \frac{(25 \text{ m/s})^2}{2(9.8 \text{ m/s}^2)} = \boxed{32 \text{ m}}$$

The roller coaster car can therefore reach a height of only 32 m, not 40 m, even if we neglect the effect of friction and air resistance as it climbs the hill. It will not be able to get to the top of the second hill.

Example 6.8

Approximately what running speed would an 80-kg pole-vaulter have to acquire to clear a pole-vault bar 17 ft off the ground? Assume that the pole-vaulter's center of gravity is 1.0 m off the ground, and that the vaulter clears the bar if this center of gravity just reaches the height of the bar.

SOLUTION

We will simplify this problem somewhat by neglecting any additional energy supplied by the vaulter's arms, any resistance of the air to the jumper's body, the slight forward velocity of the vaulter at the top of the vault, and other secondary effects, and concentrate solely on the conservation of energy involved in the process.

When the vaulter shoves the pole into the slot provided for it at the end of the runway, he or she has a certain kinetic energy. At the top of the vault all this kinetic energy has been converted into gravitational potential energy. Hence we have:

$$\text{KE}_i + \text{PE}_i = \text{KE}_f + \text{PE}_f$$

or $\text{KE}_i + 0 = 0 + mgh$

where $h = (17 \text{ ft})(0.31 \text{ m/ft}) - 1.0 \text{ m} = 4.3 \text{ m}$. Hence

$$\text{KE}_i = (80 \text{ kg})(9.8 \text{ m/s}^2)(4.3 \text{ m}) = 3.4 \times 10^3 \text{ J}$$

and so

$$v = \left(\frac{2\text{KE}_i}{m}\right)^{1/2} = \left[\frac{2(3.4 \times 10^3 \text{ J})}{80 \text{ kg}}\right]^{1/2} = \boxed{9.2 \text{ m/s}}$$

More General Conservation-of-Energy Principle

The principle of conservation of mechanical energy can be expanded to include nonmechanical forms of energy such as thermal energy and electric energy. This leads us to a much more general and inviolable principle of present-day physics, the *principle of conservation of energy*, which is usually stated as follows:

Principle of Conservation of Energy

> Energy is neither created nor destroyed; it merely changes from one form to another.

This is one of the most important principles in all of physics. We have already seen many examples of its usefulness in mechanics and will see many more throughout the rest of this course, especially in our discussion of heat. The validity of this principle is based on the fact that no experiment has ever been done which has been found to contradict it.

If this principle is correct, you may wonder why we are constantly urged to "conserve energy," since we have just said that energy is always conserved and never lost. The drive to conserve energy is more accurately a push to conserve *useful* forms of energy like the fossil fuels rather than to waste them by converting them needlessly into less useful forms of energy like heat. It is still true that energy is never destroyed or lost, but it may be converted from useful forms to relatively useless forms. (We will see this in detail in our discussion of the laws of thermodynamics in Chap. 15.)

6.8 Simple Machines

One very practical application of the principle of conservation of mechanical energy is to simple machines. A *machine* is a device that does useful work when it is provided with energy. If energy is to be conserved, it is clear that such a machine cannot do more work than the energy it receives. In other words, the work output of the machine cannot exceed the work input. At best, under ideal conditions the work output may be equal to the work or energy input; if friction or other dissipative forces are present, the work output will, of necessity, be even smaller than the input.

A *simple machine* is one so uncomplicated that it cannot be made any simpler. A good example of a simple machine is the *inclined plane* already discussed in Sec. 5.2.

Inclined Plane

Suppose we want to lift a 50-kg trunk from the street to a truck's floor which is 2.0 m off the ground. This requires the lifting of a weight $w = mg = (50 \text{ kg})(9.80 \text{ m/s}^2) = 490 \text{ N}$, which only a very strong person can do without risk.

An easier way to accomplish the same goal is to lay planks from the street to the truck, say, at an angle of 20° with respect to the street, as in Fig. 6.16, and push the trunk up the planks. Then the force which must be exerted up the plane, if there is no friction, is only $w \sin 20° = (490 \text{ N})(0.342) = 168 \text{ N}$, which is much easier for a person to provide without strain.

Of course, no energy is created in this process. The amount of work required to lift the trunk directly up from the street to the truck is

$$\mathcal{W} = Fs = (490 \text{ N})(2.0 \text{ m}) = 980 \text{ J}$$

The amount of work required to slide the trunk up the plank is, in the absence of friction,

$$\mathcal{W}' = F's' = (168 \text{ N})s'$$

where s' can be obtained from (see Fig. 6.16)

$$\sin 20° = \frac{2.0 \text{ m}}{s'}$$

or $\quad s' = \dfrac{2.0 \text{ m}}{\sin 20°} = \dfrac{2.0 \text{ m}}{0.342} = 5.85 \text{ m}$

and so $\quad \mathcal{W}' = (168 \text{ N})(5.85 \text{ m}) = 980 \text{ J}$

Hence the work done is the same in the two cases, but the force required with the inclined plane is far less than the force required without it. This is the way all simple machines work: *they multiply forces without changing the amount of work done.* In physics we never get something for nothing; in this case we are able to exert smaller forces to do a certain amount of work, but we must exert them over greater distances. Examples of other simple machines are levers, pulleys, the wheel and axle, and the screw.

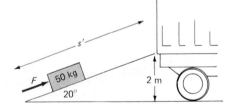

FIGURE 6.16 An inclined plane used to load a trunk onto a truck.

Levers

A lever is a device such as a wooden plank resting on a sawhorse (which serves as a fulcrum) which can be used to multiply forces in a way similar to the way an inclined plane can multiply forces.

Consider, for example, the lever in Fig. 6.17. If the lever arm L is twice

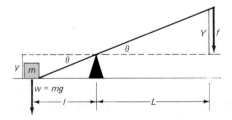

FIGURE 6.17 A lever used to lift a mass m. If the lever arm L is twice l, then the force f required to lift the weight w is only $w/2$.

as long as l, and we apply a force f in an attempt to lift the weight w, then, from the figure,

$$\tan \theta = \frac{Y}{L} = \frac{y}{l}$$

or

$$\frac{Y}{y} = \frac{L}{l} = 2$$

But if the lever is 100 percent efficient, the work input must be equal to the work output that lifts the weight, and so

$$\mathscr{W} = fY = wy$$

or

$$f = \frac{wy}{Y} = \frac{w}{2}$$

Hence a weight w can be lifted by exerting a force f which is only half as large as the weight. If the ratio of the lever arms is increased even more, the force that must be applied to lift the weight can be decreased even further. If the ratio of the lever arms becomes very large, it is possible to lift a very large weight by exerting a very small force. This is the meaning of Archimedes' celebrated claim that, if he had a lever long and rigid enough (and a fulcrum to support it), he could have moved the earth out of its orbit.

It cannot be overemphasized that we are multiplying the applied *force* here, not the *work* done. Since the large force works through a small distance, the work output is equal to the work input, but the input can be in the form of a small force working through a large distance. This is the advantage of a lever.

Pulleys

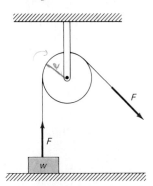

FIGURE 6.18 A simple pulley used to change the direction of a force. The applied force is equal to the lifting force.

A simple pulley is often used to make it more convenient for a person to do some useful work like lift a weight. For example, the simple pulley in Fig. 6.18 enables a person to pull down on the rope at a convenient angle and thus lift the weight w. In this case the force exerted by the person doing the lifting must be equal to w, and there is no advantage other than convenience to the pulley system.

Compound pulleys enable weights to be lifted by the exertion of forces smaller than the weights lifted. Thus consider a compound pulley system like the one in Fig. 6.19, in which two pulleys A and B are fixed, and one pulley C is free to move with the weight to be lifted. Here when we pull down on the rope with a force f, the weight w is lifted only one-third of the distance L through which the end of the rope is moved. This is because three lengths of rope are being shortened, and each one shortens only by $l = L/3$. This is also the distance the free pulley and the weight move up. Here again we have

$$\mathscr{W} = fL = wl = w\frac{L}{3} \quad \text{Torque}$$

or

$$f = \frac{w}{3}$$

Hence we can lift a weight w by exerting a force only one-third as large as w if we use a pulley system of the kind shown here.

In general it can be shown that, in the ideal case, the force required to lift a weight w is reduced to w/n, where n is the number of ropes pulling up on the

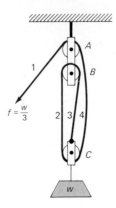

FIGURE 6.19 A compound pulley system. The force required to lift the weight w is only $f = w/3$, since there are three ropes pulling up on the free pulley attached to the weight w. When rope 1 is pulled down a distance L, ropes 2, 3, and 4 are shortened only by a distance $L/3$.

free pulley attached to the weight w. If we consider in Fig. 6.19 the static situation in which the weight is at rest, the weight w is being supported by three ropes, so that in equilibrium the tension in each rope must be only $w/3$. Since the tension in rope 4 is the same as that in rope 1 and hence the same as the force exerted by the person holding the rope, the force that must be exerted to lift the weight is also only $w/3$ in this case, or w/n for n ropes.

Mechanical Advantage

In the above examples the ratio of the distance through which the force is exerted to the distance the heavy object is moved is, in the ideal order, the same as the ratio of the weight which is moved to the force needed to move it. We call the ratio of distances the *ideal mechanical advantage*, or IMA, where

Ideal Mechanical Advantage

$$\text{IMA} = \frac{L}{l} \tag{6.15}$$

In actual fact there are, of course, losses in the system due to friction and the weight of the machine itself, and the *actual mechanical advantage* (AMA) is less than the IMA, and is given by the ratio of the output force to the input force.

Actual Mechanical Advantage

$$\text{AMA} = \frac{w}{f} \tag{6.16}$$

The *efficiency* of any simple machine can be obtained as the ratio of the work output of the machine to the work input. This is:

Efficiency of Machine

$$\text{Efficiency} = \frac{\text{work output}}{\text{work input}} \times 100\% = \frac{wl}{fL} \times 100\% = \frac{w/f}{L/l} \times 100\%$$

or

$$\boxed{\text{Efficiency} = \frac{\text{AMA}}{\text{IMA}} \times 100\%} \tag{6.17}$$

Hence the efficiency of a simple machine is the ratio of the actual to the ideal mechanical advantage. This ratio must be multiplied by 100 to give the result as a percent.

Example 6.9

A pulley system consists of three ropes pulling up on a weight of 10 N, as in Fig. 6.19. To lift the weight it is found that it is necessary to exert a force of 4.0 N.
(a) What is the ideal mechanical advantage of this pulley system?

(b) What is the actual mechanical advantage of this pulley system?
(c) What is the efficiency of this pulley system?

SOLUTION

(a) The ideal mechanical advantage of the pulley system is the ratio of the distance through which the force is exerted to the distance the 10-N weight is moved. For a pulley system with three ropes doing the lifting, this is

$$IMA = \frac{L}{l} = \boxed{3}$$

(b) The actual mechanical advantage is the ratio of the weight moved to the force exerted to move the weight, or

$$AMA = \frac{w}{f} = \frac{10\,N}{4.0\,N} = \boxed{2.5}$$

(c) The efficiency of the pulley system is then

$$Efficiency = \frac{AMA}{IMA} = \frac{2.5}{3} = 0.83 = \boxed{83\%}$$

Hence the compound pulley system is 83 percent as efficient in lifting the weight as a completely frictionless, ideal pulley system would be.

6.9 Power

The same amount of work is required to lift a 10-kg mass through 2.0 m whether it takes 2.0 s or 2.0 h to do the job. The work required is $\mathcal{W} = mgh$, and time appears nowhere in this equation. The *power* input in these two cases would be very different, however.

Definition

Power: The rate at which work is done or energy is transformed.

In the form of an equation

$$\boxed{P = \frac{\Delta \mathcal{W}}{\Delta t}}$$

(6.18)

where $\Delta \mathcal{W}$ is the amount of work done in the time Δt.

Thus in the case just mentioned, if it takes 2.0 s to move the 10-kg mass, we have

$$P = \frac{mgh}{\Delta t} = \frac{(10\ kg)(9.8\ m/s^2)(2.0\ m)}{2.0\ s} = 98\ J/s$$

If, on the other hand, it takes 2 h,

$$P = \frac{mgh}{\Delta t} = \frac{(10\ kg)(9.8\ m/s^2)(2.0\ m)}{2.0\ h(3600\ s/h)} = 0.027\ J/s$$

This unit, the joule per second (J/s), we call the watt (W), where

$$1\ W = 1\ J/s$$

This is the same unit used in electricity. A 100-W bulb, when burning, consumes electric energy at the rate of 100 J/s, that is, 100 W, and converts it into heat and light.

In the British engineering system power is measured in units of foot-pounds per second (ft·lb/s), where 1 ft·lb/s = 1.356 W.

The watt is named for Sir James Watt (1736–1819), whose steam engine was the predecessor of today's more powerful engines. Watt himself suggested the rate at which a horse works as the unit of power. This he called a *horsepower*, but horses would have short lives indeed if forced to work at this rate for very long. One horsepower (hp) is defined to be equal to 550 ft·lb/s, but since 1 ft·lb/s is equivalent to 1.356 W, we have

$$1 \text{ hp} = \left(\frac{550 \text{ ft·lb}}{s}\right)\left(\frac{1.356 \text{ W}}{1 \text{ ft·lb/s}}\right) = 746 \text{ W} \simeq 0.75 \text{ kW}$$

Thus if a car's motor has a power output of 100 hp, it is doing work at the rate of about 75 kW, or 75 kJ/s.

It must be stressed that while a watt or kilowatt is a unit of power, a kilowatthour is a unit of energy. The actual energy used, for example, by a 100-W light bulb depends on how long the bulb is left on. Hence, to find the actual energy consumed by the bulb, we must multiply the power (100 W) by the time the bulb burns. This is merely to restate Eq. (6.18) in the equivalent form

$$\Delta W = P \, \Delta t \tag{6.19}$$

Hence
$$1 \text{ J} = \left(\frac{1 \text{ J}}{s}\right)\left(1 \text{ s}\right) = 1 \text{ Ws}$$

or one joule is equal to one watt-second. Therefore one kilowatthour, the unit in terms of which we pay our electric bills, is

$$1 \text{ kWh} = 10^3 \text{ Wh} = \left(10^3 \frac{J}{s}\right)\left(\frac{3600 \text{ s}}{1 \text{ h}}\right)(1 \text{ h}) = 3.6 \times 10^6 \text{ J}$$

The kilowatthour is therefore a unit of *energy*, not of power, since the joule is an energy unit.

For the same reason you cannot pay your electric bills in terms of kilowatts, since a kilowatt merely tells you the *rate* at which energy was used; it says nothing about for how long a time the energy was used.

Power and Speed

It is important, particularly in dealing with airplanes and automobiles, to be able to express the power generated in terms of the speed of the object being moved by the applied force. In these cases the engines providing the needed force are rated in horsepower, and it is easier to see what this means if we transform Eq. (6.18) into a slightly different form. Suppose a force **F** is applied by the motor of a car and that as a result the car travels a distance **Δs** in the direction of the force in time Δt, as in Fig. 6.20. Then

$$P = \frac{\Delta W}{\Delta t} = \frac{\mathbf{F}\cdot\Delta\mathbf{s}}{\Delta t} = \mathbf{F} \cdot \bar{\mathbf{v}}$$

since Δs/Δt is the average velocity $\bar{\mathbf{v}}$ of the car during time *t*. Hence

Power
$$P = F\bar{v} \tag{6.20}$$

FIGURE 6.20 Relationship of power to average speed. Power is the product of the applied force and the average speed of the object in the direction of that force: $P = F \bar{v}$, where $\bar{v} = \Delta s/\Delta t$.

since **F** is in the direction of $\bar{v}$. Power can therefore be defined as the product of the applied force and the average speed of the object in the direction of that force.

To show the usefulness of this expression for power, let us consider its application to an automobile. The work done by a car's engine goes into overcoming friction and air resistance (as when the car is cruising at constant speed on a flat road); into lifting the weight of the car up an inclined plane (when the road is rising), thus increasing its PE; and into accelerating the car (e.g., to pass other cars), and thus increasing its KE. In all three cases it is easier to obtain the average speed of the car from the car's speedometer than it is to make the two independent measurements of distance and time needed to obtain the power from Eq. (6.18).

The following examples show how useful Eq. (6.20) is in such cases.

Example 6.10

A Pratt and Whitney JT9D jet engine on a Boeing 747 develops a thrust of 1.9×10^5 N (about 4.3×10^4 lb).
(a) If the plane is flying at a constant speed of 300 m/s (about 670 mi/h), what is the power produced by this engine?
(b) What horsepower does the engine develop?

SOLUTION

(a) Since the power is the rate at which work is done, we have, using Eq. (6.20),

$$P = \frac{\Delta W}{\Delta t} = F\bar{v}$$

$$= (1.9 \times 10^5 \text{ N})(300 \text{ m/s}) = 5.7 \times 10^7 \text{ J/s}$$

$$= \boxed{5.7 \times 10^7 \text{ W}}$$

(b) Since 1 hp = 746 $\mathcal{W}$, we have

$$P = (5.7 \times 10^7 \text{ }\mathcal{W})\left(\frac{1 \text{ hp}}{746 \text{ }\mathcal{W}}\right) = \boxed{7.6 \times 10^4 \text{ hp}}$$

Example 6.11

A 1500-kg car is traveling at a constant speed of 25 m/s (55 mi/h) on a level road. The forces of friction and air resistance on the car add up to 800 N.
(a) What power must be generated by the car's engine to keep the car moving at this speed on level ground?

(b) If the car has to climb a 15° hill at 25 m/s, what additional power is needed?
(c) If the driver accelerates to 30 m/s over a 10-s interval on a level road to pass another vehicle, what additional power must the engine provide?

SOLUTION

(a) To keep the car moving on level ground requires a power

$$P_1 = F\bar{v} = \left(800 \text{ N}\right)(25 \text{ m/s}) = \boxed{2.0 \times 10^4 \text{ W}}$$

Since automobile engines are usually rated in horsepower, this is

$$P_1 = \frac{2.0 \times 10^4 \text{ W}}{746 \text{ W/hp}} = \boxed{27 \text{ hp}}$$

(b) To climb the 15° hill the car's engine must exert a force

$$F = mg \sin \theta = (1500)(9.8 \text{ m/s}^2)(\sin 15°)$$

$$= (1500 \text{ kg})(9.8 \text{ m/s}^2)(0.26) = 3.8 \times 10^3 \text{ N}$$

Then

$$P_2 = F\bar{v} = (3.8 \times 10^3 \text{ N})(25 \text{ m/s}) = 9.5 \times 10^4 \text{ W} = \boxed{127 \text{ hp}}$$

(c) The acceleration in this case is

$$a = \frac{v_f - v_i}{t} = \frac{30 \text{ m/s} - 25 \text{ m/s}}{10 \text{ s}} = 0.50 \text{ m/s}^2$$

The force required is then

$$F = ma = (1500 \text{ kg})(0.50 \text{ m/s}^2) = 750 \text{ N}$$

The average speed is

$$\bar{v} = \frac{25 \text{ m/s} + 30 \text{ m/s}}{2} = 27.5 \text{ m/s}$$

and so the power is, in this case,

$$P_3 = F\bar{v} = (750 \text{ N})(27.5 \text{ m/s}) = 2.1 \times 10^4 \text{ W} = \boxed{28 \text{ hp}}$$

Hence we see that the greatest strain is put on the car's engine when it is climbing hills, and that about 150 hp is adequate even for climbing a 15° hill without reducing speed, allowing also for the effects of air resistance and friction.

Summary: Important Definitions and Equations

Work: The magnitude of the force applied to an object times the distance the object moves in the direction of the force.

$$\mathcal{W} = \mathbf{F} \cdot \mathbf{s} = Fs \cos \theta$$

The joule is the unit for both work and energy, where 1 joule (J) = 1 N·m.

Energy: That property of an object which enables it to do work.

    Kinetic energy: The energy possessed by an object by virtue of its motion: $\text{KE} = \frac{1}{2}mv^2$.

    Gravitational potential energy: The energy possessed by an object by virtue of its position in a gravitational field: $\text{PE}_{grav} = mgh$.

    Elastic potential energy: The energy possessed by an object by virtue of its distortion or change in size or shape: $\text{PE}_{elast} = \frac{1}{2}kx^2$.

Conservative forces: Forces which conserve the total mechanical energy of a system. Examples are gravitational and elastic forces.

Nonconservative forces: Forces which dissipate mechanical energy in the form of heat. Examples are friction and air resistance.

Work-energy theorem:

$$\mathcal{W} = \text{KE}_f - \text{KE}_i \quad \text{(if all work goes into kinetic energy)}$$

Extended work-energy theorem:

$$\mathcal{W} - \Delta Q = \Delta\text{KE} + \Delta\text{PE}$$

Principle of conservation of mechanical energy:

$$\begin{aligned} \text{Total energy} &= \text{PE} + \text{KE} \\ &= \text{constant} \quad \text{(if } \Delta Q = 0 \text{ and } \mathcal{W} = 0\text{)} \end{aligned}$$

General principle of conservation of energy (including thermal energy): Energy is neither created nor destroyed; it merely changes from one form to another.

Simple machine: A simple device which does work, like a lever, a jack, or a pulley system.

    Mechanical advantage:
        Actual: $\text{AMA} = w/f$
        Ideal: $\text{IMA} = L/l$

$$\text{Efficiency} = \frac{\text{work out}}{\text{work in}} \times 100\% = \frac{\text{AMA}}{\text{IMA}} \times 100\%$$

Power: The rate at which work is done or energy transformed.

$$P = \frac{\Delta \mathcal{W}}{\Delta t} = \mathbf{F} \cdot \bar{\mathbf{v}}$$

1 watt (W) = 1 J/s

Questions

1 Does the sun lose any energy by doing work on the earth as it moves in its orbit around the sun?

2 The dimensions of torque and of work are the same: $[M][L^2]/[T^2]$. Does this mean that torque and work are the same kind of physical quantity? Point out clearly the distinctions between them.

3 Experienced hikers, rather than step up on a log in their way on a hiking trail and then step down off it on the other side, instead step over the log to the ground on the other side (if the size of the log permits). Does this make sense from an energy point of view?

4 Figure 6.5 actually represents a highly idealized case, since it will be impossible in practice to lift by a purely vertical motion the mass m to the shelf. Rather the mass will have to be moved horizontally to the right at floor level, lifted to the height of the shelf, and then moved back to the left to its final position on the shelf. In the absence of friction and air resistance, will this change the analysis in Sec. 6.3 in any way? Why?

5 Describe the energy transformations taking place as a skier starts from the top of a ski slope, picks up speed, tries to slow down at the bottom, and finally plows into a snowbank at the bottom of the slope.

6 Discuss similarities and differences between the hydrologic cycle as a source of electric energy and the use, for the same purpose, of pumped-water storage at high elevations.

7 Discuss the energy transformations taking place from the time a pole-vaulter starts the run to the time the vaulter lands in the pit after leaping over (or hitting) the bar. Discuss also the energy transformations undergone by the pole.

8 In the compression and extension of a real spring, would you expect any energy to be lost in the form of heat? What would be the source of this heat?

9 Why cannot the kinetic energy given to the molecules of two objects when they are rubbed together be converted back into useful work at some later time?

10 We know that in any practical energy conversion scheme such as a hydroelectric plant or a steam-electric generator some mechanical energy is lost in the form of heat. Why then does it make sense to talk about a principle of conservation of energy?

11 Can you suggest a simple machine whose ideal mechanical advantage is exactly equal to 1? What is the usefulness of such a machine?

12 Can you suggest a simple machine whose ideal mechanical advantage is deliberately designed to be less than 1? Give an example of the use of such a machine (one such machine should be very familiar to most students).

13 A new electric power company places an ad in the newspaper saying that it will charge only 7 cents per kilowatt of electricity. Is this reasonable? Why?

Problems

A 1 A shopper pushes down on a 12-kg shopping cart at an angle of $20°$ with respect to the horizontal. She or he exerts a force of 6.0 N and moves the cart a distance of 20 m down an aisle of the supermarket. The work done is:
(a) 1.1×10^2 J (b) 2.3×10^3 J (c) 41 J
(d) 2.4×10^2 J (e) 0

A 2 A shopper moves a 20-kg box of groceries from the ground to a truck flatbed which is 70 cm above the ground, by sliding the box along a plank between truck and ground, where the plank makes an angle of $25°$ with respect to the ground. Neglecting friction, the work done is:
(a) 14 J (b) 1.4×10^2 J (c) 58 J (d) 124 J
(e) 5.9 J

A 3 A man pushes his daughter in a toy truck down the level driveway of their home. He exerts a constant force of 3.0 N parallel to the ground and pushes the truck 10 m in an 8.0-s interval. The mass of child and truck together is 30 kg. If we neglect friction, the kinetic energy of the truck and child together at the end of the 8.0-s interval is:
(a) 30 J (b) 30 W (c) 300 J (d) 38 J
(e) 2.9×10^3 J

A 4 The height of Victoria Falls in Rhodesia is 365 ft. A drop of water starting from rest at the top of the falls would have, at the bottom, a speed of:
(a) 85 m/s (b) 47 m/s (c) 60 m/s
(d) 33 m/s (e) None of the above

A5 A spring of mass 4.0 kg has a force constant of 20 N/m. If the spring is compressed a distance of 15 cm, its stored potential energy is:
(a) 5.9 J (b) 0.23 J (c) 0.45 J (d) 0.60 J
(e) 2.3 J

A 6 A 1500-kg car is moving along a level road with a kinetic energy of 7.5×10^4 J. The driver accelerates the car, and the engine provides an additional 5.0×10^4 J of energy during the next 10 s. If 1.0×10^4 J of this energy must be used to overcome friction, the final kinetic energy of the car is:
(a) 7.9×10^4 J (b) 13.5×10^4 J
(c) 11.5×10^4 J (d) 7.5×10^4 J
(e) None of the above

A 7 A roller-coaster car of mass 100 kg has a speed of 20 m/s at the bottom of one of the hills on the roller-coaster track. The highest hill it will be able to climb is, in the absence of friction,
(a) 20 m (b) 1.0 m (c) 200 m (d) 6.3 m
(e) 160 m

A 8 A lever is being used to lift a rock. The rock of mass 100 kg is 1.0 m from the fulcrum of the lever. The person using the lever exerts a force downward at a distance 2.5 m from the fulcrum and finds that lifting the rock requires a force of 500 N. The efficiency of the lever is:
(a) 100% (b) 250% (c) 200%
(d) 78% (e) 22%

A 9 If it takes a crane 100 s to lift a 1000-kg girder to a height of 100 m, the rate at which work is done by the crane is:
(a) 9.8×10^5 J (b) 9.8×10^5 W
(c) 9.8×10^3 W (d) 9.8×10^3 J
(e) 1.0×10^3 W

A10 One joule is equal to:
(a) 3.6×10^6 kWh (b) 2.8×10^{-7} kWh
(c) 2.8×10^{-7} kW (d) 2.8×10^{-7} kWs
(e) 3.6×10^6 kW

A11 As to order of magnitude, 1 hp is equal to:
(a) 1 W (b) 1 J/s (c) 1 kJ/s
(d) 10 kW (e) 10 W

B 1 Work has a precise meaning in physics which is somewhat different from common usage. Use the precise physics definition of work to answer the following questions.

(a) How much work is done by a 70-kg man who carries a 30-kg sack of dog biscuits 100 m along a level corridor?

(b) How much work does the same man do in carrying his load up 100 steps, each with a riser 25 cm high and a tread 50 cm wide?

(c) How much work does he do carrying his load up a ramp 100 m long, whose angle with the horizontal has a tangent of 0.10?

B 2 A 70-kg jogger moving at a speed of 2.0 m/s has a kinetic energy of 140 J. If she accelerates to a speed of 3.5 m/s, how much work must her body do to produce this acceleration?

B 3 What is the kinetic energy of an electron of mass 9.1×10^{-31} kg moving at a speed of 3.0×10^7 m/s?

B 4 An 85-kg skier starts at a height of 50 m above the bottom of a ski slope. At what speed will the skier be traveling at the bottom of the slope, in the absence of friction and air resistance?

B 5 What is the gravitational potential energy of 2.0×10^6 kg of water stored in a mountain lake at a height of 1000 m above sea level?

B 6 A spring is hung from a ceiling and a 1.0-kg mass hung from it. This increases the length of the spring by 10 cm. What is the force constant of the spring?

B 7 A 100-g ball is dropped out a window at a height of 30 m. When the ball reaches the ground, its speed is 22 m/s. How much energy has been lost in the fall because of air resistance?

B 8 A spring of force constant 20 N/m rests on a frictionless table. A 5.0-kg mass is attached to the spring, and it is displaced 20 cm from its equilibrium position and then released. What is the maximum kinetic energy of the system?

B 9 It is desired to use a lever to lift a 100-kg mass. If this mass is 1.0 m from the fulcrum of the lever, and the person trying to lift the mass can exert a force of at most 300 N, how long must be the lever arm on the lifter's side of the fulcrum?

B10 What must be the horsepower of a motor used to drive a winch to lift girders into position on a building under construction, if the girders have masses of 100 kg and are to be lifted 25 m in 10 s?

B11 A jet engine on an airplane develops a thrust of 15,000 N. If the plane is moving at a constant speed of 250 m/s, how many horsepower does this engine produce?

C 1 A person applies a force of 250 N to a car that is rolling slowly along a level track. Find the work done on the car if it moves 10 m while the person pushes: (a) in the direction of the car's motion; (b) in a direction 45° with respect to the car's displacement; (c) in a direction opposite to the car's displacement; (d) in a direction at right angles to the displacement.

C 2 A force **F** is applied at an angle of 30° above the horizontal to move a 100-kg block at constant velocity along a flat surface. The block moves 20 m, and the coefficient of sliding friction between the block and the flat surface is $\mu_k = 0.30$.

(a) Find the magnitude of the force **F**.

(b) Find the work done by the force **F**.

(c) Where did this work go?

C 3 A 10-kg box is pushed a distance of 10 m along a rough floor with $\mu_k = 0.50$ by a force of 100 N in the direction of the motion.

(a) What is the work done by this applied force?

(b) What is the work done by the force of friction on the box?

(c) What is the total work done by all forces acting on the box?

(d) What change in KE is produced by the forces acting on the box?

(e) If the initial speed of the box is 0, what is its final speed?

C 4 A 1500-kg car is moving along a level road at a speed of 15 m/s. The driver accelerates the car, and the engine provides 1.8×10^4 J of energy during the next 10 s. If 0.3×10^4 J of this energy must be used to overcome friction, what is the final speed of the car?

C 5 A boy pulls a 30-kg sled, which is initially at rest, along a level surface of ice. The coefficient of friction between the sled and the ice is 0.10. The boy exerts a force of 50 N over a 10-m stretch. Find: (a) the work done by the boy; (b) the work done against friction; (c) the final kinetic energy of the sled.

C 6 A pole vaulter clears 15 ft. If he was able to convert 90 percent of his kinetic energy into gravitational potential energy, how fast was he running when he slammed his pole into the jumping slot?

C 7 A water slide at an amusement park has the shape shown in Fig. 6.21. A girl of mass 60 kg on the slide has a speed of 2.0 m/s at height $y_2 = 15$ m. What is her speed at height $y_i = 5.0$ m?

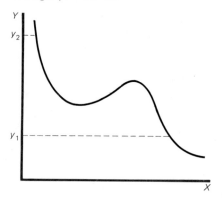

FIGURE 6.21
Diagram for Prob. C7.

C 8 A 70-kg skier starts from rest and skis for 200 m down a straight 10° slope. The skier's speed at the bottom is 25 m/s.

(a) What fraction of the skier's energy is lost to friction?

(b) What is the coefficient of friction on the slope?

C 9 A 150-g arrow is shot straight up into the air with a speed of 20 m/s. It reaches a maximum height of 19 m. Find:

(*a*)   The initial kinetic energy of the arrow.

(*b*)   The potential energy of the arrow at its topmost position.

(*c*)   The energy lost due to air resistance.

(*d*)   If the same frictional force retards the arrow as it drops back to earth, find the speed with which the arrow returns to earth.

C10   What is the average force needed to stop a bullet of mass 20 g and speed 500 m/s as it penetrates a wooden block to a distance of 20 cm?

C11   (*a*)   A spring is attached to a support and a 100-g mass hung from the bottom of the spring. The length of the spring increases by 25 cm. What is the force constant of the spring?

(*b*)   If the spring is placed on a frictionless table and attached to a wall, and the mass is then displaced through 50 cm from its equilibrium position, what is the stored potential energy in the spring?

C12   A spring of force constant 20 N/m is attached to a wall and rests on a frictionless table. A mass of 5.0 kg is attached to the spring and is displaced 25 cm from its equilibrium position and then released. Find:

(*a*)   The maximum PE in the system.

(*b*)   The maximum KE in the system.

(*c*)   The total energy of the system.

(*d*)   The maximum speed of the mass.

C13   A crane lifts a 2500-kg car to a height of 10 m in a time of 20 s. The motor driving the crane provides 18 hp. Find:

(*a*)   The work done by the crane.

(*b*)   The power output of the device.

(*c*)   The efficiency of the engine and crane system.

C14   The automobile jack shown in Fig. 6.22 has a lever arm of 0.40 m and a pitch of 4.0 mm; i.e., the top of the jack rises 4.0 mm for each full turn of the lever arm. If the efficiency of the jack is 35 percent, what force is required to lift a 1000-kg car with the jack?

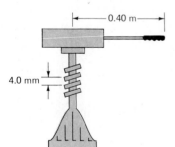

─── 0.40 m ───

4.0 mm

FIGURE 6.22
Diagram for Prob. C14.

C15   By using the wheel and axle shown in Fig. 6.23, a 500-N load can be lifted by the application of a 60-N force to the rim of the wheel. The radii are R = 90 cm and r = 8.0 cm. Determine (*a*)   the IMA; (*b*)   the AMA; and (*c*)   the efficiency of this wheel and axle.

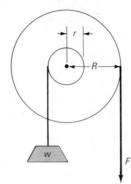

w F

FIGURE 6.23
Diagram for Prob. C15.

C16   Determine the force F needed to lift a weight of 400 N, using each of the three pulley systems shown in Fig. 6.24.

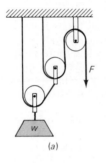

F

w

(*a*)

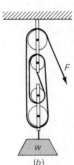

F

w

(*b*)

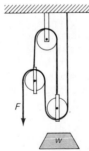

F

w

(*c*)

FIGURE 6.24
Diagram for Prob. C16.

C17   A Volkswagen, of mass 1000 kg, makes the climb from the foot of Pikes Peak, Colorado, to the top of the mountain, which is 14,100 ft above the base. If we neglect friction and air resistance, and the climb is made in $\frac{1}{2}$ h, what is the average power generated during that half hour?

C18   A rock of mass 4.0 kg falls under the force of gravity through a distance of 20 m. Plot as a function of time:

(*a*)   The work done by the force of gravity on the rock.

(*b*)   The power, or the rate at which the force of gravity does work on the rock.

C19 Montmorency Falls east of Quebec City in Canada is 90 m high and has a water flow rate over the falls of 35 m^3/s. If the density of water is 10^3 kg/m^3, what is the maximum electric power that can be generated by these falls at 100 percent efficiency?

C20 An advertisement for a car claims that a 1500-kg car can accelerate from rest to a speed of 25 m/s in 8.0 s. What average power must the motor produce to cause this acceleration, if we ignore friction and air resistance?

C21 A 90-kg jogger is moving along at a speed of 2.0 m/s and accelerates to a speed of 3.5 m/s over a 3.0-s interval.

(a) How much energy must the jogger's body provide to produce this acceleration?

(b) What is the power provided during the 30-s interval?

C22 A 2000-kg car accelerates uniformly from 15 to 24 m/s in 5.0 s. Find:

(a) The work done by the engine.

(b) The power delivered by the engine.

C23 If a horse could really work at the rate of 1 hp, how much work could it do in one 8-h day? Express your answer in both joules and kilowatthours.

C24 A rattlesnake can strike with an acceleration equivalent to an increase in speed from 0 to 60 mi/h in $\frac{1}{2}$ s. If the rattlesnake's head has a mass of 0.15 kg, what average power does the rattlesnake generate during its strike?

D1 A child on a bicycle wants to climb a 10° hill which is 200 m high. The mass of the child plus bike is 75 kg.

(a) Calculate how much work must be done by the child in climbing the hill at constant speed.

(b) If each complete revolution of the pedals moves the bike 5.0 m along the road, and if the pedals move in a circle of radius 18 cm, what average force must the child apply to the pedals? (Neglect friction and air resistance.)

D2 A trapeze artist swings on a rope 4.0 m long. He starts from rest 6.0 m above the ground, swings along an arc of a circle, and then lets go when the rope is exactly vertical. Find:

(a) The trapeze artist's speed as he lets go.

(b) His speed as he lands on the ground.

D3 What is the escape velocity of a rocket fired from the earth's surface; i.e., what is the minimum velocity needed by the rocket to escape from the earth's gravitational field?

D4 (a) From what height h above the bottom of the loop in Fig. 6.25 must a car start in order to just make it around the loop without falling off?

(b) What is the speed of the car at points A, B, and C?

D5 A pendulum ball of weight w hangs at the end of a string of length L. A variable horizontal force F_X, which starts at zero and gradually increases, pulls the ball very slowly outward until finally the string makes an angle θ with its original position. Find the work done by the force F_X.

D6 A 250-g wooden block is attached to a horizontal spring, as in Fig. 6.26. The block rests on a table for which the coefficient of sliding friction between table and block is 0.42. If the spring is first compressed 20 cm by a force of 15

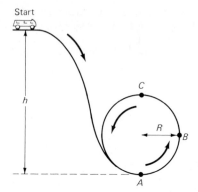

Start

FIGURE 6.25
Diagram for Prob. D4.

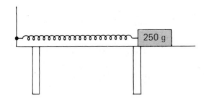

FIGURE 6.26
Diagram for Prob. D6.

N and then released, what will be its maximum displacement to the other side of its original position?

D7 A block of mass m, initially at rest, is dropped from a height h onto a vertical spring whose force constant is k, as in Fig. 6.27. Find the maximum distance y that the spring will be compressed.

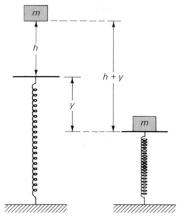

FIGURE 6.27
Diagram for Prob. D7.

D8 A car dealer is trying to find the actual power output of a car. The dealer first accelerates the 1000-kg car on level ground and finds that it will go uniformly from rest to 20 m/s in 10 s, and then finds that the car will coast to a stop from 20 m/s in 700 m, once the motor is turned off. If we make the assumption that the frictional and air resistance forces are the same for both the starting and stopping processes, what is the average power output of the car?

D9 Water flows over a dam at the rate of 1000 kg/s and falls 100 m vertically down before striking the turbine blades used to produce electricity. Find:

(*a*) The velocity of the water just before it hits the turbine blades.

(*b*) The electric power produced by the electric generator if 85 percent of the energy in the water is converted into electric energy.

Additional Readings

Brown, Sanford C.: *Count Rumford: Physicist Extraordinary*, Doubleday Anchor, Garden City, N.Y., 1962. A lively account of a most unusual physicist.

————: *Benjamin Thompson, Count Rumford*, MIT Press, Cambridge, Mass., 1981. A longer and more scholarly biography than the previous reference.

Chalmers, Bruce: *Energy*, Academic Press, New York, 1963. A very practical approach which stresses applications.

Dyson, Freeman J.: "Energy in the Universe," *Scientific American*, vol. 224, no. 3, September 1971, pp. 50–59. This is one of the best of a group of outstanding articles in an issue of *Scientific American* devoted entirely to energy. The quotation at the beginning of this chapter is taken from page 51 of this article (with permission).

Mulligan, Joseph F.: *Practical Physics: The Production and Conservation of Energy*, McGraw-Hill, New York, 1980. This text, at a lower level than the present one, contains many applications of the ideas of this chapter to our present energy problems.

Romer, Robert H.: *Energy: An Introduction to Physics*, Freeman, San Francisco, Calif., 1976. This book, at about the same level as the present text, contains both a great deal of excellent physics and many good applications of physics to the nation's energy problems.

*E*xperiment and observation provide the only way of arriving at the knowledge of the causes of all that one sees in Nature.

Christian Huygens (1629–1695)

Chapter 7

Impulse and Momentum

In the preceding chapter we introduced the first of the important conservation principles of physics, the principle of conservation of energy. In this chapter and the next we will consider two additional conservation laws for physical systems, the principle of conservation of linear momentum and the principle of conservation of angular momentum. These fundamental principles retain their validity and importance in all fields of physics, and give to physics the unity and coherence that make it so intellectually satisfying. As we will see, the principle of conservation of linear momentum explains phenomena as disparate as the recoil of a gun, the collision of two automobiles, or the interaction of two nuclear particles.

7.1 Linear Momentum

We begin by defining the *linear momentum* of a small object which we call a *particle*.

Definition

Linear momentum: The product of the mass of a particle and its velocity.

$$\mathbf{p} = m\mathbf{v}$$

(7.1)

Since the velocity $\mathbf{v}$ is a vector, so too is the momentum $\mathbf{p}$, and its direction is the same as that of the velocity. The SI unit of momentum is clearly kilogram-meters per second (kg·m/s), or newton-seconds (N·s). Just as a reference frame must always be specified with respect to which the velocity $\mathbf{v}$ is measured, the same is true for the momentum $\mathbf{p}$. For example, in Fig. 7.1 the reference frame for both the velocity and momentum of the football player is the football field.

For a large object, or for a system of particles considered as a single entity, the linear momentum of the system is the total mass of the system times the velocity of its center of mass, where the center of mass of the system is defined as in Sec. 4.9.

The word *momentum* has become almost a buzzword in newspaper articles about sporting events, but, if used carefully, its meaning corresponds well with the above definition. In football a heavy, fast running back is hard to stop because, since both m and $\mathbf{v}$ are very large, the player has a great deal of momentum. It therefore requires a very large force to stop the runner in a short time. Similarly a truck going 30 m/s can do much more serious damage

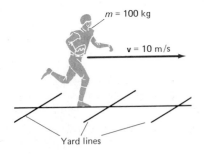

m = 100 kg

v = 10 m/s

Yard lines

FIGURE 7.1 Linear momentum: The momentum of a football player is the product of the player's mass and velocity, $\mathbf{p} = m\mathbf{v}$. If $m = 100$ kg and $v = 10$ m/s, the magnitude of the football player's momentum is 10^3 kg·m/s.

than a Volkswagen moving at 15 m/s because its momentum is so much greater. (Although, as in these cases, objects with large momenta usually also have large kinetic energies, in collisions momentum is more important because of its directional properties.)

Newton's Second Law and Momentum

Newton originally expressed his second law in terms of momentum rather than acceleration, although he called $m\mathbf{v}$ the "quantity of motion" rather than the momentum. In modern terminology we would state his second law as follows:

Newton's Second Law of Motion

> The time rate of change of the momentum of a system is equal to the net unbalanced external force applied to the system.

In the form of an equation, we can express this as

$$\mathbf{f}_{net} = \frac{\Delta \mathbf{p}}{\Delta t} \tag{7.2}$$

where $\mathbf{f}_{net}$ is the net unbalanced force acting, and $\Delta \mathbf{p}$ is the change in momentum in time Δt. We shall now show that this equation contains Newton's second law in the more usual form discussed in Chap. 3.

Consider an object of mass m undergoing acceleration from an initial velocity $\mathbf{v}_0$ to a final velocity $\mathbf{v}$ in a time interval t.

$$\frac{\Delta \mathbf{p}}{\Delta t} = \frac{m\mathbf{v} - m\mathbf{v}_0}{t} = \frac{m(\mathbf{v} - \mathbf{v}_0)}{t}$$

But, from Eq. (2.10), which is valid for constant acceleration, and Eq. (2.13), which is valid for average acceleration, we have

$$\mathbf{v} = \mathbf{v}_0 + \mathbf{a}t \qquad \text{and so} \qquad \mathbf{a} = \frac{\mathbf{v} - \mathbf{v}_0}{t}$$

Hence

$$\mathbf{f}_{net} = \frac{\Delta \mathbf{p}}{\Delta t} = \frac{m(\mathbf{v} - \mathbf{v}_0)}{t} = m\mathbf{a}$$

which is Newton's second law in the form introduced in Chap. 3.

The expression of Newton's second law in the form of Eq. (7.2) is actually more general than the form considered in Chap. 3. The change in the momentum $\Delta \mathbf{p}$ is equal to $\Delta(m\mathbf{v})$, and hence this equation states that the momentum can change because either the velocity or the mass changes. This added possibility of a change in mass becomes important in relativity theory, where mass varies with velocity, and also in the motion of rockets, where mass is constantly ejected from the rocket to accelerate it forward. The relative importance of the concept of linear momentum compared with that of velocity will become clear as we discuss one of the great integrating principles of physics—the principle of conservation of linear momentum. This principle applies only to momentum; there is no similar principle which applies to velocity.

Example 7.1

How much greater is the momentum of a truck which weighs 40,000 N than that of a Volkswagen weighing 6000 N, if the truck is going 30 m/s and the Volkswagen 15 m/s?

SOLUTION

The magnitudes of the two momenta under consideration are

$$p_1 = m_1 v_1 \quad \text{and} \quad p_2 = m_2 v_2$$

and their ratio R is

$$R = \frac{p_2}{p_1} = \frac{m_2 v_2}{m_1 v_1}$$

or $\quad R = \dfrac{m_2 g v_2}{m_1 g v_1}$

where we have multiplied numerator and denominator by g. Now $m_1 g$ is the weight of the Volkswagen and $m_2 g$ the weight of the truck. Since all we are interested in is the ratio of the two momenta, the units cancel out and we obtain:

$$R = \frac{40{,}000 \text{ N}}{6000 \text{ N}} \cdot \frac{30 \text{ m/s}}{15 \text{ m/s}} = \boxed{13.3}$$

Hence the momentum of the truck is over 13 times that of the Volkswagen. This means that it will take a force 13 times larger to stop the truck in the same time it takes to stop the Volkswagen, since from Eq. (7.2),

$$\Delta \mathbf{p} = \mathbf{f}_{\text{net}} \, \Delta t$$

Hence the force required is directly proportional to the change in momentum and is also 13 times greater.

7.2 Impulse; the Impulse-Momentum Theorem

If we take Eq. (7.2) and multiply both sides by Δt, we obtain

Impulse-Momentum Theorem

$$\boxed{\mathbf{f}_{\text{net}} \, \Delta t = \Delta \mathbf{p}} \qquad (7.3)$$

The quantity on the left, which is the product of the applied force and the time interval Δt during which the force acts, is called the *impulse*. Notice that impulse, like momentum, is a *vector* quantity. Equation (7.3), which is called the *impulse-momentum theorem*, then states that *the impulse of the force acting on a body is equal to the change in momentum of the body.* The same change in momentum can be achieved by a large force acting for a short time, or by a small force acting for a longer time. Thus in a football game a linebacker can best stop the forward progress of a tailback by a fierce tackle which lasts for a very short time interval during which the runner's velocity is reduced to zero. The linebacker could also stop the tailback by applying a lesser force for a longer period of time, but during that time interval the tailback would have picked up valuable additional yardage.

Equation (7.3) is most useful when it can be applied to situations where the applied force is essentially constant over a very short time interval, as in a perfect tackle by a linebacker. Otherwise, the force will probably vary considerably over the longer time interval, and unless we know the average value of $\mathbf{f}_{\text{net}}$, we will not be able to describe the situation accurately without the use of calculus. If the time interval is short enough, as in the case of a

FIGURE 7.2 Application of the impulse-momentum theorem to a bat striking a baseball: The impulse provided by the bat is equal to the change in the ball's momentum.

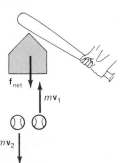

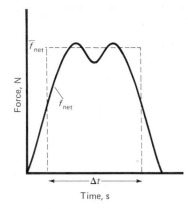

FIGURE 7.3 Variation of the force over time for a bat striking a baseball. The change in momentum is approximately equal to $\bar{f}_{net}\,\Delta t$, where $\bar{f}_{net}$ and Δt are as shown in the figure.

baseball bat striking a ball (Fig. 7.2), $\mathbf{f}_{net}$ will look something like Fig. 7.3, and the average values of the force $\bar{\mathbf{f}}_{net}$ and the time interval Δt will be good approximations to the true values to be used in the impulse-momentum equation. Figure 7.4 is a photograph of a tennis racket striking a ball at the actual instant the impulse is applied.

The dimensions of momentum are $[M][L]/[T]$, and for impulse the dimensions are

$$[F]\,[T] = [M]\,\frac{[L]}{[T^2]}\,[T] = \frac{[M]\,[L]}{[T]}$$

Hence impulse and momentum have the same dimensions, as is dictated by Eq. (7.3). They can both be expressed in the SI system as kilogram-meters per second, or, alternatively, as newton-seconds, since

$$1\ \text{N·s} = 1\ \frac{\text{kg·m·s}}{\text{s}^2} = 1\ \text{kg·m/s}$$

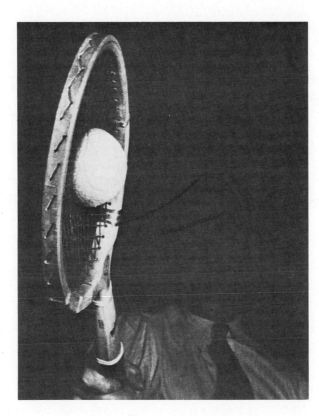

FIGURE 7.4 A high-speed flash photograph of a tennis racket striking a ball. Note how much both the ball and the racket are deformed, indicating the enormous size of the impulsive force at the instant the picture was taken. (Photo courtesy of Dr. Harold E. Edgerton.)

Example 7.2

A pitcher throws over the center of the plate a fastball which is moving at 95 mi/h when Reggie Jackson's bat collides with it. The ball, which has a mass of 5.0 kg, leaves Reggie's bat at a speed of 110 mi/h headed directly for the pitcher (who ducks!). What is the impulse imparted to the ball by the bat?

SOLUTION

From Eq. (7.3) we have

$$\mathbf{f}_{net} \, \Delta t = \Delta \mathbf{p}$$

where $\mathbf{f}_{net} \, \Delta t$ is the impulse of the force $\mathbf{f}_{net}$. In this case the change in momentum $\Delta \mathbf{p}$ is equal to $\mathbf{p}_f - \mathbf{p}_i$, where $\mathbf{p}_f$ and $\mathbf{p}_i$ are the momenta before and after the bat hits the ball. If we take the direction of $\mathbf{p}_f$ as positive, since that is the direction in which the force acts, we have

$$\Delta p = p_f - p_i$$

$$= (0.50 \text{ kg})(110 \text{ mi/h}) \left(\frac{0.447 \text{ m/s}}{1 \text{ mi/h}} \right)$$

$$- \left[(0.50 \text{ kg}) (-95 \text{ mi/h}) \left(\frac{0.447 \text{ m/s}}{1 \text{ mi/h}} \right) \right]$$

$$= (25 + 21) \text{ kg·m/s} = \boxed{46 \text{ kg·m/s}}$$

Hence the impulse $\mathbf{f}_{net} \, \Delta t = 46$ kg·m/s or 46 N·s. In this case, therefore, we have been able to find the impulse, even though we do not know either $\mathbf{f}_{net}$ or Δt.

Notice that the change in momentum here is the *vector difference* $\mathbf{p}_f - \mathbf{p}_i$ between the two momenta. Since $\mathbf{p}_f$ and $\mathbf{p}_i$ are in opposite directions, the numerical value of $\Delta \mathbf{p}$ is actually the sum of the magnitudes of $\mathbf{p}_f$ and $\mathbf{p}_i$. The vector difference is therefore $25 + 21 = 46$ kg·m/s, whereas the scalar difference between the two magnitudes is only $25 - 21 = 4$ kg·m/s. This essential vector nature of momentum must be kept in mind in all problems.

7.3 Conservation of Linear Momentum

Since $\mathbf{f}_{net} = \Delta \mathbf{p}/\Delta t$, it is clear that, if $\mathbf{f}_{net} = 0$, then $\Delta \mathbf{p}$ is equal to zero, and $\mathbf{p}$ remains constant. This is true for a single particle, for a large moving object, or for a system of particles. We can therefore express the *principle of conservation of linear momentum* as follows:

Conservation of Linear Momentum

> If no net external force acts on a physical system, the total linear momentum of the system remains constant.

The center of mass of the system will therefore continue to move with its initial velocity as long as no net unbalanced force acts from outside the

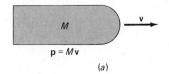

(a)

FIGURE 7.5 An exploding shell; an example of the conservation of linear momentum. (a) The linear momentum of the shell before it explodes $\mathbf{p} = M\mathbf{v}$. (b) The sum of the linear momenta of all the fragments of the shell after it explodes, $\sum_i m_i \mathbf{v}_i$, is equal to the original linear momentum $\mathbf{p} = M\mathbf{v}$.

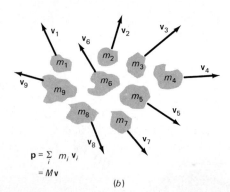

(b)

system. The momentum of the whole system cannot change, even though the momentum of individual parts of the system may change radically, as shown in Fig. 7.5.

Newton's Third Law and Momentum Conservation

The principle of conservation of linear momentum is closely related to Newton's third law of motion, as can be seen from the following discussion.

Consider two colliding pucks of masses m_1 and m_2, as in Fig. 7.6, where we assume that no external forces act on the two pucks. During this collision m_1 exerts a force $\mathbf{f}_{12}$ on m_2; m_2 exerts an equal and opposite force $\mathbf{f}_{21}$ back on m_1. Each puck exerts this force on the other puck for the same time Δt. We can therefore apply the impulse-momentum theorem in the form of Eq. (7.3) and obtain

$$\mathbf{f}_{12}\,\Delta t = \Delta \mathbf{p}_2 = \Delta(m_2 \mathbf{v}_2)$$

$$\mathbf{f}_{21}\,\Delta t = \Delta \mathbf{p}_1 = \Delta(m_1 \mathbf{v}_1)$$

But, from Newton's third law,

$$\mathbf{f}_{12} = -\mathbf{f}_{21}$$

and so

$$\mathbf{f}_{12}\,\Delta t = -\mathbf{f}_{21}\,\Delta t$$

or

$$\Delta(m_2 \mathbf{v}_2) = -\Delta(m_1 \mathbf{v}_1)$$

If m_1 and m_2 are constant, v_1 and v_2 are the velocities before the collision, and v_1' and v_2' are the velocities after the collision, we can rewrite this as

$$m_2 \mathbf{v}_2' - m_2 \mathbf{v}_2 = -(m_1 \mathbf{v}_1' - m_1 \mathbf{v}_1)$$

or

$$m_1 \mathbf{v}_1' + m_2 \mathbf{v}_2' = m_1 \mathbf{v}_1 + m_2 \mathbf{v}_2 \tag{7.4}$$

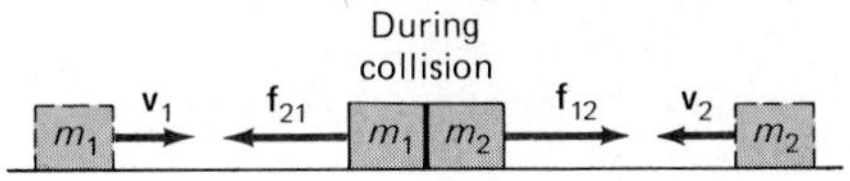

FIGURE 7.6 Head-on collision of two air-hockey pucks. Because action equals reaction during the collision, linear momentum is conserved.

This is just a statement of the principle of conservation of linear momentum: since no external force acts, the total linear momentum of the system (the two pucks) remains constant.

Since we know m_1, m_2, $\mathbf{v}_1$, and $\mathbf{v}_2$ in Eq. (7.4), we can obtain a relationship between $\mathbf{v}_1'$, and $\mathbf{v}_2'$. We cannot, however, obtain actual values for $\mathbf{v}_1'$ and $\mathbf{v}_2'$ without further information about the collision, in particular, what happens to the kinetic energy in the collision. In general, the kinetic energy is not conserved in collisions, since some of the energy of motion may be converted into heat (and perhaps sound). This is true even when no net external force acts on the system. There are, however, certain special kinds of collisions in which we can specify what happens to the kinetic energy and use this information to solve the problem fully. We will see some examples of this in the sections which follow.

It is important to observe that the principle of conservation of linear momentum can lead to solutions of problems even in cases where it would be very difficult to apply Newton's second law, since the instantaneous forces acting are not known (Example 7.3 illustrates this). Also, in dealing with many problems in atomic and nuclear physics it is impossible to specify

Christian Huygens (1629–1695)

FIGURE 7.7 Christian Huygens. (Photo courtesy of AIP Niels Bohr Library.)

Christian Huygens is one of the least well-known of the world's important physicists. This is especially odd, because he was one of the greatest scientific geniuses of all time. He made significant contributions to astronomy by discovering Saturn's ring and the satellite Titan; to mechanics by his understanding of centripetal force, of the properties of collisions, and of work and energy; to optics, where he is justly regarded as the founder of the wave theory of light; and to experimental physics, where he developed the first accurate pendulum clock for timing physical events.

Huygens was born on April 14, 1629, in The Hague, the son of the famous Constantine Huygens, who was a poet, philosopher, classical scholar, and diplomat. He was taught by a private tutor until he was 16, his education including Latin, music, and drawing. He grew to love mathematics more than all other subjects, however, and this interest dominated his later life. In 1645, Huygens entered the University of Leyden to study mathematics and law. After further studies at the University of Breda, he completed his education by traveling widely throughout Europe. Meanwhile he became involved in the scientific ferment of the day which was stimulated by the ideas of René Descartes (1596–1650), a friend of the Huygens family.

Huygens lived at home from 1650 to 1666, except for journeys to Paris and London, and these 16 years were the most productive of his scientific career. His first published work came out in 1651; it was followed by a constant stream of important books on a great variety of scientific subjects.

In 1665, Colbert, minister to Louis XIV, King of France, decided to set up the Académie Royale des Sciences in Paris, and to include Huygens as one of its charter members. The king provided financial support to enable Huygens to live in a well-appointed apartment in the Bibliothèque Royale. He remained in Paris from 1666 until 1681, except for occasional visits to his native Holland for reasons of health. In 1672, when the French armies were attacking the Dutch Republic, he stayed on in Paris. He even dedicated a celebrated book on his invention of the pendulum clock to Louis XIV, who had vowed to destroy the Dutch Republic!

In 1681 Huygens was forced to leave Paris because of ill health, and he never returned. His convalescence was slow, and in 1683 Colbert, his patron, died, and opposition mounted to Huygens' return as head of the Académie Royale.

The years after 1683 were hard ones for Huygens. He lived in The Hague, cut off from the scientific centers of the world. He remained single, living alone at his family's estate most of the time, and the income from his family's property was barely enough to support him. In 1689 he visited England, met Newton and Boyle, and returned home even more unhappy about his intellectual isolation in the Netherlands.

He died in The Hague in 1695, survived by no disciples and relatively forgotten by the swiftly moving world of science, which owed so much to his ideas.

instantaneous positions, velocities, and accelerations for the particles involved, but it is still possible to apply the law of conservation of linear momentum and obtain valuable results. For this reason the concept of linear momentum, and its associated conservation law, are among the most fruitful ideas in all of physics.

One of the physicists who first understood the importance of energy considerations in handling collision problems was Christian Huygens (1629–1695). Huygens (see Fig. 7.7 and accompanying biography), together with Galileo and Newton, helped develop mechanics into the exact science we know today.

Example 7.3

A shell is fired from a mortar with an initial speed of 200 m/s and at an angle of 45° with respect to the ground, as in Fig. 7.8. At a position a horizontal distance of 3000 m from the mortar, the shell explodes and scatters fragments in all directions.
(a) What can be said about the motion of the fragments after the shell explodes?
(b) At what position will the center of mass of all the fragments finally reach the ground?

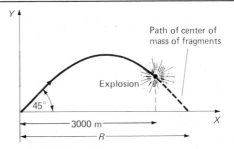

FIGURE 7.8 Trajectory of an exploding shell.

SOLUTION

(a) If we neglect air resistance, the only *external* force acting here, after the shell leaves the mortar, is the force of gravity. The shell therefore follows a parabolic trajectory, as discussed in Sec. 5.3. After the shell explodes, the center of mass of the fragments continues along the same parabolic path; i.e., if at any instant of time after the explosion the center of mass of the fragments is calculated, it will always be found along the dotted line in the trajectory of Fig. 7.8. The force of the explosion is an *internal force* and can therefore have no effect on the motion of the center of mass of the system.

(b) The center of mass of the exploded fragments will reach the ground at the same position at which the unexploded shell would have reached the ground. For an angle of 45° this is given by Eq. (5.16):

$$\text{Range} = \frac{v_0^2}{g} = \frac{(200 \text{ m/s})^2}{9.8 \text{ m/s}^2} = \boxed{4.1 \times 10^3 \text{ m}}$$

7.4 Collisions in One Dimension: Elastic Collisions

Definition

Elastic collision: A collision in which not merely the linear momentum of a system, but also its total mechanical energy, is conserved.

A good example of an elastic collision is the collision of two hockey pucks which have a coiled spring between them, so that the original kinetic energy of the pucks is first converted into elastic potential energy in the spring. Then, as the spring expands, the potential energy is converted back again into the kinetic energy of the separating pucks, and this kinetic energy is equal to the original kinetic energy. A collision between two molecules in a gas is another good example of an elastic collision.

In an elastic, head-on collision between two particles of masses m_1 and m_2, both linear momentum and kinetic energy are conserved. Hence

$$m_1\mathbf{v}_1 + m_2\mathbf{v}_2 = m_1\mathbf{v}_1' + m_2\mathbf{v}_2' \quad \text{(conservation of linear momentum)}$$

and

$$\tfrac{1}{2}m_1v_1^2 + \tfrac{1}{2}m_2v_2^2 = \tfrac{1}{2}m_1v_1'^2 + \tfrac{1}{2}m_2v_2'^2 \quad \text{(conservation of mechanical energy)}$$

Since for a head-on collision the motion is confined to a straight line, we can drop the vector notation and distinguish the direction of the velocities by + and − signs.

If the masses and initial velocities are known, these two equations can be solved for the two unknowns v_1' and v_2'. By rearranging the two equations we obtain

$$m_1(v_1 - v_1') = m_2(v_2' - v_2) \qquad (7.5)$$

and $\qquad m_1(v_1^2 - v_1'^2) = m_2(v_2'^2 - v_2^2) \qquad (7.6)$

Dividing the second equation by the first, we obtain

$$v_1 + v_1' = v_2' + v_2$$

or $\qquad v_1 - v_2 = v_2' - v_1'$

and so $\qquad v_2' - v_1' = -(v_2 - v_1) \qquad (7.7)$

Here $v_2' - v_1'$ is the velocity of m_2 relative to m_1 after the collision, while $v_2 - v_1$ is the velocity of m_2 relative to m_1 before the collision. Therefore, *the relative velocity of the two particles in a perfectly elastic collision is reversed in direction but has the same magnitude as it had before the collision.* This means that, if two balls were approaching each other before the collision, they are moving away from each other after the collision, but there is no change in the magnitude of their relative velocity.

Suppose m_2 is initially at rest, so that $v_2 = 0$. Then, from Eq. (7.5),

$$v_2' = \frac{m_1}{m_2}(v_1 - v_1')$$

and, from Eq. (7.7),

$$v_2' = \frac{m_1}{m_2}[v_1 - (v_2' - v_1)] = \frac{m_1}{m_2}(2v_1 - v_2')$$

or $\qquad v_2'\left(1 + \frac{m_1}{m_2}\right) = 2\left(\frac{m_1}{m_2}\right)v_1$

Hence $\qquad v_2' = \frac{2m_1}{m_1 + m_2}v_1 \qquad (7.8)$

Also, from Eq. (7.7),

$$v_1' = v_2' - v_1 = \frac{2m_1}{m_1 + m_2}v_1 - v_1$$

or $\qquad v_1' = \frac{m_1 - m_2}{m_1 + m_2}v_1 \qquad (7.9)$

These two equations for v_1' and v_2' in an elastic collision allow us to find these two quantities if we know the initial velocity v_1 and the masses.

An interesting example is the case in which the two masses are equal. Then, from Eqs. (7.8) and (7.9), we have

$$v_1' = 0 \qquad \text{and} \qquad v_2' = v_1$$

Hence the ball of mass m_1 stops dead and gives all its momentum and energy to the second ball, which moves away with the same velocity which ball 1 had originally, as in Fig. 7.9. Anyone who has ever played pool or billiards will recognize this as an example of what happens on a billiard table, so long as the cue ball is hit without topspin or backspin.

This complete transfer of momentum and energy in an elastic collision from a moving particle to a second particle with the same mass and at rest explains why water is used as the "moderator" in boiling-water nuclear reactors to slow down neutrons and increase the chances that they will fission

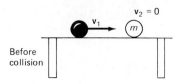

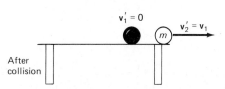

FIGURE 7.9 Collision of a moving billiard ball with a ball at rest that has the same mass. The first ball stops and the second ball moves away with the same velocity that the first ball had originally.

uranium nuclei and release useful energy. The most effective way to slow down energetic neutrons is to have them collide with particles of mass equal to the neutron mass, as we have just seen. The nuclei of the hydrogen atoms in ordinary water serve this function very well, since they consist of a single proton of mass almost identical to that of the colliding neutrons. Hence the elastic collisions between the neutrons and protons are very effective in slowing down the neutrons, as required for the efficient running of the reactor.

Example 7.4

Consider the collision of two balls, one of mass 2.0 kg moving to the right along the X axis with velocity 5.0 m/s and the other of mass 1.0 kg moving to the left along the X axis with velocity 10 m/s. If kinetic energy is conserved in the head-on collision, what are the final velocities of the two balls?

SOLUTION

As can be seen from Fig. 7.10, the total momentum before the collision is zero, since

$$p = m_1 v_1 + m_2 v_2 = (2.0 \text{ kg})(5 \text{ m/s}) - (1.0 \text{ kg})(10 \text{ m/s}) = 0$$

where we have dropped the vector notation, since all motion is along the X axis, and have used $+$ and $-$ signs to distinguish the two directions. Since momentum is conserved, the momentum after the collision is also zero, or

$$m_1 v_1' + m_2 v_2' = 0$$

Hence $\quad (2.0 \text{ kg})v_1' + (1.0 \text{ kg})v_2' = 0$

or $\quad\quad v_2' = -2v_1'$

This tells us that v_1' and v_2' will be in opposite directions, and that v_2' will be twice as large as v_1'. Hence conservation of linear momentum can be satisfied by such values as $v_1' = -4$ m/s and $v_2' = 8$ m/s, $v_1' = -2$ m/s and $v_2' = 4$ m/s, and the like. We can only find the correct value of the momenta if we specify what happens to the kinetic energy in the collision.

If, as indicated in the statement of the problem, the collision is elastic and so kinetic energy is conserved, we have

$$\tfrac{1}{2}m_1 v_1^2 + \tfrac{1}{2}m_2 v_2^2 = \tfrac{1}{2}m_1 v_1'^2 + \tfrac{1}{2}m_2 v_2'^2$$

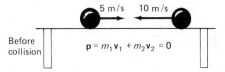

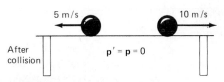

FIGURE 7.10 Elastic collision of two moving balls of different masses. In this case, the two balls reverse directions and leave the collision with the same speeds they had before the collision.

But $\quad \tfrac{1}{2}m_1 v_1^2 + \tfrac{1}{2}m_2 v_2^2$

$$= \tfrac{1}{2}(2.0 \text{ kg})(5.0 \text{ m/s})^2 + \tfrac{1}{2}(1 \text{ kg})(-10 \text{ m/s})^2$$

$$= 75 \text{ kg·m}^2/\text{s}^2$$

and so $\quad \tfrac{1}{2}(2.0 \text{ kg})v_1'^2 + \tfrac{1}{2}(1.0 \text{ kg})v_2'^2 = 75 \text{ kg·m}^2/\text{s}^2$

or

$$v_1'^2 + \tfrac{1}{2}(-2v_1')^2 = v_1'^2 + 2v_1'^2 = 75 \text{ kg·m}^2/\text{s}^2$$

so that

$$v_1'^2 = 25 \text{ m}^2/\text{s}^2 \quad \text{or} \quad v_1' = \boxed{\pm 5.0 \text{ m/s}}$$

Since the original velocity v_1 was $+5.0$ m/s, the negative sign must be chosen here since the positive sign corresponds to no collision at all. Hence $v_2' = -2v_1' = +10$ m/s, which is also the reverse of its original velocity. Thus the two balls merely reverse their original direction in this elastic collision, and leave the collision with the same speeds they had originally. This keeps the total linear momentum of the system at zero, its original value.

Example 7.5

Collisions involving atomic particles are usually good examples of elastic collisions. Compare the change in velocity experienced by a neutron in a nuclear reactor which emerges from the fission of a uranium-235 nucleus and collides head on (a) with the nucleus of a hydrogen atom (a proton) which is at rest; (b) with a uranium-235 nucleus which has a mass 235 times that of the neutron.

SOLUTION

(a)   For the head-on collision of the neutron with a proton, remembering that the proton and neutron masses are almost identical, and using Eq. (7.9), we have for the velocity of the neutron after the collision

$$v_1' = \frac{m_1 - m_2}{m_1 + m_2}v_1 = \frac{m_1 - m_1}{m_1 + m_1}v_1 = \boxed{0}$$

(b)   For the head-on collision with the ^{235}U nucleus, we have

$$v_1' = \frac{m_1 - 235m_1}{m_1 + 235m_1}v_1 = \frac{-234m_1}{236m_1}v_1 = \boxed{-0.99v_1}$$

Hence in the head-on collision with the proton, the neutron loses all its energy and comes to rest.* On the other hand, the collision with the uranium nucleus has changed the direction of the neutron but modified its speed hardly at all. Hence there is almost no change in its kinetic energy.

Since in practice few collisions are perfect head-on collisions, it takes many collisions of the neutron with hydrogen atoms before it becomes a thermal neutron.

---

*As we will see later, the neutron retains a small amount of random kinetic energy owing to the temperature of its surroundings. Because of this, such neutrons are described as being *thermalized* and are called *thermal neutrons*.

7.5 Collisions in One Dimension: Inelastic Collisions

At the opposite extreme from an elastic collision is one in which the colliding bodies stick together and move as a unit with zero relative velocity after the collision. Such a collision is called *completely inelastic*, and in such a collision much of the original kinetic energy of the two colliding particles is lost. The kinetic energy which remains is just enough to conserve the original total momentum of the system. Two freight cars which collide and are then held together by the couplers on the cars is a good example of a completely inelastic collision (see Fig. 7.11). In such a case the principle of conservation of linear momentum leads to the equation

$$m_1v_1 + m_2v_2 = (m_1 + m_2)v' \tag{7.10}$$

where v' is the final velocity of two given particles or masses, with its direction specified by a $+$ or $-$ sign. Hence the final velocity of the two masses can be obtained immediately if m_1, m_2, v_1, and v_2 are given.

The kinetic energy before the collision is

$$E_k = \tfrac{1}{2}m_1v_1^2 + \tfrac{1}{2}m_2v_2^2$$

FIGURE 7.11   Inelastic collision of two freight cars which are coupled together after the collision. If m_2 is originally at rest and the two masses are the same, half the original KE of the system is lost in the collision, while the linear momentum is, as always, conserved.

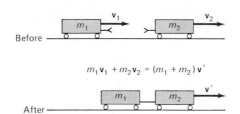

$$m_1 \mathbf{v}_1 + m_2 \mathbf{v}_2 = (m_1 + m_2)\mathbf{v}'$$

and after the collision it is

$$E_k' = \tfrac{1}{2}(m_1 + m_2)v'^2$$

Suppose that, for simplicity, we consider the case where the mass m_2 is originally at rest, so that $v_2 = 0$. Then the ratio of the kinetic energy after the collision to that before the collision is

$$\frac{E_k'}{E_k} = \frac{\tfrac{1}{2}(m_1 + m_2)v'^2}{\tfrac{1}{2}m_1 v_1^2}$$

But, from Eq. (7.10), $v' = \dfrac{m_1 v_1}{m_1 + m_2}$, and so

$$\frac{E_k'}{E_k} = \frac{(m_1 + m_2)m_1^2 v_1^2}{m_1 v_1^2 (m_1 + m_2)^2} = \frac{m_1}{m_1 + m_2} \qquad (7.11)$$

Hence the total kinetic energy *decreases* in the collision, since m_1 is less than $m_1 + m_2$. If the two masses are equal, we have

$$\frac{E_k'}{E_k} = \frac{1}{2}$$

and half the kinetic energy of the system is lost when the moving mass, say, a ball, collides with another ball of equal mass at rest, and they move away stuck together after the collision. In this case, of course, $v' = v_1/2$, as is clear from Eq. (7.10). This means that each ball after the collision has one-fourth the kinetic energy of the first ball before the collision. Hence the total kinetic energy after the collision is one-half what it was before, as we have just seen. The other half of the original kinetic energy is lost in the form of heat (and perhaps sound).

Example 7.6

A bullet of mass m is fired into a soft wooden block of mass M which is mounted as in Fig. 7.12 so that it can swing freely to a height h above its equilibrium position.
(a) Show how this so-called ballistic pendulum can be used to find the speed v of the bullet as it leaves the muzzle of the gun.
(b) How much energy is lost when the bullet collides with the block of wood? Assume that the mass of the bullet is 10 g, and that the mass of the wooden block is 10 kg.

SOLUTION

(a) If there is no net external force acting, linear momentum is conserved and we can write

$$mv = (m + M)V$$

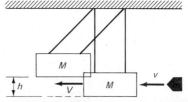

FIGURE 7.12   A "ballistic pendulum" used to find the speed of a bullet. Most of the original KE of the bullet is lost in the collision in the form of heat.

where v is the muzzle velocity of the bullet and V is the velocity of the block (and embedded bullet) as it starts its horizontal motion.

Mechanical energy is *not* conserved during this completely inelastic collision, but when the block starts to swing, the block and bullet together have a kinetic energy

$$KE = \tfrac{1}{2}(m + M)V^2$$

The block then swings upward to a new height h above its initial position, such that its increase in potential energy is exactly equal to its loss of kinetic energy. Hence

$$\tfrac{1}{2}(m + M)V^2 = (m + M)gh$$

or $\qquad V = \sqrt{2gh}$

Then, from our first equation above,

$$v = \frac{m + M}{m}V = \frac{m + M}{m}\sqrt{2gh}$$

If we know m and M and can measure h, the velocity v of the bullet can be determined. The nice feature of this setup is that a very large velocity, which is difficult to

measure directly with any accuracy, can be obtained from a simple height measurement by using this ballistic pendulum. Today much more sophisticated techniques than this are used for velocity measurements, but the principle illustrated in this example is still useful.

(b) We know that this collision is completely inelastic since the bullet sticks in the block. The ratio of the kinetic energy before and after the collision is, using the results of part (*a*), or Eq. (7.11),

$$\frac{E_k}{E_k'} = \frac{\tfrac{1}{2}mv^2}{\tfrac{1}{2}(m + M)V^2} = \frac{mv^2(m + M)^2}{(m + M)m^2v^2} = \frac{m + M}{m}$$

If $M = 10$ kg and $m = 10$ g $= 0.010$ kg, then

$$\frac{E_k}{E_k'} = \frac{m + M}{m} = \frac{10.01}{0.01} \simeq 10^3$$

Hence the final kinetic energy is only 10^{-3} or 0.1 percent of the initial kinetic energy. The other 99.9 percent of the original energy has been lost as heat when the bullet bores through the wood and finally comes to rest in the block.

Coefficient of Restitution

A quantity called the *coefficient of restitution* and designated by the symbol e is often used to categorize collision processes. We have seen in Eq. (7.7) that for an elastic collision,

$$v_2' - v_1' = -(v_2 - v_1)$$

We therefore define the coefficient of restitution in terms of this equation as follows:

$$e = \frac{-(v_2' - v_1')}{v_2 - v_1} \tag{7.12}$$

TABLE 7.1 Coefficient of Restitution e for Various Kinds of Collisions

Type of collision	Coefficient of restitution e
Elastic	1
Inelastic	$0 < e < 1$
Completely inelastic	0

For an elastic collision, for which Eq. (7.7) is valid, e is therefore equal to 1. For a completely inelastic collision, where the two particles stick together,

$$e = \frac{-(v_2' - v_1')}{v_2 - v_1} = 0$$

For all other somewhat elastic collisions, e varies between 0 and 1; the closer it comes to 1, the more nearly elastic the collision. These results are summarized in Table 7.1.

The elastic properties of a series of balls made out of different materials may be determined by dropping each ball on a thick steel plate fixed rigidly to the ground, and comparing the speed of the ball before and after it strikes the plate. We have, in this case, $e = -v_1'/v_1$, because $v_2 = v_2' = 0$. Hence, since the velocity depends on the height h' to which the ball bounces, we have

$$e = \frac{-v_1'}{v_1} = \frac{-\sqrt{2gh'}}{-\sqrt{2gh}} = \sqrt{\frac{h'}{h}} \tag{7.13}$$

TABLE 7.2 Coefficient of Restitution for Various Materials (Approximate Values for Spheres Falling on Flat Steel Plate)

Material	Coefficient of restitution, $\sqrt{\dfrac{h'}{h}}$
Glass	0.97
Steel	0.95
Rubber	0.75
Iron	0.70
Maple	0.66
Cork	0.60
Brass	0.30
Copper	0.22
Lead	0.16

where h' is the height of the ball after rebounding from the plate and h is the height from which the ball was dropped. If $h' = h$, the collision with the plate is perfectly elastic, since $e = 1$. If the ball does not rebound at all but sticks to the plate, $e = 0$ and the collision is perfectly inelastic. All other cases fall

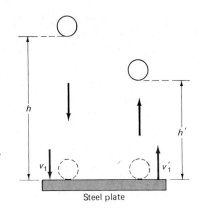

FIGURE 7.13 Method for determining the coefficient of restitution for various materials by dropping balls made of the material on a rigid steel plate. Then $e = \sqrt{h'/h}$.

between the limits 0 and 1. This is illustrated in Fig. 7.13. Table 7.2 gives values of the coefficient of restitution e obtained in this way.

Most collisions fall somewhere between elastic and completely inelastic collisions; i.e., some kinetic energy is lost in the collision. In such cases, even though we can obtain some information about the behavior of the bodies after the collision from the principle of conservation of linear momentum (which is always valid), it is impossible to obtain the final velocities of the two particles without additional information about what happens to the mechanical energy in the collision.

Example 7.7

A glass marble is dropped on a heavy steel plate resting on the floor. The marble rebounds to 0.94 of the height from which it was dropped.
(a) What is the coefficient of restitution of the glass?

(b) If the speed of the marble before it struck the plate was 10 m/s, what was its speed immediately after striking the plate?

SOLUTION

(a) Since $e = \sqrt{h'/h} = \sqrt{0.94h/h} = \sqrt{0.94}$,

$e = \boxed{0.97}$

(b) From Eq. (7.13),

$$e = \frac{-v_1'}{v_1}$$

Hence $\quad v_1' = -ev_1 = (-0.97)(10 \text{ m/s}) = \boxed{-9.7 \text{ m/s}}$

Here the minus sign merely means that the ball's direction is reversed after striking the plate.

7.6 Collisions in Two Dimensions

It is, of course, likely that two colliding particles will not hit head-on but rather at some glancing angle which will cause them to move off at angles with respect to a line originally connecting the two particles, as in Fig. 7.14. We then have to consider the collision in two (or perhaps three) dimensions. For motion in a plane, Eq. (7.4) can be broken down into X and Y components as follows:

X components: $\quad m_1 v_{1x} + m_2 v_{2x} = m_1 v_{1x}' + m_2 v_{2x}'$

Y components: $\quad m_1 v_{1y} + m_2 v_{2y} = m_1 v_{1y}' + m_2 v_{2y}'$

(7.14)

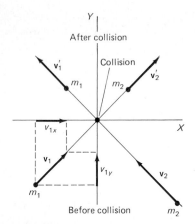

FIGURE 7.14 A collision in two dimensions. The X and Y components of the linear momentum are conserved. The two masses colliding are m_1 and m_2.

Hence, *if no external forces act on a system of particles, the components of the total linear momentum of the system along any chosen axis are the same after a collision as before the collision.*

As an example, suppose we have a high-energy nuclear particle of unknown mass m_1 and velocity v_1 colliding with a particle known to have a mass m_2 and to be at rest. Then the situation is as shown in Fig. 7.15, where the initial velocity of m_1 is taken in the X direction. After the collision, particle 1 moves off at an angle α with respect to its original direction and particle 2 moves off at an angle β with respect to the same direction. Then Eq. (7.14) becomes:

X components: $\qquad m_1v_1 + 0 = m_1v_1' \cos \alpha + m_2v_2' \cos \beta$

Y components: $\qquad 0 + 0 = m_1v_1' \sin \alpha - m_2v_2' \sin \beta \qquad$ or

$$m_1v_1' \sin \alpha = m_2v_2' \sin \beta$$

This gives us two equations containing the seven quantities m_1, m_2, v_1, v_1', v_2', α, and β. If the mass m_2 is known, and the speeds v_1' *and* v_2' and the angles α and β can be measured, we then have only two unknowns remaining, the mass m_1 and the speed v_1 of the colliding particle. Hence these can be determined from the two equations above. In this way the momentum of the particle being scattered can be measured.

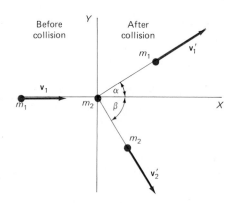

FIGURE 7.15 Collision of a high-energy nuclear particle of unknown mass m_1 and velocity v_1 with a particle of known mass m_2 that is at rest.

Example 7.8

A metal puck on a steel table has a mass of 0.25 kg and an initial speed of 2.0 m/s along the positive X axis. This puck makes a glancing collision with a stationary puck of mass 0.50 kg, and after the collision the first puck moves off at a speed of 1.0 m/s at 30° above the X axis.

(a) Find the direction of the final velocity of the second puck.
(b) Find the magnitude of the final velocity of the second puck.
(c) Is this an elastic collision?

SOLUTION

This is basically the same problem discussed in Sec. 7.6. Hence Fig. 7.15 may be used. Since linear momentum is always conserved in any collision, we have for the components of the linear momentum:

X components: $\qquad m_1v_1 = m_1v_1' \cos \alpha + m_2v_2' \cos \beta$

Y components: $\qquad m_1v_1' \sin \alpha = m_2v_2' \sin \beta$

From the first equation and the data given we have:

$$(0.25 \text{ kg})(2.0 \text{ m/s}) = (0.25 \text{ kg})(1.0 \text{ m/s})(\cos 30°) + (0.50 \text{ kg})(v_2' \cos \beta)$$

from which

$$v_2' \cos \beta = \frac{0.50 \text{ kg·m/s} - 0.22 \text{ kg·m/s}}{0.50 \text{ kg}} = 0.56 \text{ m/s}$$

(a) From the second equation we then have:

$$(0.25)(1.0 \text{ m/s})(\sin 30°) = (0.50 \text{ kg})(v_2' \sin \beta)$$

or $v_2' \sin \beta = 0.25 \text{ m/s}$

Hence $\dfrac{v_2' \sin \beta}{v_2' \cos \beta} = \dfrac{0.25 \text{ m/s}}{0.56 \text{ m/s}}$

or $\tan \beta = 0.45$ and so $\boxed{\beta = 24°}$

(b) Then, since $v_2' \sin \beta = 0.25 \text{ m/s}$,

$$v_2' = \frac{0.25 \text{ m/s}}{\sin \beta} = \frac{0.25 \text{ m/s}}{0.41} = \boxed{0.61 \text{ m/s}}$$

Hence the velocity of the second puck is 0.61 m/s, at an angle of 24° below the X axis.

(This result should be checked by substituting the answer just found back in the conservation-of-momentum equations for the X and Y components.)

(c) The kinetic energy before the collision is

$$E_k = \tfrac{1}{2}m_1 v_1^2 = \tfrac{1}{2}(0.25 \text{ kg})(0.20 \text{ m/s})^2 = 0.50 \text{ J}$$

The kinetic energy after the collision is:

$$E_k' = \tfrac{1}{2}(0.25 \text{ kg})(1.0 \text{ m/s})^2$$
$$+ \tfrac{1}{2}(0.50 \text{ kg})(0.61 \text{ m/s})^2$$
$$= 0.13 \text{ J} + 0.09 \text{ J} = 0.22 \text{ J}$$

Hence E_k' is less than E_k and energy is not conserved; so this is an *inelastic collision*.

7.7 Recoil and Rockets

In this section we discuss some practical applications of the ideas about linear momentum developed in this chapter.

Recoil

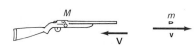

FIGURE 7.16 Recoil of a rifle when it fires a bullet.

Newton's third law, or the principle of conservation of linear momentum, can be used to explain the recoil of rifles, cannon, and even atomic nuclei which eject high-speed atomic particles.

Suppose a rifle of mass M fires a bullet of mass m, as in Fig. 7.16. Since the linear momentum of the system is zero before the firing, and the only force acting in the X direction is internal to the system, i.e., the explosion of the gunpowder, after the firing the total linear momentum of the system must also be zero. Hence we have

$$m\mathbf{v} + M\mathbf{V} = 0 \qquad \text{or} \qquad \mathbf{V} = \frac{-m}{M}\mathbf{v} \qquad (7.15)$$

where $\mathbf{v}$ is the velocity of the bullet and $\mathbf{V}$ is the velocity of the rifle. Hence the recoil velocity of the rifle is in the opposite direction to the motion of the bullet, and is reduced in size by the ratio of the two masses. For this reason, the heavier the rifle, the less the recoil is felt.

We can also obtain the ratio of the kinetic energies of rifle and bullet. We have

$$\frac{\tfrac{1}{2}mv^2}{\tfrac{1}{2}MV^2} = \frac{m}{M}\left(\frac{v}{V}\right)^2 = \frac{m}{M}\left(\frac{-M}{m}\right)^2 = \frac{M}{m} \qquad (7.16)$$

Hence the kinetic energy is inversely proportional to the masses, just as is true for the velocities from Eq. (7.15). The bullet therefore receives much more kinetic energy than the rifle, by a factor of M/m, which may be 500 or so.

Rockets

The principle of conservation of linear momentum may also be applied to the flight of rockets, as in Fig. 7.17. The operation of a rocket depends on the fact

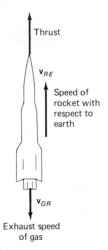

Thrust

v_{RE}

Speed of
rocket with
respect to
earth

v_{GR}

Exhaust speed
of gas

FIGURE 7.17 Rocket propulsion. The thrust of the rocket is $v_{GR} \, (\Delta m/\Delta t)$ and therefore depends on the exhaust speed of the gas with respect to the rocket (v_{GR}) and on the rate at which the gas is ejected from the rocket ($\Delta m/\Delta t$).

Thrust of a Rocket

that recoil occurs when the rocket ejects hot gases from its rear. In a rocket, fuel is burned and very hot gases are formed. The gas molecules exert an equal pressure in all directions on the inside walls of the rocket's combustion chamber. If an opening is made at one end of the chamber, the gas will stream out at a very high velocity. To conserve momentum a reaction force must be exerted on the opposite side of the chamber. It is this reaction force which propels the rocket forward, in the same way that a deflating balloon moves in the opposite direction from the emerging air. The thrust depends on the amount of gas flowing out per second and on the exhaust speed of the gas molecules.

The thrust which powers a rocket is not produced—as is sometimes erroneously supposed—by the ejected gases pushing against the air in the atmosphere, but is due to a recoil of the kind that occurs when a rifle fires a bullet. Rocket propulsion is therefore the only propulsive system which can function perfectly in a vacuum. Although aircraft jet engines use reaction forces to propel the plane, they require oxygen from the air to burn the jet-engine fuel and hence can operate only in the earth's atmosphere. The fuel in a rocket, on the other hand, is self-contained and needs no atmospheric oxygen for combustion.

Let us consider the rocket problem in a little more detail, since it ties together many important ideas developed in this chapter.

We are interested in the motion of the rocket with respect to the earth, and hence the earth will be our inertial reference frame. The exhaust gases, however, have a fixed velocity with respect to the rocket ($\mathbf{v}_{GR}$) because of the combustion process, while their velocity with respect to the earth is continually changing. Hence we can write for the velocity of the exhaust gases with respect to the earth:

$$\mathbf{v}_{GE} = \mathbf{v}_{GR} + \mathbf{v}_{RE} \tag{7.17}$$

where $\mathbf{v}_{GE}$ is the velocity of the exhaust gases with respect to the earth, $\mathbf{v}_{GR}$ is the velocity of the gases with respect to the rocket, and $\mathbf{v}_{RE}$ is the velocity of the rocket with respect to the earth. Since the motion is assumed to be in one dimension, i.e., vertically upward from the earth's surface, we can replace these velocities by their magnitudes, noting that $\mathbf{v}_{GR}$ is in the opposite direction from $\mathbf{v}_{RE}$. Hence we have

$$v_{GE} = -v_{GR} + v_{RE} \tag{7.18}$$

where the upward direction is now positive for all three velocities.

The net *external* force acting here is just the force of gravity acting downward, and so linear momentum is not conserved. Then, by the impulse-momentum theorem [Eq. (7.3)] applied to this force acting for a very brief time interval Δt, we have:

$$\mathbf{f}_{\text{net}} \, \Delta t = \Delta \mathbf{p}$$

or, again taking forces and velocities downward as negative,

$$-mg \, \Delta t = \Delta p \tag{7.19}$$

Note that the force provided by the burning of the fuel is *internal* to the system and hence cannot change the total momentum of the system.

Now, in the time Δt the total mass of the rocket, including its contents and its unburned fuel, decreases from its original value m at the beginning of the time interval to $m - \Delta m$ at the end of the time interval. During the same time Δt the speed of the rocket with respect to the ground increases from v_{RE} to $v_{RE} + \Delta v_{RE}$. Hence the change in the momentum of the *rocket* with respect to the earth is

$$(m - \Delta m)(v_{RE} + \Delta v_{RE}) - m v_{RE} = m \Delta v_{RE} - v_{RE} \Delta m - \Delta m \Delta v_{RE}$$

But the exhaust gases from the rocket also contribute to the momentum of the whole system. If an amount of gas Δm is ejected in time Δt, then the increase in the momentum of the system due to the exhaust gases is, from Eq. (7.18),

$$\Delta m \, v_{GE} = \Delta m \, (v_{RE} - v_{GR})$$

On putting these results together, Eq. (7.19) becomes:

$$-mg \, \Delta t = \Delta p = m \, \Delta v_{RE} - v_{RE} \, \Delta m - \Delta m \, \Delta v_{RE} + v_{RE} \, \Delta m - v_{GR} \, \Delta m$$

Since Δt is small, Δm is also small and so is Δv_{RE}. Hence the term $\Delta m \, \Delta v_{RE}$ is very small and so we neglect it. Then finally we have, on canceling $-v_{RE} \, \Delta m$ and $+v_{RE} \, \Delta m$:

$$-mg \, \Delta t = m \, \Delta v_{RE} - v_{GR} \, \Delta m$$

or

$$\frac{m \, \Delta v_{RE}}{\Delta t} = v_{GR} \frac{\Delta m}{\Delta t} - mg \qquad (7.20)$$

This result is simply Newton's second law of motion for the system. The left side is ma, where $a = \Delta v_{RE}/\Delta t$ is the acceleration of the rocket with respect to the earth. The right side must therefore be the net force acting on the rocket. This includes the force of gravity acting down and the upward *thrust* of the rocket due to its ejection of exhaust gases. The thrust of the rocket is therefore:

Thrust of a Rocket

$$\boxed{\text{Rocket thrust} = v_{GR} \frac{\Delta m}{\Delta t}} \qquad (7.21)$$

and depends on the exhaust speed of the gases with respect to the rocket (v_{GR}) and on the rate at which mass is ejected from the rocket, i.e., the rate at which fuel is consumed.

Since the acceleration is, from Eq. (7.20),

$$a = \frac{\Delta v_{RE}}{\Delta t} = \frac{\text{thrust}}{m} - g \qquad (7.22)$$

and the mass of the rocket plus fuel is constantly changing, the acceleration of the system depends both on g and on m, as well as on the rocket thrust.

For large thrusts v_{GR} must be large, as Eq. (7.21) shows. It is usually about 2 km/s for most chemical propellants. The time rate of change of mass $\Delta m/\Delta t$ must also be large. In most cases the fuel makes up 90 to 95 percent of the original mass of the rocket plus its contents, so that $\Delta m/m$ is about 0.95 for the whole acceleration process.

As the rocket rises, its acceleration relative to the earth increases for a number of reasons. First of all, the value of g in Eq. (7.20) decreases with altitude, while v_{GR} and $\Delta m / \Delta t$ remain constant. Hence in Eq. (7.22) the acceleration of the rocket relative to the earth increases. Secondly as time goes on, m decreases and so the acceleration a increases, according to Eq. (7.22). Finally, as the rocket picks up speed, more of the energy produced by fuel combustion goes into the velocity of the rocket with respect to the earth, and less energy goes into the velocity of the exhaust gases with respect to the earth. When $v_{GR} = v_{RE}$ in Eq. (7.18), the speed of the gases with respect to the earth (v_{GE}) is zero. When this happens the kinetic energy of the gases with respect to the earth is also zero, and so all the energy developed by the fuel goes into kinetic energy of the rocket relative to the earth.

When v_{RE} is greater than v_{GR}, v_{GE} is positive and so v_{RE}, which is equal to $v_{GE} + v_{GR}$, can be much larger than v_{GR}. Hence the rocket can attain speeds relative to the earth much greater than the speed of the exhaust gases relative to the rocket.

Since most of its fuel is consumed in powering the rocket through the earth's atmosphere, the energy which a rocket carries in its fuel supply can be utilized much more effectively if the rocket is initially given a boost by another rocket. This is the function of the booster rockets which often serve as the early stages of multistage rockets.

Example 7.9

A rocket of initial total mass 4.0×10^4 kg ejects hot gases from its combustion chamber at a rate of 1000 kg/s. The speed of the gas molecules ejected is 2.0 km/s relative to the rocket. The fuel, which makes up 90 percent of the total mass of the rocket before blast-off, is consumed in 36 s.
(a) Find the thrust of the rocket engine.

(b) Find the upward acceleration at time $t = 0$, $t = 20$ s, $t = 36$ s, and $t = 40$ s, assuming that we can neglect any change in g with altitude. (An application of the formula $g = Gm_E / r^2$ leads to the conclusion that g decreases by less than 3 percent for an increase in height of 100 km above the earth's surface.)

SOLUTION

(a) Rocket thrust $= v_{GR} \dfrac{\Delta m}{\Delta t} = (2.0 \text{ km/s})(1000 \text{ kg/s})$

$\qquad = 2.0 \times 10^6 \text{ kg·m/s}^2 = \boxed{2.0 \times 10^6 \text{ N}}$

(b) The acceleration is given by Eq. (7.22):

$$a = \frac{\text{thrust}}{m} - g$$

Now the mass of the rocket plus fuel decreases with time as the fuel is expended. If fuel is used up at the rate of 1000 kg/s, then the remaining mass of rocket plus fuel at the four times given is:

$m_0 = 4.0 \times 10^4 \text{ kg} \qquad m_{20} = 2.0 \times 10^4 \text{ kg}$

$m_{36} = 0.40 \times 10^4 \text{ kg} \qquad m_{40} = 0.40 \times 10^4 \text{ kg}$

The corresponding accelerations are:

$$a_0 = \frac{2.0 \times 10^6 \text{ N}}{4.0 \times 10^4 \text{ kg}} - 9.8 \text{ m/s}^2$$

$$= 50 \text{ m/s}^2 - 9.8 \text{ m/s}^2 = \boxed{40.2 \text{ m/s}^2}$$

$$a_{20} = \frac{2.0 \times 10^6 \text{ N}}{2.0 \times 10^4 \text{ kg}} - 9.8 \text{ m/s}^2 = \boxed{90.2 \text{ m/s}^2}$$

$$a_{36} = \frac{2.0 \times 10^6 \text{ N}}{0.4 \times 10^4 \text{ kg}} - 9.8 \text{ m/s}^2 = \boxed{490 \text{ m/s}^2}$$

Just after $t = 36$ s the fuel runs out. Hence at time $t = 40$ s there is no fuel left and hence no thrust. The acceleration of the rocket is just $\boxed{-9.8 \text{ m/s}^2}$, or the acceleration due to gravity. Notice how rapidly the acceleration increases as the mass of the fuel is reduced to zero.

Summary: Important Definitions and Equations

Linear momentum: $\mathbf{p} = m\mathbf{v}$
Units of momentum are kg·m/s or N·s.

$$\text{Newton's second law: } \mathbf{f}_{net} = \frac{\Delta \mathbf{p}}{\Delta t}$$

Impulse-momentum theorem: $\mathbf{f}_{net}\, \Delta t = \Delta \mathbf{p}$
where $\mathbf{f}_{net}\, \Delta t$ is the impulse of the force $\mathbf{f}_{net}$.

Principle of conservation of linear momentum: If no net unbalanced force acts on a physical system, the total linear momentum of the system remains constant.

Kinds of collisions:
Elastic: Both linear momentum and mechanical energy of the system are conserved.
Inelastic: Linear momentum is conserved, but mechanical energy is not.

Coefficient of restitution: $\mathbf{e} = \dfrac{-(v_2' - v_1')}{v_2 - v_1}$

In a completely elastic collision, $e = 1$.
In a completely inelastic collision, $e = 0$.

Collisions in two dimensions: Both the X and Y components of the linear momentum are conserved, if no net external force acts in the XY plane.

Rockets as examples of conservation of linear momentum:

$$\text{Rocket thrust} = v_{GR}\, \frac{\Delta m}{\Delta t}$$

$$\text{Rocket acceleration } a = \frac{\Delta v_{RE}}{\Delta t} = \frac{\text{thrust}}{m} - g$$

Questions

1 A ball is dropped from a Goodyear blimp. As it falls its linear momentum constantly increases, whereas the momentum of the blimp is unchanged. How can we talk about conservation of linear momentum in this case?

2 A meteor is traveling through the earth's atmosphere at a constant velocity $\mathbf{v}$. As it moves, it loses mass because of the intense heating caused by contact with the gases in the atmosphere. Is the momentum of the meteor changing? Why?

3 A billiard ball strikes the cushion of a billiard table at right angles to the cushion. Before it hits the cushion the ball's momentum is mv; after hitting the cushion the ball's momentum is approximately $-mv$. Hence the change in momentum is approximately $-2mv$. How can you say that linear momentum is conserved in the collision of the billiard ball with the cushion?

4 Two wooden pucks are sliding toward each other on a wooden table. The coefficient of friction between the pucks and the wooden surface of the table is 0.35.

(*a*) Is linear momentum conserved in this collision?

(*b*) Is linear momentum conserved before and after the collision?

(*c*) Discuss your answers to parts (*a*) and (*b*) so as to give a complete picture of what is happening from the viewpoint of physics.

5 Two billiard balls collide elastically on a billiard table.

(*a*) Draw a graph of linear momentum for the two balls as a function of time, from a time a few seconds before the collision to a few seconds after the collision.

(*b*) Draw a graph of kinetic energy of the two balls as a function of time for the same time interval indicated in part (*a*).

6 Two billiard balls collide inelastically. Draw the same graphs asked for in parts (*a*) and (*b*) of Question 5.

7 In Example 7.3 will the fragments of the explosion reach the ground at the same average time they would if there were no explosion?

8 Two hockey pucks of equal mass slide toward each other on a metal table. Each puck has the same speed v with respect to the table. A piece of soft wax is dropped between the two pucks just before they collide, and they stop dead and stick together.

(*a*) What is the final linear momentum of the system? Is linear momentum conserved in this collision?

(*b*) What is the final kinetic energy of the system? Is kinetic energy conserved in this collision?

9 A gold miner is stuck on a frozen lake in Nevada. The lake is so slippery that the coefficient of friction is zero and walking is impossible. The miner has a large bag of gold nuggets with him. Can you suggest how he might move himself to the edge of the lake?

10 A ball of wax is shot from a toy gun at a wall. Will the wax ball exert a larger impulse on the wall if it sticks to the wall or if it rebounds from it?

11 A baseball player is trapped in a boxcar on a deserted railroad side track and decides to move the boxcar to a more populated region by constantly throwing a baseball at full speed at the front wall of the boxcar and thus giving the car a forward velocity. Discuss the likelihood that such a technique will succeed. Will the boxcar move at all? If so, in what direction?

12 (*a*) Is it possible to have a system which has kinetic energy but no linear momentum?

(*b*) Is it possible to have a system which has linear momentum but no kinetic energy?

13 A railroad coal car is coasting along at constant speed on a frictionless track. It begins to rain and water accumulates in the bottom of the open car. Does the speed of the car change? Does it increase or decrease? Why?

14 An inventor designs a fan system for sailboats to provide a strong breeze at all times. When the powerful fan is mounted in the sailboat and directed at the sail, the inventor finds that the boat moves in the wrong (backward) direction. Is this what you would expect from the laws of physics?

Problems

A 1 A Volkswagen weighing 6000 N is traveling at 30 mi/h. A Mercedes weighing 10,000 N is traveling at a speed of 60 mi/h. The ratio of the momentum of the Volkswagen to that of the Mercedes is:
(*a*) 6/1   (*b*) 2/1   (*c*) 1/2   (*d*) 10/3
(*e*) 3/10

A 2 A sponge rubber ball of mass 0.25 kg is thrown with a speed of 30 m/s against a wall. It collides elastically with the wall and rebounds directly off the wall. The momentum of the ball after hitting the wall, with respect to its original direction, is:
(*a*) −7.5 kg·m/s   (*b*) +7.5 kg·m/s   (*c*) 225 kg·m/s
(*d*) −225 kg·m/s   (*e*) −30 kg·m/s

A 3 Two balls collide head-on in a perfectly elastic collision. The first ball has a mass of 2.0 kg and a speed of 10 m/s along the negative X axis. The second ball has a mass of 1.5 kg and a speed of 20 m/s along the positive X axis. The relative velocity of the two balls after the collision is:
(*a*) −30 m/s   (*b*) 30 m/s   (*c*) 40 m/s
(*d*) −10 m/s   (*e*) 10 m/s

A 4 A pitcher throws a baseball with a momentum of 20 kg·m/s and a batter hits the ball with a momentum of 25 kg·m/s. The impulse exerted by the hitter's bat on the ball is:
(*a*) 5 kg·m/s   (*b*) 5 N·s   (*c*) 45 N·s   (*d*) 45 J
(*e*) 45 N

A 5 A freight car of mass 4000 kg is moving with a speed of 10 m/s when it hits an identical freight car (at rest) and couples to it. The final speed of the two freight cars together is:
(*a*) 10 m/s   (*b*) 20 m/s   (*c*) 5 m/s
(*d*) 0   (*e*) None of the above

A 6 A metal ball is dropped from a height of 2.0 m on a solid steel plate. The ball rebounds to a height of 1.0 m. The coefficient of restitution for the metal out of which the ball is made is:
(*a*) 0.5   (*b*) 0.25   (*c*) 0.71   (*d*) 1.0
(*e*) None of the above

A 7 A rifle of mass 3.0 kg fires a bullet of mass 25 g at a speed of 500 m/s. The recoil momentum of the rifle is:
(*a*) 1.5 kg·m/s   (*b*) 4.2 m/s   (*c*) 4.2 kg·m/s
(*d*) 12.5 kg·m/s   (*e*) 60 kg·m/s

A 8 An 80-kg man and a 50-kg woman stand at the center of a frozen pond on ice skates. They extend their hands and push off against each other. The man acquires an initial speed of 0.25 m/s. The speed of the woman is:
(*a*) 0.25 m/s   (*b*) 0.16 m/s   (*c*) 0.40 m/s
(*d*) 0.33 m/s   (*e*) None of the above.

A 9 In Prob. A8 the ratio of the kinetic energy of the woman to that of the man is:
(*a*) 8/5   (*b*) 5.8   (*c*) 64/25   (*d*) 25/64
(*e*) (8/5)^{1/2}

A10 A rocket ejects exhaust gases at the rate of 500 kg/s at a speed with respect to the rocket of 1.7 km/s. At an instant when the rocket mass is 5.0×10^4 kg, the acceleration of the rocket is:
(*a*) −9.8 m/s²   (*b*) 9.8 m/s²   (*c*) −7.2 m/s²
(*d*) 7.2 m/s²   (*e*) 17 m/s²

B 1 A pitcher throws a baseball of mass 0.50 kg with a speed of 40 m/s. What is the linear momentum of the baseball?

B 2 A billiard ball of mass 0.60 kg and speed 0.80 m/s collides with an identical billiard ball which is at rest. The collision is perfectly elastic.
(*a*) What is the kinetic energy of the second ball after the collision?
(*b*) What is its linear momentum?

B 3 What average force is needed to stop a hammer with a momentum of 25 kg·m/s in 0.050 s?

B 4 Two identical freight cars, each of mass 4000 kg and moving with speeds of 10 m/s in opposite directions, collide. How much kinetic energy is lost in the collision?

B 5 A particle of mass 0.50 kg has a velocity of 4.0 m/s at an angle of 40° above the positive X axis. What are the X and Y components of the particle's momentum?

B 6 Two ice pucks initially at rest are driven apart by an explosion with velocities of 10 and 3.3 cm/s. What is the ratio of the masses of the two pucks?

B 7 An explosion blows a rock into three parts. Two pieces go off at right angles to each other, a 1.0-kg piece at 20 m/s and a 2.0-kg piece at 30 m/s. The third piece flies off at 40 m/s.
(*a*) Draw a vector diagram showing the direction in which the third piece of rock goes.
(*b*) What is the mass of this third piece?

B 8 A rocket ejects gas at the rate of 500 kg/s at a speed with respect to the rocket of 1.8 km/s. What is the thrust of the rocket?

B 9 A rocket is moving with a speed of 2000 m/s with respect to the earth (vertically upward). If the rocket is ejecting gas at a speed of 1.7 km/s relative to the rocket, what is the speed and direction of the exhaust gases with respect to the earth?

C 1 A 100-g marble is shot along the floor and strikes head-on a 75-g marble which is at rest. The speed of the first marble is reduced from 200 to 100 cm/s in the collision. What is the speed of the second marble after the collision?

C 2 A hockey puck of mass 0.30 kg is moving in the positive X direction with a speed of 0.20 m/s. A second, 0.15-kg puck, with a piece of wax attached to its front edge, is moving with a speed of 0.50 m/s along the negative X axis. The two pucks collide and stick together. What is the velocity of the two pucks after the collision?

C 3 A softball of mass 0.40 kg is pitched at a speed of 15 m/s. The batter hits it back directly at the pitcher at a speed of 20 m/s. The bat acts on the ball for only 0.015 s.

(a) What is the impulse imparted by the bat to the ball?

(b) What is the average force exerted by the bat on the ball?

C 4 A freight car of mass 6.4×10^4 kg is rolling slowly along a level track at a speed of 0.25 m/s. A rope is attached to the car and trails along the ground. A bystander sees the rope, runs after the freight car, picks up the rope, and tries to stop the freight car by pulling on the rope.

(a) If the maximum force the person can exert is 400 N, how long will it take to bring the freight car to rest?

(b) If an automobile is on the track 12 m from the freight car when the person begins to pull on the rope, will the freight car hit the automobile?

C 5 A ballet dancer of mass 50 kg ends a ballet by making a horizontal leap at a speed of 4.0 m/s into her partner's arms. If the male dancer is able to exert a maximum force of 300 N in catching his partner, how long will it take him to reduce her momentum to zero?

C 6 A 10-g bullet is fired horizontally into a 10-kg block of wood which is initially at rest and which has a speed of 50 cm/s after the impact. Find the initial speed of the bullet.

C 7 In a ballistic pendulum experiment (compare Example 7.6) a 12-g bullet is fired horizontally into a 5.0-kg block of wood suspended by a long cord. The bullet sticks in the wood, and the two rise together until the center of gravity of the block has been raised 15 cm from its original position. Find the initial speed of the bullet.

C 8 An astronaut whose mass is 70 kg becomes separated from her earth satellite which is moving in orbit around the earth. She finds herself 25 m away from the spaceship and moving at exactly the same speed as the ship. She has a 0.60-kg camera in her hand and decides to get back to the satellite by throwing the camera at a speed of 10 m/s in a direction away from the satellite.

(a) Will this maneuver get her back to the satellite? Why?

(b) How long will it take for the astronaut to reach the satellite?

(c) How fast will she be moving when she makes contact with the satellite?

C 9 To remove a large 10-kg rock from a building site an explosive charge is placed inside the rock. When the charge is detonated, the rock breaks up into three pieces. Two have masses of 2.0 kg each and are shot out at right angles to each other at speeds of 10 m/s.

(a) Draw a vector diagram of the momenta in this physical situation, and solve for the magnitude and direction of the velocity of the third piece of rock.

(b) Solve for the same quantity mathematically instead of graphically.

C10 Two identical balls collide head-on. One has an initial velocity of 1.0 m/s along the X axis. The other has a speed of -0.50 m/s along the X axis.

(a) Find the velocity of each of the balls after impact if the collision is elastic.

(b) Find the velocity of each of the balls if the two balls stick together.

C11 Two balls, one of mass 1.5 kg moving to the right along the X axis with a speed of 25 cm/s, and another of mass 3.0 kg moving to the left with a speed of 30 cm/s, collide head-on. If the collision is elastic, what are the final velocities of the two balls?

C12 In the first nuclear reactor constructed during World War II at Stagg Field in Chicago, graphite blocks were used as the moderator to slow down the neutrons. Graphite is made entirely of carbon atoms ($^{12}_{6}$C) with masses 12 times the neutron mass.

(a) In a head-on, elastic collision of a neutron with a carbon atom at rest, how much of the kinetic energy of the neutron is lost?

(b) How many collisions would be needed to reduce the kinetic energy of the neutron to 1 percent of its original value?

C13 Two cars, one of mass $m_1 = 1000$ kg, the other of mass $m_2 = 2000$ kg, are moving at right angles to each other when they collide and stick together. The initial velocity of m_1 was 10 m/s along the positive X axis, and that of m_2 was 8.0 m/s along the positive Y axis. What is the velocity of the wreckage of the two cars immediately after the collision?

C14 A small car of mass 2000 kg traveling at 20 m/s crashes into a truck of mass 4000 kg traveling at 25 m/s along a street at right angles to the car's path. The two vehicles lock together and the wreckage moves a distance of 15.2 m before coming to rest.

(a) Find the direction in which the wreckage moves after the collision.

(b) Find the speed of the wreckage just after the collision.

(c) Find the force of friction which brings the cars to rest.

C15 A polonium nucleus ($^{210}_{84}$Po) of mass 3.5×10^{-25} kg emits an alpha particle of mass 6.7×10^{-27} kg and speed 1.3×10^7 m/s. Find the speed with which the polonium nucleus recoils.

C16 A cannon is mounted in such a way that a constant spring force resists the recoil motion of the cannon when it fires. The cannon has a mass of 500 kg and fires a 1.0-kg shell with a muzzle velocity of 600 m/s.

(a) Find the recoil velocity of the cannon.

(b) If the resisting spring force is 4000 N, find the time it takes to stop the cannon barrel's motion.

(c) How far does the barrel move in that time?

C17 A gun mounted on a railway car can only recoil in the horizontal direction.

(a) If the gun has a mass of 50,000 kg and fires a 250-kg artillery shell at an angle of 45° and a speed of 200 m/s, what is the recoil velocity of the gun in the horizontal direction?

(b) What happens to the vertical component of the recoil momentum?

C18 An ice boat of mass 50 kg is moving along a frozen lake without friction at a constant speed of 30 m/s. A similar boat catches up with the first one and a person throws a 5.0-kg hamper of provisions into the first boat. What is the change in the first boat's velocity?

C19 Coal drops from the bottom of a coal hopper at a rate of 7.5 kg/s onto a conveyer belt moving horizontally at a speed of 2.0 m/s. What force is needed to drive the conveyer belt? (Neglect friction.)

C20 (*a*) What minimum thrust must the engines of a rocket of initial total mass 4.0×10^4 kg have to be able to lift it from the earth if the rocket is aimed straight upward?

(*b*) If the engines eject fuel gases at the rate of 100 kg/s, how fast must the gases be moving when they leave the engines?

(*c*) What is the acceleration of the rocket after 2.0 min?

C21 A rocket at rest on a launch pad is pointing directly upward. Its jet engines can eject gas molecules at a rate of 900 kg/s at a speed of 2.0 km/s. What is the greatest mass the rocket can have if it is to leave the ground?

D 1 (*a*) A head-on elastic collision occurs between two particles, one of which is initially at rest and the second of which is in motion directly toward the first particle. Find the ratio of the two masses which will lead to the transfer of maximum kinetic energy to the particle initially at rest.

(*b*) Show that if the masses of the two particles do not meet the requirement arrived at in part (*a*), then better energy transfer can be obtained by inserting a third particle between the other two, with the third particle having a mass which is the geometric mean of the masses of the other two particles; i.e., $m_3 = \sqrt{m_1 m_2}$.

D 2 Show that if a particle of mass m, traveling with a velocity v, makes an elastic collision with a second particle of the same mass m which is initially at rest, then the angle between the two velocity vectors after the collision must be 90°. (Assume that the collision is not head-on.)

D 3 An old-fashioned 500-kg cannon, mounted on wheels, fires a 5.0-kg shell horizontally at 1000 m/s.

(*a*) If the wheels turn freely, what will be the recoil speed of the gun?

(*b*) If the gun is firmly anchored in the ground, what will be the recoil speed of gun and earth together?

(*c*) If the projectile was accelerated uniformly along the 2.0-m length of the cannon, what was the average force on the projectile inside the barrel?

(*d*) What was the average force exerted back on the cannon by the expanding combustion gases in the barrel?

(*e*) How far does the gun travel in the time it takes the shell to travel the length of the gun barrel?

D 4 A projectile is fired from a gun at an angle of 45° with respect to the ground. Under normal circumstances the projectile would reach a maximum height H and its range would be R. In this case, however, when the projectile is at its maximum height H, it explodes into two exactly equal fragments. If one fragment is observed to fall vertically downward toward the ground, as if it were released from rest at height H, prove that the other fragment will strike the ground at a distance $\frac{3}{2}R$ from the gun. (*Hint*: Consider the motion of the center of mass of the projectile.)

D 5 Firefighters are spraying water on a building from the outside. A horizontal stream of water from a hose strikes a window on the outside wall and then rolls down the side of the window. If 0.10 kg of water strikes the window each second, with a speed of 2.5 m/s, find (*a*) the constant force exerted on the window by the water; (*b*) the impulse of this force over a 0.10-s interval.

D 6 A device similar to that in Fig. 7.18 is often seen in toy and novelty shops. All five balls have the same mass. When one of the balls is pulled back and released, the ball on the opposite end flies out with the same speed that the original ball had just before it collided.

(*a*) Prove that this is to be expected if the collision is elastic.

(*b*) Predict what will happen if two balls are pulled out instead of one. Justify your predictions on the basis of physical principles.

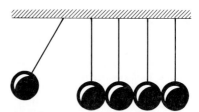

FIGURE 7.18 Diagram for Prob. D6.

D 7 A bag of jelly beans is emptied at a constant rate into the pan of a spring scale. Each jelly bean has a mass of 1.0 g and is dropped from a height of 0.70 m onto the pan. Ten jelly beans hit the scale pan each second. What is the scale reading at the end of 5.0 s, if the jelly beans all stick to the scale pan without bouncing?

D 8 A 5000-kg rocket traveling at a speed of 200 m/s is moving freely through space on a journey to the moon. The ground controllers find that the rocket has drifted off course, and that it must change direction by 12° if it is to hit the moon. By radio control the rocket's engines are fired for a brief instant in a direction perpendicular to the direction of motion of the rocket. If the rocket gases are expelled at a speed of 2000 m/s, what mass of gas must be expelled to make the needed correction in the rocket's direction?

D 9 Cloud chambers are often used to provide photographs of nuclear reactions. A proton is seen in such a photograph to have undergone an elastic collision with another particle, in which the proton's velocity was deviated by an angle of 60° from its original direction. The struck particle, which is presumed to have been at rest, goes off at an angle of 30° to the other side of the proton's original direction. What was the mass of the struck particle in terms of the proton mass? (*Hint*: Apply principles of conservation of momentum and energy.)

D10 Tarzan holds on to a vine of negligible weight and swings down from a tree limb 8.0 m above the ground. At the lowest point of his swing he picks up his mate, Jane, who is standing on the ground, and then continues his swing. He and Jane land on another tree limb located at the maximum height reached on their swing. If Tarzan has a mass of 100 kg, and Jane 70 kg, how high above the ground is the limb on which they land?

Additional Readings

Asimov, Isaac, *Asimov's Biographical Encyclopedia of Science and Technology*, 2d rev. ed., Doubleday, Garden City, N.Y., 1982. Contains a good brief account of Huygens' accomplishments.

Bos, H. J. M.: "Christian Huygens," in C. C. Gillispie (ed.), *Dictionary of Scientific Biography*, Scribner, New York, 1970–1980, vol. 6. The best available treatment of Huygens' life and work in English. The volumes of this dictionary are a gold mine of information for anyone interested in the history of science.

Feynman, Richard: *The Character of Physical Law*, MIT Press, Cambridge, Mass., 1965. Chapter 3, "The Great Conservation Principles," is particularly relevant.

Haber-Schaim, Uri, John H. Dodge, and James A. Walter: *PSSC Physics*, 5th ed., Heath, Lexington, Mass., 1981. Chapter 6 of this text includes a careful study of momentum conservation accompanied by some excellent photographs of collision processes.

Lewis, H. W.: "Ballistocardiography," *Scientific American*, vol. 198, no. 2, February 1958, pp. 89–95. An interesting attempt to study conservation of linear momentum in the pumping of blood by the heart.

*F*rom time to time in the kingdom of mankind a man arises who is of universal significance, whose work changes the current of human thought or of human experience, so that all that comes after him bears evidence of his spirit. Such a man was Shakespeare, such a man was Beethoven, such a man was Newton, and, of the three, his kingdom is the most widespread.*

E. N. da C. Andrade

Chapter 8

Rotational Motion of Rigid Bodies

The spoked wheel was first introduced about 2700 B.C. during the Bronze Age. Since that time the wheel has played a fundamental role in agriculture, industry, transportation, and recreation. For this reason the rotational motion of rigid bodies like wheels is of great importance in modern science and technology.

When an automobile wheel turns, its motion consists of a rotation about the axle of the wheel plus the translational motion of the axle along with the car (see Fig. 8.1). It can be shown likewise that the most general motion of a moving body consists of a translational motion of its center of mass with respect to some chosen frame of reference, plus the rotation of the body about its center of mass. The motion of a frisbee is a good example: the frisbee is constantly rotating about its center as it glides through the air, as in Fig. 8.2. Hence its motion is a combination of translation and rotation. Other good examples are tumbling footballs and pirouetting ballet dancers.

We have already discussed translational motion in detail in previous chapters. We have also discussed some basic ideas about rotational motion in Chap. 5. In this chapter we will amplify these ideas and apply them in particular to the motion of rigid bodies.

*See Additional Readings for source of quote.

FIGURE 8.1 The motion of a bicycle wheel: The wheel rotates about its axle with an angular speed ω while at the same time the axle executes translational motion in the direction in which the bicycle is moving at a speed v.

FIGURE 8.2 As shown here, the motion of a frisbee is a combination of translational motion and rotational motion.

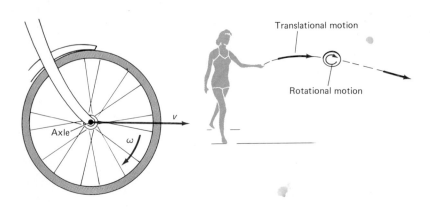

Translational motion

Rotational motion

Axle

v

ω

8.1 Parallelism Between Linear Motion and Angular Motion

In what follows we will show how the basic equations for rotational motion can be obtained by analogy from the equations we have already derived for translational motion. Even though rotational motion is in many respects analogous to translational motion, there is one very important difference. For translational motion we measure speeds in meters per second or miles per hour and assume that all portions of a solid object move with exactly this same translational speed. The taillight on a car, for example, moves at exactly the same speed as the front bumper.

For rotational motion the situation is quite different. In a given time a point on the rim of a wheel covers a greater linear distance than does a point halfway out from the center. Despite this fact the wheel rotates all in one piece, and we would therefore like to be able to describe the speed of this rotation in some convenient way. As we saw in Chap. 5, the most useful description is in terms of the number of *radians* swept out by the wheel per second. Since this is an angular measure, rotational velocities are often referred to as *angular velocities*. All points on a rotating wheel have the same angular velocity, as is clear from Fig. 8.3.

If, in the figure, point A sweeps through an arc twice as large as does point B in time Δt, it does this because its radius is twice as large. Hence the angular displacement is

$$\Delta \theta = \frac{B'B}{CB} = \frac{A'A/2}{CA/2} = \frac{A'A}{CA}$$

The angular velocity and $\Delta\theta/\Delta t$ is the same for both points A and B, and, more generally, for all points on the wheel.

The linear velocity of points A and B is, of course, quite different, for point A covers twice the distance that point B does in the same time. The relationship between the angular velocity and the linear velocity v is therefore $v = R\omega$, where R is the distance from the axis of rotation to the point where v is being calculated, as discussed previously in Sec. 5.4.

We will now summarize from previous chapters some important relations which tie together the linear and angular quantities needed in discussing rotational motion. We will write down these relationships for instantaneous values of these quantities, although in further discussions we will deal mainly with situations where either the angular velocity or angular acceleration is constant.

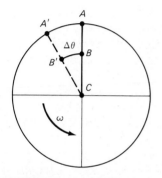

FIGURE 8.3   A rotating wheel; all points on it move with the same *angular velocity* ω but with different *linear* velocities.

Quantity	Linear motion (translation)		Angular motion (rotation)		Connection	
Displacement	Δs		$\Delta \theta$		$\Delta s = R\,\Delta\theta$	(5.19)
Velocity	$v = \lim\limits_{\Delta t \to 0} \dfrac{\Delta s}{\Delta t}$	(2.3)	$\omega = \lim\limits_{\Delta t \to 0} \dfrac{\Delta \theta}{\Delta t}$	(5.20)	$v = R\omega$	(5.21)
Acceleration	$a = \lim\limits_{\Delta t \to 0} \dfrac{\Delta v}{\Delta t}$	(2.7)	$\alpha = \lim\limits_{\Delta t \to 0} \dfrac{\Delta \omega}{\Delta t}$	(5.24)	$a = R\alpha$	(5.25)

Here the linear quantities are the linear displacement, linear velocity, and linear acceleration of a point on a rotating body a distance R from the axis of rotation, and $\Delta\theta$, ω, and α are the corresponding angular quantities for rotation about the fixed axis.

Example 8.1

A flywheel of radius 1.6 m starts from rest and rotates with constant angular acceleration until at the end of 15 s it is moving with an angular velocity of 20 rad/s.
(a) What is the angular acceleration of the wheel?
(b) What is the average angular velocity of the wheel during the 15-s interval?

(c) What is the linear velocity of a point on the rim of the flywheel at the end of 15 s?
(d) What is the linear acceleration of a point on the rim of the flywheel?
(e) What is the linear velocity of a point on the flywheel 0.80 m from its center?

SOLUTION

(a) Since the angular acceleration is constant, it is:

$$\alpha = \frac{\Delta\omega}{\Delta t} = \frac{\omega - \omega_0}{\Delta t} = \frac{20 \text{ rad/s} - 0}{15 \text{ s}} = \boxed{1.3 \text{ rad/s}^2}$$

(b) Average angular velocity:

$$\overline{\omega} = \frac{\omega_0 + \omega}{2} = \frac{0 + 20 \text{ rad/s}}{2} = \boxed{10 \text{ rad/s}}$$

(c) From Eq. (5.21),

$$v = R\omega = (1.6 \text{ m})(20 \text{ rad/s}) = \boxed{32 \text{ m/s}}$$

Again note that the radian is not a real unit, but a dimensionless ratio, and hence does not appear in the answer for v.
(d) From Eq. (5.25),

$$a = R\alpha = (1.6 \text{ m})(1.3 \text{ rad/s}^2) = \boxed{2.1 \text{ m/s}^2}$$

(e) Here

$$v = R\omega = (0.80 \text{ m})(20 \text{ rad/s}) = \boxed{16 \text{ m/s}}$$

8.2 Angular Motion with Constant Velocity

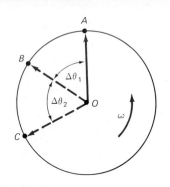

FIGURE 8.4 Rotation with constant angular velocity. Equal angles are swept out in equal times, so that $\Delta\theta_1/\Delta t = \Delta\theta_2/\Delta t = \omega = $ constant.

If a wheel is rotating about an axis with a constant angular velocity ω, then $\omega = \Delta\theta/\Delta t = $ constant, and $\Delta\theta = \omega\,\Delta t$. In this case equal angles are swept out in equal times, in the same way that a car moving at a constant speed covers equal distances in equal intervals of time. For example, if, in Fig. 8.4, it takes the same time Δt for a point on the rim of the wheel to go from A to B as it takes to go from B to C, then $AB = BC$ and so $\Delta\theta_1 = \Delta\theta_2$. Hence $\Delta\theta_1/\Delta t = \Delta\theta_2/\Delta t = \omega = $ constant. Thus, if a flywheel is rotating at a constant angular velocity of 10 rad/s, in 1 min the angular distance covered will be

$$\Delta\theta = \left(10\frac{\text{rad}}{\text{s}}\right)(1 \text{ min})\left(60\frac{\text{s}}{\text{min}}\right)$$

$$= 600 \text{ rad}$$

Since there are 2π rad in one complete revolution, the number of revolutions made by the wheel in the 1 min will be:

$$N = \frac{600 \text{ rad}}{2\pi \text{ rad/rev}} = 96 \text{ rev}$$

Example 8.2

A gear wheel on a machine is being rotated by a metal chain. The wheel makes 3 rotations each second.
(a) Through what angle does the wheel turn in 1 min?

(b) If the wheel has a radius of 10 cm, how far does a point on the rim of the wheel move each second?
(c) Through what distance will the chain move in 1 min?

SOLUTION

(a) The angular velocity of the wheel is:

$\omega = 3$ rotations/s $= 3(2\pi)$ rad/s $= 6\pi$ rad/s

Hence the angle swept out in 1 min at this angular velocity is

$\theta = \omega t = \left(\dfrac{6\pi \text{ rad}}{\text{s}}\right)(60 \text{ s}) = \boxed{360\pi \text{ rad}}$

(b) When the wheel rotates through an angle θ, a point on the rim moves through a distance

$s = R\theta = R\omega t = (0.10 \text{ m})\left(\dfrac{6\pi}{\text{s}}\right)(1 \text{ s})$

$= 0.60\pi \text{ m} = \boxed{1.9 \text{ m}}$

Another way to obtain the same result is to note that a point on the rim goes around the circumference of the wheel 3 times per second. The circumference has a length $2\pi(0.10$ m) $= 0.63$ m. Hence $s = 3(0.63$ m) $= 1.9$ m, as before.

(c) The chain must move along with a point on the rim of the wheel. Hence in 1 min the chain moves a distance:

$s = vt = R\omega t$

$= (0.10 \text{ m})\left(\dfrac{6\pi}{\text{s}}\right)(60 \text{ s}) = \boxed{113 \text{ m}}$

Example 8.3

Suppose a bicycle wheel of radius 0.33 m rotates at a constant angular velocity of 4 rev/s. **(a)** How fast does the bicycle move forward? What is the linear velocity with respect to the ground of a point on the rim that is **(b)** at the top of the wheel, **(c)** at the bottom of the rim where it makes contact with the ground?

SOLUTION

(a) We know that 4 rev/s corresponds to $4(2\pi) = 8\pi$ rad/s, since there are 2π rad in a complete circle. Hence

$\omega = \dfrac{\Delta\theta}{\Delta t} = \dfrac{8\pi \text{ rad}}{1 \text{ s}} = 25$ rad/s

The linear velocity of a point on the circumference of the wheel relative to the hub of the wheel (or to the bicycle, since the two move along together) is, then,

$v = R\omega = (0.33 \text{ m})(25 \text{ rad/s}) = 8.3$ m/s

Now, as a length Δs of the wheel's rim peels off along the ground, the wheel moves forward a distance Δs with respect to the ground, as shown in Fig. 8.5, as long as friction keeps the wheel from slipping. Hence the velocity of the bicycle with respect to the ground has a magnitude

$V = \dfrac{\Delta s}{\Delta t} = R\omega = \boxed{8.3 \text{ m/s}}$

(b) The linear velocity of the rim of the wheel *with respect to the ground* differs with position around the rim. Thus if the bike is moving along at a velocity of 8.3 m/s, the linear velocity of the point on the top of the rim *with respect to the ground* is

$v_1 = v + V$

$= 8.3$ m/s $+ 8.3$ m/s $= \boxed{16.6 \text{ m/s}}$

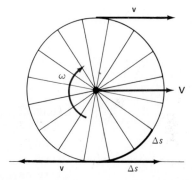

FIGURE 8.5 A bicycle wheel rotating with constant angular velocity ω. As the rim of the wheel moves through an arc Δs, the wheel moves forward a distance Δs on the ground. Hence the forward speed of the bike is equal to the linear speed of any point on the rim of the wheel with respect to the hub of the wheel.

where v is the velocity of the rim with respect to the bicycle and V is the velocity of the bicycle with respect to the ground.

(c) For the point at which the rim makes contact with the ground, on the other hand, its velocity *relative to the ground* is

$v_2 = V - v = \boxed{0}$ why!!

This is illustrated in Fig. 8.5.

8.3 Angular Motion with Constant Angular Acceleration

Let us now consider a slightly more difficult case where the angular velocity is constantly changing with time, but in such a way that the angular acceleration α remains constant. The angular motion in this case is completely equivalent to the linear motion of a ball falling under the force of gravity. All we need do is replace each linear quantity by the corresponding angular quantity, and we can obtain the equations governing uniformly accelerated *angular* motion from those already derived for uniformly accelerated *linear* motion. The equations we desire are the following, *where it is understood that all angular quantities must be expressed in radian measure*:

Equations for Angular Motion with Constant Acceleration

Linear motion with constant a:		Angular motion with constant α:	
$s = s_0 + \bar{v}t$	(2.4)	$\theta = \theta_0 + \bar{\omega}t$	(8.1)
$\bar{v} = \dfrac{v_0 + v}{2}$	(2.8)	$\bar{\omega} = \dfrac{\omega_0 + \omega}{2}$	(8.2)
$v = v_0 + at$	(2.10)	$\omega = \omega_0 + \alpha t$	(8.3)
$s = s_0 + v_0 t + \frac{1}{2}at^2$	(2.11)	$\theta = \theta_0 + \omega_0 t + \frac{1}{2}\alpha t^2$	(8.4)
$v^2 = v_0^2 + 2a(s - s_0)$	(2.12)	$\omega^2 = \omega_0^2 + 2\alpha(\theta - \theta_0)$	(8.5)

In the rotational case, since the angular acceleration is constant, the angular velocity changes from an initial value ω_0 (which may be zero) to a final value ω during the time t. Hence the average value of ω during the time interval t is $\bar{\omega} = (\omega_0 + \omega)/2$, as in Eq. (8.2) above. The total angular displacement in time t is then $\bar{\omega}t$, and if θ_0 is the initial value of θ, then at time t we have $\theta = \theta_0 + \bar{\omega}t$, which is Eq. (8.1).

The same reasoning can be used to obtain the other equations for angular motion above. But it is easier to note that, if we apply the equations for linear motion to a point on a rotating wheel of radius R so that s, v, and a are the linear displacement, velocity, and acceleration, respectively, of that point as it moves in a circle, then from Eqs. (5.19), (5.21), and (5.25), $s = R\theta$, $v = R\omega$, and $a = R\alpha$, where R is the distance of the point from the axis of rotation of the wheel. Hence, if we divide Eqs. (2.10) and (2.11) by R, we obtain

$$\frac{v}{R} = \frac{v_0}{R} + \left(\frac{a}{R}\right)t \qquad \text{or} \qquad \omega = \omega_0 + \alpha t$$

which is Eq. (8.3). Similarly, if we divide Eq. (2.11) by R, we obtain:

$$\frac{s}{R} = \frac{s_0}{R} + \left(\frac{v_0}{R}\right)t + \frac{1}{2}\left(\frac{a}{R}\right)t^2$$

or $\qquad \theta = \theta_0 + \omega_0 t + \frac{1}{2}\alpha t^2$

which is Eq. (8.4). Also, we have, on dividing Eq. (2.12) by R^2,

$$\frac{v^2}{R^2} = \frac{v_0^2}{R^2} + \frac{2a}{R^2}(s - s_0)$$

or $$\left(\frac{v}{R}\right)^2 = \left(\frac{v_0}{R}\right)^2 + 2\frac{a}{R}\left(\frac{s}{R} - \frac{s_0}{R}\right)$$

and so $$\omega^2 = \omega_0^2 + 2\alpha(\theta - \theta_0)$$

which is Eq. (8.5).

Hence solving problems for rotation with constant angular acceleration involves the use of Eqs. (8.1) to (8.5) in a manner very similar to the way we solved problems for linear motion in Chap. 2 by using Eqs. (2.4) to (2.12). These equations will allow us to solve any problem involving rotational motion with constant acceleration. Figure 8.6 gives an illustration of such motion for a wheel. As time goes on, the angular velocity of the spoke of the wheel increases and the spoke sweeps out greater angles in equal times.

FIGURE 8.6 Motion with constant angular acceleration α. Here $\omega = \omega_0 + \alpha t$. If $\omega_0 = 0$, $\omega = \alpha t$. Hence after 1 s, $\omega_1 = 1\alpha$; after 2 s, $\omega_2 = 2\alpha = 2\omega_1$; similarly, $\omega_3 = 3\omega_1$; $\omega_4 = 4\omega_1$, etc. The angles swept out in equal time intervals also increase, as is clear from the figure and the fact that $\theta = \theta_0 + \bar{\omega}t$ from Eq. (8.1).

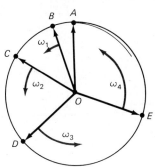

Example 8.4

The rotor of a medical centrifuge is accelerated from rest with a constant angular acceleration of 7.0 rad/s².
(a) What is its angular velocity after 5 min in radians per second? In revolutions per minute?

(b) How many revolutions has the centrifuge made in 5 min?

SOLUTION

(a) From Eq. (8.3), $\omega = \omega_0 + \alpha t$, or

$$\omega = 0 + \left(7.0\frac{rad}{s^2}\right)(5\ min)\left(60\frac{s}{min}\right) = \boxed{2.1 \times 10^3\ rad/s}$$

Then, since 1 rev = 2π rad,

$$\omega = \left(2.1 \times 10^3\frac{rad}{s}\right)\left(\frac{1\ rev}{2\pi\ rad}\right)\left(\frac{60\ s}{min}\right) = \boxed{2.0 \times 10^4\ rev/min}$$

(b) From Eq. (8.5), we have $\omega^2 = \omega_0^2 + 2\alpha(\theta - \theta_0)$. Here $\omega_0 = 0$ and $\theta_0 = 0$. Hence $\omega^2 = 2\alpha\theta$, or

$$\theta = \frac{\omega^2}{2\alpha} = \frac{(2.1 \times 10^3\ rad/s)^2}{2(7.0\ rad/s^2)} = 3.2 \times 10^5\ rad$$

$$= (3.2 \times 10^5\ rad)\left(\frac{1\ rev}{2\pi\ rad}\right) = \boxed{5.0 \times 10^4\ rev}$$

or about 50,000 rev.

The same result can be obtained by noting that the average angular velocity is, from Eq. (8.2),

$$\bar{\omega} = \frac{\omega_0 + \omega}{2} = \frac{2.0 \times 10^4\ rev/min}{2} = 1.0 \times 10^4\ rev/min$$

In 5 min the number of revolutions is then:

$$(5\ min)\left(1.0 \times 10^4\frac{rev}{min}\right) = 5.0 \times 10^4\ rev$$

as before.

8.4 Work and Rotational Kinetic Energy

Suppose we consider a wheel set into rotation about a fixed axis by pulling down on a rope wrapped around its circumference, as in Fig. 8.7. Clearly if the person pulling the rope exerts a force of magnitude F and the rope moves a distance s in the direction of the force, then the work the person does is

$$\mathcal{W} = Fs$$

But the person, by pulling on the rope, is exerting a torque setting the wheel into rotation. This torque is, as we have seen in Chap. 5, $\tau = FR$, where we assume that the force is in a direction tangent to the rim of the wheel and therefore perpendicular to R. Hence

$$\mathcal{W} = Fs = \frac{\tau s}{R} = \tau\theta$$

since $s/R = \theta$ from Eq. (5.19), and so

$$\mathcal{W} = \tau\theta \tag{8.6}$$

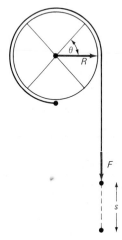

FIGURE 8.7 Torque exerted by a person pulling a rope. The person does an amount of work given by $\mathcal{W} = \tau\theta$, where the torque τ is equal to FR. The wheel rotates through an angle θ as the rope is pulled down a distance s.

Work Done in Rotating a Wheel

The work done on a wheel in setting it into rotation is the product of the torque acting and the angular displacement of the wheel.

Here the linear displacement s in the usual expression for work is replaced by the angular displacement θ, and the force F by the torque τ.

Similarly, the rate at which work is done, or the *power*, is

$$P = \frac{\Delta\mathcal{W}}{\Delta t} = \frac{\tau\Delta\theta}{\Delta t} = \tau\omega \tag{8.7}$$

Note the parallelism between this and $P = Fv$ obtained in Sec. 6.9 for linear motion.

Example 8.5

Compute the torque developed by an airplane motor which rotates at 2000 rev/min and whose power output is 3000 hp.

SOLUTION

From Eq. (8.7) for the power developed in rotational motion, we have

$$P = \tau\omega .$$

Here

$$P = 3000 \text{ hp} = (3000 \text{ hp})\left(\frac{746 \text{ W}}{1 \text{ hp}}\right) = 2.2 \times 10^6 \text{ W}$$

$$\omega = 2000 \text{ rev/min} = \frac{2000(2\pi \text{ rad})}{60 \text{ s}} = 2.1 \times 10^2 \text{ rad/s}$$

and so the torque

$$\tau = \frac{P}{\omega} = \frac{2.2 \times 10^6 \text{ N·m/s}}{2.1 \times 10^2 \text{ rad/s}} = \boxed{1.0 \times 10^4 \text{ N·m}}$$

Notice that the correct units for torque are newton-meters and not joules, and that in this case these units are *not* equivalent because of the vector nature of torque and the scalar nature of work.

Rotational Kinetic Energy

In the absence of friction the effect of the work done is to give the wheel rotational kinetic energy. This energy may be considered to be the sum of the individual kinetic energies of all the bits of mass m_i, each rotating at its own linear velocity v_i, as in Fig. 8.8. Each bit of kinetic energy is then

$$\text{KE}_i = \tfrac{1}{2}m_i v_i^2$$

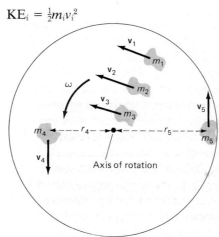

FIGURE 8.8 Rotational kinetic energy. KE_{rot} is equal to the sum of the individual kinetic energies of all the bits of mass m_i making up the wheel. As shown in the text, $\text{KE}_{\text{rot}} = \tfrac{1}{2}I\omega^2$

where the index i refers to a particular bit of mass m_i in the wheel. We also know that $v_i = \omega r_i$, where ω *is the same for all the elements of mass in the wheel since the wheel rotates as a rigid body*, and r_i is the radial distance of each element of mass from the axis of rotation.

$$\text{KE}_{\text{rot}} = \sum_{i=1}^{N} \text{KE}_i = \sum_{i=1}^{N} \tfrac{1}{2}m_i v_i^2 = \sum_{i=1}^{N} \tfrac{1}{2}m_i(\omega^2 r_i^2) = \tfrac{1}{2}\left(\sum_{i=1}^{N} m_i r_i^2\right)\omega^2$$

where the summation $\sum_{i=1}^{N}$ means that we must add together the kinetic energies of all the bits of mass from $i = 1$ to $i = N$, and we assume that there are N bits of mass making up the total mass of the wheel.

We call the sum $\sum_{i=1}^{N} m_i r_i^2$ the *rotational inertia I* (or *moment of inertia I*) of the wheel about the axis of rotation. Then the *rotational kinetic energy* is given by the following equation:

Rotational Kinetic Energy

$$\boxed{KE_{rot} = \tfrac{1}{2}I\omega^2}$$ (8.8)

Rotational Inertia

where $$\boxed{I = \sum_{i=1}^{N} m_i r_i^2}$$ (8.9)

is the *rotational inertia* of the rotating object.

Note that the rotational inertia of a body is in no way as intrinsic and unique a property of a body as is its mass. The rotational inertia *depends on the axis* about which it is calculated, whereas the mass does not.

If we compare Eq. (8.8) for rotational kinetic energy with the corresponding expression for translational kinetic energy $KE_{trans} = \tfrac{1}{2}mv^2$, we see that the linear velocity v is replaced by the angular velocity ω, and the mass m (or translational inertia) is replaced by the rotational inertia I.

I and m

The rotational inertia I therefore plays the same role in rotational motion as does the mass m in translational motion. As a result, in extending the parallelism that exists between linear and angular motion to more complicated situations, we will always have to replace the mass m by its rotational analogue, the rotational inertia I.

It should be clear that the expression for rotational kinetic energy, $\tfrac{1}{2}I\omega^2$, is correct dimensionally. Since

$$I = \sum_{i=1}^{N} m_i r_i^2 \qquad KE_{rot} \text{ has dimensions} \qquad [M][L^2]\frac{1}{[T^2]}$$

which are the same as for translational kinetic energy.

Significance of Rotational Inertia

As we have seen, for translational motion mass is a measure of inertia, that is, of how difficult it is to accelerate an object possessing that mass. In rotational motion it is not mass alone that counts, but mass times the square of the distance of that mass from the axis of rotation. Consider a rotating disk like a phonograph record, for example. Some bits of mass of the record are close to the axis and some are far away from that axis. Those bits of mass m_i close to the axis have a small r_i and therefore a very small $m_i r_i^2$, and hence contribute very little to the rotational inertia I. Bits of mass m_i far from the axis and hence with large values of $m_i r_i^2$ contribute a great deal to I. When all these bits of mass are multiplied by their proper r_i^2 and the results added, we obtain the rotational inertia I for the whole record. It is this rotational inertia which determines, for example, the torque necessary to set the record into angular motion.

The value of the rotational inertia can be changed without altering the total mass of an object. For example, suppose we had two metal disks, one of uniform density and the second constructed with most of its mass near the outside rim with just a few thin spokes to hold the disk together. If these two disks had the same mass, the second one would have a much greater rotational inertia I than the first.

In devices like flywheels and gyroscopes it is desirable to maintain as constant an angular velocity as possible, despite torques due to friction and

FIGURE 8.9 A rim-loaded flywheel, with almost all its mass concentrated at the same distance from the axis of rotation. In this case the rotational inertia is $I = MR^2$.

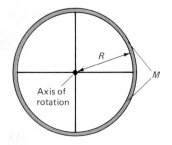

air resistance trying to slow down the rotating object. Hence such devices are constructed with massive rims and light interiors. The acceleration produced by outside torques is reduced to a minimum because the rotational inertia has been made a maximum. Figure 8.9 shows a rim-loaded flywheel, i.e., a flywheel with almost all its mass at a distance R from the axis of rotation. Its rotational inertia is therefore

$$I = \sum_{i=1}^{N} m_i r_i^2 = MR^2$$

Example 8.6

A light rope is wrapped 6 times around the rim of a wheel of radius 0.75 m. The rope is then pulled off the rim of the wheel by a person exerting a constant force of 20 N on the rope tangent to the rim.
(a) How much work is done in rotating the wheel?
(b) If there is no friction, and we neglect the KE of the rope, what is the final rotational kinetic energy of the wheel?
(c) If the rotational inertia of the wheel is 10 kg·m², what is the velocity of the wheel when it is no longer being accelerated?

SOLUTION

(a) From Eq. (8.6), $\mathcal{W} = \tau\theta$. Here the torque is

$$\tau = FR = (20 \text{ N})(0.75 \text{ m}) = 15 \text{ N·m}$$

and the angle $\theta = (6 \text{ rev})(2\pi \text{ rad/rev}) = 12\pi \text{ rad}$

Hence we have $\mathcal{W} = (15 \text{ N·m})(12\pi \text{ rad}) = \boxed{565 \text{ J}}$

(b) The final kinetic energy of the wheel is the same as the work done. i.e., $\boxed{565 \text{ J}}$

(c) From Eq. (8.8), $KE_{rot} = \frac{1}{2}I\omega^2$, and so

$$\omega = \sqrt{\frac{2KE_{rot}}{I}} = \sqrt{\frac{2(565 \text{ J})}{10 \text{ kg·m}^2}} = \boxed{10.6 \text{ rad/s}}$$

8.5 Newton's Second Law for Rotational Motion

As we have said before, perhaps the most important law in classical newtonian physics is Newton's second law of motion. Applied to translational motion, this takes the form $\mathbf{F} = m\mathbf{a}$. We now want to see what form Newton's second law takes when applied to rotational motion.

Consider a small puck of mass m on a frictionless air table. The puck is forced to rotate about a vertical axis at the end of a very light rigid rod which connects it to point C about which the rod is free to rotate, as in Fig. 8.10. If we apply a force F_t (a tangential force) on the puck at right angles to the radius R, then from Newton's second law, we have:

$$F_t = ma_t$$

where a_t is the linear tangential acceleration of the puck. Then, on multiplying by the radius R, we have:

$$RF_t = mRa_t = mR(R\alpha)$$

since $a_t = R\alpha$, where α is the angular acceleration of the puck. Now, RF_t is just the torque τ acting, since it is the product of the force and its lever arm. We then have:

$$\tau = RF_t = (mR^2)\alpha = I\alpha$$

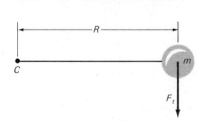

FIGURE 8.10 A small puck is forced to rotate about point C on a frictionless table. The torque needed to give the puck an angular acceleration α is $\tau = I\alpha$.

since in this case mR^2 is equal to the rotational inertia I of the puck about the axis of rotation C. This equation turns out to be true for all objects, no matter what their size or shape. Hence we can write in general:

Newton's Second Law

$$\boxed{\tau = I\alpha}$$

(8.10)

This is, indeed, the correct form of Newton's *second law for rotational motion*. This law states that *to produce an angular acceleration α, a net torque τ is needed, and the magnitude of the torque required to produce a given angular acceleration α is directly proportional to the rotational inertia I*. Note that in Eq. (8.10) the force F in Newton's second law is replaced by its rotational analogue, the torque τ, and the linear acceleration is replaced by its rotational counterpart, the angular acceleration α, as is to be expected.

A much larger torque is needed to bring to rest a 5-ton flywheel of radius 5 m, and rotating at 10 rev/s, than would be required to stop a bicycle wheel spinning at 2 rev/s. The mass of the flywheel is larger than that of the bicycle wheel, and the average distance of this mass from the axis of rotation is far greater for the flywheel than for the bicycle wheel. Hence a much larger torque is needed to produce the same change in the angular velocity.

Newton's laws also help explain why a tightrope walker in the circus often carries a long pole, which is held in such a way as to give it a large rotational inertia with respect to the rope as an axis of rotation. If the circus performer starts to lose balance and falls, he or she exerts a torque on the pole. By the angular analogue of Newton's third law, the pole then exerts a reaction torque back on the tightrope walker. Because the pole has a large rotational inertia, its initial angular acceleration about the wire is very small, but the reaction torque on the performer is much more effective in restoring balance because of the performer's smaller rotational inertia.

Example 8.7

A frictionless wheel of rotational inertia $I = 2.5 \times 10^{-3}$ kg·m² and radius 10 cm is mounted as in Fig. 8.11. A rope is wound around it and attached to a mass $m = 1.0$ kg which is then allowed to fall under gravity. Find **(a)** the linear acceleration a of the mass; **(b)** the tension in the rope (assumed massless); **(c)** the downward speed of the mass after it has fallen a distance of 1.0 m.

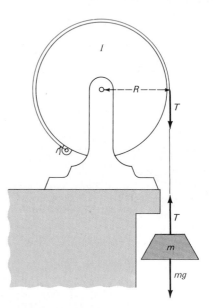

FIGURE 8.11 Diagram for Example 8.7.

SOLUTION

(a) Since the acceleration of the mass is obviously downward, we take the downward direction as positive. Then the torque τ on the wheel exerted by the tension (T) in the rope produces an angular acceleration α given by

$$\tau = TR = I\alpha \qquad \text{so that} \qquad T = \frac{I\alpha}{R}$$

or, since $\alpha = a/R$, where a is the linear acceleration,

$$T = \frac{Ia}{R^2}$$

(Note that T means *tension*, not *torque*; τ is the torque.)

Let us now apply Newton's second law to the linear motion of the mass m. We have

$$F_{\text{net}} = mg - T = ma$$

or, using the result above for T,

$$mg - \frac{Ia}{R^2} = ma$$

Hence

$$a\left(m + \frac{I}{R^2}\right) = mg$$

and so

$$a = \frac{m}{m + I/R^2}\, g$$

The acceleration is therefore less than the acceleration due to gravity, as might be expected.

Putting in numbers, we find that

$$a = \left[\frac{1.0\text{ kg}}{1.0\text{ kg} + (2.5 \times 10^{-3}\text{ kg·m}^2)/(0.10\text{ m})^2}\right]\left(9.8\,\frac{\text{m}}{\text{s}^2}\right)$$

$$= \left(\frac{1}{1 + 0.25}\right)(9.8\text{ m/s}^2) = \boxed{7.8\text{ m/s}^2}$$

(b) The tension in the rope is then

$$T = \frac{Ia}{R^2} = \frac{(2.5 \times 10^{-3}\text{ kg·m}^2)(7.8\text{ m/s}^2)}{(0.10\text{ m})^2} = \boxed{2.0\text{ N}}$$

(c) The downward speed of the mass after it has fallen a distance of 1.0 m can be found from Eq. (2.12), $v^2 = v_0^2 + 2a(s - s_0)$. Here we can take s_0 and v_0 as zero, and so we have

$$v^2 = 2as = 2(7.8\text{ m/s}^2)(1.0\text{ m}) = 15.6\text{ m}^2/\text{s}^2$$

$$v = \boxed{3.9\text{ m/s}}$$

This result for the downward speed v also could have been obtained from energy considerations. As the weight falls, it loses potential energy, but both the mass and the wheel acquire kinetic energy. To conserve energy we must have

$$mgs = \tfrac{1}{2}mv^2 + \tfrac{1}{2}I\omega^2$$

But $v = R\omega$, and so

$$mgs = \frac{1}{2}mv^2 + \frac{1}{2}\frac{Iv^2}{R^2}$$

Hence

$$v^2 = 2\left(\frac{m}{m + I/R^2}\right)gs = 2as$$

since, from part (a),

$$a = \frac{m}{m + I/R^2}\, g$$

We therefore have $v = \sqrt{2as}$, which leads to the same result as above.

8.6 Radius of Gyration; Rotational Inertias

To get a clearer idea of the meaning of the rotational inertia of an object, let us rewrite Eq. (8.9) in a slightly different form. We break up the mass of the rotating object into small pieces of *equal mass* Δm. Then

$$I = \sum_{i=1}^{N} m_i r_i^2 = \sum_{i=1}^{N} \Delta m\ r_i^2 = \Delta m \sum_{i=1}^{N} r_i^2$$

Let us multiply and divide by N, the number of pieces of mass Δm in the total mass M. We then have

$$I = N \, \Delta m \sum_{i=1}^{N} \frac{r_i^2}{N}$$

Now $N \, \Delta m$ is just the total mass M of the object, and $\sum_{i=1}^{N} (r_i^2/N)$ is the average value of r^2 for all the tiny masses making up the body. We designate this by

$$k^2 = \sum_{i=1}^{N} \frac{r_i^2}{N} \qquad (8.11)$$

The rotational inertia I can then be expressed:

$$I = Mk^2 \qquad (8.12)$$

where we call k the *radius of gyration* of the body.

Definition

The radius of gyration k is that distance from the axis of rotation at which the whole mass of the body must be considered concentrated to obtain the correct value for the rotational inertia I.

Rotational Inertias of Various Objects

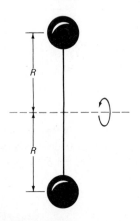

FIGURE 8.12 Rotational inertia of a dumbbell. Here $I = 2\,mR^2$.

This definition of the radius of gyration allows us to calculate the rotational inertia for some simple cases. For example, in a bicycle wheel most of the mass is located at the same distance R from the axis of rotation, and so Eq. (8.12) becomes

$$I = Mk^2 = MR^2$$

For a dumbbell of length $2R$ rotating about an axis at its center and perpendicular to the length of the dumbbell, as in Fig. 8.12, we have, if the mass of each ball at the end of the dumbbell is m,

$$I = Mk^2 = 2mR^2$$

For most objects, where the mass is irregularly distributed over a considerable region of space, it is necessary to find the rotational inertia by experiment. For some objects with symmetrical shapes calculus can be used to find the average value of r_i^2. In Table 8.1 some values of I obtained in this way for a few important symmetrical objects are given. Table 8.1 shows that the rotational inertia depends on the axis about which it is calculated, since we see that the rotational inertia of a thin rod about an axis at one end is $\frac{1}{3}ML^2$, but about an axis at the center it is only $\frac{1}{12}ML^2$.

It is also worth emphasizing that the mass of a body *cannot* be considered as concentrated at the center of mass for the purpose of computing its rotational inertia. For a bicycle wheel, for example, the center of mass of the wheel is at its center, which is on the axis of rotation. This means that, if all the mass were considered concentrated at that point, R would be zero, and hence so would the rotational inertia I. Since this is clearly wrong, we see that we cannot decide to place all the mass of an object at its center of mass for the purpose of calculating I. Rather all the mass must be considered concentrated at the *radius of gyration* if we desire to calculate the rotational inertia, as can be seen from Eq. (8.12).

TABLE 8.1   Rotational Inertia of Various Solids

Solid object	Axis	Rotational inertia I
Thin cylindrical shell, mean radius R	Axis of the cylinder	MR^2
Circular sheet of radius R	Through center, normal to sheet	$\frac{1}{2}MR^2$
Solid sphere, radius R	Any diameter	$\frac{2}{5}MR^2$
Solid cylinder, radius R	Axis of cylinder	$\frac{1}{2}MR^2$
Right circular cone, base of radius R	Axis of cone	$\frac{3}{10}MR^2$
Spherical shell, mean radius R	Any diameter	$\frac{2}{3}MR^2$
Uniform thin rod, length L	Normal to length, at one end	$\frac{1}{3}ML^2$
Uniform thin rod, length L	Normal to length, at the center	$\frac{1}{12}ML^2$

Example 8.8

Find the rotational inertia of a uniform thin rod of length 0.20 m and mass 0.40 kg **(a)** about an axis at one end normal to the length of the rod; **(b)** about an axis at the center normal to the length of the rod. **(c)** Find the radius of gyration in each case.

SOLUTION

(a) From Table 8.1, we have

$$I = \tfrac{1}{3}ML^2 = \tfrac{1}{3}(0.40 \text{ kg})(0.20 \text{ m})^2 = \boxed{5.3 \times 10^{-3} \text{ kg·m}^2}$$

(b) Here

$$I = \tfrac{1}{12}ML^2 = \tfrac{1}{12}(0.40 \text{ kg})(0.20 \text{ m})^2 = \boxed{1.3 \times 10^{-3} \text{ kg·m}^2}$$

(c) Since $I = Mk^2$, we have, for case (a),

$$k^2 = \frac{I}{M} = \frac{5.3 \times 10^{-3} \text{ kg·m}^2}{0.40 \text{ kg}} = 1.3 \times 10^{-2} \text{ m}^2$$

and so $k = \boxed{0.12 \text{ m}}$

This distance is slightly more than halfway from the axis of rotation to the end of the rod.

In case (b) we have

$$k^2 = \frac{I}{M} = \frac{1.3 \times 10^{-3} \text{ kg·m}^2}{0.40 \text{ kg}} = 3.25 \times 10^{-3} \text{ m}^2$$

and so

$$k = \boxed{5.7 \times 10^{-2} \text{ m}}$$

Again here the radius of gyration is somewhat more than halfway from the axis of rotation to the end of the rod.

Example 8.9

A solid metal sphere of radius 5.0 cm and mass 500 g starts from rest at the top of an incline of height $H = 20$ cm and rolls down the incline, as in Fig. 8.13. How fast is it moving along the ground at the bottom of the incline? Assume that it rolls without slipping and that friction is negligible.

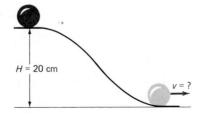

$H = 20$ cm

$v = ?$

FIGURE 8.13   Diagram for Example 8.9.

SOLUTION

This is a good example of a conservation-of-energy problem in which the initial gravitational potential energy goes into *both* translational kinetic energy and rotational kinetic energy.

The initial gravitational PE with respect to the ground is: $PE_{grav} = MgH$. The final translational KE is $\frac{1}{2}Mv^2$, and the final rotational KE is $\frac{1}{2}I\omega^2$. Hence

$$MgH = \tfrac{1}{2}Mv^2 + \tfrac{1}{2}I\omega^2$$

But $v = R\omega$, and for a solid sphere we have, from Table 8.1, $I = \tfrac{2}{5}MR^2$. Hence

$$MgH = \tfrac{1}{2}Mv^2 + \tfrac{1}{2}(\tfrac{2}{5}MR^2)\ \omega^2 = \tfrac{1}{2}Mv^2 + \tfrac{1}{2}(\tfrac{2}{5}M)v^2$$

On canceling M throughout, we have:

$$v^2 = \frac{gH}{\tfrac{1}{2} + \tfrac{1}{5}} = \tfrac{10}{7}gH$$

and so $v = \sqrt{\tfrac{10}{7}gH}$

$$= [\tfrac{10}{7}\,(9.8 \text{ m/s}^2)(0.20 \text{ m})]^{1/2} = \boxed{1.7 \text{ m/s}}$$

In this case of a solid sphere, therefore, the rotational kinetic energy is less than the translational kinetic energy by a factor of $\tfrac{2}{5}$.

Parallel-Axis Theorem

Table 8.1 gives the rotational inertias for various solids about certain special axes within the solid. Sometimes, however, we desire the rotational inertia of a body about a different axis inside or outside the solid object. In this case a very useful relationship is the *parallel-axis theorem*:

Parallel-Axis Theorem

If I_{CM} is the rotational inertia of an object of total mass M about any axis through its center of mass (CM), then the rotational inertia about a parallel axis a distance d from the center of mass is:

$$\boxed{I = I_{CM} + Md^2}$$

(8.13)

We will not prove this theorem here, but its use is illustrated in Example 8.10. The parallel-axis theorem is particularly useful in problems that combine translational motion with rotational motion.

Example 8.10

If the rotational inertia of a uniform thin rod of length L about an axis through its center and normal to its length is $\tfrac{1}{12}ML^2$, find its rotational inertia about an axis at one end and normal to its length.

SOLUTION

The parallel-axis theorem states that:

$$I = I_{CM} + Md^2$$

where d is the distance from the center of mass (CM) to the axis of rotation about which the rotational inertia is desired.

For the thin rod its rotational inertia about its center of mass is given as:

$$I_{CM} = \tfrac{1}{12}ML^2$$

Hence about an axis at one end, which is a distance $d = L/2$ from the center of mass, we have for the rotational inertia

$$I = \tfrac{1}{12}ML^2 + M\left(\frac{L}{2}\right)^2$$

$$= \tfrac{1}{12}ML^2 + \tfrac{1}{4}ML^2 = \boxed{\tfrac{1}{3}ML^2}$$

This result agrees with the value given in Table 8.1.

8.7 Rotational Impulse and Angular Momentum

We have seen in the preceding chapter that the linear momentum of a particle or system of particles is defined as $\mathbf{p} = m\mathbf{v}$, and that Newton's second law can then be written in the more general form

$$\mathbf{f}_{net} = \frac{\Delta \mathbf{p}}{\Delta t} \qquad \text{or} \qquad \mathbf{f}_{net}\, \Delta t = \Delta \mathbf{p}$$

where $\mathbf{f}_{net}\, \Delta t$ is called the *impulse* of the force $\mathbf{f}_{net}$.

A perfectly parallel treatment of rotational motion is possible. We define an *angular momentum*, by analogy with the case of linear motion, as

Angular Momentum

$$\boxed{L = I\omega} \tag{8.14}$$

Here the rotational inertia replaces the ordinary inertia (or mass), and the angular velocity replaces the linear velocity. The impulse-momentum theorem then takes the form

$$\tau = \frac{\Delta L}{\Delta t}$$

or $\qquad \tau\, \Delta t = \Delta L = \Delta(I\omega)$ $\qquad\qquad$ (8.15)

This states that the effect of a constant torque τ, exerted over a time interval Δt, is to change the angular momentum of the system by an amount ΔL. Here $\tau\, \Delta t$ is the *rotational impulse* of the torque τ.

Clearly, if the rotational inertia I does not change while the torque acts, we have

$$\tau\, \Delta t = I\, \Delta \omega \qquad \text{or} \qquad \tau = I\frac{\Delta \omega}{\Delta t} = I\alpha \tag{8.10}$$

which is Newton's second law for rotational motion, as discussed in Sec. 8.5. Here again we see the significance of I as an inertial term. The larger I, the greater the torque required to change the rotational velocity ω.

The units of angular momentum are the units of $I\omega$, that is, (kg·m^2) (1 s^{-1}), or kg·m^2/s. Notice how this differs from the units for linear momentum, which are kg·m/s.

Vector Nature of Angular Momentum

It should be pointed out that, like linear momentum, angular momentum is actually a vector quantity. In the case of the impulse-momentum theorem applied to linear motion, the force which produces the change in momentum is a vector, and can change the momentum in either its magnitude or its direction. So, too, in the rotational case, torque is a vector, and it can change the angular momentum in either magnitude or direction. As shown in Fig. 8.14, the direction of the angular momentum is taken as along the axis of rotation and in the direction in which a right-handed screw would move if turned in the direction of the rotation. The magnitude of **L** is always $L = I\omega$.

Since we are not going to stress this vector aspect of angular momentum in this book, we will usually not use vector notation for angular quantities.

FIGURE 8.14 Vector nature of angular momentum. (*a*) A right-handed screw (an ordinary carpenter's screw) moves away from the carpenter when it is turned in a clockwise direction. The angular momentum is defined to be a vector **L** along the axis of rotation. **L** for a rotating wheel is in the direction in which a right-handed screw would move when turned: (*b*) in a clockwise direction; (*c*) in a counterclockwise direction.

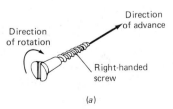

Direction of rotation

Direction of advance

Right-handed screw

(*a*)

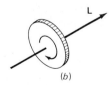

L

(*b*)

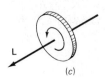

L

L

(*c*)

Example 8.11

A torque of magnitude 10^4 N·m is exerted on the rotational element of an electric generator by water falling on it from a dam for 1 min. If the rotor was initially at rest, what is its angular momentum at the end of the minute?

SOLUTION

From Eq. (8.15) we have

$$\tau \, \Delta t = \Delta L$$

Hence

$$\Delta L = (10^4 \text{ N·m})(60 \text{ s}) = 6.0 \times 10^5 \text{ N·m·s}$$

$$= \left(6.0 \times 10^5 \, \frac{\text{kg·m}}{\text{s}^2}\right)(1 \text{ m·s}) = \boxed{6.0 \times 10^5 \text{ kg·m}^2/\text{s}}$$

Since the rotor was originally at rest, this is its angular momentum at the end of the minute.

8.8 Conservation of Angular Momentum

Since from Eq. (8.15) $\tau = \Delta L / \Delta t$, if no torque acts in the time Δt we have

$$\Delta L = 0 \qquad \text{or} \qquad L = I\omega = \text{constant}$$

This is the *principle of conservation of angular momentum*, which may be stated as follows:

Conservation of Angular Momentum

> If no external torques act, the total angular momentum of an object or system of objects will remain constant.

Clearly, since $L = I\omega$, and the angular momentum **L** is a vector, this implies two things:

1 If I is constant, then the angular speed of the system remains constant; i.e., the angle swept out in equal time intervals does not change with time.

2 Since the direction of **L** is defined to be the direction of the axis of rotation, the direction of the axis of rotation will remain unchanged.

Examples of Angular Momentum Conservation

A flywheel is useful for storing energy because, if no external torques act on it, its angular momentum will remain unchanged as it rotates. Since its rotational inertia I is a constant, this means that its angular velocity will also remain unchanged and the flywheel will continue to rotate a very long time at its original angular speed. Hence the rotational kinetic energy given to it originally will remain unchanged.

A gyroscope used for navigational purposes (Fig. 8.15) is a heavy rotational wheel like a flywheel, mounted in frictionless bearings in such a way that even if the mount is moved, no torque is exerted on the wheel of the gyro. The rotating wheel therefore tends to keep its axis pointing in the same fixed direction. Hence the gyro can be used to determine the direction of an airplane or boat with respect to its initial direction.

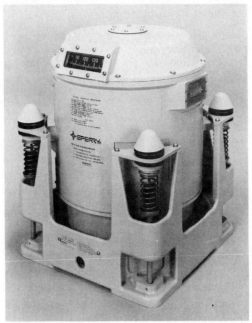

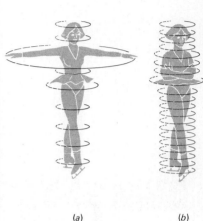

FIGURE 8.15 (left)   A gyrocompass used for navigation. (Photo courtesy of the Sperry Corporation, Electronic Systems Division.)

FIGURE 8.16 (right) Conservation of angular momentum: A figure skater doing a fast pirouette. Her rotational speed increases greatly over that in (a) when her arms are quickly brought close to her body, as in (b).

(a)   (b)

A final example is that of a figure skater doing a fast pirouette. She first starts her spin with arms extended as in Fig. 8.16a. Then she pulls her arms in close to her body, as in Fig. 8.16b, and her rotational speed increases greatly. This is because $L = I\omega$ remains constant, if we neglect the slowing torque between the skates and the ice. Hence, since I depends on the square of the distance of the skater's mass from the axis of rotation, I can be decreased by the skater pulling in her arms. Her new rotational velocity then becomes $\omega_2 = (I_1/I_2)\omega_1$. Hence if I_2 is less than I_1, ω_2 is greater than ω_1.

Example 8.12

A man stands at the center of a turntable, holding his arms extended horizontally with a 5.0-kg dumbbell in each hand. He is set rotating slowly about a vertical axis, with an angular velocity of 1 rev every 1.5 s. The rotational inertia of the turntable and the man without the dumbbells is 5.0

kg·m². The man holds the dumbbells originally at a distance of 1.0 m from the axis of rotation and then drops them quickly to his side at a distance of 20 cm from the axis of rotation. Find the new angular velocity of the man after he drops the dumbbells to his side.

SOLUTION

Since there are no external torques acting, angular momentum must be conserved. Hence $I_i\omega_i = I_f\omega_f$, where I_i and I_f are the initial and final rotational inertias of the system, and ω_i and ω_f are the initial and final angular velocities.

Now, I_i is the sum of the rotational inertias of the turntable and man, plus the rotational inertia of the dumbbells held at a distance of 1.0 m from the axis. Hence

$$I_i = 5.0 \text{ kg·m}^2 + 2(5.0 \text{ kg})(1.0 \text{ m})^2 = 15 \text{ kg·m}^2$$

Similarly,

$$I_f = 5.0 \text{ kg·m}^2 + 2(5.0 \text{ kg})(0.20 \text{ m})^2 = 5.4 \text{ kg·m}^2$$

Also

$$\omega_i = \frac{1 \text{ rev}}{1.5 \text{ s}} = \frac{2\pi \text{ rad}}{1.5 \text{ s}} = 4.2 \text{ rad/s}$$

Then

$$\omega_f = \frac{I_i}{I_f}\omega_i = \left(\frac{15 \text{ kg·m}^2}{5.4 \text{ kg·m}^2}\right)(4.2 \text{ rad/s}) = \boxed{12 \text{ rad/s}}$$

Hence the angular velocity has gone from 4.2 to 12 rad/s, or has almost tripled.

Example 8.13

When the gravitational collapse of a star occurs, its radius shrinks dramatically. Suppose that the sun undergoes such a collapse some time in the future and that its radius shrinks from its present 6.96×10^8 m to about 20 km, which is typical of some neutron stars. What would happen to the present spin rate of the sun, which is about $\frac{1}{30}$ rev per day?

SOLUTION

From the principle of conservation of angular momentum, we have $I_1 \omega_1 = I_2 \omega_2$, where for a solid sphere $I = \frac{2}{5}MR^2$. Hence

$$\omega_2 = \frac{R_1^2}{R_2^2} \omega_1 \quad \text{or} \quad f_2 = \frac{R_1^2 f_1}{R_2^2}$$

where $f = \omega/2\pi$ is the frequency of rotation. Therefore

$$f_2 = \frac{(6.96 \times 10^8 \text{ m})^2}{(20 \times 10^3 \text{ m})^2} \left(\frac{1}{30} \frac{\text{rev}}{\text{day}} \right)$$

$$= \left(4.0 \times 10^7 \frac{\text{rev}}{\text{day}} \right) \left(\frac{1 \text{ day}}{8.64 \times 10^4 \text{ s}} \right) = \boxed{4.6 \times 10^2 \text{ rev/s}}$$

Hence such a collapse would cause the sun to rotate over 400 times *a second*. Such large rotational frequencies have actually been observed in pulsars.

Summary: Important Definitions and Equations

Most of the important ideas and equations of this chapter are summarized in the accompanying table.

Comparison between Translational and Rotational Quantities

Linear motion (translation)	Physical quantity	Angular motion (rotation)
Δs, m	Displacement $\Delta s = R \, \Delta\theta$	$\Delta\theta$, rad
$v = \lim\limits_{\Delta t \to 0} \dfrac{\Delta s}{\Delta t}$	Velocity $v = R\omega$	$\omega = \lim\limits_{\Delta t \to 0} \dfrac{\Delta\theta}{\Delta t}$
$a = \lim\limits_{\Delta t \to 0} \dfrac{\Delta v}{\Delta t}$	Acceleration $a = R\alpha$	$\alpha = \lim\limits_{\Delta t \to 0} \dfrac{\Delta\omega}{\Delta t}$
m	Inertia	$I = \sum\limits_i m_i r_i^2$
F	Force; torque	$\tau = FR$
$\mathcal{W} = \mathbf{F} \cdot \mathbf{s}$	Work	$\mathcal{W} = \tau\theta$
$P = \mathbf{F} \cdot \mathbf{v}$	Power	$P = \tau\omega$
$\mathbf{F}_{\text{net}} = m\mathbf{a} = \dfrac{\Delta \mathbf{p}}{\Delta t}$	Newton's second law	$\tau_{\text{net}} = I\alpha = \dfrac{\Delta L}{\Delta t}$
$\mathbf{p} = m\mathbf{v}$	Momentum	$L = I\omega$
$\mathbf{F} \, \Delta t = \Delta \mathbf{p}$	Impulse	$\tau \, \Delta t = \Delta L$
$\frac{1}{2}mv^2$	Kinetic energy	$\frac{1}{2}I\omega^2$
If $F_{\text{net}} = 0$, $p_2 = p_1$ $m_2 v_2 = m_1 v_1$	Conservation of momentum	If $\tau_{\text{net}} = 0$, $L_2 = L_1$ $I_2 \omega_2 = I_1 \omega_1$

Equations for angular motion with constant angular acceleration:

$$\theta = \theta_0 + \bar{\omega} t \qquad \theta = \theta_0 + \omega_0 t + \tfrac{1}{2}\alpha t^2$$

$$\bar{\omega} = \frac{\omega_0 + \omega}{2} \qquad \omega = \omega_0 + \alpha t$$

$$\omega^2 = \omega_0^2 + 2\alpha(\theta - \theta_0)$$

Radius of gyration:
That distance from the axis of rotation at which the whole mass of the body must be considered concentrated to obtain the correct value for the rotational inertia I.

Questions

1 A fly sits on a phonograph record near its outer edge as it rotates at a constant speed of 33.3 rev/min.
    (*a*) Is the linear velocity of the fly constant? Why?
    (*b*) Is the angular velocity of the fly constant? Why?

2 Is there such a thing as *rotational potential energy*? Can you give an example of a situation in which this concept might have some importance?

3 Estimate your rotational inertia about a vertical axis through your center of mass when you are standing upright. Repeat for a horizontal axis through your center of mass.

4 A solid spherical ball and a solid cylinder are rolled down the same incline. If they have exactly the same mass and the same radius, which has the larger linear speed at the bottom of the incline?

5 Which bicycle wheel would be harder to stop rotating about its axle, one with an aluminum rim or one with a rim made out of lead, if the rims were the same size?

6 Can you suggest for a subway car a braking system which employs a flywheel and which would conserve energy?

7 A problem arises in the use of flywheels in automobiles and buses in that when the vehicles are going around curves torques are set up which tend to unbalance the vehicles and lead to accidents. Can you suggest a simple means to correct this problem?

8 A fly is sitting on the edge of a frisbee which is rotating with a constant angular speed about its center. The fly then begins to walk toward the center of the frisbee. What would happen to the motion of the frisbee?

9 Why do helicopters always have two horizontal rotors, either one large rotor which provides the lift and a smaller one at the tail of the plane, or two equal-size rotors which rotate in opposite directions?

10 Why do football players try to throw or kick "spirals," in which the football rotates about its long axis which is in line with the direction of the ball's motion?

11 The earth actually moves faster in its orbit around the sun in the winter than it does in the summer. Is the earth therefore closer to the sun in the winter or in the summer?

12 If hard-boiled eggs become mixed with uncooked eggs in the refrigerator, one way to separate them is to spin each egg on a table top. The hard-boiled eggs will spin faster than the uncooked ones. Explain why, using the concept of rotational inertia.

Problems

A 1 A point on the circumference of a bicycle wheel moves a distance of 3.0 m in 1 s. If the wheel has a radius of 0.33 m, the angular velocity of the wheel is:
(*a*) 1.0°/s   (*b*) 9.1 rev/s   (*c*) 1.0 rad/s
(*d*) 9.1 s^{-1} (*e*) 9.1 m/s

A 2 A point on the outer edge of a flywheel of radius 1.5 m rotates at a constant linear speed of 50 m/s. The angle swept out by a spoke on the flywheel in 10 s is:
(*a*) 333 rad   (*b*) 333 m   (*c*) 500 rad
(*d*) 100π rad   (*e*) 750 rad

A 3 A metal chain is attached to a gear rotating on an axle. The gear has a radius of 5.0 cm and turns at a constant speed of 2.5 rev/s. The linear distance moved by the chain in 1 min is:
(*a*) 47 rad   (*b*) 47 m   (*c*) 4700 m
(*d*) 300π m   (*e*) 0.78 m

A 4 The work done by a torque of 5.0 N·m in spinning a flywheel through 10 complete revolutions is:
(*a*) 100π W   (*b*) 20π J   (*c*) 50 W
(*d*) 100π J   (*e*) 50 J

A 5 The rotational kinetic energy of a 100-g steel ball of radius 1.0 cm rotated at the end of a string 0.80 m in length, if the ball makes 2 complete revolutions per second, is:
(*a*) 6.4 J   (*b*) 0.13 J   (*c*) 0.41 J   (*d*) 10.2 J
(*e*) 5.1 J

A 6 A force of 20 N is exerted on a rope wrapped around a wheel by someone pulling on the rope. The force is tangent to the rim of the wheel, which has a radius of 0.60 m. This force pulls an amount of rope off the wheel just sufficient to go around the wheel 1 time. The final rotational kinetic energy of the wheel is:
(*a*) 75 J   (*b*) 12 J   (*c*) 150 J   (*d*) 38 J
(*e*) None of the above

A 7 The torque which must be applied to a turntable with rotational inertia 5.0×10^{-3} kg·m^2 to give it an acceleration of 1.2 rad/s^2 is:
(*a*) 4.2×10^{-3} J   (*b*) 6.0×10^{-3} N·m
(*c*) 6.0×10^{-3} J   (*d*) 6.0×10^{-3} kg·m/s^2
(*e*) 4.2×10^{-3} N·m

A 8 The angular momentum of a bicycle wheel of mass 1.6 kg and radius 0.33 m rotating at an angular velocity of 2.0 rev/s is:
(*a*) 2.2 kg·m/s   (*b*) 2.2 kg·m^2/s   (*c*) 0.35 kg·m^2/s^2
(*d*) 0.35 kg·m/s   (*e*) 6.7 kg·m^2/s

A 9 A ball is rotating in a horizontal circle at the end of a string of length 1.0 m at an angular velocity of 10 rad/s. The string is gradually reduced in length to 0.50 m without any force being exerted in the direction of the ball's motion. The new angular velocity of the ball is:
(*a*) 40 rad/s   (*b*) 5.0 rad/s   (*c*) 20 rad/s
(*d*) 10 rad/s   (*e*) 2.5 rad/s

A10 In Prob. A9 the new linear speed of the ball is:
(*a*) 2.5 m/s   (*b*) 5.0 m/s   (*c*) 40 m/s
(*d*) 10 m/s   (*e*) 20 m/s

B 1 The turntable on a stereo is designed to acquire an angular velocity of 33.3 rev/min in 0.50 s, starting from rest. Find the average angular acceleration of the turntable during the 0.50 s.

B 2 A wheel is rolling along without slipping in such a way that the axle of the wheel has a translational speed of 10 m/s along the ground. If the radius of the wheel is 0.50 m, what is the angular velocity of the wheel?

B 3 The rear wheel of a bicycle is rotating at a rate of 3.0 rev/s. If the radius of the wheel is 0.33 m, how fast is the bicycle moving along the ground?

B 4 The rotor in a centrifuge is accelerated from rest with a constant angular acceleration of 6.0 rad/s^2. What is the average angular velocity of the rotor during the first minute?

B 5 An automobile engine develops a rotational torque of 475 N·m and is rotating at a speed of 3000 rev/min. What horsepower does the engine generate?

B 6 What is the rotational kinetic energy of a rim-loaded 1000-kg flywheel of radius 10 m which is rotating through 5.0 complete revolutions per second?

B 7 What is the radius of gyration of a right circular cone about the axis of the cone? (*Hint*: Use Table 8.1.)

B 8 The rotational inertia about a particular axis through its center of mass of an irregularly shaped object of mass 1.0 kg is found by experiment to be 3.0×10^{-2} kg·m². What is the rotational inertia of this same object about an axis parallel to the first axis and a distance of 1.4 m from it?

B 9 A torque of magnitude 3.0×10^{2} N·m is exerted on the rotor of an electric generator for 30 s. What is the angular momentum of the rotor at the end of the 30 s, if it was initially at rest?

B10 What would be the period of the earth in its motion around the sun, if it approached to half its present average distance from the sun while preserving the same angular momentum it now has?

B11 A diver comes off a high diving board with his body straight and with an angular velocity which would enable him to make one complete turn in 2.0 s. With his body straight his rotational inertia is 19.8 kg·m². He then tucks his body in tightly by bending at both the waist and the knees. This reduces his rotational inertia to 3.8 kg·m². How many revolutions per second can he now make in this tucked position?

C 1 A flywheel starts from rest and accelerates at a constant rate of 3.0 rad/s². When the wheel has rotated through 60π rad, what is its angular velocity?

C 2 A potter's wheel of radius 15 cm starts from rest and rotates with constant angular acceleration until at the end of 30 s it is moving with an angular velocity of 15 rad/s.

(*a*) What is the angular acceleration of the wheel?

(*b*) What is the linear velocity of a point on the rim of the potter's wheel at the end of the 30 s?

(*c*) What is the average angular velocity of the wheel during the 30-s interval?

(*d*) Through what angle did the wheel rotate in the 30-s interval?

C 3 A bicycle wheel of radius 0.30 m rolls down a hill without slipping. Its linear velocity increases constantly from 0 to 5.0 m/s in 3.0 s.

(*a*) What is the angular acceleration of the wheel?

(*b*) What is the angle through which the wheel turned in 3.0 s?

(*c*) How many revolutions did the wheel make in 3.0 s?

C 4 (*a*) A phonograph turntable of radius 15 cm and mass 0.70 kg gains an angular velocity of 33.3 rev/min in 1.5 s, starting from rest. What torque must have been exerted on the turntable during this 1.5-s interval?

(*b*) What was the rotational kinetic energy of the turntable at the end of the 1.5-s interval?

C 5 A record player has a turntable with a diameter of 28 cm and a mass of 0.50 kg.

(*a*) If it takes 2.0 s for the turntable to reach its final speed of 33.3 rev/min, what torque must be provided by the motor of the record player?

(*b*) If a record of diameter 30 cm and mass 75 g is placed on the turntable, how much must the torque increase to preserve the speed of 33.3 rev/min?

C 6 A constant force of 100 N is applied tangentially to the rim of a wheel 50 cm in radius with a rotational inertia of 2.0 kg·m² and which rotates with negligible friction.

(*a*) Find the work done on the wheel during the first 5.0 s after the force is applied.

(*b*) If the wheel starts from rest, what is its kinetic energy of rotation after 5.0 s?

(*c*) Are the answers to parts (*a*) and (*b*) the same? Why or why not?

C 7 The motor driving a grinding wheel with a rotational inertia of 0.10 kg·m² is switched off when the wheel has a rotational speed of 300 rev/min. After 10 s the wheel has slowed down to 240 rev/min.

(*a*) What is the torque exerted by friction to slow the grinding wheel down?

(*b*) If this torque remains constant, when will the grinding wheel come to rest?

C 8 A 100-g mass is suspended from the rim of a wheel of 60-cm radius, with a horizontal axis of rotation, by a light string wound around the wheel, as in Fig. 8.17. The wheel is frictionless and has a rotational inertia of 0.10 kg·m². The mass is allowed to fall freely.

(*a*) What is the linear acceleration of the 100-g mass?

(*b*) What is the tension in the string?

(*c*) What is the velocity of the falling weight after 10 s?

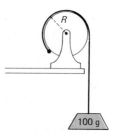

FIGURE 8.17 Diagram for Prob. C8.

C 9 A wheel with rotational inertia 0.050 kg·m² is spinning freely at 4.0 rev/s. The wheel is then connected to a cord which lifts a 4.0-kg mass from the ground. How high can the mass be lifted before the wheel stops rotating?

C10 (*a*) Find the rotational inertia of a sphere of radius 5.0 cm about an axis 50 cm from the center of the sphere. (*Hint*: Use Table 8.1 and the parallel-axis theorem.)

(*b*) How large is the error in assuming, for purposes of calculating the rotational inertia, that all the mass is concentrated at the center of the sphere?

C11 An airplane propeller has a mass of 50 kg and a radius of gyration of 0.80 m.

(*a*) Find the rotational inertia of the propeller.

(b)   How large a torque is needed to give the propeller an angular acceleration of 5.0 rev/s²?

C12   A dumbbell of length 50 cm has a small 5.0-kg mass at each end.

(a)   What is the rotational inertia of this dumbbell about an axis through its center and perpendicular to the line connecting the two masses?

(b)   What is the angular momentum of the dumbbell if it rotates about this same axis with an angular velocity of 2.0 rad/s?

C13   A 1500-kg car carries a 25-kg rim-loaded flywheel of radius 0.30 m to store energy. When the car stops, the flywheel is engaged and the car's kinetic energy is transferred to the flywheel. What will be the angular velocity of the flywheel if it absorbs all the kinetic energy of the car as it decelerates from 25 m/s to rest?

C14   A solid cylinder of mass M rolls down a ramp of height H to a level floor below. Compare the rotational kinetic energy and the translational kinetic energy of the cylinder as it rolls along the floor.

C15   A spherical ball of mass 0.20 kg and radius 4.0 cm starts from rest and rolls without slipping down a 30° incline. The center of mass of the ball falls a distance of 5.0 m in descending the incline. What is the final linear speed of the ball?

C16   In the Bohr model of the hydrogen atom an electron of mass 9.11×10^{-31} kg revolves in a circular orbit of radius 0.53×10^{-10} m about the proton at a linear speed of 2.2×10^6 m/s. What is the angular momentum of the electron in this orbit?

C17   A disk of rotational inertia 2.5×10^{-2} kg·m² is rotating with an angular velocity of 10 rad/s, when a second, nonrotating disk of rotational inertia 1.3×10^{-2} kg·m² is suddenly dropped on the first disk. If the two stick together and rotate as one disk, what is the final angular velocity of the two disks?

C18   An ice skater has a rotational inertia about a vertical axis through his body of 1.5 kg·m². If he extends his arms he can increase this rotational inertia to 1.8 kg·m². He first spins on his skates 2.0 rev/s with arms extended; then he quickly pulls his arms tightly to his side.

(a)   What is his new rotational speed?

(b)   How much kinetic energy has he gained?

(c)   Where did this energy come from?

C19   A girl of mass 50 kg jumps on a moving carousel which is rotating at a speed of 0.20 rev/s. The carousel turntable has a rotational inertia of 200 kg·m². The girl jumps toward the center of the carousel and lands at a distance 2.0 m from the center.

(a)   Find the angular velocity of the carousel after the girl jumps aboard.

(b)   Find the kinetic energy of the system both before and after the girl jumps on.

(c)   If kinetic energy is lost, where does it go?

C20   A 40-kg boy stands on the edge of a frictionless turntable of radius 4.0 m which is initially at rest. The rotational inertia of the turntable is 750 kg·m². The boy starts to run around the edge of the platform and reaches a speed of 2.0 m/s relative to the ground. What is the angular

velocity of the turntable relative to the ground, when the boy has reached this speed?

C21   A solid round platform of mass 200 kg and radius 3.0 m is mounted so that it can rotate freely around a vertical axis, but is initially at rest. A 30-kg girl comes running at 1.0 m/s in a direction tangential to the platform and jumps on. If she lands at the very edge of the platform, find the angular velocity of the platform just after the girl jumps on.

D 1   Two circular cylinders have the same radius R and mass M, but one is solid and the other is a hollow shell, with all the mass concentrated at its rim, which is a distance R from the axis of rotation of the cylinder. If the two cylinders are released together at the top of a hill which makes an angle of 20° with level ground, how far apart will they be after 5.0 s?

D 2   A mass of 500 g hangs from the rim of a wheel of radius $R = 20$ cm, as in Fig. 8.18. When released from rest, the 500-g mass falls 0.50 m in 7.5 s. Find the rotational inertia of the wheel.

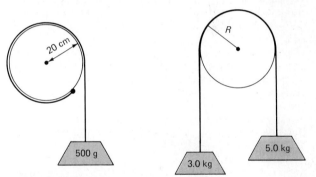

FIGURE 8.18   Diagram for Prob. D2.

FIGURE 8.19   Diagram for Prob. D3.

D 3   An Atwood's machine, of the type considered in Sec. 3.6, has a solid pulley wheel with rotational inertia 0.12 kg·m² and radius $R = 50$ cm. Two masses, one of 3.0 kg and the other of 5.0 kg, are attached to the two ends of a cord which passes over the pulley wheel, as in Fig. 8.19. The cord does not slip on the wheel but moves with the wheel.

(a)   What is the resulting linear acceleration of the two masses?

(b)   What are the tensions in the two cords?

(c)   What is the angular acceleration of the wheel?

(d)   Prove that after 1 s the work done on the system by the force of gravity is exactly equal to the kinetic energy of the whole system, so that energy is conserved.

D 4   Two pulley wheels, one of radius $R_1 = 10$ cm, the other of radius $R_2 = 30$ cm, are mounted rigidly on a common axle, as in Fig. 8.20. The rotational inertia of the two pulleys, which are clamped together, is 1.4 kg·m². Two masses of 2.0 kg and 1.5 kg are connected to cords attached to each of the pulleys, as in the figure.

(a)   Find the angular acceleration of the pulley system.

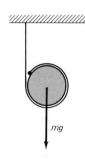

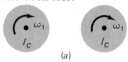

mg

FIGURE 8.20 Diagram for Prob. D4.

FIGURE 8.21 Diagram for Prob. D5.

(*b*) Find the tensions T_1 and T_2 in the two cords.

D 5 A solid cylindrical disk of mass m and radius R has a rope wrapped around its outside edge and is allowed to fall, the rope being attached to the ceiling above the disk, as in Fig. 8.21. How large is the acceleration of the disk downward compared with the acceleration due to gravity?

D 6 Two blocks are connected to a pulley of radius 20 cm and rotational inertia 0.10 kg·m², as in Fig. 8.22. The 1.0-kg block hangs freely and the 2.0-kg block slides on a frictionless plane of angle 40°. The cord does not slip on the pulley.

(*a*) What is the acceleration of the two blocks?

(*b*) What is the tension in the two cords?

(*c*) What is the linear acceleration of a point on the rim of the wheel?

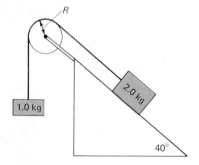

FIGURE 8.22 Diagram for Prob. D6.

D 7 Prove the parallel-axis theorem [Eq. (8.13)].

D 8 Find the rotational inertia of a hoop of radius R about a point on its rim, where the motion is confined to the plane of the hoop.

D 9 A thin meterstick of mass 0.20 kg is hinged at the floor so that it can only rotate in a vertical circle. Initially the meterstick is held in an upright position. It is then allowed to fall to the ground, rotating about the hinge at its bottom end, as in Fig. 8.23. What will be the angular speed of the meterstick when it strikes the floor?

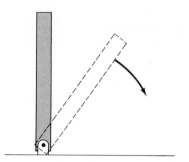

FIGURE 8.23 Diagram for Prob. D9.

D10 A metal ball of mass 2.0 kg rests on a horizontal frictionless surface, as in Fig. 8.24. A cord passes through a hole on the axis about which the ball is rotating. The ball initially moves in a circle of 0.40-m radius at an angular velocity of 10 rad/s. The radius of rotation is then slowly decreased by pulling on the cord from below the table. If the breaking strength of the cord is 200 N, what will be the radius of the circle when the cord breaks?

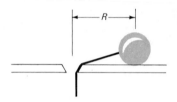

FIGURE 8.24 Diagram for Prob. D10.

D11 Two identical, flat, solid disks having moments of inertia I_C are each rotating clockwise with angular velocity ω_1, as in Fig. 8.25. They collide at point O on their rims and immediately stick firmly together.

(*a*) With what angular velocity do they rotate about the point O? (*Hint*: Use the parallel-axis theorem to find I_O.)

(*b*) Has rotational kinetic energy been conserved? If not, what fraction of the original rotational kinetic energy has been lost?

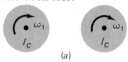

(*a*)

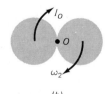

(*b*)

FIGURE 8.25 Diagram for Prob. D11.

Additional Readings

Andrade, E. N. da C.: "Isaac Newton," in *Newton Tercentenary Celebrations,* Cambridge University Press, New York, 1947. The quote at the beginning of the present chapter is taken from this essay (with permission of Cambridge University Press).

Frohlich, Cliff: "The Physics of Somersaulting and Tumbling," *Scientific American*, vol. 242, no. 3, March 1980, pp. 154–165. A very interesting account of angular momentum conservation in the motion of divers, gymnasts, astronauts, and even cats.

Kreifeldt, John G., and Ming-Chuen Chuang: "Moment of Inertia: Psychophysical Study of an Overlooked Sensation," *Science*, vol. 206, 1979, pp. 588–590. Points out the importance of rotational inertia in determining the "feel" of an object.

Post, R. F., and S. F. Post: "Flywheels," *Scientific American*, vol. 229, no. 6, December 1973, pp. 17–23. An interesting account of research on the use of flywheels to store energy.

Walker, Jearl: "The Essence of Ballet Maneuvers in Physics," *Scientific American*, vol. 246, no. 6, June 1982, pp. 146–153. A fascinating account of how dancers' leaps, turns, and pirouettes can be explained by using the basic ideas of mechanics, especially the principle of conservation of angular momentum.

*T*he ultimate aim of physical science must be to find the movements which are the real causes of all other phenomena and to determine the motive forces upon which these movements depend. In other words, its aim is to reduce all phenomena to mechanics.

Hermann von Helmholtz (1821–1894)

Chapter 9
The Structure of Atoms, Molecules, and Gases

Thus far we have discussed the motion of highly idealized particles, or point masses, whose relevance to the real world may sometimes be difficult to see. The world around us is made up of matter in a great variety of forms—gases, liquids, solids, plasmas—but none of these closely resembles the point masses of mechanics. To understand the structure of matter and its relationship to the point masses we have thus far considered, we must first understand the building blocks out of which all matter is constructed, i.e., the particles which make up atoms and molecules, and hence matter in all its forms. In this chapter we will consider the particles that combine to form atoms and molecules, and how these atoms and molecules behave to produce the observed properties of common gases like the air we breathe.

9.1 The Particles Which Make Up Atoms and Molecules

Nature has built up the material universe around us by combining a small number of elementary particles into a multitude of ingenious physical structures which we call *atoms* and *molecules*. In the past few decades experimental physicists have discovered a whole "zoo" of elementary particles, many of which are not well understood even today. Fortunately for us only three of these many particles are needed to understand the basic structure and behavior of atoms and molecules. These three are the electron, the proton, and the neutron.

The electron is a particle of very small mass, 9.11×10^{-31} kg, which carries the electric current in the wires leading to our electric toasters and TV sets. It has a negative electric charge of 1.60×10^{-19} coulomb (C), and it is this charge which gives the electron its unique importance in practical electric and electronic circuits. (The nature of charge and the use of the coulomb as the basic unit of charge will be discussed more thoroughly in Chap. 16.)

The proton has a mass 1836 times that of the electron, or 1.673×10^{-27} kg, still a small mass on any ordinary scale. The proton's charge is positive, and is exactly equal in magnitude to the charge on the electron.

The neutron has no charge, and is therefore electrically neutral. Its mass is 1.675×10^{-27} kg, which is close to that of the proton, but about 0.14 percent larger. Data for these building blocks of atoms are given in Table 9.1

TABLE 9.1   Properties of the Particles in Atoms

Particle	Symbol	Charge, C	Mass kg	Mass u*
Electron	e	-1.60219×10^{-19}	9.11×10^{-31}	5.49×10^{-4}
Nucleons:				
Proton	p	$+1.60219 \times 10^{-19}$	1.67265×10^{-27}	1.00728
Neutron	n	0	1.67495×10^{-27}	1.00866

*The symbol u signifies *unified mass units*, whose significance is clarified in Sec. 9.4.

9.2 The Structure of Atoms

~ 3×10^{-15} m

Protons   Neutrons

FIGURE 9.1 A rough sketch of the protons and neutrons in the nucleus of a carbon atom ($^{12}_{6}$C).

The three elementary particles—the positively charged proton, the negatively charged electron, and the uncharged neutron—are the basic building blocks out of which all atoms, both great and small, are constructed. Owing to the experimental work of Prof. Ernest Rutherford at Manchester in England in 1911 (see Fig. 9.2 and accompanying biography), we now know that every atom has a nucleus, or central core, which contains all the protons and neutrons, as shown in Fig. 9.1. Since the electrons are so light, the mass of the atom is basically the sum of the masses of the neutrons and protons in the nucleus. The nuclear charge is positive and numerically equal to the sum of the charges on the protons in the nucleus.

The nucleus is extremely small, with a diameter less than 10^{-14} m, and is therefore extremely dense (see Example 9.1). Since it was well known even in Rutherford's time that the diameter of atoms, as computed from the density of crystals, was about 10^{-10} m, it is clear that atoms contain a great deal of empty space. Thus, if we should attempt to draw to scale a diagram of an atom in which the atomic nucleus is shown as a round dot 1 mm in diameter, the atom would have to extend over a diameter of 10 m. For this reason, in diagrams of atoms, molecules, and nuclei in this book, the components will not be drawn to scale.

Example 9.1

The nucleus of an ordinary oxygen atom contains 8 protons and 8 neutrons, and has a radius of about 3.5×10^{-15} m. What is the density of this oxygen nucleus?

SOLUTION

The masses of protons and neutrons are approximately 1.67×10^{-27} kg. Hence the mass M of the oxygen nucleus is

$$M = 16(1.67 \times 10^{-27} \text{ kg}) = 2.67 \times 10^{-26} \text{ kg}$$

The volume V of a sphere of radius 3.5×10^{-15} m is

$$V = (\tfrac{4}{3})\pi R^3 = (\tfrac{4}{3})\pi(3.5 \times 10^{-15} \text{ m})^3 = 1.79 \times 10^{-43} \text{ m}^3$$

Hence the density d of the oxygen nucleus is

$$d = \frac{M}{V} = \frac{2.67 \times 10^{-26} \text{ kg}}{1.79 \times 10^{-43} \text{ m}^3} = \boxed{1.49 \times 10^{17} \text{ kg/m}^3}$$

Since the density of water, one of the more dense liquids, is only 10^3 kg/m^3, it is clear that this oxygen nucleus has a density about 10^{14} times that of water, and 10^{13} times that of lead, one of the most dense chemical elements found in nature.

The region of the atom outside the nucleus contains a number of electrons equal to the number of protons in the nucleus. This makes the

Ernest Rutherford (1871–1937)

FIGURE 9.2 Ernest Rutherford. (Photo Courtesy of AIP Niels Bohr Library.)

The great astronomer Sir Arthur Eddington once said that in 1911 Ernest Rutherford introduced the greatest change in our understanding of matter since the time of Democritus (about 400 B.C.). Eddington was referring to Rutherford's idea of the nuclear atom. It was this idea, and the experimental work leading up to it, which made him the acknowledged founder of modern nuclear physics.

That such a great scientist should have been born in New Zealand, at that time a remote country with few people and almost no scientific traditions, is as remarkable as was Rutherford's innate skill with scientific apparatus. Rutherford was born into a poor family. His father was a farmer who also did odd jobs to support his family of 12 children. Rutherford's simple surroundings in his youth no doubt influenced the development of his character, which re-mained open and unassuming throughout his life.

Rutherford went to Nelson College in New Zealand, where he excelled in mathematics, and later to Canterbury College in the city of Christchurch. After receiving his degree in 1893 he began some physics research in a small, drafty cellar, working on the magnetic properties of iron under the influence of electric discharges. He published his results in two scientific papers, and as a result was awarded in 1895 a scholarship to study in England at Cambridge University. When his mother came to tell him of his good luck, he was digging potatoes. He flung away his spade with a shout of glee, exclaiming: "That's the last potato I'll dig!"

At Cambridge, Rutherford worked under the great J. J. Thomson (see biography, Chap. 18) on the effect of x-rays on the conduction of electricity in gases, and later in the newly emerging field of radioactivity. From 1898 to 1907 he was professor of physics at McGill University in Montreal, Canada, where he did some of his most important work on the nature of the emissions from radioactive substances, for which he received the Nobel Prize (in chemistry) in 1908. He returned to England in 1907, and spent the years until 1919 as professor and director of the physical laboratory at Manchester. It was here that he did his renowned work on the nuclear atom. On the basis of his work on alpha particles (helium nuclei) he was led in 1911 to a description of the atom as made up of a small, very dense nucleus sur-rounded by orbital electrons. This description was taken up by Niels Bohr in 1913 and, combined with Bohr's ideas about quantum theory, provided the basic theory of the atom still accepted today.

In 1919 Rutherford succeeded Thomson as professor and director of the Cavendish Laboratory in Cambridge, and he spent his last years there directing work on artificially induced radioactivity, which eventually led to the splitting of the nucleus.

Despite his many accomplishments Rutherford was never a slave to work. He delighted in reading novels and detective stories, avidly played bridge and golf, and loved the outdoors. He would often advise his colleagues to leave their offices and laboratories and "go home and think." He was happily married to his childhood sweetheart from New Zealand. The Rutherfords' daughter, their only child, died in childbirth.

In 1931 Great Britain named Rutherford a baron, and he took the name Baron Rutherford of Nelson, after the town in New Zealand where he had been born. He immediately sent a cable to his mother in New Zealand, stating simply "Now Lord Rutherford, more your honor than mine, Ernest."

Rutherford lived an exciting and productive life until he died in 1937 at the age of 66. Although he took delight in his many accomplishments as a physicist, he would probably have taken even greater delight in the eulogy of a friend of 30 years: "Rutherford never made an enemy and never lost a friend." He was truly both a great physicist and a great man.

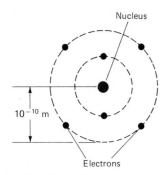

FIGURE 9.3 A very approximate model of a carbon atom.

overall atom electrically neutral, since the positive charges in the nucleus just cancel the negative charges on the electrons. These electrons are in rapid motion in the space surrounding the nucleus. It is impossible to say where any electron is at any precise instant of time, but quantum mechanics shows that it is possible to specify the regions of the atom in which the electron is most likely to be found. Despite this inherent ambiguity as to the position of the electrons, diagrams of atoms often show the electrons moving in precise circular or elliptical orbits about the nucleus. This is the way electrons were originally pictured in the atom described by Rutherford and Niels Bohr, and for some purposes this concrete picture is still useful today. Figure 9.3 shows a model of a carbon atom drawn in this approximate way.

All neutral atoms, i.e., those with zero net electric charge, contain the same number of protons as electrons. This number—1 for hydrogen, 2 for helium, 92 for uranium—distinguishes the atoms of one chemical element from those of all other elements, and accounts for most of the significant physical and chemical properties of the atom. We call this the *atomic number*, or *charge number*, of the atom, and designate it by Z.

Definition

Atomic (or charge) number Z: The number of protons in the nucleus of an atom.

FIGURE 9.4 The periodic table of the elements. The atomic number is given above the chemical symbol, and the mass number below.

The chemical elements are distinguished by subscripts that precede the chemical symbol and indicate their atomic number, e.g., $_1$H, $_2$He, $_{92}$U, or, in general, $_ZX$, where X is the chemical symbol for the particular atom. In a periodic table such as that in Fig. 9.4 the elements are listed in the order of their atomic numbers.

1 H 1.0078																	2 He 4.003
3 Li 6.942	4 Be 9.012											5 B 10.81	6 C 12.01	7 N 14.01	8 O 16.00	9 F 19.00	10 Ne 20.18
11 Na 22.99	12 Mg 24.31											13 Al 26.98	14 Si 28.09	15 P 30.97	16 S 32.06	17 Cl 35.45	18 Ar 39.95
19 K 39.10	20 Ca 40.08	21 Sc 44.96	22 Ti 47.90	23 V 50.94	24 Cr 52.00	25 Mn 54.94	26 Fe 55.85	27 Co 58.93	28 Ni 58.71	29 Cu 63.54	30 Zn 65.37	31 Ga 69.72	32 Ge 72.59	33 As 74.92	34 Se 78.96	35 Br 79.91	36 Kr 83.80
37 Rb 85.47	38 Sr 87.62	39 Y 88.91	40 Zr 91.22	41 Nb 92.91	42 Mo 95.94	43 Tc [98]	44 Ru 101.1	45 Rh 102.91	46 Pd 106.4	47 Ag 107.9	48 Cd 112.4	49 In 114.8	50 Sn 118.7	51 Sb 121.8	52 Te 127.6	53 I 126.9	54 Xe 131.3
55 Cs 132.9	56 Ba 137.3	57–71 La Series*	72 Hf 178.5	73 Ta 180.9	74 W 183.8	75 Re 186.2	76 Os 190.2	77 Ir 192.2	78 Pt 195.1	79 Au 197.0	80 Hg 200.6	81 Tl 204.4	82 Pb 207.2	83 Bi 209.0	84 Po [210]	85 At [210]	86 Rn [222]
87 Fr [223]	88 Ra [226]	89–103 Ac Series†	(104) [257]	(105) Ha [260]	(106)	(107)	(108)										

*Lanthanide series	57 La 138.9	58 Ce 140.1	59 Pr 140.9	60 Nd 144.2	61 Pm [147]	62 Sm 150.4	63 Eu 152.0	64 Gd 157.3	65 Tb 158.9	66 Dy 162.5	67 Ho 164.9	68 Er 167.3	69 Tm 168.9	70 Yb 173.0	71 Lu 175.0
†Actinide series	89 Ac [227]	90 Th 232.0	91 Pa [231]	92 U 238.0	93 Np [237]	94 Pu [242]	95 Am [243]	96 Cm [247]	97 Bk [247]	98 Cf [251]	99 Es [254]	100 Fm [253]	101 Md [256]	102 No [254]	103 Lw [257]

Ions

If an electron is removed from a neutral atom, the atom ceases to be electrically neutral and takes on a net positive charge. Such an atom is called a *positive ion*. Thus a positive lithium ion, designated as Li^+, comes into existence when a neutral Li atom loses one electron. There is also the possibility that an atom may pick up an added electron and become negatively charged. Such an atom is called a *negative ion*, such as H^-.

Definition

Ion: An atom (or molecule) which has either lost or gained electrons, and therefore has acquired a net positive or negative charge.

Since the chemical (and many physical) properties of atoms are in great part determined by the number of electrons in the atom, positive ions often behave like atoms with an atomic number 1 less than the ion, and negative ions like atoms with an atomic number 1 greater. Figure 9.5 shows a schematic picture of two important ions.

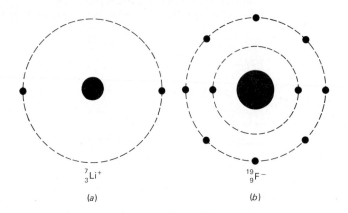

$^{7}_{3}Li^+$

(a)

$^{19}_{9}F^-$

(b)

FIGURE 9.5 Models of (a) a positive ion (Li^+); (b) a negative ion (F^-). The size of the nucleus is grossly exaggerated in these oversimplified sketches.

Isotopes

Two atoms may have the same number of protons and electrons but have different numbers of neutrons in the nucleus. Such atoms are distinguished by their mass number A, written as a superscript before the chemical symbol, and are called *isotopes*.

Definitions

Mass number A: The number of protons plus neutrons in an atomic nucleus.

Isotopes: Two atoms having the same charge number Z, but different mass numbers A.

The word *isotope* comes from two Greek words meaning the "same place," which appropriately applies to position in the periodic table. Isotopes of the same atom usually have the same chemical properties, since these depend on the electrons, but their nuclear properties may differ greatly.

The two uranium isotopes, $^{235}_{92}U$ and $^{238}_{92}U$, for example, both have 92 protons and 92 electrons. The first has $235 - 92 = 143$ neutrons in the nucleus, while the second has 146. Their chemical and physical properties, as found from chemical reactions, spectra, etc., are almost identical, but their nuclear properties are vastly different. Whereas $^{235}_{92}U$ is a fissionable material useful in nuclear reactors and bombs, $^{238}_{92}U$ is a nonfissionable and relatively useless material as found in nature.

Elements as they occur in nature are usually mixtures of various isotopes. Thus the natural abundance of uranium consists of 99.27% $^{238}_{92}U$, 0.72% $^{235}_{92}U$, and 0.006% $^{234}_{92}U$. This is of great significance for the energy problems of the United States, since the fissionable isotope $^{235}_{92}U$ needed in conventional nuclear reactors makes up less than 1 percent of the uranium occurring naturally in the earth's crust.

Because of the presence of isotopes, the average atomic mass numbers of most of the elements are not integers but decimal fractions obtained by averaging the masses and abundances of all the isotopes present in the element as found in nature. Thus chlorine consists of 75% $^{35}_{17}Cl$ and 25% $^{37}_{17}Cl$. Hence its mass number, as found in nature, is $(0.75 \times 35) + (0.25 \times 37) = 35.5$.*

The numbers of protons, neutrons, and electrons in some important isotopes are given in Table 9.2.

TABLE 9.2 Constituent Parts of Some Important Isotopes

Element	Symbol ($^A_Z X$)	Number of protons	Number of electrons	Number of neutrons	Atomic (charge) number Z	Mass number A
Hydrogen	^1_1H	1	1	0	1	1
	^2_1H	1	1	1	1	2
	^3_1H	1	1	2	1	3
Helium	^4_2He	2	2	2	2	4
Carbon	$^{12}_6C$	6	6	6	6	12
Oxygen	$^{16}_8O$	8	8	8	8	16
Iron	$^{56}_{26}Fe$	26	26	30	26	56
Uranium	$^{235}_{92}U$	92	92	143	92	235
	$^{238}_{92}U$	92	92	146	92	238

Example 9.2

Two very important chemical elements for the semi-conductor industry are (a) silicon ($^{28}_{14}Si$) and (b) germanium ($^{72}_{32}Ge$). How many electrons, protons, and neutrons does each of these contain?

SOLUTION

(a) For $^{28}_{14}Si$ the atomic number is 14, and so silicon contains 14 protons in the nucleus and 14 extranuclear electrons to make the atom electrically neutral. The number of neutrons is then obtained from the fact that

Number of neutrons $= A - Z = 28 - 14 = 14$

(b) By the same sort of analysis for $^{72}_{32}Ge$, this isotope consists of 32 protons and 40 neutrons in the nucleus, and 32 extranuclear electrons.

---

*The mass numbers of isotopes are not exact integers, for reasons which will be made clear later. To present standards of precision, the mass number of $^{35}_{17}Cl$ is 34.97867 and of $^{37}_{17}Cl$ it is 36.97750. At this stage of the course rounding these numbers off to 35 and 37 is adequate for our purposes.

Example 9.3

Copper ($_{29}$Cu) has two predominant isotopes, $_{29}^{63}$Cu, which is 69.1 percent of the copper in the earth's crust, and $_{29}^{65}$Cu, which is 30.9 percent. What is the approximate mass number A of copper as found in the earth?

SOLUTION

Mass contributed by $_{29}^{63}$Cu is $0.691 \times 63 =$ 43.5

Mass contributed by $_{29}^{65}$Cu is $0.309 \times 65 =$ 20.1

$$\text{Sum} = \boxed{63.6}$$

This value agrees reasonably well with the value given in the periodic table of the elements in Fig. 9.4.

9.3 The Combination of Atoms to Form Molecules

Very few of the objects that surround us on earth are made up of single atoms. Air, for example, consists not of atomic oxygen and atomic nitrogen, but of molecular oxygen and molecular nitrogen, in which two identical atoms are tightly bound together into a molecule. These are designated by the symbols O_2 and N_2, where the subscript after the chemical symbol indicates the number of atoms of the element present in the molecule. Similarly the sand on the beach is composed of SiO_2, silicon dioxide.

In a few cases, such as the noble gases—helium, neon, argon, krypton, xenon, radon—the structure of the atom is so stable that there is almost no tendency for the atoms to form molecules; hence the noble gases are monatomic (one-atom) gases. Most atoms, however, are more "sociable" than the "aloof" noble gases and tend to form molecules consisting of two atoms (diatomic molecules), or more than two atoms (polyatomic molecules). These molecules are held together by chemical bonds which depend on the positions and motions of the electrons in the atoms making up the molecule, and hence on the kinetic and potential energies of the electrons.

Potential-Energy Diagrams

Figure 9.6a is a potential-energy diagram for the hydrogen molecule H_2. Here we are plotting the potential energy $V(r)$ of the two-atom system against r, the separation of the two atoms. We choose an energy scale so that the potential energy is zero when the two atoms are very far apart and therefore exert no force on each other. This corresponds to the right side of the diagram, where r is very large. As the atoms approach each other, the electrons rearrange themselves so as to lower the potential energy. If the atoms come too close together, however, the two positively charged nuclei repel each other and the potential energy rises sharply. This corresponds to the left side of the diagram ($r \to 0$). In between these two extremes the potential-energy curve must have a minimum, since the right and left pieces of the curve are both descending as they move toward the center and must join together to form a continuous curve. The minimum occurs at r_0, which is the equilibrium separation of the two H atoms in H_2. At this separation the attractive and repulsive forces between the two atoms cancel out and there is no net force between them. If they are either pulled apart or pushed together, a force arises which tends to restore them to their equilibrium position as if they were connected by

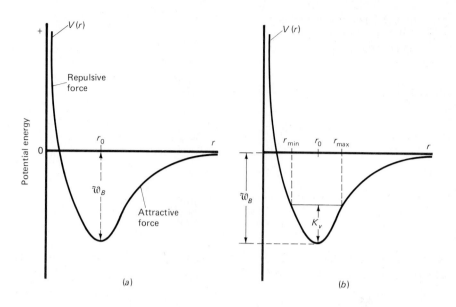

FIGURE 9.6 (a) Potential-energy diagram for the hydrogen molecule H_2: r is the distance between the two H atoms; W_B is the binding (or dissociation) energy of the molecule at its equilibrium separation r_0. (b) If a hydrogen molecule has kinetic energy of vibration K_v, its internuclear separation varies between r_{min} and r_{max}.

springs. For this reason the atoms vibrate back and forth about their equilibrium separation r_0.

The value of the potential energy at r_0 is called the *binding energy* W_B. It is negative, indicating that the system is stable, since an amount of energy W_B (called the *dissociation energy*) must be supplied to the molecule to break it up into two isolated atoms. The deeper this "potential well," i.e., the greater W_B, the greater the energy that must be added to dissociate the molecule, and hence the more stable the molecule is.

As time goes on, the molecule moves back and forth from one side of the potential-energy curve to the other, i.e., from small values of r to large values of r, as shown in Fig. 9.6b. The maximum and minimum values of r are determined by the vibrational kinetic energy K_v. In the case of a stable H_2 molecule, K_v is too small to allow the atoms to escape from the potential-energy valley in which they find themselves. If K_v is increased, say, by raising the temperature, r_{max} also increases. When $K_v = W_B$, $r_{max} \rightarrow \infty$ and the molecule breaks up into two separate atoms. This is called *dissociation*.

Chemical Bonds

There are two main kinds of chemical bonds, the ionic bond and the covalent bond. Ionic bonding occurs between atoms which easily form positive or negative ions. Li easily loses an electron and forms Li^+, and F easily picks up an electron to form F^-. The positively charged Li^+ will then attract the negatively charged F^- ion, since unlike electric charges attract each other. The two atoms therefore form a lithium fluoride molecule (LiF), with the two ions held together by an *ionic bond*, as in Fig. 9.7. Ordinary table salt, NaCl, is held together by this same kind of ionic bond.

On the other hand, in the case of H_2, as discussed above, bonding occurs because, even though the two H atoms are uncharged, the energy of the molecule is still less than the energy of the two isolated atoms. Hence if the electrons in two atoms of hydrogen can rearrange themselves into a more stable (lower-energy) configuration by forming H_2, they will do so. This is

FIGURE 9.7 The ionic bonding between the Li⁺ and the F⁻ ions to form the molecule LiF.

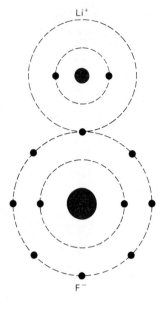

FIGURE 9.8 The formation of a hydrogen molecule from two hydrogen atoms. In this case the bonding between the two atoms is not ionic but covalent.

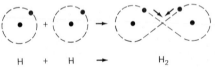

illustrated in Fig. 9.8. The bonds in molecules like H_2, O_2, and N_2 are clearly not ionic, for no ions are present. They are called *covalent*, or *homopolar*, bonds because they involve two similar atoms.

Many bonds observed in chemical compounds like CO and CO_2 are mixtures of ionic and covalent bonds.

9.4 The Mole and Avogadro's Number

The chemical equation $2H_2 + O_2 \rightarrow 2H_2O$, for the reaction shown schematically in Fig. 9.9, can be interpreted as applying to the reaction of two molecules of hydrogen with one molecule of oxygen to form two molecules of water vapor (H_2O). More commonly, however, chemical equations are interpreted as applying to "moles" of the substance involved.

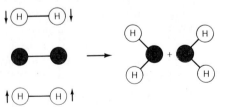

FIGURE 9.9 A simplified model of how hydrogen and oxygen molecules combine to form water vapor.

Definition

Mole (mol): The amount of a substance that contains as many elementary entities as there are atoms in 12 g of $^{12}_{6}C$. This number is called Avogadro's number N_A and is equal to 6.022×10^{23} mol⁻¹.

The mole is the SI base unit of amount of substance. In this definition the elementary entities may be atoms, molecules, ions, electrons, or other kinds of particles.

One mole of any substance contains as many elementary entities as one mole of any other substance. This can be guaranteed by taking for a mole of any substance a mass which is equal to the atomic or molecular mass number of the substance. In this book we will use the *gram-mole* (g·mol) to be consistent with what is done in chemistry courses. A gram-mole is the mass of a substance *in grams* equal to its atomic or molecular mass number. A gram-mole of any substance then contains 6.022×10^{23} atoms or molecules.

For example, to a good approximation (since the true mass numbers are not actually integers), 1 g·mol of ^4_2He consists of 4 g of ^4_2He atoms; and 1 g·mol of ordinary H_2O consists of $2(1) + 16 = 18$ g of H_2O molecules. Both samples contain Avogadro's number of atoms (for ^4_2He) or molecules (for H_2O).

When a book uses the term *mole* without further specification, it usually means gram-mole, and this will be the case in this book. The *kilogram-mole* (or kilomole) is often used in physics texts, but in this book we will stick with the gram-mole to avoid confusion. Note that if the kilogram-mole (kg·mol) is used, Avogadro's number has the value $N_A = 6.022 \times 10^{26}$ (kg·mol)$^{-1}$ since 1 kg = 10^3 g.

Avogadro's number is named for Amedeo Avogadro (1776–1856), the Italian physicist who first proposed the hypothesis that all gases at the same pressure and temperature contain the same number of particles per unit volume.

Example 9.4

If each person in the world (about 4.5 billion in 1980) were assigned the task together of counting the number of oxygen molecules in 1 mol of molecular oxygen, how long would they take, presupposing that each person can count at the rate of 5 molecules per second?

SOLUTION

One mole of molecular oxygen (O_2) contains Avogadro's number, or 6.02×10^{23} oxygen molecules. Hence 4.5 billion people counting this number of molecules at a rate of 5 molecules per second would take a time

$$t = \frac{6.02 \times 10^{23} \text{ molecules}}{5(4.5 \times 10^9) \text{ molecules/s}} = 2.68 \times 10^{13} \text{ s}$$

Converting seconds to years, we get

$$t = (2.68 \times 10^{13} \text{ s}) \left(\frac{1 \text{ h}}{3600 \text{ s}} \right) \left(\frac{1 \text{ day}}{24 \text{ h}} \right) \left(\frac{1 \text{ year}}{365 \text{ days}} \right)$$

$$= \boxed{8.5 \times 10^5 \text{ years}}$$

Hence, it would take almost 1 million years to count the molecules in just 1 mol of O_2, even with all the people in the world counting. This should give us some feeling for the staggering number of molecules which make up our universe.

Unified Mass Unit

This introduction of Avogadro's number leads us to another unit of great importance in atomic and nuclear physics, the *unified mass unit*, usually abbreviated u, and defined in terms of the mass of $^{12}_6C$. From the above discussion 1 mol of $^{12}_6C$ atoms has a mass of 12 g, and contains 6.022×10^{23} atoms. The unified mass unit (u) is then defined to be $\frac{1}{12}$ the mass of the $^{12}_6C$ atom (including electrons), where this mass is taken to be exactly 12.00 g. Then

Unified Mass Unit

$$1 \text{ u} = \frac{1}{12} \left(\frac{12.00 \text{ g}}{6.022 \times 10^{23}} \right) = 1.661 \times 10^{-24} \text{ g} = 1.661 \times 10^{-27} \text{ kg} \qquad (9.1)$$

This value of the unified mass unit is roughly equal to the mass of the proton and hence to the mass of the hydrogen atom. It is not exactly equal to the mass of the hydrogen atom, however, because of the choice of the $^{12}_6C$ atom as the standard for its definition. For this reason the mass of the hydrogen atom is, more precisely, 1.0078 u, and so

$$m(^1_1H) = (1.0078 \text{ u})(1.661 \times 10^{-27} \text{ kg/u}) = 1.674 \times 10^{-27} \text{ kg}$$

This is consistent with the fact that a hydrogen atom consists of one proton and one electron. Using the values for their masses from Table 9.1, we have

$$m(^1_1H) = (1.67265 \times 10^{-27} \text{ kg}) + (9.11 \times 10^{-31} \text{ kg}) = 1.6736 \times 10^{-27} \text{ kg}$$

which agrees to four significant figures with the result just obtained.

The atomic mass numbers found in the periodic table of the elements represent the mass, *in unified mass units*, of the atom as found in nature. Thus the atomic mass of chlorine is 35.45 u. Since 1 u = 1.661×10^{-24} g, and 1 mol of atomic Cl contains N_A atoms, the mass of 1 mol of atomic Cl is

$$m(_{17}Cl) = (35.45 \text{ u})(1.661 \times 10^{-24} \text{ g/u})(6.022 \times 10^{23}) = 35.45 \text{ g}$$

Hence *the mass in grams of 1 mol of any substance is equal to the mass in unified mass units of one atom (or molecule) of the same substance.*

The reason so many of the atoms in the periodic table have nonintegral mass numbers is therefore twofold: the presence of isotopes of different mass numbers and the fact that all mass numbers are measured relative to $^{12}_6C$, which is defined to have a mass number of exactly 12.00000 u. Another reason that mass numbers have nonintegral values will become clear later when we discuss nuclear physics and the conversion of mass into energy.

Example 9.5

The atomic mass of the oxygen isotope $^{16}_8O$ is 16.00, and oxygen as found in nature is almost 100% $^{16}_8O$. Find the mass in kilograms of an oxygen atom.

SOLUTION

Since the atomic mass is equal to the mass of one atom of any substance in atomic mass units, and 1 u = 1.66×10^{-27} kg, we have

The same method can be used for finding the mass of any atom or molecule.

$$m(^{16}_8O) = (16.00 \text{ u})(1.66 \times 10^{-27} \text{ kg/u})$$

$$= \boxed{2.66 \times 10^{-26} \text{ kg}}$$

Example 9.6

How many atoms are contained in a tiny speck of gold foil, which is in the shape of a cube 10^{-5} cm on a side? The mass number of gold is 197 and its density is 19.3×10^3 kg/m³.

SOLUTION

Since the density of any substance is given by $d = M/V$, we have for the gold speck:

Now, 1 mol of gold contains Avogadro's number N_A of atoms and has a mass of 197 g. Hence we can set up a proportion:

$$M = dV = \left(19.3 \times 10^3 \frac{\text{kg}}{\text{m}^3}\right)(10^{-7} \text{ m})^3$$

$$= 1.93 \times 10^{-17} \text{ kg} = 1.93 \times 10^{-14} \text{ g}$$

$$\frac{N \text{ atoms of gold}}{1.93 \times 10^{-14} \text{ g}} = \frac{6.02 \times 10^{23} \text{ atoms}}{197 \text{ g}}$$

and so $N = \left(\frac{1.93 \times 10^{-14}}{197}\right) (6.02 \times 10^{23} \text{ atoms})$

$$\boxed{= 5.9 \times 10^7 \text{ atoms}}$$

Hence this tiny speck of gold foil contains about 59 million atoms!

9.5 The Structure of Gases

Atoms are made up of protons, neutrons, and electrons. Molecules in turn are made up of atoms, in some cases identical atoms like the two oxygen atoms in O_2, in other cases dissimilar atoms, as in H_2O. These molecules, while preserving their separate identities, form the three states in which matter normally exists: the solid, the liquid, and the gaseous states.* In this and in the following chapter, we will see how the atomic structure of matter can explain the behavior of these three states of matter.

Gases

Definition

Gas: A substance which expands to fill uniformly the volume of any container in which it is placed. (See Fig. 9.10.)

FIGURE 9.10 (left)   A gas in a container of variable size: (*a*) Gas atoms confined to a small volume; (*b*) the piston is raised and the gas atoms move so as to fill uniformly the additional space in the container.

FIGURE 9.11 (right)   Three types of homogeneous gases: (*a*) A gas made up of atoms (He); (*b*) a gas made up of diatomic molecules (N_2); (*c*) a gas made up of polyatomic molecules (CO_2).

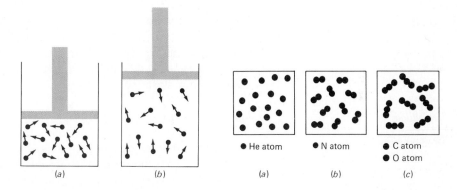

In gases the molecular speeds are very great, and the attractive forces among the molecules small. The effect of the attractive forces can be increased either by lowering the temperature or by squeezing the gas down into a very small volume by an increase in pressure. In this way it is possible to liquefy gases.

Some gases are homogeneous throughout, and are made up of either single atoms like He, diatomic molecules like nitrogen (N_2), or polyatomic molecules like carbon dioxide (CO_2). In this case every molecule (or atom) in the gas is exactly the same, as is shown in Fig. 9.11.

Other gases are mixtures of different molecules, and hence at the microscopic level are nonhomogeneous. For example, clean, dry air is 78%

**Plasmas are sometimes called "the fourth state of matter." These are highly ionized gases which exist only at very high temperatures. We shall postpone any discussion of plasmas until we take up nuclear fusion in Chap. 29.

nitrogen and 21% oxygen, with the remaining 1% consisting mostly of argon (Ar), carbon dioxide (CO_2), neon (Ne), and helium (He). Thus if we could sample a very small bit of air containing only five molecules, we would find, on average, four molecules of N_2 and one of O_2, as shown in Fig. 9.12. Note that we would never find a "molecule" of air, no matter how hard we tried, since a molecule of air does not exist. All that exists at the molecular level are molecules of a number of gases which are mixed together in the proper proportions to make the substance we colloquially call "air."

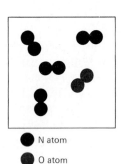

N atom

O atom

FIGURE 9.12 A sample of air containing only five molecules. On the average we would expect four of these to be N_2, and one O_2.

Important Properties of Gases

There are a number of large-scale (macroscopic) properties of gases which can be easily measured. These include:

Definition

Volume: The space occupied by a gas.

Since a gas expands to fill any container in which it is placed, the volume of a gas is equal to the interior volume of the container in which the gas finds itself. Volumes are measured in m^3.

Definition

Mass $\mathcal{M}$: The total mass of all the atoms or molecules making up the gas.

The total mass of gas is merely the number of atoms (or molecules) present in the gas times the mass of each atom (or molecule). Another way of expressing this same mass is by giving the number n of moles of gas present times the mass M of 1 mol of gas. Hence we have

$$\mathcal{M} = nM \tag{9.2}$$

Thus the mass of 5 mol of hydrogen gas (H_2) is

$$\mathcal{M} = nM = (5 \text{ mol})(2.0 \text{ g/mol}) = 10 \text{ g} = 0.010 \text{ kg}$$

Definition

Density d: The mass of gas per unit volume.

$$d = \frac{\mathcal{M}}{V} \tag{9.3}$$

In the SI system densities are expressed in kilograms per cubic meter (kg/m^3). For example, the density of H_2 is 0.0899 kg/m^3 and for radon gas it is 9.73 kg/m^3.

Definition

Pressure P: The force per unit area acting perpendicular to the surface of the container holding the gas.

$$P = \frac{F}{A}$$

(9.4)

Pressures are measured in newtons per square meter (N/m^2). (In the British engineering system pressures are given in pounds per square inch, the units in which tire pressures are still commonly measured today.) The pressure of a gas is caused by the atoms or molecules of the gas moving around in all directions inside the container and colliding with its walls. The force they exert on any given surface of the container, divided by the area of that surface, is the pressure of the gas. The pressure exerted by the gas is the same throughout the container. Thus the pressure of the air in an automobile tire is the same at any point on the wall of the tire.

9.6 Temperature

The last important property of a gas to be introduced is in some ways the most important and yet the most elusive. We all have some idea of what temperature means and could define it loosely as follows:

Definition

Temperature: A measure of the hotness or coldness of an object.

We derive our ideas about temperatures from our sense of touch. But, if we are interested in *measuring* temperatures with any scientific accuracy, our sense of touch is of little use. Although we can tell, by touching an object, whether it is hot or cold, this is a very subjective judgment that can easily lead us astray. For example, if you put your left hand under a hot water faucet and your right under a cold water faucet, let the water run for a few minutes, and then plunge your two hands at the same instant into a pot of lukewarm water, your left hand will feel cool and your right hand warm even though they are both in water at the same temperature. Hence we clearly need a more objective and quantitative way than touch to determine an object's temperature.

The scientific measurement of temperature is based on the experimental fact that two objects in contact always come to the same temperature if we wait long enough. We know from experience that if we put a TV dinner in a hot oven and wait a half hour or so, the TV dinner will become as hot as the inside of the oven. Heat constantly flows from the oven into the TV dinner, which heats up until it comes into thermal equilibrium with the oven, that is, until there is no net flow of heat into or out of the TV dinner. When this occurs, we say that the oven and the TV dinner are in *thermal equilibrium*, or "at the same temperature."

To measure temperature we use a *thermometer* which, when put into close contact with the object whose temperature is to be determined, can come into thermal equilibrium with the object. For example, when a doctor puts a thermometer into a patient's mouth to measure body temperature, the doctor expects that after a few minutes the thermometer will come into thermal equilibrium with the patient's body. Only then can the doctor be sure that the temperature as read from the thermometer actually corresponds to the patient's body temperature. This is the reason the doctor leaves the thermometer in the patient's mouth for several minutes before reading it.

The physical basis for the use of thermometers to measure temperature is an experimental fact sometimes referred to as the *zeroth law of thermodynamics* (see Fig. 9.13).

FIGURE 9.13 The zeroth law of thermodynamics: If system *A* is in thermal equilibrium simultaneously with system *B* and system *C*, then *B* and *C* must be in thermal equilibrium with each other, or they must be at exactly the same temperature.

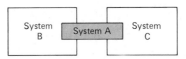

Zeroth Law of Thermodynamics

Two systems in thermal equilibrium with a third system are in thermal equilibrium with each other.

For example, if we insert a thermometer in a pot of boiling water and then into a pot of coffee, in each case allowing thermal equilibrium to be established, and if we find that in both cases the thermometer reads exactly the same value, then we say that the water and the coffee are *at the same temperature.*

Temperature is a new kind of physical quantity which cannot be defined in terms of the three fundamental quantities of mechanics—mass, length, and time. It is an *intensive* property of a substance—that is, it measures the degree of hotness or coldness—and *is independent of how much of the substance is present*, as long as the number of atoms or molecules present is statistically large. Hence the temperature of one drop of water can be the same as that of a million gallons of water. If a thermometer is first brought into thermal equilibrium with the drop of water, and then with the million gallons (assuming that the million gallons are all the same temperature), and the thermometer reads the same in the two cases, the temperature of the drop is identical with that of the million gallons, according to the zeroth law.

We will discuss practical thermometers in Sec. 13.3. For now we can consider a thermometer as simply a device that has some measurable physical property which changes with temperature, for example, the length of a column of mercury in a mercury thermometer or the electric resistance of a metal in a resistance thermometer. Such devices can be provided with temperature scales based on assigned values for important temperatures such as the boiling point of water or the melting point of ice. When such a thermometer comes into thermal equilibrium with some object whose temperature is desired, the change in this measurable property enables the temperature of the object to be read directly from the scale on the thermometer.

9.7 The Kinetic Theory of an Ideal Gas

We now want to see how much the basic ideas of the mechanics of point masses, developed in previous chapters, can tell us about the actual behavior of a very real and practical substance such as the air around us.

In doing this we will assume the validity of Newton's laws, discussed in Chap. 3. We will also make certain simplifying assumptions about the behavior of a gas that will enable us to carry out calculations relating the macroscopic (large-scale) behavior of the gas to its microscopic (atomic or molecular) structure.

We call an *ideal gas* any gas which has the following properties:

1 The gas consists of a large number (N) of identical molecules,* each of mass m. These molecules move at random in all directions with a great variety of speeds. This is the reason why a gas fills any container in which it is placed.

2 The distances between the molecules in the gas are large compared with the size of the molecules, which have diameters of about 10^{-10} m. Each molecule may therefore be considered to be a point particle of the kind previously treated in mechanics.

3 The forces on the moving molecules are negligible except when they collide with one another or with the walls of the container.

4 Collisions between two molecules or between a molecule and a wall are assumed to be perfectly elastic; i.e., no kinetic energy is converted into heat in the collision. Hence *both momentum and kinetic energy are conserved*. (See Sec. 7.4 on elastic collisions.)

Although these are the properties of an ideal gas, many real gases at low pressures and at temperatures well above the boiling point also satisfy these conditions.

Molecular Basis of Gas Pressure

We first use this model of an ideal gas to show how the pressure of a gas depends on the motion of its molecules.

Let our ideal gas, consisting of N identical molecules, be confined to a rectangular box of length L and cross-sectional area A, as in Fig. 9.14. The

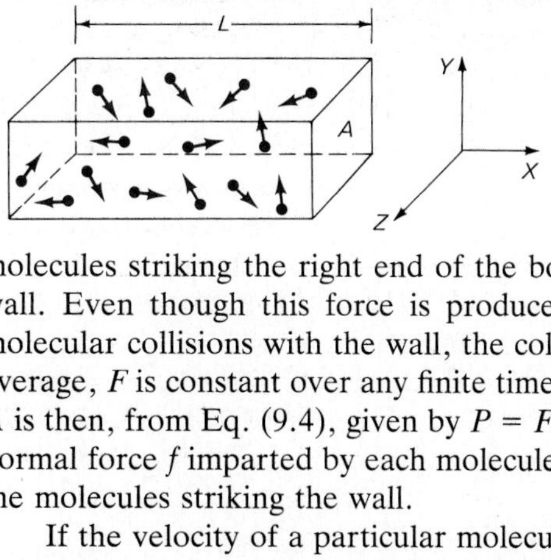

FIGURE 9.14 A box containing a gas. Its length is L, which is directed along the X axis, and its cross-sectional area (in the YZ plane) is A.

molecules striking the right end of the box exert an average force F on the wall. Even though this force is produced by large numbers of individual molecular collisions with the wall, the collisions are so frequent that, on the average, F is constant over any finite time interval. The pressure on the area A is then, from Eq. (9.4), given by $P = F/A$. To calculate F we must find the normal force f imparted by each molecule to the wall, and then sum over all the molecules striking the wall.

If the velocity of a particular molecule is v, we can resolve this velocity into components v_x, v_y, and v_z along the three edges of the box. If a molecule with velocity component v_x collides elastically with the right end of the box, the X component of its velocity is reversed, as in Fig. 9.15, so that the change in its momentum is

$$\Delta p = p_{\text{final}} - p_{\text{init}} = -mv_x - mv_x = -2mv_x$$

FIGURE 9.15 Transfer of momentum between a gas molecule and the end wall of a container. The momentum of the molecule is reversed on its hitting the rigid wall.

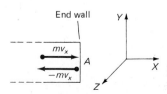

*Some gases, such as helium (He) or neon (Ne), are made up of atoms, not molecules. In what follows we will for simplicity use the term *molecule* for the basic constituents of the gas, whether they be atoms or molecules.

For our purpose, because the pressure is defined in terms of the force perpendicular to the wall and the wall is in the YZ plane, only the X component of momentum is important. Since the change in the momentum of the molecules is $-2mv_x$, the change in the momentum of the wall must be $+2mv_x$, for the total momentum of the system is conserved, according to the principle of conservation of linear momentum (Sec. 7.3).

If the molecule now moves to the left end of the box and back again without colliding with another molecule, it will do so in the time it takes to travel a distance twice the length of the box at a speed v_x, so that

$$t = \frac{2L}{v_x} \tag{9.5}$$

Hence every t s the molecule will strike the right end again and deliver up more momentum. Since the molecular speeds are so high, this may happen thousands of times a second, as we will see when we discuss the actual speeds of gas molecules. The number of collisions per second of this molecule with the wall is then $1/t = v_x/2L$.

Hence the amount of momentum transferred per second to the right end by each molecule is the amount transferred per collision times the number of collisions per second, or

$$(2mv_x)\left(\frac{v_x}{2L}\right) = \frac{mv_x^2}{L}$$

Since, by Newton's second law [see Eq. (7.2)], the force is equal to the time rate of change of momentum, i.e., to the amount of momentum transferred to the wall per second, we have for the force exerted by each molecule on the wall

$$f = \frac{mv_x^2}{L}$$

The total force is obtained by summing over all the molecules, where each molecule is assumed to have a different X component of velocity v_{xi}. Hence

$$F = \sum_i f_i = \sum_{i=1}^{N} \frac{mv_{xi}^2}{L} = \frac{m}{L} \sum_{i=1}^{N} v_{xi}^2$$

since m and L are constants. The pressure is then, from Eq. (9.4),

$$P = \frac{F}{A} = \frac{m}{LA} \sum_{i=1}^{N} v_{xi}^2 = \frac{m}{V} \sum_{i=1}^{N} v_{xi}^2$$

where V is the volume of the gas, equal to the length L of the container times its cross-sectional area A.

Hence, on multiplying both sides by V and also multiplying and dividing the right side by N, we have

$$PV = Nm \sum_{i=1}^{N} \frac{v_{xi}^2}{N}$$

But the quantity $\sum_{i=1}^{N} v_{xi}^2/N$ is simply the average value of v_x^2 for all the molecules. We write this quantity as $\overline{v_x^2}$, which is defined to be

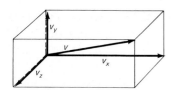

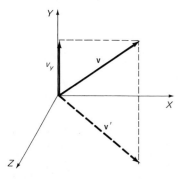

FIGURE 9.16 Relationship between the velocity of a gas molecule and the three components of that velocity.

FIGURE 9.17 Application of the pythagorean theorem to the velocity of gas molecules. **v** is the vector velocity of a gas molecule with components v_x, v_y, and v_z. In the figure **v′** is the projection of **v** on the XZ plane. Hence $v'^2 = v_x^2 + v_z^2$. But, from the diagram, $v^2 = v'^2 + v_y^2$. Hence $v^2 = v_x^2 + v_y^2 + v_z^2$.

Kinetic Energy of a Gas

$$\overline{v_x^2} = \sum_{i=1}^{N} \frac{v_{xi}^2}{N} \tag{9.6}$$

Hence we have

$$PV = Nm\overline{v_x^2} \tag{9.7}$$

Now, there is nothing about motion in the X direction to distinguish it from motion in the Y direction or Z direction. We would therefore expect, since the pressure is the same on all sides of the box, that

$$\overline{v_x^2} = \overline{v_y^2} = \overline{v_z^2}$$

From the pythagorean theorem, as shown in Figs. 9.16 and 9.17, we have, on using this last result,

$$\overline{v^2} = \overline{v_x^2} + \overline{v_y^2} + \overline{v_z^2} = 3\overline{v_x^2} \tag{9.8}$$

and so $\overline{v_x^2} = \frac{1}{3}\overline{v^2}$, where $\overline{v^2}$ is the square of the speed averaged over all the molecules, without respect to direction. Hence, from Eqs. (9.7) and (9.8),

$$PV = \tfrac{1}{3}Nm\overline{v^2}$$

We can multiply and divide the right side by 2 and obtain

$$PV = \frac{2N}{3}(\tfrac{1}{2}m\overline{v^2}) \tag{9.9}$$

Now $\tfrac{1}{2}m\overline{v^2}$ is the *average kinetic energy* of one molecule, and $N(\tfrac{1}{2}m\overline{v^2})$ is the *total kinetic energy of the gas*. We have therefore obtained the interesting and extremely important result that the product of the pressure and the volume of an ideal gas is equal to two-thirds the total kinetic energy of the gas, or

$$\boxed{PV = \tfrac{2}{3}\text{KE}} \tag{9.10}$$

This calculation, while one of the most involved in this text, is worth the effort required to understand it, since it illustrates how the classical mechanics of point masses developed in Chaps. 2 to 8 can lead to significant information about the behavior of bulk matter.

In this calculation we have neglected collisions between molecules, but the same result would be obtained even if collisions were taken into consideration. This is because all collisions are assumed to be elastic, and hence velocities are merely exchanged in the collision of two molecules (consider a billiard ball striking another billiard ball at rest). As a consequence, in our box model, there will always be some molecule returning to the right wall with a momentum mv_x corresponding to the molecule which left the wall with the same numerical value for the momentum. This must be the case since momentum in the X direction is conserved. Hence by neglecting collisions we can simplify the problem without introducing any serious error in the result. One of the secrets of theoretical physics is finding such simple models susceptible to calculation which do not radically distort the real physical situation.

The model we have used to obtain Eq. (9.10) is called the *kinetic theory of gases* model because it enables us to relate macroscopic quantities like pressure and volume to the microscopic *motion* of the gas molecules (the Greek word *kinesis* means "motion").

9.8 The Ideal Gas Law

Suppose we have a mass $\mathcal{M}$ of He gas in a container of volume V closed off by a movable piston and sitting on a hot plate, as in Fig. 9.18. Let us express the mass of the gas as $\mathcal{M} = nM$, where n is the number of moles of the gas and M is the mass in grams of 1 mol. We perform a series of experiments to see how the volume V, the pressure P, and the temperature T are related for this gas when the number of moles (n) remains unchanged.

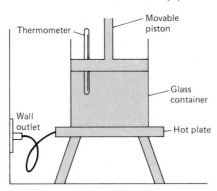

FIGURE 9.18 A glass container, filled with He gas, on a hot plate.

Boyle's Law

First we keep the temperature T fixed, and vary P and V by compressing the gas slowly with the piston. If the pressure and volume change from P_1 and V_1 before the compression to P_2 and V_2 after the compression, we find experimentally that

FIGURE 9.19 Boyle's law for an ideal gas. At constant temperature the volume of the gas varies inversely with the gas pressure. This law takes its name from the British chemist and physicist Robert Boyle (1627–1691).

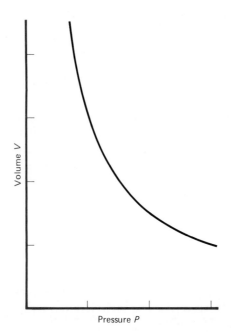

Pressure P

$$P_1V_1 = P_2V_2$$

If we repeat this for a great variety of different values of P and V, always keeping the temperature unchanged, we find that in every case

Boyle's Law

$$PV = \text{constant} \quad \text{(if } T \text{ is constant)} \tag{9.11}$$

Boyle's law simply states that, at constant temperature, the volume of a gas is inversely proportional to its pressure, since PV is equal to a constant. This is shown in Fig. 9.19.

An ideal gas is sometimes defined as a gas which obeys Boyle's law, as do most gases at reasonably low densities and not-too-low temperatures.

Charles' and Gay-Lussac's laws

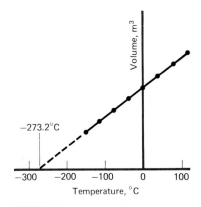

FIGURE 9.20 Variation of the volume of an ideal gas with the temperature (°C) at constant pressure.

We can perform a second experiment by keeping a fixed pressure on the gas, say, by putting some weights on the piston and allowing it to move up and down freely and thus apply a fixed force on the gas as the temperature varies. The Frenchman Jacques Charles (1746–1823) first performed this experiment and found that the volume increased linearly with temperature, at fixed pressure. This is shown in Fig. 9.20. We see that we get a straight-line graph for the volume-versus-temperature curve for helium, so long as the temperature does not come too close to −269°C, which is the temperature at which helium liquefies. Even though we cannot reduce the volume to zero, we find that, if we extrapolate the straight line in Fig. 9.20, it crosses the X axis at −273.2°C. As a matter of fact, if we repeat this experiment for a group of different ideal gases, their graphs will have different slopes, but all will cross the X axis at the same value of the temperature, −273.2°C, as shown in Fig. 9.21. This temperature, at which the volume of all ideal gases would be reduced to zero (if they did not liquefy), is called the *absolute zero of temperature.* It is the basis for the introduction of the *absolute* or *Kelvin temperature scale.*

We introduce the Kelvin temperature scale by setting its zero at −273.2°C, and choosing the size of a kelvin to be exactly the same as a Celsius (or centigrade) degree. Hence the freezing point of water is 0°C or 273.2 K,* and the boiling point of water is 100°C or 373.2 K. There are therefore 100 degrees between the freezing and boiling points of water on each of these scales. Any Celsius temperature can then be converted into the corresponding Kelvin temperature by simply adding 273.2 to the Celsius value.

When the volume of an ideal gas at constant pressure is plotted against the Kelvin temperature, we obtain a straight line which now passes through the origin, i.e., through the zero of temperature, as shown in Fig. 9.22. Hence on the Kelvin scale the volume of an ideal gas at constant pressure is directly proportional to the absolute temperature; i.e.,

Charles' Law

$$V \propto T \quad \text{(at constant } P) \tag{9.12}$$

This is *Charles' law*, sometimes also called Gay-Lussac's law.

---

*On the absolute scale an international commission has decided to use *kelvins* (symbol K) to specify temperature instead of "degrees Kelvin," or °K.

FIGURE 9.21 (left)   Variation of the volume of a group of different ideal gases with temperature, at constant pressure. Note how the extrapolations of the straight lines all cross the temperature axis at −273.2°C.

FIGURE 9.22 (right)   The same graph of volume against temperature for a group of gases as in Fig. 9.21, but with the new zero of the temperature scale taken at −273.2°C, and with temperature intervals the same as on the Celsius scale. The result is the absolute, or Kelvin, temperature scale.

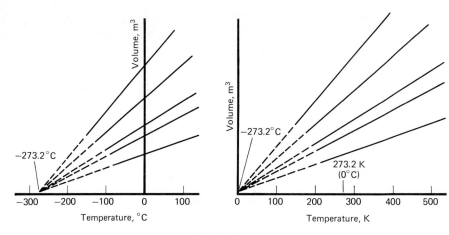

There is a third related gas law which states that, if the volume is held constant, the pressure of a gas is directly proportional to its absolute temperature, or

$$P \propto T \qquad \text{(at constant } V) \tag{9.13}$$

This is the law which governs the pressure in our automobile tires. For a fixed amount of air in a tire, the pressure increases in the summer and is reduced in the winter by amounts which depend on the change in the *Kelvin temperature*.

Example 9.7

The pressure in an automobile tire is P_1 on a hot summer day when the temperature is 38°C. What will be the tire pressure on a very cold day in winter when the temperature is −18°C? What is the percentage change in the pressure?

SOLUTION

According to Eq. (9.13),

$$P \propto T$$

or

$$\frac{P_2}{P_1} = \frac{T_2}{T_1}$$

Hence

$$P_2 = \frac{T_2}{T_1} P_1$$

Now 38°C = 38 + 273 = 311 K, and −18°C = 273 − 18 = 255 K. Hence

$$P_2 = \frac{255}{311} P_1 = \boxed{0.82P_1}$$

Hence from summer to winter the tire pressure has fallen by 18 percent. This shows why it is a good idea to add more air to automobile tires in the winter.

Note: P_1 and P_2 are here the absolute or total pressures inside the tire. An automobile pressure gauge reads the difference between the pressure inside the tire and the atmospheric pressure outside the tire. The atmospheric pressure is the pressure exerted at ground level by all the air in the atmosphere above the point on the ground. Hence the absolute pressure is the gauge pressure plus atmospheric pressure. An atmospheric pressure of 1 atmosphere (1 atm) is equal to 1.013×10^5 N/m², or 14.7 lb/in². Hence a gauge pressure of 1.73×10^5 N/m² (or 25 lb/in²) corresponds to an absolute tire pressure

$$P = P_{\text{gauge}} + P_{\text{atm}}$$

$$= 1.73 \times 10^5 \text{ N/m}^2 + 1.013 \times 10^5 \text{ N/m}^2$$

$$= 2.74 \times 10^5 \text{ N/m}^2$$

or

$$P = P_{\text{gauge}} + P_{\text{atm}}$$

$$= 25 \text{ lb/in}^2 + 14.7 \text{ lb/in}^2$$

$$= 39.7 \text{ lb/in}^2$$

Some useful factors to be employed in converting pressures are given in Table 9.3. The significance of these different pressure units, and how they are to be used, will become clearer in the chapters that follow.

TABLE 9.3   Useful Pressure Units and Their Conversion Factors

Pressure unit		Conversion factor			
		atm	N/m² (Pa)	lb/in²	mmHg (torr)
1 atm	=	1	1.013×10^5	14.7	760
1 N/m² (Pa)	=	9.87×10^{-6}	1	1.45×10^{-4}	7.50×10^{-3}
1 lb/in²	=	6.80×10^{-2}	6.90×10^3	1	51.7
1 mmHg (torr)	=	1.32×10^{-3}	133	1.93×10^{-2}	1

Ideal Gas Law

If we put these three gas laws together, we obtain the following general equation:

Ideal Gas Law

$$\boxed{PV = nRT}$$ (9.14)

where n is the number of gram-moles of gas present and R is a constant. Experimentally R turns out to be the same constant for all gases and is therefore called the *universal gas constant*. Its value in SI units is

$$R = 8.314 \text{ J/(mol·K)}$$

or $$R = 8.314 \times 10^3 \text{ J/(kg·mol·K)}$$

Equation (9.14) is called the *ideal gas law*, or the equation of state of an ideal gas. It is obeyed by most gases at pressures near or below atmospheric pressure, so long as the temperature is not too low.

For Eq. (9.14) to be applied correctly to practical problems using SI units, P should be in N/m², V in m³, and T in K. If n is in mol, then R must be in J/(mol·K); whereas if n is in kg·mol, then R must be in J/(kg·mol·K). In chemistry it is usual practice to measure P in atmospheres and V in liters. In this case R must have the value 0.0821 liter·atm/(mol·K).

Example 9.8

A metal cylinder contains 100 liters of oxygen gas at a temperature of 20°C and a pressure of 10 atm. The gas is then compressed by a piston to a volume of 50 liters, while the temperature increases to 30°C. What is the final pressure of the gas?

SOLUTION

Since, from Eq. (9.14), $PV = nRT$, we have

$$\frac{P_1 V_1}{T_1} = nR = \frac{P_2 V_2}{T_2}$$

where the subscript 1 refers to the initial situation of the gas and 2 to the final situation. Our values are therefore

$P_1 = 10$ atm $V_1 = 100$ liters $T_1 = 20°C = 293$ K

$P_2 = ?$ $V_2 = 50$ liters $T_2 = 30°C = 303$ K

We have deliberately introduced here the non-SI units liters and atmospheres to show that problems of this type can be solved no matter what units are used, so long as the units are the same on the two sides of the equation. Thus solving the above equation for P_2, we have

$$P_2 = \frac{T_2 V_1}{T_1 V_2} P_1$$

$$= \frac{(303 \text{K})(100 \text{ liters})(10 \text{ atm})}{(293 \text{ K})(50 \text{ liters})} = \boxed{20.6 \text{ atm}}$$

The temperature and volume units cancel, and the pressure is expressed in the same units as was the original pressure. Hence in problems involving changes in P, V, and T for gases, *so long as the amount of gas remains unchanged*, the units may be of any kind and may even involve two different systems and the problem can still be solved. The crucial thing to remember is that the temperature *must* be in kelvins. All gas-law problems in which the amount of gas present does not change can be solved by using the following equation:

$$\frac{P_1V_1}{T_1} = \frac{P_2V_2}{T_2}$$
(9.15)

Example 9.9

Find the volume of 1 mol of any gas at STP (standard temperature and pressure, where standard temperature is 0°C, or 273 K, and standard pressure is 1 atm, or 1.01×10^5 N/m²).

SOLUTION

Since $PV = nRT$, we have $V = nRT/P$, or

$$V = \frac{(1.0 \text{ mol}) [8.31 \text{ J/(mol·K)}](273 \text{ K})}{1.01 \times 10^5 \text{ N/m}^2}$$

$$= \boxed{22.4 \times 10^{-3} \text{ m}^3}$$

Since 1 liter is 10^3 cm³, or 10^{-3} m³, 1 mol of any gas at STP therefore occupies 22.4 liters, and 1 kg·mol occupies 22.4×10^3 liters.

Notice how in this problem, in which the gas law must be solved explicitly, a consistent set of units must be used throughout if a meaningful answer is to be obtained. Example 9.10 illustrates this same point.

Example 9.10

A tank of nitrogen gas at 0°C and a pressure of 50 atm has a volume of 0.10 m³. What is the mass of nitrogen in the tank?

SOLUTION

We have $PV = nRT$, where

$P = 50 \text{ atm} = 50(1.01 \times 10^5 \text{ N/m}^2) = 50.5 \times 10^5 \text{ N/m}^2$

$V = 0.10 \text{ m}^3$

$T = 0°C = 273 \text{ K}$

$R = 8.31 \text{ J/(mol·K)}$

Hence the number of moles of N₂ present is

$$n = \frac{PV}{RT} = \frac{(50.5 \times 10^5 \text{ N/m}^2)(0.10 \text{ m}^3)}{[8.31 \text{ J/(mol·K)}](273 \text{ K})} = 2.23 \times 10^2 \text{ mol}$$

But 1 mol of N₂ has a mass of 28 g. Hence

$$\text{Mass of N}_2 = (2.23 \times 10^2 \text{ mol})\left(\frac{28 \text{ g}}{1 \text{ mol}}\right) = 6.2 \times 10^3 \text{ g}$$

$$= \boxed{6.2 \text{ kg}}$$

Relationship between Microscopic and Macroscopic Models

We now have two equations for ideal gases derived in very different ways, one based on the molecular structure of the gas and the other on its bulk properties. They are

$$PV = \tfrac{2}{3}N(\tfrac{1}{2}m\overline{v^2})$$
(9.9)

and $PV = nRT$
(9.14)

Putting these two equations together we have

$$\tfrac{2}{3}N(\tfrac{1}{2}m\overline{v^2}) = nRT$$

Now Nm is the total mass of the gas, since N is the number of molecules and m is the mass of each molecule. This can also be expressed as nM, from Eq. (9.2), where n is the number of moles of gas and M is the mass of 1 mol. Hence $Nm = nM$ and so $\tfrac{2}{3}n(\tfrac{1}{2}M\overline{v^2}) = nRT$

or $$\tfrac{1}{2}M\overline{v^2} = \tfrac{3}{2}RT \qquad (9.16)$$

The left side is the total translational kinetic energy per mole for the random motion of the gas molecules, since M is the mass of 1 mol. *The total translational kinetic energy per mole is therefore proportional to the Kelvin temperature*, according to Eq. (9.16).

If we divide Eq. (9.16) by Avogadro's number, we obtain

$$\frac{1}{2}\frac{M\overline{v^2}}{N_A} = \frac{3}{2}\frac{RT}{N_A}$$

or $$\tfrac{1}{2}m\overline{v^2} = \tfrac{3}{2}kT \qquad (9.17)$$

Here k is a constant called *Boltzmann's constant*, where

$$k = \frac{R}{N_A} = \frac{8.314 \text{ J/(mol·K)}}{6.022 \times 10^{23} \text{ molecules/mol}}$$

$$= 1.380 \times 10^{-23} \text{ J/K per molecule} \qquad (9.18)$$

The constant k is sometimes called the *gas constant per molecule*; it plays a very important role in many branches of both classical and modern physics.

Significance of Absolute Temperature T

Equation (9.17) shows that *the Kelvin temperature T of a gas is directly proportional to the average random translational kinetic energy of the gas molecules*. This is the basic microscopic significance of temperature. At a fixed temperature all gases, no matter what their volumes, pressures, masses, or other properties, have exactly the same average kinetic energy. The temperature of a gas is a measure of this average random translational kinetic energy.

Example 9.11

Calculate (*a*) the average kinetic energy, and (*b*) the approximate average speed of oxygen (O_2) molecules in the air at room temperature (about 20°C).

SOLUTION

(**a**) From Eq. (9.17) we have for the average kinetic energy of a gas molecule:

$$\tfrac{1}{2}m\overline{v^2} = \tfrac{3}{2}kT = 1.5(1.380 \times 10^{-23} \text{ J/K})(293 \text{ K})$$

$$= \boxed{6.07 \times 10^{-21} \text{ J}}$$

since 20°C is equal to 293 K.

(**b**) Since $\tfrac{1}{2}m\overline{v^2} = 6.07 \times 10^{-21}$ J

$$\overline{v^2} = \frac{2(6.07 \times 10^{-21} \text{ J})}{m}$$

Here m is the mass of one molecule of oxygen and must be expressed in kilograms in the SI system if the speed is to be obtained in meters per second. Now

$$m = \frac{M}{N_A} = \frac{32 \text{ g/mol}}{6.022 \times 10^{23} \text{ mol}^{-1}}$$

$$= 5.3 \times 10^{-23} \text{ g} = 5.3 \times 10^{-26} \text{ kg}$$

and so $\overline{v^2} = \dfrac{2(6.07 \times 10^{-21} \text{ J})}{5.3 \times 10^{-26} \text{ kg}} = 2.29 \times 10^5 \text{ m}^2/\text{s}^2$

and $\sqrt{\overline{v^2}} = (2.29 \times 10^5 \text{ m}^2/\text{s}^2)^{1/2} = \boxed{4.8 \times 10^2 \text{ m/s}}$

The same result could have been obtained by noting that the mass of one oxygen molecule is 32 u, or 32 (1.66×10^{-27} kg) = 5.3×10^{-26} kg, as above.

This result is not the average speed in the usual sense, but rather the square root of the average of the squared speeds, a quantity called the root-mean-square speed, v_{rms}, as we shall see in the next section.

Root-Mean-Square Speed

Table 9.4 gives some useful data on molecular speeds at 0°C. In this table v_{rms}, the *root-mean-square speed*, is $\sqrt{\overline{v^2}}$. Since $\frac{1}{2}m\overline{v^2}$ is constant for all gases at any particular temperature T, the larger the mass of a molecule, the smaller v_{rms}. This is clear from the table. On the other hand, from Eq. (9.16) the translational kinetic energy per mole, $\frac{1}{2}M\overline{v^2} = \frac{1}{2}Mv_{rms}^2$, is expected to be constant for all gases at the same T, since it is equal to $\frac{3}{2}RT$. The table shows that this is approximately verified by experiment. Note that in a container 1 m long a He atom would go back and forth the length of the container about 650 times a second because of its high value for v_{rms}.

TABLE 9.4   Speeds and Kinetic Energies of Some Important Gases at 0°C

Gas	Atomic or molecular mass M, kg/(kg·mol)	Root-mean-square speed $v_{rms} = \sqrt{\overline{v^2}}$, 10^2 m/s	Translational kinetic energy $\frac{1}{2}Mv_{rms}^2 = \frac{1}{2}M\overline{v^2}$, 10^6 J/(kg·mol)
H_2	2	18.4	3.37
He	4	13.1	3.43
Ne	20	5.9	3.43
N_2	28	4.9	3.42
O_2	32	4.6	3.39
CO_2	44	3.9	3.40

Gaseous Diffusion

If we have two gases of molecular masses m_1 and m_2 at the same temperature, from Eq. (9.17) the ratio of their molecular speeds is

$$\frac{v_{rms,1}}{v_{rms,2}} = \frac{\sqrt{\overline{v_1^2}}}{\sqrt{\overline{v_2^2}}} = \frac{\sqrt{m_2}}{\sqrt{m_1}} \tag{9.19}$$

FIGURE 9.23 Gaseous diffusion plant for the separation of uranium isotopes at the Department of Energy laboratory at Oak Ridge, Tennessee. (Photo courtesy of Union Carbide.)

Thus if we have two isotopes like $^{235}_{92}U$ and $^{238}_{92}\dot{U}$ at the same temperature, the lighter isotope will have the greater average speed. Since $^{235}_{92}U$ is the fissionable isotope needed for nuclear power plants and weapons, and makes up only 0.7 percent of uranium as found in nature, its concentration needs to be increased to be useful. One way to do this is to use a process called *gaseous diffusion* in which the gas uranium hexafluoride (UF_6) is allowed to diffuse through porous barriers over long distances. The gas emerging from the barrier first has an increased concentration of the lighter isotope since the lighter atoms have higher average speeds. This emerging gas can then be passed through a second porous barrier to increase even more the concentration of the lighter isotope. Four thousand diffusion stages are required to produce the almost pure $^{235}_{92}U$ used in nuclear weapons. The gaseous diffusion plants used to enrich uranium at Oak Ridge, Tennessee, and Hanford, Washington, are among the largest engineering projects ever constructed in the United States (see Fig. 9.23).

Example 9.12

What is the percentage difference, at room temperature, between the rms speeds of UF_6 molecules containing $^{235}_{92}U$ and those containing $^{238}_{92}U$?

SOLUTION

From the periodic table (Fig. 9.4) we find that a fluorine atom has a mass of 19.00 unified mass units (u). Hence the masses of the UF_6 molecules containing the two different uranium isotopes are $235 + 6(19) = 349$ u and $238 + 6(19) = 352$ u. Hence, from Eq. (9.19),

$$\frac{v_{rms}(^{235}U)}{v_{rms}(^{238}U)} = \frac{(352)^{1/2}}{(349)^{1/2}} = \frac{18.76}{18.68}$$

and the percentage difference is

$$\frac{v_{rms}(^{235}U) - v_{rms}(^{238}U)}{v_{rms}(^{238}U)} \times 100\% = \frac{18.76 - 18.68}{18.68} \times 100\%$$

$$= \boxed{0.43\%}$$

It is because this difference in the speeds is so small that so many stages are needed to produce the pure ^{235}U used in nuclear weapons.

9.9 Distribution of Molecular Velocities in Gases

Table 9.4 gave the values for the root-mean-square speeds of some common gas molecules at 0°C. For nitrogen (N_2), which makes up 80 percent of ordinary air, the value of v_{rms} is 490 m/s. This is, of course, only a special kind of average of the actual speeds of all the N_2 molecules in the gas. These speeds actually range from near zero to speeds exceeding 1500 m/s. Figure 9.24 shows a graph of the actual distribution of speeds in N_2 at three temperatures, 0°C (273 K), 1000°C (1273 K), and 2000°C (2273 K). Notice how the peak of the curve shifts to higher speeds as the temperature rises. This is what we would expect from Eq. (9.17), which says that the average kinetic energy for a gas depends on the temperature and hence that

$$\overline{v^2} \propto T$$

and so $v_{rms} = \sqrt{\overline{v^2}} \propto \sqrt{T}$

Number of molecules with indicated speed (relative units)

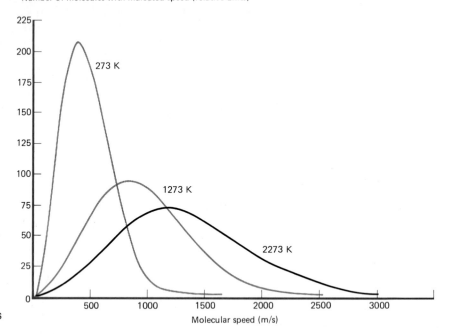

FIGURE 9.24 Maxwell distribution law for the molecular speeds of N_2 molecules at three different temperatures. Note how the peak moves to higher speeds as the temperature increases.

This is reasonably well verified by the data plotted in Fig. 9.24, if we note that v_{rms} falls somewhat to the right of the peak of each of the three curves.

The distribution of molecular speeds shown in Fig. 9.24 was first derived theoretically by the British physicist James Clerk Maxwell (1831–1879) in 1859, and is referred to as *Maxwell's distribution law*. This distribution is plotted more carefully for a sample of 10^6 oxygen molecules in Fig. 9.25. Here we plot along the Y axis the number of molecules out of the 10^6 which possess speeds in a particular range, say, from 200 to 250 m/s. When this is done for all possible speeds, from zero to very high values, we obtain the given distribution curve, which shows how the speeds are distributed in the gas. The total area under the curve corresponds to the total number of molecules (10^6), and the area of the vertical strip shown in color on the graph corresponds to the number possessing speeds between 250 and 300 m/s.

FIGURE 9.25 Distribution of molecular speeds in a sample of 10^6 oxygen (O_2) molecules at 0°C, showing the values of the most probable speed (v_p), the average speed ($\bar{v}$), and the root-mean-square speed (v_{rms}).

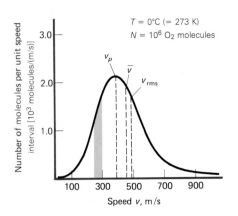

The most probable speed v_p is the speed at which the curve has its maximum. The average speed $\bar{v}$ is obtained by adding up all the speeds and dividing by the number of molecules. This is somewhat larger than v_p since the speeds must cut off at zero on the low side, while on the high side any speed can exist, in principle, up to the speed of light, 3×10^8 m/s. Of course, relatively few molecules have such high speeds (only 1 percent of O_2 molecules have speeds exceeding 1800 m/s), but enough molecules have these high speeds to produce a curve asymmetric about v_p and therefore to make $\bar{v}$ larger than v_p. The root-mean-square velocity v_{rms}, since it is obtained by averaging the squares of the speeds, emphasizes the greater speeds even more, and so $v_{rms} > \bar{v} > v_p$, as is shown in the figure.

Summary: Important Definitions and Equations

Atomic particles (see Table 9.1)
Notation for atoms: $^A_Z X$
   Z = charge number
     = number of protons in nucleus
     = number of electrons in neutral atom
   A = mass number = $Z + N$,
     where N is number of neutrons in nucleus
Ions:
   Negative (H^-): More electrons outside nucleus than protons in nucleus.
   Positive (H^+): Fewer electrons outside nucleus than protons in nucleus.
Isotopes: Atoms with the same number of protons and electrons but with a different number of neutrons in nucleus, e.g., $^{235}_{92}U$ and $^{238}_{92}U$.
Mole (or gram-mole): The amount of a substance that contains as many atoms or molecules as there are atoms in 12 g of $^{12}_6C$, that is, Avogadro's number, or 6.022×10^{23}.
Unified mass unit (u):
   1 u = 1.661×10^{-27} kg
Avogadro's number (N_A): The number of atoms or molecules in one mole of any substance. $N_A = 6.022 \times 10^{23}$ atoms (or molecules) per mole.
Gas: A substance which expands to fill uniformly the volume of any container in which it is placed.

Ideal gas law: $PV = nRT$
   where P = pressure, N/m^2
            V = volume, m^3
            n = number of gram-moles of gas
            R = the universal gas constant,
                 8.314 J/(mol·K)
            T = temperature, K
Relationship between kinetic energy and absolute temperature of a gas:

$$\tfrac{1}{2}M\overline{v^2} = \tfrac{3}{2}RT \quad \text{where } R = 8.314 \text{ J/(mol·K)}$$

$$\tfrac{1}{2}m\overline{v^2} = \tfrac{3}{2}kT \quad \text{where } k \text{ (Boltzmann's constant)}$$
$$= 1.380 \times 10^{-23} \text{J/K}$$

Different speeds of gas molecules:
   v_p (most probable speed): The speed at which the velocity distribution curve has its maximum.
   $\bar{v}$ (average speed): The sum of all speeds divided by the number of molecules

$$v_{rms} \text{ (root-mean-square speed): } v_{rms} = \sqrt{\overline{v^2}} \propto \sqrt{T}$$

$$v_{rms} > \bar{v} > v_p$$

Questions

1 (*a*) Would you expect $_4Be^+$ and $_3Li$ to have similar chemical and physical properties? Why?
   (*b*) Would you expect $_3Li^+$ and $_4Be$ to have similar properties? Why?
2 Hydrogen exists in three different isotopic forms, ^1_1H (ordinary hydrogen), ^2_1H (heavy hydrogen, or deuterium), and ^3_1H (tritium). Compare neutral atoms made of these three isotopes, and point out in what respect their properties will differ and in what respect they will be the same.

3 (*a*) Why does 1 mol of any atomic or molecular gas always contain exactly the same number of atoms or molecules?
   (*b*) Why would you expect equal volumes of any two gases at the same temperature and pressure to contain the same number of moles of these gases?
4 Which has the greater number of atoms, a gram of nitrogen (N_2) or a gram of oxygen (O_2)?
5 Two liters of hydrogen and one liter of oxygen gas are pumped into a strong steel drum and ignited by an

electric spark so that the following chemical reaction occurs: $2H_2 + O_2 \rightarrow 2H_2O$. At the end only water vapor remains. After the temperature of the drum has returned to its original value, is the pressure different than before the reaction? If so, by how much?

6 Two gas cylinders stand on a laboratory floor. One contains helium (He), the other nitrogen (N_2). Both are at exactly the same temperature and pressure. If the first cylinder contains 2 kg of $^4_2$He, and the second 8 kg of molecular $^{14}_7$N, which cylinder has the greater volume? Why?

7 Two bottles of gas sit beside each other on a laboratory shelf. One bottle contains hydrogen gas (H_2) with a total translational kinetic energy of 3.3 J. The other bottle contains nitrogen gas (N_2) with a total translational kinetic energy of 4.5 J. If the nitrogen bottle contains 1.5 times as many molecules as does the hydrogen bottle, which bottle is at the higher temperature?

8 (*a*) Why doesn't the size of gas molecules enter into either the equation for the kinetic energy of a gas or the ideal gas equation?

(*b*) Can you imagine any circumstances in which the size of the gas molecules would be important?

9 On the moon the acceleration due to gravity is only about one-sixth its value on earth. Can you use this fact and your knowledge of the kinetic theory of gases to explain why the moon has no atmosphere?

10 As one ascends directly up from the earth in a balloon, it is found that the percentage of the earth's atmosphere made up of hydrogen increases and the percentage of oxygen decreases. Explain why.

11 When a gas is compressed by rapidly pushing down on a piston to reduce its volume, the temperature of the gas rises. Can you explain this experimental fact on the basis of the kinetic theory of gases? (*Hint:* Consider what happens to the momentum of the gas molecules when they strike the moving piston.)

Problems

In all the following problems, STP, or standard temperature and pressure, means a temperature of 0°C or 273 K, and a pressure of 760 mmHg, or 1.01×10^5 N/m^2. Under these conditions 1 mol of any gas occupies 22.4 liters and 1 kg·mol occupies 22.4×10^3 liters, where 1 liter = 10^3 cm^3 = 10^{-3} m^3.

A 1 The density of an atomic nucleus exceeds that of an atom by about:
(*a*) 1 order of magnitude
(*b*) 4 orders of magnitude
(*c*) 12 orders of magnitude
(*d*) 8 orders of magnitude
(*e*) They are about the same.

A 2 The nucleus of a plutonium atom ($^{239}_{94}$Pu) contains the following number of electrons:
(*a*) 94   (*b*) 239   (*c*) 145   (*d*) 333
(*e*) None of the above

A 3 The nucleus of a radium atom ($^{226}_{88}$Ra) contains the following number of neutrons:
(*a*) 88   (*b*) 226   (*c*) 138
(*d*) 314   (*e*) None of the above

A 4 The mass of an atom of gold ($^{197}_{79}$Au) is about:
(*a*) 3.3×10^{-25} kg   (*b*) 3.3×10^{-25} g
(*c*) 1.3×10^{-25} kg   (*d*) 1.3×10^{-25} g
(*e*) 3.3×10^{-22} kg

A 5 Oxygen gas is in a rectangular container of length $L = 2.3$ m and cross-sectional area $A = 0.020$ m at a temperature of 0°C. In 1 s an oxygen atom moving along the X axis perpendicular to the area A will make approximately the following number of collisions with the end of the container:
(*a*) 1   (*b*) 10   (*c*) 100
(*d*) 1000   (*e*) None of the above

A 6 Helium gas is in a 2.0-liter container at atmospheric pressure. A piston at the end of the container is moved until the volume is reduced to 0.50 liter. If this is done slowly enough that the temperature of the gas does not change, the final pressure of the He is:
(*a*) 1 atm   (*b*) 2 atm   (*c*) $\frac{1}{2}$ atm
(*d*) 4 atm   (*e*) None of the above

A 7 The average translational kinetic energy of a He atom in a sample of gas at 20°C is about:
(*a*) 3.4×10^6 J   (*b*) 3.4×10^3 J
(*c*) 4.0×10^{-22} J   (*d*) 6.0×10^{-21} J
(*e*) None of the above

A 8 The rms speed of an argon atom ($^{40}_{18}$Ar) at room temperature is about:
(*a*) 1.8×10^5 m/s   (*b*) 4.2×10^2 m/s
(*c*) 0.9×10^5 m/s   (*d*) 3.0×10^2 m/s
(*e*) None of the above

A 9 A cylinder contains a mixture of hydrogen and oxygen gases. If the rms speed of the oxygen gas molecules is 5.0×10^2 m/s, the rms speed of the hydrogen molecules is about:
(*a*) 5.0×10^2 m/s   (*b*) 20×10^2 m/s
(*c*) 1.3×10^2 m/s   (*d*) 14×10^2 m/s
(*e*) 1.8×10^2 m/s

A10 The temperature of a gas is increased from 300 to 600 K. The rms speeds of the molecules in the gas increases by a factor of:
(*a*) 2   (*b*) 4   (*c*) $\frac{1}{2}$ (*d*) 0.71   (*e*) 1.4

B 1 A neon atom ($^{20}_{10}$Ne) has a radius of about 1.1×10^{-10} m. The nucleus of a neon atom has a radius of about 3.3×10^{-15} m.
(*a*) What is the density of a neon atom?
(*b*) What is the density of the nucleus of a neon atom?

B 2 Lithium (Li) as found in nature consists of 7.4% $^6_3$Li and 92.6% $^7_3$Li. What is the average mass number of Li as found in nature?

B 3 The element neon has the following natural abundances: $^{20}_{10}$Ne, 90.92%; $^{22}_{10}$Ne, 8.82%; $^{21}_{10}$Ne, 0.26%. What is the mass of a mole of neon as found in nature?

B 4 How many moles are contained in:

(a) 2.0 kg of $^4_2$He?   (b) 100 g of $^{235}_{92}$U?

(c) 500 g of molecular nitrogen (N_2) consisting of the isotope $^{14}_7$N?

B 5 A sample of unknown gas is contained in a tank of volume 5.0×10^{-2} m³ at a pressure of 1.0×10^5 N/m².

(a) Is it possible to obtain the kinetic energy of the gas from these data, even if we do not know the nature of the gas?

(b) If so, what is the kinetic energy of the gas?

(c) Is it possible to obtain the speed of the gas molecules from the data given?

B 6 A cylindrical container of nitrogen gas contains a piston at the top which is free to move to preserve the same atmospheric pressure throughout an experiment. The gas is slowly heated from 0 to 100°C. If the initial volume of the gas was 0.010 m³, what is its final volume?

B 7 A tank of oxygen gas at 20°C and a pressure of 10 atm has a volume of 0.015 m³. How many moles of oxygen gas are in the tank?

B 8 What is the percentage difference between the rms speeds of nitrogen molecules in a gas at 300 K and at 900 K?

B 9 What is the percentage difference between the rms speeds of $^4_2$He and $^3_2$He atoms at the same temperature?

B10 What is the rms speed of a He atom in a sample of gas at 500 K?

B11 What is the volume occupied by 5.0 mol of H_2 gas at STP?

C 1 The average molecular mass of air is about 29 g/mol, since air is 80% nitrogen. If the air in a room 3.0 m high, 5.0 m wide, and 6.0 m long is at STP, what is the mass of the air in the room?

C 2 (a) How many molecules are there in 1.0 cm³ of air at STP?

(b) On the average, how far apart are two air molecules?

(c) How does this compare with the approximate size of the atoms and molecules making up the air?

C 3 (a) The volume occupied by a neon atom is about 6.0×10^{-30} m³. In a sample of neon gas at STP, what fraction of the volume of the gas is actually occupied by neon atoms?

(b) How does this result fit in with our assumptions about ideal gases in Sec. 9.6?

C 4 In a diesel engine the cylinder compresses air from STP to $\frac{1}{20}$ of its original volume and a pressure of 60 atm. What is the final temperature of the air after it has been compressed?

C 5 A medical syringe contains air at standard atmospheric pressure. The plunger of the syringe is then pushed in and the volume of air reduced to $\frac{1}{4}$ its original volume. Find the final pressure of the air:

(a) When the compression is performed so slowly that there is no change in temperature.

(b) When the compression is performed rapidly and the temperature rises from 293 to 315 K.

C 6 In an "ultra-high" vacuum system the gas pressure can be reduced to 10^{-15} atm. How many molecules of air exist in 1.0 cm³ of this very good vacuum, if the laboratory temperature is about 20°C?

C 7 A balloon has its gas bag only partially filled with 4000 m³ of He on the ground, where the pressure is 1.0 atm and the temperature is 27°C. Find the volume of the gas bag of the balloon when it has risen to such a height that the pressure has fallen to 0.25 atm and the temperature to −18°C.

C 8 A glass bottle filled with nitrogen is sealed off at STP. If the gas is then heated to 450°C, what is the new pressure of the gas? Express your answer in both atmospheres and pounds per square inch.

C 9 An empty oil can contains 0.30 m³ of air when the cap is screwed tightly on inside a garage at 20°C. It is then set out in the sun on an asphalt driveway on a hot day, where the temperature rises to 40°C. If the original pressure was 1.0 atm, what will be the final pressure of the air in the can?

C10 (a) A tank of He gas at 27°C has a volume of 0.010 m³ at a pressure of 40 atm. How much He is in the tank?

(b) What is the density of the He gas?

C11 An automobile tire is filled to a gauge pressure of 29.4 lb/in² on a day when the temperature is 0°C. What is the tire pressure after the outside temperature has risen to 30°C? (*Hint:* Note the difference between absolute pressure and gauge pressure, as pointed out in Example 9.7.)

C12 A mass of 100 g of nitrogen (N_2) gas is contained in a glass cylinder at a pressure of 5.0×10^5 N/m² and a temperature of 27°C. What is the volume of the glass cylinder?

C13 What is the mass of the carbon dioxide (CO_2) contained in a 0.10-m³ tank at a pressure of 5.0 atm and a temperature of 25°C?

C14 In a hospital, oxygen gas for patients is stored in large cylinders in a central room which is at 15°C, and the oxygen is then pumped from there to the patients' rooms. If a cylinder is at a pressure of 50 atm and the pressure must be reduced to 1 atm before the oxygen is administered to a patient in a hospital room with a temperature of 22°C, what volume does 1.0 cm³ of oxygen from the tank occupy when it is administered to the patient?

C15 Calculate (a) the average kinetic energy and (b) the rms speed of nitrogen (N_2) molecules in the air at 400 K.

C16 The temperature of the interior of the sun is of the order of 10^7 K.

(a) Find the average translational KE of a hydrogen atom at this temperature.

(b) Find the rms speed of a hydrogen atom at this temperature.

(c) How does this speed compare with that of a hydrogen atom at 0°C?

C17 A molecule of nitrogen at the surface of the earth and at 20°C has a velocity directed upward. If it could continue to rise without colliding with any other molecules, how far from the surface of the earth would it be when it stopped?

C18 The velocity required for a projectile shot off

from the surface of the earth to escape from the earth is 11.2 km/s.

(*a*) At what temperature would hydrogen molecules reach an rms speed equal to this escape velocity?

(*b*) At what temperature would helium atoms reach this escape velocity?

C19 Find the temperature at which the rms speed of O_2 molecules is equal to the rms speed of He atoms at 27°C.

C20 A container of hydrogen gas (H_2) contains a mixture of the three isotopes of hydrogen, $_1^1H$, $_1^2H$, and $_1^3H$.

(*a*) At 0°C what is the transitional kinetic energy of molecules of these three isotopes?

(*b*) At 0°C what are the rms speeds of these three isotopes?

C21 (*a*) Why is it easier to separate two isotopes of hydrogen by gaseous diffusion than it is to separate two isotopes of uranium?

(*b*) What is the percentage difference between the rms speeds of $_1^1H$ and $_1^2H$ at room temperature, compared with the same percentage difference for $_{92}^{234}U$ and $_{92}^{235}U$?

D 1 An automobile tire whose volume is 0.025 m³ is found to have a pressure of 20 lb/in² as read from a tire gauge. If it is desired to bring this pressure up to 28 lb/in² without changing the temperature, how much air must be pumped into the tire?

D 2 A glass cylinder containing He gas at 27°C and atmospheric pressure is divided into two parts, each of volume 0.050 m³, by a flat, round metal piston of cross-sectional area 0.010 m². The gas in one part of the cylinder is then heated to 127°C. If we assume that the two parts are thermally insulated from each other so that the second part of the cylinder remains at 27°C, how far will the piston move when the temperature of the first part goes from 27 to 127°C?

D 3 A partition separates a container into two parts, one of which is 4 times the size of the other. The large-volume section holds He gas at a pressure of 3.0 atm, while the smaller volume contains O_2 gas at a pressure of 1.0 atm. The temperature is the same for both gases and does not change during the experiment. If a small door in the partition is opened by remote control, so that the gases are free to move from one side of the partition to the other, what is the final pressure of the system?

D 4 What is the average distance between two helium atoms in a tank of helium gas at STP?

D 5 Show that the pressure of an ideal gas can be written as $P = \frac{1}{3}dv^2$, where d is the density of the gas.

D 6 Suggest an experiment which would lead to a value for Avogadro's number if Boltzmann's constant is known.

D 7 Prove, using a simple numerical example, that $\overline{v^2}$ does not equal $(\overline{v})^2$ for the speeds of the molecules in a gas.

D 8 Suppose it were possible to make measurements directly of the speeds of 10 molecules from a tank of nitrogen (N_2) gas at STP and that the following speeds were obtained, with all speeds in 10^2 m/s:

9   1   3   5   6   5   8   2   5   7

For these 10 molecules, calculate (*a*) the most probable speed, (*b*) the average speed, (*c*) the rms speed.

D 9 (*a*) Calculate the kinetic energy of a mole of krypton gas ($_{36}^{84}Kr$) at 27°C.

(*b*) What is the rms speed of the krypton atoms at this temperature?

(*c*) What is the total momentum relative to the laboratory of all the krypton atoms in a cylinder of krypton gas which is sitting on the floor of the laboratory?

D10 A law of considerable importance in biological situations, for example, for the exchange of oxygen with the blood, is called *Dalton's law of partial pressures*. It states that the total pressure of a gas mixture is the sum of the partial pressures of the component gases, i.e., of the individual pressures that each gas would exert if it alone occupied the whole container. Show that Dalton's law follows directly from our model of an ideal gas.

Additional Readings

Andrade, E. N. da C.: *Rutherford and the Nature of the Atom*, Doubleday Anchor, Garden City, N.Y., 1964. A simple, brief, and interesting biography by a physicist.

Baker, Jeffrey J. W., and Garland E. Allen: *Matter, Energy, and Life*, 4th ed., Addison-Wesley, Reading, Mass., 1981. A good introduction to the physical and chemical concepts important in the life sciences.

Conant, James B. (ed.): *Robert Boyle's Experiments in Pneumatics*, Harvard University Press, Cambridge, Mass., 1950. An historical account of the pioneering work done by Robert Boyle on the gas laws.

Eve, A. S.: *Rutherford*, Macmillan, New York, 1939. This is usually considered the definitive biography of Lord Rutherford.

Nash, L. K.: *Atomic-Molecular Theory*, Harvard University Press, Cambridge, Mass., 1950. An account of the historical development of the theories of the structure of atoms.

Pauling, Linus, and Robert Hayward: *The Architecture of Molecules*, Freeman, San Francisco, 1964. A brilliant description of the structure of some key molecules by a two-time Nobel Laureate (Pauling), together with some magnificent colored drawings of their basic structure. A true work of art.

Steinherz, H. A., and P. A. Redhead: "Ultrahigh Vacuum," *Scientific American*, vol. 206, no. 3, March 1962, pp. 78–90. A very clear discussion of how vacuums as low as 10^{-12} torr could be obtained using vacuum pumps available in 1962.

Wilson, David: *Rutherford: Simple Genius*, MIT Press, Cambridge, Mass., 1983. The most recent biography of the famous physicist.

There is an astonishing imagination, even in the science of mathematics. We repeat, there was far more imagination in the head of Archimedes than in that of Homer.

Voltaire (1694–1778)

Chapter 10
Liquids and Solids

Of the three states of matter, gases are the easiest to handle quantitatively. Gas molecules are far enough apart that they can be considered to act independently of one another. Hence we can, to a good approximation, ignore the forces between gas molecules except when they collide, and derive the bulk properties of the gas from the behavior of the individual molecules. This is not true of either liquids or solids, where the intermolecular forces are much greater. Crystalline solids, however, possess an orderly molecular arrangement which repeats itself over and over again as we move through the solid. This periodic structure simplifies the calculation of many solid-state properties. But liquids possess neither the total disorder that characterizes gases nor the order that characterizes crystalline solids. Hence liquids are less well understood than either gases or solids.

10.1 The Structure of Liquids

The molecules of a liquid have widely varying speeds and kinetic energies. However, the intermolecular forces between liquid molecules, unlike those between gas molecules, are quite large and bind the liquid molecules together in a compact mass. Hence liquids do not expand to fill a container as gases do, but settle down, under the influence of gravity, to the bottom of any container in which they are placed. The intermolecular forces are not as strong in liquids as in solids and do not restrict the molecules to fixed positions in space. As a result, if external forces are applied to a liquid, it can be easily made to change shape. By applying sufficiently large forces, it is possible to split up the liquid into pieces, as when water drips from a faucet. Here the force of gravity acting on the unsupported liquid leaving the faucet pulls the water apart into small droplets which gradually change their shape as they fall, owing to the surface tension of the liquid.

Surface Tension

A molecule in the interior of a liquid is subjected to attractive forces from the other liquid molecules. Since these forces are in all directions, they cancel out, and there is no net unbalanced force on the molecule. But molecules at the surface of the liquid experience a net force pulling them back toward the interior of the liquid. As a consequence, a molecule in the surface layer has a higher potential energy than a molecule in the interior of the liquid. Therefore increasing the surface layer will increase the potential energy of the

FIGURE 10.1 A high-speed photograph, taken by Dr. Harold Edgerton of MIT, illustrating how spherical droplets are formed when a drop of liquid falls on a liquid surface. The droplets break off in the form of tiny spheres because a sphere has smaller surface area for a given volume than any other geometric shape. (Photo courtesy of Omikron/Photo Researchers.)

liquid. Since the liquid tends to minimize its potential energy, as does a roller-coaster car gliding down a hill, so it tends to minimize its surface area. As a result, the surface of a liquid behaves like a stretched membrane and produces what is called the *surface tension* of the liquid.

The concept of surface tension explains the spherical shape of liquid droplets, the "hanging" of drops of water at the end of a faucet, the behavior of liquids in narrow capillary tubes, and the fact that objects denser than water, such as a needle, can be made to float on a water surface, supported by the tight "skin" of the water. Figure 10.1 shows how surface tension produces spherical droplets when water splashes.

Evaporation

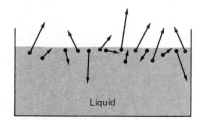

Liquid

FIGURE 10.2 The evaporation of molecules from a liquid. The colored dots indicate molecules escaping from the liquid and entering the vapor phase; the black dots indicate molecules remaining in the liquid.

Since the distribution of molecular speeds in a liquid is similar to the Maxwell distribution of Fig. 9.24, there will always be a small number of molecules possessing very large speeds, even at room temperature. At high temperatures, of course, this number increases very rapidly, as it does in gases. Some of these high-speed molecules may be near the free surface of the liquid, and may therefore be able to escape from the liquid as vapor molecules. This is the process known as *evaporation*, depicted in Fig. 10.2. To escape, a molecule must have a sufficient kinetic energy to climb the potential-energy "hill" arising from the forces binding the molecule to the liquid, in the same way that a roller-coaster car must have sufficient kinetic energy to climb to the top of a hill on the track before it can coast down the other side. Since only the fastest molecules escape by evaporation from a liquid, the average internal energy of the liquid, and therefore its temperature, is reduced by the process of evaporation. For this reason the evaporation of perspiration from our bodies in the summer tends to cool us off; thus drinking a hot liquid, even in very hot weather, is a good way to induce perspiration and cool off.

Density and Specific Gravity

Density of Liquid

As we saw for gases in Eq. (9.3), the density of a liquid is the total mass of the liquid divided by the volume of the liquid:

$$d = \frac{M}{V}$$

(10.1)

For liquids (unlike gases) the density varies very little with the pressure and temperature, and we will usually treat it as a constant. The most common liquid, water, is often taken as a standard for density measurements. The density of water at the temperature at which it is most dense, 4°C, and at atmospheric pressure, is 10^3 kg/m^3, or 1 g/cm^3. A related concept is that of *specific gravity*.

Definition

Specific gravity: The ratio of the density of any material to the density of water.

Specific gravity is a dimensionless number, since it is the ratio of two quantities which are dimensionally the same and which therefore must be expressed in the same units.

Since water has a density of 10^3 kg/m^3, or 1 g/cm^3, the density of any substance in grams per cubic centimeter is identical to its specific gravity; its density in kilograms per cubic meter is 10^3 times its specific gravity. Table 10.1 gives the specific gravity and density for a number of important liquids and gases.

TABLE 10.1 Densities and Specific Gravities for Some Liquids and Gases at STP*

Substance	Specific gravity	Density	
		g/cm^3	kg/m^3
Liquids:			
Water (H$_2$O) (at 4°C)	1.00	1.00	1.00×10^3
Mercury (Hg)	13.6	13.6	13.6×10^3
Alcohol, ethyl	0.79	0.79	0.79×10^3
Gasoline	0.68	0.68	0.68×10^3
Blood, whole	1.05	1.05	1.05×10^3
Gases:			
Hydrogen (H$_2$)	0.90×10^{-4}	0.90×10^{-4}	0.090
Helium (He)	1.79×10^{-4}	1.79×10^{-4}	0.179
Nitrogen (N$_2$)	1.26×10^{-3}	1.26×10^{-3}	1.26
Oxygen (O$_2$)	1.43×10^{-3}	1.43×10^{-3}	1.43
Carbon dioxide (CO$_2$)	1.98×10^{-3}	1.98×10^{-3}	1.98
Air	1.3×10^{-3}	1.3×10^{-3}	1.3

*STP (standard temperature and pressure) means a temperature of 0°C and a pressure corresponding to a barometric height of 760 mmHg (sometimes called 1 atm).

Example 10.1

An aqueous solution of 50% albumin at 15.5°C has a specific gravity of 1.135.
(a) What volume (in cubic meters) would be occupied by 100 g of this solution?

(b) What would the volume be in liters?

SOLUTION

Since the solution has a specific gravity of 1.135, its density is 1.135×10^3 kg/m^3, or 1.135 g/cm^3.
(a) Hence the volume occupied by 100 g, or 0.100 kg, is

$$V = \frac{M}{d} = \frac{0.100 \text{ kg}}{1.135 \times 10^3 \text{ kg/m}^3} = \boxed{8.81 \times 10^{-5} \text{ m}^3}$$

(b) Since 1 liter $= 10^3$ cm^3, we have

$$V = \frac{100 \text{ g}}{1.135 \text{ g/cm}^3} = 88.1 \text{ cm}^3 = \boxed{0.0881 \text{ liter}}$$

Since 1 liter $= 10^{-3}$ m^3, these two results are clearly consistent.

Pressure

We have already seen the meaning of pressure in the case of a gas. For a liquid we use the same definition as in Eq. (9.4),

$$P = \frac{F}{A}$$

where F is the force, *perpendicular* to the area A, exerted by the liquid. Both gases and liquids, which are sometimes lumped together and called "fluids," possess the property that any force parallel to the walls of the container will simply cause one layer of fluid to slide over another and relieve the applied stress. It is this inability of fluids to resist such *tangential* forces, i.e., forces parallel to the walls of the container, that enables them to flow so easily. Hence the only force that a gas or liquid can sustain is one at right angles to a boundary surface, and for this reason pressure is defined in terms of the force perpendicular to the boundary surface.

The SI unit of pressure is the newton per square meter (N/m²), a unit which is now officially called the pascal (Pa), after the French scientist, philosopher, and writer Blaise Pascal (1623–1662), who did some very important work in the field of hydrostatics, i.e., on the properties of fluids at rest. As we saw in the preceding chapter, atmospheric pressure is equal to 1.01×10^5 N/m² (or Pa), or to 14.7 lb/in². Atmospheric pressure is, of course, the force exerted on unit area of the ground by the weight of a column of air of the same unit cross section and extending from the surface of the earth to the top of the earth's atmosphere, as shown in Fig. 10.3.

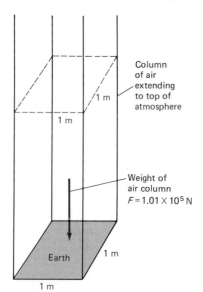

FIGURE 10.3 Atmospheric pressure: The force exerted on unit area of the ground by the weight of a column of air of the same unit cross section and extending from the earth to the top of the earth's atmosphere.

10.2 Hydrostatic Pressure

One of the most important principles governing the behavior of liquids was first proposed by Pascal. *Pascal's principle* can be stated as follows:

Pascal's Principle

The pressure applied to a confined liquid is transmitted undiminished to every part of the liquid and acts at right angles to any surface it touches.

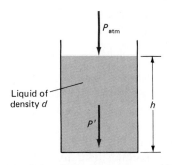

FIGURE 10.4 The pressure P' exerted on the bottom of a pail by liquid filling the pail to a depth h. Here $P' = P_{atm} + P = P_{atm} + dgh$.

Thus, if a pail of liquid has a free surface at the top, as in Fig. 10.4, the pressure of the atmosphere pressing down on the liquid surface is transmitted undiminished to every point in the liquid.

Let us try to find the total pressure produced at the bottom of this pail of liquid, if the liquid column is of cross-sectional area A and height h. The pressure on the container at its bottom produced by the liquid is $P = F/A$, where F is the weight of the column of liquid. Hence

$$F = Mg = dVg$$

where d is the density of the liquid. Since the volume V of the column is equal to Ah,

$$F = dAhg \qquad \text{and so} \qquad P = \frac{F}{A} = \frac{dAhg}{A} = dgh$$

The pressure produced by a column of liquid of height h and density d is therefore

Pressure Produced by a Liquid

$$\boxed{P = dgh} \tag{10.2}$$

Note that this is the pressure due to the liquid alone. Since atmospheric pressure is also exerted on the top of the liquid column by the air in the atmosphere, and this pressure is transmitted undiminished throughout the liquid, the total pressure at the bottom of the liquid column is actually

$$P' = P_{atm} + P = P_{atm} + dgh \tag{10.3}$$

as shown in Fig. 10.4.

The pressure exerted by the earth's atmosphere can, in principle, also be found from Eq. (10.2). This is complicated by the fact that d, the density of air at various heights above the earth, is not a constant, and g also varies with altitude. The air tends to settle toward the earth, so that its density near the earth is greater than it is in the upper atmosphere. For this reason calculus is needed to determine the pressure produced by the earth's atmosphere.

Example 10.2

Find the height of a column of water which would produce the same pressure on the earth as is produced by the earth's atmosphere.

SOLUTION

We have seen that atmospheric pressure is 1.01×10^5 Pa (or N/m²). Hence we must have

$$P = dgh = 1.01 \times 10^5 \text{ N/m}^2$$

Now, for water $d = 10^3$ kg/m³, and so

$$h = \frac{P}{dg} = \frac{1.01 \times 10^5 \text{ N/m}^2}{(10^3 \text{ kg/m}^3)(9.8 \text{ m/s}^2)} = 10.3 \text{ N·s}^2/\text{kg} = \boxed{10.3 \text{ m}}$$

since 1 N = 1 kg·m/s².

Hence the height of the water column must be 10.3 m, or 33.8 ft.

If the column consisted of mercury rather than water, its height would be reduced by a factor of 13.6, since the specific gravity of mercury is 13.6. Hence the height of the column would be 0.76 m, or 76 cm, which is a much more convenient height for a liquid column than is 10.3 m. It is for this reason that atmospheric pressure is often measured in centimeters (or millimeters) of mercury, even though it is produced by the weight of the air mass at the bottom of which we live.

The pressure varies as we go from the top to the bottom of a column of liquid like that in Fig. 10.4, because the amount of liquid above different heights varies. What about the hydrostatic pressure (i.e., the pressure due to the liquid) at some fixed depth below the surface of the liquid? Is it the same at all points? Is it the same in all directions? Clearly it must be, for if we immerse in the liquid a very light piece of plastic wrap at the end of a string, it is not pushed in one direction or the other by the fluid. Hence the pressure on its front face must be the same as that on its back face, and this is true regardless of the orientation of the probe. The hydrostatic pressure at the same depth in the fluid is then the same at all places and in all directions; and depends only on the depth below the surface of the fluid.

This remains true even for different amounts of liquid above a fixed point, so long as the height of the column remains the same. In Fig. 10.5 the pressures at A, B, and C, which are at the same distance below the liquid surface, must be the same; otherwise the fluid would flow from one place to the other. Even though there is more liquid above B and less above C than there is above A, the pressures remain the same because in B the glass walls exert upward forces, and in C downward forces, to balance out the different weights of liquid.

Another clear indication that pressures are the same everywhere at the same depth in a fluid comes from the human body, which is constantly subjected to large pressures from the weight of the air in the atmosphere. The body normally does not feel these pressures because breathing fills our lungs and tissues with air, which produces a pressure inside the body that is equal in magnitude and opposite in direction to the pressure outside. Even the very delicate human ear does not detect any pressure difference. If, however, we are in an airplane which is losing altitude rapidly, and the pressurizing system in the cabin does not work perfectly, our ears will feel pressure differences because the pressure in the middle ear on the inside of the eardrum will not be able to adjust rapidly enough to the new pressure on the outside of the eardrum. Swallowing and yawning in such circumstances can help to adjust the internal ear pressure to the external pressure.

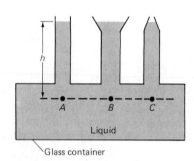

FIGURE 10.5 The pressure at the bottom of three liquid columns of the same height but of different shapes and sizes. This pressure is the same at the same depth for all three columns ($P_A = P_B = P_C$) and is equal to dgh.

Example 10.3

(a) Calculate the order of magnitude of the force exerted by the earth's atmosphere on the outside of a person's body.
(b) How large is this force in tons, i.e., in units of 2000 lb?

(c) Why does such a force cause no discomfort to the person and produce no acceleration?

SOLUTION

(a) Let us make the very crude assumption (which is quite adequate for an order-of-magnitude calculation) that the person's body has about the same surface area as a square column 1 ft on a side and 6 ft tall. The surface area in this case would be 4 times an area 1 by 6 ft, or

$$A = 4(1 \text{ ft})(6 \text{ ft}) = 24 \text{ ft}^2 = (24 \text{ ft}^2)\left(\frac{0.305 \text{ m}}{1 \text{ ft}}\right)^2 = 2.2 \text{ m}^2$$

Since atmospheric pressure is 1.01×10^5 N/m², the total force on the outside of the person's body is then

$$F = PA = \left(1.01 \times 10^5 \, \frac{\text{N}}{\text{m}^2}\right)(2.2 \text{ m}^2) = \boxed{2.2 \times 10^5 \text{ N}}$$

(b) Since 1 lb = 4.45 N, we have

$$F = (2.2 \times 10^5 \text{ N})\left(\frac{1 \text{ lb}}{4.45 \text{ N}}\right) = 5.0 \times 10^4 \text{ lb}$$

or, since 1 ton = 2000 lb,

$$F = (5.0 \times 10^4 \text{ lb})\left(\frac{1 \text{ ton}}{2000 \text{ lb}}\right) = \boxed{25 \text{ tons}}$$

(c) Although this is a huge force, it produces no discomfort even though it is pushing in on the outside of the person's body in all directions normal to the body surface, because there is an equal and opposite force exerted on the inside of the body in an outward direction. These two forces therefore cancel at every point. The net force acting on the person's body is therefore zero, and no acceleration results.

Hydraulic Devices

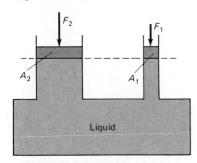

FIGURE 10.6 A hydraulic lift. Here F_2 must be equal to $(A_2/A_1)F_1$ to maintain equilibrium, since the pressure applied is transmitted undiminished to every point in the liquid. The liquid therefore exerts a large upward force on the piston at A_2, which can do useful work.

Let us consider a container of liquid with two entry tubes at the top, one of large area A_2 and the other of small area A_1, as in Fig. 10.6. Two pistons close off the entry ports. The liquid will assume the same height in the two tubes for the reasons just discussed. Suppose now that with piston 1 we apply an added force F_1 to A_1, so that an added pressure is exerted on the liquid equal to $P = F_1/A_1$. This pressure is added to the pressure at the bottom of the liquid, according to Pascal's principle, and the resulting pressure is the same at all points at the same depth below the surface. This can only be true if the pressure produced by piston 2 on A_2 is the same as the pressure on A_1. We therefore have

$$P = \frac{F_1}{A_1} = \frac{F_2}{A_2} \quad \text{or} \quad F_2 = \frac{A_2}{A_1}F_1 \tag{10.4}$$

Since A_2 can be made much larger than A_1, a small downward force F_1 can produce a larger upward force F_2. This is the principle of the hydraulic lift, which is a device that enables us to move heavy objects by applying relatively small forces to the lift.

Other examples of the usefulness of Pascal's principle are hydraulic presses and hydraulic brakes on cars.

Example 10.4

If a hydraulic lift is used to do work, say, in lifting an automobile, prove that the work output is equal to the work input, and hence that mechanical energy is conserved in the process.

SOLUTION

We have, from Eq. (10.4), $F_2 = \dfrac{A_2}{A_1}F_1$

Now, suppose we move the small piston in Fig. 10.7 down through a distance s_1 and thus cause the large piston to move upward through a distance s_2. Then the work done is F_1s_1, and the volume of the liquid in tube 1 is reduced by an amount $V_1 = s_1A_1$. The volume of liquid in tube 2 must be increased by the same amount, since the liquid is incompressible. Hence

$$V_2 = s_2A_2 = V_1 = s_1A_1 \quad \text{or} \quad s_1 = \frac{A_2}{A_1}s_2$$

and so the work done is

$$W_1 = F_1s_1 = \frac{F_1(A_2s_2)}{A_1} = F_2s_2 = W_2$$

from Eq. (10.4). Hence the work output is exactly the same as the work input, and no energy is created in the process.

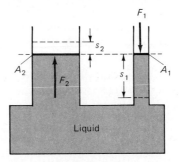

FIGURE 10.7 Diagram for Example 10.4.

A hydraulic device of this sort is equivalent to a mechanical lever which allows us to exert a small force over a large distance, and with the help of the lever produce a large force working over a short distance. In both the hydraulic lift and the lever no energy is created, but mechanical energy is conserved (so long as no mechanical energy is lost as heat).

The Barometer

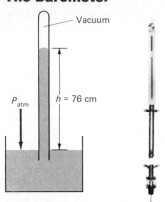

FIGURE 10.8 A mercury barometer. If a vacuum exists in the tube above the mercury column, the column will stand to a height of approximately 76 cm.

A barometer is an instrument used to measure the pressure exerted by the earth's atmosphere. Slight fluctuations in that pressure can then be used to predict changing weather patterns, since a rising pressure usually corresponds to clear weather, and falling pressure to stormy or inclement weather. The simplest way to make a barometer is to take a piece of glass tubing about a meter long, close it off at one end, fill it with mercury, and plunge it into a container of mercury without allowing any air to enter, as in Fig. 10.8. It is found that the mercury column does not fall but stays at a height of about 76 cm above the pool of mercury. Since the pressure in the pool of mercury is the same at all points at the same depth below the surface, the pressure due to 76 cmHg (about 30 inHg) must be the same as the pressure produced by the air above the mercury pool. Hence a pressure of 1 atm must correspond to the pressure produced by a 76-cm column of Hg. From Eq. (10.2) this is

$$P = dgh = (13.6 \times 10^3 \text{ kg/m}^3)(9.8 \text{ m/s}^2)(0.76 \text{ m}) = 1.01 \times 10^5 \text{ N/m}^2$$

This is the value given for standard atmospheric pressure in Sec. 10.1.

Note that the pressure at the bottom of the column of mercury is due to the mercury alone, because the top of the tube is closed off. Hence there is a very good vacuum between the top of the mercury column and the closed-off end of the tube, and there is not enough air present in that space to add notably to the pressure at the bottom of the column. A pressure of 1 mmHg is often referred to as a *torr* after Evangelista Torricelli (1608–1647), the Italian physicist who invented the mercury barometer. The best vacuum, obtainable by using powerful vacuum pumps, on earth is about 10^{-12} torr. In intergalactic space, on the other hand, the pressure is about 10^{-17} torr. This corresponds to less than one molecule per cubic centimeter of space. Because of its usefulness the millimeter of mercury (or torr) was included among the pressure units given in Table 9.3.

Manometers

FIGURE 10.9 An open-ended mercury manometer. The pressure in tube 2 is $P_2 = P_1 + dgh$. If P_1 is atmospheric pressure, then $P_2 = P_{atm} + dgh$.

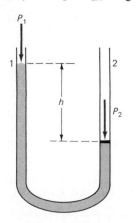

A *manometer* is another device used to measure pressure, usually the difference between the pressure on a confined gas and atmospheric pressure. A manometer consists of a U-shaped tube partially filled with a liquid such as mercury, as in Fig. 10.9. If a pressure is exerted on the top of the liquid in tube 2, say, by a piston pushing down on the liquid, the heights of the two columns will change, as shown in the figure. The difference in pressure at the top of the two columns will now be

$$\Delta P = P_2 - P_1$$

But $\Delta P = dgh$, where h is the difference in height of the two columns of Hg. Hence

$$P_2 = P_1 + dgh$$

If P_1 is the pressure of the atmosphere,

$$P_2 = P_{atm} + dgh \tag{10.5}$$

This is the pressure on the liquid produced by the piston. Obviously tube 2 could be connected to a container of gas and used in the same way to determine gas pressures both above and below atmospheric pressure.

A manometer can be used to measure the pressure of automobile tires. In this case *dgh* is the so-called gauge pressure, the difference in pressure inside and outside the tires. The absolute pressure inside the tire is the gauge pressure plus the pressure of the atmosphere. In the same way pressure on gases in metal cylinders can be measured.

In practical work, manometers are usually replaced by mechanical pressure gauges, which are simpler to use and which do not present the dangers associated with the breathing of mercury vapor. These gauges are calibrated (i.e., checked) against high-precision mercury manometers.

Example 10.5

A mercury manometer is connected to a gas cylinder. The mercury column connected to the cylinder falls to a point 18 cm below the top of the other mercury column.

(a) What is the gauge pressure of the gas in the cylinder?
(b) What is its absolute pressure?

SOLUTION

(a) Figure 10.9 illustrates this situation. The manometer reads the gauge pressure $\Delta P = dgh$, where h is the difference in height of the two mercury columns. Hence

$$\Delta P = (13.6 \times 10^3 \text{ kg/m}^3)(9.8 \text{ m/s}^2)(0.18 \text{ m})$$

$$= 24 \times 10^3 \text{ N/m}^2 = \boxed{0.24 \times 10^5 \text{ N/m}^2}$$

(b) The absolute pressure is then

$$P = P_{atm} + \Delta P$$

$$= 1.01 \times 10^5 \text{ N/m}^2 + 0.24 \times 10^5 \text{ N/m}^2$$

$$= \boxed{1.25 \times 10^5 \text{ N/m}^2}$$

Blood Pressure

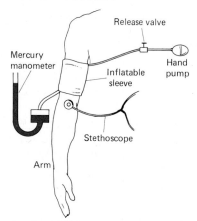

FIGURE 10.10 Apparatus for measuring blood pressure in the human body.

Mercury manometers are often used to measure blood pressure in the human body. When taking a patient's blood pressure, the doctor or nurse measures the maximum and minimum pressure of the blood as it flows through an artery in the arm, i.e., the force per unit area exerted by the blood normal to the artery wall. An inflatable sleeve is placed on the patient's arm, as in Fig. 10.10, and the sleeve is filled with air until the blood circulation is completely stopped. The doctor or nurse then gradually lowers the pressure in the sleeve until the sound of the blood pulsating through the arm can again be heard in the stethoscope. At this point the pressure of the blood outward on the artery wall is just equal to the inward force exerted by the inflated sleeve on the artery. This pressure, the *systolic* blood pressure, can then be obtained by measuring the pressure of the air in the sleeve with a manometer or other type of pressure gauge. As the pressure in the sleeve is reduced even farther, a point is reached at which the pulsing of the blood is no longer heard in the stethoscope. This pressure is called the *diastolic* pressure, and can be measured in a similar fashion.

Since blood pressure is measured with a manometer of some sort, what is measured is the gauge pressure, i.e., the difference between the pressure of the blood and atmospheric pressure. This is the only quantity of physiological interest, since it is the pressure maintained by the human circulatory system.

Normal blood pressure is about 120/80, i.e., a systolic pressure of 120 mmHg and a diastolic pressure of 80 mmHg. Too high a systolic pressure indicates the dangerous possibility that the arteries could rupture because of the high pressure exerted by the blood on the artery walls. If a blood vessel in the brain ruptures, for example, the result is a stroke.

Example 10.6

A patient's systolic pressure is 130 mmHg, and the diastolic pressure is 90 mmHg.
(a) What are these gauge pressures in SI units?
(b) What are the corresponding absolute pressures?

(c) What is the average pressure of the blood as it leaves the heart, if atmospheric pressure is taken as a reference level?

SOLUTION

(a) Since atmospheric pressure is 76 cmHg, or 1.01×10^5 N/m^2 (Pa),

$$1 \text{ cmHg} = \frac{1.01 \times 10^5 \text{ N/m}^2}{76} = 1.33 \times 10^3 \text{ N/m}^2$$

Hence 130 mmHg = 13 cmHg = $13(1.33 \times 10^3$ N/m$^2)$

$$= \boxed{1.73 \times 10^4 \text{ N/m}^2}$$

90 mmHg = $9(1.33 \times 10^3$ N/m$^2) = \boxed{1.20 \times 10^4 \text{ N/m}^2}$

(b) The corresponding absolute pressures are:

$P_S = P_{atm} + 1.73 \times 10^4$ N/m^2

$\quad = 1.01 \times 10^5$ N/m$^2 + 1.73 \times 10^4$ N/m^2

$= \boxed{1.18 \times 10^5 \text{ N/m}^2}$

$P_D = 1.01 \times 10^5$ N/m$^2 + 1.20 \times 10^4$ N/m^2

$= \boxed{1.13 \times 10^5 \text{ N/m}^2}$

It can be seen that the pressures generated by the pumping heart are relatively small, being only about 10 percent of atmospheric pressure.
(c) For this patient, the average gauge pressure of the blood is

$$\frac{90 + 130}{2} \text{ mmHg} = 110 \text{ mmHg} = 11(1.33 \times 10^3 \text{ N/m}^2)$$

$$= \boxed{1.46 \times 10^4 \text{ N/m}^2}$$

For purposes of discussion the average pressure of the blood as it enters the aorta from the heart is often taken to be 100 mmHg, or 1.33×10^4 N/m^2.

10.3 Density Determinations

You may have heard the story of how Archimedes (287–212 B.C.), the famous Greek mathematician and scientist, ran through the streets of Athens (naked, according to Vitruvius, since Archimedes had been taking a bath), shouting "*Eureka*," which in Greek means "I have found it." What he had found was a way to determine the density of the material in the king's crown.

King Hiero II of Syracuse was suspicious that a goldsmith had substituted silver for gold in his crown, and asked Archimedes to determine whether this was indeed the case. The insight Archimedes had during his bath was that, if the tub were filled to the brim, then the overflow when he stepped into the tub would be exactly equal to the volume of his body. This gave him a way to determine accurately the volume of an irregularly shaped body like Hiero's crown. He first determined the mass of the crown on a balance, and then constructed two other crowns of exactly the same mass, one of gold and the other of silver. He next immersed each one in a bucket of water, and measured the overflow in each case as exactly as possible by using a pint measure.

King Hiero's crown displaced more water than an equal mass of gold but less than an equal mass of silver. Hence, its density was lower than that of pure gold. Archimedes therefore concluded that the goldsmith had stolen some gold from the crown and replaced it with silver.

This method of Archimedes is still the simplest, although not the most accurate, way to obtain the density of an unknown solid object. We simply find the mass of the object on a balance, find its volume by measuring the overflow when it is immersed in a beaker of water, and then find its density from $d = M/V$.

Archimedes' Principle

Archimedes' Principle

Archimedes is also given credit for a related principle of hydrostatics, which is stated as follows:

An object partially or totally submerged in a fluid (liquid or gas) is buoyed up by a force equal to the weight of the fluid which the object displaces.

Hence, if the king's crown displaced $1.5 \times 10^{-4} \text{ m}^3$ of water when submerged, it would be buoyed up or supported by a force

$$Mg = dVg = (10^3 \text{ kg/m}^3)(1.5 \times 10^{-4} \text{ m}^3)(9.8 \text{ m/s}^2) = 1.5 \text{ N}$$

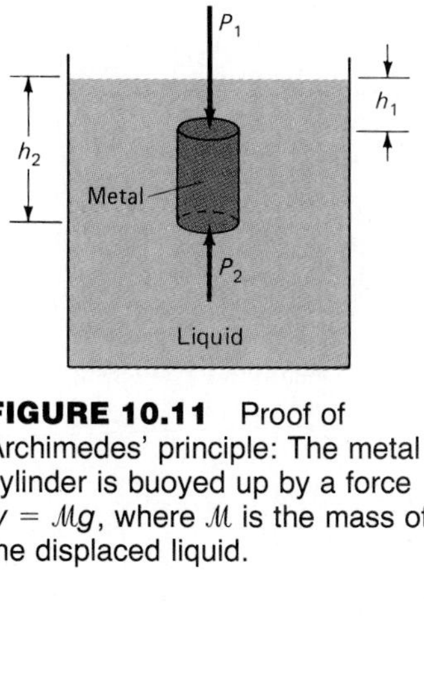

The proof of Archimedes' principle is straightforward for a regularly shaped object like a cyclinder. Thus, if a metal cylinder B is immersed in water, as in Fig. 10.11, the pressure at its top surface due to the liquid above it is

$$P_1 = dgh_1$$

and the force acting downward on the top surface of the cylinder is therefore

$$F_1 = P_1 A = dgh_1 A$$

The pressure on the bottom surface is $P_2 = dgh_2$, since the pressure is the same at any point at the same depth in the liquid. Hence the upward force on the bottom of the cylinder is

$$F_2 = dgh_2 A$$

The net upward force acting on the cylinder is then

$$F_B = F_2 - F_1 = dgA(h_2 - h_1)$$

But $A(h_2 - h_1)$ is just the volume V of the cylinder, and so

$$F_B = dgV = Mg$$

where M is the mass of the displaced liquid. This force is clearly upward, and is therefore a buoyant force. Hence, the cylinder is buoyed up by a force equal to the weight Mg of the displaced liquid.

Although derived here for a simple case, Archimedes' principle can be shown to be true in general for any object, no matter what its shape.

FIGURE 10.11 Proof of Archimedes' principle: The metal cylinder is buoyed up by a force $w = Mg$, where M is the mass of the displaced liquid.

Example 10.7

A log floats on water with $\frac{4}{5}$ its volume submerged, as in Fig. 10.12. What is the density of the log?

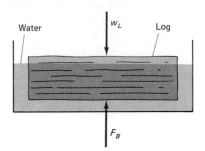

FIGURE 10.12 Diagram for Example 10.7.

SOLUTION

The weight of the log is $w_L = Mg = d_L V_L g$. The volume of water displaced is given as

$$V_W = \tfrac{4}{5}V_L$$

Hence by Archimedes' principle the buoyant force is

$$F_B = d_W V_W g = d_W \tfrac{4}{5}V_L g$$

Since the log is floating in equilibrium, the weight of the log is equal to the buoyant force exerted by the water. Hence

$$w_L = F_B$$

or $\quad d_L V_L g = \tfrac{4}{5}d_W V_L g$

or $\quad d_L = \tfrac{4}{5}d_W = \boxed{0.80 \times 10^3 \text{ kg/m}^3}$

Notice that the fraction of the log submerged is just the ratio of the log's density to the density of water; i.e., it is the specific gravity of the log.

If any object floats beneath the surface of a liquid and neither rises to the top nor sinks to the bottom, its density is the same as that of the liquid in which it is submerged. If the liquid is water, the specific gravity of the object is 1 and its density is 10^3 kg/m³.

Use of Archimedes' Principle to Determine Densities of Solids

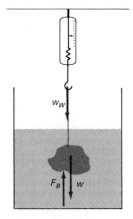

FIGURE 10.13 Determination of the density of a solid object, using Archimedes' principle.

We can also determine densities of solids by using Archimedes' principle. Suppose we first weigh the object on a spring scale in air. (A spring scale measures weights $w = mg$ directly, not masses.) Its weight is $w = Mg$. We then submerge the object in water at the end of a supporting string, as in Fig. 10.13, and measure its weight, which has now been reduced by the buoyant force of the water. Hence we have

$$w_W = w - F_B \quad \text{or} \quad F_B = w - w_W$$

where w is its weight in air, and w_W its weight in water. Now, by Archimedes' principle,

$$F_B = d_W V g = w - w_W$$

or $\quad V = \dfrac{w - w_W}{d_W g}$

Now $\quad V = \dfrac{M}{d}$

and so $\quad \dfrac{d}{M} = \dfrac{1}{V} = \dfrac{d_W g}{w - w_W}$

and $\quad \dfrac{d}{d_W} = \dfrac{Mg}{w - w_W} = \dfrac{w}{w - w_W}$ $\qquad$ (10.6)

Hence the specific gravity of a solid object (d/d_W) can be found from two weight measurements of the object—one in air, the other in water. If a liquid other than water is used, the density of the object can still be found from Eq. (10.6), so long as the density of the liquid is known.

Example 10.8

Suppose a crown like the one Archimedes tested has a weight of 30 N in air, and only 28 N when submerged in water.

(a) Is it made out of pure gold or some other material?
(b) If the crown is made partially of lead and partially of gold, what is the percentage of lead in it?

SOLUTION

(a) The density of the crown's material can be obtained from Eq. (10.6):

$$\frac{d}{d_W} = \frac{w}{w - w_W}$$

Here $\quad \dfrac{d}{d_W} = \dfrac{30 \text{ N}}{30 \text{ N} - 28 \text{ N}} = \dfrac{30 \text{ N}}{2 \text{ N}} = 15$

Hence the material has a specific gravity of 15, or a density of 15×10^3 kg/m³. It certainly is not pure gold, which has a specific gravity of 19.3, not 15. Since silver has a density of 10.5 kg/m³, and lead a density of 11.3 kg/m³, it is likely that the crown is a mixture of gold with silver or lead (see Table 10.3).

(b) Let x be the fraction of lead in the crown. Then $1 - x$ is the fraction of gold. These two fractions, each multiplied by the proper specific gravity, must yield the measured specific gravity of 15. Hence we must have:

$$x(11.3) + (1 - x)(19.3) = 15$$

Solving for x, we obtain $x = 0.56$. Hence the crown is

$$\boxed{56\% \text{ lead}}$$

and 44% gold.

10.4 Fluids in Motion

We depend on the flow of fluids throughout much of our lives. Oil and natural gas are pumped hundreds of miles through pipes to satisfy heating and transportation needs. It was a break in the flow of cooling water which started the chain of events that led to a near-catastrophe at the Three Mile Island nuclear power plant. Gigantic pipes bring fresh drinking water from mountain reservoirs to large cities. Deadly strokes occur when blood clots cut off the flow of blood to the human brain. Such phenomena are examples of fluids in motion.

We will consider in this section only the simplest kind of fluid flow—the flow along streamlines, without turbulence and without viscosity. By *streamline flow* we mean flow similar to that in Fig. 10.14. Any particle which finds itself at point A will be carried along a particular streamline and arrive at point B and later at point C. The flow is smooth and completely predictable. There are no whirlpools or sudden changes of the kind which occur, for example, in white-water rapids, where much of the energy of the water is consumed in turbulent motion and eventually converted into heat.

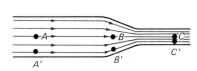

FIGURE 10.14 Streamline flow. Any liquid particle arriving at point A will move along a streamline from A to B to C. Similarly any particle at A' will move from A' to B' to C'. This behavior remains the same over time.

Equation of Continuity

If we consider a small tube of liquid whose boundaries are set by streamlines, as in Fig. 10.15, the velocity of the fluid may change as it flows along the tube. Suppose that at Q the velocity of the fluid is v_1 and the cross-sectional area A_1, and that at R the velocity is v_2 and the area A_2. Then for smooth, continuous

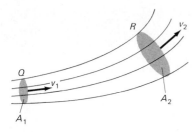

FIGURE 10.15 Equation of continuity. For streamline flow $A_1v_1 = A_2v_2$.

flow the mass passing the point Q must be equal to the mass passing the point R in the same time interval Δt. In time Δt the mass passing Q is

$$\Delta M_1 = d_1 \, \Delta V_1 = d_1 A_1 v_1 \, \Delta t$$

where $v_1 \, \Delta t$ is the length of the fluid element passing Q in time Δt, and d_1 is the density of the fluid at Q. Similarly, at R

$$\Delta M_2 = d_2 A_2 v_2 \, \Delta t$$

Then, since the amount of fluid passing Q and R must be the same in equal times,

$$\frac{\Delta M_1}{\Delta t} = \frac{\Delta M_2}{\Delta t} = d_1 A_1 v_1 = d_2 A_2 v_2$$

If we assume that the fluid is incompressible (an assumption that is *not* valid for most kinds of gas flow), so that its density does not change, d_1 is equal to d_2, and so

Equation of Continuity

$$\boxed{A_1 v_1 = A_2 v_2} \tag{10.7}$$

This is the *equation of continuity*, which says that, for streamline flow, the speed of flow of a fluid varies inversely with the cross-sectional area of the tube. Hence, if there are constrictions in blood vessels in the human body, the blood must move more rapidly to get through these constrictions.

Example 10.9

In a normal adult at rest the average speed with which the blood flows through the aorta, the main artery leading from the heart, is $\bar{v} = 0.33$ m/s.
(a) If the aorta has a cross-sectional radius of about 1.0 cm, how much blood flows per second through the aorta?
(b) If the total cross-sectional area of all the other arteries fed by the aorta is 20×10^{-4} m^2, at what speed does blood flow in these arteries?

SOLUTION

(a) The amount of blood flowing per second through the aorta is that occupying a volume of cross-sectional area $A_1 = \pi(1.0 \times 10^{-2} \text{ m})^2 = 3.1 \times 10^{-4}$ m^2 and of length 0.33 m, since all the blood in such a length will pass through the aorta at a speed of 0.33 m/s. Hence the volume of blood flowing per second through the aorta is

$$A_1 \bar{v} = (3.1 \times 10^{-4} \text{ m}^2)(0.33 \text{ m/s}) = \boxed{1.0 \times 10^{-4} \text{ m}^3/\text{s}}$$

(b) From the equation of continuity, we have

$$A_1 v_1 = A_2 v_2$$

or

$$v_2 = \frac{A_1}{A_2} v_1 = \frac{A_1 \bar{v}}{A_2} = \frac{1.0 \times 10^{-4} \text{ m}^3/\text{s}}{20 \times 10^{-4} \text{ m}^2} = 0.05 \text{ m/s}$$

The blood therefore moves about 6 times more slowly through the other arteries than it does through the aorta.

Bernoulli's Equation

Bernoulli's famous equation, named for the Swiss mathematician Daniel Bernoulli (1700–1782), is the most important equation in fluid dynamics. It describes the streamline flow in a nonviscous, incompressible fluid, as shown in Fig. 10.16. The fluid (say, water) is assumed to be pumped uphill, and the pipe is also assumed to widen as it rises. We concentrate on the fluid between points Q and R in Fig. 10.16a. In time Δt the left end of this fluid has moved from Q to Q' and the right end from R to R', as shown in Fig. 10.16b.

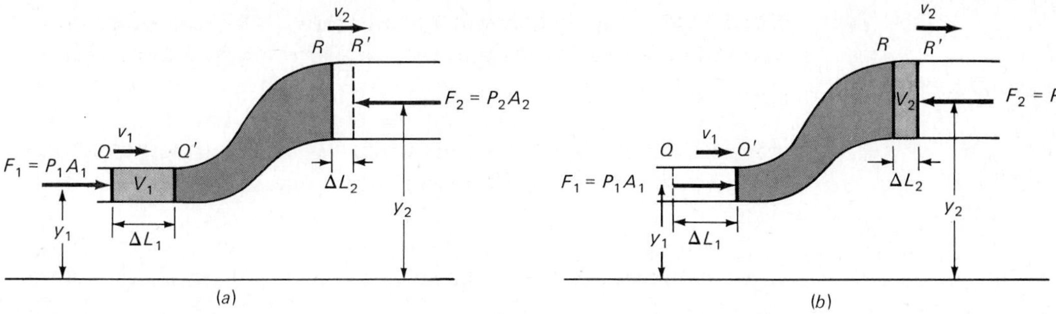

(a)

(b)

FIGURE 10.16 Bernoulli's equation for streamline flow in a nonviscous, incompressible liquid: (a) Initial position of a chosen volume of fluid. (b) Final position of the same volume of fluid.

We now apply our extended work-energy theorem to this system. We have, from Eq. (6.11),

Work done = $\Delta KE + \Delta PE$

Here the work is done by the pressures at the two ends moving the fluid through the pipe. The work done by the force F_1 at Q is $W_1 = F_1 \, \Delta L_1 = P_1 A_1 \, \Delta L_1$. Similarly, the work done by the force F_2 at R is $W_2 = -P_2 A_2 \, \Delta L_2$, since F_2 and ΔL_2 are in opposite directions. The total work done is then

$$W = W_1 + W_2 = P_1 A_1 \, \Delta L_1 - P_2 A_2 \, \Delta L_2$$

But, since from the equation of continuity $A_1 v_1 = A_2 v_2$, we have

$$A_1 \frac{\Delta L_1}{\Delta t} = A_2 \frac{\Delta L_2}{\Delta t} \quad \text{or} \quad A_1 \, \Delta L_1 = A_2 \Delta L_2$$

Now $A_1 \, \Delta L_1$ and $A_2 \, \Delta L_2$ are the volumes V_1 and V_2, and so

$$V_1 = V_2 = \frac{m}{d}$$

where m is the mass of the piece of liquid which is colored in the diagram. The rest of the liquid is unchanged in the process. Hence, if we set $V_1 = V_2 = V$, we have

$$W = P_1 A_1 \, \Delta L_1 - P_2 A_2 \, \Delta L_2 = (P_1 - P_2)V = (P_1 - P_2)\frac{m}{d}$$

and our work-energy theorem becomes

$$(P_1 - P_2)\frac{m}{d} = \Delta KE + \Delta PE = \tfrac{1}{2}mv_2^2 - \tfrac{1}{2}mv_1^2 + mgy_2 - mgy_1$$

Dividing through by m and multiplying by d, we have

$$P_1 - P_2 = \tfrac{1}{2}dv_2^2 - \tfrac{1}{2}dv_1^2 + dgy_2 - dgy_1$$

and so $\quad P_1 + \tfrac{1}{2}dv_1^2 + dgy_1 = P_2 + \tfrac{1}{2}dv_2^2 + dgy_2$

Since this is true for any two points in the flow, we have in general

Bernoulli's Equation

$$\boxed{P + \tfrac{1}{2}dv^2 + dgy = \text{constant}}$$

(10.8)

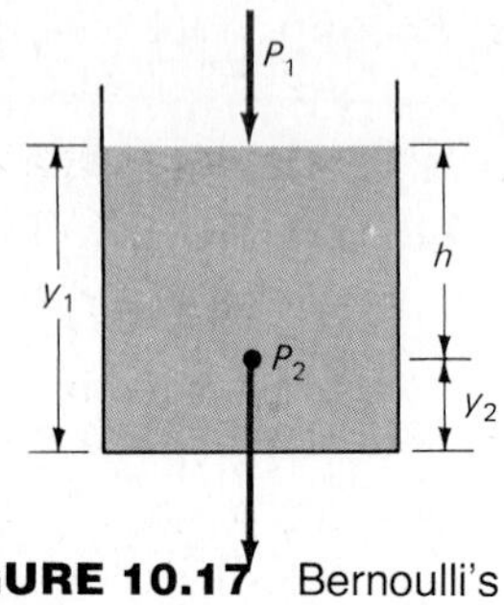

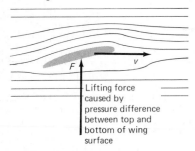

FIGURE 10.17 Bernoulli's equation applied to the pressure exerted by a fluid at rest: $P_2 = P_1 + dgh$.

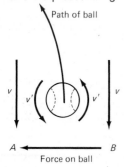

Lifting force caused by pressure difference between top and bottom of wing surface

FIGURE 10.18 Lifting force on an airplane, as predicted by Bernoulli's equation. On the top wing surface the air must travel a longer distance. Hence the air speed is higher and the pressure lower. The air travels a shorter distance along the bottom of the wing. Hence the air speed is lower and the pressure higher.

Path of ball

Force on ball

FIGURE 10.19 Why a baseball curves: An application of Bernoulli's equation. v here is the speed of the air relative to the ball.

This is Bernoulli's equation for fluid flow. All three terms in it have the dimensions of energy per unit volume (can you prove this?). It can therefore be looked at as a statement of the principle of conservation of energy applied to fluid flow. It states that in the streamline flow of an incompressible, nonviscous fluid, the pressure, speed of flow, and height of the fluid will always adjust themselves in such a way as to keep the total energy of the fluid constant.

Bernoulli's equation contains most other equations of hydrostatics and hydrodynamics as special cases. Thus it tells us that for a fluid at rest, that is $(v_1 = v_2 = 0)$

$$P_1 + dgy_1 = P_2 + dgy_2$$

Hence, if P_1 is the pressure at the surface of a liquid and P_2 is the pressure a distance h below the surface,

$$P_2 = P_1 + dg(y_1 - y_2)$$

where y_1 and y_2 are measured from a point at the bottom of the fluid, as in Fig. 10.17. Hence $y_1 - y_2 = h$, and so $P_2 = P_1 + dgh$, which is Eq. (10.5).

Also, if we consider steady horizontal flow, where y remains constant, we have

$$P + \tfrac{1}{2}dv^2 = \text{constant}$$

Hence the pressure is reduced at those points where the velocity increases and vice versa. The wings of an airplane have a curved shape which forces the air to flow more rapidly above the wing and less rapidly below it, as in Fig. 10.18. As a result, the pressure above the wing is reduced, while below the wing it is increased, according to the above equation. This provides the "lift" to keep the airplane in the air. If a plane stalls in flight so that this pattern of streamline air flow is interrupted, the lifting force disappears and the pilot must put the plane into a shallow dive to restore the air flow and hence the lift.

The curves thrown by baseball pitchers and the slices and hooks of golfers also find their explanations in the above equation. Figure 10.19 shows what happens when a pitcher throws a curve. As the ball moves forward, the air rushes past it at a velocity v. Since the ball is spinning, however, it drags some air with it with a velocity v'. Hence the resultant speed of the air at A is $v + v'$ and at B it is $v - v'$. The pressure is, therefore, by Bernoulli's principle, lower at A than at B, since the velocity is greater at A. A force is therefore set up which continuously deflects the ball in the direction of A. The rougher the surface of the ball, the greater the effect. This is the reason major-league pitchers resort to all kinds of tricks to make the surface of the ball as rough as possible.

Example 10.10

Cold water circulates through a house from the main water pipe in the basement. The water enters the house through an 8.0-cm-diameter pipe at a flow speed of 0.60 m/s under a pressure of 4.0×10^5 N/m² (i.e., 4 times atmospheric pressure).

(a) What is the flow speed of the water in a 5.0-cm-diameter pipe on the third floor, which is 9.0 m above the main water pipe in the basement?

SOLUTION

(a) From the equation of continuity for fluid flow [Eq. (10.7)] we know that the flow speed and the cross-sectional area of the pipe at any two points along the pipe are related by the equation $A_1v_1 = A_2v_2$. Hence

$$v_2 = \frac{A_1}{A_2}v_1 = \frac{(0.040 \text{ m})^2}{(0.025 \text{ m})^2}(0.60 \text{ m/s}) = \boxed{1.5 \text{ m/s}}$$

(b) From Bernoulli's equation we have

$$P_1 + \tfrac{1}{2}dv_1^2 + dgy_1 = P_2 + \tfrac{1}{2}dv_2^2 + dgy_2$$

or

(b) What is the water pressure in the third-floor pipe?

$$P_2 = P_1 + \tfrac{1}{2}d(v_1^2 - v_2^2) + dg(y_1 - y_2)$$
$$= 4.0 \times 10^5 \text{ N/m}^2 + \tfrac{1}{2}(10^3 \text{ kg/m}^3)\,[(0.60 \text{ m/s})^2 - (1.5 \text{ m/s})^2]$$
$$+ (10^3 \text{ kg/m}^3)(9.8 \text{ m/s}^2)(0 - 9.0 \text{ m})$$

Hence, since $1 \text{ kg·m/s}^2 = 1 \text{ N}$, we have

$$P_2 = 4.0 \times 10^5 \text{ N/m}^2 - 9.5 \times 10^2 \text{ N/m}^2 - 8.8 \times 10^4 \text{ N/m}^2$$
$$= 3.1 \times 10^5 \text{ N/m}^2$$

The pressure has fallen, as it must when the water flows uphill.

Example 10.11

A water cooler which is open to the atmosphere at the top contains water to a height h above the exit hole for the water at the bottom, as in Fig. 10.20. With what speed will the water flow out through the hole?

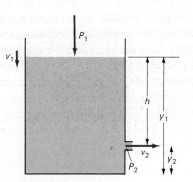

FIGURE 10.20 Speed of water flow from a water cooler. Here $P_1 = P_2$ and $v_2 = \sqrt{2gh}$.

SOLUTION

This is a good illustration of Bernoulli's principle. We have

$$P_1 + \tfrac{1}{2}dv_1^2 + dgy_1 = P_2 + dv_2^2 + dgy_2$$

Both the top of the cooler and the exit hole are assumed to be exposed to the atmosphere, so that $P_1 = P_2$. Also v_1, which is the speed with which the water level falls, is so small that we can take it to be zero to a good approximation. Hence we have

$$\tfrac{1}{2}dv_2^2 = dg(y_1 - y_2) \qquad \text{or} \qquad v_2 = \sqrt{2g(y_1 - y_2)}$$

Since $y_1 - y_2$ is here equal to h, the distance of the hole below the level of the water in the cooler, this becomes

$$v_2 = \sqrt{2gh}$$

a result sometimes called *Torricelli's theorem*.

It is interesting to note that the speed of the water in this case is the same as would be given by point mechanics for a mass m of water falling freely under the force of gravity through a distance h.

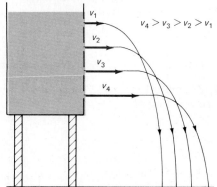

FIGURE 10.21 Flow of water from a water cooler as a function of height of water level above exit hole.

Figure 10.21 shows how the flow of the water from holes at different heights below the surface of the water would look.

10.5 Viscosity

In our discussion of Bernoulli's principle thus far we have completely neglected one very important property of fluids—viscosity. *Viscosity* is a measure of how difficult it is to cause a fluid to flow, or of how resistive the fluid is to motion through it, as we saw in Sec. 3.8, where we discussed the motion of falling objects in viscous fluids. Thus water has a higher viscosity than air; we can easily walk through the air, but it is much more difficult to walk through water. If water did not resist the motion of our bodies, swimming would be impossible, because it is the reactive force of the water on our bodies which pushes us forward when we swim, just as it is the reactive force of the ground on our feet that pushes us forward when we walk. In general the viscosity of gases is very small and that of liquids considerably greater.

Let us restrict our attention here to viscosity in liquids because this is where the effects of viscosity show up most clearly. If a liquid is flowing through a pipe, the walls of the pipe rub against the flowing liquid and tend to slow it down. Hence the liquid flows more slowly at the walls of the pipe than it does at the center. A distribution of speeds is therefore set up across the width of the pipe, similar to that in Fig. 10.22.

We can approximate this speed distribution by a model in which we consider thin cylindrical shells of liquid, each moving with a slightly different speed through the pipe. One shell rubs against the next shell in the same way that a book rubs against a desktop when it is dragged over it. We have seen that friction between the book and the desktop generates heat and wastes mechanical energy. The same thing happens with liquids. Mechanical energy is lost and heat is generated by the internal sliding of one layer of liquid over another. Hence viscosity for liquids is analogous to the frictional force between solids and depends on the force required to slide one layer of the fluid over another at a given speed.

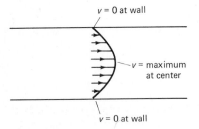

FIGURE 10.22 Distribution of speeds for the viscous flow of liquid in a pipe.

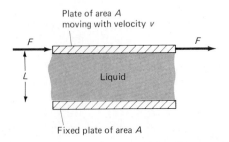

FIGURE 10.23 A way to measure the viscosity of a liquid: The greater the viscosity, the larger the force F required to give the upper plate a speed v.

One way to measure viscosities is to confine a liquid between two flat plates each of area A and then to keep the bottom plate fixed and apply a force F to the top plate to move it at a constant speed v with respect to the bottom plate. If the distance between the plates is L, as in Fig. 10.23, then we define the coefficient of viscosity η (Greek lowercase letter *eta*) as follows:

Coefficient of Viscosity

$$\eta = \frac{F/A}{v/L}$$

(10.9)

For fixed A and L, the larger the force needed to produce a velocity v, the larger η, and hence the greater the viscosity of the fluid.

Equation (10.9) shows that the SI units for the coefficient of viscosity η must be

$$\frac{N/m^2}{(m/s)/m} = \frac{N \cdot s}{m^2} \quad \text{or} \quad Pa \cdot s \quad \text{since} \quad 1\ Pa = 1\ N/m^2$$

One pascal-second is equal to 1000 centipoise (cP), a unit frequently used in tabulating viscosities. The viscosity of water at 20.20°C is taken as a reference and is exactly equal to 1.0000 cP.

The viscosities of some important liquids are shown in Table 10.2. Note that large variations occur from liquid to liquid. The viscosities of gases are smaller than those of liquids by about two orders of magnitude. Thus the viscosity of oxygen gas at 20°C is 2.0×10^{-5} Pa·s compared with 1.0×10^{-3} Pa·s for water. The coefficient of viscosity depends on temperature, falling as the temperature rises. This is what we would expect, since liquids approach the gaseous state as their temperature goes up.

TABLE 10.2 Coefficient of Viscosity of Some Important Liquids (at 20°C)

Liquid	Coefficient of viscosity η	
	Pa·s, or N·s/m²	cP
Water	1.00×10^{-3}	1.00
Benzene	0.65×10^{-3}	0.65
Carbon tetrachloride	0.97×10^{-3}	0.97
Ethyl alcohol	1.20×10^{-3}	1.20
Glycerin	1.49	1490
Mercury	1.55×10^{-3}	1.55
Methyl alcohol	0.597×10^{-3}	0.597
Sulfuric acid	25.4×10^{-3}	25.4
Turpentine	1.49×10^{-3}	1.49
Whole blood (37°C)	4×10^{-3}	4

Poiseuille's Law

It is often important to know the rate at which a liquid flows—e.g., blood through the human body or oil through the Alaskan pipeline. For example, we would expect that the greater the pressure difference between the two ends of a pipe ($\Delta P = P_1 - P_2$) and the shorter the pipeline, the greater the flow rate. Hence we would expect that the volume of fluid flowing out of the pipe per second would vary as the *pressure gradient*, $(P_1 - P_2)/L$. Also, since increased viscosity tends to slow down the flow, we would expect an inverse dependence on the coefficient of viscosity η. The dependence on the radius of the tube R is more difficult to predict. To find this dependence we can resort to an experiment on the flow of water through tubes of radius 0.01, 0.02, 0.04, and 1.0 m for the same values of $P_1 - P_2$, L, and η. The results are shown in Figs. 10.24 and 10.25. In Fig. 10.24 the volume flowing through in unit time, V/t, is plotted as a function of R. The dependence is clearly not linear, but it is hard to tell the exact functional dependence. In Fig. 10.25 we plot log (V/t) versus log R. We find a straight line with a slope of 4. Hence, as shown in Appendix 2C, V/t varies as R^4.

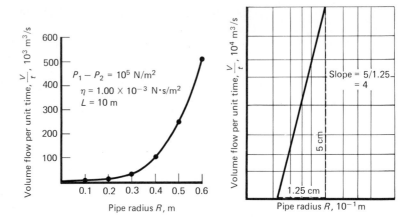

FIGURE 10.24 The volume flow of water (V/t) through a round pipe as a function of the radius (R) of the pipe, for constant pressure difference between the ends of the pipe.

FIGURE 10.25 Data of Fig. 10.24 plotted on log-log graph paper (see Appendix 2C). The slope is 4, indicating that V/t varies as R^4.

The Frenchman Jean L. M. Poiseuille* (1797–1896) combined this R^4 dependence with the other factors described above to form a complete equation for fluid flow. This is now called Poiseuille's law:

Poiseuille's Law

$$\frac{V}{t} = \frac{\pi R^4 (P_1 - P_2)}{8 \eta L}$$

(10.10)

The most remarkable term in this equation is the R^4 term, which tells us that, if the pipe's radius is increased by a factor of 2, the amount of fluid flowing out of the tube per second increases by a factor of 16. An oversimplified explanation for this fourth-power dependence on R is that since both the cross-sectional area of the pipe and the velocity of flow at the center depend on R^2, the flow rate depends on the product of the two, or on R^4 (a rigorous proof requires the use of calculus).

It is worth noting that the dependence of the flow rate on the fourth power of R can also be obtained from dimensional analysis. Since the units of η are N·s/m^2, we have for the dimensions in Eq. (10.10):

$$\frac{[L^3]}{[T]} = \frac{[L^n] \, [F]/[L^2]}{[L][F] \, [T]/[L^2]} \quad \text{or} \quad [L^n] = [L^4]$$

from which R^n must be R^4.

Blood Flow

Poiseuille's major research interest was the physiology of the circulation of the blood through arteries in the human body. It was this work which led him to his equation, which has a much wider applicability in physics and engineering than he first realized. The blood distribution system in the human body is very complicated, and it is amazing that Poiseuille's law applies to it as well as it does.

Figure 10.26 shows how the pressure varies in the circulatory system of the human body. As the figure shows, the pressure drop is very small across the large arteries and veins because of their relatively large diameters. It is in

---

*The centipoise and the poise (10^2 cP) are both named after Poiseuille.

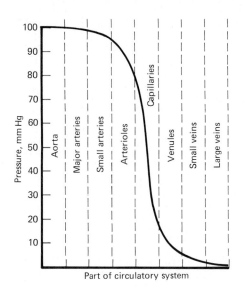

FIGURE 10.26 Pressure in the circulating system of the human body.

the small-diameter capillaries that most of the drop in pressure occurs between the blood leaving the heart through the aorta at high pressure and that returning to it at low pressure.

Poiseuille's law also points up problems which can result from hardening of the arteries (arteriosclerosis) or from the filling in of the arteries with deposits of cholesterol or other debris. According to Eq. (10.10), if in some way R is reduced by a factor of 2, with η and L remaining unchanged, then the pumping pressure of the heart must increase by a factor of 16 to keep the blood flowing at its proper rate. Usually the heart cannot completely provide this tremendous increase in pressure, and, as the artery fills up more and more, the blood circulation drops. Hence the clogging of arteries leads to two quite undesirable consequences, high blood pressure and poor blood circulation. This stenosis, or narrowing of the arteries, is often a contributing factor in chronic heart disease.

Example 10.12

Blood flows through a thin capillary in a person's leg. The capillary is 2.0 mm long and 3.0 μm in radius. The speed of the blood through the center of this capillary is 1.0 mm/s. What is the drop in the blood pressure as the blood passes through this capillary?

SOLUTION

From Table 10.2 the coefficient of viscosity for blood at 37°C (body temperature) is $\eta = 4 \times 10^{-3}$ Pa·s $= 4 \times 10^{-3}$ N·s/m^2.

From Poiseuille's equation for the viscous flow of a liquid, which is the case here, we have [Eq. (10.10)]

$$P_1 - P_2 = \frac{8\eta LV}{\pi R^4 t}$$

where V/t is the amount of blood flowing through the capillary per second. As we saw in Example 10.9, or from the equation of continuity [Eq. (10.7)], this flow rate $V/t = A\bar{v}$, where $\bar{v}$ is the average speed of flow across the diameter of the tube, if v is not constant, as is the case here. The speed of flow varies from 1.0 mm/s at the center of the capillary to 0 at the walls of the capillary, where the viscosity prevents the fluid from moving at all. Hence the average flow speed is:

$$\bar{v} = \frac{1.0 \text{ mm/s} + 0}{2} = 0.50 \text{ mm/s}$$

Then

$$P_1 - P_2 = \frac{8\eta L}{\pi R^4}(\pi R^2 \bar{v}) = \frac{8\eta L}{R^2} \bar{v}$$

$$= \frac{8(4 \times 10^{-3} \text{ N·s/m}^2)(2.0 \times 10^{-3} \text{ m})(0.50 \times 10^{-3} \text{ m/s})}{(3.0 \times 10^{-6} \text{ m})^2}$$

$$= 3.6 \times 10^3 \text{ N/m}^2 = \boxed{3.6 \times 10^3 \text{ Pa}}$$

Hence the pressure drop in this one capillary is about 27 mmHg compared to an average blood pressure of about 100 mmHg.

10.6 Other Properties of Liquids

Diffusion

By the *diffusion* of fluids (i.e., gases or liquids) we mean the following:

Definition

Diffusion: The process by which molecules of one kind pass into and intermingle with molecules of another kind in a fluid.

When two gases like hydrogen and nitrogen are mixed by opening a valve connecting containers of the two gases, they mingle with each other, and in a short time samples of the gas reveal the same relative numbers of H_2 and N_2 molecules at any place in the gas mixture. This process of diffusion arises because of the random thermal motion of the gas molecules, which carries them rapidly throughout the two connected gas containers. Diffusion also explains why, when a bottle of perfume is opened, the fragrance can be detected across a room in a few seconds' time.

Something similar happens with liquids. The molecules are also in random motion and therefore also move through the whole volume occupied by the liquid, although not as rapidly as gas molecules move. For example, if a test tube is filled with water, and then with an eyedropper some drops of bright blue copper sulfate are introduced at the bottom of the test tube, an upward diffusion of the copper sulfate molecules occurs despite the opposing force of gravity. The process is very slow, but after a few days a uniform, pale blue color can be seen throughout the liquid.

Diffusion plays an important role in biological processes. For example, muscle fibers require oxygen, which is supplied by diffusion from surrounding tissues into the muscle fibers. If an insufficient supply of oxygen is present, the muscles will not respond properly to signals given to them by the brain.

Osmosis

There are some membranes, called *semipermeable membranes*, which have pores that can pass small molecules but not large ones. For example, in a sugar solution a semipermeable membrane would be one which would pass the water molecules (the solvent) but not the sugar molecules (the solute). The movement of the molecules in this preferential diffusion process across such membranes is called *osmosis*. For example, if a red blood cell is immersed in water, it will absorb extra water through its outer membrane and grow in size, but no hemoglobin or other large molecules will pass through the membrane from the blood cell into the water.

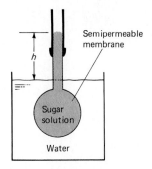

FIGURE 10.27 Osmotic pressure. The osmotic pressure is equal to the height of the sugar solution in the tube over the level of the water in which it is immersed: $P_{osm} = dgh$.

To illustrate osmosis, suppose we take a semipermeable membrane in the shape of a round balloon, with a narrow tube inserted into the mouth of the balloon, as in Fig. 10.27. We fill the membrane with a sugar solution and then immerse it in a tank of distilled water. It is found that water molecules diffuse preferentially into the sugar solution and that as a result the height of the water in the central tube rises.

This process of osmosis can be explained by assuming that both the sugar and the water molecules behave much like the molecules of an ideal gas. Since there are more water molecules per unit volume outside the membrane than inside, because the inside also contains sugar molecules, initially more water molecules attempt to pass inward through the membrane than try to pass outward per unit time. Since water molecules can pass in both ways through the membrane, there is a net flow of water to the inside of the membrane, and the water level rises in the attached tube. This flow of water continues until the pressure caused by the raised water level is just sufficient to equalize on the two sides the frequency of collision of water molecules with the membrane. When this happens, equilibrium has been established and no further net flow of water occurs. Then the water level remains fixed at some new height h above the water level in the tank.

Osmotic Pressure

The pressure that prevents further flow of the water is called the *osmotic pressure* of the liquid. This pressure is equal to the pressure produced by the central column of liquid of added height h; i.e., according to Eq. (10.2),

$$P_{osm} = dgh \tag{10.11}$$

where d is the density of the solution inside the membrane.

The basic concepts pertaining to ideal gases can be shown to apply to molecules dissolved in a liquid to form a dilute solution like the sugar molecules in our case. The sugar molecules act like an enclosed gas, and the osmotic pressure may be thought of as the pressure of this "gas." We can therefore write the ideal gas law in this case as

$$P_{osm}V = nRT \qquad \text{or} \qquad P_{osm} = \frac{n}{V}RT \tag{10.12}$$

Now, $n = m/M$, where m is the mass of the solute (sugar) in the volume V, and M is the molecular mass of the sugar. Hence we have

$$P_{osm} = \frac{m}{V}\frac{RT}{M} = c\frac{RT}{M} \tag{10.13}$$

where $c = m/V$ is the concentration, in kilograms per cubic meter, of the sugar in the solution.

Hence by measuring the osmotic pressure, the molecular mass M can be determined from the equation $M = cRT/P_{osm}$. Note that the osmotic pressure is larger for smaller values of M, and that Eq. (10.13) assumes that the concentration c is small enough that the solute can be considered to behave like an ideal gas.

The walls of living cells are semipermeable, and so some chemical substances can diffuse in and out of the cell while others are prevented from doing so. It is this osmotic behavior that preserves the proper chemical environment within the cell.

Example 10.13

Albumin is a protein molecule found in blood plasma. If a water solution containing 4.0 g of albumin per liter (1000 cm³) produces a difference in height of the liquids in Fig. 10.27 equal to 1.48 cm at 20°C, what is the molecular mass of the albumin molecule?

SOLUTION

In this case the osmotic pressure, i.e., the pressure required to stop any net flow of water, is given by Eq. (10.11):

$$P_{osm} = dgh = \left(1000 \frac{kg}{m^3}\right)\left(9.8\frac{m}{s^2}\right)(1.48 \times 10^{-2}\ m)$$

$$= 145\ N/m^2 \quad \text{(or 145 Pa)}$$

Then, on the assumption that the albumin molecules behave like an ideal gas, we have, from Eq. (10.13),

$$M = \frac{cRT}{P_{osm}} = \frac{\left(\dfrac{4.0 \times 10^{-3}\ kg}{10^{-3}\ m^3}\right)\left(8.31\dfrac{J}{mol \cdot K}\right)(293\ K)}{145\ N/m^2}$$

$$\text{or} \quad M = \boxed{67\ kg/mol}$$

This is the approximate molecular mass of the albumin. Note that the mass of an albumin molecule is about 1000 times the mass of an iron atom ($^{56}_{26}$Fe). This shows how large and complicated organic molecules like albumin are.

10.7 Solids: Structure and Density

In atomic solids the atoms vibrate back and forth about fixed positions in space in the way they would if springs connected each atom to its neighbors. In molecular solids, in addition to this motion of the molecule as a whole about a fixed position in the crystal, the atoms in the individual molecules can vibrate back and forth with respect to the center of mass of the molecule, and the molecule as a whole may also rotate about an axis passing through its fixed place in the crystal. Since the actual motion of each atom is a combination of all the motions just described, the internal motions of solids are very complicated. The only motion not possible is the movement of the molecules from one position in the crystal to another; that is, no net translational motion of the molecules in the solid can occur, even though they are in constant motion.

If the temperature of a solid is raised, the molecules vibrate much more violently. This increases the average distance of each molecule from its neighbors, just as a crowd of people swinging their arms wildly in all directions wind up having larger average separations than if they stood meekly with their hands at their sides. As a result the density of solids, defined in our usual way as $d = M/V$, decreases as the temperature increases, since the average volume per molecule increases with temperature. This density change is, however, quite small at normal temperatures, and for this reason the density of ordinary solids is usually stated without specifying the temperature at which the density was measured.

If the temperature of a solid is increased a large amount (the actual amount depending on its structure), the molecules may move so far apart while vibrating that the "springs" holding the solid together break, and the solid begins to flow; i.e., one layer of atoms begins to move with respect to another. When this happens, the substance makes a "phase transition" from the solid phase to the liquid phase, as when ice melts. The temperature at which this transition occurs is very precise in actuality, and is called the *melting point* of the solid.

TABLE 10.3 Densities and Melting Points of Some Important Solid Substances

Substance	Density, $10^3\ kg/m^3$	Melting point, °C
Ice	0.92	0
Aluminum	2.70	660
Cast iron	7.20	1230
Copper	8.89	1083
Gold	19.3	1063
Lead	11.3	327
Nickel	8.85	1455
Silver	10.5	961
Steel	7.8	1430
Tungsten	19.0	3370
Platinum	21.4	1774
Uranium	18.7	1133

Table 10.3 gives the densities and melting points of some important metals and alloys. Notice that there is no direct connection between these two properties. Thus tungsten has a melting point higher by a factor of 3 than that of either gold or uranium, even though these three elements have approximately the same densities.

10.8 Elastic Properties of Solids

If we consider a solid to be made up of molecules at fixed positions in space, with the individual molecules bound together by forces which behave somewhat as if the molecules were connected by tiny springs, we have a microscopic picture which explains much of the macroscopic behavior of solids when they are subjected to external forces. There are three basic changes which can occur when a solid is subjected to forces: (1) it can change its length in one direction, that in which the force is applied, this change being accompanied by smaller changes in the other two directions; (2) it can change its volume when subjected to a constant pressure from all sides; (3) it can change its shape when subjected to a shearing force, i.e., a force which tends to slide one layer of molecules over another.

Stress and Strain

To explain these three kinds of changes in solids, we must introduce the concepts of stress and strain. If a force is applied to a solid, the *stress* is defined as the force per unit area applied to the solid; i.e.,

Stress

$$\text{Stress} = \frac{\text{force}}{\text{area}} = \frac{F}{A} \tag{10.14}$$

For example, if we take a piece of aluminum wire and pull it at both ends, as in Fig. 10.28, the effect of this stress is to separate one plane of atoms in the aluminum from the next plane. In other words, the effect is to increase the lengths of the imaginary springs holding the solid together. It is possible, however, that a particular plane of molecules in the aluminum crystal is not perpendicular to F but is at some oblique angle with respect to F, as in Fig. 10.29. Then the stress F can be resolved into two components—F_n, a force normal to the molecular plane, and F_t, a force tangential or parallel to the

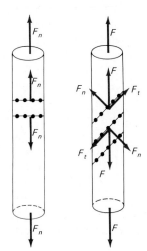

FIGURE 10.28   The stretching of an aluminum wire under a force F, which in this case is the normal force F_n.

FIGURE 10.29   A crystal atomic plane in a solid which is not normal to the applied force F. F can be broken up into two components: F_n, which tries to pull the atomic planes apart, and F_t, which tries to shear (or slide) one atomic plane over the other.

crystal plane. In this case the normal force F_n attempts to pull planes of aluminum atoms apart, whereas the tangential force F_t tends to slide one plane of atoms over another and in this way to *shear* the body apart.

Stresses produce changes in either the length, volume, or shape of a solid. Such a change is called a *strain* and is expressed as the ratio of a change in a particular physical quantity to a quantity with the same physical dimensions, usually the original value of the quantity. Thus we have the following types of change:

1 *Change in length:* If an applied force stretches a wire, we have

$$\text{Tensile strain} = \frac{L - L_0}{L_0} = \frac{\Delta L}{L_0} \tag{10.15}$$

where ΔL is the change in the original length L_0 under the applied stress. If the rod or wire is not stretched, but compressed, we have

$$\text{Compressive strain} = \frac{L_0 - L}{L_0} = \frac{\Delta L}{L_0}$$

2 *Change in volume:* Here we have

$$\text{Volume strain} = \frac{\Delta V}{V_0} \tag{10.16}$$

where ΔV is the change in the original volume V_0. In this case it is assumed that the solid is submerged in a liquid, or in some other way subjected to equal pressures on all sides, as in Fig. 10.30.

3 *Change in shape*: The shearing strain is defined in terms of the angle through which the solid is distorted. For example, consider a block *abcd* of cross-sectional area A as shown in Fig. 10.31, to which we apply a shearing force on the side *bc* to move it to the right with respect to side *ad*. If the new position side *bc* assumes is $b'c'$, then we define the shearing strain as

$$\text{Shearing strain} = \tan \phi = \frac{cc'}{cd} \tag{10.17}$$

Since for small angles (and ϕ is very small) $\tan \phi = \phi$ (see Table 11.1), the shearing strain is just equal to the angle ϕ in radians. A good example of a shear is the deformation of a book, as shown in Fig. 10.32.

As can easily be seen, all strains defined in this way are pure numbers without dimensions and without units.

Any solid will be changed somewhat in size and shape when a stress is applied to it. *An elastic body is one that returns to its original size and shape when the stress is removed*. For such elastic bodies the strain is always directly proportional to the stress so long as the stress is not too great, and so the plot of strain versus stress is a straight line, as in Fig. 10.33. The slope of this straight line is called the *elastic modulus* of the solid. There are three separate elastic moduli corresponding to the three possible types of change which can occur. These are:

Strain

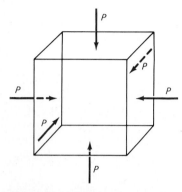

FIGURE 10.30 Stresses which reduce the volume of a solid cube: The same pressure is exerted on all six sides of the cube.

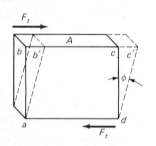

FIGURE 10.31 Change of shape of a solid under a shearing stress.

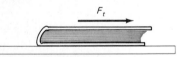

FIGURE 10.32 The deformation of a book by a shearing stress.

Elastic Moduli
Definition

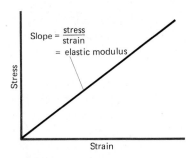

FIGURE 10.33 Plot of stress versus strain. The elastic modulus is the slope of the straight line which results.

1 *Length modulus, or Young's modulus*

$$Y = \frac{\text{tensile stress}}{\text{tensile strain}} = \frac{\text{compressive stress}}{\text{compressive strain}} = \frac{F_n/A}{\Delta L/L_0} \tag{10.18}$$

2 *Torsion, or shear, modulus, or modulus of rigidity*

$$S = \frac{\text{shearing stress}}{\text{shearing strain}} = \frac{F_t/A}{\phi} \tag{10.19}$$

3 *Volume, or bulk, modulus*

$$B = \frac{-F_n/A}{\Delta V/V_0} = \frac{-P}{\Delta V/V_0} \tag{10.20}$$

Here P is the pressure, equal to F_n/A, which is assumed to be the same over the whole external surface of the solid. The negative sign arises because an increase in pressure produces a decrease in volume.

In all three cases the dimensions of the elastic modulus are those of the term F/A, since the strain is always a dimensionless quantity.

Compressibility

The bulk modulus is a measure of how difficult it is to compress a solid. Its inverse, the compressibility, is a measure of how easy it is to compress a solid. We define the compressibility as $K = 1/B$, so that we have

$$K = \frac{1}{B} = \frac{-\Delta V/V_0}{P} \tag{10.21}$$

Definition

or the *compressibility is the fractional decrease in volume per unit increase in pressure*. Hence we have

$$\Delta V = -KPV_0 \tag{10.22}$$

so that the decrease in volume is directly proportional both to the original volume and to the applied pressure. These equations for compressibility turn out to be as valid for fluids as they are for solids.

Table 10.4 gives values for the elastic moduli of some important solids. Notice that tungsten is the hardest of all the materials listed, since it requires the largest stress to produce a given strain. Lead, on the other hand, is the softest. Notice also that the order of magnitude of all three elastic moduli is approximately the same for any one substance, with the shear modulus being roughly one-third the size of Young's modulus.

TABLE 10.4   Elastic Moduli of Metals

Metal	Young's modulus Y, 10^{10} N/m^2	Shear modulus S, 10^{10} N/m^2	Bulk modulus B, 10^{10} N/m^2
Aluminum	7.0	3.0	7.0
Brass	9.1	3.6	6.1
Copper	11.0	4.2	14.0
Iron	19.0	7.0	10.0
Lead	1.6	0.56	0.77
Steel	20.0	8.4	16.0
Tungsten	36.0	15.0	20.0

Example 10.14

A 2.0-m piece of no. 12 gauge copper wire (diameter = 0.081 in) is used to hang a large pot of flowers from the ceiling of a store. If the flower pot has a mass of 5.0 kg, how much does the wire stretch?

SOLUTION

Since $Y = \dfrac{F_n/A}{\Delta L/L_0} = \dfrac{Mg/\pi R^2}{\Delta L/L_0}$

we have $\Delta L = \dfrac{Mg/\pi R^2}{Y} L_0 = \dfrac{Mg L_0}{\pi R^2 Y}$

Here $M = 5.0$ kg, $L_0 = 2.0$ m, and

$R = \dfrac{D}{2} = (0.0405 \text{ in}) \left(\dfrac{0.305 \text{ m}}{12 \text{ in}} \right) = 1.03 \times 10^{-3}$ m

Also, from Table 10.4, $Y = 11.0 \times 10^{10}$ N/m^2 for copper. Hence

$$\Delta L = \dfrac{(5.0 \text{ kg})(9.8 \text{ m/s}^2)(2.0 \text{ m})}{\pi(1.03 \times 10^{-3} \text{ m})^2 \, (11.0 \times 10^{10} \text{ N/m}^2)}$$

$$= 2.67 \times 10^{-4} \text{ m} = 2.67 \times 10^{-1} \text{ mm} = \boxed{0.267 \text{ mm}}$$

It is clear from this that a wire stretches very little even under sizable forces.

10.9 Hooke's Law

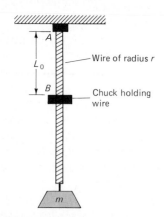

FIGURE 10.34 Young's modulus experiment for measuring the increase in the length L_0 of a wire between A and B when a stretching force is applied to it.

Hooke's Law

A standard laboratory experiment in college physics is the measurement of the stretching of a long, thin, metal wire when heavy weights are applied to it. This method for measuring Young's modulus is worth discussing here because of what it can tell us about the properties of solids.

In Fig. 10.34 we consider a wire of radius r and length L_0 between points A and B, which is stretched by attaching weights to the bottom of the wire. We desire to obtain Young's modulus for the wire. By definition this is

$$Y = \dfrac{F_n/A}{\Delta L/L_0} = \dfrac{Mg/\pi R^2}{\Delta L/L_0} \tag{10.23}$$

where M is the sum of the masses applied to lengthen the wire. All the quantities here can be easily measured except for ΔL, the very small increase in length. ΔL can be measured with a micrometer caliper or, as is usually done, with a device called an *optical lever* which essentially magnifies the small increase in the wire's length into a larger distance which can be more accurately measured.

Equation (10.23) can be turned around to read

$$F_n = \dfrac{YA \, \Delta L}{L_0}$$

or $\quad Mg = F_n = k \, \Delta L \tag{10.24}$

where $k = YA/L_0$. Since Y, A, and L_0 are all constants, we would expect that ΔL would increase linearly with the applied force F_n. We can verify this by adding weights of increasing size to the wire and measuring the corresponding increases in length ΔL. The result is shown in Fig. 10.35. As long as the weights are not too large, the strain and stress are indeed linearly related, and k is a true constant in the equation $F_n = k \, \Delta L$. This is an expression of *Hooke's law*, which says that *the strain produced in an elastic body is directly proportional to the applied stress.* This law was first discovered by Robert Hooke (1635–1703), a contemporary and in some cases a rival of Sir Isaac Newton.

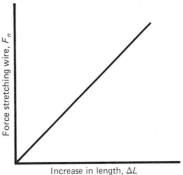

FIGURE 10.35 Hooke's law for the stretching of a wire: The increase in length is directly proportional to the force applied.

The same result is obtained for springs, even though here the stress is almost a perfect shear rather than a tensile stress. In this case k is called the *force constant* of the spring, and has the same dimensions of $[F]/[L]$ as it does in Eq. (10.24). In the SI system the units are newtons per meter. When we remove the weights, the wire will revert to its original length, as long as Hooke's law remains valid. There is a point, however, beyond which Hooke's law is not applicable.

Elastic Limit

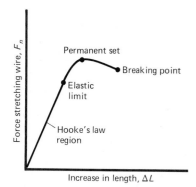

FIGURE 10.36 The elastic limit and breaking point of a metal wire.

If we try to extend further the straight line in Fig. 10.35 by applying even greater weights, we will eventually come to a point where k is no longer a constant. Instead we observe a larger-than-expected increase in L accompanying a normal increase in F_n, as in Fig. 10.36. This point, at which Hooke's law breaks down, is called the *elastic limit*. Beyond the elastic limit the wire assumes a permanent set; i.e., it will not return to its original length when the weights are removed. The wire is no longer hard and elastic but somewhat softer and more pliable, since some of the bonds holding the molecules together in the metal have been broken. If we go even further, the breaking point of the wire will be reached and the wire will separate into two pieces. The maximum force which can be applied before the wire breaks is called the *ultimate strength* of the wire. The breaking point is indicated on the graph in Fig. 10.36.

Such experimental behavior is consistent with our previously proposed model of a crystal made up of molecules held together by the equivalent of tiny springs. It is the stretching of these springs which we observe in experiments such as the Young's modulus experiment just discussed. In Chap. 29 we will discuss more in depth the structure of solids.

Summary: Important Definitions and Equations

Liquids at rest:
  Density: $d = M/V$
  Density of water = 1 g/cm^3 = 10^3 kg/m^3
  Specific gravity: The ratio of the density of a substance to the density of water.
  Pressure: $P = F_n/A$
  Atmospheric pressure: On earth, the pressure due to the weight of the air in the earth's atmosphere.
  1 atm = 1.01×10^5 N/m^2 (or Pa) = 14.7 lb/in^2

Pressure at a depth h below the surface of a liquid:
  $P = P_{atm} + dgh$
Pascal's principle: The pressure applied to a confined liquid is transmitted undiminished to every point in the liquid. As a consequence

$$F_2 = \frac{A_2}{A_1}F_1$$

Barometer: An instrument used to measure the pressure of the earth's atmosphere.

Manometer: A device to measure the difference between the pressure of a substance and that of the earth's atmosphere.

Archimedes' principle: An object partially or totally submerged in a fluid (liquid or gas) is buoyed up by a force equal to the weight of the fluid which the object displaces.

Liquids in motion:

Equation of continuity: $A_1v_1 = A_2v_2$

Bernoulli's equation: $P + \frac{1}{2}dv^2 + dgy = $ constant

Viscosity η: The property of a fluid which offers resistance to its flow and causes loss of energy in the form of heat.

Centipoise (cP): A unit of viscosity equal to 10^{-3} N·s/m^2 = 10^{-3} Pa·s.

The viscosity of water at 20.20°C is 1.0000 cP.

Poiseuille's law:

$$\frac{V}{t} = \frac{\pi R^4(P_2 - P_1)}{8\eta L}$$

Solids:

Stress: A force applied to a solid to change its size or shape; $S = F/A$.

Strain: The change in the size or shape of a solid produced by an applied stress.

Elastic modulus: The ratio of stress to strain.

Length, or Young's, modulus: $Y = \dfrac{F_n/A}{\Delta L/L_0}$

Torsion, or shear, modulus: $S = \dfrac{F_t/A}{\phi}$

Volume, or bulk, modulus: $B = \dfrac{-P}{\Delta V/V_0}$

Compressibility: $K = \dfrac{1}{B} = \dfrac{-\Delta V/V_0}{P}$

Hooke's law: The strain produced in a solid is directly proportional to the applied stress, so long as the stress does not exceed the elastic limit.

Elastic limit: That value of the stress beyond which Hooke's law is no longer valid.

Questions

1 The density of steel is approximately 8 times that of water. How then can a steel ship possibly float?

2 Why are bed sores less likely to develop on the body of a hospital patient if the hospital provides the patient with a water bed rather than a regular bed? (*Hint:* What does Pascal's principle indicate about the water in the water bed?)

3 (*a*) When we say that a normal person's blood pressure is 120/80, what precisely do we mean?

(*b*) Are we talking about gauge pressures or absolute pressures?

4 Discuss how you would determine the density of an irregularly shaped object in water, using Archimedes' principle, (*a*) when the object is more dense than water; (*b*) when the object is less dense than water.

5 Does a submarine on the ocean floor experience exactly the same buoyant force as it would at a depth of only 10 m below the ocean surface? Why?

6 Explain how a siphon works. A siphon is a tube used to transfer a liquid from a higher to a lower container, as in Fig. 10.37.

7 Explain why "water always seeks its own level," using principles discussed in this chapter.

8 When measuring blood pressure, why does a doctor always put the inflatable sleeve on the patient's arm at the level of the heart?

9 Why are the roofs of houses frequently ripped off during hurricanes? (*Hint:* Use Bernoulli's principle.)

10 Prove that the three terms in Bernoulli's equation all have the same dimensions, that is, energy per unit volume.

11 (*a*) Is it possible to pitch a baseball so that it will break upward as it approaches the batter? How?

(*b*) Is it possible to throw a pitch that will break down (a "sinker") when it approaches the batter? How?

(*c*) Is it possible, according to the laws of physics, to throw a pitch that will first rise and then almost immediately sink? Why?

12 Why does the speed of the wind increase with height above the earth's surface? What does this tell us about the nature of the fluid flow of the wind?

13 Using what you know about osmotic pressure, can you suggest what happens when you soak an infected cut in Epsom salts?

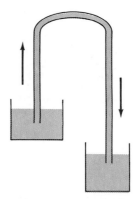

FIGURE 10.37 Diagram for Question 6: A siphon.

Problems

A 1   The density of chlorobenzene is 1.523×10^3 kg/m^3. Its specific gravity is:
(*a*) 1.523×10^3 (*b*) 1.523 g/cm^3
(*c*) 1.523 kg/m^3 (*d*) 1.523 (*e*) 1.523×10^{-3}

A 2   A pressure of 30 lb/in^2 is equal to:
(*a*)   2.0 atm   (*b*)   2.0 Pa   (*c*) 1.6×10^3 cmHg
(*d*)   2.1 N/m^2 (*e*) 2.1×10^{-1} lb/ft^2

A 3   A cylindrical jar contains mercury to a depth of 15 cm. The top is open to the atmosphere. The pressure at the bottom of the jar is:
(*a*) 2.0×10^4 Pa   (*b*) 2.0×10^4 N/m^2
(*c*) 1.2×10^5 N   (*d*)   1.0 atm   (*e*)   1.2 atm

A 4   The deepest point in the Pacific Ocean is the Marianas Trench, where the ocean depth is about 11,000 m. The pressure to which a sunken ship at the bottom of the Marianas Trench is exposed is:
(*a*) 1.1×10^8 atm   (*b*) 1.1×10^3 atm
(*c*) 1.1×10^3 N/m^2 (*d*) 1.1×10^5 atm
(*e*) 1.1×10^5 N/m^2

A 5   A hydraulic lift has one piston of radius 50 cm and another piston of radius 10 cm. If we neglect friction, the ratio of the work input of the lift to the work output is:
(*a*)   1   (*b*)   5/1   (*c*)   25/1   (*d*)   1/5   (*e*)   1/25

A 6   A log floats on water with $\frac{2}{3}$ of its volume submerged. The specific gravity of the wood in the log is:
(*a*) $\frac{2}{3} \times 10^3$ kg/m^3 (*b*) $\frac{2}{3}$ (*c*) $\frac{1}{3}$
(*d*) $\frac{3}{2}$ (*e*)   1

A 7   Water enters the basement of a house from the main water line through a pipe 15 cm in diameter. In the basement the water pipe narrows down to one with a 7.5-cm diameter. If the speed of flow of the water entering the house is 0.50 m/s, the speed of flow of the water in the narrower pipe is:
(*a*)   0.50 m/s   (*b*)   0.25 m/s
(*c*)   0.13 m/s   (*d*)   1.0 m/s   (*e*)   2.0 m/s

A 8   A water cooler open to the air at the top contains water to a height of 50 cm above the exit spout at the bottom. The water will emerge from the spout at a speed of:
(*a*)   9.8 m/s   (*b*)   9.8 m/s^2 (*c*)   3.1 m/s
(*d*)   4.9 m/s   (*e*)   2.2 m/s

A 9   In an experiment on osmosis with a sugar solution, 1.0 g of sugar of molecular mass 342 is dissolved in 2.0×10^{-3} m^3 of water at 20°C. The osmotic pressure produced is:
(*a*) 3.6×10^6 N/m^2 (*b*) 3.6×10^3 N/m^2
(*c*)   3.6 N/m^2 (*d*) 2.5×10^2 N/m^2 (*e*) 2.5×10^5 N/m^2

A10   A cube of steel 0.80 m on a side is subjected to a shearing force of 10^4 N. The deformation which results corresponds to a shear angle of:
(*a*) 1.9×10^{-7} rad   (*b*) 1.9×10^{-7} degrees
(*c*) 7.8×10^{-8} rad   (*d*) 7.8×10^{-8} degrees
(*e*)   None of the above

B 1   Prove that a density of 10 g/cm^3 is equal to a density of 10^4 kg/m^3.

B 2   A glass jar has a mass of 0.20 kg when it is empty, and has a volume of 1.4×10^{-3} m^3. A liquid is poured into the jar so that it is full. The new mass is measured to be 1.80 kg. What is the density of the liquid?

B 3   What is the pressure at the bottom of a column of gasoline 1.0 m high, if the top of the column is open to the atmosphere?

B 4   What is the force exerted by the earth's atmosphere on an area of 1.0 km^2 on the surface of the earth?

B 5   A hydraulic lift is being used to move a skid filled with mailbags. The small piston has a radius of 5.0 cm and the large piston a radius of 75 cm. If someone applies a force of 140 N to the small piston, what weight can the large piston lift, if we assume 100 percent efficiency?

B 6   A mercury manometer like the one in Fig. 10.9 is connected to a flask containing N$_2$ gas. The mercury column connected to the flask falls to a point 25 cm below the top of the other mercury column, which is exposed to the air. What is the absolute pressure of the N$_2$ gas?

B 7   A piece of metal has a weight of 100 N when weighed in air but only 91 N when weighed in water. What is the density of the metal?

B 8   Water enters the basement of an apartment house through an 8.0-cm-diameter pipe at a flow speed of 0.45 m/s and a pressure of 3.0×10^5 N/m^2. The water flows through the same pipe to the sixth floor, which is 20 m above the basement. What is the water pressure on the sixth floor?

B 9   A steel wire of length 2.5 m and diameter 3.6 mm is stretched by clamping one end in a vise and exerting a force of 200 N on the other end. How much does the wire stretch?

B10   A round, solid brass ball of radius 5.0 cm is on a ship which sinks and comes to rest at the bottom of the ocean, where the pressure is 500 atm. How much does the volume of the ball change?

B11   The compressibility of lead is 1.3×10^{-10} m^2/N. What change in volume of a 10-cm lead cube is produced by a pressure change of 5 atm?

C 1   A child of mass 36 kg is floating so that the body is just submerged beneath the surface water of a swimming pool. What is the volume of the child's body?

C 2   A glass stirring rod weighs 0.100 N in air, 0.075 N in water, and 0.050 N in an unknown liquid.
(*a*)   What is the density of the unknown liquid?
(*b*)   What is the specific gravity of the liquid?

C 3   A dam is constructed to form an artificial lake of surface area 10 km^2. At the dam itself the lake is 20 m deep.
(*a*)   What is the pressure on the dam at its base, i.e., at a depth of 20 m?
(*b*)   What is the pressure in the middle of the lake at a depth of 20 m?
(*c*)   What is the pressure on the dam at a depth of 10 m?

C 4   In hospitals patients often have to be fed intravenously; i.e., a needle is inserted in a vein in the arm and the pressure produced on the fluid by gravity forces the fluid into the patient's vein. If the fluid has the same density as water, at what height must the liquid container be placed so that the excess pressure at the vein is 50 mmHg? Assume that the blood pressure in the vein is 20 mmHg above atmospheric pressure, and that the flow rate of the blood is very small.

C 5 A block of copper has a mass of 100 g.
(*a*) What is its volume?
(*b*) If the block is suspended at the end of a string in a water tank and is completely submerged, what is the tension in the string?

C 6 A piece of metal has a weight of 2.0 N in air and 1.7 N in water.
(*a*) What is the specific gravity of the metal?
(*b*) What is its density?
(*c*) What is its volume?

C 7 A piece of nickel is suspected to be hollow. It has a weight of 3.65 N in air and 3.14 N in water. How large is the hole, if any, in the nickel.

C 8 What fraction of the total volume of an iceberg is above the water? Assume that the density of ice is 0.92×10^3 kg/m³ and that the density of sea water is 1.03×10^3 kg/m³.

C 9 A hydraulic jack of the kind shown in Fig. 10.38 is used to lift an automobile in a repair shop. The small piston of area A_1 has a radius of 1.0 cm and the large piston a radius of 25 cm. The mechanic lifts the car by exerting force on the small piston with a lever, which has the lever arms shown in the diagram. If no energy is lost in the process, what force must be exerted by the mechanic to lift a 1000-kg car?

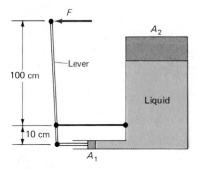

FIGURE 10.38 Diagram for Prob. C9: A hydraulic jack.

C10 An emergency pontoon bridge is laid down to carry provisions to survivors of a flood who have been isolated on a newly created island. The bridge, which rests on the water, is 3.3 m wide and 60 m long. When eight identical trucks are on the bridge, it sinks 20 cm into the water. What is the weight of one of the trucks?

C11 At the point where the main water line enters the basement of a five-story apartment house, the water gauge reads a pressure of only 1.2×10^5 N/m².
(*a*) Will this water be able to reach the fifth floor, where the water pipes are 15 m above the basement?

(*b*) What is the greatest height to which the water can reach?

C12 An 8.0-cm-diameter water pipe enters the basement of a house. The water in the pipe has a speed of 0.50 m/s. The water pipe splits into two other pipes at the same level to carry cold water to the two sides of the house. If each of these pipes has a radius of 4.0 cm, what is the speed of the water in the the two pipes?

C13 In an office building a vertical water pipe 6.0 cm in diameter contains water which is not moving. The pressure in the basement of the office building is 5.6×10^5 N/m². Fourteen floors up the pressure is only 1.4×10^5 N/m². What is the length of pipe between the basement and the fourteenth floor?

C14 A large water tank has a small hole 1.0 cm² in area punched in its bottom. Water flows in at the top of the tank at a rate of 2.0×10^{-3} m³/s. In equilibrium, when the amount of water in the tank remains constant, what will be the water level in the tank?

C15 Air flows past the upper surface of a horizontal airplane wing in streamline flow at 200 m/s and past the lower surface of the wing at 150 m/s. If the density of the air is 0.90 kg/m³ at the altitude of the plane and the wing area is 25 m²:
(*a*) What is the pressure difference between the top and bottom surfaces of the wing?
(*b*) What is the lifting force on the plane?

C16 The amount of water flowing out of a water pipe in 1 min is 0.54 m³. If the pipe has a radius of 6.0 cm, what is the average speed of the water flowing in the tube?

C17 A horizontal pipe has a constricted area where the diameter is reduced from 8.0 to 2.0 cm. In the 8.0-cm portion of the pipe the speed of water flow is 2.0 m/s and the pressure is 6.5×10^5 Pa.
(*a*) What is the speed of flow in the 2.0-cm portion of the pipe?
(*b*) What is the pressure in the 2.0-cm portion of the pipe?

C18 What pressure would be required to accelerate a column of blood 50 cm long in a rigid tube 1.0 cm in radius from zero velocity to 100 cm/s in 0.1 s, if we neglect viscosity and gravity effects? (Although the aorta is not rigid, this is roughly what the left ventricle must do with every heartbeat.)

C19 At a certain point along a pipeline the speed of water flow is 1.2 m/s and the gauge pressure is 3.0×10^5 N/m². What is the gauge pressure in the line at a point 10 m lower than the first point, where the cross-sectional area of the pipe is one-half that at the first point?

C20 How much methyl alcohol will flow through a 1.0-m length of tubing of 1.0 mm radius in 1 min, if the pressure difference across the two ends of the tubing is 0.12 atm?

C21 What must be the pressure difference across the two ends of a section of the Alaskan pipeline which is 1.0 km in length, if it is desired to pump oil through the pipeline at a rate of 0.0010 m³/s. Assume that the diameter of the pipe is 50 cm, the density of the oil 0.90×10^3 kg/m³, and the viscosity of the oil 0.50 Pa·s.

C22 A tugboat is using a steel cable to pull a ship at constant speed. The drag exerted on the ship by the water is 1.2×10^7 N. If any strain above $\Delta L/L_0 = 0.050$ will break the cable, what is the smallest-radius steel cable that can be used to pull the tug?

C23 A hydraulic press contains 0.15 m³ of oil. If the compressibility of the oil is 6.3×10^{-10} m²/N, what is the decrease in volume of the oil when it is subjected to a pressure of 100 atm?

C24 A wooden platform is suspended from a ceiling by four steel wires at its four corners. The wires are 4.0 m long and have diameters of 2.5 mm. Two people, each weighing 900 N, stand at the center of the platform. How much will the wires supporting the platform stretch?

D 1 Prove that the total force exerted on the wall of a dam of length L, behind which water stands to a height H, is $\frac{1}{2}dgH^2L$, where d is the density of the water behind the dam.

D 2 A tank of water rests on a spring scale which reads 200 N. Determine the scale reading when a steel object of volume 0.010 m³ is submerged in the water and supported by a cord connected to it and to a support above the tank.

D 3 An engineer is designing life preservers for an ocean liner. They must be able to keep one-fifth of a 90-kg body above water in an emergency. If the average density of a person is about 1.05×10^3 kg/m³, and if it is desired to use a foam material for the life preservers with a density of 0.55×10^3 kg/m³, what must be the volume of the life preservers?

D 4 A solid block of foam plastic of density 500 kg/m³ and mass 1.0 kg is to be weighed down with lead foil so that it will just sink in water.
(*a*) What mass of lead is required?
(*b*) What is the volume of the lead?

D 5 A Venturi meter of the kind shown in Fig. 10.39 can be used to measure the flow velocity of a liquid of density d in a pipe. A constriction is introduced in the pipe,

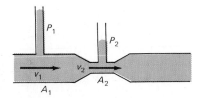

FIGURE 10.39 Diagram for Prob. D5.

and the pressures P_1 and P_2 are measured in terms of the height of the liquid in two small glass tubes attached to the top of the pipe, one over the constriction, the other at some point nearby. Prove that the velocity of the fluid in the unconstricted pipe is:

$$v_1 = A_2 \left[\frac{2(P_2 - P_1)}{d(A_2^2 - A_1^2)} \right]^{1/2}$$

D 6 A water cooler tank is resting on the floor. In addition to the regular outlet, which is 4.0 cm from the floor, the tank has another hole directly above the regular outlet and 30.0 cm above the floor. How high must the water stand in the tank for the streams of water from the two holes to hit the floor at the same place?

D 7 Suppose a sphere of radius R falls at a constant terminal speed v through a viscous material of viscosity η. Show, using dimensional analysis, that the resistive (drag) force acting on the sphere must be

$$F = k\eta vR \qquad \text{where } k = \text{constant}$$

(*Hint*: Assume that $F \propto \eta^x v^y R^z$, and evaluate x, y, and z.)

D 8 A blood transfusion is being administered to a patient by using gravity flow to supply blood at a rate of 5.0 cm³/min. Assume that the blood pressure in the vein into which the blood is being injected is 20 mmHg, and that the density and viscosity of the blood have the values given in Tables 10.1 and 10.2. The needle which is inserted into the vein has a length of 5.0 mm and a diameter of 0.40 mm. How high must the bottle containing the blood be placed above the patient's arm to provide the desired blood flow for the transfusion?

D 9 A technician is setting up a laboratory experiment and suspends a 20-kg uniform horizontal bar from the roof of the laboratory by using three steel wires each of which has a radius of 0.60 mm. The two wires at the ends of the rod are exactly 3.000 m long, but the wire at the center has a length of 3.001 m.
(*a*) By how much is each wire stretched?
(*b*) How much of the weight of the bar is supported by each wire?

D10 A 10-kg pendulum bob of radius 5.0 cm is suspended from a point 2.00 m above the ground by a steel wire of unstretched length 1.89 m and of diameter 0.10 cm. The ball is then set swinging so that its center is moving at a speed of 4.0 m/s when it passes through its lowest point. How close to the floor is the bottom of the pendulum bob at its lowest point?

Additional Readings

Allman, William F.: "Pitching Rainbows," *Science 82*, vol. 3, no. 8, October 1982, pp. 29–32. An unusual cooperative project between scientists, photographers, and the Baltimore Orioles to determine whether a baseball really curves.

Holden, Alan, and Phylis Singer: *Crystals and Crystal Growing*, Doubleday Anchor, Garden City, N.Y., 1960. A fascinating account of the growing and use of crystals.

McDonald, J. E.: "The Shape of Raindrops," *Scientific*

American, vol. 190, no. 2, February 1954, pp. 64–68. An interesting article showing how the shape of liquid drops can be explained by the forces acting on them.

Schrier, Eric W., and William F. Allman (eds.): *Newton at the Bat: The Science in Sports,* Charles Scribner's Sons, New York, 1984. The first section of this book is devoted to "Balls and Other Flying Objects."

Shortley, George, and Dudley Williams: *Principles of College Physics*, vol. 1, Prentice-Hall, Englewood Cliffs, N.J., 1959, chaps. 10, 11. This textbook contains a more detailed treatment of the structure of liquids and solids than most books at an introductory level.

Wannier, Gregory H.: "The Nature of Solids," *Scientific American*, vol. 187, no. 6, December 1952, pp. 39–48. A somewhat dated, but still very useful, account of the structure of solids by an eminent solid-state physicist.

M otion appears in many aspects—but there are two obvious kinds, one which appears in astronomy and another which is the echo of that. As the eyes are made for astronomy so are the ears made for the motion which produces harmony.

Plato (427–347 B.C.)

Chapter 11
Periodic Motion and Waves

In Chap. 6 we discussed a stretched horizontal spring. If there is no loss of mechanical energy in this system, once the spring is stretched and released the mass will repeat its back-and-forth motion (usually called a *vibration*, or *oscillation*) for an indefinite period of time. This vibration is an example of a larger class of motions which we call *periodic*.

Periodic motions include a host of important physical phenomena. For example, violin strings and air columns in organ pipes vibrate periodically, and hence the joy of music is really a delight in a special kind of periodic motion. Sound waves and the various kinds of electromagnetic waves—radio, TV, radar, light—are all periodic. A properly functioning heart beats in periodic motion. All kinds of practical machinery from jackhammers to internal-combustion engines execute periodic motion. For this reason an understanding of periodic motion is basic to an understanding of the universe in which we live.

11.1 Periodic Motion

Definitions

Periodic motion: Motion which repeats itself regularly after a fixed time interval.
Period T: The time during which a physical system completes one full cycle of its motion.

The word *period* comes from the Greek words meaning "round path," or circle. It was coined to describe the first observed periodic motion—the motion of the planets—and it referred to the time a planet took to make one complete orbit and return to its original position. The motion of a child on a swing, or of the pendulum in a grandfather's clock, is periodic. If we stretch a mass at the end of a horizontal spring a distance X_0 from its equilibrium position, as in Fig. 11.1, and then release it, the mass will speed up, pass through its original rest position at maximum speed, and then slow down as the spring begins to be compressed. The spring will compress farther until the mass reaches a position $-X_0$; then the mass stops instantaneously, speeds up

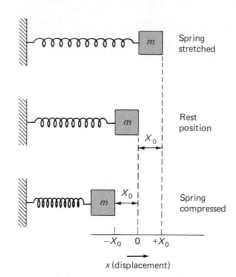

FIGURE 11.1 Motion of a mass at the end of a horizontal spring on a frictionless table. The amplitude of the motion is X_0.

again until it passes its equilibrium position with maximum speed in the opposite direction, then slows down again and stops instantaneously at its original position $+X_0$. If we assume that there is no friction or other resistive forces acting, the mass will continue to move back and forth between $-X_0$ and $+X_0$ indefinitely. Such a motion is clearly periodic, since it repeats itself over and over again in the same fixed time interval.

In this case the period T is the time it takes for the mass to go from $+X_0$ to $-X_0$ and then back again to $+X_0$, i.e., the time for a complete *cycle*, or round trip. We can also define some other important quantities for this periodic motion:

Definitions

Frequency f: The number of complete cycles (or periods) per second: $f = 1/T$.

Displacement x: The distance of the mass from its equilibrium position at any instant of time.

FIGURE 11.2 Energy transformations for a mass moving at the end of a horizontal spring. The energy oscillates between kinetic and elastic potential energy, as shown.

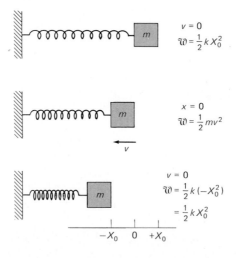

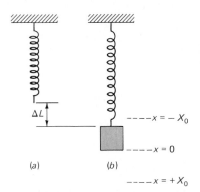

(a) (b)

$----x = -X_0$

$----x = 0$

$----x = +X_0$

FIGURE 11.3 Mass at the end of a vertical spring. The motion of the mass is the same as that for a mass at the end of a horizontal spring, except that the end of the spring is displaced a distance ΔL from its position in (a) to its new position in (b) when the mass is attached to the spring.

Amplitude X_0: The maximum value of the displacement x.

These technical terms, together with more familiar ideas like the velocity and acceleration of the moving mass, provide us with all we need to discuss periodic motion quantitatively.

In the motion of a mass at the end of a horizontal spring it is not merely the position of the mass which repeats itself over and over again in time. There is also a continuous conversion of energy from the elastic potential energy in the spring to the kinetic energy of the mass, and then back to potential energy in the spring as the mass goes from maximum displacement in one direction to maximum displacement in the opposite direction. This energy conversion repeats itself as the mass returns to its original position. This is illustrated in Fig. 11.2.

The motion of a mass at the end of a vertical spring is the same as if the spring were horizontal except that the pull of gravity displaces the mass initially to a new rest position, as shown in Fig. 11.3. The vertical motion of the mass about this new rest, or equilibrium, position is then identical with the motion of a mass at the end of a horizontal spring.

Example 11.1

A mass at the end of a horizontal spring is stretched 0.10 m from its rest position and then released. It takes 0.75 s for the mass to move through its rest position to its extreme position in the opposite direction.
(a) What is the amplitude of the motion?

(b) What is its period?
(c) What is its frequency?
(d) If the force constant of the spring is 4.0 N/m, what is the initial energy of the system?

SOLUTION

(a) Since the amplitude is the maximum displacement,

$X_0 = \boxed{0.10 \text{ m}}$

(b) Since it takes 0.75 s for the mass to make a one-way trip from one side to the other, the time for a round trip will be 2×0.75 s $= 1.5$ s. Hence $\boxed{T = 1.5 \text{ s}}$

(c) $f = \dfrac{1}{T} = \dfrac{1}{1.5 \text{ s}} = 0.67 \text{ s}^{-1}$ or $\boxed{0.67 \text{ Hz}}$

(d) As we have seen in Chap. 6, the energy stored in the spring is, from Eq. (6.8),

$\frac{1}{2}kX_0^2 = \frac{1}{2}(4 \text{ N/m})(0.10 \text{ m})^2 = 0.020 \text{ N·m} = \boxed{0.020 \text{ J}}$

11.2 Simple Harmonic Motion

The first scientists to study periodic motion were the Greeks, who were mainly interested in the role played by periodic motions in music. They therefore called periodic motion *harmonic motion* (see the chapter-opening quote from Plato). The motion of a violin string, which is usually bowed in a way which produces not a pure note but many overtones (or harmonics) of different intensities, is a very complicated kind of harmonic motion and difficult to analyze quantitatively. But there are vibrations of a single frequency which we can analyze more easily. These less complicated (or simple) vibrations constitute what is called *simple harmonic motion* (SHM).

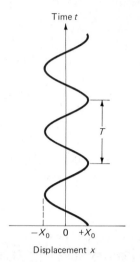

Time t

$-X_0$ 0 $+X_0$

Displacement x

FIGURE 11.4 A cosine curve for the displacement of a mass at the end of a horizontal spring as a function of time. The initial displacement of the mass is $+X_0$, and the mass vibrates back and forth between the extreme positions $-X_0$ and $+X_0$ as time goes on. (In this diagram time is plotted along the vertical axis.) The motion can be described by an equation of the form $x = X_0 \cos \omega t$.

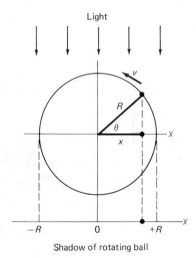

Light

$-R$ 0 $+R$

Shadow of rotating ball

FIGURE 11.5 The projection of the circular motion of a small ball on the X axis. The projected shadow executes SHM along the X axis in a way described by the equation $x = R \cos \omega t$.

A particle or system executes simple harmonic motion when its position as a function of time can be described by a single sine or cosine curve, as in Fig. 11.4, which shows the displacement of the mass at the end of a horizontal spring as a function of time. Here the amplitude of the motion is X_0 and the period is T, as shown in the diagram. The motion repeats itself over and over with a period T, which remains fixed throughout the motion. Such a motion is both periodic and simple harmonic.

Simple harmonic motion of the kind shown in Fig. 11.4 is directly related to the motion of a point in a circle, as we now proceed to show.

Consider a small ball rotating at constant speed in a circle (sometimes called the *reference circle*) of radius R, and consider what happens to the shadow of the ball along the X axis. (We might imagine a large spotlight shining down on the rotating ball from the top of the page, and projecting on the X axis a shadow of the ball.) As the ball rotates through 360°, or 2π rad, its shadow goes from $+R$ to $-R$ and back again to $+R$, as in Fig. 11.5. If we plot its position x carefully as a function of the angle θ swept out by the rotating ball, we find that x executes a perfect cosine curve similar to the one in Fig. 11.4. This has to be the case, for the cosine of θ is the ratio of the adjacent side (x) to the hypotenuse R in the triangle shown in Fig. 11.5. Since the hypotenuse is constant, as the ball rotates its shadow along the X axis (which is the adjacent side in the triangle) is always proportional to $\cos \theta$. Hence we have

$$\cos \theta = \frac{x}{R} \qquad \text{or} \qquad x = R \cos \theta$$

In this equation x is expressed in terms of the angle θ. For the SHM of a mass at the end of a spring, we are interested in how the displacement of the mass m varies with time, as in Fig. 11.4. To tie these two ideas together we note that if the ball rotates around a circle of radius R at a fixed linear speed v, then its angular velocity is, from Eq. (5.21), $\omega = v/R$, and the angle swept out in time t is, from Eqs. (5.22) and (5.23),

$$\theta = \omega t = 2\pi f t$$

where f is the frequency, i.e., the number of complete rotations per second.

Hence, since $x = R \cos \theta$, and $\cos \theta = \cos \omega t$, we have for the motion of the shadow along the X axis

$$x = R \cos \omega t \tag{11.1}$$

where R is the radius of the circle. The period of the projected motion along the X axis will be the same as the period of the circular motion, since they both complete one cycle in the same time. Hence

$$T = \frac{1}{f} = \frac{2\pi}{\omega} \tag{11.2}$$

The motion of the shadow in Fig. 11.5 is identical with the motion of a mass m vibrating along the X axis at the end of a horizontal spring of force constant k, as in Fig. 11.4. If we equate the amplitudes X_0 and R, both motions can be described by the same equation

$$x = X_0 \cos \omega t = X_0 \cos 2\pi \frac{t}{T} = X_0 \cos 2\pi ft \qquad (11.3)$$

A more formal way to establish this identity is to show that the projection of the ball's motion on the X axis has an acceleration a which bears the same relationship to its displacement x as does the acceleration of a mass at the end of a spring to the displacement of the mass. This we now proceed to show.

Acceleration in SHM and Hooke's Law

A frequently used definition of SHM is that an object executes SHM if the force controlling its motion satisfies Hooke's law. For example, we have seen in Sec. 10.8 that for a horizontal spring the increase in length x of the spring is proportional to the applied force, or

$$f' = kx$$

Now, according to Newton's third law, if a person applies the force f' by pulling on the mass at the end of the spring and thus stretching the spring, then the spring must exert an equal and opposite force $f = -f'$ back on the mass and, through it, on the person's hand. Hence the force exerted by the spring on the mass, the so-called restoring force, is

$$f = -f' = -kx \qquad (11.4)$$

Any system for which the restoring force is proportional to the displacement and in the opposite direction (i.e., the sign is minus) *is said to obey Hooke's law.* Any such system, when released, will execute SHM. The acceleration of the mass m will then be given by Newton's second law as

$$a = \frac{f}{m} = \frac{-kx}{m} \qquad \text{or} \qquad \boxed{a = -\frac{k}{m}x} \qquad (11.5)$$

This is often taken as the defining equation for SHM.

Definition

Simple harmonic motion: Any motion for which the restoring force is proportional to the displacement and in the opposite direction [as in the case of a mass at the end of a spring, where $a = -(k/m)x$].

As x varies between its maximum values of $-X_0$ and $+X_0$, the acceleration varies also but in the opposite direction. Hence when $x = X_0$, the acceleration takes on its maximum negative value $-(k/m)X_0$; when $x = -X_0$, the acceleration takes on its maximum positive value $+(k/m)X_0$.

We now want to see if the motion of the shadow along the X axis of a ball rotating in a circle satisfies Eq. (11.5). For motion at constant speed in a circle of radius R, the only acceleration which exists is the centripetal acceleration, given by Eq. (5.28) as $a = v^2/R$. The projection of this acceleration on the X axis is the acceleration of the shadow of the rotating ball along the axis. From Fig. 11.6 this projection is clearly

$$a = \frac{v^2}{R} \cos \theta = \frac{v^2}{R}\frac{x}{R} = \frac{v^2}{R^2} x = \omega^2 x \qquad (11.6)$$

where x is the displacement of the shadow along the X axis. Thus far we have neglected the fact that when x is positive, a is in the negative X direction and vice versa, as is clear from the figure. Hence we should more properly write

$$a = -\omega^2 x \tag{11.7}$$

If we compare this with Eq. (11.5) we see that in the case of the projection of the rotational motion on the X axis the acceleration is proportional to the displacement and in the opposite direction, just as was the case for motion under Hooke's law. Hence the definition of SHM in terms of Hooke's law and in terms of the projection of rotational motion on the X axis are completely equivalent. We will now use this equivalence to obtain the period T for a mass moving in SHM at the end of a spring.

Light

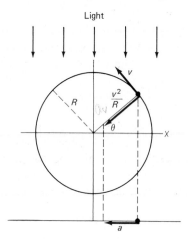

Shadow of rotating ball

FIGURE 11.6 The projection of the acceleration vector on the X axis for circular motion. Note that when x is positive the acceleration a is negative, and vice versa.

Period of Simple Harmonic Motion

Since from Eq. (11.5) we have $a = -(k/m)x$, and from Eq. (11.7) we have $a = -\omega^2 x$, we must have

$$\omega^2 = \frac{k}{m} \quad \text{or} \quad (2\pi f)^2 = \frac{k}{m}$$

and so

$$\boxed{f = \frac{1}{2\pi}\sqrt{\frac{k}{m}}} \tag{11.8}$$

The period of the motion is, then,

$$\boxed{T = \frac{1}{f} = 2\pi\sqrt{\frac{m}{k}}} \tag{11.9}$$

This is a remarkable result, since it tells us that, to obtain the period T, all we need to know is the mass of the moving object and the force constant k of the spring. Even the amplitude X_0 is of no importance, so long as Hooke's law stays valid and the motion remains simple harmonic.

Speed of a Mass Executing Simple Harmonic Motion

The total energy of a mass moving at the end of a spring of force constant k is the initial energy stored in the spring, and we have seen in Eq. (6.8) that this is

$$\mathcal{W} = \tfrac{1}{2}kX_0^2 \tag{11.10}$$

where X_0 is the initial extension of the spring and therefore the amplitude of the vibration. When the mass is in motion, its energy is just the sum of a kinetic-energy term and a potential-energy term. Hence we have

$$\mathcal{W} = \text{KE} + \text{PE} = \tfrac{1}{2}mv^2 + \tfrac{1}{2}kx^2 = \tfrac{1}{2}kX_0^2$$

and so $\quad mv^2 = k(X_0^2 - x^2)$

or $\qquad v = \sqrt{\dfrac{k}{m}} \sqrt{X_0^2 - x^2} \tag{11.11}$

This equation correctly predicts that, for $x = \pm X_0$, $v = 0$, so that the mass stops instantaneously at the two ends of its vibration. Also, when $x = 0$,

$$v^2 = \frac{k}{m} X_0^2 \qquad \text{or} \qquad v = \sqrt{\frac{k}{m}} X_0$$

This is the maximum value of the speed v.

Hence we see that all important properties of a mass moving in SHM at the end of a spring can be obtained from these four basic equations:

$$x = X_0 \cos 2\pi\frac{t}{T} = X_0 \cos \omega t \tag{11.3} \qquad\qquad T = 2\pi\sqrt{\frac{m}{k}} \tag{11.9}$$

$$v = \sqrt{\frac{k}{m}} \sqrt{X_0^2 - x^2} \tag{11.11} \qquad\qquad a = -\frac{k}{m}x \tag{11.5}$$

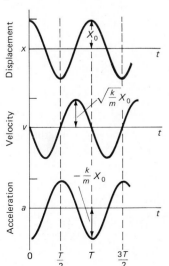

FIGURE 11.7 Graphs of the displacement x, the velocity v, and the acceleration a of a mass moving in SHM. Here T is the period of the motion.

Figure 11.7 shows graphs of the displacement, velocity, and acceleration in SHM. The acceleration curve is the negative of the displacement curve except for a change in amplitude, in accordance with Eq. (11.5). The velocity takes the form of a sine curve if x and a are represented by cosine curves. This can be shown analytically since the speed of the mass is given by

$$v = \sqrt{\frac{k}{m}} \sqrt{X_0^2 - x^2} = \sqrt{\frac{k}{m}} \sqrt{X_0^2 - X_0^2 \cos^2 \omega t}$$

$$= X_0 \sqrt{\frac{k}{m}} \sqrt{1 - \cos^2 \omega t} = X_0 \sqrt{\frac{k}{m}} \sqrt{\sin^2 \omega t} = X_0 \sqrt{\frac{k}{m}} \sin \omega t$$

This is the magnitude of the velocity. The square root yields both $+$ and $-$ signs, but the sign here must be negative, since it is clear that as x decreases from X_0 to 0, the velocity of the mass is in the *negative X* direction. Hence

$$v = -\sqrt{\frac{k}{m}} X_0 \sin \omega t \qquad \text{or} \qquad v = -\sqrt{\frac{k}{m}} X_0 \sin \frac{2\pi t}{T} \tag{11.12}$$

To summarize, then, the displacement, velocity, and acceleration of a mass moving in SHM at the end of a horizontal spring are given by:

$$x = X_0 \cos \omega t \qquad (11.3) \qquad\qquad v = -\omega X_0 \sin \omega t \qquad (11.12)$$

$$a = -\omega^2 X_0 \cos \omega t \qquad \text{where} \qquad \omega = 2\pi f = \left(\frac{k}{m}\right)^{1/2} = \frac{2\pi}{T} \qquad (11.7)$$

These values are consistent with the graphs in Fig. 11.7.

Example 11.2

A mass of 0.50 kg moves in simple harmonic motion at the end of a horizontal spring of force constant 2.0 N/m.
(a) What is the period of the motion?
(b) If the amplitude of the motion is 0.10 m, what is the equation for the displacement at any instant of time t?

(c) What is the equation for the acceleration of the mass as a function of time?

SOLUTION

(a) From Eq. (11.9),

$$T = 2\pi\sqrt{\frac{m}{k}} = 2\pi\sqrt{\frac{0.50 \text{ kg}}{2.0 \text{ N/m}}} = \boxed{3.14 \text{ s}}$$

(b) From Eq. (11.1) we have

$$x = X_0 \cos \omega t = 0.10 \cos \sqrt{\frac{k}{m}}\, t = \boxed{0.10 \cos 2t}$$

where the amplitude is in meters.

(c) From Eq. (11.7), the acceleration is

$$a = -\omega^2 x = -\left(\frac{k}{m}\right)x$$

$$= \frac{-2.0 \text{ N/m}}{0.50 \text{ kg}}(0.10 \cos 2t) = \boxed{-0.40 \cos 2t}$$

where the acceleration is in meters per second squared.

Example 11.3

A mass of 0.40 kg is displaced 0.20 m and then released, and moves at the end of a horizontal spring of force constant 2.0 N/m.
(a) What are the values of the displacement, velocity, and acceleration when the vibrating mass is at point $x = X_0$, i.e., the initial displacement of the system?

(b) What are the values of the displacement, velocity, and acceleration when the vibrating mass is at point $x = 0$, i.e., its equilibrium position?
(c) Verify that the total energy of the system is the same at $x = 0$ and at $x = X_0$.

SOLUTION

(a) The displacement is given as $x = X_0 = \boxed{0.20 \text{ m}}$

From Eq. (11.11) we have for the speed v

$$v = \left(\frac{k}{m}\right)^{1/2}(X_0^2 - x^2)^{1/2} = \boxed{0} \qquad \text{since} \qquad x = X_0$$

From Eq. (11.5) we have for the acceleration:

$$a = -\frac{k}{m}x = -\frac{k}{m}X_0 = -\frac{2.0 \text{ N/m}}{0.40 \text{ kg}}(0.20 \text{ m}) = \boxed{-1.0 \text{ m/s}^2}$$

(b) When the mass is at its equilibrium position,

$$x = \boxed{0}$$

From Eq. (11.11), the speed is

$$v = \left(\frac{k}{m}\right)^{1/2}(X_0^2 - x^2)^{1/2} = \left(\frac{k}{m}\right)^{1/2}X_0 = \left(\frac{2.0 \text{ N/m}}{0.40 \text{ kg}}\right)^{1/2}(0.20 \text{ m})$$

$$= (5.0 \text{ s}^{-2})^{1/2}(0.20 \text{ m}) = \boxed{0.45 \text{ m/s}}$$

The sign of the velocity is $+$ or $-$ depending on whether the mass is moving in the positive or negative X direction.

From Eq. (11.5), $a = -\frac{k}{m}x = \boxed{0}$

(c) The total mechanical energy is

$$\mathcal{W} = \text{PE} + \text{KE} = \tfrac{1}{2}kx^2 + \tfrac{1}{2}mv^2$$

For $x = X_0$:

$W = \frac{1}{2}(2.0 \text{ N/m})(0.20 \text{ m})^2 + 0 = 0.040 \text{ N·m} = \boxed{0.040 \text{ J}}$

Hence the total energy of the system is the same at $x = 0$ and at $x = X_0$.

For $x = 0$:

$W = \frac{1}{2}k(0)^2 + \frac{1}{2}(0.40 \text{ kg})(0.45 \text{ m/s})^2 = \boxed{0.040 \text{ J}}$

11.3 The Simple Pendulum

A simple pendulum consists of a heavy spherical bob at the end of a very light supporting string or rod, as in Fig. 11.8. The length L of such a pendulum is the distance from the point of suspension about which the pendulum pivots to the center of mass of the bob. The period T of the pendulum is the time it takes for the pendulum to execute a complete vibration, from one side to the other and back again.

Finding the period of a simple pendulum would be relatively simple if we were sure that it executed simple harmonic motion. We must first see if this is actually the case. The situation here is quite different from that of a mass at the end of a spring, where the force producing the motion is internal to the moving system; i.e., the force depends on the elastic properties of the spring itself. A pendulum, however, has no elastic properties and its motion is caused by an external force, the force of gravity pulling on the bob of the pendulum. When the pendulum is at rest, the force of gravity downward is perfectly balanced by the upward force F produced by the tension in the string, as shown in Fig. 11.9. For any other position, however, the force of gravity can be broken up into two components, one parallel to the string and one at right angles to it, as in Fig. 11.10. The component parallel to the string, $mg \cos \theta$, is smaller than the tension F in the string, since a net unbalanced force directed along the string toward 0 is required to keep the ball moving in a circle (i.e., a centripetal force). The component normal to the string, $mg \sin \theta$, is the force tending to restore the bob to its equilibrium position. We now want to see whether this restoring force is proportional to the displacement. If it is, pendulum motion is an example of SHM and we easily will be able to find the pendulum's period.

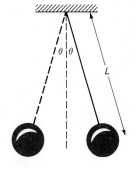

FIGURE 11.8 A simple pendulum.

FIGURE 11.9 Forces on a pendulum bob at rest.

FIGURE 11.10 Forces on a pendulum bob when displaced from its equilibrium position.

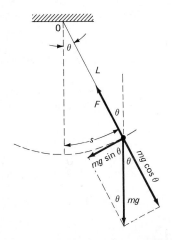

From Fig. 11.10 the linear displacement of the bob from its rest position at any time is

$$s = L\theta \tag{11.13}$$

where θ is the angular displacement (in radians) of the string and s is the arc of the circle in which the bob moves. The force acting to restore the bob to its rest position is

$$f = mg \sin \theta \tag{11.14}$$

Of course, as the pendulum swings, θ is constantly changing and so is f. For example, when $\theta = 0$, $f = 0$ and there is no restoring force at the equilibrium position of the pendulum bob.

If the motion is to be SHM, then the restoring force must be proportional to the displacement, or f must equal $-ks$. Let us see if this is true. We have, from Eqs. (11.13) and (11.14),

$$\frac{f}{s} = \frac{-mg \sin \theta}{L\theta}$$

where we have introduced the minus sign to indicate that the restoring force is always in the opposite direction from the displacement; i.e., when the displacement is to the right, the force is to the left, and vice versa. We can write this as

$$f = \frac{-mgs}{L}\left(\frac{\sin \theta}{\theta}\right) \quad \text{or} \quad f = -ks\left(\frac{\sin \theta}{\theta}\right)$$

where
$$k = \frac{mg}{L} \tag{11.15}$$

is equivalent to a force constant in this case.

Hence Hooke's law is valid and we have SHM only if $(\sin \theta)/\theta$ is a constant. Now this is certainly not true in general. Thus the sine of 30° is $\frac{1}{2}$ and the sine of 90° is 1, so that an increase in the angle by a factor 3 increases the value of the sine only by a factor of 2. Hence, *in general, pendulum motion is not SHM*, for f is not equal to $-ks$ and Hooke's law is not valid.

It is well known, however, that, for small values of θ, $\sin \theta$ and θ (measured in radians) are approximately equal. In such cases where the amplitude of vibration is small, $(\sin \theta)/\theta \simeq 1$, $f \simeq -ks$, and we have a very good approximation to SHM. To see how good the approximation is, we list in Table 11.1 the values of θ, $\sin \theta$, and $(\sin \theta)/\theta$ for angles from 0 to 90°. Note that for angles less than 15° the error introduced by assuming that $\sin \theta$ and θ are equal is only about 1 percent. We can therefore conclude:

For pendulum vibrations in which the amplitude of the motion does not exceed 15°, pendulum motion is a very good approximation (but still only an approximation) to SHM.

In this case, since the period for SHM is $T = 2\pi\sqrt{m/k}$ from Eq. (11.9), and since from Eq. (11.15) $k = mg/L$, we have for the period of the pendulum

TABLE 11.1   Table of Sines and Tangents for Small Angles

Angle θ,°	Angle θ, rad	sin θ	$\dfrac{\sin \theta}{\theta}$	tan θ
1.0	0.0175	0.0175	1.000	0.0175
2.0	0.0349	0.0349	1.000	0.0349
3.0	0.0523	0.0523	1.000	0.0524
4.0	0.0698	0.0698	1.000	0.0699
5.0	0.0873	0.0872	0.999	0.0875
6.0	0.1047	0.1045	0.998	0.1051
7.0	0.1221	0.1219	0.998	0.1228
8.0	0.1396	0.1392	0.997	0.1405
9.0	0.1571	0.1564	0.996	0.1584
10.0	0.1745	0.1736	0.995	0.1763
12.0	0.2094	0.2079	0.993	0.2126
15.0	0.2618	0.2588	0.989	0.2679
20.0	0.3490	0.3420	0.980	0.3640
30.0	0.5236	0.500	0.955	0.5774
45.0	0.7853	0.707	0.900	1.000
60.0	1.047	0.866	0.827	1.732

$$T \simeq 2\pi \sqrt{\frac{m}{mg/L}}$$

Period of a Pendulum   or $$T \simeq 2\pi \sqrt{\frac{L}{g}}$$ (11.16)

Hence to an approximation that is good for small angles, the period of a simple pendulum depends only on the length of the pendulum and on the acceleration due to gravity at the place where the pendulum is located. It does not depend on the mass of the bob, nor on the material out of which the bob is made.

Practical Applications

Equation (11.16) has a number of very interesting consequences. First of all, if we have a pendulum of fixed length L and we measure its period T, we can find the acceleration due to gravity g, since $g = 4\pi^2 L/T^2$ from Eq. (11.16). This gives us a much more accurate way to determine g than by timing the fall of a swiftly moving object. A very accurate pendulum can be used in this way to determine variations in the earth's gravity field with altitude and latitude. Such variations sometimes provide clues to the possible location of mineral or fossil-fuel deposits. Pendulum clocks can also be used to tell time, since their period provides us with a convenient time interval (see Fig. 11.11).

Example 11.4

An astronaut sets up a pendulum of length 1.0 m on the surface of the moon and finds that its measured period is 4.9 s.

(a) What is the acceleration due to gravity on the surface of the moon?

(b) Show that this result is consistent with what you would expect from the mass and radius of the moon given in Table C1.

SOLUTION

(a) We have, from Eq. (11.16), $T = 2\pi \sqrt{L/g}$. Hence

$$g = \frac{4\pi^2 L}{T^2} = \frac{4\pi^2(1.0 \text{ m})}{(4.9 \text{ s})^2} = \boxed{1.6 \text{ m/s}^2}$$

The acceleration due to gravity on the surface of the moon is therefore about one-sixth what it is on earth, where it is 9.8 m/s².

(b) From Newton's law of universal gravitation, $f_G = Gmm'/r^2$. Hence the ratio of g for the moon to g for the earth is:

$$\frac{g_M}{g_E} = \frac{Gmm_M/r_M^2}{Gmm_E/r_E^2} = \frac{m_M r_E^2}{m_E r_M^2}$$

$$= \left(\frac{7.35 \times 10^{22} \text{ kg}}{5.98 \times 10^{24} \text{ kg}}\right)\left[\frac{(6.37 \times 10^6 \text{ m})^2}{(1.72 \times 10^6 \text{ m})^2}\right]$$

$$= 0.17 \simeq \tfrac{1}{6}$$

which agrees with the result in (a).

FIGURE 11.11 Compensation device to keep the period of a pendulum constant when the temperature changes. Rods A, B, and C are made of a metal like steel, rods D and E of a different metal like brass. An increase in temperature lengthens all the rods. The expansion of rods A, B, and C increases the length of the pendulum. The expansion of D and E raises the top of rod C and thus tends to decrease the pendulum length. The lengths of the rods and their thermal expansion coefficients can be chosen so as to leave the pendulum length unaltered as temperature changes. The period of the pendulum is therefore unchanged.

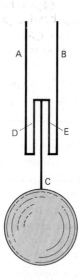

11.4 Damped and Forced Oscillatory Motion

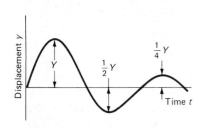

FIGURE 11.12 Damped motion of a mass at the end of a vertical spring. The amplitude of the vibration decreases as time goes on.

This far we have assumed that mechanical energy is conserved when an object oscillates in SHM. In the real world, however, this is seldom the case. A bowed violin string will only vibrate and emit sound for a few seconds before the sound becomes too weak to be heard. A swinging pendulum, if left to itself, will swing in arcs of smaller and smaller amplitude and gradually come to a full stop. These are examples of *damped* oscillatory motion. By *damping* we mean the conversion of mechanical energy into thermal energy by friction, air resistance, and other processes which dissipate or waste energy. These lead to a gradual decrease in the total mechanical energy of the system, and hence to a reduction in the motion of the system which eventually brings it to rest.

Figure 11.12 shows what happens in a damped mechanical system consisting of a mass vibrating at the end of a vertical spring. The amplitude of the motion is gradually reduced, and since the total mechanical energy of the system is proportional to the square of the amplitude, as is clear from Eq. (11.10), the mechanical energy of the system grows smaller as time goes on.

Forced Vibrations

Thus far we have been considering the vibrations of mechanical systems like springs and pendulums when they are given an initial amount of energy and then left alone to vibrate at their own natural frequencies. Such an oscillation is called a *free oscillation*. Some of the most interesting physical effects occur, however, when we attempt to drive a physical system like a pendulum at a frequency which may or may not be equal to that of the freely vibrating system. Oscillations produced by external periodic applied forces are called *forced oscillations*.

The pushing of a child on a swing is a good example of a forced oscillation. The swing is a good approximation to a simple pendulum, and has a period which depends on the length of the swing's ropes. Anyone who has ever pushed a swing knows that, unless the period of the pushes are timed to the motion of the swing, the swing will never build up a large amplitude of oscillation. On the other hand, it is possible to build up very large amplitudes in the swing by pushing it gently, but timing the pushes to conform exactly to the motion of the swing, in other words, matching the period of the applied force to the natural period of the swing.

Definition

Resonance: The large-amplitude vibrations of an object or system which result from its being driven at a frequency equal to its natural frequency of oscillation.

Examples of Resonance

When we tune our radios or TV sets, we are changing the natural frequency of the circuit which picks up the transmitted electromagnetic waves and are matching it to the frequency of the wave emitted by the desired station. In this case the transmitted signal is the driving force and the detecting circuit is the system being driven. If the driving frequency and that of the system being driven are the same, resonance occurs, and the detected signal becomes large enough to be amplified further and converted to the sounds and sights which entertain us.

Many of the most important properties of matter depend on resonance effects. It is resonance between the natural frequencies of the molecules in an object and the frequencies of the light incident on the object which determines its color. It is resonance between the frequencies of the ozone molecules in the upper atmosphere and the frequencies of ultraviolet light which absorbs most of the ultraviolet from the sun and protects us from its detrimental effects. It is the repeated resonance interactions between atoms and radiation which produce the pure, intense light characteristic of a laser.

Resonance can, of course, also have bad effects. Soldiers marching across a bridge have to break step for fear that resonance between their marching frequency and some natural frequency of the bridge could cause the bridge to oscillate with such large amplitudes that it would collapse. An opera singer, hitting a very loud, high note which resonates with a natural frequency of the molecules in a glass, can shatter the glass, even from across a room.

Resonance can also have deleterious effects on living cells. There is evidence that, at certain frequencies related to the size and shape of the cell, the rate at which cells are destroyed by microwaves and very high frequency sound waves is greatly increased.

Some of the most important experiments in physics done in recent years are based on resonance effects. Here natural frequencies of atoms and molecules, which can tell us a great deal about the structure of matter (since these frequencies are determined by the bonding forces holding such structures together), are measured by bringing an external applied frequency into resonance with the built-in, natural frequencies of the system being studied. Since the external frequencies can be measured to extremely high accuracy, the natural frequencies which determine the behavior of matter can be obtained to the same accuracy. This is the basic idea behind such experimental techniques as nuclear magnetic resonance (NMR), which is becoming increasingly important as a diagnostic tool in medicine, and electron spin resonance (ESR).

11.5 Vibrating Bodies as the Source of Waves

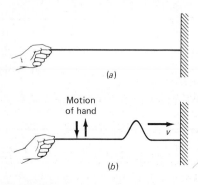

FIGURE 11.13 A pulse produced on a stretched string: (a) String stretched tight by person's hand. (b) Hand moves quickly up and down once and produces a pulse along the string.

FIGURE 11.14 (a) Vibration of a strip of spring steel held rigidly in a vise. (b) Displacement of the tip of the steel strip as a function of time.

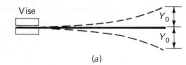

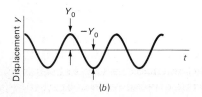

Suppose we attach a string to a wall and then hold on to the other end of the string, as shown in Fig. 11.13. If the free end of the string is moved rapidly upward and then back down to its original position, as in the figure, a "pulse" will travel along the string toward the wall. This pulse carries energy, since the molecules of the string are set in motion by it, but it carries no mass, since the molecules of the string merely vibrate back and forth at right angles to the direction of the pulse. Once the pulse has gone by, each molecule is back at rest at its original position in the string. Some phenomena which are loosely referred to as "waves" actually involve pulses similar to this one.

The name *wave* is, however, more properly restricted to changes produced in a medium (in this case string) by a *periodic* force applied to it. Suppose, for example, that we take a thin strip of spring steel and mount it rigidly between the jaws of a vise, as in Fig. 11.14a. If the free end of the steel strip is displaced from its equilibrium position and released, it will vibrate in SHM about this equilibrium position, as shown in the figure, since the restoring force will be proportional to the displacement and in the opposite direction. The equation for this vibration will be $y = Y_0 \cos \omega t$, as diagramed in Fig. 11.14b.

If now we attach the free end of this steel strip to the string discussed above and set the steel strip into SHM, it will send down the string a true wave, i.e., a periodic change in the displacement of the particles of the string. Suppose the steel strip is struck by a rubber hammer and thus set into SHM. Then, as the strip executes SHM, the particles in the end of the string near the strip are also forced into SHM. They, in turn, interact with their neighboring particles and set them into SHM. As a result, a wave travels down the string toward the right. The shape of the string at the end of quarter-period time intervals is shown in Fig. 11.15a to e, where the period is that of the steel strip. As long as the strip continues to vibrate, the particles of the string will also vibrate in SHM with the same frequency as the steel strip and pass on their motion to their neighbors. Finally, after a certain period of time, almost the entire string will be in sinusoidal motion, as shown in Fig. 11.15f.

The distance marked in Fig. 11.15e and f is called the *wavelength* of the wave and is designated by the symbol λ (Greek lowercase *lambda*, the equivalent of an English ell).

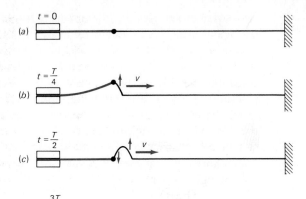

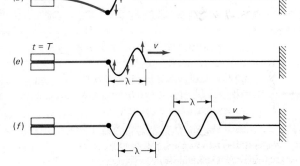

FIGURE 11.15 A wave on a string generated by the vibration of a strip of spring steel at one end of the string. The displacement of the string is shown at intervals of *T*/4, where *T* is the period of the vibrating strip. In (*e*) the time is equal to one full period, and one complete wavelength is moving down the string with a speed *v*. In (*f*) the wave fills almost the whole string.

Definition

Wavelength λ: The distance between any two consecutive equivalent points on a wave pattern.

Let us now try to find the relationship between the period T of the vibrating steel strip, the wavelength λ of the wave on the string, and the speed of the wave along the string. We see from Fig. 11.15*a* to *e* that in the time T it takes for the strip to execute one full period of its motion in the Y direction, the wave has moved a distance λ in the X direction, for when the strip goes through one complete cycle, the wave pattern also traces out one complete cycle and starts to repeat itself again. Hence the speed of the wave is the distance covered divided by the time, or

$$v = \frac{\lambda}{T}$$

Since the frequency f is equal to $1/T$, we have

$$\boxed{v = \lambda f}$$

(11.17)

Speed of a Wave

The speed v of a wave is the product of its frequency f and its wavelength λ.

This is one of the most important equations governing wave motion. We have derived it for a particular kind of wave, but it applies as well to water waves, to sound waves, and to the many different varieties of electromagnetic waves. Frequently physicists determine the speed of a wave indirectly by measuring the frequency and wavelength and using Eq. (11.17) to calculate the speed. In this equation, if λ is in meters and f in hertz, then v is in meters per second, since a hertz (or cycle per second) has the units s^{-1}, for a cycle is a pure number without dimensions or units. The reciprocal relationship between frequency and wavelength predicted by Eq. (11.17) when the speed of propagation is constant is illustrated in Fig. 11.16.

Notice that in the above case the speed of the wave is not determined by λ or by f, but by the properties of the medium through which the wave travels. This is true of all waves. Once this speed is known, for any given applied frequency f, the wavelength is then determined by the relationship $\lambda = v/f$.

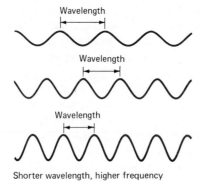

FIGURE 11.16   Reciprocal relationship between wavelength and frequency: For waves moving with a constant velocity v, the longer the wavelength, the lower the frequency, and vice versa, since $v = \lambda f$.

Example 11.5

An FM radio station sends out its programs on a wavelength of 3.3 m. To what frequency must a home radio receiver be tuned if these programs are to come in loud and clear in the home? (Radio waves travel at 3.0×10^8 m/s.)

SOLUTION

This is an example where resonance between two frequencies, the broadcast frequency of the radio station and the natural frequency of the home radio picking up the signal, must be the same. The frequency of the FM station is:

$$f = \frac{c}{\lambda} = \frac{3.0 \times 10^8 \text{ m/s}}{3.3 \text{ m}} = 91 \times 10^6 \text{ s}^{-1} = 91 \text{ MHz}$$

Hence the receiver must be tuned to this same frequency.

Properties of Waves

We can summarize the preceding discussion in the following definition of a wave:

Definition

Wave: A method of energy propagation through space which involves periodic variations in the medium through which the energy is transported.

For example, if a wave is moving down a stretched string, the displacement of particles along the string varies periodically in both time and space. If the string is vibrating back and forth in the Y direction while the wave moves down the string in the X direction, the periodicity of the displacement of one string particle in time can be expressed by an equation of the form:

$$y = Y_0 \cos \omega t = Y_0 \cos 2\pi f t = Y_0 \cos \frac{2\pi t}{T} \tag{11.18}$$

Similarly, the periodicity of the displacement of string particles along the X axis at one instant of time can be expressed by

$$y = Y_0 \cos \frac{2\pi x}{\lambda} \tag{11.19}$$

for this equation states that the value of y for $x = \lambda, 2\lambda, 3\lambda, \ldots, n\lambda$ is the same as its value for $x = 0$, that is, $y = Y$. Hence the wave is also periodic in x; the spatial period after which the motion repeats itself is the wavelength λ.

Energy Propagation

A wave transports energy from one place to another. It does this in a very different way than would a bowling ball moving down the alley and knocking over bowling pins at the other end. In this latter case mass is moved from one place to another, and kinetic energy goes with the moving mass. In a wave there is no transport of mass from one place to another; there is only a transport of *energy*. In a sound wave, for example, the molecules of the air move back and forth at fixed positions in space, and still the wave transmits enough energy to our eardrum to enable us to hear the sound. The light from the sun passes through vast regions of space where no mass at all is present to carry the energy, and still the sun provides the energy to heat and light the earth and fuel the process of photosynthesis. In the case of both sound and light, therefore, energy is transmitted in the form of waves, with no associated net transport of mass. The same is true for radio and TV transmission and for all other forms of wave motion.

As an example consider the situation in Fig. 11.17 for a wave moving along a string. In (a) the wave is moving to the right at a speed v, but has not reached point P as yet. A short time later, at (b), the wave has passed by point P and the particles in the string at P are vibrating back and forth in SHM along a line perpendicular to the direction of wave propagation. Hence, while in (a) the particles at P had no vibrational kinetic energy, once the wave reaches them they have some kinetic energy of vibration. We therefore see that energy has been transported from point A to point P, even though there has been no transport of mass.

FIGURE 11.17 The transport of energy from point A to point P by a wave on a string: (a) Here the particles in the string at P have no vibrational kinetic energy. (b) Here the particles at P have vibrational kinetic energy delivered to them by the wave.

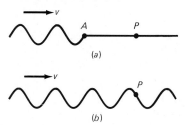

Example 11.6

(a) A wave travels down a stretched string at a speed of 2.0 m/s. The frequency of the wave is 60 Hz and its amplitude is 0.020 m. Write an expression indicating the displacement of a particle on the string as a function of time at a fixed distance x from the end of the string.

(b) Write a similar equation for the variation of the displacement as a function of distance along the string.

SOLUTION

(a) Our basic equation here is Eq. (11.18), $y = Y_0 \cos (2\pi t/T)$. The amplitude Y_0 is given as 0.020 m. The frequency $f = 60$ Hz, and so

$$T = \frac{1}{f} = \frac{1}{60} \text{ s. Hence our equation is}$$

$$y = 0.020 \cos \frac{2\pi t}{\frac{1}{60}} = \boxed{0.020 \cos 120\pi t}$$

where the amplitude is in meters.

(b) The equation giving the variation of y with x is Eq. (11.19). In this case the wavelength is given by

$$\lambda = \frac{v}{f} = \frac{2.0 \text{ m/s}}{60 \text{ Hz}} = \frac{1}{30} \text{ m}$$

Hence our equation is

$$y = Y_0 \cos \frac{2\pi x}{\lambda} = 0.020 \cos \frac{2\pi x}{\frac{1}{30}} = \boxed{0.020 \cos 60\pi x}$$

11.6 Transverse and Longitudinal Waves

Nature provides us with a great variety of different kinds of waves—sound waves, water waves, earthquake (seismic) waves, light waves, and many more. These are all examples of two basic kinds of waves—*transverse* and *longitudinal*.

Definitions

Longitudinal waves: Waves in which the vibrations in the medium are parallel to the direction of energy transport.

Transverse waves: Waves in which the vibrations in the medium are perpendicular to the direction of energy transport.

Transverse Waves

The simplest example of a transverse wave is a wave traveling down a stretched string, as discussed in the preceding section. A water wave is a reasonable approximation to a transverse wave, since the water molecules move up and down at right angles to the direction of the wave. Water waves are not perfect examples of transverse waves, however, since the molecules also have some motion back and forth parallel to the surface of the liquid as the wave passes along the surface.

One of the best examples of a transverse wave is an electromagnetic wave. In this case no molecules at all are needed for the propagation of the wave. Variations in the electric and magnetic fields in a direction perpendicular to the direction of propagation constitute the wave, as we shall see in Chap. 21.

Speed of Transverse Waves on a String

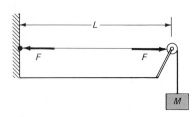

FIGURE 11.18 A string stretched between a wall and a weight at the other end.

As is true of all waves, the speed of transverse waves on a string depends on the properties of the medium through which the wave travels. Clearly this speed will depend in some way on the inertia of the string (i.e., its mass), and on the tension F in the string, since this determines how tightly the motion of one molecule in the string is coupled to that of its neighbors. The tension is the same at any point along a string attached to a wall at one end, and stretched by weights at the other end, as in Fig. 11.18. The total mass of the string will clearly vary with its length (L), but the length of the string should not change the speed of the waves, so long as the tension remains fixed, since the wave speed depends on the properties of the medium at a particular point in space. Hence the important factor must be not the total mass m, but rather the mass per unit length, or the linear density $\mu = m/L$ (μ is the Greek lowercase letter *mu*, the equivalent of the letter m). Hence it seems reasonable to expect that the speed of transverse waves on a string will depend on F and on μ, and as a matter of fact it is difficult to imagine any other property of the string which could affect the speed. (Can you suggest one?) A possibility might be the amplitude Y_0 in the expression $y = Y_0 \cos \omega t$ for the displacement of the string at any time t [Eq. (11.18)], and we will include it in our analysis just to make sure that we are not missing something important.

We want to use dimensional analysis (discussed briefly in Chap. 1) to obtain the functional dependence of the speed v on the tension F, the linear density μ, and perhaps the amplitude Y_0 for a transverse wave on a string. Since speed has dimensions $[L]/[T]$, and we are assuming that it is a function of only F, μ, and Y_0, we can write the dimensional equation:

$$\frac{[L]}{[T]} = [F]^p \, [\mu]^q \, [Y_0]^r \tag{11.20}$$

where p, q, and r are powers to be determined to make the above equation dimensionally correct. Since a force has dimensions $[M][L]/[T]^2$, μ has dimensions $[M]/[L]$, and Y_0 is simply a distance, this equation becomes

$$[L][T]^{-1} = \frac{[M]^p[L]^p}{[T]^{2p}} \frac{[M]^q}{[L]^q}[L]^r$$

or $[L]^1[T]^{-1} = [M]^{p+q}[L]^{p-q+r}[T]^{-2p}$

For this dimensional equation to be an identity, the powers of $[L]$, $[T]$, and $[M]$ must be the same on the two sides. Hence we must have $p + q = 0$, since M does not occur on the left side. Also we must have $p - q + r = 1$ and $-2p = -1$. Hence $p = \frac{1}{2}$, $q = -\frac{1}{2}$, and therefore $r = 0$, since $p - q + r = 1$. We have therefore proved by dimensional analysis that the speed of the wave does not depend on the amplitude Y_0, so long as the tension F and the linear density μ remain constant.

Hence we have for the speed, from Eq. (11.20),

$$v \propto F^p \mu^q Y_0^r \propto F^{1/2}\mu^{-1/2} \quad \text{or} \quad v \propto \left(\frac{F}{\mu}\right)^{1/2}$$

We have used a proportionality sign instead of an equals sign here because dimensional analysis cannot tell us what the constant of proportionality is. By considering the actual forces acting and applying Newton's second law it is possible to prove that the constant of proportionality in this case is exactly 1. We will therefore assume that the following equation is correct without attempting to prove it rigorously:

$$v = \sqrt{\frac{F}{\mu}} \tag{11.21}$$

We have found, therefore, that the speed of a transverse wave on a string depends directly on the square root of the tension in the string, and inversely on the square root of the linear density of the string. For any given frequency f, the wavelength of a wave on the string is therefore

$$\lambda = \frac{v}{f} = \frac{1}{f}\sqrt{\frac{F}{\mu}} \tag{11.22}$$

Example 11.7

(a) The string of a violin has a mass of 3.0 g and a length of 40 cm. What must be the tension in the string if the speed of a wave on the violin string is to be 200 m/s?

(b) For a frequency of 440 Hz, what would be the wavelength produced by this string?

SOLUTION

Here $\mu = \dfrac{m}{L} = \dfrac{3.0 \times 10^{-3} \text{ kg}}{0.40 \text{ m}} = 7.5 \times 10^{-3} \text{ kg/m}$

and $v = 200$ m/s

Since $v = \sqrt{F/\mu}$, $F = \mu v^2 = (7.5 \times 10^{-3} \text{ kg/m})(200 \text{ m/s})^2$

$$= 300 \text{ kg·m/s}^2 = \boxed{300 \text{ N}}$$

This is quite a large force, since it is the tension which would be produced in the string if we attached the string to a hook on the ceiling and hung about 30 kg from its free end.

(b) $\lambda = \dfrac{v}{f} = \dfrac{200 \text{ m/s}}{440/\text{s}} = \boxed{0.45 \text{ m}}$

Longitudinal Waves

If we attach a tuning fork to a long, narrow spring in such a way that the prongs of the fork move parallel to the length of the spring, we can send a longitudinal wave along the spring, as shown in Fig. 11.19. Here the elements of the spring move back and forth parallel to the length of the spring. Such waves are sometimes called *compressional waves*, since they consist of a series of compressions and rarefactions (i.e., stretchings of the spring) moving along the spring in a periodic fashion.

Sound waves are another good example of longitudinal waves. In a sound coming to our ears from a bandstand, for example, the air molecules vibrate back and forth about fixed positions in space along a line drawn from the bandstand to our ears. This causes periodic changes in the density and pressure of the air along this line.

Since we will be devoting the next chapter to sound waves, we will postpone until then any further discussion of the speed of longitudinal waves.

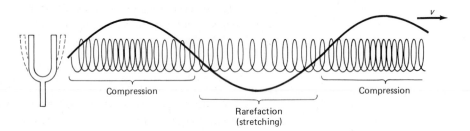

FIGURE 11.19 Longitudinal wave on a spring. The sine curve shows the periodic nature of the compressional wave moving along the spring.

Compression

Rarefaction (stretching)

Compression

11.7 Mathematical Description of Wave Motion

We now want to find a mathematical description of a transverse wave moving along a string in the positive X direction. Let us first consider a function of the form $y = 2^{-x^2}$. This function is shown in Fig. 11.20a and has a form resembling the wave pulse shown in Fig. 11.13. It can be seen that y is centered along the

FIGURE 11.20 A pulse of the form $y = 2^{-(x-b)^2}$ moving along the positive X axis. As b increases, the pulse is centered at higher and higher values of x.

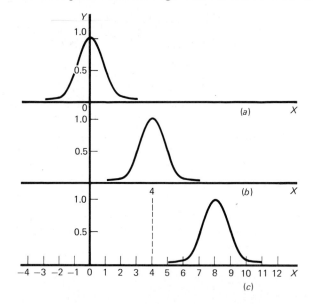

X axis at the origin, where its value is $y = 1$, and that it rapidly falls to zero to either side of $x = 0$. Consider next the similar function $y = 2^{-(x-4)^2}$. This has a maximum value of 1 when $x = 4$ and falls to zero rapidly for lower and higher values of x, as in Fig. 11.20b. Similarly, for $y = 2^{-(x-8)^2}$ we obtain the graph of y, centered on $x = 8$, shown in Fig. 11.20c. From these three diagrams it is clear that $y = 2^{-(x-b)^2}$ is a symmetric function centered at $x = b$, and that as b increases, the pulse described by y moves to the right along the positive X axis.

Suppose now we let b vary in proportion to the time t, so that $b \propto t$, or $b = vt$, where v is a constant. Then we have $y = 2^{-(x-vt)^2}$. Then as time goes on, $b = vt$ increases, and, as we saw above, as b increases, the pulse moves to the right. In this case the speed with which it moves is simply $v = b/t$, so that v is the wave speed.

The same is true of many functions containing the argument $x - vt$. For such functions, as v increases, y moves along the positive X axis at a speed v.

Let us now return to Eq. (11.19), which describes the displacement of the particles of a string at right angles to the X axis at one instant of time:

$$y = Y_0 \cos \frac{2\pi x}{\lambda}$$

If we desire this displacement to be communicated from one particle to another along the positive X axis at a speed v, we need to modify this expression in the same way we modified 2^{-x^2} above, by changing x to $x - vt$. We then have:

$$y = Y_0 \cos \left[\frac{2\pi}{\lambda}(x - vt) \right]$$

which describes the motion of a wave along the positive X axis. This can be written as follows:

$$y = Y_0 \cos \left[2\pi \left(\frac{x}{\lambda} - \frac{vt}{\lambda} \right) \right] = Y_0 \cos \left[2\pi \left(\frac{x}{\lambda} - ft \right) \right]$$

Equation for Wave Motion: or $$\boxed{y = Y_0 \cos \left[2\pi \left(\frac{x}{\lambda} - \frac{t}{T} \right) \right]}$$ (11.23)

This equation represents the motion of a transverse wave, periodic in both x and t, moving along the positive X axis.

To see this consider that when t becomes equal to T, $2T$, $3T$, . . . , for a fixed value of x, y assumes the same value as at $t = 0$, since a change of 2π in an angle leaves its cosine unchanged. The same is true at places where x becomes equal to λ, 2λ, 3λ, . . . , for a fixed value of t. When $t = 0$ and $x = 0$, $\cos 0 = 1$ and $y = Y_0$. This same value of y occurs at a distance x along the positive X axis from the origin at a later time t when $x/\lambda - t/T$ becomes equal to zero, for then $\cos [2\pi(x/\lambda - t/T)] = 1$ and y is again equal to Y_0. Hence the peak of the wave travels a distance x in time t, where these two variables are related by the equation

$$\frac{x}{\lambda} - \frac{t}{T} = 0 \quad \text{and so} \quad v = \frac{x}{t}$$

where, from above, $x/t = \lambda/T = \lambda f$. This means that Eq. (11.23) correctly describes a wave moving along the positive X axis with a speed $v = \lambda f$, as given by Eq. (11.17).

If the wave should be moving in the negative X direction, the corresponding equation would be

$$y = Y_0 \cos \left[2\pi \left(\frac{x}{\lambda} + \frac{t}{T} \right) \right] \qquad (11.24)$$

In this case, as t increases, the peak of the wave determined by $y = Y_0$ moves along the negative X axis, since only for negative values of x can $(x/\lambda + t/T)$ be equal to zero, for t is necessarily positive. Hence, as time goes on, the wave moves to the left along the negative X axis (this can also be seen by considering the function 2^{-x^2} again for the case where this function changes to $2^{-(x+b)^2}$).

Example 11.8

The equation of a transverse wave traveling on a string is

$$y = 2.0 \cos \left[\pi (0.50x - 200t) \right]$$

where all distances are in centimeters.

(a) Find the amplitude, wavelength, frequency, period, and speed of the wave.
(b) If the mass per unit length of the string is 4.0 g/cm, find the tension F in the string.

SOLUTION

(a) The general expression for a transverse wave of this kind is given by Eq. (11.23):

$$y = Y_0 \cos \left[2\pi \left(\frac{x}{\lambda} - \frac{t}{T} \right) \right]$$

To find the designated quantities, we must put our equation in a similar form and compare the quantities in the two equations. We have:

$$y = 2.0 \cos \left[2\pi (0.25x - 100t) \right]$$

On comparing these two equations we see that we have:

Amplitude $Y_0 = \boxed{2.0 \text{ cm}}$

Wavelength $\lambda = \dfrac{1 \text{ cm}}{0.25} = \boxed{4.0 \text{ cm}}$

Period $T = \dfrac{1}{100} \text{ s} = \boxed{0.010 \text{ s}}$

Frequency $f = \dfrac{1}{T} = \dfrac{1}{0.010 \text{ s}} = 100 \text{ s}^{-1} = \boxed{100 \text{ Hz}}$

Speed $v = \lambda f = (4.0 \text{ cm})(100 \text{ Hz}) = \boxed{4.0 \text{ m/s}}$

(b) Since $v = (F/\mu)^{1/2}$, and
$\mu = 4.0$ g/cm $= (4.0 \times 10^{-3} \text{ kg})/(10^{-2} \text{ m}) = 0.40$ kg/m, we have

$$F = \mu v^2 = (0.40 \text{ kg/m})(4.0 \text{ m/s})^2 = \boxed{6.4 \text{ N}}$$

11.8 The Superposition Principle; Phase

Two or more waves can pass through the same region of space independently of one another. Thus in a concert hall the sound of a violin and of an oboe can be clearly distinguished by the members of the audience even though the sound waves from these two instruments are traveling to their ears through the same region of space at the same instant of time. The only way that such behavior is possible is if the two waves act independently on the particles of the air and on the eardrums of the members of the audience and produce a

FIGURE 11.21 Superposition of two waves traveling in the same direction along the X axis with the same speed, but different frequencies and amplitudes. The displacement of the resultant wave at any position x is the algebraic sum of the displacements produced by the individual waves in (a) and (b), as shown in (c).

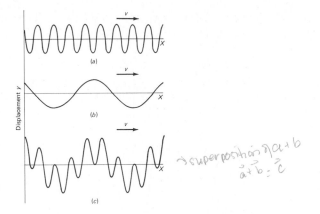

motion which is the *superposition* of the motions due to the two separate waves. This means that at any instant of time the displacement of an air molecule is simply the algebraic sum of the displacements caused by the individual waves, as in Fig. 11.21. This is true for situations in which Hooke's law, in the form of Eq. (11.4), is valid, i.e., where the restoring force is proportional to the displacement. This is the case for most small-amplitude waves, but is not true for large-amplitude waves, such as shock waves, where the superposition principle breaks down.

Interference of Waves and Phase Differences

By the *interference* of two waves we mean the result of superimposing the two waves on the same medium. Thus if we have two transverse waves traveling along a string in the same direction, with the same frequency and the same amplitude, the resulting motion of the string will depend on the algebraic addition (i.e., with the proper sign) of the displacements due to the two waves at every point along the string. How the displacements add will depend on the relative phases of the two waves. By *phase* we mean the following:

Definition

Phase: The stage in a cycle that a wave (or other periodic system) has reached at a particular time or at a particular place.

Phase becomes important only when we are considering the superposition of two or more waves of different phases, and hence we will focus our attention on the *relative phase* of one wave with respect to a second wave. For example, in Fig. 11.22 waves *a* and *b* are *in phase* because they reach their maximum values in the positive direction for exactly the same values of *x*. In this case the resulting motion of the string will be exactly the same as that due to the motion of either wave alone, but with an amplitude twice as large. This is called *constructive interference*, since at every point the displacements of the string produced by the two waves add constructively.

On the other hand, Fig. 11.23 shows *destructive interference*, where the two waves are 180°, or π rad, *out of phase*. Since their amplitudes are the same, the negative displacements of one wave cancel the positive displacements of the other, and the result is no motion of the string.

Another example of a phase difference is that between sine and cosine functions of the form:

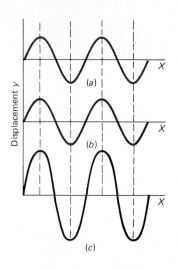

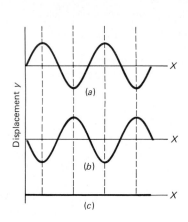

FIGURE 11.22 Constructive interference of two waves of the same frequency, amplitude, and phase. The wave shown in (c) is the sum of the waves in (a) and (b).

FIGURE 11.23 Destructive interference of two waves of the same frequency and amplitude, but 180°, or π rad, out of phase. This leads to complete cancellation of the two waves for all values of x, as shown in (c).

$$y_1 = Y_0 \cos \left[2\pi \left(\frac{x}{\lambda} - \frac{t}{T} \right) \right]$$

and

$$y_2 = Y_0 \sin \left[2\pi \left(\frac{x}{\lambda} - \frac{t}{T} \right) \right]$$

Since, from Appendix 1C, $\sin \theta = \cos (\theta - 90°) = \cos (\theta - \pi/2)$, we can write

$$y_2 = Y_0 \cos \left[2\pi \left(\frac{x}{\lambda} - \frac{t}{T} \right) - \frac{\pi}{2} \right]$$

and we see that y_1 and y_2 are 90°, or $\pi/2$ rad, out of phase. This can be seen from Fig. 11.24, where wave 2 has to be shifted through $\pi/2$ rad, which corresponds to a distance of $\lambda/4$ along the X axis, to be in phase with wave 1.

In the general expression for a wave the phase angle is usually written as an angle ϕ in the equation:

$$y = Y_0 \cos \left[2\pi \left(\frac{x}{\lambda} - \frac{t}{T} \right) + \phi \right] \tag{11.25}$$

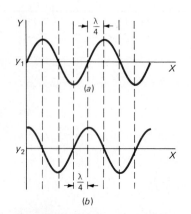

FIGURE 11.24 The phase difference between (a) $y = \sin x$ and (b) $y = \cos x$. If the cosine function is shifted by $\lambda/4$ along the +X axis, it becomes identical with the sine function.

It can be seen that a phase difference of ϕ between two waves means that one wave must be shifted through a fraction of a wavelength equal to $\phi/2\pi$ to be perfectly superimposed on the other wave. This is the case for the sine and cosine functions above, where the phase difference is $\pi/2$. This means that the sine wave had to be shifted by an amount $(\phi/2\pi)\lambda = [(\pi/2)/2\pi]\lambda = \lambda/4$ to be coherent with the cosine wave. Two waves of the same wavelength differing in phase by ϕ rad can always be superimposed by shifting one of them by an amount $(\phi/2\pi)\lambda$ in the direction of the wave's motion.

11.9 Standing Waves on a String

Suppose we consider a transverse wave which is produced by the periodic up-and-down motion of a string at its left end and which moves along the string in the positive X direction, as in Fig. 11.25. If the right end of the string is attached to a wall at a distance L from the source, the wave is reflected back

FIGURE 11.25 Transverse wave on a string. The wave is moving toward the wall at the right.

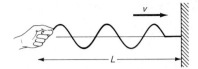

FIGURE 11.26 Transverse wave reflected from a rigid wall at one end of a string. A phase change of π rad occurs on reflection.

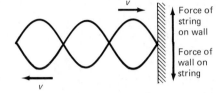

from the rigid wall and returns in the negative X direction. The reflected wave will be 180° out of phase with the original wave, since the two must add to give zero displacement at the rigid wall. Hence we say that there is a phase change of 180°, or π rad, on reflection at the wall.

This phase change can also be explained by Newton's third law. When the wave reaches the wall and the string pulls up on the wall at the point of contact, the rigid wall exerts an equal and opposite force down on the string, as in Fig. 11.26. It is this force which starts the reflected wave back down the string. The reflected wave has had its phase changed from up to down, i.e., by π rad, on reflection by the rigid wall. It is possible to show that, if the end of the string is not attached to an unmovable wall but to a bolt free to slide in a slot on the wall, the bolt will move in the same direction as the string, and will start a reflected wave which is *in phase* with the incident wave at the wall. Hence:

Phase Change on Reflection

Reflection from a fixed end causes a phase change of 180°, or π rad; reflection from a free end causes no change of phase.

We will see many applications of these ideas when we discuss sound and light.

After the original wave is reflected, we will have two waves of the same amplitude, frequency, and speed traveling in opposite directions along the string, since we presuppose that the source continues to produce waves. We now want to find the result of the superposition of these two waves.

We will represent the two waves by the equations [see Eqs. (11.23) and (11.24)]:

$$y_1 = Y_0 \cos\left[2\pi\left(\frac{x}{\lambda} - \frac{t}{T}\right)\right] \quad \text{and} \quad y_2 = Y_0 \cos\left[2\pi\left(\frac{x}{\lambda} + \frac{t}{T}\right)\right]$$

Then the superposition of these two waves leads to:

$$y = y_1 + y_2 = Y_0\left[\cos 2\pi\left(\frac{x}{\lambda} - \frac{t}{T}\right) + \cos 2\pi\left(\frac{x}{\lambda} + \frac{t}{T}\right)\right]$$

But, from Appendix 1C, we have for two angles A and B,

$$\cos(A + B) = \cos A \cos B - \sin A \sin B$$

and $\quad \cos(A - B) = \cos A \cos B + \sin A \sin B$

so that $\cos (A + B) + \cos (A - B) = 2 \cos A \cos B$

and, on setting $A = 2\pi x/\lambda$ and $B = 2\pi t/T$, our expression above for y becomes:

$$y = 2 Y_0 \cos \frac{2\pi x}{\lambda} \cos \frac{2\pi t}{T} = Y' \cos \frac{2\pi t}{T} \tag{11.26}$$

The resultant therefore is a wave periodic in time t, but with an amplitude $Y' = 2Y_0 \cos (2\pi x/\lambda)$, which is a periodic function of the distance x along the string from the source.

Thus the amplitude Y' has a minimum value (namely, zero) when $x = \lambda/4, 3\lambda/4, 5\lambda/4$, or, in general $(2n + 1)\lambda/4$, with n an integer. These points of no motion we call *nodes*. The amplitude has a maximum value of $2Y_0$ when $x = 0, \lambda/2, \lambda, 3\lambda/2$, or, in general, $n\lambda/2$, with n an integer. These points are called *antinodes*, or *loops*, since the string has its maximum motion at these points. The nodes and loops are shown in Fig. 11.27. Adjacent pairs of nodes are one-half wavelength apart, as are adjacent pairs of antinodes. In the case being considered here the left end of the string is an antinode, for we have assumed that motion is being imparted to the string at that point. If the string were, on the other hand, held by clamps at both ends, and then forced to vibrate at its center, we would again have the superposition of waves moving in both directions, but in this case we would necessarily have nodes at the two ends of the string. Since $x = 0$ at the left end of the string, in order to satisfy the condition that there be a node at the left end the above equation would have to take the form

$$y = 2Y_0 \sin \frac{2\pi x}{\lambda} \cos \frac{2\pi t}{T}$$

This corresponds to a shift in phase of 90°, or $\pi/2$ rad, for the standing wave.

Because of the looks of the string described by Eq. (11.26), with some points at rest and others in violent motion, and with no appearance of energy flow from one part of the string to the other, the resultant wave is called a *standing wave*. Hence *two superimposed traveling waves of the same frequency, amplitude, and speed, moving in opposite directions in the same medium, produce a standing wave.*

How two traveling waves can be combined graphically to produce such a standing wave is shown in Fig. 11.28.

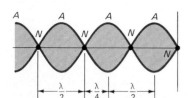

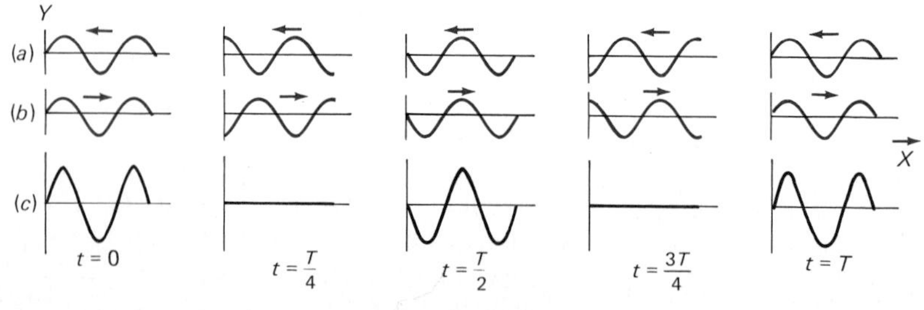

FIGURE 11.27 Nodes (N) and loops (or antinodes A) for a standing wave on a string. Adjacent pairs of nodes are $\lambda/2$ apart, as are adjacent pairs of antinodes.

Standing Wave

FIGURE 11.28 Graphical combination of two waves moving in opposite directions to form a standing wave. The wave in (a) moving toward the left and the wave in (b) moving toward the right are shown in (c) superimposed at intervals of $T/4$. Notice that the nodes remain fixed at positions separated by $\lambda/2$, and the antinodes also remain fixed at separations of $\lambda/2$. Hence the name *standing wave*.

Example 11.9

A standing wave on a string can be described by Eq. (11.26), with Y_0 equal to 0.050 m and λ equal to 0.10 m. What is the value of the amplitude of the standing wave at the following distances x along the string?

(a) 0 **(b)** 0.10 m **(c)** 0.025 m
(d) 0.050 m **(e)** 0.10 m **(f)** 0.35 m

SOLUTION

The amplitude of the standing wave is, from Eq. (11.26),

$$Y' = 2Y_0 \cos \frac{2\pi x}{\lambda} = 0.10 \cos \frac{2\pi x}{0.10} = 0.10 \cos 20\pi x$$

(a) When $x = 0$, $\cos 20\pi x = 1$, and so $Y' = \boxed{0.10 \text{ m}}$

(b) When $x = 0.010$, we have $Y' = 0.10 \cos [20\pi(0.01)]$ $= 0.10 \cos 0.2\pi$. But $\cos 0.2\pi = \cos 36° = 0.81$, and so

$$Y' = 0.10(0.81) = \boxed{0.081 \text{ m}}$$

(c) $Y' = 0.10 \cos [20\pi(0.025)] = 0.10 \cos 0.5\pi = \boxed{0}$

since $\cos (\pi/2) = \cos 90° = 0$.

(d) $Y' = 0.10 \cos [20\pi(0.05)] = 0.10 \cos \pi = \boxed{-0.10 \text{ m}}$

(e) $Y' = 0.10 \cos [20\pi(0.10)] = 0.10 \cos 2\pi = \boxed{0.10 \text{ m}}$

(f) $Y' = 0.10 \cos [20\pi(0.35)] = 0.10 \cos 7\pi = \boxed{-0.10 \text{ m}}$

Note that $\cos 7\pi$ is the same as $\cos \pi$, since after 2π rad all trigonometric functions repeat themselves.

The minus signs in the amplitudes make no difference here, since as time goes on, the displacement y varies rapidly between $+Y'$ and $-Y'$. Hence we have maximum displacements for the standing wave at intervals of $\lambda/2$, or 0.05 m, that is, for $x = 0, 0.05, 0.10, 0.15, 0.20, 0.25, 0.30, 0.35$ m. Between two consecutive loops we have nodes where $Y' = 0$, for example, at $x = 0.025$ m.

11.10 Forced Vibrations of a String

If the length of a string fixed at both ends is chosen to be an integral multiple of half wavelengths of the wave generated by a mechanical vibrator attached to the string, then large-amplitude standing waves may be set up in the string by the superposition of the waves traveling back and forth along the string.

Resonance for Waves on a String

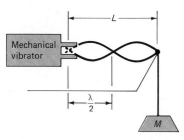

FIGURE 11.29 String being driven by a mechanical vibrator. Resonance can be produced either by keeping the mass M fixed and varying the frequency of the mechanical vibrator, or by keeping the frequency fixed and varying M. In the case shown $f_2 = (1/L) \sqrt{F/\mu}$.

Suppose we have a string driven by a mechanical vibrator at one end and passing over a pulley to a hanging weight at the other end, as in Fig. 11.29. Even though the end tied to the vibrator must be in motion, its amplitude is so small that we can, to a good approximation, consider both ends of the string to be nodes. When such a string is driven by the vibrator, the waves running back and forth along the string in most cases cancel one another out, and nothing is observed except a slight blur to indicate that the string is actually moving. If, however, the tension in the string is kept fixed by the hanging weight and the frequency of the vibrator is varied, large standing waves will be observed in the string for certain discrete frequencies. This is an example of *resonance*, in which the reflected waves add together constructively to produce large-amplitude oscillations of the string at the antinodes, because the applied frequency is equal to one of the natural frequencies of vibration of the string.

In this case the speed of the transverse waves is $v = (F/\mu)^{1/2}$, where μ is the linear density of the string and F the tension [see Eq. (11.21)]. Since we require nodes at the two ends of the string, for standing waves the length L of the string must be an integral multiple (n) of half wavelengths, i.e.,

$$L = \frac{n\lambda}{2} \quad \text{or} \quad \lambda_n = \frac{2L}{n} \tag{11.27}$$

with $n = 1, 2, 3, 4,. \ldots$ Here λ_n signifies the wavelength corresponding to a particular value of n. The natural frequencies of the string are, therefore, from Eqs. (11.21) and (11.27),

$$f_n = \frac{v}{\lambda_n} = \frac{\sqrt{F/\mu}}{2L/n}$$

or $\quad f_n = \frac{n}{2L} \sqrt{\frac{F}{\mu}}, \quad$ with $n = 1, 2, 3, \ldots$ (11.28)

These are the natural frequencies of a string of length L, with tension F and linear density μ. Whenever the string is driven at one of these frequencies, large-amplitude standing waves result, because the frequency of the driver (the vibrator) is in resonance with one of the natural frequencies of the string. The vibration shown in Fig. 11.29, for example, shows resonance for $n = 2$, since the wavelength is equal to the length of the string.

Resonance can also be achieved by keeping the vibrator's frequency fixed and varying the tension in the string by adding weights until one of the natural frequencies f_n of the string corresponding to some integral value of n becomes exactly equal to the applied frequency. This is accomplished by changing the speed (v) of the wave (by changing F) to a value such that the wavelength (v/f) satisfies Eq. (11.27) for some integral value of n.

In Eq. (11.28) the frequency with $n = 1$ is called the *fundamental frequency*, or the *first harmonic*, that with $n = 2$ the *second harmonic*, and so forth. The second harmonic therefore has twice the frequency of the fundamental. Sometimes the frequency with $n = 2$ is called the *first overtone*, with $n = 3$ the second overtone, etc. Hence the first overtone means the same as the second harmonic, the second overtone corresponds to the third harmonic, and so on. In what follows we will always use harmonics instead of overtones because the numerical relationship of a harmonic to the fundamental frequency is clear once it is stated that it is the second, third, or nth harmonic.

Figure 11.29 therefore corresponds to the string vibrating in its second harmonic ($n = 2$). The fundamental frequency of the string would correspond to no nodes except the two at the ends. In that case the length of the string would be $L = \lambda/2$.

Example 11.10

In Fig. 11.29 the vibrator has a frequency of $f = 30$ Hz, while the string is 3.0 m long and has a linear density $\mu = 0.40$ g/cm. The tension F is varied by adding weights to the string passing over the pulley.

(a) What value of the tension is needed to cause the string to resonate with the vibrator in one loop? (Assume that the points where the string meets the tuning fork and the pulley are approximately nodal points of the motion.)

(b) What is the speed of the waves on the string in this case?

SOLUTION

From Eq. (11.28) we have

$$f_n = \frac{n}{2L}\sqrt{\frac{F}{\mu}} \quad \text{from which} \quad F = \frac{4L^2 f_n^2 \mu}{n^2}$$

In this case the frequency does not change but is kept fixed at $f = 30$ Hz, while F must vary with n, the number of loops in the standing wave. Also $\mu = 0.40$ g/cm $= (0.40 \times 10^{-3}$ kg$)/(10^{-2}$ m$) = 0.040$ kg/m.

For one loop, $n = 1$ and so

$$F_1 = \frac{4L^2 f^2 \mu}{1^2} = 4(3.0 \text{ m})^2(30 \text{ s}^{-1})^2(0.040 \text{ kg/m})$$

$$= 1.3 \times 10^3 \text{ kg m/s}^2 = \boxed{1.3 \times 10^3 \text{ N}}$$

(b) For one loop, $F_1 = 1.3 \times 10^3$ N, and since $\mu = 0.040$ kg/m, we have

$$v_1 = \left(\frac{1.3 \times 10^3 \text{ N}}{0.040 \text{ kg/m}}\right)^{1/2} = \boxed{180 \text{ m/s}}.$$

Summary: Important Definitions and Equations

Periodic motion:
Motion which repeats itself after a fixed time interval.
Period (T):
The time interval during which a physical system completes one full cycle of motion.
Frequency (f):
The number of complete cycles per second; $f = 1/T$.
Displacement (x):
The distance of a moving system from its equilibrium position at some instant of time t.
Amplitude (X_0):
The maximum value of the displacement.
Simple harmonic motion (SHM):
   1 A periodic motion for which the restoring force is proportional to the displacement and in the opposite direction: $F = -kx$.
   2 The projection on a diameter of the motion of a particle moving with constant speed in a circle.
Equations for SHM of a mass at end of a spring:

$$x = X_0 \cos\frac{2\pi t}{T} \qquad v = \sqrt{\frac{k}{m}}\sqrt{X_0^2 - x^2}$$

$$a = -\frac{k}{m}x \qquad T = 2\pi\sqrt{\frac{m}{k}}$$

Simple pendulum:
A heavy bob at the end of a weightless string.
Motion is SHM only for small amplitudes for which $(\sin\theta)/\theta \simeq 1$.
For small amplitudes, $T = 2\pi\sqrt{L/g}$.
Damping:
The decrease in amplitude of a periodic mechanical system by the conversion of mechanical energy into heat.
Wave:
A method of energy propagation through space which involves periodic variations in the medium through which the wave travels.
Wavelength (λ):
The distance between any two consecutive equivalent points on a wave pattern.

Speed of a wave:
$v = \lambda f$
Longitudinal waves:
The particle motion is parallel to the direction of the wave.
Transverse waves:
The particle motion is perpendicular to the direction of the wave.
Speed of transverse wave on a string:
$v = \sqrt{F/\mu}$, where F is the tension in the string, and $\mu = m/L$ is the linear density.

General equation for a wave:

$$y = Y_0 \cos\left[2\pi\left(\frac{x}{\lambda} - \frac{t}{T}\right) + \phi\right]$$

Superposition principle:
Two or more waves can pass through the same region of space independently of one another. The resulting displacement of the medium is the sum of the displacements caused by the individual waves.
Standing wave:
A wave pattern produced by two superimposed traveling waves of the same frequency, amplitude, and speed, moving in opposite directions in the same medium. At some positions the amplitude is zero (nodes); at other positions the amplitude is a maximum (loops, antinodes).

$$y = 2Y_0 \cos\frac{2\pi x}{\lambda}\cos\frac{2\pi t}{T}$$

Resonance:
Large oscillations of a system which result when the frequency applied to the system is the same as one of the natural frequencies of the system.
Resonant frequencies of a string of length L with nodes at both ends:

$$f_n = \frac{n}{2L}\sqrt{\frac{F}{\mu}} \quad \text{where} \quad n = 1, 2, 3, \ldots$$

Questions THINK ABOUT

1 Give a few examples of motions which are periodic but not simple harmonic. Is it possible for a motion to be simple harmonic without being periodic?

2 A spring is so constructed that, when stretched, the restoring force is proportional to the square of the displacement and in the opposite direction. Is the motion of a mass at the end of this spring SHM?

3 (a) Draw a graph of the kinetic energy of the vibrating horizontal spring as a function of time, starting at $t = 0$ when the displacement is $x = X_0$, and ending at $t = T$ when the displacement is again $x = X_0$.

(b) Repeat part (a) for the potential energy of the spring.

4 A mass is vibrating up and down at the end of a vertical spring. When the mass is at its lowest position ($y = -Y_0$) in its vibration, what are the values of the displacement, the speed, and the acceleration of the mass?

5 Provide a convincing explanation of why the period of a simple pendulum does not depend on the mass of the pendulum bob.

6 At what point of the motion of a mass at the end of a spring is the acceleration zero? Compare this situation with that in which a ball is thrown vertically upward, stops, and falls back to earth again. When is the acceleration zero in this case?

7 Why would you expect to obtain a more accurate value for g by timing the vibrations of a simple pendulum than you could obtain by timing the free-fall of an object?

8 Will a pendulum clock adjusted to keep perfect time at the foot of Mount Whitney in California also keep perfect time at the top of Mount Whitney?

9 Two perfectly tuned violins are resting on a table. A musician bows a string on one violin very strongly and then stops the string with a finger, but the sound of the same note continues to come from the second violin. Explain what has happened. Does the table play any role in this phenomenon?

10 When a car is stuck in a rut, a recommended method of getting it moving again is to rock it back and forth where it is stuck, each time changing from forward speed to reverse and then back again to forward. Why do you think such a method might work? Indicate what happens to the kinetic energy and the potential energy of the car with time, for the case where this method succeeds in extricating the car?

11 Suggest two different ways in which resonance can be achieved between the natural frequency of a child on a swing and the frequency of the pushes applied to the swing.

12 (a) Is the wave described by the equation

$$y = 0.20 \cos (60\pi t + 120\pi x)$$

moving in the positive or negative X direction? Why?

(b) What is the speed of such a wave?

13 A standing wave is set up in a string with a mechanical vibrator. It is found that a screwdriver can be touched to the exact center of the string without damping the vibration in any way. What are some possible wavelengths for the wave moving up and down the string in terms of the length of the string?

14 If a tunnel could be drilled completely through the earth along a diameter, and a rock were dropped into the tunnel, what would happen to the rock? Why?

Problems

A 1 A mass at the end of a horizontal spring is stretched 0.050 m from its rest position and then released. If it takes 1.0 s for the mass to move through its rest position to an extreme position in the opposite direction, the frequency of the vibration is:
(a) 1.0 s   (b) 2.0 s   (c) 1.0 Hz   (d) 0.50 Hz
(e) 2.0 Hz

A 2 A ball is rotating in a circle of radius 0.10 m, making 2.5 rev/s. An equation which correctly describes the projection of this motion on the X coordinate axis is:
(a) $x = 0.20 \sin 5\pi t$ (b) $x = 0.10 \cos 5t$
(c) $x = 0.05 \cos 5\pi t$ (d) $x = 0.10 \cos 5\pi t$
(e) $x = 0.20 \sin 5\pi t$

A 3 A mass of 2.0 kg at the end of a spring of force constant 5.0 N/m moves in SHM. The frequency of this motion is:
(a) 4.0 s   (b) 4.0 Hz   (c) 0.25 Hz
(d) 1.25 Hz   (e) 0.80 Hz

A 4 A mass of 0.40 kg at the end of a horizontal spring is displaced 0.20 m and released, and moves in SHM at the end of a horizontal spring of force constant 2.0 N/m. The KE of the system when $x = 0.10$ m is equal to:
(a) 0.040 J   (b) 0.020 J   (c) 0.010 J
(d) 0.045 J   (e) 0.030 J

A 5 For a simple pendulum of amplitude 10°, the error involved in assuming that the motion is SHM is about:
(a) 5%   (b) 0.5%   (c) 0.17%   (d) 1%
(e) Exactly 0; pendulum motion is SHM

A 6 On the surface of the moon the period of a simple pendulum is 2.0 s. The length of the pendulum is:
(a) 16 cm   (b) 16 m   (c) 4.0 m
(d) 0.99 m   (e) 1.98 m

A 7 A small length of a string stretched along the X axis is vibrating up and down in SHM at a distance x from the free end, which is being driven with an amplitude of 0.010 m at a frequency of 120 Hz. The wave travels along the string at a speed of 6.0 m/s. The equation describing the motion of the string at this point x is:

(a) $y = 0.010 \cos 120\pi t$ (b) $y = 0.010 \cos \dfrac{2\pi x}{120}$

(c) $y = 0.010 \cos \dfrac{2\pi x}{20}$ (d) $y = 0.010 \cos 240\pi t$

(e) $y = 0.010 \cos 20\pi x$

A 8 A stretched string is 2.0 m long. If standing waves are produced on this string by varying the applied frequency, the following *cannot* be a possible value for the wavelength obtained:

(a) 4.0 m   (b) 2.0 m   (c) 1.33 m
(d) 3.0 m   (e) 1.0 m

A 9 An applied frequency which would produce resonance in a stretched string of length 2.0 m, with a tension of 1.0×10^3 N and linear density 0.025 kg/m, is:

(a) 25 Hz   (b) 125 Hz   (c) 150 Hz
(d) 175 Hz   (e) 225 Hz

A10 A string has a length 2.0 m and a linear density $\mu = 0.045$ kg/m. A mechanical vibrator of frequency 60 Hz is applied to the string, and the tension is varied to achieve resonance. The tension in the string required to produce a large-amplitude vibration of the string in three loops is:

(a) 288 N   (b) 4.5 N   (c) 40.5 N
(d) 864 N   (e) 2.6×10^3 N

B 1 A horizontal spring has a force constant of 5.0 N/m. It is stretched through a distance of 0.060 m and then released. What is the total mechanical energy of the system?

B 2 A vertical spring has a force constant of 3.0 N/m. A mass of 2.0 kg is attached to the spring, and it is stretched a distance of 2.0 cm and then released. What is the numerical relationship between the acceleration of the mass and its displacement?

B 3 A mass at the end of a spring moves in SHM with an amplitude of 2.0 cm and at a frequency of 2.5 Hz. Write a general equation for the displacement of the mass as a function of time.

B 4 A mass at the end of a spring moves in SHM of amplitude 0.15 m and period 2.0 s. Write a general equation for the velocity of this mass as a function of time t.

B 5 What is the frequency of a simple pendulum of length 2.0 m at a place where the acceleration due to gravity is 9.79 m/s²?

B 6 A wave moving along a stretched string has a period of $\frac{1}{100}$ s and a wavelength of 20 cm. What is the speed of the transverse wave on the string?

B 7 A cello string has a mass of 2.0 g and a length of 0.68 m. What must be the tension in the string if the speed of a wave traveling along the cello string is to be 50 m/s?

B 8 A traveling wave on a string is described by the equation $y = 0.020 \cos (50\pi x - 100\pi t)$ where all distances are in meters. What is the speed of such a wave along the string?

B 9 A standing wave is set up on a string by two traveling waves, each of amplitude 0.020 m and wavelength 0.80 m, moving in opposite directions along the string. The resultant standing wave may be described by the equation:

$$y = \left(2Y_0 \cos \frac{2\pi x}{\lambda}\right)\left(\cos \frac{2\pi t}{T}\right)$$

What is the amplitude of this standing wave at point $x = 0.010$ along the string?

B10 A string is 10 m long; weights are tied to it at one end to produce a tension of 500 N. What must be the linear density of the string if it is to resonate to a frequency of 60 Hz and vibrate in three segments?

C 1 An object whose mass is 0.50 kg is hung from a spring, and it is observed that the spring stretches 0.020 m under this load. The body is now pulled down an additional 0.010 m from this equilibrium position and released. Find (a) the amplitude, (b) the period, (c) the frequency of the resulting motion. (d) Write an equation describing the position of the object as a function of time.

C 2 A mass of 0.50 kg moves in SHM at the end of a horizontal spring of force constant 2.0 N/m. The amplitude of the motion is 0.10 m.

(a) What is the value of the velocity when the mass passes through its equilibrium position?

(b) Prove that the total energy of the system is the same at $x = 0$ as it is at $x = X_0$.

C 3 The SHM of a mass at the end of a vertical spring is described by the equation $y = Y_0 \cos (2\pi t/T)$, where $Y_0 = 0.050$ m and $T = 2.0$ s. Draw graphs of the displacement of the mass, its velocity, and its acceleration as a function of time. Use the correct amplitudes and relative phases for the three graphs.

C 4 A 200-g mass vibrates in SHM at the end of a spring. The amplitude of the motion is 0.20 m and the period is 2.4 s. Find (a) the frequency of the motion, (b) the force constant of the spring, (c) the maximum speed of the mass, (d) the speed when the displacement is 0.10 m, (e) the acceleration when the displacement is 0.10 m.

C 5 An unknown mass m is hung from a spring, and it is observed that the increase in length of the spring is 0.10 m. The mass is then pulled down farther and released. What is the frequency of the resulting SHM?

C 6 A horizontal string 4.0 m long has a mass of 1.60 g.

(a) What must be the tension in the string so that a 100-Hz wave moving along the string has a wavelength of 50 cm?

(b) What mass must be hung from the end of the string to produce this tension?

C 7 Four students jump into a car and notice that the springs of the car sink down 6.0 cm under the 350-kg mass of the four students. If the total load supported by the springs (i.e., car plus students) is 1000 kg, what is the period of vibration of the loaded automobile?

C 8 What is the acceleration due to gravity on a planet where an astronaut sets up a simple pendulum of length 1.0 m and finds that it makes 100 complete vibrations of small amplitude in 380 s?

C 9 A simple pendulum has a period of 1.98 s at a place where $g = 9.80$ m/s². What is the value of g at a different place where this same pendulum has a period of only 1.97 s?

C10   A pendulum which has a period of exactly 2.00 s at the Greenwich Observatory in England, where $g = 9.812$ m/s^2, is taken to Paris, where it is found to lose 20 s a day. What is the acceleration due to gravity in Paris?

C11   A pendulum bob of mass 1.0 kg hangs at the end of a string of length 2.0 m. The bob is displaced until the string makes an angle of 15° with the vertical and is then released. What is the KE of the bob when it passes through its equilibrium position?

C12   A transverse wave of amplitude 0.12 m and wavelength 1.5 m travels from left to right along a long, horizontally stretched string with a speed of 1.5 m/s. At time $t = 0$, the left end of the string has a displacement $y = 0.12$ m.

(a)   Write an equation describing the motion of a particle at the left end of the string as a function of time.

(b)   Write an equation describing the motion of a particle 1.0 m from the left end of the string as a function of time.

(c)   What is the maximum transverse velocity of any particle in the string?

C13   A wave of frequency 60 Hz and amplitude 0.020 m travels from left to right along a stretched string at a speed of 2.0 m/s. At $t = 0$ the displacement at $x = 0$ is 0.020 m. Draw a sketch of the shape of the string at a time $t = 3.0$ s, clearly indicating distance along the string from the point $x = 0$.

C14   The equation of a transverse traveling wave on a string is $y = 0.10 \cos (40\pi x - 100\pi t)$, where all distances are in meters. Find (a) the amplitude, (b) the wavelength, (c) the frequency, (d) the period, (e) the speed of the wave. (f) What is the tension in the string, if its linear density is 0.40 kg/m?

C15   A metal wire is under a tension of 200 N. The length of the wire is 1.0 m and its mass is 2.0 g.

(a)   What is the speed of transverse waves along the wire?

(b)   What is the frequency of the fundamental vibration of this wave?

(c)   What is the frequency of its second and third harmonics?

C16   A banjo string is 30 cm long and has a fundamental frequency of 256 Hz.

(a)   What is the speed of waves along the banjo string?

(b)   What is the tension in the string if the linear density of the string is 0.20 g/cm?

C17   A wire stretched under tension has a fundamental frequency of 440 Hz. What would be its new fundamental frequency if the wire were made of the same material but had twice the cross-sectional radius, twice the tension, and was twice as long as the original wire?

C18   Two wires of the same cross section are stretched with equal tensions between the same two supports. One wire is made of steel, with a density of 7.8×10^3 kg/m^3, and the other of nickel, with a density of 8.9×10^3 kg/m^3. If the fundamental frequency of the steel wire is 220 Hz, what is the fundamental frequency of the nickel wire?

C19   A stretched wire on a resonating box is under tension at both ends.

(a)   At what point should the wire be plucked to make its fundamental note most prominent?

(b)   If it is desired to make the second harmonic more prominent, where should the wire be plucked? Would touching the wire at its center after it has been plucked help in making the second harmonic more prominent than the fundamental?

(c)   What would you suggest to make the third harmonic most prominent?

C20   A string attached at both ends to supports vibrates at a frequency of 440 Hz with four nodes between its two ends. At what frequency will the number of nodes be reduced to one?

C21   A steel piano wire 0.40 m long, of mass 6.0 g, is stretched with a tension of 500 N.

(a)   What is the fundamental frequency of this piano wire?

(b)   What is the highest harmonic of this piano wire that could be heard by a person able to hear frequencies up to 15,000 Hz?

C22   A string 2.0 m long, with a linear density of 0.030 kg/m, is attached to a mechanical vibrator of variable frequency. The other end of the string passes over a pulley to a 20-kg mass.

(a)   What is the lowest frequency of the vibrator at which this string will resonate?

(b)   What frequencies correspond to the second, third, and fourth harmonics of the string?

D 1   A metal block of mass $m = 1.0$ kg rests on a frictionless surface, as in Fig. 11.30. It is attached to two walls in a room by two springs, one on each side of the mass. One spring has a force constant $k_1 = 40$ N/m and the other a force constant 25 N/m. The block is pushed a small distance to one side of its equilibrium position and released. With what frequency will it vibrate?

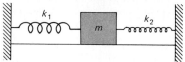

FIGURE 11.30   Diagram for Prob. D1.

D 2   The same two springs as in Prob. D1 are connected to a mass of 1.0 kg, but now the two springs are on the same side of the mass, as in Fig. 11.31. If the mass is

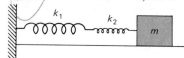

FIGURE 11.31   Diagram for Prob. D2.

displaced a small distance x and released, what is the frequency of the resulting SHM? (*Hint*: The tension must be the same in the two springs at any instant of time.)

D 3 Prove that the natural frequency of a mass oscillating at the end of a spring which obeys Hooke's law is the same as that of a simple pendulum of length equal to the increase in length the mass produces when hung on the spring.

D 4 A 5.0-g bullet is shot with a speed of 200 m/s into a 1.0-kg block which is attached to a wall by a spring of force constant $k = 200$ N/m, as in Fig. 11.32. The block is originally at rest. If we ignore friction forces between the block and the table, what will be the amplitude of vibration of the mass after the bullet has come to rest?

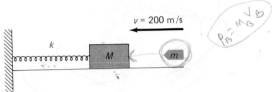

FIGURE 11.32 Diagram for Prob. D4.

D 5 Assume that the fundamental frequency of a stretched string must depend on the mass, length, and tension in the string, according to the equation

$$f_1 \propto m^x L^y F^z$$

or $\quad f_1 = Cm^x L^y F^z$

Use dimensional analysis to determine x, y, and z. Then compare your result with the more complex result in Eq. (11.28) and find the value of C needed in the above equation.

D 6 A pendulum vibrates in SHM with an amplitude of 12°. Compare the fraction of time spent by the pendulum bob between -6 and $+6°$ with the sum of the times spent between -12 and $-6°$ and between $+6$ and $+12°$.

D 7 Show that Eq. (11.23) can also be written as

$$y = Y_0 \cos\left[\frac{2\pi}{\lambda}(x - vt)\right]$$

where v is the wave velocity.

D 8 Two transverse waves are traveling down a string in the same direction. They have a speed of 4.0 m/s, a frequency of 240 Hz, and an amplitude of 1.0 cm. The two waves can be described by the equations

$$y_1 = Y \cos\left[2\pi\left(\frac{t}{T} - \frac{x}{\lambda}\right)\right]$$

$$y_2 = Y \cos\left[2\pi\left(\frac{t}{T} - \frac{x}{\lambda}\right) + \phi\right]$$

When $t = 0$, the resultant displacement of the string at the point we designate as $x = 0$ is 0.50 cm.

(*a*) What is the value of the phase angle difference ϕ between the two waves?

(*b*) What is the resultant displacement of the string at a distance $x = 1.5$ m along the string at a time $t = 3.00$ s?

D 9 A standing wave is set up on a string by combining two waves traveling in opposite directions with identical amplitudes of 0.015 m and the same wavelength of 0.12 m. What is the amplitude of the standing wave for the following values of x along the string?

(*a*) 0   (*b*) 0.030 m   (*c*) 0.060 m
(*d*) 0.090 m   (*e*) 0.18 m   (*f*) 0.20 m

D10 A transverse wave travels down a string with an amplitude of 0.010 m, a frequency of 120 Hz, and a wavelength of 0.10 m.

(*a*) What is the speed of such a wave?

(*b*) If you are told that when $t = 0$ and $x = 0$, the displacement $y = Y/4$, what is the phase angle ϕ?

(*c*) Write a general expression for such a traveling wave.

(*d*) Find the displacement of the string at a time $t = 5.0$ s at a position $x = 2.5$ m.

D11 (*a*) Use the results of Example 11.6 to write a general expression for the wave discussed there.

(*b*) If, when $t = 0$ and $x = 0$, we find that $y = Y_0/2$, what is the value of the phase angle ϕ?

(*c*) Prove from the general expression for this wave that its speed is 2.0 m/s.

(*d*) Find the value of the displacement of the string at a time $t = 10$ s, at a position $x = 2.0$ m.

Additional Readings

Bascom, Willard: "Ocean Waves," *Scientific American*, vol. 201, no. 2, August 1959, pp. 74–84. Some interesting data on one of the most complicated kind of waves—those on the oceans of the world.

Boorse, D. M.: "Motion of the Ground in Earthquakes," *Scientific American*, vol. 237, no. 6, December 1977, pp. 68–78. Discusses how earthquakes are produced and detected, and the nature of earthquake waves.

Feather, N.: *Vibrations and Waves*, Penquin, London, 1964.

An extended essay which includes some very perceptive comments.

French, A. P.: *Vibrations and Waves*, Norton, New York, 1971. A textbook at a more advanced level mathematically than this one; it contains some very good illustrations and experimental results.

Kock, Winston E.: *Sound Waves and Light Waves*, Doubleday Anchor, Garden City, N.Y., 1965. A very well illustrated account of the similarities between sound and light waves, with particular emphasis on applications to the communications industry.

For this was the first time that I came in closer contact with the world leaders in scientific research in those days—Helmholtz, above all the others. But I learned to know Helmholtz also as a human being, and to respect him as a man no less than I had always respected him as a scientist. For with his entire personality, integrity of convictions and modesty of character, he was the very incarnation of the dignity and probity of science. These traits of character were supplemented by a true human kindness, which touched my heart deeply.

Max Planck (1858–1947)

Chapter 12
Sound Waves

Many of our greatest pleasures—friendly conversation around the dinner table, the rustle of autumn leaves, the boom of the ocean surf, a song by a favorite recording artist, a Beethoven symphony—come to us by sound waves striking our ears. In this chapter we will apply the ideas about waves developed in the preceding chapter to sound waves, to see how they are produced, transmitted, and detected. We will also discuss the miracle of the human ear and its exquisite sensitivity to the sounds which surround us.

12.1 The Nature of Sound Waves

Definition

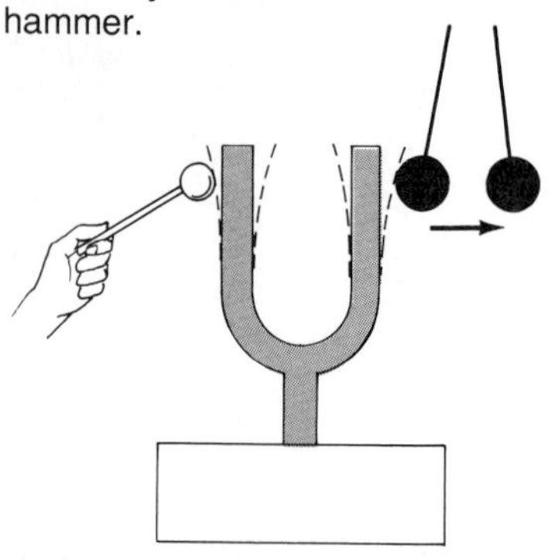

FIGURE 12.1 A tuning fork excited by a blow from a rubber hammer.

Sound wave: A longitudinal wave consisting of periodic variations in the displacement of the particles of the material and in the pressure, parallel to the direction in which the wave is traveling.

Sound waves travel through a physical medium such as air, water, or steel, and the properties of the medium determine the speed of the sound waves. Let us consider what happens when a rubber hammer strikes a tuning fork, as in Fig. 12.1. The prongs of the tuning fork move back and forth in SHM at a frequency determined by the material and shape of the tuning fork. The amplitude of the vibration is very small, but the fact that the fork is moving is clear from its effect on a light cork ball, as in the figure. Suppose we put a small tube open at both ends next to one prong of the tuning fork so that the prong moves parallel to the length of the tube. As the tuning fork prong moves toward the tube, it compresses the air ahead of it, and this compression is passed on from one group of particles* to another along the length of the tube. When the tuning fork prong moves away from the tube, it produces a partial vacuum and reduces the density of the air. This is called a *rarefaction*, and this rarefaction follows the compression down the length of the tube, as in Fig. 12.2a. As the tuning fork continues to vibrate in SHM, it imparts a succession of compressions and rarefactions to the air. These disturbances travel down the tube and make up the sound wave. Just as the tuning fork executes SHM, so too the particles of the air execute SHM and pass on the sound to their neighbors by collisions.

The result is a wave in every way equivalent to the transverse wave of Fig. 11.15 except that, since this is a longitudinal wave, the motion of the air particles is parallel to the direction of motion of the wave instead of at right angles to it.

This makes it a bit more difficult to diagram a compressional wave, even though the physics of its motion is basically the same as that of a transverse wave. In Fig. 12.2b we show as peaks the compressions, in which the particles are pushed closer together and the pressure and density take on their maximum values. The rarefactions, in which the particles are pulled farther apart and the pressure and density are lowered, are shown as troughs. Halfway between a peak and a trough is a point on the wave pattern where the air pressure and density retain their equilibrium values. (Don't be confused by the fact that here again we are plotting changes in pressure along the Y direction even though they actually occur in the X direction. On a graph the significance of the X and Y axes is what we choose to make it, and hence the Y axis can be used to graph changes which actually occur along the horizontal direction in space.)

The wavelength of a sound wave is the distance between any two successive points of maximum compression or any two successive points of

---

*Here the term *particle* is deliberately used to allow for the possibility that these particles are not individual molecules but clusters of molecules. For a gas the periodic motion of these particles is superimposed on the random translational motion of the individual gas molecules.

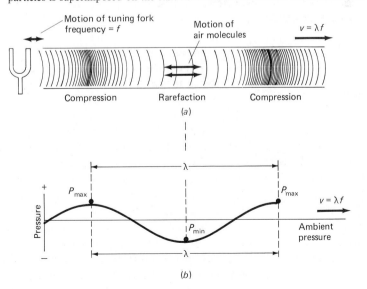

FIGURE 12.2 A sound wave moving through a gas: (a) The compressions and rarefactions caused by the wave moving through the gas. (b) The variations in pressure in the gas as a function of position.

maximum rarefaction in the wave, as shown in Fig. 12.2. The wavelength is related to the speed of the wave and to its frequency by the basic equation for all wave motion, $v = \lambda f$.

As the compressions and rarefactions move down the tube, each group of particles executes SHM along the wave direction. The displacement from its equilibrium position of any particle in the path of the wave can be represented by an equation similar to Eq. (11.23). For the displacement of the particle from its equilibrium position we will use the symbol X (to distinguish it from x, the distance along the horizontal axis), and for the amplitude or maximum value of X we will use the symbol X_{max}. For sound waves the amplitude X_{max} is very small, varying from about 10^{-5} to 10^{-11} m. Our wave equation for a sound wave moving in the positive X direction is then

$$X = X_{max} \cos\left[2\pi\left(\frac{t}{T} - \frac{x}{\lambda}\right)\right]$$ (12.1)

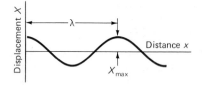

FIGURE 12.3 Graph of the displacement of the particles in a gas against distance for a sound wave passing through the gas.

where, as usual, T is the period of the SHM, λ is the wavelength of the wave, and x is the distance along the X axis. This is shown in Fig. 12.3. A compressional wave moving in the opposite direction would then be represented by the equation

$$X = X_{max} \cos\left[2\pi\left(\frac{t}{T} + \frac{x}{\lambda}\right)\right]$$ (12.2)

In both these cases we have assumed that the phase angle ϕ is zero. Sound waves are produced by just about any form of oscillatory motion—a plucked violin string, a rotary lawn motor, a power saw—since they all impart a periodic motion to the air molecules which carry the sound to our ears. Whether the sound is perceived as pleasant or unpleasant, musical or unmusical, depends on the frequencies and harmonics present, as we shall see.

Propagation of Sound Waves

Sound waves propagate readily through all forms of matter—solids, liquids, and gases alike. In this way they differ from transverse mechanical waves, which cannot pass through gases and liquids since fluids possess no elasticity in a direction perpendicular to the direction of wave propagation. Sound waves cannot propagate through a vacuum, since there are no molecules present to carry the compressional wave. This can easily be seen by putting an alarm clock inside a vacuum chamber, setting it to alarm and then withdrawing the air, as in Fig. 12.4. As the air pressure decreases, the sound grows weaker and weaker and finally becomes inaudible. This need for molecules to carry sound waves is confirmed by the fact that sound propagates better through solids than through gases, for in solids there are more molecules per unit volume to carry the sound.

On the moon, which has no atmosphere, the astronauts could hear no sound except that communicated through the solid surface of the moon. For the same reason sounds appear louder under water than they do in air, since liquids conduct sound better because of their greater density. This same fact explains why American Indians put their ears to the ground to hear the sounds of another's approach, since sounds transmitted through the ground are stronger than those transmitted through air. This technique has the added

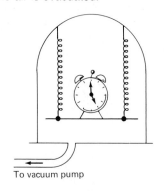

FIGURE 12.4 An alarm clock ringing in a chamber from which the air is evacuated.

To vacuum pump

advantage that the sounds of intruders reached their ears more quickly through the ground than through the air. For, as we shall see, sound travels more rapidly through solids than through gases.

Example 12.1

Compressional waves have a speed of about 340 m/s in air, and audible sound has a frequency range from about 20 to 20,000 Hz.

(a) What is the wavelength of a sound wave of frequency 10^4 Hz?

SOLUTION

(a) Since for any wave $\lambda = v/f$, here we have

$$\lambda = \frac{340 \text{ m/s}}{10^4 \text{ s}^{-1}} = 0.034 \text{ m} = \boxed{3.4 \text{ cm}}$$

It is a worthwhile exercise to compare this wavelength with the average distance between the molecules in a gas at STP.

(b) Our general equation for the motion of the particles in the path of a sound wave is Eq. (12.1):

$$X = X_{max} \cos \left[2\pi \left(\frac{t}{T} - \frac{x}{\lambda} \right) \right]$$

(b) Write an equation describing the longitudinal motion of the particles of the air through which this sound passes, as a function of time and distance along the positive X axis, which is the direction in which the sound is moving.

Here $T = \dfrac{1}{f} = \dfrac{1}{10^4 \text{ s}^{-1}} = 10^{-4} \text{ s}$ and $\lambda = 0.034$ m

Hence $X = X_{max} \cos \left[2\pi \left(\dfrac{t}{10^{-4}} - \dfrac{x}{0.034} \right) \right]$

$$= \boxed{X_{max} \cos \left[2\pi \left(10^4 t - 29x \right) \right]}$$

where times are in seconds and distances in meters.

Again we have assumed here that the phase angle ϕ is equal to zero.

12.2 The Speed of Sound Waves

In the preceding chapter we saw that the speed of a transverse wave on a string is $v = (F/\mu)^{1/2}$. Here μ is the mass per unit length, an *inertial* factor, and F is the tension in the string, an *elastic* factor, since this tension determines how tightly coupled one molecule of the string is to its neighbors and hence how elastically the string rebounds when it is displaced at right angles to its length. We might, for these reasons, expect that the speed of sound waves would depend in the same way on the elastic and inertial properties of the medium through which the sound passes. This indeed turns out to be the case, but we will not take the time to prove it here (see Prob. D1).

If the medium is a fluid, i.e., a liquid or a gas, the most important inertial property is the density of the undisturbed fluid d_0, for this is an inertial property which is independent of the volume of the fluid. Similarly, the only elastic property a fluid has is its resistance to compression, and this is measured by its bulk modulus, given by Eq. (10.20) as $B = -P/(\Delta V/V_0)$. Then by analogy with $v = (F/\mu)^{1/2}$ for waves on a string, we might expect that the speed of sound in a fluid would take the form:

$$v = \left(\frac{B}{d_0} \right)^{1/2} \tag{12.3}$$

This indeed is the result of a more careful derivation of the speed of sound in fluids.

Speed of Sound in Fluids

The speed of longitudinal waves in fluids depends directly on the square root of the bulk modulus and inversely on the square root of the equilibrium density of the fluid.

As might be expected, for a solid metal rod struck at one end, we must replace the bulk modulus by the elastic modulus for the length change in the solid, i.e., Young's modulus. Then we obtain in the same way

$$v = \left(\frac{Y}{d_0}\right)^{1/2} \tag{12.4}$$

where Y is Young's modulus.

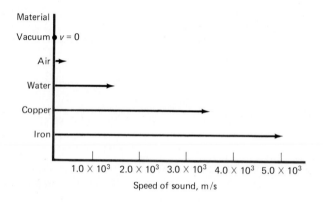

FIGURE 12.5 The speed of sound in a variety of materials.

TABLE 12.1 Elastic and Inertial Properties of Various Materials and the Speed of Sound

Material	Density d_0, kg/m^3	Bulk modulus B, N/m^2	Speed of sound, m/s Calculated value $v = (B/d_0)^{1/2}$	Measured value
		Gases and liquids		
Air (STP)	1.3	1.41×10^5	330	332
Carbon dioxide (STP)	1.98	1.41×10^5	266	258
Hydrogen (STP)	9.0×10^{-2}	1.41×10^5	1.25×10^3	1.27×10^3
Water	1.00×10^3	2.0×10^9	1.41×10^3	1.43×10^3
Mercury	1.35×10^4	2.5×10^{10}	1.36×10^3	1.45×10^3

Material	Density d_0, 10^3 kg/m^3	Young's modulus Y, 10^{10} N/m^2	Speed of sound, m/s Calculated value $v = (Y/d_0)^{1/2}$, $\times 10^3$	Measured value $\times 10^3$
Solids				
Aluminum	2.70	7.0	5.09	5.10
Copper	8.89	11.0	3.52	3.56
Iron	7.20	19.0	5.14	5.13
Nickel	8.90	21.4	4.90	4.97
Tungsten	19.0	36.0	4.35	4.32

Figure 12.5 shows the speed of sound in a variety of materials. We see that sound travels more rapidly in water than in air, and more rapidly in metals than in water, because, as Table 12.1 shows, the elastic properties of solids increase more rapidly than do their densities compared with liquids, and the same is true for liquids compared with gases. The fact that the experimental values for the sound velocity in a great variety of materials agree so well with the values calculated from the simple equations $v = (B/d_0)^{1/2}$ and $v = (Y/d_0)^{1/2}$ gives us confidence in the correctness of these equations.

Dependence on Temperature

The velocity of sound (or of nonaudible longitudinal waves) varies little with temperature for liquids or solids. In gases, however, the dependence is considerable. It can be shown (see Prob. D4) that the speed of sound v_1 at a temperature T_1 is related to the speed of sound v_2 at a temperature T_2 by the equation

$$\frac{v_1}{v_2} = \left(\frac{T_1}{T_2}\right)^{1/2}$$

where both T_1 and T_2 are absolute temperatures, i.e., Celsius temperatures plus 273°.

Hence if the speed of sound in air at STP is 330 m/s, its speed at room temperature (20°C) is $v = (293/273)^{1/2} (330 \text{ m/s}) = 342 \text{ m/s}$.

Example 12.2

Steel has a density 6000 times that of air, and an elasticity about 2×10^6 greater than air. If the speed of sound in air at STP is 330 m/s, what is its speed in steel?

SOLUTION

Since, from Eq. (12.4), $v = (Y/d_0)^{1/2}$, we have

$$\frac{v_{\text{steel}}}{v_{\text{air}}} = \left[\frac{Y_{\text{steel}}/d_{0,\text{steel}}}{Y_{\text{air}}/d_{0,\text{air}}}\right]^{1/2} = \left[\frac{Y_{\text{steel}}d_{0,\text{air}}}{Y_{\text{air}}d_{0,\text{steel}}}\right]^{1/2}$$

$$= \left[(2 \times 10^6)\left(\frac{1}{6 \times 10^3}\right)\right]^{1/2} = [0.33 \times 10^3]^{1/2} = 18$$

Hence

$$v_{\text{steel}} = 18 v_{\text{air}} = 18 \times 330 \text{ m/s} = \boxed{5.9 \times 10^3 \text{ m/s}}$$

This is not too different than the value given for iron in Table 12.1, as we would expect.

Example 12.3

At the beginning of a 100-m dash the starting signal is a gunshot fired by a starter in line with the runners. There are eight lanes, each 1.22 m wide.

(a) What is the time delay between the starting signal heard by the sprinter in lane 1 and by the sprinter in lane 8?
(b) Could this have any effect on the outcome of the race?

SOLUTION

(a) The distance between lane 1 and lane 8 is $7 \times 1.22 \text{ m} = 8.54 \text{ m}$. Since the speed of sound in air at 20°C is about 340 m/s, the time delay between the signals reaching the two runners is

$$\Delta t = \frac{8.54 \text{ m}}{340 \text{ m/s}} = \boxed{0.025 \text{ s}}$$

(b) Since 100 m can be covered in about 10 s by good sprinters, their speed averages 10 m/s. Hence the distance corresponding to a time delay of 0.025 s is:

$$\Delta s = (0.025 \text{ s})(10 \text{ m/s}) = 0.25 \text{ m}$$

or $\quad \Delta s = (0.25 \text{ m})(3.28 \text{ ft/m}) = 0.82 \text{ ft}$

In a very close race this distance of almost a foot could well be decisive.

12.3 Standing Sound Waves; Resonance

As we saw in the preceding chapter, a stretched string has certain fundamental or natural frequencies which depend on the length of the string and the speed of the transverse waves moving along the string. Standing waves can be set up in the string by waves of the same frequency and speed moving in opposite directions. The form of these standing waves was given by Eq. (11.26):

$$y = 2Y_0 \cos \frac{2\pi x}{\lambda} \cos \frac{2\pi t}{T}$$

The same analysis can be applied to two sound waves traveling in opposite directions along the length of an organ pipe or some similar tube. Standing waves again result, in which the displacement of a particle at a position x and at time t can be described by a similar equation:

$$X = 2X_{max} \cos \frac{2\pi x}{\lambda} \cos \frac{2\pi t}{T} \tag{12.5}$$

Here X_{max} is the displacement amplitude of each of the two original waves, i.e., the maximum displacement along the X axis of a particle from its equilibrium position when a sound wave passes through. X is the displacement resulting from the superposition of the two waves traveling in opposite directions. We see that it is periodic in time, but that it has an amplitude which varies with position x along the organ pipe. At some places this amplitude is zero; at other places it takes on its maximum value $2X_{max}$. Hence we have a standing compressional wave in the organ pipe.

It is more difficult to visualize standing waves in this case because the nodes and loops are for motion in the X direction, which is also the direction in which the wave is traveling. For clarity, therefore, we will diagram standing compressional waves in the same fashion in which we diagramed standing transverse waves in Fig. 11.27; i.e., we will plot on the vertical axis displacements which actually take place along the X axis. Thus, in Fig. 12.6 we show the distance along the organ pipe as the X axis. Along the Y axis we plot the displacements of the particles in the X direction. The result is a graph showing points of no motion of the air, or nodes, at intervals of $\lambda/2$ along the pipe, and other points where the displacement has a maximum value of $2X_{max}$. These antinodes, or loops, are halfway between the nodes.

FIGURE 12.6 Standing sound waves. Here N are the nodes and A the antinodes, or loops. Compare the standing wave pattern in this case with that shown for transverse waves in Fig. 11.27.

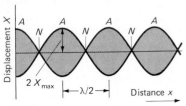

Reflection of Sound Waves

Sound waves are reflected from the end of the organ pipe in the same way that transverse waves on a string are reflected from the end of the string. If the organ pipe is closed at the top end away from the source, then no displacement of the particles is possible at the closed end, and it must be a displacement node. Put a bit differently, this means that, on reflection from a closed end, the reflected wave suffers a phase change of 180°, or π rad. The displacement of the particles at the end is produced, therefore, by two equal forces 180° out of phase. Hence the two forces cancel out and a node results. If the organ pipe is open at the top end, on the other hand, the air moves with maximum displacement there and we have an antinode, or loop. This sets the boundary conditions for the top end of the organ pipe. Since the bottom end is presumed to be where the organ pipe is excited, that end will always be a loop, or antinode.

Resonant Frequencies

An organ pipe, therefore, has a number of natural frequencies which are determined by the length of the pipe and the boundary conditions at the two ends. If a sound is injected into the bottom of the pipe at one of these natural frequencies, a standing wave pattern of large amplitude is set up, because all the reflected waves add together at the antinodes to give a very large amplitude there. This, of course, will result in a loud sound.

The natural frequencies of open and closed organ pipes differ greatly. For an open organ pipe, with displacement antinodes at the two ends, the longest wavelength possible will be $\lambda_1 = 2L$, where L is the length of the pipe (see Fig. 12.7). This corresponds to the fundamental frequency $f_1 = v/\lambda_1 = v/2L$. The second harmonic will have $\lambda_2 = L$, or $f_2 = v/L = 2f_1$. Similarly, the third harmonic has $f_3 = 3f_1 = 3v/2L$, and so forth. In this case of an open organ pipe all harmonics can be present, and the fundamental frequency is $v/2L$. The arrangement of nodes and loops for an open organ pipe vibrating at these frequencies is shown in Fig. 12.7.

An organ pipe closed at one end is quite different. Here we still must have a displacement loop at the open end to enable the pipe to be excited, but there is a node at the closed end. Since nodes and loops are only $\lambda/4$ apart, in this case the fundamental vibration must have a wavelength $\lambda_1 = 4L$, or $f_1 = v/\lambda_1 = v/4L$, which is half the fundamental for an open pipe. The next harmonic which can occur is shown in Fig. 12.8. The length of the pipe in this case must be $\frac{3}{4}\lambda$, or λ_3, and $f_3 = v/\lambda_3 = 3v/4L$. We have designated this frequency by f_3 since it is 3 times the fundamental and hence is the third harmonic. Similarly, the next possible harmonic is $\lambda_5 = 4L/5$, or $f_5 = 5v/4L = 5f_1$, which is the fifth harmonic.

For a closed organ pipe, therefore, the fundamental frequency is $v/4L$, where L is the length of the pipe, and only odd harmonics of this frequency occur. No even harmonics can be produced, since they cannot satisfy the boundary conditions set by the actual physical situation.

FIGURE 12.7 (left) Standing sound waves in an open organ pipe.

FIGURE 12.8 (right) Standing sound waves in an organ pipe closed at the top end.

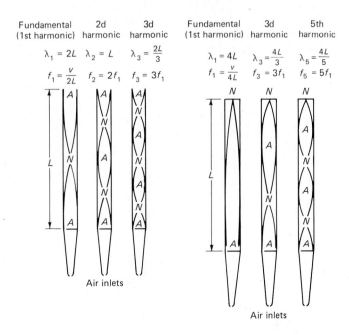

When the organ pipes are excited by an organist, normally not merely the fundamental but many of the lower harmonics of that fundamental are excited. It is clear that two organ pipes of the same length, one open and one closed, will emit very different sounds because both the fundamental note and the harmonic content will be different.

The Stethoscope

A stethoscope (Fig. 12.9) is a device doctors use to listen to the sound produced by the heart and other organs of the human body. It is a simple device which merely conducts the desired sound to the doctor's ears and excludes all other irrelevant sounds. It consists of a body-contact piece connected by hollow rubber tubing to earpieces. Different contact pieces are used depending on the frequency of the sound to be detected; just as different-sized organ pipes resonate at different frequencies. If the contact piece has dimensions close to those needed to resonate at the frequency being picked up, the stethoscope is more sensitive and hence more useful to the doctor. When invented in 1816, the stethoscope inaugurated a new era in diagnostic medicine, since for the first time a doctor could obtain information from inside the human body without using surgery.

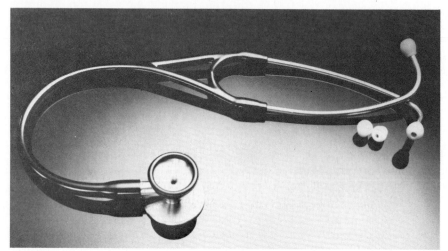

FIGURE 12.9 A modern stethoscope. This 3M Littman stethoscope is intended for heart examinations. Many modern stethoscopes come with interchangeable chestpieces to achieve maximum sensitivity in the examination of different parts of the body. *(Photo courtesy of 3M Medical Products Division.)*

Pressure Nodes and Antinodes

Figure 12.10 is a diagram of a *standing wave* in a closed tube filled with a gas, caused by a sound wave passing down the tube to the right and then being reflected from the closed end and returning to the left. The nodes and antinodes for the displacement of the particles are clearly marked. Associated with the variation in the displacement of the particles as a function of distance in the standing wave are variations in the gas pressure. We now consider these in more detail.

FIGURE 12.10 Nodes (*N*) and antinodes (*A*) for the displacement and pressure in a standing sound wave. Note that pressure nodes occur at the same positions as displacement antinodes, and vice versa.

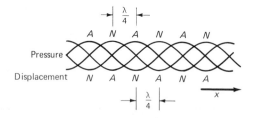

The particles at a displacement node N in the figure are not moving at all (that is what we mean by a displacement node). The particles to the left of any node (N), however, are moving toward N at the same time that the particles to the right of the node are also moving toward N. This convergence of particles leads to an increase in pressure at N. A half cycle later, i.e., $T/2$ s later, the particles near N on its two sides are all moving away from N and thus lowering the pressure. Hence at N, the displacement node, there are large changes in the gas pressure in the standing wave, and N is a pressure antinode, or loop.

At the displacement antinode A, on the other hand, the particles near A on both sides are all moving in the same direction as the particles at A. Hence there is no resultant pressure variation at A, and A is a node for the pressure variations. *The resultant pressure nodes are therefore displaced by a distance of $\lambda/4$ from the displacement nodes*, and the pressure antinodes coincide with the displacement nodes, and vice versa. This is shown in Fig. 12.10. For this reason, if the displacement wave is represented by a cosine curve, the pressure wave is represented by a sine curve.

Example 12.4

The apparatus shown in Fig. 12.11 can be used to obtain the speed of sound in air. A glass tube is filled with water, the level of which can be adjusted by moving the water container up and down. A tuning fork, of fixed frequency 1000 Hz, is set into vibration above the open top of the glass tube. It is found that the sound intensity is large when the water level is at the following distances below the open end of the tube: 8.3, 24.8, and 41.3 cm. Find the speed of sound in air.

SOLUTION

The sound intensity is large when resonance occurs and standing waves are set up in the air column. The standing waves set up must have a displacement node at the air-water interface, and a displacement antinode at the top of the tube, as shown in Fig. 12.12 (in an actual experiment, the antinode occurs slightly above the top of the tube). Hence we have, from the diagram,

$$\frac{\lambda}{4} = 8.3 \text{ cm} \qquad \text{or} \qquad \lambda = 33.2 \text{ cm}$$

$$\frac{3\lambda}{4} = 24.8 \text{ cm} \qquad \text{or} \qquad \lambda = 33.1 \text{ cm}$$

$$\frac{5\lambda}{4} = 41.3 \text{ cm} \qquad \text{or} \qquad \lambda = 33.0 \text{ cm}$$

Average $\lambda = 33.1 \text{ cm} = 0.331 \text{ m}$

Then the speed of sound is

$$v = \lambda f = (0.331 \text{ m})(1000 \text{ s}^{-1}) = \boxed{331 \text{ m/s}}$$

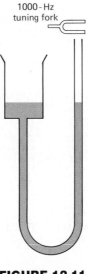

FIGURE 12.11

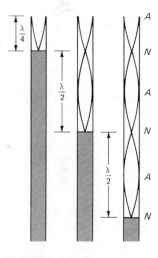

FIGURE 12.12

Example 12.5

A 1.0-m-long aluminum rod is mounted by a clamp at its exact center, and set into an audible longitudinal vibration by rubbing it with resin. One end of the rod is attached to a diaphragm which protrudes into a glass pipe which has been dusted with talcum powder, as in Fig. 12.13. It is found

FIGURE 12.13 Kundt's tube experiment. The vibrating rod produces sound waves in the air of the tube. For the correct tube length, standing waves are produced and a light powder accumulates in mounds at the displacement nodes, which are $\lambda/2$ apart.

that the position of the piston at the other end of the tube can be adjusted so that the talcum powder forms little mounds 6.3 cm apart. What is the measured speed of sound in the air of the tube?

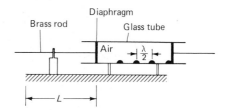

SOLUTION

This is an example of two different standing longitudinal waves, one in the aluminum rod and another in the air in the tube. The frequency of these two standing waves must be the same if resonance is to occur, as it does when the powder is sufficiently agitated to form the little mounds.

For the aluminum rod, which is supported at its center, the fundamental vibration will be one with a node at the center and antinodes at the two ends. Hence the length L of the rod will be $\lambda_R/2$, and so

$$L = 1.0 \text{ m} = \frac{\lambda_R}{2} \quad \text{or} \quad \lambda_R = 2.0 \text{ m}$$

Since no other information about the rod is given except the fact that it is aluminum, it can be assumed that we must obtain the speed of sound in aluminum from the density and elastic modulus of aluminum, which are given in Table 12.1. We then have

$$f = \frac{v}{\lambda_R} = \frac{5.10 \times 10^3 \text{ m/s}}{2.0 \text{ m}} = 2.55 \times 10^3 \text{ s}^{-1} = 2.55 \times 10^3 \text{ Hz}$$

In the glass tube the powder accumulates at the displacement nodes to which it is driven by the agitated molecular motion at the displacement loops. Since these nodes are $\lambda/2$ apart, we have

$$\frac{\lambda}{2} = 6.3 \text{ cm} \quad \text{or} \quad \lambda = 12.6 \text{ cm} = 0.126 \text{ m}$$

The speed of sound in the air of the tube is then

$$v = \lambda f = (0.126 \text{ m})(2.55 \times 10^3 \text{ s}^{-1}) = \boxed{3.2 \times 10^2 \text{ m/s}}$$

12.4 Interference of Sound Waves of Different Frequencies

In the preceding section we discussed the production of standing waves in organ pipes by the superposition of sound waves of the same frequency traveling in opposite directions inside the organ pipe. Here displacement nodes and loops occurred as a function of *distance* along the length of the pipe. It is also possible to produce interference between two waves which results in amplitude variations not as a function of distance but as a function of *time*.

What we need to produce this effect are two sound waves of slightly different frequencies traveling through space in the same direction. An example might be the sound waves produced by two tuning forks, of very close frequencies, both traveling along the same path to a listener's ear. Here we concentrate on the motion of the air particles at some fixed distance x along the path of the two waves. The air is being subjected to the displacements from the two waves at the same time, and its resultant motion as a function of time is just the superposition of these two motions. The displacement of the air at the desired position due to the first wave of frequency f_1 is then

$$X_1 = X_{max} \cos \left(\frac{2\pi t}{T_1} \right) = X_{max} \cos 2\pi f_1 t$$

Similarly, for the second wave, which we assume has the same amplitude X_{max} but a different frequency f_2, we have

$$X_2 = X_{max} \cos 2\pi f_2 t$$

From the principle of superposition the resultant motion of the air particles is then

$$X = X_1 + X_2 = X_{max} (\cos 2\pi f_1 t + \cos 2\pi f_2 t) \tag{12.7}$$

Now, from trigonometry (Appendix 1C), we have

$$\cos (A + B) + \cos (A - B) = 2 \cos A \cos B \tag{12.8}$$

Let $A + B = C$ and $A - B = D$. Then we have, on adding,

$$2A = C + D \quad \text{or} \quad A = \frac{C + D}{2} \tag{12.9a}$$

and, on subtracting, $2B = C - D \quad \text{or} \quad B = \frac{C - D}{2} \tag{12.9b}$

Hence, we have, on making these substitutions in Eq. (12.8) above,

$$\cos C + \cos D = 2 \cos \frac{C + D}{2} \cos \frac{C - D}{2}$$

Then, if we let $C = 2\pi f_1 t$ and $D = 2\pi f_2 t$, Eq. (12.8) becomes

$$X = X_{max} \left[2 \cos \frac{2\pi(f_1 + f_2)t}{2} \cos \frac{2\pi(f_1 - f_2)t}{2} \right]$$

or $\quad X = \left[2X_{max} \cos \frac{2\pi(f_1 - f_2)t}{2} \right] \cos \frac{2\pi(f_1 + f_2)t}{2} \tag{12.10}$

Hence the resultant displacement of the air at the position under consideration occurs at a frequency $(f_1 + f_2)/2$, which is just the average of the two frequencies, but the expression in the brackets is an amplitude factor which varies in time at a frequency $(f_1 - f_2)/2$. This results in what are called *beats*.

Definition

Beats: The regular increase and decrease in the amplitude of sound (or other) waves caused by two waves of slightly different frequencies being superimposed in the same medium.

Although we have derived Eq. (12.10) for only one distance from the source, it should be clear that it holds for every distance from the source. How the amplitude of the resultant wave varies at one position x as a function of *time* is shown in Fig. 12.14. In looking at this figure, keep in mind that beats occur because we are constantly going from a situation where the two waves interfere constructively to one where they interfere destructively at a later time. This leads to the rise and fall in the amplitude of the resultant wave, and hence in the loudness of the sound heard, as time goes on.

FIGURE 12.14 Beats produced by two sound waves of slightly different frequencies at a point in space: (a) Displacements of the two waves of different frequencies traveling in the same direction as a function of time. (b) Resultant displacement obtained by superimposing the displacement of the two traveling waves.

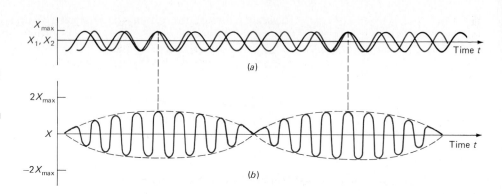

When we speak of a beat, we mean a maximum amplitude for the resulting motion, which results in a large pulse of sound. This occurs whenever in Eq. (12.10) $\cos\left[2\pi(f_1 - f_2)t/2\right]$ equals either 1 or -1. Hence the amplitude is a maximum twice in each cycle, and so the number of beats is $2(f_1 - f_2)/2 = f_1 - f_2$.

Beat Frequency

The beat frequency is equal to the difference in the frequencies of the two sound sources. This gives us a convenient way of tuning musical instruments or radio transmitters to a desired frequency. That frequency is generated by a frequency standard of some sort, and then the instrument or oscillator is tuned to "zero beat" with it, i.e., tuned until the number of beats has been reduced exactly to zero. Then $f_1 - f_2 = 0$, or $f_2 = f_1$.

A piano tuner uses a zero-beat technique to adjust the frequency of a piano string to that of a standard tuning fork. The members of a symphony orchestra tune their instruments by listening for beats between their instruments and the standard tone produced by a violin or oboe at 440 Hz.

The human ear can detect beats only up to about 7 per second. Beyond this, however, various kinds of electronic detectors and frequency counters can be used to measure the beat frequency. This is extremely useful because it enables any unknown frequency to be compared with a known standard by using the beat-frequency method. Since the unknown can be measured in this way to better than ± 1 Hz irrespective of the value of the frequency, it is possible to measure high electromagnetic frequencies to 1 part in 10^{12} or better by using beat-frequency techniques.

Example 12.6

The fourth harmonic of a standard 440-Hz tuning fork is compared with an unknown frequency produced by a plucked string. Three beats are heard per second when the two are sounded at the same time.
(a) What are the possible frequencies of the plucked string?

(b) If the frequency of the tuning fork is lowered by sticking a bit of wax on the vibrating prongs, the beat frequency increases. What is the actual frequency of the plucked string?

SOLUTION

(a) The frequency of the fourth harmonic of the tuning fork is 4×440 Hz = 1760 Hz. If 3 beats are heard when this is compared with the unknown, the two frequencies differ by 3 Hz. Hence the frequency of the unknown is either

1757 or 1763 Hz

(b) If the frequency of the tuning fork is lowered, we are told that the beat frequency increases. Hence the frequency of the tuning fork must have initially been *lower* than that of the plucked string, and so the frequency of the plucked string must be

If it were 1757 Hz, attaching wax to the tuning fork would lower the tuning fork frequency and decrease the beat frequency.

1763 Hz

12.5 The Detection of Sound

Since sound waves in air consist of small variations in the displacement of the air particles and in the pressure of the air, any device which is sensitive to these changes can be used as a sound detector. In a microphone, for example, a thin diaphragm is set into vibration by the incident sound wave. The motion of the diaphragm is then converted into an electric signal by a *transducer*, which converts mechanical impulses into electric impulses. These electric impulses are then carried over telephone lines to a telephone receiver, or transmitted through space as electromagnetic waves to radio and television sets.

Carbon microphones contain granules in the space behind the diaphragm exposed to the sound. The changes in air pressure in the sound wave cause slight changes in the density and hence in the electric resistance of the carbon. These in turn cause changes in an electric current flowing through the carbon granules. In a telephone this current is passed through a type of transformer called a *repeating coil* and sent over wires to a telephone receiver. The receiver then converts the current variations back into motion of a diaphragm and hence into sound waves. A rough sketch of a carbon microphone and telephone is shown in Fig. 12.15.

A condenser microphone is a more sensitive type which converts sound into electric impulses by a process depending on the behavior of an electric condenser (capacitor) in an electric circuit.

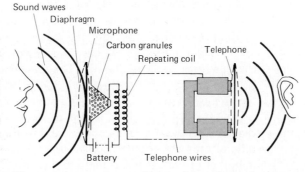

FIGURE 12.15 A carbon microphone and telephone system.

The Human Ear

The ear is an extremely sensitive detector of sound, as good as the most sensitive microphones made. Thus a displacement of the eardrum by only 10^{-11} m (smaller than the average diameter of an atom, which is about 2×10^{-10} m) can give rise to a sound which the normal human ear can hear.

Figure 12.16 is a drawing of the human ear, which performs the same function as a microphone, but in a much more complicated and delicate fashion. The eardrum, which divides the outer ear from the middle ear, is set into vibration by an incident sound wave. The ear then amplifies this motion mechanically and converts it into electric signals which proceed to the brain.

The outer ear contains the ear canal, which is about 2.7 cm long and ends at the eardrum. If we consider this canal to be a closed pipe vibrating in its fundamental mode, we have, from Fig. 12.8, $\lambda_1 = 4L = 4(0.027 \text{ m}) = 0.11 \text{ m}$. For sound waves with a speed of 330 m/s in air, this corresponds to a frequency $f = v/\lambda = (330 \text{ m/s})/(0.11 \text{ m}) \simeq 3000 \text{ Hz}$. Hence the ear is most sensitive to frequencies around 3000 Hz because at that frequency resonance occurring in the outer ear increases the pressure fluctuations at the eardrum. This resonance is rather broad, and so the ear is most sensitive over a range of a few thousand hertz on either side of 3000 Hz.

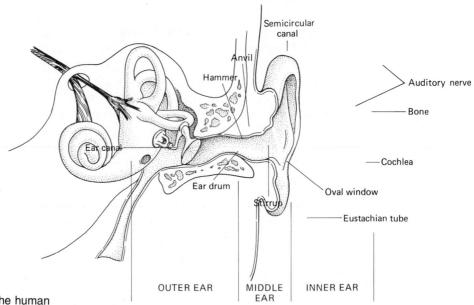

FIGURE 12.16 The human ear.

The middle ear transmits the sound from the eardrum to the oval window which separates the middle ear from the inner ear. The sound is transmitted by three tiny bones (ossicles), the hammer, anvil, and stirrup, which act as a lever system to multiply in the transmission process the force applied to the eardrum by a factor of about 2. The *pressure* at the oval window is increased by a factor of about 40 with respect to that on the eardrum, however, because the area of the oval window is 20 times smaller than the area of the eardrum. From Eq. (9.4) $P = F/A$, and here the force is transmitted unchanged (or slightly increased) by the ossicles, and so the pressure varies inversely as the ratio of the two areas. Since the area of the oval window is 20 times smaller than that of the eardrum, the pressure increases by the same factor.

The inner ear contains the semicircular canals, which are important for maintaining balance, and the cochlea, a snail-shaped cavity filled with a liquid called lymph. The stirrup presses on the oval window and sets up sound vibrations in the lymph. Thirty thousand sensory cells located on the partition which divides the cochlea into two parts detect the sound vibrations in the lymph and pass on the signal to the auditory nerve fibers, which carry them to the brain where they are perceived as sounds.

Because this process is rather complex and incompletely understood, we will not dwell further on it here.

Defective hearing can be due to a breakdown at any point in the ear's complicated sound-transmission system. Hearing aids can sometimes help those hard of hearing. These are basically combinations of a microphone and an amplifier which boost the sound level at the eardrum and thus compensate for defects in the ear structure itself.

12.6 Intensity and Loudness

By the intensity of a sound wave (or any other kind of wave) we mean the following:

Definition

Intensity of a wave: The amount of energy transported per second through unit area normal to the direction of the wave.

$$\text{Intensity } I = \frac{\mathcal{W}}{tA} \tag{12.11}$$

Since *power* is energy per unit time, the dimensions of intensity are those of *power per unit area*, and the units of intensity are watts per square meter (W/m²). (Note that $\mathcal{W}$ signifies *energy* and W signifies *watts*.) For example, if the sound generated by a rock group causes 1 J of sound energy per second to pass through a 1-m² area between them and their audience, the sound intensity would be 1 W/m².

We have seen that for the SHM of a mass at the end of a spring the total mechanical energy of the system is $\mathcal{W} = \frac{1}{2}kX_0^2$, where X_0 is the amplitude in the equation of motion $x = X_0 \cos 2\pi ft$. In a sound wave each particle of the medium oscillates back and forth in SHM in a way described by the equation

$$X = X_{\max} \cos 2\pi ft \tag{12.12}$$

where $f = (1/2\pi)\sqrt{k/m}$ and the displacement X is parallel to the direction of the wave. Since $k = 4\pi^2 f^2 m$, the mechanical energy possessed by each vibrating particle is $\mathcal{W}_1 = \frac{1}{2}kX_{\max}^2 = 2\pi^2 f^2 m X_{\max}^2$. As the sound wave moves through the material, it imparts this amount of energy to each particle it passes. The total energy imparted per second to all the particles in a volume V is then

$$\frac{\mathcal{W}}{t} = \frac{\mathcal{W}_1 Vn}{t} = \frac{\mathcal{W}_1 ALn}{t}$$

where n is the number of particles per unit volume, and the volume V is equal to AL, where L is the length of the volume. But L/t is the speed of the wave along the X axis, and so

$$\frac{\mathcal{W}}{t} = \mathcal{W}_1 Avn \tag{12.13}$$

Hence the total energy imparted to the particles in a volume of unit cross-sectional area per second is:

$$I = \frac{\mathcal{W}}{tA} = \frac{\mathcal{W}_1 Avn}{A} = 2\pi^2 f^2 X_{\max}^2 mvn \tag{12.14}$$

We therefore see that:

The intensity of a wave is proportional to the square of the amplitude of the vibrating particles in the medium and to the square of the frequency.

The intensity also depends on the density of the material, since n is the number of particles per unit volume.

The presence of the area A in the expression for the intensity [Eq. (12.11)] means that the intensity falls off with distance from the source, unless the wave is a plane wave carrying energy only in one direction, as in Fig. 12.17. If a sound is radiated in all directions from a point source such as an exploding shell, however, its intensity falls off as $1/A$, or $1/(4\pi R^2)$, where $4\pi R^2$ is the area of the surface of a sphere of radius R. Hence as R increases, the intensity falls off as $1/R^2$, as shown in Fig. 12.18. Here the same amount of energy ($\mathcal{W}$) passes through the areas A_1 and A_2 per second. Since A_2 is larger than A_1, and $I = \mathcal{W}/tA$, the intensity at A_2 is *lower* than at A_1 by a factor A_1/A_2, and hence by a factor of R_1^2/R_2^2.

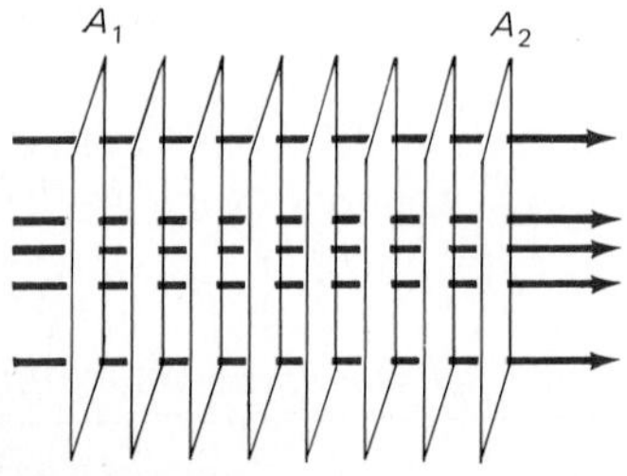

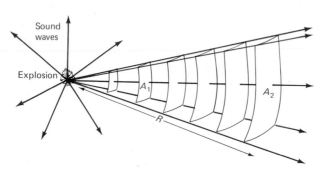

FIGURE 12.17 A plane sound wave. The intensity remains constant with distance, and the cross-sectional area of the wave is the same for all distances.

FIGURE 12.8 A spherical sound wave. The intensity falls off as $1/R^2$, since the area through which the wave passes increases as R^2.

Loudness

The intensity of a sound wave is an objective property which can be measured quantitatively by instruments. The loudness of a sound, on the other hand, is a *subjective property* which depends on the nature and condition of the ear doing the hearing. The frequency of a sound will, for example, change the perceived loudness, even for the same intensity. A dog can hear an intense sound of 20,000 Hz, even though most human beings would hear nothing at that frequency. Also, the loudness of a sound may vary markedly from individual to individual depending on age, sharpness of hearing, amount of wax in the ears, and other subjective factors. Hence loudness is a less quantitative and more subjective characteristic than is sound intensity, although it is possible to make the general statement that the more intense a sound of a given frequency, the louder it will appear to any listener.

Decibel Scale

An individual listener does not perceive a sound B, which has 10 times the intensity of A, to be 10 times as loud as A. Rather, the human ear, like other

TABLE 12.2   Intensity Levels of Some Common Sounds

Sound	Intensity,* W/m^2	Intensity level β (compared to $I_0 = 10^{-12}$ W/m^2), dB
Threshold of hearing	10^{-12}	0
Rustle of leaves	10^{-11}	10
Whisper	10^{-10}	20
Radio playing softly	10^{-8}	40
Ordinary conversation	10^{-6}	60
Busy traffic	10^{-5}	70
Screaming child	10^{-3}	90
Jackhammer	10^{-2}	100
Motorcycle	10^{-1}	110
Drums of symphony orchestra at climax of *1812 Overture*	1	120
Loud indoor rock concert	1	120
Threshold of pain	10	130
Jet engine	10^2	140
Cap pistol	10^4	160

*At a place very near to the sound source.

sense organs, has a logarithmic response to changes in intensity.* By this we mean that if B has 10 times the intensity of A, and C has 100 times the intensity of A, then the ear perceives C to be as much louder than B as B is than A, because the ratios 100/10 and 10/1 are equal.

When B has 10 times the intensity of A, the ratio of their intensities is 10, and the logarithm of 10 is 1. The difference in sound intensity between A and B is then said to be 1 bel, named in honor of Alexander Graham Bell (1847–1922), the inventor of the telephone. Similarly, if C is 10 times more intense than B, it is 1 bel more intense than B, and therefore 2 bels more intense than A. A sound 1000 times more intense than A has an intensity 3 bels above that of A, and so forth. This reduces multiplicative factors of 10 to additive factors of 1, and creates a *logarithmic scale*. This kind of scale is useful because it corresponds to the actual way the human ear responds to sound, i.e., logarithmically.

The bel is too large a unit to be convenient. Hence another unit, one-tenth as large and called the *decibel* (dB), has been introduced. We therefore define the *intensity level* of a sound, designated by the Greek letter β (*beta*, the equivalent of an English b), as follows:

Intensity Level

$$\beta \text{ (in dB)} = 10 \log \frac{I}{I_0} \qquad (12.15)$$

The factor 10 arises simply because of our converting from bels to decibels. I is the intensity of the sound for which β is being calculated; and I_0 is some reference intensity, usually taken as the minimum audible intensity, or *threshold of hearing*, which is $I_0 = 1.0 \times 10^{-12}$ W/m^2.

---

*This is an important psychological law, known as the *Weber-Fechner law*, named in honor of Ernst Weber (1795–1878), who discovered the law, and Gustav Fechner (1801–1887), who popularized its use.

For example, the intensity of a jet plane just leaving the runway, which is about 100 W/m², has an intensity level

$$\beta = 10 \log \frac{I}{I_0} = 10 \log \frac{100 \text{ W/m}^2}{1.0 \times 10^{-12} \text{ W/m}^2}$$

$$= 10 \log 10^{14} = 10(14) = 140 \text{ dB}$$

Every increase in intensity by a factor of 10 therefore corresponds to an increase of 10 dB. Hence a sound of 100 W/m² is 140 dB, or 10^{14} times as intense, as the threshold of hearing.

Some intensities and intensity levels for commonly heard sounds are given in Table 12.2.

Frequency Response of the Human Ear

Figure 12.19 shows how the threshold of hearing of the human ear varies with frequency. It is most sensitive to frequencies between about 2000 and 5000 Hz, and has almost no response below 30 Hz or above 16,000 Hz. Sound waves are often defined as compressional waves with frequencies between 20 and 20,000 Hz, but there are few, if any, ears which can hear this full range of frequencies. As a matter of fact, for many people expensive stereo systems are a bad investment, since they are not able to hear the very low and very high frequencies in the "full-frequency response" of the electronic components. Figure 12.20 shows the frequency response of a variety of recording systems.

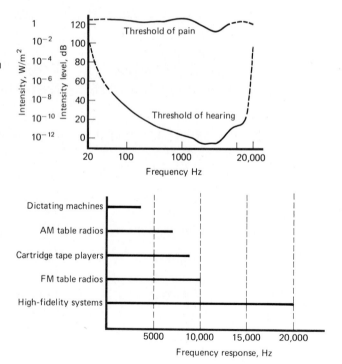

FIGURE 12.19 Response of human ear to sound at various frequencies. The intensity is given in both watts per square meter and decibels. The upper curve indicates the intensity which produces pain in the ear; the lower curve the minimum audible intensity at various frequencies. Note that the frequency scale is logarithmic.

FIGURE 12.20 Frequency response of various recording systems.

The *threshold of pain* for the human ear occurs at about 130 dB, and varies very little with frequency, as is clear from Fig. 12.19. There is a slight dip in the curve at the frequencies where the ear is most sensitive, as might be expected. Clearly the 140-dB sound of a jet plane 30 m above the ground is about 10 times greater than the threshold of pain for the ear, and continued exposure to such intense sounds can damage the ear.

Figure 12.19 points up clearly the difference between intensity and loudness. A 40-dB sound would appear rather loud to a listener if its frequency were 3000 Hz, but would be barely audible at 80 Hz.

Example 12.7

An amplified steel guitar has an intensity level 5 dB higher than the same unamplified sound. How much more intense is the amplified sound than the unamplified sound?

SOLUTION

From Eq. (12.15) we have

$$\beta = 10 \log \frac{I_2}{I_1} = 5 \text{ dB}$$

where I_2 is the intensity of the amplified sound and I_1 that of the unamplified sound. Therefore

$$\log \frac{I_2}{I_1} = \frac{5}{10} = 0.5$$

We can then find from a hand calculator, or from tables, that

$$\log 3.2 = 0.5$$

Hence $\qquad \log \frac{I_2}{I_1} = 0.5 = \log 3.2$

from which $\qquad \dfrac{I_2}{I_1} = 3.2$

or $\qquad I_2 = 3.2 I_1$

Hence the amplified sound is $\boxed{3.2 \text{ times}}$ more intense than the unamplified sound.

Example 12.8

A hi-fi set's power output is 50 W at 1000 Hz. The specifications for the set state that this output falls off by 8 dB at 35 Hz. What is the power output at 35 Hz?

SOLUTION

From Eq. (12.11) the sound intensity is equal to $I = \mathcal{W}/tA$, where $\mathcal{W}/t$ is the power P. Hence the power output is directly proportional to the intensity of the sound, and we can write

$$\frac{P_2}{P_1} = \frac{I_2}{I_1}$$

(Note that P here means *power*, not *pressure*.) But the intensity level is, from Eq. (12.15),

$$\beta = 10 \log \frac{I_2}{I_1} = 10 \log \frac{P_2}{P_1}$$

Now $\beta = 8$ dB and $P_2 = 50$ W, as stated in the problem, with P_1 to be determined. Then we have

$$8 \text{ dB} = 10 \log \frac{50 \text{ W}}{P_1}$$

or $\qquad 0.8 = \log \dfrac{50 \text{ W}}{P_1}$

From a hand calculator we find that $\log 6.3 = 0.8$, and so

$$\log 6.3 = \log \frac{50 \text{ W}}{P_1} = 0.8$$

Hence, taking antilogs,

$$6.3 = \frac{50 \text{ W}}{P_1}$$

or $\qquad P_1 = \dfrac{50 \text{ W}}{6.3} = \boxed{7.9 \text{ W}}$

Hence the fall-off in power of 8 dB reduces the power by a factor of $(50 \text{ W})/(7.9 \text{ W}) = 6.3$.

12.7 Quality of Sound and Pitch

Frequency is the fundamental physical property which indicates the number of oscillations of a string or air column per second. From a physiological point of view the corresponding attribute of sound is called the *pitch*.

Definition

Pitch: The perceived physiological response to the frequency of a sound wave.

Frequency and pitch are closely related but are not identical, since the loudness of a note of a particular frequency may change its perceived frequency, or its pitch, as heard by a listener. Thus a loud foghorn sounds lower in pitch to a hearer than a sound of the same frequency but of less intensity.

Musical instruments make use of a variety of vibrating parts to produce sound. Many instruments like the violin, piano, cello, and guitar use vibrating strings for this purpose. The length of the string, its linear density, and its tension can then be chosen to produce the desired fundamental note, as discussed previously. Other instruments like flutes, French horns, trumpets, and pipe organs set up vibrating air columns inside the instruments, and vary the length of these air columns and the method of excitation to select the desired note. How the particular fundamental is excited also determines the number and relative intensity of the higher harmonics generated.

Sounds produced by musical instruments are often referred to by subjective terms like *reedy, round, mellow*. The *quality* of sound, or *timbre*, is determined by the number of harmonics present, their relative intensities, and how quickly they decay or die out. The timbre of an instrument is determined not merely by the size of the resonant cavity in the instrument, but also by the shape of the instrument and the material from which it is made. A sound "spectrum" for a piano is shown in Fig. 12.21. Note how all the harmonics are integral multiples of one fundamental frequency. The relative intensities of the harmonics depend on the loudness of the note struck

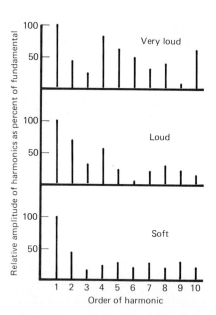

FIGURE 12.21 Sound spectra of a piano for a fundamental at 256 Hz emitted at different intensities: very loud, loud, and soft.

Hermann von Helmholtz (1821–1894)

FIGURE 12.22   Hermann von Helmholtz. *(Photo courtesy of AIP Niels Bohr Library.)*

Hermann von Helmholtz was one of the most versatile scientists who ever lived. He was born in Potsdam, Germany, in 1821, the son of a high school teacher. From an early age he wanted to be a physicist but his family could not afford the money required for his studies. Instead, his father persuaded him to take up medicine, since he could receive a free medical education on condition that he serve as a doctor in the Prussian army after receiving his medical degree. This he did, studying at the Institute for Medicine and Surgery in Berlin from 1838 to 1842, and fulfilling his obligation as an army surgeon from 1843 to 1848. Helmholtz's real interest is clear when we read that even in the army barracks he insisted on setting up a small laboratory for research in physiology and physics.

In 1847 Helmholtz published a mathematical formulation of the principle of conservation of energy which caused quite a stir in scientific circles. On the basis of it and his other researches he was appointed professor of physiology at the University of Königsberg. There he devoted himself to the physiology of the eye, explaining the mechanism of lens accommodation in the eye, and in 1851 he invented the ophthalmoscope (see Sec. 24.8). This instrument, still the basic tool used by eye doctors to peer into the eye's interior, immediately made Helmholtz famous. In 1852 he also became the first experimentalist to measure the speed of nerve impulses in the human body. He was particularly happy about this, because one of his former professors, an eminent biologist, had said that it could not be done.

In 1858 he was appointed to a chair in anatomy at Heidelberg, and while there devoted most of his time to studies in optics and sound. In 1867 he published his famous *Treatise on Physiological Optics* in three volumes. His days in Heidelberg, from 1858 to 1871, enabled him to work with other great scientists like the physicist Kirchhoff (see Sec. 17.7) and the chemist Bunsen in what has been called an era of brillance "such as has seldom existed for any university and will not readily be seen again."

In 1871 Helmholtz abandoned physiology in favor of physics. He accepted the chair in Physics at the University of Berlin, and spent the rest of his life there.

As he grew older Helmholtz became more and more interested in the mathematical side of physics, and made noteworthy contributions to fluid mechanics, thermodynamics, and electrodynamics. He devoted the last decade of his life to an attempt to unify all of physics under one fundamental principle, the principle of least action. This attempt, while evidence of Helmholtz's lifelong interest in philosophy, was no more successful than Einstein's quest for a unified field theory. Helmholtz died in 1894 as the result of a fall suffered on board ship while on his way back to Germany from the United States.

It is difficult to exaggerate the influence Helmholtz had on nineteenth-century science, not merely in Germany but throughout the world. It was during his lifetime that Germany gained its preeminence in science, which it was not to lose until the Second World War. His own research contributions, together with the impetus he gave to other scientists and would-be scientists by his teaching, research administration, and popular lectures, had much to do with the scientific Renaissance which Germany experienced during his lifetime. At that time Helmholtz was, next to Bismarck and the old emperor, the most illustrious man in the German empire.

Helmholtz was a sensitive and sickly man all his life, plagued by severe migraine headaches and fainting spells. He sought relief from his pain in music and the arts, and in mountain climbing in the Alps. It is intriguing to imagine what he might have accomplished if he had been in good health for all his 73 years! On Helmholtz's death Lord Kelvin summed up his accomplishments: "In the historical record of science the name of Helmholtz stands unique in grandeur, as a master and leader in mathematics, and in biology, and in physics."

on the piano, and hence on the force applied to the fundamental string, as the figure shows. When the fundamental is very loud, the fourth harmonic is almost as loud as the fundamental. We can show by a simple experiment that harmonics are present in any note struck on the piano. If we release the dampers on the strings of the piano, and then strike one key, we will find that a number of other strings will also vibrate in addition to the one struck. These are strings corresponding to harmonics of the note struck. The struck note must therefore contain these frequencies to an extent sufficient to excite the other strings by resonance.

Analysis of Musical Sounds

The Greeks were the first scientists to understand something of the nature of music. After the Greeks the greatest progress in this field was made by Hermann von Helmholtz (see Fig. 12.22 and accompanying biography), who devoted much of his career to a study of sound and light, and particularly of their physiological effects. He designed a set of spherical cavities which resonated at certain fixed frequencies determined by their dimensions. By playing a complex musical sound at a frequency whose harmonics correspond to the resonant frequencies of these cavities, he was able to set them into oscillation and so determine the harmonic content of the musical note. Today this can be done much more easily by using oscilloscopes and electronic filters to filter out all harmonics but the one of interest. That one can then be observed on an oscilloscope screen (similar to a TV screen), and its amplitude measured. Since the intensity of the harmonic is proportional to the square of the amplitude, the percentage of the sound made up by this particular harmonic can be determined. In this way sound spectra like those in Fig. 12.21 can be plotted.

Helmholtz also succeeded in synthesizing musical sounds by combining harmonics in different proportions. He set up electrically driven tuning forks of the proper frequency and amplitude to duplicate any desired complex sound. The modern electronic music synthesizers are improved versions of Helmholtz's device.

12.8 The Doppler Effect in Sound

We have all probably observed that if a fire engine is approaching our street the frequency of its siren appears higher than when the same engine is leaving our neighborhood. This is an example of how the frequency of a sound changes depending on the relative motion of the source of the sound and the hearer of the sound. This effect has been named the *Doppler effect*, after Christian Doppler (1803–1853), the Austrian physicist who first explained it in 1842.

Source at Rest; Observer in Motion

Suppose a tuning fork at rest emits a pure tone of frequency f. This sound moves through the air in all directions with a speed v determined by the properties of the air, and a wavelength determined by the fundamental equation $v = \lambda f$. If, now, an observer moves toward the tuning fork at a speed v_0, as in Fig. 12.23, he or she will hear a higher frequency, as the following argument will show.

We assume that the air is at rest with respect to the tuning fork. The number of complete waves passing the observer per second if he or she were

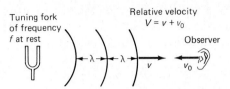

FIGURE 12.23 The Doppler effect: Observer in motion toward a sound source. The frequency of the observed sound is increased.

at rest would be $f = v/\lambda$. If the observer is in motion at a speed v_0 toward the source of sound, the observer's relative speed with respect to the sound wave becomes $V = v + v_0$. The wavelength is not affected by the observer's motion, but the frequency heard is. The apparent frequency observed is

$$f' = \frac{V}{\lambda} = \frac{v + v_0}{\lambda} = f + \frac{v_0}{\lambda} = f\left(1 + \frac{v_0}{f\lambda}\right) = f\left(1 + \frac{v_0}{v}\right)$$

Hence the frequency heard by the observer is not f but a higher frequency

$$f' = f\left(1 + \frac{v_0}{v}\right) \tag{12.16}$$

which depends on the ratio of the observer's speed (v_0) to the speed of sound (v). Since the velocity of sound in air is about 340 m/s, the observer would have to travel at a speed of 3.4 m/s to increase the frequency heard by 1 percent.

If the observer is moving away from the sound source, we have

$$f' = \frac{v - v_0}{\lambda} = f\left(1 - \frac{v_0}{f\lambda}\right) = f\left(1 - \frac{v_0}{v}\right) \tag{12.17}$$

In this case the frequency heard would be lower than f.

In the case of a moving observer, then, the shift in frequency is due to the fact that the number of waves reaching the observer's ear per second changes, and hence so does the *frequency* that is heard.

Source in Motion; Observer at Rest

This is a very different case from the one just discussed. Here the motion of the source with respect to the air changes the *wavelength* of the sound wave in the air. This sound wave still travels through the air at a speed v determined by the physical properties of the air. When it reaches the observer, however, a different frequency is heard since the wavelength has been changed.

Consider, as in Fig. 12.24, a tuning fork initially at S_1 emitting sound at a frequency f and moving through the air at a velocity v_s toward an observer who is at rest. In this case the motion of the source pushes together the waves it emits in the direction in which it is moving, as shown in the figure. In the periodic time $T = 1/f$ the source moves a distance $\Delta\lambda = v_s/f$, while the sound wave travels a distance λ, where $\lambda = v/f$. Hence the actual wavelength produced in the air by the moving source is no longer λ, but

$$\lambda' = \lambda - \Delta\lambda = \frac{v}{f} - \frac{v_s}{f} = \frac{v - v_s}{f}$$

This sound wave then reaches the observer at its original velocity v with respect to the air, since he or she is at rest with respect to the air. The frequency heard is therefore

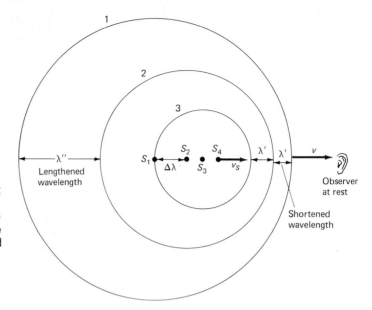

FIGURE 12.24 The Doppler effect: Source in motion toward observer. Wavefront 1 is emitted when the source is at S_1, wavefront 2 when the source is at S_2, and so forth. This leads to a decrease of the wavelength in the forward direction, and an increase in the wavelength in the backward direction. The result is an increase in the frequency heard by the observer when the sound source is approaching.

$$f' = \frac{v}{\lambda'} = \frac{v}{(v - v_s)/f} = \frac{1}{1 - v_s/v} f \qquad (12.18)$$

Hence the frequency that is heard is again increased by the motion of the source with respect to the observer. If the source moves away from the observer, the wavelength is increased as shown by λ'' in the figure, and we can show in the same way that the new (lower) frequency is:

$$f' = \frac{1}{1 + v_s/v} f \qquad (12.19)$$

The results of Eqs. (12.16) to (12.19) can be combined into the following single equation:

Doppler Effect Equation

$$f' = \frac{1 \pm v_0/v}{1 \mp v_s/v} f \qquad (12.20)$$

where the upper signs in numerator and denominator apply to motion of source and observer toward each other, and the lower signs apply to motion away from each other. In the case where both source and observer are moving with respect to the air, Eq. (12.20) can also be conveniently used.

Notice that, even though relative motion of source and observer toward each other always raises the frequency, and relative motion away from each other always lowers the frequency, the actual change differs depending on whether the tuning fork or the observer does the moving. This is because $1 + v_0/v$ is not exactly equal to $1/(1 - v_s/v)$, even when $v_0 = v_s$.

Example 12.9

Show that the observed frequency of a tuning fork sounding a note of 500 Hz is different in the following two cases:

(a) The tuning fork moves away from a stationary observer at a speed of 34 m/s.

(b) The observer moves away from a stationary tuning fork at a speed of 34 m/s.

(c) What would happen if the speeds were 340 m/s in the two cases? Assume that the speed of sound in air at 20°C is 340 m/s.

SOLUTION

(a) Here, from Eq. (12.19) we have

$$f' = \frac{1}{1 + v_s/v} f = \frac{1}{1 + (34 \text{ m/s})/(340 \text{ m/s})} f = \frac{500 \text{ Hz}}{1.10}$$

$$= \boxed{455 \text{ Hz}}$$

(b) From Eq. (12.17) we have

$$f' = \left(1 - \frac{v_0}{v}\right)f = \left(1 - \frac{34 \text{ m/s}}{340 \text{ m/s}}\right)(500 \text{ Hz})$$

$$= (1 - 0.10)(500 \text{ Hz}) = \boxed{450 \text{ Hz}}$$

Hence there is a difference of 5 Hz in the two observed frequencies. This is very slight (about 1 percent), but the difference becomes much more significant as v_0 or v_s increases.

(c) If $v' = 340$ m/s, we have, in case (a),

$$f' = \frac{1}{1 + v_s/v} f = \frac{1}{1 + (340 \text{ m/s})/(340 \text{ m/s})}(500 \text{ Hz})$$

$$= \boxed{250 \text{ Hz}}$$

On the other hand, in case (b) we would have

$$f' = \left(1 - \frac{340 \text{ m/s}}{340 \text{ m/s}}\right)(500 \text{ Hz}) = \boxed{0}$$

Hence when the tuning fork is moving away from the observer at the speed of sound, the sound frequency which eventually reaches the stationary observer is cut in half. When the observer is moving away at the speed of sound, however, no sound at all is heard, since the speed of the observer is the same as that of the sound wave. Hence no changes in the air's density or pressure strike the observer's ears, and no sound is heard.

12.9 Supersonic and Ultrasonic Waves

Although the terms *supersonic* and *ultrasonic* might at first glance appear to have similar meanings, technically they have quite different significances. Supersonic is used for *speeds* greater than the speed of sound; ultrasonic refers to *frequencies* of sound waves which are higher than those of audible sound, i.e., frequencies above 20,000 Hz.

Supersonic Speeds

The elastic and inertial properties of the molecules in the air determine the speed of sound waves in air. The faster the air molecules rebound after being compressed, the faster the sound wave is propagated. The speed with which air molecules can rebound from a flying aircraft also determines whether they are able to get out of the way of the plane. As the plane approaches the speed of sound, it is increasingly more difficult for the air molecules to get out of its way, since the plane is going almost as fast as the speed at which the molecules can rebound. At the speed of sound the plane piles up all the air molecules in front of it into one large compressional crest, since they cannot move fast enough to escape. This creates what was once called a "sound barrier," since it was believed that any airplane would be destroyed by turbulence if it tried to fly through the layer of molecules it had piled up in front of itself.

Finally on October 24, 1947, a piloted plane broke through the sound barrier by traveling at a speed greater than the speed of sound in the surrounding air. This is referred to as "exceeding Mach 1," where by Mach 1 is meant flight at the speed of sound, i.e., about 340 m/s near the earth's surface. Similarly, Mach 2 means twice the speed of sound, or about 680 m/s. Some supersonic jets now fly routinely at speeds greater than Mach 3.

Definition

Mach number: The ratio of the speed of a flying object (v_P) to the speed of sound (v_S) at the flight altitude.

$$\text{Mach number} = \frac{v_P}{v_S} \tag{12.21}$$

When a plane is flying at supersonic speeds (i.e., greater than Mach 1) it compresses the air ahead of it and sets up a sharp dividing line between this region of strong compression and the surrounding air. This compression region extends in a cone behind the plane, as shown in Fig. 12.25, and is called a shock wave.* The angle of the shock wave cone depends on the Mach number. From Fig. 12.25 it can be seen that the cone angle θ is an angle in the right triangle $S_1 S_4 R$ and has a sine given by

$$\sin \theta = \frac{v_S t}{v_P t} = \frac{v_S}{v_P} = \frac{1}{\text{Mach number}} \tag{12.22}$$

Hence the higher the Mach number, the smaller the angle of the cone produced by the shock wave.

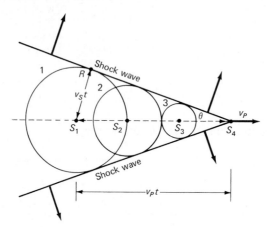

FIGURE 12.25 Shock wave generated by a plane in supersonic flight at approximately Mach 3.

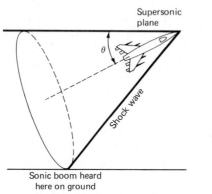

FIGURE 12.26 Sonic boom connected with a shock wave.

When a plane passes overhead traveling at supersonic speeds, the piled-up compressions of the air spread out from the shock wave produced by the plane. The shock wave gradually changes into an ordinary sound wave as it moves away from the plane. When this wave reaches the ground, as in Fig. 12.26, an observer hears a *sonic boom*. This boom is heard after the plane has

---

*In the case of light the analogue of the shock wave produced in sound is called *Cerenkov radiation* and is emitted by charged particles which move through a medium at speeds greater than the speed of light in that medium.

FIGURE 12.27 An SST (supersonic transport). *(Photo courtesy of British Airways.)*

passed overhead, since the plane is traveling more rapidly than are the sound waves reaching the ears of the observer. A sonic boom is therefore due to the constructive interference of a large number of compressional pulses at the ground which can shatter windows and create noise levels harmful to human beings and animals. Much of the destructive power of nuclear or conventional explosives comes from the shock waves they generate.

Sonic booms associated with the SST (see Fig. 12.27), the supersonic transport which Air France and British Airways fly across the Atlantic to the United States, have led to the banning of the supersonic flight by SST's over landmasses in the United States. Hence the SSTs fly at supersonic speeds only over the oceans.

Ultrasonic Frequencies

As noted before, the maximum frequency range for the human ear is roughly from 20 to 20,000 Hz. Compressional waves of frequencies below and above this range can be produced and detected by a variety of electronic techniques. "Sound" waves with frequencies below 20 Hz are called *infrasonic* waves (i.e., below the audible range). They are produced by earthquakes, thunderstorms, and very heavy vibrating machinery. Sound waves with frequencies above 20,000 Hz are called *ultrasonic* waves (i.e., above the audible range), and are of great importance both from a theoretical and practical point of view.

Ultrasonic waves can be generated by using quartz crystals as sources. Such crystals are *piezoelectric*, which means that if an alternating electric voltage is applied to the crystal, it will vibrate mechanically and generate sound waves at the same frequency as the applied voltage. In this way sound waves of very short wavelengths can be generated. For example, at a frequency of 100 kHz, or 10^5 Hz, the wavelength is only about 3 mm for sound waves in air, and at 1 GHz, or 10^9 Hz, it is only 3×10^{-4} mm. Ultrasonic waves have been produced at frequencies as high as 600 GHz, at which frequency the wavelength is only 5×10^{-10} m, or approximately the size of an atom.

Normal audible sound waves have such long wavelengths that they simply bend around small objects without much reflection occurring, in the same way water waves flow around the support pillars of a pier. But if the

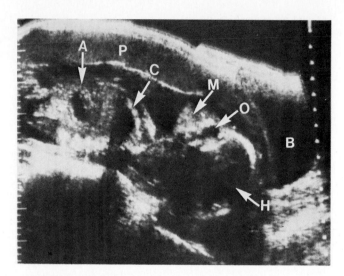

FIGURE 12.28 Ultrasonic photograph of a child in the mother's womb. The letters on the photo have the following significance: H=head; O=eye socket; M=mouth; C=chest; A=abdomen; P=placenta; B=maternal bladder. *(Photo courtesy of Steven A. Mervis, Division of Ultrasound and Radiologic Imaging, Thomas Jefferson University Hospital, Philadelphia, Pennsylvania.)*

wavelength is short enough, sound waves behave more like light waves and are reflected from objects larger than their wavelengths. For this reason ultrasonic waves can be used to find flaws or imperfections in metals, to locate tumors in the human body, and to detect intruders in homes or suspicious articles in baggage. These are examples of nondestructive testing using ultrasound. In the form of what is now called *sonar* (*so*und *na*vigation *a*nd *r*anging) ultrasonic waves can also be used underwater to locate submarines or schools of fish.

Because they are much less dangerous to living tissue than x-rays, ultrasonics have gradually replaced x-rays for many medical applications. Ultrasonics can even be used to look at the development of a fetus without risk to the mother or the child-to-be (see Fig. 12.28).

Example 12.10

A plane is flying at supersonic speed through air at a height where the speed of sound is 320 m/s. The shock wave generated has cone angle of 20°.

(a) What is the Mach number?
(b) How fast is the plane going?

SOLUTION

(a) From Eq. (12.22)

$$\text{Mach number} = \frac{1}{\sin \theta} = \frac{1}{\sin 20°} = \frac{1}{0.34} = \boxed{2.9}$$

Hence the speed of the plane is 2.9 times the speed of sound.

(b) Since the speed of sound is given as 320 m/s, the plane's speed is

$$2.9 \times 320 \text{ m/s} = \boxed{936 \text{ m/s}}$$

or about 2000 mi/h.

Summary: Important Definitions and Equations

Sound wave:

A longitudinal wave consisting of periodic variations in the displacement of the particles of the material, and in the pressure, along the direction in which the wave is traveling.

Equation for the displacement of particles in a sound wave moving along the positive X axis:

$$X = X_{max} \cos \left[2\pi \left(\frac{t}{T} - \frac{x}{\lambda} \right) \right]$$

Speed of sound in gases or liquids: $v = \sqrt{\dfrac{B}{d_0}}$

Speed of sound in solids: $v = \sqrt{\dfrac{Y}{d_0}}$

Standing sound waves: $X = 2X_{\max} \cos \dfrac{2\pi x}{\lambda} \cos \dfrac{2\pi t}{T}$

Reflection of sound waves:
  1 At closed end of a pipe: Phase change of π rad, or 180°
  2 At open end of a pipe: No phase change

Resonant frequencies of organ pipes:
  1 Open: $f_1 = \dfrac{v}{2L}$; $f_2 = 2f_1$; $f_3 = 3f_1$; . . . (All harmonics present)

  2 Closed: $f_1 = \dfrac{v}{4L}$; $f_3 = 3f_1$; $f_5 = 5f_1$; . . . (No even harmonics)

Pressure variations in a sound wave moving in positive X direction:

$$P = P_{\max} \sin\left[2\pi\left(\dfrac{t}{T} - \dfrac{x}{\lambda}\right)\right]$$

Interference of two waves of different frequencies traveling in the same direction:

$$X = \left[2X_{\max} \cos \dfrac{2\pi(f_1 - f_2)t}{2}\right] \cos \dfrac{2\pi(f_1 + f_2)t}{2}$$

Beats:
  The regular increase and decrease in sound intensity caused by two waves of slightly different frequencies being superimposed on the same medium. The beat frequency is the difference in the frequencies of the two sound sources which produce the waves.

Intensity of a sound wave:
  The amount of energy transported per second through a 1-m² area normal to the direction of the wave:

$$I = \dfrac{\mathcal{W}}{tA} \propto P_{\max}^2 \propto X_{\max}^2$$

  The intensity of a wave is also proportional to the square of the frequency.

Loudness:
  The perceived strength of a sound.

  Intensity level of a sound: β (in dB) $= 10 \log \dfrac{I}{I_0}$

Decibel (db):
  A unit on a logarithmic scale used to measure sound intensities. Every increase in intensity by a factor of 10 corresponds to an increase of 10 dB (or 1 bel).
Threshold of hearing:
  $I_0 = 1.0 \times 10^{-12}$ W/m² (weakest sound that can be heard by human ear).
Frequency response of human ear:
  From 30 to 16,000 Hz (often quoted as 20 to 20,000 hz).
Pitch:
  The perceived physiological response to the frequency of a sound wave.
Physical and physiological properties of sounds:

Physical property	Corresponding physiological property
Intensity	Loudness
Frequency	Pitch
Harmonic content	Quality, timbre

Doppler effect for sound:

$$f' = \dfrac{1 \pm v_0/v}{1 \mp v_s/v} f$$

Supersonic speeds:
  Velocities greater than the speed of sound in the medium.
  Mach number $= \dfrac{v_P}{v_S}$

Shock wave:
  An intense compressional pulse produced, for example, by planes flying at supersonic speeds.

  Angle of shock wave cone: $\sin\theta = \dfrac{v_S}{v_P} = \dfrac{1}{\text{Mach number}}$

Ultrasonic waves:
  Sound waves with frequencies above 20,000 Hz.

Questions

1 Why are transverse mechanical waves unable to pass through gases and liquids? What is lacking which is essential for the propagation of such transverse waves?

2 Can you suggest a way that waves could be used to test whether the core of the earth (i.e., its deep interior) is a liquid or a solid?

3 Can you suggest how a sound wave could be used as a crude vacuum gauge to determine the gas pressure within a closed tube?

4 Why are steel wires rather than copper wires used for stringed musical instruments?

5 (a) In a traveling sound wave are there any particles which are always at rest as the wave passes through?

(b) Are there any particles which are sometimes at rest? Indicate which particles, and for how long they remain at rest.

6 (a) In a standing sound wave are there any particles which are always at rest?

(b) Where are these particles located?

7 Equation (12.5) describes the displacement of particles in a standing sound wave in a glass tube filled with air:

$$X = 2X_{max} \cos \frac{2\pi x}{\lambda} \cos \frac{2\pi t}{T}$$

(a) On the basis of this equation describe in words the motion of a particle at the fixed position $x = x_1$ as a function of time.

(b) In the same way describe, at a fixed instant of time $t = t_1$, the displacements of the particles as x varies along the length of the tube.

8 Workers inside caissons, i.e., watertight boxes inside which construction workers do work on tunnels and bridges under water, are observed to have high-pitched voices. The air pressure in these caissons must be higher than atmospheric pressure in order to keep water out. Can you explain why this leads to the raising of the pitch of human voices?

9 Perhaps you have experienced the fact that some people have a harder time than others in carrying an ice-cube tray filled almost to the top with water without spilling the water. If we presume that all the people involved exercise equal care to avoid spilling the water, what might be a reasonable physical explanation of the different results?

10 Would it be correct to say that the intensity of a sound wave is proportional to the square of the amplitude of the displacement variations in the wave? Why?

Problems

(Unless otherwise stated, use 340 m/s for the speed of sound in all problems.)

A 1 The speed of sound in metals exceeds its speed in air by about:
(a) One order of magnitude (b) Two orders of magnitude (c) Three orders of magnitude (d) Four orders of magnitude (e) They are about equal

A 2 The wavelength of a 5000-Hz sound wave in steel is about:
(a) 1 μm (b) 1 mm (c) 1 cm (d) 1 m (e) 1 km

A 3 A sound wave travels down a piece of solid copper tubing which is supported only at its center, and is reflected at the end. The phase change on reflection for the sound wave is:
(a) 0 (b) $\pi/4$ (c) $3\pi/2$ (d) $\pi/2$ (e) None of the above

A 4 The ratio of the frequency of the third harmonic of a closed organ pipe to the second harmonic of an open organ pipe of the same length is:
(a) 2/3 (b) 3/2 (c) 4/3 (d) 3/4 (e) 2/1

A 5 A standing sound wave is set up in a closed organ pipe at a frequency of 3400 Hz. At the position $x_1 = 0.050$ m there is a displacement node. A pressure node occurs at:
(a) 0.10 m (b) 0.075 m (c) 0.20 m (d) 0.15 m (e) 0.25m

A 6 A sound is produced by the explosion of a firecracker. Very near the explosion the sound intensity is I_1. The sound intensity at a distance 10 times farther away from the firecracker is:
(a) $I_1/10$ (b) $I_1/2$ (c) $I_1/100$ (d) I_1 (e) 10 I_1

A 7 The intensity of a 50-dB sound exceeds that of a 10-dB sound by a factor of:
(a) 40 (b) 10^5 (c) 5 (d) 10^4 (e) 4

A 8 An airplane is approaching a ship at sea at a speed of 25 m/s. The ship sounds a warning whistle at a frequency of 10.0×10^3 Hz. If we assume that the speed of sound along the plane's path is 320 m/s, the frequency heard by the pilot of the airplane is:
(a) 4.1×10^3 Hz (b) 10.0×10^3 Hz (c) 15.8×10^3 Hz (d) 17.0×10^3 Hz (e) None of the above; the frequency will be too high to be heard

A 9 A fire engine passes an intersection at a speed of 20 m/s, while sounding its siren at a frequency of 5.0×10^3 Hz. After the fire engine has passed, a person in a car stopped at the intersection hears a frequency of:
(a) 5.0×10^3 Hz (b) 4.7×10^3 Hz (c) 5.3×10^3 Hz (d) 2.9×10^2 Hz (e) None of the above

A10 A plane is flying at Mach 2.5 through air at a height above the ground where the speed of sound is 300 m/s. The speed of the plane is:
(a) 120 m/s (b) 300 m/s (c) 250 m/s (d) 750 m/s (e) None of the above

B 1 A sound wave of displacement amplitude 10^{-12} m has a wavelength of 6.8 cm and a frequency of 5000 Hz.
(a) What is the speed of this sound wave?
(b) Write an equation describing the longitudinal motion of a molecule of the air through which the wave travels (presume that the wave is moving along the positive X axis).

B 2 The density of silver is 10.5×10^3 kg/m^3, and Young's modulus for silver is 7.75×10^{10} N/m^2. What is the speed of sound in silver?

B 3 What is the frequency of the third harmonic of a closed organ pipe 2.4 m long?

B 4 A nickel rod 1.6 m long is mounted at a point

0.40 m from one end. It is then set into longitudinal vibration by rubbing it with resin. At what frequency will the rod vibrate?

B 5   Two violins are badly out of tune. When one plays a note of 440 Hz, a beat note of 3 Hz is heard between the two violins. What are the possible frequencies of the second violin?

B 6   The sound emitted by a small radio carries 10^{-12} J of energy per second through an area of 1.0 cm^2 at right angles to the direction of the sound wave. What is the intensity of the sound?

B 7   What is the absolute intensity of a sound which is 80 dB above the threshold of hearing?

B 8   A ship is approaching a dock at a speed of 6.8 m/s. The ship sounds its warning whistle at a frequency of 10.0×10^3 Hz. What is the frequency heard by a person on the dock?

B 9   A plane is flying at supersonic speed through air at a height where the speed of sound is 310 m/s. The shock wave generated by the plane makes a cone angle of 35°. What is the Mach number corresponding to the plane's speed?

B10   A plane is flying at Mach 0.85. Its engines produce a whining sound at 3000 Hz. What frequency will be heard by an observer in the tower of the airport as the plane approaches?

C 1   (*a*)   A sound wave in water has a speed of 1.4×10^3 m/s. What must be the frequency of this sound wave for it to have a wavelength of 30 cm in water?

(*b*)   Write an equation describing the longitudinal motion of a particle of the water as a function of time and distance, as the sound wave passes through the water (assume that the wave is moving in the positive X direction).

C 2   The speed of sound in tin is half the speed of sound in aluminum, and the density of tin is 7.2×10^3 kg/m^3. Use these data and Table 12.1 to find Young's modulus for tin.

C 3   The speed of sound in zinc is about 3700 m/s. If the density of zinc is 7.1×10^3 kg/m^3, what is Young's modulus for zinc?

C 4   Two identical sound sources A and B of frequency 440 Hz and equal amplitude are located at distances of exactly 25 m to either side of a microphone M, as in Fig. 12.29. The two sources are turned on at the same time and are in perfect phase.

(*a*)   Describe what happens as the sound source at B is moved toward M. In particular, at what distances from M is the sound most intense, and at what distances from M is the sound least intense?

(*b*)   Is there any place where there will be complete cancellation of the sounds received by the microphone?

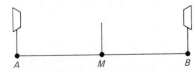

FIGURE 12.29   Diagram for Prob. C4.

C 5   (*a*)   An open organ pipe of length 1.6 m is emitting a sound which includes a frequency of 425 Hz. To what harmonic does this frequency correspond?

(*b*)   If the air in the organ pipe is replaced with carbon dioxide, what will be the frequency of this same harmonic?

C 6   An open organ pipe emits a fifth harmonic at 700 Hz.

(*a*)   What must be the length of a closed organ pipe which will have this same frequency as a fifth harmonic?

(*b*)   If the closed organ pipe has exactly the same length as the open organ pipe, what harmonic of the closed pipe will correspond in frequency to the fifth harmonic of the open pipe?

C 7   The lowest frequency that can be heard by the human ear is about 20 Hz. What is the smallest size room in which standing sound waves at this frequency can be set up?

C 8   A glass tube is filled with water whose height can be adjusted, as in Fig. 12.11. The tube above the water is kept filled with helium (He) gas. A tuning fork, of fixed frequency 1760 Hz, is set into vibration above the open top of the tube. It is found that the sound intensity is large when the water level is the following distances from the open end of the tube: 13.7, 41.5, and 69.2 cm. Find the speed of sound in He.

C 9   A copper rod of length 1.6 m is mounted in a support clamp at a position 0.40 m from one end of the rod.

(*a*)   At what frequency will such a rod vibrate?

(*b*)   If this rod is used to agitate the air in a Kundt's tube like that in Fig. 12.13, how far apart will be the mounds of talcum powder in the tube?

C10   A Kundt's tube (Fig. 12.13) is filled with air and set into vibration by a diaphragm at one end. It is found that the small mounds of powder are 15 cm apart on the average. The air is then replaced by an unknown gas, and it is found that the distance between the mounds now averages 22 cm.

(*a*)   What is the speed of sound in the unknown gas?

(*b*)   Is this gas more or less dense than air?

C11   Two pure notes are sounded, one at 440 Hz and the other at 446 Hz.

(*a*)   What is the frequency of the resultant note which is heard?

(*b*)   At what rate does the amplitude of this note change in time?

C12   The frequency of a rotating mirror is measured by beating the whistling frequency it produces when it rotates against an audio-frequency oscillator generating a fixed frequency of 100 Hz. It is found that 5 beats are produced between these two sources, and that if the frequency of the rotating mirror is slightly increased, the number of beats is reduced. What is the frequency of the rotating mirror?

C13   A noise-level meter indicates that the sound intensity level in rush-hour traffic is 75 dB. What is the corresponding sound intensity?

C14   (*a*)   A baby is crying loudly and the sound reaches an intensity level of 10^{-3} W/m^2. How much energy passes in 1 min through an area 1 m on a side near where the baby is crying?

(*b*)   How high could this amount of energy lift a 1.0-g mass?

C15 A woman is adjusting the balance of the speakers in her stereo set and decides that she must double the sound output from one speaker. By how many decibels must she increase the intensity level of that speaker?

C16 An audio amplifier is guaranteed to have a gain of 25 dB. How much can such an amplifier increase the intensity of a sound?

C17 One speaker of a stereo set is putting out sound with an intensity of 5.0×10^{-8} W/m². The second speaker is putting out sound of intensity 8.0×10^{-9} W/m². What is the intensity level of the first sound with respect to the second?

C18 The intensity level 40 m from a loudspeaker is 90 dB. What is the power output in watts of the speaker?

C19 Jennifer and Gregory each have tuning forks which emit a pure tone at 440 Hz. If Gregory stands still and Jennifer runs toward him while holding her tuning fork, how fast must Jennifer run if Gregory is to hear 2 beats per second between the two tuning forks?

C20 A firefighter is riding on the back of a fire engine moving at 25 m/s. The truck's siren, which is at the front, is sounding a note of 3000 Hz.

(*a*) What is the wavelength of the sound wave in the air near the middle of the fire engine?

(*b*) What frequency does the firefighter hear?

C21 A sound source emits waves at a frequency of 5000 Hz in air.

(*a*) Find the wavelength observed by a listener at rest with respect to the source.

(*b*) Find the wavelength a listener observes if the source moves with a speed of 34 m/s toward the listener.

(*c*) What frequency does the listener hear?

(*d*) Find the wavelength observed by a listener moving toward the stationary source at a speed of 34 m/s.

(*e*) What frequency is heard in this case?

C22 A police car moving at 36 m/s is chasing a speeding motorist who is traveling at 30 m/s. The police car has a siren going at 3000 Hz. What frequency does the driver of the speeding car hear?

C23 A physics teacher is demonstrating the Doppler effect in class by taking a tuning fork of frequency 440 Hz and moving it very rapidly toward a blackboard at the front of the classroom, thus producing beats between the sound produced by the moving tuning fork and that reflected from the blackboard. If the students are to hear a beat frequency of 2 Hz, how rapidly must the teacher move the tuning fork?

C24 Two trains are moving along in opposite directions on parallel tracks and are approaching a train station. The trains are each sounding their horns at 500 Hz. One train is moving at 35 m/s. What is the speed of the other train if a passenger on the platform hears a beat frequency of 4 Hz?

C25 The flight of a plane is being simulated in a wind tunnel by blowing air at very high speeds past the stationary model of an airplane. How fast would the air have to travel for a shock wave of cone angle 45° to develop?

C26 A plane is flying at Mach 0.90 and carries a sound source that emits a loud 2000-Hz signal. After the plane passes an observer on the ground, what frequency does the observer hear?

C27 An SST is flying at Mach 1 and is producing a vibration at a frequency of 1000 Hz. What frequency would be heard by an observer almost directly in the path of the plane?

D 1 Use dimensional analysis to show that the dependence of the speed of sound on the density and bulk modulus of a fluid must take the form: $v \propto (B/d_0)^{1/2}$, where d_0 is the density of the undisturbed fluid and B is the bulk modulus $[B = -P/(\Delta V/V_0)]$.

D 2 What stress would be required in a stretched nickel wire if the speed of transverse waves is to be 1/50 the speed of longitudinal waves in the same wire? (*Hint:* First solve algebraically to cancel out unknown quantities.)

D 3 A child drops a rock down into a well. The sound of the rock hitting water is heard by the child 3.2 s after the rock is dropped. How deep is the well?

D 4 Make use of the assumption that the speed of sound in a gas is proportional to the rms speed of the gas molecules to prove that the speed of sound v_1 at a temperature T_1 is related to the speed of sound v_2 at a temperature T_2 by

$$\frac{v_1}{v_2} = \left(\frac{T_1}{T_2}\right)^{1/2}$$

D 5 Make an order-of-magnitude estimate of the highest-frequency "sound wave" that can possibly be produced in a piece of solid material.

D 6 Two small loudspeakers are facing each other at the two ends of a 5.0-m bench. Loudspeaker A emits a sound of pressure amplitude twice that of loudspeaker B, where both sounds are at 1000 Hz and are in phase at the speakers.

(*a*) At what point on the line joining the two speakers is the intensity a minimum?

(*b*) Write an expression for the intensity of the sound at that point.

(*c*) At what other points along the line joining the two speakers will destructive interference occur?

D 7 Two small loudspeakers A and B face each other at the end of a 4.0-m laboratory bench. A small microphone is located at C, a point 1.0 m from A on the line joining A and B. A variable-frequency audio oscillator feeds the two loudspeakers and they produce sounds of the same frequency, amplitude, and phase. At what frequencies will minimum sound be heard at C?

D 8 There is one intensity level for a sound which has the property that, if it is quadrupled (i.e., increased by a factor of 4), the intensity of the sound is also quadrupled. What is the intensity level?

D 9 Derive an expression for the frequency heard by a listener when both sound source and listener are in motion toward each other through the air along a straight line connecting source to observer.

D10 Derive an expression for the Doppler effect for the two cases of moving source and moving observer in the situation where the air is also moving with a speed of u m/s with respect to the ground.

Additional Readings

Benade, Arthur H.: *Horns, Strings and Harmony*, Doubleday Anchor, Garden City, N.Y., 1960. An interesting account of the behavior of the human ear, of vibrating systems, and of musical instruments.

Devey, Gilbert B., and Peter N. T. Wells: "Ultrasound in Medical Diagnosis," *Scientific American*, vol. 238, no. 5, May 1978, pp. 98–112. An excellent article on the use of ultrasound for medical diagnosis, containing many ultrasonic pictures of human organs, and even of triplets in their mother's womb. Included is an interesting account of how the Doppler effect can be used with ultrasonic waves in medical diagnosis.

Fletcher, Neville H., and Suzanne Thwaites: "The Physics of Organ Pipes," *Scientific American*, vol. 242, no. 1, January 1983, pp. 94–103. An interesting article on how organ pipes produce their majestic sounds.

Hudspeth, A. J.: "The Hair Cells of the Inner Ear," *Scientific American*, vol. 248, no. 1, January 1983, pp. 54–64. An account, including scanning electron micrographs, of the exquisitely sensitive transducers which enable us to hear.

Jeans, Sir James: *Science and Music*, Cambridge University Press, New York, 1953. An old book, but still a classic in the field, based in great part on Helmholtz's original work in this field.

Kahl, Russell (ed.): *Selected Writings of Hermann von Helmholtz*, Wesleyan University Press, Middletown, Conn., 1971. Probably the two most accessible short accounts of Helmholtz's life are Kahl's introduction to this collection of Helmholtz's writings (pp. xii–xiv) and Helmholtz's own autobiographical sketch (pp. 466–478).

Miller, Dayton C.: *The Science of Musical Sounds*, 2d ed., Macmillan, New York, 1926. Another classic in this field. It is much more experimentally based than is Jeans' book.

The Physics of Music (readings from *Scientific American*, with an introduction by Carleen Maley Hutchins), Freeman, San Francisco, 1978. An excellent collection of articles by experts in the field of sound and acoustics.

Pierce, John R.: *The Science of Musical Sound*, Freeman, San Francisco, 1983. A *Scientific American* book on acoustics for the nonspecialist, written from the viewpoint of contemporary electronic and computer-generated music.

Rigden, John S.: *Physics and the Sound of Music*, Wiley, New York, 1977. A more up-to-date version of this interesting subject than can be found in the works of Jeans and Miller.

Taylor, Charles: *Sounds of Music*, British Broadcasting Corporation, London, 1976. A fascinating popular account based on the Christmas lectures at the Royal Institution in London in 1971.

Van Bergeijk, Willem A., John R. Pierce, and Edward E. David, Jr.: *Waves and the Ear*, Doubleday Anchor, Garden City, N.Y., 1960. A discussion of the subjective and objective aspects of sound and hearing.

*A*nd, *in reasoning on this subject, we must not forget to consider that more remarkable circumstance, that the source of Heat generated by friction, in these experiments, appeared evidently to be inexhaustible. . . . And it appears to me to be extremely difficult, if not quite impossible, to form any distinct idea of anything capable of being excited and communicated in the manner the Heat was excited and communicated in these experiments, except it be* motion.

Count Rumford (1753–1814)

Chapter 13

Temperature and Heat

In the next three chapters we will be discussing the basic laws governing the conversion of heat into other forms of energy, and the conversion of other forms of energy into heat. These laws are crucial to our own health and comfort, which depend so sensitively on our bodies being at the right temperature. An understanding of the world's energy problems also requires clear ideas about the nature of temperature and heat.

There are two very different approaches to this important subject. The first, the science of *thermodynamics*, deals only with measurable properties of large-scale matter, such as the volume, pressure, and temperature of gases, liquids, and solids. The second approach relates these macroscopic (large-scale) properties to the behavior of the atoms or molecules in the substance. This approach, called *statistical mechanics*, uses the basic quantities of mechanics—such as velocities, kinetic energies, and momenta —but, since the number of molecules is so large, employs statistical techniques to obtain average values for them. One of the greatest triumphs of physics has been its ability to tie together in a meaningful way these two very different approaches to heat phenomena.

13.1 Temperature

We have already seen in Sec. 9.6 a rough definition of temperature as a measure of the hotness or coldness of an object. Hotness and coldness are very relative concepts, however, and we need a much more objective and precise definition of temperature for scientific work. This we found from kinetic theory in Eq. (9.17), where $\frac{1}{2}m\bar{v}^2 = \frac{3}{2}kT$, which indicates:

Kelvin Temperature

The Kelvin temperature T of a gas is directly proportional to the average random translational kinetic energy of the gas molecules.

This is the basic significance of temperature at the atomic or molecular level.

At a fixed temperature T, all gases have exactly the same average kinetic energy, no matter what their masses, pressures, volumes, or other properties.

We now proceed to discuss how to measure temperatures and how to set up useful temperature scales.

13.2 Gas Thermometers

A thermometer is a device used to measure temperature. It consists of a solid, liquid, or gas with some physical property which varies with temperature, as, for example, the volume of a gas at constant pressure, the pressure of a gas at constant volume, the volume of a liquid, the length of a solid rod, or the electrical resistance of a metal. It is very difficult to measure the average kinetic energy of the molecules of a substance directly. Hence we obtain the temperature *indirectly* by first measuring some property of the substance which changes with the temperature, and then using the measured value of that property to obtain the temperature.

Common gases have the great advantage for use in thermometers that they all have the same coefficient of volume expansion, as long as the gas pressure is low. All gases change in volume by 1/273.15 of the volume at 0°C for every Celsius degree (or kelvin) change in temperature. We can write this as

$$\Delta V = \frac{V}{273.15} \Delta T = 0.0036610\ V\ \Delta T \tag{13.1}$$

If, then, we measure the volume of a gas at 0°C and some given pressure, we can determine the change in temperature ΔT at constant pressure in terms of the measured volume change ΔV for the gas. This is the idea used in a constant-pressure gas thermometer.

In practice, constant-pressure gas thermometers are not often used because the gas volume changes are too large for convenient measurement. Instead constant-volume gas thermometers are used, in which the volume is

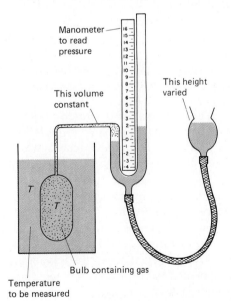

FIGURE 13.1 A constant-volume gas thermometer. By raising and lowering the bulb at the right the volume of the gas in the bulb at the left can be kept constant. The manometer is then used to measure the pressure at constant volume. Since $PV=nRT$, the absolute temperature is proportional to the pressure.

held constant and the pressure measured as a function of temperature. The ideal gas law, $PV = nRT$, predicts that, for V constant, P is directly proportional to the absolute temperature T. Hence the absolute temperature can be obtained from measurements of the pressure.

Even with a constant-volume gas thermometer different gases will give slightly different results for the temperature of an object. But it is found that if the density of the gas is sufficiently low, all gas thermometers will yield exactly the same temperature. For this reason the constant-volume gas thermometer is used to establish the temperature scale employed for scientific work throughout the world.

A drawing of a constant-volume gas thermometer is shown in Fig. 13.1. Such a thermometer is far from convenient to use, since the temperature probe is a large bulb containing gas. For this reason simpler thermometers, such as the common mercury-in-glass thermometer, are more frequently used in routine temperature measurements. They are all ultimately checked (or *calibrated*) against a standard constant-volume gas thermometer.

Absolute Zero

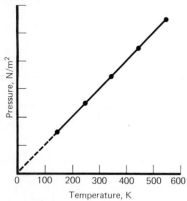

FIGURE 13.2 Graph of pressure against absolute temperature for a constant-volume gas thermometer.

Definition

A gas thermometer gives a precise physical meaning to the zero of temperature. Since the volume of an ideal gas changes by $V/273.15$ for each degree change in the absolute temperature, the change in volume for a decrease in temperature of $273.15°$ from an initial temperature of $0°C$ is:

$$\Delta V = \frac{V}{273.15} \Delta T = -\frac{273.15}{273.15} V = -V$$

and the resulting volume is $V + \Delta V = V - V = 0$. Hence the absolute zero is the temperature at which the volume of the gas would be reduced to zero. In practice, the gas would actually liquefy long before absolute zero was reached.

Something similar happens if we use a constant-volume gas thermometer and gradually reduce the temperature. We obtain the experimental curve for the pressure shown in Fig. 13.2. If the temperature could be lowered far enough, the pressure would also go to zero at $T = 0$. Hence we can define absolute zero as follows:

Absolute zero: That temperature at which the pressure of an ideal gas at constant volume would fall to zero.

This temperature is chosen as the zero on the absolute (or Kelvin) scale of temperature and has an "absolute" significance not shared by the zeros on the Celsius or Fahrenheit scales, which are arbitrarily defined points and represent no limit to the lowest temperature attainable.

The Quest for Absolute Zero

Over the years physicists have devoted a great deal of effort to produce lower and lower temperatures, both to see how close they can come to absolute zero and to investigate the properties of matter at these very low temperatures.

Temperatures as low as 4.2 K are easily attainable, since ordinary helium ($_2^4\mathrm{He}$) can be liquefied at this temperature by using specialized refrigeration equipment. Hence every low-temperature physics laboratory (Fig. 13.3) contains a supply of liquid helium to enable experiments to be performed at

FIGURE 13.3 Physicists at General Electric studying the behavior of electric transmission cables at low temperatures. Here a 10-ft cable sample is being lowered into a bath of liquid nitrogen at 77 K (−320°F) to simulate actual operating conditions. *(Photo by G.E. Research and Development Center; courtesy of U.S. Department of Energy.)*

4.2 K. The temperature can be lowered even further by pumping away the helium vapor which constantly "boils" off the liquid helium. As we noted previously, the evaporation of high-speed molecules from a liquid lowers the temperature of the liquid. In this way temperatures as low as 0.7 K have been produced. Using $^3_2$He, which liquefies at 3 K, even lower temperatures of 0.3 K have been reached. Still lower temperatures can be attained using the interaction of paramagnetic salts with large magnetic fields.

What physicists find in their quest for absolute zero, however, is that the closer they get to absolute zero the more difficult it becomes to go further. They can always reduce the temperature somewhat below the lowest temperature previously attained, but this never brings them to absolute zero, no matter how many times they repeat the process. A mathematician would say that these experiments approach absolute zero "asymptotically," which means that they get very close but never actually arrive at the desired goal. For this reason it is generally accepted as a basic law of nature that zero on the absolute scale is unattainable in a finite number of operations. This is sometimes referred to as *the third law of thermodynamics*.

13.3 Other Kinds of Thermometers

In addition to gas thermometers, which are the ultimate standards for all temperature measurements, there are a variety of other, more convenient types of thermometers.

Mercury-in-Glass Thermometers

Suppose we fill a small glass bulb with mercury metal, which is a liquid at ordinary temperatures, and attach to the bulb a long, thin, hollow tube with a uniform bore, as in Fig. 13.4. If we heat the mercury in the bulb with a flame, the mercury will gradually come to the flame temperature. Then the mercury will expand on heating and rise up the glass tube (usually called a *capillary tube*). The glass will also expand, but this effect will be small, since the

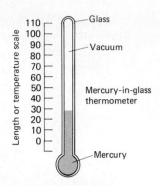

FIGURE 13.4 A mercury-in-glass thermometer.

coefficient of volume expansion for Pyrex glass is only about one-twentieth that for mercury. If we presuppose that the volume changes linearly with the temperature, we have:

$$\Delta V = k\ \Delta T \qquad \text{or} \qquad \pi R^2\ \Delta L = k\ \Delta T$$

where R is the radius of the capillary tube and ΔL is the change of the height of the mercury column for a change in temperature ΔT. Since ΔT is proportional to ΔL, the temperature can be determined from a measurement of the height of the mercury column.

A mercury-in-glass thermometer with an extremely narrow bore can be a very accurate and sensitive temperature-measuring instrument in the range from somewhat below 0°C to a few hundred degrees Celsius. Its range is limited by the fact that mercury becomes a solid at −39°C and a vapor at 357°C.

Resistance Thermometers

Resistance thermometers consist of a fine, high-resistance wire enclosed in a thin-walled tube for protection. Since the electrical resistance of the wire depends on the temperature (as we will see in Sec. 17.4), it can be used to measure that temperature. Electrical resistance can be measured very precisely, and a resistance thermometer is therefore a very useful instrument for the measurement of temperature.

Thermocouples

A thermocouple is an electrical device which converts temperature differences into electric voltages. It consists of two pieces of different kinds of metal, say, one of pure platinum and the other of a platinum-rhodium alloy. These wires are joined together to form a complete loop. If one junction (the place where the two wires are joined) is kept in an ice-water bath at 0°C, and the other junction is placed in contact with an object whose temperature is desired, a voltage will be produced between the two junctions. This voltage is proportional to the temperature difference between the junctions. Hence a voltage measurement will yield the unknown temperature.

Example 13.1

The volume of mercury increases by 18×10^{-5} times its original volume for each 1°C increase in temperature. A mercury-in-glass thermometer has a quartz bulb of volume 0.500 cm³ and an attached quartz tube with a bore of 0.10 mm. Initially the bulb is completely filled with mercury. If the temperature in a room changes from 20 to 25°C, what is the increase in the height of the mercury in the tube? (We can neglect the expansion of the quartz because it is so small, only 0.12×10^{-5} times its original volume, and less than 1/100 the value for mercury.)

SOLUTION

The increase in volume of the mercury is

$$\Delta V = \frac{18 \times 10^{-5}}{°C} V\ \Delta T$$

$$= \frac{18 \times 10^{-5}}{°C}(0.500\ \text{cm}^3)(25°C - 20°C) = 4.5 \times 10^{-4}\ \text{cm}^3$$

Now, the increased volume of mercury must raise the height of the mercury column since the bulb is completely filled.

The change in height ΔL of a mercury column of cross-sectional area πR^2 is related to the increased volume ΔV by:

$$\Delta V = \Delta L\ \pi R^2$$

Hence

$$\Delta L = \frac{\Delta V}{\pi R^2} = \frac{4.5 \times 10^{-4}\ \text{cm}^3}{3.14(0.0050\ \text{cm})^2} = \boxed{5.7\ \text{cm}}$$

13.4 Temperature Scales

The temperature of any system in thermal equilibrium can be represented by a number. We now want to see how to assign such numbers or, in other words, how to set up a workable temperature scale. This is done by selecting certain *fixed points* which correspond to easily reproducible temperatures and assigning numbers to them on a chosen temperature scale. Good examples of fixed points are the boiling point and the freezing point of water at standard atmospheric pressure.

Absolute (Kelvin) Temperature Scale

On the absolute scale the zero is fixed by the properties of ideal gases, as previously discussed. It is the temperature at which the pressure of an ideal gas goes to zero at constant volume. On this scale the freezing point of water, as determined by a gas thermometer, is 273.15 K, and the boiling point of water is 373.15 K. The difference in temperature between the freezing and boiling points is exactly 100 kelvins, just as it is 100 Celsius degrees on the Celsius scale. Hence the size of a degree on the absolute scale is exactly the same as the size of a degree on the Celsius scale. Consequently *temperature differences* have the same numerical values on these two scales, as shown in Fig. 13.5.

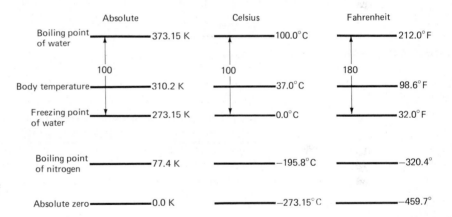

FIGURE 13.5 A comparison of three temperature scales.

Celsius Temperature Scale

This temperature scale, first proposed in 1742 by the Swedish astronomer Anders Celsius (1701–1744), is much used both for scientific and practical purposes. On this scale the freezing point of water is assigned the number 0°C, and the boiling point of water, 100°C. This scale is sometimes called the centigrade scale (*centi-* $= \frac{1}{100}$) because there are 100 equal divisions between these two fixed points on this scale.

Fahrenheit Temperature Scale

The German physicist Gabriel Daniel Fahrenheit (1686–1736) first used mercury as a thermometric liquid. He set 100°F as the approximate temperature of the human body, and 0°F as the lowest temperature obtained with an ice-salt mixture. This made the boiling point of water 212°F, and the temperature of an ice-water mixture (the freezing point of pure water) 32°F. He then divided the distance between these two fixed points on a mercury-in-glass thermometer into 180 equal divisions, or Fahrenheit degrees, and thus created the Fahrenheit temperature scale. We now know that on the Fahrenheit scale the normal temperature of the human body is closer to 98.6°F than it is to 100°F.

Relations between Temperature Scales

An interval of one degree on the absolute scale is of exactly the same size as on the Celsius scale. Hence to obtain the absolute temperature from a given Celsius temperature all we must do is add 273.15 to the Celsius value, i.e.,

$$T(\text{K}) = T(°\text{C}) + 273.15 \tag{13.2}$$

In the range between the freezing and boiling points of water there are 180 Fahrenheit degrees but only 100 Celsius degrees. Hence the Celsius degree is *larger* than the Fahrenheit degree by the factor 180/100 = 9/5. There are therefore *fewer* Celsius degrees in any given temperature interval, because of the larger size of the Celsius degree.

For example, suppose we have a Celsius temperature of 45°. This is 45 Celsius degrees above the freezing point of water, which occurs at 0°C. On the Fahrenheit scale it is $\frac{9}{5}(45) = 81$ Fahrenheit degrees above the freezing point of water. The freezing point of water is, however, not 0°F but 32°F. Hence 45°C corresponds to a Fahrenheit temperature of 81°F + 32°F = 113°F.

This result can be put in the form of an equation:

$$T'(°\text{F}) = \frac{9}{5}T(°\text{C}) + 32 \tag{13.3}$$

This yields: $T(°\text{C}) = \frac{5}{9}\left[T'(°\text{F}) - 32\right]$ (13.4)

There is only one temperature at which Fahrenheit and Celsius scales yield exactly the same temperature. This is −40°C, which is equal to −40°F.

TABLE 13.1 Important Physical Temperatures on the Fahrenheit, Celsius, and Kelvin Temperature Scales

Physical quantity	Temperature		
	°F	°C	K
Deuterium-deuterium fusion	4×10^8	2×10^8	2×10^8
Deuterium-tritium fusion	7×10^7	4×10^7	4×10^7
Surface of hottest known star	4×10^5	2×10^5	2×10^5
Surface of sun	10,000	5500	5800
Melting point of tungsten	6,170	3410	3683
Light-bulb filament	4,600	2500	2800
Uranium fuel rod (in operating reactor)	4,000	2200	2500
Bunsen burner flame	3,400	1870	2150
Steam temperature in a typical fossil-fuel power plant	930	500	770
Steam temperature in a typical nuclear-fuel power plant	660	350	620
Boiling point of water	212.0	100.0	373.15
Human body temperature	98.6	37.0	310.2
Room temperature	70	21	294
Triple point of water	32.02	0.01	273.16
Freezing point of water	32.0	0	273.15
Solid CO_2 (dry ice)	−110	−74	194
Liquid nitrogen (boiling point)	−320	−196	77
Liquid helium (boiling point)	−452	−269	4.2
Absolute zero	−459.67	−273.15	0

This is sometimes helpful in remembering the conversion formulas given above.

The relationship of these three temperature scales is shown in Fig. 13.5. Values of some important temperatures on the Celsius, Fahrenheit, and absolute scales are given in Table 13.1. Note that they range all the way from absolute zero to hundreds of millions of degrees.

Example 13.2

A sticker on the windshield of a Volkswagen Rabbit indicates that the antifreeze in the radiator is good to temperatures of −35°C, or −31°F. Prove that these two temperatures are the same.

SOLUTION

From our rule for converting Celsius temperatures to Fahrenheit we have

$$T'(°F) = (\tfrac{9}{5})T(°C) + 32 = (\tfrac{9}{5})(-35°C) + 32 = \boxed{-31°F}$$

Hence −35°C is the same as −31°F.

Example 13.3

What are the (*a*) Celsius and (*b*) absolute temperatures corresponding to the normal human body temperature of 98.6°F?

SOLUTION

(a) We have

$$T(°C) = (\tfrac{5}{9})[T'(°F) - 32.0] = (\tfrac{5}{9})(98.6 - 32.0)$$

$$= (\tfrac{5}{9})(66.6) = \boxed{37.0°C}$$

(b) Then, using the relationship between Celsius and absolute temperatures,

$$T(K) = T(°C) + 273.2 = 37.0 + 273.2 = \boxed{310.2 \text{ K}}$$

Hence normal body temperature is 310.2 K.

13.5 Expansion of Solids and Liquids with Temperature

Many thermometers for the scientific measurement of temperature rely on the fact that a solid changes its length and a liquid its volume *linearly* with temperature and are therefore convenient to use in thermometers.

As we saw in Chap. 10, the molecules of a metal vibrate about fixed positions as if they were connected by springs. Heating a metal causes the molecules to vibrate more vigorously and therefore to move farther apart on the average. If the metal object is a long, thin rod of original length L_0 at some temperature T_0, then we can denote by ΔL the change in length due to a change in temperature ΔT, as in Fig. 13.6. From experiment we find that, for most materials at ordinary temperatures, the increase in length is proportional both to the original length of the rod and to the change in temperature,

$$\Delta L \propto L_0\,\Delta T \qquad \text{or} \qquad \Delta L = \alpha L_0\,\Delta T \qquad (13.5)$$

Here the coefficient α is called the *coefficient of linear expansion*, and is defined as follows:

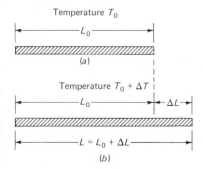

FIGURE 13.6 Increase in length of a metal rod with temperature: (*a*) Length at T_0; (*b*) length at $T = T_0 + \Delta T$.

Definition

Coefficient of linear expansion α: The increase in length of a solid per unit length per unit change in temperature:

$$\alpha = \frac{\Delta L}{L_0 \Delta T} \tag{13.6}$$

FIGURE 13.7 Expansion of a hole in a metal block when the block is heated: (*a*) Before heating. (*b*) After heating, the hole expands in the same proportion as the rest of the block.

The new length of the rod is then:

$$L = L_0 + \Delta L = L_0 + \alpha L_0 \, \Delta T$$

or $\boxed{L = L_0(1 + \alpha \, \Delta T)} \tag{13.7}$

Hence *the length increases linearly with the change in temperature* ΔT, and this change in length can therefore be used to measure an unknown temperature by relating it to some known temperature.

The coefficient of linear expansion varies from substance to substance, and its value must be obtained by experiment. Table 13.2 gives values of α for a variety of useful substances. Note that α has units $(^\circ C)^{-1}$, since $\Delta L / L_0$ is dimensionless, and so $\alpha \, \Delta T$ must also be dimensionless.

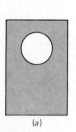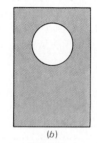

(a)   (b)

FIGURE 13.8 A way of demonstrating that holes in metal blocks increase in size on heating. The rods *A* and *B* are made of the same metal. The radius of the knob at the end of *B* and of the hole in *A* are precisely the same, so that at room temperature the knob can just be pushed through the hole. If the metal knob is now heated in a Bunsen flame, it expands so much that it can no longer pass through the hole in *A*. If, while the knob is maintained at this higher temperature, the metal surrounding the hole is heated to the same temperature as that of the knob, the knob will again pass through the hole, indicating that the hole has increased its area on being heated.

TABLE 13.2 Values of Thermal Coefficients of Linear and Volume Expansion

Substance	Coefficient of linear expansion α, $10^{-6} \ (^\circ C)^{-1}$	Coefficient of volume expansion β, $10^{-5} \ (^\circ C)^{-1}$
Solids:		
Aluminum	23	7.2
Brass	19	6.0
Copper	17	4.2
Pyrex glass	3.2	0.96
Invar	0.7	0.27
Lead	29	8.7
Steel	11	3.6
Fused quartz	0.4	0.12
Liquids:		
Ethyl alcohol		112
Glycerin		51
Mercury		18
Turpentine		97

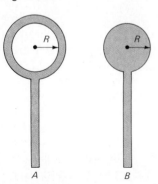

A   B

If we consider a two-dimensional object like a flat rectangular plate, of original length L_0 and width W_0, then both L_0 and W_0 increase with temperature to new values $L_0 + \Delta L$, and $W_0 + \Delta W$. Hence the new area is

$$A = (L_0 + \Delta L)(W_0 + \Delta W) = (L_0 + \alpha L_0 \, \Delta T)(W_0 + \alpha W_0 \, \Delta T)$$

$$= L_0 W_0(1 + 2\alpha \, \Delta T + \alpha^2 \, \Delta T^2) \simeq A_0(1 + 2\alpha \, \Delta T)$$

where we have dropped $\alpha^2 \, \Delta T^2$ because it is usually small compared with unity (for brass, if ΔT is 100°C, $\alpha \, \Delta T = 1.9 \times 10^{-3}$ and $\alpha^2 \, \Delta T^2 = 3.6 \times 10^{-6}$). Hence we obtain, to a good approximation,

$$A = A_0(1 + 2\alpha \, \Delta T) \tag{13.8}$$

The area therefore also increases linearly with the change in temperature, the proportionality factor being in this case 2α rather than α.

A similar analysis for volumes (try to derive it yourself) shows that

$$V = V_0(1 + 3\alpha \, \Delta T) \tag{13.9}$$

It is also possible to define a *coefficient of volume expansion* for solids in a similar fashion to the way we defined a coefficient of linear expansion in Eq. (13.7):

$$V = V_0(1 + \beta \, \Delta T) \tag{13.10}$$

where β is the coefficient of volume expansion of the solid. Hence we would expect from comparing Eq. (13.9) with (13.10) that for solids β would be equal to 3α. Some values of α and β for important solids are given in Table 13.2. Note that for these solids β is indeed approximately equal to 3α.

If we take a metal object with a circular hole in it, as in Fig. 13.7, we might think that the hole would grow smaller with increased temperature. What actually happens, however, is the reverse: the hole grows *larger*. Heating the metal increases the average distance between the molecules in the metal. Hence the molecules around the edge of the hole in Fig. 13.7 move farther apart. This increases the circumference of the hole, and hence also its area. In other words the effect of increasing the temperature is similar to making a photographic enlargement of the metal block. Everything is blown up in the same ratio, including the hole. This can be verified experimentally using the apparatus of Fig. 13.8.

Bimetallic Strips and Thermostats

A bimetallic strip consists of two thin strips of different metals cemented together. Since the two metals expand at different rates as the temperature increases, the bimetallic strip bends more and more as the temperature increases, as shown in Fig. 13.9, since this is the only way the two metals can expand without cracking the cement holding them together. Such strips can be wound into a coil and attached to a pointer which indicates the temperature on a circular scale. This makes a relatively crude, but simple, kind of thermometer, as illustrated in Fig. 13.10.

FIGURE 13.9 A bimetallic strip. Heating the strip causes it to bend because of the uneven expansion of the two metals.

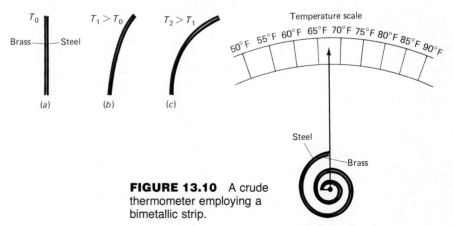

FIGURE 13.10 A crude thermometer employing a bimetallic strip.

A *thermostat* is a bimetallic strip which controls the electric power to keep the temperature of a region of space (a room or an oven, for example) within preset limits. Suppose we install a straight bimetallic strip at 70°F, a comfortable room temperature, as in Fig. 13.11. We then set electric relays in such a way as to keep the temperature always between 67 and 73°F. As the house warms up, the bimetallic strip bends to the right until at 73°F it touches an electric contact. This throws a switch which turns the furnace off. As the house cools down, the two metals contract and the strip gradually straightens out and then bends to the left. When the temperature reaches 67°F, the bimetallic strip closes another electric contact, which turns the furnace on once again. A thermostat like this is the simplest kind of *feedback mechanism*, since it constantly turns the furnance on and off in response to a signal indicating what the house temperature is.

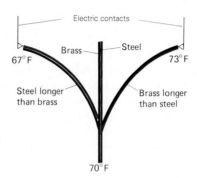

FIGURE 13.11 A thermostat consisting of a bimetallic strip and two electric contacts.

Liquids

Because the shape of a liquid is so easily changed, only volume changes are significant for liquids. Here again we find that the volume increases approximately linearly with the change in temperature, the change being given by

$$V = V_0(1 + \beta \, \Delta T) \tag{13.11}$$

where β is the coefficient of volume expansion of the liquid, as for solids. Some values for β for important liquids are included in Table 13.2.

Most liquids expand in volume linearly with temperature in this way. Water, our most common liquid, is an exception, and is therefore not very useful as a thermometer liquid. The volume of a fixed amount of water is smallest at 4°C, as shown in Fig. 13.12, and increases on both sides of 4°C. Above 4°C the expansion of water is not linear, although it is reasonably close to it. But from 4 down to 0°C, where it freezes, water again expands. Hence ice at 0°C is *less dense* than water is at 4°C. It is for this reason that ice floats at the top of a lake or river. If this were not the case and ice were *more dense* than cold water, lakes and rivers would freeze from the bottom up and never completely unfreeze even in summer.

The volume expansion of solids, liquids, and gases are the same in principle, but very different quantitatively. As Table 13.2 shows, the coeffi-

FIGURE 13.12 Density of water in the vicinity of 0°C. As can be seen, water is most dense and hence has its smallest volume at 4°C. Water expands both above and below 4°C.

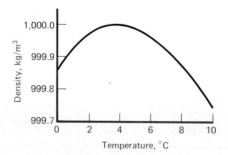

cient of volume expansion is about 10 times larger for liquids than for solids. For gases the coefficient of volume expansion is about 300 times that for solids.

Example 13.4

Invar (an alloy of iron and nickel in a 5/3 ratio) is a material with a very low coefficient of linear expansion and is therefore used in clocks and precise scientific instruments where slight changes in temperature might otherwise intro-duce errors. Calculate the increase in length of an Invar pendulum rod 1 m long, as the temperature changes from 15 to 25°C.

SOLUTION

The increase in length is given by Eq. (13.6):

$$\Delta L = \alpha L_0\, \Delta T$$

where, from Table 13.2, $\alpha = 0.7 \times 10^{-6}$ $(°C)^{-1}$, and $\Delta T = 25°C - 15°C = 10°C$. Therefore

$$\Delta L = \frac{0.7 \times 10^{-6}}{°C}(1\ \text{m})(10°C) = 0.7 \times 10^{-5}\ \text{m}$$

$$= \boxed{0.7 \times 10^{-2}\ \text{mm}}$$

The increase in length of the 1-m Invar rod is then less than $\frac{1}{100}$ mm.

Example 13.5

A solid brass block at 20°C has a circular hole in it of radius 2.000 cm. If the brass block is thrown into a pot of boiling water at 100°C, what is the area of the hole in the block when its temperature reaches 100°C?

SOLUTION

From the preceding discussion, the hole must get *larger*. The original area of the hole was $A_0 = \pi R_0^2 = \pi(2.000\ \text{cm})^2 = 12.57\ \text{cm}^2$. The radius changes, on heating, from R_0 to R, where, from Eq. (13.7),

$$R = R_0(1 + \alpha\, \Delta T)$$

$$= (2.000\ \text{cm})\left[1 + \left(\frac{19 \times 10^{-6}}{°C}\right)(80°C)\right]$$

$$= 2.000(1 + 152 \times 10^{-5})\ \text{cm} = 2.003\ \text{cm}$$

Then the new area is $\pi(2.003\ \text{cm})^2 = \boxed{12.61\ \text{cm}^2}$

13.6 Heat as a Form of Energy

We have seen that at the atomic level the temperature of a gas is a measure of the average random translational kinetic energy of the gas molecules. The higher the absolute temperature, the greater the kinetic energy, and therefore the greater the average molecular speed. Temperature is independent of how much gas is present or of what kind of gas it is. When we measure the temperature (in kelvins) of the He gas in Fig. 13.13, we measure the average random translational kinetic energy of the He atoms, and nothing else.

FIGURE 13.13 Helium gas in a container on a hot plate.

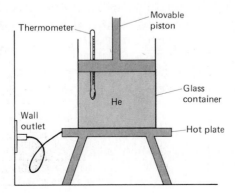

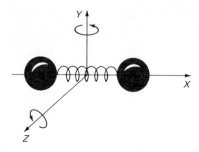

FIGURE 13.14 Vibration and rotation of a nitrogen molecule (N_2). The two N atoms can vibrate back and forth as if connected by springs, and the whole molecule can rotate like a dumbbell around the Y and Z axes.

What then do we mean by *heat*, and how do we distinguish the concept of heat from that of temperature? To answer this question we must first clarify what we mean by the thermal energy of a gas like He at a temperature T. By *thermal energy* we mean the sum of all the energies associated with the random motions of the atoms in the substance. Thermal energy depends on the temperature, but it also depends on the number of atoms (or molecules) present. If each group of atoms has an average kinetic energy determined by the Kelvin temperature T, then the more atoms present, the greater the total thermal energy of the gas, even if T remains fixed.

For more complicated systems there are new kinds of motions which make additional contributions to the thermal energy even for the same values of the temperature and the same number of molecules. Thus in a sample of nitrogen gas a nitrogen molecule N_2 can rotate like a dumbbell in space, or the two N atoms can vibrate back and forth as if they were connected by springs (see Fig. 13.14). Hence, while the temperature depends only on the translational kinetic energy of the gas molecules, the thermal energy depends not only on the temperature but also on the number and kinds of atoms and molecules in the gas. In other words:

Definition

Thermal energy: The sum of all the random mechanical energies of the atoms and molecules in a substance.

What then is heat? The term *heat* is usually reserved for the *transfer* of thermal energy from one substance to another.

Definition

Heat: The thermal energy transferred between a system and its surroundings, as a result of temperature differences only.

Thermal energy and heat are the same kind of physical quantities, just as mechanical energy and work are in mechanics. Just as we use the term *work* to refer to the *transfer* of mechanical energy from one part of a system to another, so too we use the term *heat* for the *transfer* of thermal energy from one part of a system to another. This distinction will become important when we discuss the first law of thermodynamics.

To clarify the distinction between heat and temperature a bit more, consider what happens when we turn up the hot plate in Fig. 13.13 to a temperature T_2 higher than the initial temperature T_1 of the gas. Heat flows from the hot plate through the glass into the He in the container. On a molecular level this means that the agitated molecules in the hot plate pass on some of their random molecular motion to the glass molecules, which pass it on to the He atoms inside. We say that a *flow of heat* has increased the translational kinetic energy of the gas molecules and thus has increased the thermal energy and hence the temperature of the gas. The amount of heat ΔQ required to change the temperature by an amount $\Delta T = T_2 - T_1$ will obviously depend on how large ΔT is. Hence we can write

$$\Delta Q \propto \Delta T$$

where ΔQ is the heat required to raise the temperature of the He gas from T_1 to T_2. But the amount of heat, that is, the amount of disordered energy, required to raise the temperature by an amount ΔT will depend also on how many gas atoms are present: the amount of heat required to change the temperature of 10^{12} atoms from T_1 to T_2 will be 10 times greater than the amount of heat required to raise the temperature of 10^{11} identical atoms by the same amount ΔT. In other words, the amount of heat will also be proportional to the mass of gas present, and we can write, for ΔT constant, $\Delta Q \propto m$. We then combine these two results and obtain

$$\Delta Q \propto m \, \Delta T \qquad \text{or} \qquad \boxed{\Delta Q = cm \, \Delta T} \tag{13.12}$$

where c is a constant of proportionality which differs from gas to gas or from substance to substance. It depends on the kinds of atoms or molecules present in the substance, and is called the *specific heat capacity*. The specific heat capacity of a substance is usually obtained from measurements made in the laboratory, although it is possible to calculate theoretically specific heat capacities in certain simple cases.

Analogy between Heat Flow and Water Flow

Heat naturally flows of its own accord from a higher temperature to a lower one, and never the reverse, as we know from our own experience. There is a very good analogy between the flow of water over a dam from a higher level of gravitational potential energy to a lower one, and the flow of heat from a hotter temperature to a colder temperature. Just as water always flows *down* a gravitational gradient, so heat always flows *down* a temperature gradient. Neither water nor heat ever flows "uphill." This is illustrated in Fig. 13.15.

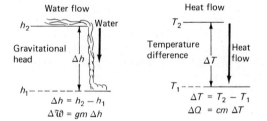

FIGURE 13.15 Analogy between heat flow and the flow of water over a dam.

It was Julius Robert Mayer (see Fig. 13.16, and accompanying biography) who first put forth the thesis that heat was nothing but another form of energy (disordered molecular energy), although Count Rumford (see biography, Chap. 6) had a vague grasp of this crucial idea many years before. This is clear from Rumford's statement quoted at the beginning of this chapter in which he equated heat with motion (i.e., the motion of molecules). It is interesting that Mayer was led to his conclusion that heat was merely another form of energy by his study of the oxidation of blood in the human body.

13.7 Specific Heat Capacities

Now that we have discussed heat and temperature and set up our temperature scales, the only thing remaining before we take up some very practical problems of heat transfer is the introduction of the units in terms of which heat is measured. Since heat is a form of energy, it could well be measured in joules, the unit of energy in the SI system. Initially, however, the science of

Julius Robert Mayer (1814–1878)

FIGURE 13.16 Julius Robert Mayer. *(Courtesy of AIP Niels Bohr Library.)*

It is unusual to find an important person in the history of physics who did not understand vectors, who constantly confused the concepts of mass and weight, and who was terrified by calculus (perhaps this should give us all hope!). Such a man was Julius Robert Mayer. Despite his shortcomings he is recognized today as the first person to emphasize the significance of the concept of *energy* in physics and the biological and health sciences.

Mayer was born in Heilbronn in Bavaria in 1814. Interested in science as a young man, he decided to make a career for himself in medicine. He entered the University of Tübingen in 1832, took physics for only one se-mester, and received his M.D. degree in 1838. He then shipped out as a doctor on a freighter for Java in the East Indies. There he made the observation that changed his life. He noticed that the blood in the sailor's veins was a much brighter red than he had observed in patients back in Germany. He surmised that the redness of the venous blood meant less oxidation of the food consumed and less heat produced, since less heat was necessary in tropical climates. This made him ponder how the human body derived heat from food and how energy is transformed in the process. This interest in energy became the driving passion of his life.

From 1841 until his death Mayer practiced medicine in Heilbronn, and became chief surgeon of the town. In his spare time he did some experiments and struggled with difficult theoretical concepts in an attempt to understand the nature of energy and its transformations. He knew so little physics that many of his articles were rejected as incompetent by the scientific journals of the day, and he was forced to publish them privately at his own expense.

In a paper published in 1842 Mayer stated clearly his conviction that there is a definite quantitative relationship between the height from which a mass m falls to the ground and the heat generated when it strikes the ground. In modern terms he was saying, however obscurely, that the potential energy (mgh) of the mass is equal to the heat generated when the mass strikes the ground, and hence that energy is conserved if heat is considered a form of energy. He also made a calculation of the ratio of the unit for mechanical energy (the joule) to the unit for heat (the calorie), on the basis of the thermal properties of gases, and found that it was 3.59 J/cal. This differs considerably from today's accepted value of 4.18 J/cal, but this was because of the inaccurate values he used for some constants in his calculations.

The years from 1846 to 1850 were dismal ones for Mayer. Three of his children died in a 2-year period. He was accused by local scientists of being more a mad philosopher than a competent scientist. And, finally, he became embroiled in a controversy with the great British physicist Sir James Joule (see Fig. 14.11) about the priority of their discoveries in the energy field.

Finally, in May 1850, in a fit of despair, Mayer threw himself out of his bedroom window to the street 30 ft below but escaped without serious injury. He spent 3 years in a mental hospital and did little scientific work after his release in 1853. Physicists around the world gradually came to appreciate his scientific work, but by this time were unsure whether he was dead or alive.

In his later years Mayer finally reaped some of the fruits of his scientific labors. In 1871 he received the Copley Medal of the Royal Society of London, and later the Prix Poncelet of the Paris Academy of Sciences and an honorary Ph.D. degree from his own University of Tübingen.

heat grew up separately from the science of mechanics and developed its own units, the kilocalorie in countries using the metric system and the British thermal unit in the British engineering system. Both these units are still much used today, and we must therefore have clear ideas of their meanings and magnitudes before we can treat heat transfer problems correctly.

The definitions of these two units are based on the fact that the amount of heat added to a fixed mass of water can be deduced from the change in temperature produced, as is clear from Eq. (13.12).

Kilocalorie (kcal or Cal): The amount of heat required to raise the temperature of one kilogram of water one degree Celsius (from 14.5 to 15.5°C).

British thermal unit (Btu): The amount of heat required to raise the temperature of one pound of water one degree Fahrenheit (from 63 to 64°F).

There is a smaller unit, the calorie (cal), which is just 1/1000 of the kilocalorie. When we talk about the number of calories in a potato or a candy bar, we are really talking about the kilocalorie (the "big calorie"), not the calorie (or "small calorie"). As we shall see, 1 kcal is equal to 4.186×10^3 J.

Note that in both the above definitions the kilocalorie and the British thermal unit are so defined that the specific heat of water is unity in both systems. Hence for water $c = 1$ kcal/(kg·°C) in the metric system, and $c = 1$ Btu/(lb·°F) in the British system. This leads to the result that 1 kcal = 3.96 Btu $\simeq$ 4 Btu.

Since it takes 1 kcal to raise the temperature of 1 kg of water 1°C (or 1 K), the amount of heat required to raise more than 1 kg of a substance more than 1°C may be obtained easily from Eq. (13.12):

$$\Delta Q(\text{kcal}) = cm(\text{kg})\ \Delta T(°C) \tag{13.12}$$

where, for water, $c = 1$ kcal/(kg·°C). Hence to raise the temperature of 5 kg of water 2°C requires 10 kcal. The units and dimensions of the specific heat capacity are clear from its definition:

$$c = \frac{\Delta Q}{m\ \Delta T}\ \text{kcal/(kg·°C)} \tag{13.13}$$

Specific heat capacity c: The quantity of heat, in kilocalories, which must be supplied to a unit mass (1 kg) of a substance to raise its temperature by one degree (1°C or 1 K).

For example, we find from experiment that

$c_{\text{copper}} = 0.093$ kcal/(kg·°C)

Then, to raise the temperature of 2.0 kg of copper 10°C requires

$$\Delta Q = cm\ \Delta T = \left(0.093\ \frac{\text{kcal}}{\text{kg·°C}}\right)(2.0\ \text{kg})(10°C) = 1.86\ \text{kcal}$$

In many practical problems involving specific heat capacities, two bodies at different temperatures come in contact, as when we pour cold milk into hot coffee in the morning. The coffee loses heat and the milk gains heat until the two come to an equilibrium temperature. The equilibrium temperature can be found fairly easily if we know the amounts of each substance present and their specific heat capacities. A similar problem is solved in Example 13.6.

In Table 13.3 the specific heat capacities of a number of important substances are shown. Since they vary somewhat with temperature, the values given are for temperatures in the immediate vicinity of room temperature (about 20°C).

Definitions

TABLE 13.3 Specific Heat Capacities

Substance	Specific heat capacity c, kcal/(kg·°C)*
Metals:	
Aluminum	0.22
Iron	0.113
Copper	0.093
Mercury	0.033
Silver	0.056
Lead	0.031
Beryllium	0.47
Nonmetals:	
Water	1.00
Ice	0.55
Wood	0.42
Glass	0.12
Body tissue	0.80
Human blood	0.92
Human body (average)	0.85
Urea	0.32
Sugar	0.27

*Or cal/(g·°C).

Definition

Example 13.6

Suppose we drop a 1.0-kg piece of aluminum at 200°C into 5.0 kg of water at room temperature (20°C). If no heat is exchanged with the surroundings, what is the final temperature of the water and the aluminum?

SOLUTION

The aluminum will lose heat and the water will gain heat until they reach a final equilibrium temperature T_f. The amounts of heat gained by the water and lost by the aluminum must be equal, according to the principle of conservation of energy, and can be calculated from Eq. (13.12). Thus the heat given out by the aluminum is

$$\Delta Q_{Al} = c_{Al} m_{Al} \Delta T_{Al}$$

$$= \left(0.22\frac{\text{kcal}}{\text{kg·°C}}\right)(1.0 \text{ kg})[(200 - T_f)°C]$$

$$= 0.22(200 - T_f) \text{ kcal} = (44 - 0.22T_f) \text{ kcal}$$

Similarly, the heat taken in by the water is

$$\Delta Q_w = c_w m_w \Delta T_w$$

$$= \left(1.0\frac{\text{kcal}}{\text{kg·°C}}\right)(5.0 \text{ kg})[(T_f - 20)°C]$$

$$= 5.0(T_f - 20) \text{ kcal} = (5T_f - 100) \text{ kcal}$$

If no heat is lost or gained by the system, $\Delta Q_{Al} = \Delta Q_w$, and so

$$44 - 0.22T_f = 5T_f - 100 \quad \text{or} \quad 5.22T_f = 144$$

$$\text{and} \quad T_f = \boxed{28°C}$$

Note that the final temperature of the mixture is within 8°C of the initial temperature of the water, even though the aluminum was at 200°C. This is because there is 5 times as much water as aluminum, and also because the water has a specific heat capacity 5 times that of the aluminum. The water, because of its large specific heat capacity, "hides" the effect of the heat given to it by the aluminum.

In general, all problems of this kind can be solved by setting up equations in the form

$$c_1 m_1 (T_1 - T_f) = c_2 m_2 (T_f - T_2)$$

where T_f is the final equilibrium temperature. The units used must be consistent. In the SI system these are kilocalories, kilograms, and degrees Celsius. In this system absolute temperatures may also be used, since temperature *differences* like $T_1 - T_f$ are the same on the Celsius and absolute temperature scales.

13.8 Molar Heat Capacities of an Ideal Gas

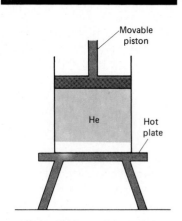

FIGURE 13.17 Helium gas in a container closed by a movable piston and resting on a hot plate.

Let us consider an ideal, monatomic gas like He in a closed container with a piston at one end, as shown in Fig. 13.17. We assume that the collisions of the gas molecules among themselves and with the walls of the container are perfectly elastic, and that no other forces act except when collisions occur. From Eq. (9.16) we know that, under such assumptions, the kinetic energy per mole of gas is

$$KE = \tfrac{1}{2}M\overline{v^2} = \tfrac{3}{2}RT \tag{13.14}$$

where R is the universal gas constant [in cal/(mol·K)]. Since the only energy the gas possesses is kinetic, its thermal energy is:

$$U = \tfrac{3}{2}nRT \tag{13.15}$$

where n is the number of moles of gas. Equations (13.14) and (13.15) show that the thermal energy of an ideal gas depends only on its temperature and not on other variables like pressure and volume.

We now want to heat the gas shown in Fig. 13.17 and in this way determine its specific heat capacity. Since we are here dealing with n mol of the gas, we will use the molar heat capacity C. This is the amount of heat required to raise the temperature of 1 mol of the gas 1°C (or 1 K). We can, however, add heat to the gas in many different ways, two of the most important of which are at constant volume and at constant pressure.

1 *Constant volume:* Here we do not allow the piston in Fig. 13.17 to move. The amount of heat ΔQ required to change the temperature by ΔT is then

$$\Delta Q = nC_V \, \Delta T \qquad\qquad (13.16)$$

where C_V is the molar heat capacity *at constant volume*. Since the volume of the gas does not change, the heat added must go into the thermal energy U of the gas. The increase in thermal energy of the gas is therefore

$$\Delta U = \Delta Q = nC_V \, \Delta T \qquad\qquad (13.17)$$

2 *Constant pressure:* Here we allow the piston to move up, *maintaining the same pressure P* (say, that of the atmosphere) on the gas. The gas does an amount of work ΔW in moving the piston of area A through a distance Δs, where

$$\Delta W = F \, \Delta s = PA \, \Delta s = P \, \Delta V$$

since $\Delta V = A \, \Delta s$ is the increase in volume of the gas. In this case the heat input is

$$\Delta Q = nC_P \, \Delta T \qquad\qquad (13.18)$$

where C_P is the molar heat capacity at constant pressure. Part of the heat added is used to perform the work ΔW, the rest is stored in the gas as thermal energy ΔU. Hence we have

$$\Delta Q = \Delta W + \Delta U \qquad \text{or} \qquad nC_P \, \Delta T = P \, \Delta V + \Delta U \qquad (13.19)$$

But the thermal energy of the gas depends only on the temperature, as Eq. (13.15) shows. The change in thermal energy is given by Eq. (13.17) as $\Delta U = nC_V \, \Delta T$, and *has the same numerical value for any particular change in temperature ΔT*, irrespective of whether the change was carried out at constant pressure, constant volume, or in some more complicated fashion. Hence, on putting together Eqs. (13.17) and (13.19), we have

$$nC_P \, \Delta T = P \, \Delta V + nC_V \, \Delta T \qquad\qquad (13.20)$$

But, from the ideal gas law [Eq. (9.14)], $PV = nRT$, and so, at constant pressure, $P \, \Delta V = nR \, \Delta T$. Hence we have, on substituting in Eq. (13.20),

$$nC_P \, \Delta T = nR \, \Delta T + nC_V \, \Delta T$$

or
$$\boxed{C_P - C_V = R} \qquad\qquad (13.21)$$

Since $R = 8.31$ J/(mol·K) $= 1.99$ cal/(mol·K), we would expect that the difference between the molar heat capacities of gases at constant pressure and constant volume would be about 2 cal/(mol·K). This is indeed what we find from the experimental values in Table 13.4, not merely for monatomic gases like He and Ne, but also for a variety of other, more complicated, molecular gases.

TABLE 13.4 Experimental Values of Molar Heat Capacities and γ for Selected Gases

Gas	Molar heat capacities, cal/(mol·K)		$C_P - C_V$, cal/(mol·K)	$\gamma = C_P/C_V$
	C_P	C_V		
Monatomic:				
He	4.97	2.98	1.99	1.67
Ne	4.97	2.98	1.99	1.67
Ar	4.97	2.98	1.99	1.67
Diatomic:				
H_2	6.87	4.88	1.99	1.41
N_2	6.95	4.96	1.99	1.40
O_2	7.03	5.03	2.00	1.40
Cl_2	8.29	6.15	2.14	1.35
Polyatomic:				
CO_2	8.83	6.80	2.03	1.30
H_2O	8.20	6.20	2.00	1.32
C_2H_6	12.35	10.30	2.05	1.20

Value of Molar Heat Capacities C_P and C_V

Since for an ideal gas, from Eq. (13.15), $U = \frac{3}{2}nRT$, if the temperature changes by an amount ΔT, we have $\Delta U = \frac{3}{2}nR\,\Delta T$.

But, from Eq. (13.17) for an expansion at constant volume, $\Delta U = nC_V\,\Delta T$. Hence

$$\tfrac{3}{2}nR\,\Delta T = nC_V\,\Delta T$$

or $\quad C_V = \tfrac{3}{2}R$ (13.22)

The molar specific heat capacity at constant volume should therefore be, for a monatomic gas, $\frac{3}{2}R$ or about 3 cal/(mol·K).

Since, from Eq. (13.21), $C_P = C_V + R$, $C_P = \frac{3}{2}R + R = \frac{5}{2}R$, and, for a monatomic gas the specific heat capacity at constant pressure should be about 5 cal/(mol·K).

A very important and useful quantity in thermodynamics is the ratio of the two molar heat capacities. This ratio is universally designated by the symbol γ (the Greek *gamma*, the equivalent of an English g), where

$$\gamma = \frac{C_P}{C_V}$$ (13.23)

For a monatomic gas $\gamma = C_P/C_V = (\frac{5}{2})R/(\frac{3}{2})R = \frac{5}{3} = 1.67$. The data in Table 13.4 show that these theoretical values for C_V, C_P, and γ are all indeed verified by experiment for monatomic gases.

It can be seen that for the diatomic gases listed in Table 13.4 γ is *not* equal to 1.67 but is close to 1.40. This can be explained by the fact that a diatomic molecule resembles a dumbbell which can rotate rigidly in space. Such a "dumbbell" molecule can rotate freely about two perpendicular axes of rotation. It is therefore said to have two additional *degrees of freedom*, each one having associated with it a kinetic energy of $\frac{1}{2}RT$ per mole. Hence, instead of C_V being equal to $\frac{3}{2}R$, it is equal to $\frac{5}{2}R$, and $C_P = \frac{5}{2}R + R = \frac{7}{2}R$. Then $\gamma = C_P/C_V = 1.40$. This is an example of the classical principle of the equipartition of energy, which states that each degree of freedom has associated with it an energy of $\frac{1}{2}RT$ per mole. The breakdown of this principle contributed significantly to the development of modern quantum physics.

Example 13.7

One mole of He gas is heated at a constant volume from 27 to 97°C.
(a) What is the change in thermal energy of the gas?
(b) What is the molar heat capacity at constant volume (C_V) of the gas?

(c) What is the molar heat capacity at constant pressure (C_P) of the gas?
(d) What is the ratio $\gamma = C_P/C_V$?

SOLUTION

(a) From Eq. (13.15) we have for a constant-volume process: $\Delta U = \frac{3}{2}nR\,\Delta T$

Here $n = 1$ mol, and $\Delta T = (370 - 300)$ K = 70 K. Hence

$$\Delta U = \frac{3}{2}(1\text{ mol})\left(8.31\frac{\text{J}}{\text{mol·K}}\right)(70\text{ K}) = \boxed{8.75 \times 10^2\text{ J}}$$

(b) From Eq. (13.17) we have

$$C_V = \frac{1}{n}\frac{\Delta U}{\Delta T} = \frac{8.75 \times 10^2\text{ J}}{(1\text{ mol})(70\text{ K})} = \boxed{1.25 \times 10^1\text{ J/(mol·K)}}$$

The same result would, of course, have been obtained using

$$C_V = \frac{3}{2}R$$

(c) From Eq. (13.21),

$$C_P = C_V + R = \frac{3}{2}R + R = \frac{5}{2}R$$

Hence

$$C_P = \frac{5}{2}[8.31\text{ J/(mol·K)}] = \boxed{2.08 \times 10^1\text{ J/(mol·K)}}$$

C_P is greater than C_V since some of the heat goes into the work required to increase the volume. The values obtained here for C_V and C_P agree with those in Table 13.4 when the conversion is made from joules to calories.

(d) $\gamma = \dfrac{C_P}{C_V} = \dfrac{2.08 \times 10^1}{1.25 \times 10^1} = \boxed{1.67}$

13.9 Change of Phase: Heats of Fusion and Vaporization

By the term *phase* we mean the state in which a substance exists, i.e., solid, liquid, or gas. Thus the chemical substance H_2O can exist in the gaseous phase as steam or water vapor, in the liquid phase as ordinary water, and in the solid phase as ice. All substances can normally exist in any one of these three phases under the proper conditions of pressure and temperature. For example, water at atmospheric pressure becomes steam at temperatures above 100°C and ice at temperatures below 0°C.

From our previous discussion of specific heat capacities we might think that whenever we add heat to a substance, we necessarily raise its temperature. A simple experiment shows us that this is not always the case. Consider a glass beaker containing water and ice cubes, as in Fig. 13.18. We insert a thermometer and find that the temperature of the water-ice mixture is 0°C (or 32°F). Then we turn on the hot plate and begin to heat the ice-water mixture. We might expect the temperature to begin to rise immediately, but it does not! Instead the temperature remains fixed at 0°C until all the ice has melted; only then does the water begin to behave as we might expect from our previous discussion of specific heats (see Fig. 13.19).

The explanation of this unexpected behavior is that we are dealing here not with a simple substance like water or oxygen gas, but with a mixture of ice

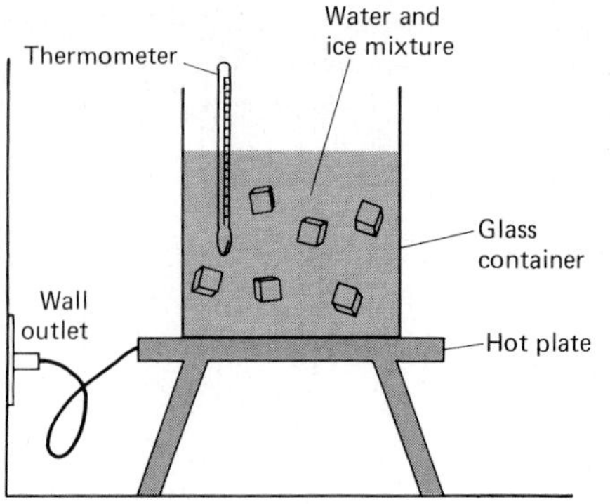

FIGURE 13.18 A beaker filled with ice and water at 0°C and resting on a hot plate.

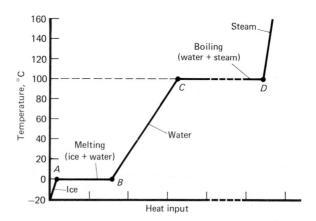

FIGURE 13.19 Variation of temperature of an ice-water mixture with heat input. Note the break in the scale for the boiling part of the curve. Since the heat of vaporization is 540 kcal/kg and the heat of fusion is about 80 kcal/kg, the boiling part of the curve should be about 7 times as long as the melting part. Note also that, since the specific heats of ice and of steam are less than the specific heat of water, the rising straight lines for ice and steam have steeper slopes than does the line for water.

and water, i.e., with H_2O in *two different phases*. To convert ice to water at 0°C, i.e., to produce a *phase change*, requires an input of energy to break the bonds holding the water molecules together in the ice crystal. Hence so long as there is any ice present, all the heat we put in is used in breaking bonds and converting ice into water. Only after all the ice has been changed to water does the applied heat begin once again to increase the random kinetic energies of the water molecules, and to show up as a change in temperature of the water. Again we see from this example that temperature and heat are two very different concepts. At the melting point the heat input to the ice is increasing as we melt it, but the temperature remains unchanged.

Definition

Heat of fusion: The amount of heat required to melt 1 kg of a substance from solid to liquid.

For water the heat of fusion is 80 kcal, since, when ice melts, each kilogram absorbs 80 kcal and, when ice freezes, each kilogram releases 80 kcal. The numerical value of the heat of fusion is different for different substances. We may, then, write in general

$$Q = L_f m \tag{13.24}$$

where Q is the heat in kilocalories, m is the mass in kilograms, and L_f is the heat of fusion in kilocalories per kilogram.

If, after all the ice is melted, we continue to heat the water, its temperature will increase constantly, as shown in Fig. 13.19, until 100°C is reached. Then boiling occurs, and all the applied heat is used to break the H_2O molecules free from the liquid and allow them to escape as vapor molecules. To do this requires a "heat of vaporization" of 540 kcal/kg. While boiling is taking place, the temperature remains fixed at 100°C (or 212°F). It is this constancy of temperature at the point where a change of phase occurs that makes melting and boiling points so useful in defining temperature scales.

No matter how vigorously water boils, it remains at the same temperature (100°C at atmospheric pressure). This is why double boilers are useful for melting substances which scorch easily. In a double boiler such substances are subjected only to the boiling temperature of the water regardless of how hot the burner on the stove may be.

For vaporization we can write an equation similar to Eq. (13.24) above:

$$Q = L_V m \tag{13.25}$$

where L_V is the *heat of vaporization*.

Heat of vaporization: The heat required to convert 1 kg of a substance from liquid to vapor.

After all the water has been converted into steam, we can collect the steam in a closed vessel and continue to heat it. Its temperature will increase above 100°C according to Eq. (13.12), where c is now the specific heat capacity of *steam*, 0.48 kcal/(kg·°C). This is shown in Fig. 13.19. Since the specific heat of steam is about half that of water, the slope of the steam part of the curve (>100°) is steeper than the slope of the water part (between 0 and 100°C).

Values of the heats of fusion and vaporization for various substances are given in Table 13.5. Notice how much larger the heats of fusion and vaporization are for water than for other common substances. The same is true of the specific heat of water. This explains the moderating effect on the climate of large bodies of water like the Atlantic and Pacific oceans, since large amounts of heat are required to change the temperature of the oceans by small amounts.

Definition

TABLE 13.5 Heats of Fusion and Vaporization for Typical Substances

Substance	Heat of fusion, kcal/kg	Heat of vaporization, kcal/kg
Water	79.7	539.6
Mercury	2.82	70.6
Methyl alcohol	16.4	262.8
Ethyl alcohol	24.9	204
Nitrogen	6.09	47.6
Oxygen	3.30	50.9
Hydrogen	14.0	108
Helium	1.25	5

Example 13.8

How much heat is required to convert 0.50 kg of liquid mercury at 356.6°C to mercury vapor?

SOLUTION

We have seen that the heat required to vaporize a liquid is $Q = mL_V$, where in this case $L_V = 70.6$ kcal/kg, from Table 13.5.

Hence

$$Q = 0.50 \text{ kg } (70.6 \text{ kcal/kg}) = \boxed{35.3 \text{ kcal}}$$

Example 13.9

An ice storm in Eau Claire, Wisconsin, leaves a 2-in layer of ice on the driveway of a professor's house, which has an area of 80 m². The next day the temperature is exactly 0°C, and the professor wants to know whether to break up the ice or let the sun melt it away. If the driveway is directly exposed to the sun, and if the average solar energy falling on the driveway is 200 W/m², how long will it take to melt all the ice? (The density of ice was given in Table 10.3 as 0.92×10^3 kg/m³.)

SOLUTION

Consider a 1.0-m² slab of ice 2.0 in thick. Now

$$2.0 \text{ in} = (2.0 \text{ in})\left(2.54 \frac{\text{cm}}{\text{in}}\right) = 5.08 \text{ cm} = 0.051 \text{ m}$$

Hence the volume of ice in an area 1.0 m² is

$$V = (0.051 \text{ m})(1.0 \text{ m}^2) = 0.051 \text{ m}^3$$

and the mass of this much ice is $m = dV = (0.92 \times 10^3$ kg/m³)(0.051 m³) = 47 kg.

The amount of heat required to melt this ice to water at 0°C is

$$Q = mL_f = (47 \text{ kg})(80 \text{ kcal/kg}) = 3.8 \times 10^3 \text{ kcal}$$

or, converting to joules,

$$Q = (3.8 \times 10^3 \text{ kcal})(4.18 \times 10^3 \text{ J/kcal}) = 1.6 \times 10^7 \text{ J}$$

Now, 200 W/m² is equal to 200 J/s falling on 1.0 m², and so the time required to melt all the ice is

$$t = \frac{Q}{P} = \frac{1.6 \times 10^7 \text{ J}}{200 \text{ J/s}} = 8.0 \times 10^4 \text{ s}$$

or $\quad t = (8.0 \times 10^4 \text{ s})\left(\frac{1 \text{ h}}{3600 \text{ s}}\right) = \boxed{22 \text{ h}}$

Hence if the professor wants the ice removed, he or she had better start chopping away at it, since it will be dark before half the ice melts.

Note: The area of the driveway is of no importance in this problem, since the ice is assumed to have the same thickness over the whole driveway, and the energy input from the sun is the same over the whole driveway. Hence the result would be the same, whether calculated for 1 m² or 80 m².

Vapor Pressure and Boiling

The heat of vaporization is usually given for the normal boiling temperature of liquids, e.g., 100°C for water at atmospheric pressure. At different pressures, however, liquids boil at different temperatures and also have different heats of vaporization at these temperatures. To see why this is so, consider the following.

If a liquid is placed in a closed container from which some of the air has been evacuated, some molecules will evaporate from the liquid into the gaseous state, and other vapor molecules will return from the gaseous to the liquid state. When the number of molecules leaving the liquid is exactly equal to the number returning to the liquid, the vapor is said to be *saturated*, and the vapor is said to be in *equilibrium* with the liquid.

Definition

Vapor pressure: The pressure of a vapor which is in equilibrium with a liquid (or solid), i.e., the pressure of the saturated vapor.

As the temperature is increased, the number of molecules leaving the liquid increases, and as a result the vapor pressure also increases rapidly.

If a liquid is heated continuously, some evaporation occurs even at low temperatures and increases as the temperature increases. Bubbles of vapor are formed within the liquid, but if the external pressure is greater than the vapor pressure of the liquid at the given temperature, the bubbles collapse as they are formed. If the temperature is increased to the point where the vapor pressure of the liquid becomes equal to the external pressure, then these bubbles rise up through the liquid and *boiling* occurs. For water at atmospheric pressure this occurs at 100°C. We can therefore define the *boiling point* as follows:

Definition

Boiling point: That temperature at which the vapor pressure of a liquid is equal to the external pressure.

At the boiling point the liquid is rapidly converted to a vapor at the temperature of the boiling point, while for each kilogram boiled off an amount of energy equal to the heat of vaporization must be provided by some source external to the liquid.

For water under atmospheric pressure (760 mmHg, or 760 torr) boiling occurs at 100°C with a heat of vaporization of 540 kcal/kg. At Denver, Colorado, which is approximately 1 mile above sea level, atmospheric pressure is reduced to about 610 torr, and water boils at about 94°C with a heat of vaporization of 543 kcal/kg. In a pressure cooker, on the other hand, the boiling point of water is raised by increasing the pressure on the water above that of the atmosphere. Hence in this way it is possible to cook food at a higher temperature.

Changes in pressure also affect the freezing point of liquids but to a much more limited extent. Thus an increase in pressure of 1 atm lowers the freezing point of water by only about 0.007°C.

Sublimation

Under conditions of reduced pressure, ice may be made to change directly to steam without going through the liquid state. This is called *sublimation*, and the heat required to cause this transformation is called the *heat of sublimation*. Sublimation is not normally observed for water, because the required pressure is so low, below 4.58 torr.

A pressure-temperature phase diagram for water is shown in Fig. 13.20*a*. It includes the *triple point* of water at 0.01°C and 4.58 mmHg.

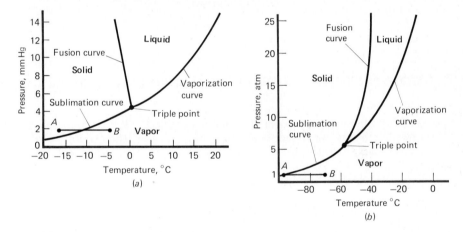

FIGURE 13.20 Phase diagrams for (*a*) water (H_2O) and (*b*) carbon dioxide (CO_2).

Definition

Triple point: That point on a pressure-temperature phase diagram at which a substance can exist in all three phases (solid, liquid, and vapor) simultaneously.

Note on the phase diagram for water that, by keeping the pressure constant below 4.58 mmHg and increasing the temperature, it is possible to have the substance *sublime*, as indicated by the line $A \rightarrow B$ on the diagram.

Figure 13.20*b* shows a similar phase diagram for carbon dioxide (CO_2), on which the triple point occurs at a pressure of 5.11 atm and a temperature of -56.6°C. Hence at atmospheric pressure CO_2 sublimes, passing directly from the solid phase ("dry ice") to the vapor phase. The fusion curve for CO_2 has a positive slope, which is typical of most substances and unlike the negative slope shown for H_2O (a most untypical substance) in Fig. 13.20*a*. Notice the similar forms of the curves describing the phase transitions for these two substances, even though the pressures and temperatures shown differ greatly.

Example 13.10

How much heat is required to boil 0.50 kg of water originally at room temperature (20°C) and to raise the temperature of the steam to 120°C? This can be done by collecting all the steam in a closed vessel and continuing to heat it until it reaches 120°C. [The specific heat capacity of steam is 0.48 kcal/(kg·°C).]

SOLUTION

Let us break this problem down into three parts:

1 The heat required to raise the temperature of 0.5 kg of water from 20 to 100°C is

$$Q_1 = cM \, \Delta T = \left(1 \frac{\text{kcal}}{\text{kg·°C}}\right)(0.50 \text{ kg})[(100 - 20)°\text{C}] = 40 \text{ kcal}$$

2 The heat required to boil 0.50 kg of water at 100°C is

$$Q_2 = mL_V = (0.5 \text{ kg})\left(540 \frac{\text{kcal}}{\text{kg}}\right) = 270 \text{ kcal}$$

3 The heat required to heat 0.50 kg of steam from 100 to 120°C is

$$Q_3 = cm \, \Delta T = \left(0.48 \frac{\text{kcal}}{\text{kg·°C}}\right)(0.5 \text{ kg})[(120 - 100)°\text{C}]$$

$$= 4.8 \text{ kcal}$$

The total heat required is then

$$Q = Q_1 + Q_2 + Q_3$$

$$= (40 + 270 + 4.8) \text{ kcal} = \boxed{315 \text{ kcal}}$$

Note the very large contribution made to the total by the process of vaporization.

Summary: Important Definitions and Equations

Temperature:
A measure of the average random translational kinetic energy of the atoms or molecules in a substance.
Temperature scales (Kelvin, Celsius, and Fahrenheit):

$$T(\text{K}) = T(°\text{C}) + 273.15$$

$$T'(°\text{F}) = (\tfrac{9}{5})T(°\text{C}) + 32.0$$

Absolute zero:
That temperature at which the pressure of an ideal gas falls to zero at constant volume; 0 K = −273.15°C.
Temperature expansion of a rod: $\quad L = L_0 (1 + \alpha \, \Delta T)$
where α is the coefficient of linear expansion.

Volume expansion of a solid or liquid: $\quad V = V_0(1 + \beta \, \Delta T)$
where β is the coefficient of volume expansion.

Thermal energy:
The sum of all the random mechanical energies of the atoms on molecules in a substance.
Heat:
The thermal energy transferred between a system and its surroundings as a result of temperature differences only.

$$\Delta Q = cm \, \Delta T = cm(T_2 - T_1)$$

Specific heat capacity (c):
The quantity of heat (in kilocalories) which must be supplied to a unit mass (1 kg) of a substance to raise its temperature by one degree (1°C or 1 K).
Molar heat capacity:
The specific heat capacity for 1 mol rather than 1 kg of a substance.
Heat units:
One kilocalorie (kcal): The amount of heat required to raise the temperature of one kilogram of water one degree Celsius (from 14.5 to 15.5°C).
One British thermal unit (Btu): The amount of heat required to raise the temperature of one pound of water one degree Fahrenheit (from 63 to 64°F).
Internal energy of an ideal monatomic gas:
$$U = \tfrac{3}{2}nRT$$
where n = number of moles
R = 8.31 J/(mol·K)
T = temperature, K
Molar heat capacity of gases: For all gases
$$C_P - C_V = R$$
where C_P = molar heat capacity at constant pressure
C_V = molar heat capacity at constant volume

Ratio of the two molar heat capacities: $\gamma = \dfrac{C_P}{C_V}$

Phase:
The state in which a substance exists, i.e., solid, liquid, or gas.
Phase equilibrium:
A physical situation in which two different phases of the same substance coexist at the same temperature.
Phase transition:
A change in the phase of a substance accompanied by the absorption or liberation of heat.

$$Q = Lm$$

where L = heat absorbed or liberated, kcal/kg
Heat of fusion of ice:
The amount of heat (80 kcal) required to melt 1 kg of ice to water at 0°C.
Heat of vaporization of water:
The amount of heat (540 kcal) required to vaporize 1 kg of water to steam at 100°C.
Sublimation:
The conversion of the solid phase of a substance to the vapor phase without passing through the liquid phase.

Questions

1 Does the following question, often asked by a parent of a sick child, make sense physically? "Do you have a temperature?" What is really meant?

2 Name three other properties of substances, in addition to the thermal expansion of gases, liquids, and solids, which might be used to measure temperature?

3 Why in practice cannot a single thermometer be used to measure temperatures all the way from 0 to 10^8 K?

4 Is there any temperature which is the same in degrees Fahrenheit and in kelvins?

5 If it proves difficult to open a jar with a screw-type metal lid, the conventional wisdom is that you should put the lid of the jar under hot running water. Why would you expect such a technique to help in opening the jar?

6 How is the specific heat capacity of a gas related to the structure of the molecules in the gas?

7 Show, by analogy with the flow of water, why heat always flows of its own accord from a higher to a lower temperature and never the reverse.

8 Heats of fusion and vaporization are sometimes referred to as "latent heats." Can you explain why this is an apt name for them?

9 Explain why in a coastal state like Maine less snow and higher temperatures occur in winter along the coast than occur farther inland.

10 The vapor pressure of a liquid does not depend on the external pressure. Why, then, does the boiling point of a liquid depend on the external pressure?

11 You want to make water boil, but have no way to heat it. You do have available, however, a vacuum chamber from which the air can be pumped out. Explain how you might be able to make water boil at room temperature (about 20°C) using this apparatus.

12 Why is water a particularly good substance to use in home heating systems, whether they be hot-water or steam heating systems?

13 Explain why in a house whose furnace breaks down in the middle of winter there is danger that the water pipes will burst. Can you suggest a way to prevent this from happening?

14 Count Rumford once remarked that he was amazed at how long apple pies retain their heat after being taken out of the oven. What explanation can you provide for this behavior?

Problems

A 1 The order-of-magnitude difference between the temperature required for deuterium-deuterium (D-D) fusion, (2.0×10^8)°C, and room temperature (on the Kelvin scale) is:
(a) 10^6 (b) 10^7 (c) 10^{-6} (d) 10^{-7}
(e) 10^5

A 2 When an ordinary metal is heated, the order of magnitude of its change in length per unit length per Celsius degree is:
(a) 10^{-3} (b) 10^{-5} (c) 10^{-7} (d) 10^{-9}
(e) 10^{-12}

A 3 The order of magnitude of the change in area per unit area per Celsius degree for an average metal when it is heated is:
(a) 10^{-3} (b) 10^{-5} (c) 10^{-6} (d) 10^{-10}
(e) 10^{-12}

A 4 The order of magnitude of the change in volume per unit volume per Celsius degree of an average liquid when it is heated is:
(a) 10^{-3} (b) 10^{-5} (c) 10^{-6} (d) 10^{-8}
(e) 10^{-10}

A 5 It takes 10 kcal to raise the temperature of a certain mass of water 10°C. The amount of heat required to raise the temperature of the same mass of ice 10°C will be about:
(a) 10 kcal   (b) 1 kcal   (c) 20 kcal
(d) 5 kcal   (e) 2 kcal

A 6 The quantity of heat required to raise the temperature of 5 kg of iron from 20 to 300°C is:
(a) 1400 kcal   (b) 158 cal   (c) 158 kcal
(d) 1400 Btu   (e) 1.24×10^4 kcal

A 7 A mixture of 100 g of ice and 200 g of water is

heated by applying 5 kcal of heat to the mixture. The final temperature of the ice-water mixture will be:
(a) 50°C (b) 100°C (c) 16.7°C (d) 0°C
(e) 25°C

A 8 The specific heat capacity of iron is larger than the specific heat capacity of water by about:
(a) One order of magnitude
(b) Zero orders of magnitude
(c) Two orders of magnitude
(d) Three orders of magnitude
(e) None of the above

A 9 The internal energy of 1 mol of an ideal monatomic gas at 300 K is about:
(a) 4×10^3 J (b) 4×10^3 kcal (c) 4×10^6 J
(d) 4×10^6 kcal (e) 2×10^6 J

A10 The molar heat capacity at constant volume of krypton gas is 2.98 cal/(mol·K). Its molar heat capacity at constant pressure [in cal/(mol·K)] is:
(a) 2.98 (b) 3.98 (c) 4.98 (d) 1.98
(e) 0

A11 The heat of vaporization of water is larger than the heat of vaporization of helium by about:
(a) One order of magnitude
(b) Zero orders of magnitude
(c) Two orders of magnitude
(d) Three orders of magnitude
(e) None of the above

B 1 Express the temperatures 0 K, 300 K, and 600 K in degrees Fahrenheit.

B 2 Express the temperatures −40°F, 68°F, and 98.6°F in degrees Celsius.

B 3 A nurse takes a patient's temperature and finds that it is 102°F. What is this temperature on the Celsius scale?

B 4 Waterford crystal is made by first melting a mixture of silica, potash, and red lead oxide in a furnace at 1200°C. What is this temperature in degrees Fahrenheit and in kelvins?

B 5 In kelvins, what is the ratio of the melting point of steel to that of lead? What is this ratio in degrees Celsius? (See Table 10.3.)

B 6 A steel bridge is 500 m long at 20°C. How much will it shorten when the temperature falls to 0°C?

B 7 If the earth were a steel ball, what temperature change would make its radius increase 35 m?

B 8 How much heat is required to produce a 10°C temperature change in 2.0 kg of (a) aluminum, (b) ice, and (c) water?

B 9 How many kilocalories are required to heat 1.0 kg of water in a 0.10-kg aluminum container from 5 to 50°C? From 50 to 95°C?

B10 What is the specific heat of mercury in J/(kg·K)?

B11 How much heat is required to convert 2.0 kg of solid mercury at −38.87°C to liquid mercury at the same temperature?

B12 How much heat must be extracted from 0.5 kg of hydrogen gas at 20.4 K to convert it to liquid at the same temperature?

B13 If the "heat output" of a normal person (120 W) could all be used to melt ice, how many kilograms per second would be melted?

C 1 An aluminum meterstick of mass 0.40 kg reads correctly at 20°C. How much longer will the meterstick be if it absorbs 7.0 kcal of heat?

C 2 A Pyrex flask in a refrigerator at −10°C holds exactly 1000 cm³ of ethyl alcohol. If the refrigerator fails, how much alcohol spills by the time the temperature reaches 22°C?

C 3 A 13.6-kg sample of liquid mercury occupies 10^3 cm³ at its melting point (−38.9°C). (a) What is its volume just as it reaches its boiling point (356.9°C)? (b) How many kilocalories of heat does this take?

C 4 Suppose a 75-kg person consumes 2500 kcal in a day. If all that energy went into heat, what would the person's final temperature be? [Assume $c = 0.83$ kcal/(kg·°C).] (b) How many kilocalories should be taken in to get the person to the boiling point of water?

C 5 One kilogram of water gives up 80 kcal upon freezing. If another 80 kcal is removed, what is the final temperature of the ice?

C 6 Suppose air has density 1.3 kg/m³ and specific heat capacity 0.20 kcal/(kg·°C). If the volume of air in the passenger compartment of an automobile is 2.0 m³, how long would it take the body heat of one person (120 W) to raise the air temperature from 15 to 37°C?

C 7 If the amount of radiant solar energy reaching the car of Prob. C6 each second is 200 J/m², and if the effective area of the car windows is 1.5 m², how long would it take the sun alone to raise the temperature by the same amount?

C 8 A 0.70-kg cup [$c = 0.22$ kcal/(kg·°C)] is at 20°C. If 0.25 kg of water at 95°C is poured into the cup, what is the final temperature?

C 9 Suppose we drop a 1.0-kg piece of iron at 200°C into 5.0 kg of water in a 0.20-kg aluminum container at room temperature (20°C). If no heat is exchanged with the surroundings, what is the final temperature of the iron and the water?

C10 A 5.0×10^{-2} kg iron vessel and the 0.20 kg of water in it are at 20°C. If 8.0×10^{-2} kg of metal shot at 100°C is dropped into it, the final temperature is 24°C. What is the specific heat of the metal?

C11 A 0.21-kg brass calorimeter contains 0.10 kg of water. If 1.20 kcal is required to raise the temperature of the water and container 10°C, what is the specific heat of brass?

C12 According to the Law of Dulong and Petit, most metals have a molar heat capacity of 6 cal/(mol·°C) at room temperature. A metal has molecular mass 197. If it obeys the Dulong-Petit law, what is its specific heat?

C13 One mole of CO_2 is heated at constant pressure from −10 to 37°C. What is the ratio of the final volume to the initial volume?

C14 How many kilocalories are required to change 2.0 kg of ice at −20°C to water at 100°C?

C15 If 0.20 kg of ice at −5°C is put into water at 0°C, how much water turns to ice?

C16 One-fifth of a kilogram of ice at $-10°C$ is put into 0.50 kg of soft drink $[c = 0.85$ kcal/(kg·°C)] at 22°C. What is the final temperature?

C17 One hundred grams of ice is in an insulated container.

(*a*) How much water at 100°C must be admitted in order to just melt all the ice?

(*b*) What mass of steam at 100°C would be just sufficient to melt the ice?

C18 A 5.0-kg iron ball at 1200°C is put into a 2.0-kg iron pail containing 12 kg of water at 20°C. How much steam is produced?

C19 A 1.0-mm-thick crust of ice covers a windshield of area 0.50 m². (*a*) How much energy is required to melt the ice? (*b*) If the job is to take 3 min, what must be the output of the heater (in watts)?

C20 Air conditioners are sometimes rated in tons. Thus a 2-ton air conditioner removes heat at the same rate as would the melting of 2 tons (1 ton = 2000 lb) of ice per day. What is the rate (in watts) at which a 2-ton air conditioner removes heat from a house?

C21 When a person is talking, about 10^{-6} J/m² of energy is being radiated each second. If 1 g of ice absorbs energy at this rate through one face of a 1.0-cm² cube, how long would it take to melt the ice by "talking at it"?

C22 To compare how much energy per particle is involved:

(*a*) Calculate the molar heat of fusion for water, mercury, hydrogen, and helium.

(*b*) Do the same for the heats of vaporization. Comment on trends or abnormalities.

C23 From what height should an ice cube fall in order to acquire sufficient energy to melt itself? Express your answer in kilometers and in miles.

C24 For simplicity, assume that the heat of vaporization of water is independent of temperature. If so, how many kilograms of perspiration (water) are needed to change the temperature of a 75-kg person by one Celsius degree?

C25 A marathon runner generates heat at a rate of 10^3 W while in the middle of a race (as compared with 120 W for a person at rest).

(*a*) How many kilograms of water per hour must the runner lose by evaporation, with the skin at a temperature of 35°, to keep the runner's body temperature from rising? (Assume that the heat of vaporization of water at 35°C is 577 kcal/kg.)

(*b*) What percentage of the runner's total body mass of 80 kg does this loss represent?

C26 Compare the amount of steam that must be circulated in a steam-heating system to produce 10^3 kcal of heat, with the amount of hot water that would have to be circulated to achieve the same heating effect. Assume that the hot water leaves the furnace at 60°C and returns at 35°C.

D 1 Show that the volume coefficient of expansion for an ideal gas is $1/T$ if the gas is heated at constant

pressure. At what temperature does this equal the volume coefficient of expansion for copper?

D 2 Consider a mercury-in-glass thermometer. If the cross section of the capillary is a constant A, and if the mercury just fills the bulb of volume V_0 at 0°C, show that the length of the mercury in the capillary at temperature t°C is $L = (V_0/A)(\beta - 3\alpha)t$, where β is the volume coefficient of expansion of mercury and α is the linear coefficient of expansion of the glass.

D 3 A steel tape measure gives the length of a brass rod to be 125.00 cm when both are at 20°C. What would the tape measure read when the temperature increases to 40°C? (*Note:* Five significant figures are required in your answer.)

D 4 Two rods, each of 5.0-cm² cross-sectional area and 15 cm long, are placed end to end and held rigidly in clamps at the two free ends. One rod is made of steel and the other of brass. If, after the rods are put in place, they are heated by a flame until their temperature increases by 100°C, what is the stress set up in each rod?

D 5 When the outside temperature is 10°C, a steel beam of cross-sectional area 120 cm² is installed in a building with its ends bolted securely to two immovable pillars. In the summer the temperature of the beam rises to 35°C. What is the compressional force on the beam?

D 6 Find the numerical relationship between the kilocalorie and the British thermal unit.

D 7 The temperatures of three different liquids are maintained at 10, 20, and 30°C, respectively. It is found that mixing equal masses of the first two liquids leads to a final temperature of 16°C, and that mixing equal masses of the second and third liquids leads to a final temperature of 24°C. What will be the final temperature if equal masses of the first and third liquids are mixed together?

D 8 An aluminum can of mass 60 g is used as a "calorimeter" in which heat experiments are performed. The calorimeter contains 120 g of water-ice mixture. An aluminum bar of mass 120 g is taken out of a steam bath, where it has reached a temperature of 100°C, and dropped into the calorimeter. The temperature of the calorimeter's contents rises to 6.0°C. How much ice was originally in the calorimeter?

D 9 A closed vessel contains liquid water in equilibrium with its vapor at 100°C and 1 atm. Compare the mean kinetic energy of a vapor molecule with the energy required to transfer one molecule from the liquid to the vapor phase.

D10 Substance X has the following properties:

$$c_{solid} = 0.4 \text{ kcal/(kg·°C)} \qquad c_{liquid} = 1.5 \text{ kcal/(kg·°C)}$$

Melting point = 35°C   Boiling point = 210°C

Heat of fusion = 105 kcal/kg

Substance X and water are immiscible and do not react chemically with each other. If 2 kg of substance X and 5 kg of ice start out together in a container at $-15°C$, how long does it take the combination to reach 60°C if heat is put in at the rate of 1 kcal/s?

Additional Readings

Holton, Gerald: *Introduction to Concepts and Theories in Physical Science*, 2d ed. (revised and with new material by Stephen G. Brush), Addison-Wesley, Reading, Mass., 1973. Chapter 17 in particular is excellent with respect to the historical and philosophical aspects of the material in this and the next chapter.

Lindsay, Robert Bruce: *Julius Robert Mayer, Prophet of Energy*, Pergamon Press, New York, 1973. The introductory biographical sketch in this book is probably the best account in English of Mayer's life. This book also contains reprints of all Mayer's important scientific papers.

MacDonald, D. K. C.: *Near Zero: An Introduction to Low Temperature Physics*, Doubleday/Anchor, Garden City, N.Y., 1961. A popular book which includes discussions of magnetic cooling, superconductivity, the properties of liquid helium, and some interesting applications.

Scientific American, vol. 191, no. 3, September 1954. This issue contains nine articles on heat, thermal energy, and temperature. Especially good is the one entitled "What Is Heat?," by Freeman J. Dyson.

Zemansky, M. W.: *Temperatures Very Low and Very High*, Van Nostrand, New York, 1964. An interesting account of how matter behaves at extreme temperatures.

The two books on Count Rumford by Sanford Brown referred to in Chap. 6 are also relevant here.

I shall lose no time in repeating and extending these experiments, being satisfied that the grand agents of nature are, by the Creator's fiat, indestructible; and that, whenever mechanical force [i.e., work] is expended, an exact equivalent of heat is always obtained.

Sir James Joule (1818–1889)

Chapter 14

The Transfer of Heat and the First Law of Thermo-dynamics

Many of us have used a hand pump to put air into a bicycle or automobile tire and noticed that both the tire and the pump became hot in the process. Where did the heat needed to raise the temperature come from? No obvious source of heat was present, since the air in the tire and the air outside the tire were probably initially at the same temperature. We did work, however, to operate the pump and hence we might suspect that there was some relationship between the work we did and the thermal energy that raised the temperature of the air in the tire. In this chapter, after first discussing the heat transfer processes of conduction, convection, and radiation, we will consider how mechanical work can increase the thermal energy of a substance. Then, by putting together these ideas on the transfer of heat and the performance of work, we will arrive at the first general principle of the science of heat, the first law of thermodynamics.

14.1 The Conduction of Heat

The molecules in a metal hot plate are in violent, agitated motion, since they are at a high temperature. These molecules collide with the molecules in the outermost molecular layer of a glass flask set on top of the hot plate and transfer some of their energy to the glass molecules. These in turn increase their motion, and, even though they cannot move freely through the glass, they collide with and pass on some of their excess energy to these neighboring molecules, which in turn pass energy to their neighbors, etc., until finally the gas molecules within the container receive the heat which has passed through the glass.

Definition

This process, in which heat is carried from one molecule to the next through a series of collisions, is called *conduction. In the conduction of heat, molecules, while remaining at relatively fixed positions in space, absorb and pass on heat in the form of random mechanical motion.* We describe this process by saying that a certain amount of heat Q is conducted through the glass from a high temperature T_2 to a lower temperature T_1. The amount of heat conducted per second depends on both the number and kind of molecules involved.

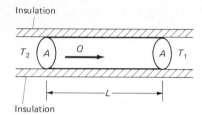

Insulation

T_2 A Q A T_1

Insulation

L

FIGURE 14.1 The conduction of heat along a piece of solid glass pipe of length L and cross-sectional area A.

In Fig. 14.1 we consider the conduction of heat through a piece of solid glass pipe whose ends are kept at temperatures T_2 and T_1, and whose sides are insulated so that no heat can escape. We might expect that the amount of heat transferred would be directly proportional to the cross-sectional area A of the glass normal to the direction of heat flow (for this determines the number of molecules involved) and to the time t the heat flows. We might also expect the amount of heat transferred to be proportional to the temperature gradient $(T_2 - T_1)/L$, just as the flow of water in a pipe is proportional to the pressure gradient $(P_2 - P_1)/L$, according to Poiseuille's law, Eq. (10.10). Hence we can write

$$Q \propto \frac{At(T_2 - T_1)}{L}$$

or

$$P = \frac{Q}{t} \propto \frac{A(T_2 - T_1)}{L}$$

Here P is the rate at which heat (a form of energy) flows, and therefore has the dimensions of *power*. We can convert this proportionality into an equality by introducing a constant of proportionality which we call the *thermal conductivity k*. We then have

$$\boxed{P = \frac{Q}{t} = k\frac{A(T_2 - T_1)}{L}}$$

(14.1)

In this equation, which has been well confirmed by experiment, the rate of heat conduction Q/t is usually expressed in kilocalories per second, A is in meters squared, L in meters, and $T_2 - T_1$ in degrees Celsius or kelvins. The units for the thermal conductivity k are therefore

$$\frac{\text{kcal}}{\text{s}} \cdot \frac{\text{m}}{\text{m}^2} \cdot \frac{1}{°C} \quad \text{or} \quad \text{kcal/(s·m·°C)}$$

The rate of heat flow can also be expressed in joules per second (or watts), in which case k would have to be in W/(m·°C).

The thermal conductivity k varies greatly from material to material, as Table 14.1 shows, being large for a good heat conductor and small for a poor one. Metals have large thermal conductivities and are very good conductors of heat just as they are good conductors of electricity. Gases are poor heat conductors, or good heat insulators. Many materials like Styrofoam or rock wool are good insulators because they are porous and contain large amounts of air in their interstices. Similarly, clothing insulates the body by trapping a layer of warm air next to the skin which prevents heat flow from the body to the cold air outside.

Carpeting in a bathroom is more comfortable than tile to bare feet on a winter morning because the carpeting conducts heat poorly. As a consequence, it heats up quickly to the temperature of our bodies and feels relatively warm to the touch. A bare tile bathroom floor, on the other hand, quickly conducts heat away from our feet and for this reason feels much colder.

TABLE 14.1 Thermal Conductivities of Some Common Materials

Material	Thermal conductivity k	
	kcal/(s·m·°C)	W/(m·°C)
Metals:		
Copper	9.2×10^{-2}	3.8×10^2
Aluminum	4.9×10^{-2}	2.1×10^2
Brass	2.6×10^{-2}	1.1×10^2
Steel	1.2×10^{-2}	5.0×10
Silver	1.0×10^{-1}	4.2×10^2
Gold	7.0×10^{-2}	2.9×10^2
Nonmetals:		
Granite	$\sim5.0 \times 10^{-4}$	~2.1
Glass	$\sim2.0 \times 10^{-4}$	~0.84
Ice	4.0×10^{-4}	1.7
Red brick	$\sim1.5 \times 10^{-4}$	~0.63
Concrete	$\sim2.0 \times 10^{-4}$	~0.84
Water	1.4×10^{-4}	0.58
Wood (across grain)	$0.2\text{–}0.4 \times 10^{-4}$	$0.084\text{–}0.168$
Masonite	0.1×10^{-4}	0.04
Glass wool	0.1×10^{-4}	0.04
Rock wool	0.09×10^{-4}	0.038
Air	0.057×10^{-4}	0.024
Styrofoam	0.024×10^{-4}	0.010

Example 14.1

(a) Calculate the rate of heat flow through a glass window 1.0 m wide and 1.5 m high if the glass is 2.0 mm thick. Assume that the temperatures of the inner and outer surfaces of the glass are 10 and 5°C, respectively.
(b) Suppose that the window is replaced by a regular wall with rock wool insulation 15 cm thick. Assume that the inner and outer surfaces of the wall remain at the same temperatures as the inner and outer surfaces of the window, and calculate the rate of heat flow through the 1.0×1.5 m area of wall.

SOLUTION

(a) From Eq. (14.1),

$$\frac{Q}{t} = k\frac{A(T_2 - T_1)}{L}$$

For the window,

$$\frac{Q}{t} = \left(2.0 \times 10^{-4}\,\frac{\text{kcal}}{\text{s·m·°C}}\right)(1.0\text{ m})(1.5\text{ m})\frac{(10-5)°\text{C}}{2.0 \times 10^{-3}\text{ m}}$$

$$= \boxed{0.75\text{ kcal/s}}$$

(b) For the wall,

$$\frac{Q}{t} = \left(0.09 \times 10^{-4}\,\frac{\text{kcal}}{\text{s·m·°C}}\right)(1.0\text{ m})(1.5\text{ m})\frac{(10-5)°\text{C}}{0.15\text{ m}}$$

$$= \boxed{4.5 \times 10^{-4}\text{ kcal/s}}$$

Hence the loss of heat is reduced by a factor of $0.75/(4.5 \times 10^{-4}) = 1.7 \times 10^3$ when the window is replaced by a wall, under the conditions stated in the problem.

This very rough calculation shows that a major source of heat loss in houses is by conduction through windows. Note that we have not said that the temperatures inside and outside the house were 10 and 5°C. It is likely that in this case the inside temperature might be close to 20°C and the outside air temperature well below freezing. Much of the fall in temperature actually occurs in the thin layers of air that are immediately adjacent to the glass on the two sides. How effective these layers are in reducing the conduction of heat depends on how much they are stirred up by the motion of the surrounding air. A strong wind, for example, will constantly replace the layer of air on the outside of the window pane with colder air and hence increase the heat loss through the window. Storm windows and thermal-pane windows reduce this heat loss by introducing a layer of trapped air, or a vacuum, between the inner and outer windows. This trapped air layer reduces the heat losses by both conduction and convection.

Heat Conduction in Houses

Understanding Eq. (14.1) can help us save energy in our homes. Since most heat losses in a house are by conduction, insulating our houses against conductive heat losses is very important. This means making k, A, and $T_2 - T_1$ as small as possible and L as large as possible. We have little control over T_1, which in this case is the outside temperature, but lowering the house thermostat by a few degrees and thus lowering T_2 can provide real savings. The surface area A in contact with the outside is, in general, a function of the kind of house we live in. Townhouses and apartments are much easier to keep warm in winter than detached houses because the latter have so much more surface area in direct contact with the elements. The crucial factor in insulating a house (in addition to caulking cracks and holes) is to use materials with low thermal conductivities and to make L as large as possible by having the walls as thick as possible. Modern building practice calls for relatively thin outer and inner walls, with the space between them filled with large thicknesses of insulating materials of low k. This is in contrast to older building styles, in which very thick stone walls were used.

14.2 The Convection of Heat

If we set a beaker of water on a hot plate, we find that once the heat is conducted through the glass wall of the beaker, a new kind of heat transfer process occurs. The glass molecules conduct heat to the water molecules touching the glass. The heated water molecules then rise to the surface of the water and are replaced by cooler water molecules, which are then heated and repeat the cycle, as in Fig. 14.2.

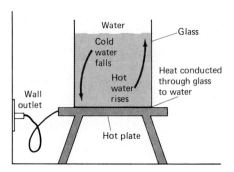

FIGURE 14.2 The convection of heat in a beaker of water.

Definition

The process whereby heated molecules carry thermal energy with them physically as they move from one part of a fluid to another is called convection.

In this case it occurs because hot water, being lighter than cold water, spontaneously rises to the surface. A similar process occurs in a room heated by a radiator. Air molecules are constantly being heated by contact with the radiator. They rise and are replaced by cooler air molecules, which are in turn heated, rise, and repeat the cycle.

Theory of Convection

Convection is a difficult topic to handle quantitatively, because it involves both a hydrodynamic problem (the complicated flow of a fluid) and a heat transfer problem. In fact the situation is so involved that it is impossible to derive any general equation to describe convection. The best we can do is to define a convection coefficient h by analogy with the coefficient of thermal conductivity k previously introduced. We therefore write for the rate at which heat is transferred by convection

$$P = \frac{Q}{t} = hA(T_2 - T_1) \tag{14.2}$$

Here A is the area of the hot surface (at temperature T_2) with which the fluid is in contact, and T_1 is the temperature of the main body of the fluid. The convection constant h, which must have units of W/(m²·°C) or kcal/(s·m²·°C) to make Eq. (14.2) dimensionally correct, has to be determined from experiment in each case, since it depends on the properties of the fluid, including its rate of flow, and on the size, shape, and orientation of the hot object.

14.3 Heat Transfer by Radiation

The third method by which heat can be transferred from one object to another is by *radiation*. Even though in nuclear physics the term *nuclear radiation* includes particles like electrons and alpha particles, in discussing heat we restrict the term *radiation* to the following meaning:

Definition

Radiation: The transfer of electromagnetic energy from one object to another without any transfer of matter and without the need for a material medium to accomplish this transfer.

For example, when we are stretched out on the beach in the summertime, we are absorbing energy from the sun, but we can detect no material substance being transported from the sun to our bodies. Similarly, the warmth we derive from a fireplace is mostly from radiation, since most of the heated air is convected up the chimney and not into the room.

Electromagnetic radiation (which will be discussed in detail in Chap. 21) includes a great variety of different kinds of electromagnetic waves—such as radio waves, light waves, and x-rays—which are produced and detected in different ways depending on their wavelengths. In this chapter we are concentrating on one particular kind of electromagnetic radiation which may be defined as follows:

Definition

Thermal radiation: The electromagnetic radiation emitted by hot objects because of their temperature.

Any object which is hotter than its surroundings loses heat to its surroundings. Some of this loss occurs by conduction and convection. But, especially at high temperatures, much of this loss occurs by means of thermal radiation.

Thermal radiation is produced by the random molecular motions of the atoms and molecules in a hot object like the sun. The motion of charged particles produces electromagnetic waves which are radiated out into space in the same way that the motion of electric charges in the transmitting antenna of a radio station sends radio waves out into space. Since the sun is so hot (6000 K), the electromagnetic waves radiated are partly in the visible region of the spectrum, where we perceive them as light, and partly in the infrared region of the spectrum, where we perceive them by their heating effects. For objects at temperatures below 1000 K most of the electromagnetic radiation emitted is in the infrared region of the spectrum, with wavelengths from about 0.7×10^{-6} m (the red end of the visible region) to about 1 mm in length. For this reason infrared radiation produced by hot objects is often referred to as thermal radiation, or heat radiation, although it is nothing but electromagnetic radiation in a particular wavelength range.

The frequency range spanned by infrared radiation includes the vibrational and rotational frequencies of many molecules in the human body. Hence when radiation from the sun or from an infrared lamp strikes our bodies, resonance occurs between frequencies in the infrared radiation and the natural frequencies of the molecules in our bodies. This sets the molecules into forced motion, increases their random kinetic energies, and hence increases our body temperature. This transfer of heat to our bodies is accomplished without the transfer of any matter from the sun or infrared lamp. The warm glow we experience when in front of a blazing fire is produced by this absorption of thermal radiation by our bodies.

Laws Governing Thermal Radiation

Stefan's radiation law, first proposed by the Austrian physicist Josef Stefan (1835–1893) to explain experimental data he had collected, states that the rate at which energy is radiated by an object at an *absolute temperature T* is proportional to the surface area A of the object and to the *fourth* power of the absolute temperature.

$$P = \frac{Q}{t} = e\sigma A T^4 \tag{14.3}$$

where σ (Greek *sigma*, the equivalent of the letter s) is a universal constant called the *Stefan-Boltzmann constant*, with a value $\sigma = 5.67 \times 10^{-8}$ W/(m^2·K^4). The emissivity e is a number between 0 and 1 which depends on the substance emitting the radiation and its surface condition. Dark, rough surfaces have emissivities close to 1, the value for a perfect radiator (called a *blackbody*), while white, shiny surfaces have values closer to 0. For example, the value of e for copper is about 0.3, for light human skin about 0.6, and for dark human skin about 0.8.

A second law governing thermal radiation was proposed by the German physicist Wilhelm Wien (1864–1928) in 1893, and is called the *Wien displacement law*. Wien showed that the wavelength at which the emitted radiation had maximum intensity could be related to the absolute temperature of the source by the following simple equation:

$$\lambda_{max}T = k = 2.898 \times 10^{-3} \text{ m·K} \tag{14.4}$$

The Wien displacement law fits well the experimental data which are plotted in Fig. 14.3 and which show that the maximum intensity radiated by a hot object occurs at shorter wavelengths (and hence higher frequencies) as the temperature increases. The constant k is called *Wien's constant*. This law predicts, for example, that if the sun emits radiation as a blackbody at 6000 K (which is approximately correct), then the wavelength at which maximum emission occurs should be

$$\lambda_{max} = \frac{k}{T} = \frac{2.898 \times 10^{-3} \text{ m·K}}{6000 \text{ K}} = 0.48 \times 10^{-6} \text{ m} = 0.48 \text{ } \mu\text{m}$$

FIGURE 14.3 Experimental radiation curves for different temperatures of the emitter. These graphs show the amount of energy radiated per unit time as a function of the wavelength at which the energy is emitted. Note how rapidly the total amount of energy emitted per unit time increases with the temperature, and how the peak of the curves moves to shorter wavelengths as the temperature increases.

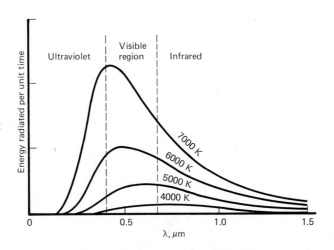

TABLE 14.2 Color Temperature of Stars

Color	Temperature, K
Reddish stars	2000–3000
Orange stars	3000–5000
Yellow stars (sun)	5000–8000
White stars	8000–12,000
Bluish stars	> 12,000

This wavelength is in the green region of the visible spectrum, which is exactly where solar radiation has its greatest intensity.

Wien's displacement law also explains why a piece of steel in a blast furnace will, as it heats up, first glow a dull red, then turn bright red, then appear yellow, and finally look almost pure white. The steel appears white because most of the radiation being emitted is in the visible region of the spectrum, and all the visible colors are present to a sufficient extent to produce an overall white effect. To make this description a bit more quantitative, Table 14.2 shows how the colors of stars can be used to estimate their approximate temperatures.

14.4 Radiation Balance

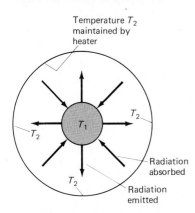

FIGURE 14.4 A metal sphere surrounded by an outer spherical shell maintained at a temperature T_2. As time goes on, the temperature of the metal sphere increases from T_1 to T_2.

Suppose we put a metal sphere at an initial temperature T_1 inside an evacuated spherical shell which is maintained at a higher temperature T_2 by an electric heater, as in Fig. 14.4. Both the sphere and the shell will be radiating energy at a rate determined by Stefan's law, but since the outside shell is hotter it will radiate at a faster rate. As time goes on, the temperature of the metal sphere gradually increases and comes to a final temperature T_2 which is the same as that of the outer shell. The sphere is then in thermal equilibrium with the outer shell. But it is now radiating energy at a rate $P = e\sigma A T_2^4$, according to Stefan's law, since its new temperature is T_2. If the sphere's temperature is not changing, it must also be *absorbing* radiation from the outer shell at exactly the same rate, $P = e\sigma A T_2^4$, as it is *radiating* energy. Hence the rate of absorption of energy by the sphere is the same as its rate of emission at the same temperature. A good emitter is therefore a good absorber, and a poor emitter is a poor absorber, the emissivity e being the same for both processes.

If a hot object with emissivity e at an initial temperature T_2 is surrounded by a closed metal shell initially at room temperature T_1 (about 300 K), as in Fig. 14.5, then the hot object initially radiates energy at a rate $e\sigma A T_2^4$ and absorbs energy from the shell at a rate $e\sigma A T_1^4$. Hence the net rate of energy flow from the hot object to the cooler object is

$$P = \frac{Q}{t} = e\sigma A(T_2^4 - T_1^4) \tag{14.5}$$

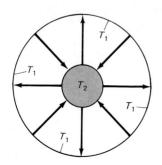

FIGURE 14.5 A metal sphere and a surrounding shell at a lower temperature and with no built-in sources of heat. The hot sphere loses energy and the colder shell absorbs energy until they both come to the same temperature. Finally a condition of radiation balance is reached in which each object absorbs exactly the same amount of energy it emits.

If the hot object has no built-in source of heat, its temperature will gradually fall from T_2 and the temperature of the surrounding shell will increase from T_1 until both come to the same final temperature T_f. At that time both continue to radiate and absorb energy, but no net exchange of energy occurs, since each object radiates just as much energy as it absorbs. This final situation is an example of *radiation balance*.

You may wonder why all hot objects do not eventually radiate all their energy away as heat and cool down to absolute zero, for only then will the radiated power P be equal to zero in Eq. (14.3). This is exactly what would happen if energy were not continuously supplied to these objects to enable them to continue to radiate without cooling down. The sun, for example, has fusion processes going on in its interior which constantly convert mass into energy. This is the source of the energy the sun radiates. The earth, for its part, is constantly receiving some of the sun's energy to replenish the energy the earth radiates into space. It is the radiation balance between the energy falling on the earth from the sun and the energy the earth radiates into space that determines the earth's temperature at any time. That temperature will constantly adjust itself until the earth is emitting exactly the same amount of energy that it receives from the sun. Any drastic change in the energy radiated by the sun or in its distance from the earth could destroy this precious balance and bring an end to life on earth by making the earth's temperature either too high or too low to sustain human life.

Something similar occurs with the human body. Since the skin temperature of the human body is about 33°C, and the walls of a room on a cold day may be at 18°C, the human body is constantly losing thermal energy by radiation to the walls as in Fig. 14.6. This, together with other sources of heat loss like conduction and convection, would be expected to make the temperature of the body fall. The body responds, however, by increasing the rate at which it converts food or body fat into thermal energy to keep the body temperature at its normal interior value of 37°C. This heat is then carried to the surface of the body by *convection,* with the heart acting as the pump and the blood as the circulating fluid. It can be shown from the food intake of the average person that the human body must generate heat at a rate of about 120 J/s, or 120 W on the average. One person therefore produces as much heat as that given out by one 120-W light bulb burning continuously.

FIGURE 14.6 Radiation of energy by human body to the walls of a room on a cold day. Both the body and the walls are continually absorbing and emitting radiation, but since the temperature of the body is about 33°C and that of the walls about 18°C, there is a net flow of energy from the body to the walls.

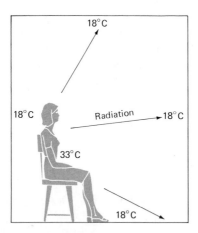

Example 14.2

(a) Compare the rate at which energy is radiated by 1 m²
of the sun's surface, if this is assumed to be at 6000 K, with
the rate at which energy is radiated by 1 m² of the earth's
surface, assumed to be at 290 K. Assume that the emissivity
e is the same for the earth and the sun.

(b) If the radius of the sun is 6.96×10^8 m, and its
emissivity is about 0.93, find the rate at which the sun
radiates energy at 6000 K.

SOLUTION

(a) From Eq. (14.3), we have for the sun and earth,
respectively,

$$P_S = \left(\frac{Q}{t}\right)_S = e\sigma A T_S^4 \quad \text{and} \quad P_E = \left(\frac{Q}{t}\right)_E = e\sigma A T_E^4$$

Hence the ratio of these two rates, since in both cases $A = 1$
m², is

$$R = \frac{P_S}{P_E} = \frac{e\sigma A T_S^4}{e\sigma A T_E^4} = \frac{T_S^4}{T_E^4}$$

and so $\quad R = \dfrac{(6000 \text{ K})^4}{(290 \text{ K})^4} = \boxed{1.8 \times 10^5}$

Hence in 1 s, 1 m² of the sun's surface radiates 1.8×10^5 times the energy radiated by the same surface area of the

earth in the same time. Since the surface area of the sun is
1.2×10^4 times that of the earth, the overall ratio of the
rates of radiation is 2.2×10^9. The sun therefore radiates
energy at a rate approximately 2 billion times that of the
earth.

(b) The rate at which the sun radiates energy is:

$$P_S = \left(\frac{Q}{t}\right)_S = e\sigma A T_S^4$$

$$= 0.93 \left(5.67 \times 10^{-8} \frac{\text{W}}{\text{m}^2 \cdot \text{K}^4}\right)(4\pi)\,(6.9 \times 10^8 \text{ m})^2\,(6000 \text{ K})^4$$

$$= \boxed{3.9 \times 10^{26} \text{ W}}$$

Example 14.3

A swimmer clad in a tight-fitting bathing suit is standing near an indoor pool. The surface temperature of the
swimmer's body is 33°C, and the temperature of the walls of
the room is 20°C.
(a) If the surface area of the swimmer's body is 1.3 m² and

its average emissivity is 0.65, find the net rate at which
energy is radiated by the swimmer's body.
(b) At what wavelength is the energy radiated a maximum?

SOLUTION

(a) From Eq. (14.5) we have $P = Q/t = e\sigma A(T_2^4 - T_1^4)$.
Here T_2 is 306 K and T_1 is 293 K. Hence

$$P = (0.65)\left(5.67 \times 10^{-8} \frac{\text{W}}{\text{m}^2 \cdot \text{K}^4}\right)(1.3 \text{ m}^2)(306^4 - 293^4) \text{ K}^4$$

$$= (4.8 \times 10^{-8})(1.4 \times 10^9) \text{ W} = \boxed{67 \text{ W}}$$

Hence the heat radiation from the swimmer's body is
roughly equivalent to the power consumed by a 60-W light
bulb.

(b) From Eq. (14.4) we have

$$\lambda_{max} = \frac{k}{T} = \frac{2.898 \times 10^{-3} \text{ m} \cdot \text{K}}{306 \text{ K}}$$

$$= 9.47 \times 10^{-6} \text{ m} = \boxed{9.47 \; \mu\text{m}}$$

This is in the infrared region of the spectrum, since it is
longer than 0.7 μm and shorter than 1 mm, which represent
the approximate limits of the infrared spectrum.

14.5 Some Applications of Heat Transfer Processes

Let us consider a few applications of the heat transfer processes already
discussed in this chapter in order to get some feeling for their practical
importance.

Newton's Law of Cooling

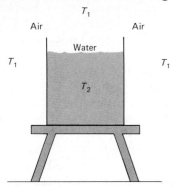

FIGURE 14.7 Newton's law of cooling: Water at a temperature T_2 in a room where the air temperature is T_1 cools down at a rate $P = K(T_2 - T_1)$, if $T_2 > T_1$.

Thermos Bottles (Dewars)

Among the many discoveries in physics made by Sir Isaac Newton was a simple law obeyed by objects on cooling down from a higher temperature to room temperature, so long as their initial temperature was not too great. If the initial temperature of the object was T_2 and the temperature of its surroundings was T_1, as in Fig. 14.7, Newton found that the rate of heat loss was

$$P = \frac{Q}{t} = K(T_2 - T_1) \tag{14.6}$$

Here K is an empirical constant, i.e., one which must be obtained from experiment for each particular physical situation. Equation (14.6) is called *Newton's law of cooling*. It predicts that the rate of cooling will slow down as T_2 is reduced, and go to zero when T_2 becomes equal to T_1.

It can be shown that Newton's law of cooling is an immediate consequence of the laws we have already discussed for conduction, convection, and radiation, so long as T_2 is not too much larger than T_1 (see Prob. D7).

A thermos bottle [or *dewar*, after its inventor, the British physicist Sir James Dewar (1842–1923)] is able to preserve the temperature of liquid gases like N_2 or He, or of hot and cold drinks, for long periods of time. Why does not the coffee in a thermos cool down quickly to room temperature? The thermos makes heat transfer from the hot coffee to the cooler air as difficult as possible by blocking off the three channels of conduction, convection, and radiation, thus drastically reducing the rate at which cooling (or heating) occurs.

A typical thermos bottle is shown in Fig. 14.8. Evacuating the space between the inner and outer walls leaves few molecules in this space to convey heat by conduction or convection. Hence the only way heat can be lost (except for inevitable small losses near the neck of the bottle) is through radiation. Radiation loss can be greatly reduced by silver-plating the two inner surfaces of the vacuum jacket. These silvered surfaces reflect back any radiation trying to escape from the liquid in the thermos and any radiation trying to enter it from outside.

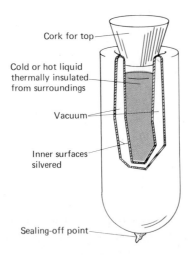

Cork for top

Cold or hot liquid thermally insulated from surroundings

Vacuum

Inner surfaces silvered

Sealing-off point

FIGURE 14.8 A thermos bottle, or dewar.

Thermography

Infrared radiation from an object can be detected using a camera equipped with infrared-sensitive film, or by a more sensitive type of infrared scanner that converts small variations in the temperature of radiating objects into pictures called *thermographs* in which each scanned region is displayed with a color shading characteristic of its temperature.

Thermography is very useful in medicine. The human body, with a skin temperature of about 33°C (306 K), radiates in the infrared region of the spectrum, according to the Wien displacement law. Temperature differences in the body can be recorded on thermographs, as in Fig. 14.9. This can be quite helpful in detecting breast cancers, vascular disorders, arthritis, and other diseases. Tumors are often found in areas where the temperature is higher due to increased metabolic activity in the vicinity. This leads to more intense infrared radiation. Such a technique is especially useful in finding breast cancers, since thermographs can find temperature differences of less than 0.07°C, and the temperature differences associated with breast cancers are usually at least a few tenths of a Celsius degree. Growths as small as 1 cm in diameter can be detected with this valuable technique, which also avoids the problems associated with use of x-rays.

Figure 14.9 Serial thermograms of the hands of a 25-year-old woman who smokes a pack of cigarettes a day: (*a*) The presmoking thermogram demonstrates normal warm (white) hands; (*b*) 1 min after she smokes a single filtered cigarette, her hands become 6°C cooler; (*c*) 30 min later the temperature is starting to rise again (the fingers are becoming white in the thermogram). *(Photo courtesy of American Cancer Society.)*

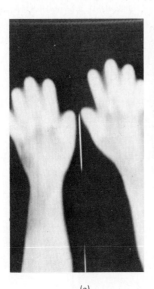
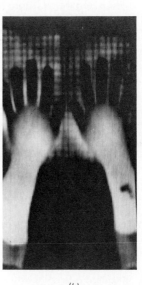

(*a*)   (*b*)   (*c*)

Solar Energy

The sun is the ultimate source of 99.98 percent of all the energy we use on earth. It radiates energy into space at a rate of 3.9×10^{26} J/s, or 3.9×10^{26} W, as we found in Example 14.2. The source of this enormous energy supply is the conversion of the mass of the sun into energy at a rate of 4.3×10^9 kg/s.* This is a fantastic rate of decrease in mass, but since the sun's total mass is 2×10^{30} kg, the sun could continue to exist for some 10^{17} years even at the rate that its mass is presently being converted into energy.

Figure 14.10 shows that the sun radiates a continuous spectrum with maximum intensity in the green region at a wavelength of about 0.5 μm.

---

*This conversion is governed by Einstein's famous equation $E = mc^2$ (see Sec. 25.7).

About one-half the solar radiation is concentrated in the visible range, from 0.4 to 0.7 μm, with the remainder mostly in the infrared, as the figure shows.

Although the sun radiates energy at a rate of 3.9×10^{26} W, this radiation is spread out in all directions over a full sphere. Only one two-billionths of it falls on the earth, but this still amounts to 1.73×10^{17} W, which is about 500,000 times the electric power generating capacity of the United States!

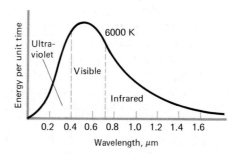

FIGURE 14.10 The distribution of the energy radiated by the sun at 6000 K as a function of wavelength. Most of the radiation is in the visible and infrared regions of the spectrum.

When we consider that this radiant power is spread out over the whole earth, that much of it is absorbed or reflected by the earth's atmosphere, that cloud cover obscures the sun on many days of the year, and that half the earth is in darkness at any one time, the average intensity of solar energy falling on a house in the central United States turns out to be only about 200 W/m². In other words, if all the solar energy falling on 1 m² of the earth's surface could be collected and fully converted into electric energy, on the average it would just suffice to light two 100-W electric light bulbs on a continuous basis. This shows one of the main problems associated with using solar energy to meet our energy needs.

Example 14.4

Assume that the average rate at which solar energy falls on a house in the central United States is 200 W/m². Consider a one-story house of 1500-ft² floor space, and assume that it requires 10^9 J of energy per day to provide heat and hot water on a cold day in January.
(a) How many joules of solar energy are incident per day on the flat roof of the house?
(b) If this energy could be collected and used with 100 percent efficiency (which is not possible), what fraction of the roof area of the house would be required for heat collectors?
(c) If a solar-electric power plant capable of producing electric energy at a rate of 1 GW (or 10^9 W) were to be built in the central United States, what is the *minimum* land area such a power plant would require?

SOLUTION

(a) The incident power is

$$P = \left(200 \, \frac{W}{m^2}\right)(1500 \text{ ft}^2)\left[\frac{1 \text{ m}^2}{(3.28 \text{ ft})^2}\right] = 28 \text{ kW}$$

Hence the energy incident in 1 day is

$$\mathcal{W} = Pt = \left(28 \, \frac{kJ}{s}\right)\left(\frac{3600 \text{ s}}{1 \text{ h}}\right)\left(\frac{24 \text{ h}}{1 \text{ day}}\right)(1 \text{ day})$$

$$= \boxed{2.4 \times 10^9 \text{ J}}$$

This is more than twice the energy needs of the house, even in January.
(b) The fraction of the roof area needed for collectors at 100 percent efficiency is:

$$\frac{1 \times 10^9 \text{ J}}{2.4 \times 10^9 \text{ J}} = 0.42 = \boxed{42\%}$$

(c) To produce 10^9 W when the incident power is 200 W/m² would require an area

$$A = \frac{10^9 \text{ W}}{200 \text{ W/m}^2} = 5 \times 10^6 \text{ m}^2 = \boxed{5.0 \text{ km}^2}$$

This would be a square area about 7 km (or 4.3 mi) on a side. This is the *minimum* area needed, because solar energy cannot be collected and converted into usable heat with 100 percent efficiency. Fossil-fuel or nuclear plants providing this same amount of power, which is the output of a typical large electric power plant, occupy much smaller land areas.

14.6 The Mechanical Equivalent of Heat

The fields of mechanics and heat developed independently, and as a result work was for many years measured in mechanical units (joules) while heat was measured in thermal units (kilocalories). The conclusive experimental work that established the exact quantitative relationship between a quantity of work (in joules) and a quantity of heat (in kilocalories) was performed by James Joule (see Fig. 14.12 and accompanying biography). In his most famous experiment Joule used the paddle-wheel apparatus of Fig. 14.11.

Joule's Experiment

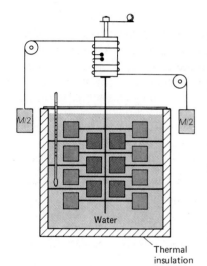

Joule's apparatus consisted of two heavy masses attached to a rope wound around a pulley. When the masses were allowed to fall under gravity, they unwound the rope from the pulley and turned a set of paddles immersed in a tank of water which was thermally insulated from its surroundings. The churning paddles did work on the water and heated it up. In this way Joule could compare the work done by the falling mass with the heat required to produce the measured temperature change in the water.

If the masses are each $M/2$ and if each is raised to a height h off the ground, its gravitational potential energy with respect to the ground is $Mgh/2$. As the two masses fall slowly through a distance h back to the ground, their total potential energy is converted into an amount of work

$$\mathcal{W} = Mgh$$

where $\mathcal{W}$ is in joules.*

As the rope unwinds, the moving paddles stir up the water and increase its temperature by an amount ΔT which can be measured with a thermometer. The amount of heat absorbed by the water can then be calculated from Eq. (13.12),

$$Q = cm\ \Delta T$$

where m is the mass of the water. Hence

$$Q = \left(1\frac{\text{kcal}}{\text{kg·°C}}\right)(m\ \text{kg})(\Delta t\ \text{°C})$$

is the amount of heat added in *kilocalories*.

Joule compared the values of these two quantities obtained in his paddle-wheel experiment and in this way found a numerical value for the mechanical equivalent of heat,† or *Joule's equivalent*, where

$$\text{Joule's equivalent} = \frac{\mathcal{W}}{Q} \qquad \left(\frac{\text{J}}{\text{kcal}}\right) \tag{14.7}$$

FIGURE 14.11   Paddle-wheel apparatus used by Joule in measuring the mechanical equivalent of heat (Joule's equivalent).

---

*Throughout this discussion we will use joules and kilocalories as our units, even though the early workers in this field used different units for work and heat.

†When we consider the way Joule and other physicists have measured this quantity, it appears that a more apt title for it would be "the heat equivalent of work."

James Prescott Joule (1818–1889)

FIGURE 14.12 Sir James Joule. *(NBS Archives; courtesy of AIP Niels Bohr Library.)*

Of the four men whose names are usually associated with the development of the first law of thermodynamics—Mayer, Rumford, Helmholtz, and Joule—Joule was the true experimental physicist. It was his lifelong dedication to confirming the conservation-of-energy principle in the laboratory that finally convinced his contemporaries of that principle's universal validity.

Joule was born in 1818 at Salford, near Manchester, England, the son of a wealthy owner of a brewery. He was educated at home, and for 3 years had the eminent chemist John Dalton, then 70 years of age, as his tutor. Dalton imparted to Joule his great love for science and his passion for sound numerical data on which to base scientific theories and laws. Unfortunately Joule had little mathematical training, and this prevented him from making even more significant contributions to physics in his later life.

Joule had no real profession and no job except for his involvement in running his father's brewery. Up until 1854, when the brewery was sold, Joule worked there and did his experiments before or after work. After 1854 he had the time and the funds to continue his physics experiments in a laboratory he built at his home. Later in his life Joule suffered financial difficulties and needed a government subsidy from Queen Victoria to continue his research.

During the years from 1837 to 1847 Joule devoted all his available time to a variety of experiments on the conversion of various forms of energy—mechanical, electric, chemical—into heat. He developed thermometers capable of measuring temperatures to 1/200 of a Fahrenheit degree, and this gave his work a precision unattainable by other physicists up to that time. In 1840 he proposed for the first time that the rate at which heat was generated by an electric current I passing through a wire of resistance R was $Q/t = I^2R$. This is now called *Joule's law*.

In June 1847 Joule presented a paper to the British Association meeting in Oxford which contained the most accurate experimental value for the mechanical equivalent of heat (Joule's equivalent) obtained up to that time. Joule's peers were sleepy and unimpressed until a young man in the audience, William Thomson (later Lord Kelvin), pointed out to his fellow scientists the significance of Joule's work. This was the turning point in Joule's career.

In 1850 Joule was elected to the Royal Society on the basis of his work on energy conversion, and he became very influential in physics circles. He received many scientific honors, serving as president of the British Association for the Advancement of Science in 1872 and again in 1887, and having the unit of energy (the joule) named after him.

Joule had excellent laboratory skills and a passion for careful, precise work. Even on his honeymoon he took time out to measure the temperature of the water at the top and bottom of a scenic waterfall to see if the difference agreed with the value predicted by the conservation-of-energy principle! He believed that nature was simple, and strove to find the simple relationships (like Joule's law in electricity) which he was convinced must exist between important physical quantities. His discovery of just two such relationships made a major contribution to the development of the concept of energy as we know it today.

Joule's efforts to determine the value of the mechanical equivalent of heat culminated in a paper he presented before the Royal Society in June 1847. It contained an overwhelming amount of experimental data from five series of experiments using water, sperm oil, and mercury as fluids and the churning by paddle wheels, friction, and electricity as heat sources. Joule obtained results with 5 percent agreement from all five series. Translated into SI units, his result corresponds to a value of 4.15×10^3 J/kcal for the mechanical equivalent of heat. The best modern value of Joule's equivalent is 4.186×10^3 J/kcal, and so Joule's work was accurate to better than 1 percent, a significant achievement even by contemporary standards.

Joule's Equivalent

$$1 \text{ kcal} = 4.186 \times 10^3 \text{ J}$$

14.7 Heat and Work

FIGURE 14.13 An expanding gas doing work on a movable piston. The work done is $\Delta \mathcal{W} = F \, \Delta s = P \, \Delta V = P_0(V_2 - V_1)$, if the pressure remains constant at P_0.

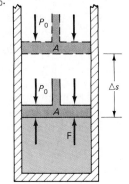

Joule's experiments showed conclusively that work can be converted into heat and that the amount of heat produced from a given amount of work is always the same. We also know that heat can be converted into work. For example, if a gas is heated it will expand at constant temperature and do work on a piston, as in Fig. 14.13. In this case, the heat transferred to the gas is converted into work.

It would therefore appear that heat and work are both forms of *energy*, and that this is the reason one can be converted into the other. These two forms of energy differ in that *heat* is *disordered energy* that flows from one object to another because of the temperature difference between the two objects, whereas *work* is *ordered energy* which is transferred when one object exerts a force on another object and moves it through a distance in the direction of the force.

Work Done by a Gas

Let us consider a gas contained in a cylinder and maintained at constant pressure by a piston at one end, as in Fig. 14.13. If the gas expands a distance Δs against the pressure P exerted by the piston, then the work done by the gas on expanding is

$$\Delta \mathcal{W} = \mathbf{F} \cdot \Delta \mathbf{s} = PA \, \Delta s = P \, \Delta V \qquad (14.8)$$

since $F = PA$ and $\Delta V = A \, \Delta s$.

The work done by the gas is considered positive if the volume increases and negative if the volume decreases. The work will be in joules if the pressure is in newtons per square meter and the volume in cubic meters, since $1 \text{ N·m} = 1 \text{ J}$.

If the pressure P is constant and equal to P_0, then the work done by the gas in a change of volume $\Delta V = V_2 - V_1$ is:

$$\Delta \mathcal{W} = P_0 \, (V_2 - V_1)$$

FIGURE 14.14 The work done by an expanding gas at constant pressure P_0. The work is the area shown: $\Delta \mathcal{W} = P_0(V_2 - V_1)$.

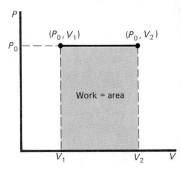

In Fig. 14.14, $P_0(V_2 - V_1)$ is simply the area under the *PV* curve between the two volumes V_1 and V_2.

PV Work

Hence we say that *the PV work done by a gas on expanding is equal to the area under its PV curve.*

Suppose we have a gas in an initial state specified by its pressure and volume, P_0 and V_0, and we want to change the gas to a new state specified by P_f and V_f. There are a great variety of ways of doing this, as indicated in Fig. 14.15. We can simply let the gas expand at constant temperature according to Boyle's law from point A to point C. We can also let the gas expand at constant pressure from point A to point B, and then reduce the pressure at constant volume to bring the gas to point C, where the pressure and volume are the desired P_f and V_f. Since the amount of gas remains constant, and n is therefore constant in the ideal gas equation $PV = nRT$, and since $P_0V_0 = P_fV_f$ for the two processes, the temperature at the end of the two processes is the same and is equal to T_0. We could also carry out this expansion process in a variety of other ways. Clearly in each case the work done would be different, since the area under the PV curve would be different, as Fig. 14.15 shows.

Hence we see that the work done by a gas on expanding depends not merely on the initial and final states, but also on the *path* (on a *PV* diagram) the gas takes in going from its initial to its final state. *The work done on or by the gas is dependent on the path.* For this reason the *work content* of a gas is a meaningless phrase, for knowing the pressure and volume of a gas tells us nothing about how much work has been done on or by the gas in getting to that state. Thus in Fig. 14.16, if we take the gas from the initial state to the final state by path ABC, and then back to the initial state by the path CA, the gas has done a net amount of work in the process, and still the final state of the system is the same as it was initially. Hence the amount of work done in going from one state to another can be determined only if the path taken is fully specified.

FIGURE 14.15 The work done by a gas in going from point A (P_0, V_0) to point C (P_f, V_f) depends on the path taken. The work done along the path $A{\rightarrow}B{\rightarrow}C$ is greater than along the path $A{\rightarrow}C$, since the area under the curve is greater.

FIGURE 14.16 The work done in a cyclic process. The work done by a gas in going from A to B to C and then back to A again, by the path shown, is the area indicated on the diagram.

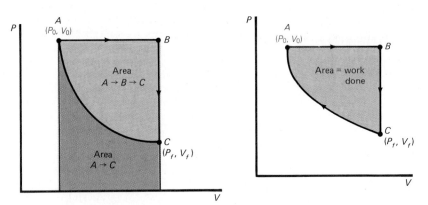

Heat Flow

Heat flow is similarly dependent on path. In Fig. 14.15 a greater amount of heat must be supplied to take the system from the initial to the final state by path ABC than directly by path AC, if the initial and final temperatures T_0 and T_f are the same for the two processes. Hence the flow of heat into or out of the gas is also dependent on the path taken and not merely on the initial and final states. For this reason the phrase *the heat content of a system* is as meaningless as is *the work content of a system*. Both are basically unclear concepts that do not lend themselves to precise definition. For example, if we could assign an arbitrary value to the heat content of a gas in some standard state, the heat content in some other state would be the heat content in the standard state plus whatever heat was added in going from that standard state to the new state. But this added heat depends entirely on the path taken. Since there are an infinite number of paths, there are an infinite number of values for the heat content, and hence this term is meaningless.

Both heat and work therefore depend on path and are not conserved quantities in the sense that the energy of a particle in a gravitational field is conserved. In the gravitational case the work done in moving the particle from one point to another depends only on the two endpoints and not on the path taken. Neither heat nor work share this conservative property; both are dependent not merely on the coordinates (i.e., for a gas, P, V, and T) of the endpoints, but also on the physical *path* taken to go from one endpoint to the other.

Example 14.5

In Fig. 14.17 a gas expands from point A to point B on the PV diagram at constant pressure, and the pressure is then changed at constant volume to take the gas from B to C. The gas is finally compressed back to point A along the path CA.

(a) How much work has been done *by* the gas in going from A to B?

(b) How much work has been done by the gas in going from B to C?

(c) How much work has been done by the gas in returning from C to A?

(d) What is the net work done by the gas in the complete process?

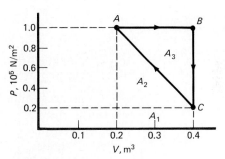

FIGURE 14.17 Diagram for Example 14.5.

SOLUTION

(a) The work done here is the constant pressure times the change in volume, or

$$\Delta W_{AB} = P(V_B - V_A)$$

$$= (1.0 \times 10^5 \text{ N/m}^2)(0.40 \text{ m}^3 - 0.20 \text{ m}^3)$$

$$= 0.20 \times 10^5 \text{ N·m} = \boxed{0.20 \times 10^5 \text{ J}}$$

which is the area between the X axis and the line AB.

(b) $\Delta W_{BC} = P(V_C - V_B) = \boxed{0}$ since $V_B = V_C$.

(c) $\Delta W_{CA} = P(V_A - V_C)$

Here the pressure varies along with the volume, but the work done is still the area under the PV curve. This is the sum of the two areas A_1 and A_2, with the latter being the area of a triangle. Hence the total area A under the curve is

$$A = A_1 + A_2$$

$$= (0.20 \times 10^5 \text{ N/m}^2)(0.20 \text{ m}^3 - 0.40 \text{ m}^3)$$

$$+ \tfrac{1}{2}[(1.0 - 0.20) \times 10^5 \text{ N/m}^2](0.20 \text{ m}^3 - 0.40 \text{ m}^3)$$

Hence

$$\Delta W_{CA} = -0.04 \times 10^5 \text{ J} - 0.08 \times 10^5 \text{ J} = \boxed{-0.12 \times 10^5 \text{ J}}$$

This work is *negative*, since work is being done *on* the gas to compress it.

(d) The net work done by the gas is then

$$\Delta W = \Delta W_{AB} + \Delta W_{BC} + \Delta W_{CA}$$

$$= 0.20 \times 10^5 \text{ J} + 0 - 0.12 \times 10^5 \text{ J}$$

$$= 0.080 \times 10^5 \text{ J} = \boxed{8.0 \times 10^3 \text{ J}}$$

Hence the net work is positive, indicating that net work is done *by* the gas. The same result can be obtained directly by calculating the area A_3 inside the triangle ABC on the PV diagram. Here

$$\Delta W = \tfrac{1}{2}[(1.0 - 0.20) \times 10^5 \text{ N/m}^2](0.40 \text{ m}^3 - 0.20 \text{ m}^3)$$

$$= 0.080 \times 10^5 \text{ J} = 8.0 \times 10^3 \text{ J}$$

as before.

14.8 The First Law of Thermodynamics

Thus far we have used the phrase *thermal energy* to describe the random motion of the atoms and molecules in a substance. This thermal energy can be increased by doing work on the gas or decreased when the gas does work on an external object like a piston. But a physical system has other forms of energy—such as the energy in chemical bonds, the binding energy of atoms, and the energy stored in the atomic nucleus—which cannot under normal circumstances be released to do work. Hence to state the first law of thermodynamics precisely, we need to introduce a broader term to include all the forms of energy which a physical system like a gas can have. This we call *internal energy*.

Definition

Internal energy (U): The sum of the kinetic and potential energies of all the individual particles making up a physical system.

If a system with internal energy U interacts with its surroundings, it can do so in two ways. First the system can have *work done on it*, in which case U *increases*, or it can *do work* on its surroundings, in which case U *decreases*. Secondly, *heat* can flow *into* the system and thus *increase* the internal energy U, or heat can flow *out of* the system and thus *decrease* U.

The great difference between internal energy and work or heat is that internal energy is a *state function*. By this we mean that U depends only on the values of certain physical quantities (for a gas, P, V, and T), and that once these quantities are determined, the value of U is completely determined. P, V, and T are called *thermodynamic state variables*. No matter how complicated the interaction of the gas with its surroundings, once it returns to the state specified by its original values of P, V, and T, its internal energy is exactly what it was originally.* This is true of all thermodynamic state functions.

---

*Note that if P, V, and the amount of gas present are determined for a gas, then its temperature is fully determined by the gas law $PV = nRT$.

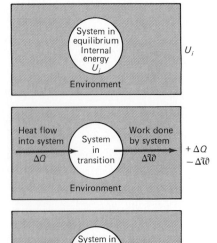

FIGURE 14.18 Diagram to show the content of the first law of thermodynamics. A system in equilibrium with its environment and with an internal energy U_i interacts with its environment by absorbing heat ΔQ from the environment and doing work ΔW on the environment. Its final internal energy is $U_f = U_i + \Delta Q - \Delta W$.

Internal energy and other state functions are therefore quantities similar to potential energy in mechanics. Their values are independent of the processes (or paths) by which they arrive at a particular state, in the same way that potential energy in a gravitational field is a function only of a particle's position in the field and not of the path taken to get there.

We can now state the first law of thermodynamics in terms of the internal energy of a system. Suppose that the initial internal energy of the system is U_i, that an amount of heat ΔQ flows into the system, and an amount of work ΔW is done by the system, as in Fig. 14.18. Then, on the basis of the work of Rumford, Mayer, Joule, Helmholtz, and others, we are led to the following conclusion. If the internal energy of a system changes from an initial value U_i to a final value U_f, then

First Law of Thermodynamics

$$\boxed{U_f - U_i = \Delta U = \Delta Q - \Delta W}$$ (14.9)

where ΔU = the *increase* in internal energy of the system
  ΔQ = the heat flow *into* the system
  ΔW = the work done *by* the system

The minus sign in Eq. (14.9) is needed because when work is done *by* the system, energy flows *out* of the system.

Equation (14.9) states that, if heat ΔQ flows into a system and the system performs an amount of work ΔW, then the change in the internal energy of the system is simply $\Delta U = \Delta Q - \Delta W$. This equation makes clear the fact that internal energy, heat, and work are all forms of energy, and that the *total energy is conserved* in any process so long as all forms of energy are taken into consideration.

All three quantities in Eq. (14.9) must be expressed in the same units, either joules or kilocalories. If a problem is stated in inconsistent units, you can use Joule's equivalent to convert joules to kilocalories, or vice versa, as Example 14.6 shows.

Example 14.6

Determine the value of Joule's equivalent from the following data: 4.00 kcal of heat is supplied to a system; the system does 6.70×10^3 J of external work; at the end of the process the internal energy of the system has increased by 1.00×10^4 J.

SOLUTION

From the first law of thermodynamics,

$\Delta U = \Delta Q - \Delta W$ and so

1.00×10^4 J $= 4.00$ kcal $- 6.70 \times 10^3$ J   or

4.00 kcal $= 1.00 \times 10^4$ J $+ 6.70 \times 10^3$ J $= 1.67 \times 10^4$ J

Hence

Joule's equivalent $= \dfrac{1.67 \times 10^4 \text{ J}}{4.00 \text{ kcal}} = \boxed{4.18 \times 10^3 \text{ J/kcal}}$

Conservation of Energy

When we are dealing with an isolated system, i.e., a system which does not exchange either heat or work with its surroundings, both ΔQ and ΔW are zero in Eq. (14.9). Hence

$$U_f - U_i = \Delta U = \Delta Q - \Delta W = 0 \quad \text{or} \quad U_f = U_i$$

We can therefore state the first law of thermodynamics in the following form:

First Law of Thermodynamics

> In any isolated system the total internal energy remains constant, even though it may be transformed from one kind of energy to another.

This is the most general form of the principle of conservation of energy. It generalizes from the principle of conservation of mechanical energy to include all other forms of energy, in particular, thermal energy. The internal energy of an isolated system cannot be changed by any process (including a chemical or biological process) taking place *inside* the system. If the system is not isolated, then heat can flow into or out of the system, and the system can do work or have work done on it. In this case, as we have seen, $\Delta U = \Delta Q - \Delta W$.

Example 14.7

In Fig. 14.19, when an ideal gas is taken from state A to state B along the path ACB, 1.91×10^{-2} kcal of heat flows into the system and 30 J of work is done by the system.
(a) What is the change in the internal energy of the gas when it goes from A to C to B?
(b) What is the change in internal energy of the gas along path ADB?
(c) How much heat flows into the system along path ADB if the work done by the system along ADB is 10 J?
(d) When the system is returned from B to A along the curved path, the work done is 20 J. Does the system absorb or liberate heat, and how much?

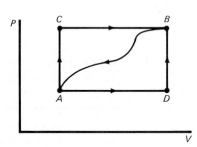

FIGURE 14.19 Diagram for Example 14.7.

SOLUTION

The units must be consistent throughout the problem. We choose to use joules throughout. The heat flowing into the system is then:

$$\Delta Q = (1.91 \times 10^{-2} \text{ kcal})(4.18 \times 10^3 \text{ J/kcal}) = 80 \text{ J}$$

(a) Then, from Eq. (14.9), we have for the change in the internal energy of the gas in this case:

$$\Delta U = \Delta Q - \Delta W$$

For path ACB, $\quad \Delta Q = 80 \text{ J} \quad$ and $\quad \Delta W = 30 \text{ J}$

Hence $\quad \Delta U_{AB} = \Delta Q - \Delta W = 80 \text{ J} - 30 \text{ J} = \boxed{50 \text{ J}}$

(b) For path ADB the change in internal energy must be the same, since U *is independent of the path*. Hence here also

$$\Delta U_{AB} = \boxed{50 \text{ J}}$$

(c) Since $\Delta U_{AB} = \Delta Q - \Delta W = 50 \text{ J}$,

$$\Delta Q = \Delta U_{AB} + \Delta W = 50 \text{ J} + 10 \text{ J} = \boxed{60 \text{ J}}$$

(d) On going from B to A along the curved path, the work done *by the gas* is *negative*, since the volume decreases and work must therefore be done on the gas. Hence

$$\Delta U_{BA} = \Delta Q - \Delta W$$

where here $\Delta U_{BA} = -\Delta U_{AB} = -50 \text{ J}$, and $\Delta W = -20 \text{ J}$, and so

$$\Delta Q = \Delta U_{BA} + \Delta W = -50 \text{ J} - 20 \text{ J} = \boxed{-70 \text{ J}}$$

Hence 70 J of heat is liberated when the system goes from B to A along the curved path.

Perpetual-Motion Machines of the First Kind

Because no energy transfer process is possible in which more work is done or energy produced than is put in, as Eq. (14.9) makes clear, a logical consequence of the first law is that a device known as a perpetual-motion machine of the first kind cannot work.

Definition

A perpetual-motion machine of the first kind is a machine which is designed to do work continually either with no input of energy or with an input of energy that is less than the work output. Although many perpetual-motion machines, such as the one in Fig. 14.20, have been and continue to be proposed by inventors, we are assured by the first law of thermodynamics that such machines cannot possibly work.

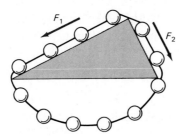

FIGURE 14.20 A proposed perpetual-motion machine of the first kind. The chain of balls is supposed to keep rotating, since there are always more balls on the left slope than on the right slope, and hence F_1 is greater than F_2. Thus the chain is supposed to rotate in a counterclockwise direction forever, without the need of any energy input. (What is the fallacy in this argument?)

Alternative Statement of First Law

For this reason *an alternative statement of the first law of thermodynamics is that a perpetual-motion machine of the first kind is impossible*, since it would violate the principle of conservation of energy. We cannot emphasize too strongly that this limitation is *not* due to engineering problems in designing or building such machines; it is due to the nature of the physical universe and is therefore not subject to change. We cannot get more work out of a machine than the energy we put in.

First-Law Efficiency

The *first-law efficiency* (e_1) of a physical process is defined as follows:

$$e_1 = \frac{\text{work done (or energy transferred) by a system}}{\text{energy input to the system}} \tag{14.10}$$

The maximum possible value of e_1 is unity, since we can never get more work out of a system than we put in. Thus an oil furnace in a home is said to be 60 percent efficient ($e_1 = 0.60$) if 60 percent of the oil's heat of combustion is delivered to the house as heat. (We will see in the next chapter that there is an even more meaningful way than this first-law efficiency to assign efficiency ratings to energy conversion processes.)

Example 14.8

In a coal-burning steam-electric power plant generating electricity at a rate of 1.0 GW (10^9 W), coal is burned to provide steam for the turbines at a rate of 7.2×10^5 kcal/s. What is the first-law efficiency of such a plant?

SOLUTION

In this case by the first-law efficiency we mean

$$e_1 = \frac{\text{electric energy out}}{\text{thermal energy in}} = \frac{W_{\text{el}}}{W_{\text{th}}}$$

Since we are given rates of energy production rather than absolute amounts of energy, we need to select a certain time interval, say, 1 s, over which to compare the energies. The electric energy produced in 1 s is then

$$\mathcal{W}_{el} = (1.0 \text{ GW})(1.0 \text{ s}) = (10^9 \text{ J/s})(1.0 \text{ s}) = 1.0 \times 10^9 \text{ J}$$

The thermal energy put in is

$$\mathcal{W}_{th} = \left(7.2 \times 10^5 \frac{\text{kcal}}{\text{s}}\right)\left(4.18 \times 10^3 \frac{\text{J}}{\text{kcal}}\right)(1.0 \text{ s})$$

$$= 3.0 \times 10^9 \text{ J}$$

Then the first-law efficiency is

$$e_1 = \frac{\mathcal{W}_{el}}{\mathcal{W}_{th}} = \frac{1.0 \times 10^9 \text{ J}}{3.0 \times 10^9 \text{ J}} = 0.33 = \boxed{33\%}$$

This is a typical efficiency for large steam-electric power plants.

14.9 Applications of the First Law of Thermodynamics

Let us consider the implications of the first law of thermodynamics in the form of Eq. (14.9) for a number of important physical processes. For simplicity we will confine our applications to ideal gases.

Isothermal Process

An *isothermal (equal-temperature) process is one which takes place at constant temperature*. In order for the temperature to remain constant when a gas expands, the changes in the pressure and volume of the gas must be carried out very slowly so that every state in the process is an equilibrium state. For an ideal gas that obeys Boyle's law, the internal energy of the gas does not change during an isothermal expansion, ΔU is 0, and so, from Eq. (14.9), ΔQ equals $\Delta \mathcal{W}$. Hence an amount of heat flows into the gas which is just sufficient to enable the gas to do the work required for the expansion without changing its internal energy.

Adiabatic Process

An *adiabatic* (from the Greek word meaning "impassable") *process is one in which no heat enters or leaves the system*. Here $\Delta Q = 0$, and so from Eq. (14.9), $\Delta U = -\Delta \mathcal{W}$. If work is done by the system, then the system's internal energy *decreases* by an amount equal to the work done.

An adiabatic process can be achieved either by thermally insulating the system against heat transfers, or by carrying out the process so rapidly that heat has no chance to flow into or out of the system. Thus the compression and rarefactions of a gas in a sound wave are adiabatic processes, since they take place so rapidly that heat cannot flow away from the compressions or into the rarefactions.

Adiabatic processes play an important role in mechanical engineering, because many processes in internal-combustion engines take place adiabatically.

Throttling Process A special, and very practical, adiabatic process is a throttling process, in which a fluid, originally at high pressure, flows through a needle valve or other tiny opening into a region of lower pressure. Since the process is assumed to be adiabatic, $\Delta Q = 0$, and so $\Delta U = -\Delta \mathcal{W}$, where $\Delta \mathcal{W}$ is the work done by the fluid in expanding.

If the gas is an ideal gas, and if it expands into a vacuum, then $\Delta \mathcal{W} = P \Delta V$. Since for a vacuum $P = 0$, we must have $\Delta \mathcal{W} = 0$. Hence the internal energy of the gas does not change in this case, and the temperature also does not change.

For most real gases, however, the gas molecules exert small forces on one another. Therefore, on expanding, the gas does work against these forces, and ΔW is finite and positive. Hence $\Delta U = -\Delta W$ is negative, and the temperature falls as the internal energy falls.

When liquids like ammonia and freon are subjected to a throttling process, the liquid vaporizes when it expands into the region of lower pressure because the reduced pressure lowers the boiling point below the actual temperature of the liquid, as in Fig. 14.21. An amount of heat equal to the heat of vaporization of the liquid must be supplied to evaporate each kilogram of liquid. Since $\Delta Q = 0$, this energy can only come from the internal energy of the fluid, and so the temperature will fall, as it does when the liquid in a spray-type can of deodorant or hairspray is converted into vapor. Throttling like this can be used to provide cooling in refrigeration systems. Such systems almost always involve changes of state because the associated large heats of vaporization make the cooling more effective.

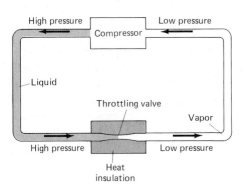

FIGURE 14.21 A throttling process. Such a process causes the temperature to fall. Throttling processes play an essential role in working refrigerators and air conditioners since they bring about the drop in temperature needed for refrigeration or air conditioning.

Isochoric Process

An *isochoric (equal-volume) process is one in which the volume of the system remains unchanged.* If heat is added to a gas and the volume is unchanged, then no work is done and $\Delta W = 0$. Hence $\Delta U = \Delta Q$ and all the heat added goes into the internal energy of the gas. The very sudden increase in temperature and pressure when a gasoline-air mixture ignites in an automobile engine is close to an isochoric process. Here the heat added to the system increases the pressure and the internal energy, but does not change the volume of the mixture. This isochoric process is immediately followed by the power stroke of the engine, which is basically an adiabatic process. This performs useful work (by turning the car's wheels) at the expense of the internal energy of the system, which had just been increased by the isochoric combustion process.

Isobaric Process

An *isobaric (equal-pressure) process is one that takes place at constant pressure.* Here the work done is $\Delta W = P \, \Delta V$, and so

$$\Delta U = \Delta Q - P \, \Delta V \tag{14.11}$$

When water is heated and boiled in a pot at atmospheric pressure, the heating of the water and the conversion of water to steam all take place isobarically. In this case,

$$\Delta Q = cm \, \Delta T + mL_v$$

where m is the mass of the water and L_v is the heat of vaporization. At the boiling point the volume changes from V_{liq} to V_{vap}. Hence

$$P \, \Delta V = P(V_{\text{vap}} - V_{\text{liq}})$$

where P is usually atmospheric pressure. Then

$$\Delta U = U_{\text{vap}} - U_{\text{liq}} = \Delta Q - P \, \Delta V = cm \, \Delta T + mL_v - P(V_{\text{vap}} - V_{\text{liq}}) \qquad (14.12)$$

Because for an isobaric process $\Delta Q = \Delta U + P \, \Delta V$, from Eq. (14.11), any heat added to a gas at constant pressure goes partly into the internal energy of the gas and partly into PV work.

Living Systems and the First Law

A living body is *not* an isolated system and energy is *not* transferred into the system only in the form of heat or work. Rather the supply of internal energy in the human body is constantly being replenished by the food we eat. For a healthy adult the energy taken in as food minus the energy lost in the form of waste products is exactly equal to the sum of the work done by the individual and the heat lost to the surroundings. Hence the first law of thermodynamics remains valid so long as we consider food as an additional, and very important, source of internal energy. Of course, if the intake of food exceeds the sum of the work done by the individual and the heat lost to the surroundings, the excess energy is stored in the body. If the person is physically active, this energy will be stored as muscle tissue; if not, it will be stored as fat. This excess energy can then be made available to the body at a later time when food may be lacking.

Example 14.9

One kilogram of iron is heated from room temperature (20°C) to 60°C. For purposes of this problem the expansion of the iron can be considered negligible over this temperature range.
(a) How much heat is added to the system?

(b) What is the increase in internal energy of the iron?
(c) If all the increased internal energy could be converted into useful work, say, to lift another 1-kg iron mass, how high could the 1-kg mass be lifted by the work produced?

SOLUTION

(a) From Eq. (13.12), $\Delta Q = cm \, \Delta T$, or

$$\Delta Q = \left(\frac{0.113 \text{ kcal}}{\text{kg} \cdot {}^\circ\text{C}}\right)(1.0 \text{ kg})(60 - 20){}^\circ\text{C} = 4.52 \text{ kcal}$$

Hence, in energy units,

$$\Delta Q = (4.52 \text{ kcal})(4.18 \times 10^3 \text{ J/kcal}) = \boxed{18.9 \times 10^3 \text{ J}}$$

(b) From the first law of thermodynamics we have $\Delta U = \Delta Q - \Delta W$. Here no work is done, and so

$$\Delta U = \Delta Q = \boxed{18.9 \times 10^3 \text{ J}}$$

This is the increase in the internal energy of the iron at 60°C from what it was at 20°C.

(c) If this increase in internal energy could be converted completely into the work required to lift a 1-kg iron mass, we would have

$$mgh = 18.9 \times 10^3 \text{ J}$$

and so

$$h = \frac{18.9 \times 10^3 \text{ J}}{(1.0 \text{ kg})(9.8 \text{ m/s}^2)} = 1.93 \times 10^3 \text{ m} = \boxed{1.93 \text{ km}}$$

Since 1 mi = 1.61 km, this is enough energy to lift the 1-kg mass to a height of more than a mile! This shows that remarkably large amounts of energy are stored in matter in the form of the random thermal motions of molecules. The problem, as we will see in the next chapter, is that it is difficult to convert this thermal energy into useful work.

Summary: Important Definitions and Equations

The transfer of heat:
1. Conduction: A process in which molecules, while remaining at relatively fixed positions in space, absorb and pass on heat in the form of random mechanical motion.

$$\frac{Q}{t} = \frac{kA(T_2 - T_1)}{L}$$

2. Convection: A process in which heated molecules carry thermal energy with them physically as they move from one part of a fluid to another.

$$\frac{Q}{t} = hA(T_2 - T_1)$$

3. Radiation: The transfer of electromagnetic energy from one object to another without any transfer of matter or the need for a material medium to accomplish this transfer.

Thermal radiation:
The electromagnetic radiation emitted by hot objects because of their temperature.

Stefan's radiation law:
A law relating the rate P at which energy is radiated by a body of surface area A and emissivity e to its absolute temperature T:

$$P = \frac{Q}{t} = e\sigma AT^4$$

where $\sigma = 5.67 \times 10^{-8}$ W/(m²·K⁴)

Wien displacement law:

$$\lambda_{max}T = k = 2.898 \times 10^{-3} \text{ m·K}$$

Newton's law of cooling:

$$P = \frac{Q}{t} = K(T_2 - T_1)$$

Mechanical equivalent of heat (Joule's equivalent):

$$\frac{W}{Q} = 4.186 \times 10^3 \text{ J/kcal}$$

where W is in joules and Q in kilocalories.

Heat and work
Heat (Q): Disordered energy that flows from one object to another because of a temperature difference between them.

Work (W): Ordered energy that is transferred when one object exerts a force on another object and moves it through a distance in the direction of the force.

Work done by a gas: The work done by a gas on expanding is equal to the area under its pressure-volume curve:

$$\Delta W = P \, \Delta V$$

Both the work done and the heat transferred in any physical process are dependent on the path followed by that process.

Internal energy (U):
The sum of the kinetic and potential energies of all the individual particles making up the substance. The internal energy is independent of the path followed by the physical process and depends only on the end-points.

First law of thermodynamics:
If the internal energy of a system changes from U_i to U_f, then $U_f - U_i = \Delta U = \Delta Q - \Delta W$, where ΔU is the increase in internal energy of the system, ΔQ is the heat flow into the system, and ΔW is the work done by the system.

Principle of conservation of energy:
In any isolated system the total internal energy remains constant, even though it may be transformed from one kind to another, *or a perpetual-motion machine of the first kind is impossible.*

First-law efficiency:

$$e_1 = \frac{\text{work done (or energy transferred) by a process}}{\text{energy input to system}}$$

Thermodynamic processes:
Isothermal (same temperature): For an isothermal process involving an ideal gas,

$$\Delta U = 0 \qquad \text{and so} \qquad \Delta Q = \Delta W$$

Adiabatic (no heat enters or leaves the system):

$$\Delta Q = 0 \qquad \text{and so} \qquad \Delta U = -\Delta W$$

Isochoric (same volume):

$$\Delta W = 0 \qquad \text{and so} \qquad \Delta U = \Delta Q$$

Isobaric (same pressure):

$$\Delta U = \Delta Q - \Delta W = \Delta Q - P \, \Delta V$$

Questions

1 Why is it more difficult to keep a house warm on a winter's day if a very strong wind is blowing?

2 Why is the term *radiator* for the heating elements used in houses somewhat inappropriate?

3 What is the basic difference between the radiation emitted by atomic nuclei in the form of alpha particles (helium nuclei) and beta rays (electrons), and the radiation emitted by the sun?

4 It is sometimes said that, if you want to keep your coffee warm for a longer period of time, it is advisable to put the cream in immediately after pouring the coffee and not wait to do so. Can you justify such an opinion from the viewpoint of physics?

5 What are the basic problems associated with the use of solar energy to produce large amounts of electric energy in a convenient and economical way?

6 If both ΔW and ΔQ are dependent on the path taken in any thermodynamic process, how can the change in internal energy ΔU be independent of path?

7 Does the first law of thermodynamics apply to living systems? Does it have to be modified in any way to be so applied?

8 Modern refrigerators have in some cases been made more attractive in appearance by reducing the thickness of their walls. If the thickness of all the insulating walls of a refrigerator is cut in half, what will be the effect on the cost of running the refrigerator for a day, all other factors remaining the same? Why?

9 Argon atoms (Ar) and oxygen molecules (O_2) have roughly the same masses. The specific heat capacity of O_2 is, however, about twice that of Ar. Explain from the structure of the two gases why this is so.

10 Why are white instead of black robes used by missionaries in Africa, where the climate is hot?

11 Does the earth absorb exactly the same amount of energy from the sun as the earth radiates into space? What role is played by the heat flowing from the hot interior of the earth to its surface?

12 The plunger of a bicycle pump is pushed down quickly, compressing the air inside the cylinder. Why do the air and the pump's cylinder become warm?

13 Distinguish as carefully as you possibly can between *heat* and *temperature*.

14 Why do sleeping animals curl up to keep warm in winter, and stretch out to keep cool in summer?

15 Doctors have reported that a number of people have died of heat stroke after remaining in hot tubs at about 45°C for several hours. Can you suggest what happens physically to raise the body temperature excessively under such circumstances?

Problems

A 1 The amount of heat conducted per second through a cylindrical copper rod of cross-sectional area 0.10 m^2, whose ends are 10 m apart and kept at temperatures of 50 and 30°C, is:
(*a*) 1.8×10^{-2} kcal (*b*) 7.6×10^2 kcal (*c*) 7.6×10^2 J (*d*) 1.8×10^{-2} J (*e*) None of the above

A 2 The thermal conductivity of metals is greater than that of common nonmetals like glass and brick by about:
(*a*) Two orders of magnitude
(*b*) One order of magnitude
(*c*) Four orders of magnitude
(*d*) Zero orders of magnitude
(*e*) None of the above

A 3 A piece of metal is at a temperature of 27°C. This temperature is then increased to 327°C. The radiation emitted by the hot metal increases by a factor of about:
(*a*) 2.0×10^4 (*b*) 12 (*c*) 4 (*d*) 8 (*e*) 16

A 4 The absolute temperature of a blast furnace is increased by one order of magnitude. The amount of radiation emitted per second by the furnace increases by:
(*a*) Four orders of magnitude
(*b*) Three orders of magnitude
(*c*) Two orders of magnitude
(*d*) One order of magnitude
(*e*) None of the above

A 5 The earth radiates like a blackbody at about 290 K. Blackbody radiation from the earth therefore has a maximum intensity at a wavelength of about:
(*a*) 10 m (*b*) 10 μm (*c*) 1 μm (*d*) 0.5 μm (*e*) 10^{-7} m

A 6 Water flows over a waterfall which is 50 m high. The increase in temperature of the water at the bottom of the fall compared with the top should be about:
(*a*) 0.01°C (*b*) 0.1°C (*c*) 1.0°C (*d*) 10°C (*e*) 50°C

A 7 Five joules of energy is equal to:
(*a*) 1.2×10^{-3} cal (*b*) 1.2×10^{-3} kcal (*c*) 21 cal (*d*) 21 kcal (*e*) 1.2 kcal

A 8 The first-law efficiency of an electric power plant that takes in 1.5×10^9 J of heat in the form of steam each second, and produces electricity at the rate of 0.40×10^9 W, is:
(*a*) 2.7% (*b*) 27 (*c*) 0.27 (*d*) 37% (*e*) 3.7

A 9 A gas expands isothermally and does an amount of work equal to 50 J. The heat ΔQ added to the gas during the expansion is:
(*a*) -50 J (*b*) 100 J (*c*) -100 J (*d*) 50 J (*e*) 0

A10 A gas is compressed adiabatically from a volume of 2×10^{-3} to 1×10^{-3} m^3. The work done on the gas during the compression is 50 J. The change in internal energy of the gas is:
(*a*) 0 (*b*) $+50$ J (*c*) -50J (*d*) $+100$ J (*e*) -100 J

A11 A gas is compressed at a constant pressure of 10^5 N/m^2 from a volume of 2×10^{-3} to 1×10^{-3} m^3. The work done by the gas is:
(*a*) -10^2 J (*b*) 10^2 J (*c*) 0 (*d*) 10^2 kcal (*e*) -10^2

A12 In an isochoric (equal-volume) process 20 J of heat is added to a gas whose pressure is 10^5 N/m^2 and whose volume is 10^{-4} m^3. The change in internal energy of the gas is:
(*a*) $+20$ J (*b*) -20 J (*c*) 0 (*d*) $+10$ J (*e*) -10 J

A13 In the situation described in Prob. A12, the work done by the gas is:
(a) +20 J (b) −20 J (c) 0 (d) +10 J
(e) −10 J

B 1 Consider a window of area 1.0 m² and 10^{-3} m thick. If the temperatures of the inner and outer surfaces are 10 and 5°C, respectively, how much heat per second flows through such a "window" of (a) copper, (b) glass, and (c) Styrofoam?

B 2 The inside of the steel wall of a boiler is at 102°C and the outside is at 62°C. If the wall is 10^{-2} m thick, how much heat per second comes through a 1.0-m² section of the wall?

B 3 The air temperature at the upper surface of the ice layer on a pond is −10°C. The temperature of the bottom surface is 0°C.
(a) If the ice layer is 10^{-2} m thick, how much heat per second comes through 1 m² of surface?
(b) Does ice form more rapidly when the layer is thick or thin?

B 4 How much heat per second do you lose by conduction from your hand if it is against a window when the outside temperature is (a) 0°C, (b) −10°C? (Assume $T = 33$°C, glass thickness = 2.0 mm, and $A = 4.0 \times 10^{-3}$ m² for your hand.)

B 5 How much heat per second is radiated by a 1-m² surface with emissivity $e = 1$ at 10^4 K?

B 6 At what wavelength is the radiation highest for a radiating object at 0°C?

B 7 What would the temperature of a body have to be if its peak radiation intensity were at $\lambda = 10^{-10}$ m (i.e., an x-ray)?

B 8 If all the energy radiated by the sun on the earth could be used to heat the oceans of the earth, how long would it take to change the temperature of the oceans from 0 to 100°C, i.e., from freezing to boiling? Assume that the mass of the oceans is about 1.5×10^{21} kg.

B 9 A gas expands adiabatically, doing 100 J of work.
(a) How much heat is added?
(b) Calculate ΔU for the gas.

B10 A gas expands isothermally, doing 100 J of work.
(a) How much heat must be added?
(b) Calculate ΔU for the gas.

B11 During a constant-volume process, the internal energy of a gas increases by 100 J.
(a) How much heat must be added?
(b) How much work is done by the gas?

B12 You blow up a balloon to a volume 10^{-1} m³. How much work does the surface of the balloon do against the atmosphere? How high would this amount of work lift a 75-kg person?

B13 During a physics demonstration, a metal can is crushed by removing the inside air with a pump. The initial volume is 9×10^{-3} m³ and the final volume is 10^{-4} m³. How much work does the atmosphere do while crushing the can?

C 1 The submerged area of a 1.5×10^{-2} m thick wooden boat is 2.0 m². If the temperature drop through the wood is 7.0°C, what power goes from the boat to the water?

C 2 A wooden shack is in the form of a cube 2.0 m on a side. The planks are 1.5×10^{-2} m thick. The temperature difference across the wall is 5.0°C. At 120 W per person, at least how many people are in the shack?

C 3 (a) What is the temperature gradient across the window in Example 14.1?
(b) Thirty-five miles below the earth's surface the temperature of the melting rock is about 1400°C. If the surface temperature is 20°C, what is the temperature gradient?
(c) Since the thermal conductivity of glass and granite are roughly the same, what conclusion do you draw?

C 4 The effective area of a Styrofoam ice chest is 2.0 m². How thick are its walls if it takes 24 h to melt 5.0 kg of ice inside the chest on a day when the temperature is 35°C?

C 5 In Prob. C4, if the Styrofoam were replaced by steel of the same thickness, how long would the ice last?

C 6 The inside wall temperature of a steel oven is 150°C and the outside wall temperature is 120°C.
(a) If the surface area of the oven is 1.5 m² and the walls are 5.0×10^{-3} m thick, how much heat per second is given off by conduction?
(b) What thermal conductivity would make the result 5.0×10^3 W?

C 7 A copper bar is 2.0 m long. The cross-sectional area is 4.0×10^{-4} m². It is insulated so that heat may flow only from one end to the other. If one end of the bar is in boiling water and the other is in ice, what power flows along the bar?

C 8 Since the specific heat of the human body is close to that of water, let us assume that the body's thermal conductivity equals that of water, 0.58 W/(m·°C). Using $T_{internal} - T_{surface} = 4$°C, and an effective thickness of 10^{-1} m, calculate how much heat per second the swimmer of Example 14.3 loses by conduction.

C 9 Using the results of Example 14.3 and Prob. C8, calculate the convection coefficient for the swimmer if the total heat loss is 120 W.

C10 (a) What power, in watts, is radiated by a 10^{-4}-m² object ($e = 1$) at 20°C? What would this be if (b) the Celsius temperature were doubled, (c) the Kelvin temperature were doubled?

C11 Molten iron is in a bucket-shaped container at 1600°C. If the surface area of the iron is 10 m², how much heat is lost by radiation each second? (Take $e = 0.83$.)

C12 A light-bulb filament is 2.5×10^{-1} m long and has a cross-sectional diameter 5×10^{-4} m. (a) If the filament ($e = 0.3$) runs at 1100°C, what power is radiated?
(b) Is all this in the visible region?

C13 A 10^{-3} kg piece of ice ($e = 0.1$) in the form of a cubic block at 0°C is ejected from a satellite. If the temperature in orbit is 3 K, how much heat per second does the ice radiate initially?

C14 The flame in a fireplace has area 0.5 m². The room temperature is 20°C. The net radiation from the fire is 10^3 W. Calculate the temperature of the flame if (a) $e = 0.1$, (b) $e = 0.5$, and (c) $e = 1$.

C15 A person sits perfectly still in a bathtub. The body's surface temperature is 33°C and the body's surface area is 1.3 m². What should the water temperature be if the bather is to receive 0.20 W of radiation from the water? (Take $e = 0.6$.)

C16 From our discussion of sound we know that the threshold of pain and barely audible sound correspond to intensities of 1 W/m² and 10^{-12} W/m², respectively. At what temperatures would a perfect radiator ($e = 1$) radiate with these intensities?

C17 A furnace in a private home in Minneapolis breaks down at 9 p.m., when the outside temperature is 0°C (32°F). During the 3 h from 9 p.m. to midnight the house temperature falls from 20 to 17°C. If the outside temperature remains right at the freezing point all night, how long will it take for the house temperature to fall to 14°C, 11°C, 8°C?

C18 The cycle shown in Fig. 14.22 is for an ideal gas.

(a) Find the work done along each of the processes AB, BC, CD, and DA, specifying whether work is done by the gas or on the gas.

(b) What is the total work done by the gas on going around the path $ABCDA$?

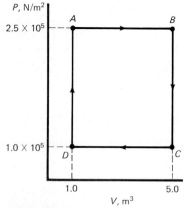

FIGURE 14.22 Diagram for Prob. C18.

C19 Repeat Prob. C18 if all the pressures and volumes are doubled; i.e., $P_A = 5 \times 10^5$ N/m², $V_A = 2.0$ m³, etc.

C20 If the cycle in Fig. 14.22 is for helium and the temperature at D is 27°C:

(a) Find the temperature at points A, B, and C.

(b) Calculate ΔQ for AB, BC, CD, and DA, using $\Delta Q = mc \, \Delta T$.

C21 Two moles of N_2 gas are at 100°C.

(a) What is v_{rms} for a nitrogen molecule?

(b) If the gas does 10^3 J of work during an adiabatic expansion, what is v_{rms} at the end of the process?

C22 In Prob. C21, what is the final v_{rms} if 10^3 J of work is done on the gas during an adiabatic compression?

C23 One kilogram of copper is heated from 20 to 60°C.

(a) How much work does the copper do by pushing back the atmosphere? Take $P = 1$ atm.

(b) Calculate $\Delta W / \Delta Q$ for this process. Does the approximation $\Delta W = 0$ seem justified?

D 1 The thickness and thermal conductivities of a two-layer wall are denoted by L_1, L_2, k_1, and k_2. For a temperature difference $T_2 - T_1$ across the combination, show that the heat per second transferred across an area A is

$$P = \frac{A(T_2 - T_1)}{L_1/k_1 + L_2/k_2}$$

(For three layers, a term L_3/k_3 would appear in the denominator, and similarly for more layers.)

D 2 (a) Calculate P for a 1-m² glass window 0.66 cm thick.

(b) Calculate P for a 1-m² window composed of two 0.33-cm-thick glass panes sandwiching a 0.33-cm-thick layer of air. Assume $T_2 - T_1 = 5$°C. (See Prob. D1.)

D 3 A cooking utensil is to be 3 mm thick, one layer being 2 mm and the other 1 mm thick. If aluminum and brass are used, which combination will pass more heat per second? (See Prob. D1.)

D 4 Two flat metal plates are firmly soldered together. Each plate has a surface area of 400 cm² and a thickness of 5.0 mm. One plate is made of copper and the other of brass. The plates are laid flat on a hot metal plate whose surface temperature is 150°C. The upper surface of the metal plates is at room temperature (20°C).

(a) Find the rate at which heat flows through the plates to the air. (Neglect any heat loss at the edges.)

(b) Find the temperature of the surface where the copper and brass plates are soldered together.

D 5 Three large slabs of the same material and with the temperature gradients shown in Fig. 14.23 are suddenly stuck together.

(a) Show that the increase in the average temperature of the middle slab for a time interval Δt is

$$\frac{\Delta T_{av}}{\Delta t} = \frac{k}{cd} \frac{(\Delta T/\Delta X)_R - (\Delta T/\Delta X)_L}{l}$$

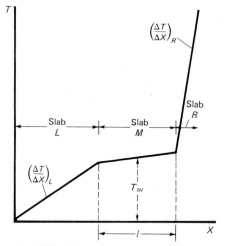

FIGURE 14.23 Diagram for Prob. D5.

where k is the thermal conductivity, c is the specific heat, and d is the density. (The second factor is just a geometric factor, different for each problem, but k/cd determines how rapidly temperature *changes* propagate in a given substance.)

(*b*) Let $\alpha = k/cd$ be the diffusivity for a given substance. Calculate α for copper, water, and air.

D 6 A kilogram of water at 0°C is ejected from an orbiting satellite surroundings which are at a temperature of 3 K.

(*a*) How much time elapses before it turns into ice?

(*b*) Why doesn't 2 kg take exactly twice as long to turn to ice? (Assume the water takes on a spherical shape.)

D 7 Show that Newton's law of cooling follows directly from the equations governing the conduction, convection, and radiation of heat so long as the two temperatures involved are not too far apart. [*Hint*: Show that for radiation $T_2^4 - T_1^4 \simeq K(T_2 - T_1)$.]

D 8 When a person sleeps in pajamas, covered only by a sheet, 53 percent of the heat loss from the person's body to the room is by way of radiation, 28 percent by evaporation, and 19 percent by convection, with conduction contributing an insignificant amount. If the person's skin temperature is 33°C and room temperature is 20°C, what is the rate at which the person is losing heat to the room by all possible processes? Assume that the body surface area is 1.3 m^2 and that the emissivity of the skin is 0.65 in this case.

D 9 The temperature of 5.0 mol of nitrogen (N_2) gas is raised from 20 to 200°C at constant atmospheric pressure.

(*a*) Find the increase in internal energy of the gas.

(*b*) Find the external work done by the gas.

(*c*) Find the heat added to the gas by an external source.

(*Hint*: Use the values of C_P and C_V given in Table 13.4.)

D10 A cylinder containing He gas is closed off at the top by a 10-kg movable piston of surface area 100 cm^2, which is subjected to atmospheric pressure at its top surface. The He is first heated from 20 to 120°C, and the piston rises 30 cm. Then the piston is held fast in its new position while the gas is cooled back down to 20°C.

(*a*) How much heat was added to the gas during the heating process?

(*b*) How much heat did the gas lose as it cooled back down to 20°C?

(*c*) What was the net gain or loss of heat by the gas in the complete cycle?

D11 Consider the same situation as in Example 14.7 and Fig. 14.19. We now assume that the internal energy of the gas is 20 J at point A and 60 J at point D, that the work done by the system along path ADB is 10 J, and that the difference in internal energy between A and B is $\Delta U_{AB} = 50$ J, as found in Example 14.7.

(*a*) Find the heat absorbed in the expansion from A to D.

(*b*) Find the heat absorbed in the constant-volume process in which the gas goes from D to B.

D12 In an isobaric process at atmospheric pressure, 2.0 kg of water is converted into steam at 100°C. The steam occupies a volume 1670 times that of the same mass of water. How is the heat that is required to produce the vaporization distributed between the work done by the system and the internal energy of the system?

Additional Readings

Angrist, Stanley W.: "Perpetual Motion Machines," *Scientific American*, vol. 218, no. 1, January 1968, pp. 114–122. A fascinating article relating attempts made over the past 400 years to escape the implications of the first and second laws of thermodynamics. It contains many interesting photographs and sketches of proposed perpetual-motion machines.

Behrman, Daniel: *Solar Energy: The Awakening Science*, Little, Brown, Boston, 1976. A good popular account of solar-energy research in all parts of the world.

Feynman, Richard: *The Character of Physical Law*, MIT Press, Cambridge, Mass., 1965. Chapter 3, "The Great Conservation Principles," is particularly relevant.

Holton, Gerald: *Introduction to Concepts and Theories in Physical Science*, Addison-Wesley, Reading, Mass., 1973. Chapter 17 is very good on the first law of thermodynamics.

Kelly, James B.: "Heat, Cold, and Clothing," *Scientific American*, vol. 194, no. 2, February 1956, pp. 109–116.

A brief article on the function of clothing in keeping us warm and cool.

Shamos, Morris H. (ed.): *Great Experiments in Physics*, Holt, Rinehart and Winston, New York, 1959. Each chapter of this book is devoted to an important experiment in the history of physics, as described by the originator of the experiment and annotated by the editor. Chapter 12 is devoted to Joule's work on the mechanical equivalent of heat.

Steffens, Henry John: *James Prescott Joule and the Concept of Energy*, Science History Publications, New York, 1979. A useful account not merely of the work of Joule but also of the contributions of Mayer, Kelvin, Helmholtz, and Clausius to the development of thermodynamics.

Tierney, John: "Perpetual Commotion," *Science 83*, vol. 4, no. 4, May 1983, pp. 30–39. A good account, accompanied by photographs, of attempts to evade the implications of the laws of thermodynamics.

In order to go further it is necessary to talk quantitatively. We must measure heat precisely in terms of numbers. . . . First it is clear that to specify heat we must use at least two numbers: one to measure the quantity of energy, the other to measure the quantity of disorder. The quantity of energy is measured in terms of a practical unit called the calorie. . . . The quantity of disorder is measured in terms of the mathematical concept called entropy.*

Freeman J. Dyson (1923–   )

Chapter 15

The Second Law of Thermodynamics

Science and technology move forward together. Sometimes advances in a field like physics lead to giant steps forward in technology. The development of the transistor, for example, revolutionized the electronics industry. At other times technological developments lead to great progress in pure science. One noteworthy case was the development of radar electronics during World War II. The radar equipment developed for the detection of enemy aircraft turned out to be almost perfectly adapted to studies of atomic and molecular structure. Thus the years immediately following the war, when physicists returned to their laboratories and brought their radar gear with them, saw the rapid development of new fields such as microwave spectroscopy, electron spin resonance, and atomic and molecular beam spectroscopy.

*See Additional Readings for source of quote.

FIGURE 15.1 An early form of steam engine: The Corliss Engine. (*Photo courtesy of the Granger Collection.*)

In the field of thermodynamics physics has profited from technology in a special way. Around 1800 the first truly practical steam engine was put into operation in Great Britain, marking the beginning of the industrial revolution. Throughout the nineteenth century the steam engine (see Fig. 15.1) took over the operation of pumps and industrial machines, the propulsion of boats and trains, and a variety of onerous tasks previously performed by draft animals or slaves. Physicists were anxious to develop the steam engine to its maximum possible efficiency. This was the basis for much of the progress in thermodynamics made in the nineteenth century, and in particular for Carnot's pioneering work on the theory of heat engines, which led directly to the second law of thermodynamics.

15.1 Reversible and Irreversible Processes

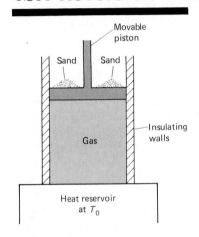

FIGURE 15.2 A gas in a container with insulating walls and a movable piston at the top. The gas is in contact with a heat reservoir at temperature T_0.

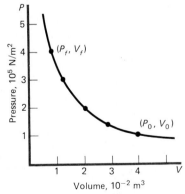

FIGURE 15.3 PV diagram for an isothermal (reversible) compression of the gas in Fig. 15.2.

Before discussing heat engines and the second law of thermodynamics we must clarify the difference between reversible and irreversible processes. Consider a gas in thermal equilibrium, like the one in Fig. 15.2, with its thermodynamic state defined by its volume V_0, pressure P_0, and temperature T_0. We surround the container with perfectly insulating walls at the sides and top, and set the container on a heat reservoir from which heat can flow into or out of the gas to keep its temperature at its initial value T_0. We can reduce the volume of the gas to $V_f = V_0/4$ by carrying out either of the following processes:

1 *Reversible compression.* We can gradually increase the pressure P by dropping sand, a few grains at a time, on the piston and giving the gas time to maintain constant thermal equilibrium with the reservoir as it contracts. In this way the volume can be slowly decreased to $V_0/4$, while the pressure increases to $4P_0$ and the temperature remains constant at T_0 throughout every step of the process. Hence each intermediate state is an equilibrium state, and the change can be graphed on a PV diagram, as in Fig. 15.3. This slow process through a very large number of equilibrium states is called a *reversible process*. Such a process can be made to go in the reverse direction by an infinitesimal change in the operating conditions, say, by removing a few grains of sand at a time from the piston instead of dropping them on the piston.

2 *Irreversible compression.* We can also push the piston in Fig. 15.2 down very suddenly and change the pressure from P_0 to $4P_0$. The temperature will initially increase, then decrease, and the pressure and volume will fluctuate in unpredictable ways during the actual compression. Finally, however, the temperature will return to T_0 because the gas must come into thermal equilibrium with the reservoir which is maintained at T_0. At the end of the process the pressure will therefore be $P_f = 4P_0$, the volume $V_f = V_0/4$, and temperature $T_f = T_0$, just as in the case of a reversible compression. During the actual process of compression, however, we have no way of knowing the values of P, V, and T, since the temperature has fluctuated, and Boyle's law for an isothermal process does not apply. *The intermediate states in the process do not correspond to states in which the gas is in thermal equilibrium.* Hence the system passes from one equilibrium state specified by (P_0, V_0, T_0) to another equilibrium state specified by ($4P_0$, $V_0/4$, T_0) through a series of

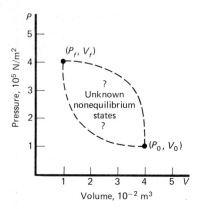

FIGURE 15.4 *PV* diagram for an irreversible compression of the gas in Fig. 15.2.

unknown nonequilibrium states, as shown in Fig. 15.4. Such a process is called *irreversible* because it cannot be made to go in the reverse direction by a small change in the operating conditions. The sudden compression discussed above is irreversible because we do not know even the path followed by the gas in going from V_0 to V_f and hence there is no way to reverse it. There are other cases where a compression may be carried out slowly, but the process is still irreversible. For example, suppose the walls of the container exerted a frictional force on the piston in the example above. Then, even in a slow compression, energy would be lost in the form of heat and the process would again be irreversible. This is always true when friction or air resistance is present. Any process that is not *completely* reversible is said to be irreversible.

Any macroscopic physical process in the real world is irreversible. A reversible process is an idealized case which is extremely important from a theoretical point of view but can never quite be achieved in practice. The slow expansion or compression of an ideal gas is, however, a very close approximation to a reversible process, and we will use it as our model of a reversible process in the discussion that follows.

Isothermal and Adiabatic Processes

All isothermal processes are reversible, because the only way the temperature can be kept constant is to carry out the process very slowly. In the case of an isothermal expansion or compression of a gas, we have already seen that the product of the pressure and the volume must remain constant as long as the temperature is unchanged; i.e.,

Isothermal Process

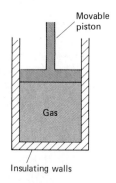

Movable piston

Gas

Insulating walls

FIGURE 15.5 A reversible adiabatic process, in which a gas is surrounded on all sides by insulating walls and is then slowly compressed.

$$\boxed{PV = \text{constant}} \quad \text{(Boyle's law)}$$

The graph of such a process on a *PV* diagram is an *isotherm*. It takes the form of the hyperbola shown in Fig. 15.3.

Adiabatic processes can be either reversible or irreversible. Here we will consider a *reversible adiabatic process*, such as that in Fig. 15.5, in which a gas is surrounded by insulating walls on all sides and then compressed. In this case no heat flows out of the gas ($\Delta Q = 0$), and so the temperature must rise since the work done on the gas in the compression increases its internal energy U. Hence $PV (= nRT)$ can no longer be a constant since T is not a constant. Rather we find experimentally that if the compression is carried out slowly so that the gas is in thermal equilibrium at every stage of the process, the *PV* graph is as shown in Fig. 15.6. Such a graph of a reversible adiabatic process is called an *adiabat. This has a steeper slope than an isotherm* and crosses a series of isotherms as the gas is compressed, since the temperature continually rises in such an adiabatic compression. To see why the slope is steeper for an adiabat than for an isotherm, note that if the pressure is doubled in an isothermal process the volume is cut in half. In an adiabatic compression, however, if the pressure is doubled, the volume is not quite cut in half since the increase in temperature acts to expand the gas. This prevents the volume from falling to $V_0/2$, and the slope of the graph is therefore steeper.

It is found that the adiabat in Fig. 15.6 satisfies the equation

Adiabatic Process

$$\boxed{PV^\gamma = \text{constant}} \tag{15.1}$$

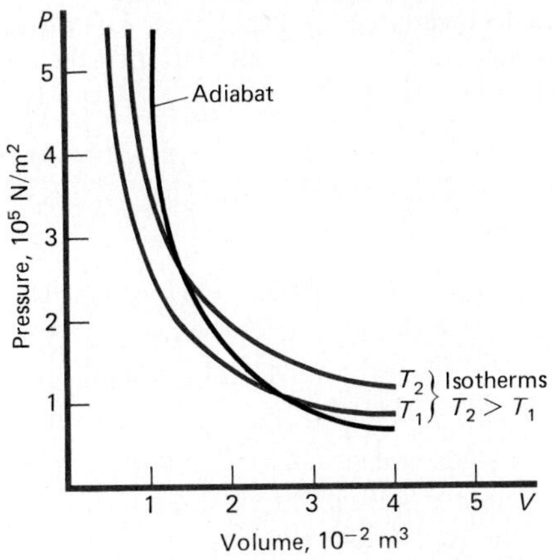

FIGURE 15.6 *PV* diagram for a reversible adiabatic process; an adiabat. Notice how the adiabat crosses the isotherms, since adiabats have steeper slopes.

where γ is the ratio of the molar heat capacities (C_P/C_V) defined in Sec. 13.8, and where the constant on the right side of the equation depends on the amount of gas present. Equation (15.1) then describes the behavior of the pressure and volume in a reversible adiabatic process, just as Boyle's law, $PV = $ constant, describes the same behavior for an isothermal process.

Since, for any ideal gas, $PV = nRT$, we have from Eq. (15.1)

Adiabatic Process

$$\frac{nRT}{V} V^\gamma = \text{constant} \qquad \text{or} \qquad \boxed{TV^{\gamma-1} = \text{constant}} \qquad (15.2)$$

for n and R are both constants for any sample of gas, and have no effect on the functional form of the equation. Hence the two basic equations governing reversible adiabatic processes in an ideal gas are

$$P_1 V_1^\gamma = P_2 V_2^\gamma \qquad (15.3)$$

and

$$T_1 V_1^{\gamma-1} = T_2 V_2^{\gamma-1} \qquad (15.4)$$

Note that the temperatures must be expressed on the absolute scale, since the ideal gas law requires absolute temperatures. These equations can be used to find how any one of the three fundamental properties of an ideal gas—P, V, and T—varies when the other properties vary in an adiabatic process, just as $PV/T = $ constant can be used in the isothermal case.

Example 15.1

Suppose that 10^{-3} m^3 of neon gas at room temperature and atmospheric pressure is compressed adiabatically to twice atmospheric pressure.

(a) What is the final volume of the gas?
(b) What is its final temperature?

SOLUTION

(a) For an adiabatic process, we have $P_1 V_1^\gamma = P_2 V_2^\gamma$. Here $V_1 = 10^{-3}$ m^3, and $P_2 = 2P_1$. Also, since neon is a monatomic gas, $\gamma = 1.67$. Hence we have

$$\left(\frac{V_2}{V_1}\right)^\gamma = \frac{P_1}{P_2} = \frac{1}{2}$$

or

$$\frac{V_2}{V_1} = (\tfrac{1}{2})^{1/\gamma} = (\tfrac{1}{2})^{1/1.67} = (\tfrac{1}{2})^{(0.60)} = 0.66$$

and so $V_2 = 0.66 V_1 = \boxed{0.66 \times 10^{-3} \text{ m}^3}$

(b) Here we have $T_1 V_1^{\gamma-1} = T_2 V_2^{\gamma-1}$. Hence

$$\frac{T_2}{T_1} = \left(\frac{V_1}{V_2}\right)^{\gamma-1} = \left(\frac{10^{-3} \text{ m}^3}{0.66 \times 10^{-3} \text{ m}^3}\right)^{0.67} = (1.5)^{0.67} = 1.3$$

and so $T_2 = 1.3 T_1 = 1.3(300 \text{ K}) = \boxed{390 \text{ K}}$

Note on the numerical calculation: The unusual powers to which numbers are raised in a problem of this sort, e.g., $(1.5)^{0.67}$, can be calculated by use of logarithms. However, it is much easier to use the y^x function found on many hand calculators. The procedure consists of selecting the function y^x and entering the numerical values of x and y according to the rules specified for the particular calculator being used.

Work Done in Isothermal and Adiabatic Processes

We saw in Sec. 14.7 that the work done by a gas in expanding its volume by an amount ΔV at a constant pressure P is $\Delta W = P\,\Delta V$. For an *isobaric* process this leads to the total work done during the expansion, $W = P_0(V_f - V_0)$. This does not help us to find the work done in isothermal and adiabatic processes, however, since the pressure P is *not constant* in such processes. To obtain the work in these cases calculus is needed, but useful results can also be obtained graphically, as we will now show for the work done by a gas in an isothermal expansion.

The graph of an isothermal expansion, in which the volume of the gas increases from $V_0 = 1.0 \times 10^{-2}\ \mathrm{m}^3$ to $V_f = 4.0 \times 10^{-2}\ \mathrm{m}^3$, is shown in Fig. 15.7. Since the temperature remains constant in this isothermal process, the pressure and the volume must satisfy the equation:

$$PV = P_0V_0 = (4.0 \times 10^5\ \mathrm{N/m^2})(1.0 \times 10^{-2}\ \mathrm{m^3}) = 4.0 \times 10^3\ \mathrm{J}$$

and so $P = (4.0 \times 10^3\ \mathrm{J})/V$. The values of P obtained from this equation are given with the corresponding values of V in Table 15.1.

The work done in this expansion is the area under the curve between V_0 and V_f. To obtain an approximate value for this area graphically, we break up the area into six vertical strips, each of width $\Delta V = 0.50 \times 10^{-2}\ \mathrm{m}^3$. We then calculate the average value of the pressure $(\overline{P})$ for each strip by simply averaging the values of P at the two edges of each strip. The area of each strip is then, to a reasonable approximation, the work done by the gas during the volume change covered by the strip. The total work done by the gas is the sum of the areas of the six strips:

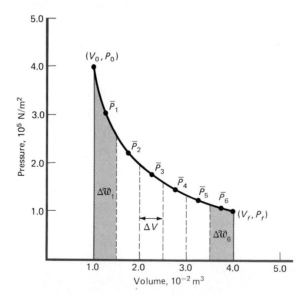

FIGURE 15.7 Calculation of the work done during an isothermal expansion.

TABLE 15.1 Work Done by a Gas in an Isothermal Expansion (Fig. 15.7)

Volume V, 10^{-2} m³	Pressure P, 10^5 N/m²	Average pressure for vertical strip $\bar{P}$, 10^5 N/m²	Width of strip ΔV, 10^{-2} m³	Work done, $\Delta W = \bar{P} \, \Delta V$, 10^3 J
1.0	4.0			
		3.35	0.50	1.68
1.5	2.7			
		2.35	0.50	1.18
2.0	2.0			
		1.80	0.50	0.90
2.5	1.6			
		1.45	0.50	0.73
3.0	1.3			
		1.20	0.50	0.60
3.5	1.1			
		1.05	0.50	0.53
4.0	1.0			

Total work done $W = 5.62 \times 10^3$ J

$$W = \sum_{i=1}^{6} \Delta W_i = \sum_{i=1}^{6} \bar{P}_i \, \Delta V$$

This summation is carried out in Table 15.1 and leads to the result $W = 5.62 \times 10^3$ J, as shown there.*

For an adiabatic process, since $PV^\gamma = P_0 V_0^\gamma$, we have

$$P = \frac{\text{constant}}{V^\gamma}$$

This equation can then be used to plot P versus V, and the work done obtained from the area under the curve, as in the above example.

This graphical method can be used to find the work done by a gas in any kind of physical process, as long as it is possible to construct a continuous PV graph for the process.

15.2 Entropy

Consider Fig. 15.8. We have two blocks of metal, one at a high temperature T_1, the other at a lower temperature T_2. The blocks are initially insulated from each other. In such a condition they represent an *ordered system*, for we can clearly distinguish two kinds of molecules, those at temperature T_1, with a higher average kinetic energy, and those at T_2, with a lower average kinetic energy. These two blocks could be used as the hot and cold reservoirs in a heat engine to produce useful work, as we will see.

Suppose, now, that the two blocks are put in contact. Heat conduction will soon bring them both to the same intermediate temperature T_f. Then we

---

*From calculus it can be found that the exact value for the work done by the gas in this isothermal expansion from V_0 to V_f is

$$W = P_0 V_0 \ln \frac{V_f}{V_0} = (4.0 \times 10^5 \text{ N/m}^2)(1.0 \times 10^{-2} \text{ m}^3) \ln \frac{4}{1} = (4.0 \times 10^3 \text{ J})(1.39) = 5.55 \times 10^3 \text{ J}$$

Hence our graphical result agrees with this exact result to two significant figures, which is quite good considering the crudeness of the averaging process used to obtain $\bar{P}$.

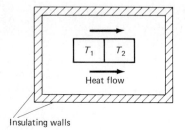

FIGURE 15.8 Flow of heat between two bodies at different temperatures within a thermally insulated box. Since $T_1 > T_2$, heat flows from 1 to 2. The process is an irreversible adiabatic one, and the entropy increases.

will no longer be able to distinguish one group of molecules from the other, nor to use the blocks to do useful work. Order has thus given way to *disorder* which is irreversible. To separate the two groups of hot and cold molecules from each other once they have come to the same average temperature would require the flow of heat from a colder temperature to a hotter one—an impossibility.

Why does the first process, the flow of heat from a hot to a cold temperature, occur so easily, whereas the reverse process never occurs? The answer given by physics is that in the first process a quantity called the *entropy* increases, and therefore the process can occur in practice, whereas the second process never occurs because it would require a *decrease* in entropy.*

What is this mysterious quantity called entropy? The above example indicates that entropy is a measure of *disorder*. An increase in entropy means an increase in disorder. Since, as we all know from experience, things spontaneously become more disordered (consider the clothes in your room or the records in your record collection), in any spontaneous process the entropy always increases.

Disorder and Probability

The reason why all physical systems tend to move from an ordered state to a disordered state is simply that disordered states are more probable because there are more of them. For example, there is only *one* correct alphabetical order for the cards in a library card catalog, a highly ordered state. There are, however, thousands or perhaps millions of incorrect orders for the cards, some with one card out of place, some with two out of place, etc. Hence if anyone randomly mixes up the cards in a library file drawer, a disordered state is much more likely to result than the original ordered state, since there is only one ordered state but a very large number of disordered states. We say that disordered states are highly probable but that the one ordered state is highly improbable.

The entropy of any isolated system therefore always tends to increase, because disorder (which entropy measures) is always more probable than order.[†] This is the reason our record collections are usually out of order and our rooms often a mess. Disorder has a way of winning out over order in many facets of the physical universe, since entropy must continually increase.

Quantitative Definition of Entropy

To make any further progress we must quantify the concept of entropy, as suggested by Freeman Dyson at the beginning of this chapter, by defining entropy in a measurable way. Suppose a gas expands isothermally against a

---

*The German physicist Rudolf Clausius (1822–1888) introduced the word *entropy* in 1865, basing it on a Greek word signifying *transformation*, and deliberately making it as similar as possible to the word *energy*.

[†]This is usually expressed in the form of an equation, $S = k \ln w$, where S is the entropy, k is Boltzmann's constant, w is the so-called thermodynamic probability, and ln is the natural logarithm (base e). This equation connects a thermodynamic, or macroscopic, quantity, the *entropy*, with a statistical, or microscopic, quantity, the *thermodynamic probability*. As an example, for a system of 100 coins the thermodynamic probability of all 100 coming up heads when tossed is 1, since there is only 1 way this configuration can be achieved. Hence for this highly improbable state $S = k \ln (1) = 0$. For 50 heads and 50 tails, on the other hand, $w \simeq 10^{29}$, and $\ln w \simeq 67$. For this highly probable and therefore disordered state, the entropy S is much larger.

piston from a volume V_0 to a volume V_f. It does work ΔW in the process, and, to keep its internal energy and therefore its temperature constant, an amount of heat ΔQ must flow into the gas from the environment to balance out the work done, since by the first law of thermodynamics, $\Delta U = \Delta Q - \Delta W$. This heat flows into the gas slowly and reversibly in such a way as to keep the temperature T constant. We define the change in entropy (ΔS) in this process as the heat added to the gas divided by the absolute temperature, i.e.,

Definition of Entropy

$$\boxed{\Delta S = \frac{\Delta Q}{T}}$$ (15.5)

where ΔQ is positive if heat flows into the system and negative if heat flows out of the system. Common units for entropy are joules per kelvin (J/K), or kilocalories per kelvin (kcal/K).

We have already seen that Q is not a state variable. On the other hand, it turns out that entropy, as defined above, *is* a state variable (like the internal energy U), since it always has the same value for the same values of P, V, and T. Because entropy is a state variable, the change in entropy in any closed cyclical process is zero, just as the change in the internal energy of a thermodynamic system in a cyclic process is zero, as discussed in the preceding chapter. In the case of the entropy the cyclic process must be a reversible one. This means that every intermediate state in the process must be an equilibrium state, so that it is possible to specify the temperature at which the heat ΔQ is transferred at any stage of the process. Hence we have, for any reversible process,

$$\sum_{\text{cycle}} \Delta S = \sum_{\text{cycle}} \frac{\Delta Q}{T} = 0$$

15.3 Entropy Changes

We now want to consider the changes in entropy which occur in some important physical processes, where we include entropy changes in both the physical system and its environment.

Reversible Processes

Suppose an ideal gas expands isothermally from a volume V_0 to a volume V_f, as in Fig. 15.7. In so doing the gas must take in an amount of heat ΔQ from a heat reservoir to keep the temperature constant at its initial value T_0. Hence the change in the entropy of the gas (which we will call the *system*) is $\Delta S_{\text{sys}} = \Delta Q/T_0$. The heat reservoir (which we will refer to by the general term *environment*) must lose an equal amount of heat ΔQ at the same temperature. Therefore the change in the entropy of the environment is $\Delta S_{\text{env}} = -\Delta Q/T_0$. The total change in the entropy of the system plus environment is therefore:

$$\Delta S = \Delta S_{\text{sys}} + \Delta S_{\text{env}} = \frac{\Delta Q}{T_0} - \frac{\Delta Q}{T_0} = 0$$

The entropy of the system plus that of its environment remains unchanged. This turns out to be true for all *reversible processes*.

Irreversible Processes

As an example of an irreversible process consider the preceding example of the flow of heat between two blocks of metal, one at a high temperature T_1, the other at a lower temperature T_2. We put these two blocks in contact inside a thermally insulated box, as in Fig. 15.8, and they eventually come to a final intermediate temperature T_f. This is an irreversible adiabatic process. We want to calculate the change in entropy of the system plus its environment for this process.

We will not attempt a rigorous calculation of the entropy in this case. But it can be seen that entropy must increase in this irreversible process. The hot metal's temperature falls from T_1 to T_f, and hence its average temperature T_H during this process is certainly greater than T_f. Similarly, the cold metal's temperature rises from T_2 to T_f, and so its average value T_C is certainly less than T_f. If the heat lost by the hot metal and gained by the cold metal is ΔQ, then the change in entropy of the system is

$$\Delta S = -\frac{\Delta Q}{T_H} + \frac{\Delta Q}{T_C}$$

Since $T_H > T_f > T_C$, we must have $\dfrac{\Delta Q}{T_C} > \dfrac{\Delta Q}{T_H}$

and $\qquad \Delta S = -\dfrac{\Delta Q}{T_H} + \dfrac{\Delta Q}{T_C} > 0 \qquad$ or $\qquad \Delta S > 0$

In this irreversible process the entropy of the system has increased. The entropy of the environment is unchanged, since the system is thermally insulated from the environment, i.e., it is an *isolated system*. This result turns out to be valid for all irreversible processes. Hence we can conclude:

In an irreversible process the entropy of any physical system plus its environment always increases.

We can now put our two results together in the following summary statement:

Entropy Statement of Second Law

> The entropy of any physical system plus its environment increases in all irreversible processes and remains constant in all reversible processes. Hence *the total entropy of any isolated system never decreases.*

Since, as we have seen, all real macroscopic processes are irreversible, this necessarily implies that the entropy of any isolated physical system is constantly increasing. This is one statement of the *second law of thermodynamics*, whose significance will become clearer when we apply it to some practical heat engines in what follows.

The entropy statement of the second law implies that any isolated system will spontaneously go from an ordered state to a disordered state, since this increases its entropy. The greater the disorder, the less work the system will be able to do, as in the case of heat conduction discussed above. Hence, even though energy is always conserved, an increase of entropy in a system means that the system will be able to do less work than if the system had a lower entropy. Degradation of energy therefore goes hand in hand with increase in entropy—and entropy is always increasing in any isolated system. This is important for understanding the world's energy problems.

Example 15.2

Five kilograms of water at 0°C is frozen to ice at 0°C. **(a)** Calculate the change in entropy of the ice-water system.

(b) Show that the result is consistent with the entropy statement of the second law of thermodynamics.

SOLUTION

(a) Since the temperature is constant and equal to 0°C or 273 K, we have:

$$\Delta S = S_{ice} - S_{water} = -\frac{\Delta Q}{T}$$

for the ice-water system loses heat when the water freezes. Hence

$$\Delta S = -\frac{\Delta Q}{T} = -\frac{(80 \text{ kcal/kg})(5.0 \text{ kg})}{273 \text{ K}} = -1.5 \text{ kcal/K}$$

or

$$\Delta S = \left(-1.5 \frac{\text{kcal}}{\text{K}}\right)\left(4.18 \times 10^3 \frac{\text{J}}{\text{kcal}}\right) = \boxed{-6.3 \times 10^3 \text{ J/K}}$$

(b) In this case the entropy of the ice-water system *decreases*. This does not violate the second law of thermodynamics, however, since the entropy of the environment increases by exactly the same amount when the ice gives off heat on freezing. Hence the entropy change of system plus environment is *zero*. This is what we would expect, since freezing is an isothermal process in which water at 0°C is converted to ice at 0°C. Therefore it is necessarily a reversible process, and the entropy of system plus environment remains unchanged in all reversible processes, as we have seen.

15.4 Heat Engines; the Carnot Engine

One of the basic functions of our technological civilization is to convert the energy in fossil and nuclear fuels into useful mechanical or electric energy efficiently and economically. At the present time the process followed in doing this always includes the following steps:

Internal energy (in fuel) → heat → mechanical or electric energy

Even though the first step of this process can be carried out with nearly 100 percent efficiency, the conversion of heat into useful energy is a very inefficient process (5 to 40 percent efficient), as we will see. The rest of the energy is lost in the form of heat exhausted at a low temperature by the machine that converts the heat into useful work.

Definition

A device that converts heat into work is called a heat engine.

A heat engine takes a *working substance* (often H_2O) through a series of operations known as a *cycle*. In all heat engines the basic process is the same. The internal energy of the fuel is used to heat the working substance to a high temperature. The working substance then does some useful work, and discharges the remainder of the original heat into another reservoir at a lower temperature. Such a heat engine can be represented by the flow diagram in Fig. 15.9.

In this process the working substance is used over and over again. It is always restored to its original condition at the end of the cycle by contact with the heat reservoir at temperature T_H, which receives its energy from the fuel being burned. Since the working substance is unchanged in a complete cycle, its internal energy is also unchanged, and so, from the first law,

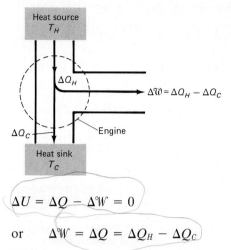

FIGURE 15.9 A generalized heat engine, in which the engine takes in heat ΔQ_H at a temperature T_H, converts part of it into useful work ΔW and discards the remainder (ΔQ_C) as heat to a cold reservoir at temperature T_C.

$$\Delta U = \Delta Q - \Delta W = 0$$

or $\quad \Delta W = \Delta Q = \Delta Q_H - \Delta Q_C$

Hence the total work done by the engine is the difference between the heat ΔQ_H taken in at the hot reservoir and the heat ΔQ_C given out at the cold reservoir.

The first-law efficiency e_1 of such an idealized heat engine is then defined as the ratio of the work it produces to the energy it takes in as heat. Therefore

$$e_1 = \frac{\text{work output during cycle}}{\text{heat input during cycle}} = \frac{\Delta W}{\Delta Q_H} \tag{15.6}$$

or $\quad e_1 = \dfrac{\Delta Q_H - \Delta Q_C}{\Delta Q_H} = 1 - \dfrac{\Delta Q_C}{\Delta Q_H}$ $\tag{15.7}$

Hence for maximum efficiency ΔQ_C must be as small as possible and ΔQ_H must be as large as possible. No engine has ever been built that makes ΔQ_C zero, and no heat engine therefore is 100 percent efficient.

Carnot Cycle

A *Carnot engine* is a particular type of idealized heat engine which Sadi Carnot (see Fig. 15.10 and accompanying biography) proved was the most efficient possible heat engine operating between any two given temperatures. Carnot proved that all engines of this type operating between the same two temperatures have the same efficiency regardless of the nature of the working substance. In our case we will, for simplicity, choose an ideal gas as the working substance.

The distinguishing feature of a Carnot engine is that the cycle in which the engine works is defined by two reversible isothermal processes and two reversible adiabatic processes. The gas is confined to a cylinder with insulating walls and a movable piston at one end. At the other end the cylinder can be put in contact with heat reservoirs or insulating stands as needed. The Carnot cycle then consists of the four steps illustrated in Fig. 15.11. These are:

(a) *Isothermal expansion:* The gas is put in contact with the reservoir at T_H and, after achieving temperature equilibrium, is expanded isothermally. Its initial volume V_1 changes to a new volume V_2, with the pressure falling from P_1 to P_2. The gas takes in an amount of heat ΔQ_H from the reservoir at T_H which is just sufficient to keep the temperature constant at T_H throughout the

Nicolas Léonard Sadi Carnot (1796–1832)

FIGURE 15.10 Nicolas Léonard Sadi Carnot. (*Photo courtesy of AIP Niels Bohr Library*.)

Carnot was a remarkable member of a very distinguished French family. He was born in Paris, the eldest son of Lazare Carnot, who served as minister of war under Napoleon and wrote treatises on military strategy, mechanics, geometry, and calculus. The young Carnot was educated at home by his father in mathematics, physics, languages, and music. At the age of 16 he entered the Ecole Polytechnique, of which his father was a founder, and continued his studies until 1814, when he fought for Napoleon in the futile defense of Paris against Napoleon's European enemies. He then took a 2-year course in military engineering at Metz, and during 1816 to 1818 served in the French army as a second lieutenant in charge of planning fortifications.

Carnot did not have the temperament of a soldier and in 1818 took a permanent leave of absence from the military to pursue his real interest, the application of physics and mathematics to the improvement of the steam engines which were then becoming popular in France as part of the industrial revolution. This passion occupied him until his premature death from cholera in 1832 at the age of 36, except for two interruptions: a trip to Magdeburg in the summer of 1821 to visit his then-exiled father, and another tour of active duty with the military from 1826 to 1828.

In 1824 Carnot published his great and only work, *Reflections on the Motive Power of Heat*. In it he addressed questions such as whether there is a limit to the amount of work that can be produced from a given quantity of heat, and whether steam is the best substance to use in engines, and, if so, what can be done to improve the efficiency of steam engines. His approach to these problems was in great part derived from his father. Although he used the caloric theory of heat in his book, it seems clear that he broke away from it in his later years and embraced the kinetic theory of heat, which was ultimately to win out over the caloric theory.

Carnot was the first physicist to treat quantitatively the interconversion of work and heat, and thus he deserves to be called the father of thermodynamics. He showed that what is important in producing work from heat are the high and low temperatures between which the heat engine operates. It is possible to use Carnot's results to obtain the second law of thermodynamics, even though Carnot never made this clear in his work. Kelvin and Clausius did so years later, and in so doing paid tribute to Carnot's pioneering work, Kelvin referring to him as "the profoundest thinker in thermodynamic philosophy in the first thirty years of the nineteenth century."

Carnot was a quiet, scholarly individual, who lived by the avowed principle: "Say little about what you know, nothing about what you do not know." He was by no means a narrow scientist; throughout his short life he continued to study economics, French literature, and music, and to cultivate gymnastics, fencing, and dancing as relaxations from his scientific work. He was also very practical, a prime motivation in his work being the hope that he could improve people's lives by developing cheaper and more efficient steam engines to reduce the burden of work.

expansion. On the *PV* diagram in Fig. 15.12 the gas moves from point *a* to point *b*.

(b) *Adiabatic expansion:* The heat reservoir is replaced by an insulating stand and the gas is expanded adiabatically and reversibly. P_2, V_2, T_H change to P_3, V_3, T_C, with $T_C < T_H$; that is, the temperature falls. In Fig. 15.12 the gas goes from *b* to *c*.

(c) *Isothermal compression:* The insulating stand is replaced by another heat reservoir at a temperature T_C. The gas is then compressed isothermally

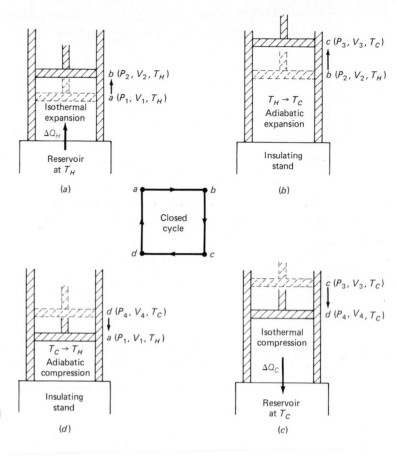

FIGURE 15.11 The four steps in a Carnot cycle: *(a)* Isothermal expansion; *(b)* adiabatic expansion; *(c)* isothermal compression; and *(d)* adiabatic compression. In a Carnot cycle all four processes are reversible.

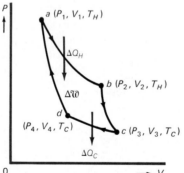

FIGURE 15.12 *PV* diagram for the Carnot cycle of Fig. 15.11. The enclosed area ($= \Delta \mathcal{W}$) is bounded by two isotherms and two adiabats.

at T_C. An amount of heat ΔQ_C is given off to the reservoir at the temperature T_C so as to keep the gas temperature constant. P_3, V_3 change to P_4, V_4, and the gas goes from c to d on the PV diagram.

(d) *Adiabatic compression:* An insulating stand again replaces the heat reservoir. The gas is compressed adiabatically back to its initial state. P_4, V_4, T_C change to P_1, V_1, T_H, which are the original values of the pressure, volume, and temperature of the gas, which goes from d to a in Fig. 15.12.

The gas has therefore been taken around a closed, reversible cycle. As Fig. 15.12 shows, heat ΔQ_H has flowed into the gas at temperature T_H, and a lesser amount of heat ΔQ_C has flowed out of the gas at T_C. Since the internal energy of the gas is unchanged in the cycle, the difference $\Delta Q_H - \Delta Q_C$ has been converted into work, where the work is equal to the area enclosed by the two isotherms and the two adiabats in Fig. 15.12.

The entropy change of the gas in this Carnot cycle is

$$\Delta S = \frac{\Delta Q_H}{T_H} - \frac{\Delta Q_C}{T_C}$$

Every part of this cycle is, however, reversible and hence the change in entropy in the complete cycle is zero. Hence

$$\frac{\Delta Q_H}{T_H} - \frac{\Delta Q_C}{T_C} = 0 \qquad \text{or} \qquad \frac{\Delta Q_C}{\Delta Q_H} = \frac{T_C}{T_H} \tag{15.8}$$

The overall first-law efficiency of the Carnot cycle is therefore, from Eq. (15.7),

First-Law Efficiency of a Carnot Engine
$$e_1 = 1 - \frac{\Delta Q_C}{\Delta Q_H} = 1 - \frac{T_C}{T_H} \qquad \boxed{e_1 = 1 - \frac{T_C}{T_H}} \tag{15.9}$$

This is the maximum possible efficiency for a heat engine operating between temperatures T_H and T_C.

For a conventional steam engine operating between 600 and 100°C this efficiency is

$$e_1 = 1 - \frac{273 + 100}{273 + 600} = 1 - 0.43 = 0.57 = 57\%$$

Any heat engine which is not a Carnot engine has a lower efficiency than this. One reason is that, since all processes in a Carnot cycle are reversible, it represents an idealized case. All real heat engines involve some irreversible processes, like friction or turbulence, which dissipate energy and reduce the efficiency of the engine below the Carnot value. This led Carnot to the following conclusion:

Carnot Result for Heat Engines

> The efficiency of all reversible heat engines operating between the same two temperatures is the same, and no irreversible engine working between the same two temperatures can have as great an efficiency.

Hence thermodynamics determines the maximum possible efficiency of all heat engines, and there is nothing that can be done by human ingenuity to improve this efficiency except to make T_H as high as possible and T_C as low as possible in Eq. (15.9).

A Carnot engine is not a very practical engine, since the cyclic process must be carried out so slowly that the area surrounded by the two isotherms and two adiabats in Fig. 15.12 must be very small. Hence the work output of such an engine is also too small to be practical.

Example 15.3

A nuclear power plant on a submarine produces steam at 350°C to drive a steam-electric turbine. The temperature on the low side of the turbine is 27°C.
(a) What is the maximum possible efficiency of the steam turbine?

(b) If 10^{12} J of heat is supplied by the nuclear fuel, how much of this is converted into useful work (again assuming maximum possible efficiency)?
(c) How much energy is dissipated as waste heat?

SOLUTION

(a) The maximum possible efficiency is that of a Carnot engine operating between the two given temperatures, i.e., 623 and 300 K. This efficiency is

$$e_1 = 1 - \frac{T_C}{T_H} = 1 - \frac{300}{623} = 0.52 = \boxed{52\%}$$

(b) From Eq. (15.6) we have

$$e_1 = \frac{\Delta W}{\Delta Q_H} = 0.52$$

Hence $\Delta W = e_1 \Delta Q_H = (0.52)(10^{12} \text{ J}) = \boxed{5.2 \times 10^{11} \text{ J}}$

(c) The waste heat is, from the principle of conservation of energy,

$$\Delta Q_C = \Delta Q_H - \Delta W = 10^{12} \text{ J} - 0.52 \times 10^{12} \text{ J} = 0.48 \times 10^{12} \text{ J}$$

$$= \boxed{4.8 \times 10^{11} \text{ J}}$$

15.5 Practical Heat Engines

One of the first practical heat engines was the steam engine on which the industrial revolution was built.

Steam Engines

The steam engine utilizes the energy in high-pressure steam to do useful work. A steam engine consists of a cylinder with two valves A and B at either end, as in Fig. 15.13. These valves serve alternately as input and exhaust valves for the steam. First, hot steam is injected into the cylinder through valve A. As the steam drives the piston to the right, valve A is closed and valve B is opened to exhaust the gases to the right of the piston. When the steam has fallen in temperature because of the work it has done in moving the piston, valve A is again opened, but now to serve as an exhaust valve. At the same time hot steam is injected through valve B. This drives the piston back to the left and the remaining spent steam is exhausted from valve A. The process continues in this way, with successive bursts of high-pressure steam driving the piston first to the right and then to the left. The piston's motion is converted into the rotary motion of a flywheel by the action of the connecting rods and the "crosshead" shown in the figure.

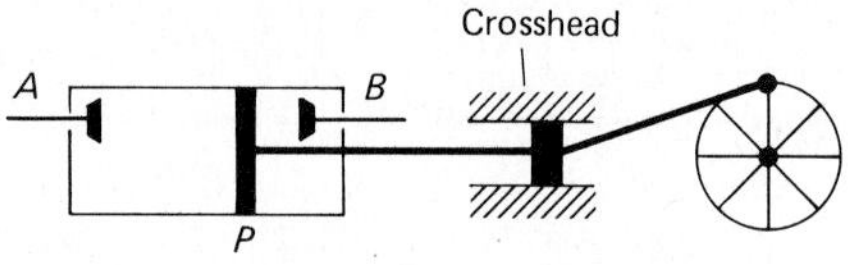

FIGURE 15.13 A model of a steam engine. Hot gases drive the piston P back and forth. The crosshead converts this back-and-forth motion into the rotary motion of a flywheel.

Steam Turbines

In a steam turbine the steam produced in a boiler impinges on a set of blades attached to a rotating wheel. Some of the internal energy of the steam is converted into rotational energy of the wheel.

A prime example of a practical steam turbine is the typical fossil-fuel steam-electric power plant shown in Fig. 15.14. The nature of the electric generator will be discussed in Chap. 19. The boiler converts water into steam,

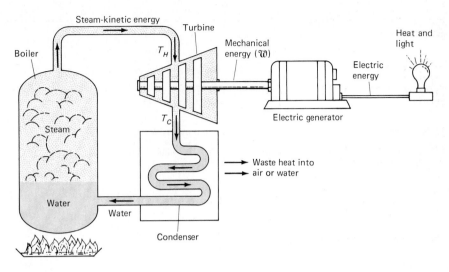

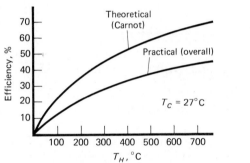

FIGURE 15.14 A fossil-fuel steam-electric power plant.

FIGURE 15.15 Theoretical and practical efficiencies of a typical fossil-fuel power plant.

which impinges on the blades of a turbine and causes the turbine to rotate. This rotating turbine then produces electricity. The steam loses some of its energy in driving the turbine and is then condensed back to water in a set of pipes cooled by flowing river water or blown air. Hence the overall function of the steam engine is to take heat from the steam at temperature T_H, convert part of it into useful work ΔW, and deliver the rest of it as waste heat to the condenser at the temperature of the cooling water T_C.

The theoretical Carnot efficiency of such an engine is $e_1 = 1 - T_C/T_H$, and the actual efficiency even less than this. Figure 15.15 shows the theoretical and practical efficiencies of a typical fossil-fuel power plant as a function of T_H, when $T_C = 27°C = 300$ K. We can see that even for $T_H = 700°C$ (= 973 K), the actual efficiency is only about 50 percent, and has leveled off so much as a function of T_H that higher temperatures would not be justified in terms of the increased efficiency to be expected. For this reason most fossil-fuel plants today use steam at about 1000°F (538°C), and hence have theoretical efficiencies of about 60 percent and overall practical efficiencies of about 40 percent. This means that 60 percent of the input energy is thrown away as waste heat (resulting in what is called *thermal pollution*).

Nuclear-powered steam-electric plants are even less efficient than fossil-fuel powered plants, because safety considerations require that the steam be produced at a lower temperature (~350°C). Hence the practical efficiency of such nuclear plants is only 30 percent or so.

Internal-Combustion Engine

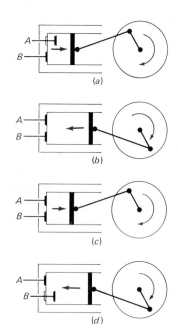

FIGURE 15.16 An internal-combustion engine. A complete cycle involves the following processes: (a) intake stroke, (b) compression stroke, (c) power stroke, and (d) exhaust stroke.

Internal-combustion engines differ from steam engines (external-combustion engines) in that the heat is produced by combustion inside the engine itself and not in a separate boiler outside as in a steam engine. A good example is the gasoline engine used in automobiles all over the world. In each cylinder of such an engine gasoline is burned and the expanding hot gases from the combustion do work on the piston to drive the wheels of the car. Only part of the energy from the gasoline is converted into useful work; the remainder is lost to the environment in the form of heated air and exhaust products.

The four processes which make up one complete cycle of the piston are illustrated in Fig. 15.16. On the *intake stroke* the piston is pulled back by the rotating crankshaft. The intake valve A opens and a mixture of gasoline vapor and air is drawn into the cylinder. This is followed by the *compression stroke* in which the piston compresses this fuel mixture to about $\frac{1}{7}$ of its original volume (called the *compression ratio*). A spark from the spark plug ignites the fuel mixture, and the temperature and pressure in the cylinder increase greatly. This increased pressure drives the piston outward in the *power stroke*. When the volume of gas in the cylinder reaches its maximum value, determined by the position of the piston, the *exhaust stroke* starts. Valve B opens and the exhaust products of the combustion are forced out. The piston is now back where it started and the cycle begins again.

Diesel engines are the most efficient of all internal-combustion engines because the combustion chamber becomes hot enough to ignite the fuel without the need for a spark. As a result, T_H is higher and the efficiency improves. The prices paid for this increased efficiency of diesel engines are the added initial cost of the sturdier metals required to stand the higher temperatures and the greater weight of the engine per horsepower produced. Some typical efficiencies of various kinds of engines are given in Table 15.2.

TABLE 15.2 First-Law Efficiencies of Various Kinds of Engines

Kind of engine	Practical efficiency, %
Fossil-fuel-powered steam turbine	40
Nuclear-powered steam turbine	35
Automobile engine (internal-combustion)	20–25
Aircraft engine	20–23
Marine engine	30–35
Diesel engine	26–38
Gas turbine, single cycle	20–30
Gas turbine, combined cycle	40–50

Example 15.4

Consider two electric power plants, one a fossil-fuel plant with an efficiency of 40 percent, the other a nuclear plant with an efficiency of 30 percent. If each generates the same amount of energy, how much more waste heat is produced by the second plant than by the first?

SOLUTION

Since the work done by either plant is $\Delta W = e_1 \, \Delta Q_H$, and the efficiency $e_1 = 1 - \Delta Q_C / \Delta Q_H$, we have

$$\frac{\Delta Q_C}{\Delta Q_H} = 1 - e_1 \quad \text{or} \quad \Delta Q_H = \frac{\Delta Q_C}{1 - e_1}$$

Hence
$$\Delta W = \frac{e_1}{1 - e_1} \Delta Q_C$$

Now, ΔW is the energy generated by the plants in some fixed time interval and is the same for the two plants; hence we have, using subscripts n and f for the nuclear and fossil-fuel plants, respectively,

$$\Delta W = \frac{e_{1,n}}{1 - e_{1,n}} \Delta Q_{C,n} = \frac{e_{1,f}}{1 - e_{1,f}} \Delta Q_{C,f}$$

But $e_{1,n} = 0.30$ and $e_{1,f} = 0.40$, and so

$$\frac{0.30}{0.70} \Delta Q_{C,n} = \frac{0.40}{0.60} \Delta Q_{C,f}$$

or

$$\frac{\Delta Q_{C,n}}{\Delta Q_{C,f}} = \frac{0.70(0.40)}{0.60(0.30)} = \frac{0.28}{0.18} = \boxed{1.55}$$

Hence 55 percent more waste heat is produced by the nuclear plant, even though its efficiency is only 10 percent lower than the fossil-fuel plant.

15.6 The Absolute Thermodynamic Temperature Scale

Equation (15.8), $\Delta Q_C/\Delta Q_H = T_C/T_H$, is at the core of thermodynamics in the same way that Newton's second law is at the core of mechanics. In 1848 the British physicist William Thomson [Lord Kelvin (1824–1907)] proposed that the heat ΔQ_H taken in and the heat ΔQ_C given out by a Carnot engine be used to define a new thermodynamic scale of temperature. This is now called the *absolute thermodynamic scale*, since it is independent of the working substance. It is also called the *Kelvin scale* after its founder. The ratio of two temperatures on this scale is defined as follows:

$$\frac{T_C}{T_H} = \frac{\Delta Q_C}{\Delta Q_H} \tag{15.10}$$

Even though ΔQ_C and ΔQ_H depend on the substance used in the Carnot engine, their ratio does not. It depends only on the two temperatures involved and can therefore be used to define a temperature scale. Thus if ΔQ_C is $\frac{1}{10}$ of ΔQ_H, as measured in a Carnot cycle, and T_H is known to be 50 K, then T_C is equal to 5 K.

This absolute scale is still not complete, for it yields only ratios of temperatures. We must also fix one point on the scale by assigning a particular numerical temperature to a well-defined physical system. We choose to take 273.16 K as the temperature of the triple point of water (see Fig. 13.20). Since this corresponds to 0.01°C above the freezing point of water, this makes the freezing point of water 273.15 K, and absolute zero −273.15°C.

Such a scale has advantages over the similar scale previously defined in terms of measurements with a constant-volume gas thermometer. The problem with such a thermometer is that measured temperatures differ slightly with the gas used, and even with the density of a particular gas, as we have seen. If a perfect, low-density, ideal gas is used in a constant-volume thermometer, then the temperature measured with such a thermometer agrees perfectly with the absolute thermodynamic temperature. Hence in the future we will consider these two scales as equivalent and use kelvins to denote the temperature on both. The scientific importance of the Kelvin temperature scale is that, although for convenience we must measure temperature with a practical thermometer, the Kelvin scale is completely independent of the physical properties of any such thermometer. It is for this reason that it is truly an *absolute* temperature scale.

In practice no gas thermometer will work below 1 K. However, Eq. (15.10) can still be used. If we carry out a Carnot cycle between temperatures of, say, 2 K and some unknown temperature below 1 K, as in Fig. 15.17, then

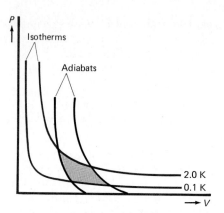

FIGURE 15.17   A Carnot cycle used to measure low temperatures. Measurements of the heat taken in and given off in the Carnot cycle can be used to obtain the lower temperature, if the higher temperature is known.

$$T_C = \frac{\Delta Q_C}{\Delta Q_H} \, T_H = \frac{\Delta Q_C}{\Delta Q_H} \,(2 \text{ K})$$

Hence measurements of the heat taken in and the heat given off in the Carnot cycle can be used to determine the absolute temperature T_C, even if it is much below 1 K.

As the temperature falls, the amount of heat ΔQ_C given off in a Carnot cycle at the lower temperature T_C gets smaller and smaller, according to Eq. (15.10). In the limit of absolute zero, ΔQ_C will also have to be zero. Hence another definition of absolute zero is the following:

Definition

Absolute zero is the temperature of a reservoir into which a Carnot engine would eject no heat.

Example 15.5

In a low-temperature physics experiment a Carnot cycle is carried out between a warmer reservoir which is at the temperature of liquid helium (4.2 K) and a colder reservoir whose temperature is unknown. In the cycle the heat taken from the upper reservoir is found to be 2.0 J and the heat delivered to the lower reservoir is 0.10 J. What is the temperature of the colder reservoir?

SOLUTION

Since $T_H/T_H = \Delta Q_C/\Delta Q_H$,

$$T_C = \frac{\Delta Q_C}{\Delta Q_H} T_H = \frac{0.10 \text{ J}}{2.0 \text{ J}}(4.2 \text{ K}) = 0.050(4.2 \text{ K}) = \boxed{0.21 \text{ K}}$$

15.7 The Second Law of Thermodynamics

A major difficulty in understanding the second law of thermodynamics is that it is unique in physics in that it is not a statement of what is, or what can be, but rather of what *cannot* be. It is a *statement of impossibility*. The many different forms in which this impossibility can be expressed make the second law appear more complex than it really is. In this section we will try to tie together the results of our previous discussions into a coherent account of the second law. Keep in mind the spontaneous tendency in nature to proceed from a state of greater order to one of lesser order (or comparative disorder).

Gases expand to fill any available space; two objects at different temperatures when put in contact finally reach the same temperature. The explanation for this tendency is ultimately found in a statistical analysis of the behavior of the huge number of molecules which constitute matter as we know it. Although this analysis constitutes an argument from statistical mechanics rather than thermodynamics, this tendency for nature to move from order to disorder is basic to understanding the second law of thermodynamics.

Entropy Statement of Second Law

We have seen that entropy is a measure of disorder. Hence if nature spontaneously moves in the direction of disorder, the entropy of any isolated system tends to increase, or, as stated in Sec. 15.3, "The entropy of any isolated physical system increases in all irreversible processes and remains constant in all reversible processes. Hence the total entropy of any isolated system never decreases." All physical processes in the real world are irreversible, and therefore entropy must increase in any isolated physical system. This entropy statement is the most basic and far-reaching statement of the second law of thermodynamics.

Kelvin Statement of Second Law

Another statement of the second law is based on the extension of Carnot's work by Lord Kelvin. According to Carnot, the efficiency of a completely reversible engine is $e_1 = 1 - T_C/T_H$. In the ideal case, if T_C could be zero on the Kelvin scale, the maximum efficiency would be 100 percent for a Carnot engine. The Kelvin statement of the second law merely states that such a 100 percent efficient heat engine is impossible:

Kelvin Statement

It is impossible to construct a heat engine which, operating in a cycle, produces no effect except the extraction of heat from a source and the conversion of this heat completely into work.

Perpetual-Motion Machine Statement

Kelvin's statement of the second law leads immediately to the conclusion that a perpetual-motion machine of the second kind is impossible.

Definition

Perpetual-motion machine of the second kind: A machine that would take heat from a reservoir and convert it entirely into work.

A perpetual-motion machine of the *first kind* is one that could run forever without the input of any energy. Such a machine would violate the *first law* of thermodynamics. By analogy a machine that would violate the *second law* of thermodynamics is called a perpetual-motion machine of the *second kind*. Such a machine would be a 100 percent efficient heat engine, which the Kelvin statement of the second law does not allow. Hence we can state:

Perpetual-Motion Statement

Since a heat engine with an efficiency of 100 percent is impossible, a perpetual-motion machine of the second kind is also impossible.

An example of such a perpetual-motion machine of the second kind would be a ship which could take in water from the surface of the ocean, remove heat from it, use the heat to power an engine to move the ship, and then dump the water back into the surface waters of the ocean from which it originally came. Carnot's work shows that this is impossible because the efficiency of a Carnot engine working between two identical temperatures T_0 is

$$E = 1 - \frac{T_0}{T_0} = 0$$

Hence such an engine could perform no work. Note that if the intake and output waters are from different depths and hence have different temperatures, in principle an engine could be driven by this temperature difference. The efficiency of such an engine would be very low, however, because of the small temperature differences available in the oceans.

Clausius' Summary of Thermodynamics

Probably the most terse statement of both laws of thermodynamics has been provided by Rudolf Clausius:

1 The *energy* of the universe is constant.

2 The *entropy* of the universe tends always toward a maximum.

In more colloquial terms the first law tells us that the best we can do is break even, since we can never get something for nothing in the energy business. The second law tells us that we can never actually break even, because we must always lose some useful energy in any energy-conversion process.

The Clausius statements of the two laws of thermodynamics, while succinct, are not as practical as the statement of these laws given above for isolated systems. Clausius' statement of the first law would allow, for example, the sudden disappearance of a quantity of energy in the midst of an experiment, so long as the same amount of energy reappeared at the same instant in some other part of the universe. Such occurrences would make physics experiments impossible.

Because of the crucial importance of the second law of thermodynamics to the universe as we know it, C. P. Snow seems justified to remark in his book *The Two Cultures* that for a humanist to be ignorant of the second law of thermodynamics is as shocking as for a scientist to be ignorant of the writings of Shakespeare.

Example 15.6

In recent years there has been increased interest in a process for generating energy from the oceans first proposed in 1881 by Jacques d'Arsonval (1851–1940) and now usually called OTEC (ocean thermal energy conversion). This plan would use the temperature difference between the surface waters and the deep waters of the ocean to generate electric energy. In the tropics the surface temperature of the ocean is about 25°C, and at depths of 1000 m the temperature falls to about 5°C. What is the maximum possible efficiency of a Carnot engine working between these two temperatures?

SOLUTION

Since $e_1 = 1 - T_C/T_H$, we have

$$e_1 = 1 - \frac{278 \text{ K}}{298 \text{ K}} = 0.067 = \boxed{6.7\%}$$

This is, of course, the *maximum* efficiency. The practical efficiency would be closer to 2 to 3 percent. Hence very large flow rates of water through the engine will be required to produce the same power output provided by more efficient engines operating with large temperature differences. As a result the pumps and pipes required are very large and costly, and it is unclear at this time whether OTEC will ever be a commercially viable system. While it does not violate the second law of thermodynamics, its efficiency is very low.

15.8 Some Applications: Heat Pumps and Refrigerators

Let us now apply the principles developed in the preceding section to some practical devices commonly used in homes and businesses.

Heat Pumps

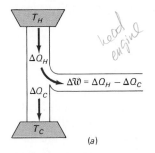

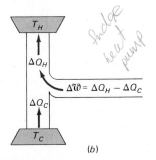

FIGURE 15.18 Comparison of (a) a heat engine with (b) a heat pump.

According to the second law of thermodynamics, heat will never flow spontaneously from a low temperature to a high temperature, since that would result in a decrease in the entropy of the system and its environment. Hence we can never expect to heat a house by opening a window on a cold day and hoping that heat will flow into the house from the outside cold air. The reverse is obviously what will happen. The second law, however, in no way prevents use from *pumping* heat from the air outside into the house to warm it. In the process the pump must do work and consume energy, but *the heat transferred can be much greater than the energy consumed by the pump*. This is why the *heat pump* is becoming increasingly popular for the heating of houses in moderate climates.

In Fig. 15.18 we compare a heat pump with a heat engine from a thermodynamic viewpoint. In the heat engine shown in Fig. 15.18a an amount of heat ΔQ_H is taken from a hot reservoir at temperature T_H. Some of this heat is converted into useful work ($\Delta W = \Delta Q_H - \Delta Q_C$), and the remaining ΔQ_C is exhausted to the cold reservoir at temperature T_C. In a heat pump we reverse this process. As shown in Fig. 15.18b, we take an amount of heat ΔQ_C from a cold reservoir at temperature T_C and pump it up, by doing work on it, to a higher-temperature reservoir at temperature T_H. The amount of heat delivered to the hot reservoir is then $\Delta Q_H = \Delta Q_C + \Delta W$, since we assume that no energy is lost in the process. In a practical heat pump T_C is the temperature of the outside air, T_H is the temperature of the inside of the house, and ΔW is the energy contributed by the pump, which is usually a compressor run by electric power.

As stated previously, the practical value of a heat pump lies in the fact that the amount of heat transferred to the interior of the house can be much greater than the amount of energy consumed in the transfer process. As an example of this, let us suppose that in Fig. 15.18 an amount of heat ΔQ_C is taken from the colder temperature T_C outside the house and an amount ΔQ_H is delivered to the house at the warmer temperature T_H inside. Then, from the first law, $\Delta Q_H = \Delta W + \Delta Q_C$, and so

$$\frac{\Delta Q_H}{\Delta W} = 1 + \frac{\Delta Q_C}{\Delta W} = 1 + \frac{\Delta Q_C}{\Delta Q_H - \Delta Q_C} = \frac{\Delta Q_H}{\Delta Q_H - \Delta Q_C} = \frac{1}{1 - \Delta Q_C/\Delta Q_H}$$

But, since for a perfect Carnot engine, from Eq. (15.10), $\Delta Q_C/\Delta Q_H = T_C/T_H$, we have

$$\frac{\Delta Q_H}{\Delta W} = \frac{1}{1 - T_C/T_H} = \frac{T_H}{T_H - T_C}$$

The amount of heat transferred to the inside of the house is therefore

$$\Delta Q_H = \frac{T_H}{T_H - T_C} \Delta W \tag{15.11}$$

Note the remarkable result here. If T_H and T_C are close together (i.e., if it is not too cold outside), then Q_H can be much greater than ΔW; that is, much more heat can be transferred than the energy input required by the pump (ΔW). Of course, this does not mean that energy is being created; the heat brought into the room is the sum of the work done and the heat extracted from the outside air. (Remember that air even at $-10°C$ is far above absolute zero and hence has considerable thermal energy, even though we may think of it as being cold.) Hence the first law of thermodynamics is satisfied. The secret of the heat pump's success is that it enables us to convert relatively useless energy from the outside air into useful heat to warm our houses, and to do this with an expenditure of work that is small compared with the heat transferred to the inside of the house.

Coefficient of Performance of a Heat Pump Since $\Delta Q_H/\Delta W$ for a heat pump can be much greater than unity (or 100 percent efficiency), and since efficiencies are usually defined so as to have a maximum value of 100 percent, the quantity $\Delta Q_H/\Delta W$ is not in this case called the efficiency, but rather the *coefficient of performance* (CP) of the system. For a heat pump:

Coefficient of Performance of Heat Pump

$$\text{CP} = \frac{\Delta Q_H}{\Delta W} \leq \frac{T_H}{T_H - T_C} \tag{15.12}$$

where the equal sign applies in the ideal order, whereas the inequality sign applies to real devices. This is just the reciprocal of the expression for efficiency of a Carnot engine. Example 15.7 shows how sensitive is the coefficient of performance of a heat pump to the difference in temperature $T_H - T_C$.

Note that the equality sign in Eq. (15.12) refers to the ideal case, where we have a completely reversible (Carnot) cycle. Hence practical CPs are considerably less than those predicted by Eq. (15.11) but are still greater than unity in most cases. Values of 3 to 5 are typical coefficients of performance for heat pumps. The great advantage of a heat pump is evident when we pay our electric bills. We pay only for ΔW, the work done, and not for the heat extracted from the outside air, which is free.

Figure 15.19 shows how a practical heat pump works. The outdoor source of heat for the pump can be a water reservoir buried in the ground or can be the earth itself, but in practice the outside air is most frequently used. This makes for a simpler, less costly system, but reduces the efficiency of the heat pump when it is very cold outside. For this reason an auxiliary heating system must be used to supplement the heat pump when the outside

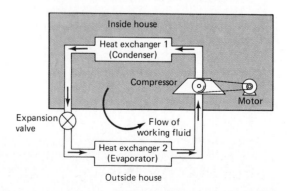

FIGURE 15.19 A practical heat pump. The working fluid flows around in a closed cycle, constantly taking heat from the outside air and carrying it inside the house.

temperature falls much below freezing. One way to avoid this problem is to use solar heating to increase the temperature of the air providing the thermal input to the heat pump.

Example 15.7

Suppose that we want to keep the inside of a house at 70°F (21°C). Compare the theoretical coefficient of performance of a heat pump working to do this:

(a) When the outside temperature is 10°C.
(b) When the outside temperature is −10°C, that is, 10 degrees below freezing on the Celsius scale.

SOLUTION

(a) When $T_C = 10°C = 283$ K, we have

$$CP_{10} = \frac{(273 + 21)\ K}{(294 - 283)\ K} = \frac{294}{11} = 27$$

Note that in such calculations we could use Celsius degrees *in the denominator* since the difference between two temperatures is the same whether they be expressed in kelvins or Celsius degrees, but we *must* use *Kelvin* temperatures in the numerator.

(b) When $T_C = -10°C = 263$ K, we have

$$CP_{-10} = \frac{294\ K}{(294 - 263)\ K} = \frac{294}{31} = 9.5$$

Hence the efficiency of the heat pump is reduced by a factor of 3 when the temperature falls 20 Celsius degrees outside the house.

Note that even when the temperature is below freezing outside, a heat pump can, in the ideal order, transfer to the interior of a house an amount of heat which is almost 10 times greater than the work performed by the pump.

Air Conditioners

An air conditioner is nothing but a heat pump run backward; i.e., the outside and inside of the house are simply exchanged in Fig. 15.19. The working fluid in this case (usually freon) is cooled by a throttling process as it enters the house, picks up heat inside the house, is compressed as it leaves the house to a temperature above that of the outside air, and then delivers heat to the outside air before starting the cycle all over again.

The coefficient of performance of an air conditioner is slightly different than for a heat pump, because the important thing here is not the amount of heat (ΔQ_H) delivered to the outside air, but rather the heat (ΔQ_C) removed from inside the house. Hence we define the ideal coefficient of performance in this case as

Coefficient of Performance of Air Conditioner & Fridge

$$CP = \frac{\Delta Q_C}{\Delta W} = \frac{\Delta Q_C}{\Delta Q_H - \Delta Q_C} \le \frac{T_C}{T_H - T_C} \tag{15.13}$$

Real CPs are somewhat less than the ideal value corresponding to the equals sign, because of friction and other irreversible factors in the operation of the unit. The amount of heat removed from the house when an amount of work ΔW is performed is then

$$\Delta Q_C = CP\ \Delta W \le \frac{T_C}{T_H - T_C}\ \Delta W \tag{15.14}$$

Again the heat removed can be much larger than the work done.

Refrigerators

A refrigerator is basically an air-conditioning unit of a special kind which removes heat from the inside of the refrigerator box and deposits it in the kitchen. (Feel the coils at the rear of your refrigerator and notice how warm they are. They are the condensing coils which deliver heat to the room.) The coefficient of performance for a refrigerator is given by the same expression found for an air conditioner, i.e., Eq. (15.13).

Second-Law Efficiency

There is an important distinction between the efficiencies of energy conversion processes as calculated using the first and second laws of thermodynamics.

We discussed the first-law efficiency in Sec. 14.8. This determines the efficiency of any particular machine such as a heat engine by calculating the ratio of the work output of the machine to the energy input. The second-law efficiency compares the energy consumed by a machine in doing a particular task with that of the most efficient machine capable of performing the *same task*. We define the second-law efficiency as follows:

Second-Law Efficiency

$$e_2 = \frac{\text{least amount of energy needed for a task}}{\text{actual energy used in performing the task}}$$

(15.15)

For example, if we assume that using a heat pump with a CP of 3 is the most efficient way to heat a house, then to provide an amount of heat $\Delta\mathcal{W}$ to the house requires an input of $\Delta\mathcal{W}/3$ of electric energy. If we decide to heat the house using electric-resistance heating, the amount of electric energy required is $\Delta\mathcal{W}$, even at 100 percent efficiency. Hence the second-law efficiency for electric-resistance heating in this case is:

$$e_2 = \frac{\text{energy required by heat pump}}{\text{energy required by resistance heating}} = \frac{\mathcal{W}/3}{\mathcal{W}} = \frac{1}{3} = 33\%$$

Hence we would expect from this simple example that a savings of 60 to 70 percent should be possible in some cases in heating houses by heat pumps rather than by electric-resistance heating.

To conserve our energy resources we must use the energy device best adapted to any particular task, i.e., the device that uses the least amount of energy. This means that we must find the source with the highest *second-law efficiency* e_2. Otherwise, we will waste energy because the energy conversion devices we choose will be poorly adapted to the task at hand.

Summary: Important Definitions and Equations

Reversible process:
A thermodynamic process carried out so slowly that every intermediate state is an equilibrium state; a process that can be reversed by an infinitesimal change in the operating conditions.

Irreversible process:
A thermodynamic process for which the intermediate states are not equilibrium states; any process that is not completely reversible.

Equations for reversible adiabatic process:

$$P_1 V_1^{\gamma} = P_2 V_2^{\gamma} \qquad T_1 V_1^{\gamma-1} = T_2 V_2^{\gamma-1}$$

Entropy:
A state variable that measures the disorder of a physical system. For any reversible process the change in entropy is:

$$\Delta S = \frac{\Delta Q}{T}$$

Entropy statement of second law of thermodynamics:
The entropy of any physical system plus its environment increases in all irreversible processes and remains constant in all reversible processes. Hence the total entropy of a closed system never decreases.

Heat engine:
A device for converting heat into useful work.

Carnot engine:
The most efficient possible heat engine operating between any two given temperatures. The first-law efficiency of a Carnot engine is:

$$e_1 = \frac{\Delta\mathcal{W}}{\Delta Q_H} = 1 - \frac{\Delta Q_C}{\Delta Q_H} = 1 - \frac{T_C}{T_H}$$

Absolute thermodynamic (Kelvin) temperature scale:
A temperature scale based on the fact that for a Carnot cycle, $\Delta Q_C/\Delta Q_H = T_C/T_H$, and the definition of the triple point of water as 273.16 K.

Absolute zero (thermodynamic definition):

The temperature of a reservoir into which a Carnot engine would eject no heat.
Perpetual-motion machine of the second kind:
A 100 percent efficient engine that takes heat from a reservoir and converts it entirely into work (such a machine is impossible by the second law of thermodynamics).
Coefficient of performance of a heat pump:

$$CP = \frac{\Delta Q_H}{\Delta W} \leq \frac{T_H}{T_H - T_C}$$

Coefficient of performance of air conditioner (or refrigerator):

$$CP = \frac{\Delta Q_C}{\Delta W} \leq \frac{T_C}{T_H - T_C}$$

Second-law efficiency:

$$e_2 = \frac{\text{least amount of energy needed for task}}{\text{actual energy used in performing the task}}$$

Questions

1 Distinguish clearly between a perpetual-motion machine of the first kind and a perpetual-motion machine of the second kind. Give a few examples of each.

2 In discussions of the world's energy crisis, why is the second-law efficiency a more revealing criterion of the worth of an energy conversion process than is the first-law efficiency?

3 Discuss all the energy transformations that take place in a steam-electric turbine (see Fig. 15.14), beginning with the energy in the fossil or nuclear fuel and ending with the waste heat delivered to the cooling system.

4 Consider Table 15.2, which lists the practical efficiencies of various kinds of engines. Present arguments to show that the efficiencies listed for the first six kinds of engines are reasonable, given the temperatures and other conditions under which the engines operate.

5 Why cannot the theoretical efficiency of a Carnot engine be achieved in any practical engine?

6 One way to help solve the problem of shortages developing in our supplies of metals like aluminum and tin is to recycle these metals, i.e., collect old tin cans and aluminum containers and convert them back into useful forms in a reprocessing plant. Why cannot we solve our present energy problems in a similar fashion by recycling the energy we have available, since we know from the first law of thermodynamics that energy is never destroyed but is merely converted to a different form?

7 Is it possible to cool down the kitchen on a hot summer's day by opening the door of the refrigerator? Why? What will be the actual result? Would the result be the same if a window air conditioner were placed in the middle of a room and turned on?

8 Why cannot the huge amount of energy stored in the core of the earth and in the oceans be used to solve our energy problems? Can these sources of energy contribute in some way to our energy needs? How?

9 Name a few spontaneous processes in which an ordered state changes to a disordered (or less-ordered) state.

10 Explain what is meant by the following statement: "The flow of heat down a temperature gradient is always associated with an increase of entropy and a degradation of energy."

11 If you compare two states of a system, one a state of low entropy and the other a state of high entropy, then the high-energy state has a greater probability of occurring. From your knowledge of the physical significance of entropy, explain why you would expect this to be the case.

12 Which has the greater entropy, 1 mol of liquid mercury, or 1 mol of solid mercury?

13 (*a*) In an isothermal expansion of an ideal gas, does the entropy increase, decrease, or stay the same?
(*b*) In an adiabatic expansion of an ideal gas, does the entropy increase, decrease, or stay the same?

14 Which is the stronger statement of the first and seconds laws of thermodynamics, that in terms of an isolated system, or Clausius' statement in terms of the energy and entropy of the universe?

15 Since the earth remains at a reasonably constant temperature, it must radiate into space roughly the same amount of energy it receives from the sun. Compare the entropy changes in the earth due to the thermal radiation it receives from the sun and the thermal radiation it emits into space. Does this help explain why the entropy of the earth also remains roughly constant over time?

16 In living organisms order increases and hence entropy would appear to decrease. Explain why living organisms do not involve a violation of the law of increasing entropy.

Problems

(For some of the problems in this chapter the values of γ given for selected gases in Table 13.4 are needed.)

A 1 A gas is expanded irreversibly from a volume V to a volume $3V$. At the end of the process, the temperature is the same as it was at the beginning and the pressure of the gas has changed from P to $P/3$. When the volume was equal to $2V$ during the expansion, the pressure was:
(*a*) $P/2$　(*b*) $2P$　(*c*) P　(*d*) $P/3$
(*e*) It is impossible to say

A 2 If 0.50 m³ of argon gas at room temperature and atmospheric pressure is compressed adiabatically to a pressure 5 times atmospheric, the final volume of the gas is:
(a) 2.5 m³ (b) 0.1 m³ (c) 0.38 m³
(d) 0.19 m³ (e) 2.5 m³

A 3 The temperature at the end of the process in Prob. A2 will be:
(a) 570 K (b) 570°C (c) 300 K
(d) 300°C (e) 157 K

A 4 If 1 kg of ice at 0°C is melted to water at the same temperature, the entropy change of the ice-water system is:
(a) −0.29 kcal/K (b) +0.29 kcal/K (c) 0
(d) −0.29 J/K (e) +0.29 J/K

A 5 The Carnot efficiency of a heat engine operating between temperatures of 27 and 627°C is:
(a) 0 (b) 67% (c) 33% (d) 4.3%
(e) 95.7%

A 6 The ratio of the energy thrown away as waste heat to the energy consumed as fuel in a practical heat engine is about:
(a) 0 (b) 1.0 (c) 0.20 (d) 0.40 (e) 0.70

A 7 The Carnot efficiency of a fossil-fuel, steam-electric plant exceeds the efficiency of proposed OTEC (ocean thermal energy conversion) plants by about:
(a) 10 orders of magnitude
(b) 5 orders of magnitude
(c) 2 orders of magnitude
(d) 1 order of magnitude (e) None of the above

A 8 A heat pump operates between a temperature of 0°C outside the house and 20°C inside the house. If it requires 10^9 J of heat to keep the house warm for a day, in the ideal order the electrical input to the heat pump would have to be:
(a) 10^9 J (b) 6.8×10^7 J (c) 15×10^9 J
(d) 5×10^8 J (e) 0

A 9 The coefficient of performance of a heat pump working between a temperature of −10°C outside and 10°C inside a house is:
(a) 0.92 (b) 0.64 (c) 1.55 (d) 14.2 (e) 1.08

A10 The coefficient of performance of an electric freezer working between the temperature of −10°C and room temperature (20°C) is:
(a) 8.8 (b) 9.8 (c) 1.0 (d) 0.10
(e) 0.33

B 1 A sample of nitrogen gas is at $T = 500$ K and $V = 1.0$ m³. After an adiabatic expansion, the final volume is 10 m³. What is the final temperature?

B 2 A sample of argon gas is at $P = 3.0 \times 10^5$ N/m² and $V = 1.0$ m³. After an adiabatic expansion, the final pressure is 1.0×10^5 N/m². What is the final volume?

B 3 Calculate the change in entropy when 1.0 kg of ice melts (see Table 13.5).

B 4 Calculate the change in entropy for the melting of 1.0 kg of solid hydrogen at −259°C (see Table 13.5).

B 5 An engine does 10^4 J of work for each 10^5 J of heat added. What is its efficiency?

B 6 During every cycle 10^5 J of heat is added to an engine and 10^4 J of heat is thrown away by the engine. What is its efficiency?

B 7 What is the efficiency of a Carnot engine which runs between
(a) 100 and 200 K, (b) 200 and 300 K,
(c) 300 and 400 K?

B 8 What would the temperature of the low-temperature reservoir of a Carnot engine have to be in order that the efficiency be 100%?

B 9 A Carnot device operates between 0 and 37°C.
(a) Calculate the efficiency if it is run as an engine.
(b) Calculate the coefficient of performance if it is run as a heat pump.

B10 A Carnot device operates between 0 and 300°C.
(a) Calculate the efficiency if it is run as an engine.
(b) Calculate the coefficient of performance if it is run as a heat pump.

B11 A Carnot air conditioner operates between 20 and 37°C. What is its coefficient of performance?

B12 A bathysphere is 1000 m below the ocean surface, a depth where the water temperature is 5°C. The interior of the bathysphere is to be at 20°C. What is, in the ideal order, the coefficient of performance of a Carnot heat pump which will maintain this temperature difference?

C 1 A 10^{-3}-m³ sample of neon gas at 300 K and atmospheric pressure is allowed to expand adiabatically to a final pressure of 0.50 atm.
(a) What is the final volume of the gas?
(b) What is the final temperature?
(c) What is the change in the internal energy?

C 2 A monatomic gas for which $P_0 = 10^6$ N/m² and $V_0 = 1.0$ m³ undergoes an expansion so that the final pressure is 10^5 N/m².
(a) What is the final volume if the expansion is isothermal?
(b) What is the final volume if the expansion is adiabatic?
(c) Calculate the work done in both the isothermal expansion and the adiabatic expansion (do this graphically by drawing the process on a PV diagram).
(d) In which process does the gas do more work?

C 3 At the base of a mountain the air pressure is 1.0 atm and the temperature is 300 K.
(a) If air rises adiabatically to the top of the mountain ($P = 0.95$ atm), what is the temperature? Use $\gamma = 1.4$ since air is mostly oxygen and nitrogen.
(b) If some of the water vapor condenses, will the temperature tend to go back up?

C 4 (a) If 700 J of heat is extracted from a gas during an isothermal compression, how much work is done?
(b) Is the work done on the gas or by the gas?
(c) What is the entropy change for the gas if the temperature is 27°C?

C 5 A gas does 10^4 J of work during an isothermal expansion. What is the increase in entropy of the gas if the expansion takes place at (a) 0.0°C, (b) −130°C?

C 6 With the data given in Chap. 14, calculate the entropy change per second for the sun.

C 7 In Prob. C7 of Chap. 14:
(a) What is the change in entropy per second for the boiling water?
(b) What is this change for the ice-water mixture?

(c)   What is the total change of entropy for the system and its environment?

C 8   An engine operates between 500 and 300 K. If it takes in 5×10^5 J of heat and does 1.5×10^5 J of work each cycle, how does its efficiency compare with the corresponding Carnot engine?

C 9   Assume that the turbines in a power plant are as efficient as Carnot engines. They operate between 350 and 27°C.

(a)   Calculate the efficiency.

(b)   If it is required that 10^{12} J of work be done, how much heat must be added at the high temperature and rejected at the low temperature?

C10   For a Carnot cycle similar to that of Fig. 15.12, the entropy increase of the gas from point a to point b is 10^5 J/K. If $T_a = T_b = 400$ K,

(a)   How much heat is given to the gas along ab?

(b)   How much work is done by the gas along ab?

(c)   How much heat is given to the gas along bc?

C11   A 70-kg person consumes 2.0×10^8 kcal. If this person were a Carnot engine operating between 37 and 33°C, how high could he or she be lifted with the work produced?

C12   The cycle shown in Fig. 15.20 consists of two isotherms and two constant-volume processes. The numbers of joules listed indicate how much heat is taken in or expelled in each process.

(a)   How much work is done per cycle?

(b)   What is the efficiency?

(c)   What is the efficiency of a Carnot engine operating between these isotherms?

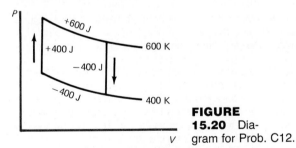

FIGURE 15.20 Diagram for Prob. C12.

C13   A Carnot heat pump runs between 20 and 0°C.

(a)   If 2000 J of heat is absorbed from the outside air, how many joules of heat is supplied to the inside of the house?

(b)   How much work must the compressor do?

(c)   What is the coefficient of performance?

C14   A Carnot air conditioner runs between 20 and 37°C.

(a)   If 2000 J of heat is extracted from the inside, how many joules of heat is given to the outside?

(b)   How much work must the compressor do?

(c)   What is the coefficient of performance?

C15   Before the ideal Carnot heat pump is turned on, the temperature inside and outside a cabin is 4°C.

(a)   If 200 J per cycle is delivered by the compressor, how much heat per cycle does the cabin receive when the inside temperature has reached 5°C?

(b)   How much when it is 32°C?

(c)   What is the coefficient of performance in the ideal order in each case?

C16   (a)   A Carnot engine operates between a hot reservoir at 320 K and a cold reservoir at 260 K. If it absorbs 1000 J of heat at the hot reservoir, how much work does it deliver?

(b)   If the same engine, working in reverse, functions as a refrigerator between the same two reservoirs, how much work must be supplied to remove 2000 J of heat from the cold reservoir?

D 1   A cylinder contains He gas of volume 0.10 m^3 at a pressure of 1.0×10^5 N/m^2 and a temperature of 27°C. The gas is first heated at constant volume to a pressure of 2.0×10^5 N/m^2, and then at constant pressure to a temperature of 427°C, as in Fig. 15.21.

(a)   Calculate the total heat input during these processes (a to b and b to c). The gas is next cooled at constant volume to its original pressure and then at constant pressure to its original volume.

(b)   Find the total heat output during these processes (c to d and d to a).

(c)   Find the total work done by the gas in the entire cyclic process (abcda). (Hint: Use Table 13.4.)

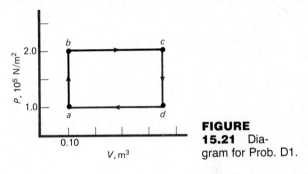

FIGURE 15.21 Diagram for Prob. D1.

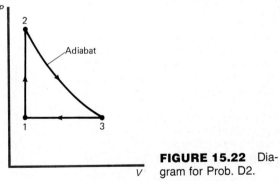

FIGURE 15.22 Diagram for Prob. D2.

D 2   A reversible heat engine carries 1 mol of an ideal monatomic gas around the cycle shown in Fig. 15.22. Process 1-2 takes place at constant volume, process 2-3 is adiabatic, and process 3-1 takes place at constant pressure. The temperatures at the indicated points are $T_1 = 300$ K, $T_2 = 600$ K, $T_3 = 355$ K.

(a)   Compute the heat exchange ΔQ, the change in internal energy ΔU, and the work done ΔW for each of the three processes and for the cycle as a whole.

(*b*) If the initial pressure at point 1 is 10^5 N/m², find the pressure and the volume at points 2 and 3.

D 3 A mole of ideal gas goes from temperature T_1 to temperature T_2 during an adiabatic process.

(*a*) Show that the change in internal energy of the gas is given by

$$\frac{3}{2}RT_1\left[\left(\frac{V_1}{V_2}\right)^{\gamma-1} - 1\right]$$

(*b*) If $2V_2 = V_1$, would the energy increase be larger for Ne or for CO_2?

D 4 In Fig. 15.23, *ab* is an isotherm, *bc* a constant-volume curve, and *ca* an adiabat; and $P_a = 2.0 \times 10^5$ N/m². Because the pressure of a gas is a state variable, it is true that $\Delta P_{ac} = \Delta P_{ab} + \Delta P_{bc}$. Because the entropy is a state variable, a completely analogous result holds: $\Delta S_{ac} = \Delta S_{ab} + \Delta S_{bc}$. (Assume 1 mol of an ideal monatomic gas.)

(*a*) Calculate ΔS_{ab} for 1 mol of an ideal gas.

(*b*) Find ΔS_{ac}, noting that path *ac* is an adiabat.

(*c*) Find ΔS_{bc}.

D 5 Starting with $\Delta Q = \Delta U + P\,\Delta V$, show that

$$\Delta S = C_V\frac{\Delta T}{T} + R\frac{\Delta V}{V} = R\left(\frac{3}{2}\frac{\Delta T}{T} + \frac{\Delta V}{V}\right)$$

for 1 mol of a monatomic ideal gas. (Note that the universal gas constant has dimensions of entropy per mole. Note also that we can have $\Delta S \neq 0$ even if $\Delta T = 0$ so long as there is a change in volume.)

D 6 One kilogram of water at 10°C is mixed with one kilogram of water at 0°C.

(*a*) What is the final temperature?

(*b*) Calculate the entropy decrease of the warm water between 10 and 9°C, using 9.5°C for the average temperature. Do the same for 9 to 8°C, etc., down to 5°C.

(*c*) Calculate the entropy increase of the cold water between 0 and 1°C, using 0.5°C for the average temperature. Do the same for 1 to 2°C, etc., up to 5°C.

(*d*) Add the five entropy increases and the five entropy decreases together. Is the result positive, negative, or zero? Does this show that mixing is an irreversible process?

D 7 In Fig. 15.24 curves *AB*, *FC*, and *ED* are isotherms. *AFE* and *BCD* are adiabats.

(*a*) What is the efficiency of cycle *ABDEA*?

(*b*) In an attempt to improve the efficiency, cycles *ABCFA* and *FCDEF* are operated in such a way that the heat rejected by the former is the heat input to the latter.

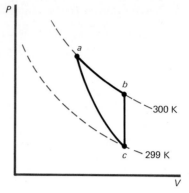

FIGURE 15.23 Diagram for Prob. D4.

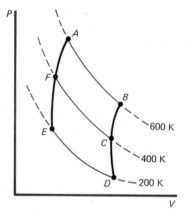

FIGURE 15.24 Diagram for Prob. D7.

What is the efficiency of this tandem setup?

D 8 Prove that a 100 percent efficient heat engine would lead to a *decrease* in entropy of the heat engine plus the environment, which would violate the entropy statement of the second law given in the text.

D 9 Rudolph Clausius has stated the second law of thermodynamics as follows:

There is no such thing as a perfect refrigerator, i.e., a machine which transfers heat from a cold temperature to a hot temperature in a closed cyclical process without any other change in the system.

Prove that, if a perfect refrigerator existed, the entropy of the refrigerator plus the environment would *decrease* in such a refrigeration process, which would violate the entropy statement of the second law given in the text.

Additional Readings

Dyson, Freeman J.: "What Is Heat?," *Scientific American*, vol. 191, no. 3, September 1954, pp. 58–63. The quotation at the beginning of this chapter is from page 59 of this thoughtful article (with permission).

Fenn, John B.: *Engines, Energy, and Entropy: A Thermodynamics Primer*, Freeman, San Francisco, 1982. An attempt to make thermodynamics clear and enjoyable to readers with minimal technical backgrounds.

Mendelssohn, Kurt: *The Quest for Absolute Zero*, 2d ed., Wiley, New York, 1977. A more technical discussion of the subject than MacDonald's book (see Chap. 13).

Saperstein, Alvin M.: *Physics: Energy in the Environment*, Little, Brown, Boston, 1975. Chapters 10 and 11 of this book contain a detailed discussion of entropy.

Wilson, S. S.: "Sadi Carnot," *Scientific American*, vol. 245, no. 2, August 1981, pp. 134–145. One of the best short accounts of the little-known life of Carnot.

Zener, Clarence: "Solar Sea Power," *Physics Today*, vol. 26, no. 1, 1973, pp. 48–53. A good article on the prospects for extracting useful energy from the oceans.

If we accept the hypothesis that the elementary substances are composed of atoms, we cannot avoid concluding that electricity also is divided into definite elementary portions, which behave like atoms of electricity.

Hermann von Helmholtz (1821–1894)

Chapter 16
Electrostatics

We live in an electrical age. Most of us use electrical devices—radios, television sets, toasters, refrigerators—every day, and would be lost without them. Modern industry and business rely to an ever-increasing extent on electric power to drive machines and process data. In the next six chapters we will examine the basic facts and laws of electricity and magnetism in order to understand better the electrical world in which we live. We will start with the basic ideas governing charges at rest (electrostatics), and build up through the study of charges in motion (electrodynamics) and the magnetic fields they produce (electromagnetism) to the grand synthesis of electromagnetic theory contained in Maxwell's equations.

16.1 Elementary Electric Charges and Their Interaction

We know from experience that electric charges build up on our bodies when we walk on a wool rug on a cold winter's day and that these charges then produce sparks when we touch a metal doorknob or a water pipe; charges also accumulate on clouds and produce lightning and thunder when they are released with explosive force. A great many electrostatic phenomena like these can be explained once we understand the structure of atoms. In the eighteenth century, however, when most of the original work on electrostatics was done, scientists not only did not understand the structure of atoms, they were not sure whether atoms even existed.

Franklin's Contributions

The scientist who did much of the original work to clarify our understanding of the behavior of electric charges at rest was the great American statesman Benjamin Franklin (see Fig. 16.1 and accompanying biography). In the early 1740s Franklin carried out a variety of experiments with static charges which were produced by frictional rubbing or by electrostatic machines (Fig. 16.2). He introduced the idea of positive and negative charges, and concluded that the total charge was always exactly the same after the rubbing as it was before the rubbing. This principle is now known as the *principle of conservation of electric charge*. Franklin's introduction of this principle and the concepts of positive and negative charge was his most significant contribution to electrostatics.

Benjamin Franklin (1706–1790)

FIGURE 16.1 Benjamin Franklin (1706-1790) at his workbench. Notice the Leyden jar used for electrostatic experiments near Franklin's right arm. *(Photo courtesy of the Burndy Library, Norwalk, Connecticut.)*

Benjamin Franklin is best known today as one of the founding fathers of the United States and as an author of the Declaration of Independence. Long before the United States was founded, however, Franklin was world-famous as the man who had tamed and explained lightning, introduced the idea of positive and negative electric charges, and made sense of the many puzzling phenomena of electrostatics.

Franklin was born in Boston in 1706, the fifteenth of 17 children.

When he was 17 years old, he went to Philadelphia to work as a printer. Franklin had only 2 years of formal schooling, but he loved books and studied foreign languages, philosophy, and science on his own. He became so involved in scientific study that in 1748 he turned over to his foreman what was then his own printing business, intending to devote the rest of his life to science.

During the years from 1747 to 1753 Franklin carried out his most important work on the nature of electricity (see Sec. 16.1), the conservation of charge, and the properties of Leyden jars (Sec. 16.8). This work culminated in the publication in 1651 of his *Experiments and Observations on Electricity*, which went through five editions in English, three in French, one in Italian, and one in German.

One of Franklin's discoveries in electricity was the important role played by pointed metal objects. He found that a grounded metal rod with a pointed end could cause a charged conductor to lose its charge if the point was brought within 6 in of the charged object, but that a flat or rounded conductor had no such effect. Franklin used this discovery in the design of lightning rods to protect buildings. A lightning rod is simply a sharp-pointed rod which is connected by a very good conducting path from the roof of a building to the ground. A lightning rod allows charge to leak off thunder clouds before it builds up sufficiently to create destructive lightning bolts. Also, if lightning should strike, it would hit the lightning rod, not the house, and the charge would pass through the lightning rod to the ground without damaging the house.

For at least 50 years before Franklin's time European scientists had speculated that lightning was electrical in nature. Franklin was the first scientist with the insight and the courage to prove this by his famous lightning-kite experiment.

By 1753 Franklin had become so involved in the political life of colonial America that he was forced to abandon most of his scientific work. His later success in winning the support of France for the American Revolution was as much due to his stature in Europe as a scientist and an intellectual as it was to his ability to charm both the men and women of France by his simplicity, urbanity, and wit. His good humor is reflected in the comment he made after trying to kill a turkey with an electric shock, and nearly killing himself in the process: "I meant to kill a turkey, and instead I nearly killed a goose."

Among Franklin's many inventions were the rocking chair, the Franklin stove, and bifocal glasses. Franklin had an attitude toward life which reflected the youthfulness of his spirit. In 1787, at the age of 81, after 50 years of public service, Franklin wrote to a Dutch friend asking him to come to America, where "in the little remainder of my life . . . we will make plenty of experiments together."

Benjamin Franklin was, without doubt, the first great American scientist, and has been called the "wisest American," a well-deserved title.

Modern Version of Franklin's Theory

A combination of Franklin's theory with our more modern understanding of atomic structure leads to the following summary of the behavior of electric charges.

FIGURE 16.2 An electrostatic machine which produces large electrostatic charges by frictional rubbing. This machine, built by Dumotiers Frères, Paris, for Professor Nicholas T. de Saussure of the University of Geneva, Switzerland, is now in the Burndy Library, Norwalk, Connecticut. *(Photo courtesy of the Burndy Library.)*

Basic Facts of Electrostatics

There are two different kinds of electric charge: the charge on the proton; which is designated as positive (+), and the charge on the electron, which is designated as negative (−). Charges of like sign repel each other, and charges of unlike sign attract each other.

Many electrostatic phenomena require only these simple laws for an explanation. For example, if two amber rods are rubbed with fur, one amber rod repels the other. This can be shown by mounting one rod as shown in Fig. 16.3a and bringing the other rod near it. Also, if two glass rods are rubbed with silk, one rod will repel the other in the same way. If, however, a charged glass rod is brought near a charged amber rod, they will attract each other, as in Fig. 16.3b. The reason is that the amber rods are both charged negatively,

FIGURE 16.3 *(a)* The repulsion of two amber rods which contain excess negative charge. *(b)* The attraction of a glass rod for an amber rod; the glass rod is charged positively and the amber rod negatively.

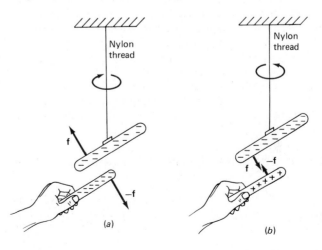

the glass rods positively: like charges repel each other, and unlike charges attract.

A modern statement of Franklin's law of conservation of charge would take the following form:

Conservation of Charge

In any closed system the algebraic sum of all the electric charges remains constant.

Even in the case of pair annihilation, in which masses of an electron and a positron (a positively charged electron) are converted completely into radiant energy, charge is still conserved, since the charge on the positron is exactly equal to the charge on the electron, but has the opposite sign. Hence the *algebraic sum* of the two charges before and after the annihilation process is the same, namely, zero.

16.2 Conductors and Insulators

Suppose two metal spheres are connected together by a long wooden rod, as in Fig. 16.4a. If one sphere is then charged by touching it with an amber rod rubbed with fur, it is found that the second sphere exerts no force on a charged amber rod brought near it. If, on the other hand, the two spheres are connected by a metal rod, as in Fig. 16.4b, then the second sphere will exert a repulsive force on another charged amber rod. It appears that charge flowed from the first sphere to the second through the metal rod but was unable to flow through the wood. Thus metals are called electric *conductors*, while substances like wood, glass, wax, and plastics are called electric *insulators*.

Metals are good conductors of electricity because they contain many free electrons. In metals at least one electron from each atom is free to move through the crystal lattice. Since 1 mol of a metal contains 6.02×10^{23} atoms, there are in 1 mol of a metal at least 6.02×10^{23} electrons free to carry charge from one place to another. The lattice is still electrically neutral, since there are just as many positive charges as negative charges in it. In wood or glass, on the other hand, almost all the electrons are tightly bound to atoms or

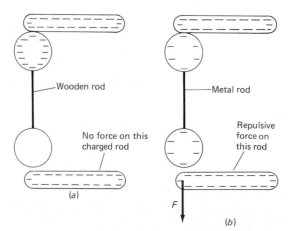

FIGURE 16.4
(a) Demonstration that a wooden rod is an electric insulator.
(b) Demonstration that a metal rod is an electric conductor.

molecules and cannot move about freely. Hence in a perfect insulator there can be no electric current, i.e., no flow of charge through the material.

At room temperature the best conductors known are silver and copper. Since silver is a precious metal, and therefore expensive, most electric wiring is made of copper. Good insulators are quartz, glass, and some synthetic materials such as Teflon. Note that to produce static electricity you need an insulator, like glass or wax, so that no flow of charge can occur to neutralize the static charges produced by friction or by an electrostatic machine.

16.3 The Gold-Leaf Electroscope

Thus far our exploration of electrostatics has been descriptive and qualitative, as it was in the eighteenth century. But physics is a quantitative science, and eighteenth-century physicists badly needed an instrument to measure charges quantitatively. Such an instrument is the gold-leaf electroscope, first developed by Abraham Bennet in 1787. This consists of a metal shaft with a round metal knob on top and two pieces of very light gold foil at the bottom, as in Fig. 16.5. The shaft of the electroscope passes through an insulating support in the case, which is enclosed in glass to protect the delicate gold leaves from air currents. When the electroscope is charged by touching its knob with a charged rod, excess charge flows to the gold leaves. The two leaves acquire like charges and therefore repel each other, and their deflection can be used as a rough measure of the charge on the electroscope. For more quantitative measurements the two gold leaves can be replaced by a flat brass plate and a single gold leaf mounted on the shaft in such a way that the mutual repulsion between the two causes the gold leaf to pull away from the plate, as in Fig. 16.6. The angular deflection of the gold leaf with respect to the brass plate is then a measure of the charge on the instrument, which is now called an *electrometer*, i.e., an instrument for measuring electric charge.

FIGURE 16.5 A gold-leaf electroscope.

FIGURE 16.6 A simple gold-leaf electrometer.

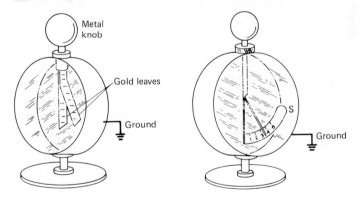

Metal knob

Gold leaves

Ground

S

Ground

Forces Exerted by Electric Charges at a Distance

Suppose we take an electroscope, as in Fig. 16.7, and first touch the knob with a charged amber rod. Some of the excess electrons from the amber will flow to the two pieces of gold foil, and they will stand apart. If we now remove the amber rod, charge it up again, and bring it close to the knob without touching it, we find that the gold leaves separate even more. The excess electrons on the charged rod repel the free electrons in the knob and shaft of the electroscope. The free electrons move as far away from the charged rod as

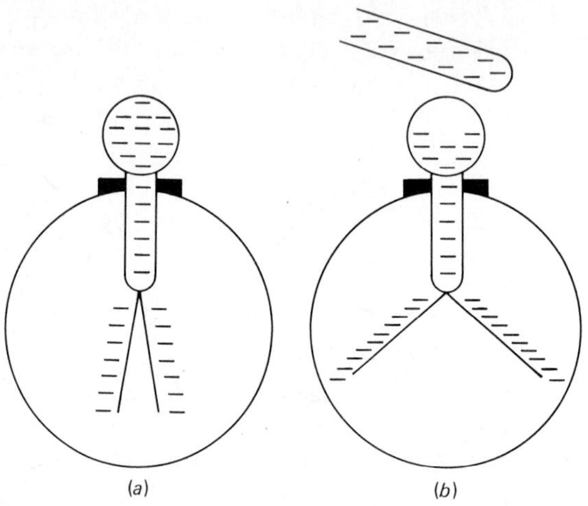

FIGURE 16.7 Effect of a negatively charged rod on a charged electroscope: (*a*) Electroscope charged negatively. (*b*) A negatively charged rod increases the separation of the leaves of the electroscope.

possible, and end up on the gold leaves. The excess electrons on the gold leaves cause an increased deflection of the electroscope, and the size of this increased deflection depends on how close the charged rod is to the knob. If the rod is removed, the excess deflection completely disappears. On the other hand, if a *positively* charged glass rod is brought near the knob of the electroscope, as in Fig. 16.8, electrons are drawn *away* from the gold leaves by the attractive force of the positive charges and the electroscope deflection is *decreased* as long as the glass rod is held near the electroscope knob. Hence an electroscope containing a charge of known sign can be used to distinguish positive from negative charges.

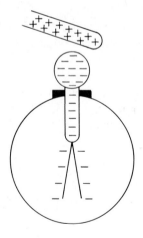

FIGURE 16.8 Effect of a positively charged rod on a negatively charged electroscope. The separation of the leaves is reduced.

**Charging an
Electroscope
by Induction**

It is possible to charge an electroscope without touching it with a charged rod or other source of charge, as in Fig. 16.9. Suppose we bring a negatively charged amber rod near the knob of an uncharged electroscope. The gold leaves diverge because electrons are being repelled by the charged rod and flow into the leaves. If we now touch the knob with our free hand and then remove our free hand while still keeping the rod in place, the excess electrons in the gold leaf, in their effort to get as far away as possible from the charged

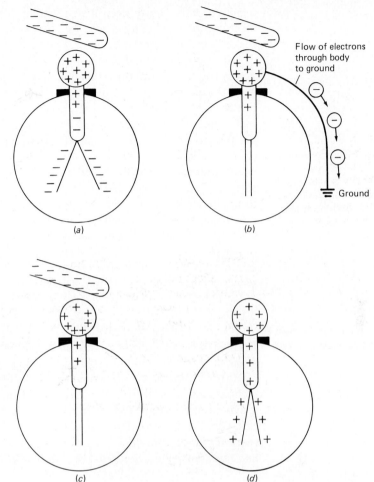

FIGURE 16.9 Charging an electroscope by induction: (*a*) A negatively charged amber rod is brought near the knob of an uncharged electroscope. (*b*) The knob is touched with the free hand. In an effort to get as far away as possible from the charged rod, electrons flow from the electroscope to ground through the body. (*c*) The free hand is then removed, breaking the connection to ground and leaving the electroscope positively charged. (*d*) The charged rod is removed, and the separation of the leaves indicates that the electroscope contains a net positive charge.

rod, flow through our hand and body to the ground,* and the electroscope leaves collapse. If, finally, we remove the rod, the gold leaves again diverge, and we find that the electroscope is *positively* charged, since we have taken

FIGURE 16.10 Grounding a charged sphere. The excess negative charge flows off the sphere to ground, and leaves the sphere uncharged.

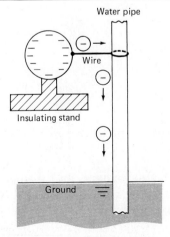

*The word *ground* is used here in the same sense that we talk about "grounding a TV set," or, as the British would say, "earthing a telly." The earth is considered as an almost infinite source of electric charge and is also a reasonably good conductor. Hence if any charged object is connected to the earth by a conducting path, any net negative charge on the object will flow from the object to ground, and any net positive charge will be canceled by a flow of electrons from ground to the object. This is illustrated in Fig. 16.10. The symbol $\underline{\overline{}}$ is often used to designate a ground.

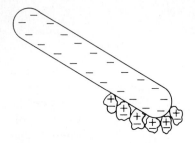

FIGURE 16.11 Attraction of a charged amber rod for small pieces of paper. The positive and negative charges of the paper are slightly separated, and the result is an attractive force.

negative charge away from it by touching it with our hand. We say that the electroscope has been *charged by induction*, i.e., without any charged body making direct contact with it.

Charges can be induced even on good insulators. For example, a charged amber rod will attract small bits of paper. The negative charges on the rod will repel the electrons in the atoms of the paper and attract the positively charged nuclei. Even though paper is an insulator and the electrons are not free to move through the paper, the atoms near the surface become *polarized*; i.e., the positive charges are displaced slightly in one direction and the negative charges in the opposite direction. This results in a layer of positive charge near the surface of the paper, with the electrons pushed back slightly from the surface. The attraction of the electrons in the amber for the positive charges in the paper is greater than the repulsion between the electrons on the rod and the electrons in the paper, since the latter electrons are farther away from the rod. Hence there is a net attractive force, and the amber picks up the bits of paper, as in Fig. 16.11. The electrostatic precipitators used to remove coal ash, sulfur oxides, and other pollutants from the chimneys of coal-burning power plants work in a similar fashion.

16.4 Coulomb's Law

In 1788 Charles-Augustin Coulomb (1736–1806), a French engineer and scientist, performed some experiments which put electrostatics on a quantitative basis for the first time. Coulomb used a torsion balance (Fig. 16.12) and a technique similar to the one used by Henry Cavendish to measure the gravitational constant (see Sec. 5.1; note that Coulomb's work *preceded* that of Cavendish by about 10 years). Coulomb concluded that *the force between two charges at rest is directly proportional to the product of the two charges and inversely proportional to the square of the distance between them*. In the

FIGURE 16.12 Coulomb's electrostatic torsion balance. The charged sphere *b* exerts a force on charged sphere *a*, which twists the wire support about a vertical axis. The force between the two spheres can be measured in terms of the twist of the suspension.

FIGURE 16.13 The Coulomb's law force between two charged spheres: $f_{el} = \dfrac{k_e q q'}{r^2}$.

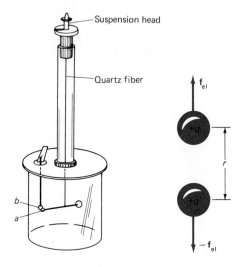

form of an equation, we have, for the situation in Fig. 16.13, where q is the symbol conventionally used for an amount of charge, and f_{el} is the magnitude of the electrostatic force:

Coulomb's Law:

$$f_{\text{el}} \propto \frac{qq'}{r^2} \quad \text{or} \quad \boxed{f_{\text{el}} = k_e \frac{qq'}{r^2}} \tag{16.1}$$

Coulomb's law is analogous to Newton's law of universal gravitation, both in the inverse-square dependence on distance and in the direct dependence on the magnitude of the charges (or masses, in the gravitational case). The major difference is that the electric force is repulsive when q and q' are of the same sign, and attractive when they are of opposite signs, whereas the gravitational force between two masses is always attractive, since masses do not have $+$ and $-$ signs. The electric force between two electrons is also larger than the gravitational force by 42 orders of magnitude! (See Prob. C20.) For this reason, in problems in electrostatics we completely ignore the presence of gravitational forces, which are negligible compared with the electrostatic forces between even very small charges.

Let us now evaluate the constant k_e in Coulomb's law [Eq. (16.1)]. Its value depends on the units we choose for the other quantities in the equation. In SI units f_{el} is measured in newtons, r in meters, and q in a unit called (after the French physicist) the *coulomb* (C) and defined as follows.*

Definition

A coulomb (C) is that charge which, when placed one meter away from an equal and like charge in a vacuum, repels it with a force of 9.0×10^9 N.

It is then possible to obtain the value of k_e from this definition. If, in Eq. (16.1) we let $f_{\text{el}} = 9.0 \times 10^9$ N, $r = 1$ m, and $q = q' = 1$ C, then Eq. (16.1) becomes

$$f_{\text{el}} = 9.0 \times 10^9 \text{ N} = k_e \frac{(1 \text{ C})(1 \text{ C})}{(1 \text{ m})^2}$$

and so $k_e = 9.0 \times 10^9$ N·m²/C²

Once k_e is determined in this way, it can be used in Coulomb's law to find unknown charges in terms of measured forces and distances or, if the charges are known, to find unknown electrostatic forces.

The coulomb is very large compared with the charges on the electron and proton. In fact a coulomb is the charge carried by 6.25×10^{18} electrons, as was shown in a famous measurement of the charge on the electron by the American physicist Robert A. Millikan in 1906 (see Fig. 16.14). Millikan measured the charge on the electron by balancing the electrostatic force on charged droplets of oil against the gravitational force acting on the same droplets. The apparatus he used for this purpose is shown in Fig. 16.15.

Millikan and his coworkers made measurements on thousands of oil drops, sometimes observing through the telescope the same oil drop for days at a time. They found that the measured charge q was always an integral multiple of an elementary charge e; that is, $q = e, 2e, 3e, 4e, \ldots, ne$ with n an

---

*In the SI system of units the coulomb is not a fundamental unit but a derived unit more properly defined in terms of the unit of electric current, the *ampere*. This will be discussed more fully in subsequent chapters, where we will see that a more precise value of k_e is 8.988×10^9 N·m²/C².

FIGURE 16.14 Robert A. Millikan (1868–1953), the American physicist famous for his measurement of the charge on the electron and for his experimental verification of Einstein's theory of the photoelectric effect (Sec. 26.3). For these accomplishments he received the 1923 Nobel Prize in physics. (*Photo courtesy of California Institute of Technology Archives.*)

FIGURE 16.15 Millikan's apparatus for measuring the charge on the electron. An atomizer sprays tiny oil droplets into the region between the charged plates P and P'. The motion of the oil droplets under the combined gravitational and electrostatic fields is then observed through the telescope.

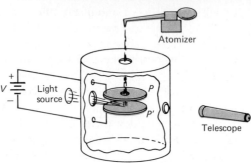

integer. Even when an x-ray source was used to change the charge on an oil drop by ionization, *the change in charge was always found to be an integral multiple of e*. Also, no charge was ever observed to be less than e, which was found to be approximately 1.60×10^{-19} C.

Millikan's interpretation of these results was that e was the charge on the electron, and that each oil drop measured had either one, or two, or three, or n extra electrons, and hence that the charge was an integral multiple of the so-called elementary charge—the smallest charge ever experimentally observed, or the charge on the electron.

Elementary Electronic Charge:

$$e = 1.602 \times 10^{-19} \text{ C}$$

Example 16.1

A comb drawn through a person's hair on a dry day causes 10^{12} electrons to leave the person's hair and stick to the comb.

(a) Is the force between the comb and the hair attractive or repulsive?

(b) What is the magnitude of this force when the comb is 1.0 m from the person's hair?

SOLUTION

(a) The force is *attractive* because the charge on the comb is negative and that on the hair is positive. Unlike charges attract.

(b) From Coulomb's law we have

$$f_{el} = \frac{k_e q_1 q_2}{r^2} = (9.0 \times 10^9 \text{ N·m}^2/\text{C}^2)(10^{12}) \,(+1.6 \times 10^{-19} \text{ C})$$

$$\times (10^{12}) \,(-1.6 \times 10^{-19} \text{ C})/(1.0 \text{ m})^2 = \boxed{-2.3 \times 10^{-4} \text{ N}}$$

where the minus sign means that the force is attractive.

Electrostatic Problems Involving Point Charges

Coulomb's law can be used to solve a variety of problems involving point charges at rest. In so doing, three basic ideas, developed in previous chapters, should be kept in mind:

1 *Newton's third law.* To every action there is an equal and opposite reaction. Hence if a charge q_A exerts an electric force $\mathbf{f}_{AB}$ on q_B, then q_B exerts an equal and opposite force $\mathbf{f}_{BA}$ back on q_A, as in Fig. 16.16. Here $\mathbf{f}_{BA} = -\mathbf{f}_{AB}$. Note that this equality is true even if q_A and q_B are very different in magnitude. Thus q_A could be 100 μC and q_B only 2.0 μC, and it would still be true that $\mathbf{f}_{BA} = -\mathbf{f}_{AB}$; that is, the two forces are equal in magnitude but opposite in direction. This must be true since the equation for Coulomb's law is unchanged when q and q' are interchanged.

2 *The principle of superposition.* Each electrostatic force acts independently of all other forces present. Hence the resultant electrostatic force on a charge is the vector sum of the individual forces produced by each of the charges present.

FIGURE 16.16 Newton's third law in the electrostatic case: $\mathbf{f}_{AB}$ is equal in magnitude to $\mathbf{f}_{BA}$, but is in the opposite direction.

3 *Vector addition.* Electrostatic forces are, like all forces, *vectors*. Hence if two charges q_A and q_B exert forces on q_C, as in Fig. 16.17, then the forces $\mathbf{f}_{AC}$ and $\mathbf{f}_{BC}$ must be added *vectorially* to obtain the resultant force $\mathbf{f}_{el}$ on q_C. In so doing, the direction of each force depends on whether the interaction of the charges is attractive or repulsive, and this is determined by the signs of the two charges.

Using these basic ideas and Coulomb's law, it is possible to solve all electrostatic problems involving point charges, no matter how many charges are involved. Some illustrations are given in Examples 16.2 and 16.3.

FIGURE 16.17 Application of Coulomb's law to a configuration of three charges, indicating how the electrostatic forces must be added as *vectors* to obtain correct results.

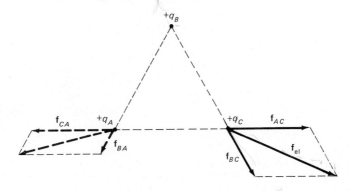

Example 16.2

(a) The force required to give a 1000-kg car an acceleration of 2.5 m/s^2 is 2.50×10^3 N. What would the charge have to be on two small spheres 1.0 m apart to produce the same force between the spheres, if we assume that both spheres have the same charge?

(b) If the charge on each sphere is negative, how many excess electrons are there on each sphere?

SOLUTION

(a) From Coulomb's law, with $q' = q$, we have

$$f_{el} = \frac{k_e q q'}{r^2} = \frac{k_e q^2}{r^2}$$

Hence $q^2 = \dfrac{r^2 f_{el}}{k_e} = \dfrac{(1.0 \text{ m})^2 \, (2.50 \times 10^3 \text{ N})}{9.0 \times 10^9 \text{ N·m}^2/\text{C}^2} = 2.77 \times 10^{-7} \text{ C}^2$

and so $q = 5.26 \times 10^{-4} \text{ C} = \boxed{526 \ \mu\text{C}}$

Two charges of 526 μC at a distance of 1 m apart would therefore produce this relatively large force of 2.5×10^3 N.
(b) The charge on the electron is 1.60×10^{-19} C. Hence the number N of electrons in 526 μC is:

$$N = \frac{526 \ \mu\text{C}}{1.60 \times 10^{-19} \text{ C/electron}} = \frac{526 \times 10^{-6} \text{ C}}{1.60 \times 10^{-19} \text{ C/electron}}$$

$$= \boxed{3.29 \times 10^{15} \text{ electrons}}$$

Example 16.3

Three charges are arranged at the corners of a right triangle, as shown in Fig. 16.18, where q_B and q_C are located on the X axis. Find the magnitude and direction of the resultant force acting on q_C if the charges are $q_A = -6.0 \ \mu\text{C}$, $q_B = +4.0 \ \mu\text{C}$, and $q_C = +2.0 \ \mu\text{C}$.

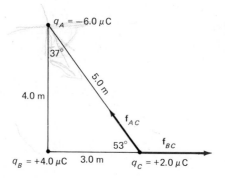

FIGURE 16.18 Diagram for Example 16.3.

SOLUTION

The force exerted on q_C by q_A is

$$f_{AC} = \frac{k_e q_A q_C}{r^2}$$

$$= \left(9.0 \times 10^9 \ \frac{\text{N·m}^2}{\text{C}^2}\right) \frac{(-6.0 \times 10^{-6} \text{ C}) \, (+2.0 \times 10^{-6} \text{ C})}{(5.0 \text{ m})^2}$$

$$= -4.3 \times 10^{-3} \text{ N}$$

This is an attractive force (the sign is −) and is directed from q_C toward q_A along the line joining q_A and q_C. Also

$$f_{BC} = \left(9.0 \times 10^9 \ \frac{\text{N·m}^2}{\text{C}^2}\right)\frac{(+4.0 \times 10^{-6} \text{ C}) \, (+2.0 \times 10^{-6} \text{ C})}{(3.0 \text{ m})^2}$$

$$= +8.0 \times 10^{-3} \text{ N}$$

This is a repulsive force directed along the positive X axis and away from q_B at q_C.

The two forces on q_C are as shown in Fig. 16.19. To find the resultant we first find the X and Y components of the two force vectors at C. The angles in the triangle can be found from our knowledge of the three sides of the triangle:

$$f_X = 8.0 \times 10^{-3} \text{ N} - (4.3 \times 10^{-3} \text{ N}) \cos 53°$$

$$= (8.0 - 2.6) \times 10^{-3} \text{ N} = 5.4 \times 10^{-3} \text{ N}$$

$$f_Y = (4.3 \times 10^{-3} \text{ N}) \sin 53° = 3.4 \times 10^{-3} \text{ N}$$

Then we have

$$f_{el} = \sqrt{f_X^2 + f_Y^2} = \sqrt{(5.4)^2 + (3.4)^2} \times 10^{-3} \text{ N}$$

$$= \boxed{6.4 \times 10^{-3} \text{ N}}$$

The direction of the force is given by

$$\tan \theta = \frac{f_Y}{f_X} = \frac{3.4 \text{ N}}{5.4 \text{ N}} = 0.63$$

and so $\theta = \boxed{32°}$

Hence the resultant force on q_C is of magnitude 6.4×10^{-3} N at an angle of 32° with respect to the positive X axis, as shown in the figure.

The forces on q_A and q_B can be found in a similar fashion.

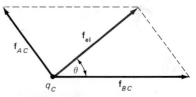

FIGURE 16.19 Forces on charge q_C in Example 16.3.

16.5 The Electric Field; Gauss' Law

If we have a negative point charge $-q$ at some point in space and bring up a test charge $+q'$ to a point A a distance r from the charge $-q$, as in Fig. 16.20, then q' will experience a force given by Coulomb's law as $f_{el} = -k_e\, qq'/r^2$,

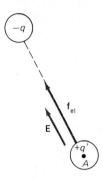

FIGURE 16.20 A charge q' at point A experiences a force $\mathbf{f}_{el}$ due to the charge $-q$. The electric field at point A is then in the direction of $\mathbf{f}_{el}$ and of magnitude $E = f_{el}/q'$.

where the negative sign indicates an attractive force. We now introduce a vector quantity called the *electric field strength* (or *electric intensity*) defined as follows:

Definition

Electric field strength: The electric force $\mathbf{f}_{el}$ experienced by a positive test charge q' at some position in space, divided by the magnitude of the charge. That is:

$$\mathbf{E} = \frac{\mathbf{f}_{el}}{q'}$$

(16.2)

In the case of the point charge above, the magnitude of the electric field strength is $k_e qq'/q'r^2 = k_e q/r^2$, and its direction is toward the charge $-q$ which is the source of the field. If the point charge producing the field were positive, the field would be of the same magnitude but directed *away from* the positive point charge.

There is one important restriction on this definition: the test charge q' must be so small that it does not change in any way the field being measured. The units of electric field strength are those of force per unit charge, or newtons per coulomb (N/C).

Example 16.4

In Example 16.3 what is the electric field strength at the position occupied by charge q_C in Fig. 16.19?

SOLUTION

We found that the resultant force on q_C was $f_{el} = 6.4 \times 10^{-3}$ N at an angle of 32° with respect to the positive X axis. Hence the electric field at q_C is in this same direction, and its magnitude is:

$$E = \frac{f_{el}}{q_C} = \frac{6.4 \times 10^{-3}\ \text{N}}{2.0 \times 10^{-6}\ \text{C}} = \boxed{3.2 \times 10^3\ \text{N/C}}$$

Electric Lines of Force

By plotting the electric field at various points in space, the nature of the field produced by various arrangements of charges can be visualized.

The resultant electric field produced by any set of electric charges can be obtained mathematically from Coulomb's law or can be measured experimentally in the laboratory. Here we will merely show the results for a few important types of fields. Figures 16.21 and 16.22 show the fields of single negative and positive charges. Since the test charge is positive by convention, the lines of force are always directed toward negative charges and away from positive charges, for that is the direction in which a positive charge would move if free to do so.

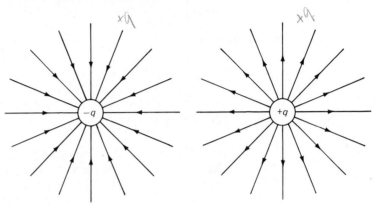

FIGURE 16.21 (left) The electric field of a negative point charge. These lines actually fill up an entire sphere in three dimensions, but in the diagram only a two-dimensional cross section is shown.

FIGURE 16.22 (right) The electric field of a positive point charge.

A very important charge distribution is that of an *electric dipole*, shown in Fig. 16.23. This consists of two charges of equal magnitude but opposite sign placed very close together. Some molecules like HCl are natural electric dipoles. Other molecules can be polarized in strong electric fields, leading to a separation of the centers of positive and negative charge, and to what we call an *induced dipole*. The field shown in Fig. 16.23 is appropriate to either kind of electric dipole.

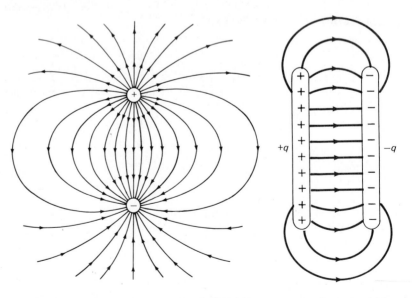

FIGURE 16.23 (left) Field of an electric dipole. Again these lines fill up a three-dimensional space. The complete picture can be obtained by rotating this figure through 360° about the line connecting the two charges.

FIGURE 16.24 (right) Cross-sectional view of the electric field between two flat plates containing equal and opposite electric charges.

Figure 16.24 shows the electric field between two flat metal plates, with charge $+q$ on the left plate and $-q$ on the right plate. Here the charge is distributed uniformly over each of the plates. In this case the lines of force are everywhere normal* (i.e., perpendicular) to the surfaces of the plates except very near the edges. Since the number of lines of force passing through a unit area perpendicular to the lines of force remains unchanged as we move from one plate to the other, the electric field strength **E** is *constant* at any point between the plates, so long as we do not get too close to the edges.

It is important to observe that, as Figs. 16.21 to 16.24 show, the lines of the electric field always begin on positive charges and end on negative charges. It is therefore impossible in an electrostatic field for a line of force to close on itself, for to do this it would have to point toward a positive charge or away from a negative charge, which is opposite to the direction in which a positive test charge would move. In many cases, as in Fig. 16.22, the negative charges on which the field lines end may be a very large distance away from the positive charges originating the field. Despite this, we can be confident that the field lines for static charges actually end on negative charges somewhere in space.

Gauss' Law

One of the great contributions to physics made by the eminent German mathematician Karl Friedrich Gauss (1777–1855) describes an important property of electrostatic fields, and is called *Gauss' law*. Gauss' law provides great insight into the solution of many practical problems in electrostatics.

Consider a positive point charge q whose electric field is that shown in Fig. 16.25, with N lines of force drawn from the charge to represent its magnitude q (for our purposes N can be any arbitrarily chosen number). Suppose we draw two spherical surfaces around the charge as a center, one with radius R, the other with radius $2R$. These two surfaces are everywhere normal to the radial lines of the electric field **E**. The surface area of the first sphere is $4\pi R^2$, of the second $4\pi(2R)^2 = 16\pi R^2$. The magnitude of the electric field at R is $E_R = k_e q/R^2$, and at $2R$ it is $E_{2R} = k_e q/(2R)^2 = k_e q/4R^2$. Hence, on multiplying the magnitude of the electric field strength by the surface area in each case, we obtain

---

*In physics *normal* means "perpendicular to," a meaning that is derived from the use of the word in the construction trade. In the building of houses, beams must be at right angles to each other, in which case they are referred to as "squared," or "normal." Hence normal came to mean "at right angles," and this is the meaning of the word as it is used in physics.

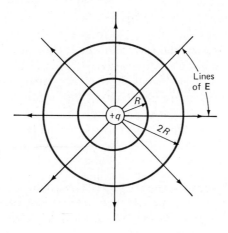

FIGURE 16.25 Gauss' law: The electric field of a positive point charge. Note that the lines drawn here as circles really represent spheres.

$$E_R(4\pi R^2) = \frac{k_e q}{R^2}(4\pi R^2) = 4\pi k_e q$$

and
$$E_{2R}(16\pi R^2) = \frac{k_e q}{4R^2}(16\pi R^2) = 4\pi k_e q$$

The product of the electric field strength and the area normal to the electric field vector, when summed over a complete sphere, is the same for $2R$ as it is for R. In fact, it is the same no matter what the distance is from q, the source of the electric field.

This is Gauss' law, which we have proved only for the simple case of a point charge and spherical surfaces. It states that if we multiply each element of the area of any closed surface by the component of **E** normal to that element of the area, and then sum over the entire surface, the result is a constant ($4\pi k_e$) times the total charge q inside the surface, or, in symbols,

$$\sum E_\perp \Delta A = 4\pi k_e q$$

The quantity $\sum E_\perp \Delta A$ is called the *electric flux* through the surface. It can be shown that this result remains true in general for any number of charges inside any closed surface. This more general form of Gauss' law can be written as follows:

Gauss' law

$$\boxed{\sum E_\perp \Delta A = 4\pi k_e \sum q} \tag{16.3}$$

where $E_\perp$ is the component of the electric field perpendicular to an element of area ΔA, and the sum is taken over the total area of the surface and the total charge within the surface.

The surface to be used in applying Gauss' law is completely arbitrary; it may be any surface that is useful in solving a particular problem. Such a surface is called a *gaussian surface*; in solving electrostatic problems using Gauss' law we need to find the proper gaussian surface for the problem. Usually we choose a surface that has the same symmetry as the charge distribution, and which therefore enables us to evaluate $\sum E_\perp \Delta A$ easily. We will show just a few examples here to indicate the power of Gauss' law.

Charge on a Spherical Metal Surface

If a hollow metal sphere is charged negatively with a charge $-Q$, the excess electrons will repel one another and assume a uniform distribution over the whole surface of the sphere. The lines of electric field intensity in this case point toward the surface of the metal and end on the excess electrons. They are also normal to the surface of the metal, as in Fig. 16.26. If they were not, the electric field would have a component parallel to the metal surface. Since the electrons are free to move, they would therefore move to a position where that parallel component would be reduced to zero. Hence *the only components of the electric field which remain are perpendicular to the surface of the metal*.

Let us construct a spherical gaussian surface of radius R just outside the metal sphere. Since the lines of electric field intensity are everywhere perpendicular to the gaussian surface, Gauss' law tells us that

$$EA = 4\pi k_e \sum q = 4\pi k_e(-Q)$$

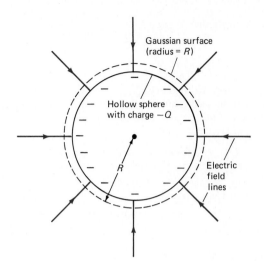

FIGURE 16.26 Field of a negatively charged hollow metal sphere.

Hence $\qquad E = \dfrac{-4\pi k_e Q}{A} = \dfrac{-4\pi k_e Q}{4\pi R^2} = \dfrac{-k_e Q}{R^2}$

But this is just the result which would be obtained from Coulomb's law for a point charge of magnitude $-Q$ at the center of a sphere of radius R. Hence *the electric field of a charged metal sphere at any point outside the sphere is the same as if all the charge were concentrated at its center.*

Charge on Metal Objects

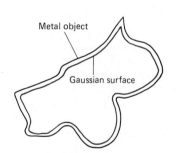

FIGURE 16.27 Charge distribution on an irregularly shaped solid metal object.

Suppose we have an irregularly shaped solid metal object like the one in Fig. 16.27, and charge it negatively. Where does the excess negative charge end up after the very brief time required for the charges to move to equilibrium positions and come to rest?

To answer this question we can construct a gaussian surface just inside the outer boundary of the surface, as in the figure. Then we have

$$\sum E_\perp \, \Delta A = 4\pi k_e \sum q \qquad\qquad (16.3)$$

But the electric field strength inside any metal must be zero in the static case, since the charges are free to flow when the field is applied and will continue to flow until no net field remains. Hence the left side of Eq. (16.3) is zero, and therefore the right side must also be zero. The only way this can be true is if $\sum q = 0$. Therefore there can be no excess electric charge in the interior of the metal. *All the excess charge on a metal is on the outside.*

We have here obtained some very important insights into the behavior of metals by using Gauss' law. To obtain the same results by using Coulomb's law would require considerably more work.

Example 16.5

Find the electric field E at a distance R from a long, straight wire which contains a total charge q spread uniformly over a length L. Assume that R is very small compared with L.

SOLUTION

The charge per unit length on the wire is $\lambda = q/L$. Since a wire can be considered to be a very thin cylinder, the wire has cylindrical symmetry. This gives us a hint that, if we apply Gauss' law, the gaussian surface we use should be a cylindrical surface, as in Fig. 16.28. We choose the gaussian surface to have a radius R and a length ΔL. The surface area of the curved sides of this cylinder is then $2\pi R\ \Delta L$.

We have seen that the field near a metal will be everywhere perpendicular to the surface of the metal, in this case, the long wire. The lines of the electric field will also be perpendicular to the curved surface of this cylinder at every point on it. For the same reason no lines of the electric field will pass through the two flat ends of the cylinder, and the electric field will be radial and equal to E_R. Hence we have, from Gauss' law,

$$\sum E_\perp\ \Delta A = 4\pi k_e \sum q \qquad \text{or} \qquad E_R(2\pi R\ \Delta L) = 4\pi k_e \lambda\ \Delta L$$

since the amount of charge within the gaussian surface is $\lambda\Delta L$. Hence

$$E_R = \frac{4\pi k_e \lambda\ \Delta L}{2\pi R\ \Delta L} = \boxed{\frac{2k_e\lambda}{R}}$$

Hence the field of a long, straight wire falls off as $1/R$, that is, as the first power of the distance from the wire. This

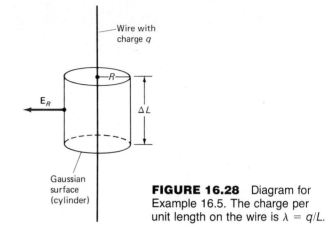

FIGURE 16.28 Diagram for Example 16.5. The charge per unit length on the wire is $\lambda = q/L$.

example shows the power of Gauss' law in solving problems which otherwise would be very complicated.

16.6 Electric Potential Difference; Voltage

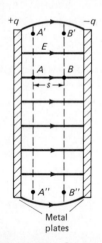

FIGURE 16.29 Cross-sectional view of the electric field between two charged flat metal plates. Here $A'A''$ and $B'B''$ are two flat surfaces parallel to the metal plates. (As we will see later in this section, these two surfaces are equipotential surfaces.)

Let us consider the field between two flat metal plates, one charged positively, the other negatively, as in Fig. 16.29, where the two charges are the same in magnitude. The electric field in this case is uniform between the plates, as we have seen. (We will derive the explicit mathematical expression for this field in Sec. 16.7.) Suppose the distance between the points A and B is s, and that we introduce a test charge $+q'$ and move it from B to A. Since we are moving a positive charge in the opposite direction to the force ($\mathbf{f}_{el}$) exerted on it by the field $\mathbf{E}$, we must do work against this force. The amount of work done is calculated just as it was in mechanics from

$$\mathcal{W} = f_{el}s = q'Es \tag{16.4}$$

Now the electric field is a conservative field of a type similar to the gravitational field discussed in Sec. 6.3. This might be expected because of the similarity of Newton's law of universal gravitation to Coulomb's law. Hence the work done in moving the charge from B to A must increase the potential energy of the charge, just as lifting a rock increases the gravitational potential energy of the rock. Since the charge is moved so slowly that it can be presumed no work goes into kinetic energy, and since the field is conservative so that no energy is lost as heat, our work-energy theorem becomes simply

$$\mathcal{W} = \Delta\text{PE} = \text{PE}_A - \text{PE}_B \tag{16.5}$$

i.e., all the work goes into the increase in the electric potential energy of the charge q. Here PE_A means the electric potential energy of the charge q at point A in the field, and PE_B is the same at point B. On dividing Eq. (16.5) by q', we have

$$\frac{\mathcal{W}}{q'} = \frac{PE_A}{q'} - \frac{PE_B}{q'}$$

Definition

The potential energy per unit charge of a charge q' in an electric field is called the electric potential V.

$$V = \frac{PE}{q'} \tag{16.6}$$

In our case the potential at A is $V_A = PE_A/q'$, and at B it is $V_B = PE_B/q'$. We then have:

Definition of Voltage

$$\boxed{\frac{\mathcal{W}}{q'} = V_A - V_B = V_{AB}} \tag{16.7}$$

where $V_{AB} = V_A - V_B$, and is called the *potential difference*, or the *voltage*, between point A and point B. The voltage between A and B is equal to the work done by an outside force in moving a unit charge from B to A. It is also equal to the work done by the field in moving a unit positive charge from A to B in the direction of the field.

The fact that the voltage between two points in an electrostatic field is equal to the work done in moving a unit charge from one place in the field to another is a completely general result, even though we have derived it for an especially simple case.

Since, in Fig. 16.29, the positive charge q has a higher potential energy at A than at B, we say that there is a *voltage drop* between A and B, and that a positive charge will, if released, of its own accord, flow "downhill" from A to B. Similarly, if we want to move the charge "uphill" from B to A, we must do an amount of work $\mathcal{W} = q'V_{AB}$ against the field, which is directed from A to B. On the other hand, a negative charge will spontaneously flow from B to A; that is, it will flow *up* a potential hill.

The Unit of Voltage

In the SI system $\mathcal{W}$ is measured in joules, q in coulombs, and V in *volts* (V). Hence, since $\mathcal{W} = q'V_{AB}$,

$$1\ J = 1\ C \cdot 1\ V \qquad \text{or} \qquad 1\ V = 1\ J/C$$

Definition

If one joule of work is done to move one coulomb of charge from one point in an electric field to another, then the voltage (or potential difference) between these two points is one volt.

The voltage between the positive ($+$) and negative ($-$) terminals of an automobile storage battery is usually 12 V. This means that there is an electric field in the region around the battery terminals, and that if we move a positive charge from the negative to the positive terminal of the battery, the work done would be 12 J for each coulomb of positive charge carried from the negative to the positive terminal. Since the electrostatic field is conservative, this amount of work is the same no matter what path is taken.

Equipotential Surfaces

A useful concept in dealing with electric fields is that of an *equipotential surface*.

Definition

Equipotential surface: A surface all parts of which are at the same electric potential.

Since there is no potential difference (or voltage) between any two points on such a surface, no work need be done in moving a charge from one place to another on that surface. We must therefore conclude that the electric field has no components parallel to the equipotential surface, for if it did, work would have to be done in moving a charge along that surface against the field. Hence the *electric field* **E** *must be everywhere perpendicular to an equipotential surface*.

For example, in Fig. 16.29 $A'A''$ and $B'B''$ are flat surfaces parallel to the two charged flat plates. Any two points on the surface $A'A''$ are at the same electric potential because they are the same distance from the plates. $A'A''$ is therefore an equipotential surface, as is $B'B''$. These surfaces, as can be seen from the diagram, are everywhere perpendicular to the lines of the electric field **E**. This remains true no matter how complicated the charge distribution.

The Electronvolt: A Unit of Energy

We have just seen that, if the voltage between two points in an electric field is 1 V, then it takes 1 J of work to move 1 C of charge between these two points. Suppose, however, that we do not want to move 1 C of charge, but only a single electron with a charge 1.60×10^{-19} C through a voltage of 1 V. How much work must be done? The work in this case is

$$\mathcal{W} = (1.602 \times 10^{-19} \text{ C})(1 \text{ V}) = 1.602 \times 10^{-19} \text{ J}$$

This amount of work (or energy) is called an *electronvolt*.

Definition

Electronvolt (eV): The work required to move one electronic charge through a potential difference of one volt.

$$1 \text{ eV} = 1.60 \times 10^{-19} \text{ J}$$

The electronvolt is much used as an energy unit in atomic and nuclear physics, as we will see.

Relationship between Electric Field and Change in Potential

We now want to establish a relationship between the electric field **E** and the change in the potential ΔV. In doing this we must be very careful about *signs*. If, in Fig. 16.29, we move a charge q' a distance Δs in the direction from B to A, we must exert a force just equal to the electric force $\mathbf{f}_{el} = q'\mathbf{E}$, but in the opposite direction. Hence the force we exert is: $\mathbf{f} = -q'\mathbf{E}$. In moving the charge through a distance Δs we do an amount of work

$$\Delta \mathcal{W} = \mathbf{f} \cdot \Delta\mathbf{s} = f \, \Delta s = -q'E \, \Delta s$$

This produces a change in the potential energy of the charge equal to

$$\Delta \text{PE} = \Delta \mathcal{W} = -q'E \, \Delta s$$

But, by the definition of potential difference, ΔV is equal to the work done per unit charge in moving the charge q' through the distance Δs. Hence, from Eq. (16.7) we have

$$\Delta V = \frac{\Delta \mathcal{W}}{q'} \quad \text{or} \quad \Delta \mathcal{W} = q' \, \Delta V$$

On combining these two results for the work $\Delta \mathcal{W}$, we have

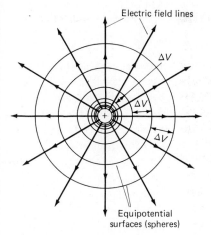

Electric field lines

ΔV

ΔV

ΔV

Equipotential
surfaces (spheres)

FIGURE 16.30 Cross-sectional
drawing of the field lines and
equipotential surfaces for a
positive point charge.

$$\Delta W = q' \, \Delta V = -q'E \, \Delta s$$

and so
$$E = \frac{-\Delta V}{\Delta s} \tag{16.8}$$

where ΔV is the change in electric potential along the distance Δs, and Δs is normal to the equipotential surfaces at every point. Here the minus sign indicates that E is positive when $\Delta V/\Delta s$ is negative, since E always points in the direction in which V *decreases*. Hence when ΔV has a minus sign, E must have a plus sign. We therefore say that *the electric field strength is the negative gradient of the potential*, where the potential gradient and therefore the field strength is measured in volts per meter.*

This expression for the relationship between the electric field strength and the gradient of the potential can be shown to be true for all electrostatic situations, no matter how complicated.

As an example, consider the field of the point charge shown in Fig. 16.30. The field lines and the equipotential surfaces are drawn on the figure. Suppose successive pairs of equipotential lines vary in potential by ΔV, which in this case we take to be 10 V. Then the closer together the equipotential surfaces are, the smaller the value of Δs, and therefore the greater the strength of the field E, since $E = -\Delta V/\Delta s$. Hence the electric field strength is large where the equipotential surfaces are crowded together, and weak where the equipotential surfaces are far apart. This can be compared to the way contour maps show elevations on the surface of the earth. A crowding together of gravitational equipotential surfaces means a large change in height over a small distance along the earth's surface and thus indicates the presence of mountains or cliffs.

Metals

In an electrostatic field, every point of a metal must be at the same potential, since, if it were not, an electric field would exist because of the potential gradient. This field would cause the free electrons to move until the electric field was reduced to zero. This means that the surface of any conductor is an equipotential surface, regardless of its shape. Since the field lines are everywhere perpendicular to the equipotentials, the electric field at the surface must be everywhere perpendicular to the surface of the metal (Fig. 16.31).

---

*The units volts per meter and newtons per coulomb can be used interchangeably for the electric field strength, since

$$1 \frac{V}{m} = \frac{1 \text{ J/C}}{1 \text{ m}} = \frac{1 \text{ N·m}}{1 \text{ C·m}} = \frac{1 \text{ N}}{C}$$

FIGURE 16.31 Electric field
lines for an irregularly shaped
metal object. The lines of **E** are
perpendicular to the surface at
every point.

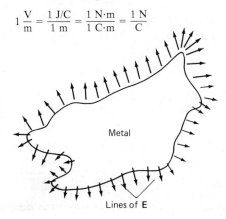

Metal

Lines of **E**

Example 16.6

A 12-V storage battery drives 50 C of charge through a light bulb in 1 min.
(a) How much work has been done by the battery?

(b) What is the power, i.e., the rate at which work is done by the battery?

SOLUTION

(a) Since the voltage, or potential difference, is the work done per unit charge, or

$$\frac{W}{q'} = V_{AB}$$

we have $W = q'V_{AB} = (50 \text{ C})(12 \text{ V}) = \boxed{600 \text{ J}}$

(b) Since power is the rate at which work is done,

$$P = \frac{W}{t} \quad \text{and so} \quad P = \frac{600 \text{ J}}{60 \text{ s}} = 10 \text{ J/s} = \boxed{10 \text{ W}}$$

Hence the power delivered by the storage battery is the same as that consumed by a 10-W light bulb.

Example 16.7

An electron in a TV set is accelerated toward the screen by a voltage of 1000 V. The screen is 35 mm from the electron source.
(a) How much work is done by the voltage in accelerating the electron?

(b) What is the speed of the electron when it strikes the screen?

SOLUTION

(a) Since the work done per unit charge by the field is $W/q' = V$, we have

$$W = q'V = (1.6 \times 10^{-19} \text{ C})(1000 \text{ V}) = \boxed{1.6 \times 10^{-16} \text{ J}}$$

(b) By the principle of conservation of energy all this work must be converted into kinetic energy of the electron, Hence

$$KE = \tfrac{1}{2}mv^2 = 1.6 \times 10^{-16} \text{ J}$$

and so $v = \left[\dfrac{2(1.6 \times 10^{-16} \text{ J})}{9.1 \times 10^{-31} \text{ kg}} \right]^{1/2} = \boxed{1.9 \times 10^7 \text{ m/s}}$

Note that the distance between source and screen is of no importance here. The crucial factor is the voltage between source and screen; this completely determines the kinetic energy of the electron and hence its speed.

Absolute Potential

In the case of the gravitational field we have seen that the zero of potential energy is completely arbitrary, and that the only factor of interest is the difference in potential energy between two points at different distances from the earth's center. For the electric field we are usually interested only in differences of electrostatic potential energy, but in certain cases it is useful to choose a particular zero for the potential energy in an experimentally reproducible fashion. Thus for electric circuits it is customary to "ground" one part of the circuit and take the electric potential energy of a charge at ground potential to be zero. In this way the so-called *absolute electric potential*, or the potential energy per unit charge with respect to ground, can be obtained.

Another way to choose a zero for the electric potential energy is particularly useful in dealing with atoms and molecules. A charge q' in the field of a point charge q has an electric potential energy which depends on its distance from q. We can define the *absolute potential energy* of q' as *the work*

Definition

required to bring q' from a very large distance away (which we usually call infinity) to the point at which it is located. This is equivalent to choosing $r = \infty$ as the place of zero potential.

The *absolute potential V_a* at a point a distance r_a from point charge q is the work done in bringing a *unit* test charge from infinity up to that point in the field, as in Fig. 16.32. This turns out to be:

Potential of a Point Charge

$$V_a = \frac{k_e q}{r_a} \tag{16.9}$$

Hence, whereas the *field* of a point charge varies as $1/r^2$, the absolute *potential* of a point charge varies as $1/r$. Dimensionally this must be the case, since from Eq. (16.8), $E = -\Delta V / \Delta s$, and so the field must vary as V_a / r_a, or as $1/r_a^2$.

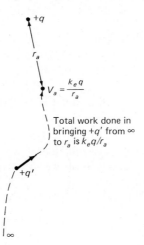

FIGURE 16.32 Absolute potential at a point in the field of a point charge $+q$. The absolute potential at a distance r_a from the charge is $V_a = k_e q / r_a$.

Example 16.8

Consider two point charges, $q_1 = +1.0 \ \mu C$ and $q_2 = +2.0 \ \mu C$ separated by a distance of 10 m, as in Fig. 16.33.
(a) At what point on the line (or on its extension) joining the charges is the electric field zero?
(b) At what point on the same line is the absolute electric potential zero?
(c) Is there any point in the space surrounding these two charges at which the absolute electric potential is zero? Why?

SOLUTION

(a) Since the electric field strength is a *vector* and the two charges are positive, the only place where the fields due to the two charges can cancel is on the line between the two charges, for only here are the two electric field vectors in opposite directions.

Let r_1 be the distance from q_1 to the point where the field vanishes, and let $r_2 = 10 - r_1$ be the distance from q_2 to the same point. Then we have, from the definition of the electric field,

$$E_1 = \frac{k_e q_1}{r_1^2} \qquad E_2 = \frac{k_e q_2}{r_2^2} = \frac{k_e q_2}{(10 - r_1)^2}$$

At the point where the field vanishes, $E_1 = E_2$ in magnitude, and they are in the opposite direction. Hence we have

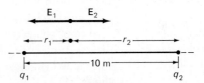

FIGURE 16.33 Diagram for Example 16.8.

$$\frac{k_e (1.0 \ \mu C)}{r_1^2} = \frac{k_e (2.0 \ \mu C)}{(10 - r_1)^2}$$

or

$$(10 - r_1)^2 = 2r_1^2$$
$$10 - r_1 = \sqrt{2} r_1$$

from which

$$2.41 r_1 = 10 \ \text{m} \qquad \text{or}$$

$$r_1 = \boxed{4.1 \ \text{m from charge } q_1}$$

Hence the point where the electric field vanishes is on the line between the two charges, and 4.1 m from the smaller charge. There is no other point on the line joining the two charges where the field vanishes, since outside the charges the two fields are in the same direction. Of course, as $r \to \infty$, $E \to 0$.

(b)   The electric potential is a *scalar*. In this case it is equal to $k_e q_1/r_1$ for the first charge and $k_e q_2/r_2$ for the second charge. Since both charges are positive, their sum will never be zero. It is only because of the *vector* nature of the electric field that there was one point, found in part (*a*), where it vanished. This points up the clear distinction between the *vector* electric field strength and the *scalar* electric potential.

(c)   From the argument just presented there is no point in space at which the electric potential due to two positive charges is zero, except for a point infinitely far from both charges, since in that case both potentials go to zero.

16.7 The Storage of Electric Charge

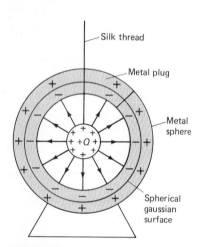

— Silk thread

— Metal plug

— Metal sphere

— Spherical gaussian surface

FIGURE 16.34 The introduction of a charge $+Q$ into the interior of a metal sphere.

Consider a hollow spherical conductor which is mounted on an insulating stand. We initially ground the metal to make sure no excess charge remains on it, and then we remove the ground. Next we introduce into the center of the sphere a silk thread with a metal ball at the end, and then fill in the sphere with a metal plug, as in Fig. 16.34. The ball has been previously given a charge $+Q$. What happens to the metal sphere as a result of the introduction of the positive charge into the cavity?

To answer this question we construct a spherical gaussian surface confined everywhere to the interior of the metal, as in the figure. Then we have, from Gauss' law,

$$\sum E_\perp \Delta A = 4\pi k_e \sum q$$

But the electric field inside a metal, as we have seen, is everywhere zero, and so the left side of this equation is zero. Hence the right side is also zero, and so $\sum q = 0$. Since we know we have a charge $+Q$ on the small sphere inside the gaussian surface, we must have a charge $-Q$ distributed over the inner surface of the large sphere to make the total charge within the gaussian surface zero. Hence electrons are pulled away from the outside of the metal surface, which must therefore be left positively charged, with a charge $+Q$, as shown. The effect of introducing the charge $+Q$ into the interior of the hollow sphere has then been to induce a charge $+Q$ on the outside surface of the large sphere and a charge $-Q$ on the inner surface.

This interpretation is verified by attaching the outer surface of the sphere to an electroscope. When the charge Q is introduced, the leaves of the electroscope diverge. Moving the charge Q around does not affect the electroscope deflection. Finally, if we touch the small sphere to the interior of the large sphere, there is still no change in the electroscope's response. This is because the charge $+Q$ on the ball cancels the charge $-Q$ on the interior of the large sphere, leaving a charge $+Q$ still on the outside of the sphere. Since the electroscope responds to this charge on the outer surface, and this charge does not change, there is no change in the electroscope deflection.

Hence the effect of touching the charge $+Q$ to the inside of the sphere is to transfer that charge completely to the outside of the sphere. If we repeat this process over and over again, we can transfer large amounts of charge to the sphere, and hence build up a very large voltage on the sphere with respect to ground, since the absolute potential of the sphere is proportional to its charge, according to Eq. (16.9).

This idea of transferring charge from one conductor to another by internal contact is put to good use in the *Van de Graaff generator*, which in this way builds up a high potential on a large sphere. This high potential can then be used to accelerate atomic particles.

16.8 Capacitors and Dielectrics ᴅᴹᴵᵀ

FIGURE 16.35 A Leyden jar. (*Photo courtesy of Smithsonian Institution.*)

In the early days of electricity, before batteries had been developed, a crucial piece of apparatus for electrical experiments was the Leyden jar. This is a glass jar like those used for preserves, with both inner and outer surfaces covered with metal foil, as in Fig. 16.35. A metal knob at the top is attached, through a metal shaft mounted on a nonconducting plug in the neck of the jar, to a metal chain making contact with the inside metal foil at the bottom of the jar. The Leyden jar therefore consisted of two electric conductors separated by the insulating glass of the jar. It was the first example of what we now call a *capacitor* (often called a *condenser*), which is a device used for storing electric charge. The name *Leyden jar* comes from the University of Leyden in Holland, at which the first capacitor of this kind was used for electrical experimentation.

If the knob of a Leyden jar is repeatedly touched with a negatively charged rod, the inner conducting surface will build up a large negative charge on the inside metal surface. If the outside of the jar is grounded, electrons will flow to ground and leave a positive charge on the inner surface of the outer conductor. This positive charge will be exactly equal to the negative charge on the inside of the jar. Such a device, in which we have equal amounts of positive and negative charge separated by an insulator (glass in this case) is a typical capacitor. Capacitors are very useful devices for a variety of purposes, including the tuning of radios and TV sets.

Parallel-Plate Capacitor

Let us consider a very simple type of capacitor, a parallel-plate capacitor in a vacuum. In this case we have two metal plates, one with a charge $+Q$, the other with a charge $-Q$, at a distance s apart in a vacuum, as in Fig. 16.36. If we vary the amount of charge on the two plates, we find that the voltage V_0 between the plates is directly proportional to the charge Q; that is,

$$Q \propto V_0 \qquad \text{or} \qquad Q = C_0 V_0$$

FIGURE 16.36 A parallel-plate capacitor.

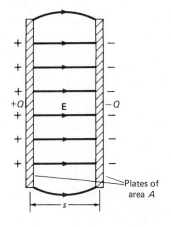

The constant of proportionality C_0 is called the *capacitance*.

Capacitance: A measure of the ability of a system to store electric charge; the ratio of the charge on a capacitor to the voltage across the capacitor.

$$C_0 = \frac{Q}{V_0} \tag{16.10}$$

In Eq. (16.10) the capacitance C_0 depends on the dimensions of the capacitor and the material between the plates (in this case a vacuum). Since $V_0 = Q/C_0$, a large C_0 indicates the ability of the system to store large charges Q without the voltage becoming too great.

Farad

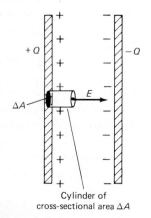

FIGURE 16.37 Gaussian surface for a parallel-plate capacitor. A cylinder of cross-sectional area ΔA, with one end inside the left plate and the other end between the plates, contains an amount of charge $\Delta Q = \sigma \Delta A$.

Since $C_0 = Q/V_0$, the units of capacitance are coulombs per volt, now called *farads* (F) after the British physicist Michael Faraday (1791–1867). Hence 1 F = 1 C/V.

To find the relationship of the capacitance to the size and material of the capacitor, we again resort to Gauss' law. Since the electric field between the plates is uniform, we have only to find the field strength at any one point, for it will be the same at all points between the plates. To find the field at one point we construct a cylindrical gaussian surface of cross-sectional area ΔA which starts within the left plate and ends at some point between the plates, as in Fig. 16.37. Since there is no electric field within the metal plate, and since $E_\perp \Delta A$ vanishes at the curved sides of the cylinder, the only place $E_\perp \Delta A$ is not zero is at the right end of the cylinder. Hence we have

$$\sum E_\perp \Delta A = E \Delta A = 4\pi k_e \sum q$$

where E is the electric field strength between the plates. But the charge within the gaussian cylinder is $\sigma \Delta A$, where σ is the surface density of charge (in coulombs per square meter) on the inside of the left plate. Hence

$$E \Delta A = 4\pi k_e \sigma \Delta A \qquad \text{or} \qquad E = 4\pi k_e \sigma$$

If we replace the constant $4\pi k_e$ by a new constant $1/\epsilon_0$, we have

$$E = \frac{\sigma}{\epsilon_0} \tag{16.11}$$

Here the constant ϵ_0 is called the *permittivity* of a vacuum. Since Q is the total charge on the plate of area A, σ must equal Q/A, and so

$$E = \frac{Q}{\epsilon_0 A}$$

Also, since the field between the plates is uniform, $V_0 = Es$, from Eq. (16.8). On substituting this value for V_0 we obtain

$$C_0 = \frac{Q}{V_0} = \frac{Q}{Es} = \frac{Q}{Qs/\epsilon_0 A} = \frac{\epsilon_0 A}{s} \tag{16.12}$$

The capacitance of a parallel-plate capacitor is therefore directly proportional to the area of the plates and inversely proportional to the distance between the plates. In SI units A must be in square meters and s in meters to obtain C_0 in farads. Note that C_0 is completely independent of the charge on the capacitor and depends only on the "hardware," i.e., on the size and shape of the device.

The constant ϵ_0, the permittivity of a vacuum (sometimes called the *permittivity of free space*), has a value:

$$\epsilon_0 = \frac{1}{4\pi k_e} = \frac{1}{4\pi(9.0 \times 10^9 \text{ N·m}^2/\text{C}^2)} = 8.85 \times 10^{-12} \frac{\text{C}^2}{\text{N·m}^2}$$

Since $\epsilon_0 = C_0 s/A$, equivalent units for the permittivity are farads per meter. Hence the value of ϵ_0 in tables is usually given as $\epsilon_0 = 8.85 \times 10^{-12}$ F/m.

Dielectrics

Suppose we have the parallel-plate capacitor of Fig. 16.36 with its plates charged but disconnected from any battery or other source of charge. We first measure the voltage between the plates with an electrometer or other kind of voltmeter. Then we fill the region between the plates with a nonconducting material, usually called a *dielectric*. The voltage is observed to decrease, although the charge on the plates remains exactly what it was before. The effect of the dielectric has been to *increase* the capacitance of the capacitor over its value for a vacuum. Whereas in a vacuum we had $C_0 = Q/V_0$, we now have a new capacitance

$$C = \frac{Q}{V} = KC_0 \tag{16.13}$$

where $K = C/C_0$, *the ratio of the capacitance with the dielectric in place to the capacitance in a vacuum, is called the dielectric constant of the material*. The process just outlined gives us an operational way of defining exactly what we mean by the dielectric constant. For a vacuum $K = 1$, and for air it is very close to 1. Values for some common dielectrics are given in Table 16.1.

The capacitance of our new capacitor with dielectric inserted is then

$$C = KC_0 = \frac{K\epsilon_0 A}{s} = \frac{\epsilon A}{s} \tag{16.14}$$

where ϵ is called the *permittivity of the dielectric*, and is equal to $K\epsilon_0$. Since K is always greater than 1, the presence of the dielectric *increases* the capacitance.

We see that *the effect of a dielectric is to increase the amount of charge that can be stored in a capacitor without increasing the capacitor's voltage*, since $C = Q/V$. This is the reason the Leyden jar worked so well as a charge-storage device. The glass in the jar is thin and has a dielectric constant of about 5; hence it is a good way to store charge.

In addition to increasing capacitance, dielectrics also play a very practical role in capacitors by making it possible to bring the plates very close together without touching, thus increasing the capacitance, and by making it more difficult for high voltages to produce ionization of the air and an electric discharge from one plate to the other.

Practical capacitors often consist of sheets of metal foil separated by thin sheets of plastic or other dielectric material. These can be rolled into cylinders which have large surface areas A, with plates very close together so that s is small. They therefore have large capacitances, although they usually remain in the microfarad (μF) range. The tuning capacitors on radios vary the area of the capacitor plates by allowing one set of plates to be moved with respect to a second, stationary set of plates, as in Fig. 16.38. In this way the capacitance

TABLE 16.1 Dielectric Constants of Various Materials*

Material	Dielectric constant $K = \epsilon/\epsilon_0$
Vacuum	1.00000
Air	1.00059
Water	80
Paper	3.5
Porcelain	6.5
Bakelite	4.8
Quartz	4.3
Pyrex glass	4.5
Amber	2.7
Polystyrene	2.6
Teflon	2.1
Transformer oil	2.2
Titanium dioxide	~100
Barium titanate	~10,000

*For 20°C and unvarying or very low frequency electric fields.

FIGURE 16.38 A tuning capacitor for a radio.

can be varied and the radio tuned. Capacitors in radios usually run from the microfarad range (10^{-6} F) to the picofarad range (10^{-12} F).

Example 16.9

A parallel-plate capacitor has a plate area of 50 cm² and a plate separation of 1.0 cm. A potential difference $V_0 = 200$ V is applied across the plates with no dielectric present. The battery is then disconnected and a piece of bakelite inserted which fills the complete volume between the plates.

What is (a) the capacitance, (b) the charge on the plates, and (c) the voltage across the plates both before and after the dielectric is inserted?

SOLUTION

(a) Before the dielectric is inserted, air fills the space between the plates. Since the dielectric constant of air is very close to unity, we have for the capacitance before the bakelite is inserted

$$C_0 = \frac{\epsilon_0 A}{s} = \frac{(8.85 \times 10^{-12} \text{ C}^2/\text{N·m}^2)(50 \times 10^{-4} \text{ m}^2)}{0.010 \text{ m}}$$

$$= 4.4 \times 10^{-12} \text{ F} = \boxed{4.4 \text{ pF}}$$

After the dielectric is inserted, using the value $K = 4.8$ from Table 16.1, we have

$$C = \frac{\epsilon A}{s} = \frac{K\epsilon_0 A}{s} = KC_0 = 4.8(4.4 \text{ pF}) = \boxed{21 \text{ pF}}$$

(b) Before the dielectric is inserted,

$$Q_0 = C_0 V_0 = (4.4 \times 10^{-12} \text{ F})(200 \text{ V}) = 880 \times 10^{-12} \text{ C}$$

$$= \boxed{8.8 \times 10^{-10} \text{ C}}$$

After the dielectric is inserted the charge *must remain the same*, since the battery is disconnected and the charge cannot leave the plates. Hence

$$Q = Q_0 = \boxed{8.8 \times 10^{-10} \text{ C}}$$

(c) The voltage before the dielectric is inserted is given as

$$V_0 = \boxed{200 \text{ V}}$$

The voltage after the bakelite is inserted is then

$$V = \frac{Q}{C} = \frac{8.8 \times 10^{-10} \text{ C}}{21 \times 10^{-12} \text{ F}} = \boxed{42 \text{ V}}$$

Hence the effect of the dielectric in this case is to increase the capacitance and decrease the voltage, while the charge remains unchanged.

Summary: Important Definitions and Equations

Electric charge:
The two different kinds of electric charge are called positive ($+$) and negative ($-$). Charges of like sign repel each other and charges of unlike sign attract each other.

Principle of conservation of charge:
In any closed system the algebraic sum of all the electric charges present remains constant.

Coulomb's law:

$$f_{el} = \frac{k_e q q'}{r^2} \qquad \text{where } k_e = 9.0 \times 10^9 \text{ N·m}^2/\text{C}^2$$

Coulomb (the unit of electric charge):
That charge which, when placed one meter away from an equal and like charge, repels it with a force of 9.0×10^9 N.

Elementary charge:
The charge on the electron, equal to 1.602×10^{-19} C.

Electric field strength:
The vector force $\mathbf{f}_{el}$ experienced by a test charge q' at some position in space, divided by the magnitude of the test charge.

$$\mathbf{E} = \frac{\mathbf{f}_{el}}{q'}$$

Gauss' law: $\sum E_\perp \Delta A = 4\pi k_e \sum q$

where $E_\perp$ is the component of the electric field perpendicular to the element of surface ΔA, and the summations are taken over the surface area A of the gaussian surface and over the total charge within that surface.

Electric potential (V):
The potential energy per unit charge of a charge q' at a point A in an electric field:

$$V_A = \frac{PE_A}{q'}$$

Voltage:
The potential difference (or potential energy difference per unit charge) between two points in an electric field. If $\mathcal{W}$ is the work done by the field in moving a charge q from A to B in the field, then the voltage between A and B is

$$V_{AB} = V_A - V_B = \frac{PE_A}{q} - \frac{PE_B}{q} = \frac{\mathcal{W}}{q}$$

Volt (unit of potential difference or voltage):
If 1 J of work is done to move 1 C of charge from one point to another in an electric field, then the voltage between these two points is 1 V, that is, $1 \text{ V} = 1 \text{ J/ C}$.

Electronvolt:
An amount of energy equal to 1.6×10^{-19} J.

Equipotential surfaces:
Imaginary surfaces in space which are at every point perpendicular to the electric field, and all points of which are therefore at the same potential.

Absolute potential:
The difference between the electric potential of a charged particle at some point in an electric field and its value at infinity. For the field of a point charge q:

$$V_a = \frac{k_e q}{r_a}$$

Capacitor:
A device for storing electric charge, consisting of two charged metal plates separated by an insulator, or dielectric.

Capacitance:
The ratio of the charge on a capacitor to the voltage across its plates. The unit of capacitance is the farad (F), where $1 \text{ F} = 1 \text{ C/V}$.

Field between the plates of a parallel-plate capacitor of charge Q and plate area A:

$$E = \frac{Q}{\epsilon_0 A}$$

where $\epsilon_0 = 1/4\pi k_e$ is the permittivity of a vacuum (or of free space) and is equal to 8.85×10^{-12} F/m.

Capacitance of a parallel-plate capacitor:

$$\text{In vacuum: } C_0 = \frac{\epsilon_0 A}{s}$$

$$\text{With dielectric: } C = \frac{\epsilon A}{s}$$

Dielectric constant:
The ratio of the permittivity of a material to the permittivity of a vacuum:

$$K = \frac{\epsilon}{\epsilon_0}$$

Questions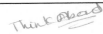

1   Explain why distilled water is a very poor conductor of electricity, but if ordinary table salt (NaCl) is poured into the water, it becomes a good conductor.

2   Suggest some ways in which the electrostatic field resembles the gravitational field, and some ways in which they differ.

3   Draw the lines of electric force surrounding two small spherical charges, each of magnitude $+Q$, separated by a distance R which is much larger than the radius of the charged spheres.

4   How could you test to see if an electric field existed at some given point in space?

5   Is it possible for two lines of an electric field to cross each other? Why?

6   Can two different equipotential surfaces intersect? Why?

7   If you comb your hair repeatedly on a dry winter day, the comb becomes electrically charged. Indicate how you would determine whether this charge is positive or negative. What would be the resultant charge on your hair?

8   Why do woolen socks coming out of a clothes dryer tend to stick together?

9   Why do electrostatic experiments not work well on damp, humid days?

10   An uncharged cork sphere at the end of a string is attracted by a charged amber rod, collides with the rod, and then jumps violently away from it. From that time on the sphere is repelled by the amber rod. Explain exactly what has happened.

11   Is the surface of the earth an equipotential surface for the gravitational force? Is this exactly true, or is it an approximation of the truth?

12   A physicist decides to call the potential of the earth $+1000$ V instead of zero. Is this possible without invalidating the basic laws of electrostatics? What effect would such an assumption have on measured values of absolute potentials and of potential differences?

13   If a capacitor is connected directly across a battery, does each of the two plates of the capacitor receive exactly the same charge? Why? Is this true even if the plates are of greatly different sizes?

Problems

A 1 The ratio of the electric force between two protons at rest to the gravitational force between them is about:
(a) 10^{42} (b) 10^{39} (c) 10^{36} (d) 10^{45}
(e) 10^{48}

A 2 The electrostatic force between a proton and an electron separated by a distance of 10^{-10} m is:
(a) 2.3×10^{-8} N (b) 2.3×10^{-4} N
(c) 2.3×10^{8} N (d) 2.3×10^{4} N (e) 2.3 N

A 3 The electric field intensity at a point 0.30 m to the right of a charge $Q = 3.0 \times 10^{-4}$ C is (where a minus sign on e means that the field is directed toward Q, a plus sign means away from Q):
(a) -3.0×10^{7} N/C (b) $+3.0 \times 10^{7}$ N/C
(c) -3.0×10^{7} N (d) $+3.0 \times 10^{7}$ N
(e) -3.0×10^{7} J

A 4 The radius of a gold nucleus is about 6.6×10^{-15} m, and the nucleus contains 79 protons. The electric potential at the surface of the gold nucleus is:
(a) 1.7×10^{7} J (b) 1.7×10^{7} C
(c) 2.6×10^{21} V (d) 1.7×10^{7} V
(e) 1.9×10^{-3} V

A 5 To increase the capacitance of a parallel-plate capacitor by two orders of magnitude, the diameter of its round plates must be increased by
(a) One order of magnitude
(b) Two orders of magnitude
(c) Four orders of magnitude
(d) Three orders of magnitude
(e) None of the above

A 6 A long, straight wire carries a static electric charge of 10^{-4} C/m. The electric field at a distance 2.0 cm from the wire is:
(a) 9.0×10^{7} N/C (b) 4.5×10^{9} J/C
(c) 9.0×10^{9} N/C (d) 9.0×10^{7} N/m
(e) 4.5×10^{9} N/C

A 7 Two equipotential surfaces between the charged parallel plates of a capacitor correspond to the values $V_A = 30$ V and $V_B = 20$ V, respectively. The distance between these equipotential surfaces is 1.0 cm. The electric field E_{AB} in the region between the plates has a value:
(a) $+25$ V/M (b) $+10^{3}$ V/m (c) -10^{3} V
(d) $+10^{3}$ V (e) -10^{3} V/m

A 8 A voltage of 120 V drives 10 C of electric charge through an electric heater each second. The work done by the voltage source in 1 s is:
(a) 1.2×10^{3} J (b) 1.2×10^{3} N (c) 12 J
(d) 12 N (e) 1.2×10^{4} J

A 9 The capacitance of an air capacitor with plates of area 5.0 cm² separated by a distance of 1 mm is:
(a) 4.4 pF (b) 4.4 μF (c) 4.4×10^{-10} F
(d) 1.8 pF (e) 1.8 μF

A10 A capacitor of capacitance 100 pF is charged with a voltage source of 5000 V. The energy stored in the capacitor is:
(a) 2.5×10^{-7} J (b) 1.25×10^{-3} V
(c) 2.5×10^{-3} J (d) 1.25×10^{3} J
(e) 1.25×10^{-3} J

B 1 Two point charges, one of $+20$ μC, the other of -10 μC, are separated by a distance of 10 cm in air.
(a) What is the magnitude and direction of the force on the 20-μC charge?
(b) If this positive charge is on a small plastic sphere of mass 1.0 g, what is the acceleration of this sphere caused by the electrostatic force?

B 2 A point charge of 100 μC repels another point charge with a force of 2.0 N when placed 1.0 m away from the second charge. Find the sign and magnitude of the second charge.

B 3 What are the magnitude and direction of the electric field **E**:
(a) At a distance of 20 cm from a positive point charge of magnitude 30 μC?
(b) At a distance of 20 cm from a negative point charge of magnitude 30 μC?

B 4 What are the magnitude and direction of the electric field at a distance 1.0 m from a long, thin copper wire containing a charge of 10 μC per meter of length?

B 5 (a) If 10 J of work is done by a person to move 0.050 C of positive charge from point A to point B in an electrostatic field, what is the difference in potential between points A and B?
(b) Which point, A or B, is at the higher potential?

B 6 The positive terminal of a "dry-cell" battery is $+1.5$ V with respect to the negative terminal. The two terminals are separated by a distance of 4.0 cm. What are the magnitude and direction of the electric field along a straight line connecting the two terminals?

B 7 How many joules are there in 1 MeV?

B 8 A radio capacitor of capacitance 20 pF is charged by a voltage of 400 V. What is the charge on the capacitor?

B 9 What is the capacitance of a parallal-plate capacitor with air between the plates, a plate area equal to 0.010 m², and a plate separation of 1.0 mm?

B10 (a) What is the capacitance of the moon?
(b) What would the voltage of the moon be if it were charged to 100 μC?

B11 Prove that 1 F = 1 C²/(N·m).

C 1 Three positive charges are at the corners of an equilateral triangle of side 20 cm: $q_A = 50$ μC, $q_B = 25$ μC, and $q_C = 100$ μC
(a) What is the magnitude of the resultant force on q_C?
(b) What angle does this force make with respect to the line joining q_A and q_B?

C 2 Consider the problem of Example 16.3, which is diagramed in Fig. 16.18. Solve this problem for the magnitude and direction of the forces on (a) the charge q_A; (b) the charge q_B.

C 3 For the three charges in Fig. 16.39 calculate the electric field E:
(a) At the charge q_B.
(b) At a point 5.0 cm above the charge q_B and equidistant from the charges q_A and q_C.

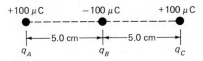

FIGURE 16.39 Diagram for Prob. C3.

C 4 (a) Determine the electric field midway between two positive charges of 10 μC each, if the charges are 5.0 cm apart and the charges are immersed in transformer oil of dielectric constant 2.2.

(b) What is the electric field if one charge is positive and the other negative?

C 5 A square $ABCD$ has sides 20 cm in length. Positive charges of 50 μC are placed at corners A and B, and negative charges of 50 μC are placed at corners C and D. Calculate the magnitude and direction of the electric field at the center of the square.

C 6 Two charged spheres, one of charge $+100$ μC and the other of charge -100 μC, are held in fixed positions with their centers separated by 20 cm.

(a) What is the electric potential energy of the -100-μC charge?

(b) What is the absolute electric potential of the -100-μC charge?

C 7 Two point charges, $q_1 = +2.0$ μC and $q_2 = -5.0$ μC, are separated by a distance of 2.0 m.

(a) At what point on the line joining the charges, or on its extension, is the electric field zero?

(b) At what point on the same line is the absolute electric potential zero?

(c) If both charges were positive, would there be a point on the line joining the charges where the electric field would be zero? Where would it be located?

(d) If both charges were positive, would there be a point where the absolute electric potential would be zero?

C 8 In a modern version of Millikan's experiment, polystyrene spheres of mass 1.60×10^{-14} kg were used. It was found that a voltage of 4000 V across two horizontal plates 4.00 mm apart was able to balance out the force of gravity on a sphere. If we assume that the sphere contained one electronic charge, what is the value for the electronic charge?

C 9 In a Cottrell electrostatic precipitator used to remove soot particles from a factory chimney, two electrodes are 1.0 cm apart and the voltage between them is 10^5 V.

(a) If the net charge on each soot particle is about 200 elementary charges, how large is the Coulomb force on each soot particle?

(b) If each soot particle has a mass of about 10^{-5} kg, what is the acceleration of each particle?

C10 (a) How much energy is required to separate completely the two ions in the potassium iodide molecule, K^+I^-, where the ions are initially 3.0×10^{-10} m apart in air, and the K ion has one net positive charge and the I ion has one net negative charge?

(b) How would this energy differ if the potassium iodide were immersed in water?

C11 In Prob. C5, what is the absolute potential at the center of the square?

C12 Two charges of $+6.0$ μC and -5.0 μC are 80 cm apart.

(a) Find the absolute potential at a point midway between them.

(b) How much work is required to carry a $+10.0$-C charge from infinity to that point?

C13 An electron is located midway between two parallel metal plates 2.0 cm apart. One of the plates is at a potential 300 V higher than the other.

(a) What is the potential gradient between the plates?

(b) What is the magnitude of the force on the electron?

(c) What is the direction of the force on the electron?

C14 A Van de Graaff accelerator which produces voltages of 10^6 V is used to accelerate a beam of protons.

(a) What is the kinetic energy of the protons after they have been accelerated by this potential difference?

(b) What is their speed?

C15 The plates of a parallel-plate capacitor are of area 0.50 m² and are 4.0 mm apart. If a voltage of 10^5 V is applied across the capacitor, which has air between the plates:

(a) What is the capacitance?

(b) What is the charge on each plate?

(c) What is the electric field between the plates?

C16 A parallel-plate capacitor has two plates, each of area 0.030 m², separated by a 0.30-cm air gap.

(a) Find the capacitance of this capacitor.

(b) If the capacitor is connected across a 600-V source, find the charge on its plates.

(c) Find the electric field $\mathbf{E}$ between the plates.

C17 (a) Calculate the capacitance of a parallel-plate capacitor with a 0.50-cm-thick layer of bakelite between the plates, each of which has an area of 0.010 m².

(b) If this capacitor is connected to a 600-V source, calculate the charge on each plate of the capacitor.

C18 A parallel-plate capacitor has a plate area of 100 cm² and a plate separation of 4 mm. A potential difference of 1000 V is applied across the plates with only air between the plates. The battery is then disconnected, and a piece of Pyrex glass inserted to fill completely the space between the plates. Both before and after the dielectric is inserted:

(a) What is the capacitance?

(b) What is the charge on the plates?

(c) What is the voltage across the plates?

C19 In the Bohr theory of the hydrogen atom an electron circles the hydrogen nucleus (a proton) in an orbit of radius 0.53×10^{-10} m. The electrostatic attraction of the proton for the electron furnishes the centripetal force needed to hold the electron in its circular orbit. Find:

(a) The force of electrical attraction between the proton and the electron.

(b) The speed of the electron.

C20 Calculate the ratio of the electric force between two electrons to the gravitational force between them.

D 1 Suppose that all the electrons in a gram of carbon ($^{12}_{6}$C) could be moved to a position 1.0 m away from the carbon nuclei. What would be the force of attraction between the electrons and the nuclei?

D 2 Two metal spheres, one of radius 10 cm and charge $+2.0$ μC, and the other of radius 20 cm and charge $+10.0$ μC, are initially some distance apart. If the two spheres are brought together and allowed to touch, what will be the final charge on each sphere?

D 3 Two small conducting balls, each of mass 1.0 g, hang from a stand on parallel insulating threads of length 75 cm, so that the two balls just touch. A positively charged rod then touches the two balls and gives them equal positive charges. They repel each other and take up new positions in which each hangs at an angle of 30° with respect to the vertical. What is the charge on each ball?

D 4 A $+10.0$-μC charge is placed 50 cm away from a -2.5-μC charge.

(*a*) Where can a third charge (of either sign) be placed so that it experiences no net force?

(*b*) Does this point correspond to stable or unstable equilibrium?

D 5 The sphere of a Van de Graaff generator is to be charged to a potential of 10^6 V. The dielectric strength of air is 3.0×10^6 V/m, which means that an electric field greater than 3.0×10^6 V/m will cause the air to ionize and discharge the sphere. What is the minimum radius the sphere must have to retain its charge?

D 6 A proton is released from rest in a uniform electric field and moves a distance of 1 m in 1.0×10^{-3} s.

(*a*) What is the electric field strength?

(*b*) What is the voltage between the two points?

D 7 An alpha particle has a charge $+2e$, where e is the electronic charge.

(*a*) Find the electric flux through a sphere of radius 2.0×10^{-10} m surrounding the alpha particle.

(*b*) What is the electric flux through a sphere of radius 2.0×10^{-9} m surrounding the alpha particle?

D 8 Positive charge is uniformly distributed throughout a spherical volume of radius R, the volume charge density being ρ in units of C/m³.

(*a*) Use Gauss' law to prove that the electric field inside the volume, at a distance r from the center, is $E = \rho r/3\epsilon_0$.

(*b*) Prove that the electric field at the surface of the sphere is the same as that produced by a point charge at the center of the sphere with the same total charge as the sphere.

D 9 A 10.0-μF capacitor is charged to 100 V and a 5.0-μF capacitor is charged to 150 V. The capacitors are then disconnected from the voltage source. The positive plates are connected together, and the negative plates are also connected together.

(*a*) What is the voltage across each capacitor?

(*b*) What is the charge on each capacitor?

D10 A parallel-plate capacitor with plates separated by air receives 10 μC of charge when connected to a 300-V battery. The plates, still connected to the battery, are then immersed in transformer oil. How much additional charge must flow from the battery to the plates of the capacitor?

D11 A spherical capacitor consists of an inner metal sphere of radius R_a supported on an insulating stand at the center of a hollow metal sphere of inner radius R_b. The inner sphere has a charge $+q$ and the outer sphere a charge $-q$. Prove that the capacitance of the two spheres is:

$$C = \frac{1}{k_e} \frac{R_b R_a}{R_b - R_a}$$

Additional Readings

Cohen, I. Bernard: "Benjamin Franklin," in *Lives in Science*, a *Scientific American* book, Simon and Schuster, New York, 1957. An excellent brief article by the man who has done much to restore Franklin to his rightful place in the history of physics. Cohen sees Franklin as "a major transitional figure between the physics of Newton and the physics of Faraday and Clerk Maxwell."

————: *Benjamin Franklin, Scientist and Statesman*. Scribner, New York, 1975. A well-illustrated account based on Cohen's article on Franklin in the *Dictionary of Scientific Biography*.

Fletcher, Harvey: "My Work with Millikan on the Oil-Drop Experiment," *Physics Today*, vol. 35, no. 6, June 1982, pp. 43–47. An interesting article suggesting that many of the best ideas in Millikan's work were provided by Fletcher when he was a graduate student.

Heilbron, J. L.: "Franklin's Physics," *Physics Today*, vol. 29, no. 7, July 1976, pp. 32–37. A short, well-illustrated discussion of Franklin's contributions to physics.

Moore, A. D.: *Electrostatics: Exploring, Controlling and Using Static Electricity*, Doubleday Anchor, Garden City, N.Y., 1968. A fascinating account of practical electrostatics by an electrical engineer who has devoted his life to the subject.

Roller, Duane, and Duane H. D. Roller: *The Development of the Concept of Electric Charge*, Harvard University Press, Cambridge, Mass., 1954. This is one of the best of the Harvard Case Studies in Experimental Science, and considers the development of electricity from the Greeks to Coulomb.

Shamos, Morris H.: *Great Experiments in Physics*, Holt, Rinehart and Winston, New York, 1959. Coulomb's work is described in his own words in Chap. 5 of this very useful book.

In going on with these experiments, how many pretty systems do we build, which we soon find ourselves obliged to destroy! If there is no other use discovered of electricity, this, however, is something considerable, that it may help to make a vain man humble.

Benjamin Franklin (1706–1790)

Chapter 17

Direct Current Electricity

The phenomena of electrostatics are interesting and basic to our understanding of the physical universe, but electric currents are more important for practical purposes. (Important exceptions are the Xerox dry copying process and the Cottrell dust precipitator, which depend completely on electrostatics.) To power our machines and computers, to heat our toasters and stoves, charges at rest are of little value. Instead, we need a continuous flow of electric charge, or an electric current, to accomplish such tasks. In this chapter we begin with a discussion of direct currents (dc). In practice most electrical machines run on alternating current (ac), but many of the ideas developed for direct current will be applicable to alternating current, which we will discuss in Chap. 20.

17.1 Direct Currents and EMF's

We can define *electric current* and *direct current* as follows:

Definitions

Electric current is the rate of flow of electric charge.
A direct current (dc) is the rate of flow of electric charge that is always in the same direction in the wire carrying the charge

Suppose, for example, that we have a metal wire in which electric charge is flowing, as in Fig. 17.1. Consider a point B along the wire. The magnitude of the direct current I is defined to be the amount of charge in coulombs which passes point B in one second. Thus, if an amount of charge Q passes point B in t s, then by definition the current I is

$$I = \frac{Q}{t} \qquad (17.1)$$

In the SI system of units a current of *one coulomb per second* is called an *ampere* (A). Hence

Definition of Ampere

$$1 \text{ A} = 1 \text{ C/s}$$

The direction of the current is taken to be the direction in which a *positive* charge would flow if free to move around a closed loop, or *electric*

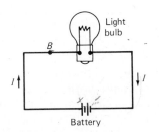

FIGURE 17.1 An electric current *I* in a wire. The current is the amount of charge (in coulombs) passing point *B* in the wire in 1 s.

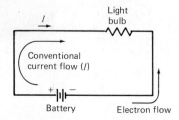

FIGURE 17.2 The conventional current in a wire. Its direction is that in which a positive charge would flow, and is therefore in the opposite direction to that of the electrons which actually carry the current.

Batteries

Definition

circuit. The reason for this choice of direction is that we have defined the electric field in terms of the movement of positive charges. Hence for the electric current to be in the direction of the electric field in a wire, we must take the direction of the current as that in which a positive charge would flow. This is only a "convention," or arbitrary decision, and for this reason such a current is sometimes called the *conventional current.* In metals all the current is carried by electrons moving in a direction opposite to that of the conventional current. This should cause no difficulty since, from an electrical point of view, a positive charge moving in one direction is completely equivalent to a negative charge moving in the opposite direction. In this book, when we speak of the direction of a current we will always mean the direction of the conventional current, as in Fig. 17.2.

One of the most useful sources of direct current is the chemical *battery.*

A battery is a device which converts chemical energy into electric energy on a continuous basis, and which tries to maintain a constant potential difference across its terminals.

Chemical reactions go on inside a battery and separate the positive from the negative charges. Electrons are removed from the positive terminal of the battery and deposited at the negative terminal. This process proceeds until the electrostatic potential energy of the separated charges is exactly equal to the potential-energy difference produced by the chemical reactions taking place in the battery. When this state is reached, no further separation of charge occurs. If, however, the two terminals of the battery are connected to a light bulb or other electrical device, charge flows through the light bulb and depletes the excess positive and negative charges built up on the battery terminals. Chemical reactions then resume in the battery in order to maintain the charge separation at its original value and hence a constant potential difference across its terminals. This allows a current to be drawn from the battery on a continuous basis.

EMF of a Battery

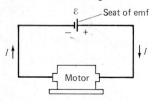

FIGURE 17.3 A seat of emf (a battery in this case) delivering current to an electric motor.

Definition

In a battery a positive charge has a higher potential energy at the positive terminal than it would have at the negative terminal. This difference in potential energy is exactly equal to the work done by the molecular forces which move the positive charge from the negative to the positive terminal inside the battery. Hence if the two terminals are connected to an electric motor, as in Fig. 17.3, a current will flow in the external circuit from the + to the − terminal of the battery and drive the motor, which can then do useful work. This work has its ultimate source in the chemical energy provided by the battery.

A device like a battery which transforms nonelectric energy into electric energy is called a seat of emf.

The numerical value of an emf is denoted by $\mathcal{E}$ (*not E*, which we will continue to use for the magnitude of the electric field strength), and is measured in terms of the work done per unit charge, or in units of joules per coulomb, or volts. Thus in a 1.5-V dry cell, each coulomb of charge which flows through the battery gains 1.5 J of electric potential energy at the expense of the chemical energy in the cell.

Since, by definition, $\mathcal{E} = W/Q$, the work done by a battery in moving a charge Q through the battery is $W = Q\mathcal{E}$. The value of the emf $\mathcal{E}$ is determined by the chemical processes going on inside the battery, and has a different numerical value for each different type of battery.

The concept of emf is similar to that of voltage, and both have the same dimensions and units (usually volts), but the term *emf* is applied only to parts of a circuit in which forms of energy other than electric energy are continuously converted into electric energy (or vice versa).

Practical Batteries

There are many varieties of batteries in common use. The most popular are the *dry cells* used in flashlights, automobile storage batteries, and the small batteries used in cameras, hand calculators, and watches. Storage batteries and some of the newer batteries have the advantage over dry cells that they can be recharged, where *recharged* means that after some of the chemical energy in the battery has been converted into electric energy, an electric current can be forced through the battery in the reverse direction to convert electric energy back into chemical energy. In an automobile both charging and discharging of the car's storage batteries go on automatically while the car is running. Figure 17.4 shows several kinds of commonly used batteries.*

In electric circuits a battery is designated by the symbol ⊣⊢ , where the long vertical line always represents the positive terminal of the battery. Sometimes a repeated symbol like ⊣⊦⊢ is used to indicate that the battery contains more than one simple cell. If the cells are connected *in series*, i.e., one after the other in a line, with the positive terminal of one cell connected

---

*An amazing feature of batteries is that they depend on built-in potential differences that nature freely provides. These differences can be compared to the natural differences in gravitational potential energy that can be used to generate hydroelectric power.

FIGURE 17.4 Some practical batteries. (*Photo courtesy of Union Carbide Corporation.*)

to the negative terminal of the next cell, then the voltages of the two cells add and the cells are said to be *in series aiding*. Thus two 1.5-V dry cells connected in this way yield a 3.0-V battery, as in Fig. 17.5. If two cells are connected in series but with two positive (or two negative) terminals together, then the cells are said to be *in series opposing* and the effective emf is the difference of the two individual emf's, as in Fig. 17.6.

FIGURE 17.5 Two batteries connected in a series-aiding circuit (in the same direction).

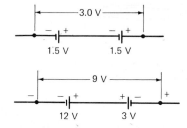

FIGURE 17.6 Two batteries connected in a series-opposing circuit (in opposite directions).

Example 17.1

A dry cell with an emf of 1.5 V is connected to the two flat plates of a capacitor inside a vacuum chamber, as in Fig. 17.7.
(a) How much mechanical energy would a proton gain in moving from plate D to plate C?
(b) How much mechanical energy would an electron gain in moving from plate C to plate D?
(c) Where does this mechanical energy come from?

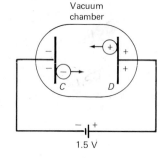

FIGURE 17.7 Circuit for Example 17.1.

SOLUTION

(a) Here plate D is kept at a potential 1.5 V higher than plate C by the emf of the battery. Hence the electric potential energy of a proton, with charge $Q = +1.6 \times 10^{-19}$ C, is higher at D than at C by an amount

$$\Delta PE = QV_D - QV_C = Q(V_D - V_C) = (1.6 \times 10^{-19} \text{ C})(1.5 \text{ V})$$

$$= \boxed{2.4 \times 10^{-19} \text{ J}}$$

This electric energy is converted into mechanical energy when the proton moves from D to C. Hence the proton gains 2.4×10^{-19} J of mechanical energy. Its velocity when it strikes plate C can be obtained from the fact that $\frac{1}{2}mv^2 = 2.4 \times 10^{-19}$ J, if there is no resistive force opposing the proton's motion.

(b) An electron will gain just as much mechanical (kinetic) energy in moving from C to D as a proton does in moving from D to C, since the charge on the electron is negative and of the same magnitude as the proton's charge. Its kinetic energy will also be the same, although its speed will be quite different because of the difference between the two masses. (What will be the difference in the final velocities of the two particles?)

(c) Ultimately this kinetic energy comes from the chemical energy in the battery.

17.2 Electric Currents in Electrolytes

Distilled water, like many other pure liquids, is a poor electric conductor, for there are no free electrons or ions to move through the liquid. If, however, some ordinary table salt is added, the water becomes a good conductor. Such solutions of salts in water are termed *electrolytes*. If a battery is connected to two electrodes immersed in the salt solution, as in Fig. 17.8, the Na^+ ions will migrate to the negative terminal of the battery, and the Cl^- ions will migrate

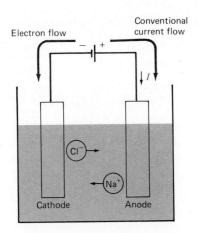

FIGURE 17.8 An electrolytic cell containing salt (NaCl) in solution. Inside the cell the current is carried both by positive Na ions and negative Cl ions. Outside the cell the current is carried by electrons in the wire.

to its positive terminal, thus carrying an electric current. In this case the current is carried by both positive and negative charges moving in opposite directions through the liquid, although in the wire connected to the battery only electrons carry the current.

Faraday's Laws of Electrolysis

In 1833 Michael Faraday (see biography, Chap. 19) showed that the study of currents in electrolytes can cast much light on the ultimate structure of matter and electricity. Faraday found in his experiments on what we now call *electrolysis* that the passage of a current through a NaCl solution liberates chlorine gas at the anode and deposits solid sodium on the cathode. Careful experimental study of this effect led Faraday to the following conclusions:

Faraday's Laws

1 *The mass of substance deposited or liberated from a solution undergoing electrolysis is directly proportional to the amount of charge flowing through the solution.*

2 *One faraday of charge ($Q_F = 96,500$ C) deposits or liberates an amount of substance equal to the mass of one mole of the substance divided by its valence.*

Atomicity of Charge

Faraday's laws are susceptible to a very simple atomic interpretation. For monovalent substances like Na and Cl each ion in solution carries a net electric charge of either plus or minus one elementary charge. Every time a Cl^- ion reaches the anode, it gives up its excess electron to the anode, becomes a neutral atom, and bubbles up to the surface as a gas. Similarly, a Na^+ ion picks up an electron from the cathode and attaches itself to the cathode as neutral sodium metal. Suppose, for purposes of illustration, that the electron picked up by the Na^+ is the same one deposited by the Cl^- at the anode. For each molecule of NaCl coming out of solution, one electron has therefore moved through the circuit outside the electrolytic cell. The more electrons that flow, the larger the current in the circuit and the more Na and Cl liberated.

If this atomic interpretation is correct, then 96,500 C corresponds to a number of electrons equal to Avogadro's number (N_A) flowing in the circuit, for this amount of charge releases 1 mol of Na^+ and 1 mol of Cl^- or 1 mol of NaCl in all. Corresponding to each molecule released there is one electron flowing through the external circuit, as we have just seen. Hence we can write:

$$Q_F = 96,500 \text{ C} = N_A e \qquad (17.2)$$

where Q_F is the faraday, e is the charge on the electron, and N_A is Avogadro's number. We therefore obtain

$$e = \frac{96,500 \text{ C/mol}}{(6.03 \times 10^{23})/\text{mol}} = 1.60 \times 10^{-19} \text{ C}$$

If we assume that Avogadro's number is known, electrolysis experiments therefore yield a value for the charge on the electron in good agreement with the one found by Millikan in the oil-drop experiment discussed in the preceding chapter. Hence we have a physical picture, or a *model*, of atomic structure, in which the pieces fit together extremely well, not merely qualitatively but also quantitatively.

Example 17.2

A current of 15 A is maintained for 30 min through an electrolytic cell containing $CuSO_4$. The salt breaks up into Cu^{2+} and SO_4^{2-} in solution. How much copper is deposited at the cathode?

SOLUTION

The amount of charge flowing through the cell is

$$Q = It = (15 \text{ A})(30 \text{ min}) = \left(15\frac{\text{C}}{\text{s}}\right)(30 \text{ min})\left(60\frac{\text{s}}{\text{min}}\right)$$

$$= 2.7 \times 10^4 \text{ C}$$

Now, from Faraday's laws, 1 faraday (96,500 C) will plate out 63.5/2 g of Cu^{2+}, since each ion carries two electronic charges. Hence we can set up a proportion:

$$\frac{x}{2.7 \times 10^4 \text{ C}} = \frac{\frac{1}{2}(63.5)}{9.65 \times 10^4 \text{ C}}$$

or

$$x = \frac{2.7 \times 10^4 \text{ C}}{9.65 \times 10^4 \text{ C}}(31.8 \text{ g}) = \boxed{8.9 \text{ g}}$$

Hence 8.9 g of copper is deposited on the cathode.

17.3 Electric Currents in Metals

Unlike the situation with electrolytes, where the current is carried by both positive and negative ions, in a metal wire the current is carried solely by negatively charged electrons. The metal atoms, which lose electrons and become positive ions, are bound to fixed positions in the crystal lattice and cannot move. Only the outermost (or valence) electrons from these atoms can move freely through the wire. They move in a direction opposite to that of the conventional current, as we have seen.

Ohm's Law

Suppose we take a piece of thin copper wire of length L and attach a voltage source across its two ends. What happens? If the left end of the wire is at a potential V_A and the right end at a potential V_B, as in Fig. 17.9, then the work done by the voltage in moving a charge q' from A to B in the wire is, from Eq. (16.7),

$$\mathcal{W} = (V_A - V_B)q'$$

The work done per unit charge is therefore

$$\frac{\mathcal{W}}{q'} = \frac{f_{el}L}{q'} = V_A - V_B$$

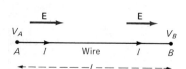

FIGURE 17.9 Electric field **E** and current in a wire when a voltage is applied across it.

where f_{el} is the electric force on the charge q'. But f_{el}/q' is the magnitude of the electric field strength **E** in the wire, and so

$$E = \frac{V_A - V_B}{L} \qquad (17.3)$$

In a wire with a voltage across its two ends there is an electric field strength of magnitude $E = (V_A - V_B)/L$, which is directed parallel to the wire at any point along its length. Note that, unlike the situation in electrostatics, an electric field *can* exist inside a conductor. This is because the applied external voltage does not allow the charges in the metal to come to rest at positions where they could cancel the field.

Let $V = V_A - V_B$ be the voltage across the wire. If we now increase V, the field inside the wire increases, and experiment shows that the current through the wire also increases in direct proportion to V, as shown in Fig. 17.10. Hence

$$I \propto V \qquad \text{or} \qquad V \propto I$$

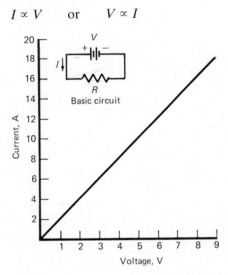

FIGURE 17.10 Ohm's law for the relationship of the current to the applied voltage in a simple circuit containing only a resistance R.

and we can write, on introducing a constant of proportionality R,

Ohm's Law

$$\boxed{V = IR} \qquad (17.4)$$

The larger R is, the smaller I is for any applied voltage V. Hence R measures the resistance of the wire to the flow of current and is therefore called the *resistance* of the wire. It is denoted by the symbol ⌇⌇⌇ on circuit diagrams.

Definition

The resistance is the ratio of the voltage applied across any element of a dc circuit to the current through that element.

Equation (17.4) is called *Ohm's law* after the German physicist Georg Simon Ohm (1787–1854) who discovered it. For this reason we call the unit of resistance the *ohm*.

Definition

One ohm (Ω) is the resistance of a wire across which the application of a voltage of one volt produces a current of one ampere in the wire.

Hence $1\ \Omega = 1$ V/A

The symbol for the ohm is the Greek capital *omega*, the last letter in the Greek alphabet.

Note that there is a voltage *drop* across a resistor in a dc circuit, so that $V_B < V_A$, if the current is flowing from A to B. This is because the battery must do work to move the charge from A to B. Energy is therefore lost in the process and appears as heat in the resistor R.

Example 17.3

The resistance of a graphite rod is $3.0 \times 10^{-1}\ \Omega$.
(a) What voltage must be applied across the two ends of the rod to produce a current of 10 A through the rod?

(b) If the required voltage is supplied by a battery which delivers current for 10 min, how much work does the battery do?

SOLUTION

(a) From Ohm's law,

$V = IR = (10\ \text{A})(3.0 \times 10^{-1}\ \Omega) = \boxed{3.0\ \text{V}}$

(b) Since $\mathscr{W} = (V_A - V_B)q' = (3.0\ \text{V})q'$, and

$q' = It = (10\ \text{A})(10\ \text{min}) = (10\ \text{A})(600\ \text{s})$

$= \left(10\dfrac{\text{C}}{\text{s}}\right)(600\ \text{s}) = 6.0 \times 10^3\ \text{C}$

we have $\mathscr{W} = (3.0\ \text{V})(6.0 \times 10^3\ \text{C}) = \boxed{1.8 \times 10^4\ \text{J}}$

Simple Electric Circuits

A simple kind of electric circuit is a flashlight containing a small bulb lighted by two D cells, each with an emf of 1.5 V, as in Fig. 17.11a. In this case the conventional electric current flows from the positive terminal of the top battery through wires to the filament of the light bulb and then back to the negative terminal of the lower battery by way of the metal case of the flashlight.

The circuit diagram corresponding to this physical situation is shown in Fig. 17.11b. The filament of the light bulb is represented by the resistance R. For a flashlight R is about 10 Ω. The connecting wires do have some resistance, but this resistance is so small that it can be neglected or can be lumped in with that of the filament wire to make up the resistance R. In circuit

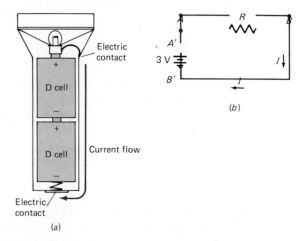

FIGURE 17.11 *(a)* A flashlight: A simple electric circuit. The return circuit is through the metal case of the flashlight. *(b)* Circuit diagram for the flashlight. The resistance R includes the resistance of the flashlight bulb and all other resistances in the circuit.

diagrams straight lines like those connecting the battery to the resistance R are therefore assumed to have zero resistance. As a result, in the diagram points like A and A', or B and B', are at exactly the same potential, since, by Ohm's law, $V_A - V_{A'} = IR' = 0$, if $R' = 0$; and so $V_A = V_{A'}$. *Only wires in which an electric current encounters sufficient resistance to generate a measurable amount of heat are called resistors and indicated by the resistance symbol on a circuit diagram.* The electric current is the same at every point in the circuit of Fig. 17.11b. The electrons act as an incompressible fluid which flows around the circuit in the same way that water flows through a pipe. An equation similar to the hydrodynamic equation of continuity is therefore valid. For a closed loop the amount of charge passing any point in the circuit per second is exactly the same as the amount passing any other point in the circuit, since electric charge cannot be created or destroyed.

17.4 Resistance and Resistivity

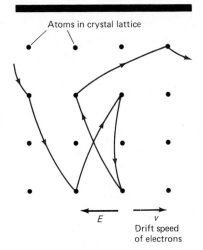

Atoms in crystal lattice

E v
Drift speed
of electrons

FIGURE 17.12 Drift speed of electrons through a wire. The electrons move in a direction opposite to the electric field in the wire at a rate determined by the number of collisions they make with the atoms in the metal crystal.

Factors Determining Electric Resistance

When a voltage is applied across a wire, the electrons do not move very rapidly along the length of the wire. An electron is accelerated by the voltage, since it experiences a force of magnitude eE, where e is the magnitude of the charge on the electron and E is the electric field strength in the wire. The acceleration of the electron is then, by Newton's second law, $a = eE/m$. Because the electron mass m is so small, this acceleration is very large even for a small applied voltage. The electron therefore picks up speed very rapidly, but almost immediately collides with one of the positively charged metal ions in the wire. It loses a little of its energy in the collision, bounces off the atom, is accelerated again, collides again, and thus makes its way by a very irregular path along the length of the wire, as in Fig. 17.12. As a result the *drift speed* of the electrons through the wire (that is, the average speed at which they progress along the length of the wire) is only on the order of 1 mm/s. This is in contrast to the very large speeds the electrons can acquire during the brief periods of time between collisions.

The current that flows through a wire is proportional to the drift speed of the electrons through the wire. Hence the effect of the collisions that the electrons undergo is to reduce the current, much in the same way that rocks, fallen trees, and other debris in a river reduce the flow rate of the river water. The smaller the current for any given applied voltage, the larger the resistance.

To find the factors which determine the electric resistance of a wire, let us do a simple experiment. Suppose we take a piece of copper wire 1 m long, apply a voltage of 6 V across it with a storage battery, and measure the current which flows. (This measurement can be made with an *ammeter*, to be discussed in the next chapter.) Suppose we find that the current is 1 A. Then, from Ohm's law, the resistance of the wire is $R = V/I = 6$ V/1 A $= 6\ \Omega$. If we repeat this experiment with a piece of the same wire 2 m long, we find that this time R is 12 Ω; if the wire is 3 m long, its resistance is 18 Ω; and so on. Hence the resistance of the wire is found by experiment to be proportional to its length. The reason is that the electric field inside the wire is given by $E = V/L$ [Eq. (17.3)], where V is the applied voltage and L is the length of the wire. Hence increasing L decreases E and therefore decreases the current, since the electric field E provides the force to move the charges through the wire.

Similarly, if we use copper wires of the same length but different cross-sectional areas, we find by experiment that the resistance varies inversely with the area of the wire. This is reasonable, since the thicker the wire is, the more free electrons there are in any given length to carry the current.

Putting these experimental observations together, we may write for the resistance

$$R \propto \frac{L}{A} \quad \text{or} \quad \boxed{R = \rho \frac{L}{A}} \tag{17.5}$$

Here L and A tell us the shape of the wire used; they are geometric factors which are independent of the material out of which the wire is made. The dependence of the resistance on the material of the wire is contained in the proportionality constant ρ (Greek *rho*, the equivalent of the English r), which is called the *resistivity* of the material. The value of the resistivity must be obtained by laboratory measurement; it is small for good conductors and large for poor conductors, as Table 17.1 shows. The units of resistivity are, from Eq. (17.5), $\Omega \cdot m^2/m = \Omega \cdot m$ (ohm-meters). Thus copper has a resistivity of 1.7×10^{-8} $\Omega \cdot m$ at 0°C.

TABLE 17.1 Electric Resistivity and Temperature Coefficient of Resistivity for Various Materials at 0°C

Material	Resistivity, $\Omega \cdot m$	Temperature coefficient of resistivity, 10^{-3} (°C)$^{-1}$
Conductors:		
Silver	1.5×10^{-8}	4.1
Copper	1.7×10^{-8}	4.1
Gold	2.4×10^{-8}	4.0
Aluminum	2.6×10^{-8}	3.9
Tungsten	5.5×10^{-8}	4.7
Iron	8.85×10^{-8}	6.2
Platinum	9.83×10^{-8}	3.7
Lead	22×10^{-8}	3.9
Mercury	94×10^{-8}	0.88
Nichrome (60% Ni, 24% Re, 16% Cr)	100×10^{-8}	0.4
Semiconductors:		
Germanium	~0.5*	−50
Silicon	20–2000*	−70
Nonconductors (insulators):		
Glass	~10^{12}	
Amber	~5×10^{14}	
Sulfur	~10^{14}	
Fused quartz	~10^{17}	

*Depending on purity.

Resistance of a Wire

The resistance of a wire is directly proportional to the resistivity of the metal and to the length of the wire, and inversely proportional to the cross-sectional area of the wire.

Temperature Coefficient of Resistivity

Like many other physical properties, resistivity depends on temperature. The resistivity of pure metals increases linearly with temperature. This is because raising the temperature increases the amplitude of vibration of the atoms in the metal lattice and thus increases the likelihood of collisions and decreases the current through the wire. The expression for the increase of resistivity with temperature is similar to that for the increase in length of a metal rod with temperature [Eq. (13.7)]. We have

$$\rho_T = \rho_0 \left(1 + \alpha \, \Delta T\right) \tag{17.6}$$

Definition

where ρ_T is the resistivity of the metal at a temperature $T°C$, ρ_0 is its resistivity at $0°C$, and α is called the *temperature coefficient of resistivity* and is a measure of *the fractional increase in resistivity per 1°C rise in temperature*. Since $\alpha \, \Delta T$ must be dimensionless, α has units $(°C)^{-1}$. Values of α for some important metals are given in Table 17.1. The dependence of the resistance of a copper

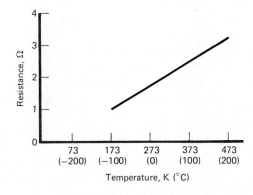

FIGURE 17.13 The resistance of a piece of copper wire, 100 m long and 1 mm² in cross-sectional area, as a function of temperature.

From Tables 13.2 and 17.1 it can be seen that the temperature coefficients of resistivity for metals are at least 100 times larger than the coefficients of linear expansion for the same metals; hence temperature effects on L and A can be reasonably ignored in calculating the resistance R as a function of temperature.

Since the resistance R for a metal is proportional to its resistivity, Eq. (17.6) is equivalent to:

$$R_T = R_0(1 + \alpha \, \Delta T) \tag{17.7}$$

Because the resistance of a metal is a function of temperature, *resistance thermometers* are frequently used to measure high temperatures, at which ordinary thermometers will not function. (How would you expect a resistance thermometer to work?) A platinum resistance thermometer is particularly useful for this purpose, since platinum melts only at 1774°C and is readily available in highly purified form.

Measuring Resistance: The Ohmmeter

An *ohmmeter* is a useful device for measuring electric resistance. It consists of a battery connected to a series resistor and a meter which measures current (an ammeter), as in Fig. 17.14. When an unknown resistor is connected in series with this circuit, the constant voltage V drives a current I through the unknown and the series resistances. Hence

$$I = \frac{V}{R_X + R_{ser}}$$

When $R_X \to \infty$, $I \to 0$; when $R_X = 0$, $I = V/R_{ser}$. Hence the deflection of the meter, which measures the current, is roughly inversely proportional to the resistance R_X. The scale on the meter can then be calibrated to read resistance directly. Such a scale is highly nonlinear and not very precise. Despite this, an ohmmeter is a convenient, fast way to determine approximate values of resistance.

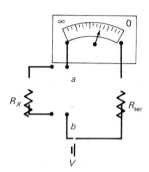

FIGURE 17.14   An ohmmeter. A rough value of the resistance R_X can be obtained by a measurement of the current through the meter, since

$$I = \frac{V}{(R_X + R_{ser})}$$

Resistance of the Human Body

An instrument similar to an ohmmeter can be used to determine the resistance of the human body. The tissues and fluids beneath the skin contain a large number of ions and hence conduct electricity almost as well as many metals. By making a small current flow between two points on the surface of the body it is therefore possible to measure the electric resistance between these points. Often one electrode is attached to a patient's leg and the other is moved over the body, with a voltage of 60 V or so applied between the two electrodes. Since, for constant voltage, the current flowing varies inversely with the resistance of the path beneath the skin, the electric resistance of the body can be determined in this way.

It is found that nerve damage or tumors beneath the skin greatly increase the electric resistance near the site of the abnormality. This technique has been found to be very effective in detecting cancers. The measured increase in resistance even provides a good indication of how advanced the cancer is.

Example 17.4

(a)   What is the resistance at 0°C of a 1.0-m-long piece of no. 5 gauge copper wire which has a cross-sectional area of 16.8 mm²?

(b)   What would be the resistance of the same piece of wire at room temperature (20°C)?

SOLUTION

(a)   Using the basic equation $R = \rho L/A$ and the value of ρ for copper from Table 17.1, we have

$$R = \frac{(1.7 \times 10^{-8}\ \Omega\cdot m)(1.0\ m)}{(16.8\ mm^2)(10^{-3}\ m/mm)^2} = \boxed{1.02 \times 10^{-3}\ \Omega}$$

(b)   Since, from Table 17.1, $\alpha = 4.1 \times 10^{-3}$, we have $R_T = R_0 (1 + \alpha T)$, or

$R_{20°} = R_{0°} [1 + (4.1 \times 10^{-3})(°C)^{-1}(20°C)] = R_{0°} (1 + 0.082)$

$$= (1.02 \times 10^{-3}\ \Omega)(1.082) = \boxed{1.10 \times 10^{-3}\ \Omega}$$

Hence the resistance changes by about 8 percent for a 20°C change in temperature.

Example 17.5

A platinum resistance thermometer is used to determine the melting point of lead. The thermometer has a resistance of 100 Ω at 0°C and 228 Ω when it is immersed in a mixture of solid and liquid lead at thermal equilibrium. What is the melting point of lead?

SOLUTION

Since $R_T = R_0(1 + \alpha T)$, we have

$$T = \frac{R_T - R_0}{\alpha R_0} = \frac{228\ \Omega - 100\ \Omega}{[3.9 \times 10^{-3}\ (°C)^{-1}](100\ \Omega)} = \boxed{328°C}$$

Hence the melting point of lead is about 328°C.

17.5 Kirchhoff's Rules

In 1845 Gustav Kirchhoff (see Fig. 17.16 and accompanying biography) proposed two rules which are extremely useful in solving problems relating to electric circuits and are now known as *Kirchhoff's rules*.

Kirchhoff's first rule, called the *junction rule*, is based on the law of conservation of electric charge.

Junction Rule

The sum of all the currents entering any junction point is equal to the sum of all the currents leaving that junction point.

Here a *junction point* is a point in the circuit at which three or more wires come together.

For example, as applied to the junction point A in Fig. 17.15a, Kirchhoff's first rule states that, if the currents are assumed to be in the directions shown in the figure, then $I_1 = I_2 + I_3$. In writing down this equation we are forced to guess the directions of the three currents, and our assumptions may be incorrect. If so, we find on solving the circuit problem that one or more of the currents is negative. This means that our original assumption about the direction of that particular current was wrong, and that the actual direction is the reverse of the assumed direction.

If we consider water flowing in a pipe instead of charges flowing in a wire, as in Fig. 17.15b, then clearly all the water that flows into point A in one pipe must flow away from A in the other two pipes. The same is true of electric charge and hence of electric current. The amount of charge flowing into A per second (I_1) must be equal to the amount of charge leaving A per second ($I_2 + I_3$), if charge is to be conserved. Kirchhoff's junction rule is a statement of this fact.

Kirchhoff's second rule, called the *loop rule*, is based on the principle of conservation of energy.

FIGURE 17.15 (a) Kirchhoff's junction rule for three currents: $I_1 = I_2 + I_3$. (b) The behavior of the current is the same as that of a fluid which flows from one pipe into two separate pipes.

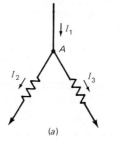

(a)

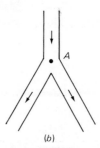

(b)

Gustav Robert Kirchhoff (1824–1887)

FIGURE 17.16 Gustav Robert Kirchhoff. (*Photo courtesy of AIP Niels Bohr Library; Meggers Gallery of Nobel Laureates.*)

Kirchhoff was one of the group of university professors who led Germany to preeminence in physics at the end of the nineteenth century. He was born in Königsberg, the son of a law counselor who had many friends on the faculty at the University of Königsberg; he married the daughter of one of his professors; and he spent his life doing research and teaching at the universities of Breslau, Heidelberg, and Berlin. While in Heidelberg, he worked in close association with R. Bunsen, the famous chemist, and H. Helmholtz (see biography, Chap. 12), the physiologist-physicist, during what has been called Heidelberg's golden age of science.

Kirchhoff's first important contribution to physics, the analysis of electrical networks (discussed in Sec. 17.5), was made while he was still a university student and only 21 years old.

When he went to Heidelberg in 1854, Kirchhoff found Bunsen trying to analyze salts on the basis of the distinctive colors they imparted to the flame of what we now call a Bunsen burner. Kirchhoff pointed out that a much more reliable test for the presence of various elements would be the line spectrum they emit, which could be detected using the spectroscope which he proceeded to construct with Bunsen's help. Together they determined the line spectra of many of the elements, and in the process discovered the elements cesium and rubidium.

Kirchhoff's work on the emission spectra of the elements led him to consider the relationship of their spectral lines to the dark lines observed by Fraunhofer in the sun's spectrum. Kirchhoff showed that the famous yellow D lines in the emission spectrum of sodium corresponded perfectly to two dark lines in the solar spectrum. This made him conclude that the sun's atmosphere contained sodium and that this sodium absorbed characteristic wavelengths out of the continuous radiation emitted by the sun. He verified that these were indeed the sodium lines by means of a laboratory experiment involving a sodium light source and NaCl salt ignited in a Bunsen burner.

His study of the relationship between the emission and absorption of radiation then led Kirchhoff to investigate the radiation from hot objects and to propose a hollow metal cavity (a blackbody) as an ideal source for such radiation. His radiation law (Sec. 14.4) became the key to the study of radiation. Max Planck, who succeeded Kirchhoff to the chair of theoretical physics at the University of Berlin, took up Kirchhoff's ideas and used them as the basis for his quantum theory, one of the truly great developments in the history of physics (see Chap. 26).

Despite his enormous research ability, Kirchhoff was a somewhat dull lecturer. Still, he worked assiduously on his lectures on theoretical physics at Berlin in the years after 1875, when his poor health prevented further experimental work. Many of the physicists who made Germany the mecca of theoretical physics in the years from 1900 to 1935 learned physics from Kirchhoff's published lectures.

For many years, as the result of an accident, Kirchhoff was confined to a wheelchair or crutches. Yet he retained his cheerful disposition and great enthusiasm for his work. One day his banker, unimpressed by Kirchhoff's ability to locate elements in the sun with his spectroscope, asked him, "Of what use is gold in the sun if I cannot bring it down to earth?" Some years later, when Great Britain presented Kirchhoff with a gold medal and a large cash prize in gold sovereigns for his research, Kirchhoff handed it over to the same banker, with the sly remark: "Here is your gold from the sun."

Loop Rule

The algebraic sum of the changes in electric potential encountered in a complete traversal of any closed circuit is equal to zero.

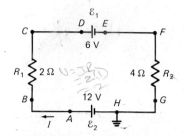

FIGURE 17.17 Kirchhoff's loop rule: The sum of the emf's is equal to the sum of the *IR* drops around the circuit.

When there is a steady current in a closed circuit there is a fixed value for the electric potential at every point in that circuit, i.e., for the potential energy per unit charge at each point. If we take a charge at point H in Fig. 17.17 and follow it completely around the circuit, returning finally to point H, then its potential energy must be the same at the end as it was at the beginning. Hence the algebraic sum of the changes in potential it encounters as it goes around the complete circuit must be zero—which is the loop rule.

Solving the circuit of Fig. 17.17 for the single current which flows is not difficult. The net emf is 12 V − 6 V = 6 V, since the two are in opposite directions, and the resultant emf is in the direction of $\mathcal{E}_2$ (we assume here that the emf's and terminal voltages are identical for each battery). The resistance is $R = 2\,\Omega + 4\,\Omega = 6\,\Omega$. Hence $I = (6\ \text{V})/(6\ \Omega) = 1$ A. A current of 1 A therefore flows around the circuit in a clockwise direction.

To consider the changes in potential as we go around this circuit, let us unfold the circuit into a straight line, as in Fig. 17.18, where points H at the beginning and end of the unfolded circuit are identical electrically. This will enable us to indicate below each circuit element exactly how the potential varies as we move through that element. For simplicity, we choose to ground point H to put it at zero potential.

As we move through the emf $\mathcal{E}_2$ from H to A, the potential increases by 12 V, the emf of the battery. Between A and B there is no change in potential, since there is no resistance between A and B. In passing through the resistor R_1 the potential falls by 2 V, since 1 A × 2 Ω = 2 V, and C is at a lower potential than B. Again there is no change in potential between C and D. In going through the emf $\mathcal{E}_1$, however, the potential *decreases* by another 6 V, since we are going from a high potential to a low potential. On passing through the resistor R_2 the potential drops again from 4 V to zero, and we return to point H at the same (zero) potential at which we started.

FIGURE 17.18 Circuit of Fig. 17.17 unfolded into a straight line, with values of the potential at various points throughout the circuit indicated on the vertical scale. Since point H is at ground potential, the circuit is completed through ground.

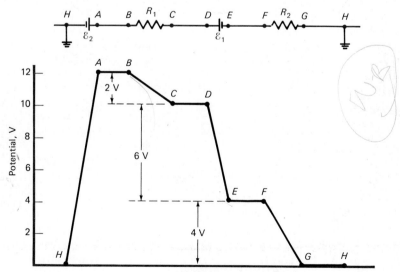

Hence the changes in potential in going around the circuit are $12\,V - 2\,V - 6\,V - 4\,V = 0$, as required by the loop rule. This analysis helps make clear two points of great importance in assigning signs to the changes in potential in applying the loop rule.

Sign Convention

1 If a resistor is traversed in the same direction as the current in any branch of a circuit, the change in potential is $-IR$ because there is a *drop* in potential in moving through a resistor in the direction of the current; if a resistor is traversed in the opposite direction to that of the current, the change in potential is $+IR$.

2 If a seat of emf is traversed in the direction of the emf, the change in potential is $+\mathscr{E}$; if in the opposite direction it is $-\mathscr{E}$. (Look carefully at Fig. 17.18 and be sure you understand the sign convention being used.)

In applying the loop rule the assumed directions of the currents must be the same as those assumed in applying the junction rule. These may turn out to be the wrong directions, but the right answer will still be obtained so long as the current directions are taken consistently throughout the problem, i.e., in all applications of both the junction rule and the loop rule.

The application of Kirchhoff's two rules to a circuit in which all the emf's and resistances are given leads to a number of equations in which all terms are known except the currents. In general, there will be just a sufficient number of independent equations to solve algebraically for all the unknown currents. This is illustrated in the following examples. In some cases the currents may be known, and Kirchhoff's rules can be applied to find unknown resistances.

Example 17.6

Solve the circuit of Fig. 17.19 for the three currents I_1, I_2, and I_3. Assume that the terminal voltage is equal to the emf for each of the two batteries.

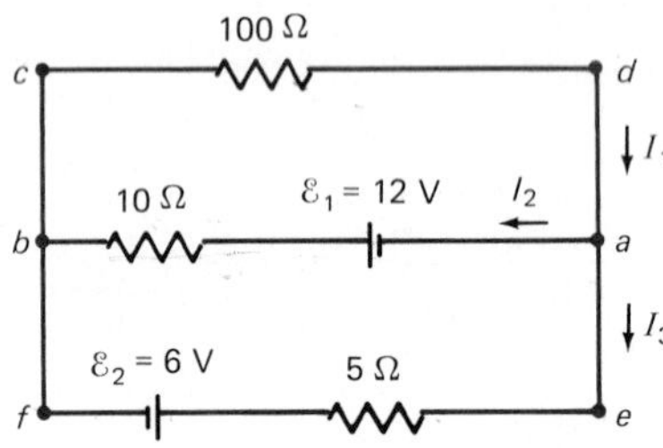

FIGURE 17.19 Circuit diagram for Example 17.6.

SOLUTION

Applying the junction rule to point a and assuming the currents to be in the directions shown in Fig. 17.19, we have

$$I_1 = I_2 + I_3 \qquad [1]$$

We now apply the loop rule, going around each of the three possible loops in a clockwise direction, and noting that there are different currents in each of the three branches. For the upper loop *abcd* we have, using the sign convention previously discussed,

$$\mathscr{E}_1 - I_2(10\ \Omega) - I_1(100\ \Omega) = 0 \qquad [2]$$

For the lower loop *aefb* we have

$$-I_3(5\ \Omega) - \mathscr{E}_2 + I_2(10\ \Omega) - \mathscr{E}_1 = 0 \qquad [3]$$

For the full loop *defc* we have

$$-I_3(5\ \Omega) - \mathscr{E}_2 - I_1(100\ \Omega) = 0 \qquad [4]$$

Then [4] becomes, on replacing I_1 by $I_2 + I_3$ from [1],

$$-5I_3 - 6 - (I_2 + I_3)100 = 0$$

and so $\qquad 100I_2 = -105I_3 - 6 \qquad [5]$

From [3] we have

$$-5I_3 - 6 + 10I_2 - 12 = 0 \qquad \text{or} \qquad 10I_2 = 5I_3 + 18 \qquad [6]$$

On multiplying [6] by 10 and combining with [5], we have

$$100I_2 = 50I_3 + 180 = -105I_3 - 6$$

which leads to $\qquad I_3 = \dfrac{-186}{155} = \boxed{-1.20 \text{ A}}$

Then, from [6], $10I_2 = 5(-1.20 \text{ A}) + 18$

or $\qquad I_2 = \boxed{1.20 \text{ A}}$

and so $\qquad I_1 = I_2 + I_3 = 1.20 \text{ A} - 1.20 \text{ A} = \boxed{0}$

Hence $I_1 = 0$, $I_2 = 1.20$ A and is in the originally assumed direction, and $I_3 = 1.20$ A but is in a direction opposite to that shown in the diagram. Hence a current of 1.20 A flows in a counterclockwise direction in the lower loop, and no current at all flows in the upper loop. The reason for this is that the potential difference between points a and b due to $\mathscr{E}_1$ is 12 V $-$ (10 Ω)(1.20 A) = 0. Since $V_{ab} = V_{ac}$, there is no voltage across d and c to drive current through the 100-Ω resistor. Hence $I_1 = 0$, as found from Kirchhoff's rules.

Note that we did not use Eq. [2] at all in this solution. Actually it is redundant, since only two of the three loop equations are independent. Hence the two loop equations and the junction equation must be used to solve the problem.

17.6 Some Applications of Kirchhoff's Rules

EMF and Terminal Voltage of a Battery

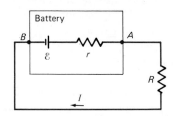

FIGURE 17.20 EMF and terminal voltage for a battery: $V_{AB} = \mathscr{E} - Ir$.

In a battery chemical energy is continually being converted into electric energy. No battery is perfect, however, and some energy is always lost in the form of heat inside the battery. This loss depends on the amount of current through the battery. The battery therefore behaves as if it had an *internal resistance r* which functions in exactly the same way an ordinary resistance would function in the circuit external to the battery.

If we allow for the battery's internal resistance, the correct circuit diagram for a seat of emf delivering a current I to an external resistance R is shown in Fig. 17.20. Applying Kirchhoff's loop rule, we find, on going around the circuit in the direction of the conventional current,

$$\mathscr{E} - Ir - IR = 0 \qquad \text{or} \qquad \mathscr{E} - Ir = IR$$

Now, $\mathscr{E} - Ir$ is the *terminal voltage* of the battery, or the voltage V_{AB} appearing between points A and B in the circuit, that is, the voltage which would be measured by attaching a voltmeter to the two terminals of the battery while it is delivering current. Hence

$$\boxed{V_{AB} = \mathscr{E} - Ir} \qquad\qquad (17.8)$$

Definition

The terminal voltage of a battery is its emf reduced by the Ir drop across its internal resistance.

Note that the terminal voltage is equal to the emf only when $I = 0$, that is, when no current is being drawn from the battery. The more current drawn, the lower the terminal voltage of the battery, according to Eq. (17.8). If the battery were a storage battery being charged, the terminal voltage would be *larger* than the emf (prove this).

In starting a car very large currents ($\sim$100 A) must be provided by the car's battery for a short period of time. Thus if you start your car with the headlights on, the lights dim while the battery is delivering this large starting current, since the terminal voltage is reduced by Ir.

The internal resistance of a fresh battery is very small, perhaps about 0.05 Ω for a small D cell. However, as the battery ages, its internal resistance increases to many ohms. Finally the internal resistance becomes so large that the battery ceases to be useful, for as soon as it delivers any measurable current its terminal voltage falls almost to zero.

Example 17.7

Consider the circuit of Fig. 17.21. Here a battery with emf of 12 V has an internal resistance of 0.10 Ω. If the battery is connected to two light bulbs whose resistances when hot are 50 and 100 Ω, respectively, as shown in the figure, what are the currents in the three branches of the circuit?

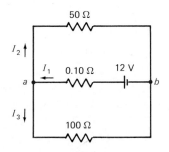

FIGURE 17.21   Circuit diagram for Example 17.7.

SOLUTION

Let us solve the problem using Kirchhoff's rules. At the junction point a we have $I_1 = I_2 + I_3$ from the junction rule.

From the loop rule we have, on going around the loops in a clockwise direction:

For the upper loop: $12 \text{ V} - (0.10 \text{ Ω})I_1 - (50 \text{ Ω})I_2 = 0$

For the lower loop: $-12 \text{ V} + (0.10 \text{ Ω})I_1 + (100 \text{ Ω})I_3 = 0$

On adding these two equations, we obtain

$$100 I_3 = 50 I_2 \quad \text{or} \quad I_2 = 2I_3$$

Hence $I_1 = I_2 + I_3 = 2I_3 + I_3 = 3I_3$

On substituting this in our second loop equation, we have

$$-12 \text{ V} + (0.10 \text{ Ω})(3I_3) + (100 \text{ Ω})I_3 = 0$$

from which $I_3 = \dfrac{12 \text{ V}}{100.3 \text{ Ω}} = \boxed{0.12 \text{ A}}$

Hence $I_2 = 2I_3 = 2(0.12 \text{ A}) = \boxed{0.24 \text{ A}}$

and $I_1 = 3I_3 = 3(0.12 \text{ A}) = \boxed{0.36 \text{ A}}$

In this case all the currents come out positive, which means that our original assumption about directions was correct.

Resistors in Series

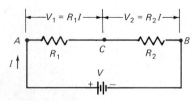

FIGURE 17.22   Two resistors in series: $R = R_1 + R_2$.

Kirchhoff's rules can be used to obtain some very practical equations for combining resistors in electric circuits.

There are two ways to combine resistors. The first way is to connect them *in series*, as in Fig. 17.22. Here the same current flows through the two resistors in succession. By Kirchhoff's loop rule we have:

$$V - IR_1 - IR_2 = 0$$

where V is the terminal voltage of the battery. Hence

$$V = I(R_1 + R_2)$$

But the equivalent resistance of a circuit which would draw the same current I when the same voltage V was applied to it is

$$R = \frac{V}{I} \quad \text{or} \quad V = IR$$

Hence we must have

$$IR = I(R_1 + R_2) \qquad \text{or}$$

$$\boxed{R = R_1 + R_2} \qquad\qquad (17.9)$$

For *three* resistors in series,

$$R = R_1 + R_2 + R_3 \qquad \text{or, in general} \qquad R = \sum_n R_n$$

Hence we conclude:

Resistors in Series

For resistors in series, the equivalent resistance is the sum of the individual resistances.

There is a practical problem with any series circuit of this kind. For example, if R_1 is the filament resistance of a light bulb and R_2 is the resistance of a toaster's heating element, then if the light bulb burns out, the toaster will not work when the two are connected together in series, for there is no longer a complete circuit by which current can flow from A to B. This was a great nuisance in the series-wired sets of Christmas-tree lights used a few decades ago. Every time one bulb burned out, the whole set went dark and considerable time was wasted in finding the burnt-out bulb. For this reason series connections now are rarely used in house wiring. The one exception is a fuse put in series with an appliance precisely to cut off the current to the appliance if a malfunction blows the fuse.

Resistors in Parallel

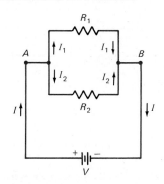

FIGURE 17.23 Two resistors in parallel: $1/R = 1/R_1 + 1/R_2$.

The second way to combine resistors is to connect R_1 and R_2 in parallel, as in Fig. 17.23. Here the current I splits into two parts, a current I_1 passing through R_1 and a current I_2 passing through R_2.

By Kirchhoff's loop rule applied to the upper branch,

$$V - I_1R_1 = 0 \qquad \text{or} \qquad V = I_1R_1$$

Similarly, for the lower branch we have:

$$V - I_2R_2 = 0 \qquad \text{or} \qquad V = I_2R_2$$

Hence $\qquad I_1 = \dfrac{V}{R_1} \qquad$ and $\qquad I_2 = \dfrac{V}{R_2}$

But, from Kirchhoff's junction rule, $I = I_1 + I_2$, and so

$$I = \frac{V}{R} = \frac{V}{R_1} + \frac{V}{R_2}$$

where R is the equivalent resistance of the circuit, i.e., the resistance which will lead to the actual current I when the voltage V is applied across that resistance. If we now cancel V on both sides of the preceding equation, we obtain:

$$\boxed{\frac{1}{R} = \frac{1}{R_1} + \frac{1}{R_2}} \qquad\qquad (17.10)$$

For three resistors in parallel,

$$\frac{1}{R} = \frac{1}{R_1} + \frac{1}{R_2} + \frac{1}{R_3} \qquad \text{or, in general,} \qquad \frac{1}{R} = \sum_n \frac{1}{R_n}$$

Resistors in Parallel

For resistors in parallel, the reciprocal of the equivalent resistance is the sum of the reciprocals of the individual resistances.

Note that in this case the equivalent resistance is *less* than any of the individual resistances. The reason is that by providing alternate, parallel paths we are in a certain sense increasing the cross-sectional area of the conductor and thus decreasing the overall resistance.

In most household circuits, appliances are connected in parallel. Hence each appliance has the full 120 V applied across it, and the current it draws is determined by its resistance. Parallel circuits have the great advantage that the burnout of one light bulb or resistor does not interrupt the current to the other appliances connected in parallel with it.

Example 17.8

Find the current delivered by a 6-V battery when it is connected to two light bulbs, each of resistance 50 Ω and connected (*a*) in parallel; (*b*) in series.

SOLUTION

(**a**) The circuit is as shown in Fig. 17.24*a*, with $R_1 = R_2 = 50\ \Omega$. Since the two resistors are in parallel, we have

$$\frac{1}{R} = \frac{1}{50\ \Omega} + \frac{1}{50\ \Omega} = \frac{2}{50\ \Omega} \quad \text{or} \quad R = 25\ \Omega$$

Hence

$$I = \frac{V}{R} = \frac{6\ \text{V}}{25\ \Omega} = \boxed{0.24\ \text{A}}$$

(**b**) In this case, the circuit is as in Fig. 17.24*b*, and $R = R_1 + R_2 = 100\ \Omega$. Then

$$I = \frac{V}{R} = \frac{6\ \text{V}}{100\ \Omega} = \boxed{0.060\ \text{A}}$$

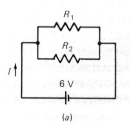

(*a*)

(*b*)

FIGURE 17.24 Circuit diagrams for Example 17.8.

Example 17.9

(**a**) In the circuit of Fig. 17.25 find the current I in the 18-Ω resistor. The terminal voltage of the battery is 6.0 V.
(**b**) Find the currents flowing in the individual branches of the circuit.

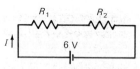

FIGURE 17.25 Circuit diagram for Example 17.9.

SOLUTION

(**a**) Here we must first find the equivalent resistance of the circuit. In such complicated circuits the best approach in applying the laws for combining resistances in series and in parallel is to start with the innermost combination of resistances. This is the combination of the 6-Ω and the 12-Ω resistors in parallel. We find

$$\frac{1}{R_{\text{eq}}} = \frac{1}{6\ \Omega} + \frac{1}{12\ \Omega} = \frac{3}{12\ \Omega} \quad \text{or} \quad R_{\text{eq}} = 4\ \Omega$$

We can then add this in series with the 8-Ω resistor to obtain $R'_{\text{eq}} = 8\ \Omega + R_{\text{eq}} = 8\ \Omega + 4\ \Omega = 12\ \Omega$.

Next we must combine R'_{eq} with the 12-Ω resistor in parallel to obtain R''_{eq}. We have

$$\frac{1}{R''_{eq}} = \frac{1}{R'_{eq}} + \frac{1}{12\ \Omega} = \frac{1}{12\ \Omega} + \frac{1}{12\ \Omega} = \frac{1}{6\ \Omega}$$

and so $R''_{eq} = 6\ \Omega$. This must then be combined in series with the 18-Ω resistor to yield a final equivalent resistance for the circuit of $R'''_{eq} = 24\ \Omega$.

Our circuit has now been reduced to a voltage of 6.0 V driving a current through an equivalent resistance of 24 Ω. Hence, from Ohm's law,

$$I = \frac{V}{R} = \frac{6.0\ \text{V}}{24\ \Omega} = \boxed{0.25\ \text{A}}$$

The steps in this process of combining resistances are illustrated in Fig. 17.26.

(b) Since $R'_{eq} = 12\ \Omega$ and is in parallel with another 12-Ω resistor, the 0.25-A current must split in two, with half going through the 12-Ω resistor and half going through the other branch. The 0.125 A going through the upper branch must also split so that twice as much current goes through the 6-Ω resistor as through the 12-Ω resistor, since the voltage across the two is the same. Hence $\boxed{0.042\ \text{A}}$ goes through the 12-Ω resistor and $\boxed{0.084\ \text{A}}$ through the 6-Ω resistor. (As a check calculate the voltages across all elements of the circuit, and show that they all add correctly to the 6 V provided by the battery, as they must according to Kirchhoff's loop rule. The results are shown in Fig. 17.27.)

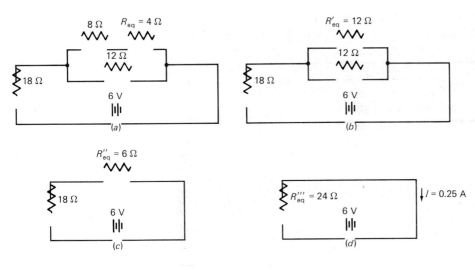

FIGURE 17.26 Steps in combining resistors in Example 17.9.

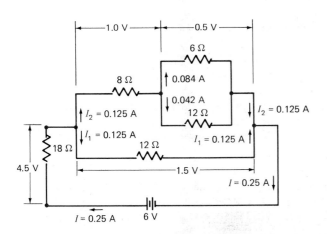

FIGURE 17.27 Voltages across circuit elements in Example 17.9.

17.7 Capacitors in DC Circuits

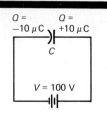

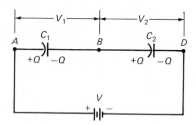

FIGURE 17.28 A capacitor connected across a battery.

We saw in Sec. 16.8 that capacitors are devices used to store electric charge. The capacitance of such a device is defined as $C = Q/V$, where Q is the charge on each of the two plates of the capacitor and V is the voltage across the capacitor. For example, in Fig. 17.28 the capacitor, indicated by the circuit symbol $\dashv\vdash$, has a capacitance

$$C = \frac{Q}{V} = \frac{10\ \mu C}{100\ V} = 0.10 \times 10^{-6}\ F = 0.10\ \mu F$$

Capacitors in Series

Just as resistors can be combined in series and in parallel to yield desired equivalent resistances for use in dc circuits, so too can capacitors.

Figure 17.29 shows two capacitors, of capacitance C_1 and C_2, combined in series. We desire to find the equivalent capacitance of this series combination, i.e., the value of C in the expression $C = Q/V$, where V is the applied voltage and Q is the charge on the capacitor plates.

Since, from Kirchhoff's loop rule, the algebraic sum of the changes in electric potential in going around the circuit must add to zero, we have

$$V - V_1 - V_2 = V - \frac{Q_1}{C_1} - \frac{Q_2}{C_2} = 0$$

FIGURE 17.29 Two capacitors in series: $1/C = 1/C_1 + 1/C_2$.

But the charges on the two capacitors must be the same, since current flows until the charges on the left plate of C_1 and on the right plate of C_2 are equal in magnitude. These charged plates then induce equal and opposite charges on the interior plates of the two capacitors, so that each capacitor ends up with an identical charge $Q = Q_1 = Q_2$.

Since, from above, $V = Q\ (1/C_1 + 1/C_2)$, and the equivalent capacitance of the series combination is $C = Q/V$, we have

$$\frac{V}{Q} = \frac{1}{C} = \frac{1}{C_1} + \frac{1}{C_2} \qquad \text{so that} \qquad \boxed{\frac{1}{C} = \frac{1}{C_1} + \frac{1}{C_2}} \tag{17.11}$$

Capacitors in Series

For capacitors in series, the reciprocal of the equivalent capacitance is the sum of the reciprocals of the individual capacitances.

For n capacitors in series we would therefore have

$$\frac{1}{C} = \sum_n \frac{1}{C_n}$$

Capacitors in Parallel

Figure 17.30 shows two capacitors connected in parallel. In this case the voltage across the two capacitors is the same, as is clear from the figure, but the charge Q is divided between the two so that $Q = Q_1 + Q_2$. Then the equivalent capacitance is:

$$C = \frac{Q}{V} = \frac{Q_1 + Q_2}{V} = \frac{Q_1}{V} + \frac{Q_2}{V} = C_1 + C_2$$

or $\qquad \boxed{C = C_1 + C_2} \tag{17.12}$

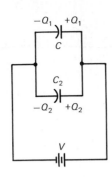

FIGURE 17.30 Two capacitors in parallel: $C = C_1 + C_2$.

Capacitors in Parallel

For capacitors in parallel, the equivalent capacitance is the sum of the individual capacitances.

For n capacitors in parallel we would have $C = \sum_{n} C_n$. Note that these results show that capacitors in series add in the same way that resistors add in parallel; and capacitors in parallel add in the same way that resistors add in series.

DC Circuits Containing Resistors and Capacitors

Capacitors do not usually play very important roles in dc circuits because no dc current can flow through a capacitor. An applied voltage will merely charge up a capacitor in a dc circuit, and then the current will stop. For example, consider the circuit of Fig. 17.31. Suppose the key K is closed and current flows. Charge will build up on the plates of the capacitor very quickly until the voltage between points A and B is equal to the applied voltage V, that is, until $V_{AB} = Q/C = V$. After the charge reaches a value Q, determined by $Q = CV$, the capacitor behaves as if it were no longer in the circuit. All the current now flows through the resistor R, with the current being given by Ohm's law, $I = V/R$.

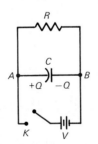

FIGURE 17.31 A capacitor and a resistor in parallel.

Example 17.10

A 6.0-V battery is connected across points A and B in Fig. 17.32.
(a) What is the equivalent capacitance between points A and B?
(b) What is the charge on the 4.0-μF capacitor?
(c) What is the charge on each of the 1.0-μF capacitors?

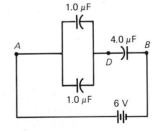

FIGURE 17.32 Diagram for Example 17.10.

SOLUTION

(a) The equivalent capacitance of the two 1.0-μF capacitors in parallel is $C_{eq} = C_1 + C_2 = 1.0\ \mu\text{F} + 1.0\ \mu\text{F} = 2.0\ \mu\text{F}$. When C_{eq} is combined in series with the 4.0-μF capacitor, we have for the equivalent capacitance between points A and B:

$$\frac{1}{C} = \frac{1}{2.0\ \mu\text{F}} + \frac{1}{4.0\ \mu\text{F}} = \frac{3}{4.0\ \mu\text{F}}$$

and so $C = \dfrac{4.0}{3}\ \mu\text{F} = \boxed{1.33\ \mu\text{F}}$

(b) Since $Q = CV$, we have

$Q = (1.33\ \mu\text{F})(6.0\ \text{V}) = \boxed{8.0\ \mu\text{C}}$

This is the charge on the 4.0-μF capacitor, since it receives the total negative charge delivered by the battery.

(c) The charge on the left plates of the two 1.0-μF capacitors must also be 8.0 μC. Hence each 1.0-μF capacitor

has a charge of $\boxed{4.0\ \mu\text{C}}$

To check this result, let us apply Kirchhoff's loop rule to a path from A to D to B through the top branch. We have

$$V - V_{AD} - V_{DB} = 0$$

or

$$V = V_{AD} + V_{DB} = \frac{4.0\ \mu\text{C}}{1.0\ \mu\text{F}} + \frac{8.0\ \mu\text{C}}{4.0\ \mu\text{F}} = 6.0\ \text{V}$$

This is indeed equal to the applied voltage, and it serves as a good check on our results.

Example 17.11

Consider the circuit of Fig. 17.33, which includes two batteries whose terminal voltages are as indicated, three resistors, and a capacitor, whose values are all given in the diagram.
(a) Find the currents I_1, I_2, and I_3 in this circuit.
(b) Find the voltage across the capacitor.
(c) Find the charge on the capacitor.

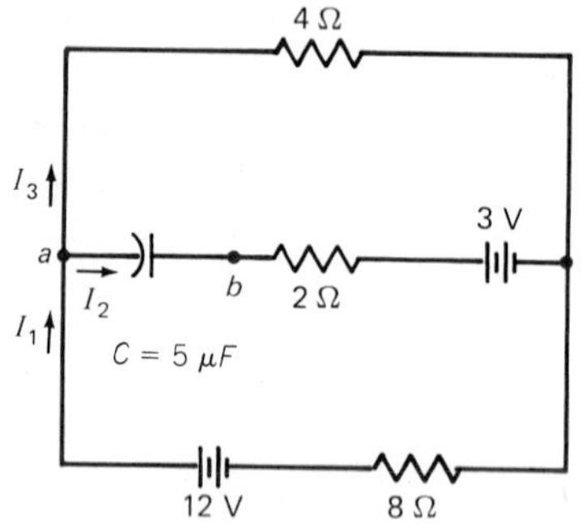

FIGURE 17.33 Diagram for Example 17.11.

SOLUTION

(a) No dc current can flow through a capacitor. Hence once the capacitor is charged, no current flows through it or through the whole middle branch. Hence

$\boxed{I_2 = 0}$

The current therefore flows in a simple series circuit of only one branch, with a voltage of 12 V driving a current through two resistors. Hence

$$I_1 = \frac{\sum \mathscr{E}}{\sum R} = \frac{12\ \text{V}}{4\ \Omega + 8\ \Omega} = \frac{12\ \text{V}}{12\ \Omega} = \boxed{1\ \text{A}}$$

Since $I_2 = 0$,

$I_3 = I_1 = \boxed{1\ \text{A}}$

(b) Let us apply Kirchhoff's loop rule to the upper branch of the circuit, going around the circuit in a clockwise direction starting at c. Then we have, since no current flows through the 2-Ω resistor,

$$3\ \text{V} + V_{ba} - (4\ \Omega)(1\ \text{A}) = 0$$

or $V_{ba} = 4\ \text{V} - 3\ \text{V} = \boxed{1\ \text{V}}$

where side a of the capacitor is 1 V more positive than side b.
(c) Since $C = Q/V$, we have

$$Q = CV = (5 \times 10^{-6}\ \text{F})(1\ \text{V}) = 5 \times 10^{-6}\ \text{C} = \boxed{5\ \mu\text{C}}$$

17.8 Energy and Power in DC Circuits

Electric energy is useful to us because it is so easily converted into other forms of energy—heat in a heating element, light in a light bulb, mechanical motion in a motor. In all these cases the electric energy produced in a seat of emf is eventually converted into some other form of energy.

Electric Power

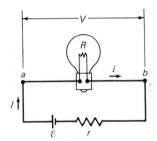

FIGURE 17.34 Electric light bulb connected to a battery.

Definition

In dc circuits like that in Fig. 17.34, when a charge q flows from point a to point b through a light bulb of resistance R, the work done by the battery in moving the charge from a to b is

$$\mathcal{W} = PE_a - PE_b = qV$$

where V is the voltage drop between a and b. If this work $\mathcal{W}$ is done in time t, then the power delivered by the battery to the light bulb is

$$P = \frac{\mathcal{W}}{t} = \frac{qV}{t} = IV \quad \text{or} \quad \boxed{P = VI} \qquad P = I^2R \qquad (17.13)$$

Power in a dc electric circuit is the product of the voltage and the current.

This is true for the circuit as a whole or for any individual element of the circuit, so long as the current used is the current through that element and the voltage is the voltage across that element. In SI units, in which voltage is measured in volts and current in amperes, the unit of power is the watt (W), or joule per second. Hence

$$\boxed{1 \text{ W} = 1 \text{ V} \cdot 1 \text{ A}}$$

This equation shows the advantage of the SI system of units. Using this system, electric power can be expressed in the same unit as mechanical power, the watt, if the voltage is measured in volts and the current in amperes, as is commonly done today.

Home Usage of Electricity

The *rate* at which a home uses electric energy is therefore 120 V (the voltage) times the current drawn at any particular time.* This is the *power*, or *the rate at which energy is consumed*, and is measured in watts or kilowatts (10^3 W). If we multiply the average rate of electric energy consumption during a month by the number of hours in the month, we obtain the number of kilowatthours (kWh) of electric energy used in that month. Here

$$\boxed{1 \text{ kWh} = 3.6 \times 10^6 \text{ J}}$$

$$1 \text{ kWh} = \left(10^3 \frac{\text{J}}{\text{s}}\right) (1 \text{ h}) \left(3600 \frac{\text{s}}{\text{h}}\right) = 3.6 \times 10^6 \text{ J}$$

---

*Even though almost all house wiring now carries alternating current (ac) rather than direct current (dc), Ohm's law and other basic equations of this chapter remain valid as long as the circuits contain only resistances. This applies to many practical appliances like electric heaters, electric stoves, toasters, light bulbs, etc. In such cases, I refers to the *effective value* of the ac current, i.e., the dc current which would have the same heating effect as the ac being considered. Here V also refers to the *effective value* of the ac voltage. These ideas will be clarified and expanded in Chap. 20.

Monthly electric bills are paid in terms of kilowatthours. Note that the kilowatt is a unit of *power*, while the *kilowatthour* is a unit of *energy*. Kilowatts and kilowatthours are frequently confused in newspaper and magazine articles. Monthly electric bills are based on the number of kilowatt-hours used, not on the number of kilowatts used, which indicates only the *rate* at which electric energy is used and not the *amount* of energy used. When you fill up your gasoline tank at a service station, you do not pay for the rate at which gasoline is pumped into your car but for the amount of gasoline put into the tank. Similarly, you do not pay the electric utility company for the rate (in kilowatts) at which they deliver electric energy to your house, but for the total electric energy (in kilowatthours) delivered.

Example 17.12

A 150-W light bulb is connected to a 120-V line.
(a) What is the current drawn from the line?
(b) What is the resistance of the light bulb while it is burning?

(c) How much energy is consumed if the light is kept on for 6.0 h?
(d) What would be the cost of this energy at 8 cents per kilowatthour?

SOLUTION

(a) Since the power $P = IV$, we have

$$I = \frac{P}{V} = \frac{150 \text{ W}}{120 \text{ V}} = \boxed{1.25 \text{ A}}$$

(b) The resistance of the light bulb is

$$R = \frac{V}{I} = \frac{120 \text{ V}}{1.25 \text{ A}} = \boxed{96 \text{ } \Omega}$$

(c) $\mathcal{W} = Pt = (150 \text{ W})(6.0 \text{ h})\left(3600 \frac{\text{s}}{\text{h}}\right)$

$= \left(150 \frac{\text{J}}{\text{s}}\right)(2.16 \times 10^4 \text{ s}) = 3.2 \times 10^6 \text{ J}$

Since 1 kWh = 3.6×10^6 J,

$$\mathcal{W} = (3.2 \times 10^6 \text{ J})\left(\frac{1 \text{ kWh}}{3.6 \times 10^6 \text{ J}}\right) = \boxed{0.90 \text{ kWh}}$$

The same result could be obtained by using kilowatthours as the energy unit from the beginning. Then

$$\mathcal{W} = Pt = (150 \text{ W})(6.0 \text{ h}) = 900 \text{ Wh} = \boxed{0.90 \text{ kWh}}$$

(d) $\text{Cost} = \left(\frac{8.0 \text{ cents}}{1 \text{ kWh}}\right)(0.90 \text{ kWh}) = \boxed{7.2 \text{ cents}}$

17.9 Joule Heating

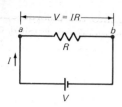

FIGURE 17.35 Joule heating in a resistor: $P = I^2R$.

We can now combine two previous results into an equation which Joule used to help establish the first law of thermodynamics. Consider the simple circuit of Fig. 17.35, in which a battery of terminal voltage V drives a current I through a resistance R. In the battery chemical energy is being converted into electric energy. In the resistor this electric energy is being converted into thermal energy, since the moving electrons collide with the atoms in the metal and give up some of their kinetic energy to these atoms. The random motion of the metal atoms is increased, and hence their temperature (and therefore their thermal energy) increases. The energy imparted to the electrons by the battery is ultimately used to heat the metal. This is called *Joule heating*.

Since the voltage across the resistance R is V, the power delivered to the resistor is $P = VI$. But, since, from Ohm's law, $V = IR$, we have

$$P = (IR)I = I^2R \quad \text{or} \quad \boxed{P = I^2R} \qquad (17.14)$$

Joule Heating

The rate at which thermal energy is generated in a resistance R is proportional to the resistance and to the square of the current through the resistance.

The total energy converted into heat in time t in the resistor is then

$$\mathcal{W} = Pt = I^2Rt \qquad (17.15)$$

Electrical Equivalent of Heat

To show the validity of the first law of thermodynamics we can perform an experiment which is the electrical counterpart of Joule's famous paddle-wheel experiment discussed in Sec. 14.6. Suppose we take the resistor of Fig. 17.35 and immerse it in water contained in a thermos bottle, as in Fig. 17.36. In time t the amount of electric energy delivered to the resistor is I^2Rt. This energy is all converted into heat in the resistor, and the temperature of the water increases. If the mass of the water is m, then the increase in thermal energy of the water is $cm\,\Delta T$, where c is the specific heat capacity of the water. In this way we can measure an amount of electric energy in joules (I^2Rt) and the equivalent amount of thermal energy in kilocalories ($cm\,\Delta T$).*

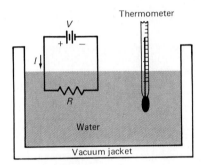

FIGURE 17.36 Experiment to determine the electrical equivalent of heat. From a measurement of V and I, the electric energy delivered in time t can be determined. From a measurement of the increase in temperature of the water of mass m, the heat input to the water can be determined.

When this experiment is carefully performed we find (as did Joule) that we always obtain the same relationship between the joule and the kilocalorie:

$1 \text{ kcal} = 4.18 \times 10^3 \text{ J}$

This is the same basic result Joule obtained from his paddle-wheel experiment, in which mechanical energy was converted into heat.

Hence we have another proof that energy is neither created nor destroyed, but only changes form. In this case energy is conserved but is degraded in the conversion process, since the ordered chemical energy stored in the battery has ultimately been converted into the disordered, random motion of the water molecules. Thus in this electrical version of Joule's experiment we find confirmation of both the first and second laws of thermodynamics: energy is conserved and entropy increases.

Example 17.13

A 50-Ω resistor is attached to a 120-V circuit and immersed in 1.5 kg of water. If current is drawn from the line for 15 min, what is the increase in temperature of the water, presupposing that all the electric energy goes into heating water?

---

*In a careful experiment the temperature increase of the inside of the thermos bottle, of the thermometer, and of any stirrer used must also be included in the calculation.

SOLUTION

The current drawn from the circuit is

$$I = \frac{V}{R} = \frac{120 \text{ V}}{50 \text{ } \Omega} = 2.4 \text{ A}$$

The electric energy going into heating the water is then

$$\mathcal{W} = I^2Rt = (2.4 \text{ A})^2(50 \text{ } \Omega)(15 \text{ min})\left(60 \frac{\text{s}}{\text{min}}\right) = 2.6 \times 10^5 \text{ J}$$

$$\mathcal{W} = (2.6 \times 10^5 \text{ J})\left(\frac{1 \text{ kcal}}{4.18 \times 10^3 \text{ J}}\right) = 62 \text{ kcal}$$

Then, from our equation for the specific heat capacity (from Chap. 13), we have $\mathcal{W} = cm \, \Delta T$, or

$$\Delta T = \frac{\mathcal{W}}{cm} = \frac{62 \text{ kcal}}{[1 \text{ kcal/(kg·°C)}](1.5 \text{ kg})} = \boxed{41°C}$$

Summary: Important Definitions and Equations

Electric current:
The rate of flow of electric charge.

$$I = \frac{Q}{t} \qquad 1 \text{ ampere (A)} = 1 \text{ C/s}$$

Conventional current: A current in the direction in which positive charges would flow in an electric circuit.

Seat of emf ($\mathcal{E}$):
A device (such as a battery or generator) which transforms nonelectric energy into electric energy on a continuous basis.

$$\mathcal{E} = \frac{\mathcal{W}}{Q}$$

where $\mathcal{E}$ is in volts, if $\mathcal{W}$ is in joules and Q in coulombs.

Battery:
A device which converts chemical energy into electric energy on a continuous basis.
Terminal voltage of battery: $V_{AB} = \mathcal{E} - Ir$, where r is the internal resistance of the battery.

Faraday (Q_F):
The amount of charge (96,500 C) which deposits or liberates an amount of substance from an electrolytic solution equal to the mass of one mole of the substance divided by its valence.

$$Q_F = N_A e$$

Ohm's law:
$V = IR$, where the resistance R (in ohms) is equal to $\rho L/A$, and ρ is the resistivity (in ohm-meters) of the material making up the wire.

$$1 \text{ ohm } (\Omega) = 1 \text{ V/A}$$

Temperature dependence of resistance:
$R_T = R_0(1 + \alpha \, \Delta T)$, where α is the temperature coefficient of resistivity, or the fractional increase in resistivity per degree Celsius rise in temperature.

Combinations of resistors:

$$\text{In series: } R = R_1 + R_2 \qquad \text{In parallel: } \frac{1}{R} = \frac{1}{R_1} + \frac{1}{R_2}$$

Combinations of capacitors:

$$\text{In series: } \frac{1}{C} = \frac{1}{C_1} + \frac{1}{C_2} \qquad \text{In parallel: } C = C_1 + C_2$$

Kirchhoff's rules:
1 The sum of all the currents entering any junction point is equal to the sum of all the currents leaving that junction point.
2 The algebraic sum of the changes in potential encountered in a complete traversal of any closed circuit is equal to zero. (If a resistor is traversed in the same direction as the current through the resistor, the change in potential is $-IR$; if in the opposite direction, the change in potential is $+IR$. If a seat of emf is traversed in the direction of the emf, the change in potential is $+\mathcal{E}$; if in the opposite direction, it is $-\mathcal{E}$.)

Electric power:

$$P = VI \qquad 1 \text{ W} = 1 \text{ V·1 A}$$

Joule heating:

$$P = I^2R \qquad \mathcal{W} = I^2Rt$$

Questions

1 When a dry cell is driving current through an external circuit, electrons inside the battery are moving from the + to the − terminal of the battery. How can this happen, since the electrons are attracted to the + terminal and repelled by the − terminal?

2 Give some examples of situations in which currents

flow, but Ohm's law is not obeyed. Is it proper to call a relationship like $V = IR$ a law of physics if it has only limited applicability?

3 A 6-V battery is connected in series with a capacitor.

(a) If the wires leading to the capacitor have a very low resistance, what is the voltage across the capacitor plates after all the charge has ceased to flow?

(b) If the wires leading to the capacitor had a very high resistance, how would the situation change? Would the battery have expended the same amount of chemical energy as in part (a)?

4 A student tries to turn on a light in a classroom and observes that the switch box is quite hot. Can you suggest a possible cause?

5 As the temperature of a wire is increased, the average kinetic energy of the electrons increases in proportion to the absolute temperature. Would you not expect therefore that the drift velocity of the electrons would increase with the temperature? Why, then, does the resistance of a wire increase with increasing temperature?

6 Would you expect the internal resistance of a dry cell battery to be a constant? Why?

7 Can the terminal voltage of a battery be zero? Why?

8 In a flashlight battery containing zinc and carbon electrodes, both the zinc and the carbon, if left by themselves, are electrically neutral. How then can a flashlight produce a potential difference between the zinc and the carbon electrodes?

9 If you measure the resistance of a 100-W light bulb, you will find that it is smaller than for a 60-W bulb. Is this not inconsistent with the fact that $P = I^2R$?

10 Why is it dangerous to replace a burnt-out 20-A fuse which is constantly blowing with a 30-A fuse in the hope that the fuse will no longer blow?

11 What happens when a light bulb burns out? Why is there a momentary flash of light just before the bulb goes dark?

12 Car batteries are often rated in ampere-hours (A·h). Is this a unit of current, power, voltage, energy, or charge? Why?

13 The voltage across a connecting wire in a electric circuit is usually assumed to be zero, since the wire has no resistance, and so $V = IR = 0$. How then can current flow through the wire? Can you suggest a good analogy between this situation and the flow of a liquid through a pipe?

Problems

A 1 The work done by a 6.0-V battery in moving a single electron through the battery from its positive to its negative terminal is:
(a) 6.0 J   (b) 6.0 V   (c) 1.6×10^{-19} J
(d) 9.6×10^{-19} J   (e) 1.6×10^{-19} W

A 2 A 6.0-V battery has an internal resistance of 0.50 Ω. To reduce the terminal voltage of the battery to 3.0 V, the battery will have to deliver to the external circuit a current of:
(a) 3.0 A   (b) 6.0 A   (c) 0.50 A   (d) 1.0 A
(e) 12.0 A

A 3 A storage battery delivered a 10-A current for a time of 10 min to a heating coil. The amount of charge which flowed through the heating coil in the 10 min was:
(a) 6.0×10^3 C   (b) 100 C   (c) 10 C
(d) 6.0×10^3 J   (e) 1 C

A 4 The amount of sodium deposited on the cathode of an electrolytic cell by the flow of 1 C of charge through the cell is about:
(a) 1 mol   (b) 10^{-6} mol   (c) 10^{-5} mol
(d) 0.5 mol   (e) 10^{-3} mol

A 5 A piece of metal wire is cut into three equal lengths, and the pieces are then put together side by side to form a thicker wire. The resistance of the new wire is less than that of the original wire by a factor of:
(a) 3   (b) 6   (c) 9   (d) 27   (e) 81

A 6 The resistivity of a metal like silver is less than that of an insulator like sulfur by a factor of:
(a) 10^8 (b) 10^{14} (c) 10^6 (d) 10^{16} (e) 10^{22}

A 7 If the temperature of a copper wire increases by 250°C, the resistance of the wire increases by a factor of about:
(a) 2   (b) 10^2 (c) 250   (d) 10^3 (e) 10

A 8 The fractional increase in resistance of a metal for a change in temperature of 1°C exceeds its fractional increase in length for the same temperature change by about:
(a) One order of magnitude
(b) Two orders of magnitude
(c) Three orders of magnitude
(d) Four orders of magnitude
(e) They are about the same

A 9 A 200-Ω and a 100-Ω resistor are connected in parallel. The equivalent resistance of the combination is:
(a) 300 Ω   (b) 20 Ω   (c) 150 Ω   (d) 67 Ω
(e) 100 Ω

A10 A power line carries 1000 A and has a resistance of 0.10 Ω. The rate at which electric energy is converted into heat in the line is:
(a) 100 W   (b) 10^5 W   (c) 10^5 J
(d) 10^4 W   (e) 10 W

B 1 A 12-V automobile battery is being charged. A current of 20 A flows into the battery for 2 h. How much electric charge has been delivered to the battery?

B 2 A current of 10 A is maintained for 15 min through an electrolytic cell containing Cu^{2+} ions. How much copper is deposited?

B 3 A voltage of 120 V is applied to a light bulb of resistance 200 Ω.

(*a*) What is the current through the light bulb?

(*b*) How much charge flows through the light bulb every hour?

B 4 A sparrow stands on a 240,000-V high-tension wire carrying 1000 A. The line has a resistance of 2.0×10^{-5} Ω/m. What voltage does the bird feel across its body if its feet are 5 cm apart?

B 5 What is the resistance, at 0°C, of a piece of nichrome wire 0.50 m long and of cross-sectional area 1.0 mm²?

B 6 Two resistors, one of resistance $R_1 = 20.0$ Ω and one of resistance $R_2 = 60.0$ Ω are connected in parallel to a voltage source of 120 V.

(*a*) What is the equivalent resistance of the circuit?

(*b*) What is the current delivered by the battery?

(*c*) What is the current through R_1 and R_2 individually?

B 7 A 6.0-V battery which has an internal resistance of 0.50 Ω is delivering 2.0 A to the external circuit. What is the terminal voltage of the battery?

B 8 A 60-W bulb is connected to a 120-V line.

(*a*) What is the current which flows?

(*b*) What is the resistance of the light bulb?

B 9 A power line of resistance 1.0 Ω carries 100 A of current for a day. How much energy has the power company lost in the form of heat in the line?

B10 If the current through a 100-W light bulb is 0.83 A:

(*a*) What is the resistance of the light bulb?

(*b*) What is the line voltage?

C 1 How much work is done by a storage battery, of emf equal to 12 V and negligible internal resistance, if it delivers a 50-A current for 10 s?

C 2 In an x-ray tube an electron is accelerated by a voltage of 15×10^3 V between the cathode and the anode of the tube.

(*a*) When the electron crashes into the anode, what energy does it possess?

(*b*) What is the electron's speed?

C 3 In the video tube of a TV set the electron beam is accelerated toward the screen by a voltage of 2.0×10^4 V. With what speed does an electron strike the screen of the tube?

C 4 Chlorine gas (Cl_2) can be produced from a solution containing Cl^- ions, e.g., from Na^+Cl^-, by electrolysis.

(*a*) What must be the current through an electrolytic cell to produce 5 kg of Cl_2 gas in 10 h?

(*b*) What is the volume of this gas, stored as Cl_2 at 0°C and a pressure of 50 atm?

C 5 Objects made of iron are often plated with cadmium to make them rustproof. If the cadmium is present in an electrolytic cell in the form of Cd^{2+} ions, how much cadmium will be plated on the iron object in 5 h by a 20-A electric current?

C 6 A voltage of 100 V is applied across the ends of a piece of silver wire 1 m long.

(*a*) What is the electric field in the wire?

(*b*) What is the force acting on the electron?

(*c*) What is the electron's acceleration?

C 7 In starting a car a 12-V battery is required to deliver an extremely large current, perhaps as much as 200 A, for a few seconds. There is always some resistance in the wires leading to the motor from the battery and in the lugs attaching the wires to the battery terminals. What is the maximum allowable resistance in each of the two leads if a 200-A current is to be delivered by the battery? (This provides a good reason for keeping the battery lugs on a car clean to reduce the contact resistances as much as possible.)

C 8 What is the resistance at the temperature of boiling water (100°C) of a piece of no. 5 gauge copper wire, of length 4.5 m and of cross-sectional area 16.8 mm²?

C 9 A platinum resistance thermometer has a resistance of 50 Ω at 0°C. It is then immersed in a mixture of solid and liquid tin at 2260°C. What resistance will the thermometer now show?

C10 A 100-W incandescent bulb has a resistance of 16 Ω when it is at room temperature, and 144 Ω when it is hot and delivering light to the room. What is the approximate temperature of the bulb when in use, if the filament of the bulb is made out of tungsten?

C11 We have two resistors, one of 4.0 Ω and the other 8.0 Ω.

(*a*) If the two resistors are connected in series in a circuit, what is the equivalent resistance of the circuit?

(*b*) If the same two resistors are connected in parallel, what is the equivalent resistance?

C12 Calculate the equivalent resistance of the following combinations of resistors:

(*a*) A 3.0-Ω resistor, a 4.0-Ω resistor, and two 5.0-Ω resistors in series.

(*b*) A 4.0-Ω resistor, a 2.0-Ω resistor, and a 6.0-Ω resistor in parallel.

C13 In the circuit of Fig. 17.37 calculate:

(*a*) The equivalent resistance of the circuit.

(*b*) The current through each of the resistors in the circuit.

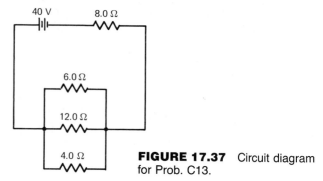

FIGURE 17.37 Circuit diagram for Prob. C13.

C14 In the circuit of Fig. 17.38, in which the internal resistance of the battery is 0.40 Ω, calculate:

(*a*) The equivalent resistance of the circuit.

(*b*) The current through each of the resistors in the circuit.

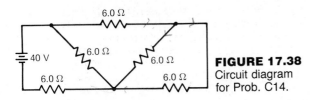

FIGURE 17.38
Circuit diagram
for Prob. C14.

C15 (*a*) What is the equivalent capacitance of the following three capacitors in series: $C_1 = 0.50 \ \mu F$; $C_2 = 0.20 \ \mu F$; $C_3 = 0.10 \ \mu F$?
(*b*) What is the equivalent capacitance of the same three capacitors in parallel?

C16 A voltage of 50 V is connected across three capacitors in series. The total charge on the three capacitors is 5.0 μC. If two capacitors have capacitances of $C_1 = 0.50 \ \mu F$ and $C_2 = 0.25 \ \mu F$, what is the value of C_3?

C17 A 0.30-μF capacitor is connected in parallel with a 0.15-μF capacitor. This combination is then connected in parallel with a 0.40-μF capacitor. What is the equivalent capacitance of the combination?

C18 In the circuit of Fig. 17.39, what is the charge on the 10-μF capacitor?

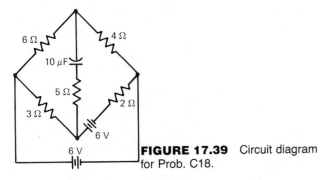

FIGURE 17.39 Circuit diagram for Prob. C18.

C19 Find the currents I_1, I_2, and I_3 in the three branches of the circuit of Fig. 17.40. Assume that the batteries have zero internal resistance.

C20 Solve the circuit of Fig. 17.40 for the currents I_1, I_2, and I_3, under the assumption that each battery has an internal resistance of 1.0 Ω.

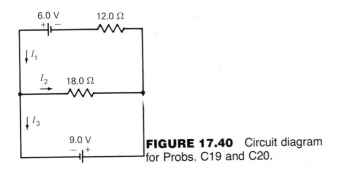

FIGURE 17.40 Circuit diagram for Probs. C19 and C20.

C21 If an electric air-conditioning unit draws 15 A from a 120-V line and is used 24 h a day during July, how much will the electricity to run this unit cost for the month of July, if the local electrical rate is 8 cents per kilowatthour?

C22 Suppose you burn five 100-W bulbs, each for 6 h a day.
(*a*) How much electric energy do you consume each day?
(*b*) What is the cost of this energy at 8 cents per kilowatthour?

C23 In the circuit of Fig. 17.40, calculate:
(*a*) The power produced by the two batteries.
(*b*) The power dissipated as heat in the two resistors.
(*c*) Do your results indicate that energy is conserved? (Assume that the batteries have zero internal resistance.)

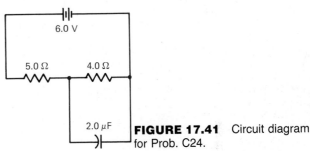

FIGURE 17.41 Circuit diagram for Prob. C24.

C24 In the circuit of Fig. 17.41, find the charge on the 2-μF capacitor if the battery has an internal resistance of 1.0 Ω.

C25 Some light bulbs possess two separate filaments and allow three different wattages depending on how the filaments are connected, for example, 50-100-150 W.
(*a*) What resistances would be required in the two filaments for the bulb to consume 50 W and 100 W of electric power when connected to a 120-V line?
(*b*) Show that the resistance needed to make the bulb function as a 150-W bulb is merely the sum of the two filament resistances in parallel. Assume that Ohm's law remains valid, even though the current and voltage here are alternating rather than direct.

C26 Batteries are often rated in terms of ampere-hours, i.e., the number of hours the battery would last if current were drawn from it at the rate of 1 A. Consider a simple 1.5-V D cell used in flashlights. If it is rated for 3 A·h and the current drawn by the flashlight is 0.15 A:
(*a*) How long will the battery last?
(*b*) What will be the total energy delivered by the battery during its lifetime?
(*c*) If the battery costs 50 cents, what is the cost per kilowatthour for the electric energy used?
(*d*) How does this compare with the rate of about $0.08 per kilowatthour at which utilities sell electric energy?

C27 A student leaves her car in the parking lot of the university at 8 a.m. She forgets to turn off the car's lights. Assume that each of the two front lights have 40-W bulbs and the two rear lights 6-W bulbs. The battery is rated at 100 A·h. How long will it take to completely run down the car's fresh 12-V storage battery?

C28 An electric space heater draws 7.5 A from a 120-V line.

(a) If this heater is kept on 14 h per day, how much electric energy is used?

(b) What is the cost of this energy at 8 cents per kilowatthour?

(c) If all the electric energy is converted into heat, how much heat is produced in kilocalories?

D 1 Objects can be gold- or silver-plated using electrolysis. It is desired to gold-plate a metal object to a thickness of 20 μm (that is, 20×10^{-6} m). The surface area of the metal object is 0.010 m^2 and it is placed in an electrolytic cell containing gold (Au^{3+}) ions. If a current of 5.0 A is driven through the cell, how long will it take to complete the gold plating?

D 2 If you are given some nichrome wire, with $R = 24$ Ω, what is the best way to get more heat from this wire—by winding one heating coil out of the wire, or by cutting it in half, winding two separate coils, and connecting them in parallel? In each case the coils are connected across a 120-V line.

D 3 Compute the current through the 12-Ω resistor in Fig. 17.42.

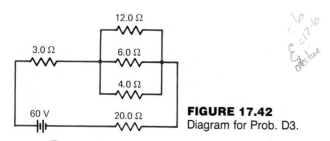

FIGURE 17.42
Diagram for Prob. D3.

D 4 Prove that when a storage battery with internal resistance r is being charged, the voltage which must be applied to charge it is $V = \mathcal{E} + Ir$, where I is the charging current and $\mathcal{E}$ is the emf of the battery.

D 5 Calculate the equivalent resistance of the circuit of Fig. 17.43.

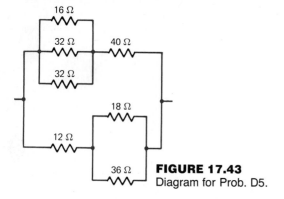

FIGURE 17.43
Diagram for Prob. D5.

D 6 Copper is a slightly better conductor than aluminum, but its density is greater. Show that for two electric power lines 1000 m in length an aluminum line will have a lower resistance than a copper line of the same weight.

D 7 (a) Determine the currents I_1, I_2, and I_3 in Fig. 17.44, under the assumption that the internal resistance (r) of each of the three batteries is 1.0 Ω.

(b) What is the terminal voltage of the 18-V battery?

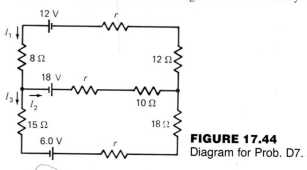

FIGURE 17.44
Diagram for Prob. D7.

D 8 In the circuit of Fig. 17.45, $\mathcal{E}_1 = 12$ V, $\mathcal{E}_2 = 6$ V, and both have zero internal resistance. Also $R_1 = 24$ Ω, $R_2 = 12$ Ω, and $R_3 = 6$ Ω. Find I_1, I_2, and I_3.

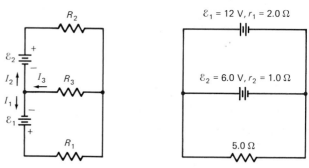

FIGURE 17.45
Diagram for Prob. D8.

FIGURE 17.46
Diagram for Prob. D9.

D 9 (a) Calculate the current delivered by each emf—$\mathcal{E}_1$ (internal resistance, 2 Ω) and $\mathcal{E}_2$ (internal resistance, 1 Ω)—in Fig. 17.46.

(b) Calculate the current through the 5-Ω resistor.

(c) What is the terminal voltage of each battery?

D10 Calculate the voltage between the points P and P' in the circuit of Fig. 17.47. Assume that the three emf's have negligible internal resistance and that $\mathcal{E}_1 = 40$ V, $\mathcal{E}_2 = 20$ V, and $\mathcal{E}_3 = 60$ V.

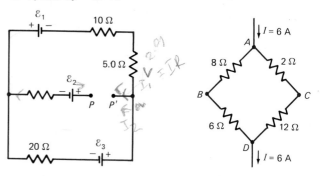

FIGURE 17.47
Diagram for Prob. D10.

FIGURE 17.48
Diagram for Prob. D11.

D11 (a) What is the potential difference between points B and C in Fig. 17.48?

(b) Which point is at the higher potential?

D12 Figure 17.49 is the diagram of a Wheatstone bridge, which is used to measure unknown resistances to high accuracy by a "null" method that compares them with standard resistors of precisely known values. In the diagram R_1 and R_2 are fixed resistances, R is a variable standard resistor, and X is the unknown. Prove, using Kirchhoff's rules, that, if R is adjusted until no current flows through the meter M when the key K is closed, then $X = (R_2/R_1)R$.

D13 Two cells, one of emf 1.2 V and internal resistance 0.50 Ω, and the other of emf 2.0 V and internal resistance 0.10 Ω, are connected in parallel, as shown in Fig. 17.50. This combination of two emf's is then connected in series with an external resistance of 5.0 Ω.

(a) What is the current through the external resistor?

(b) Prove that the power generated by the two emf's is equal to the power dissipated as heat in the circuit.

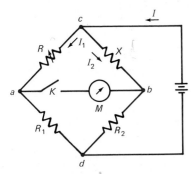

FIGURE 17.49 Circuit diagram for Prob. D12: A Wheatstone bridge.

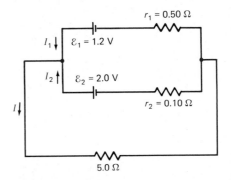

FIGURE 17.50 Circuit diagram for Prob. D13.

Additional Readings

Asimov, Isaac: *Asimov's Biographical Encyclopedia of Science and Technology*, 2d rev. ed., Doubleday, Garden City, N.Y., 1982. This book contains brief biographies of some 1510 famous scientists, written in Asimov's lucid and entertaining style. Among the physicists included are Kirchhoff, Ohm, Volta, Galvani, and others who developed the physics presented in this chapter.

Healy, Timothy J.: *Energy, Electric Power and Man*, Boyd and Fraser, San Francisco, 1974. Included are good discussions of electric power, electric automobiles, and alternative energy sources.

Magie, William Francis: *A Source Book in Physics*, Harvard University Press, Cambridge, Mass., 1963. In this useful volume are a brief account of Ohm's life and a long excerpt from the paper in which he first proposed his famous law. It also contains material on the discovery of the battery by Galvani and Volta.

Rosenfeld, L.: "Gustav Robert Kirchhoff," in *Dictionary of Scientific Biography*, Scribner, New York, 1970, vol. 7, pp. 379–382. One of the few accounts available in English of Kirchhoff's life and contributions to physics.

Shepard, G. M.: "Microcircuits in the Nervous System," *Scientific American*, vol. 238, no. 2, February 1978, pp. 92–103. An interesting discussion of the microcircuits in the brain which provide a deeper understanding of the neural mechanisms underlying behavior.

*T*he experimental investigation by which Ampère established the laws of the mechanical action between electric currents is one of the most brilliant achievements in science. The whole, theory and experiment, seems as if it had leaped full grown and full armed from the brain of the "Newton of Electricity."

James Clerk Maxwell (1831–1879)

Chapter 18
Magnetism

The year 1820 was a notable one in the history of physics. In that year Oersted discovered that an electric current deflected a compass needle—the first clear evidence that magnetism and electricity were interrelated. A week after Oersted's discovery was reported to the French Academy of Sciences in Paris, Ampère had extended Oersted's work to show that two current-carrying conductors exerted forces on each other through the magnetic fields they produced. Thus began the branch of physics called *electrodynamics*, a name given to the field by Ampère himself.

Immediately scientists all over the world rushed to their laboratories to carry out experiments involving magnets and electric currents, driven not only by a desire to understand nature better but also by the sense that such remarkable discoveries might have very important practical consequences, as they indeed did. Witness the motors, generators, electric meters, telephones, and other more sophisticated electromagnetic devices which surround us today. As Isaac Asimov once remarked, nothing like the discoveries of Oersted and Ampère were seen again "until the announcement of nuclear fission a century later."

18.1 Magnets and the Magnetic Field

Most of us played with bar magnets and compass needles when we were young. Observations made during such experimentation lead to the following basic facts about the behavior of magnets:

1 *There are two different kinds of magnetic poles*, or places on a magnet where the magnetic effects are particularly strong. These are called north (N) and south (S) poles because, when placed in the earth's magnetic field, the north pole of a magnetic needle will point north and the south pole will point south. The two kinds of magnetic pole (N and S) are analogous to the two kinds of electric charge (+ and −).

2   *Like magnetic poles repel each other and unlike poles attract each other.* This again is similar to the behavior of electric charges. (See Fig. 18.1.)

3   *It is impossible to produce an isolated north or south pole in a magnet.** The production of one kind of magnetic pole is always accompanied by the simultaneous production of an opposite pole on the other end of the magnet. As an example, north and south poles cannot be separated by cutting a bar magnet in half. All that happens is that two smaller bar magnets result, each with N and S poles at the two ends, as shown in Fig. 18.2.

4   *Either pole of a magnet will attract previously unmagnetized materials containing iron, cobalt, or nickel, which are the primary magnetic substances.* (See Fig. 18.3.)

The rest of this chapter will be devoted to elucidating these experimental facts.

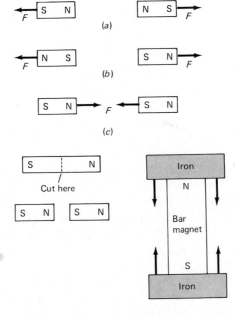

FIGURE 18.1   The interaction between magnetic poles: *(a)* Two north poles repel each other. *(b)* Two south poles repel each other. *(c)* A north pole attracts a south pole.

FIGURE 18.2   Cutting a bar magnet in two: The result is two smaller bar magnets.

FIGURE 18.3   Either pole of a magnet will attract unmagnetized materials which contain iron, cobalt, or nickel.

The Magnetic Field

We have seen that surrounding any stationary electric charge is an electric field, which can be represented by lines of electric force, whose density measures the electric field strength **E** and whose direction indicates the direction of the electric field at any point in space.

In a similar way, surrounding any magnetic pole is a magnetic field, which can be represented by lines of magnetic force, whose density measures the magnetic field strength **B** and whose direction indicates the direction of the magnetic field at any point in space. The direction of the field is conventionally taken to be the direction of the force experienced by an

---

*The history of physics is marked by an almost continuous effort to isolate magnetic "monopoles" in the laboratory, but it is at present unclear whether such odd "beasts" will ever be captured and tamed.

isolated N pole in the field and hence the direction in which this N pole would move in the field.

According to this convention, the lines of a magnetic field always begin at the N pole of a magnet and end at a S pole, as can be seen in the diagrams of Fig. 18.4 for a variety of permanent magnets. An isolated N pole will always be pushed by the magnetic field from the N pole of the magnet to the S pole and hence will move along the field lines in the diagram in the direction shown by the arrows. The lines of the magnetic field are illustrated by the use of iron filings in Fig. 18.5.

FIGURE 18.4 Some permanent magnets, with magnetic field lines drawn to show that the lines of the field start on north poles and end on south poles: (a) A bar magnet; (b) a horseshoe magnet; (c) a laboratory magnet with large, flat pole pieces.

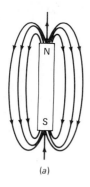
(a)

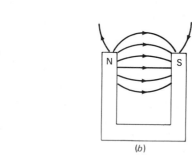
(b)

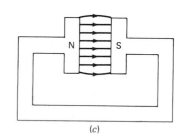
(c)

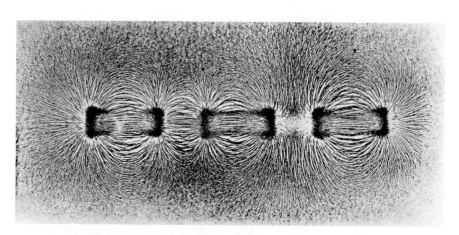

FIGURE 18.5 Magnetic field lines made visible by the use of iron filings which align themselves in the direction of the magnetic field. The adjacent poles of the first two magnets attract each other, while those of the second and third magnets repel each other. Notice the similarity of these magnetic field patterns to those of an electric dipole shown in Fig. 16.23. (*Photo courtesy of Omikron/Photo Researchers.*)

The magnitude of the magnetic field strength could, in principle, be mapped by measuring the force experienced by an isolated N pole when placed at different points in the field. In practice, however, this cannot be done, since we cannot obtain the isolated pole we need to probe the field. For this reason we will restrict ourselves for the present to the following somewhat incomplete definition of the *magnetic field strength* **B**.

Definition

*The magnetic field strength **B** is a vector whose direction at any point in space is that in which a small, isolated N pole would be moved by the magnetic force at that point, and whose magnitude is proportional to the force experienced by that N pole.*

18.2 Force on a Current-Carrying Conductor in a Magnetic Field

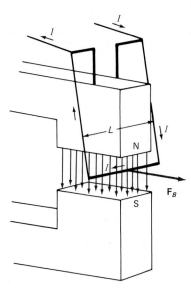

A magnetic field has no effect on *stationary* electric charges. Experimentally it is found, however, that a magnetic field does exert a force on *moving* electric charges and therefore on electric currents.

One of the easiest ways to obtain the magnetic field strength **B** is to measure the force exerted on a wire carrying a current through a magnetic field. Suppose we have a magnetic field **B** in the vertical direction provided by a permanent magnet, as in Fig. 18.6. We suspend a light rectangular loop of wire from the magnet, as in the figure, with a straight length (L) of wire in the horizontal plane and at right angles to the magnetic field. When a current is passed through the wire of length L, the wire is pushed out in the direction shown in the figure. Hence an added force must occur—a magnetic force—on the current-carrying wire in the magnetic field. This force can be measured in terms of the angle the wire makes with the vertical when it comes to rest, for this angle depends on the relative sizes of the magnetic and gravitational forces acting on the wire (see Example 18.2).

Further experimentation leads to the conclusion that the magnitude of the force F_B depends directly on the length of the wire in the magnetic field, L, on the magnitude of the current I through the wire, and on the strength of the magnetic field at right angles to the wire, which we will designate by $\mathbf{B}_\perp$. Hence we can write

FIGURE 18.6 Apparatus to measure the force on a current-carrying conductor in a magnetic field. When a current I passes through the wire of length L in the magnetic field **B**, the wire experiences a force $F_B = BIL$ in the direction shown.

$$F_B \propto B_\perp IL$$

As usual, we can convert this proportion into an equality by introducing a constant of proportionality, but in this case we prefer to define the units for magnetic field strength in such a way as to make this proportionality constant equal to unity, so that

Force Produced by a Magnetic Field

$$F_B = B_\perp IL \qquad (18.1)$$

Since in the SI system F_B will be in newtons (N), I in amperes (A), and L in meters (m), the unit for $B_\perp$ must be N/(A·m) to make Eq. (18.1) dimensionally correct. In the SI system 1 N/(A·m) is called a *tesla* (T).

Definition

One tesla, which is equal to one newton per ampere-meter, is the strength of that magnetic field in which a current-carrying conductor of length one meter oriented perpendicular to the magnetic field and carrying a current of one ampere will experience a force of one newton.

An older, non-SI unit still frequently used for the magnetic field strength is the *gauss* (G) where 1 G = 10^{-4} N/(A·m) = 10^{-4} T. A field of 1 T, or 10^4 G, is a rather large magnetic field. The magnetic field of the earth is on the order of 0.5 G, or 5×10^{-5} T.

The direction of the force on the current-carrying conductor can be found by experiment. The results conform to the following rule, called the *right-hand force rule*. (See Fig. 18.7; also check the direction of deflection shown in Fig. 18.6 against this rule.)

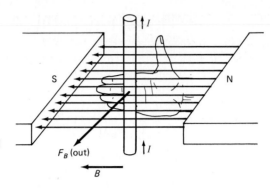

FIGURE 18.7 The right-hand force rule. If the fingers are in the direction of the magnetic field and the thumb in the direction of the current, then the palm of the hand faces in the direction of the force produced.

Right-Hand Force Rule

If the fingers of the outstretched right hand are placed in the direction of the magnetic field, and the thumb in the direction of the conventional current I, then the palm of the hand faces in the direction of the force on the current-carrying conductor.

FIGURE 18.8 How the magnetic force on a wire varies with the angle between the current direction and the magnetic field: *(a)* Current is parallel to lines of **B**; no force results. *(b)* Current makes an angle θ with **B**; force is $BIL \sin \theta$. *(c)* Current is perpendicular to **B**; force is BIL.

Note that **B** must have a component normal to the wire or no force arises. If the angle between **B** and the wire is θ, then $B_\perp$ equals $B \sin \theta$. Hence the force varies as the sine of the angle between B and the wire, which goes from 0 to 1 as θ goes from 0 to 90°. This is illustrated in Fig. 18.8.

Note the clear distinction Eq. (18.1) makes between the magnetic force F_B and the magnetic field $B_\perp$. In this case F_B differs from $B_\perp$ not only in magnitude and units but also in *direction*.

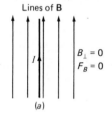

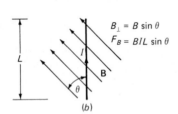

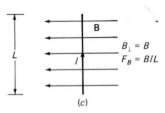

Example 18.1

A straight wire carries a 1.5-A current at right angles to a magnetic field of strength $B = 1000$ G, as in Fig. 18.9. If the length of the wire in the magnetic field is 5.0 cm:
(a) What is the magnitude of the force on the wire?
(b) What is the direction of this force?

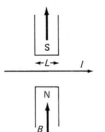

FIGURE 18.9 Diagram for Example 18.1.

SOLUTION

(a) The magnitude of the magnetic force is $F_B = B_\perp IL$. Since 10^3 G $= 10^{-1}$ T, and the field is perpendicular to the direction of the current, we have

$$F_B = (10^{-1} \text{ T})(1.5 \text{ A})(5.0 \times 10^{-2} \text{ m}) = \boxed{7.5 \times 10^{-3} \text{ N}}$$

(b) If the situation is as in Fig. 18.9 and we apply the right-hand rule, we find that the force is directed *out of the plane of the paper*.

Example 18.2

In the setup of Fig. 18.6, the horizontal wire has a mass of 100 g and is 10 cm long. The magnetic field strength is 0.50 T. What angle do the supporting wires make with the vertical when a current of 4.0 A passes through the wire?

SOLUTION

The force on the current-carrying strip of wire of length L is, since the field is perpendicular to the wire,

$$F_B = B_\perp IL = (0.50 \text{ T})(4.0 \text{ A})(0.10 \text{ m})$$

$$= \left(0.50 \frac{\text{N}}{\text{A·m}}\right)(0.40 \text{ A·m}) = 0.20 \text{ N}$$

The gravitational force on the wire is:

$$w = mg = (0.10 \text{ kg})(9.8 \text{ m/s}^2) = 0.98 \text{ N}$$

Hence $\tan \theta = \dfrac{F_B}{w} = \dfrac{0.20 \text{ N}}{0.98 \text{ N}} = 0.20$

and $\theta = \boxed{11°}$

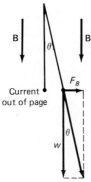

FIGURE 18.10   Diagram for Example 18.2.

where θ is the angle shown in Fig. 18.10.

18.3 Force on a Charged Particle in a Magnetic Field

In atomic and nuclear physics we are more interested in the force on individual charged particles in motion through a magnetic field than we are in forces on current-carrying wires.

Let us consider positive charges moving in a beam with a velocity **v** at right angles to a uniform magnetic field **B**, as in Fig. 18.11. Since moving charges constitute a current, we expect that the magnitude of the magnetic force on a length L of the beam will be $F_B = B_\perp IL$, as given by Eq. (18.1). The current I is the amount of charge passing any point along the path of the beam in 1 s. Hence

$$I = \frac{nq}{t} \tag{18.2}$$

where n is the number of charges passing that point in time t, and q is the magnitude of each charge. We assume that the charge density in the beam is constant, so that I is the same at any point along the beam.

FIGURE 18.11   Forces exerted by a magnetic field on charged particles moving in a beam through a magnetic field. The force is at right angles to both **B** and **v**, and in this case is given by the right-hand force rule as directed into the paper.

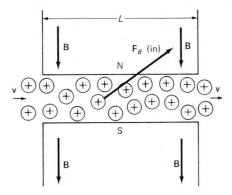

For convenience we choose t to be the time it takes for a charge to move the distance L at a speed v (assumed the same for all charges in the beam), where L is the width of the magnetic field. Hence

$$L = vt \tag{18.3}$$

But for the time t, as chosen in this way, n is equal to the total number of charges in the length L, for the charges at the beginning of L will just reach the end of L in time t, since they move at a speed $v = L/t$. Hence the magnetic force on the beam, which is the force on n individual charges, is, from Eqs. (18.1), (18.2), and (18.3),

$$F_B = B_\perp IL = B_\perp \frac{nq}{t} vt = Bnqv^*$$

The force on a single charge q is then:

$$\boxed{f_B = \frac{F_B}{n} = B_\perp qv} \tag{18.4}$$

Force on a Moving Charge

A charge q moving with a velocity $\mathbf{v}$ at right angles to a magnetic field $\mathbf{B}$ experiences a force at right angles to both $\mathbf{B}$ and $\mathbf{v}$ of magnitude $B_\perp qv$.

For a positive charge the right-hand force rule again gives us the direction of the force f_B.

Direction of Force

If the fingers of the right hand are placed in the direction of $\mathbf{B}$ and the thumb in the direction of the velocity $\mathbf{v}$, then the palm of the hand faces in the direction of the force on a moving positive charge.

If the charge q is negative, as it would be for an electron, q in Eq. (18.4) is negative, and hence f_B will be in the *opposite direction* to that predicted by the right-hand rule for a positive charge. The directions of the forces in these two cases are shown in Fig. 18.12.

---

*The result of this derivation has been to show that in the expression $F_B = B_\perp IL$ we can replace IL by nqv. This same kind of substitution turns up in many other fields of physics, e.g., in the kinetic theory of gases in Sec. 9.7. As another example, if a stretch of highway 1000 m in length contains n automobiles (q here stands for an automobile), each moving at a speed of $v = 25$ m/s, and the number of cars passing any point on the road per second is 10 (this is equivalent to I, being the "car current"), then the total number of cars on the 1000-m stretch is, from the above equation:

$$n = \frac{IL}{qv} = \frac{(10 \text{ cars/s})(1000 \text{ m})}{(1 \text{ car})(25 \text{ m/s})} = 400$$

Hence there are 400 cars on this stretch of road.

FIGURE 18.12 The direction of the forces on *(a)* positive and *(b)* negative charges moving through a magnetic field. The effect of these forces is to deflect the charged particles in the way shown in this figure.

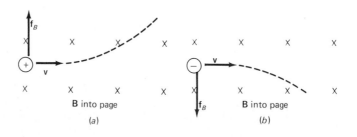

B into page

(a)

B into page

(b)

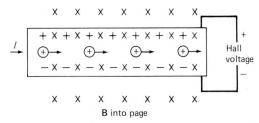

FIGURE 18.13 The Hall effect: Current carried by positive charges moving in the direction of the conventional current *I*.

Hall Effect

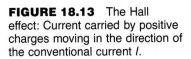

Suppose we take a thin, flat ribbon of metal with the flat side perpendicular to a magnetic field **B** (into the paper in Fig. 18.13), and pass a current *I* along the length of the wire ribbon.

As we have seen, a positive charge flowing in one direction is completely equivalent to a negative charge flowing in the opposite direction. Therefore, when we establish a current in the wire, there is no immediate way of knowing whether positive charges or negative charges carry the current. The Hall effect enables us to decide unambiguously between these two possibilities. Suppose in Fig. 18.13 the conventional current is driven to the right by the battery. Then, if the current were carried by positive charges, the magnetic field would deflect the charges upward, and the upper edge of the ribbon would become positively charged and the lower edge negatively charged, according to our right-hand force rule. Hence a *Hall voltage* would appear across the ribbon, with the voltage being directed from the top to the bottom of the ribbon.

If, on the other hand, the current is carried by *electrons* moving to the *left*, then, because of their negative signs the electrons would also be deflected *upward* in the magnetic field. In this case the upper edge of the wire would become negatively charged and the lower edge positively charged, so that the Hall voltage would be from the bottom to the top of the ribbon, as in Fig. 18.14.

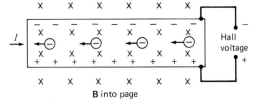

FIGURE 18.14 The Hall effect: Current carried by negative charges moving in the opposite direction to that of the conventional current.

When the Hall voltage is measured, it is found that for metal conductors the voltage is from the bottom to the top of the ribbon in the case described above. Hence *the current in metals is carried by electrons moving in a direction opposite to that of the conventional current.* This is not the case in some semiconductors, where the Hall voltage shows that the current is apparently carried by positive charges. The nature of these charges will be discussed in Chap. 29.

The Hall voltage can be shown to be given by

$$V_H = Bwv_D \tag{18.5}$$

where *B* is the strength of the magnetic field, *w* is the width of the strip (from top to bottom in Fig. 18.14), and v_D is the drift velocity of the charges. Hence the magnetic field strength is directly proportional to the magnitude of the Hall voltage. This is used in an instrument called a Hall-effect gaussmeter* for the measurement of magnetic fields. These are very convenient instruments calibrated to convert measurements of the Hall voltage directly into magnetic field strengths.

Paths of Charged Particles in a Magnetic Field

If an atomic particle of charge q is shot into a magnetic field with a velocity **v** at right angles to the direction of the field **B**, as in Fig. 18.15, it is deflected at right angles to its original direction by the magnetic field. As it changes its direction slightly, it continues to be deflected at right angles to its new direction, and so on. Since the force **F** continues to be at right angles to the velocity **v**, it acts like a string keeping a ball rotating in a circle. As a result the path of the particle is a *circle*. The radius of the circle can be found by noting that the centripetal force needed for the circular motion is here provided by the magnetic field, so that, from Eqs. (18.4) and (5.29),

$$Bqv = \frac{mv^2}{r} \quad \text{or} \quad r = \frac{mv}{Bq} \tag{18.6}$$

(In most of the following examples the arrangement of the apparatus ensures that **B** is perpendicular to **v**. Thus we will write simply B instead of $B_\perp$.) Hence if the charge of the particle is known (it is usually one or two elementary charges), together with the magnitude of the magnetic field, the momentum mv of the particle can be obtained from a measurement of the radius of the circle in which the particle moves, since $mv = Bqr$. For electrons the circles are very small because of the electron's small mass. This makes it easy to identify electrons in the photographs of particle tracks taken in cloud

---

*Since the gauss is a unit of magnetic field strength, the name *gaussmeter* is given to instruments designed to measure magnetic fields. Similarly, a *degausser* is a device for removing magnetic fields induced in submarines, watches, or other steel or iron objects after they have been in a magnetic field like that of the earth.

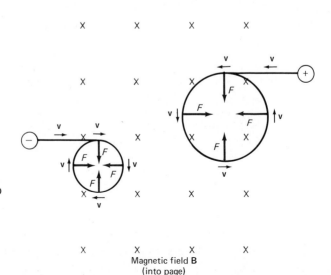

FIGURE 18.15 Charged particles moving at right angles to a magnetic field. They are bent by the magnetic field into closed circles, with (in this case) electrons rotating in clockwise circles and protons in counterclockwise circles.

Magnetic field **B**
(into page)

chambers and other kinds of nuclear detectors. From the direction in which the particle is bent in the magnetic field it is also possible to tell whether its charge is positive or negative, as is clear from Fig. 18.15.

If the speed v of the particle is known, for example, from the voltage used to accelerate it to that speed, then from Eq. (18.6) we have

$$\frac{q}{m} = \frac{v}{Br} \tag{18.7}$$

and the charge-to-mass ratio can also be obtained from a measurement of r.

The charge-to-mass ratio is a very important quantity for any atomic particle, since, once q/m is known, a measurement of the charge q leads directly to a value for the particle's mass m.

Example 18.3

An electron in a vacuum tube is accelerated by a voltage of 2000 V and enters a region in which there is a uniform magnetic field of 1.0 T at right angles to the direction of the electron's motion.

(a) What is the force on the electron due to the magnetic field?

(b) In what path does the electron move in the magnetic field? Be as specific as possible.

SOLUTION

(a) To find the magnetic force on the electron, we first need to know its speed v. Since the electron has been accelerated by 2000 V, the work done on it is $\mathcal{W} = qV = \frac{1}{2}mv^2$, for in this case all the work done must go into kinetic energy. Hence

$$v = \sqrt{\frac{2qV}{m}} = \left[\frac{2(1.6 \times 10^{-19}\ \text{C})(2.0 \times 10^3\ \text{V})}{9.1 \times 10^{-31}\ \text{kg}}\right]^{1/2}$$

$$= 2.7 \times 10^7\ \text{m/s}$$

Then the magnetic force is

$$f_B = Bqv$$

$$= \left(1.0\ \frac{\text{N}}{\text{A·m}}\right)(1.6 \times 10^{-19}\ \text{C})\left(2.7 \times 10^7\ \frac{\text{m}}{\text{s}}\right)$$

$$= \boxed{4.3 \times 10^{-12}\ \text{N}}$$

(b) The electron's direction will constantly change as it is continuously deflected at right angles to its original direction of motion. Hence it will move in a circle at a constant speed v.

The radius of the circle is given by Eq. (18.6):

$$r = \frac{mv}{Bq} = \frac{(9.1 \times 10^{-31}\ \text{kg})(2.7 \times 10^7\ \text{m/s})}{(1.0\ \text{T})(1.6 \times 10^{-19}\ \text{C})}$$

$$= 1.5 \times 10^{-4}\ \text{m} = \boxed{0.15\ \text{mm}}$$

Mass Spectrometer A mass spectrometer is an instrument used to separate ions in a magnetic field according to their masses, and hence to measure atomic and molecular masses.

Charged ions are first produced by bombarding with electrons the gases being studied. Positive ions emerge in a beam through slit S_1 in Fig. 18.16 and are accelerated through slit S_2 by a voltage applied between the two slits. An ion with a charge $+q$ will have a kinetic energy qV, where V is the voltage accelerating the ion. Hence $\frac{1}{2}mv^2 = qV$. After passing through slit S_2, the ion moves at constant speed v into a magnetic field perpendicular to v. Hence the ion is deflected in a circular path of radius r, given again by Eq. (18.6).

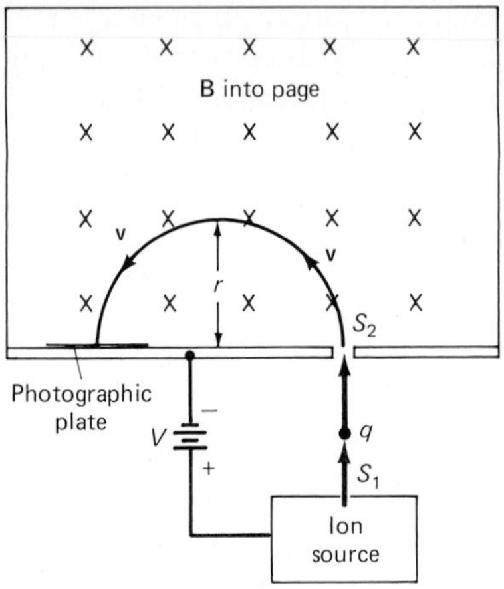

The figure shows a mass spectrometer with magnetic field B into the page (marked with ×'s), velocity v, radius r, entry points S_1 and S_2, charge q, potential V, an ion source, and a photographic plate.

FIGURE 18.16   A mass spectrometer. The mass of the ion can be determined from the position at which it strikes the photographic plate.

$$r = \frac{mv}{Bq} = \frac{m}{Bq}\sqrt{\frac{2qV}{m}} = \sqrt{\frac{2Vm}{B^2q}}$$

and so $m = \dfrac{B^2qr^2}{2V}$ (18.8)

The ion strikes a photographic plate or electronic detector after moving through a semicircle, as shown in the figure, and in this way r can be measured. Since q, V, and B are presumed known, the mass m of the ion can be determined.

Results obtained with mass spectrometers make it clear that many purified elements consist of *isotopes* which have identical chemical properties but slightly different masses due to the presence of different numbers of neutrons in their nuclei.

Cyclotron   Many of the particle accelerators used in national laboratories like Brookhaven on Long Island and Fermilab near Chicago use magnetic fields to control the paths of the particles. The *cyclotron*, one of the earliest particle accelerators, was developed in 1930 by Ernest O. Lawrence (see Fig. 18.17) at the University of California at Berkeley. It used a large magnetic field to keep the protons or other charged particles moving in circles.

FIGURE 18.17   Ernest O. Lawrence (1901–1958). Lawrence supplied the key idea needed to accelerate charged particles to very high speeds around circular paths. Without his insight many of the advances made in nuclear and elementary-particle physics in the last generation would have been delayed. For his development of the cyclotron Lawrence received the Nobel Prize in physics in 1939. Element 103, lawrencium, is named after him. (*Photo courtesy of Lawrence Radiation Laboratory, Berkeley, California.*)

A cyclotron has two D-shaped cavities called *dees*, as in Fig. 18.18. Each time the protons cross the region between the dees, a voltage is applied to speed them up. Hence they move with increasing speed and in paths of increasing radius as time goes on. Finally they acquire the desired high energy and can be directed out of the cyclotron to an external target where their interaction with other nuclei is observed and studied.

The voltage applied to the dees must be in resonance with the frequency of rotation of the protons if large energies are to be developed, for every time the protons pass through the gap the voltage must be such as to accelerate them again. The voltage must reverse itself, therefore, in the time it takes for the protons to make half a revolution. This is accomplished by using an alternating (ac) voltage across the dees which is in perfect synchronism with the particles' motion. The protons are therefore accelerated twice in each revolution by the electric field **E**.

The magnetic field supplies the force to keep the protons moving in a circle of radius r, given by Eq. (18.6), $r = mv/Bq$. The speed v changes each time the protons are accelerated by the electric field in the gap. The radius r therefore changes along with v.

The time required for the protons to make a complete revolution is the period, which is, using Eq. (18.6),

$$T = \frac{\text{distance}}{\text{speed}} = \frac{2\pi r}{v} = \frac{2\pi r}{Bqr/m} = \frac{2\pi m}{qB}$$

and so the frequency of revolution is

$$f = \frac{1}{T} = \frac{qB}{2\pi m} \tag{18.9}$$

What makes the cyclotron work is the fact that *the frequency f is independent of either v or r*, as is clear from Eq. (18.9). Hence from a knowledge of the field B and q/m for the particle it is possible to find the frequency f which must be used for the accelerating voltage. This frequency is unchanged throughout the experiment. Even though v and r increase rapidly, they increase in such a way that v/r remains constant and so the frequency also remains constant.

FIGURE 18.18 Schematic diagram of a cyclotron. Protons are accelerated by electric fields between the dees and kept moving in a circle by the strong magnetic field. The protons spiral outward, moving more rapidly as the radius of their orbit increases, and finally crash into a target outside the cyclotron.

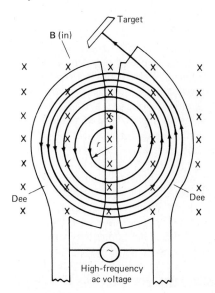

Joseph John Thomson (1856–1940)

FIGURE 18.19 Sir Joseph John (J. J.) Thomson. (*Photo courtesy of AIP Niels Bohr Library.*)

J. J. Thomson is famous both as the discoverer of the electron and as the teacher and mentor of many of this century's greatest physicists. He became a physicist almost by accident. His father wanted him to be an engineer, but when the time came, the family did not have enough money to pay for his engineering apprenticeship. A friend suggested to his father that he send his son to Owens College in nearby Manchester. At Owens young Thomson had excellent instructors in mathematics and physics who kindled his interest in a physics career.

In 1876 J. J. (as he was always called by everyone, including his son, Nobel Laureate George Thomson) went to Trinity College at Cambridge University to study mathematics and science—and spent the next 64 years of his life there! In 1884, at the age of 28, he was elected to the Cav-

endish Professorship at Cambridge and became director of the Cavendish Laboratory, the outstanding physics laboratory in England at that time. He was elected Master of Trinity College at Cambridge in 1918 and gave up the Cavendish Chair in 1919 in favor of Ernest Rutherford. He remained Master of Trinity until shortly before his death in 1940.

Thomson's most famous work in physics was the clarification of the nature of cathode rays and the proof that they were what we now call electrons (see Sec. 18.4). He did other important work on the behavior of positive ions in electric and magnetic fields, and on the structure of atoms. He was neither a high-powered theoretical physicist nor a gifted experimentalist, but rather an unusually adept "broker" between the two. In 1906 Thomson received the Nobel Prize in physics "for his theoretical and experimental investigation of the conduction of electricity in gases," and 2 years later was knighted by the British Crown.

Thomson did a magnificent job in stimulating and encouraging the research of the outstanding group of physicists who flocked from all over the world to work with him at the Cavendish. He trained eight Nobel Laureates, 27 Fellows of the Royal Society, and many other highly successful physics teachers and researchers. In the 1920s it was said that all physics professors in the United States were either Thomson's students or students of his students. He was a good, if somewhat penurious, administrator, and much beloved by his students. Their respect and admiration for him are reflected in the parodies of popular songs they wrote in his honor to be sung at the annual

meetings of the Cavendish Physical Society, rather raucous occasions when old stories were retold, much wine consumed, and songs sung about the Cavendish and its professors.

No one has captured J. J.'s boyish enthusiasm and remarkable genius better than his former Cavendish student, Nobel Laureate F. W. Aston, in this statement on the occasion of Thomson's death in 1940:

> Working under him never lacked thrills. When results were coming out well his boundless, indeed childlike, enthusiasm was contagious and occasionally embarrassing. Negatives just developed had actually to be hidden away for fear he would handle them while they were still wet. Yet when hitches occurred, and the exasperating vagaries of an apparatus had reduced the man who had designed, built and worked with it to baffled despair, along would shuffle this remarkable being, who, after cogitating in a characteristic attitude over his funny old desk in the corner, and jotting down a few figures and formulae in his tiny tidy handwriting, on the back of somebody's Fellowship thesis, or on an old envelope, or even the laboratory cheque book, would produce a luminous suggestion, like a rabbit out of a hat, not only revealing the cause of trouble, but also the means of cure: This intuitive ability to comprehend the inner working of intricate apparatus without the trouble of handling it appeared to me then, and still appears to me now, something verging on the miraculous, the hall-mark of a great genius.[*]

[*]*The London Times*, Sept. 4, 1940.

18.4 J. J. Thomson's Determination of *e/m* for the Electron

Sir Joseph John Thomson usually receives credit for the discovery of the electron. (See Fig. 18.19 and accompanying biography.) Although there had been electron theories before Thomson, it was Thomson's work in 1896 to 1897 that pinned down the nature of the electron. Thomson performed many careful and ingenious experiments on the deflection of cathode rays in electric and magnetic fields. (*Cathode ray* is the name given to the glow which spreads out from the cathode of an evacuated tube when a high voltage is applied between the anode, or positive terminal, and the cathode, or negative terminal.)

Figure 18.20 shows the evacuated glass tube used by Thomson in his most famous experiment. Thomson showed that the cathode rays were deflected by both electric and magnetic fields, which proved that they were material particles. Also, the direction of their deflection was consistent with the idea that their charge was negative. He then clarified the nature of these particles by measuring the ratio of their charge *e* to their mass *m*.

FIGURE 18.20   The evacuated glass tube used by Thomson in his work on the charge-to-mass ratio *e/m* of the electron. The electrons originate at *C*, pass through the slits *A* and *B* and the deflecting plates *D* and *E*, and strike the screen at the end of the tube.

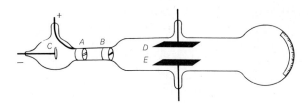

Measurement of *e/m*

To measure *e/m* Thomson first determined the position of the beam on the screen in the absence of electric and magnetic fields. Then he applied a voltage *V* between plates *D* and *E*, so that an electric field was established between the plates of magnitude $E = V/d$, where *d* is the distance between the plates. This field exerted a force on the electrons equal to $F_Y = Ee$, which produced an upward deflection of the electron beam from its original position on the screen.

Next Thomson established a magnetic field at right angles to the electric field, and uniform in the region between the plates. He chose a magnetic field which would deflect the electrons in a downward direction, and varied the electric field until the "crossed" electric and magnetic fields restored the beam to its original, undeflected position. Then the magnetic force was equal in magnitude and opposite in direction to the electric force. Thus according to Eq. (18.4),

$$Ee = Bev_X \quad \text{or} \quad v_X = \frac{E}{B} \tag{18.10}$$

Using this value of v_X and the fact that the electron was acted on by the electric field only while it was between the plates, i.e., during a time *t* equal to L/v_X where *L* is the length of the deflecting plates, Thomson obtained the following result (see Prob. D6):

$$\boxed{\frac{e}{m} = \frac{E \tan \theta}{LB^2}} \tag{18.11}$$

Here tan θ is just the ratio Y/X of the deflection on the screen in the absence of the magnetic field to the distance from the center of the plates to the screen, as shown in Fig. 18.21. Hence it can be measured, as can B, E, and L, and thus a value of e/m for the electron can be obtained.

Thomson found that e/m for the electron was three orders of magnitude larger than for the lightest ion (H^+) studied by electrolysis! Presumably the charge on the electron did not differ much in magnitude from the charge on H^+, and hence *the mass of the electron must be less than $\frac{1}{1000}$ of the mass of* H^+. As a result Thomson's work provided clear evidence that the electron was a real particle, of negative charge and very small mass, and that, no matter what its source, it had the same charge-to-mass ratio e/m. Electrons clearly seemed to be one of the prime constituents of all matter.

The presently accepted value of 1.76×10^{11} C/kg for e/m can be combined with the charge on the electron, as determined by the Millikan oil-drop experiment, $e = 1.60 \times 10^{-19}$ C, to yield a value for the electron mass given by

$$m = \frac{e}{e/m} = \frac{1.60 \times 10^{-19} \text{ C}}{1.76 \times 10^{11} \text{ C/kg}} = 9.11 \times 10^{-31} \text{ kg}$$

FIGURE 18.21 Diagram showing how the angle θ in Thomson's experiment may be obtained by measurements of the deflection Y of the beam on the screen, and the distance X between the center of the deflecting plates and the screen.

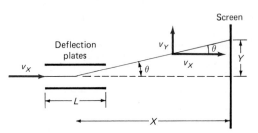

18.5 Magnetic Field of a Straight, Current-Carrying Wire

Thus far we have considered the way currents and moving charges react to the presence of magnetic fields, without asking where these magnetic fields came from. We now turn our attention to understanding how magnetic fields are produced.

Let us consider the simple, but revolutionary, experiment first performed in 1820 by Hans Christian Oersted (1777–1851), a professor at the University of Copenhagen. According to Oersted's own account he was giving a course of lectures on electricity and magnetism and had set up a current-carrying wire above a compass needle to see if there was any interaction between the two. During the lecture he passed current through the wire and observed a very feeble force acting to line up the needle in a direction *at right angles* to the wire. The effect was so weak that his audience was not impressed, but in July 1820 Oersted repeated his demonstration with a more powerful battery and with a group of notables present to witness his discovery. When Oersted reversed the direction of the current, the compass needle reversed its direction, but still lined up perpendicular to the wire, as in Fig. 18.22. As Oersted moved the needle around the wire, it always pointed at right angles to the wire. Hence the magnetic field formed closed circles around the wire,

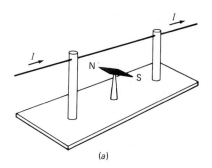
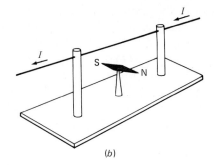
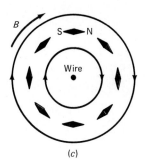

(a)   (b)   (c)

FIGURE 18.22 Oersted's experiment: *(a)* The magnetic field near a long, straight wire carrying a current *I*. *(b)* When the current reverses direction, the magnetic field also reverses direction. *(c)* The magnetic field is everywhere at right angles to the wire, as can be seen from the direction of the compass needles which line up in the direction of the magnetic field. Here the current is flowing into the paper.

as in Fig. 18.23. In this way Oersted first demonstrated the connection between electricity and magnetism which many physicists had suspected for years. Oersted's discovery marked the beginning of the study of electromagnetism.

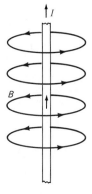

FIGURE 18.23 Magnetic field lines around a current-carrying wire. The lines form closed circles in a plane, at right angles to the direction of the current.

News of Oersted's discovery reached Paris in September 1820, and André-Marie Ampère (see Fig. 18.24) immediately set out to verify and extend Oersted's discovery. Within a week he presented important new

FIGURE 18.24 André-Marie Ampère (1775–1836). Ampère extended Oersted's discovery by showing the interaction of two current-carrying wires, even when no permanent magnetic field was present. He also developed the solenoid and showed that it had many properties like those of a permanent magnet. He was the founder of "electrodynamics," a name he coined himself. The unit of electric current is named after him. (*From a painting by Edgar Maxence now in the Institute of Electrical Engineers, London. Courtesy of AIP Niels Bohr Library.*)

results to the French Academy. Not only do magnets and electric currents exert forces on each other, electric currents exert forces on other electric currents. We would say that a steady direct current in a straight wire produces a magnetic field about the wire, which in turn exerts a force on another current-carrying wire placed in that field. Ampère found that the direction of the magnetic field is circular about the wire and could be specified by the following right-hand field rule (see Fig. 18.25):

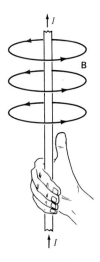

FIGURE 18.25 Right-hand field rule for the magnetic field produced by a current-carrying wire.

Right-Hand Field Rule

If a wire is grasped with the right hand, with the thumb pointed in the direction of the current, the magnetic field is circular about the wire and in the direction in which the fingers curl.

Ampère concluded from his experiments that all magnetism is produced by electric currents and that even in permanent magnets it is some kind of tiny electric current which produces the magnetism.

With a Hall-effect gaussmeter or other magnetic field measuring device we can measure **B** as a function of the current I in a straight wire and the distance r from the wire to the point at which the field is measured. We find that the magnitude of B varies as follows:

$$B \propto \frac{I}{r} \quad \text{or} \quad B = \frac{k'I}{r} \tag{18.12}$$

Here k' is still to be determined. We can determine k' in terms of the force between two current-carrying wires.

Force between Two Long, Straight Current-Carrying Conductors

If we have two long parallel wires, one carrying a current I and the other I' in the same direction, as in Fig. 18.26, we find, as Ampère did, that they attract each other. The reason is that the wire with current I produces a magnetic field which is circular around the wire and of magnitude $B = k'I/r$, from Eq. (18.12). This field is perpendicular to the current I' in the other wire at any point along the length L of the wire. Hence, from Eq. (18.1) we have

$$F_B = B_\perp I'L = k'\frac{II'L}{r} \quad \text{or} \quad \frac{F_B}{L} = k'\frac{II'}{r} \tag{18.13}$$

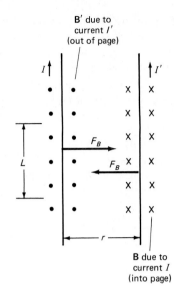

B' due to
current I'
(out of page)

B due to
current I
(into page)

FIGURE 18.26 Force between two wires carrying electric current in the same direction. The two wires attract each other.

This is the magnitude of the force per unit length between the two wires. By first applying the right-hand field rule and then the right-hand force rule you should be able to convince yourself that the two wires attract each other. Notice how Newton's third law is obeyed in this situation. On the other hand, if I and I' are in opposite directions, the two wires repel each other (prove this also).

Definition of the Ampere

If, in Eq. (18.13), we set $I = I' = 1$ A, and $r = 1$ m, then

$$\frac{F_B}{L} = k' \frac{(1 \text{ A})(1 \text{ A})}{1 \text{ m}}$$

We now define the *ampere* as follows:

Definition

One ampere is that current which, flowing in each of two long parallel wires one meter apart, results in a force between the wires of exactly 2×10^{-7} N per meter of length.

In terms of this definition, then, k' must have the value

$$k' = 2 \times 10^{-7} \text{ N/A}^2 = 2 \times 10^{-7} \text{ T·m/A}$$

since
$$1 \text{ T·m/A} = \left(1 \frac{\text{N}}{\text{A·m}}\right)\left(1 \frac{\text{m}}{\text{A}}\right) = 1 \text{ N/A}^2$$

Frequently k' is written as $\mu_0/2\pi$, where $\mu_0 = 4\pi \times 10^{-7}$ T·m/A. μ_0 is called the *permeability of free space* (or of a vacuum), and plays a similar role in magnetism to that played by ϵ_0, the permittivity of free space, in electrostatics. Equation (18.12) for the magnetic field at a distance r from a long, straight wire carrying a current I then becomes:

$$B = k'\frac{I}{r} = \frac{\mu_0 I}{2\pi r} = 2 \times 10^{-7} \frac{I}{r} \qquad (18.14)$$

To make this equation dimensionally correct μ_0 must have the units T·m/A, as found above.

The preceding definition of the ampere is an *operational definition* which allows us to obtain 1 A precisely in terms of *operations* to be performed using a *current balance*. This is an instrument in which the repulsive force between two equal currents can be balanced against the gravitational force required to restore the wires to their original separation. Figure 18.27 shows a current balance of the kind used to determine the standard ampere at the National Bureau of Standards (NBS) in Gaithersburg, Maryland.

Once the ampere is defined in this way, the coulomb can be defined as the amount of charge passing any point in a wire per second when the wire is carrying a current of one ampere; i.e.,

$$1 \text{ C} = (1\text{A})(1 \text{ s}) = 1 \text{ A·s}$$

(The tentative definition of the coulomb given in Sec. 16.4 is therefore no longer needed.)

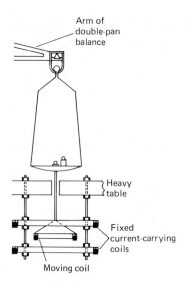

FIGURE 18.27 A current balance of the type used to determine the standard ampere at the National Bureau of Standards in Gaithersburg, Maryland. The same current is made to flow around three circular loops, two of which are fixed rigidly to the table, and the middle one which is hung from one arm of an equal-arm balance, like that in Fig. 2.5. In the case shown here, because of the direction of the current, the moving coil is attracted by the lower coil and repelled by the upper coil. This force can be balanced by adding weights to the other pan of the balance. The standard ampere is then defined in terms of the weights added and the coil dimensions.

Arm of double-pan balance

Heavy table

Fixed current-carrying coils

Moving coil

Example 18.4

Two long, straight wires, each of length 0.50 m and separated by a distance of 1.0 cm, are connected to a battery, as in Fig. 18.28. The resistance of the circuit external to the battery is 3.0 Ω, and the applied voltage is 6.0 V.
(a) What is the current in the circuit?
(b) What is the magnitude of the force between the wires?
(c) Is this force one of attraction or repulsion?

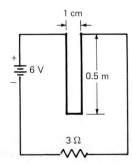

1 cm

6 V

0.5 m

3 Ω

FIGURE 18.28 Diagram for Example 18.4.

SOLUTION

(a) From Ohm's law, $I = \dfrac{V}{R} = \dfrac{6.0 \text{ V}}{3.0 \ \Omega} = \boxed{2.0 \text{ A}}$

(b) As we saw in Sec. 18.5, the force between two wires of length L separated by a distance r is:

$$F_B = \frac{\mu_0 I I' L}{2\pi r}$$

$$= \left(2 \times 10^{-7} \frac{\text{T·m}}{\text{A}}\right)\left[\frac{(2.0 \text{ A})(2.0 \text{ A})(0.50 \text{ m})}{(0.010 \text{ m})}\right]$$

$$F_B = \left(2 \times 10^{-7} \frac{\text{N}}{\text{A·m}}\right)\left(1 \frac{\text{m}}{\text{A}}\right)(4.0 \text{ A}^2)(50)$$

$$= 400 \times 10^{-7} \text{ N} = \boxed{4.0 \times 10^{-5} \text{ N}}$$

(c) Since the currents are in opposite directions in the two wires, the force is one of *repulsion*.

18.6 Ampère's Circuital Law

Equation (18.12) resulted from an experimental determination of the magnetic field **B** produced by the current in a long, straight wire. What about the fields produced by more complicated configurations of current-carrying wires? In 1826 Ampère generalized Eq. (18.12) to obtain an expression for the magnetic field produced by any configuration of currents, an expression now called *Ampère's circuital law*.

The circuital law is based on the idea that work must be done to carry a magnetic pole in a closed path about any wire or group of wires carrying electric current. Because the amount of work done depends on the strength of the magnetic field produced by the current, a calculation of the work can, in principle, lead to a value for the magnetic field.

Suppose we have a number of wires carrying currents, as in Fig. 18.29. If we make a complete circuit, or path, around the wires, and along each short piece Δl of this path evaluate the product of Δl and the component of **B** parallel to Δl (we label this parallel component $B_{\parallel}$), then the following equation is valid:

Ampère's Circuital Law

$$\boxed{\sum B_{\parallel}\,\Delta l = \sum \mu_0 I} \qquad (18.15)$$

The first sum is taken around any closed path, and the second sum over all the currents (taken with the correct sign) passing through the surface bounded by this closed path. Ampère proved that his circuital law is valid for *any* closed path taken around the wires.

FIGURE 18.29 Ampère's circuital law. Here three wires carry currents through the plane *ABCD*. If any continuous path is taken around these three wires in this plane, with Δl a distance along the path, and $B_{\parallel}$ the value of the magnetic field **B** parallel to Δl, then $\sum B_{\parallel}\,\Delta l = \sum \mu_0 I$, according to Ampère's law. Here $\sum I = I_2 + I_3 - I_1$, since I_1 is in the opposite direction to I_2 and I_3.

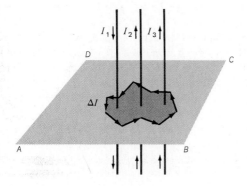

Let us apply Ampère's law to the example already discussed, the magnetic field at a distance r from a long, straight wire.

In this case we choose as our path a circle of radius r around the wire. Then **B** is circular around the wire and is everywhere parallel to Δl and has the same magnitude at every point on the circle, since each point is the same distance from the wire. Also there is only one current I through the plane bounded by the path we have chosen. Hence

$$\sum B_{\parallel}\,\Delta l = B \sum \Delta l = B(2\pi r) \qquad \text{and} \qquad \sum \mu_0 I = \mu_0 I$$

and so Eq. (18.15) becomes

$$2\pi r B = \mu_0 I \qquad \text{or} \qquad B = \frac{\mu_0 I}{2\pi r}$$

which is Eq. (18.14).

Ampère's law plays the same role in magnetism that Gauss' law plays in electrostatics. In both cases considerable symmetry must exist in the physical situation if the application of the law is to be of much practical use. For this reason we will make limited use of Ampère's law in solving problems in this chapter, but its theoretical importance will become clear when we discuss electromagnetic waves in Chap. 21.

18.7 Circular Coils and Solenoids

A very important kind of magnetic field is that produced by a single circular coil of wire, through which a current I flows, as in Fig. 18.30. We can apply the right-hand field rule to find the direction of the magnetic field **B** shown in the figure. At the center of the coil the field is perpendicular to the plane of the coil over a significant distance. Since the center of the coil is at the same distance r from every point on the coil, we might expect that the magnitude of the field **B** would vary directly with the current through the coil and inversely with the distance r from the wire to the center of the coil, as it does in Eq. (18.14) for the field of a long, straight wire. This is found from theory and by experiment to be correct, except that the constant of proportionality is different in this case. The magnetic field at the center of a current loop of radius r turns out to be:

Magnetic Field of a Current Loop $B = \dfrac{\mu_0 I}{2r}$ (18.16)

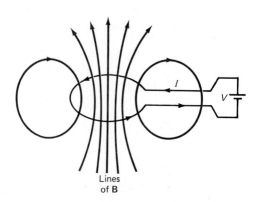

FIGURE 18.30 Magnetic field of a current loop.

Lines of B

The magnetic field produced by a single loop of wire, as shown in Fig. 18.30, is very similar to the electric field of an electric dipole, as shown in Fig. 16.23. A current-carrying loop of wire is therefore called a *magnetic dipole*, and the quantity IA, where A is the area of the circular loop, is called the *dipole moment* of the current loop. If there are N turns of wire instead of 1 turn, the magnetic field becomes $B = \mu_0 NI/2r$, and the dipole moment becomes NIA. When placed in a magnetic field, a current loop tends to line up so that its dipole moment is in the direction of the applied magnetic field.

The Solenoid

A *solenoid* is a tight cylindrical coil of wire with a diameter small compared with its length, through which a current passes, as in Fig. 18.31. Let us apply the circuital law to such a solenoid containing N turns of wire in a length L, each turn carrying a current I. The magnetic field inside the solenoid will be directed along the axis of the tube inside the tube, as can be seen from the diagram for a single loop in Fig. 18.30. Outside the solenoid the magnetic field will be very much weaker, as the rough sketch of the magnetic field in Fig. 18.31 indicates. This is because outside the solenoid the magnetic fields produced by the currents flowing into and out of the paper tend to cancel. We choose as the path for our application of Ampère's law the rectangular path *abcda* in Fig. 18.32. Along *ab* the magnetic field is very weak and $B_\parallel \, \Delta l$ is nearly zero. Along *bc* and *da* the field is very weak outside the solenoid and is at right angles to Δl inside the solenoid, and so $\sum B_\parallel \, \Delta l = 0$ for these pieces of the path. This leaves the path *cd*, which is of length L, along which **B** has a constant magnitude B parallel to L. Hence Ampère's law in the form $\sum B_\parallel \, \Delta l = \sum \mu_0 I$ becomes

FIGURE 18.31 A solenoid, through which a current I flows, and the magnetic field produced.

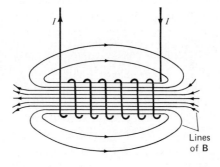

Lines of B

FIGURE 18.32 Application of Ampère's law to obtain the field of a solenoid near its center and along its axis.

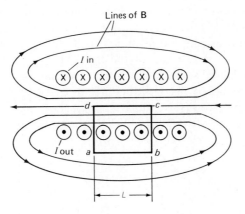

Lines of B

I in

I out

$$BL = \mu_0 NI$$

since there are N turns of wire in the distance L which carry current I through the surface bounded by the path *abcd*. Hence

$$B = \mu_0 \frac{NI}{L} = \mu_0 nI$$

The magnetic field produced by a solenoid near its center and along its axis is therefore

Field of a Solenoid

$$\boxed{B = \mu_0 nI}$$

(18.17)

where n is the number of turns per unit length of the solenoid (N/L) and I is the current flowing through each of the turns. B is constant inside the solenoid except near the ends where it begins to "fringe," or bend away from the solenoid axis, as shown in Fig. 18.32.

Example 18.5

An air-core solenoid is wound with 50 turns of wire for every 10 cm of length. A current of 1.0 A passes through the windings of the solenoid. What is the magnetic field strength inside the solenoid at a point on its axis?

SOLUTION

$$B_0 = \mu_0 nI$$
$$= \left(4\pi \times 10^{-7} \frac{\text{T·m}}{\text{A}}\right)\left(\frac{50}{0.10 \text{ m}}\right)(1.0 \text{ A})$$

$$B_0 = \boxed{6.3 \times 10^{-4} \text{ T}} \qquad \text{or} \qquad B_0 = 6.3 \text{ G}$$

Electromagnets

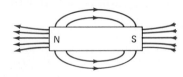

FIGURE 18.33 The magnetic field produced by a bar magnet.

The magnetic field of the solenoid in Fig. 18.31 is almost identical to that of the bar magnet shown in Fig. 18.33. The solenoid is therefore an *electromagnet*, which behaves basically like a bar magnet when current flows through it, but has little or no magnetic field remaining when the current is turned off. To increase the magnetic field strength of electromagnets, the central core of the solenoid is often filled with an iron alloy. The dipole moments of the iron atoms line up with the applied field and add greatly to the magnetic field.

Both permanent magnets and electromagnets attract objects containing iron, cobalt, or nickel by inducing magnetic dipole moments in the material, in the same way that an electric field induces electric dipole moments. For example, if the solenoid is brought near a piece of iron, as in Fig. 18.34, the field of the solenoid will so modify the iron that a south magnetic pole will appear near the solenoid's north pole, and a north magnetic pole will appear at the other end of the iron. Since the iron's south pole is closer to the solenoid's north pole, a net attractive force results. Electromagnets are frequently used on the ends of derrick cables to lift large pieces of steel or scrap iron, since the magnetic field can be turned on or off as desired merely by turning the electric current on or off.

FIGURE 18.34 The use of a solenoid with an iron core to induce a magnetic moment in a piece of iron, resulting in an attractive force between the solenoid and the iron. The magnetic field of the solenoid lines up the atomic magnets in the iron parallel to the magnetic field. Since this brings the north poles in the iron closer to the south pole of the solenoid, an attractive force results.

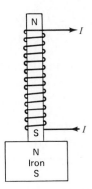

18.8 Magnetic Properties of Materials

An air-core solenoid makes a very weak electromagnet, but by placing a soft-iron* core inside the solenoid the magnetic field strength both inside and outside the solenoid can be increased by a factor of a thousand or more. The iron greatly enhances the magnetic field and changes the expression for B from $B_0 = \mu_0 nI$, in which μ_0 is the magnetic permeability of a vacuum (or of air, which is very close to a vacuum magnetically), to $B = \mu nI$, where μ is the magnetic permeability of the iron or other material filling the core of the solenoid. The ratio B/B_0 is called the *relative magnetic permeability* and is represented by the symbol K_m, where

$$K_m = \frac{B}{B_0} = \frac{\mu nI}{\mu_0 nI} = \frac{\mu}{\mu_0} \tag{18.18}$$

Hence $K_m = \dfrac{\mu}{\mu_0}$ or $\mu = K_m \mu_0$ $\tag{18.19}$

Depending on the value of K_m we can distinguish three kinds of materials with very different magnetic properties:

FIGURE 18.35 Orientation of needlelike samples of diamagnetic, paramagnetic, and ferromagnetic materials in a magnetic field.

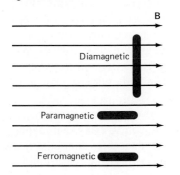

1 *Diamagnetic materials.* For some materials like glass, sulfur, and bismuth, μ is slightly less than μ_0, and K_m is slightly less than unity. These are called diamagnetic materials. If suspended near a strong magnet, a diamagnetic material has the peculiar property that it will move from a region in which the magnetic field **B** is strong into a region in which **B** is weak, so that as few lines of **B** as possible pass through the material. For the same reason, a needlelike sample of the material will orient itself with its long axis *perpendicular* to the lines of **B** if free to do so, as shown in Fig. 18.35.

2 *Paramagnetic materials.* Many ordinary materials like aluminum are paramagnetic. For such materials μ is slightly greater than μ_0, and K_m is greater than unity. If placed in a magnetic field a paramagnetic material

---

*"Soft" iron refers to iron that is easily magnetized but which will not hold a permanent magnetic field. "Hard" iron is an alloy which is difficult to magnetize but which makes a good permanent magnet.

experiences small forces moving it toward the strong points of the field. A needlelike sample of the material will also orient itself parallel to an applied magnetic field, so as to increase the magnetic field already present, as in Fig. 18.35. Values for the permeabilities of some diamagnetic and paramagnetic materials are given in Table 18.1. None of them differs much from unity.

TABLE 18.1   Values of the Relative Permeability for Various Materials at Room Temperature

Nature of material	Material	Relative permeability $K_m = \dfrac{\mu}{\mu_0}$
Ferromagnetic	Iron	5,500*
	Permalloy (55% Fe, 45% Ni)	25,000*
	Mumetal (77% Ni, 16% Fe, 5% Cu, 2% Cr)	100,000*
Paramagnetic	Aluminum	1.000023
	Iron ammonium alum	1.00066
	Magnesium	1.000012
	Titanium	1.000071
	Uranium	1.00040
Diamagnetic	Bismuth	0.999834
	Copper	0.999990
	Gold	0.999964
	Mercury	0.999971

*Maximum values; actual value depends on the applied field.

3 *Ferromagnetic materials*. Ferromagnetic materials are mainly the elements iron, cobalt, and nickel, and their alloys. They have very large values of μ and hence relative permeabilities of 1000 to 100,000, as shown in Table 18.1. As we know from experience, ferromagnetic materials like iron are subject to large forces pulling them into the strongest part of a magnetic field. Similarly, iron needles and bar magnets line themselves up in the direction of an applied magnetic field. Ferromagnetic materials are much used in permanent magnets. They have the unusual property that μ is not a constant but depends both on the current through the solenoid and on what currents have passed through the solenoid previously. For large currents *saturation* occurs, and the magnetic field strength increases very little with increased current in the solenoid.

Atomic Explanation of Magnetism

Since the time of Ampère investigators have shown conclusively that magnetic effects are all produced by *currents*. Ampère first suggested that all objects contain a multitude of small current loops in which currents flow continuously without loss of energy. We now believe that Ampère's idea was basically correct. The elementary current loops are thought to consist of the motion of electrons in orbits around the atomic nucleus, and of the spin of individual electrons about an axis through their center, as in Fig. 18.36, just as the earth rotates on its axis while moving in its orbit around the sun. This leads to magnetic dipole moments for the electrons and to associated magnetic fields. In diamagnetic and paramagnetic materials these dipole moments tend to

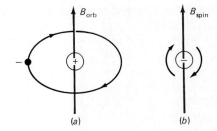

FIGURE 18.36 Current loops in atoms which produce magnetic moments: *(a)* A current loop due to the rotation of an electron in an orbit about the nucleus. *(b)* Spin of an electron about an axis through its center.

cancel each other out and produce almost zero net magnetic dipole moment. In fact, in diamagnetic materials the net dipole moment in a magnetic field is negative, leading to μ being less than μ_0; whereas in paramagnetic materials the net dipole moment is slightly greater than zero, and μ a little greater than μ_0, as can be seen in Table 18.1.

Ferromagnetic Materials

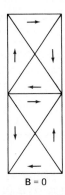

FIGURE 18.37 Magnetic domains oriented in all directions with no preferred orientation in absence of a magnetic field. (The situation is oversimplified to make comparisons with Fig. 18.38 clearer.)

Ferromagnetic materials such as iron, cobalt, and nickel have electrons with magnetic dipole moments which do not cancel out. These dipole moments can be aligned in an applied magnetic field and can increase that field greatly. This is the phenomenon of *ferromagnetism* (the name comes from the Latin word for iron, *ferrum*).

In samples of iron there are regions in which the adjacent ions interact so that all the magnetic dipoles align themselves in the same direction, even in the absence of an external magnetic field. Such regions are called *magnetic domains*, and they have dimensions of less than 1 mm on a side. The domains are oriented randomly with respect to one another in the absence of an applied magnetic field, as in Fig. 18.37. When an external field is applied, however, whole domains rotate and line up in the direction of the field (see Fig. 18.38) and produce a greatly enhanced field **B**. Also, the size of some domains is increased by the action of the field. This makes μ greater than μ_0 by orders of magnitude, and the substance is therefore ferromagnetic. The value of μ is not constant, but depends on the applied field and the previous magnetic history of the sample.

FIGURE 18.38 Orientation of magnetic domains in the direction of an applied magnetic field: *(a) Domain growth*: Domains in the direction of the magnetic field grow, while domains in the opposite direction decrease in size. *(b) Domain rotation*: Individual domains rotate in such a way that they have a larger component of their magnetic moments in the direction of the applied field.

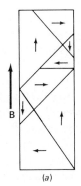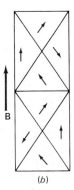

Domains can be made visible by putting some finely divided iron oxide powder on the iron. The powder tends to collect on the boundaries between adjacent domains and make them visible, as in Fig. 18.39.

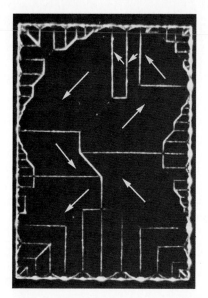

FIGURE 18.39 A photograph of actual ferromagnetic domain boundaries in a single-crystal nickel platelet of dimensions 71 by 99 μm. The arrows indicate the direction of the magnetic field in each domain. (*Photo courtesy of Ralph W. DeBlois, General Electric Research Center.*)

Permanent Magnets The existence of domains also explains the existence of permanent magnets. In ferromagnetic substances some of the domains remain aligned in the original field direction even after the external field is removed. They therefore have become permanent magnets.

18.9 The Earth's Magnetic Field

The earth acts as if it had a relatively short bar magnet, some few hundred miles long, buried near its center, as shown in Fig. 18.40. This produces a dipole field similar to that of a bar magnet. The axis of this magnet does not correspond to the axis of rotation of the earth, but is tilted about 15° with respect to it, as shown in the figure. Since the north pole of a compass needle is, by definition, the end of the needle which points *north* when free to rotate in the earth's field, it must be attracted by a *south* pole. Hence the earth's south magnetic pole is situated near its geographic north pole, and vice versa, as the figure shows.

The angle made by a compass needle with true geographic north is called the *magnetic declination*, and varies from about 20°E to 20°W for most of the United States. Figure 18.40 shows that the direction of **B** at most places on the earth is not parallel to the earth's axis. In addition, the lines of the earth's magnetic field are not horizontal at any place on the earth's surface except near the equator. This means that in the northern hemisphere a compass needle will dip below the horizontal if pivoted around a horizontal axis at the earth's surface. The angle of dip below the horizontal is called the *magnetic inclination*. In the United States the dip angle varies from a low of 39° for Hawaii to 75° for the state of Maine.

In practice the strength of the earth's magnetic field **B** is usually measured in two steps. First a compass needle, free to move only in a horizontal plane, is used to find the direction of the earth's magnetic field at the point of interest. The magnitude of the horizontal component of **B** is then measured with a gaussmeter. Then a magnetic dip needle (see Fig. 18.41),

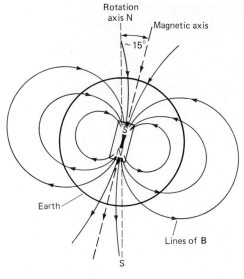

FIGURE 18.40 Magnetic field of the earth, showing how it behaves as if it had a bar magnet some few hundred miles long buried near its center.

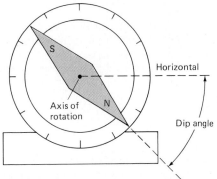

FIGURE 18.41 A magnetic dip needle. Its axis of rotation is at its center and is horizontal.

whose axis of rotation in the horizontal plane is perpendicular to the horizontal component of the earth's field, is used to measure the angle made by the earth's field with the horizontal.

Example 18.6

The magnetic inclination at Ponce in Puerto Rico is 49.5°. If the *vertical* component of the earth's magnetic field is measured with a gaussmeter and found to be 0.35 G, what is the magnitude of the earth's magnetic field at Ponce?

SOLUTION

Since the magnetic inclination is the angle between the horizontal and the direction of the earth's magnetic field, the situation is as shown in Fig. 18.42. Then $\sin \theta = B_Y/B$, or $B = B_Y/\sin \theta$, and

$$B = \frac{0.35 \text{ G}}{\sin 49.5°} = \frac{0.35 \text{ G}}{0.76} = \boxed{0.46 \text{ G}}$$

This value, close to $\frac{1}{2}$ G, is typical of magnetic field values on the earth's surface. This is a very small field relative to the tesla fields ($1 \text{ T} = 10^4 \text{ G}$) which can be easily produced in the laboratory.

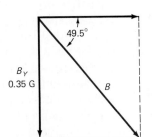

FIGURE 18.42 Diagram for Example 18.8.

18.10 Applications of Magnetism

A very important application of the ideas of this chapter is the development of electrical meters to measure currents and voltages. Suppose we suspend a rectangular coil of wire in a magnetic field **B**, as in Fig. 18.43. The two sides of the coil around which current flows are of length L and are at right angles to the magnetic field **B**. When a current is established in the coil, as in the figure, side 1 experiences a force F_B out of the plane of the paper, according to our right-hand force rule. Similarly side 2 experiences a force equal in magnitude but opposite in direction, i.e., into the plane of the paper. Hence two torques

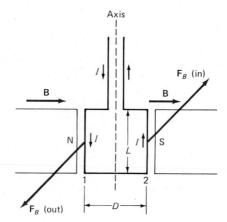

FIGURE 18.43 A rectangular coil of wire in a magnetic field. The current I passing around the rectangular coil produces a couple which rotates the coil about the vertical axis shown.

(see Sec. 4.4) are set up to twist the coil about its axis, the magnitude of the total torque (or couple) being the sum of the force times the lever arm for each force. This is

$$\tau = 2\left(F_B \frac{D}{2}\right) = F_B D = (B_\perp IL)D = B_\perp IA \tag{18.20}$$

where A is the area of the coil. If the coil consists of N turns of wire, then $\tau = B_\perp NIA$. Note that the quantity NIA is the *magnetic dipole moment* of the loop, and that the field tries to orient this dipole moment in the direction of the field, just as it would a compass needle.

This torque twists the wire to an equilibrium position where the restoring torque of the twisted suspension just balances out the torques on the current-carrying conductors in the magnetic field. Since the angular displacement of the coil is proportional to the torque τ, this displacement is also proportional to the current I, from Eq. (18.20). By putting known currents through the coil, the instrument can be calibrated; i.e., the readings corresponding to these known currents can be marked on the scale. By interpolation the values of unknown currents can then be found. Such an instrument is called a *galvanometer*, a very sensitive instrument for measuring small currents. Sturdy galvanometers have pointers directly attached to their coils. These pointers can move over the scale and indicate the value of the current, as shown in Fig. 18.44.

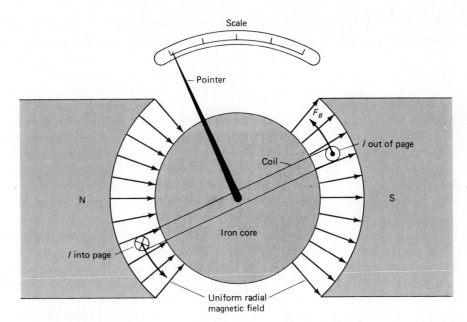

FIGURE 18.44 A galvanometer with a pointer attached to the coil of Fig. 18.43. The sides of the coil which experience the torque are normal to the plane of the paper in the diagram.

Ammeters

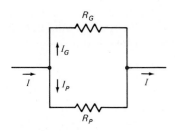

FIGURE 18.45 Conversion of a galvanometer into an ammeter by putting a shunt resistor R_P in parallel with the galvanometer (R_G).

A delicate galvanometer can be used to measure very small currents directly, but cannot carry large currents without sustaining deflections large enough to damage the meter. It is possible, however, to convert any galvanometer into an ammeter of desired range by passing only a tiny, precisely known fraction of the total circuit current through the galvanometer coil.

Suppose we have a galvanometer which will give a full-scale deflection for a current of 100 μA, but with which we want to measure currents up to 5 A. We first measure the resistance of the coil and find that it is, say, 25 Ω. Then we set up the circuit of Fig. 18.45, in which an external resistor R_P (called a *shunt* resistor) is put in parallel with the galvanometer. We want to find the value of the shunt resistor R_P which will convert the galvanometer into an ammeter with 5-A full-scale deflection.

From Kirchhoff's junction rule applied to the circuit in Fig. 18.45, we have $I = I_G + I_P$; and from the loop rule we have

$$-I_G R_G + I_P R_P = 0 \qquad \text{or} \qquad R_P = \frac{I_G}{I_P} R_G$$

Now, the current through the galvanometer must be 100 μA for full-scale deflection, and the current I which is being measured must be 5 A for full-scale deflection. Hence

$$I_P = I - I_G = 5 \text{ A} - 0.0001 \text{ A} = 4.9999 \text{ A}$$

and so $\qquad R_P = \dfrac{0.0001 \text{ A}}{4.9999 \text{ A}}(25 \ \Omega) = 0.00050 \ \Omega = 5.0 \times 10^{-4} \ \Omega$

The shunt resistor required is therefore quite small. When the 5.0×10^{-4} Ω resistor (usually just a short length of wire) is inserted in parallel with the galvanometer, the galvanometer becomes an ammeter which is deflected full-scale by a 5-A current. We refer to it as a "5-A ammeter." The very small equivalent resistance (less than R_P) of the meter has the advantage that the meter does not change the current in the circuit significantly, since the resistance of ordinary circuits is orders of magnitude larger than the 5.0×10^{-4} Ω meter resistance.

Voltmeters

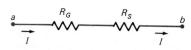

FIGURE 18.46 Conversion of a galvanometer into a voltmeter by putting a resistor R_S in series with the galvanometer (R_G).

The same galvanometer can be converted into a *voltmeter* of any desired range. Suppose we want to use this galvanometer with a current sensitivity of 100 μA and resistance $R_G = 25$ Ω to measure voltages up to 300 V. To achieve this we connect a resistor R_S *in series* with the galvanometer, as in Fig. 18.46.

Here we want 300 V connected across a and b to produce a current of 100 μA. From Ohm's law, we must have

$$V_{ab} = (R_S + R_G)I$$

or $\qquad$ 300 V = $(R_S + 25 \ \Omega)(100 \times 10^{-6} \text{ A})$

Hence $\qquad R_S = \dfrac{300 \text{ V}}{10^{-4} \text{ A}} - 25 \ \Omega \approx 3.0 \times 10^{6} \ \Omega$

If we put a 3.0×10^{6} Ω resistor in series with the galvanometer, it will therefore function as a 300-V full-scale voltmeter.

When connected directly across (i.e., in parallel with) the voltage to be measured, a voltmeter's high resistance draws very little current, and hence does not change the properties of the circuit it is being used to measure. Electrometers (Sec. 16.3), with resistances as high as 10^{13} Ω, are often used as high-resistance voltmeters.

The proper way to connect ammeters and voltmeters in an electric circuit is shown in Fig. 18.47. Ammeters must always be connected in series with the other elements of the circuit; voltmeters must always be connected in parallel with the voltage to be measured (do you see the reason for this?).

FIGURE 18.47 Use of ammeters and voltmeters in electric circuits: *(a)* A simple electric circuit consisting of a voltage source V and a resistor R. *(b)* Connection of the ammeter in series with the resistor, and of the voltmeter in parallel with the resistor.

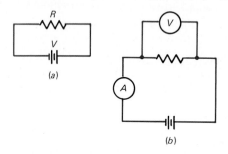

Example 18.7

A galvanometer has a current sensitivity of 10 μA and a resistance of 40 Ω.

(a) What resistance must be connected in parallel with the galvanometer to make it into a 1.0-A ammeter? What is the equivalent resistance of the new ammeter?

(b) What resistance must be connected in series with the galvanometer to convert it into a 1000-V voltmeter? What is the equivalent resistance of the new voltmeter?

SOLUTION

(a) From the junction rule $I_P = I - I_G = 1.0 \text{ A} - 10^{-5}$ A = 0.99999 A. Then

$$R_P = \frac{I_G}{I_P}R_G = \left(\frac{10^{-5} \text{ A}}{0.99999 \text{ A}}\right)(40 \ \Omega) = 40 \times 10^{-5} \ \Omega$$

$$= \boxed{4.0 \times 10^{-4} \ \Omega}$$

The equivalent resistance of the new ammeter is given by

$$\frac{1}{R} = \frac{1}{R_G} + \frac{1}{R_P} = \frac{1}{40 \ \Omega} + \frac{1}{4.0 \times 10^{-4} \ \Omega}$$

The second term on the right is very much larger than the first, which can be neglected by comparison. Hence

$$\frac{1}{R} = \frac{1}{4.0 \times 10^{-4} \ \Omega} \qquad \text{or} \qquad R = \boxed{4.0 \times 10^{-4} \ \Omega}$$

(b) From Ohm's law we have $V = (R_S + R_G)I$, or

$$1000 \text{ V} = (R_S + 40 \ \Omega)(10^{-5} \text{ A})$$

and so $R_S = \dfrac{10^3 \text{ V}}{10^{-5} \text{ A}} - 40 \ \Omega = 10^8 \ \Omega - 40 \ \Omega \approx \boxed{10^8 \ \Omega}$

Again in this case R_S is so large that the equivalent resistance of the two resistors in series is almost identical to $10^8 \ \Omega$.

Summary: Important Definitions and Equations

Magnetic field strength (B):
A vector whose direction at any point in space is that in which a small, isolated north pole would be moved by the magnetic force if placed at that point, and whose magnitude is proportional to the force experienced by that north pole.

Force on a current-carrying conductor of length L in a magnetic field:
Magnitude: $F_B = B_\perp IL$
  where $B_\perp$ is in teslas (T); 1 tesla = 1 N/(A·m) = 10^4 G
Direction (right-hand force rule): If the fingers of the outstretched right hand are placed in the direction of the magnetic field and the thumb in the direction of the conventional current I, then the palm of the hand faces in the direction of the force on the current-carrying conductor.

Force on a charged particle in a magnetic field:
Magnitude: $f_B = B_\perp qv$
Direction: For a *positive charge* given by right-hand force rule.

Magnetic field produced by a straight, current-carrying wire:
Magnitude: $B = \dfrac{\mu_0 I}{2\pi r} = 2 \times 10^{-7}\dfrac{I}{r}$
  where r is the distance from the wire, and $\mu_0 = 4\pi \times 10^{-7}$ T·m/A is the permeability of free space
Direction (right-hand field rule): If a wire is grasped with the right hand, with the thumb pointed in the direction of the current, the magnetic field is circular about the wire and in the direction in which the fingers curl.

Force between two long straight, current-carrying conductors:
Magnitude: $\dfrac{F_B}{L} = \dfrac{\mu_0 II'}{2\pi \ r}$

Direction: Attraction if two currents are in same direction; repulsion if in opposite directions.

Ampere (official definition):
That current which, flowing in each of two long parallel wires one meter apart, results in a force between the wires of exactly 2×10^{-7} N per meter of length.

Ampère's circuital law:
$$\sum B_\parallel \Delta l = \sum \mu_0 I$$

Magnetic field of a circular current loop:
$$B = \frac{\mu_0 I}{2r}$$

Magnetic field of a solenoid:
In vacuum (or air): $B = \mu_0 nI$, where n is the number of turns per unit length.
  In material of magnetic permeability μ: $B = \mu nI$

Relative magnetic permeability:
$$K_m = \frac{\mu}{\mu_0}$$

Magnetic materials:

Diamagnetic: μ is slightly less than μ_0; K_m is slightly less than 1.

Paramagnetic: μ is slightly greater than μ_0; K_m is slightly greater than 1.

Ferromagnetic: μ is very much greater than μ_0; K_m is very much greater than 1; also μ is not constant but varies with both the magnetizing current and the previous history of the material.

Atomic explanation of magnetism:

Magnetism is caused by current loops at the atomic level. These current loops act as magnetic dipoles whose behavior in magnetic fields determines the magnetic behavior of the material.

Earth's magnetic field:

The earth has a magnetic field of about 5×10^{-5} T (or 0.5 G), with a south magnetic pole situated near its geographic north pole, and vice versa.

Questions

1   What would happen if the north pole of a magnet were brought near a positively charged light plastic ball? What would happen if a south pole of the same magnet were brought near the same positively charged light plastic ball?

2   If a particle with charge q enters a magnetic field **B** with a velocity **v** which has a component in the direction of **B**, describe the path of the particle in the magnetic field.

3   In the interaction of two wires carrying current in the same direction, how can you be sure that the interaction is not an electrostatic one? (*Hint*: What would happen if you had three wires all carrying current in the same direction?)

4   Does the magnetic field do any work on a charged particle moving at right angles to the magnetic field? If not, how can it deflect the particle? If so, where does the energy come from?

5   What conclusions can you draw from the fact that J. J. Thomson repeated his *e/m* determination for vacuum tubes with electrodes of different metals like aluminum, platinum, and iron, and containing different gases like hydrogen and carbon dioxide in addition to air, and found, within the limited accuracy of his experiment, always the same value for *e/m*?

6   J. J. Thomson performed another measurement of *e/m* for the electron; the experiment involved the collision of a beam of electrons with the inside of a metal can. Thomson measured with an electrometer the total charge on the can, and obtained the total kinetic energy of the electron beam by measuring the increase in temperature of the can. Would you expect this method to be as accurate as the method for *e/m* discussed in the text? Why?

7   Would you expect the earth's magnetic field to introduce a sizable error in J. J. Thomson's measurement of *e/m*? How could you determine the approximate size of the error introduced?

8   What is the evidence that cathode rays are really electrons and not some kind of electromagnetic wave?

9   Why does either pole of a permanent magnet attract a piece of unmagnetized iron?

10   Why can you form a chain of paper clips at the end of a small bar magnet? What has happened to each paper clip?

11   Two ions have the same mass, but one has a single positive charge and the other a double positive charge. How will their final positions vary as measured by a mass spectrometer like that in Fig. 18.16?

12   You have an electron at rest and want to accelerate it to a high speed. Can you do this with a static electric field? With a static magnetic field? Explain your answer.

13   A bright student suggests that, since a long wire experiences a force in a magnetic field, we could have airplanes land and take off vertically by passing large currents through metal rods running the length of the plane. The force provided by the earth's magnetic field on the metal rod would then provide the needed upward or downward force. Discuss this suggestion from the point of view of the physics involved.

14   The picture on your TV screen is rather blurry, and a friend suggests that it is probably due to the deflection of the electron beam by the earth's magnetic field. How would you go about deciding whether this could possibly be the cause of the poor picture?

15   Since two electrons at rest repel each other, do two electrons moving along parallel lines attract each other? Why or why not?

16   Discuss all the laws of physics which are illustrated in the operation of a cyclotron.

17   What would you expect the direction of the magnetic inclination on the earth to be in the southern hemisphere?

Problems

A 1   Consider a three-dimensional rectangular coordinate system with X, Y, and Z axes as shown in Fig. 18.48. A uniform magnetic field **B** exists along Y, directed downward into the XZ plane. A wire is stretched along the X axis and carries a current I in the negative X direction. The force on the wire is in the direction of:

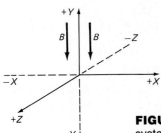

FIGURE 18.48 A coordinate system for Prob. A1.

(a) $-Y$ (b) $-Z$ (c) $+Y$ (d) $+Z$
(e) $-X$

A 2 The coordinate system is the same as in Prob. A1. An electron moves in the negative Z direction, i.e., from $+Z$ to $-Z$. The force on the electron is in the direction of:
(a) $+X$ (b) $-X$ (c) $+Z$ (d) $-Z$
(e) $+Y$

A 3 If a current of 1 A is carried by a wire of length 10 cm at right angles to a magnetic field **B**, and the wire experiences a force of 10^{-2} N, the strength of the magnetic field is:
(a) 1 T   (b) 10 T   (c) 10^{-1} T
(d) 10^{-3} T   (e) 10^{-2} T

A 4 A proton and an electron each enter the same magnetic field with identical velocities at right angles to the field. The ratio of the radius of the circular orbit of the proton to the radius of the electron's orbit will be about:
(a) 1   (b) 1/2000   (c) 2000/1
(d) $(4 \times 10^6)/1$ (e) $1/(4 \times 10^6)$

A 5 An electron of speed 2.0×10^6 m/s enters a region of space in which there are crossed electric and magnetic fields. If the electric field produced by the deflecting plates is 20,000 V/m, and the electron emerges from the fields without being deflected from its original path, the magnetic field **B** must be equal to:
(a) 4×10^{10} T   (b) 2×10^4 T   (c) 100 T
(d) 10^{-1} T   (e) 10^{-2} T

A 6 A particle of velocity 2×10^6 m/s is shot into a magnetic field of strength 2.0 T at right angles to **v**. The particle moves in a circle of radius 1.0 cm. The particle is a:
(a) Electron   (b) Neutron   (c) ^{3}He nucleus
(d) ^{4}He nucleus   (e) Proton

A 7 The magnetic field at a distance of 1 m from a wire carrying a current of 1 A is:
(a) 2×10^{-7} G   (b) 4×10^{-7} T   (c) $\dfrac{\mu_0}{2\pi}$ T

(d) 1 T   (e) $\dfrac{1}{2\pi}$ T

A 8 A solenoid with air core has 10 turns per centimeter which carry 1 A of current. The magnetic field inside the solenoid is about:
(a) 1 T   (b) 10^{-2} T   (c) 10^2 T
(d) 10^{-3} T   (e) 10^{-5} T

A 9 The permeability of a particular sample of iron is measured to be 2.5×10^{-3} T·m/A. The relative magnetic permeability (K_m) of the iron sample is about:
(a) 2.5×10^{-3} T   (b) 2.0×10^3 T·m/A

(c) 2.0×10^3 (d) 2.5×10^{-3} (e) 5.0×10^{-4}

A10 In a laboratory experiment on the optical pumping of rubidium a magnetic field of 50 G is used. The error introduced in such an experiment by the neglect of the earth's magnetic field would be about one part in:
(a) 10^4 (b) 10   (c) 10^2 (d) 10^3 (e) 10^5

B 1 In the situation shown in Fig. 18.9 the length L of the wire is 10 cm, the current is 5.0 A, and the magnetic field **B** has a strength of 0.50 T in the direction shown.
(a) What is the magnitude of the force of the wire?
(b) What is its direction?

B 2 A rectangular galvanometer coil of area 5.0 cm² is suspended in a magnetic field of strength 0.10 T with the plane of the coil parallel to the magnetic field. What is the torque on the coil when a current of 0.10 mA flows through the coil?

B 3 What is the magnitude of the force on an electron moving with a speed of 3.0×10^5 m/s at right angles to a magnetic field of strength 0.55 T?

B 4 How many complete revolutions does an electron make per second in a cyclotron in which the magnetic field **B** is 50×10^{-4} T?

B 5 An electric power line carries a current of 500 A in a straight line at a height of 20 m above the ground.
(a) What is the strength of the magnetic field produced at the ground?
(b) How does field compare with the magnetic field of the earth?

B 6 Two long parallel wires are 10.0 cm apart and carry current of 2.0 A and 5.0 A in the same direction. Find the magnitude and direction of the force between the wires.

B 7 What is the magnetic field of a solenoid of length 50 cm, and consisting of 500 turns, when there is a current of 2.0 A through the solenoid?

B 8 Approximately what is the maximum magnetic permeability of a sample of permalloy?

B 9 The horizontal component of the earth's magnetic field at Tucson, Arizona, is 0.26 G, and the magnetic inclination is 59°. Find the strength of the earth's magnetic field at Tucson.

C 1 A long electric cable is suspended above the earth and carries a 100-A current parallel to the surface of the earth. The earth's magnetic field has a value of 0.56×10^{-4} T and makes an angle of 70° with respect to the cable. What is the force exerted on a 10-m length of the cable due to the earth's magnetic field?

C 2 A galvanometer coil is rectangular in shape and of length 4.0 cm and width 2.0 cm. It is mounted with its two long sides perpendicular to a magnetic field **B** of 0.010 T, with its two short sides initially parallel to the magnetic field. If a current of 100 μA passes through the coil:
(a) What is the force on one of the long sides?
(b) What is the torque on one of the long sides about an axis parallel to the long side and passing through the middle of the short sides?
(c) If the coil rotates through 90° so that its plane is

now perpendicular to the magnetic field, what is the force on one of its long sides?

(*d*) What is now the torque on one of the long sides?

C 3 A rectangular galvanometer coil of length 3.0 cm and width 1.5 cm, containing 10 turns of light wire, is suspended in a magnetic field with the plane of the coil parallel to the direction of the magnetic field. If the torque on the coil is 1.8×10^{-7} N·m when a current of 0.20 mA passes through the coil, what is the strength of the magnetic field **B**?

C 4 A rectangular coil of 20 turns is suspended in a magnetic field of magnitude 0.20 T, with its plane parallel to the direction of the field. The coil has a height of 12 cm and a width of 6 cm. What must be the current in the coil to produce a torque of 5.0×10^{-6} N·m?

C 5 An electron in a vacuum is first accelerated by a voltage of 10,000 V and then enters a region in which there is a uniform magnetic field of 0.60 T at right angles to the direction of the electron's motion.

(*a*) What is the force on the electron due to the magnetic field?

(*b*) What will be the speed of the electron after it has been in the magnetic field for 10 s? Why?

C 6 Crossed electric and magnetic fields can be used as a "velocity selector" to select particles of a particular speed from a beam containing charged particles with a variety of speeds. If the electric field in a particular region of space is 6.0×10^7 V/m and the magnetic field at right angles to the electric field is 1.5 T, what will be the speed of protons which can pass perpendicular to these fields without being deflected?

C 7 A beam of electrons passes undeflected through "crossed" electric and magnetic fields, where $E = 8.0 \times 10^3$ V/m and $B = 2.0 \times 10^{-3}$ T. If the electric field is turned off, the electrons move in the magnetic field in circular paths of radius 1.2 cm. Determine the e/m ratio for the electrons from these data.

C 8 Protons in a cyclotron are moving with a speed of 2.0×10^7 m/s in a circle of radius 0.80 m.

(*a*) What is the strength of the magnetic field required to keep them moving in this circle?

(*b*) What is the magnitude of the force exerted on the protons by this magnetic field?

C 9 A proton is moving with a speed of 2.0×10^6 m/s in a magnetic field whose direction is unknown. When moving along the positive Z axis in Fig. 18.49, the proton experiences no force. When moving along the positive X axis, it feels a force of 1.0×10^{-13} N in the positive Y direction. What are the magnitude and direction of the magnetic field?

C10 An electron of speed 3.0×10^6 m/s is shot into a magnetic field of 50×10^{-4} T (50 G). It then moves in a circle of 3.5-mm radius. What is the value of e/m for the electron obtained from these data?

C11 A 12-V storage battery is connected across two long wires in parallel, one of resistance 4.0 Ω and the other of resistance 3.0 Ω. A 1.0-m length of one wire is arranged to be parallel to a 1.0-m length of the second wire, and at a

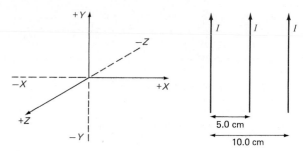

FIGURE 18.49 **FIGURE 18.50**

distance of 10 cm from it. What is the force between the two 1.0-m lengths of wire?

C12 Three long, straight wires carry identical parallel currents of $I = 10$ A, as shown in Fig. 18.50.

(*a*) What is the resultant force on the center wire?

(*b*) What is the force on the wire at the left?

C13 What current would have to flow in an air-core solenoid with 100 turns per meter in order to produce a magnetic field equal to that of the earth, i.e., about 5×10^{-5} T?

C14 What is the maximum possible value of the magnetic field produced by an iron-core solenoid of 100 turns and length 20 cm when a current of 10 A is established in the solenoid?

C15 If the horizontal component of the earth's magnetic field at Cambridge, Massachusetts, is 0.17 G and the magnetic inclination is 73°, find the vertical component of the earth's magnetic field at Cambridge.

C16 A galvanometer has a current sensitivity of 100 μA and a resistance of 15 Ω. It is desired to use this galvanometer as a 30-A full-scale ammeter.

(*a*) How can the galvanometer be modified to accomplish this goal?

(*b*) What is the equivalent resistance of the galvanometer after it is thus modified?

C17 A galvanometer of current sensitivity 50 μA and resistance 50 Ω is to be used as a high-voltage voltmeter to measure voltages up to 10,000 V.

(*a*) How can the galvanometer be modified to accomplish this goal?

(*b*) What is the equivalent resistance of the galvanometer after it is thus modified?

D 1 A proton is accelerated by a voltage of 50,000 V and then enters a region in which a uniform magnetic field of 1.0 T exists.

(*a*) What is the force on the proton?

(*b*) What is the radius of the circle in which the proton moves?

D 2 In a repetition of J. J. Thomson's experiment a beam of electrons is first accelerated through a voltage of 10^4 V. The beam then enters into a region between two deflection plates, as in Fig. 18.21, where a voltage of 10^3 V is applied to the two plates separated by a distance of 2.0 cm. What must be the value of the magnetic field which, when applied at right angles to the electric field, will allow the electrons to pass through the plates without being deflected?

D 3 Singly charged uranium ions, some of mass number 235 and the others of mass number 238, make up the beam in a mass spectrometer.

(*a*) What would be the percentage difference in the radii of the two uranium isotopes in the mass spectrometer?

(*b*) What conclusions can you draw from this about the feasibility of using electromagnetic separation to produce the ^{235}U needed for nuclear power plants and nuclear weapons?

D 4 Prove that, for a charged particle moving at right angles to a uniform magnetic field, the momentum of the particle is directly proportional to the radius of its circular orbit.

D 5 A proton is shot at a speed of 5.0×10^6 m/s at an angle of 30° with respect to a magnetic field of 0.30 T directed along the positive *X* axis. Describe the path followed by the proton. (*Hint*: Resolve the motion of the proton into its *X* and *Y* components.)

D 6 Derive J. J. Thomson's result for *e/m* for the electron [Eq. (18.11)]. (*Hint*: Use Fig. 18.21 and the calculated values for the electric and magnetic fields on the electron.)

D 7 Two long parallel wires are hung by cords of 3.0-cm length from a common axis, as in Fig. 18.51. The wires have a mass of 0.010 kg/m and carry the same current in opposite directions. What is the current if the two cords each make an angle of 30° with the vertical?

D 8 A long, straight wire carries a current of 2.0 A. An electron travels with a speed of 5.0×10^4 m/s parallel to the wire, 20 cm from the wire, and in the same direction as

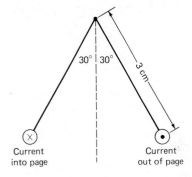

FIGURE 18.51
Diagram for Prob. D7.

the current. What force does the magnetic field of the current exert on the moving electron?

D 9 A toroid is a long solenoid which has been bent into a circle of radius *r*. The average length of the toroid is then $2\pi r$. If the toroid contains *N* turns in all, use Ampère's law to find the field inside the coils of the toroid when it carries a current *I*.

D10 A coaxial cable consists of a small, solid central conductor, separated from a cylindrical outside conductor by insulating disks. If the two conductors carry equal currents in opposite directions, use Ampère's circuital law to find the magnetic field in the space outside the coaxial cable.

D11 A long, straight wire carries a current of 25 A along the axis of a long solenoid. The solenoid has 200 turns per meter and carries a current of 10 A. Find the magnitude of the resultant field at a point inside the solenoid and 4.0 mm from its axis.

Additional Readings

Anderson, David L.: *The Discovery of the Electron*, Van Nostrand, Princeton, N.J., 1964. A narration of the experimental and theoretical developments which led to our present ideas about the nature of electrons and of electricity.

Bitter, Francis: *Magnets: The Education of a Physicist*, Doubleday Anchor, Garden City, N.Y., 1959. An unusual book which combines an abundance of useful information about magnetism with the story of the author's career as a distinguished physicist.

Carrigan, Richard A., Jr., and W. Peter Trower: "Super-heavy Magnetic Monopoles," *Scientific American*, vol. 246, no. 4, April 1982, pp. 106–118. Many research laboratories are hunting for the type of monopole discussed in this article.

Dibner, Bern: *Oersted and the Discovery of Electromagnetism*, Blaisdell, New York, 1962. A historical account of Oersted's life and work.

Furth, H. P., M. A. Levine, and R. W. Waniek: "Strong Magnetic Fields," *Scientific American*, vol. 198, no. 2, February 1958, pp. 28–33. An article on the production

of magnetic fields as high as 100 T, using huge pulses of electric energy to power electromagnets.

Kunzler, J. E., and M. Tannenbaum: "Superconducting Magnets," *Scientific American*, vol. 206, no. 6, June 1962, pp. 60–67. Describes the development of the first generation of superconducting magnets.

Rayleigh, Lord, *The Life of Sir J. J. Thomson*, Dawsons of Pall Mall, London, 1969. An interesting biography by a physicist who knew Thomson well.

Runcorn, S. K.: "The Earth's Magnetism," *Scientific American*, vol. 193, no. 3, September 1955, pp. 152–162. Geophysical explanation of the earth's magnetic field and its changes over time.

Tipler, Paul A.: *Modern Physics*, Worth, New York, 1978. Section 3.2 contains a good discussion of the measurement of *e/m* by J. J. Thomson.

Weinberg, Steven: *The Discovery of Subatomic Particles*, *Scientific American* Library, Freeman, San Francisco, 1983. Contains a good account of Thomson's measurements and of the work of Oersted and Ampère.

*A*nd if the path indicated was a false one, warning could only come from an intellect of great freshness—from a man who looked at phenomena with an open mind and without preconceived notions, who started from what he saw, not from what he had heard, learned, or read. Such a man was Faraday.

Heinrich Hertz (1857–1894)

Chapter 19

Electro-magnetic Induction

Just as the cathedrals at Chartres and Mont-Saint-Michel were the symbols of medieval western culture, so the dynamo—the electric generator—was the symbol of the industrial revolution and the technological age which followed it in Europe and the Americas. Although a more appropriate symbol for present-day civilization might be the computer chip or the industrial robot, electric generators and motors remain the workhorses of today's world. In this chapter we will develop the basic theory that enables us to understand these industrial "movers and shakers."

19.1 Motional EMF's

In the preceding chapter we saw that a wire carrying a current through a magnetic field experiences a force which tends to set the wire in motion. Physicists like Michael Faraday (1791–1867) suspected that the reverse process might also occur, i.e., that the motion of a conductor through a magnetic field might cause a current to flow in the conductor.

To see that this does indeed happen, consider the circuit of Fig. 19.1, in which a metal rod of length *l* between points *a* and *b* slides along a U-shaped wire frame in a magnetic field **B**, which is directed into the paper in the diagram. Suppose we move the rod to the right at a constant speed *v* at right angles to the lines of **B**. Positive charges in the rod will be subjected to forces

FIGURE 19.1 A motional emf produced in a rod moving to the right in a magnetic field which is directed into the paper.

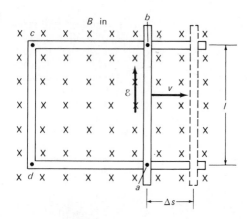

in the direction from a to b, as given by the right-hand force rule, and negative charges will experience forces in the direction from b to a. Since there is a complete circuit, electrons will flow in a clockwise direction around the circuit as long as the rod continues to move. We have therefore produced a *conventional current* in the *counterclockwise* direction *abcda*. The work we have done in moving the rod has produced an emf and in this case also a current. For this reason the emf produced is called a *motional emf*, which is one kind of *induced emf*.

The *magnitude* of the motional emf can be found from the fact that the magnitude of the force on each charge is, from Eq. (18.4),

$$f_B = B_\perp qv$$

and the work done in moving the charge from a to b is

$$\mathcal{W} = f_B l = B_\perp qvl$$

The emf $\mathcal{E}$ is then the work done per unit charge, or

$$\mathcal{E} = \frac{\mathcal{W}}{q} = B_\perp vl \qquad (19.1)$$

The magnitude of the motional emf is thus the product of the magnetic field strength, the length of the rod, and the speed with which the rod moves at right angles to the magnetic field. As usual, $\mathcal{E}$ is in volts if B is in teslas [N/(A·m)], v is in meters per second, and l is in meters.

If B is not perpendicular to the plane of the circuit, then $B_\perp = B \cos\theta$, and Eq. (19.1) becomes:

$$\mathcal{E} = Bvl \cos\theta \qquad (19.1a)$$

where θ is the angle between B and the normal to the plane.

The *direction* of the induced emf and hence of the conventional current is in this case *counterclockwise*, as we have just found by applying the right-hand force rule. Example 19.1 shows that the current must flow in this direction if energy is to be conserved.

Example 19.1

The length of the rod resting on the metal frame in Fig. 19.1 is 0.20 m, and its speed v to the right is 0.10 m/s at right angles to the magnetic field. If the resistance of the loop is 0.20 Ω, and if $B = 1.0$ T [= 1.0 N/(A·m)], find:

(a) The motional emf.
(b) The current which flows.
(c) The magnitude and direction of the force on the rod because of the current flowing through it.

(d) The power needed to move the rod against this force.
(e) Is this power equal to the rate at which electric energy is being produced?
(f) What is the rate at which energy is consumed in the resistance of the circuit?

SOLUTION

(a) The motional emf is

$$\mathcal{E} = B_\perp vl = \left(1.0 \, \frac{N}{A \cdot m}\right)\left(0.10 \, \frac{m}{s}\right)(0.20 \, m)$$

$$= 0.020 \, N \cdot m/C = 0.020 \, J/C = \boxed{0.020 \, V}$$

(b) The current is, from Ohm's law,

$$I = \frac{\mathcal{E}}{R} = \frac{0.020 \, V}{0.020 \, \Omega} = \boxed{1.0 \, A}$$

Goofed oops!

(c) Since the rod is carrying a current of 1.0 A in a magnetic field **B**, it experiences a force equal, from Eq. (18.1), to

$$F_B = B_\perp Il = \left(1.0\ \frac{N}{A \cdot m}\right)(1.0\ A)(0.20\ m) = \boxed{0.20\ N}$$

The direction of this force can be found from the right-hand force rule and is to the left in the diagram. Hence the current produces a force (in the opposite direction to **v**) which tries to prevent the motion that induces the emf.

(d) To move the rod requires work against this opposing force. The rate at which this work is done is

$$P = \frac{\mathcal{W}}{t} = \frac{Fs}{t}$$

where s is the distance that the rod moves to the right in time t, so that $s/t = v$. Hence we have

$$P = Fv = (0.20\ N)(0.10\ m/s) = \boxed{0.020\ W}$$

(e) The rate at which electric energy is produced is

$$P = \mathcal{E}I = (0.020\ V)(1.0\ A) = \boxed{0.020\ W}$$

Hence the mechanical power put into the circuit by the force moving the rod is equal to the electric power generated. Energy is therefore conserved, as it must be.

(f) Since $P = I^2R$, we have

$$P = (1.0\ A)^2(0.020\ \Omega) = \boxed{0.020\ W}$$

The work done therefore ends up as heat in the resistance of the loop.

19.2 Faraday's Induction Law

Equation (19.1) is one possible expression of a more general law called *Faraday's induction law* after the great British physicist Michael Faraday (see Fig. 19.2 and accompanying biography), who first discovered electromagnetic induction.*

Magnetic Flux

In Fig. 19.1 the lines of the magnetic field **B** are perpendicular to the area A bounded by the four sides of the electric circuit. We call the product of the magnitude of **B** and A the *magnetic flux* Φ (Greek capital *phi*), so that $\Phi = BA$. In the general case **B** may not be perpendicular to A, and we write instead

---

*Joseph Henry (1797–1878; see Fig. 19.3) the first head of the Smithsonian Institution in Washington, D.C., actually did some of the key experiments on electromagnetic induction in 1830 before Faraday. But heavy teaching responsibilities at the Albany Academy kept Henry from completing and publishing his work.

FIGURE 19.3   Joseph Henry, the first American scientist to do any serious work in electricity after Benjamin Franklin's great contributions made around 1750. Henry's life and scientific discoveries paralleled those of Faraday in many ways. Henry is credited with developing the principles of the telegraph (before Morse), of electromagnetic induction (independently of Faraday), of self-induction, and of the electric motor. The unit of inductance, the henry, is named after him. (*Photo courtesy of National Portrait Gallery, Smithsonian Institution.*)

Michael Faraday (1791–1867)

FIGURE 19.2 Michael Faraday. *(Photo courtesy of AIP Niels Bohr Library.)*

Michael Faraday is a unique figure in the history of science. Born into a poor family which could not afford to pay for his education, he was almost completely self-taught. Despite this initial handicap, so great were Faraday's scientific curiosity and laboratory skills that he made outstanding contributions to both chemistry and physics, and is considered by many to be the greatest experimental physicist who ever lived.

Faraday was saved from oblivion by a felicitous combination of hard work and good fortune. As a youth he was apprenticed to a bookbinder and spent his lunch hours reading some of the books sent to the shop for binding. An article on electricity in the *Encyclopedia Britannica* particularly captured his fancy, and he decided to devote his life to science. He applied to Sir Humphrey Davy, the director of the Royal Institution in London, for a position and eventually was taken on as a valet and laboratory assistant to Davy.

Faraday's laboratory skills developed so quickly that he soon outshone his scientific sponsor—a source of lasting tension between them. Faraday became known as one of the best analytical chemists in England, much sought after by the growing chemical industry of the day, especially after he succeeded in isolating benzene. Later he moved on to a study of ions in solution and discovered the laws governing electrolysis (see Sec. 17.2). His most famous contributions to science, however, were his induction law (Sec. 19.2) and his development of a prototype electric generator.

Faraday believed that, because a current-carrying wire experiences a force and tends to move in a magnetic field, the reverse should also be true—a magnet should be able to produce a current in a wire. For many years he was unable to prove this experimentally, but one day he plunged a magnet down into a coil and observed that a current did indeed flow momentarily. The essential thing was the *relative motion* of magnet and coil, something which had escaped him before.

Faraday was the first physicist to develop the idea of electric and magnetic *fields*. He did not have the mathematical tools needed to treat field problems theoretically, and his monumental work, *Experimental Researches in Electricity*, contains not a single equation! James Clerk Maxwell (see biography, Chap. 21) took Faraday's ideas and used them as the basis for his famous equations describing the electromagnetic field. For this reason Albert Einstein once said that Faraday had the same relationship to Maxwell in the development of electromagnetism that Galileo had to Newton in the development of mechanics.

Despite his many successes as a scientist, Faraday always remained "as simple, charming and unaffected as a child," in the words of H. Helmholtz. He rejected worldly honors and possible wealth because of his devotion to science and to the religious ideals of the obscure Sandemanian sect to which he remained faithful throughout his life. He turned down the presidency of both the Royal Institution and the Royal Society of London and also refused to be knighted. He requested a "gravestone of the most ordinary kind" when he died, and that is exactly what he received. His grave in Highgate Cemetery, London, contains a small tombstone inscribed simply: "Michael Faraday. Born 22 September 1791. Died August 25 1867."

The true memorials to Faraday's genius are the laws of electrolysis and of electromagnetic induction which now bear his name, as do the unit for capacitance, the farad, and the unit for a mole of electric charge, the faraday.

$$\Phi = B_\perp A = BA \cos \theta \tag{19.2}$$

as is clear from Fig. 19.4.

If we have a small area ΔA over which B is constant, and if the lines of B are all perpendicular to ΔA, then the magnetic flux is $\Delta\Phi = B\,\Delta A$, and the magnitude of B is

$$B = \frac{\Delta\Phi}{\Delta A} \tag{19.3}$$

The magnetic field strength is therefore sometimes called the *magnetic flux density*, since it can be measured in terms of the number of lines of magnetic flux passing through a given area at normal incidence.

The unit of magnetic flux is the weber (Wb), named in honor of Wilhelm Weber (1804–1890), a German physicist and collaborator with Gauss. Hence, since $\Phi = BA$,

$$1\ \text{Wb} = \left(1\frac{\text{N}}{\text{A·m}}\right)(1\ \text{m}^2) = 1\ \text{N·m/A} \tag{19.4}$$

Because of Eq. (19.4) the units of the magnetic field strength B are often given as webers per square meter (Wb/m²), where

$$1\frac{\text{Wb}}{\text{m}^2} = 1\frac{\text{N}}{\text{A·m}} = 1\ \text{T} = 10^4\ \text{G}$$

Here again different units are used to describe the same physical quantity. In what follows we will try to confine ourselves to the use of the tesla and the weber per square meter, and will choose between these two to simplify calculations as much as possible.

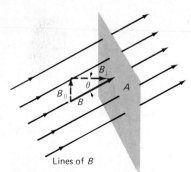

FIGURE 19.4 Definition of magnetic flux. The lines of the magnetic field **B** here pass through the area A in a direction that is not perpendicular to the area. The magnetic flux Φ is defined as $\Phi = AB_\perp = AB \cos \theta$, where $B_\perp = B \cos \theta$ is the component of **B** perpendicular to A.

Example 19.2

If the distance da in Fig. 19.1 is 0.30 m, and if all other quantities are as in Example 19.1, what is the magnetic flux through the circuit $abcd$ produced by a permanent magnet?

SOLUTION

We are told that $B = 1.0\ \text{T} = 1.0\ \text{Wb/m}^2$. Also the length da is equal to 0.30 m, and the length ab is 0.20 m. Hence the area enclosed by the circuit $abcd$ is

$$A = (0.30\ \text{m})(0.20\ \text{m}) = 0.060\ \text{m}^2$$

Since B is perpendicular to A, we have

$$\Phi = BA = \left(1.0\frac{\text{Wb}}{\text{m}^2}\right)(0.060\ \text{m}^2) = \boxed{0.060\ \text{Wb}}$$

Faraday's Law

Now that we have introduced the concept of magnetic flux, let us return to Fig. 19.1 and look at this circuit from a different point of view. We found in Eq. (19.1) that the induced emf is $\mathscr{E} = B_\perp vl$. But if the rod moves a distance Δs in time Δt, then $v = \Delta s/\Delta t$. Now, as the rod moves, $l\,\Delta s$ is simply the change in area (ΔA) of the closed loop in time Δt. Hence, using these results in Eq. (19.1), we have

$$\mathscr{E} = B_\perp \frac{\Delta s}{\Delta t} l = B_\perp \frac{\Delta A}{\Delta t} = \frac{\Delta\Phi}{\Delta t} \tag{19.5}$$

The induced emf in the circuit is numerically equal to the rate of change of magnetic flux through the circuit.

This is *Faraday's induction law* except for one important omission. We need to find the correct sign in Eq. (19.5) to reproduce the result found experimentally for the situation in Fig. 19.1.

In Fig. 19.1 when we moved the rod to the right in the magnetic field B, we *increased* the flux through the circuit, since the area A was increased in the expression for the flux $\Phi = B_\perp A$. We have seen that the induced emf is counterclockwise around the loop. Using our right-hand field rule we therefore find that the induced emf produces a magnetic field out of the plane of the paper which reduces B and hence *decreases* the flux through the circuit. If, in Eq. (19.5), the flux is *increasing* and has a plus sign, then the induced emf $\mathscr{E}$ acts to *decrease* the flux and hence must have a minus sign. The correct assignment of signs therefore changes Eq. (19.5) to:

$$\mathscr{E} = -\frac{\Delta\Phi}{\Delta t}$$

If, instead of a single turn of wire, we have N turns connected in series, an emf is induced in each of them. The total emf induced in the complete circuit is then

Faraday's Induction Law

$$\boxed{\mathscr{E} = -N\frac{\Delta\Phi}{\Delta t}} \qquad (19.6)$$

This is *Faraday's induction law*, the most important equation in all electromagnetism, and the basis for the operation of generators, motors, transformers, and a great variety of other electrical devices. As we shall see, this is a completely general law. It is not necessary that the flux Φ be changed by moving part of a circuit, as in the example above. No matter how the flux through a circuit changes, there is an induced emf in the circuit. For example, if a bar magnet is pushed into or pulled out of the coil of wire in Fig. 19.5, an induced emf is produced as long as the magnet is moving, since it is changing the magnetic flux through the circuit. Once the magnet stops moving, the induced emf vanishes. Hence in this case a moving magnetic field produces an electric field, just as in the case of a moving charge a moving electric field (attached to the charge) produced a magnetic field. This is the "symmetry" between electric and magnetic fields that physicists like Faraday expected to find in nature.

FIGURE 19.5 A change in magnetic flux through a circuit induces an emf and, if the circuit is complete, a current. The current in *(b)* is opposite in direction to that in *(a)*, since in *(a)* the flux through the circuit is increasing while in *(b)* it is decreasing.

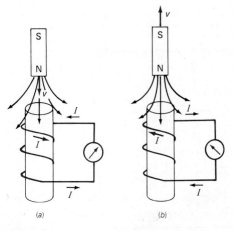

(a)   (b)

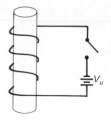

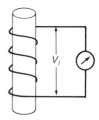

FIGURE 19.6 Emf induced in one coil by the changing current in a second coil whose flux links the first coil.

Another example of the Faraday induction law is shown in Fig. 19.6. Here we have two solenoids, one connected to a battery, the other to an electric meter. If we throw the switch and pass a current through the upper coil, there is an induced emf in the lower coil while the magnetic field in the upper coil is increasing from zero to its final value. During this time $\Delta\Phi/\Delta t$ is not zero in the lower coil, because some field lines from the upper coil pass through the lower coil thus linking the two coils. Once the magnetic field in the first coil reaches its final value, the induced emf vanishes. If, now, we open the switch and allow the magnetic field in the first coil to fall to zero, an emf is again induced in the second coil, but in the opposite direction from the first case, as shown in Fig. 19.7. If we change the flux through the second coil by first increasing the current through the first coil and then decreasing it, and we do this repeatedly, an emf and a changing current will be induced in the second coil as long as the current in the first coil is changing. Hence we have a way of inducing a current in a coil that has *no direct physical connection* with a power source or with a coil through which current flows. This is how a *transformer* (such as one used with model railroads) works. Its performance can be understood on the basis of Faraday's induction law, as we will see in the next chapter.

FIGURE 19.7 *(a)* The applied voltage V_u as a function of time in the upper coil of Fig. 19.6. *(b)* The induced emf in the lower coil of Fig. 19.6 as a function of time. Note than an induced emf is only present in the lower coil when the current through the upper coil is *changing*.

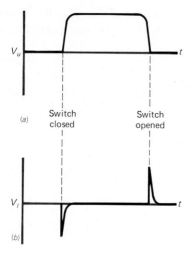

Example 19.3

In the circuit of Fig. 19.1 the rod *ab* is at rest at a position where the length *da* is 0.30 m, as in Example 19.2. The magnetic field *B*, which is being produced by an electromagnet, is suddenly turned off and falls to zero in 0.25 s.

(a) What is the magnitude of the induced emf?
(b) What is the direction of the induced emf?

SOLUTION

(a)   As found in Example 19.2, the flux through the coil in this case is initially 0.060 Wb. Since there is only one turn, $N = 1$, and Faraday's law becomes

$$\mathscr{E} = -N\frac{\Delta\Phi}{\Delta t} = -N\frac{\Phi_{\text{final}} - \Phi_{\text{init}}}{t_{\text{final}} - t_{\text{init}}}$$

$$= -\frac{0 - 0.060 \text{ Wb}}{0.25 \text{ s}} = 0.24 \frac{\text{Wb}}{\text{s}} = \boxed{0.24 \text{ V}}$$

(b)   In this case $\Delta\Phi/\Delta t$ is negative since the flux is *decreasing* to zero. The induced emf $\mathscr{E}$ therefore acts to drive a current which will *increase* the flux. Hence $\mathscr{E}$ is in the opposite direction from that in Fig. 19.1, which means that it is directed in a *clockwise direction*.

It may not be immediately clear that 1 Wb/s is the same as 1 V. But, from Eq. (19.4), we have 1 Wb = 1 N·m/A, and so

$$\frac{1 \text{ Wb}}{1 \text{ s}} = \frac{1 \text{ N·m}}{1 \text{ A·s}} = \frac{1 \text{ J}}{1 \text{ C}} = 1 \text{ V}$$

as required.

19.3 Lenz' Law

The Russian physicist Heinrich Lenz (1804–1865) duplicated many of the results of Faraday and Henry without any prior knowledge of their work. The law which goes by his name merely accounts for the negative sign in Faraday's law.

Lenz' Law

The direction of an induced emf is always such that it opposes the change producing it.

The cause of an induced emf and current may be the motion of a conductor in a magnetic field. If we move the rod in Fig. 19.1 to the right, the current that flows sets up a force toward the left to oppose the motion, as we found in Example 19.1, part (*c*). This is in accordance with Lenz' law. It is also in accordance with the principle of conservation of energy, for if the force produced by the induced current were to the right, we would be able to generate an electric current without doing any work, which would violate the first law of thermodynamics.

Another illustration of Lenz' law is shown in Fig. 19.5. If we push the north pole of the magnet into the coil, the induced emf must cause a current in such a direction as to make the end of the solenoid near that north pole also a north pole, so that the two will repel each other. This means that we must do work to push the magnet into the coil, and it is this work which is converted into electric energy in the coil. Similarly, if we pull the north pole away from the coil, the current flows in such a direction as to make the near end of the coil a south pole, so that the two poles attract each other and require that work be done to separate them. In this case the induced current flows in the opposite direction from that current produced when the north pole was pushed into the coil. The minus sign in Eq. (19.6) is therefore required both by Lenz' law and the principle of conservation of energy.

In any particular case Faraday's induction law will yield the same result as Lenz' law, but Lenz' law is easier to visualize and apply. Note that Lenz' law does not state that the induced current opposes the flux itself; it opposes the *change in flux* which produces the induced emf and current.

Example 19.4

A long solenoid with an air core has 400 turns per meter and a cross-sectional area A of 10 cm². The current through the solenoid increases from 0 to 50 A in 2 s in the direction shown in Fig. 19.8. A loop of wire consisting of 10 turns and of cross-sectional area 100 cm² and resistance 0.050 is put in place around the solenoid near its center.
(a) What is the magnitude of the induced emf in the loop of wire?
(b) What is the magnitude of the current in the wire loop?
(c) What is the direction of the induced emf and current?

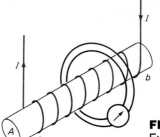

FIGURE 19.8 Diagram for Example 19.4.

SOLUTION

(a) We must first find the flux through the solenoid, which is, from Eqs. (19.2) and (18.17),

$$\Phi = BA = \mu_0 nIA$$

The rate of change of the flux with time is then

$$\frac{\Delta\Phi}{\Delta t} = \mu_0 nA \frac{\Delta I}{\Delta t}$$

$$= \left(4\pi \times 10^{-7} \frac{\text{Wb}}{\text{A·m}}\right)\left(\frac{400}{1 \text{ m}}\right)(10 \text{ cm}^2)\left(\frac{10^{-4} \text{ m}^2}{1 \text{ cm}^2}\right)\left(\frac{50 \text{ A}}{2 \text{ s}}\right)$$

$$= 1.3 \times 10^{-5} \text{ Wb/s}$$

Then, by Faraday's induction law, the induced emf is

$$\mathscr{E} = -N\frac{\Delta\Phi}{\Delta t} = -10(1.3 \times 10^{-5} \text{ Wb/s}) = \boxed{-1.3 \times 10^{-4} \text{ V}}$$

In this case all the flux is confined to the inside of the solenoid, and so the area of the pickup coil is of no importance since the flux through it is always the same, no matter what its area.

(b) The induced current is then

$$I_s = \frac{\mathscr{E}}{R} = \frac{-1.3 \times 10^{-4} \text{ V}}{0.050} = -2.6 \times 10^{-3} \text{ A} = \boxed{-2.6 \text{ mA}}$$

(c) The current through the solenoid is clockwise and increasing as we look along its axis from its lower end in the figure; $\Delta\Phi/\Delta t$ is therefore increasing from a to b. Hence the induced current must produce flux in the opposite direction, i.e., from b to a. A *counterclockwise* induced current accomplishes this. The direction of the induced current is therefore the opposite of the current increasing the flux, as predicted by Lenz' law.

Eddy Currents

FIGURE 19.9 Eddy currents in a rotating disk.

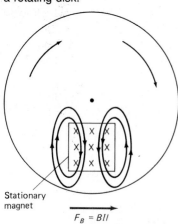

Stationary magnet

$F_B = BIl$

An interesting illustration of Lenz' law is found in the case of *eddy currents*, or currents flowing through metals in regions of space where some magnetic flux is changing. Consider Fig. 19.9, in which a metal disk rotates in a magnetic field produced by a horseshoe magnet, whose poles only cover a small rectangular portion at the bottom of the disk. According to our right-hand force rule, positive charges moving with the disk to the left through the magnetic field will experience a downward force. This causes a conventional current to flow toward the bottom of the disk. This current must flow through a complete circuit and hence must flow upward in regions of the disk outside the magnetic field, as shown in the figure. According to Lenz' law the current flowing downward produces a force $F_B = BIl$ to the right that acts to slow down the wheel (apply the right-hand force rule). The currents flowing upward, on the other hand, are outside the magnetic field and create no force on the disk. As a consequence the net effect of the eddy currents is to *retard* the motion of the wheel. This is an example of *eddy-current damping*, often used to damp galvanometers and other oscillatory systems, and to provide braking action on subway cars.

19.4 Mutual Induction and Self-Induction

Mutual Induction

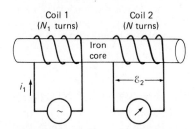

FIGURE 19.10 Two coils wound together on the same iron core, so that any magnetic flux through coil 1 produces a flux through coil 2.

In Fig. 19.10 two coils are wound together on the same iron core. An induced emf is produced in coil 2 by a change in the current in coil 1. It is often useful to express the induced emf in terms of this change in current rather than the change in the magnetic flux. We will indicate the value of the changing current in coil 1 at any instant of time by the symbol i.

In Fig. 19.10 the iron core links the two coils, and hence any flux through coil 1 leads to a flux through coil 2. Φ is the flux produced in coil 2 by a current i in coil 1. This flux is proportional to i, and so if i changes by Δi, then the flux will change by $\Delta\Phi$, where

$$\Delta\Phi \propto \Delta i$$

Since there are N turns of wire on coil 2 and N is a constant, we can also write:

$$N\,\Delta\Phi \propto \Delta i$$

If we call the constant of proportionality M, this becomes

$$N\,\Delta\Phi = M\,\Delta i$$

where
$$M = N\frac{\Delta\Phi}{\Delta i} \tag{19.7}$$

M is called the *mutual inductance*, and is a constant which depends on geometrical factors such as the size of the coils, their proximity, the number of turns on coil 1, and the mutual orientation of the two; and also on the nature of the core material. M is presumed to be a constant for any set of coils, since the flux Φ through coil 2 increases in direct proportion to the current in coil 1, and so $\Delta\Phi/\Delta i$ is a constant.

Now, from Faraday's law [Eq. (19.6)], we have for the induced emf

$$\mathscr{E} = -N\frac{\Delta\Phi}{\Delta t} = -M\frac{\Delta i}{\Delta t} \tag{19.8}$$

Hence we see that *the mutual inductance M is the induced emf in coil 2 per unit rate of change of current in coil 1*. M is therefore a measure of the coupling of the coils, i.e., of how large an emf is induced in coil 2 by a current change in coil 1. Notice that there is no *direct* coupling, either mechanical or electrical, of the two coils. Rather, they are coupled through the magnetic field, whose reality is strongly confirmed by induction experiments of this kind.

The mutual inductance M has units of V/(A/s), or *henrys* (H), where:

$$1\text{ H} = \frac{1\text{ V}}{1\text{ A/s}} = 1\text{ V·s/A}$$

This unit takes its name from the American physicist Joseph Henry as noted above.

Example 19.5

Two coils of wire are placed next to each other on a common axis. The mutual inductance of the pair is 50 mH. If the current in one coil increases from 0 to 10 A in 0.10 s, what is the induced emf in the other coil?

SOLUTION

As we have just seen,

$$\mathscr{E} = -M\frac{\Delta i}{\Delta t}$$

$$= -(50 \times 10^{-3} \text{ H})\left(\frac{10 \text{ A}}{0.10 \text{ s}}\right) = -5.0 \ \frac{\text{V·s}}{\text{A}} \cdot \frac{\text{A}}{\text{s}} = \boxed{-5.0 \text{ V}}$$

The minus sign indicates that the induced emf will drive current in the *opposite* direction from that of the increasing current which induces the emf, according to Lenz' law. If the current in the first coil were *decreasing* from 10 A to 0, then the induced emf would drive current in the *same* direction as the original current, again according to Lenz' law.

Self-Induction

We really do not need two coils to produce inductive effects. If we have a single coil and vary the current in the coil, the magnetic field produced by the coil will change in time. Since this field links through the coil itself, the magnetic flux through the coil will therefore change. This results in a *self-induced emf* opposing the change in the current, according to Faraday's induction law. This induced emf delays the change in current, whether the current is increasing or decreasing, and the amount of delay depends on the self-inductance of the circuit. By analogy with mutual induction, we define the *coefficient of self-inductance* of the circuit as follows:

$$L = N\frac{\Delta\Phi}{\Delta i}$$

so that $\quad L\,\Delta i = N\,\Delta\Phi \qquad\qquad$ (19.9)

Note that Φ is the flux produced by the current i in the same coil. Again the self-inductance L (usually called simply the *inductance*) depends on geometric factors like the size and shape of the coil, as well as on the permeability of the material filling the core of the coil. It is much greater for an iron core, for example, than it is for air. For ferromagnetic materials it is also not constant, but varies with the current i. (For the rest of this section we will, however, assume that L may be considered constant.)

Since $L\,\Delta i = N\,\Delta\Phi$, Faraday's law becomes

$$\mathscr{E} = -N\frac{\Delta\Phi}{\Delta t} = -L\frac{\Delta i}{\Delta t} \qquad\qquad (19.10)$$

where the unit for L is again the henry.

Any element of a circuit having a self-inductance is called an *inductor* and is represented by the symbol ⟋⟋⟋ on circuit diagrams. Since coils are made out of wire, they also have a resistance R associated with the inductance L. Such coils can therefore be considered as a resistance R in series with an inductance L.

Metal Detectors in Airports

Many of the metal detectors at airports use magnetic induction. Two inductance coils are parts of two electric circuits which produce electric signals at the same frequency. One coil is completely shielded, but the other is located near the person or baggage being scanned for the presence of metal objects. Any metal object will have a current induced in it by the changing magnetic field produced by the oscillator. This changes the inductance of the coil by adding mutual inductance. This change in inductance shifts the frequency of that coil with respect to the other (we will see in Sec. 20.4 how the frequency depends on the inductance), and a beat note is produced. The sound of this beat note is then a clear indication of the presence of a metal object.

We will see more applications of inductors when we discuss ac circuits in the next chapter.

19.5 Energy Stored in an Inductor

Suppose we have an inductor of inductance L through which we establish a current by connecting a variable dc voltage across it. We increase the voltage and thus increase the current i from zero to its final value I. An induced emf $\mathscr{E} = -L\,\Delta i/\Delta t$ is produced across the inductor, and this emf is in the opposite direction to the applied voltage, by Lenz' law. Hence work must be done against this emf by the battery to establish the current. The work done by the battery during time Δt is, using Eq. (19.10),

$$\Delta \mathscr{W} = P\,\Delta t = \mathscr{E}\,i\,\Delta t = L\,\frac{\Delta i}{\Delta t}\,i\,\Delta t$$

or $\quad \Delta \mathscr{W} = Li\,\Delta i$

To find the total work done in establishing the current I, we must sum Δi over all the values of i from zero to I. [Note the similarity here to the equation for the work done by the expansion of an ideal gas, $\Delta \mathscr{W} = P\,\Delta V$ (Eq. 14.8).] Here the work done is the area under the curve when we plot Li against i, as in Fig. 19.11, just as in the PV case the work was the area under the curve when we plotted P versus V.

On the straight-line graph of Li versus i in the figure, the work done in changing the current from i to $i + \Delta i$ is $\Delta \mathscr{W} = Li\,\Delta i$, which is the area of the small vertical rectangle shown in color. Just as for the ideal gas, the total work done is the sum of all such small rectangles, or the total area under the curve. Since this total area is triangular, its area is one-half the base of the triangle times its altitude, or

$$\mathscr{W} = \tfrac{1}{2}I(LI) = \tfrac{1}{2}LI^2$$

Hence the energy stored in an inductor, which is equal to the energy needed to establish a current I through the inductor, is

$$\boxed{\mathscr{W} = \tfrac{1}{2}LI^2} \qquad (19.11)$$

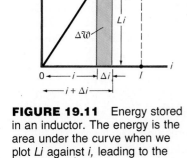

FIGURE 19.11 Energy stored in an inductor. The energy is the area under the curve when we plot Li against i, leading to the result $\mathscr{W} = \frac{1}{2}LI^2$.

Once the current is established, no further work is required to keep the current flowing through a "pure" inductor. If the inductor has resistance, of course, work must be done to replenish the energy lost as heat in the resistance.

If the circuit is interrupted after the current I is established, then the energy $\mathscr{W} = \frac{1}{2}LI^2$ is returned to the circuit when the magnetic field collapses.

Energy in the Magnetic Field

The magnetic field of a typical inductor is the same as that for a solenoid and is given, for an air-core solenoid carrying a current i, by Eq. (18.17),

$$B = \mu_0 ni = \mu_0 \frac{N}{l} i$$

where we assume that there are N turns in length l of the solenoid. Now, the magnetic flux Φ is, from Eqs. (19.2) and (18.17),

$$\Phi = BA = \mu_0 \frac{N}{l} iA \quad \text{and so} \quad \Delta\Phi = \mu_0 \frac{N}{l} A \, \Delta i$$

Then, from Eq. (19.9),

$$L = N\frac{\Delta\Phi}{\Delta i} = \mu_0 \frac{N^2}{l} A \tag{19.12}$$

This is the equation for the self-inductance of a solenoid of length l and cross-sectional area A, containing N turns of wire.

Using this value for L in our expression for the energy in an inductor, we have, from Eqs. (19.11) and (18.17),

$$\mathcal{W} = \tfrac{1}{2}Li^2 = \tfrac{1}{2}\mu_0 \frac{N^2}{l} A \left(\frac{Bl}{\mu_0 N}\right)^2 \quad \text{or} \quad \mathcal{W} = \frac{1}{2}\frac{B^2}{\mu_0} lA$$

This represents the energy stored in the magnetic field of the inductor. But the interior volume of the solenoid is lA, and so the energy density, or the energy per unit volume (J/m³), is

$$w = \frac{\mathcal{W}}{V} = \frac{\mathcal{W}}{lA} = \frac{1}{2}\frac{B^2}{\mu_0}$$

The energy density for the energy stored in a magnetic field in free space (or in air) is then

$$\boxed{w = \frac{B^2}{2\mu_0}} \tag{19.13}$$

If the inductor has a ferromagnetic core with permeability μ, then Eq. (19.13) becomes $w = B^2/2\mu$.

We therefore see that it is possible to have energy stored in free space where there is no matter present. Whenever a magnetic field **B** exists, energy is present. This will become important in Chap. 21 when we discuss the propagation of energy through space in the form of electromagnetic waves.

19.6 A Circuit with Inductance and Resistance: The *RL* Circuit

Figure 19.12 shows an inductor in series with a resistor. The resistance R may include the resistance of the wire coils of the inductor in addition to any other resistors in the circuit. This combination of a noninductive resistor and a nonresistive inductor may be connected to a source of constant terminal voltage V by throwing the switch shown in the figure to cd, thus connecting a to c and b to d.

Once a constant current I is established in the coil, this current can be reduced to zero by throwing the switch to ef, so that a is connected to e and b to f, and e and f are shorted together by a short, low-resistance wire. Then the current in the circuit has no applied voltage to sustain it, and it decreases to zero, not immediately, but in a time determined by the inductance and resistance in the circuit.

FIGURE 19.12 An *RL* circuit.

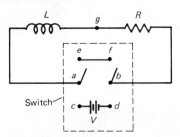

Decreasing Current

When the switch is thrown to *ef*, an emf is induced in the inductor and acts to maintain the current. We can apply Kirchhoff's loop rule by traversing the circuit from *a* back to *a* in a clockwise direction and obtain, since there is no applied voltage,

$$V_{ag} + V_{gb} = 0 \qquad \text{or} \qquad -L\frac{\Delta i}{\Delta t} - Ri = 0$$

and so

$$\frac{\Delta i}{\Delta t} = -\frac{Ri}{L} \qquad\qquad (19.14)$$

Here $\Delta i/\Delta t$ is the rate at which the current is changing, or the slope of the graph of *i* against *t*. In this case the current is decreasing, but the rate of decrease becomes less as the current becomes smaller, since $\Delta i/\Delta t$ is proportional to *i*. If we plot an experimental curve of *i* against *t* in this case, we obtain a *decreasing exponential* curve of the type shown in Fig. 19.13. The actual rate of decrease depends on the values of *R* and *L*, being rapid for *R* large and *L* small, and more gradual for *R* small and *L* large.

FIGURE 19.13 Decay of current in the circuit of Fig. 19.12 after the switch is thrown to *ef*, removing the battery from the circuit.

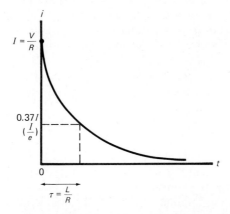

The solution to the above equation can be obtained by methods beyond the scope of this book and takes the form:

$$i = Ie^{-Rt/L} \qquad\qquad (19.15)$$

where *e* is the *natural base* of the exponential function and has a value $e \simeq 2.718$. (When this base is used, the value of the slope of a graph of e^x against *x* is equal to the value of e^x itself at any point, i.e., $\Delta e^x/\Delta x = e^x$.*)

In Eq. (19.15), when $t = L/R$, $i = I/e = I/2.718 = 0.37\,I$. Hence the current has fallen from *I* to $0.37\,I$ in L/R s. For this reason L/R is called the *time constant* of the circuit.

Definition

The *time constant determines the rate at which a physical quantity (the current in this case) approaches its final, steady-state value.* Time constants turn out to be of great practical value when dealing with complicated electric circuits.

Increasing Current

If we now throw the switch back to position *cd*, the current rises but does not immediately reach its final value because of the effect of the inductor. At any

---

*For a further discussion of exponential functions see Appendixes 1E and 2C, and Clifford Schwartz, *Prelude to Physics,* Wiley, New York, 1983.

instant of time the potential difference across the inductor is $V_{ag} = L\,\Delta i/\Delta t$, where a is at a higher potential than g because the induced emf is opposing the increase of current. Also the voltage across the resistor is $V_{gb} = iR$. By Kirchhoff's loop rule we therefore have

$$V = V_{ag} + V_{gb} = L\frac{\Delta i}{\Delta t} + iR \qquad \text{and so} \qquad \frac{\Delta i}{\Delta t} = \frac{V - iR}{L} = \frac{V}{L} - \frac{iR}{L} \qquad (19.15a)$$

As the current i increases, the right side of this equation gets smaller and so $\Delta i/\Delta t$ also becomes smaller.

Hence the graph of i versus t will be a line starting with a slope V/L when $t = 0$, but that slope will get smaller and approach zero as i approaches its final value I. The resulting experimental curve is shown in Fig. 19.14. When the current reaches its final value, $\Delta i/\Delta t$ is zero, and so

$$\frac{\Delta i}{\Delta t} = 0 = \frac{V}{L} - \frac{R}{L}I \qquad \text{or} \qquad I = \frac{VL}{LR} = \frac{V}{R}$$

which is what we would expect from Ohm's law if there were no inductor in the circuit. *An inductance has an effect on a circuit only when the current through it is changing.*

Equation (19.15a) can be solved for i as a function of time and the solution takes the form:

$$i = \frac{V}{R}(1 - e^{-Rt/L})$$

Here again the solution contains a decreasing exponential term $e^{-Rt/L}$, which falls from an initial value 1 to 0 as t goes from 0 to ∞. As a result the current i increases gradually from 0 to its final value $I = V/R$. The time constant of the circuit is again that value of t for which the exponential term is equal to 1/e, that is, $t = L/R$, as in the decreasing-current case. Here, when $t = L/R$, $i = (V/R)(1 - 1/e) = (V/R)(1 - 0.37)$, or $i = 0.63V/R = 0.63\,I$, and the current has reached 63 percent of its final steady-state value when $t = L/R$.

The time constant τ is therefore the same for LR circuits whether the current is increasing or decreasing. It is:

$$\tau = \frac{L}{R} \qquad\qquad (19.16)$$

RL circuits with short time constants allow the current to reach its final value very quickly after the voltage is applied. Circuits with large time constants, i.e., large inductances and small resistances, prevent the current from reaching its final value for considerable times. An inductor therefore acts as an inertial term in an electric circuit.

Note that the units of the time constant L/R are

$$\frac{\text{H}}{\Omega} = \frac{\text{V·s}}{\text{A}}\frac{\text{A}}{\text{V}} = \text{s}$$

which is correct, since τ is dimensionally a time.

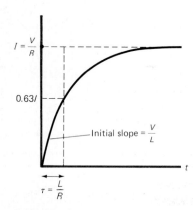

FIGURE 19.14 Current through the *RL* circuit of Fig. 9.12 as a function of the time after the switch is thrown to *cd* connecting the battery to the circuit.

Example 19.6

A coil has a resistance of 5.0 Ω and an inductance of 100 mH. At a particular instant of time after a battery is connected to the coil, the current is 2.0 A and is increasing at a rate of 20 A/s.

(a) What is the voltage applied to the coil?
(b) What is the time constant of the circuit?
(c) What is the final value of the current?

SOLUTION

(a) From Kirchhoff's loop rule

$$V = L\frac{\Delta i}{\Delta t} + iR$$

where V is the applied voltage and i is the current at some particular instant of time. When $i = 2.0$ A,

$$V = (100 \text{ mH})\left(20\,\frac{A}{s}\right) + (2.0 \text{ A})(5.0\,\Omega) = 2\text{ V} + 10\text{ V}$$

$$= \boxed{12\text{ V}}$$

(b) The time constant of the circuit is L/R. Hence

$$\tau = \frac{L}{R} = \frac{100 \times 10^{-3}\text{ H}}{5.0\,\Omega} = 20 \times 10^{-3}\text{ s} = \boxed{0.020\text{ s}}$$

(c) When $\Delta i/\Delta t = 0$, $V = IR$. Hence

$$I = \frac{V}{R} = \frac{12\text{ V}}{5.0\,\Omega} = \boxed{2.4\text{ A}}$$

19.7 Energy Stored in a Capacitor

Because of the parallels between RC and RL circuits, we will now consider RC circuits and the energy stored in a capacitor.

Work must be done in charging a capacitor, since charges must be moved from one plate to the other against the force set up by the charges already in place on the two capacitor plates. This work winds up as stored electric energy in the capacitor. We want to find how much energy is stored in a capacitor when its final charge is Q (that is, $+Q$ on one plate and $-Q$ on the other) and when the voltage between the plates is V.

The work done in charging the capacitor is the work required to move the total charge Q from one plate to the other against the potential difference between the plates, which starts at zero and ends up at V. Hence, since the voltage is the work done per unit charge, the work done in moving a charge Q is the product of Q and the average voltage on the plates. Since V varies linearly with Q (for $V = Q/C$, with C constant), as the charge on the plates increases from 0 to Q, the voltage across the plates changes from 0 to V. The average voltage is therefore $V/2$, and the work done is:

$$\mathcal{W} = Q(\tfrac{1}{2}V) = \tfrac{1}{2}QV \tag{19.17}$$

This then is the energy stored in the capacitor. Equation (19.17) is true in general for any capacitor.

Since $C = Q/V$, we may also write *the energy stored in a capacitor* in the following equivalent ways:

$$\boxed{\mathcal{W} = \tfrac{1}{2}QV = \tfrac{1}{2}CV^2 = \frac{1}{2}\frac{Q^2}{C}} \tag{19.17}$$

Example 19.7

A parallel-plate capacitor in air has a capacitance of 5 pF. A voltage of 100 V is applied by a storage battery across the plates, which are 1.0 cm apart.

(a) What is the energy stored in the capacitor?

(b) The battery is disconnected and one plate is moved until it is 2.0 cm from the other plate. What is the energy stored in the capacitor now?

(c) Everything is the same as stated above except that the battery is left connected to the plates when one plate is again moved until it is 2.0 cm from the other plate. What is the energy stored in the capacitor in this case?

SOLUTION

(a) The energy stored in the original capacitor is

$$\mathcal{W} = \tfrac{1}{2}CV^2 = \tfrac{1}{2}(5 \times 10^{-12}\ \text{F})(100\ \text{V})^2 = \boxed{2.5 \times 10^{-8}\ \text{J}}$$

(b) The energy stored in a capacitor is also given by $Q^2/2C$. Hence if the battery is disconnected so that the charge remains unchanged, the energy varies inversely with the capacitance. In this case C is reduced by a factor of 2 [since s in Eq. (16.17) is doubled when the plate is moved]. We therefore have

$$\mathcal{W}' = \frac{Q^2}{2C'} = \frac{Q^2}{2C/2} = \frac{Q^2}{C} = 2\mathcal{W}$$

Hence the stored energy has been doubled by pulling the plates apart, and

$$\mathcal{W}' = 2(2.5 \times 10^{-8}\ \text{J}) = \boxed{5.0 \times 10^{-8}\ \text{J}}$$

The increase in the energy is equal to the work done in separating the plates.

(c) If the battery is left connected, $V' = V$, and so

$$\mathcal{W}' = \tfrac{1}{2}C'V^2 = \frac{1}{2}\left(\frac{C}{2}\right)V^2 = \tfrac{1}{2}(\tfrac{1}{2}CV^2) = \tfrac{1}{2}\mathcal{W}$$

Hence in this case the stored energy is *reduced* to

$$\tfrac{1}{2}(2.5 \times 10^{-8}\ \text{J}) = \boxed{1.25 \times 10^{-8}\ \text{J}}$$

Even though the stored energy in the capacitor has been reduced, positive work has been done to pull the plates apart against the attractive force holding them together. Where does this work go? Both this work and half the original stored energy in the capacitor must be returned to the battery if energy is to be conserved in this case. This energy is consumed in charging the storage battery, since charge must be returned from the capacitor to the battery when the capacitance is reduced.

Energy in the Electric Field

It is possible to use Eq. (19.17) to find the energy stored in an electric field, whether it be between the plates of a capacitor or in some region of outer space.

Let us concentrate on a parallel-plate capacitor with a substance of dielectric constant K between the plates. Then its capacitance is, from Eq. (16.14),

$$C = \frac{K\epsilon_0 A}{s} = \frac{\epsilon A}{s}$$

Also for a parallel-plate capacitor the voltage between the plates is $V = Es$, where E is the magnitude of the electric field and s is the distance between the plates. Hence the energy stored in the capacitor is

$$\mathcal{W} = \tfrac{1}{2}CV^2 = \tfrac{1}{2}CE^2s^2 = \frac{1}{2}\frac{\epsilon A}{s}E^2s^2 = \tfrac{1}{2}\epsilon E^2 As$$

But As is the volume between the plates. Hence the energy stored per unit volume, or the *energy density* w, is

$$w = \frac{\mathcal{W}}{As} = \tfrac{1}{2}\epsilon E^2 \qquad\qquad (19.18)$$

Hence the electric energy stored per unit volume in a region of space is $\tfrac{1}{2}\epsilon E^2$, where $\epsilon = K\epsilon_0$ is the permittivity of the material in which the field is located. In a vacuum $w = \tfrac{1}{2}\epsilon_0 E^2$.

This expression for the energy density, while derived here for a special case, can be shown to be valid for any region of space in which an electric field exists.

If electric and magnetic fields exist together in space, then the total energy stored in unit volume of the combined electric and magnetic fields is

$$w = \tfrac{1}{2}\epsilon_0 E^2 + \frac{1}{2}\frac{1}{\mu_0}B^2 \qquad\qquad (19.19)$$

and depends on the square of the magnitude of both the electric and magnetic fields. Note the symmetry that exists in these two expressions; that is, E and B both appear squared in the expression for w, and the constants multiplying E^2 and B^2 are of similar form. The existence of such fields in free space, and the possibility that energy can be transferred back and forth between the electric and magnetic fields, form the basis for electromagnetic waves, as we will see in Chap. 21.

Example 19.8

In a certain region of space the magnetic field has a value of 1.0×10^{-2} T, and the electric field a value of 2.0×10^6 V/m. What is the density of the energy stored in the combined electric and magnetic field?

SOLUTION

We have for the energy density of the electric field, from Eq. (19.18), with $\epsilon = \epsilon_0$,

$$\tfrac{1}{2}\epsilon_0 E^2 = \tfrac{1}{2}\left(8.85 \times 10^{-12}\ \frac{C^2}{N\cdot m^2}\right)(2.0 \times 10^6\ V/m)^2$$

$$= (8.85)\left(2.0\ \frac{C^2}{N\cdot m^2}\right)\left(1\ \frac{J}{C\cdot m}\right)^2 = \boxed{18\ J/m^3}$$

For the magnetic field,

$$\frac{1}{2}\frac{1}{\mu_0}B^2 = \frac{1}{2}\left(\frac{1}{4\pi \times 10^{-7}}\ \frac{A}{T\cdot m}\right)(1.0 \times 10^{-2}\ T)^2$$

$$= \frac{1}{8\pi} \times 10^3\ \frac{A}{m}\cdot T = \left(40\ \frac{A}{m}\right)\left(1\ \frac{N}{A\cdot m}\right)$$

$$= 40\ N/m^2 = \boxed{40\ J/m^3}$$

Hence the energy density for the two fields together is

$$w = 18\ J/m^3 + 40\ J/m^3 = \boxed{58\ J/m^3}$$

19.8 A Circuit with Capacitance and Resistance: The *RC* Circuit

Consider the *RC* circuit of Fig. 19.15, with a switch to allow us to connect a voltage across the *RC* combination or to complete the circuit with no applied voltage in the circuit.

Charging the Capacitor

If we throw the switch to *cd*, the battery causes a current to flow and charge up the capacitor C. The initial current is just $I = V/R$, but it decreases to zero as times goes on. We have, from Kirchhoff's loop rule,

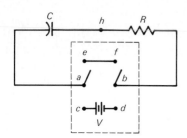

FIGURE 19.15 An RC circuit.

$$V = V_{ah} + V_{hb} = \frac{q}{c} + Ri$$

where i is the current through R at some instant of time t, and q is the charge on the capacitor at the same instant of time. Hence

$$i = \frac{V - q/C}{R} = \frac{V}{R} - \frac{q}{RC}$$

or $\quad i = \frac{V}{R}\left(1 - \frac{q}{CV}\right) = \frac{V}{R}\left(1 - \frac{q}{Q}\right)$

As time goes on, q increases until it reaches its final value Q, when the voltage V is completely across the capacitor. When this happens,

$$i = \frac{V}{R}(1 - 1) = 0$$

Hence i decreases exponentially in time from I to zero, and q increases from zero to Q, as shown in Figs. 19.16 and 19.17. The exponential decrease in i may be written

$$i = \frac{V}{R}e^{-t/RC} = Ie^{-t/RC} \tag{19.20}$$

In this case the *time constant* τ is

$$\tau = RC \tag{19.21}$$

for when $t = RC$ the current has fallen to $I/e = 0.37I$, and the charge has risen to $0.63Q$.

Note that the units of the time constant RC are

$$(\Omega)(F) = \left(\frac{V}{A}\right)\left(\frac{C}{V}\right) = \left(\frac{C}{C/s}\right) = s$$

as must be the case for a time constant.

FIGURE 19.16 Current in the RC circuit of Fig. 19.15 as a function of time. After the switch is thrown to cd, the current falls exponentially to zero with a time constant $\tau = RC$.

FIGURE 19.17 The charge on the capacitor in the RC circuit of Fig. 19.15 as a function of time. After the switch is thrown to cd, the charge approaches its dc value with a time constant $\tau = RC$.

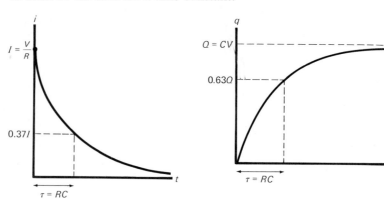

Discharging the Capacitor

Once the capacitor has its full charge Q, we can throw the switch to position ef in Fig. 19.15 and let the capacitor discharge through the resistor. Here both q and i fall gradually to zero. We have, from Kirchhoff's loop rule,

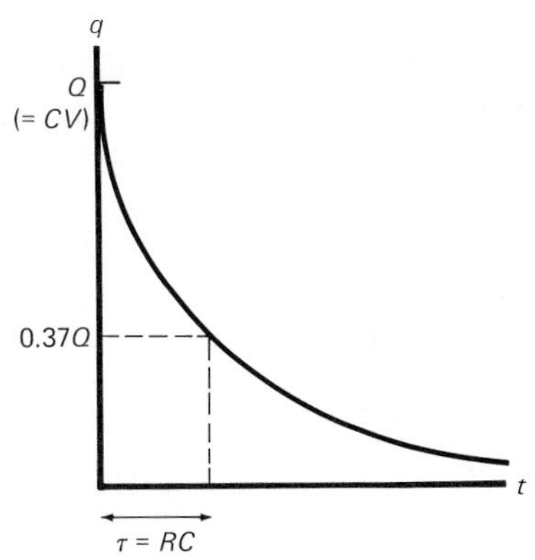

$$0 = V_{ah} + V_{hb} \quad \text{or} \quad 0 = +\frac{q}{C} - iR$$

since the direction of i is now from h to a. Hence $i = q/RC$, and initially, when $i = I$ and $q = Q$, we have

$$I = \frac{Q}{RC} = \frac{V}{R}$$

where V is the original voltage used to charge the capacitor. Here

$$q = Qe^{-t/RC} \quad \text{and} \quad i = Ie^{-t/RC} \tag{19.22}$$

The charge q decreases to $0.37Q$ within one time constant $\tau = RC$, as shown in Fig. 19.18, and the same is true for the current.

Again, the shorter the time constant, the more rapidly a capacitor can be charged or discharged. The longer the time constant, the more slowly the capacitor charges or discharges. This can have disastrous results if someone, feeling confident because the voltage source has been disconnected, touches both ends of a large capacitor which has been charged to a high voltage (see Example 19.9).

FIGURE 19.18 The variation with time of the charge on the capacitor of Fig. 19.15 when it discharges through a resistor R. The charge decays exponentially with a time constant $\tau = RC$.

Example 19.9

A discharge of more than 10 J of energy through the human body is extremely hazardous and often lethal.
(a) To what voltage would an 80-μF capacitor have to be charged to store 10 J of energy?
(b) Suppose this same capacitor was charged to 1000 V and that then the voltage source was disconnected and the capacitor allowed to discharge through a resistance of 1 MΩ. If a laboratory technician made the mistake of touching both terminals of the capacitor 1 min after the voltage had been disconnected, how much energy would be discharged through the technician's body?

SOLUTION

(a) We have $\mathcal{W} = \frac{1}{2}CV^2 = 10$ J, and so

$$V = \sqrt{\frac{2(10 \text{ J})}{80 \times 10^{-6} \text{ F}}} = \boxed{5.0 \times 10^2 \text{ V}}$$

(b) In this case we have an RC series circuit with $R = 10^6 \, \Omega$ and $C = 80 \times 10^{-6}$ F. Hence the time constant of the circuit is

$$\tau = RC = (10^6 \, \Omega)(80 \times 10^{-6} \text{ F}) = 80 \text{ s}$$

The charge q on the discharging capacitor as a function of time, with t taken as zero when the voltage is disconnected, is given by Eq. (19.22):

$$q = QE^{-t/RC}$$

where in this case $Q = CV = (80 \times 10^{-6} \text{ F})(1000 \text{ V}) = 8.0 \times$ 10^{-2} C. After 1 min, or 60 s, the charge on the capacitor is therefore:

$$q = (8.0 \times 10^{-2} \text{ C})(e^{-(60 \text{ s}/80 \text{ s})}) = (8.0 \times 10^{-2} \text{ C}) \, e^{-0.75}$$

$$= (8.0 \times 10^{-2} \text{ C})(0.47) = 3.8 \times 10^{-2} \text{ C}$$

The energy still stored in the capacitor is therefore:

$$\mathcal{W} = \frac{q^2}{2C} = \frac{(3.8 \times 10^{-2} \text{ C})^2}{2(80 \times 10^{-6} \text{ F})} = \boxed{9.0 \text{ J}}$$

Hence the energy the technician receives through the body is still almost a lethal 10-J dose, even though 1 min has elapsed since the voltage source was disconnected from the capacitor. In other words, always short-circuit large capacitors before handling them. Otherwise you may live (or not live!) to regret it.

Thyratron Sweep Circuit

An interesting and practical circuit using a capacitor and a resistor in series is the sweep circuit shown in Fig. 19.19, which is of the kind used to sweep the

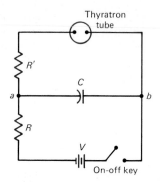

FIGURE 19.19   A thyratron sweep circuit.

electron beam back and forth across TV screens. In the figure R is a much larger resistor than R'. The voltage V charges up the capacitor C through the resistor R, with the charging process being confined to the lower portion of the charging curve shown in Fig. 19.17, so that the charging curve approximates a straight line. The voltage V_{ab} across the capacitor is also present across the gas-filled thyratron tube shown in the figure, since there is no voltage drop across R' as long as no current flows. When the voltage V_{ab} reaches a preselected value V', this is sufficient to ionize the gas in the thyratron tube so that it will conduct. The capacitor therefore discharges very rapidly through the ionized gas, because R' is so small that the time constant is also very small. The voltage V then takes over and starts to charge up the capacitor again, and this process repeats itself continuously.

The resulting *sawtooth voltage* across the capacitor is shown in Fig. 19.20. In a TV set this voltage is used to deflect the electron beam. The relatively slow rise of the V_{ab}-versus-t curve corresponds to the sweep of the electron beam from left to right on the TV screen. The much more rapid fall of the voltage from V' back to zero corresponds to the almost instantaneous flyback of the beam from right to left on the screen. In oscilloscopes and other instruments that display electric signals, such as an electrocardiograph, the proper selection of C and R fixes the appropriate time constant for any desired application.

FIGURE 19.20   The sawtooth voltage produced by a thyratron sweep circuit of the kind shown in Fig. 19.19. V_{ab} is the voltage across the capacitor C. This is the type of voltage used to sweep an electron beam back and forth across a TV or oscilloscope screen.

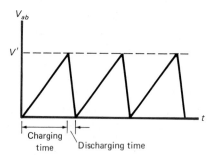

Example 19.10

The electron beam in a TV set sweeps across the screen at a rate of 30 lines per second. If the sweep mechanism is a thyratron circuit, provided by charging and discharging a 10-μF capacitor through a resistor, what must be the approximate values of R and R' in Fig. 19.19 in order for the sweep circuit to function properly?

SOLUTION

Since the time of one sweep is 1/30 s, the time constant for the charging of the capacitor must be approximately 1/30 s. Hence, since the time constant $\tau = RC$, we have

$$\tau = RC = \frac{1}{30} \text{ s}$$

or $R = \dfrac{1 \text{ s}}{30C} = \dfrac{1 \text{ s}}{30(10 \times 10^{-6} \text{ F})} = \boxed{3.3 \times 10^3 \ \Omega}$

For the flyback of the electron beam, a time constant much shorter than this is needed, certainly 1/100 times smaller. Let us take 10^{-4} s as a reasonable value. Then we have

$$\tau = R'C = 10^{-4} \text{ s}$$

and $R' = \dfrac{10^{-4} \text{ s}}{10 \times 10^{-6} \text{ F}} = \boxed{10 \ \Omega}$

19.9 Electric Generators

Faraday's induction law is the basis for the operation of all electric generators. A simple model of an electric generator, or dynamo, is shown in Fig. 19.21. A rectangular coil *abcd* is mechanically forced to rotate about an axis OO'

FIGURE 19.21 A simple electric generator (dynamo). Mechanical motion of the coil in a magnetic field produces a current through the meter.

FIGURE 19.22 End-on view of the generator of Fig. 19.21. The coil is rotating with an angular velocity ω, and the sides of the coil *ad* and *bc* cut the magnetic field lines and produce an emf.

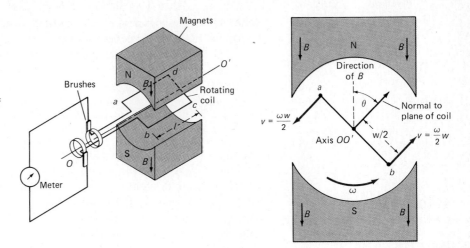

perpendicular to a uniform magnetic field of strength B. This rotation converts mechanical energy into electric energy and drives current through the meter in the external circuit.

Figure 19.22 gives an end-on view along the axis OO' shown in the preceding figure. The width of the coil is w, and its sides extend inward (into the paper) to a depth l. The two sides ad and bc have motional emf's and currents induced in them as they rotate through the magnetic field.

If the coil rotates with an angular velocity ω, then the linear speed of the side ad is $v = \omega(w/2)$. The motional emf induced in side ad is $\mathcal{E} = B_\perp vl$, where l is the length ad. The direction of $\mathcal{E}$ is from d to a, by the right-hand rule. But

$$B_\perp v = Bv \sin \theta$$

where θ is the angle between the lines of B and the normal to the plane of the coil (see Fig. 19.22). When v is parallel to B, $\theta = 0$, $\sin \theta = 0$, and no emf is induced. When v is at right angles to B, $\theta = 90°$, $\sin \theta = 1$, and the induced emf is a maximum. Hence $Bv \sin \theta$ correctly represents the physical situation. Thus we have

$$\mathcal{E} = Bvl \sin \theta = B\omega\frac{w}{2} l \sin \theta = \frac{BA\omega}{2} \sin \theta$$

since the area of the coil $A = wl$. (Note that this expression for $\mathcal{E}$ shows that the flux $\Phi = BA$ through the coil is changing as the coil rotates, in accordance with Faraday's induction law.)

Now this is the induced emf only for the side ad. For side bc, an identical expression is obtained. Since this second emf is in the direction from b to c, these two emf's reinforce each other. Also, let us assume that there are N turns on the coil. Then the total emf is

$$\mathcal{E} = 2N \frac{BA\omega}{2} \sin \theta = NBA\omega \sin \theta$$

or $\quad \mathcal{E} = NBA\omega \sin \omega t = NBA\omega \sin 2\pi ft$

since $\theta = \omega t$ and $\omega = 2\pi f$, where f is the frequency, or the number of complete rotations made per second.

The maximum emf occurs when $\theta = \omega t = 90°$, and $\mathcal{E}_{max} = NBA\omega$. Hence we have

$$\mathcal{E} = \mathcal{E}_{max} \sin \omega t = \mathcal{E}_{max} \sin 2\pi ft \qquad (19.23)$$

As time goes on, therefore, the rotating coil produces a sinusoidal voltage of the kind shown in Fig. 19.23. As the coil rotates through one complete cycle, or 360°, the emf goes from a maximum to zero, to a maximum in the opposite direction, to zero, and finally back to maximum in the original direction, as shown. If the circuit is complete, a current results, which is constantly changing in time according to Eq. (19.23). Hence this generator is called an *alternating current* (or ac) *generator*.

The output of the generator can be fed to an external circuit by attaching the two ends of the coil to rotating rings which make contact with metal brushes, as in Fig. 19.21.

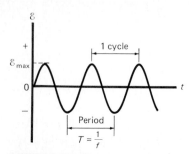

FIGURE 19.23 The emf produced by the generator of Fig. 19.21 as a function of time. Notice that the emf is "alternating" in nature and will produce an alternating current (ac) in a complete circuit.

DC Generators

The current in Fig. 19.23 is obviously positive in direction half the time and negative the other half. If we want direct current, we can obtain it from the same generator by attaching the two ends of the coil to a single ring which is split in half and makes contact with two metal brushes, as shown in Fig. 19.24. In this case the split ring, called a *commutator*, reverses the external contacts just at the instant the current being generated reverses. As a result, the current through the load is always in one direction but is pulsating in magnitude, as shown in Fig. 19.25. Such a pulsating current can be smoothed out by using more than one coil rotating about the same axis, or by well-known electronic techniques, to give a very good imitation of the direct current delivered by a battery.

FIGURE 19.24 A generator using a split-ring commutator to produce direct current. The commutator reverses the current every half cycle so as to keep it always flowing in one direction, although changing in magnitude.

FIGURE 19.25 The pulsating dc current produced by the generator of Fig. 19.24.

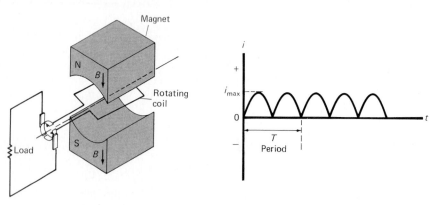

Practical Electric Generators

In a typical fossil-fuel steam-electric power plant, coal, oil, or natural gas is burned to heat water and produce steam. The high-energy steam molecules then strike the blades of a turbine (a modern version of an old-fashioned water wheel) and set it into rotational motion (see Fig. 15.14). This rotational motion is conveyed through a moving shaft to the rotor of an electric generator which converts mechanical motion into electric energy and feeds it into transmission lines for the use of consumers.

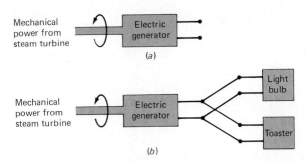

FIGURE 19.26 Loading the generator: *(a)* No load—open circuit. *(b)* Closed circuit—load attached. The addition of a light bulb and a toaster to the generator output increases the required energy input to the generator.

The above discussion has not indicated how the load (e.g., the light bulb or toaster in Fig. 19.26) affects the operation of the generator. If a generator is running with no load attached, as in Fig. 19.26a, once the rotor is moving at full speed the only mechanical power the turbines must provide to keep the shaft turning is that needed to overcome friction and other mechanical losses in the system. If, now, we connect a light bulb (or a whole city of light bulbs) to the generator, the bulb draws current from the generator. That current flows in such a direction that, by Lenz' law, it sets up a magnetic force which acts to keep the shaft from turning. To overcome this turning resistance, additional mechanical power must be provided by the steam turbine, which means that more fuel must be burned to produce the steam to drive the turbine. If we now add a toaster (or another city) in parallel with the light bulb, the current increases even more, as does the resisting force, and still more mechanical power is needed.

Hence the greater the load, the more mechanical energy required and the more fuel consumed. This is another example of the *principle of conservation of energy*, since the energy consumed by the light bulb and the toaster must ultimately come from the energy provided by the fossil fuel or other energy source.

Example 19.11

The armature of a 60-Hz ac generator has 200 coils of wire wound on it. The area of each coil is 0.25 m². What must be the strength of the magnetic field in which this armature rotates if the peak voltage output of the generator is to be 170 V?

SOLUTION

From the above discussion the maximum emf produced by the rotating armature is

$$\mathscr{E}_{max} = NBA\omega$$

and so

$$B = \frac{\mathscr{E}_{max}}{NA\omega} = \frac{\mathscr{E}_{max}}{NA(2\pi f)}$$

$$= \frac{170 \text{ V}}{200(0.25 \text{ m}^2)(120\pi/\text{s})} = \boxed{0.0090 \text{ T}} = 90 \text{ G}$$

19.10 Electric Motors

An electric motor is a device that converts electric energy into continuous rotational motion.

As we saw in our discussion of the galvanometer in Sec. 18.10, a current-carrying coil in a magnetic field experiences a rotational torque. If the current is dc and the plane of the coil is originally parallel to the lines of **B**, the

FIGURE 19.27 A coil carrying a dc current in a fixed magnetic field: *(a)* The torques on the current-carrying wires normal to the paper at positions 1 and 2 are in opposite directions and drive the coil to position 3, where it freezes. *(b)* Current through the coil reverses every half cycle, so that the torques at positions 1 and 2 are in the same direction and the coil is set into rapid rotation about the axis at 0.

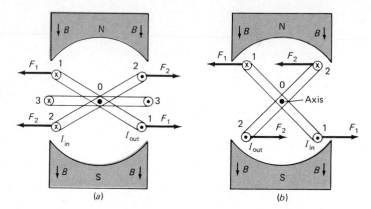

torque will be a maximum and will rotate the coil to a position where its plane is at right angles to the magnetic field. There, however, the coil will stop. This can be seen in Fig. 19.27*a*, where it is clear that the torques at positions 1 and 2 for the coil are in opposite directions and will tend to drive the coil to position 3, where it will "freeze."

To make the coil rotate continuously we must make the current through the coil reverse every time the coil moves through position 3. Then the torque will be a continuous one and the coil will pick up speed as it rotates, as in Fig. 19.27*b*. This can be accomplished by using an ac voltage to drive current through the coil. An ac motor is therefore nothing but an ac generator run in reverse.

Motors can also be made to operate off dc voltages using split-ring commutators to reverse the current, as in Fig. 19.28.

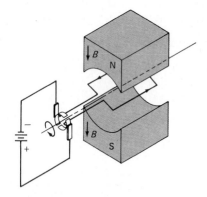

FIGURE 19.28 A motor operating off a dc source using a split-ring commutator to reverse the current each half cycle.

Back EMF in a Motor

The constant torque on the coil of a motor does not accelerate it to greater and greater rotational speeds. This is because, as the coil turns, an emf is generated in it which opposes the coil's rotation according to Lenz' law. This is called a *back emf*. As the rotor speeds up, the back emf increases until a constant terminal speed is reached. This situation is quite similar to a ball falling through a viscous fluid. The resistive force is proportional to the ball's velocity, and when this force equals the force of gravity, the ball ceases to be accelerated and falls at a constant terminal velocity. Something similar happens with an electric motor. The back emf depends on $B_\perp vl$, and thus the faster the coil rotates, the larger the back emf and the larger the force opposing the motion of the armature.

Since the back emf serves to limit the current flowing through the motor, overloading a motor has the effect of slowing down the rotor and reducing the back emf. Hence more current is drawn from the source. If this current becomes too large, it is possible to burn out the motor.

When a motor is just starting, there is little back emf. Hence the motor draws a larger current when it is starting than when it is up to full speed. This explains why house lights dim when a large motor like that in an air conditioner is first turned on. The large starting current drawn by the air conditioner causes an increased IR drop in the house wiring which reduces the voltage across light bulbs in the same circuit as the air conditioning unit.

Example 19.12

A dc motor has a resistance of 4.0 Ω. When running at full speed on a 120-V line, it draws a current of 5.0 A.
(a) What is the back emf produced by the motor?
(b) What power is delivered to the motor by the line?
(c) At what rate is energy lost as heat in the motor?
(d) How much useful power is produced by the motor?
(e) What is the efficiency of such a motor?
(f) What current does such a motor draw when it is just starting up?

SOLUTION

(a) If the applied voltage is V and the back emf $\mathcal{E}$, then the applied voltage must be equal to the sum of the voltage drops across the motor, or $V = \mathcal{E} + IR$. Hence

$$\mathcal{E} = V - IR = 120 \text{ V} - (5.0 \text{ A})(4.0 \text{ Ω}) = \boxed{100 \text{ V}}$$

(b) The power delivered by the line is

$$P_1 = VI = (5.0 \text{ A})(120 \text{ V}) = \boxed{600 \text{ W}}$$

(c) Energy is lost as heat at a rate

$$P_2 = I_2 R = (5.0 \text{ A})^2(4.0 \text{ Ω}) = \boxed{100 \text{ W}}$$

(d) The useful power produced by the motor is the difference between (b) and (c):

$$P = P_1 - P_2 = 600 \text{ W} - 100 \text{ W} = \boxed{500 \text{ W}}$$

The same result could have been obtained by taking the product of the back emf and the current which flows.

$$P = \mathcal{E}I = (100 \text{ V})(5.0 \text{ A}) = \boxed{500 \text{ W}}$$

as just found.

(e) The efficiency is merely the ratio of the useful power out to the total power in, or

$$\text{Efficiency} = \frac{P}{P_1} = \frac{500 \text{ W}}{600 \text{ W}} = 0.83 = \boxed{83\%}$$

(f) When the motor is just starting up, the back emf is zero since v is zero in the expression $\mathcal{E} = B_\perp vl$. Hence the starting current is

$$I_S = \frac{V}{R} = \frac{120 \text{ V}}{4.0 \text{ Ω}} = \boxed{30 \text{ A}}$$

Note how much larger this current is than the current when the motor is running at full speed.

Summary: Important Definitions and Equations

Induced emf's
   Motional emf: The emf produced by the motion of a conductor through a magnetic field: $\mathcal{E} = B_\perp vl$

   Faraday's induction law: $\mathcal{E} = -N\dfrac{\Delta\Phi}{\Delta t}$

   where Φ = magnetic flux = $B_\perp A$.

Lenz's law: The direction of an induced emf is always such that it opposes the change producing it.
EMF due to mutual induction:

$$\mathcal{E} = -M\frac{\Delta i}{\Delta t} \qquad Mi = N\Phi$$

EMF due to self-induction:

$$\mathcal{E} = -L\frac{\Delta i}{\Delta t} \qquad Li = N\Phi$$

Inductors

Unit of inductance (M or L): 1 henry (H) = 1 V·s/A

Energy stored in an inductor: $\mathcal{W} = \frac{1}{2}LI^2$

Energy density in a magnetic field B (energy stored per unit volume):

$$w = \frac{B^2}{2\mu_0}$$

RL circuit: Time constant $\tau = L/R$

Time constant: A time which determines the rate at which a physical quantity approaches its final, steady-state value.

Capacitors

Energy stored in a capacitor:

$$\mathcal{W} = \tfrac{1}{2}QV = \tfrac{1}{2}CV^2 = \frac{1}{2}\frac{Q^2}{C}$$

Energy density in an electric field E (energy stored per unit volume):

$$w = \tfrac{1}{2}\epsilon_0 E^2$$

RC circuit: Time constant $\tau = RC$

Electric generators:

Machines that convert mechanical energy of rotation into electric energy.

EMF produced by an ac generator: $\mathcal{E} = \mathcal{E}_{max} \sin \omega t = \mathcal{E}_{max} \sin 2\pi ft$ where $\mathcal{E}_{max} = NBA\omega$.

Electric motors:

Machines equivalent to electric generators run in reverse which convert electric energy into mechanical energy of rotation.

Questions

1 What would happen if a straight rod that is not part of a complete circuit moved at right angles to a magnetic field? Would an emf be produced? Would a current flow? If so, how long would the current last?

2 Is it possible for a stationary, unchanging magnetic field to produce an electric current without a violation of the principle of conservation of energy? Explain your answer.

3 If, in Fig. 19.1, the rod *ab* is at rest and the magnetic field is suddenly turned off, is the induced emf clockwise or counterclockwise? Indicate how you arrived at your answer.

4 Discuss the relationship between Faraday's induction law and the operation of an electric generator.

5 A copper ring is mounted on a ring stand 1 m above the floor. A bar magnet is held above the ring with its north pole toward the floor. If the magnet is then dropped through the ring to the floor, what are the forces acting on the magnet (*a*) before it reaches the ring; (*b*) while it is passing through the ring; (*c*) after it has passed through the ring?

6 Sensitive galvanometers use closed coils of wire mounted adjacent to the deflecting coils of the meter to damp the motion of the galvanometer movement after it is displaced. Explain how such a magnetic damping mechanism would operate.

7 A galvanometer has an eddy-current damper attached to it in a way similar to that in Fig. 19.9. Does this damping system reduce the energy supplied to the galvanometer by the current through it? If so, where does the energy go?

8 Can you suggest a way to decrease inductive effects in the winding of resistance coils for use in standard resistance boxes?

9 How could you decide whether the earth is moving through a magnetic field that fills all space, or is carrying its own magnetic field along with it as it moves?

10 Why is any work at all required to keep a steam-electric generator turning when no load is attached to the generator terminals? What happens when a load is attached?

11 Are nuclear steam-electric power plants any more efficient than coal-burning plants? Why?

12 A motor which is being forced to do too much work will frequently blow a fuse or circuit breaker before any damage is done to the motor itself. Explain what happens in this case. When is the fuse most likely to blow?

13 Explain how a combination of strong magnetic fields and eddy-current effects could be used to separate ground-up solid waste into three groups: ferrous metals, nonferrous metals, and nonmetallic substances.

Problems

A 1 In Fig. 19.1 the rod *ab*, which is 0.20 m long, is moved to the right with a speed of 3 m/s. The magnetic field strength is 1.5 T and is directed at right angles to the plane of the current loop. If the current loop has a break in it, the induced emf is:

(*a*) 0   (*b*) 0.90 V   (*c*) 0.90 T   (*d*) 1.0 V
(*e*) 0.10 V

A 2 The Wb/m² is a unit of:

(*a*) Magnetic flux   (*b*) Area
(*c*) Magnetic field strength   (*d*) Force   (*e*) Pressure

A 3 A circular coil of wire consists of 20 turns, and has an area equal to $0.10 \, m^2$. The magnetic field through the coil changes by 2.0 T in 0.10 s. The magnitude of the induced emf in the coil is:
(a) 40 T   (b) 4.0 V   (c) 400 V   (d) 40 J
(e) 40 V

A 4 The unit 1 Wb/s is the same as:
(a) 1 J   (b) 1 V   (c) 1 T   (d) 1 N
(e) 1 Wb/m²

A 5 A coil has a self-inductance of 100 mH. If the current through it is increasing at the rate of 5.0 A/s, the induced voltage in the coil is:
(a) −0.50 V   (b) + 50 V   (c) + 0.50 V
(d) −50 V   (e) −0.020 V

A 6 A coil of length 0.10 m has a cross-sectional area of $0.025 \, m^2$ and contains 100 turns. The inductance of such a coil with an air core is:
(a) 0.25 H   (b) 0.25 T   (c) $3.14 \times 10^{-7} \, T$
(d) $3.14 \times 10^{-7} \, H$ (e) $3.14 \times 10^{-3} \, H$

A 7 Two induction coils are identical in size and shape, but coil 1 contains 100 times as many turns of wire as coil 2. The ratio of the inductances L_1/L_2 is:
(a) 10^2 (b) 10^{-2} (c) 10^4 (d) 10^{-4} (e) 1

A 8 A circuit contains an inductance coil with $L = 50$ mH and $R = 100 \, \Omega$. A voltage of 100 V is placed across the coil. The time it takes for the current to reach 0.63 A is:
(a) $5.0 \times 10^3 \, s$ (b) $5.0 \times 10^{-3} \, s$ (c) 5.0 s
(d) $5.0 \times 10^{-5} \, s$ (e) $5.0 \times 10^{-4} \, s$

A 9 In the previous question the final current through the coil will be:
(a) 1.0 A   (b) 2.0 A   (c) 0.50 A
(d) 10^2 A   (e) 2×10^{-3} A

A10 A capacitor has a capacitance of 2.0 μF and is charged to a voltage of 1000 V. The energy stored in the capacitor is:
(a) $2.0 \times 10^3 \, J$ (b) $2.0 \times 10^{-3} \, J$ (c) 1.0 J
(d) 2.0 J   (e) $1.0 \times 10^{-3} \, J$

A11 A circuit consists of a 10-μF capacitor in series with a 500-Ω resistor. In 5.0×10^{-3} s after a voltage of 100 V is connected across this combination, the current will be:
(a) 0.20 A   (b) 0.063 A   (c) 0.13 A
(d) 0.037 A   (e) 0.074 A

A12 A coil consists of 50 turns and has a cross-sectional area of $5.0 \times 10^{-3} \, m^2$. It rotates at a frequency of 60 Hz in a magnetic field of 2.0 T. The maximum induced emf is:
(a) 30 V   (b) 189 V   (c) $1.89 \times 10^4 \, V$
(d) 60 V   (e) 378 V

B 1 A rod slides along a metal frame similar to that in Fig. 19.1. If the length of the rod is 0.50 m, its speed is 0.30 m/s, and a magnetic field of 0.10 T is directed at right angles to the plane of the metal frame, find the induced emf in the circuit made up of the fixed metal frame and the sliding rod.

B 2 What is the flux produced by a magnetic field B directed perpendicular to a coil of area $0.10 \, m^2$, if the magnitude of B is $0.50 \, Wb/m^2$?

B 3 (a) The magnetic field at right angles to a coil of 50 turns and area $0.025 \, m^2$ increases at a rate of 0.10 T/s. What is the induced emf in the coil?

(b) How would the result change if the magnetic field were decreasing at a rate of 0.10 T/s?

B 4 Two coils are wound on a common iron core, as in Fig. 19.10. If a current of 4.0 A in coil 1 produces a magnetic flux of $1.2 \times 10^{-3} \, Wb/m^2$ in coil 2, which has 150 turns, what is the mutual inductance between the two coils?

B 5 The current through the primary coil of a transformer is increasing at a rate of 5.0 A/s.
(a) If the self-inductance of the coil is 200 mH, what is the magnitude of the induced emf in the primary coil?
(b) What is the direction of the induced emf?

B 6 How much energy is stored in the magnetic field of a 200-mH inductor when the current through the inductor is 5.0 A?

B 7 What is the time constant of a circuit consisting of an inductance of 50 mH in series with a 2000-Ω resistor?

B 8 What is the energy density between the plates of a parallel-plate capacitor when the voltage across the plates is 500 V and the distance between the plates 1.0 mm (in air)?

B 9 What is the time constant of a circuit consisting of a 1.0-μF capacitor and a 10-MΩ $(10 \times 10^6 \, \Omega)$ resistor?

B10 It is desired to produce an alternating voltage of maximum value 5000 V with a generator containing a coil of 200 turns and cross-sectional area $1.0 \, m^2$ rotating at 60 Hz. What must be the strength of the magnetic field in which the coil rotates?

B11 A dc motor has a resistance of 5.0 Ω and draws a current of 10 A when running at full speed on a 120-V line. What is the back emf produced by the motor?

C 1 A rod slides along a metal frame of the shape shown in Fig. 19.29. The earth's field has a strength of $0.35 \times 10^{-4} \, T$ at right angles to the plane of the frame. If the rod is of length 0.20 m inside the frame, and moves with a speed of 1.0 m/s:
(a) What is the motional emf produced by the rod?
(b) If the resistance of the frame and rod combined is 0.050 Ω, what current flows?
(c) What are the magnitude and direction of the force on the rod?
(d) Show that the mechanical power needed to move the rod is exactly equal to the electric power consumed in the resistance of the frame and rod.

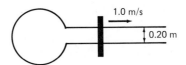

FIGURE 19.29 Diagram for Prob. C1.

C 2 An air-core solenoid contains 100 turns of wire per meter and has a cross-sectional area of $0.12 \, m^2$. A current of 75 A is established in the solenoid and is then reduced to zero in 3.0 s by opening a switch. A pickup coil of 50 turns circles the solenoid near its center.
(a) What are the magnitude and direction of the induced emf in the pickup coil?

(b)   If the resistance of the loop is 0.0080 Ω, what is the induced current?

C 3   A train is moving with a speed of 10 m/s in a northerly direction. If the vertical component of the earth's magnetic field in the vicinity is 0.50 G, find the magnitude and direction of the emf induced in an axle of the train, if the axle is of length 2.0 m.

C 4   A transformer consists of two coils of wire wound on a common toroidal iron core. If the mutual inductance of the pair is 100 mH, and the current in the first coil decreases from 20 A to 0 in 0.50 s, what is the induced emf in the second coil?

C 5   An inductance coil of self-inductance 100 mH is connected to a circuit in which the current is changing with time in a sawtooth pattern, as in Fig. 19.30. Draw a graph indicating how the induced emf in the circuit varies in time. On this graph show the actual numerical value of $\mathcal{E}$ at each instant of time.

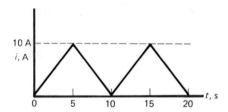

FIGURE 19.30   Diagram for Prob. C5.

C 6   A solenoid has a length of 20 cm, a cross-sectional area of 0.020 m², and contains 100 turns.
(a)   What is its inductance?
(b)   If the current through the coil increases by 5.0 A, how does the magnetic flux through the coil change?

C 7   (a)   How large would a magnetic field B have to be to store as much energy per unit volume as is stored in an electric field E of strength 10^4 V/m?
(b)   What is the numerical ratio of E (in V/m or N/C) to B (in T) in this case?

C 8   Two coils of wire C and C' have 500 and 1000 turns, respectively. A current of 2.0 A in C produces a flux of 2.0×10^{-4} Wb through C and a flux of 1.5×10^{-4} Wb through C'. Find:
(a)   The self-inductance of coil C.
(b)   The mutual inductance of C and C'.
(c)   The average induced emf in C' when the current in C falls to zero in 0.5 s.

C 9   A circuit consists of a 10-Ω resistor and a 500-mH inductor connected in series to a 12-V battery. When the current is increasing at the rate of 0.20 A/s, what is the value of the current?

C10   A current of 10 A flows in a circuit containing an inductor with $L = 10$ mH and a resistor with $R = 100$ Ω. A key in the circuit is opened at $t = 0$. How long will it take for the current to fall to 3.7 A?

C11   A long iron-core solenoid is connected across a 12-V battery and the current rises to 0.63 of its final value in 0.80 s. This procedure is then repeated with the iron ore removed, and the time required for the current to rise to 0.63 of its final value is only 0.0010 s.

(a)   Calculate the relative permeability of the iron.
(b)   Find the inductance of the air-core solenoid if the maximum current is 0.60 A.

C12   An average lightning stroke contains about 5×10^8 J of energy. If we assume that this stroke is produced by the flow of charge from one plate of a capacitor to another (analogous to a flow from one cloud to another), that the distance between the plates is 1000 m, and that the voltage across the plates 10^8 V, how large would the plates have to be to store the 5×10^8 J released in the lightning bolt?

C13   A circuit consisting of a 100-μF capacitor in series with a 1.0-MΩ resistor is connected across a 6.0-V battery.
(a)   What is the time constant of the circuit?
(b)   What is the initial value of the current when a key in the circuit is closed?
(c)   What is the charge on the capacitor when the current i is 2.0×10^{-6} A?

C14   A 5.0-μF capacitor is attached to a 100-V battery and a 4.0×10^4 Ω resistor.
(a)   What is the final charge on the capacitor?
(b)   If the battery is shorted out of the circuit so as to allow the capacitor to discharge, how long does it take for the charge on the capacitor to fall to 37 percent of its initial value?
(c)   Prove that at the instant of time when $q = 0.37Q$, where Q is the initial charge on the capacitor, the voltages across C and R add to give the 100 V provided by the battery.

C15   A 5.0-μF capacitor is charged to a potential of 100 kV. After being disconnected from the power source, it is connected across a 10-MΩ resistor.
(a)   What rate does the capacitor discharge initially?
(b)   How long will it take for the voltage across the capacitor to fall to 37 kV?

C16   A generator coil rotates in a magnetic field in a fashion similar to that in Fig. 19.21. If the coil contains 100 turns, has an area of 1.5 m², and makes 60 rev/s in a magnetic field of 0.10 T:
(a)   What is the maximum emf induced in the coil?
(b)   Draw a graph of the emf in the coil as a function of time t for one complete cycle.

C17   An ac motor develops an emf of 120 V when rotating at 1500 rev/min. At what speed must it rotate to generate 240 V?

C18   A dc motor with a resistance of 4.0 Ω draws a current of 12 A when running at full speed off a 60-V battery.
(a)   What is the starting current for this motor?
(b)   What is the back emf when the motor is up to full speed?

C19   For the motor of Prob. C18, prove that the electric power delivered by the battery is equal to the sum of the power delivered by the motor and the power lost as heat in the motor resistance.
(a)   Show this at the instant the motor is starting up.
(b)   Show this when the motor is running at full speed.

D 1   A 6.0-cm-diameter circular coil of wire has a resistance of 2.0 Ω. It is located in a magnetic field of 0.30 T directed at right angles to the plane of the coil. If the coil is quickly removed from the field in a time 0.10 s:

(a)   What is the induced emf in the coil?

(b)   How much electric energy is lost in the process?

(c)   Where does this energy go?

D 2   The moving rod in Fig. 19.1 is assumed to have negligible resistance, but the resistance of the U-shaped conductor is assumed to be R.

(a)   Prove that the rate at which mechanical work must be done to move the rod is $P = B^2 l^2 v^2 / R$.

(b)   Prove that this is equal to the electric power dissipated as heat in the circuit.

D 3   A rod 0.50 m long is pivoted so that it can rotate about a fixed point P, as in Fig. 19.31. If it rotates at 4 rev/s, and a magnetic field of 0.20 T exists parallel to its axis of rotation and covering the whole area swept out by the rod, what is the potential difference between the two ends of the rod?

FIGURE 19.31   Diagram for Prob. D3.

D 4   A "search coil" combined with a ballistic galvanometer (whose deflection is proportional to the charge flowing through it) is often used to measure magnetic fields. If a search coil in a magnetic field of strength B is suddenly pulled completely out of the field, show that the magnetic field B is given by $B = RQ/NA$, where N is the number of turns and A is the area of the coil, R is the resistance of the search coil and galvanometer combined, and Q is the total charge which flows through the galvanometer.

D 5   One meter of no. 18 copper wire (diameter = 0.10 cm) is formed into a circular loop. It is then placed in a uniform magnetic field directed at right angles to the plane of the loop. The magnetic field increases in time at a constant rate of 0.020 T/s. At what rate is heat generated in the loop?

D 6   An inductance coil of resistance 20 Ω and inductance 0.50 H is connected to a 120-V dc source. At what rate will the current in the coil rise:

(a)   At the instant the coil is connected to the power source?

(b)   At the instant the current reaches 75 percent of its maximum value?

D 7   Prove that the expression $\sqrt{L/C}$ has the units of ohms. Here L is an inductance and C is a capacitance.

D 8   The magnetic field B perpendicular to a single-turn loop of wire of radius 0.20 m and resistance 0.10 Ω changes with time as shown in Fig. 19.32.

(a)   Plot the induced emf as a function of time.

(b)   Plot the rate of energy dissipation as a function of time.

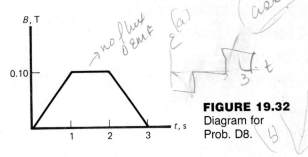

FIGURE 19.32 Diagram for Prob. D8.

D 9   A motor has a resistance of 4.0 Ω and is connected to a 120-V power line. The motor draws a current of 5.0 A when running at full speed. What current will the motor draw if the speed is reduced to 90 percent of full speed by application of a load?

Additional Readings

Coulson, Thomas: *Joseph Henry, His Life and Work*, Princeton University Press, Princeton, N.J., 1950. The standard biography of a great American physicist.

Kaplan, Joseph: "Michael Faraday (1791–1896)," in *Great Men of Physics: The Humanistic Element in Scientific Work*, Tinnon-Brown, Los Angeles, Calif., 1969. A popular lecture on Faraday's contributions to physics.

Kondo, Herbert: "Michael Faraday," in *Lives in Science, Scientific American* book, Simon and Schuster, New York, 1957, pp. 127–140. A brief and useful account of Faraday's life and contributions to physics.

McDonald, D. K. C.: *Faraday, Maxwell and Kelvin*, Doubleday Anchor, Garden City, N.Y., 1964. This book includes excellent short accounts of the lives and work of three great physicists, but is marred by the constant intrusion of the author's personal asides on physics and life in general.

Shiers, George: "The Induction Coil," *Scientific American*, vol. 224, no. 5, May 1971, pp. 80–87. A fascinating history of the development of the induction coil as a source of high voltage and of the role it played in important physics discoveries by Roentgen, Hertz, and J. J. Thomson.

Williams, L. Pearce: "Michael Faraday and the Physics of 100 Years Ago," in Jerry B. Marion, *A Universe of Physics: A Book of Readings*, Wiley, New York, 1970, pp. 85–97. An account of how Faraday broke with the mechanistic concepts of his day. Williams has also written the definitive biography of Faraday: *Michael Faraday*, Basic Books, New York, 1964.

Wilson, Mitchell: "Joseph Henry," in *Lives in Science*, op. cit., pp. 141–153. A good short description of the life of Joseph Henry.

A ny notice about the progress of physics in the latter part of the last century would be like the play of Hamlet *without the Prince of Denmark if it did not deal with the part played in it by Lord Kelvin, who for more than forty years before his death in 1907 had been the most potent influence in British physics.*

J. J. Thomson (1856–1940)

Chapter 20

Alternating Current Circuits

At the end of the nineteenth century a heated controversy erupted over whether the electric power to be delivered to consumers in the United States should be in the form of alternating or direct current. George Westinghouse embraced the ideas of Nikola Tesla and strenuously supported alternating current. Thomas Edison, on the other hand, fought furiously and often unscrupulously for direct current, first lobbying for the New York State Legislature to adopt alternating current for the electric chair and then using the electric chair as an example of the deadly nature of alternating current to advocate the adoption of the less dangerous direct current for commercial use! In 1893 the proponents of alternating current won two crucial victories: Tesla's ac system was chosen to light the World Columbian Exposition in Chicago, and Westinghouse received the contract to deliver hydroelectric power in ac form from Niagara Falls to American homes and factories. Since that time ac power has assumed almost complete dominance in the electric utility industry in the United States.

20.1 AC Circuit Containing Only Resistance

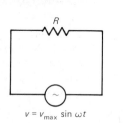

FIGURE 20.1 An ac voltage of form $v = v_{max} \sin \omega t$ applied to a simple resistor, with $\omega = 2\pi f$.

In the preceding chapter we saw that the voltage output of an ac generator takes the form $v = v_{max} \sin 2\pi ft$, where f is the frequency, or the number of complete rotations of the generator coil per second, and v is the voltage at any instant of time t. If this voltage is applied across a simple resistor, as in Fig. 20.1, it is found by experiment that the current i through the resistor is also sinusoidal and is in step with the voltage variations. It takes the form $i = i_{max} \sin 2\pi ft$. The current reaches its maximum value i_{max} at the same time as the voltage, and passes through zero at the same instant as the voltage, as shown in Fig. 20.2. We can therefore conclude:

In an ac circuit containing only resistance the current and the voltage are in phase.

(To refresh your mind on phase and phase relationships, which we will be referring to frequently in what follows, review Sec. 11.7.)

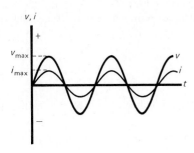

FIGURE 20.2 The ac voltage and current as a function of time for a purely resistive circuit. The current is in phase with the voltage.

In dealing with ac currents and voltages it is important to distinguish a number of different terms used to describe ac quantities:

Definitions

i (or v): The *instantaneous value* of an ac quantity, i.e., its value at some particular instant of time.

i_{max} (or v_{max}): The *amplitude* in the expression for an ac quantity, for example, $i = i_{max} \sin \omega t$. The amplitude is also referred to as the *peak value* of the quantity.

$i_{p\text{-to-}p}$: The *peak-to-peak value* of an ac quantity, i.e., the difference between its maximum positive and its maximum negative value; $i_{p\text{-to-}p} = 2i_{max}$ (see Fig. 20.2).

For a purely resistive circuit Ohm's law applies, and, hence, for the instantaneous current and voltage we have:

$$i = \frac{v}{R}$$

Since the peak values i_{max} and v_{max} are just the values of i and v at particular instants of time, and since the current and the voltage reach their peaks at the same instant of time, we also have:

$$i_{max} = \frac{v_{max}}{R}$$

Effective, or RMS, Values of Voltage and Current

From Fig. 20.2 we can see that the *average value* of an ac voltage or current is zero, since it is positive just as much as it is negative. In many cases, however, such as in supplying power to light a lamp, an alternating current can be of considerable use even though its average value is zero. The effectiveness of the current in producing heat in the filament of a lamp bulb does not depend on the direction of the current. The rate at which heat is generated is determined by Joule's law, i^2R, and this quantity is *not* zero when averaged over a cycle, even though the current i is.

To find the *effective value* of an ac current, we average the heating effect of the current over one cycle and equate it to the constant current which would produce the same effect. We have for the power,

$$P = I^2R = \overline{i^2R} = \overline{i^2}R$$

Definition

where I is the *effective current*, or the *constant current that would produce the same heating effect, on the average, as does i*. Here $\overline{i^2}$ is the mean-square current averaged over one cycle. Then, since $i = i_{max} \sin 2\pi ft = i_{max} \sin \omega t$, we have

$$I^2 R = \overline{i_{max}^2 \sin^2 \omega t}\, R$$

or $\quad I^2 = \overline{i_{max}^2 \sin^2 \omega t} = i_{max}^2\, \overline{\sin^2 \omega t}$

Now the average value of $\sin^2 \omega t$ is $\frac{1}{2}$,* and so

$$I^2 = \frac{i_{max}^2}{2} \quad \text{and} \quad I = \frac{i_{max}}{\sqrt{2}} = 0.707 i_{max} \tag{20.1}$$

Similarly, $\quad V = \dfrac{v_{max}}{\sqrt{2}} = 0.707 v_{max} \tag{20.2}$

The effective values of ac currents or voltages are thus 0.707 times their peak, or maximum, values. This is shown in Fig. 20.3. Since $i_{max} = v_{max}/R$,

$$0.707 i_{max} = \frac{0.707 v_{max}}{R}$$

and so $\quad I = \dfrac{V}{R}$

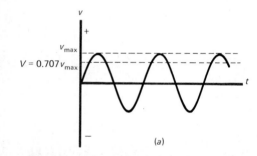

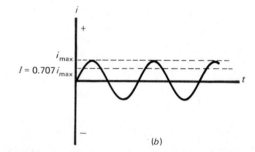

FIGURE 20.3 The effective, or rms, value (V or I), and the peak, or maximum, value (v_{max} or i_{max}), of: (a) an ac voltage; (b) an ac current.

In a purely resistive ac circuit the effective current and the effective voltage satisfy Ohm's law.

Effective currents and voltages are also called rms currents and voltages since they are obtained by first *squaring* the quantity, then taking the *mean* or average of the squared quantity over a period, and finally taking the square *root* of this average, as we did above. Hence we can define an rms current as follows:

Definition

i_{rms}: *That constant current which would produce the same heating effect as a given ac current.* (It is identical with the effective current.)

---

*To show that the average value of $\sin^2 \omega t$ is $\frac{1}{2}$, we can use a trigonometric identity from Appendix 1C, with $x = \omega t$. We have

$$\cos 2\omega t = 1 - 2 \sin^2 \omega t \quad \text{or} \quad 2 \sin^2 \omega t = 1 - \cos 2\omega t$$

Taking time averages of all quantities, we have:

$$\overline{2 \sin^2 \omega t} = 1 - \overline{\cos 2\omega t}$$

But the average value of any simple sine or cosine function over one period is zero, since it is positive just as much as it is negative. Hence

$$\overline{2 \sin^2 \omega t} = 1 \quad \text{or} \quad \overline{\sin^2 \omega t} = \tfrac{1}{2}$$

$$i_{\text{rms}} \equiv I = 0.707i_{\text{max}}$$

Similarly, $v_{\text{rms}} \equiv V = 0.707v_{\text{max}}$

For ordinary ac house voltages which have effective values of about 120 V, the peak value v_{max} is $V/0.707 = 170$ V. Most ac voltmeters and ammeters are calibrated to read effective, or rms, values directly.

In what follows we will always, unless the contrary is stated, use the symbols I and V to denote the *effective, or rms, values* of ac currents and voltages.

Example 20.1

An ac voltage of effective value 240 V is connected across a 60-Ω resistor. What are (*a*) the effective current, (*b*) rms current, (*c*) peak current, (*d*) maximum current, (*e*) peak-to-peak current?

SOLUTION

(a) The effective current is

$$I = \frac{V}{R} = \frac{240 \text{ V}}{60 \text{ }\Omega} = \boxed{4.0 \text{ A}}$$

(b) The rms current is the same as the effective current.

(c) $i_{\text{peak}} = 1.41I = 1.41(4.0 \text{ A}) = \boxed{5.6 \text{ A}}$

(d) The peak current and the maximum current are the same.
(e) The peak-to-peak current is, as its name implies, twice the peak current. Hence

$$i_{\text{p-to-p}} = 2i_{\text{peak}} = \boxed{11.2 \text{ A}}$$

20.2 AC Circuit Containing Only Inductance

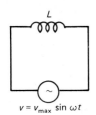

FIGURE 20.4 An ac voltage applied to a purely inductive circuit.

$v = v_{\text{max}} \sin \omega t$

Next we consider what happens when we apply an ac voltage across an inductance coil, as in Fig. 20.4. Initially we presume that the resistance of the coil is negligible, but in a later section we will consider more realistic coils having both resistance and inductance.

Even if the coil has no resistance, the ac current is limited in value by the inductor in a way similar to that in which a resistor limited the current in Fig. 20.2. An equation similar to Ohm's law for a pure-resistance circuit applies in this case, and takes the form

$$V = IX_L \tag{20.3}$$

where X_L is called the *inductive reactance*, and V and I are the *effective* (or rms) values of the voltage and current. X_L has the dimensions of resistance and is measured in ohms, just like an ordinary resistance.

Phase of the Current

When an ac voltage is applied across a coil, the changing current in the coil produces a changing magnetic flux through the coil. This changing flux produces an induced emf, which must, at any instant of time, be equal to the applied voltage. Hence Kirchhoff's loop rule gives:

$$v - L\frac{\Delta i}{\Delta t} = 0 \qquad \text{or} \qquad v = L\frac{\Delta i}{\Delta t} \tag{20.4}$$

The applied voltage is therefore proportional to the time rate of change of the current in the circuit, or to the slope of the current curve. Since the applied voltage is represented by a sine curve of the form $v = v_{max} \sin \omega t$, we want to find a function to represent the current i, which has the property that its slope is a sine curve. To find this function, let us first plot a few important values of i.

When $\sin \omega t = 1$, $v = v_{max}$, $\Delta i/\Delta t$ has its maximum positive value, and so i is increasing most rapidly in time. This corresponds to point 1 on the graph of Fig. 20.5, at which point $i = 0$ and $\Delta i/\Delta t = v_{max}/L$. Again, from Eq. (20.4), when $v = 0$, $\Delta i/\Delta t = 0$, which means that there is no change in i with time and so i must be either at its maximum positive value or at its maximum negative value. Since we already know that the slope of i versus t is upward at point 1, points 2 and 3 must be as shown on the graph.

If we plot a whole series of points in this way, the current curve shown in Fig. 20.5 results. The graph for i is clearly not a sine curve, since i is not equal to 0 when $t = 0$. Rather it turns out to be the negative of a cosine curve, since its value for $t = 0$ is not i_{max}, as it would be for a cosine function, but $-i_{max}$, as expected for the negative of a cosine curve. Hence we can write the current through the inductor as a function of time in the following form:

$$i = -i_{max} \cos \omega t$$

Both v and i therefore follow sinusoidal curves, but *the current lags behind the voltage in time*. The voltage, for example, reaches its maximum value at point 4 at an *earlier time* than the current reaches its maximum value at point 5. The time lag between v and i is seen from the graph to be 90°, or one-fourth of a cycle.

The current in an ac circuit containing only inductance lags the applied voltage by 90°, or $\pi/2$ rad.

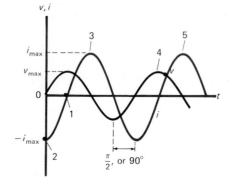

FIGURE 20.5 Current and voltage for an ac circuit containing only inductance (Fig. 20.4). The current lags the voltage by 90°.

Dependence of X_L on L and f

The inductive reactance X_L limits the current in an inductive circuit just as the resistance R limits the current in a resistive circuit. Since the induced emf in the coil is $-L\,\Delta i/\Delta t$, the larger L is, the greater the induced emf which prevents the current from building up. Hence the larger L, the smaller the effective current which flows through the inductor. Also, the higher the frequency, the larger $\Delta i/\Delta t$, since the current changes more rapidly with time.* This again produces a larger induced emf and a smaller effective

---

*It can be shown, using calculus, that $\Delta i/\Delta t$ is proportional to the angular frequency ω of the applied voltage. This leads directly to Eq. (20.5).

current. Hence the inductive reactance, which varies inversely with the current [from Eq. (20.3)], must be proportional to both L and f, and so

$$X_L \propto fL$$

or, putting in 2π as the constant of proportionality (a value obtained from a more rigorous analysis), we have

$$X_L = 2\pi fL = \omega L \tag{20.5}$$

where $\omega = 2\pi f$ is the angular frequency, and the effective voltage and current are related by the equation $V = IX_L$ [Eq. (20.3)].

In an inductive ac circuit, the factor which is analogous to the resistance in a purely resistive circuit is the inductive reactance $X_L = 2\pi fL = \omega L$.

Note that if we connect an inductance coil with no resistance to a dc voltage, $f = 0$, $X_L = 0$, and $I = V/X_L \to \infty$. In this case there is no induced emf and hence no opposition to the flow of current. Any practical coil usually has sufficient resistance to keep the current within finite limits.

At the other extreme, if the product of f and L is large, then the effective current I is small, no matter how large the applied voltage. The fact that X_L varies directly with f and with L is consistent with the fact noted above that X_L must have the dimensions of resistance and be measured in ohms. The units of fL are:

$$\frac{1}{s} \cdot H = \frac{1}{s} \cdot \frac{V \cdot s}{A} = \frac{V}{A} = \Omega$$

as required.

Example 20.2

An inductance coil has an inductance of 400 mH and a resistance of 2.0 Ω. Find the current in the coil if the applied voltage is:

(a) 120 V dc.
(b) 120 V ac at 60 Hz.
(c) What is the peak current in case (b)?

SOLUTION

(a) For direct current the inductance has no effect, since $\Delta i/\Delta t = 0$. Hence all we need is Ohm's law:

$$I = \frac{V}{R} = \frac{120\ V}{2.0\ \Omega} = \boxed{60\ A}$$

(b) For alternating current the inductive reactance is

$$X_L = 2\pi fL = 2\pi(60\ s^{-1})(0.40\ H) = 151\ \Omega$$

Since this is almost 100 times larger than the 2.0-Ω resistance, we omit the resistance for now (we will see shortly how to include it), and obtain for the rms ac current

$$I = \frac{V}{X_L} = \frac{120\ V}{151\ \Omega} = \boxed{0.79\ A}$$

Notice the tremendous difference between the ac and dc currents in these two cases, even though the circuits are identical and the effective ac voltage applied in (b) is of the same magnitude as the dc voltage in (a).

(c) $i_{peak} = 1.41I = 1.41(0.79\ A) = \boxed{1.1\ A}$

20.3 AC Circuit Containing Only Capacitance

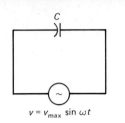

FIGURE 20.6 An ac circuit containing only a capacitor.

$v = v_{max} \sin \omega t$

Let us consider the circuit of Fig. 20.6, in which an ac voltage is connected to a capacitor C. In this case the ac current is limited by the capacitor in a way similar to that in which a resistor limits the current in a pure-resistance circuit. Here we have:

$$V = IX_C \qquad (20.6)$$

where X_C is called the *capacitive reactance.*

Phase of the Current

If the applied voltage is given by $v = v_{max} \sin \omega t$, then the charge on the capacitor at any instant of time is $q = Cv = Cv_{max} \sin \omega t$. Hence the *charge* on the capacitor is *in phase* with the applied voltage.

The current in the circuit, however, is $i = \Delta q/\Delta t = C \, \Delta v/\Delta t$, and so the current is not in phase with the applied voltage; rather i is proportional to the slope of the curve for $v = v_{max} \sin \omega t$.

We therefore desire to find a mathematical function which represents the slope of a sine curve. To find this function, we will again plot a few values of the slope of the voltage curve in Fig. 20.7. At point 1, $v = v_{max}$, its slope is zero (the graph is horizontal), and hence i must be equal to zero. At point 2, $v = -v_{max}$, the slope of v is again 0, and so i at point 2 is also equal to zero. If we plot a whole series of points in this way, the current curve shown in Fig. 20.7 results. It is a cosine curve, and so we have found that the slope of a sine curve is a cosine curve. Hence we can write the current in the capacitive circuit as a function of time in the following form:

$$i = i_{max} \cos \omega t$$

In this case it can be seen that *the current leads the voltage by 90°.* By this we mean that the current attains any particular value like that at point 3 in the diagram one-fourth cycle, or 90°, *ahead of* the voltage (point 4).

In an ac circuit containing only capacitance, the current leads the voltage by 90°, or π/2 rad.

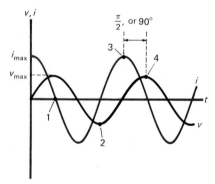

FIGURE 20.7 Current and voltage for an ac circuit containing only capacitance (Fig. 20.6). The current leads the voltage by 90°.

Dependence of X_C on C and f

Since $i = \Delta q/\Delta t$, the more rapidly the charge on the capacitor changes in time, the greater the current i. Hence the higher the frequency f, the larger the current i.* A capacitance thus acts something like a large resistance at low frequencies and like a small resistance at high frequencies. At $f = 0$, or dc, the equivalent resistance is infinite and no current at all flows.

Also, since $i = C \, \Delta v / \Delta t$, the larger C, the larger the current. The current produced by any given voltage is therefore proportional to the capacitance C. Since, from Eq. (20.6), $I = V/X_C$, and since I is proportional to both f and C, as we have just seen, we have:

$$I = \frac{V}{X_C} \propto fC \qquad \text{or} \qquad X_C \propto \frac{1}{fC}$$

and so
$$X_C = \frac{1}{2\pi fC} = \frac{1}{\omega C} \qquad\qquad (20.7)$$

where we have again chosen the constant of proportionality to be 2π in accordance with a more complete analysis than given here.

In a capacitive ac circuit the factor which is analogous to the resistance in a purely resistive circuit is the capacitive reactance $X_C = 1/(2\pi fC)$.

Again here the units of X_C are ohms, since $1/(2\pi fC)$ has units

$$\frac{1}{(1 \ \text{s}^{-1})(\text{F})} = \frac{1}{(1 \ \text{s}^{-1})(\text{C/V})} = \frac{\text{V}}{\text{A}} = \Omega$$

A summary of the major results of the preceding three sections is given in Table 20.1.

TABLE 20.1 Comparison of AC Circuits Containing R, L, and C

Circuit	$\dfrac{V}{I}$	Phase relations	Power loss in circuit (from Sec. 20.5)
R only	R	Current and voltage are *in phase*	I^2R
L only	$X_L = 2\pi fL$	Current *lags* voltage by 90°, or $\pi/2$ rad	0
C only	$X_C = \dfrac{1}{2\pi fC}$	Current *leads* voltage by 90°, or $\pi/2$ rad	0

Example 20.3

A 10-μF capacitor is connected to a 120-V, 60-Hz ac source.
(a) What is the effective value of the current?

(b) How does this current change if the frequency is increased to 6.0 MHz (6.0×10^6 Hz)?

SOLUTION

(a) $X_C = \dfrac{1}{2\pi fC} = \dfrac{1}{2\pi(60 \ \text{s}^{-1})(10^{-5} \ \text{F})} = 265 \ \Omega$

$I = \dfrac{V}{X_C} = \dfrac{120 \ \text{V}}{265 \ \Omega} = \boxed{0.45 \ \text{A}}$

(b) In this case, where $f = 6.0 \times 10^6$ Hz,

$X_C = \dfrac{1}{2\pi fC} = \dfrac{1}{2\pi(6.0 \times 10^6 \ \text{s}^{-1})(10^{-5} \ \text{F})} = 2.65 \times 10^{-3} \ \Omega$

---

*Here again a more rigorous treatment leads to the result that $i = \Delta q / \Delta t \propto \omega = 2\pi f$.

$$I = \frac{V}{X_C} = \frac{120 \text{ V}}{2.65 \times 10^{-3} \ \Omega} = \boxed{4.5 \times 10^4 \text{ A}}$$

capacitive circuit at high frequencies, even if the applied voltage is relatively low, as in this case.

Notice the very large ac currents which can flow in a

20.4 The *LC* Circuit

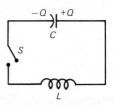

FIGURE 20.8 An *LC* circuit in series with a switch *S*. The capacitor is initially charged with a charge *Q*, and then the switch is closed.

Let us consider a circuit containing both an inductor *L* and a capacitor *C* in series with a switch (*S*), as in Fig. 20.8, and therefore referred to as an *LC* circuit. We presume initially that there is no resistance and no battery or other source of emf in the circuit. The capacitor is first charged to a charge *Q*, and in this way energy is stored in the electric field of the capacitor. We want to know the behavior of this circuit once the switch *S* is closed.

The best approach to this problem is to note that electric energy is conserved at any instant of time, since there is no resistance present to dissipate energy. We can therefore write an expression for the constant energy in the system at any instant of time. Once current flows, the energy initially stored in the capacitor is partially converted to energy in the magnetic field of the inductor, and we have, using Eqs. (19.17) and (19.11),

$$\frac{Q^2}{2C} = \frac{q^2}{2C} + \tfrac{1}{2}Li^2$$

where *q* and *i* are the *instantaneous* values of the charge on the capacitor and the current. Then, solving for *i*, we have

$$i = \frac{\pm 1}{\sqrt{LC}} \sqrt{Q^2 - q^2} \qquad (20.8)$$

This equation is very similar in form to the equation for the velocity of a particle of mass *m* moving in simple harmonic motion at the end of a spring, as given by Eq. (11.11):

$$v = \pm \sqrt{\frac{k}{m}} \sqrt{X_0^2 - x^2}$$

Just as $v = \Delta x / \Delta t$ in the mechanical case, so too $i = \Delta q / \Delta t$ in the electrical case. Also, just as X_0 is the maximum value of *x*, or the amplitude of the mechanical motion, so too *Q* is the maximum value of the charge *q* on the capacitor.

Hence, just as for the mechanical case the displacement was $x = X \cos \omega t$, with $\omega = \sqrt{k/m}$, so too here, on comparing Eq. (20.8) with Eq. (11.3), we obtain

$$q = Q \cos \omega t = Q \cos 2\pi f t$$

where $\quad \omega = 2\pi f = \sqrt{\dfrac{1}{LC}} \quad$ or $\quad f = \dfrac{1}{2\pi} \sqrt{\dfrac{1}{LC}} \qquad (20.9)$

Note that here the inertial factor *m* corresponds to a similar inertial factor *L* in the electrical case, since the inductance acts to oppose any change in the current through the circuit. Also the elastic constant *k*, which determines the

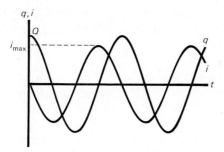

FIGURE 20.9 Graphs of the charge q on the capacitor and the current i in the *LC* circuit of Fig. 20.8 for the idealized case in which no electric energy is converted into heat as the current flows.

restoring force in the system, is replaced by $1/C$ in the electrical case, since the smaller C is, the more rapidly the capacitor must charge and discharge and hence the higher the frequency is. Graphs of q and i in Fig. 20.9 parallel perfectly those for the displacement x and the velocity v in Fig. 11.7.

Hence we have found that an *LC* circuit will oscillate with a natural frequency determined by the inductance L and the capacitance C in the circuit.

The natural angular frequency for an LC circuit is $\omega = \sqrt{1/LC}$. *The number of complete oscillations per second is* $f = \omega/2\pi = (1/2\pi)(1/\sqrt{LC})$.

Energy Transfer in an *LC* Circuit

In such an *LC* circuit the energy moves back and forth between the electric field $(\frac{1}{2}q^2/C)$ and the magnetic field $(\frac{1}{2}Li^2)$ in the same way that the energy changes back and forth from elastic potential $(\frac{1}{2}kx^2)$ to kinetic energy $(\frac{1}{2}mv^2)$ in the mechanical case. What happens is shown in Fig. 20.10. Initially the capacitor is charged to a voltage v_{max} and has a charge Q, as in (a). When the switch S is closed, a current i flows which discharges the capacitor. The flow of charge increases until $q = 0$ and i takes its maximum value $i_{max} = \sqrt{1/LC}\sqrt{Q^2 - 0} = \sqrt{1/LC}\,Q$, as in (b). At this point all the energy is in the magnetic field of the inductor. The current then continues to flow but decreases in magnitude as the charge builds up until it has the same maximum value but the opposite sign as what it had originally, as in (c). The capacitor then discharges again and the current builds up to a maximum value in the opposite direction, as in (d). One complete cycle corresponds to q returning

FIGURE 20.10 Oscillation of energy between electric and magnetic fields in the *LC* circuit of Fig. 20.8. In (e), at the end of one complete cycle or period, the circuit is back in its original condition (a).

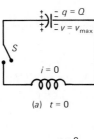

(a) $t = 0$

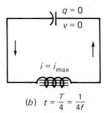

(b) $t = \dfrac{T}{4} = \dfrac{1}{4f}$

(c) $t = \dfrac{T}{2} = \dfrac{1}{2f}$

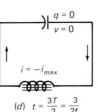

(d) $t = \dfrac{3T}{2} = \dfrac{3}{2f}$

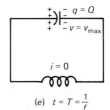

(e) $t = T = \dfrac{1}{f}$

to its original value of $+Q$, and i returning to its initial value of zero, as shown in Fig. 20.10e. In parts (a), (c), and (e) of Fig. 20.10 the energy is all in the electric field of the capacitor; in parts (b) and (d) it is all in the magnetic field of the inductor.

Hence we see that there is an almost perfect analogy between the oscillations of an mk mechanical circuit and of an LC electric circuit. The same oscillation of energy at a frequency determined by L and C (or m and k) is found, and there is a perfect correspondence between mechanical quantities in one case and electrical quantities in the other. This is evident from Table 20.2, in which these correspondences are explicitly shown.

TABLE 20.2 Comparison of Mechanical and Electrical Oscillations

Mechanical: SHM of a mass on a spring (mk circuit)	Physical quantity	Electrical: LC circuit
m	Inertial factor	L
k	Elastic factor	$\dfrac{1}{C}$
$x(x = X \cos \omega t)$	Displacement—Charge	$q(q = Q \cos \omega t)$
$v = \dfrac{\Delta x}{\Delta t}$	Velocity—Current	$i = \dfrac{\Delta q}{\Delta t}$
$\left(v = \pm \sqrt{\dfrac{k}{m}} \sqrt{X^2 - x^2}\right)$		$\left(i = \pm \sqrt{\dfrac{1}{LC}} \sqrt{Q^2 - q^2}\right)$
$\omega = \sqrt{\dfrac{k}{m}}$	Angular frequency of oscillation	$\omega = \sqrt{\dfrac{1}{LC}}$
$f = \dfrac{1}{2\pi} \sqrt{\dfrac{k}{m}}$	Number of oscillations per second (linear frequency)	$f = \dfrac{1}{2\pi} \sqrt{\dfrac{1}{LC}}$
$\frac{1}{2}mv^2$	Kinetic energy—Magnetic energy	$\frac{1}{2}Li^2$
$\frac{1}{2}kx^2$	Elastic potential energy—Electric energy	$\dfrac{1}{2}\dfrac{1}{C}q^2$
$\frac{1}{2}kx^2 + \frac{1}{2}mv^2 = \frac{1}{2}kX^2$	Conservation of energy	$\dfrac{1}{2}\dfrac{1}{C}q^2 + \dfrac{1}{2}Li^2 = \dfrac{1}{2}\dfrac{1}{C}Q^2$

The striking parallelism between a mechanical system on the one hand and a very-different-looking electrical system on the other hand, as revealed in Table 20.2, shows the symmetry that pervades physics. In this case the symmetry is a mathematical one, contained in the equations describing the motion of the system in the two cases. Once we understand the behavior of the mechanical system, the behavior of the corresponding electrical system becomes much easier to understand.

Effect of Resistance

If we add to an LC circuit a resistance R, we obtain the RLC circuit of Fig. 20.11. This is equivalent to a *damped* simple harmonic oscillator. Just as frictional damping dissipates mechanical energy, so here resistance dissipates electric energy in the form of heat. Energy still oscillates back and forth between the electric field of the capacitor and the magnetic field of the

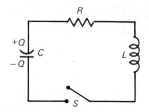

FIGURE 20.11 A circuit containing a resistance *R* in series with an inductance *L* and a capacitance *C*, that is, an *RLC* series circuit. The capacitor is charged with a charge *Q* and then the switch *S* is closed.

FIGURE 20.12 Current in an *RLC* series circuit as a function of time.

Lord Kelvin's Contribution

inductor, but the total amount of useful energy decreases with each cycle. Hence the amplitude of the current decreases with time, as shown in Fig. 20.12.

Any real *LC* circuit has resistance, since the inductor is made of wire which must have at least some resistance. Hence any isolated *RLC* circuit will oscillate only for a finite amount of time. To produce an electrical oscillation of constant amplitude over a long period of time, energy must be constantly fed into the circuit to compensate for the energy being lost as heat in the resistor.

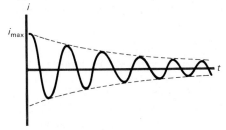

In 1853 the British physicist William Thomson, or Lord Kelvin as he was later called (see Fig. 20.13 and accompanying biography), published in the *Philosophical Magazine* a famous paper entitled "On Transient Electric Currents." This paper for the first time derived Eq. (20.9) above:

$$f = \frac{1}{2\pi} \frac{1}{\sqrt{LC}}$$

This simple equation enables the frequency of different electrical systems to be calculated, and hence makes it possible to tune transmitters and receivers to the same frequency by varying *L* and *C*. J. J. Thomson once referred to it as the basic equation of "modern wireless telegraphy, telephony, and broadcasting." For this reason Kelvin's 1853 paper has been of great importance in the development of modern communications systems.

Example 20.4

A 10-μF capacitor is connected in series with an inductance of 100 mH and a switch. The capacitor is first charged to a voltage of 100 V. The charging unit is then removed, and the switch is closed.
(a) What is the natural frequency of oscillation of this circuit?

(b) What is the maximum charge on the capacitor?
(c) What is the maximum energy stored in the capacitor?
(d) What is the maximum energy stored in the inductor?
(e) What is the maximum current in the circuit?

SOLUTION

(a) $f = \dfrac{1}{2\pi} \sqrt{\dfrac{1}{LC}} = \dfrac{1}{2\pi} \dfrac{1}{\sqrt{(10^{-5}\ \text{F})(0.10\ \text{H})}} = \boxed{159\ \text{Hz}}$

(b) The maximum charge on the capacitor is the initial charge

$Q = CV = (10^{-5}\ \text{F})(100\ \text{V}) = \boxed{10^{-3}\ \text{C}}$

(c) The maximum energy stored in the capacitor is

$W = \frac{1}{2}CV^2 = \frac{1}{2}(10^{-5}\ \text{F})(100\ \text{V})^2 = \boxed{0.050\ \text{J}}$

(d) From the energy-conservation principle the maximum energy stored in the inductor is also 0.050 J, since the energy oscillates back and forth between the electric field of the capacitor and the magnetic field of the inductor.
(e) The maximum energy in the inductor is $\frac{1}{2}Li_{max}^2 = 0.050$ J. Hence

$$i_{max}^2 = \frac{2(0.050\ \text{J})}{L} = \frac{0.10\ \text{J}}{0.10\ \text{H}} = 1.0\ \frac{\text{V·A·s}}{\text{V·s/A}} = 1.0\ \text{A}^2$$

and so

$i_{max} = \boxed{1.0\ \text{A}}$

Sir William Thomson, Baron Kelvin of Largs (1824–1907)

FIGURE 20.13 William Thomson, Lord Kelvin. *(Photo courtesy of AIP Niels Bohr Library; Zeleny collection.)*

William Thomson was a remarkably versatile scientist. He was an excellent mathematician who at the same time had a love of precision measurements, and he developed instruments such as mirror galvanometers, electrometers, and ac bridge circuits for accurate electrical measurements. Carrying this interest in apparatus even further, he developed highly practical engineering instruments like depth-sounders, marine compasses, and tide predictors. The culmination of his practical engineering accomplishments was his work as a consultant on the 1900-mi underwater transatlantic cable put in place between Ireland and Newfoundland during the years 1850 to 1866.

It was his work on the transatlantic cable that made Thomson the toast of Victorian society, and led to his knighthood (and the name Lord Kelvin), and ultimately to a large fortune in royalties from the cable.

William Thomson was born in Belfast in 1824, the son of a mathematics professor who moved to the University of Glasgow when William was 8 years old. William was taught at home by his father until he was 10, at which time he began to attend classes at the university. There he upset some of his fellow students by always volunteering the answers to his father's questions in mathematics class. He went to Cambridge University at the age of 17 and remained there for 4 years, studying mathematics and physics, graduating second in his class. He then did some postgraduate work in Paris under Regnault. In 1845, at the age of 21, he was appointed Professor of Natural Philosophy at Glasgow, a position he retained for 53 years! Although he loved to travel, he found Glasgow the perfect setting for his genius and never considered leaving it for a better academic post. On three occasions he even turned down the prestigious Cavendish Professorship at Cambridge.

Lord Kelvin is most famous for his work in heat and thermodynamics, particularly for his development of the absolute scale of temperature (Sec. 15.6) and his formulation of the second law of thermodynamics. He also served as a bridge between the experimentally oriented Faraday and the mathematically inclined James Clerk Maxwell in the development of electromagnetism.

Kelvin was a lively, but not very clear, lecturer, who tended to digress on topics of great interest to himself but unrelated to the subject matter at hand. In 1884, during one of his many visits to the United States, he gave a series of 20 lectures at the Johns Hopkins University in Baltimore on recent developments in physics. Yet even in this formal setting it appeared that some of his lectures developed out of questions raised over breakfast on the morning of the lecture!

Despite his many triumphs Kelvin is sometimes passed over when lists of the world's great physicists are drawn up. This is partially because he never developed a group of disciples around him, and partially because he resisted new ideas, such as relativity and radioactivity. Also, some people felt that he was not really interested in discussing their ideas but only in expounding his own. J. J. Thomson (no relation) summed up Lord Kelvin's attitude very well by an allusion to the theory of blackbody radiation (see Sec. 26.2): "In fact he [Kelvin] was a remarkable exception to that very important physical principle that good radiators are good absorbers."

Kelvin died in 1907 and was buried next to Newton in Westminster Abbey, an honor which he richly deserved for his many contributions to physics.

20.5 *RLC* Series Circuit with Applied AC Voltage

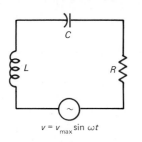

We now want to consider the "driven" *RLC* series circuit of Fig. 20.14. An ac voltage of form $v = v_{max} \sin \omega t$ drives current through a resistor, inductor, and capacitor in series. We presume that the effective current I in the circuit is proportional to the effective voltage V applied, so that

$$\boxed{V = ZI}$$ (20.10)

where Z is the *impedance* of the circuit, and is some combination, still to be determined, of the resistances and reactances in the circuit. Equation (20.10) is the generalization of Ohm's law for ac circuits.

To find both the numerical and the phase relationships between the current and the voltage, we use the vector diagram of Fig. 20.15. We make use of the fact that the current is the same in all three elements of the circuit, and hence we take the effective current I, drawn along the positive X axis, as the reference line with respect to which the phases of all the voltages are determined. V_R, V_L, and V_C are the effective voltages across R, L, and C, respectively. $V_R = IR$ is in phase with the current and hence is plotted along the positive X axis in the direction of the current I. On the other hand, the current I lags the voltage V_L across the inductor by 90°, and hence V_L is drawn along the positive Y axis; and the current leads the voltage across the capacitor by 90°, and hence V_C is drawn along the negative Y axis, as shown in the figure. (Note that all angles are being measured in a *counterclockwise* direction.)

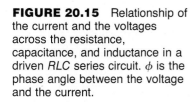

FIGURE 20.14 An *RLC* series circuit with an applied ac voltage.

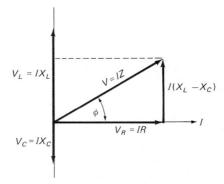

FIGURE 20.15 Relationship of the current and the voltages across the resistance, capacitance, and inductance in a driven *RLC* series circuit. ϕ is the phase angle between the voltage and the current.

The effective voltage V applied to the circuit is then the sum of V_R, V_L, and V_C. Since these three have different phases, they must be added as *vectors* to take account of their phase relationships. From Fig. 20.15 we have for the magnitude of the resultant of these three vectors,

$$V^2 = V_R^2 + (V_L - V_C)^2$$

or $$I^2Z^2 = I^2R^2 + (IX_L - IX_C)^2$$

from which $$Z^2 = R^2 + (X_L - X_C)^2$$

and so $$Z = \sqrt{R^2 + (X_L - X_C)^2}$$ (20.11)

Hence the impedance Z in any series *RLC* circuit is the square root of the sum of the squares of the total resistance and the total reactance in the circuit, as

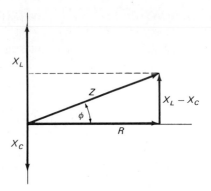

FIGURE 20.16 Impedance diagram for *RLC* circuit. Here $Z^2 = R^2 + (X_L - X_C)^2$, where Z is the impedance of the circuit, and X_L and X_C are the inductive and capacitive reactances, respectively.

shown in Fig. 20.16. The units of Z are ohms, since X_L, X_C, and R are all in ohms.

The effective (rms) voltage and current in an RLC series circuit are related by the equation

$$\boxed{V = ZI}$$

where the impedance $Z = \sqrt{R^2 + (X_L - X_C)^2}$.

Power Factor

$$\cos \phi = \frac{R}{Z} \qquad (20.12)$$

where ϕ is the *phase angle* between the applied voltage V and the current I. The rate at which energy is converted into joule heat is then:

$$P = I^2R = I^2(Z \cos \phi) = I(IZ \cos \phi)$$

or $\qquad P = IV \cos \phi \qquad (20.13)$

Here $\cos \phi$ is called the *power factor* of the *RLC* circuit.

For a pure resistance R, we have $Z = R$, $\cos \phi = R/Z = 1$, and $P = IV$, as in the dc case. For a pure inductance or a pure capacitance, on the other hand, $\phi = \pm 90°$, $\cos \phi = 0$, and so $P = IV(0) = 0$. Hence *no power is dissipated as heat in a pure inductance or capacitance*. Electric energy simply flows in and out of the capacitor as it charges and discharges, and magnetic energy flows in and out of the coil as the current builds up and decays, as discussed previously.

In an RLC series circuit all the power delivered is dissipated as I^2R heat in the resistor R.

These results are included in the last column of Table 20.1 (Sec. 20.3).

Example 20.5

A 1000-Ω resistor is connected in series with a 0.20-H inductor and a 0.30-μF capacitor. The applied voltage is $V = 170 \sin 2\pi ft$, where $f = 2000$ Hz. Find:

(a) The impedance of the circuit.
(b) The effective current in the circuit.

(c) The effective voltages across the resistor, capacitor, and inductor.
(d) The power factor of the circuit.

SOLUTION

(a) $Z^2 = R^2 + (X_L - X_C)^2$

$$= (1.0 \times 10^3 \ \Omega)^2 + \left[2\pi(2000)(0.20) \right.$$

$$\left. - \frac{1}{2\pi(2000)(0.30 \times 10^{-6})} \right] \Omega^2$$

$$= [1.00 \times 10^6 + (2.51 \times 10^3 - 0.26 \times 10^3)^2] \ \Omega^2$$

$$= 6.06 \times 10^6 \ \Omega^2$$

and so $\quad Z = \boxed{2.46 \times 10^3 \ \Omega}$

(b) Since the maximum voltage is 170 V, the effective voltage is 0.707(170 V) = 120 V. Then

$$I = \frac{V}{Z} = \frac{120 \ \text{V}}{2.46 \times 10^3 \ \Omega} = \boxed{49 \ \text{mA}}$$

(c) $V_R = IR = (0.049 \ \text{A})(1000 \ \Omega) = \boxed{49 \ \text{V}}$

$V_L = IX_L = I(2\pi fL)$

$$= (0.049 \ \text{A})(2\pi)(2000 \ \text{s}^{-1})(0.20 \ \text{H}) = \boxed{123 \ \text{V}}$$

$V_C = IX_C = \dfrac{I}{2\pi fC} = \dfrac{0.049 \ \text{A}}{2\pi(2000 \ \text{s}^{-1})(0.30 \times 10^{-6} \ \text{F})} = \boxed{13 \ \text{V}}$

(d) $\cos \phi = \dfrac{R}{Z} = \dfrac{1000 \ \Omega}{2.46 \times 10^3 \ \Omega} = \boxed{0.41}$

ϕ is therefore 66°

(e) The power dissipated by the circuit.
(f) Explain how the voltages across R, L, and C add to give the voltage applied to the circuit.

(e) $P = VI \cos \phi = (120 \ \text{V})(0.049 \ \text{A})(0.41) = \boxed{2.4 \ \text{W}}$

(f) If added directly, we would have $V = (49 + 123 + 13)$ V = 185 V, which is more than the effective voltage applied. But we must add the voltages by *taking account of their phase differences*. This yields

$$V^2 = V_R^2 + (V_L - V_C)^2 = (49 \ \text{V})^2 + (123 - 13)^2 \ \text{V}^2$$

$$= 14.5 \times 10^3 \ \text{V}^2$$

and so $\quad V = 120 \ \text{V}$

which is equal to the applied voltage, as it must be. This is shown in Fig. 20.17.

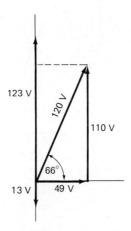

FIGURE 20.17 Diagram for Example 20.5.

20.6 Series Resonance

We have seen that the phase angle ϕ is the angle between the current and the voltage in an ac circuit, where from Fig. 20.15,

$$\tan \phi = I\frac{X_L - X_C}{IR} = \frac{X_L - X_C}{R} \tag{20.14}$$

If $X_L = X_C$, $\tan \phi = 0$, $\phi = 0$, and the current and the voltage are *in phase*, as in a purely resistive circuit. When $X_L = X_C$,

$$Z = \sqrt{R^2 + (X_L - X_C)^2} = R$$

Hence $V = IR$ and the current takes on its maximum value limited only by the resistance in the circuit. In this case the circuit is said to be in *series resonance*.

Series Resonance

Resonance in an RLC series circuit occurs when the inductive reactance X_L and the capacitance reactance X_C are equal; the current is then a maximum and in phase with the applied voltage.

Figure 20.18 shows the behavior of X_L, X_C, and I as a function of the frequency of the voltage applied to the circuit.

Since the resonant frequency is that frequency at which X_L is equal to X_C, we have

$$2\pi f L = \frac{1}{2\pi f C} \quad \text{or} \quad f_{\text{res}} = \boxed{\frac{1}{2\pi}\frac{1}{\sqrt{LC}}} \tag{20.15}$$

This is, of course, the same basic equation previously given for the natural frequency of a simple circuit [Eq. (20.9)]. *Resonance occurs when the applied frequency is equal to this natural frequency.*

FIGURE 20.18 Graphs of the capacitive reactance X_C, the inductive reactance X_L, and the current I in a driven *RLC* series circuit, as a function of the frequency of the applied voltage. Notice how X_C and X_L are equal and therefore cancel (since they have opposite phases) at the resonant frequency f_{res}, at which the current attains its maximum value.

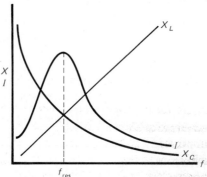

Importance of Resonance

Resonance phenomena occur in all branches of physics from mechanics and classical electromagnetism to lasers and elementary particle physics.

The mechanical analogue of an *RLC* circuit is a mass vibrating at the end of a spring and damped by friction or some other resistive force. If this mass is driven by an external force at a varying frequency, there will be one frequency at which the motion of the mass will build up to very large amplitudes. This is the natural mechanical frequency of the system. A common example of such behavior is the pushing of a child on a swing. The pushing frequency must be equal to the natural frequency of the swing if large amplitudes are to develop.

A very important practical use of resonance occurs in radio and TV sets, where series resonance is used to select a desired station or channel. The capacitance in an *RLC* series circuit is varied until the antenna circuit is in series resonance with the carrier frequency of the desired broadcasting station. At this frequency the emf induced in the antenna causes a large current to flow in the antenna circuit. This current can then be made to produce an audio or video signal. Other stations or channels have frequencies not in resonance with the antenna circuit and hence produce negligibly small currents in that circuit.

Dangers of Series Resonance Circuits

It is important to realize that in a series circuit at resonance the voltage across the capacitor and the inductor individually can be many times larger than the 120-V line voltage or other applied voltage, even though the total voltage across the combination of L and C together is zero because of the opposite phases of the voltages across each. This can present dangers both to the circuit elements and to anyone who is foolhardy enough to touch the leads to either the inductor or the capacitor. That very large voltages can arise in this way when a circuit is at resonance is shown in Example 20.6.

Example 20.6

An RLC series circuit is connected to a variable-frequency ac generator whose effective (or rms) voltage output is maintained at 120 V. The values of the circuit elements are $L = 100$ mH, $C = 20 \times 10^{-9}$ F, $R = 50\ \Omega$.
(a) What is the resonant frequency of the circuit?

(b) How much current flows in the circuit at resonance?
(c) What is the voltage across the capacitor at resonance?
(d) What is the voltage across the inductor at resonance?
(e) What is the voltage across the resistor at resonance?

SOLUTION

(a) For resonance,

$$f_{res} = \frac{1}{2\pi}\frac{1}{\sqrt{LC}} = \frac{1}{2\pi}\frac{1}{\sqrt{(0.10\ \text{H})(20 \times 10^{-9}\ \text{F})}}$$

$$= \boxed{3.56 \times 10^3\ \text{Hz}}$$

(b) The effective current at resonance is

$$I = \frac{V}{R} = \frac{120\ \text{V}}{50\ \Omega} = \boxed{2.4\ \text{A}}$$

(c) $V_C = IX_C = \dfrac{1}{2\pi fC}$

$$= \frac{2.4\ \text{A}}{2\pi(3.56 \times 10^3\ \text{s}^{-1})(20 \times 10^{-9}\ \text{F})}$$

$$= \boxed{5.36 \times 10^3\ \text{V}}$$

(d) $V_L = IX_L = I(2\pi fL)$

$$= (2.4\ \text{A})(2\pi)(3.56 \times 10^3\ \text{s}^{-1})(0.10\ \text{H})$$

$$= \boxed{5.36 \times 10^3\ \text{V}}$$

(e) $V_R = IR = (2.4\ \text{A})(50\ \Omega) = 120\ \text{V}$

as expected.

Note in the answers to parts (c) and (d) that the voltages across the capacitor and inductor are each over 5000 V, which is almost 50 times the applied voltage! Since these voltages are always opposite in phase, and since they have the same value at resonance, the voltage across the capacitor and inductor in series at resonance is zero, and all the 120 V applied is across the resistor. Despite this fact the 5000 V across either capacitor or inductor by itself is enough to kill a person under the right (or wrong!) circumstances.

20.7 Transformers

Over 300,000 mi of overhead high-voltage electric cables line the United States, and they occupy over 7 million acres of land. Any transmission of electric power over electric cables loses energy in the form of Joule heat, $W = I^2Rt$, where I is the effective ac current and R is the resistance of the line. Since resistance is directly proportional to the length of the line, the longer the power line, the greater the energy loss. Consequently electric power transmission over long distances can be very wasteful; in some cases 10 percent of the energy transmitted is lost. This loss can be avoided by transmitting the power at a high voltage and thus reducing the current I and hence the I^2R losses. If the voltage is increased by a factor of 10^3, for example, the power loss is decreased by a factor of 10^6. Hence, many recently installed power lines in the United States carry power at voltages as high as 765,000 V.

The advantage of ac over dc in power transmission is mainly the ease with which ac voltages can be "stepped up" or "stepped down." In electric power distribution systems it is desirable at both the generating and the receiving end to deal with relatively low voltages for reasons of safety and convenience. No one wants a child's electric train, for example, to operate off 100,000 V. On the other hand, power companies, for economic reasons, want to transmit power at high voltages. They therefore step up (i.e., increase) the voltage at the generator, transmit this high-voltage power over the electric lines, and then step down the voltage at the house or factory where the power is to be used. The piece of electrical equipment that does this is a *transformer*.

Definition

A transformer is an electrical device which uses electromagnetic induction to step up voltages and simultaneously to step down currents, or vice versa, in an ac circuit.

A simplified version of a transformer is shown in Fig. 20.19. It consists of two coils of wire, a primary coil (C_1) and a secondary coil (C_2), wound on the same iron core. The iron core concentrates the magnetic flux and guarantees that it is the same in the primary coil, with N_1 turns of wire, as it is in the secondary coil, with N_2 turns of wire.

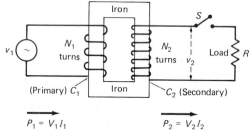

FIGURE 20.19 A simplified version of a transformer. For such a transformer $V_2/V_1 = N_2/N_1 = I_1/I_2$.

In Fig. 20.19 the primary coil of the transformer is connected to an ac generator whose voltage output is $v_1 = v_{max} \sin \omega t$. Initially we assume that switch S is open, so that no current flows in the secondary circuit. The varying flux through the primary due to the changing current induces an emf in the primary because of its self-inductance. We have, on equating the applied voltage to the induced emf,

$$v_1 = L \frac{\Delta i}{\Delta t} = N_1 \frac{\Delta \Phi}{\Delta t}$$

But $\Delta \Phi$ is the same, by assumption, in both the primary and secondary coils. Hence the induced emf in the secondary, which has N_2 turns, is

$$v_2 = N_2 \frac{\Delta \Phi}{\Delta t}$$

We have therefore

$$\frac{\Delta \Phi}{\Delta t} = \frac{v_1}{N_1} = \frac{v_2}{N_2} \quad \text{or} \quad \frac{v_2}{v_1} = \frac{N_2}{N_1}$$

If we take rms averages of v_1 and v_2, this becomes

$$\boxed{\frac{V_2}{V_1} = \frac{N_2}{N_1}}$$

(20.16)

N_2/N_1 is called the *turns ratio* of the transformer. When $N_2 > N_1$, the transformer is a *step-up transformer*. When $N_2 < N_1$, it is a *step-down transformer*, where *step-up* and *step-down* refer to the *voltage* change. The great advantage of using a transformer is that, without any moving parts, we can transform voltages merely by selecting the right values for N_1 and N_2. Transformers also consume very little energy and hence can be kept running even if they are not delivering power to consumers. The ease with which transformers operate on ac, and the lack of any equivalent device for dc operation, are prime reasons for the use of alternating rather than direct current.

Delivery of Power

We now close the switch S in Fig. 20.19 and deliver power to a load resistance R. Then the power dissipated in the load is $I_2^2 R$. An equal amount of power must be fed into the primary from the generator so that, if there are no losses in the system, we must have

Transformer Equation

$$P_1 = P_2$$

and so $\qquad V_1 I_1 = V_2 I_2 \qquad$ and $\qquad \boxed{\dfrac{I_1}{I_2} = \dfrac{V_2}{V_1} = \dfrac{N_2}{N_1}} \qquad$ (20.17)

The transformer therefore *steps down the current* at the same time and by the same ratio as it *steps up the voltage*, and vice versa.

The way in which the primary draws the exact amount of power needed to supply the output of the secondary is as follows. When the secondary circuit is open, the flux in the core is produced by the changing primary current alone. When the switch S is closed, however, both primary and secondary currents contribute to the flux. The secondary current, according to Lenz' law, tends to weaken the flux and hence to decrease the induced emf. Since the applied voltage is fixed and must be equal to the induced emf, additional current flows in the primary to increase the flux and the induced emf. Hence the power input $P_1 = V_1 I_1$ increases until it equals the power output $P_2 = V_2 I_2$.

Transformers are used not merely in the transmission of electric power, but for a variety of other useful purposes. They are, for example, used to step down 120-V power to 6 V for doorbells in homes, and to step up 120-V power to 10^5 V to accelerate the electron beam in television sets.

Impedance Matching

Another very important use of transformers is as impedance-matching devices. It can be shown that for maximum power delivery from a source to a resistive load, the load must have the same resistance as the internal resistance of the source. This can be accomplished by using a transformer to match the two resistances. If the load resistance is R_2, then $R_2 = V_2/I_2$. But, if a transformer is inserted between the source and the load,

$$V_2 = \frac{N_2 V_1}{N_1} \qquad \text{and} \qquad I_2 = \frac{N_1 I_1}{N_2}$$

so that
$$R_2 = \frac{(N_2/N_1)V_1}{(N_1/N_2)I_1} = \frac{V_1}{(N_1/N_2)^2 I_1}$$

and so
$$I_1 = \frac{V_1}{(N_1/N_2)^2 R_2} = \frac{V_1}{R'}$$

Hence the load resistance R_2 appears to the source like a resistance

$$R' = \left(\frac{N_1}{N_2}\right)^2 R_2 \tag{20.18}$$

where N_2/N_1 is the turns ratio of the transformer. By choosing the proper turns ratio, therefore, R' can be made equal to the internal resistance R_1 of the source. This is called *impedance matching*, since in more complicated cases the load may include both reactances and resistances.

Figure 20.20 shows an audio amplifier, which is a high-impedance ac source, connected to a loudspeaker, which is a low-impedance device. The impedance of the loudspeaker is increased to match that of the amplifier by using a transformer with the proper turns ratio. Some impedance-matching transformers on stereo amplifiers come with many different output connections, so that the proper turns ratio can be chosen to fit the speaker which the amplifier is to drive.

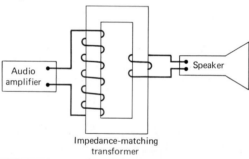

FIGURE 20.20 Using an impedance-matching transformer to match the output of a high-impedance audio amplifier to a low-impedance speaker system.

Example 20.7

An electric generating plant produces electric energy at a rate of 1 GW and a voltage of 12,000 V.
(a) If it is desired to transmit this power at 720,000 V, what must be the turns ratio of the step-up transformer used?
(b) What current would be sent out over the power lines if the transmission were at 12,000 V?

(c) What current would be transmitted if the voltage were stepped up to 720,000 V?
(d) What would be the ratio of the power lost as heat in (b) and (c)?

SOLUTION

(a) $\dfrac{N_2}{N_1} = \dfrac{V_2}{V_1} = \dfrac{720{,}000 \text{ V}}{12{,}000 \text{ V}} = \boxed{60/1}$

(b) $I_1 = \dfrac{P}{V_1} = \dfrac{10^9 \text{ W}}{12{,}000 \text{ V}} = \boxed{8.3 \times 10^4 \text{ A}}$

(c) Since the power remains unchanged, if there are no losses in the transformer,

$I_2 = \dfrac{P}{V_2} = \dfrac{10^9 \text{ W}}{7.2 \times 10^5 \text{ V}} = \boxed{1.4 \times 10^3 \text{ A}}$

(d) The ratio of the power lost in the two cases would be:

$$\frac{P_1}{P_2} = \frac{I_1^2 R}{I_2^2 R} = \frac{(8.3 \times 10^4 \text{ A})^2}{(1.4 \times 10^3 \text{ A})^2} = (60)^2 = \boxed{3.6 \times 10^3}$$

Hence more than 3000 times the power lost as heat at 720,000 V would be lost at 12,000 V.

Example 20.8

An audio amplifier, with an internal impedance of 2000 Ω, is used to drive a loudspeaker with an impedance of 5.0 Ω. What must be the turns ratio of the transformer used to match the impedances?

SOLUTION

We want the transformer to convert the impedance of the loudspeaker from $R_2 = 5.0$ Ω to $R' = 2000$ Ω. Hence, since

$$R' = \left(\frac{N_1}{N_2}\right)^2 R_2$$

we have

$$\left(\frac{N_1}{N_2}\right)^2 = \frac{R'}{R_2} = \frac{2000 \text{ Ω}}{5 \text{ Ω}} = 400$$

and so
$$\frac{N_1}{N_2} = \sqrt{400} = \boxed{20}$$

Hence the primary of the transformer must have 20 turns for each turn of the secondary to match impedances and deliver maximum power to the loudspeaker.

20.8 Practical Aspects of Electric Power

Figure 20.21 shows how transformers are typically used in the transmission of electric power. A generating plant produces ac power at 12,000 V. Before leaving the plant, this voltage is stepped up to 240,000 V by using a 20/1 step-up transformer. The power is then transmitted at 240,000 V over high-voltage power lines to a substation where a second transformer steps down the voltage to 12,000 V. The power is then distributed at this voltage to sites where it is needed. At a large factory or in residential areas additional smaller transformers step down the voltage again by a ratio of 50/1 to yield the 240 V normally supplied to homes and factories. This transmission process avoids the large power losses which would be incurred in the transmission of electric power over large distances at low voltage (and hence high current) levels, and at the same time provides the large amount of current needed at 120 or 240 V ac to run most electric machines and appliances. Another advantage of such a procedure is that it keeps excessively high voltages out of homes and factories, where they could be dangerous to human life.

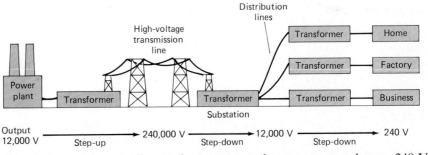

FIGURE 20.21 Use of transformers in electric power transmission.

Home Electricity

Electric companies usually supply ac power to homes at two voltages, 240 V for large appliances like central air-conditioning units and clothes dryers, and 120 V for smaller home appliances and lights. In modern homes a number of independent 120-V circuits each carry a maximum of 20 A of current. An individual circuit may include three or four double outlets in the walls of a room. Each time we plug in a new light or appliance we draw more current from that particular circuit, for the appliances are all connected in parallel. In most cases little danger exists of exceeding the 20-A capacity of a circuit, since we would have to plug twenty-four 100-W bulbs into the same 120-V circuit to

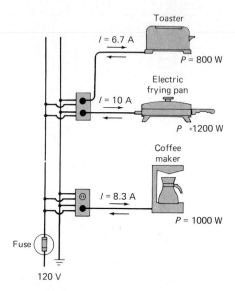

FIGURE 20.22 Plugging a number of different kitchen appliances into one house circuit fused at 20 A.

exceed the capacity of the line. When a number of electric heating appliances are being used, however, as in the kitchen, it is quite easy to exceed this 20-A capacity.

Suppose, for example, that we have the situation shown in Fig. 20.22. We have two double outlets above a kitchen counter, both part of the same circuit. We plug into the outlets an electric toaster which draws a power of 800 W, an electric frying pan rated at 1200 W, and an automatic coffee maker rated at 1000 W. The toaster draws a current of $I = P/V = (800 \text{ W})/(120 \text{ V}) = 6.7$ A; the frying pan $(1200 \text{ W})/(120 \text{ V}) = 10$ A; and the coffeemaker $(1000 \text{ W})/(120 \text{ V}) = 8.3$ A. Since the wiring in the circuit is only meant to carry 20 A, we have overloaded the circuit, and there is danger of overheating the wires and producing a fire. To prevent this, safety devices must be included in the circuit.

Safety Devices

Fuses and Circuit Breakers Building codes now require protection of each house circuit by a fuse or circuit breaker, usually in a junction box in the garage, furnace room, or basement. A *fuse* is merely a screw-in device which puts a piece of metal ribbon in series with the circuit, as in Fig. 20.23. If more than 20 A flows in the circuit, the ribbon melts and cuts off all further current. If one of the appliances is removed from the line, and a new fuse put in place, everything should again function normally.

In most modern homes *circuit breakers* are used in place of fuses. A circuit breaker is also wired in series with the rest of the circuit. Two kinds are in common use—a thermal and a magnetic device. The thermal circuit breaker (Fig. 20.24) is a bimetallic strip (see Sec. 13.3) which becomes so hot

FIGURE 20.23 An electric fuse. The current must flow from the bottom tip of the fuse through a fine metal ribbon to the metal side of the fuse body to complete the circuit. An overload causes the ribbon to melt and open the circuit.

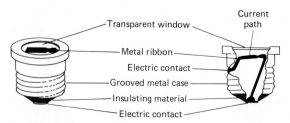

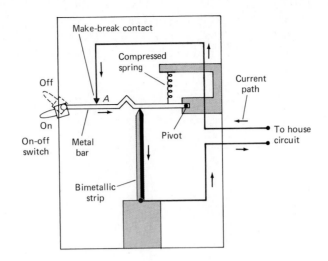

FIGURE 20.24 A thermal (bimetallic strip) circuit breaker. The bending of the overheated bimetallic strip to the left causes the pivoted horizontal rod to fall until the notch engages the metallic strip. This opens the circuit at point A.

when it carries excessive current that it bends enough to open the circuit. In a magnetic circuit breaker, excessive current produces a magnetic field large enough to pull open a switch and interrupt the current. The advantage of circuit breakers over fuses is that they can be reset easily and need not be replaced.

Another way that a fuse or circuit breaker may be blown is if a *short circuit* occurs. A short circuit occurs when two wires, one at a high potential and the other at ground potential, cross or touch. As a result the current has a very low resistance path to ground, since the wires have very low resistance, and a very large current is drawn from the line. A short circuit may occur either in the house wiring itself or in an appliance plugged into the house wiring.

A short circuit is indicated if fuses or circuit breakers are continually being blown on the same circuit. All appliances should then be removed from the line and plugged in again one at a time. The one that blows the fuse is clearly the culprit and needs repair. If the circuit itself is defective, the fuse will blow even when nothing is plugged into the line.

Grounding of Appliances Most 120-V appliances now have a third (round) prong on the power plug. This connects the metal case of the appliance to ground. Hence, if the high-voltage wire inside the appliance should by accident touch the case of the appliance, a short circuit will result and a fuse will blow (as in Fig. 20.25). If the ground wire were absent, as in

FIGURE 20.25 The use of a three-pronged plug on an electric appliance to prevent electric shocks. The plug grounds the case of the appliance so that no shock will be received from the case even if high-voltage wires inside the appliance should make contact with the case.

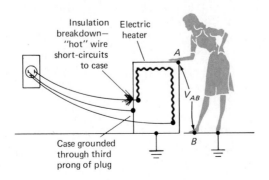

older two-pronged plugs, then after the high-voltage wire touched the case of the appliance the case would be at 120 V above ground. If a person touched it, the current would flow through the person's body, producing an "electric shock."

Example 20.9

(a) How many 150-W light bulbs can be plugged into a 120-V circuit fused for 20 A before the fuse is blown? Assume that the bulbs are connected in parallel.

(b) What would happen if the same number of bulbs were connected in series?

SOLUTION

(a) For a parallel connection the voltage across each bulb is the same 120 V. Hence each light bulb draws a current

$$I = \frac{P}{V} = \frac{150 \text{ W}}{120 \text{ V}} = 1.25 \text{ A}$$

Hence the number of light bulbs required to draw 20 A is

$$N = \frac{20 \text{ A}}{1.25 \text{ A/bulb}} = \boxed{16 \text{ bulbs}}$$

(b) If the 16 bulbs were connected in series, the resistance of the circuit would be the sum of the resistances of the individual bulbs. Since each bulb is rated at 150 W for a voltage of 120 V across it, the resistance of each bulb is:

$$R_1 = \frac{V}{I} = \frac{120 \text{ V}}{1.25 \text{ A}} = 96 \text{ }\Omega$$

The total series resistance of the circuit would then be

$$R = 16(96 \text{ }\Omega) = 1.5 \times 10^3 \text{ }\Omega$$

and the current flowing would be only

$$I = \frac{V}{R} = \frac{120 \text{ V}}{1.5 \times 10^3 \text{ }\Omega} = 8.0 \times 10^{-2} \text{ A} = 0.08 \text{ A}$$

Hence the circuit would not be overloaded and the fuse would not be blown, but the light bulbs would give very little light.

Summary: Important Definitions and Equations

AC currents and voltages:
   Instantaneous currents and voltages of the form:

$$i = i_{max} \sin \omega t = i_{max} \sin 2\pi f t$$

$$v = v_{max} \sin \omega t = v_{max} \sin 2\pi f t$$

where f is the frequency of the current or voltage.
   Effective (or rms) values: $I = 0.707 i_{max}$; $V = 0.707 v_{max}$
   Resistive circuit (R only): $V = IR$; I and V are in phase.
   Inductive circuit (L only); $V = IX_L$; I lags V by 90°, or $\pi/2$ rad.
   Inductive reactance: $X_L = \omega L = 2\pi f L$
   Capacitive circuit (C only): $V = IX_C$; I leads V by 90°, or $\pi/2$ rad.
   Capacitive reactance: $X_C = \dfrac{1}{\omega C} = \dfrac{1}{2\pi f C}$

LC circuit:

   Natural frequency: $f = \dfrac{\omega}{2\pi} = \dfrac{1}{2\pi} \dfrac{1}{\sqrt{LC}}$

RLC series circuit with applied ac voltage:
   $V = ZI$, where the impedance $Z = \sqrt{R^2 + (X_L - X_C)^2}$.
   Power factor: $P = IV \cos \phi$, where $\cos \phi = R/Z$ is called the power factor of the RLC circuit. In an RLC series circuit all the power delivered is dissipated as $I^2 R$ heat in the resistance, none in the inductor or capacitor.
   Series resonance: That condition in an RLC series circuit when X_L and X_C are equal; the current is then a maximum and in phase with the applied voltage. This occurs then the applied frequency is equal to the natural frequency of the RLC circuit, $f = (1/2\pi)(1/\sqrt{LC})$.

Transformer:
   An electrical device which uses electromagnetic induction to step up voltages and simultaneously step down currents, or vice versa, in an ac circuit.

   Transformer equation: $\dfrac{I_1}{I_2} = \dfrac{V_2}{V_1} = \dfrac{N_2}{N_1}$

Questions

1 Indicate some reasons why you would expect ac voltages and currents to be more dangerous than dc voltages and currents of the same effective value.

2 Can you suggest how to wind an electrical resistor so that it has no inductance?

3 Why would you expect a coil with inductance L and no resistance to be an idealization, never achievable in the real world?

4 Why does a capacitor behave like a high resistance at low frequencies and like a low resistance at high frequencies?

5 An ac circuit contains an inductor which has a resistance R. Is there any electric power lost in the inductor? If so, how much power is lost?

6 What are the maximum and minimum possible values of the power factor for an ac circuit? Is the power factor characteristic of the applied voltage, or of the load, or of some combination of the two?

7 If you want to reduce your electric bill, do you want a large or a small power factor at 60 Hz? Can you suggest a way to change the power factor while still allowing the appliances in your house to operate?

8 A doorbell transformer is designed for a primary voltage of 120 V and a secondary voltage of 6 V. Suppose by mistake you mix up the primary and secondary connections when you install the transformer. What would you expect to happen?

9 Why does not a transformer violate the first law of thermodynamics, since it enables a voltage to be increased to any desired amount?

10 What kind of transformer is needed for the door-bell in a home? For the electron beam in a TV set?

11 Will the length of a copper cable carrying power-line currents at high voltages increase or decrease as the temperature rises? Will its resistance increase or decrease as the temperature rises? What conclusions can you draw from these facts?

12 Why is 240 V ac instead of 120 V used for clothes dryers, electric stoves, and other appliances which draw large electric currents?

13 Point out some practical jobs for which dc is absolutely essential and others for which either dc or ac will do.

14 Some people have maintained that dc should be favored over ac for electric power transmission, because dc "is getting somewhere," since it flows in one direction and is therefore useful. On the other hand, ac "never gets anywhere," since it constantly flows back and forth and hence cannot be useful. Comment on such an opinion from what you know of the nature and uses of electric current.

15 The advice is often given to technicians or physicists working with high-voltage electronic circuits that they should always keep one hand in their pockets when they are taking measurements in the vicinity of high-voltage wires. Why is this good advice?

Problems

A 1 A circuit contains only a 50-Ω resistor. An ac voltage is applied to this circuit with peak value 340 V. The rms current which flows is:
(*a*) 6.8 A   (*b*) 3.4 A   (*c*) 0.15 A
(*d*) 0.21 A   (*e*) 4.8 A

A 2 A circuit contains only an inductance L, which has a value of 0.20 H. An ac voltage of peak value 170 V and frequency 60 Hz is applied to the circuit. When the ac voltage is equal to 170 V, the ac current is:
(*a*) 0   (*b*) 1.6 A   (*c*) 2.3 A   (*d*) -1.6 A
(*e*) -2.3 A

A 3 An ac voltage of peak value 85 V and frequency 50 Hz is applied to a capacitor C of capacitance 20 μF. The rms current which flows is:
(*a*) 0.54 A   (*b*) 0.38 A
(*c*) -0.54 A   (*d*) -0.38 A
(*e*) 0

A 4 In the same situation given in Prob. A3, the current in the circuit when $v = -85$ V is:
(*a*) 0   (*b*) -1.7×10^{-3} A   (*c*) 1.7×10^{-3} A
(*d*) 0.53 A   (*e*) -0.53 A

A 5 If the frequency applied to a purely capacitive circuit is increased by 10 orders of magnitude, the current through the circuit:

(*a*) Increases by 10^5 (*b*) Decreases by 10^5
(*c*) Increases by 10^{10} (*d*) Decreases by 10^{10}
(*e*) Remains the same

A 6 An LC circuit oscillates at a frequency of 1000 Hz. If it is desired to raise this frequency by three orders of magnitude (i.e., by a factor of 1000), the values of the inductance L and the capacitance C must be changed as follows:
(*a*) Both L and C decreased by three orders of magnitude   (*b*) Both L and C decreased by six orders of magnitude   (*c*) Either L or C decreased by three orders of magnitude   (*d*) Either L or C increased by six orders of magnitude   (*e*) None of the above

A 7 An LC circuit with $L = 0.50$ H and $C = 0.20$ μF has an initial charge on the capacitor of 100 μC. The maximum current which flows in this circuit is:
(*a*) 100 μA   (*b*) 0.32 A   (*c*) 50 A
(*d*) 1.0 A   (*e*) 0.01 A

A 8 An RLC series circuit has $L = 0.10$ H, $C = 100$ μF, $R = 20$ Ω. A voltage is applied at a frequency of 1000 Hz. The power factor of the circuit is:
(*a*) 31   (*b*) 0.032   (*c*) 0.97   (*d*) 0.50
(*e*) None of the above

A 9 It is desired to step up a voltage of 120 V to

96,000 V. The turns ratio N_2/N_1 of the transformer used must be:

(a) 28/1 (b) 800/1 (c) 1/800 (d) 1/28
(e) $(64 \times 10^4)/1$

A10 What must be the turns ratio N_2/N_1 of a transformer used to match a source impedance $R_1 = 4500 \ \Omega$ to an output impedance $R_2 = 5.0 \ \Omega$?

(a) 900/1 (b) 1/900 (c) 150/1 (d) 1/30
(e) 30/1

B 1 An ac voltage of rms value 120 V is connected across a light bulb with a resistance of 100 Ω.

(a) What is the rms current which flows?
(b) What is the maximum, or peak, current which flows?

B 2 A 60-Hz ac voltage of effective value 120 V is applied across a pure inductor of inductance 100 mH.

(a) What is the effective current which flows?
(b) When the voltage has its maximum value of 170 V, what is the value of the current?

B 3 A radio capacitor of capacitance 0.101 μF is connected to an ac voltage source of rms value 6.0 V at 60 Hz.

(a) What is the rms current which flows?
(b) What is the maximum, or peak, current which flows through the capacitor?

B 4 A circuit consists of a resistor with $R = 200 \ \Omega$, a capacitor with $C = 0.15 \ \mu$F, and an inductor with $L = 100$ mH and a resistance $R_L = 50 \ \Omega$.

(a) What is the inductive reactance of the circuit for an applied 60-Hz voltage?
(b) What is the capacitive reactance of the circuit for an applied 60-Hz voltage?

B 5 What is the natural frequency of an LC circuit with $L = 0.20$ mH and $C = 1.0 \ \mu$F?

B 6 What is the resonant frequency of an RLC circuit with $L = 0.20$ mH, $C = 100 \ \mu$F, and $R = 0.10 \ \Omega$?

B 7 An RLC circuit is in resonance with a 60-Hz applied voltage. If the inductance in the circuit is 60 mH, what is the capacitance in the circuit?

B 8 An electric-train transformer steps down the 120-V house current to 6 V. If the current supplied to the transformer by the house circuit is 0.50 A, what is the current output of the transformer?

B 9 An electric generator produces ac power at an effective voltage of 19,125 V. It is desired to transmit this power over high-tension lines at a voltage of 765,000 V by using a step-up transformer. If the primary coil of the transformer consists of 100 turns, how many turns must the secondary have?

C 1 An ac voltage of the form $v = v_{max} \sin 2\pi ft$, with $f = 60$ Hz and $V_{max} = 170$ V, is applied across a 60-W light bulb.

(a) When $t = 0$, i.e., when the voltage is first applied, what is the current through the circuit?
(b) When $t = T/4$, where T is the period (equal to $1/f$) of the applied voltage, what is the current in the circuit?
(c) What is the rms current in the circuit?

C 2 An ac voltage of the form $v = V_{max} \sin 2\pi ft$, with

$f = 60$ Hz and $V_{max} = 340$ V, is applied to a light bulb of resistance 100 Ω. Draw graphs of the applied voltage and the current as a function of time t for one complete voltage cycle. Indicate the numerical values of the maximum voltages and currents.

C 3 A voltage of the form $v = V_{max} \sin 2\pi ft$, with $f = 60$ Hz and $V_{max} = 170$ V, is applied to a pure inductor with inductance 0.20 H.

(a) What is the current through the circuit when $t = T/4$, where T is the period?
(b) What is the current when $t = T/2$?
(c) What is the current when $t = T$?

C 4 An ac voltage of form $v = V_{max} \sin 2\pi ft$, with $f = 2.5 \times 10^6$ Hz and $V_{max} = 0.10$ V, is applied to a pure inductor with $L = 0.10$ H.

(a) Draw on the same sheet of paper graphs of the voltage v and the current i in the circuit as a function of time.
(b) What is the effective current in the circuit?

C 5 A radio capacitor of capacitance 0.020 μF is attached to a voltage source of the form $v = V_{max} \sin 2\pi ft$, where $f = 60$ Hz and $V_{max} = 170$ V.

(a) What is the current through the circuit at the instant the voltage is applied ($t = 0$)?
(b) What is the current when $v = 170$ V?
(c) What is the current when $t = T/2$, where T is the period?

C 6 For the situation of Prob. C5:

(a) Draw on the same sheet of paper graphs of the voltage v and the current i in the circuit as a function of time.
(b) What is the effective value of the current?
(c) What is the maximum charge on the capacitor?

C 7 A 60-μF capacitor is connected in series with an inductance of 10 mH and a switch. The capacitor is first charged to a voltage of 170 V. The charging battery is then removed, and the switch is closed.

(a) What is the maximum charge on the capacitor?
(b) What is the maximum current in the circuit?
(c) What is the resonant frequency of the circuit?

C 8 In the situation given in Prob. C7, plot on the same sheet of graph paper the energy stored in the magnetic field of the inductor, and the energy stored in the electric field of the capacitor, as a function of time. Show that the total energy in the circuit remains constant in time.

C 9 A capacitor with $C = 0.15 \ \mu$F is connected in series with a 50-Ω resistor and connected to a 120-V, 60-Hz power line. Find:

(a) The current in the circuit.
(b) The phase angle between the current and the applied voltage.
(c) The power factor.
(d) The power loss in the capacitor.

C10 What is the resistance of a coil if its impedance is 50 Ω and its inductive reactance is 25 Ω?

C11 Find the impedance at 60 Hz of an RLC circuit with $L = 100$ mH, $C = 10 \ \mu$F, and $R = 50 \ \Omega$.

C12 If an effective ac voltage of 120 V at 60 Hz is applied to an RLC circuit with $L = 10$ mH, $C = 0.10 \ \mu$F, and $R = 100 \ \Omega$, what effective current flows in the circuit?

C13   In the circuit of Prob. C12:

(*a*)   What is the power factor of the circuit?

(*b*)   What is the power consumed in the circuit? In what element of the circuit is the power consumed?

C14   An inductance coil with $L = 0.15$ H and resistance 20 Ω is connected across a 120-V, 60-Hz line. Find:

(*a*)   The current through the coil.

(*b*)   The phase angle between the current and the applied voltage.

(*c*)   The power factor.

(*d*)   The power lost in the coil in the form of heat.

C15   A radio is tuned to a chosen station by a circuit consisting of a fixed 0.25-mH inductance and a variable capacitor which is set to 20 pF (20×10^{-12} F).

(*a*)   What is the frequency of this station?

(*b*)   What is the wavelength of the waves it transmits?

C16   A 120-V voltage is applied at the resonant frequency of a circuit consisting of a 10-Ω resistor, a 20-mH inductor, and a 10-μF capacitor in series.

(*a*)   What are the voltages across the inductor and the capacitor at resonance?

(*b*)   What is the voltage across the resistor at resonance?

(*c*)   What is the total voltage across the inductor, capacitor, and resistor in series at resonance?

C17   An RLC series circuit is connected to a variable-frequency ac generator of rms voltage 120 V. In the circuit $L = 100$ mH, $C = 2.0 \times 10^{-6}$ F, and $R = 150$ Ω.

(*a*)   What is the resonant frequency of the circuit?

(*b*)   What is the current at resonance?

(*c*)   What is the voltage across the capacitor, inductor, and resistor at resonance?

(*d*)   What is the power delivered to the circuit at resonance?

C18   A voltage of 120 V is applied across a circuit consisting of a 10-Ω resistor, a 50-mH inductor, and a 30-μF capacitor in series.

(*a*)   What is the resonant frequency of this circuit?

(*b*)   What is the power factor of the circuit at resonance?

C19   An electric generating plant produces electric energy at a rate of 0.50 GW and a voltage of 10,000 V.

(*a*)   If it is desired to transmit this power at 400,000 V, what must be the turns ratio of the transformer used?

(*b*)   What current would be sent out over the power lines if it were transmitted at 10,000 V?

(*c*)   What current would be transmitted at 400,000 V?

C20   A power company is transmitting power to a city at a rate of 0.50 GW over a power line with a resistance of 0.50 Ω. Compare the power losses in such a system if the power is transmitted at 10,000 and at 400,000 V.

C21   A stereo amplifier has an impedance of 1500 Ω. What must be the turns ratio of the transformer used to match this impedance to:

(*a*)   A loudspeaker of impedance 3.0 Ω?

(*b*)   A public-address system of impedance 50 Ω?

C22   An ac source of internal resistance 10,000 Ω is to feed current to a load of resistance 20 Ω.

(*a*)   How should the source be matched to the load?

(*b*)   When the impedances are matched, what is the ratio of the current through the source to the current through the load?

D 1   An inductance coil has an inductance of 300 mH and a resistance of 20 Ω. Find the current in the coil if the applied voltage is:

(*a*)   120 V dc

(*b*)   120 V ac at 10^6 Hz.

D 2   A 100-Ω resistor is connected in series with a 0.10-H inductor and a 30-μF capacitor. The applied voltage has the form $v = 170 \sin 2\pi f t$, where $f = 1000$ Hz. Find:

(*a*)   The impedance of the circuit.

(*b*)   The effective current.

(*c*)   The effective voltages across R, L, and C.

(*d*)   Show by means of a vector diagram how the voltages across R, L, and C add to give the voltage applied to the circuit.

D 3   A capacitor is placed in parallel across a resistive load R in the circuit of Fig. 20.26. Suppose that R is 250 Ω and that C is 0.10 μF. What percent of the incoming current will pass through the capacitor to ground rather than through the load when:

(*a*)   The frequency is 100 Hz?

(*b*)   The frequency is 1.0 MHz?

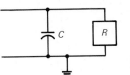

FIGURE 20.26   Diagram for Prob. D3.

D 4   A "black box" contains two circuit elements in series, but it is not known whether they are capacitors, inductors, or resistors. When connected to a 120-V, 60-Hz circuit, the current which flows is 6.0 A, and this current lags the voltage by 40°. What are the two elements in the black box and what are their numerical values?

D 5   An inductance coil draws 2.0 A dc when connected to a 12-V battery. When connected to a 120-V, 60-Hz power source the effective current drawn is 5.0 A. What is (*a*) the resistance, (*b*) the inductance of the coil?

D 6   A coil having resistance and inductance is connected across a 120-V dc line. The current is 1.3 A. The coil is then connected across a 120-V, 60-Hz line and the effective ac current is 0.65 A. What are (*a*) the resistance, (*b*) the impedance, (*c*) the reactance, (*d*) the inductance of the coil?

D 7   A 6-V battery is connected in series with two unknown electric components, and the current is 600 mA. A 60-Hz, 6-V ac source then replaces the battery, and the current becomes 300 mA.

(*a*)   What are the circuit elements and what are their numerical values?

(*b*)   How does the current change if the frequency is increased to 1.0 MHz?

D 8   A step-down transformer at the end of a transmission line reduces the voltage from 2400 to 120 V. The power output of the transformer is 10.0 kW and the overall efficiency of the transformer is 92 percent. The primary transformer winding has 1000 turns.

(a) How many turns does the secondary winding have?

(b) What is the power input to the transformer?

(c) What is the current in each of the transformer coils?

D 9 When a cathode-ray tube has its electrostatic deflection plates connected across a 90-V battery, as in Fig. 20.27a, the spot on the fluorescent screen is deflected through 10 cm. If the plates are now connected across a resistance of 20 Ω in parallel with an ac voltmeter and in series with a 60-Hz ac generator, as in Fig. 20.27b, the voltmeter reads 45 V and the length of the trace on the screen becomes 14 cm.

(a) Can you explain these seemingly contradictory data? (*Hint:* What is the beam displacement per volt in the two cases?)

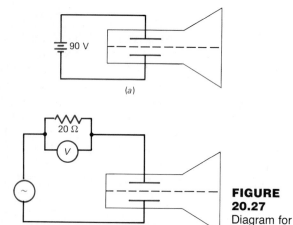

(a)

(b)

FIGURE 20.27
Diagram for Prob. D9.

(b) What is the rms current through the 20-Ω resistor?

D10 An inductance coil with $L = 5.0$ H and resistance 200 Ω is connected to a 120-V, 60-Hz source. A capacitor is connected in series with the coil to bring the power factor up to 1.0.

(a) What is the current in the circuit after the capacitor has been inserted in the circuit?

(b) What value of capacitance is needed to bring the power factor to 1.0?

(c) What is the peak voltage that the capacitor must be able to sustain?

D11 Figure 20.28 shows an ac bridge which is similar to a Wheatstone bridge for direct current (see Fig. 17.50), but which in this case is used to measure frequencies. When the bridge is balanced, a is at the same potential as b and no current flows through the earphones connected between a and b. If $R_1 = R_2$, $L = 100$ mH, and $C = 1.0$ μF, at what frequency will the bridge be balanced? (Assume that the resistance of the coil is very small.)

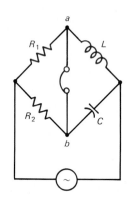

FIGURE 20.28   Diagram for Prob. D11.

Additional Readings

Asimov, Isaac: *Understanding Physics,* vol. II: *Light, Magnetism and Electricity*, New American Library, New York, 1966. Chapter 13 of this popular book contains a good discussion of alternating current circuits.

Barthold, L. O., and H. G. Pfeiffer: "High-Voltage Power Transmission," *Scientific American*, vol. 210, no. 5, May 1964, pp. 39–47. Includes a discussion of dc lines, very-high-voltage lines and new approaches to the transmission of electric power.

Cheney, Margaret: *Tesla: Man Out of Time*, Laurel Dell, New York, 1983. A very interesting and more complete life than O'Neill's, but a bit weak on the scientific side.

Gray, Andrew: *Lord Kelvin: An Account of His Scientific Life and Work*, Chelsea, New York, 1973. A delightful book, a reprint of the original 1908 edition by a former student and assistant of Kelvin at the University of Glasgow.

Hammond, Allen L., William D. Metz, and Thomas H. Maugh II: *Energy and the Future*, American Association for the Advancement of Science, Washington, D.C., 1973. This book contains four brief but excellent

chapters on new approaches to the transmission of electric power.

O'Neill, John J.: *Prodigal Genius: The Life of Nikola Tesla*, Ives Washburn, New York, 1944. A lively study of the life and mind of an eccentric scientific genius by a personal friend.

Schwartz, Brian B., and Simon Foner: "Large-Scale Applications of Superconductivity," *Physics Today*, vol. 30, no. 7, July 1977, pp. 34–43. Includes applications of superconductivity to power transmission, and to electric motors and generators, superconducting magnets, and magnetic levitation of trains.

Sharlin, Harold Issadore: *Lord Kelvin, The Dynamic Victorian*, Pennsylvania State University Press, University Park, Pa., 1979. The most recent account of Kelvin's life and scientific accomplishments.

Snowden, Donald P.: "Superconductors for Power Transmission," *Scientific American*, vol. 226, no. 4, April 1972, pp. 84–91. Discusses the advantages and problems associated with underground superconducting power lines for both ac and dc power transmission.

A Clerk Maxwell was required, a second man of the same depth and independence of insight, to build up in the normal forms of our systematic thinking the great structure whose plan was present to Faraday's mind, which he saw clear before him, and endeavored to render apparent to his contemporaries.

Hermann von Helmholtz (1821–1894)

Chapter 21

Electro-magnetic Waves

In the preceding five chapters we have introduced the basic physical laws governing the behavior of charges at rest and in motion, and electric and magnetic fields and their interactions. Now we can put all these pieces together into a complete picture of electromagnetic theory. This was first done in the years between 1864 and 1873 by James Clerk Maxwell (see Fig. 21.1 and accompanying biography). The four equations he put forward, now called simply Maxwell's equations, unified elements of electricity and magnetism once thought entirely distinct and independent. In addition they integrated electromagnetic theory with light by predicting the existence of electromagnetic waves traveling at a speed which was identical to the speed of light. As a consequence, Maxwell's equations provide the theoretical foundation on which today's radio, radar, and television communication systems are built. In the realm of electromagnetics Maxwell's equations play a role similar in importance to that of Newton's laws of motion in mechanics.

21.1 Maxwell's Equations

Maxwell showed that all the theories of electromagnetism can be summarized by four basic equations which describe the behavior of electric and magnetic fields in space. Of these four equations three were restatements of results previously obtained by Coulomb, Gauss, Oersted, Faraday, Henry, and others, which we have already seen in previous chapters. The fourth equation, while based on the work of Ampère, introduced Maxwell's key idea of displacement current, which was the last piece needed to complete the intricate mosaic of electromagnetic theory.

Because precise statements of Maxwell's equations require calculus, we will confine ourselves here to stating the physical content of each of the four laws and to illustrating their use in a few simple cases. This will provide both a useful review of important aspects of electricity and magnetism discussed in the preceding five chapters, and an introduction to the application of Maxwell's equations to electromagnetic waves.

Maxwell's First Equation   Maxwell's first equation is a consequence of Coulomb's law in electrostatics, but is more conveniently stated in terms of Gauss' law as given by Eq. (16.3):

James Clerk Maxwell (1831–1879)

FIGURE 21.1 James Clerk Maxwell. *(Photo courtesy of AIP Niels Bohr Library.)*

James Clerk Maxwell, the greatest theoretical physicist of the nineteenth century, was born in 1831, the year in which Faraday discovered electromagnetic induction, and died in 1879, the year in which Einstein was born. Maxwell was a true link between these two great physicists, for it was he who put Faraday's experimental results and far-reaching insights into useful mathematical equations, and it was on the foundation of Maxwell's equations that Einstein built his theory of relativity.

Maxwell was born in Edinburgh of a well-known Scottish family. His mother died when he ws 9 years old,

and he was raised by his father and an aunt. As a youth he revealed great mathematical genius, which earned him the nickname "dafty" from his schoolmates. Maxwell had not only the ability to handle mathematical abstractions, but also a remarkable curiosity about how things worked and a fascination with constructing working models of practical devices. He constantly bombarded his elders with questions such as "What does it do?" or "What's the particular go of that?"

In 1847 Maxwell enrolled at the University of Edinburgh, and in 1850 moved on to Cambridge University, from which he graduated in 1854, second in his class in mathematics. In 1856 he accepted an appointment as a professor at Marischal College in Aberdeen, Scotland, mainly to be near his father, who unfortunately died just before his son took over his new post. After 5 years (1860 to 1865) at King's College in London, Maxwell went into semiretirement at his family estate, Glenlair, near the village of Parton in Scotland to devote as much time as possible to research.

In 1871 Maxwell agreed to come out of retirement to organize the newly created Cavendish Laboratory at Cambridge, the first laboratory for experimental physics at that university. There Maxwell began the great tradition which has marked physics at Cambridge ever since.

Maxwell made great contributions to many fields of physics. Among these were his work on kinetic theory and the statistical theory of gases, his theory of color vision, his work in thermodynamics and astrophysics, and in particular his development of the equations of the electromagnetic field (discussed in this chapter). Heinrich Hertz once referred to Maxwell's equations as behaving as if they had "an independent life and an intelligence of their own, as if they were wiser than ourselves, indeed wiser than their discoverer, as if they gave forth more than he had put into them." Today, even after the revolutionary discoveries in this century of relativity and quantum mechanics, Maxwell's equations remain as valid and useful as ever.

This remarkable scientist was at the same time a gentle, religious, selfless man who loved children (he and his wife had none of their own), family, and good friends. He once wrote: "Work is good, and reading is good but friends are better." He also had a fine sense of humor, once delivering a lecture on the telephone, then newly developed, in which he referred to "the perfect symmetry of the whole apparatus—the wire in the middle, the two telephones at the ends of the wire, and the two gossips at the ends of the telephones."

Maxwell's career came to a sudden, sad end when he succumbed to cancer at the age of 48.

$$\sum E_{\perp}\Delta A = 4\pi k_e \sum q$$

Here $E_{\perp}$ is the component of the electric field perpendicular to the element of area ΔA of a gaussian surface enclosing a sum of charges $\sum q$. In applying this first equation of Maxwell we sum over the total area of the surface and all the charges within the surface.

FIGURE 21.2 (left)   A spherical gaussian surface used to calculate the electric field of a point charge at a distance *r* from the charge.

FIGURE 21.3 (right)   Lines of the electric field in the neighborhood of charges. For the spherical gaussian surface *a*, $\Sigma\, E_\perp\, \Delta A$ is positive, since the charge within the gaussian surface is positive. For the gaussian surface *c*, $\Sigma\, E_\perp\, \Delta A$ is negative, because the charge within is negative. For the gaussian surface *b*, $\Sigma\, E_\perp\, \Delta A$ is zero, because there is no charge within. This means that as many lines of **E** must enter the gaussian surface *b* as leave it, as shown.

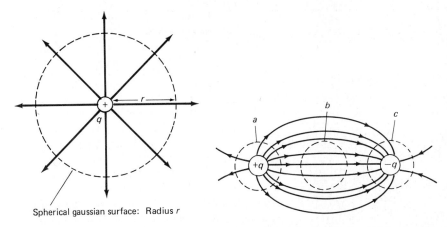

Spherical gaussian surface: Radius *r*

As an example, recall how Eq. (16.3) was used to find the field of a point charge *q*. In Fig. 21.2 we construct a spherical gaussian surface about a single point charge at the center of the sphere. Equation (16.3) then becomes

$$E(4\pi r^2) \;=\; 4\pi k_e q \qquad \text{or} \qquad E \;=\; -\,\frac{k_e q}{r^2}$$

which is the same result obtained from Coulomb's law for the field of a point charge.

Implicit in Gauss' law is the idea that all the lines of a static electric field must begin and end on electric charges. By convention these lines are taken to begin on positive charges and end on negative charges, as in Fig. 21.3. The electric charges are therefore the "poles" of the electric field.

Maxwell's Second Equation

Maxwell's second equation points out the essential difference between magnetic field lines and electric field lines. Magnetic field lines, which are due to electric currents, do not start or stop in space but form closed continuous curves. This is shown for a current loop in Fig. 21.4. The reason the lines of the magnetic field always close on themselves is that there is no magnetic equivalent to an isolated electric charge; magnetic monopoles do not exist. (The poles we think we observe in permanent magnets are due to a concentration of the field lines in the region around the poles, but the magnetic field lines themselves are continuous, as shown in Fig. 21.5.)

FIGURE 21.4 (left)   The magnetic field lines produced by a current loop. All the lines of the magnetic field **B** form closed loops in space.

FIGURE 21.5 (right)   A permanent magnet. Even in this case the lines of **B** are closed, since inside the magnet they continue on from the S to the N pole.

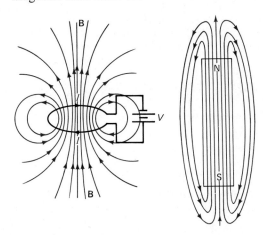

As a result, if an equation equivalent to Eq. (16.3) in electrostatics is written for the magnetic field, the right side is always zero, since there are no isolated magnetic poles, and so as many lines of **B** pass inward through the area A of any closed surface as pass outward through it. Hence

$$\sum B_\perp \, \Delta A = 0 \tag{21.1}$$

This is Maxwell's second equation.

Maxwell's Third Equation

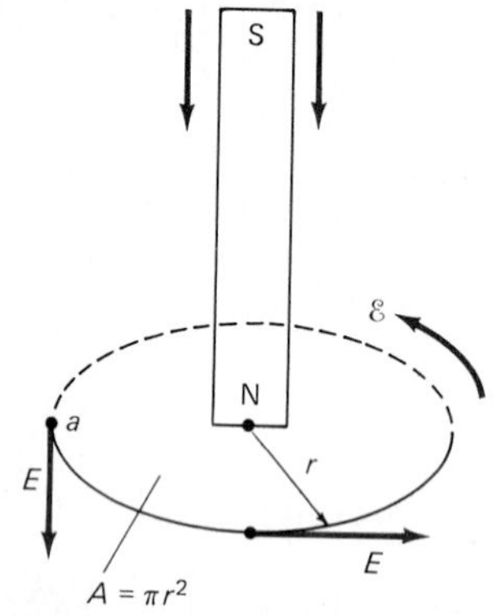

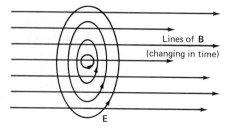

FIGURE 21.6 Induced emf in a circular coil of one turn when a permanent magnet is moved through the plane of the coil. The induced emf causes an electric field in the wire. Hence a magnetic field changing in time produces an electric field.

Faraday's induction law, as given in Eq. (19.6),

$$\mathscr{E} = -N \frac{\Delta \Phi_B}{\Delta t}$$

states that a magnetic flux Φ_B changing in time produces an induced emf (we introduce here the symbol Φ_B for magnetic flux, instead of merely Φ, since we want to distinguish it from a similar symbol to be used later for *electric* flux).

Figure 21.6 shows a single circular coil of wire of radius r through which the magnetic flux is changing at a rate $\Delta \Phi_B / \Delta t$, where $\Phi_B = B_\perp A$ and where the area A remains constant while B varies in time. The emf $\mathscr{E}$ is then, from the definition of emf, the work done in moving a unit charge around the loop starting at point a and returning back to point a. Hence we have, if $\mathscr{W}$ is the work done in moving a charge q around a complete loop,

$$\mathscr{E} = \frac{\mathscr{W}}{q} = \frac{\mathbf{F} \cdot \mathbf{s}}{q} = \mathbf{E} \cdot \mathbf{s} = E(2\pi r) \tag{21.2}$$

where **E**, the electric field in the wire, is equal to the force exerted on a unit positive charge when placed at any point in the wire. Hence Eq. (19.6) becomes

$$\mathscr{E} = E(2\pi r) = -N \frac{\Delta \Phi_B}{\Delta t}$$

Since here $N = 1$, we have

$$E = -\frac{1}{2\pi r} \frac{\Delta \Phi_B}{\Delta t} = -\frac{A}{2\pi r} \frac{\Delta B_\perp}{\Delta t} \tag{21.3}$$

This is Maxwell's third equation, which states simply that a magnetic field **B** changing in time produces an electric field **E**, as in Fig. 21.7.

FIGURE 21.7 A magnetic field **B** changing in time produces an electric field **E** even in free space.

Maxwell's Fourth Equation

Maxwell's fourth equation is based on Ampére's circuital law, as given in Eq. (18.15),

$$\sum B_\parallel \, \Delta l = \mu_0 I$$

Maxwell's great achievement was to recognize that the circuital law was incomplete. Since a changing magnetic field produces an electric field,

Maxwell expected to find that a changing electric field would produce a magnetic field. There was no experimental evidence to justify this idea in Maxwell's day. Maxwell, however, was able to show that, since a changing electric field **E** normal to an area A leads to a change in the electric flux $\Phi_E = E_\perp A$ through the area, this changing electric flux is equivalent to an electric current I_D, called by Maxwell the *displacement current* and defined as:

$$I_D = \epsilon_0 \frac{\Delta \Phi_E}{\Delta t} = \epsilon_0 A \frac{\Delta E_\perp}{\Delta t} \tag{21.4}$$

Any change in the electric flux therefore produces a displacement current which, like all currents, has a magnetic field associated with it. In the next section we will see how Maxwell arrived at this important result.

When Ampére's circuital law is extended to include the possibility of a displacement current I_D in addition to a conventional current I, we have

$$\sum B_\parallel \, \Delta l = \mu_0 (I + I_D)$$

or, from Eq. (21.4),

$$\sum B_\parallel \, \Delta l = \mu_0 I + \mu_0 \epsilon_0 \frac{\Delta \Phi_E}{\Delta t} = \mu_0 I + \mu_0 \epsilon_0 \, A \, \frac{\Delta E_\perp}{\Delta t} \tag{21.5}$$

In this equation $B_\parallel$ is the component of **B** parallel to the closed path taken around the area A. If a real current I flows through this area *or* if the electric field perpendicular to this area changes in time, then a magnetic field is produced which can be calculated from Eq. (21.5). Figure 21.8 illustrates these two possibilities.

These then are Maxwell's basic equations for the electromagnetic field. These four equations, together with two equations that provide operational definitions of the electric and magnetic fields, enable us to solve all problems

FIGURE 21.8   Magnetic field **B** produced by: *(a)* a real electric current I in a wire; *(b)* a change in the electric field lines **E** between the plates of a capacitor, i.e., a displacement current. For the same value of the current, these two fields cannot be distinguished.

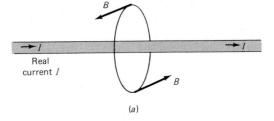

(a)

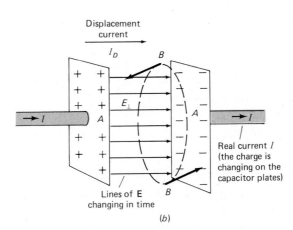

(b)

in electromagnetism. We have already seen these two other equations:

$$f_{el} = qE \tag{16.2}$$

and $\qquad f_B = B_\perp qv \tag{18.4}$

These merely tell us how to obtain the magnitude of the electric field **E** and the magnetic field **B** by measuring the force these fields exert on a charge q which is at rest in Eq. (16.2), and on a charge q which moves with a velocity **v** perpendicular to **B** in Eq. (18.4).

21.2 Derivation of Maxwell's Displacement Current

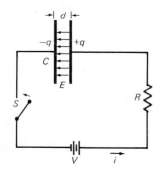

FIGURE 21.9 Maxwell's displacement current. When a capacitor C is being charged through a resistor R, the electric field between the plates changes at the rate $\Delta E/\Delta t$. Maxwell showed that this was equivalent to a displacement current between the plates of magnitude $I_D = \epsilon_0 A \, \Delta E/\Delta t$, where A is the area of the capacitor plates.

In this section we will present a somewhat simplified version of Maxwell's argument that a varying electric field is equivalent to an electric current and hence produces a magnetic field.

We consider a capacitor of capacitance C in series with a resistance R, as in Fig. 21.9, with the distance between the capacitor plates being d. A battery is connected across the circuit. At time $t = 0$ the switch S is closed and the battery delivers a current i, which falls from its initial value I to zero in a time determined by the time constant of the circuit. At the same time, the charge q on the capacitor increases gradually from zero to its final value Q, as in Fig. 19.17. As the charge q varies, so does the voltage across the capacitor, since $v = q/C$.

The electric field in the region between the plates is $E = v/d$, and so

$$q = Cv = CEd \tag{21.6}$$

Now C and d are both constants. Hence as the charge on the plates changes in time, the electric field must change according to the equation:

$$\frac{\Delta q}{\Delta t} = Cd \frac{\Delta E}{\Delta t} \tag{21.7}$$

Here $\Delta q/\Delta t$ is just the instantaneous current i in the wire, and from Eq. (16.12) the capacitance of a parallel-plate capacitor is just $C = \epsilon_0 A/d$. Hence

$$i = \frac{\Delta q}{\Delta t} = Cd \frac{\Delta E}{\Delta t} = \epsilon_0 A \frac{\Delta E}{\Delta t}$$

Since the field E is here assumed to be normal to the area A, this equation can be written as:

$$i = \epsilon_0 A \frac{\Delta E_\perp}{\Delta t} \tag{21.8}$$

Equation (21.8) relates the change in the electric field between the plates of the capacitor to the actual current (i) in the rest of the circuit. If we assume that a current equal to $\epsilon_0 A \, \Delta E_\perp/\Delta t$ flows through the capacitor when the field $E_\perp$ changes with time, then we have a continuous flow of current through the whole circuit, including the region between the capacitor plates. This explains why ac currents can flow in a circuit that contains a capacitor. Since the electric field is always changing in the region between the plates, an equivalent current is always flowing through the capacitor, and this current is numerically equal, from Eq. (21.8), to the real current in the rest of the circuit.

The current defined in this way in terms of the change in the electric field was called by Maxwell the *displacement current*, and is defined in terms of Eq. (21.8) as:

Maxwell's Displacement Current

$$I_D = \epsilon_0 A \frac{\Delta E_\perp}{\Delta t}$$

(21.9)

Hence just as Faraday's induction law states that a changing magnetic field produces an electric field, so Eq. (21.9) states that a changing electric field produces a displacement current, which in turn creates a magnetic field.

Even though Eq. (21.9) has been derived for the field between the plates of a capacitor, this equation is valid for all situations in which an electric field changes with time. If we write for the electric flux $\Phi_E = E_\perp A$, by analogy with the magnetic flux $\Phi_B = B_\perp A$, then the displacement current can be written as:

$$I_D = \epsilon_0 \frac{\Delta \Phi_E}{\Delta t}$$

If this displacement current is treated as an added current in Ampére's circuital law, we then obtain Eq. (21.5):

$$\sum B_\parallel \, \Delta l = \mu_0 I + \mu_0 \epsilon_0 \frac{\Delta \Phi_E}{\Delta t}$$

(21.5)

This is Maxwell's fourth equation, based on his brilliant insight that a magnetic field can be produced not only by an ordinary electric current but also by a changing electric field. This was the crucial step needed to introduce symmetry between the electric and magnetic fields, thus rounding out electromagnetic theory.

The great utility of Maxwell's displacement current is that it helps us both to predict and to understand the propagation of electromagnetic waves through space, as we will see in the sections which follow.

Example 21.1

A capacitor with plates of area 400 cm² and an air gap between the plates of 1.0 mm is being charged by a dc power supply at a rate of 5.0 nA.
(a) At what rate is the electric field between the plates changing?
(b) What is the displacement current while the capacitor is being charged?

(c) What is the magnetic field a distance 50 cm from the wire which carries the charge to the capacitor?
(d) What is the magnetic field produced by the displacement current at a distance 50 cm from the center of the capacitor?

SOLUTION

(a) To find the rate at which the electric field is changing, we need to know the rate at which the voltage between the plates is changing. Since $v = q/C$, this is

$$\frac{\Delta v}{\Delta t} = \frac{1}{C} \frac{\Delta q}{\Delta t}$$

Here $\Delta q/\Delta t = i = 5.0 \times 10^{-9}$ A $= 5.0 \times 10^{-9}$ C/s, and the capacitance is

$$C = \frac{\epsilon_0 A}{d} = \left(8.85 \times 10^{-12} \, \frac{C^2}{N \cdot m^2}\right)\left(\frac{0.040 \, m^2}{0.0010 \, m}\right)$$
$$= 3.5 \times 10^{-10} \, F$$

and so $\dfrac{\Delta v}{\Delta t} = \dfrac{5.0 \times 10^{-9} \, C/s}{3.5 \times 10^{-10} \, F} = 14$ V/s

Then, since $E = v/d$, we have

$$\frac{\Delta E}{\Delta t} = \frac{1}{d} \frac{\Delta v}{\Delta t} = \frac{14 \, V/s}{0.0010 \, m} = \boxed{1.4 \times 10^4 \, V/(m \cdot s)}$$

(b) The displacement current I_D is, from Eq. (21.9),

$$I_D = \epsilon_0 A \frac{\Delta E}{\Delta t}$$

$$= \left(8.85 \times 10^{-12} \frac{C^2}{N \cdot m^2}\right)(0.040 \ m^2)\left(1.4 \times 10^4 \frac{V}{m \cdot s}\right)$$

$$= \boxed{5.0 \times 10^{-9} \ A}$$

Note that this is numerically equal to the current in the external circuit.

(c) Since for the magnetic field around a wire, from Eq. (18.14), $B = \mu_0 I/2\pi r$ and $I = 5.0 \times 10^{-9}$ A, we have

$$B = \left(\frac{4\pi \times 10^{-7}}{2\pi} \frac{N}{A^2}\right)\left(\frac{5.0 \times 10^{-9} \ A}{0.50 \ m}\right)$$

$$= \boxed{2.0 \times 10^{-15} \ T}$$

(d) Since the displacement current I_D is numerically equal in magnitude to the real current I, from parts (b) and (c), both being 5.0×10^{-9} A, the magnetic field B around the capacitor at a distance of 50 cm from its center is the same as calculated for the field around the wire. Hence B is equal to 2.0×10^{-15} T, as in part (c). This is quite a small magnetic field because the rate at which the capacitor is being charged is very low.

21.3 The Nature of Electromagnetic Waves

Once we understand the content of Maxwell's equations, the nature of electromagnetic waves becomes clearer. Time-varying electric fields moving through space produce magnetic fields which also change in time. These time-varying magnetic fields in turn produce electric fields. Hence even in regions of space in which there are no charges, no magnets, and no real currents, electromagnetic waves can exist. A "leap-frogging" effect of changing electric fields producing magnetic fields, which in turn produce electric fields, occurs continuously, as in Fig. 21.10. As time goes on, the whole package of electric and magnetic fields moves through space, for each field produced is in a different region of space from the changing field producing it. Hence the energy in the electric and magnetic fields is transported through space by electromagnetic waves, even through regions in which no matter exists. Maxwell's work suggested that this is the way that light also propagates through space.

FIGURE 21.10 The propagation of an electromagnetic wave: A changing magnetic field **B** produces a changing electric field **E**, which in turn produces a changing magnetic field **B**, and so forth. This leap-frogging effect carries electromagnetic energy through space.

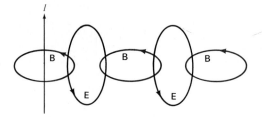

Properties of Electromagnetic Waves

Solving Maxwell's equations for the electric and magnetic fields in a region of space where there are no charges leads to the following important results:

1 Electromagnetic waves in a vacuum (or in "free space") are all *transverse*. The magnetic field **B** and the electric field **E** are at right angles to each other and to the direction of propagation. This is illustrated in Fig. 21.11 for the usual case where both **E** and **B** are of sinusoidal form, with **E** being confined to the XY plane and **B** to the XZ plane, while the wave travels along the positive X axis.

2 All electromagnetic waves travel with the same speed in a vacuum. (We will show in the next section that this speed is $c = 1/\sqrt{\mu_0 \epsilon_0}$.)

3 Various kinds of electromagnetic waves differ in their frequency f (the number of complete variations of **B** and **E** per second) and their wavelength λ [the distance between any two consecutive points along the direction of the wave at which the phases of **E** (or of **B**) are the same]. The wavelength and frequency are connected by the usual equation for waves, $\lambda f = c$, where the speed c in a vacuum is the same for all frequencies and wavelengths.

FIGURE 21.11 A plane electromagnetic wave moving along the X axis. The electric field **E** is confined to the XY plane and the magnetic field **B** to the XZ plane, so that the wave is transverse.

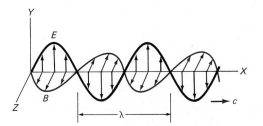

In the next section we will use Maxwell's equations to obtain a formula that will yield a numerical value for c. The agreement of such a value with experimental measurements of the speed of various kinds of electromagnetic waves will be a strong indication of the validity of Maxwell's theory.

21.4 The Predicted Speed of Electromagnetic Waves

We apply Maxwell's equations to a somewhat artificial model of electric and magnetic fields moving through space, to see at what speed these fields must travel if they are to reproduce themselves as the wave moves along. Our model will be closer to a "pulse" than to a conventional wave. (When a stone is dropped into a swimming pool, the outward-traveling ring of water is a pulse.) We consider a constant electric field in the Y direction and an associated constant magnetic field in the Z direction. These two fields are moving along the X axis, as shown in Fig. 21.12. At some instant of time the fields exist only to the left of the line 1-2 in the figure, whereas to the right of this line both fields are zero. We are interested in how rapidly the boundary between the two regions of field and no-field must move in the X direction to satisfy Maxwell's equations. Note that even though our fields do not correspond to a conventional sinusoidal wave, **B** and **E** are indeed transverse

FIGURE 21.12 An electromagnetic pulse traveling along the X axis. It consists of a constant electric field **E** in the Y direction and a constant magnetic field **B** in the Z direction. Here we show the magnetic field lines passing through an imaginary rectangle $defg$ in the XY plane.

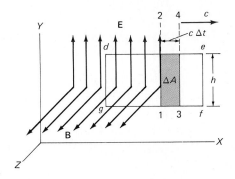

to each other and to the direction of propagation, as required by Maxwell's equations.

Relationship between Electric and Magnetic Fields

We construct an imaginary rectangle *defg* in the *XY* plane and consider the change in the magnetic flux $\Phi_B = B_\perp A$ through the rectangle as the wave moves along. Suppose the boundary moves from position 1-2 to position 3-4 in time Δt. The change in the area pierced at right angles by the magnetic flux in time Δt is then $\Delta A = hc\,\Delta t$, where h is the height of the rectangle and c is the speed of the boundary. Then , since **B** is everywhere perpendicular to A, we have

$$\frac{\Delta \Phi_B}{\Delta t} = B\,\frac{\Delta A}{\Delta t} = \frac{Bhc\,\Delta t}{\Delta t} = Bhc$$

By Faraday's induction law this change in flux will induce an emf around the perimeter of the rectangle, of magnitude

$$\mathscr{E} = -N\frac{\Delta \Phi_B}{\Delta t} = Bhc \qquad (21.10)$$

where the loop is assumed to have one turn in this case so that $N = 1$, and where we neglect the sign since we are only interested in the magnitude of the result.

Now, from Eq. (21.2) the induced emf can also be written as $\mathscr{E} = \mathbf{E} \cdot \mathbf{s}$, and so, from Eq. (21.10),

$$\mathbf{E} \cdot \mathbf{s} = Bhc$$

Here **s** is a distance along the imaginary loop *defg*.

We desire to see if an electric field **E** exists which will satisfy this equation and at the same time have the form assumed in Fig. 21.12. At the top and bottom of the rectangle **E** is perpendicular to **s** and so $\mathbf{E} \cdot \mathbf{s} = 0$. At the right end of the rectangle $\mathbf{E} = 0$, since the boundary has not yet reached that point. Hence the only contribution to $\mathbf{E} \cdot \mathbf{s}$ comes from the left end of the rectangle and is there equal to Eh. Thus we have

$$Eh = Bhc \qquad \text{or} \qquad \boxed{E = cB} \qquad (21.11)$$

Example 21.2

An electromagnetic wave in free space consists of a sinusoidal electric field whose rms value is 5.00×10^3 N/C and a magnetic field perpendicular to this electric field. What is the rms value of the magnetic field?

SOLUTION

Since for all electromagnetic waves the relationship between the magnitudes of the electric field **E** and the magnetic field **B** is $E = cB$, we have

$$B = \frac{E}{c} = \frac{5.00 \times 10^3 \text{ N/C}}{3.00 \times 10^8 \text{ m/s}} = \boxed{1.67 \times 10^{-5} \text{ T}} = 0.167 \text{ G}$$

The units are correct because

$$\frac{\text{N/C}}{\text{m/s}} = \left(\frac{\text{N}}{\text{C}}\right)\left(\frac{\text{s}}{\text{m}}\right) = \text{N/(A·m)}$$

But, since $F = BIL$, 1 T = 1 N/(A·m). Hence the answer is correctly given in teslas.

**Speed of
Electromagnetic Waves
in Free Space**

We continue to use this same model of an electromagnetic wave moving along the positive X axis, but now focus on the electric flux through an imaginary rectangle *defg* constructed in the XZ plane, as in Fig. 21.13. Using Maxwell's fourth equation [Eq. (21.5)] and noting that no real currents exist in this case, we have

$$\sum B_{\parallel} \, \Delta l = \mu_0 \epsilon_0 \frac{\Delta \Phi_E}{\Delta t}$$

where Δl is to be taken around the perimeter of the rectangle. In this case E is constant and everywhere perpendicular to A, but A changes by an amount ΔA in time Δt. Hence

$$\sum B_{\parallel} \, \Delta l = \mu_0 \epsilon_0 E \frac{\Delta A}{\Delta t}$$

Also $\Delta A = wc \, \Delta t$, where w is the width of the rectangle, and so

$$\sum B_{\parallel} \, \Delta l = \mu_0 \epsilon_0 E w c$$

Here the magnetic field **B** is at right angles to the top and bottom of the rectangle, so that $\sum B_{\parallel} \, \Delta l$ is zero at the top and bottom. $\sum B_{\parallel} \, \Delta l$ also vanishes at the right side, since **B** $= 0$ there. Hence we have, using the nonvanishing contribution from the left side,

$$\sum B_{\parallel} \, \Delta l = Bw = \mu_0 \epsilon_0 E w c$$

or $\quad B = \mu_0 \epsilon_0 E c$ $\hfill$ (21.12)

But $E = cB$ from Eq. (21.11). Also the **B** field, whose changes produced the **E** field, must in turn be generated by changes in the **E** field. Hence the **B** field in Eqs. (21.11) and (21.12) must be the same, and so

$$B = \mu_0 \epsilon_0 c^2 B \qquad \text{or} \qquad c^2 = \frac{1}{\mu_0 \epsilon_0}$$

Thus we have finally

*Speed of Electromagnetic
Waves in Space*

$$\boxed{c = \frac{1}{\sqrt{\mu_0 \epsilon_0}}}$$
$\hfill$ (21.13)

The postulated configuration of electric and magnetic fields is therefore consistent with the laws of electromagnetism only if the boundary moves with

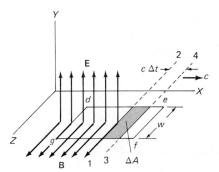

FIGURE 21.13 The same electromagnetic pulse shown in Fig. 21.12. Here we focus on the electric field lines passing through an imaginary rectangle *defg* in the XZ plane.

the speed $c = 1/\sqrt{\mu_0\epsilon_0}$. This turns out to be a completely general result, not merely for pulses but for all electromagnetic waves.

The numerical value of c is then

$$c = \frac{1}{\sqrt{[8.85 \times 10^{-12} \text{ C}^2/(\text{N}\cdot\text{m}^2)](4\pi \times 10^{-7} \text{ N}\cdot\text{s}^2/\text{C}^2)}}$$

or $c = 2.998 \times 10^8$ m/s.*

At the time Maxwell did his work (1864) it had already been determined by the optical experiments of Fizeau (1849) and Foucault (1850) that light traveled at a speed very close to 3.0×10^8 m/s. The agreement of this experimental result with the result we have just calculated using Maxwell's equations suggests strongly that light is an electromagnetic wave in a particular frequency range to which the human eye is sensitive. Subsequent work has confirmed this suggestion. Maxwell had thus achieved a true synthesis of light and electromagnetism.

The Experiments of Hertz

Maxwell predicted, on the basis of his equations, that electromagnetic radiation with essentially the same properties as light should exist at frequencies far below and far above that of light. Almost 10 years after Maxwell's death, the German physicist Heinrich Hertz (see Fig. 21.14) produced low-frequency electromagnetic radiation from a spark coil and detected it by observing sparks jumping across the gap of a detector coil some meters away, as in Fig. 21.15. These waves, which we now call *radio waves*,

---

*The symbol c is universally used for the speed of electromagnetic waves in a vacuum. The c stands for *celeritas*, a Latin word meaning "speed."

FIGURE 21.14   Heinrich Hertz (1857–1894). Hertz was a protégé of Helmholtz at the University of Berlin. Although he died very young, at the age of only 37, he had already made many outstanding contributions to physics, in particular, his experimental confirmation of the predictions of Maxwell's electromagnetic theory. His work foreshadowed the development of radio by Marconi. *(Photo from Deutsches Museum, Munich; courtesy of AIP Niels Bohr Library.)*

FIGURE 21.15   Hertz's apparatus for generating and detecting radio waves. The generator is sometimes called a *hertzian dipole.*

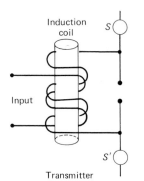

Induction coil

Input

Transmitter

Detector

fully supported Maxwell's predictions and led to the acceptance of his theory of electromagentism, which had been previously regarded as highly speculative.

In 1887, at the University of Karlsruhe, Hertz produced stationary electromagnetic waves in air and measured the distance between adjacent nodes to obtain their wavelength. Using this and the frequency of the waves, which he obtained from the measured inductance and capacitance of the oscillating circuit, he calculated the speed of the waves to be 3.2×10^8 m/s. He cautiously stated that his value "only holds good as far as the order of magnitude is concerned," but it was sufficiently accurate to show that light and radio waves traveled with essentially the same speed in air.

Hertz also proved that the electric field was transverse to the direction of propagation of the radio waves, and that all the electric vectors were in the same direction. In addition he was able to show that radio waves could be reflected and refracted in the same way as is light. Hence he proved conclusively that electromagnetic waves of wavelengths near 1 m and light waves with wavelengths 10^6 times shorter had the same fundamental properties. Thus Hertz provided the experimental confirmation needed to establish Maxwell's theory.

Speed of Electromagnetic Waves in Dielectric Materials

Electromagnetic waves will not propagate very far into conductors, but they will propagate through insulators, or dielectrics. In this case the speed of the wave is

$$v = \frac{1}{\sqrt{\epsilon\mu}} \tag{21.14}$$

where ϵ, the permittivity of the dielectric, replaces the permittivity of free space ϵ_0, and μ, the magnetic permeability of the dielectric, replaces the permeability of free space μ_0 in Eq. (21.13). Now, the ratio ϵ/ϵ_0 is the dielectric constant K [see Eq. (16.14)] and the ratio μ/μ_0 is the relative permeability K_m [see Eq. (18.18)]. Hence

$$v = \frac{1}{\sqrt{\epsilon\mu}} = \frac{1}{\sqrt{KK_m\ \epsilon_0\mu_0}}$$

or
$$v = \frac{c}{\sqrt{KK_m}}$$

For all but ferromagnetic materials K_m is very close to unity, and so*

$$v = \frac{c}{\sqrt{K}} \tag{21.15}$$

The speed of electromagnetic waves in a material with dielectric constant K is given therefore by $v = c/\sqrt{K}$, where c is the speed of light in a vacuum. (As we will see when we discuss light, this is sometimes written as $v = c/n$, where $n = \sqrt{K}$ is called the *index* of refraction of the material.)

---

*The dielectric constant K is frequency-dependent. Its value in Eq. (21.15) must therefore be chosen to fit the frequency involved. For example, at very low frequencies K for water is about 80, but at visible frequencies it is 1.78, so that $n = 1.33$.

Example 21.3

What is the frequency of 3.0-cm microwaves, i.e., of electromagnetic waves of 3.0-cm wavelength in free space?

SOLUTION

Since $c = f\lambda$, we have

$$f = \frac{c}{\lambda} = \frac{3.0 \times 10^8 \text{ m/s}}{0.030 \text{ m}} = \boxed{10^{10} \text{ Hz}}$$

Example 21.4

A low-frequency radio wave of wavelength 30 m passes from air into a block of Pyrex glass.
(a) What is the speed of the radio wave in the Pyrex?

(b) What is the new value of the frequency? Of the wavelength?

SOLUTION

(a) From Table 16.1 the dielectric constant of Pyrex glass at low frequencies is 4.5, while the dielectric constant of air is very close to that of free space (it is 1.00059 compared with 1.00000). Hence we have

$$v = \frac{c}{\sqrt{K}} = \frac{3.0 \times 10^8 \text{ m/s}}{\sqrt{4.5}} = \boxed{1.4 \times 10^8 \text{ m/s}}$$

(b) The frequency of an electromagnetic wave is determined by the source producing it and does not change when the medium changes. What does change is the wavelength, which varies in proportion to the speed in the medium. Here

$$f = \frac{c}{\lambda_{\text{air}}} = \frac{v}{\lambda_{\text{Pyrex}}}$$

and so $\quad \lambda_{\text{Pyrex}} = \frac{v}{c}\lambda_{\text{air}} = \frac{1.4}{3.0}(30 \text{ m}) = \boxed{14 \text{ m}}$

Hence the wavelength is shortened in the Pyrex in the same ratio that the speed is decreased.

21.5 Energy in Electromagnetic Waves; Intensity

A television transmitter sends a signal through space which is absorbed by the antennas on our houses and converted into useful sound and pictures on our TV sets. No *matter* is transported through space in this case; what is transported is *energy*. In this section we will investigate how the amount of energy transported by a wave depends on the strength of the electric and magnetic fields making up the wave.

In Eq. (19.19) we found that the total energy stored in unit volume of the combined electric and magnetic fields was

$$w = \tfrac{1}{2}\epsilon_0 E^2 + \tfrac{1}{2}\frac{1}{\mu_0}B^2 \tag{21.16}$$

Since for an electromagnetic wave $B = E/c$ from Eq. (21.11), and since $c = 1/\sqrt{\epsilon_0\mu_0}$, we can write Eq. (21.16) as

$$w = \tfrac{1}{2}\epsilon_0 E^2 + \frac{1}{2}\frac{E^2}{c^2\mu_0} = \tfrac{1}{2}\,\epsilon_0 E^2 + \tfrac{1}{2}\epsilon_0 E^2$$

or $\quad w = \epsilon_0 E^2 \tag{21.17}$

This shows that half the energy in the wave is stored in the electric field and half in the magnetic field.

As an electromagnetic wave moves through space, it carries along with it the energy in the electromagnetic field. Let us denote by S the *incident power per unit area*, i.e., *the energy transported per unit time across the unit cross-sectional area perpendicular to the direction in which the wave is moving.*

Definition

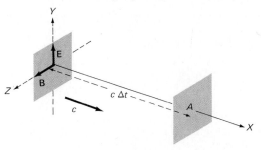

FIGURE 21.16 The transport of energy through the area A by a plane electromagnetic wave. The intensity of the wave is the amount of energy carried by the wave per second through a 1-m² area perpendicular to the direction of propagation of the wave.

In Fig. 21.16 a plane wave front which is part of a continuous wave moving along the positive X axis travels a distance $c\,\Delta t$ in time Δt. If we consider a cross-sectional area A at right angles to the direction of the wave, then in time Δt the wave occupies an additional volume ΔV of space, where $\Delta V = Ac\,\Delta t$. The amount of energy filling this volume is

$$\Delta\,W = w\,\Delta V = \epsilon_0 E^2 Ac\,\Delta t$$

The incident power per unit area is, then, from the definition given above:

$$S = \frac{\Delta W}{A\,\Delta t} = \frac{\epsilon_0 E^2 Ac\,\Delta t}{A\,\Delta t}$$

so that

$$\boxed{S = \epsilon_0 E^2 c} \tag{21.18}$$

This is the rate at which energy is transported across a unit area at right angles to the direction of propagation of the wave. It is measured in $J/(s{\cdot}m^2)$, or W/m^2.

Since half the energy is in the electric field and half is in the magnetic field, it is useful to use the identities previously derived to convert this to a more symmetric form:

$$S = \epsilon_0 cE(cB) = \epsilon_0 c^2 EB = \frac{EB}{\mu_0}$$

so that

$$\boxed{S = \frac{1}{\mu_0} EB} \tag{21.19}$$

This equation shows how the power delivered by the electromagnetic wave depends symmetrically on E and B.

Equation (21.19) gives the intensity of the electromagnetic wave at any instant of time when E and B are the *instantaneous* magnitudes of the electric and magnetic fields. To find the *average* value of S we must take averages of E and B over time. If the electric field takes the form $E = E_0 \sin \omega t$, the average value of E^2 is $\overline{E^2} = E_0^2/2$ (as for the effective value of an ac current or voltage; see Sec. 20.1). Hence

$$\boxed{\overline{S} = \epsilon_0 \frac{E_0^2}{2}\,c = \frac{E_0 B_0}{2\mu_0} = \frac{1}{\mu_0}\frac{B_0^2}{2}\,c} \tag{21.20}$$

The average value of S at a point is called the *intensity* of the radiation at that point. The intensity $\bar{S}$ of an electromagnetic wave therefore depends directly on the amplitudes of the E and B fields. The electric and magnetic fields each contribute exactly the same amount to the energy transported.

Example 21.5

For the wave described in Example 21.2 find the intensity of the wave using:
(a) The value given there for the rms value of the electric field.

(b) The result calculated there for the rms value of the magnetic field.

SOLUTION

(a) Since, from Eq. (21.18), $S = \epsilon_0 E^2 c$, the intensity is $\bar{S} = \epsilon_0 \overline{E^2} c$. But, by definition,

$$E_{rms}^2 = \overline{E^2} \quad \text{and so} \quad \bar{S} = \epsilon_0 E_{rms}^2 c$$

Hence we have

$$\bar{S} = \left(8.85 \times 10^{-12} \frac{C^2}{N \cdot m^2}\right)\left(5.00 \times 10^3 \frac{N}{C}\right)^2 (3.00 \times 10^8 \text{ m/s})$$

$$= 6.6 \times 10^4 \frac{N}{s \cdot m} = 6.6 \times 10^4 \frac{J}{s \cdot m^2} = \boxed{6.6 \times 10^4 \text{ W/m}^2}$$

The units are clearly correct, for a watt per square meter means that 1 J of energy per second is transported across a cross-sectional area of 1 m² perpendicular to the direction of the wave. This agrees with our definition of intensity.
(b) For all electromagnetic waves we must have $E = cB$, and so

$$S = \epsilon_0 c^2 \, B^2 c = \epsilon_0 \frac{1}{\mu_0 \epsilon_0} B^2 c = \frac{1}{\mu_0} B^2 c$$

Then $\quad \bar{S} = \frac{c}{\mu_0} \overline{B^2} = \frac{c}{\mu_0} B_{rms}^2$

and so $\quad \bar{S} = \left(\frac{3.00 \times 10^8 \text{ m/s}}{4\pi \times 10^{-7} \text{ N/A}^2}\right)(1.67 \times 10^{-5} \text{ T})^2$

$$= \left(6.6 \times 10^4 \frac{m/s}{N/A^2}\right)\left(1 \frac{N}{A \cdot m}\right)^2 = 6.6 \times 10^4 \frac{m}{s} \cdot \frac{N}{m^2}$$

$$= 6.6 \times 10^4 \frac{J}{s \cdot m^2} = \boxed{6.6 \times 10^4 \text{ W/m}^2}$$

This is the same result obtained in part (*a*), as must be the case.

21.6 The Complete Electromagnetic Spectrum

All electromagnetic waves have the same basic nature and travel at the same speed c in a vacuum, but different kinds of electromagnetic waves have different frequencies and wavelengths, with the wavelength and frequency related by the equation $\lambda f = c$. Electromagnetic waves in different frequency ranges are produced in different ways, are detected in different ways, and interact with matter in different ways. This is one example of a frequently occurring situation in physics in which quantitative changes (in this case in frequency and wavelength) lead to important qualitative differences. Figure 21.17 shows the *complete electromagnetic spectrum*, i.e., the complete range of electromagnetic waves. It gives the wavelengths, frequencies, and energies* corresponding to each part of the spectrum, and some indication of how electromagnetic waves of a particular frequency are produced and detected.

We will discuss the production and detection of radio waves later in this chapter, and visible radiation, x-rays, and gamma rays in subsequent chapters.

---

*The energies given here are the *photon* energies associated with particular frequencies. Their significance will be discussed in Chap. 26.

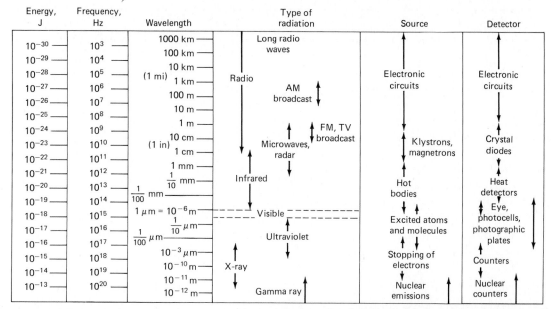

Energy, J	Frequency, Hz	Wavelength	Type of radiation	Source	Detector
10^{-30}	10^{3}	1000 km	Long radio waves	Electronic circuits	Electronic circuits
10^{-29}	10^{4}	100 km			
10^{-28}	10^{5}	10 km	Radio		
10^{-27}	10^{6}	(1 mi) 1 km	AM broadcast		
10^{-26}	10^{7}	100 m			
10^{-25}	10^{8}	10 m			
10^{-24}	10^{9}	1 m	FM, TV broadcast		
10^{-23}	10^{10}	(1 in) 10 cm	Microwaves, radar	Klystrons, magnetrons	Crystal diodes
10^{-22}	10^{11}	1 cm			
10^{-21}	10^{12}	1 mm	Infrared		
10^{-20}	10^{13}	$\frac{1}{10}$ mm		Hot bodies	Heat detectors
10^{-19}	10^{14}	$\frac{1}{100}$ mm			Eye, photocells, photographic plates
10^{-18}	10^{15}	$1\ \mu m = 10^{-6}$ m	Visible	Excited atoms and molecules	
10^{-17}	10^{16}	$\frac{1}{10}\ \mu m$	Ultraviolet		
10^{-16}	10^{17}	$\frac{1}{100}\ \mu m$			
10^{-15}	10^{18}	$10^{-3}\ \mu m$	X-ray	Stopping of electrons	Counters
10^{-14}	10^{19}	10^{-10} m			
10^{-13}	10^{20}	10^{-11} m	Gamma ray	Nuclear emissions	Nuclear counters
		10^{-12} m			

FIGURE 21.17 The complete electromagnetic spectrum. The energies associated with different frequencies are calculated from the equation $E = hf$, to be discussed in Chap. 26.

The complete electromagnetic spectrum stretches all the way from long radio waves with frequencies below 10 Hz to gamma rays with frequencies above 10^{22} Hz. Hence the high and low frequencies differ by a factor of about 10^{21}, or 2^{70}. In music, the stretch of frequencies between a fundamental note, or first harmonic, and the second harmonic is called an *octave*, since on a piano this stretch corresponds to the eight (octo) notes do, re, mi, fa, so, la, ti, do. For example, middle C on a piano (or *do*) has a frequency of 264 Hz. Its second harmonic has a frequency of 528 Hz, and the octave is said to include all frequencies from 264 to 528 Hz. If we apply this idea of octaves to the electromagnetic spectrum, there are approximately 70 octaves between a frequency of 10 Hz and a frequency of 10^{22} Hz, as we saw above. Of these 70 octaves, visible light, with frequencies from 3.9×10^{14} to 7.7×10^{14} Hz, occupies less than one octave. Still this one octave is of unique importance for life on earth, since human and animal eyes are sensitive only to this part of the complete electromagnetic spectrum. Electromagnetic radiation from the sun in this same frequency range is also crucial to the synthesis of carbohydrates from carbon dioxide and water by plants in the process we call *photosynthesis*.

The names given to the various regions of the electromagnetic spectrum are based on the way a particular kind of wave is produced and detected. There is considerable overlap in the designations used at the boundaries between the various regions of the spectrum. For example, short-wavelength radio waves with wavelengths below about 1 m are usually called *microwaves* or *radar* waves, and are produced in electron tubes known as *klystron* or *magnetron* tubes. In the region below 1 cm they blend into the *infrared* region of the spectrum, where the electromagnetic radiation is produced by the vibration and rotation of molecules inside very hot objects.

Similarly, electromagnetic waves with wavelengths of about 10^{-11} m are called either x-rays or gamma rays depending on how they are produced. If such radiation is produced by the sudden stopping of an electron beam, it is called x-radiation (or x-rays); if it is emitted from the nucleus of an atom, it is called gamma radiation (or gamma rays).

Example 21.6

Helium-neon lasers emit very intense red light at a wavelength of 0.633 μm.
(a) What is the frequency of this light?

(b) What is the frequency of the second harmonic of this radiation? In what region of the spectrum does it occur?

SOLUTION

(a) $f = \dfrac{c}{\lambda} = \dfrac{3.00 \times 10^8 \text{ m}}{0.633 \times 10^{-6} \text{ m}} = \boxed{4.74 \times 10^{14} \text{ Hz}}$

$\lambda_2 = \dfrac{3.00 \times 10^8 \text{ m/s}}{9.48 \times 10^{14} \text{ m/s}} = 0.316 \ \mu\text{m} = \boxed{316 \text{ nm}}$

(b) The second harmonic of this frequency is

$f_2 = 2(4.74 \times 10^{14} \text{ Hz}) = \boxed{9.48 \times 10^{14} \text{ Hz}}$

The wavelength of the second harmonic is then

This is in the *near-ultraviolet* range, since the visible region of the spectrum usually is considered to include wavelengths from about 390 nm (violet) to 770 nm (red).

21.7 Measurements of the Speed of Electromagnetic Waves

The precise measurement of the speed of electromagnetic waves constitutes one of the most interesting and important chapters in the history of experimental physics. The first successful measurement of the speed of light over a light path confined to the earth was carried out in 1849 by the French physicist Armand Fizeau (1819–1896). Fizeau set up a rapidly rotating toothed wheel on one hilltop and a mirror on another 5 miles away. Light from a strong source passed through a narrow gap between two adjacent teeth of the wheel, as in Fig. 21.18, traveled the 5 miles to the mirror, and was

FIGURE 21.18 Fizeau's rotating toothed-wheel apparatus for measuring the speed of light c.

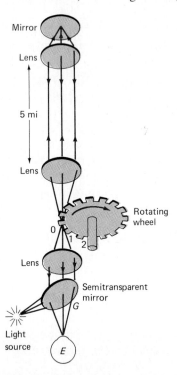

reflected back. If the wheel was turning at the right speed, the reflected light pulse returned just in time to pass through another gap (1 or 2 in the figure) between the teeth of the disk. From the wheel's speed of rotation Fizeau was able to calculate the time required for light to travel the 10 miles. His value for c was some 5 percent too high, but his success encouraged other physicists to improve on his basic method.

Measurements like Fizeau's are called *time-of-flight measurements*, i.e., measurements of the time required for a beam of electromagnetic radiation to travel a known distance. More accurate values of c can be obtained by *standing wave measurements*, i.e., by measuring the wavelength (λ) and the frequency (f) of electromagnetic waves, and then calculating c from the equation $c = \lambda f$. Almost all determinations of c are made in one of these two ways.

Time-of-Flight Measurements

The most famous measurements of the speed of light in the United States were those carried out by the American physicist A. A. Michelson (see Fig. 24.28 and accompanying biography) in the years 1878 to 1930, using a rotating mirror in place of Fizeau's toothed wheel. In 1923 Michelson had a 35-km path between two California mountain peaks, Mount Wilson and Mount San Antonio, surveyed to an accuracy of less than 1 in. He used an eight-sided revolving mirror, shown in Fig. 21.19, to time the travel of a light pulse over the 70-km round-trip path.

Michelson directed a beam of light at the revolving mirror as it rotated. Some of this light struck one face of the mirror at an angle so that it was reflected toward the distant mirror and then reflected back to the rotating mirror. If, in the time the pulse of light had traveled the distance $2d$, the mirror had rotated one-eighth of a turn, another mirror face would be in the right position to reflect light into an observing telescope. The speed of rotation of the mirror was varied until bright light was observed in the telescope. Then the time for light to travel $2d$ was 1/8 of the period of rotation of the eight-sided mirror (or perhaps $n/8$, with n an integer). Michelson's 1926 result for c was 299,796 ± 4 km/s, a value which agrees within his stated accuracy with the value accepted today.

An improvement on Michelson's method was introduced by E. Bergstrand in 1950. Bergstrand used a *geodimeter*, which replaced Michelson's rotating mirror with an electronic device that pulsed on and off at a variable frequency both the light beam and the detector. By varying this on-off frequency until maximum light intensity was obtained at the detector,

FIGURE 21.19 Michelson's eight-sided rotating mirror for measuring c.

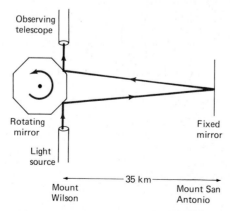

Bergstrand was able to time the passage of light to a distant reflector and back. He improved on both the accuracy and the precision with which c was known, his final result being $c = 299,792.7 \pm 0.25$ km/s. The geodimeter is now used routinely in land surveys, since, once the speed of light is known accurately, measuring the time required for light to travel an unknown distance leads to a precise value for that distance.

Standing Wave Measurements

The technique used for standing wave measurements of c is similar to that used to measure the speed of transverse waves on strings. We set up a standing wave pattern, measure the distance (d) between nodes to obtain the wavelength ($\lambda = 2d$), and from that and the known frequency obtain c, since $c = \lambda f$.

Before the developement of lasers it was not possible to apply this technique to infrared or visible radiation, since it was impossible to measure directly the frequency of such radiation. Important measurements were, however, made in the microwave and radio wave regions of the spectrum, where frequencies could be measured more easily.

For example, in 1958 K. D. Froome in Great Britain measured c using a microwave interferometer. His apparatus, which is shown in Fig. 21.20, consisted of a four-horn* microwave interferometer operating at 72 GHz ($\lambda = 4.2$ mm). Froome generated a beam of microwaves whose frequency could be measured to a very high accuracy using well-known microwave or radar techniques. This radiation was split into two beams and radiated from two horns toward a detector mounted on a cart which could be moved along a track. As the cart was moved along, positions of constructive and destructive interference were observed between the two beams. The distance moved by the cart between nulls was $\lambda/2$, since the motion of the cart through a distance of one-half wavelength decreased one path by $\lambda/2$ and increased the other path by the same amount. Since many nulls could be detected over the 8-m movement of the cart, a very precise value of λ could be obtained. From $c = \lambda f$ Froome found that c was equal to $299,792.50 \pm 0.10$ km/s, which improved on both the accuracy and the precision of Bergstrand's work.

The most accurate measurements of c ever made were those carried out at the National Bureau of Standards in the United States and at the National Physical Laboratory in England in the early 1970s. After Froome's measure-

FIGURE 21.20 Froome's microwave interferometer for measuring c.

---

*A horn is merely a piece of flanged hollow metal for guiding the microwaves out into the air from inside a metal pipe called a *waveguide*.

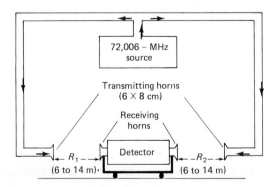

ment it became clear to many physicists that the ideal way to determine c was to extend Froome's method from the microwave to the infrared and visible regions of the spectrum and simultaneously measure frequency and wavelength for a monochromatic light source in a vacuum. A high-precision wavelength measurement could be made in this region, using well-established interferometric techniques (to be discussed in Sec. 24.8).

The real scientific breakthrough came when it became possible to measure the *frequencies* of infrared and visible laser lines to very high accuracy. This is done by taking microwave frequencies (at about 10^{11} Hz), which are accurately known in terms of the Cs (cesium) frequency standard (see Sec. 2.1), and multiplying them into the infrared region of the spectrum, using a variety of advanced electronic techniques. These infrared frequencies (at about 10^{13} Hz) are then compared with higher infrared frequencies by similar techniques, resulting in the measurement of the frequency of near-infrared lines (at about 10^{14} Hz) to almost the accuracy of the Cs frequency standard.

The value obtained from these measurements, $c = 299,792.458 \pm 0.001$ km/s, shows an improvement in accuracy of two orders of magnitude over Froome's work and is one of the most precise determinations of a physical constant ever made.

Some of these results for the speed of light were summarized in Table 1.1.

Example 21.7

In the measurement of c using an eight-sided rotating mirror, Michelson found that the mirror rotated 1/8 rev in the time it took light to travel to a mirror 35 km away and return to the rotating mirror. From the known value of c, what was the frequency of revolution of the mirror in Michelson's experiment?

SOLUTION

The time for one complete revolution of the eight-sided mirror is $T = 1/f$, where f is the number of revolutions per second of the mirror. The time for a rotation of 1/8 rev is therefore $t = 1/8f$. Hence

$$c = \frac{d}{t} = \frac{70 \times 10^3 \text{ m}}{1/8f} = (70 \times 10^3 \text{ m})\,(8f)$$

Hence

$$f = \frac{2.998 \times 10^8 \text{ m/s}}{560 \times 10^3 \text{ m}} = \boxed{535 \text{ rev/s}}$$

21.8 The Generation of Electromagnetic Waves: Radio Waves

Let us now consider how electromagnetic waves are generated and detected. These processes differ from one region of the electromagnetic spectrum to another, but as a useful example we will here consider the generation of waves in the *radio-frequency* region of the spectrum, i.e., waves with wavelengths of about 100 m.

We have already seen that an *RLC* circuit has a resonant frequency $f = (1/2\pi)(1/\sqrt{LC})$. If we feed energy into such a circuit at this frequency, we build up large oscillations of energy between the electric field of the capacitor and the magnetic field of the inductor. These oscillations still do not constitute an electromagnetic wave, since the electric and magnetic fields are largely confined to separate regions of space (the regions around the capacitor and the inductor) and do not overlap to any extent. To produce electromagnetic waves we need overlapping electric and magnetic fields to allow a changing electric field to produce a magnetic field, and a changing magnetic field in turn to produce an electric field, as required to sustain the waves.

The Electric Field

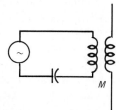

FIGURE 21.21 A dipole antenna.

To produce overlapping electric and magnetic fields we use an antenna. The type shown in Fig. 21.21 is called a *dipole antenna*, because of its similarity to the hertzian dipole of Fig. 21.15. It consists of a long rod which is connected through a mutual inductance M to an oscillating circuit feeding in energy at a frequency f.

The net distribution of electric charge on such an antenna will vary periodically in time with a period $T = 1/f$. Figure 21.22 shows the charge distribution at intervals of $T/4$ from $t = 0$ to $t = T$. At $t = 0$ the entire antenna is assumed to be electrically neutral and the electric field **E** at point P along the X axis is then zero. A time $T/4$ later, however, the oscillator has moved electrons from the bottom to the top of the antenna, so that the top becomes electrically negative and the bottom electrically positive, as in Fig. 21.22b. The electric field **E** at point P is therefore the vector sum of the forces exerted by the net positive and negative charges on a unit positive charge at point P, and is directed vertically upward, as shown in the figure. A time $T/4$ later the oscillator has driven a sufficient number of electrons to the bottom of the rod to make top and bottom both electrically neutral again, and the field zero, as in Fig. 21.22c. This process then repeats itself in the opposite direction, and at time $3T/4$, **E** has the same magnitude as in (b), but is in the opposite direction.

FIGURE 21.22 Charge distribution on a dipole antenna at intervals of $t = T/4$. The electric field **E** at point P oscillates in time as shown.

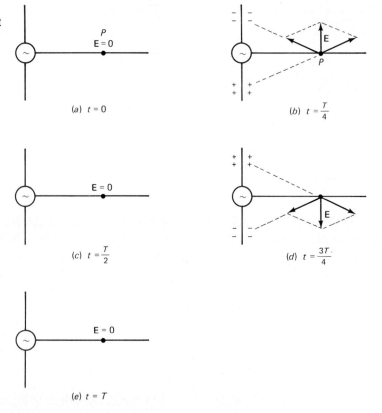

When $t = T$, the oscillator is as it was initially, and the electric field is again zero.

The crucial point here is that these electric field lines are not static. As the oscillator drives charge up and down the antenna, these electric field lines move away from the antenna with a speed c. Hence the electric field configuration along the positive X axis will look like that in Fig. 21.23, where the five points marked correspond to the five times discussed in relation to Fig. 21.22. It is clear that such an electric field distribution is identical to that previously shown in Fig. 21.11. If the wavelength is 100 m, then this is the distance between points a and e in Fig. 21.23.

Note that the electric field lines no longer begin and end on charges once the wave is radiated into space. Rather the field lines form continuous curves in space, as indicated in Fig. 21.24. The electric field shown in Fig. 21.23 is only the strength of the electric field along the positive X axis. The dotted lines in Fig. 21.24 attempt to show the complete lines of the electric field. Notice how all lines of **E** form closed curves.

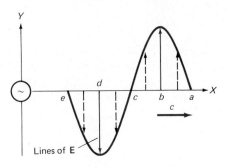

FIGURE 21.23 Electric field **E** produced by a dipole antenna as a function of distance from the antenna along the X axis.

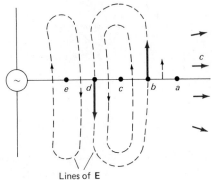

FIGURE 21.24 Complete lines of the electric field for a traveling electromagnetic wave. The lines of **E** form closed loops in space.

The Magnetic Field

While the oscillator is moving charge up and down the antenna in the Y direction, it is also producing a changing magnetic field in the Z direction, since, as the oscillator goes through one period T, the conventional current goes from a maximum in one direction to zero to a maximum in the other direction and then back to zero again. What happens is shown in Fig. 21.25, where the right-hand rule can be used to obtain the direction of **B** from the indicated direction of the conventional current I. The lines of the magnetic field **B** form closed circles around the antenna rod and are always at right angles to the lines of the electric field **E**.

We see then that the magnetic field **B** also goes through a complete cycle

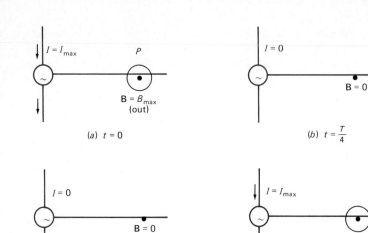

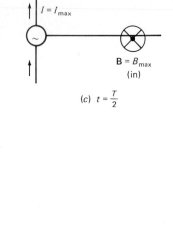

FIGURE 21.25 Magnetic field **B** produced by a dipole antenna as a function of time. The magnetic field at P oscillates in time as shown.

FIGURE 21.26 Magnetic field **B** produced by a dipole antenna as a function of distance from the antenna along the X axis.

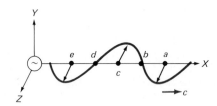

in one period T. Again this magnetic field moves through space at a speed c, and is diagramed in Fig. 21.26. Since the electric field is in the Y direction and the magnetic field in the Z direction when the wave is moving along the positive X axis, the electric field **E**, the magnetic field **B**, and the direction of propagation are mutually perpendicular to one another, as must be the case for electromagnetic waves in free space.

Phase Relations

There is one problem with the oversimplified model we have just used. Since the current and the charge flowing in the antenna are 90° out of phase, the electric and magnetic fields near the antenna are also 90° out of phase, as is clear from Figs. 21.23 and 21.26. As the wave spreads away from the antenna, however, **B** and **E** gradually come into phase with each other, because the fields due to the oscillating charges and alternating currents have a different dependence on distance from their source. Putting Figs. 21.23 and 21.26 together, we then obtain Fig. 21.27, which is identical with our previous Fig. 21.11.

We therefore conclude:

Radio waves are electromagnetic waves produced by the motion of accelerated electric charges.

The accelerated motion of electrons up and down a straight-rod dipole antenna therefore produces long-wavelength electromagnetic waves which satisfy the predictions of Maxwell's equations: the electric and magnetic fields are periodic in both time and space, are transverse and mutually perpendicular, and move with the speed of light c.

Electromagnetic radiations in the microwave and x-ray regions of the spectrum are also produced by the acceleration of electrons. It is the *acceleration* which is the real source of the radiated electromagnetic energy. To accelerate the electrons requires a force, and this force does work in moving the electrons. For radio waves this work is provided by the oscillator, and the energy input to the oscillator ultimately becomes the energy radiated.

FIGURE 21.27 Combination of the electric and magnetic fields shown for an electromagnetic wave moving along the positive X axis in Figs. 21.23 and 21.26. The result is the electromagnetic wave of Fig. 21.11.

21.9 The Detection and Modulation of Radio Waves

Electromagnetic waves in the radio region can be detected using an antenna similar to the transmitting antenna of Fig. 21.21. We set up a straight-line dipole antenna in the direction of the electric field radiated by the antenna, as in Fig. 21.28. The incident ac field sets electrons oscillating in the antenna, which generates a current in the tunable LC circuit connected to the antenna. By varying the capacitance C the circuit can be tuned to resonance and large currents produced. These can then be amplified and made to produce the audio or video signals required for radio and television.

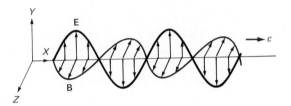

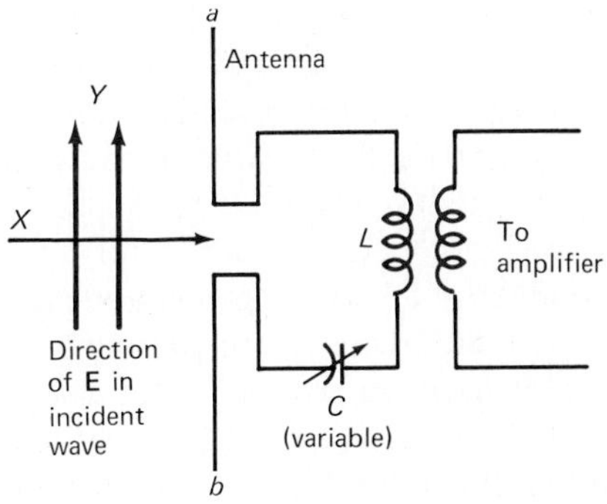

FIGURE 21.28 A straight-line dipole receiving antenna. The dipole ab is oriented in the direction of the electric field **E,** which in this case is in the Y direction.

Magnetic Detection

Radio waves can also be detected by using a loop antenna to pick up the signal from the oscillating magnetic field of the wave, as in Fig. 21.29. If the magnetic field is in the XZ plane and the loop is in the XY plane, so as to be

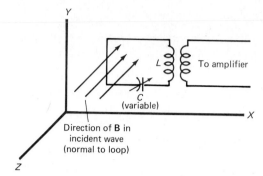

FIGURE 21.29 A loop antenna for receiving radio waves. The plane of the loop is at right angles to the direction of the magnetic field **B**, which in this case is in the Z direction.

normal to the lines of B, then as the wave passes by the loop, the magnetic flux through the loop changes with time and induces an emf in the loop. Again in this case a tunable ac circuit connected to the loop can lead to sizable currents, which can then be amplified further and used to produce sound or visible signals.

Both the dipole and loop antennas are highly directional because, for maximum signal reception, the dipole antenna must be lined up parallel to the electric field and the loop antenna must be at right angles to the magnetic field. A dipole antenna oriented perpendicular to the electric field of the oncoming electromagnetic wave, or a loop antenna with its plane parallel to the magnetic field of the wave, will detect no signal.

Modulation

Much of the information transmitted by radio consists of voices and music at audio frequencies in the range 20 to 20,000 Hz. The latter frequency corresponds to a wavelength for electromagnetic waves of

$$\lambda = \frac{3 \times 10^8 \text{ m/s}}{2 \times 10^4/\text{s}^{-1}} = 1.5 \times 10^4 \text{ m}$$

For good transmission and reception dipole antennas must be about $\lambda/2$ in length, which means that, for audio frequencies of 20,000 Hz, we would need antenna lengths of about 7,500 m. This is clearly impractical. To overcome this difficulty information at audio frequencies is used to *modulate* higher-frequency electromagnetic waves at the transmitting end.

Definition

Modulation: The superposition of an audio (or video) signal on a higher-frequency carrier wave so that the information contained in the signal can be transmitted by the carrier wave at a speed c.

In practice the carrier frequency for an FM radio station is about 10^8 Hz, corresponding to a wavelength of about 3 m. At the receiving end the audio signal is extracted from the higher-frequency carrier wave and the information is converted to useful form. This means that the transmitting and receiving antennas can be designed for wavelengths much shorter than those in the audio range, making them more compact and practical.

There are two different kinds of modulation used to transmit information at low frequencies using a higher-frequency carrier wave; amplitude modulation and frequency modulation.

Amplitude Modulation   With amplitude modulation (AM) the *amplitude* of a fixed-frequency carrier wave is modified, or "modulated," to transmit the

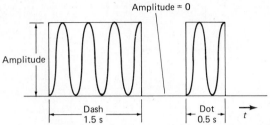

FIGURE 21.30 Amplitude modulation (AM). The amplitude of the carrier wave is varied over time to carry the desired information. Here amplitude modulation is being used to transmit the dots and dashes of Morse code.

desired information. For example, the dots and dashes of Morse code can be sent in this way by transmitting pulses of electromagnetic radiation of different lengths, as in Fig. 21.30. In this case the amplitude of the modulation varies between zero and some constant value.

Frequency Modulation   With frequency modulation (FM) the amplitude of the carrier wave is not changed, but its *frequency* is expanded or contracted to reflect the information being transmitted. To transmit dots and dashes in this way, the carrier frequency might be switched from 1 to 1.5 MHz and kept there for 1.5 s to indicate a dash, or for 0.5 s to indicate a dot, as in Fig. 21.31. The same technique can be used, with some modifications, to transmit music or the human voice.

FIGURE 21.31 Frequency modulation (FM). The frequency of the carrier wave is varied over time to carry the desired information. Here frequency modulation is being used to transmit the dots and dashes of Morse code.

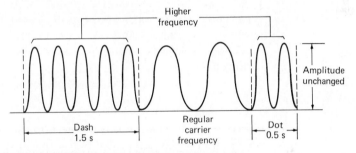

21.10 Practical Radio and Television

Figure 21.32 shows the various steps required to transmit and receive voice or music by means of radio. A microphone first converts the sound into an audio-frequency (AF) electric signal, which is then amplified and fed into a mixer. The mixer takes the fixed-frequency, fixed-amplitude output of a radio-frequency (RF) oscillator and superimposes the AF signal on it. This modulation may be either AM or FM. The modulated RF signal is then amplified further and fed to an antenna from which it radiates outward at speed c.

At the receiving end the modulated RF signal is picked up by an antenna and fed to an RF tuner and amplifier. The tuner is a variable-frequency RLC circuit whose frequency can be tuned to resonance with the incoming RF frequency. This RF signal is then amplified and fed to a demodulator, which functions in a reverse fashion from the mixer. The demodulator first rectifies the signal, and then passes low frequencies (the AF information) on to an AF amplifier, but bypasses the high-frequency RF to ground, since it is no longer needed. The AF signal is then further amplified and fed to a speaker which converts it back into audible sound.

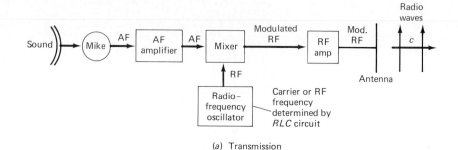

(a) Transmission

FIGURE 21.32 Schematic of a system for transmitting and receiving radio signals. Here AF signifies *audio frequency* and RF means *radio frequency*. The carrier wave operates at a radio frequency; the information being carried is at audio frequencies.

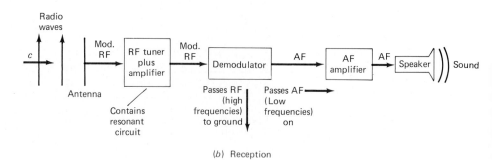

(b) Reception

The process for picture transmission by television is basically the same, with a TV camera and TV receiver replacing the radio microphone and speaker. Television sound is usually transmitted by FM, while the video is transmitted by AM.

The approximate frequency ranges of radio and TV bands are shown in Table 21.1.

TABLE 21.1 Frequencies Used in Radio and Television

Type of electromagnetic wave	Frequency range
Audio frequencies (AF)	20–20,000 Hz
AM (amplitude-modulation) radio	540–1600 kHz
CB (citizen's band)	About 27 MHz
TV: Channels 2 to 6	54–88 MHz
FM (frequency-modulation) radio	88–108 MHz
TV: Channels 7 to 13	174–216 MHz
UHF (Ultrahigh frequency)	470–890 MHz

Expansion of Communication Channels

Every television channel has its own distinctive carrier frequency which, to avoid interference, must be well separated from all other allowed TV frequencies. For this reason the assigned width of a standard TV channel is 4 MHz. If we assume that this channel width remains fixed throughout the electromagnetic spectrum, then as we go to higher and higher frequencies, one octave will contain a larger and larger number of possible transmission frequencies. Thus in the octave from 60 to 120 MHz there is room for 15 TV channels, each 4 MHz wide. Three octaves up the scale, in the frequency range 480 to 960 MHz, one octave can contain 120 conventional TV channels, or 8 times as many, of the same 4-MHz width.

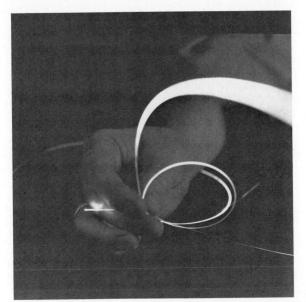

FIGURE 21.33 Optical fibers, illuminated by laser light, being used to transmit information. Twelve such fibers are embedded in plastic to form a flat ribbon. Twelve such ribbons stacked in a cable can transmit over 40,000 voice messages simultaneously. *(Photo courtesy of Bell Telephone Laboratories.)*

The number of TV channels included in one octave of the electromagnetic spectrum thus doubles with the doubling of frequency which occurs as we go from one octave to the next. Since the frequencies of visible light (about 10^{14} Hz) are about a million times higher than those now used for TV transmission (about 10^8 Hz), this means that a million times more information channels are available in one octave of the visible electromagnetic spectrum than in one octave of the TV frequency range.

With the advent of lasers to provide the needed carrier waves, the possibility of using visible light to transmit voices and other information is fast becoming a reality. This opens up vast numbers of new channels of communication, so vast that interference between channels will become virtually impossible.

However, even laser light cannot penetrate buildings, heavy clouds, or other obstacles. But this sort of interference can be avoided by using optical fibers or light pipes (see Sec. 23.10 and Fig. 21.33) to transmit laser light through cables either above or below ground. The huge increase in the number of messages that can be sent simultaneously over such cables promises to revolutionize the communications industry.

Summary: Important Definitions and Equations

Maxwell's equations:

Four equations containing most of the important laws of electricity and magnetism, and leading to the prediction that electromagnetic waves exist and travel at a speed c.

$$1 \quad \sum E_\perp \, \Delta A = 4\pi k_e \sum q$$

$$2 \quad \sum B_\perp \, \Delta A = 0$$

$$3 \quad E = -\frac{1}{2\pi r} \frac{\Delta \Phi_B}{\Delta t}$$

$$4 \quad \sum B_\parallel \, \Delta l = \mu_0 I + \mu_0 \epsilon_0 \frac{\Delta \Phi_E}{\Delta t}$$

Maxwell's displacement current:

$$I_D = \epsilon_0 \frac{\Delta \Phi_E}{\Delta t} = \epsilon_0 A \frac{\Delta E_\perp}{\Delta t}$$

where $E_\perp$ is the component of the electric field strength **E** perpendicular to the area A.

Electromagnetic waves:

Transverse waves in which the sinusoidal electric and magnetic fields are at right angles to each other and to the direction of propagation, and which travel through free space at a speed $c = 1/\sqrt{\mu_0 \epsilon_0} = 2.998 \times 10^8$ m/s. For all electromagnetic waves in a vacuum $c = \lambda f$, where λ is the wavelength in meters and f is the frequency in hertz.

Speed of electromagnetic waves in dielectric materials:

$$v = \frac{c}{\sqrt{K}}$$

where K is the dielectric constant.

Relationship between magnitudes of **E** field and **B** field:

$$E = cB$$

Intensity of an electromagnetic wave: The energy transported per unit time across unit cross-sectional area perpendicular to the direction in which the wave is moving.

Average intensity: $\bar{S} = \epsilon_0 \dfrac{E_0^2}{2} c = \dfrac{E_0 B_0}{2\mu_0} = \dfrac{1}{\mu_0}\dfrac{B_0^2}{2} c$

where E_0 and B_0 are the amplitudes of the periodic **E** and **B** fields.

Complete electromagnetic spectrum

All the electromagnetic waves from radio waves (low frequency, long wavelength) to gamma rays (high frequency, short wavelength).

Radio waves:

Electromagnetic waves of frequencies between 10^1 and 10^8 Hz produced by the accelerated motion of electrons up and down a straight-rod dipole antenna.

Modulation: The superposition of an audio (or video) signal on a higher-frequency carrier wave so that the information contained in the signal can be transmitted by the carrier wave at a speed c.

Amplitude modulation (AM): The amplitude of the carrier wave is varied to carry the desired information.

Frequency modulation (FM): The frequency of the carrier wave is varied to carry the desired information.

Questions

1 A very long, thin bar magnet has its north pole 1.0 m from its south pole. A spherical surface is constructed which surrounds the north pole and intersects the magnet 0.30 m from its end, as in Fig. 21.34. What is the net flux of the magnetic field through this gaussian surface? Why?

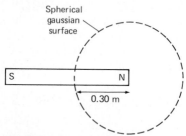

FIGURE 21.34 Diagram for Question 1.

2 What was the key idea first introduced by Maxwell in putting forth his equations to describe electricity, magnetism, and their interactions? Why was this idea so important?

3 Point out three ways in which sound waves differ from electromagnetic waves and three ways in which a wave traveling along a stretched string differs from electromagnetic waves.

4 Electromagnetic waves transport energy through space. Would you expect that they would also transport momentum? Give reasons for your answer.

5 Can you suggest why gamma rays differ so radically from radio waves in their properties, even though they are both part of the same electromagnetic spectrum?

6 X-rays are produced when electrons moving at high speeds collide with a metal target. Can you suggest the cause of the continuous x-ray spectrum produced?

7 Why should you expect standing wave measurement to be more accurate than time-of-flight measurement of the speed of light in a vacuum?

8 Why would you expect less "static" and other kinds of interference on FM radio than on AM radio?

9 A radio transmitter has a vertical antenna, i.e., one directed along the Y axis, as in Fig. 21.21. At the receiving end it is desired to detect the signal magnetically by placing a loop of wire in the direction which will yield maximum signal. If the radio wave is being propagated along the positive X axis, should the loop of wire be placed in the XY plane, the XZ plane, or the YZ plane?

10 It is possible to send smoke signals using a Morse code of dots and dashes. Is this equivalent to AM or to FM in the case of electromagnetic waves? Why?

11 Why are the "rabbit ears" on some television sets about a meter long when fully extended?

Problems

A 1 The electric field between the plates of an air capacitor of plate area 0.050 m^2 is changing at a rate of 10^4 V/(m·s). Maxwell's displacement current is, in this case:
(a) 0   (b) 10^4 A   (c) 4.4×10^{-11} A
(d) 4.4×10^{-10} A   (e) 4.4×10^{-9} A

A 2 An electromagnetic wave is traveling through free space. At a particular instant of time the electric field at a point along the path of the wave is 4.2×10^3 V/m. At the same time and at the same place the magnetic field is:
(a) 1.4×10^{-5} G   (b) 1.4×10^{-5} T
(c) 1.4×10^{-7} T   (d) 4.2×10^{-7} T
(e) 0

A 3 At low frequencies the dielectric constant of Pyrex glass is 4.5. The speed of low-frequency radio waves in Pyrex should be about:
(a) 3.0×10^8 m/s   (b) 14×10^8 m/s
(c) 0.67×10^8 m/s   (d) 6.4×10^8 m/s
(e) 1.4×10^8 m/s

A 4 The speed of light in a piece of light flint glass is 1.9×10^8 m/s. The wavelength of the red light ($\lambda = 633$ nm in a vacuum) emitted by a He–Ne laser and passing through the flint glass is:
(a) 633 nm   (b) 999 nm   (c) 401 nm
(d) 78 nm   (e) 5142 nm

A 5 At the time and place given in Prob. A2, the intensity of the electromagnetic wave has a magnitude:
(a) 4.2×10^1 V/m   (b) 17.6×10^6 V^2/m^2
(c) 4.7×10^4 W/m^2 (d) 4.7×10^4 J/m^2 (e) 0

A 6 The intensity of the electromagnetic wave described in Prob. A5, if calculated using the magnetic field instead of the electric field, will be:
(a) The same as in A5
(b) The same as in A5, on the average, but not at any instant of time
(c) Smaller than in A5
(d) 0
(e) Greater than in A5

A 7 Compared with radio waves of frequency 10 Hz, gamma rays of 10^{22} Hz have frequencies higher by:
(a) A factor of 21   (b) 21 orders of magnitude
(c) 2^{21} (d) 10^{-21} (e) None of the above

A 8 The number of octaves between radio waves of frequency 10 Hz and gamma rays of frequency 10^{22} Hz is:
(a) 21   (b) 70   (c) 10^{70} (d) 23
(e) None of the above

A 9 The sodium atom has a spectrum with two very close spectral lines (a doublet) in the yellow region with a wavelength at about 589 nm. The third harmonic of these two lines occurs in the following region of the spectrum:
(a) Yellow   (b) Infrared   (c) Microwave
(d) Ultraviolet   (e) X-ray

A10 A state trooper's car sends out a radar signal at a frequency of 10.5 GHz. The wavelength of this signal is about:
(a) 3 m   (b) 30 m   (c) 0.3 m   (d) 0.003 m
(e) None of the above

B 1 An air capacitor with plates of area 200 cm^2 and a gap of 2.0 mm between the plates is being charged by a battery at a rate of 1.0×10^{-8} C/s. What is the rate at which the electric field between the plates is changing?

B 2 In the situation of Prob. B1 what is the displacement current while the capacitor is being charged?

B 3 Heinrich Hertz produced what we now call radio waves, with a wavelength of about 3.0 m. What was the frequency of the oscillating electric charges producing these radio waves?

B 4 A radar wave has associated with it a magnetic field of rms value 2.0×10^{-6} T. What is the rms value of the electric field associated with the same radar wave?

B 5 For light waves the index of refraction $n \, (= \sqrt{K})$ of flint glass is 1.58. What is the speed of light in this glass? Is it greater or less than in air?

B 6 A radio wave traveling through air consists of a sinusoidal electric field of rms value 6.0×10^2 V/m and a magnetic field perpendicular to this electric field. How much energy is stored on the average in a unit volume of space as this wave passes through it?

B 7 WGMS-FM in Washington, D.C., transmits at a frequency of 103.5 MHz.
(a) To what wavelength does this correspond?
(b) What is the wavelength corresponding to the second harmonic of this frequency?

B 8 What is the frequency band corresponding to the microwave region of the spectrum, extending from wavelengths of about 1 mm to 1 m?

B 9 How long did it take light to travel the 70-km round trip between Mount Wilson and Mount San Antonio in Michelson's 1923 measurement of the speed of light?

B10 A geodimeter is used to measure the distance between two points A and B in the Mohave Desert. It is found that it takes light 2.4016×10^{-4} s to make the trip from A and B and back again. What is the distance between A and B?

C 1 An air capacitor with plates of area 100 cm^2 and an air gap of 0.50 mm between the plates is being charged by a battery at a rate of 2.0×10^{-9} C/s.
(a) What is the capacitance of the capacitor?
(b) What is the current in the circuit while the capacitor is being charged?
(c) What is the displacement current?
(d) What is the magnetic field at a distance 20 cm from a wire carrying current from the battery?
(e) What is the magnetic field at a distance 20 cm from the center of the capacitor plates?

C 2 A straight wire 2.0 m long is oriented in the direction of the electric field vector **E** in an electromagnetic wave.
(a) To produce a voltage difference between the ends of the wire of maximum value 10^{-2} V, how large must the maximum electric field of the wave be?

(*b*) For the same wave what will be the maximum magnetic field?

C 3 A 60-Hz radio wave passes from air into a block of quartz of dielectric constant 4.3.

(*a*) What is the speed of the radio wave in the quartz?

(*b*) What is the wavelength of the radio wave in the quartz?

(*c*) What is the frequency of the radio wave in the quartz?

C 4 An electromagnetic wave is moving through a material for which $K = 8$ and $K_m = 1200$ at the frequency of the wave.

(*a*) Find the speed of propagation of the wave through the material.

(*b*) If the frequency is 2.0 kHz, find the wavelength of the wave.

C 5 A light wave traveling through air consists of a sinusoidal electric field of rms value 30 V/m and a magnetic field perpendicular to this electric field. Find the intensity of this wave, using:

(*a*) The rms value of the electric field.

(*b*) The rms value of the magnetic field.

C 6 How much energy is transported across a 10-cm^2 area in 1 min by an electromagnetic wave whose electric field **E** has an rms value of 100 V/m?

C 7 About how many octaves are occupied:

(*a*) By the radio region of the complete electromagnetic spectrum?

(*b*) By the x-ray region of the spectrum?

C 8 A ruby laser emits a strong red light at a wavelength of 0.694 μm. What is the wavelength of the second harmonic of this radiation?

C 9 An amateur ("ham") radio operator desires to construct a tuner to receive wavelengths in the neighborhood of 20 m. She has available a capacitor with a capacitance variable between 20 and 80 pF. If she decides to wind an inductance coil to use in the tuning circuit, what must be the inductance of such a coil?

C10 An *RLC* circuit with $L = 1.0$ mH, $C = 10$ μF, and $R = 0.50$ Ω, resonates at its natural frequency. If this circuit is attached to a dipole antenna, what is the wavelength of the radiation produced?

C11 An FM receiver has a dial marked with frequencies from 88 to 108 MHz. If the fixed inductance in the circuit is 1.0 μH, what range of values must the variable capacitor cover to tune the receiver over the entire frequency range?

C12 What would be the difference in the wavelengths of the electromagnetic waves used to transmit information at audio frequencies (about 10^3 Hz) and at radio frequencies (about 10^6 Hz)?

C13 What is the wavelength range covered by TV channels 7 to 13?

C14 On the Fourth of July a large fireworks display is set off at a distance of 1.5 km from an observer. How long after the observer sees the visual display will the sound of the initial explosion be heard? (Assume that the speed of sound is 340 m/s.)

C15 Who will hear the voice of a radio announcer first, a woman in the studio audience 40 m away fom the announcer whose voice reaches her directly, or her husband who is listening to the radio in their home 160 km from the radio studio? By how much?

C16 In Bohr's original model of the hydrogen atom the electron circled the proton 3.6×10^{15} times per second.

(*a*) If this were the source of the radiation emitted by the hydrogen atom, what would be the wavelength emitted by the atom?

(*b*) In what region of the spectrum would this wavelength occur?

(*c*) How could such a model explain the fact that the hydrogen atom emits not one wavelength, but many different ones?

D1 (*a*) From the expression $c = 1/\sqrt{\mu_0\epsilon_0}$, prove that c has the dimensions of a velocity.

(*b*) From the expression $c = E/B$, prove that c has the dimensions of a velocity.

D2 (*b*) Show that the expression ϵ_0E^2 has the dimensions of energy per unit volume.

(*b*) Show that the expression ϵ_0E^2c has the proper dimensions for the intensity of an electromagnetic wave.

D3 A circular loop antenna has a cross-sectional area of 0.050 m^2 and consists of 200 turns. An electromagnetic wave of frequency 10 MHz passes through the loop antenna. If the maximum value of the electric field in the wave is 5.0×10^{-2} V/m, what is the approximate amplitude of the emf generated in the loop antenna?

D4 A parallel-plate capacitor consisting of two circular plates of area $A = \pi R^2$ separated by air is being charged by a constant current I.

(*a*) Show that the magnetic field between the capacitor plates at a distance r from the line connecting the centers of the two plates is given by $B = \mu_0Ir/2A$ for r less than R. (*Hint*: Use Ampère's circuital law and Maxwell's displacement current.)

(*b*) Show that for r equal to R this expression reduces to the usual expression for the magnetic field at a distance R from a long, straight wire.

D5 The solar constant is the rate at which electromagnetic energy radiated by the sun falls on the earth's atmosphere. Its value, at high noon, for radiation falling on the atmosphere at right angles to the earth's surface, is 1.35 kW/m^2.

(*a*) Find the rms values of the electric **E** and the magnetic field **B** needed to produce this energy flow from the sun to the earth.

(*b*) If the distance from the sun to the earth is 1.5×10^{11} m, find the total power radiated by the sun.

D6 A 100-W light bulb emits about 10 percent of its 100-W input as visible light which is radiated uniformly in all directions. Calculate the rms values of the electric field **E** and the magnetic field **B** for the light wave at a distance of 3.0 m from the bulb.

D7 It has been suggested that, to reduce energy losses in electric power cables, electricity should be transmitted through space in the form of electromagnetic waves. Suppose that we wanted to transmit an amount of electric power comparable to that handled by modern transmission lines, say, 1000 A at 750 kV, and to do this using a beam of 50-m^2 cross-sectional area.

(*a*) What would the rms values of E and B in the electromagnetic wave have to be?

(*b*) Does this strike you as a realistic scheme?

Additional Readings

Hertz, Johanna (ed.): *Heinrich Hertz: Memoirs, Letters, Diaries*, San Francisco Press, San Francisco, 1977. A fascinating account of the education and development of a great physicist, as told mostly in his own words.

Jaffe, Bernard: *Michelson and the Speed of Light*, Doubleday Anchor, Garden City, N.Y., 1960. A brief biography of Michelson, emphasizing his life-long obsession with the measurement of *c*.

MacDonald, D. K. C.: *Faraday, Maxwell and Kelvin*, Doubleday Anchor, Garden City, N.Y., 1964. Contains some very interesting biographical material and photographs relating to Maxwell's life.

Morrison, Philip, and Emily Morrison: "Heinrich Hertz," *Scientific American*, vol. 197, no. 6, December 1957, pp. 98–106. A brief biography of a man who accomplished a great deal in a short time.

Newman, James R.: "James Clerk Maxwell," in *Lives in Science*, a *Scientific American* book, Simon and Schuster, New York, 1957, pp. 155–180. A brief account of Maxwell and his contributions to physics.

Shamos, Morris H.: *Great Experiments in Physics*, Holt, Rinehart and Winston, New York, 1965. Contains excerpts both from Maxwell's papers on electromagnetic theory and Hertz's papers on the measurement of the properties of electromagnetic waves.

Tolstoy, Ivan: *James Clerk Maxwell—A Biography*, University of Chicago Press, Chicago, 1982. A brief, nontechnical biography of Maxwell.

Chapter 22

The Wave Nature of Light: Diffraction, Interference, and Polarization

Before the middle of the nineteenth century heated controversies about the nature of light had marked the history of physics. Some scientists, like Newton, thought that light was the motion of particles in a beam; others believed that it was a wave motion. When Maxwell proposed his electromagnetic theory in 1865 and predicted a speed $c = 2.998 \times 10^8$ m/s for all electromagnetic waves, a value equal to the then-known speed of visible light, it appeared that light had finally been proved to be an electromagnetic *wave* similar in most respects to radio waves but of a much higher frequency.

The place of visible light in the complete electromagnetic spectrum is shown in Fig. 22.1. The wavelengths of visible light range from about 3.9×10^{-7} to 7.6×10^{-7} m. Often light wavelengths are given in units of micrometers (1 μm = 10^{-6} m), nanometers (1 nm = 10^{-9} m), or angstroms (1 Å = 10^{-10} m). Thus the visible spectrum runs from 390 to 760 nm or from 3900 to 7600 Å. (It is helpful to remember that most atoms are 1 to 2 Å in diameter.)

Although Maxwell's work was far from the last shot in the battle over the nature of light, as we will see in Chap. 26, a wave theory does explain extremely well many observed optical phenomena including reflection, refraction, diffraction, interference, and polarization. In this chapter we will apply a wave theory of light to these phenomena.

FIGURE 22.1 Place of visible light in the complete electromagnetic spectrum. It occupies less than one octave out of 70, and is bounded by the infrared on the low-frequency (long-wavelength) side, and by the ultraviolet on the high-frequency (short-wavelength) side.

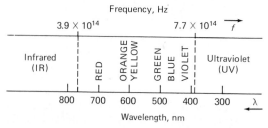

22.1 Huygens' Principle

In 1670, long before Maxwell's day, Christian Huygens (see Fig. 7.7 and accompanying biography) had proposed a wave theory of light. Although his work has now been both justified and replaced by Maxwell's more comprehensive theory, it contains one key idea which is very useful in predicting how light will behave in particular physical situations. This is Huygens' principle:

Huygens' Principle

Every point on a wave front may be considered as a source of secondary spherical wavelets which spread out in the forward direction at the speed of light in the medium. The new wave front is then the surface tangent to all these secondary wavelets.

Definition

A wave front is a surface joining all adjacent points on a wave which have the same phase. *

Thus if we have a plane electromagnetic wave moving to the right along the X axis, as in Fig. 22.2, we can choose as our wave fronts those positions on the wave which have the same phase and where the electric field **E** has its maximum value. Consecutive wave fronts like AA' and BB' are then planes normal to the X axis and separated by one wavelength λ, since λ is by definition the distance between consecutive points of equal phase.

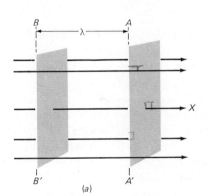

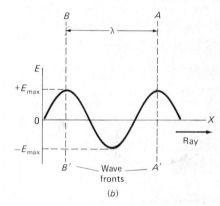

FIGURE 22.2 A plane electromagnetic wave: (*a*) The planes represent wave fronts spaced a wavelength λ apart, and the arrows indicate rays. (*b*) The variation of the electric field E for such a plane wave. The wave fronts are shown separated by the distance between the maximum positive values of E.

For a spherical electromagnetic wave the wave fronts AA' and BB' traveling in any given direction are parts of spherical surfaces, again separated by a distance λ, as shown in Fig. 22.3. In both these cases *rays* can be drawn at right angles to the wave fronts to indicate the direction in which the wave is actually traveling. For plane waves the rays are directed along the positive X axis; for spherical waves they are directed radially outward from the point source of the wave.

Huygens' principle provides a technique for predicting how a wave will move in particular circumstances. Thus for the plane wave shown in Fig. 22.4, if we take one plane wave front and consider that all points on it produce

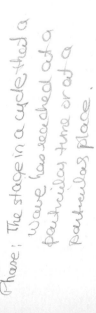

---

*By having "the same phase" we mean that all points on a wave front "are doing the same thing at the same time." For a more mathematical discussion of phase see Sec. 11.8.

FIGURE 22.3 A spherical electromagnetic wave. The wave fronts are spherical in shape and are spaced a wavelength apart, while the rays are radial. Far away from the source, however, small sections of the wave fronts become approximately planar.

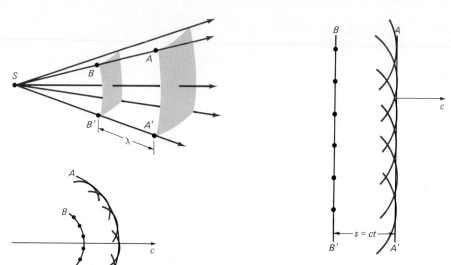

FIGURE 22.5 Propagation of a spherical wave front according to Huygens' principle. Again, the Huygens' wavelets produced by the spherical wave front *BB'* combine to form the spherical wave front *AA'* in the direction in which the wave is moving.

FIGURE 22.4 Propagation of a plane wave front according to Huygens' principle. The Huygens' wavelets produced by the plane wave front *BB'* combine to form the plane wave front *AA'* in the direction in which the wave is moving.

Huygens' wavelets in the forward direction, we obtain the result shown in the figure. All the spherical wavelets travel exactly the same distance in the same time interval. Therefore at some later time the tangents to all the little wavelets form another plane wave front which has moved a distance s along the X axis, where $s = ct$. Hence in this case Huygens' principle correctly predicts the observed motion of the plane wave along the positive X axis.

The same consistent results are obtained when Huygens' principle is applied to a spherical wave front, as in Fig. 22.5. In this case the tangents to the wavelets arising at a spherical wave front form a spherical surface similar to the original wave front but of larger radius. The spreading of the wave from the original point source is again correctly predicted.

Huygens' principle will be particularly useful throughout this chapter when we have to deal with light waves moving from one medium to another or falling on various obstacles and slits.

Example 22.1

What is the frequency range of visible light?

SOLUTION

In Sec. 21.6 we saw that the wavelength and frequency of electromagnetic radiation are related by the equation $\lambda f = c$.

Assuming that the wavelengths of visible light run from 390 nm for violet to 760 nm for red, we have for the frequency limits of the visible spectrum:

$$f_V = \frac{c}{\lambda_V} = \frac{3.0 \times 10^8 \text{ m/s}}{390 \times 10^{-9} \text{ m}} = 7.7 \times 10^{14} \text{ s}^{-1} = \boxed{7.7 \times 10^{14} \text{ Hz}}$$

$$f_R = \frac{c}{\lambda_R} = \frac{3.0 \times 10^8 \text{ m/s}}{760 \times 10^{-9} \text{ m}} = 3.9 \times 10^{14} \text{ s}^{-1} = \boxed{3.9 \times 10^{14} \text{ Hz}}$$

Hence the frequency range of visible light is from about 3.9×10^{14} Hz (red) to 7.7×10^{14} Hz (violet).

22.2 Reflection and Refraction on the Basis of a Wave Theory

Huygens' principle immediately leads to a satisfactory explanation of the reflection and refraction of light.

Reflection of Light

Consider a plane wave front incident on a mirror resting on a flat table, as in Fig. 22.6. The planes of the mirror and of the incident wave fronts are all drawn normal to the plane of the paper. A ray, which is by definition perpendicular to the wave front, is then in the plane of the paper. So too is the normal to the mirror. Here the reflection of light can be represented either by the wave front diagram of Fig. 22.6 or by the ray diagram of Fig. 22.7.

FIGURE 22.6 (left) Reflection of a plane wave front. The wave fronts shown are at right angles to the direction of the wave.

FIGURE 22.7 (right) Ray diagram for the reflection of light by a mirror. Experimentally the incident ray, the reflected ray, and the normal to the mirror are all in the same plane.

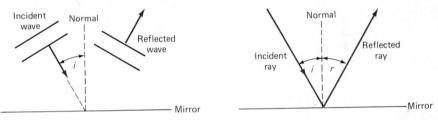

Experimentally we find that the reflection of light is governed by two rules:

1 *The incident ray, the reflected ray, and the normal to the surface all lie in the same plane* (the plane of the paper in Fig. 22.7).

2 *The angle of reflection r is equal to the angle of incidence i, where both angles are measured with respect to the normal to the mirror.*

Laws of Reflection

To show that angle r is equal to i in Fig. 22.7, we use Huygens' principle and the experimental fact that the incident and reflected rays and the normal are all in the same plane. In Fig. 22.8 the wave front AA' is perpendicular to the three rays shown. If Huygens' wavelets are drawn from AA', then in the time it takes one wavelet to go from A' to R', a wavelet at A'' travels from A'' to B and is then reflected back to B'. A wavelet originating at A travels the distance AR in the same time. Hence the new reflected wave front is the surface tangent to these three wavelets and is RR' in the diagram.

Now $A'R' = AR$, since these are distances traveled by light in equal times in the same medium. The right triangles $AA'R'$ and ARR' then have two equal sides, for the hypotenuse AR' is the same in both, and they are therefore congruent. Hence angle $A'R'A$ is equal to angle RAR'. But, from Figs. 22.8 and 22.9,

$$\angle A'R'A = 90° - i \quad \text{and} \quad \angle RAR' = 90° - r$$

Hence $\quad 90° - i = 90° - r \quad$ and so $\quad r = i \quad$ (22.1)

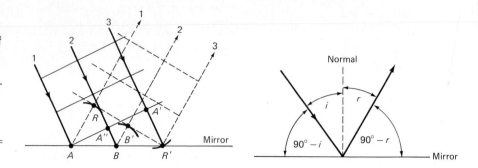

FIGURE 22.8 (left) Diagram used to prove that the angle of reflection is equal to the angle of incidence when light is reflected from a mirror. Here the distance AR is equal to the distance $A'R'$.

FIGURE 22.9 (right) Relationship between angles for the reflection of light. If $90° - i = 90° - r$, then $i = r$.

Hence, *on reflection the angle of reflection is equal to the angle of incidence*, as stated above and as confirmed by experiment. This applies only to a perfectly flat, mirrorlike surface. For rough surfaces, the law of reflection still holds, but different rays are reflected in different directions because they strike parts of the surface oriented at different angles with respect to the incident ray. This we call *diffuse reflection*, which is illustrated in Fig. 22.10.

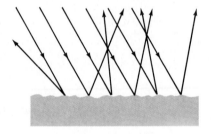

FIGURE 22.10 The diffuse reflection of light. The angle of reflection is still equal to the angle of incidence for each ray, but since the surface is rough, the reflected rays go off in many different directions.

Example 22.2

A plane wave from a spotlight strikes a mirror at an angle of 37° with respect to the normal and is reflected by the mirror. What is the angle between the incident and the reflected waves?

SOLUTION

Since the angle of reflection is equal to the angle of incidence, we have for the angle between the incident and reflected waves

$$i + r = i + i = 2i = 2(37°) = \boxed{74°}$$

Refraction of Light

Refraction is the bending of a wave front of light when it travels from one medium to another.

The crucial point to remember about refraction is that the speed of light is different in different materials. This speed is determined by Eq. (21.15), $v = c/n$, where c is the speed of light in a vacuum and n is the index of refraction of the material for light of the particular frequency used. Hence the speed of light is highest in a vacuum, somewhat smaller in gases, and much smaller in solids, as is clear from Table 22.1.

In Fig. 22.11 we consider a plane wave front PQ incident at an angle on a boundary between medium 1 (say, air) and medium 2 (perhaps glass). Points P and Q on the wave front can be considered sources of Huygens' wavelets.

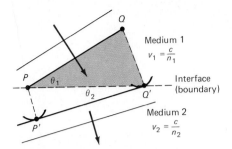

FIGURE 22.11 Refraction of light according to Huygens' principle. Since v_1 is greater than v_2 (by assumption), the incident ray 1 is refracted toward the normal and assumes the direction of ray 2.

TABLE 22.1 Indices of Refraction* for Important Materials

Material	Index of refraction n $(= c/v)$
Air (STP)	1.00029
Hydrogen (STP)	1.00013
Water	1.33
Benzene	1.50
Ice	1.31
Glass:	
Borosilicate crown	1.52
Light flint	1.58
Heavy flint	1.65
Heaviest flint	1.89
Diamond	2.42

*For yellow light, with $\lambda = 590$ nm.

Let us suppose that in time t the wavelet from Q traveling perpendicular to PQ just reaches Q' at the surface of the glass. In the same time the wavelet from P has traveled a shorter distance PP' in the glass, because its speed is slower in glass than in air. Hence the new wave front $P'Q'$ can be found by drawing a straight line from Q' which is tangent to the Huygens' wavelet at P'. Then

$$\frac{QQ'}{PP'} = \frac{v_1 t}{v_2 t} = \frac{v_1}{v_2} \tag{22.2}$$

But, from the figure,

$$\sin \theta_1 = \frac{QQ'}{PQ'} \quad \text{and} \quad \sin \theta_2 = \frac{PP'}{PQ'}$$

and so

$$\frac{QQ'}{PP'} = \frac{PQ' \sin \theta_1}{PQ' \sin \theta_2} = \frac{\sin \theta_1}{\sin \theta_2}. \tag{22.3}$$

On combining Eqs. (22.2) and (22.3) we obtain

$$\frac{v_1}{v_2} = \frac{\sin \theta_1}{\sin \theta_2} \quad \text{or} \quad \frac{c/n_1}{c/n_2} = \frac{\sin \theta_1}{\sin \theta_2}$$

Snell's Law

and so finally

$$\boxed{n_2 \sin \theta_2 = n_1 \sin \theta_1} \tag{22.4}$$

From Fig. 22.12 it can be seen that θ_1 is the angle between the incident ray and the normal to the surface, and θ_2 is the angle between the refracted ray and the normal to the surface. Equation (22.4) is called *Snell's law of refraction* after W. Snell (1591–1626), a Dutch mathematician. It predicts that a light ray will always be bent toward the normal in the more optically dense material, i.e., the material with larger index of refraction, for the larger n_2, the smaller θ_2.

It is important to note that light is bent toward the normal in the material in which light travels *less* rapidly. Sir Isaac Newton had proposed a particle theory of light which explained the refraction of light by the difference in the forces exerted on the particles by the two media, the more dense medium exerting a larger force and causing light to move more rapidly. A measure-

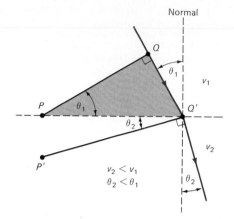

FIGURE 22.12 Relationship between angles for the refraction of light. Since the wave front PQ is perpendicular to the ray QQ', and the normal is perpendicular to the boundary between the two media, the two angles marked θ_1 are equal; and the angles θ_2 are also equal.

ment of the speed of light in water, made by Foucault in 1850, clearly showed that light has a lower speed in water than in air, and that Newton's theory must therefore be wrong. There had been increasing evidence that a wave theory was correct but physicists had been unwilling to accept it because of Newton's monumental contributions to mechanics. This example points up the danger of bowing to authority in an experimental field like physics.

Example 22.3

A searchlight is being used at the edge of a swimming pool to locate a lost key at the bottom of the pool. The key is 4.0 m below the surface and 4.0 m from the edge of the pool. At what angle will the searchlight have to point with respect to the normal to locate the key, if the light enters the water 1.0 m from the edge of the pool?

SOLUTION

The situation is as in Fig. 22.13. Light is bent toward the normal in the water, and so angle θ_1 is greater than θ_2. By Snell's law we then have

$$n_1 \sin \theta_1 = n_2 \sin \theta_2$$

For air $n_1 \simeq 1$; for water $n_2 = 1.33$. Hence

$$\sin \theta_1 = \frac{n_2}{n_1} \sin \theta_2 = \frac{1.33}{1} \sin \theta_2$$

Now, in the figure AC is the hypotenuse of the 3-4-5 right triangle ABC and therefore has a length of 5 m.

Hence $\qquad \sin \theta_2 = \dfrac{BC}{AC} = \dfrac{3}{5} = 0.60$

and $\qquad \sin \theta_1 = 1.33(0.60) = 0.80$

so that $\qquad \theta_1 = \boxed{53°}$

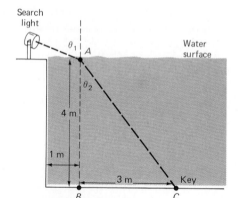

FIGURE 22.13 Figure for Example 22.3.

This is the angle at which the searchlight would have to point with respect to the normal to the surface of the pool to locate the key at the bottom of the pool. Someone looking along the direction of the searchlight beam would think that the pool was shallower than it really is, since the eye (and brain) would assume that the light reaching the eyes from the key had traveled in a straight-line path.

Dispersion

The index of refraction of any material varies with wavelength, as shown in Fig. 22.14 for three different kinds of glass. Since the index of refraction is in this case smaller for longer wavelengths, red light is refracted less than blue light in going from air to glass.

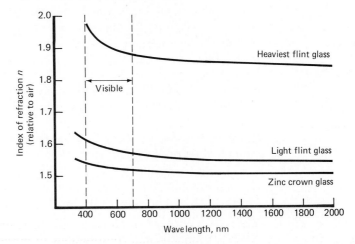

FIGURE 22.14 Dependence of index of refraction on wavelength for three kinds of glass.

Definition

The dependence of the index of refraction of a material on wavelength is called dispersion.

It is dispersion that causes a glass prism to break up white light into all the colors of the rainbow, as Newton first observed. Since the different colors correspond to different wavelengths, they are refracted by different amounts and emerge from the prism going in different directions. This will be discussed in more detail in Sec. 23.3.

Medium with Varying Index of Refraction

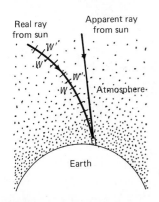

FIGURE 22.15 Refraction of sunlight by the atmosphere. The refraction makes the sun appear to be at a higher point in the sky than it really is.

If the index of refraction of a medium through which light passes is gradually changing, the light passing through it is gradually deflected from a straight path by refraction. The density of the earth's atmosphere and hence its index of refraction decrease as we go up in the atmosphere. The deflection of sunlight which results can be understood from Fig. 22.15. Plane wave fronts of light coming from the sun have their upper portions W' traveling through air of lower density than their lower portions W. Greater density means a higher index of refraction and hence a lower velocity. Thus the part of the wave front at W' travels more rapidly than the part at W and the wave front is gradually bent around as shown. When the wave front reaches an observer on the earth, the observer assumes that the light is coming from a direction normal to the wave front, and hence sees the sun at a higher position in the sky than it actually occupies, as the figure shows.

Light waves entering the earth's atmosphere in a direction parallel to the surface of the earth are bent in such a way that stellar objects appear about 1/2° higher in the sky than they actually are. This is approximately the angle subtended by the width of the solar disk on the earth. Hence when the sun appears to be just above the horizon at sunrise and sunset, in reality it is just below it.

22.3 Diffraction of Light

One of the common objections once raised against a wave theory of light was that better-known kinds of waves like water waves and sound waves bend around obstacles, whereas light appears to travel in perfect straight lines and to cast sharp shadows, as would a beam of particles. But in the middle of the seventeenth century the Italian Jesuit priest Francesco Grimaldi (1618–1663) showed that this was not the whole story. Grimaldi demonstrated that when a beam of light passes through two small holes, one behind the other, and then falls on a dark surface, the band of light on that surface is a little wider than the original beam. Grimaldi surmised that the beam had been bent outward by a slight amount at the edges of the hole. This bending he called *diffraction*.

Definition

Diffraction is the bending of waves into the shadow region when they pass through holes or slits comparable in size to the wavelength.

The fact that diffraction is not more evident for light in everyday life can be explained by Huygens' principle. Suppose a plane beam of light falls on a very wide slit, as in Fig. 22.16a. Using Huygens' construction we find that the wave fronts remain planar across the slit, but that there is a slight bending of the wave front at the edges, since there are no additional Huygens' wavelets beyond the edges to create the same kind of plane wave front which is produced in the interior of the slit. As long as the wavelength of light (5×10^{-7} m) is small compared with the width of the slit (say, 10 cm), this bending or diffraction is not easily noticeable. If we narrow the slit down considerably, however, as in Fig. 22.16b, there is a more noticeable bending of light into the shadows at the two edges of the slit. Finally, if the slit becomes of the same relative size as the wavelength of the light, for example, 10^{-6} m, or 10^{-3} mm, then there is room only for a few Huygens' wavelets across the slit width and the plane wave is converted into a cylindrical wave, as shown in Fig. 22.16c. The corresponding behavior for water waves is shown in Fig. 22.17.

FIGURE 22.16 Diffraction of light by a slit of varying width: (a) very wide slit; (b) narrower slit; (c) very narrow slit. The designations of width are all relative to the wavelength of the light.

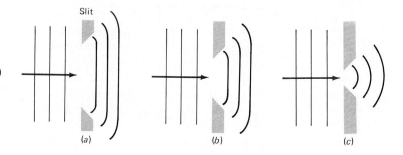

Slit

(a)   (b)   (c)

These observations will be made more quantitative in the next section. Here we merely want to stress that *observable diffraction only occurs if we are dealing with slits having openings of the same order of magnitude as the wavelength of the light used*. Since this wavelength is very small compared with common objects and openings, the diffraction of light is not easily observed in everyday life. But for water waves, sound waves, and radio waves, where the wavelengths may be measured in centimeters, meters, or (for radio waves) even in kilometers, diffraction effects are much larger, and

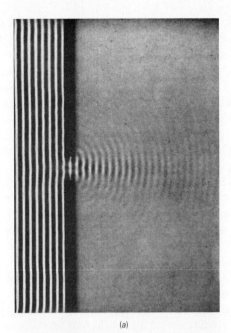

(a)

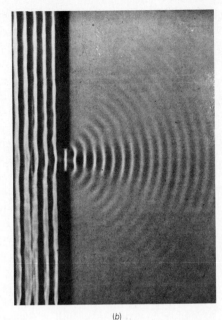

(b)

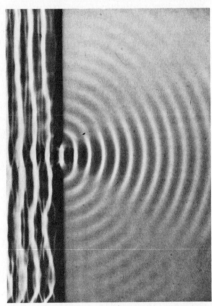

(c)

FIGURE 22.17 Diffraction of water waves of different wavelengths by a slit. In (a) the wavelength of the water waves λ is 0.3w, where w is the slit width. A very clear shadow of the slit can be seen to the right of the slit. In (b) λ has been increased to 0.5w, and some bending into the shadow region can be seen. In (c) λ is 0.7w, and the wave passing through the opening has been almost completely converted into a cylindrical wave. No shadow of the slit can be seen. (*Photo courtesy of Educational Development Center, Newton, Massachusetts.*)

we get accustomed to the fact that sound travels around corners, and that radio waves bend into the shadows of large buildings to reach our home radios.

Coherent and Incoherent Light Sources

In the following sections we will be discussing the superposition and interference of light waves. In such discussions the distinction between coherent and incoherent light sources is crucial.

In many cases electromagnetic radiation is produced by oscillating electric charges with the frequency of oscillation determining the region of the spectrum in which the radiation is emitted. If, in a given source, the charges all oscillate *in step*, or in unison, the source is said to be *coherent*. If the charges oscillate independently and randomly, the source is said to be *incoherent*.

Ordinary sources of visible light—incandescent light bulbs, fluorescent tubes, carbon arcs—are incoherent. Laboratory sources of radio waves and microwaves are normally coherent, because they amplify the output of coherent oscillators such as *LC* circuits. The development of the *laser*—which is basically an optical amplifier and which will be discussed more in detail in Chap. 27—has extended the range of coherent sources into the optical region of the electromagnetic spectrum.

Diffraction and interference effects are easily demonstrated with lasers, because the superimposed waves are all from the same coherent source. Hence if two light waves leave the laser in phase but travel different distances to a screen or to the eye, they will produce constructive or destructive interference depending on the phase difference between the two waves introduced by the different paths traveled.

With a conventional light source the interference of light can be demonstrated only by selecting a very small region of the source. The light emitted from such a small region has sufficient coherence to produce interference. A broad, incoherent light source will not, however, produce an observable interference pattern because waves from different parts of the source will have random phase relationships, and the superposition of such waves of random phase produces no observable constructive or destructive interference.

22.4 Single-Slit Diffraction

When Grimaldi first made his observations on diffraction, he thought that he saw colored fringes on the screen in the region outside the enlarged image of the hole through which the light passed. We now know that these colored fringes were due to the superposition of Huygens' wavelets coming from different positions across the width of the hole, or from what are usually called different "zones." Grimaldi succeeded in producing fringes with sunlight because he used two small apertures, one to select a very small region of the incident sunlight and provide sufficient coherence for the second aperture to produce fringes by diffraction and the resulting interference between diffracted waves.

For simplicity let us use a He–Ne laser as a light source, as in Fig. 22.18. The laser produces intense, highly monochromatic (one-color), coherent red light of wavelength $\lambda = 633$ nm. This light falls on a narrow slit of width w and then on a screen a large distance from the slit. The observed pattern on the screen is referred to as the *diffraction pattern* of the single slit. It is found experimentally to consist of alternating bright and dark fringes parallel to the slit.

To explain the observed diffraction pattern we need to apply the ideas of the superposition and interference of waves developed in Sec. 11.8. (It may be a good idea to review these ideas before proceeding.) Let the distance between slit and screen be L. The Huygens' wavelets that go straight through the slit will arrive at position P on the screen in phase and produce a bright central image of the slit. To find the width of this central image we must find

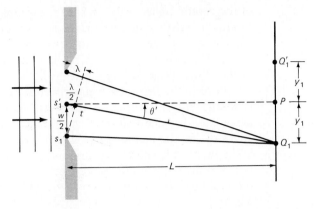

FIGURE 22.18 Calculation of a single-slit diffraction pattern. Light from the two halves of the slit are 180° out of phase and cancel at Q_1.

FIGURE 22.19 Angles used in calculating single-slit diffraction. From geometry, $\theta = \theta'$.

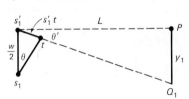

the positions on either side of P on the screen at which the Huygens' wavelets from different parts of the slit cancel out because they travel *different distances* and arrive at the screen 180° out of phase. These are the positions at which each Huygens' wavelet has superimposed on it another wavelet exactly $\lambda/2$ out of phase with it. This leads to cancellation and therefore darkness at the points Q_1 and Q_1' on either side of P on the screen.

To find the distance y_1 from P to Q_1 we consider light traveling in a direction from slit to screen so that the path difference between a Huygens' wavelet arising at the upper edge of the slit in Fig. 22.18 and the wavelet coming from the lower edge of the slit is $\Delta = \lambda$, where λ is the wavelength of the light being used. Then we divide the slit in Fig. 22.18 into two halves (or zones), a top half and a bottom half, and consider the superposition of pairs of Huygens' wavelets, one from the top half and one from the bottom half, arising at points separated by a distance $w/2$ across the slit. (Remember that we are looking down on this slit from above. It actually extends some distance into the plane of the paper.) If point Q_1 is chosen as indicated above, then a Huygens' wavelet from s_1 will cancel a Huygens' wavelet from s_1' at Q_1, and so on, as we take pairs of Huygens' wavelets from points a distance $w/2$ apart, since the two paths will always differ by $\lambda/2$ for these wavelets. The difference in the paths of these pairs is $s_1't$ in Fig. 22.18. Now, from Fig. 22.19,

$$s_1't = \frac{w}{2}\sin\theta = \frac{w}{2}\sin\theta'$$

Also for $L \gg y_1$, θ' is a very small angle, and so

$$\sin\theta' \simeq \tan\theta' = \frac{y_1}{L}$$

For dark fringes we must have

$$s_1't = \frac{\lambda}{2} = \frac{w}{2}\sin\theta' \qquad \text{or} \qquad \sin\theta' = \frac{\lambda}{w}$$

and so $\qquad y_1 = L\sin\theta' \qquad$ or $\qquad \boxed{y_1 = \frac{L\lambda}{w}} \qquad$ (22.5)

The first dark fringe therefore occurs at Q_1, a distance $\lambda L/w$ from P. A similar dark fringe is found at Q_1' on the other side of P. Note that the width of of the central image is $2y_1 = 2\lambda L/w$ and that it gets wider as the wavelength increases and the slit gets narrower. In other words, the narrower the slit compared with the wavelength being used, the wider the central diffraction image. Since $\sin \theta' = \lambda/w$, when λ is equal to w, $\sin \theta' = 1$, $\theta' = 90°$, and the image of the slit spreads to cover the entire hemisphere to the right of the slit.

Beyond the first dark fringe on either side of the central maximum there is another much less intense bright fringe. This is followed by a second dark fringe at a distance y_2 from P. To find y_2 we consider wavelets from the two edges of the slit in Fig. 22.18 which have path differences of 2λ. Then we break up the slit into four zones, and consider the superposition of Huygens' wavelets arising from the points on the slit $w/4$ apart. In a fashion similar to that used for the first dark fringe we find that in this case dark fringes occur when

$$\frac{wy_2}{4L} = \frac{\lambda}{2} \quad \text{or} \quad y_2 = \frac{2\lambda L}{w}$$

If this process is repeated we find that *dark fringes* occur at distances

Dark Fringes

$$y_m = \frac{m\lambda L}{w} \quad (m \text{ integral}) \tag{22.6}$$

The distance between dark fringes is then $\lambda L/w$, which is half the width of the central maximum fringe. Between adjacent dark fringes, bright fringes of decreasing brightness occur. Since Eq. (22.6) contains the wavelength, the single-slit diffraction pattern varies with the wavelength of the light used. If white light is used, a series of colored fringes are obtained, since y_m varies with λ. These are the fringes first observed by Grimaldi.

FIGURE 22.20 Theoretically predicted diffraction pattern for a single slit.

FIGURE 22.21 Photograph of an experimental diffraction pattern for a single slit. Note that the width of the central maximum is indeed twice the width of the secondary maxima, as predicted by theory. (*Photo from Cagnet, Françon, and Thrierr, Atlas of Optical Phenomena; reprinted by permission of Springer-Verlag, Heidelberg.*)

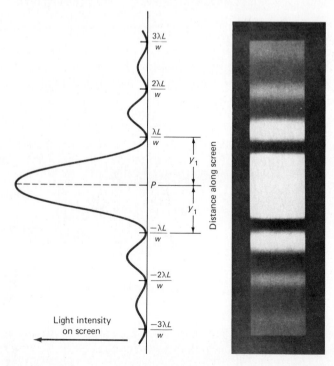

The expected diffraction pattern on the screen is shown in Fig. 22.20, and a photograph of an actual pattern is shown in Fig. 22.21. For $w >> \lambda$, $y_1 \to 0$ and we have almost perfect straight-line propagation of light. This predicts the validity of a *ray theory of light* (see Chap. 23), which holds as long as there are no obstacles or openings of a size comparable to the wavelength of the light used.

Diffraction around a Circular Obstacle

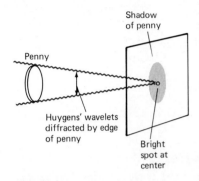

FIGURE 22.22 Prediction of bright spot at the exact center of the diffraction pattern of a penny.

If a plane light wave from a laser is incident on a penny, the penny will throw a shadow on a screen behind it, but there will be weak fringes around the shadow arising from the superposition of Huygens' wavelets diffracted around the edge of the penny. In addition there will be a very tiny bright spot at the exact center of the shadow, as shown in Fig. 22.22. This is what would be expected according to Huygens' principle. Suppose the light in the plane wave hitting the penny is coherent; i.e., all the waves are perfectly in phase. Then the Huygens' wavelets produced at the edge of the penny will be in phase at the screen at a point directly opposite the center of the penny, for all the wavelets will travel exactly the same distance to reach that point. Hence a bright spot of light should appear at the center. A photograph of an actual diffraction pattern obtained in this way is shown in Fig. 22.23.

When the French physicist Augustin Fresnel (1788–1827) first presented his wave theory of light in 1816, it was ridiculed by many French scientists. One of Fresnel's strongest critics was S. Poisson (1781–1840), who pointed out that Fresnel's theory led to the absurd conclusion that a bright spot should appear at the center of the diffraction pattern of an opaque circular object. Much to Poisson's chagrin, in 1818 another French physicist. D. F. Arago (1786–1853), actually carried out the experiment and found the bright spot. To commemorate Poisson's unintentional contribution to this landmark proof of Fresnel's wave theory of light, the bright spot is now called *Poisson's spot*.

FIGURE 22.23 Actual photograph of the diffraction pattern of a circular opaque object in laser light, indicating the predicted bright spot at the center of the shadow. (*Photo from Cagnet, Françon, and Thrierr, Atlas of Optical Phenomena; reprinted by permission of Springer-Verlag, Heidelberg.*)

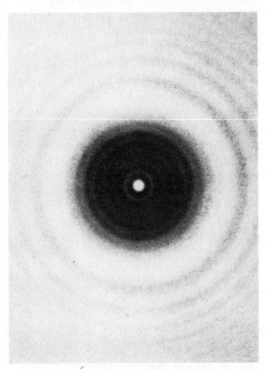

Example 22.4

Light from a He–Ne laser ($\lambda = 633$ nm) is incident on a single slit of width 0.10 mm.
(a) Describe the diffraction pattern formed on a screen 2.0 m away from the slit.

(b) Describe what happens if the slit width is reduced to 0.01 mm.

SOLUTION

(a) By Eq. (22.6) the dark fringes occur at distances $y_m = m\lambda L/w$ from the center of the pattern. Hence the first dark fringe is at

$$y_1 = \frac{1(633 \times 10^{-9} \text{ m})(2.0 \text{ m})}{0.10 \times 10^{-3} \text{ m}} = 1.3 \times 10^{-2} \text{ m} = \boxed{1.3 \text{ cm}}$$

Hence the central image is 2(1.3 cm) = 2.6 cm wide, and the dark fringes on either side of the central image are spaced 1.3 cm apart.

(b) If the slit is narrowed to 0.010 mm, we have

$$y_1 = \frac{1(633 \times 10^{-9} \text{ m})(2.0 \text{ m})}{0.01 \times 10^{-3} \text{ m}} = 1.3 \times 10^{-1} \text{ m} = \boxed{13 \text{ cm}}$$

In this case the central image is 10 times as wide, and the fringes are spaced 10 times as far apart as in (a), that is, 13 cm apart.

22.5 Interference: Young's Double-Slit Experiment

One of the crucial experiments in establishing the wave nature of light was first performed in 1803 by a remarkable British medical doctor Thomas Young (see Fig. 22.24 and accompanying biography). As a physician, Young was interested in the workings of human ears and eyes. This led him to dedicate a great deal of his time to a study of sound and light. He knew that if two sound waves reach the ear 180° out of phase, they cancel each other out and no sound is heard. It occurred to him that a similar interference effect should result with two beams of light if, like sound, light consisted of waves. This led Young to an experiment now commonly referred to as *Young's double-slit experiment*.

Consider the physical situation shown in Fig. 22.25, in which we again use a laser for convenience. Two very narrow vertical slits separated by a distance d between their centers are cut in a piece of metal. Light from the laser is directed at the two slits, passes through, and strikes a screen at a distance L from the slits. Huygens' wavelets produced by diffraction at the slits combine to form cylindrical wave fronts (because of the long, narrow slits), and these wave fronts from the two slits are superimposed on the screen. This leads to positions of constructive and destructive *interference*

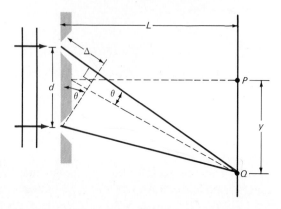

FIGURE 22.25 Young's double-slit experiment. The intensity at point Q depends on the size of Δ and hence on the angle θ.

Thomas Young (1773–1829)

FIGURE 22.24 Thomas Young. (*Photo courtesy of AIP Niels Bohr Library.*)

Few men or women have ever contributed so much first-rate scholarly work to so many different fields of learning as did Thomas Young.

Born in Somerset, England, in 1773, Young learned to read at the age of 2 and by the age of 6 is said to have read the complete Bible twice. By 19 he had learned on his own to read Latin, Greek, Italian, Hebrew, Arabic, Persian, Turkish, and Ethiopian, and had time to master calculus, chemistry, and physics on the side! No wonder his fellow students at Cambridge nicknamed him 'Phenomenon Young."

From 1792 to 1799 Young was engaged in his medical studies at London, Edinburgh, Göttingen, and Cambridge, receiving his M.D. degree from Göttingen in 1796. For most of his life he practiced medicine, but without great success. Constantly torn between the practice of medicine and his love for pure research, Young lacked sufficient interest in the routine problems of patients to build up a successful medical practice. Fortunately he had inherited a large fortune from his uncle, and did not have to rely on medicine for a livelihood.

During his medical studies Young became interested in the physiology of the human eye. He was the first to explain how the human eye can focus on objects at different distances (accommodation). He also first explained, by experimenting with his own eyes, the nature of astigmatism (see Sec. 24.2). Finally, research on the nature of color blindness led him to put forth his three-color theory of color vision, which proposed that only three different kinds of color receptors are present in the human eye, one for blue, one for green, and one for red light. All other colors are merely combinations of these three basic colors. This theory, now called the Young-Helmholtz theory, is the basis of modern color photography and color television.

Young's work on vision naturally led him to explore the nature of light. His experiments with double-slit interference, Newton's rings, and thin films convinced him that light was a wave motion of some kind. Young never developed the mathematics to explain his wave model of light properly, and hence the French physicist Fresnel, who did, is often given more credit than Young for the wave theory.

Young also made contributions to other fields of physics, e.g., to the stretching of metals (Young's modulus), to the understanding of the concept of energy, and to the theory of the tides.

In his latter years Young devoted most of his time to his first love, languages. He devoted years of intense study to deciphering the Egyptian hieroglyphics found on the Rosetta stone discovered in the Nile Delta in 1799. He never achieved complete success, but his ideas were the basis for our first understanding of this important ancient form of communication.

Young was a brilliant intellectual, but had little understanding of human sensitivities. He was honest— sometimes too honest—in appraising the scholarly work of his contemporaries, and made enemies as a result. Much of his scientific work was published anonymously because he did not want his patients to think that his heart was not in his medical practice.

Thomas Young was never a modest man and liked to boast: "I never spent an idle day in my life." Given the wealth of his contributions to knowledge, we can well believe him.

depending on the *path difference* between the wavelets from the two slits, and hence to bright and dark vertical fringes on the screen.

To find the positions of the bright and dark fringes we proceed in a manner similar to that used with the single-slit case. Suppose we consider the fringes produced at a distance y to either side of a line P on the screen parallel to the slits and directly opposite a point midway between the slits. Constructive interference will occur when the wavelets from the two slits travel distances to the screen which differ by $\Delta = m\lambda$, with m integral. From Fig. 22.25 we have

$$\sin\theta = \frac{\Delta}{d}$$

and so the condition for constructive interference is

$$\Delta = d\sin\theta = m\lambda \qquad (m = 0, 1, 2, \ldots) \tag{22.7}$$

But, from the figure, $\tan\theta = y/L$, and θ is usually very small, since y is much smaller than L. For small angles we can set $\sin\theta = \tan\theta = y/L$, and so Eq. (22.7) becomes

$$d\sin\theta = \frac{dy}{L} = m\lambda \quad . \quad \text{or} \quad \boxed{\sin\theta = \frac{m\lambda}{d}} \tag{22.8}$$

Bright Fringes

and so $\qquad \boxed{y_m = \frac{m\lambda L}{d}} \qquad (m \text{ integral}) \tag{22.9a}$

where y_m is the position of the mth-order bright fringe.

For destructive interference, on the other hand, the distance between the paths of the two wavelets must be $(m + \frac{1}{2})\lambda$, with $m = 0, 1, 2, \ldots$, or

Dark Fringes

$$\boxed{y'_m = \frac{(m + \frac{1}{2})\lambda L}{d}} \qquad (m \text{ integral}) \tag{22.9b}$$

Figure 22.26 shows how the waves from the two slits lead to constructive and destructive interference.

Hence we expect to find a series of alternating bright and dark fringes, if Young's wave interpretation of his double-slit experiment is valid. Figure 22.27 provides ample confirmation of Young's ideas. In this case it can be seen that the fringes are equally spaced as we go out from the central image, exactly as predicted by theory.*

Equation (22.9a) or (22.9b) can be used to obtain a value for the wavelength λ of the light used, since d can be measured with a traveling microscope, and y_m and L with a measuring tape or meterstick. The results, of course, confirm the fact that the wavelength of visible light is about 500 nm. As a result d must be very small or the fringes would be so close together that it would be impossible to distinguish them.

Note that here again, as the slits are brought closer together, the interference pattern spreads out. Also the fringe separations depend directly

---

*There is also a diffraction pattern for each single slit, but we assume that the width of each slit is much smaller than d, so that the single-slit diffraction pattern is much wider than the double-slit interference pattern and produces only a gradual variation in the intensity of the sharper double-slit fringes.

FIGURE 22.26 An illustration of how the constructive interference of Huygens' wavelets leads to bright lines at positions like y_0 and y_1, and how destructive interference leads to the cancellation of the two incident wavelets and hence darkness at positions like y_0'.

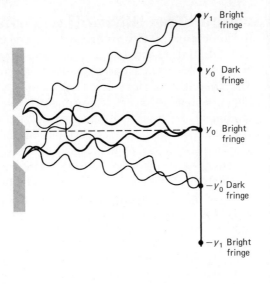

FIGURE 22.27 An experimental double-slit diffraction pattern. The fringes are equally spaced as predicted by theory. (*Photo from Cagnet, Françon, and Thrierr, Atlas of Optical Phenomena; reprinted by permission of Springer-Verlag, Heidelberg.*)

on the wavelength λ. For microwaves, for example, for which $\lambda \simeq 10$ cm, double-slit interference patterns are easily obtained for slit separations of 20 cm.

Interference and Diffraction

Interference and diffraction are similar in that they are both basically *superposition* effects which depend on the addition of wave disturbances at a given point, taking account of the phase differences between the superimposed waves. The distinction usually made between interference and diffraction is that *interference* arises from the superposition of a finite number of waves coming from *different* coherent sources, as in Young's double-slit experiment. *Diffraction*, on the other hand, comes from the superposition of an infinite number of wavelets arising from different places on the *same* original wave front (different zones), as in single-slit diffraction.

Example 22.5

Coherent light of wavelength 589 nm from a small region of a sodium arc lamp falls on a double slit with a slit separation of 0.10 mm. The interference pattern is produced on a screen 2.5 m from the slits. Calculate the separation on the screen of the two fourth-order bright fringes on either side of the central image.

SOLUTION

The easiest way to handle this problem is to calculate the distance y_4 of the fourth-order bright fringe from the central image, and then to double this value to obtain the distance between the two fourth-order fringes.

From Eq. (22.8) we have
$$y_m = \frac{m\lambda L}{d}$$

$$= \frac{4(589 \times 10^{-9}\ \text{m})(2.5\ \text{m})}{0.10 \times 10^{-3}\ \text{m}} = 5.9 \times 10^{-2}\ \text{m} = 5.9\ \text{cm}$$

Hence the distance between the two fourth-order fringes on either side is

$$2y_m = 2(5.9\ \text{cm}) = \boxed{11.8\ \text{cm}}$$

Note Phy2384

22.6 Multislit Interference: The Diffraction Grating

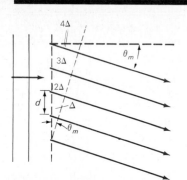

FIGURE 22.28 Interference produced by light passing through a diffraction grating. The intensity of light at any point on the screen depends on the value of Δ.

If the number of slits producing interference fringes is gradually increased from 2 to 3 to 4, and so on, up to some very large number, we find that the bright fringes become narrower and brighter as the number of slits increases. The limiting case here is the *diffraction grating*, invented by Joseph von Fraunhofer in 1820, and consisting of as many as 10,000 or more slits per centimeter. Such gratings are produced by highly specialized engines which rule with a diamond stylus very fine lines on glass at equal intervals. The unruled regions of glass then act as the slits. Such diffraction gratings produce extremely sharp interference fringes and are much used in spectrometers for the analysis of the spectra of atoms and molecules.

The theory of the diffraction grating is identical with that of the double slit. We consider a large number of Huygens' wavelets moving from the slits in a direction so that the path difference between the wavelets from two adjacent slits is Δ, as in Fig. 22.28. For constructive interference adjacent wavelets must satisfy the condition $\Delta = m\lambda$, where m is an integer. Hence, if d is the distance between two adjacent slits,

$$\sin \theta_m = \frac{\Delta}{d} = \frac{m\lambda}{d}$$

Grating Equation

or

$$\sin \theta_m = \frac{m\lambda}{d}$$

(22.10)

[Notice that this equation is identical with Eq. (22.8) for two slits.] Here θ_m is the angle corresponding to the mth-order bright fringe, where m is 0 for the central, undeviated image, $m = 1$ for the first-order fringes to either side of the central image, $m = 2$ for the second-order fringes, etc. The slit separation d, in meters, is the reciprocal of the number of lines per meter ruled on the grating.

In practice a diffraction grating is mounted as in Fig. 22.29. A *collimator* (which uses a lens of proper focal length) takes diverging light from a light source and converts it into a beam of plane waves. These plane waves strike a diffraction grating normal to its surface. The parallel waves emerging from all the slits at any particular angle are then collected and the image observed with a telescope. This telescope is free to move in a circle around an axis parallel to the grating rulings. The angles corresponding to bright fringes can then be read from a scale over which the telescope moves. In this way θ can be

FIGURE 22.29 A grating spectrometer. A diffraction grating produces images of the slit of the collimator at angular positions which depend on the wavelengths emitted by the light source.

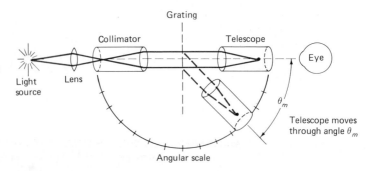

measured and the wavelength λ of the light obtained from the grating equation. This apparatus is usually called a *grating spectrometer*.

The main difference between the fringe patterns from a double slit and a grating is the remarkable increase in intensity and sharpness of the fringes with the grating. Figure 22.30 illustrates such a situation, where d, the distance between slits, is the same in the two cases. The increased intensity and sharpness occurs because, with so many sources of light provided by the slits in the grating, even a slight change in the angle θ_m produces a sufficient amount of destructive interference to cancel out the light at all angles except those satisfying Eq. (22.10). Hence all the original light is concentrated into very intense, narrow fringes in the interference pattern.

FIGURE 22.30 Comparison between interference fringes produced by a double slit and those produced by a diffraction grating. The grating fringes are much more intense and sharper.

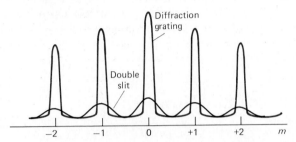

If white light is used as a light source for a grating, and the collimator contains a narrow slit through which the light enters the system, then a bright white line is seen with the telescope for light passing straight through the grating. This is the zero-order ($m = 0$) fringe. To either side of this central image we find a complete spectrum like a rainbow stretching from the violet, which has the shortest wavelength and is deviated least, to the red, which has the longest wavelength and is deviated most. This is the first-order ($m = 1$) spectrum, in this case a continuous spectrum because the white light contains all visible wavelengths. Beyond this first-order spectrum there are weaker, higher-order spectra.

If the light source is a mercury arc lamp, such as those used to light highways, we find a central image which is a bluish-purple line, and discrete lines to either side of it in the first order, ranging outward from 405 nm (violet) through 436 nm (blue) and 546 nm (green) to 579 nm (yellow). We call this a discrete spectrum, or a *line spectrum*. Beyond these first-order lines there will be higher-order images, the number depending on the value of the grating space d. Since $\sin \theta_m$ cannot exceed 1, there is a limit on the order m possible for any given wavelength and grating space.

Example 22.6

A spectrometer uses a grating ruled with 8000 lines per centimeter.
(a) How many orders of the sodium D line at 589 nm can be observed with such a spectrometer?
(b) How many orders of the mercury violet line at 405 nm can be observed?

(c) A second-order yellow line from an unknown element is observed to fall very close to the third-order mercury violet line of wavelength 405 nm. What is the wavelength of the unknown line?

SOLUTION

(a) 8000 lines per centimeter corresponds to a grating space

$$d = \frac{1}{8.0 \times 10^5 \text{ m}^{-1}} = 1.25 \times 10^{-6} \text{ m}$$

From the grating equation we have $d \sin \theta_m = m\lambda$

and so
$$\sin \theta_m = \frac{m\lambda}{d} = \frac{m(589 \times 10^{-9} \text{ m})}{1.25 \times 10^{-6} \text{ m}} = \frac{m}{2.1}$$

Now, the maximum value of $\sin \theta_m$ is 1, for which θ_m is 90°.

Hence the largest possible value of m is $\boxed{m = 2}$

(b) For the mercury 405-nm line, we find in the same way,

$$\sin \theta_m = \frac{m(405 \times 10^{-9} \text{ m})}{1.25 \times 10^{-6} \text{ m}} = \frac{m}{3.1}$$

Hence the largest possible value of m is $\boxed{m = 3}$

(c) From the grating equation, with θ_m the same for the two lines, since they fall very close together, we have

$$m_2\lambda_2 = m_1\lambda_1$$

Here $m_1 = 3$ and $\lambda_1 = 405$ nm for the mercury violet line. Also $m_2 = 2$ for the unknown line of the second order. Hence

$$2\lambda_2 = 3(405 \text{ nm})$$

or
$$\lambda_2 = \frac{3}{2}(405 \text{ nm}) = \boxed{608 \text{ nm}}$$

This method can be used to find the wavelengths of unknown lines in experimental spectra, and is referred to as the *method of overlapping orders*.

22.7 Interference in Thin Films

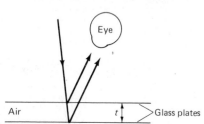

FIGURE 22.31 Interference produced by a thin air film of thickness *t*. The waves reflected by the top and bottom glass plates interfere.

A. A. Michelson once observed that the brilliant colors exhibited by the diamond beetle were caused by a natural diffraction grating built into the beetle's wings. The beautiful colors seen in soap bubbles or in the feathers of peacocks, however, are produced by the interference of two beams of light reflected by thin films.

For example, suppose that a very thin film of air is trapped between two pieces of flat glass, as in Fig. 22.31. If a beam of yellow light from a sodium arc shines down almost normal to the glass surfaces, some of the light will be reflected from the interface between the bottom of the upper plate and the air, and some will be reflected from the interface between the air and the top of the lower plate. These two beams will produce destructive or constructive interference depending on whether their path difference is equal to an odd or an even number of half wavelengths.

Let the thickness of the air film be t. Then the difference in the paths of the two beams shown in the figure is $\Delta = 2t$. We would expect that constructive interference and bright light would result if $\Delta = m\lambda$, where m is an integer, and destructive interference and darkness would result if $\Delta = (m + \frac{1}{2})\lambda$, with m again an integer.

This is not the entire picture, however, since an additional phase difference is introduced between the two beams on reflection. The first beam in this case is reflected where the optically more dense glass meets the less dense air. There is therefore no phase change on reflection, just as there is no phase change when a wave on a string is reflected from a free end of the string (see Sec. 11.9). On the other hand, the other beam which is reflected from the interface where the air film meets the top surface of the lower piece of glass suffers a phase change of 180° on reflection, just as does a transverse wave on a string when reflected from a fixed end, as in Fig. 11.26. Hence an additional phase difference of 180° is introduced between the beams on reflection, which is equivalent to an additional path difference of $\lambda/2$. When this added phase change is taken into account, the condition for *constructive interference* becomes

Constructive Interference

$$2t = (m + \tfrac{1}{2})\lambda$$

(22.11)

Here the $\lambda/2$ in Eq. (22.11) and the additional $\lambda/2$ from the phase change on reflection lead to the path difference being simply $m\lambda$, and so constructive interference and bright fringes result.

Similarly, for destructive interference and dark fringes we must have

Destructive Interference

$$2t = m\lambda$$

(22.12)

For white light this will mean constructive interference for some wavelengths and destructive interference for others, and so white light reflected back from the thin film will show only some colors. This explains the colors seen in thin films of oil spread out on wet, and therefore reflecting, surfaces.

Films of Materials Other Than Air

If the thin film consists of water, oil, or some other material between glass plates, the observed results are basically the same as those observed for air, except that the wavelength of the light in the film is reduced from λ to λ/n, since the speed of light in the film is reduced to $v = c/n$. Hence for *destructive* interference we must now have (allowing for the phase change on reflection)

$$t = \frac{m\lambda}{2n} \qquad \text{or} \qquad nt = \frac{m\lambda}{2}$$

(22.13)

The product nt of the index of refraction of the material and its actual thickness is called the *optical thickness* of the film. If optical thickness is used in place of t, Eqs. (22.11) and (22.12) remain valid, with λ the vacuum wavelength for the light, as usual.

Wedges

If the thin film is not of constant thickness, but wedge-shaped, then as the thickness increases, a series of alternating bright and dark fringes appear in monochromatic light. This is because the thickness satisfies the conditions first for constructive interference, then for destructive interference, and so forth, given by Eqs. (22.11) to (22.13). Fringes like those in Fig. 22.32b therefore appear. If both the plates are optically flat, i.e., ground flat to within a small fraction of a wavelength, the fringes will be straight. If the fringes are not perfectly straight, this is an indication of lack of optical flatness in either or both of the plates of glass. One glass plate which is known to be

FIGURE 22.32 Fringes produced by a wedge-shaped, thin air film: (*a*) Experimental setup for observing such fringes. (*b*) Observed fringes. The form of the fringes indicates that the air film varies greatly in thickness. Each dark fringe describes a region of equal thickness in the film; between two adjacent fringes the change in thickness is $\lambda/2n$. (*Photo from Cagnet, Françon, and Thrierr, Atlas of Optical Phenomena; reprinted by permission of Springer-Verlag, Heidelberg.*)

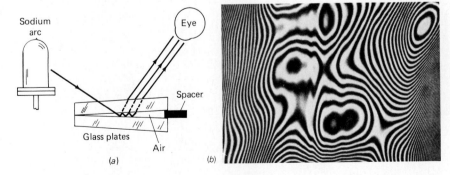

optically flat is often used to test the flatness of another glass plate. If the fringes produced by the two plates are not straight, the second plate is polished until they become straight. In this way the optically flat mirrors and glass plates needed for today's research laboratories can be produced.

Newton's Rings

A striking example of thin-film interference called *Newton's rings* occurs when the film is the air trapped between a flat glass plate and the convex surface of a lens with a very large radius-of-curvature, as in Fig. 22.33. This is basically the same situation as for the wedge discussed above except that the fringes now are circular, because the points at which the film has the same thickness form a series of concentric circles about the point of contact of lens and plate. In this case the distance between the two glass surfaces does not increase linearly, as it did for two flat plates. Rather, because of the convex curvature of the lens, the thickness of the air film increases more rapidly the farther out we go from the center of the lens. As a result the fringes are spaced farther apart near the center of the lens and get closer together as we go out from the center, as shown in Fig. 22.34. Where the lens makes perfect

FIGURE 22.33 (*a*) Experimental arrangement for producing Newton's rings. In this case the wedge has circular symmetry, leading to the circular fringes usually referred to as Newton's rings. (*b*) Diagram indicating how reflection from the top and bottom surfaces of the air film produces bright and dark fringes.

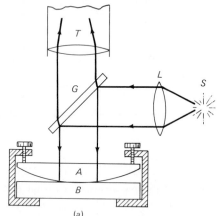

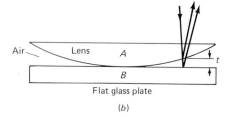

FIGURE 22.34 A Newton's rings interference pattern made by Dr. William F. Meggers of the National Bureau of Standards using a mercury source containing only the single isotope $^{198}_{80}$Hg. Such a source produces a particularly sharp fringe pattern. In the photo Dr. Meggers holds such a mercury source in his hand (*Photo courtesy of National Bureau of Standards and Omikron/Photo Researchers, Inc.*)

contact with the glass plate the center spot is not bright, but dark. This is clear evidence that there is a difference in phase of 180° produced in the two beams on reflection, as discussed previously.

Soap Bubbles

A soap bubble consists of a thin film of soapy water solution. When looked at in white light, a soap bubble shows bright colors. For some particular wavelength of light, i.e., some particular color, let us suppose that the optical thickness of the bubble wall (nt) is exactly $\lambda/2$. Then the optical path difference for light reflected from the inside and outside of the bubble wall will be exactly λ. But the beam reflected at the front surface of the film undergoes a 180° phase change, while the beam reflected at the back experiences no phase change. Hence the two beams are 180° out of phase and cancel. The light corresponding to the wavelength is therefore not reflected by the bubble. For other wavelengths, however, constructive interference will occur and brightly colored reflected light will be seen. As the bubble grows in size and begins to evaporate, its wall thins out and the colors seen by reflected light change. When the thickness of the bubble's wall becomes much less than the wavelength of any visible light, the bubble appears black just before it finally bursts.

Example 22.7

A soap bubble 250 nm thick is illuminated by white light. The index of refraction of the soap film is 1.36.
(a) What colors are *not* seen by reflected light?

(b) For what colors does constructive interference occur?
(c) What color does the soap bubble appear to be?

SOLUTION

In this case our treatment of films of materials other than air in the preceding section, and in particular Eq. (22.13), are applicable.
(a) For *destructive interference* we must have $nt = m\lambda/2$. Hence the wavelengths which are *not* reflected must satisfy the equation

$$\lambda_m = \frac{2nt}{m} \qquad m = 1, 2, 3, \ldots$$

We find

$$\lambda_1 = \frac{2nt}{1} = 2(1.36)(250 \text{ nm}) = \boxed{680 \text{ nm}}$$

$$\lambda_2 = \frac{2nt}{2} = (1.36)(250 \text{ nm}) = 340 \text{ nm}$$

These are the only wavelengths even close to the visible region of the spectrum for which destructive interference occurs: 680 nm is right in the middle of the red region of the spectrum, while 340 nm is in the ultraviolet and cannot be seen in any case. Hence the only nonreflected color is *red*.
(b) For *constructive interference* we must have, using Eq.

(22.11) modified for the proper optical thickness of the soap film,

$$2nt = (m + \tfrac{1}{2})\lambda$$

Hence

$$\lambda_m = \frac{2nt}{m + \tfrac{1}{2}}$$

and so

$$\lambda_1 = \frac{2nt}{3/2} = \frac{4}{3}(1.36)(250 \text{ nm}) = \boxed{453 \text{ nm}}$$

$$\lambda_2 = \frac{2nt}{5/2} = \frac{4}{5}(1.36)(250 \text{ nm}) = 272 \text{ nm}$$

A wavelength of 272 nm is in the ultraviolet region. Hence the only strong reflection is in the blue-violet region at 453 nm.
(c) The reflected light will be weak in the red region of the spectrum, and strong in the blue-violet region. Hence its color will be a pronounced blue.

22.8 Some Applications of Interference

Nonreflecting Films

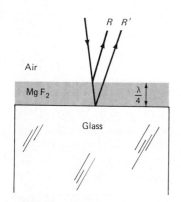

FIGURE 22.35 A nonreflective coating for glass. If the two reflected beams R and R' are 180° out of phase, they cancel and no light is reflected.

An interesting example of the practical utility of interference is the use of thin films as antireflection coatings on camera lenses and other optical components. Suppose we want all light of a particular wavelength to pass through a glass plate without any loss due to reflection. We can evaporate a thin film of a transparent substance like magnesium fluoride (MgF_2) on the outer surface of the glass and make this layer equal in thickness to $\lambda/4$, where λ is the wavelength of the light in the film. Then, as shown in Fig. 22.35, the light reflected at the film-glass interface will have traveled a distance $\lambda/2$ more than the light reflected from the air-film interface. Hence destructive interference results in the reflected light, and no net reflection occurs. We do not have to worry about phase changes in this case, since the index of refraction of the film in practice always has a value between that of the air and that of the glass, and hence the same phase change of 180° on reflection occurs at each interface.

Almost all optical parts of high quality are now coated to reduce reflection. Coated lenses and mirrors often have a purplish hue by reflected light. This is because perfect cancellation of the two beams can only occur for one wavelength, usually chosen to be in the middle of the visible spectrum in the green. Hence some red and some violet light is reflected, leading to the observed purplish color.

By using multilayer films of the proper thickness the reflection of light over a wide range of wavelengths can be controlled in a similar fashion. Such multilayer films may also be used to increase the reflection of light, or to split a light beam into two beams with any desired ratio of intensities. The "invisible glass" once in vogue in fashionable shop windows is another illustration of the use of nonreflecting films of this sort.

Example 22.8

Magnesium fluoride (MgF_2) has an index of refraction $n = 1.38$. How thick a coating of MgF_2 is needed on a store window to produce minimum reflection from the glass in the green region of the visible spectrum near 550 nm? Assume that light strikes the coating almost at normal incidence.

SOLUTION

For zero reflection we want the film thickness to be equal to $\lambda/4$ for green light *in the film*. Since the speed of light in the film is $v_f = c/n$, the wavelength in the film is:

$$\lambda_f = \frac{\lambda}{n} = \frac{550 \text{ nm}}{1.38} = 399 \text{ nm}$$

Hence the film thickness must be:

$$t = \frac{\lambda_f}{4} = \frac{399 \text{ nm}}{4} = \boxed{100 \text{ nm}}$$

Holography

A more modern application of light interference is the storage of three-dimensional images on two-dimensional photographic plates by a process called *holography*. Holography is a two-step process in which an object is first illuminated by coherent light (like that from a laser) and an interference pattern is produced on a photographic plate between the light arising from the coherent source and the light scattered by the object, as in Fig. 22.36. Then the developed photographic plate (called a *hologram*) is illuminated by light

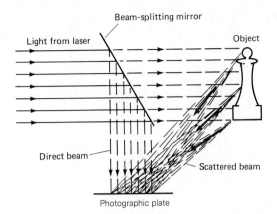

Beam-splitting mirror

Light from laser

Object

Direct beam

Scattered beam

Photographic plate

FIGURE 22.36 Making a hologram: An interference pattern is produced on the photographic plate between the light coming from a laser and the same light scattered from the object of which a hologram is desired, in this case, a chess pawn.

from the same coherent source, and a three-dimensional image of the original object results. Amazingly, the holographic image appears truly three-dimensional. A viewer changing position with respect to the image experiences the same changes in perspective experienced in walking around a real three-dimensional object.

Holography is developing at a very fast pace. Holographic microscopes and moving pictures are being developed. Not only are visible-light holograms being made, but holograms using infrared, ultraviolet, and microwave radiation, and even sound waves at ultrasonic frequencies, are being produced. Acoustic holography has been used, for example, to obtain images of the embryo in a mother's womb in which all the baby's organs and their relative sizes and positions can be seen clearly, without any known danger to the mother or the infant from the radiation.

22.9 The Polarization of Light

Diffraction and interference are clear examples of wave properties of light. They do not, however, indicate whether light is a transverse or a longitudinal wave. Sound, after all, is a longitudinal wave and still exhibits diffraction and interference effects. The most convincing experimental proof that light is a *transverse* wave comes from the phenomenon called *polarization*.

Just as the word *polarity* applied to a battery refers to the *direction* of the emf produced by the battery, so too the word *polarization* refers to the *direction* of the electric field vector **E** in a light wave or other kind of electromagnetic wave.

The electromagnetic waves discussed in Chap. 21 were all transverse waves, and if the wave was moving along the positive X axis, as in Fig. 21.10, the electric field **E** and the magnetic field **B** were perpendicular to the direction of propagation and to each other. If, for example, the **E** field were along the Y axis, the **B** field would have to be along the Z axis, as shown in the figure. Such a wave is called a *linearly polarized* wave.

Definition

Linearly polarized light*: A light wave in which all the electric field vectors are oriented in the same direction.

---

*There are other important kinds of polarized light, such as *circularly polarized* light, which we will not discuss here.

Since the light emitted by most light sources is produced by the superposition of the light from a very large number of atoms oriented at random with respect to one another, light as normally found in nature is *unpolarized*.

Definition

Unpolarized light: A light wave in which the electric field vectors are distributed in all possible directions normal to the direction of propagation of the light.

A beam of unpolarized light directed out of the paper toward us can therefore be represented with its electric field vectors oriented in all possible directions in the plane of the paper, as in Fig. 22.37. It is possible to change such an unpolarized beam of light into a beam linearly polarized along any desired direction at right angles to the direction of propagation by using a piece of *Polaroid*.*

Polaroid sheets have a preferred (or "easy") axis determined by the arrangement of the molecules in the sheet. Light polarized parallel to this axis can pass through the Polaroid sheet; light polarized at right angles to this axis is absorbed and does not pass through. Light whose **E** vector makes an angle θ with the easy axis of the Polaroid has a component of **E** parallel to the axis of the Polaroid sheet, $E \cos \theta$, which is passed by the Polaroid, whereas the component normal to the axis, $E \sin \theta$, is absorbed.

When unpolarized light such as that in Fig. 22.37 passes through a piece of Polaroid, therefore, it emerges linearly polarized along the direction determined by the easy axis of the Polaroid, as in Fig. 22.38. Used in this way, a piece of Polaroid is called a *polarizer*. Once the light has been polarized, a second piece of Polaroid, with its axis rotated through an angle of 90° with respect to the polarizer, will cut off the light completely, as in Fig. 22.39, since all the electric vectors are now at right angles to the easy axis of the Polaroid. A second piece of Polaroid, used in this way, is called an *analyzer*, for it can be used to determine (analyze) the polarization of the light incident on the Polaroid sheet. Note that a polarizer and an analyzer are identical. The different names arise from their different positions in the path of the beam of light and hence their different functions.

If polarized light is incident on an analyzer, the amplitude of the transmitted wave depends on the component of the electric field vector

---

*Polaroid is the trade name given to a plastic material capable of polarizing light, and invented in 1934 by Edwin H. Land (born 1909).

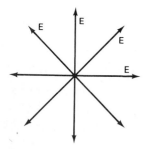

FIGURE 22.37 Unpolarized light. In this case the beam of light is directed out of the plane of the paper toward the reader.

FIGURE 22.38 The conversion of unpolarized light into linearly polarized light by a Polaroid sheet (a polarizer).

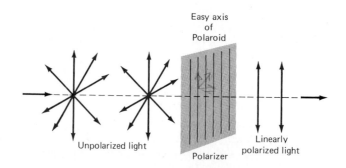

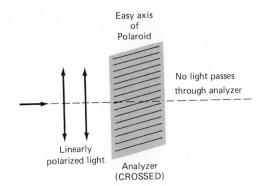

FIGURE 22.39 Use of a second sheet of Polaroid (an analyzer) to cut off a beam of polarized light.

parallel to the axis of the analyzer, and hence to the cosine of the angle θ between **E** and the easy axis of the Polaroid. For the amplitudes E'_{max} and E_{max} of the transmitted and incident waves we therefore have

$$E'_{max} = E_{max} \cos \theta$$

Since the intensity of the waves is proportional to the square of their amplitudes we have

$$\boxed{I' = I \cos^2 \theta} \tag{22.14}$$

Hence the intensity of the light transmitted by an analyzer depends on the square of the cosine of the angle between the polarization direction of the incoming wave and the axis of the Polaroid.

For example, if $\theta = 0$, $I' = I$ and the intensity is not reduced at all; if $\theta = 90°$, or the analyzer is "crossed" with respect to the polarization direction of the incoming light, $I' = 0$ and no light is transmitted.

Before the discovery of Polaroid *Nicol prisms* made of calcite were used as polarizers and analyzers for polarized light. These are still used today in preference to Polaroid in physics research laboratories.

Example 22.9

Linearly polarized light is incident on a piece of Polaroid at an angle of 60° with respect to the axis of the Polaroid. What is the intensity of the light passing through the Polaroid compared with the intensity of the incident light?

SOLUTION

Since $I' = I \cos^2 \theta$, we have

$$\frac{I'}{I} = \cos^2 60° = 0.25 = \boxed{\frac{1}{4}}$$

Hence the intensity of the light passed by the Polaroid is one-fourth that of the incident light.

Polarization by Reflection

If unpolarized light strikes a nonmetallic material like glass or water at any but normal incidence, the beam becomes partially polarized by reflection. Thus in Fig. 22.40 an unpolarized light beam is incident on a sheet of glass at an angle θ with respect to the normal. It is found by experiment that the reflected light is partially linearly polarized—i.e., the **E** vectors point in all directions normal

FIGURE 22.40 Polarization of unpolarized light by reflection from a glass plate at Brewster's angle. The polarization of the reflected light is total, with the **E** vectors parallel to the glass surface, when the reflected and refracted waves are at right angles. This occurs when $\theta + \theta_r = 90°$. The refracted light is not completely polarized, but the amplitude of the **E** vectors perpendicular to the air-glass interface is greater than for **E** vectors parallel to the glass surface.

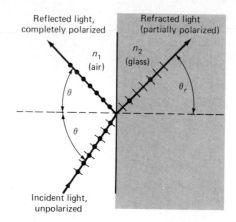

to the direction of propagation—but the amplitudes of these vibrations are greater parallel to the surface of the glass (i.e., normal to the plane of the paper) than they are at right angles to the plane.

For one particular angle of incidence, θ_B, called *Brewster's angle*, the linear polarization produced in the reflected light is complete. This occurs when all the components of the electric field not parallel to the surface of the glass are refracted through the glass, and only the parallel components are reflected. It is found by experiment that this occurs when the reflected and refracted beams are at right angles, so that from Fig. 22.40 we have

$$\theta + \theta_r = 90°$$

But, from Snell's law,

$$n_1 \sin \theta = n_2 \sin \theta_r$$

or $\quad n_1 \sin \theta = n_2 \sin (90° - \theta) = n_2 \cos \theta$

and so $\quad \tan \theta = \dfrac{n_2}{n_1}$

This can be written as

Complete Polarization by Reflection

$$\boxed{\tan \theta_B = n} \tag{22.15}$$

where θ_B is Brewster's angle, i.e., the angle for which complete polarization by reflection is achieved, and $n \ (= n_2/n_1)$ is the index of refraction of medium 2 with respect to medium 1.

Sunlight reflected from a flat horizontal reflecting surface such as a lake, river, highway, ocean, or desert is partially polarized with its electric vector horizontal, as just discussed. Polaroid sunglasses employ sheets of Polaroid with their easy axes vertical to cut out most of this reflected light. The light passing through the glasses is then the vertically polarized half of the diffusely scattered and unpolarized light from the trees, boats, hills, or houses we desire to see. The Polaroid sunglasses reduce the amount of light getting through to our eyes, but we see more clearly because much of the glare is removed.

Example 22.10

At what angle must light be incident on a piece of borosilicate crown glass, with index of refraction 1.52, to produce 100 percent linearly polarized light by reflection?

SOLUTION

From our discussion above, the light must be incident at Brewster's angle, given by

$$\tan \theta_B = n = \frac{n_2}{n_1}$$

Here n_1 is the index of refraction of air, which is equal to 1 to four significant figures. Hence we have: $\tan \theta_B = \dfrac{1.52}{1} = 1.52$

and so $\theta_B = \boxed{57°}$

For light incident on crown glass at 57°, therefore, the reflected light will be 100 percent linearly polarized.

Summary: Important Definitions and Equations

Huygens' principle:
   Every point on a wave front may be considered as the source of secondary wavelets which spread out in the forward direction at the speed of light in the medium. The new wave front is then the surface tangent to all these secondary wavelets.

Wave front:
   A surface joining all adjacent points on a wave which have the same phase.
   Ray: A line drawn perpendicular to a wave front at any point.

Reflection of light:
   1. The incident ray, the reflected ray, and the normal to the surface all are in the same plane.
   2. The angle of reflection r is equal to the angle of incidence i.

Refraction of light:
   The bending of a wave front of light when it travels from one medium to another.
   Snell's law of refraction: $n_2 \sin \theta_2 = n_1 \sin \theta_1$

Dispersion of light:
   The dependence of the index of refraction of a material on wavelength, leading to a difference in the angle of refraction.
   Diffraction: The bending of waves into the shadow region when they pass through holes or slits comparable in size to the wavelength.
   Single-slit diffraction:
   For dark fringes: $y_m = \dfrac{m\lambda L}{w}$

Interference of light:
   The superposition of two or more beams of light so that they reinforce each other (constructive interference—bright fringes) or cancel each other out (destructive interference—dark fringes), depending on their relative phases.

Young's double-slit experiment:
   For bright fringes: $y_m = \dfrac{m\lambda L}{d}$; $\sin \theta = \dfrac{m\lambda}{d}$
   For dark fringes: $y_m' = \dfrac{(m + \frac{1}{2})\lambda L}{d}$

Diffraction grating:
   Grating equation: $\sin \theta_m = \dfrac{m\lambda}{d}$

Thin films:
   1 Air film between glass plates:

   Bright fringes in reflected light: $2t = (m + \frac{1}{2})\lambda$

   Dark fringes in reflected light: $2t = m\lambda$

   2 Fluid film between glass plates (or other identical materials):

   Bright fringes in reflected light: $2nt = (m + \frac{1}{2})\lambda$

   Dark fringes in reflected light: $2nt = m\lambda$

   where nt is the optical thickness of the film, and n the index of refraction.

Holography:
   The use of the interference of light to store three-dimensional images on two-dimensional photographic plates.

Polarization of light:
   Linearly polarized light: A light wave in which all the electric vectors are oriented in the same direction.
   Polarization by reflection: For 100 percent polarization of reflected light, $\tan \theta_B = n$.

Diffraction and *interference* are convincing proofs of the *wave nature* of light.

Polarization is a convincing proof that light waves are *transverse waves*.

Questions

1 Why is a light beam reflected better by a plane mirror than by a piece of cardboard?

2 If light is incident normal to the surface of an air-glass interface, no refraction occurs. Show from Snell's law why this is to be expected.

3 Explain why an oar that is half in the water and half out of the water appears to bend at the point where it enters the water.

4 If a source of sound can be heard when it cannot be seen because of obstacles in the way, how can both sound and light be propagated as waves?

5 Explain why coherent light diffracted around a penny will produce a bright spot at the center of the penny on a screen.

6 Why does a soap bubble appear dark (and not bright) just before it bursts as it grows larger and thinner?

7 An oil film spreads out on wet pavement and produces colors by interference in reflected light. Near the edges, where the film is thinnest, the oil slick appears bright.

(*a*) Why does the oil slick appear bright, rather than dark, where it is very thin?

(*b*) What limits can you set on the value of the index of refraction of the oil from the information given?

8 For Young's double-slit experiment explain what happens to the bright fringes as:

(*a*) The wavelength of the light is increased.

(*b*) The distance to the screen is increased.

(*c*) The distance between the slits is increased.

(*d*) The width of each individual slit is increased.

9 Give a theoretical justification for the experimental fact that in a Newton's-rings interference pattern the fringes get closer together as they move out from the point of contact of lens and glass plate.

10 What is the difference between diffraction and interference?

11 (*a*) If light as it occurs in nature is "unpolarized," does this mean that the electric and magnetic vectors are not confined to a plane perpendicular to the direction of propagation?

(*b*) What then do we mean by unpolarized light?

(*c*) Can you suggest how unpolarized light might be produced?

12 Why are the diameters of some microwave transmitting antennas so large? (These are the large dish-shaped antennas you often see on ships, at airports, and at radio relay stations.) Assume that the microwaves used are from 1 to 10 cm in length.

13 Explain how one might use polarized light to measure the index of refraction of something opaque, like a lump of coal.

14 Unpolarized light is incident on a crossed polarizer and analyzer, and no light passes through the combination. A student claims to be able to make some light pass through the analyzer by placing a third piece of Polaroid between the polarizer and the analyzer. Will this work? If so, explain how a third piece of polarizing material can increase the transmission of the light.

Problems

A 1 The wavelength of visible light is shorter than the wavelength of an FM radio wave of frequency 100 MHz by about:

(*a*) 1 order of magnitude   (*b*) 3 orders of magnitude
(*c*) 7 orders of magnitude
(*d*) 10 orders of magnitude
(*e*) 14 orders of magnitude

A 2 Light enters a plate of boro silicate crown glass at an angle of 43° with respect to the normal to the air-glass surface. The angle made by the refracted beam with the normal is:

(*a*) 0°   (*b*) 26°   (*c*) 64°   (*d*) 32°   (*e*) 90°

A 3 The speed of blue light (400 nm) in light flint glass is less than the speed of red light (700 nm) in the same glass by about (see Fig. 22.14):

(*a*) 2%   (*b*) 20%   (*c*) 10%   (*d*) 0.2%
(*e*) 0.02%

A 4 Diffraction effects become important for visible light when the light encounters slits or holes with dimensions of the order of:

(*a*) 1 m   (*b*) 10^{-3} m   (*c*) 10 cm
(*d*) 1 mm   (*e*) 10^{-6} m

A 5 A single slit is 1.5 m from a screen. Monochromatic light of wavelength 633 nm falls on the slit and the third dark fringe falls at a distance of 3.0 cm from the center of the diffraction pattern. The width of the single slit is about:

(*a*) 1 mm   (*b*) 0.1 mm   (*c*) 1 cm
(*d*) 0.01 mm   (*e*) None of the above

A 6 A double-slit interference experiment uses a screen 1.0 m from the double slit, whose width is 0.50 mm. If the third-order bright fringe falls a distance of 4.0 mm from the center of the pattern, the wavelength of the light used must be:

(*a*) 670 nm   (*b*) 396 nm   (*c*) 472 nm
(*d*) 525 nm   (*e*) 590 nm

A 7 A diffraction grating is ruled with 12,000 lines to the centimeter. It is illuminated with mercury light. The angular displacement of the first-order green line at 546 nm with respect to the central image will be:

(*a*) 41°   (*b*) 3.7°   (*c*) 0°   (*d*) 49°
(*e*) None of the above

A 8 A thin film of water is trapped between two glass plates. On looking straight down on this water "sandwich," a blue color is observed, indicating that red light has been absorbed. One reasonable value for the thickness of the water film is:

(a) 0.26 μm   (b) 0.35 μm   (c) 0.15 μm
(d) 0.20 μm   (e) None of the above

A 9 A soap bubble 350 nm thick is illuminated by white light. The index of refraction of the soap film is 1.36. The most intense color in the reflected light from the bubble will be:

(a) Red   (b) Yellow   (c) Green   (d) Blue
(e) Violet

A10 Unpolarized light is incident on a piece of glass at an angle of 59° and it is found that the reflected light is 100 percent linearly polarized. The index of refraction of the glass with respect to the air is:

(a) 0.85   (b) 1.0   (c) 1.53   (d) 1.2   (e) 1.66

B 1 At time $t = 0$ a plane wave front of light is at a position 50 m from the origin along the X axis. What is the position of the same wave front 20×10^{-9} s later, if the light is moving through a vacuum?

B 2 A point source of light emits a spherical light wave in air. At time $t = 0$ the radius of the spherical wave front is 100 m. What is the radius of the wave front when $t = 3.0 \times 10^{-6}$ s?

B 3 A plane wave front is incident on a flat piece of plate glass ($n = 1.50$) so that the normal to the wave front makes an angle of 40° with respect to the normal to the glass surface.

(a) What is the angle of reflection?
(b) What is the angle of refraction in the glass?

B 4 Light from a He–Ne laser ($\lambda = 633$ nm) is incident on a narrow slit of width 0.020 mm. What is the position of the fourth-order dark fringe produced by the slit on a screen 1.8 m away from the slit?

B 5 Light of wavelength 579 nm from a mercury arc falls on a double slit, where the distance between the centers of the two slits is 0.080 mm. On a screen 2.0 m from the slit what is the location in the interference pattern:

(a) Of the first dark fringe?
(b) Of the first bright fringe?

B 6 Light of wavelength 546 nm falls on a diffraction grating normal to its surface. The grating is ruled with 7500 lines per centimeter. What is the angle corresponding to the first bright fringe produced by the grating?

B 7 Light of wavelength 589 nm from a sodium arc falls at right angles on an air film trapped between two pieces of glass. What is the minimum thickness of air film for which no visible light will be reflected back?

B 8 A soap bubble of index of refraction 1.36 is illuminated by white light. No red light in the region around 700 nm is reflected from the walls of the bubble. What is the minimum thickness of the bubble walls which will explain this observation?

B 9 A magnesium fluoride film, of index of refraction 1.38, is used to coat a prism to reduce the light reflected from the front face of the prism. What must be the thickness of the film to reduce the reflection of the red light ($\lambda = 633$ nm) from a He–Ne laser to zero?

B10 Unpolarized light falls on a glass surface at an angle of 56°. It is found that the reflected beam is 100 percent linearly polarized with its electric vectors parallel to the surface of the glass. What is the index of refraction of the glass?

B11 Linearly polarized light falls on a piece of Polaroid with the direction of polarization of the light making an angle of 45° with the easy axis of the Polaroid. If the intensity of the light striking the Polaroid is 50 mW/cm², what is the intensity of the light emerging from the Polaroid?

C 1 It is desired to use two plane mirrors to turn a beam of light around so that it returns in the direction from which it came. At what angles can the two mirrors be arranged to accomplish this reversal?

C 2 It is desired to have a beam of light enter a block of heavy flint glass at such an angle (other than zero) that it will be refracted normal to the air-glass interface. Is this possible? If so, what must be the angle of incidence? If not, why not?

C 3 A light beam is directed inward through the outer glass wall of an aquarium tank at an angle of 45° with respect to the normal.

(a) If the index of refraction of the glass is 1.53, and of the water 1.33, at what angle with respect to the normal will a fish see the beam enter the tank? (*Hint:* Apply Snell's law repeatedly.)
(b) Would the same result be obtained if the glass were not present and the light went directly from the air to the water?

C 4 White light passes from air into a block of heavy flint glass at an angle of 55° with respect to the normal. What is the difference in the angle of refraction between the red light and the blue light in the beam? (*Hint:* Use Fig. 22.14.)

C 5 Mercury light at 436 nm is incident on a narrow slit of width 0.10 mm. Find the width of the central bright fringe on a screen 2.5 m away.

C 6 Light from a hydrogen discharge tube falls on a narrow slit 0.050 mm in width. On a scale 2.0 m from the slit what is the distance between the third-order dark fringe produced by the blue-green 486-nm line and that produced by the blue 434-nm line?

C 7 Light from a sodium arc, with $\lambda = 589$ nm, is incident on a slit of width 0.15 mm.

(a) Describe the diffraction pattern on a screen 1.0 m away from the slit.
(b) Describe what happens if the slit width is reduced to 0.05 mm.

C 8 What must be the separation between two slits which will produce a fourth-order bright fringe at a distance of 2.0 cm from the center of the interference pattern on a screen 1.0 m from the slit? Assume that:

(a) The light used is red at a wavelength of 700 nm.
(b) The light used is blue at a wavelength of 400 nm.

C 9 A double slit has a slit separation of 0.010 mm. For red light at 633 nm what is the highest-order bright fringe which can be observed on a screen?

C10 Light from a tunable dye laser is incident on a double slit. If the slit separation is 0.050 mm, and the first bright fringe occurs at a distance of 2.5 cm from the center of the interference pattern on a screen 2.0 m from the slit, what is the wavelength of the light from the laser?

C11 An adjustable double slit is used in Young's experiment. Initially the slit separation is 0.10 mm and the distance between the fifth-order bright fringes to either side of the central image on the screen is 16 cm. What is the distance between the same fringes if the slit separation is reduced to 0.02 mm?

C12 Light of 546 nm from a mercury arc falls on a diffraction grating ruled with 30,000 lines per inch. What is the angular separation between the first-order images to either side of the central maximum?

C13 How many orders of fringes of the 436-nm blue line of mercury can be observed when this light is incident on a grating ruled with 5000 lines per centimeter?

C14 A third-order violet line from an unknown element is observed to fall near the second order of the mercury 579-nm yellow line. What is the wavelength of the unknown line?

C15 An oil film of index of refraction 1.3 is trapped between two pieces of glass. No light is reflected by such a film when 536-nm light falls on it at normal incidence. What is the minimum thickness of the oil film which will satisfy these conditions?

C16 A metal shim is used to separate the ends of two long pieces of flat glass, while the other two ends remain in direct contact. The plates are 50 cm long. If mercury light with $\lambda = 436$ nm is incident on the plates, the distance between two consecutive dark fringes is found to be 1.0 cm. What is the thickness of the shim?

C17 A soap bubble 400 nm thick is illuminated by white light. The index of refraction of the film is 1.36.

(*a*) What colors do not appear in the reflected light?

(*b*) What colors appear strongly in the reflected light?

C18 A magnesium fluoride film of index of refraction 1.38 reduces the reflection of green light ($\lambda = 546$ nm) almost to zero. What wavelengths will such a coated window reflect strongly?

C19 The index of refraction of a plate of flint glass is 1.66. At what angle of incidence on this plate is the reflected light 100 percent linearly polarized if the plate is immersed (*a*) in oil, (*b*) in water, (*c*) in benzene?

C20 At what angle is the sun above the horizon when the sunlight reflected from a swimming pool can be completely obscured with a piece of Polaroid?

D 1 A beam of light from an underwater flashlight is directed from the bottom of a pool at an angle of 75° with respect to the normal to the surface of the water. At what angle will the beam emerge from the water?

D 2 Prove that, when white light passes through a 60° (equiangular) prism, the violet light is deviated more than the red, leading to the continuous spectrum first observed by Newton. (For convenience assume that the light beam in the prism travels parallel to one side of the prism.)

D 3 The brilliance of diamonds arises partially from their very large index of refraction, $n = 2.42$. Hence light becomes trapped inside the diamond for some time before it emerges. What is the largest angle at which light can strike the diamond-air interface and still escape from the diamond?

D 4 A light wave passes from air into a piece of glass of index of refraction n. It is found experimentally that the angle of incidence is twice the angle of refraction.

(*a*) Find the angle of incidence in terms of n.

(*b*) What are the minimum and maximum values of n for which this situation can occur?

D 5 Light of wavelength 633 nm from a He–Ne laser falls on a double slit where each slit has a width of 0.010 mm and the distance between the centers of the two slits is 0.10 mm. Compare the widths of the single-slit diffraction pattern and the double-slit interference pattern produced on a screen 1.0 m away.

D 6 In the Young's double-slit experiment of Prob. B5 a piece of glass is placed in front of one of the slits. Then a thin film of MgF_2, with $n = 1.38$, is evaporated on the glass to a thickness of 1.0×10^{-4} cm. How far does the first bright fringe move sideways when the thin film is deposited on the glass?

D 7 A long-focal-length lens is supported horizontally a short distance above the flat end of a steel cylinder. The cylinder is 10 cm high and is rigidly clamped at its base. Newton's rings are produced between the curved surface of the lens and the flat surface of the top of the steel cylinder. The light used is that from a sodium lamp at 589 nm. A microscope is used to observe the Newton's rings. When the temperature of the cylinder is raised 20°C, 80 fringes are observed to move past the cross hairs of the microscope. What is the coefficient of linear expansion of steel?

D 8 The yellow light from a sodium arc consists of two very close wavelengths of 589.0 and 589.6 nm, respectively. If this light falls normally on a plane diffraction grating with 500 lines per centimeter:

(*a*) What is the angular separation of these two lines in the first order?

(*b*) In the second order?

(*c*) If the human eye can distinguish lines with an angular separation of at least 1 minute of arc, will it be able to see these two lines distinctly?

D 9 The wavelength limits of the visible spectrum are 390 and 760 nm. If white light falls at normal incidence on a plane diffraction grating ruled with 5000 lines per centimeter, what is the angular spread of:

(*a*) The first-order spectrum?

(*b*) The second-order spectrum?

(*c*) The third-order spectrum?

D10 The captain of a naval vessel desires to send a message to a Coast Guard lighthouse on shore by using a light beam which is 100 percent linearly polarized parallel to the ocean surface. The captain decides to polarize the beam by reflecting it from the ocean surface. If the transmitter is 20 m above the ocean surface and the receiver is on a cliff 150 m above the ocean surface, how far from the cliff should the captain locate the ship to carry out this mission? (Assume that the index of refraction of the water at the frequency used is 1.33.)

Additional Readings

Baumeister, Philip, and Gerald Pincus: "Optical Interference Coatings," *Scientific American*, vol. 233, no. 6 December 1970, pp. 58–75. Describes the making of multilayer interference films and their use for a variety of practical purposes.

Cagnet, M., M. Françon, and J. C. Thrierr: *Atlas of Optical Phenomena*, Springer-Verlag, Heidelberg, 1962. Contains a magnificent selection of photographs illustrating important optical phenomena, including diffraction and interference.

Ingalls, Albert G.: "Ruling Engines," *Scientific American*, vol. 186, no. 6, June 1952, pp. 42–54. A description of the various kinds of machines used to attain the fantastic precision required in the ruling of diffraction gratings.

Light and Its Uses: Making and Using Lasers, Holograms, Interferometers, and Instruments of Dispersion, readings from *Scientific American* (introduction by Jearl Walker), Freeman, New York, 1980. Contains useful articles on holography and a good bibliography.

Moore, A. D.: "Henry A. Rowland," *Scientific American*, vol. 246, no. 2, February 1982, pp. 150–161. An account of the success of the famous Johns Hopkins University professor in producing high-precision diffraction gratings.

Ruechardt, Edward: *Light, Visible and Invisible*, University of Michigan Press, Ann Arbor, Mich., 1958. A very well illustrated discussion of the wave properties of light and of x-rays.

Shamos, Morris H. (ed.): *Great Experiments in Physics*, Holt, Rinehart and Winston, New York, 1959. Chapter 7 is devoted to Young's work on the interference of light, and chap. 8 to Fresnel's work on diffraction.

Shurcliff, William A., and Stanley S. Ballard: *Polarized Light*, Van Nostrand, New York, 1964. The nature and applications of polarized light.

Van Heel, A. C. S., and C. H. F. Velzel: *What Is Light?*, McGraw-Hill, New York, 1968. Much of this introductory college text is devoted to well-illustrated discussions of diffraction, interference, and polarization.

Wehner, Rudiger: "Polarized Light Navigation by Insects," *Scientific American*, vol. 235, no. 1, July 1976, pp. 106–115. An account of experiments showing that bees and ants find their way home by detecting the polarization of scattered sunlight from the sky.

Wood, Alexander, and Frank Oldham: *Thomas Young, Natural Philosopher, 1773–1829*, Cambridge University Press, Cambridge, 1954. One of the few biographies of a surprisingly neglected figure in the history of physics.

Wood, Elizabeth A.: *Science for the Airplane Passenger*, Ballantine, New York, 1969. This book contains an interesting chapter on optical phenomena as observed from airplanes.

*A*ll the compliments that I have received from Arago,
Laplace and Biot never gave me as much pleasure as the
discovery of a theoretic truth, or the confirmation of a
calculation by experiment.

Augustin Fresnel (1788–1827)

Chapter 23

Geometrical Optics: Mirrors and Lenses

In the preceding chapter we presented strong evidence that light is a wave phenomenon. Why then, in prescribing eyeglasses for patients, do optometrists never take into consideration the wave theory of light? How can the makers of eyeglass lenses predict the path of light through the glasses to the retina of the eye without considering diffraction and interference effects? Optometrists and eyeglass makers use geometrical, or ray, optics in their work rather than wave, or physical, optics. In this chapter we will discuss the validity and advantages of such an approach in dealing with optical components such as mirrors and lenses.

23.1 Geometrical, or Ray, Optics

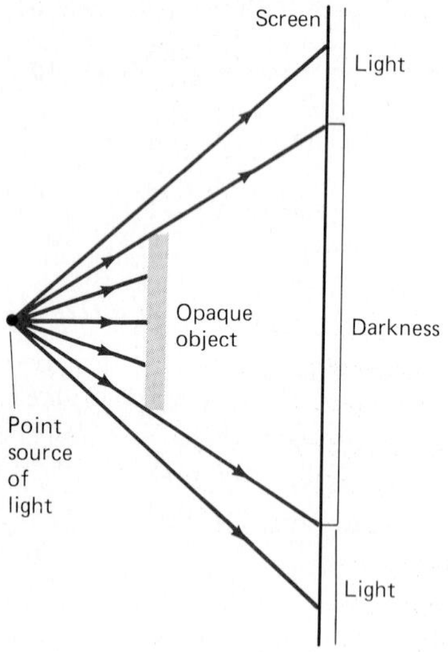

FIGURE 23.1 The shadow cast by an opaque object on a screen, on the assumption that light travels along rays in perfectly straight lines.

In the previous chapter we used a wave theory and Huygens' principle to explain the behavior of light. This *wave model* works extremely well, and with certain properties of light, such as diffraction and interference, it is absolutely necessary to achieve agreement with experiment. With regard to other properties, however, such as reflection and refraction, a simpler *ray model* works equally well.

The ray model of light assumes that light travels in straight lines along "rays." These rays are the normals to the wave fronts discussed in Chap. 22. Opaque objects therefore cast sharp shadows whose dimensions can be calculated by geometry. For this reason *geometrical optics* is sometimes used as a synonym for *ray optics*. Similarly, if light passes through an aperture large compared with the wavelength, the area of a screen illuminated by the light can be found, with the aid of a little geometry, by assuming that light travels in perfectly straight lines. These two cases are illustrated in Figs. 23.1 and 23.2. Note that we have completely ignored diffraction effects in these diagrams.

The wavelength of visible light is about 500 nm, or 5×10^{-5} cm, which is very small compared with the pupil of the eye or most common objects. As a result the diffraction and interference of light are rarely observed in our everyday lives. This is the justification for using a simpler ray model. As we saw in Chap. 22, the straight-line propagation of light is relative and depends on the ratio λ/w, where λ is the wavelength of the light and w is the smallest dimension of the object or aperture on which the light is incident. Since the first dark fringe in a single-slit diffraction pattern occurs at $y_1 = \lambda L/w$ [see Eq.

(22.5)], then if $w \gg \lambda$, y_1 approaches zero and the angle to the first minimum is so small that it is unobservable.

For visible light and objects even as small as 1 mm, $\lambda/w \simeq (10^{-7}\,\text{m}/10^{-3}\,\text{m}) \simeq 10^{-4}$, and the deviation of light from a straight-line path is only about 1 part in 10,000. For most practical purposes in our daily lives, therefore, we can neglect diffraction and interference effects and assume that light travels in straight lines. In the rest of this chapter we will assume that this is the case.

FIGURE 23.2 The illumination of a screen by light passing through a wide slit, on the assumption that light travels along rays in perfectly straight lines.

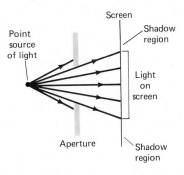

Example 23.1

Calculate the approximate ratio of wavelength to obstacle size (say, that of a typical house) for the propagation through space of:
(a) AM radio waves.
(b) FM radio waves.
(c) Microwaves (radar waves).
(d) Light waves.

Point out how you would expect these four kinds of radiation to differ in their interaction with obstacles in their path.

SOLUTION

From Table 21.2 the frequency of AM radio waves is about 10^6 Hz, and of FM radio waves about 10^8 Hz. From Fig. 21.17 the wavelength of microwaves is about 10 cm.

(a) For AM radio, therefore,

$$\lambda = \frac{c}{f} \simeq \frac{3 \times 10^8\,\text{m/s}}{10^6\,\text{s}^{-1}} \simeq 300\,\text{m}$$

and $\quad \dfrac{\lambda}{w} \simeq \dfrac{300\,\text{m}}{20\,\text{m}} \simeq \boxed{15}$

Here we have taken 20 m as one dimension of a typical house in the path of the radio wave. In this case λ/w is so large that diffraction is very important and the radio waves bend around any obstacle in their path. AM radio waves, therefore, do not travel in straight lines and do not cast sharp shadows.

(b) For FM radio,

$$\lambda = \frac{c}{f} \simeq \frac{3 \times 10^8\,\text{m/s}}{10^8\,\text{s}^{-1}} \simeq 3\,\text{m}$$

and $\quad \dfrac{\lambda}{w} \simeq \dfrac{3\,\text{m}}{20\,\text{m}} \simeq \boxed{0.15}$

The value for λ/w is still relatively large, and as a result diffraction remains important. If the obstacle is very large,

like a gigantic office building, however, diffraction may be greatly reduced and FM radio waves may not propagate completely into the shadow of the building. Because of this smaller value of λ/w, FM radio waves have paths closer to straight lines than do AM radio waves. For this reason FM radio waves do not follow the curvature of the earth, and, as a consequence, FM communication is limited to distances of 20 to 40 mi.

(c) For 10-cm radar waves, we have

$$\frac{\lambda}{w} \simeq \frac{0.10\,\text{m}}{20\,\text{m}} = 0.005$$

In this case λ/w is so small that diffraction effects can be neglected and we can use ray optics. For example, a radar beam striking airplanes or ships at sea is strongly reflected back, which explains why radar is so useful for detection purposes.

(d) For visible light, with $\lambda = 5 \times 10^{-7}$ m,

$$\frac{\lambda}{w} \simeq \frac{5 \times 10^{-7}\,\text{m}}{20\,\text{m}} \simeq 2.5 \times 10^{-8}$$

and diffraction is completely negligible for large objects such as a house at the wavelengths of visible light.

23.2 Reflection and Refraction on a Ray Model

The ray model of light is especially useful in dealing with reflection and refraction, and hence in dealing with the behavior of mirrors, lenses, and many optical instruments.

Reflection of Light

Consider a ray of light incident at an angle of incidence i with respect to the normal to a plane mirror, as in Fig. 23.3. This ray is normal to the plane wave front incident on the mirror. The reflected ray is similarly normal to the reflected wave front. The speed of any point along the ray is the same for both incident and reflected rays, since they are in the same medium. Hence the predicted behavior for the reflection of light on a ray model will be identical with the results based on a wave model. We can therefore conclude, as we did in the preceding chapter, that (1) the incident ray, the reflected ray, and the normal to the reflecting surface are all in the same plane; and (2) the angle of reflection is equal to the angle of incidence.

These results for the reflection of light are the same as would be obtained for the path of a ball making a perfectly elastic collision with the floor at an angle of incidence i, as in Fig. 23.4. The ball would be reflected at an angle $r = i$, and its motion would be in the plane defined by its initial path and the normal to the floor, since no force acts on the ball to move it out of that plane.

FIGURE 23.3 The reflection of light on a ray model. The angle of reflection r is equal to the angle of incidence i.

FIGURE 23.4 A ball bouncing off the ground at an angle. Here too the angle of reflection is equal to the angle of incidence.

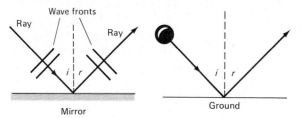

Refraction of Light

We consider light rays incident at an angle θ_1 with respect to the normal to the interface between two different media, one of index of refraction n_1 and the other of index of refraction n_2, as in Fig. 23.5. Our argument here is the same as in the case of reflection. The incident rays shown in the figure are normal to the wave front PQ. The refracted rays are also normal to the wave front $P''Q''$. Hence whatever we have proved about wave fronts will also be true for the rays normal to these wave fronts. We are therefore led to the same result found in Sec. 22.2, that is, to Snell's law:

FIGURE 23.5 The refraction of light on a ray model. This leads to Snell's law: $n_1 \sin \theta_1 = n_2 \sin \theta_2$.

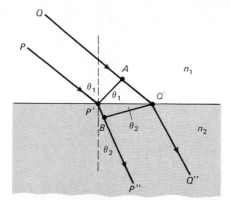

FIGURE 23.6 The reflection and refraction of a light ray by a block of glass. The light enters the glass from above, and is partially reflected and partially refracted at the upper surface. At the lower surface the ray is again partially reflected and partially refracted, as shown. Since the ray is refracted toward the normal on passing from air to glass, and away from the normal on passing from glass to air, and the two faces of the block are parallel, the emerging ray has the same direction as the incident ray, but has been displaced sideways. (*Photo courtesy of Educational Development Center, Newton, Massachusetts.*)

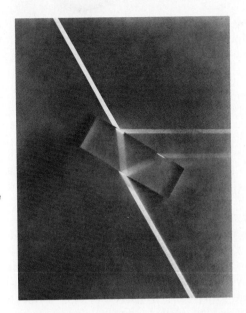

$$n_1 \sin \theta_1 = n_2 \sin \theta_2 \tag{23.1}$$

Here θ_1 is the angle made by the incident ray with the normal, and θ_2 is the angle made by the refracted ray with the normal. Figure 23.6 shows the reflected and refracted rays produced at air-glass interfaces.

Hence we see that the same results are to be expected for reflection and refraction whether we use a wave or a ray model for light. This is the reason that ray optics is adequate for most problems involving lenses and mirrors, which will be our main concern in this chapter.

Total Internal Reflection

According to Snell's law,

$$\sin \theta_2 = \frac{n_1}{n_2} \sin \theta_1$$

If light is traveling from a more dense (n_1) to a less dense (n_2) optical medium (from glass to air, for example), as in Fig. 23.7, n_1 is greater than n_2. For small angles of incidence, part of the light is refracted into the less dense medium, and part is reflected back into the more dense medium. When the angle of incidence is such that $\sin \theta_c = n_2/n_1$, we have $\sin \theta_2 = 1$, $\theta_2 = 90°$, and the light ray C incident at the angle θ_c is refracted along the interface between the two boundaries. For any angle greater than θ_c, as that for ray D, only reflection

FIGURE 23.7 Total internal reflection. Light travels from a medium of index of refraction n_1 to a medium of index of refraction n_2, where $n_1 > n_2$. For angles greater than θ_c, where $\sin \theta_c = n_2/n_1$, no light passes into the upper medium; the light is totally reflected back into the lower medium.

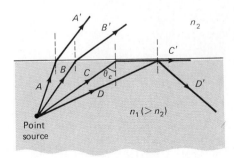

(and no refraction) occurs; i.e., the light is no longer partially but is *totally* reflected at the interface. This is called *total internal reflection* and occurs for all angles greater than the critical angle θ_c given by

Critical angle

$$\boxed{\sin \theta_c = \frac{n_2}{n_1}}$$

(23.2)

For example, consider light passing from borosilicate crown glass, with $n_1 = 1.52$, into air. Then the critical angle is given by:

$$\sin \theta_c = \frac{n_2}{n_1} = \frac{1}{1.52} = 0.66 \qquad \text{and} \qquad \theta_c = 41°$$

For all angles of incidence greater than 41°, the light is totally reflected. Figure 23.8 shows an actual light beam undergoing total internal reflection.

When total internal reflection occurs, no energy is lost by refraction or absorption, but the interface acts as a perfect reflector. This makes it convenient, for example, to use 45° prisms (i.e., solid triangular blocks of glass, with two 45° angles and one 90° angle) in place of mirrors to reflect light in binoculars. This is shown in Fig. 23.9, in which all angles of incidence on the prism sides are 45°, which is greater than the critical angle of 41°. "Corner-cube" reflectors, which reflect any light beam directly back on itself, also use total internal reflection for the same purpose.

FIGURE 23.8 Total internal reflection. Light rays enter a prism from the left at different angles with respect to the normal to the prism surface. The top four rays are partially reflected and partially refracted, as can be seen. The two lowest rays, however, are totally reflected at the glass-air interface, since their angle of incidence exceeds the critical angle. These two rays are much more intense than the other rays emerging from the bottom of the prism. (*Photo courtesy of Educational Development Center, Newton, Massachusetts.*)

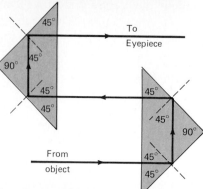

FIGURE 23.9 The use of 45° prisms to reflect light in binoculars. Since the critical angle for total internal reflection is about 41°, the light is completely reflected twice in each prism, as shown.

Waterford and other types of crystal used in chandeliers and glassware are made from very heavy lead glass, for which n_1 is about 1.89. Hence the critical angle in this case is only 32°. The crystal is therefore cut at such angles that much of the incident light is reflected back and forth inside the crystal many times by successive total internal reflections before it escapes. This gives the crystal its uniquely bright, dazzling appearance. This sparkling effect is even greater in diamonds, for which $n = 2.42$ and θ_c is only 24°.

Example 23.2

Light passes from a prism of heavy flint glass directly into a second prism of borosilicate crown glass, which is in contact with it. If the light incident on the interface makes an angle of 60° with respect to the normal, as in Fig. 23.10, what happens to the light after it strikes the interface?

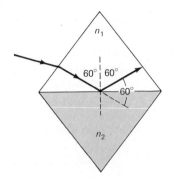

FIGURE 23.10   Diagram for Example 23.2.

SOLUTION

Usually we would expect that light would be partially reflected and partially refracted at the interface. To see if that happens in this case let us calculate the critical angle for light passing from the more dense to the less dense glass. The critical angle is given by

$$\sin \theta_c = \frac{n_2}{n_1} = \frac{1.52}{1.89} = 0.80$$

and so $\theta_c = 54°$

Since the light is incident at an angle of 60°, which is greater than the critical angle, no refraction occurs, but *the light is totally reflected.*

For the reflected light, the angle of incidence is equal to the angle of reflection. The reflected ray therefore makes an angle of 60° with the direction of the incident ray just before it hit the interface, as can be seen from Fig. 23.10.

Fiber Optics

Total internal reflection enables light to be transmitted inside thin glass fibers, since the light will be internally reflected off the sides of the fiber and will therefore follow the fiber's contour. In this way light can be carried around corners as long as the bends are not too sharp, so that the light always strikes the sides at angles greater than the critical angle. This is shown in Fig. 23.11. Such fibers can be used to look at otherwise inaccessible objects.

FIGURE 23.11   (a) Bending of light around curves by optical fibers. (b) Photo of pinpoints of light emerging from glass fibers used in light communication systems. (*Photo courtesy of Bell Telephone Laboratories.*)

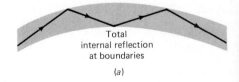

Total internal reflection at boundaries

(a)

(b)

In practice, glass fibers of uniform thickness and of diameters about 2×10^{-6} m are commonly used. These are made of glass of high refractive index coated with a thin layer of glass of lower refractive index to provide a suitable boundary. These fibers are combined in bundles which are flexible enough to bend easily without breaking. If the end of a fiber bundle of this sort is polished flat, almost all the light entering one end will emerge from the other. These fiber bundles are therefore frequently referred to as *light pipes*.

More recently, transparent plastics have been used to replace the glass fibers, but it is difficult to make them as thin as glass fibers.

Fiber optics are now used routinely to take pictures inside the human body using cystoscopes and similar medical instruments. For example, a light pipe can be inserted into a patient's stomach through the mouth and used to examine the stomach for ulcers. Light that is transmitted down the outer layer of the light pipe's optical fibers is scattered back by the stomach wall and transmitted through the central portion of the fiber bundle to produce an image of the stomach wall. This image can be either observed visually or recorded photographically. Since the fibers are very thin and numerous, excellent detail can be achieved in the final image produced.

Such light pipes can also be used to transmit high-intensity laser light inside the body for use in surgery and cancer therapy. The use of fiber optics in medicine again demonstrates that a simple physical principle can have far-reaching consequences in other fields of research.

23.3 Refraction and Dispersion of Light by a Prism

Prisms are important components in many optical systems. For example, in Fig. 23.12 we show a glass prism of index of refraction n and vertex angle A between the two glass surfaces at which refraction occurs. A light ray passes from air into the prism at an angle θ_1 with respect to the normal to its left face. The ray is refracted toward the normal at the glass surface, passes through the prism, is refracted away from the normal at the right face, and emerges at an angle δ with respect to the original ray. δ is called the *angle of deviation* and can be found by successive applications of Snell's law to the two prism faces.

To simplify our derivation we assume that the deviation produced is the minimum possible deviation for any given prism and wavelength of light. This minimum occurs when the light ray passes through the prism symmetrically or parallel to its bottom face, as in the figure. In this case $\theta_1 = \theta_1'$, and by symmetry half the total deviation occurs at each face. The *angle of minimum deviation* δ_m is then equal to $2\delta_1$.

At the first face we have, from Snell's law, with $n_1 = 1$ and $n_2 = n$,

$$\sin \theta_1 = n \sin \theta_2$$

FIGURE 23.12 Refraction of light by a prism. It is presumed that the angle of incidence is such that $\theta_1 = \theta_1'$, in which case the deviation δ_m is found experimentally to be a minimum.

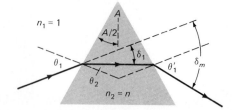

where n is the index of refraction of the glass. The deviation at the first face, δ_1, can be found from the fact that $\theta_1 = \theta_2 + \delta_1$, as can be seen from the figure. But $\theta_2 = A/2$, since the sides making the angles θ_2 and $A/2$ are mutually perpendicular. Also $\delta_1 = \delta_m/2$, from above. Hence

$$\theta_1 = \frac{A}{2} + \frac{\delta_m}{2} = \frac{A + \delta_m}{2}$$

Substituting in Snell's law we then have:

$$\sin \frac{A + \delta_m}{2} = n \sin \theta_2 = n \sin \frac{A}{2}$$

or finally, $$n = \frac{\sin\left[(A + \delta_m)/2\right]}{\sin (A/2)} \tag{23.3}$$

This is a very useful formula for finding indices of refraction of solids and liquids. For transparent solids the material can be made into a prism, and light of a single wavelength passed through it at minimum deviation. Then a measurement of the angle of the prism A and of the angle of minimum deviation δ_m yields a value for n in Eq. (23.3). A and δ_m can be measured to high accuracy with a spectrometer. With precision equipment and a little care the index of refraction may be obtained to 1 part in 10^5 or 10^6 with this method.

For liquids hollow prisms filled with the liquid can be used, and n determined from a measurement of δ_m.

Dispersion by a Prism

We saw in the preceding chapter that the index of refraction of glass varied slowly with wavelength in the visible region of the spectrum (see Fig. 22.14). From Eq. (23.3) we can therefore conclude that for larger values of n, $\sin\left[(A + \delta_m)/2\right]$ is larger and hence the deviation δ_m is also larger. Since n is larger for violet light than for red, violet light must be deviated most and red light least by a prism. Hence, when white light is incident on a prism, as in Fig. 23.13, we obtain a continuous spectrum of colors in the deviated light, running from red, which is least deviated, through orange, yellow, green, and blue to violet, which is most deviated. In 1666 Sir Isaac Newton made the first recorded observation of this dispersion of sunlight by a prism.

If the light source emits not white light but light characteristic of a particular element or molecule, then the prism produces sharp images of a slit placed in front of the incident light. These line images of the slit occur at wavelengths peculiar to each material. Hence a prism can be used in place of a grating as the dispersing element in the spectrometer of Fig. 22.29. The spectrum produced by a prism spectrometer, however, is the reverse of that produced by a grating spectrometer, since a grating deviates red light most and violet light least.

FIGURE 23.13 A prism used to break up white light into its component colors. The red light is deviated least, and the violet most, leading to a complete visible spectrum.

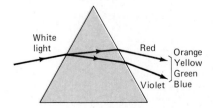

Example 23.3

In a physics laboratory experiment a student is given a prism with all three angles equal to 60°, and told to determine the exact type of glass from which the prism is made. The student finds that, when red light of wavelength 700 nm is incident on the prism at such an angle as to produce minimum deviation, the measured angle of minimum deviation is 44°. What can the student conclude about the kind of glass in the prism?

SOLUTION

The student can use Eq. (23.3) to find n for the prism, since A and δ_m are both known. The result is:

$$n = \frac{\sin\left[(A + \delta_m)/2\right]}{\sin(A/2)}$$

$$= \frac{\sin\left[(60° + 44°)/2\right]}{\sin(60°/2)} = \frac{\sin 52°}{\sin 30°} = \frac{0.79}{0.50} = \boxed{1.58}$$

From Fig. 22.14 the student can see that this value agrees reasonably well at 700 nm with the value of n for light flint glass. Hence the student might conclude that the prism is made of *light flint glass*. Obviously, other materials with indices of refraction near 1.58 are also possible, but light flint glass is a reasonable choice given the data in the problem.

23.4 Spherical Mirrors

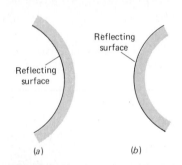

(a)   (b)

FIGURE 23.14 Types of spherical mirrors: (a) a concave mirror; (b) a convex mirror.

One of the most important applications of the laws governing the reflection of light is to mirrors. Most of us look at ourselves in a plane or concave mirror at least once a day, but probably never wonder how such a mirror works. An ordinary plane mirror is made by applying a thin metallic backing to a flat, polished glass surface. A spherical mirror has the shape of a piece cut out of a spherical surface where the radius of the sphere R is called the *radius of curvature* of the mirror. Such mirrors can be made to reflect almost 100 percent of the incident light.

If the reflecting surface of a spherical mirror corresponds to the *inside* portion of the sphere, the mirror is a *concave* mirror (the "cave" in *concave* has the same meaning as *cavity*). If the reflecting surface corresponds to the *outside* portion of the sphere, then the mirror is a *convex* mirror (see Fig. 23.14). Concave mirrors are often used as shaving or makeup mirrors, since they provide large magnifications of objects near the mirror. Convex mirrors are often found in stores and on automobiles, because a small convex mirror provides a large field of view.

In Fig. 23.15 the line AB drawn normal to the reflecting surface of a concave mirror at its center is called the *principal axis of the mirror*. (Note that, although all our diagrams of mirrors and lenses are only in two dimensions, a concave mirror is a three-dimensional object. Because of the spherical symmetry, however, ray diagrams in the plane of the paper are

FIGURE 23.15 A concave mirror. Rays parallel to the principal axis and not too far off that axis all pass through the focal point F after reflection. Here R is the radius of curvature and f the focal length of the mirror.

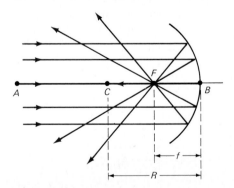

identical to those for any other plane passing through the principal axis. Hence for simplicity we will continue to draw ray diagrams in two dimensions only. If such diagrams are rotated through 360° around the principal axis, they provide a complete picture of the actual behavior of the mirror.)

It is found by experiment that rays striking a spherical concave mirror parallel to its principal axis, and not too far off that axis, are reflected by the mirror so that they all pass through the same point F, called the *focal point*, or *focus*, on the principal axis, as in Fig. 23.15.

Definitions

The point through which all rays close to and parallel to the principal axis pass after reflection from a concave mirror is called the focal point, or focus, of the mirror. The distance along the principal axis from the focus F to the mirror is called the focal length f of the mirror.

All rays from an object a large distance away, such as the sun, are parallel when they strike the mirror and are therefore brought to a focus at F.

Focal Point and Radius of Curvature

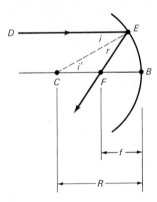

FIGURE 23.16 Relationship between the radius of curvature R and the focal length f of a spherical mirror: $f = R/2$, as derived in the text.

Figure 23.16 shows a spherical mirror of known radius of curvature R. We want to find the position of the focal point F. Suppose we draw a line CE from the center of curvature of the mirror C to the mirror surface at E. The line CE is then perpendicular to the mirror at E. Suppose a light ray DE traveling parallel to the principal axis of the mirror strikes the mirror at E. It is reflected so that the angle of reflection r is equal to the angle of incidence i. Hence i' is also equal to r, since i and i' are alternate interior angles. The triangle CEF is therefore isosceles, and so $CF = FE$.

Now, if the ray is not too far off the axis, FE is *approximately* equal to FB. Hence, to this approximation,

$$R = CB = CF + FB = FE + FB = 2FB \qquad (23.4)$$

Since F is what we have called the focal point of the mirror, FB is equal to the focal length f. Hence $R = 2f$, or

$$\boxed{f = \frac{R}{2}} \qquad (23.5)$$

The *focal length of a spherical mirror is equal to half its radius of curvature.*

Spherical Aberration

FIGURE 23.17 Spherical aberration. A spherical mirror does not focus off-axis rays at the focal point F but at points nearer to the mirror.

The result of the preceding section is only an approximation, since as the rays depart farther from the principal axis of the mirror, they focus not at F but closer to the mirror, as shown in Fig. 23.17. This lack of perfect focusing by a

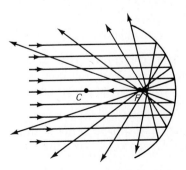

spherical mirror is called *spherical aberration*. It can be overcome by using a parabolic mirror instead of a spherical one.

Since a parabolic mirror will bring all parallel rays incident on the mirror to a focus at *F*, a light bulb at *F* will produce a beam of rays parallel to the principal axis of the mirror (this is referred to as the *principle of the reversibility of light rays*). For this reason automobile headlights have an inner surface consisting of a parabolic mirror, with a small incandescent bulb at its focus. Consequently such headlights cast very straight beams of light ahead of the car.

Solar furnaces also use large parabolic mirrors to catch the sun's rays and focus them down to a small, but very intense, real image of the sun. Temperatures of 3900°C have been reached at the focus of such furnaces. The energy concentrated at the focal point can be used to produce high-temperature steam to generate electricity in steam-electric turbines.

In what follows we will limit ourselves to spherical mirrors with diameters small compared with their radius of curvature, so that the approximations introduced above will lead to negligible errors for most practical purposes.

Image Formation by Concave Mirrors

If an *object* is placed at point *P* on the reflecting side of a concave mirror and the mirror produces an *image* at some point *Q* on the same side of the mirror, as in Fig. 23.18, then we have the following definitions of *object distance* and *image distance*.

Definitions

Object distance p: The distance of the object from the mirror along the principal axis.

Image distance q: The distance of the image from the mirror along the principal axis.

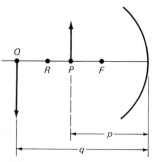

FIGURE 23.18 Image formation by a concave mirror. An object at *P* produces an image at *Q*.

There are two different ways of obtaining the image produced by a concave mirror—one graphical, the other analytical.

Graphical Method The graphical method consists of drawing the rays emanating from key points on the object and locating the points at which these rays are focused by the mirror. For this purpose four kinds of rays are of particular importance:

1 A ray that is parallel to the principal axis and which is always reflected through the focal point *F* of the mirror.

2 A ray that passes through the focal point first and which is then reflected parallel to the principal axis (the reverse of ray 1).

3 A ray which strikes the mirror at its vertex (i.e., the point at which the principal axis intersects the mirror). Such a ray is always reflected in such a way that the angle of incidence with respect to the principal axis is equal to the angle of reflection.

4 A ray that passes through the center of curvature and which is reflected back through the center of curvature, since it strikes the mirror normal to its surface.

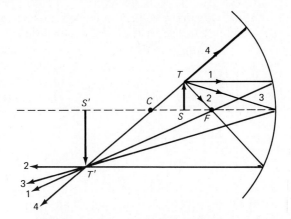

FIGURE 23.19 Rays used to locate the image formed by a concave mirror graphically. In this case the image $S'T'$ is real, inverted, and larger than the object. In this and all subsequent diagrams of image formation by mirrors and lenses the rays from the object will be drawn in black. Rays which form the image after reflection or refraction will be in color.

These four kinds of rays, all originating at the tip of an arrow, are shown in Fig. 23.19. It is seen that all four rays originating at the tip T come together (*converge*) at T' to form the tip of the image (actually only two rays are needed, but it is helpful to draw at least three as a check). Similarly, all other points of the object are imaged in the same way. An observer looking directly at the mirror from a distance greater than the image distance therefore sees an image which is in this case real, inverted, and larger than the original, as can be seen from the figure. By a *real image* we mean an image formed by light rays actually coming together (converging) to form the image, as in the figure. Whether the image is larger or smaller than the object depends on whether the object is inside or outside the center of curvature C.

Figure 23.20 shows the formation of a *virtual image* of the tip of an arrow by a concave mirror. In this case the four rays discussed above *diverge* after striking the mirror and so no real image is formed. Rather the rays *appear to come* from a point T' behind the mirror, although there are, of course, no real light rays behind the mirror. In this case of *diverging* rays, the image is said to be *virtual*. The image in the present case is upright, larger than the original, and virtual, as the figure shows.

In dealing with mirrors, it should always be kept in mind that, if the reflected (image) rays are *converging*, then the image is *real*; if they are *diverging*, then the image is *virtual*. Drawing a ray diagram at the beginning of each problem will clarify this distinction.

FIGURE 23.20 Formation of a virtual image by a concave mirror. In this and subsequent diagrams dotted lines mean that the light rays never actually pass over these paths.

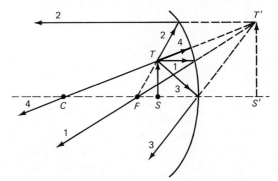

Analytical Method Using the analytical method, we can find an equation relating the image distance to the object distance and the focal length (or radius of curvature) of the mirror.

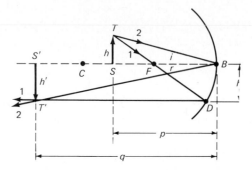

FIGURE 23.21 Obtaining the image of an object *ST* analytically. Here *p* is the object distance and *q* is the image distance, *h* is the height of the object, and *h'* is the height of the image.

The situation in Fig. 23.21 is the same as in Fig. 23.19. We draw two rays from the object *ST* and use them to form the image *S'T'*. Ray 1 passes through the focal point and is reflected parallel to the principal axis to point *T'*. Ray 2 strikes the mirror where it intersects the principal axis and is then also reflected to *T'*.

The angle *i* is equal to *r*, and so the right triangles *BST* and *BS'T'* are similar. Hence we have

$$\frac{h}{-h'} = \frac{p}{q} \tag{23.6}$$

where *p* and *q* are the object and image distances, respectively, and *h* and *h'* are the object and image sizes; and we have included a minus sign to indicate that the image is upside-down, or *inverted*. Also the right triangles *STF* and *BDF* are similar if we assume that *BD* approximates a straight line for mirrors small compared with their radius of curvature. Hence

$$\frac{h}{-h'} = \frac{SF}{FB} = \frac{p-f}{f} \tag{23.7}$$

On combining Eqs. (23.6) and (23.7) we therefore have

$$\frac{p}{q} = \frac{p-f}{f} \quad \text{or} \quad pf = pq - qf$$

On dividing through by *pqf*, we obtain

$$\frac{1}{q} = \frac{1}{f} - \frac{1}{p}$$

or, using Eq. (23.5),

Mirror Equation

$$\boxed{\frac{1}{p} + \frac{1}{q} = \frac{1}{f} = \frac{2}{R}} \tag{23.8}$$

This is the mirror equation, from which the image distance *q* can be obtained if the object distance and focal length (or radius of curvature) are known. Once the image distance is found, the magnification can be obtained from Eq. (23.6). The magnification *M* is defined as the ratio of the image size to the object size. We have, using Eq. (23.6),

$$\boxed{M = \frac{h'}{h} = \frac{-q}{p}} \tag{23.9}$$

We now want to introduce sign conventions that will apply to all cases of the formation of images by simple mirrors. The conventions we will use are the following:

Sign Conventions

When the object, image, or focal point is on the reflecting side of the mirror, the corresponding distance is positive (+). If any of these is behind the mirror, the corresponding distance is negative (−). If the magnification M (= −q/p) is positive, the image is upright; if negative, the image is inverted.*

In the situation shown in Fig. 23.19 the object is on the reflecting side of the mirror. Hence p is positive. The mirror is concave, and so f is on the reflecting side of the mirror and is positive. Graphically the image is on the reflecting side of the mirror, and so q is also positive in this case. Therefore the magnification is $M = -q/p$, which is negative, and so the image is *inverted*. If we want to solve any similar problem using Eq. (23.8), we substitute values for p and f and solve for q. If q is positive, the image is on the reflecting side of the mirror and the reflected rays *converge* to a *real* image; if q is negative, as in Fig. 23.20, the rays appear to *diverge* from an image on the nonreflecting side of the mirror and the image is *virtual*.

The analytical method is in most cases easier to use and more accurate than the graphical method, as long as we stick faithfully to the adopted sign conventions. It is always useful to draw a ray diagram to make sure that our results agree with the analytical solution with respect to the nature of the image (real or virtual, inverted or upright) and its approximate size and location. Mistakes can be caught easily in this way at an early stage in problem solving. Applying these methods of solution will be clearer after we work through Examples 23.4 and 23.5.

Example 23.4

A chess pawn 3.0 cm in height is located 40 cm from a concave mirror with radius of curvature $R = 50$ cm, as in Fig. 23.22. Find the location and size of the image of the pawn, and determine whether it is inverted or upright.

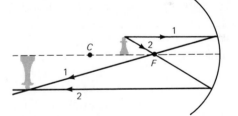

FIGURE 23.22 Diagram for Example 23.4.

SOLUTION

From the mirror equation with $p = +40$ cm, $f = R/2 = (+50 \text{ cm})/2 = +25$ cm (since the mirror is concave), and $h = 3.0$ cm, we have

$$\frac{1}{q} = \frac{1}{f} - \frac{1}{p} = \frac{1}{25 \text{ cm}} - \frac{1}{40 \text{ cm}} = \frac{1}{67 \text{ cm}}$$

or $q = \boxed{+67 \text{ cm}}$

The magnification is

$$M = -\frac{q}{p} = -\frac{67 \text{ cm}}{40 \text{ cm}} = -1.67$$

and so $h' = Mh = (-1.67)(3.0 \text{ cm}) = \boxed{-5.0 \text{ cm}}$

where the minus sign means that the image is inverted.

---

*A situation in which an object is on the nonreflecting side of a mirror occurs in complex lens and mirror systems where the image produced by one element of the system becomes the object for another element of the system.

Hence the image is real, since q is positive; inverted, since M is negative; larger than the object; and located 67 cm from the reflecting side of the mirror. Figure 23.22 shows a ray diagram for this problem, which indicates that our analytical solution is correct. Only two rays are used here to find the image, but it is clear from the diagram that these two rays *converge* to form a *real* image, as was found from the analytical solution.

Example 23.5

This example is similar to Example 23.4, but the chess pawn is now at a distance 12.5 cm from the mirror, which is inside the focus of the mirror. Find the location and size of the image, and determine whether it is inverted or upright.

SOLUTION

In this case we have

$$\frac{1}{q} = \frac{1}{f} - \frac{1}{p} = \frac{1}{25\ \text{cm}} - \frac{1}{12.5\ \text{cm}} = \frac{-12.5}{25(12.5)\ \text{cm}} = -\frac{1}{25\ \text{cm}}$$

Hence $\quad q = \boxed{-25\ \text{cm}}$

and the image is on the nonreflecting side of the mirror. It is therefore *virtual*. The magnification is:

$$M = -\frac{q}{p} = -\frac{-25\ \text{cm}}{12.5\ \text{cm}} = +2.0$$

and the image size is

$$h' = Mh = 2.0(3.0\ \text{cm}) = \boxed{6.0\ \text{cm}}$$

In this case the concave mirror produces an upright, enlarged, virtual image of the type produced by a shaving or makeup mirror. Figure 23.23 shows a ray diagram for this

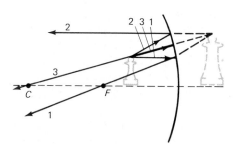

FIGURE 23.23 Diagram for Example 23.5.

case. The rays arising at the top of the pawn and striking the mirror appear to an observer to *diverge* from the top of a larger image of the pawn behind the mirror, but no light rays actually penetrate behind the mirror to that image. The image is therefore virtual. It is useful to compare the ray diagram in Fig. 23.23 with that in Fig. 23.22 to see how the situation changes when the object is moved from outside F to inside F.

Convex Mirrors

When parallel rays fall on a convex mirror, as in Fig. 23.24, they are reflected at angles which make it appear as though they came from a single point just behind the mirror. This point is called the *virtual focus* of the convex mirror. The focal length is again equal to half the radius of curvature of the mirror, as shown in the figure.

The equations and sign conventions used for concave mirrors remain valid for convex mirrors, with the one change that the focal length f (and hence the radius of curvature R) must be taken as negative, since the focal point occurs on the nonreflecting side of the mirror.

In constructing ray diagrams for convex mirrors we proceed as for concave mirrors except that the four rays of importance are the following:

1 A ray that is parallel to the axis and which is reflected as if it came from the virtual focus of the mirror

2 A ray that is heading toward the virtual focus and which is reflected parallel to the principal axis (by the principle of reversibility of rays)

3 A ray that is heading toward the vertex of the mirror and which is reflected so that the angle of reflection with respect to the principal axis is equal to the angle of incidence

4 A ray that is heading toward the center of curvature of the mirror and which is reflected directly back on itself

These four rays enable the image of any point on an object to be located for a convex mirror; they are used to locate a virtual image in Fig. 23.25.

FIGURE 23.24 A convex mirror. Rays parallel to the principal axis are reflected so that they appear to come from a virtual focus behind the mirror. In this case f is the focal length and R is the radius of curvature of the mirror.

FIGURE 23.25 Graphical method of obtaining the image formed by a convex mirror. The four rays from the tip T of the arrow all appear to come from T' after reflection in the mirror.

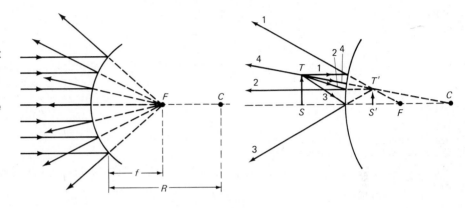

Example 23.6

A security mirror in a department store is a convex mirror of focal length 25 cm. What height does a man 1.8 m tall, and 20 m from the mirror, appear to be in the mirror, and what is the location of his image?

SOLUTION

Here $f = -25$ cm, since the mirror is convex, and $p = 20$ m. Hence the mirror equation becomes

$$\frac{1}{q} = \frac{1}{f} - \frac{1}{p} = \frac{1}{-0.25 \text{ m}} - \frac{1}{20 \text{ m}} = -\frac{20.25}{5.0 \text{ m}}$$

and

$$q = \frac{-5.0 \text{ m}}{20.25} = -0.25 \text{ m} = \boxed{-25 \text{ cm}}$$

Hence the reflected rays diverge, and the image of the man is *virtual* and approximately 25 cm behind the mirror.

In this case the magnification is

$$M = -\frac{q}{p} = -\frac{-0.25 \text{ m}}{20 \text{ m}} = +0.0125$$

and so the image is upright. The image size is:

$$h' = Mh = 0.0125(1.8 \text{ m}) = 0.023 \text{ m} = \boxed{2.3 \text{ cm}}$$

Hence the image is nearly 100 times smaller than the object in this case. It is because of this drastic reduction in size that a small convex mirror can cover such a large field of view. A graphical solution of this problem is shown in Fig. 23.26.

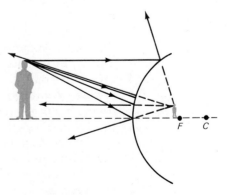

FIGURE 23.26 Graphical solution for Example 23.6.

23.5 Plane Mirrors

A plane mirror can be treated as a special case of a concave (or convex) mirror, with the radius of curvature taken as infinite, since as R approaches infinity, a piece out of a spherical surface of radius R approaches a flat plane. If $R \rightarrow \infty$, $f = R/2 \rightarrow \infty$ and $1/f \rightarrow 0$. Hence the mirror equation becomes:

$$\frac{1}{p} + \frac{1}{q} = \frac{1}{f} = 0 \qquad \text{or} \qquad q = -p$$

Hence *for a plane mirror the image distance is equal to the object distance.* The image is virtual, since q is negative, and is as far behind the mirror as the object is in front of the mirror. This is shown graphically in Fig. 23.27. All the light rays originating at P and striking the mirror appear to come from a point Q behind the mirror. Application of the law governing the reflection of light and a little geometry lead to the conclusion that the image and object distances are the same in the diagram.

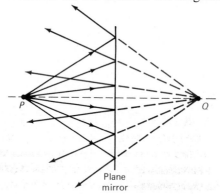

FIGURE 23.27 Image formation by a plane mirror. The image always appears as far behind the mirror as the object is in front of the mirror.

Size and Nature of the Image

Suppose we have an extended object like the arrow in Fig. 23.28, and we want to know not merely where the image is, but whether it is upright or inverted and how large it is. We have already found that $q = -p$ for a plane mirror, and hence that the image of the arrow is as far behind the mirror as the object is in front of it. The magnification of the object is then:

$$M = -\frac{q}{p} = -\frac{-p}{p} = +1$$

The image is therefore upright and the same size as the object. This agrees with our experience with plane mirrors. Since the image is behind the mirror, it is also virtual. This means that a piece of white paper put at point Q will never have an image formed on it, and a photographic plate at Q will never record an image.

Even though the virtual image in a plane mirror is upright, the right and left sides of the image are reversed. For example, if a woman is looking into a plane mirror and raises a hairbrush with her right hand, her image appears to raise the brush with the left hand. Such a lateral inversion is sometimes called a *perversion*.

An example of such a lateral inversion is shown in Fig. 23.29. In this case the object is a *right-handed* set of coordinate axes, characterized by the fact that if the X axis is turned in the direction of the Y axis, a right-handed screw (a normal screw) would move along the Z axis; that is, it would move *out* of the plane of the paper in the figure. On reflection in the plane mirror the three

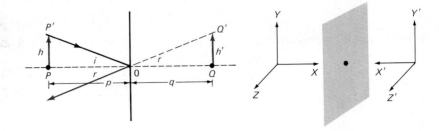

FIGURE 23.28 Formation of the image of an extended object by a plane mirror.

FIGURE 23.29 Lateral inversion by a plane mirror. Reflection in a plane mirror changes a right-handed coordinate system into a left-handed coordinate system.

axes have the images shown. Now, however, in the image the coordinate axes are *left-handed*, for if a right-handed screw were turned from X' into Y', it would no longer move in the direction of Z', but in the opposite direction, that is, *into* the plane of the paper in the figure. This inversion of right-handed systems into left-handed systems on reflection in a plane mirror turns out to have considerable theoretical importance in elementary-particle physics (see Chap. 30).

23.6 Thin Converging Lenses

Just as the laws of reflection determine the behavior of mirrors, so Snell's law of refraction determines the behavior of lenses. A glass lens can be considered to be a series of glass prisms, as in Fig. 23.30, with the light passing through the lens being refracted by the prisms as shown. In practice the surfaces of lenses are polished curved surfaces, of course, with none of the sharp breaks evident in the figure.

There are two basic kinds of lenses, *converging* and *diverging*.

Definitions

A converging lens is a lens that brings all incident light rays parallel to its principal axis together at a point called the focal point of the lens.

A diverging lens is a lens that spreads out all incident light rays parallel to its principal axis so that they appear to arise from a focal point on the object side of the lens.

The behavior of these two kinds of lenses for incident parallel rays (i.e., for an object very far from the lens) is shown in Fig. 23.31. Sketches of some

FIGURE 23.30 A glass lens considered as a series of prisms.

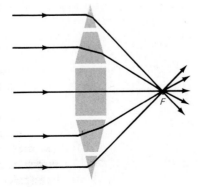

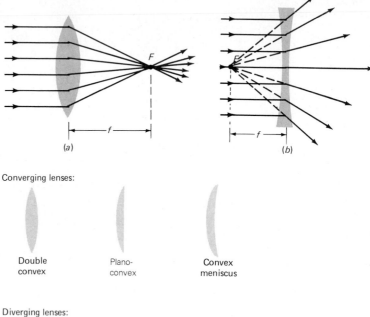

FIGURE 23.31 Types of lenses: (a) converging lens; (b) diverging lens. The focal length f of each is shown.

Converging lenses:

Double convex   Plano-convex   Convex meniscus

FIGURE 23.32 Some frequently used types of converging and diverging lenses. Note that converging lenses are always thicker at the center than at the rim of the lens, while the opposite is true for diverging lenses.

Diverging lenses:

Double concave   Plano-concave   Concave meniscus

frequently used types of converging and diverging lenses are shown in Fig. 23.32.

In what follows we will limit our discussion to *thin lenses*, i.e., to lenses whose thicknesses are small compared with their focal lengths. For such lenses we will draw ray diagrams as if the refraction produced by the lens occurred at the central plane of the lens, as in Fig. 23.31.

Converging lenses behave very much like concave mirrors, as shown in Fig. 23.33. Hence the definitions, sign conventions, and ray diagrams used are also very similar. For lenses we use the following definitions, as illustrated in Fig. 23.34:

FIGURE 23.33 Comparison of a converging lens with a concave mirror: (a) a concave mirror; (b) a converging lens. Both form inverted real images of objects outside their focal points.

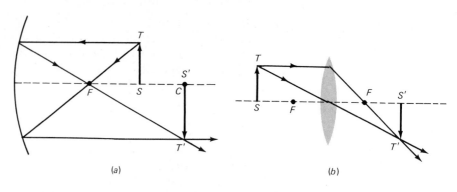

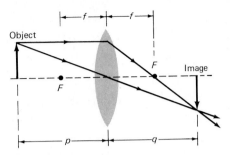

FIGURE 23.34 Definition of object distance *p*, image distance *q*, and focal length *f* for a converging lens. Note that any lens has two focal points, one on each side of the lens and equidistant from the lens center.

Definitions

Object distance p: The distance of the object from the center of the lens as measured along or parallel to the principal axis.

Image distance q: The distance of the image from the center of the lens as measured along or parallel to the principle axis.

Focal length f: The image distance for an object at infinity.

Sign Conventions

The object distance p is positive (+) if the object is on the side of the lens from which the light is coming. In this case the rays coming from the object are diverging from a real object.*

The image distance q is positive (+) if the image is on the other side of the lens from the object. In this case the rays are converging toward a real image. Otherwise q is negative (−).

The focal length is positive for a converging lens and negative for a diverging lens.

If the magnification M (= −q/p) is positive, the image is upright; if negative, the image is inverted.

Ray Diagrams

In locating graphically the images formed by lenses, it is again useful to draw rays from a point on the object which will come together to form a point on the image. The rays we will use are the following:

1 A ray that is parallel to the principal axis and which is refracted through the focus of the lens.

2 A ray that passes through the focus on the object side of the lens and which is refracted parallel to the principal axis (again, a ray must behave this way because of the principle of reversibility of rays).

3 A ray that passes through the exact center of the lens and, which is not refracted at all by the lens. (Actually such a ray, while not bent, is displaced slightly to one side. For a thin lens, however, this displacement is very small and can be neglected.)

---

*The object distance is negative only when we are dealing with systems of lenses. In that case the image produced by one lens may be on the side of a second lens other than the side from which the light is coming.

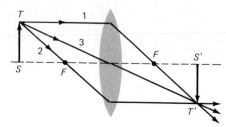

FIGURE 23.35 Graphical location of the real image formed by a converging lens.

These three rays are used in Fig. 23.35 to locate the image of the tip of an arrow which serves as the object for a converging lens. The rest of the object will be imaged to form the real, inverted image shown in the figure.

Thin-Lens Equation

The equation governing the behavior of thin lenses is found in a manner similar to that used to derive the mirror equation.

In Fig. 23.36 we see that the right triangles SOT and $S'OT'$ are similar, and hence that

$$\frac{-h'}{h} = \frac{OS'}{OS} = \frac{q}{p} \tag{23.10}$$

Hence the magnification of a thin converging lens is:

$$M = \frac{h'}{h} = \frac{-q}{p}$$

Also the right triangles OPF and $S'T'F$ are similar, and so

$$\frac{-S'T'}{OP} = \frac{FS'}{OF} \qquad \text{or} \qquad \frac{-h'}{h} = \frac{q-f}{f}$$

Note that we have taken h' as negative since it is in the opposite direction from h. But, from Eq. (23.10), $h'/h = -q/p$, and so

$$\frac{q-f}{f} = \frac{q}{p} \qquad \text{or} \qquad pq - pf = qf$$

Finally, on dividing by pqf, we have

Lens Equation

$$\frac{1}{f} - \frac{1}{q} = \frac{1}{p} \qquad \text{or} \qquad \boxed{\frac{1}{p} + \frac{1}{q} = \frac{1}{f}} \tag{23.11}$$

This is the equation for a thin converging lens. It is identical in form to the mirror equation, Eq. (23.8).

FIGURE 23.36 Diagram used for deriving the lens equation: An object ST at a distance p from the lens produces an image $S'T'$ at a distance q from the lens.

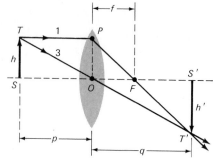

The sign conventions introduced above must be carefully followed in applying the lens equation. The use of these sign conventions will be clarified in Examples 23.7 and 23.8.

Example 23.7

A book 20 cm wide and 25 cm high is placed 1.5 m from the center of a large converging lens of 50 cm focal length and at right angles to the lens axis, as in Fig. 23.37.
(a) What is the position and nature of the image produced?
(b) What is the size of the image of the book?

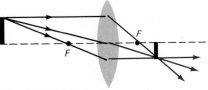

FIGURE 23.37   Diagram for Example 23.7.

SOLUTION

(a) Using the lens equation, we have

$$\frac{1}{q} = \frac{1}{f} - \frac{1}{p} = \frac{1}{50 \text{ cm}} - \frac{1}{150 \text{ cm}} = \frac{2}{150 \text{ cm}} = \frac{1}{75 \text{ cm}}$$

Hence $q = \boxed{75 \text{ cm}}$

Since this image distance is positive, the image is *real*.
(b) The magnification is

$$M = -\frac{q}{p} = -\frac{75 \text{ cm}}{150 \text{ cm}} = -\frac{1}{2}$$

Hence the image is reduced in size and inverted. The width of the image is reduced to

$$w' = Mw = -\tfrac{1}{2}(20 \text{ cm}) = \boxed{-10 \text{ cm}}$$

and the height to

$$h' = Mh = -\tfrac{1}{2}(25 \text{ cm}) = \boxed{-12.5 \text{ cm}}$$

where the minus signs merely indicate the inverted nature of the real image.

The image of the book is therefore 10 by 12.5 cm in area. It would so appear on a screen placed 75 cm from the lens.

A ray diagram showing how the image is produced is shown in Fig. 23.37. It shows only how the height is reduced in the image; the width is changed in the same proportion.

Example 23.8

The book in Example 23.7 is placed 25 cm from the same lens.
(a) What is the position and nature of the image produced?

(b) What is the size of the book's image?

SOLUTION

(a) Here

$$\frac{1}{q} = \frac{1}{f} - \frac{1}{p} = \frac{1}{50 \text{ cm}} - \frac{1}{25 \text{ cm}} = -\frac{1}{50 \text{ cm}}$$

and so $q = \boxed{-50 \text{ cm}}$

The image is therefore on the same side of the lens as the object. It is *virtual*, since the image distance is negative. The rays on the right side of the lens are therefore diverging.
(b) The magnification is:

$$M = -\frac{q}{p} = -\frac{-50 \text{ cm}}{25 \text{ cm}} = +2$$

and so the image is *upright*. The width of the image is:

$$w' = Mw = 2(20 \text{ cm}) = \boxed{40 \text{ cm}}$$

and its height is

$$h' = Mh = 2(25 \text{ cm}) = \boxed{50 \text{ cm}}$$

The image of the book therefore appears to be 40 by 50 cm to someone looking through the lens at the book. This is, of course, a virtual image, which is impossible to see on a screen or to record directly on a photographic film, because the light rays never really converge to such an image.

A ray diagram for this case is provided in Fig. 23.38. It is instructive to compare this figure with Fig. 23.37, noting how the nature of the image changes as the object is moved from a position outside the focus of the lens to inside the focus of the lens.

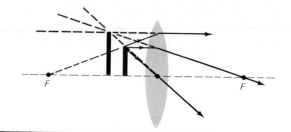

FIGURE 23.38 Diagram for Example 23.8.

23.7 Thin Diverging Lenses

A diverging lens behaves in a fashion similar to the way a convex mirror behaves, as can be seen from Fig. 23.39. A diverging lens produces an upright, virtual image of an object. This can be seen from the ray diagram in Fig. 23.40, or by applying the thin-lens equation. Remember that the focal length of a diverging lens is always negative, according to our sign conventions. In drawing ray diagrams for diverging lenses the three best rays to use are (1) a ray that is parallel to the principal axis and which is refracted so that it appears to come from the focus of the lens on the object side; (2) a ray that is directed at the focus on the other side of the lens and which is refracted parallel to the principal axis; (3) a ray that passes through the center of the lens and whose path is unchanged. These three rays are shown in Fig. 23.40.

In the case of a diverging lens we note that the image distance q can be found from the lens equation

$$\frac{1}{p} + \frac{1}{q} = \frac{1}{f} \tag{23.11}$$

For a diverging lens, since $1/q = 1/f - 1/p$, where f is negative, it can be seen that the right side of this equation is *always* negative, and so the image distance q must also be negative, as long as p is positive. Hence the image produced by a diverging lens is always virtual and upright with respect to the original object. The magnification can then be obtained from $M = h'/h = -q/p$.

FIGURE 23.39 Comparison of (a) a diverging lens with (b) a convex mirror. Both form virtual, upright images of objects.

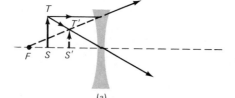

(a)

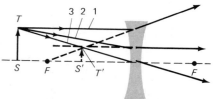

FIGURE 23.40 Ray diagram for image formation by a diverging lens. The image $S'T'$ is virtual and upright and, in this case, smaller than the object ST.

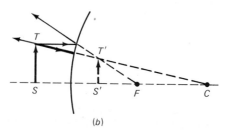

(b)

Example 23.9

An arrow is 24 cm from a double-concave (diverging) lens whose focal length is 36 cm, as in Fig. 23.41.
(a) What will be the nature and location of the image of the arrow?
(b) How large will the image be compared with the size of the original arrow?

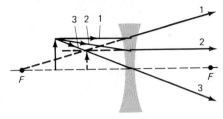

FIGURE 23.41 Diagram for Example 23.9.

SOLUTION

(a) Since a diverging lens has a negative focal length, we have $f = -36$ cm. Our lens equation therefore becomes

$$\frac{1}{q} = \frac{1}{f} - \frac{1}{p} = \frac{1}{-36 \text{ cm}} - \frac{1}{24 \text{ cm}} = -\frac{5}{72 \text{ cm}} = -\frac{1}{14.5 \text{ cm}}$$

and so $q = \boxed{-14.5 \text{ cm}}$

This means that the image is 14.5 cm from the lens on the same side of the lens as the object, and is therefore *virtual*.

(b) The magnification is:

$$M = -\frac{q}{p} = -\frac{-14.5 \text{ cm}}{24 \text{ cm}} = +0.60$$

The image is therefore upright and 3/5 the size of the object. This agrees well with the graphical result shown in Fig. 23.41.

23.8 Combinations of Lenses

Most practical optical systems such as those used in microscopes and telescopes use more than one lens or mirror. Hence it is important to be able to locate the final image produced by a system of two or more lenses. Here we will do this for a few general cases. More specific applications will be taken up in the next chapter.

In dealing with combinations of lenses, take one lens at a time, ignoring for the moment all other lenses. The image produced by that lens, as determined using the lens equation and the sign conventions already introduced, is then taken as the object for the second lens. If there is a third lens, the image produced by the second lens is used as the object for the third lens, and so on, until the final image is obtained. Example 23.10 illustrates this technique.

Example 23.10

The telephoto lenses much used in photography and television allow larger images of distant objects to be formed without requiring excessively long cameras. Figure 23.42 is a diagram of a telephoto lens combination, consisting of a converging lens L_1 followed by a diverging lens L_2. L_1 has a focal length of 125 mm and L_2 has a focal length of -70 mm (diverging), and the two lenses are separated by a distance of 75 mm in the camera.

(a) If this camera is used to take a close-up shot of a football player on the field 100 m away, what is the position of the final image formed by this lens combination?
(b) What is the magnification produced by the two lenses?
(c) In order to accomplish the same task, what would the

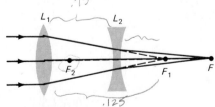

FIGURE 23.42 F_1 is the position where the image formed by the converging lens would fall if L_2 were not present. The actual image formed by the two lenses in combination falls at F.

length of a single-lens camera have to be in comparison with the length of the camera using the telephoto lens?

SOLUTION

(a) For the first lens,

$$\frac{1}{q_1} = \frac{1}{f_1} - \frac{1}{p_1} = \frac{1}{0.125 \text{ m}} - \frac{1}{100 \text{ m}}$$

and so $\quad q_1 = 0.125$ m   *image*

and $\quad M_1 = -\dfrac{q_1}{p_1} = -\dfrac{0.125 \text{ m}}{100 \text{ m}} = -0.00125$

The image would therefore be real and inverted, if the light were not intercepted by the second lens.

The image produced by the first lens then serves as the object for the second lens. The object distance is in this case

$$p_2 = 0.075 - 0.125 = -0.050 \text{ m}$$

where the negative sign means that the object is on the *opposite side* of the lens from where the light is coming, i.e., from the original object as shown in the figure. Hence the lens equation for the second lens becomes:

$$\frac{1}{q_2} = \frac{1}{f_2} - \frac{1}{p_2} = \frac{1}{-0.070 \text{ m}} - \frac{1}{-0.050 \text{ m}}$$

and so $\quad q_2 = \boxed{+0.175 \text{ m}}$

and the image is 17.5 cm to the right of lens L_2 and is a real image, since light rays actually converge to that point.

(b) The magnification produced by the second lens is

$$M_2 = -\frac{q_2}{p_2} = -\frac{0.175 \text{ m}}{-0.050 \text{ m}} = +3.5$$

The overall magnification is then

$$M = M_1 M_2 = (-0.00125)(+3.5) = \boxed{-0.0044}$$

and the final image is inverted and reduced in size.

(c) The length of the above lens system from lens L_1 to the final image is:

Length of camera = 7.5 cm + 17.5 cm = $\boxed{25 \text{ cm}}$

For a single-lens camera to produce the same image size for an object 100 m away, we would need:

$$p = 100 \text{ m}$$

$$q = Mp = 0.0044(100 \text{ m}) = 0.44 \text{ m} = \boxed{44 \text{ cm}}$$

The length of the camera from lens to film would therefore be 44 cm, which is about twice as long as required by the telephoto lens system, for the same size image of an object 100 m away.

Diopters

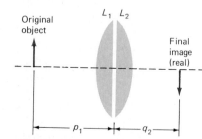

FIGURE 23.43 Combination of two plano-convex lenses placed very close together. They form a real, inverted image of the object.

When testing a patient's eye for glasses, an optometrist places several different lenses in front of the eye at the same time, and from this combination of lenses determines the single lens the patient needs to see clearly with glasses. How does the optometrist figure out the focal length of the single lens needed to produce the same effect as the correct combination of lenses?

This is a simple problem in the combination of lenses, but here the lenses being combined are all very close together, so close in fact that it is reasonable to neglect their separation compared with the object and image distances involved. The situation is then as in Fig. 23.43 for two plano-convex lenses very close together, where we measure all distances with respect to the center of the two-lens system.

We proceed as in the preceding example. We first find the image produced by lens L_1 while ignoring lens L_2. We have

$$\frac{1}{p_1} + \frac{1}{q_1} = \frac{1}{f_1} \tag{23.12}$$

In this case the image distance q_1 is clearly on the other side of the lens from the object.

We now apply the lens equation to the second lens L_2. But here its object is the image produced by the first lens, and this image is to the right of the lens. The object distance for the second lens is therefore *negative*, and is $p_2 = -q_1$. The lens equation for the second lens is then:

$$\frac{1}{p_2} + \frac{1}{q_2} = \frac{1}{f_2} \quad \text{or} \quad \frac{1}{-q_1} + \frac{1}{q_2} = \frac{1}{f_2} \tag{23.13}$$

If we add Eqs. (23.12) and (23.13), we obtain

$$\frac{1}{p_1} + \frac{1}{q_2} = \frac{1}{f_1} + \frac{1}{f_2} \tag{23.14}$$

Here p_1 is the distance of the original object from the two lenses, and q_2 is the distance of the final real image from the two lenses. Hence the two lenses together act as if they were a single lens of focal length f, where, from Eq. (23.11),

$$\frac{1}{p_1} + \frac{1}{q_2} = \frac{1}{f} \tag{23.15}$$

On comparing Eqs. (23.14) and (23.15) we see that

$$\frac{1}{f} = \frac{1}{f_1} + \frac{1}{f_2} \tag{23.16}$$

Hence the reciprocal of the focal length for the combination of two lenses located very close together is the sum of the reciprocals of the focal lengths of the individual lenses. (It might be helpful to remember that this is exactly the way two electrical resistors combine when connected in parallel.)

We then define the *power* of a lens as follows:

Definition

The power (P) of a lens in diopters (D) *is the reciprocal of its focal length in meters, or* $P = 1/f$.

$$1 \text{ D} = 1 \text{ m}^{-1}$$

Then the power of two lenses expressed in diopters can be added directly to give the power of the lens combination, as is clear from Eq. (23.16).

The power of a converging lens is positive when expressed in diopters, since the focal length is positive. Similarly, the power of a diverging lens is negative. Converging and diverging lenses are therefore often referred to as positive and negative lenses, respectively.

Example 23.11

Two converging lenses with focal lengths 0.20 and 0.30 m, respectively, are combined with a third diverging lens of focal length 0.50 m to form a single compound lens, where all three lenses are very close together. What is the focal length of this compound lens?

SOLUTION

The diverging lens must be assigned a negative focal length by our sign convention. Hence we have

$$\frac{1}{f} = \frac{1}{f_1} + \frac{1}{f_2} + \frac{1}{f_3}$$

$$= \frac{1}{0.20 \text{ m}} + \frac{1}{0.30 \text{ m}} + \frac{1}{-0.50 \text{ m}} = \frac{5.0 + 3.3 - 2.0}{1 \text{ m}}$$

$$\text{or} \quad f = \frac{1}{6.3} \text{ m} = \boxed{0.16 \text{ m}}$$

---

*Diopter is derived from two Greek words meaning "see through."

The same result could have been obtained using the power of the lenses in diopters. We have

$$P = P_1 + P_2 + P_3$$

$$= \frac{1}{0.20}D + \frac{1}{0.30}D - \frac{1}{0.50}D = 6.3 \text{ D}$$

Then, using the definition of lens power, we have

$$f = \frac{1}{P} = \frac{1}{6.3 \text{ D}} = \boxed{0.16 \text{ m}}$$

as before.

Example 23.12

One way to measure the focal length of a diverging lens is to place a converging lens of known focal length in contact with the diverging lens and to find the focal length of the combination. If parallel light from the sun is focused at a point 30 cm from the center of the lens combination, and the focal length of the converging lens is $f_1 = 15$ cm, what is the focal length of the diverging lens?

SOLUTION

Since $1/f = 1/f_1 + 1/f_2$, we have

$$\frac{1}{0.30 \text{ m}} = \frac{1}{0.15 \text{ m}} + \frac{1}{f_2}$$

where f_2 is the focal length of the diverging lens. Then

$$\frac{1}{f_2} = \frac{1}{0.30 \text{ m}} - \frac{1}{0.15 \text{ m}} = -\frac{0.15}{0.045 \text{ m}}$$

or $\quad f_2 = \boxed{-0.30 \text{ m}}$

Hence the focal length is 30 cm for the diverging lens and is negative, as required. We could have obtained the same result by adding lens powers, as in the previous example.

23.9 Lens Aberrations

The theory we have developed for image formation by lenses assumed that a simple lens would focus all rays arising at one point on an object to one point on the image regardless of the location of the point or the wavelength of the light used. This represents an oversimplification of the complicated process of image formation by a simple lens. No simple lens forms a perfect image of any given object. Imperfections, which we call *lens aberrations*, lead to lack of clarity and sharpness in the image. There are five principal kinds of such lens aberrations, but for our purposes we will discuss only three: spherical aberration, chromatic aberration, and astigmatism.

Spherical Aberration

If we use a simple spherical lens to form an image of a point source of light, we find that rays passing through the outer portion of the lens surface are focused closer to the lens than rays passing near the center of the lens (the same thing happens with a spherical mirror). As a result there is no single sharp image of the point object, but rather a tiny circular patch of light. The diameter of this circular patch will vary with distance from the lens. In practice, the best place to put a photographic film to record the image is where this circle has its smallest diameter. This is called the *circle of least confusion*, and is shown in Fig. 23.44.

FIGURE 23.44 Spherical aberration. Rays passing through the center of the lens are focused at a different distance from the lens than are rays passing through the outside portions of the lens. Hence a point object does not produce a point image.

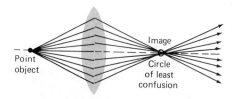

Just as with mirrors spherical aberration can be corrected by using parabolic mirrors, so too spherical aberration can be eliminated by using nonspherical lens surfaces. These are very difficult and expensive to produce, however, and hence it is better to use combinations of lenses chosen to reduce spherical aberration. Many compound lenses used in research equipment are corrected for spherical aberration in this way.

Even with a simple lens, spherical aberration can be reduced by decreasing the diameter of, or stopping down, the lens with a diaphragm and using only its central portion. This, of course, greatly reduces the light passed by the lens and hence the intensity of the image produced.

Chromatic Aberration

Spherical aberration occurs even with the perfectly monochromatic light from a laser. There is another aberration, however, which arises from the fact that the index of refraction of the glass in a lens is different for different wavelengths. We have seen that a prism refracts violet light more than it refracts red light. The same is true of a lens, since a lens can be considered to be an array of small prisms, as in Fig. 23.30. As a result a simple lens will focus violet or blue light closer to the lens than it focuses red light. Hence a white light source will produce a blurred image of an object, and this image will have colored edges, as shown in Fig. 23.45. This lens defect is called *chromatic aberration* (*chroma* is the Greek word for color). The human eye has the same problem in focusing on both red and blue colors at the same time, since the focal length of the eye for the two colors is slightly different.

FIGURE 23.45 Chromatic aberration: Because the index of refraction of glass is different for different colors, images in these colors fall at different distances from the lens. A blue image with red edges is formed near the lens, and a red image with blue edges slightly farther away from the lens.

In 1758 John Dollond (1706–1761), an English optician, discovered a way to eliminate chromatic aberration. He combined two lenses, one converging, the other diverging, to make an *achromatic doublet*, i.e., two lenses showing no color. The two lenses are made of different kinds of glass with indices of refraction such that the combination brings any two chosen colors to the same sharp focus.* The light paths through an achromatic doublet designed for blue and red light are shown in Fig. 23.46. In many cases achromatic triplets or more complicated compound lenses are used to correct chromatic aberration. Often they can be designed to correct for spherical aberration at the same time.

FIGURE 23.46 An achromatic doublet. The first element in the compound lens is made from low-dispersion glass. The second element, a diverging lens, is weaker than the first element but is made from high-dispersion glass. The combination cancels out all dispersion and focuses all colors at the same focal point. (The dispersion is greatly exaggerated for clarity.)

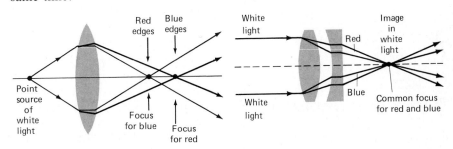

Astigmatism

A point source which is off the principal axis of a lens is not focused to a sharp point image on the other side of the lens, as shown in Fig. 23.47. Rather, rays from the off-axis point which are in vertical planes parallel to that defined by

---

*Dollond's discovery was particularly important for the development of high-grade microscopes. Newton's discovery of reflecting telescopes had eliminated the problem of chromatic aberration from astronomical work by replacing lenses with mirrors. This was, however, not possible for microscopes.

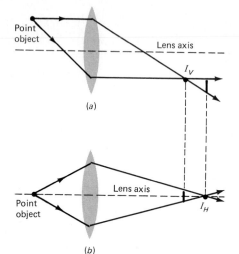

FIGURE 23.47 Astigmatism: (a) Side view of a point object above the axis of the lens, together with the image formed of that object by the lens. (b) Top view of the same point object and its image. Two focal lines result instead of a single focal point.

the principal axis and the object point, converge to one image point I_V. At this point, however, rays in horizontal planes do not converge and so the image at I_V is a horizontal line. Similarly, rays in horizontal planes normal to that defined by the principal axis and the object point converge to a second image point I_H. At this point, however, the vertical rays have already begun to diverge and so the image is a vertical line, which is also shown in the figure. In the case of an off-axis point, therefore, we have two so-called *focal lines* at right angles to each other and at different distances from the lens. The circle of least confusion which falls in between these two focal lines is then the closest approximation we have to a focal point in this case. This defect is known as *astigmatism.** Astigmatism can be reduced by using only the central portion of the lens, in the same manner in which spherical aberration is reduced, or by replacing the single lens by a complicated lens system corrected for astigmatism.

The design of compound lenses to reduce aberrations has been greatly improved in recent years by the use of high-speed computers. The usual technique is to combine two lenses, or a lens and a mirror, in such a way that the aberration introduced by one just cancels the aberration caused by the other. Even with computers not all aberrations can be drastically reduced at the same time, but it is possible to determine which aberrations will be most troublesome for a particular application and then to choose a lens system which will reduce these aberrations to within tolerable limits.

Summary: Important Definitions and Equations

Ray (or geometrical) optics:
   A model of light which assumes that light travels in perfect straight lines, or along "rays." Ray optics is only valid when $\lambda/w \ll 1$, where λ is the wavelength of the electromagnetic radiation and w is the smallest dimension of the object or aperture on which the radiation falls.

---

*The term *astigmatism* comes from Greek—the prefix *a* meaning "not," and *stigma* meaning "point"—thus indicating that a lens will not focus an off-axis point in the object plane to a point in the image plane.

Refraction of light

Snell's law: $n_1 \sin \theta_1 = n_2 \sin \theta_2$

Total internal reflection: When light passes from a more dense optical medium to a less dense optical medium, no refraction occurs for light incident at an angle greater than the critical angle θ_c, where $\sin \theta_c = n_2/n_1$. Total reflection then occurs.

Mirrors

Object distance (p): The distance of the object from the mirror along the principal axis.

Image distance (q): The distance of the image from the mirror along the principal axis.

Focal length (f): The distance of the focus F from the mirror along the principal axis.

Focus (or focal point): The point through which all rays parallel to the mirror's principal axis are reflected by a mirror; $f = R/2$, where R is the radius of curvature of the mirror.

Real image: An image formed by light rays actually coming together to form the image.

Virtual image: An image formed by light rays which appear to come from a particular point but never actually pass through that point.

Thin lenses

Converging lens: A lens which brings all incident light rays parallel to its principal axis together at the focal point of the lens.

Diverging lens: A lens which spreads out all incident light rays parallel to its principal axis so that they appear to come from a focal point on the other side of the lens.

(The definitions given for mirrors of object distance, image distance, focal length, focus, real image, and virtual image are the same for lenses except that *lens* replaces *mirror* in each definition. All important equations and sign conventions for mirrors and lenses are summarized in Table 23.1.)

Power of a lens (in diopters): The reciprocal of its focal length in meters.

$$P = \frac{1}{f} \qquad 1 \text{ diopter (D)} = 1 \text{ m}^{-1}$$

Lens aberrations:

Imperfections in the formation of images by lenses.

Spherical aberration: Lack of sharpness in focusing by a lens because off-axis rays parallel to the principal axis have a different focus than those passing through the center of the lens.

Chromatic aberration: Lack of sharpness and the presence of colored fringes in the image caused by variation of focal length of a lens with the wavelength of the light.

Astigmatism: Lack of sharpness in image because the lens does not focus off-axis objects to a point but produces two images in the form of short lines at different distances from the lens.

TABLE 23.1   Important Equations and Sign Conventions for Lenses and Mirrors

Quantity	Positive sign (+)	Negative sign (−)
$$\text{Mirrors: } \frac{1}{p} + \frac{1}{q} = \frac{1}{f} = \frac{2}{R}$$		
Object distance p	If on reflecting side of mirror	If on nonreflecting side of mirror
Image distance q	If on reflecting side of mirror (real image)	If on nonreflecting side of mirror (virtual image)
Focal length $f = \dfrac{R}{2}$	Focal point on reflecting side (concave mirror)	Focal point on nonreflecting side (convex mirror)
Magnification: $M = \dfrac{h'}{h} = -\dfrac{q}{p}$	If image is upright	If image is inverted
$$\text{Lenses: } \frac{1}{p} + \frac{1}{q} = \frac{1}{f}$$		
Object distance p	If on side of lens from which light is coming	If on opposite side of lens than side from which light is coming
Image distance q	If on opposite side of lens from object (real image)	If on same side of lens as object (virtual image)
Focal length f	Converging lens	Diverging lens
Magnification: $M = \dfrac{h'}{h} = -\dfrac{q}{p}$	If image is upright	If image is inverted

Questions

1 Give a few examples of optical observations made in everyday life which *cannot* be explained by a ray theory of light.

2 What advantage can you see in using prisms at angles greater than the critical angle to reflect light instead of using plane mirrors for the same purpose? What disadvantages can you see in the use of such prisms?

3 (*a*) Why is the index of refraction of a medium always larger than unity?

(*b*) What kind of medium would have an index of refraction less than unity?

4 Can a light ray in air be totally reflected at a glass surface if the angle of incidence is just right? Why?

5 Show by ray diagrams and from the mirror equation that the magnification of a concave mirror is less than 1 if the object is outside the center of curvature and greater than 1 if it is inside the center of curvature. What is the situation when the object is placed at the center of curvature?

6 A concave mirror produces a real image. Is the image necessarily inverted?

7 (*a*) Indicate the similarities which exist between the behavior of converging lenses and concave mirrors.

(*b*) Do the same for diverging lenses and convex mirrors.

8 Five converging lenses are arranged in a line. Their focal lengths are such that the image formed by one lens always falls at such a position that the image formed by the next lens is real. Is the final image produced by the system of lenses upright or inverted?

9 Can you suggest any circumstances in which the image formed by a diverging lens would be a real image? Draw a ray diagram to explain your answer.

10 Some newspapers in an apartment ignite and the fire department is called to put out the fire. The fire captain claims that the fire was caused by a fishbowl full of water sitting just inside a sunlit window. The bowl focused the sun's rays and set the newspapers on fire. Do you think this is a possible explanation of the fire?

11 Is it possible for the same glass lens to be either converging or diverging depending on the medium in which it is embedded? Explain your answer, using a ray diagram.

12 (*a*) How would you determine the focal length of a converging lens?

(*b*) How would you determine the focal length of a diverging lens?

Problems

A 1 Red light is incident on a slit of width 0.10 mm. The ratio λ/w in this case is:
(*a*) 7×10^{-6} (*b*) 7×10^{-3} (*c*) 7×10^{-1}
(*d*) 70 (*e*) 7×10^{3}

A 2 Light travels down a light pipe made of flint glass ($n = 1.66$) coated on the outside by borosilicate crown glass. The critical angle for total internal reflection inside the light pipe is:
(*a*) $66°$ (*b*) $24°$ (*c*) $0°$ (*d*) $41°$ (*e*) $37°$

A 3 A concave mirror has a radius of curvature of 2.0 m. A bouquet of roses is placed 1.0 m from the mirror on its reflecting side. The roses will be focused sharply on a screen which is at a distance from the mirror of:
(*a*) 2.0 m   (*b*) 1.0 m   (*c*) 0.50 m
(*d*) 4.0 m   (*e*) None of the above

A 4 In Prob. A3 the roses are placed 1.5 m from the mirror. The image appears at a distance from the mirror of:
(*a*) 1.5 m   (*b*) 1.0 m   (*c*) 2.0 m   (*d*) 3.0 m
(*e*) None of the above

A 5 A convex mirror has a focal length of 1.0 m. An upright arrow is located 0.60 m from the reflecting side of the mirror. The image produced is:
(*a*) Virtual and 0.63 times as tall as the object
(*b*) Real and 0.63 times as tall as the object
(*c*) Inverted and 4.4 times as tall as the object
(*d*) Upright and 4.4 times as tall as the object
(*e*) None of the above

A 6 The distance of the lens in the human eye from the retina, on which the image is focused, is about 1.7 cm. To focus on a book 0.50 m from the eye, the focal length of the eye must be about:
(*a*) 50 cm   (*b*) 0.59 cm   (*c*) 1.6 cm
(*d*) 1.6 m   (*e*) 3.4 cm

A 7 If the distance of the lens in the human eye from the retina is about 1.7 cm, an arrow 5.0 cm high placed 25 cm from the eye will form an image on the retina of height:
(*a*) 5.0 cm   (*b*) 1.0 cm   (*c*) 0.34 cm
(*d*) 8.5 cm   (*e*) None of the above

A 8 A vase is located 40 cm from a diverging lens of focal length 40 cm. The location of the image will be:
(*a*) 20 cm from the lens on the same side as the object
(*b*) 20 cm from the lens on the opposite side from the object
(*c*) 40 cm from the lens on the same side as the object
(*d*) 40 cm from the lens on the opposite side from the object
(*e*) None of the above

A 9 A converging lens of focal length 20 cm and a diverging lens of focal length 50 cm are combined into a compound lens. The focal length of the compound lens is:
(*a*) 100 cm   (*b*) 14 cm   (*c*) 33 cm
(*d*) 10 cm   (*e*) None of the above

A10 Two lenses are combined to form a compound lens. One of the lenses has a power of $+10$ D and the other

of −5.0 D. The focal length of the resulting lens is:
(*a*)   100 cm   (*b*)   50 cm   (*c*)   20 cm
(*d*)   10 cm   (*e*)   None of the above

B 1   How wide would be the central image of a single-slit diffraction pattern on a screen 10.0 m from the slit if the slit width is 1.0 m and the wavelength is that of a He–Ne laser, 633 nm?

B 2   How wide would be the central image of a diffraction pattern on a screen 10.0 m from the slit for a radio wave of 100-MHz frequency passing through a 1.0-mm slit?

B 3   The critical angle for total internal reflection of a light beam passing from diamond to air is 24°. What is the index of refraction of diamond?

B 4   If an object 0.60 m from a concave mirror produces a real image 0.30 m from the mirror, what is the radius of curvature of the mirror?

B 5   An object is placed 0.20 m from the reflecting surface of a convex mirror of focal length 0.50 m. What is the location, size, and nature of the image produced?

B 6   What is the magnification produced by a converging lens of focal length 0.15 m for an object placed 0.20 m from the lens?

B 7   A converging lens of focal length 0.20 m forms a virtual image of an object. The image appears to be 0.40 m from the lens on the same side as the object. Where was the object located?

B 8   A double-concave (diverging) lens of focal length 0.20 m is used to view a bowl of fruit 1.0 m from the lens.
(*a*)   What is the location of the image?
(*b*)   What is the size and nature of the image?

B 9   (*a*)   Two lenses, one of power +2.0 D and the other of power −1.5 D are combined into one compound lens. What is the power of the compound lens in diopters?
(*b*)   What is the focal length of the compound lens?

B10   Two diverging lenses with focal lengths 0.20 and 0.25 m, respectively, are combined with a converging lens of focal length 0.50 m to form a single compound lens.
(*a*)   What is the focal length of the combination?
(*b*)   What is the power of the compound lens?

C 1   A light beam falling on the mirror of a Cavendish balance (see Sec. 3.5) is initially reflected directly back on itself. When two heavy lead spheres are brought near the two small lead balls at the ends of the balance arm, the mirror is deflected through an angle of 1.5°. If a scale is mounted 20 m from the mirror, what is the deflection of the reflected light beam on the scale?

C 2   (*a*)   What is the critical angle for the total internal reflection of a light beam passing from a diamond into very heavy flint glass?
(*b*)   What is the critical angle for a light beam passing in the opposite direction, from very heavy flint glass into diamond?

C 3   A beam of red light passes symmetrically through a prism with three equal angles. The measured angle of minimum deviation is 41°. What is the index of refraction of the prism glass?

C 4   A beam of yellow light passes through a hollow prism of vertex angle $A = 60°$, and the deviation of the beam is zero, within the limits of error of the experiment. The prism is then filled with an unknown liquid and the angle of minimum deviation is found to be 28°. What is the index of refraction of the liquid?

C 5   A concave mirror has a radius of curvature of 1.5 m. Calculate the position and kind of image produced when an object is placed at the following positions:
(*a*)   2.0 m from the mirror.
(*b*)   1.0 m from the mirror.
(*c*)   0.50 m from the mirror.
(*d*)   Describe how the image changes as the object is gradually moved nearer to the mirror as in parts (*a*) to (*c*).

C 6   (*a*)   A concave mirror has a radius of curvature of 0.60 m. A vase is placed 0.50 m from the mirror and on its principal axis. Use graphical methods to locate the image and obtain its approximate size.

C 7   The real image produced by a concave mirror is found to be one-third the size of the object. If the distance from the mirror to the screen on which the image appears is 0.75 m, what is the focal length of the mirror?

C 8   A convex mirror with a 100-cm radius of curvature is used to reflect the light from an object placed 75 cm in front of the mirror. Find the location of the image and its relative size.

C 9   A concave mirror of radius of curvature 1.50 m is placed behind a chess piece which is 0.20 m from the reflecting side of the mirror along its principal axis. What is the size and nature of the image produced, and where does it occur?

C10   A convex mirror of radius of curvature 1.50 m is placed 0.20 m from a chess piece which is on the reflecting side of the mirror. What is the size and nature of the image produced, and where does it occur?

C11   A convex mirror has a radius of curvature of 1.0 m. Calculate the position and kind of image produced when an object is placed at the following points:
(*a*)   1.5 m from the mirror.
(*b*)   0.75 m from the mirror.
(*c*)   0.25 m from the mirror.
(*d*)   Describe how the image changes as the object is gradually moved nearer to the mirror as in parts (*a*) to (*c*).

C12   (*a*)   A convex mirror has a radius of curvature of 0.75 m. A vase is placed 0.60 m from the mirror on its principal axis. Use graphical methods to locate the image and obtain its approximate size.
(*b*)   Solve the problem by using the mirror equation and compare the two results.

C13   Repeat both parts of Prob. C12 for the same mirror and an object distance of 0.30 m.

C14   Two plane mirrors *A* and *B* are set up at right angles to each other, as in Fig. 23.48. A ray of light is incident on *A* at an angle of incidence of 65°, as shown. It then strikes mirror *B* and is reflected.

(*a*) At what angle of incidence does the light strike mirror *B*?

(*b*) Prove that after the reflection from *B* the ray is moving parallel to the ray incident on *A* but in the opposite direction.

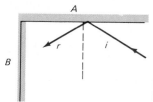

FIGURE 23.48 Diagram for Prob. C14.

C15 Two mirrors, each 1.0 m long, are separated by a distance of 22 cm and are facing each other, as in Fig. 23.49. A light ray is incident on one end of the mirrors at an angle of incidence of 25°.

(*a*) How many times is the ray reflected before it reaches the other end of the mirror system?

(*b*) What is the final direction of the ray after emerging from between the mirrors?

FIGURE 23.49
Diagram for Prob. C15.

C16 The magnification produced by a converging lens is found to be 2.5 for an object placed 0.20 m from the lens. What is the focal length of the lens?

C17 A double-convex lens (converging) has a focal length of 0.40 m. Calculate the position and kind of image (including its size) produced when an object is placed at the following points:

(*a*) 1.5 m from the lens.

(*b*) 0.40 m from the lens.

(*c*) 0.20 m from the lens.

(*d*) Describe how the image changes as the object is moved gradually nearer to the mirror as in parts (*a*) to (*c*).

C18 (*a*) A converging lens has a focal length of 0.55 m. A vase is placed 0.65 m from the lens on its principal axis. Use graphical methods to locate the image and obtain its approximate size.

(*b*) Solve the same problem analytically by using the lens equation, and compare the two results.

C19 Repeat both parts of Prob. C18 for the same lens and an object distance of 0.30 m.

C20 A man 2.0 m tall is observed through a diverging lens of focal length 1.0 m. The man is a distance of 3.0 m from the lens.

(*a*) What will be the nature and location of the image of the man?

(*b*) How large will the man appear to be?

C21 (*a*) A diverging lens has a focal length of 1.6 m. A rose in a bud vase is placed 1.9 m from the lens on its

principal axis. Use graphical methods to locate the image and to obtain its approximate size.

(*b*) Solve the same problem analytically and compare the two results.

C22 Repeat both parts of Prob. C21 for the same lens and an object distance of 1.0 m.

C23 A converging lens has a focal length of 25 cm, and a second converging lens has a focal length of 50 cm. An object is placed 25 cm from the first lens and on the side away from the second lens.

(*a*) If the lenses are 50 cm apart, where does the final image appear?

(*b*) If the lenses are 2.0 m apart, where does the image appear?

C24 A converging lens of focal length 25 cm is put into contact with a diverging lens of unknown focal length. This compound lens focuses sunlight at a point 40 cm from the center of the lens combination. What is the focal length of the diverging lens?

C25 An object is placed 24 cm in front of a converging lens with $f = 16$ cm. At a distance of 72 cm beyond the first lens is a diverging lens of focal length 12 cm.

(*a*) Find the position and size of the final image produced.

(*b*) Is the image real or virtual, erect or inverted?

C26 An object is placed 40 cm in front of a converging lens which has a focal length of 20 cm. This lens is 50 cm from a plane mirror. Find the final image produced by this system.

C27 A flint glass lens, with $n = 1.66$, has a power in air of $+2.0$ D. What would be its power if submerged in water?

C28 If a light pipe is made of very heavy flint glass fibers coated with borosilicate crown glass, what is the largest bend in the pipe which can be sustained without losing some of the light by refraction?

C29 Light is incident on one end of an optical fiber at an angle of incidence θ_1 and is refracted into the fiber at an angle θ_2, as in Fig. 23.50. It then strikes the side of the fiber at an angle i. If the index of refraction of the fiber is 1.40, what is the largest angle of incidence θ_1 that the ray can have and still be totally reflected from the wall of the fiber?

C30 An optical fiber consists of an interior glass fiber of index of refraction $n_1 = 1.65$ coated with a thin layer of glass of index of refraction $n_2 = 1.40$.

(*a*) What is the critical angle for total reflection of a ray inside the fiber?

(*b*) If light is incident on the end of the fiber, as in Fig. 23.51, what is the largest angle of incidence θ_1 that the ray can have and still be totally reflected from the fiber wall?

D 1 A small rock lies at the bottom of a large water tank, 8.0 m below the surface of the water. Determine the diameter of a circular piece of wood which, when floating on the surface of the water directly above the rock, will totally hide the rock, regardless of the angle of the line of sight.

D 2 Prove that for a convex mirror, no matter where an object is placed on the reflecting side of the mirror, the image will always be virtual and erect.

D 3 A spherical mirror has both sides reflecting. A

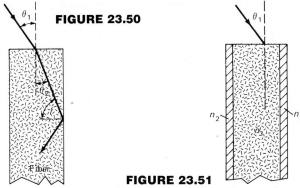

FIGURE 23.50

FIGURE 23.51

$$f' = \frac{n_l\,(n_g - 1)f}{n_g - n_l}$$

where f is the focal length in air.

D 7 Show that the formula in Prob. D6 correctly predicts the behavior of a glass lens immersed in a liquid for the two limiting cases:

(a) $n_l \to n_g$
(b) $n_l \to 1$

D 8 A light bulb and a screen are placed at a fixed distance D apart.

(a) Show that if a converging lens of focal length f, where $f < D/4$, is inserted between them, it will produce a real image of the object on the screen for two positions separated by a distance $d = \sqrt{D(D - 4f)}$.

(b) Prove that at the two positions found in part (a) the ratio of the two images sizes is $(D - d)^2/(D + d)^2$.

(c) Why have the focal lengths been limited to values such that $f < D/4$?

D 9 Consider two converging lenses, one of focal length 20 cm and a second of focal length 40 cm, with their centers 1.0 m apart on the common axis of the two lenses. If an object is placed 30 cm from the center of the first lens on the side away from the second lens, what is the nature of the final image produced and where does it fall?

D10 Consider two converging lenses, one of focal length 25 cm and a second of focal length 50 cm, with their centers 75 cm apart. If an object is placed 20 cm from the center of the first lens on the side away from the second lens, what is the nature of the final image produced and where does it fall?

woman looks into one side of the mirror and sees an image of her face 42 cm in back of the mirror. When she looks into the other side of the mirror she sees an image 14 cm in back of the mirror.

(a) How far is her face from the mirror, assuming that the distance is the same in both cases?

(b) What is the focal length of each side of the mirror?

D 4 How far apart are the object and image if a lens of focal length 30 cm forms a real image 3 times as high as the object?

D 5 If a diverging lens is to be used to form an image which is half the size of the object, where must the object be placed?

D 6 Prove that for a glass lens, with n_g as its index of refraction, immersed in a liquid of index of refraction n_l, the focal length of the lens in the liquid is not f, but

Additional Readings

Bragg, Sir William: *The Universe of Light*, Dover, New York, 1959. An expanded and well-illustrated version of a course of lectures given by the Nobel Prize winner in 1931 at the Royal Institution in London.

Fraser, Alistair B., and William H. Mach: "Mirages," *Scientific American*, vol. 234, no. 1, January 1976, pp. 102–111. A discussion of an interesting optical phenomenon caused by the refraction of light.

Greenler, Robert: *Rainbows, Halos and Glories*, Cambridge University Press, New York, 1980. A discussion of optical and meteorological phenomena in the atmosphere.

Miller, Stewart E.: "Lightwaves and Telecommunication," *American Scientist,* vol. 72, no. 1, January-February 1984, pp. 66–71. An interesting account of recent developments in the use of light pipes for long-range communication.

Minnaert, M.: *Light and Color in the Open Air*, Bell, London, 1959. A classic discussion of some commonly observed optical phenomena.

Newton, Sir Isaac: *Opticks, or a Treatise on the Reflections, Refractions, Inflections, and Colours of Light*, Dover, New York, 1979. A more accessible work to the beginning physics student than the famous *Principia*, since in it Newton describes experiments on light which he performed himself. This edition contains introductory material by I. Bernard Cohen, Albert Einstein, and Sir Edmund Whittaker.

Nussenzweig, H. Moyses: "Theory of the Rainbow," *Scientific American*, vol. 236, no. 4, April 1977, pp. 116–126. An article which indicates how complicated the physical explanation of a rainbow really is.

Smith, F. Dow: "How Images Are Formed," *Scientific American*, vol. 219, no. 3, September 1968, pp. 96–108. How wave theory and computer programs are needed to supplement ray optics in designing high-quality lenses. This issue of *Scientific American* also contains some other excellent articles on light.

*W*here the telescope ends, the microscope begins. Who
is to say of the two, which is the grander view?

Victor Hugo (1802–1885)

Chapter 24
Optical Instruments

Cameras, microscopes, telescopes, and more specialized optical instruments are among the triumphs of modern technology. Without them we would neither know as much as we do about the world in which we live, nor be able to enjoy its wonders as fully as we now can. All optical instruments consist of combinations of the lenses and mirrors previously discussed, sometimes in combination with prisms and diffraction gratings. The basic behavior of such instruments can be predicted using the theory developed in the preceding two chapters.

24.1 Cameras

The simplest optical device for viewing images on a screen or for recording images on a photographic plate is the *pinhole camera*.

Pinhole Camera

The pinhole camera* uses neither lenses nor mirrors but is still capable of producing sharp images with good *depth of field*; that is, the camera can form sharp images of objects at very different distances from the camera. It consists merely of a single small pinhole in the front of a light-tight box. If an illuminated object like the light bulb filament in Fig. 24.1 is placed opposite the pinhole, a ray from point A on the filament will pass through the pinhole and strike the screen at the rear of the box at point A'. Similarly, a ray from B will strike the screen at B'. Each point on the object will in the same way be imaged at a single point on the screen. Hence a real, inverted image of the light bulb filament will be produced on the screen. It is easily shown that the magnification M produced by this simple camera is equal to q/p, where q is the image distance and p the object distance, as for mirrors and lenses.

*The pinhole camera was originally called a *camera obscura*, the Latin words for "dark room."

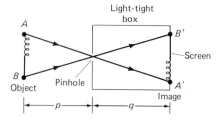

FIGURE 24.1 A pinhole camera. Its magnification is q/p.

Excellent pictures of stationary objects can be taken with a pinhole camera and photographic film. The only problem is that very little light gets through the pinhole and hence long exposure times, of the order of minutes or even hours, may be needed to collect sufficient light to form an image bright enough to expose the film. During this exposure time the object, pinhole, and film have to remain perfectly motionless. If the hole is increased in size to reduce the exposure time, the image begins to blur, because rays from different parts of the object now overlap at the same point on the screen.

Cameras with Lenses

For most practical purposes the pinhole camera has been replaced by cameras employing converging lenses in place of pinholes. In the old-fashioned box camera the photographic film was placed in the focal plane of the lens. Distant objects would then be focused sharply on the film. Since the lens had a fixed focal length, objects closer to the lens would be out of focus. This difficulty was overcome by adding a bellows to the camera that allowed the image distance to be adjusted until a sharp image of the desired object was obtained on the photographic plate or film. This process of moving the lens with respect to the photographic plate is called *focusing*. A sketch of a simple bellows camera is shown in Fig. 24.2.

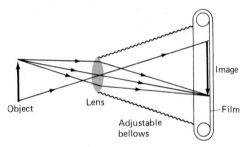

FIGURE 24.2 A bellows-type camera. The distance between the converging lens and the photographic film can be adjusted with the bellows to bring into focus objects at different distances from the lens.

f-Number

A very important characteristic of a camera is the brightness of the image it forms, because the brighter the image, the shorter the exposure time required. Short exposure times usually result in sharper images, since the image is relatively unaffected by motion of the camera or object during the exposure. The brightness of the image depends on the amount of light passing into the camera per second. This, in turn, depends on the area of the lens through which the light enters. This area is $A = \pi R^2 = \pi D^2/4$, where D is the diameter of the lens. Hence we may write:

$$\text{Image brightness} \propto D^2 \tag{24.1}$$

The brightness also depends on the focal length of the lens. For a simple lens we have for the magnification, from Eq. (23.10),*

$$M = \frac{h'}{h} = \frac{q}{p}$$

---

*In this chapter we will abandon the convention of inserting a minus sign in this equation to predict whether the image is upright or inverted. This will be clear from the ray diagrams drawn for the optical instruments discussed.

The height of the image is therefore proportional to f, and hence the area of the image is proportional to f^2, or

$$A \propto f^2 \qquad (24.2)$$

Now, the larger the area A of the image, the lower the brightness, since the same amount of light is spread out over a larger area. Hence we have:

$$\text{Image brightness} \propto \frac{1}{A} \propto \frac{1}{f^2} \qquad (24.3)$$

Putting Eqs. (24.1) and (24.3) together we obtain

$$\text{Image brightness} \propto \frac{D^2}{f^2} \propto \left(\frac{D}{f}\right)^2 \qquad (24.4)$$

Definition

The ratio of focal length to lens diameter (f/D) is called the f-number of the lens.

A lens of diameter 5 cm and f-number equal to 2 (often written as f/2) has a focal length given by

$$f = D(\text{f-number}) = (5 \text{ cm})(2) = 10 \text{ cm}$$

The smaller the f-number, the brighter the image, according to Eq. (24.4). "Fast" lenses, which have large diameters and short focal lengths, often have f-numbers as low as 1.0. Because fast lenses are large and have short focal lengths, they also have larger lens aberrations than slower lenses and hence need considerable correction before they can be used in good cameras. Since the smaller the aperture, the greater the depth of field (see the discussion of the pinhole camera), fast lenses with large apertures have greatly reduced depths of field.

A sketch of a corrected compound camera lens is shown in Fig. 24.3.

Modern cameras contain adjustments to change both the effective f-number of the lens and the shutter speed. The f-number is controlled by a "stop," or variable iris diaphragm, behind the lens, which varies the brightness of the image by varying the effective diameter D of the lens. Most

FIGURE 24.3 A typical corrected compound camera lens. The adjustable iris diaphragm removes the aberrations which would be introduced by the outer portion of the lens.

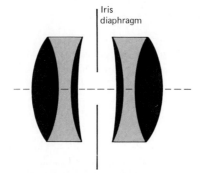

Iris diaphragm

cameras have f-stop markings of 1.0, 1.4, 2.0, 2.8, 4.0, 5.6, 8, 11, 16, 22, and 32. Each number in this series is approximately $\sqrt{2}$ times the preceding one. Since the f-number is equal to f/D, increasing the f-number from 1 to 1.4 means that the effective lens diameter D has been decreased by a factor of 1.4. Therefore the open lens area and the brightness are both decreased by a factor of 2, since $A \propto D^2$. Hence the image brightness is reduced by a factor of 2 for each successive f-stop marking.

Shutter Speed

The shutter controls the time during which light enters the camera and falls on the photographic film. This may vary from time exposures of seconds or minutes to flash exposures of 10^{-3} s or less. To photograph moving objects, very fast shutter speeds are required. Normally, to avoid the fuzziness in the image that results from inadvertent motion of the camera by the photographer, shutter speeds of 1/50 s or less are used.

The shutter speed, nature of the lens, and f-stop used determine the amount of light entering the camera during an exposure. Too little light leads to *underexposure*, where none but the brightest parts of the image can be seen; too much light leads to *overexposure*, in which everything has a washed-out appearance. Modern cameras contain automatic exposure devices to adjust the shutter and lens diaphragm for optimal exposures.

Example 24.1

(a) An f/2 camera lens is first used wide open, i.e., without a diaphragm to stop down the lens. How will the brightness of the image differ if the lens is then stopped down to f/8?

(b) How will the shutter speed have to change with the f/8 stop if the amount of light entering the camera is to remain the same?

SOLUTION

(a) The brightness depends on $(D/f)^2$ or on (1/f-number)^2. Hence the brightness of the f/8 lens is proportional to 1/64. The brightness of the wide-open f/2 lens, on the other hand, is proportional to 1/4. Hence the relative brightness is

$$\frac{1/4}{1/64} = \boxed{16}$$

The unstopped f/2 lens therefore produces an image 16 times brighter than the same lens stopped down to f/8.

(b) To compensate for the 16-fold reduction in brightness on changing to the f/8 stop, the shutter speed must be changed to increase the exposure time by a factor of 16. This will allow the same amount of light to enter the camera.

24.2 The Human Eye

The human eye in many ways resembles a simple camera, but it is exceedingly more complicated. The lens in the eye takes the place of the camera lens and forms a real, inverted image of an object on the *retina*, which is equivalent to the photographic film in a camera. Rods and cones in the retina act as transducers to convert light energy into electric signals. These signals then travel along the optic nerve to the brain, where the final image is constructed.

Just as a camera can be focused by changing the lens-film distance, so too the eye is automatically focused by a process called *accommodation*. The lens in the eye is fixed in position about 1.7 cm from the retina. Since it is made of flexible tissue, however, its shape can be modified by the ciliary muscles of the eye. In this way a normal eye can be made to focus on objects from infinity to 25 cm from the eye. Objects closer than 25 cm, called the *near point of the eye*,

FIGURE 24.4 The change in the shape of the eye lens to focus on objects at different distances: (*a*) Eye focused on an object at infinity; (*b*) eye focused on an object near the eye. The eye lens increases its thickness and, therefore its power to focus on close objects. When its thickness is increased in this way, the eye focuses objects at infinity not on the retina but considerably in front of it, as shown.

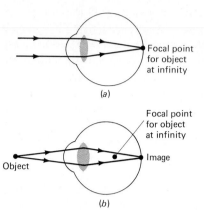

cannot be seen clearly with the unaided eye, since the lens of the normal eye cannot be made thick enough to focus at these short distances. Figure 24.4 shows how the shape of the lens changes as the object distance varies. The thicker the eye lens, the shorter its focal length and the greater its power.

Figure 24.5 is a diagram of the human eye. The *cornea* is the transparent membrane which acts as the outer window of the eye. The *iris* is the colored part—usually blue or brown—of the eye, and the *pupil* appears as the black circle at the center. The pupil is black because it is a hole and no light is reflected from it. The iris acts as a diaphragm which by a feedback mechanism adjusts the size of the pupil spontaneously to allow the proper amount of light into the eye for clear vision.

The *cornea* produces most of the refraction needed for the eye to converge the light incident on it. The *cornea* has a large index of refraction compared with air, and hence a large amount of refraction occurs at the air-cornea interface. On the other hand, the refractive indices of the cornea, the lens, and the aqueous and vitreous bodies adjacent to the lens are not too different, and relatively little refraction occurs once the light rays pass through the cornea. The optical power of the cornea is about 40 diopters (D), whereas all the other eye components, including the lens, have a total power of only 20 to 24 D (see Sec. 23.8). The overall power of the eye is therefore between 60 and 64 D. The amount of accommodation possible then amounts to only about 4 D on the average.

FIGURE 24.5 Diagram of the human eye.

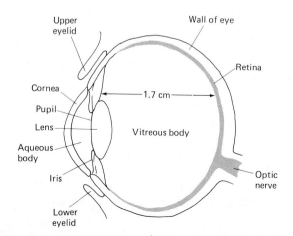

If the eye lens is completely removed, as in a cataract operation, the patient can still see. Eyeglasses can make up for the converging power of the removed lens, but accommodation is no longer possible and objects at different distances from the eye cannot all be seen clearly.

Example 24.2

(a) If the human eye has a lens power of 60 D, what is the focal length of the eye?
(b) How does the focal length of the eye change when, by accommodation, its power changes to 64 D?

(c) To what object positions do these two focal lengths correspond, if we assume that the distance from eye lens to retina is 1.70 cm?

SOLUTION

(a) Since for any lens $f = 1/P$, we have

$$f = \frac{1}{60 \text{ D}} = \frac{1}{60} \text{ m} = 0.0167 \text{ m} = \boxed{1.67 \text{ cm}}$$

Hence the eye acts like a converging lens of 1.67 cm focal length. This agrees with our previous statement that the retina is located about 1.7 cm from the center of the lens of the eye.

(b) If the power changes to 64 D, we have:

$$f = \frac{1}{64 \text{ D}} = \frac{1}{64} \text{ m} = 0.0156 \text{ m} = \boxed{1.56 \text{ cm}}$$

Hence the full range of accommodation of the average human eye changes its focal length by only about 0.1 cm, or 1 mm.

(c) For $f = 1.67$ cm, as in part (a),

$$\frac{1}{p} = \frac{1}{f} - \frac{1}{q} = \frac{1}{1.67 \text{ cm}} - \frac{1}{1.70 \text{ cm}} = \frac{0.599 - 0.588}{1 \text{ cm}}$$

$$= 0.011 \text{ cm}^{-1}$$

and so $\quad p = \dfrac{1}{0.011} \text{ cm} = \boxed{91 \text{ cm}}$

For $f = 1.56$ cm, as in part (b),

$$\frac{1}{p} = \frac{1}{f} - \frac{1}{q} = \frac{1}{1.56 \text{ cm}} - \frac{1}{1.70 \text{ cm}} = \frac{0.641 - 0.588}{1 \text{ cm}}$$

$$= 0.053 \text{ cm}^{-1}$$

and so $\quad p = \dfrac{1 \text{ cm}}{0.053} = \boxed{19 \text{ cm}}$

Hence this range of focal lengths corresponds to object distances from about 20 cm to 1 m.

Defects of the Eye

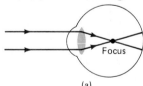

(a)

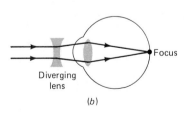

(b)

FIGURE 24.6 (a) Nearsightedness: The eye cannot focus on distant objects. (b) Correction for nearsightedness using diverging eyeglass lenses.

Nearsightedness Nearsightedness (myopia) is a defect which allows a person to see *near* objects clearly, but makes it impossible for the eye to focus on objects at a distance, even when the eye is totally relaxed. Correcting for myopia requires eyeglasses containing a diverging lens. This lens reduces the positive power of the lens in the eye and enables distant objects to be focused clearly on the retina. This is illustrated in Fig. 24.6.

Farsightedness Farsightedness (hyperopia) is a defect which allows a person to see *distant* objects clearly but makes it impossible to focus nearby objects on the retina. A perfectly normal eye in an adult cannot focus on objects closer than 25 cm from the eye. Correcting for farsightedness requires glasses with converging lenses, as shown in Fig. 24.7. Farsightedness often occurs in older people as their ciliary muscles lose the power to accommodate to near objects. The eye's power to accommodate changes from as much as 11 D at age 10 to 1 D at age 70, as shown in Fig. 24.8. Note particularly the rapid falloff during the years from 30 to 50.

Most prescriptions for eyeglasses range from 1 to 4 D, corresponding to focal lengths from 1 m to 25 cm. The wearer of *bifocal glasses* looks through

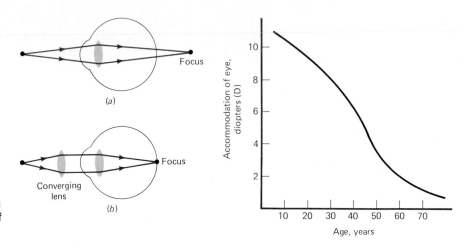

FIGURE 24.7 (*a*) Farsightedness: The eye cannot focus on near objects. (*b*) Correction for farsightedness using converging eyeglass lenses.

FIGURE 24.8 Accommodation of the human eye as a function of age.

diverging lenses at objects far away and through converging lenses at objects up close. *Trifocals* add a third lens, for middle-distance vision, between these two.

Astigmatism Astigmatism results from a lack of perfect spherical symmetry in the cornea and lens of the eye. In this case a person looking at the pattern of Fig. 24.9 sees some of the spokes of the wheel as darker and clearer than the others. This happens because the lenses of the person's eyes are not perfectly spherical but somewhat cylindrical. As a result the eye focuses rays from a point in a plane *perpendicular* to the cylinder axis to a point; but it focuses rays from a point in a place *parallel* to the cylinder axis not to a point but to a line, as shown in Fig. 24.10. For this reason, lines drawn at different angles will look different to a person with astigmatism, and this enables an oculist to locate the cylindrical axis in the astigmatic eye.

FIGURE 24.9 Pattern to test for astigmatism in the human eye.

FIGURE 24.10 Astigmatism in the human eye. In this case the eye lens is not perfectly spherical but somewhat cylindrical. As a result it focuses point objects not into a point but into a line, as shown.

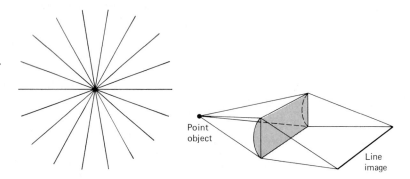

Correcting for astigmatism requires eyeglasses with cylindrical lenses which exactly cancel out the cylindrical components in the natural eye.

Off-axis astigmatism of the kind found in lenses is not a problem for the human eye, because the eye normally rotates in its socket so as to put the object directly on axis.

Chromatic aberration does exist in the human eye, but it is greatly reduced by the yellowish pigment called *macula lutea* in the fovea or color-sensitive part of the retina. This makes the eye much more sensitive in the yellow-green region of the spectrum, as shown in Fig. 24.11, and filters

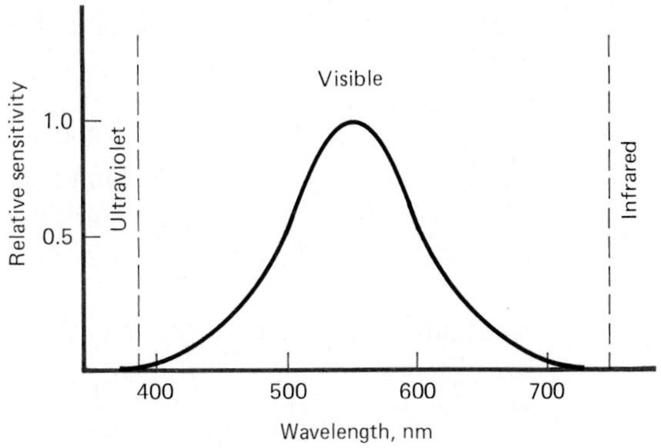

FIGURE 24.11   Sensitivity of human eye to wavelengths in the visible region of the spectrum. The eye is especially sensitive in the yellow-green region of the spectrum at about 550 nm.

out much of the red and blue light that would make chromatic aberration more noticeable.

Example 24.3

An elderly man can read a newspaper only when it is at least 60 cm from his eyes. What kind of reading lenses does he need in his glasses? Assume that the glasses will be worn very close to his eyes.

SOLUTION

The usual near point of the human eye is 25 cm. Hence we want the man to see print clearly when it is 25 cm from his eyes. His glasses therefore must produce a virtual image of the newspaper at a distance of 60 cm from his eyes when the paper is actually held at the ordinary reading distance of 25 cm. Hence we have

$$\frac{1}{p} + \frac{1}{q} = \frac{1}{f}$$

or

$$\frac{1}{f} = \frac{1}{25 \text{ cm}} + \frac{1}{-60 \text{ cm}} = \frac{+35}{1500 \text{ cm}} = \frac{+7}{300 \text{ cm}}$$

Note that the sign on the image distance is negative, since it is on the same side of the lens as the object. Hence

$$f = \frac{+300 \text{ cm}}{7} = +43 \text{ cm}$$

The power of the required lens is then

$$P = \frac{1}{f} = \frac{1}{+0.43 \text{ m}} = \boxed{+2.3 \text{ D}}$$

Hence a lens of +2.3-D power is needed, i.e., a converging lens with power 2.3 D.

24.3 The Magnifying Glass (Simple Magnifier)

The closer an object is to the eye, the larger the image on the retina and the more detail to be seen in it, as shown in Fig. 24.12. The actual size of an object is of little consequence in determining how big it appears to be. What is important is its *angular size*. An object h cm high at a distance d cm from the eye subtends an angle given by $\tan \phi = h/d$. Since h/d is usually small, we can assume that $\phi \simeq \tan \theta = h/d$, and take ϕ (in radians) as a measure of the angular size of the object.

FIGURE 24.12   The size of the image formed on the retina of the eye. The object is of the same size in both (a) and (b), but the object forms a much larger image on the retina in (b) because the object is closer to the eye.

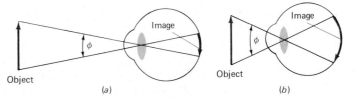

For example, a quarter (coin) of diameter 2.5 cm held at arm's length (60 cm) subtends an angle $\phi \simeq h/d = 2.5/60 = 0.042$ rad, whereas the sun, which has a diameter of 1.39×10^9 m, subtends an angle at the earth of only $\phi = (1.39 \times 10^9 \text{ m})/(1.49 \times 10^{11} \text{ m}) = 0.009$ rad, or an angle 5 times smaller. The quarter therefore appears *bigger* than the sun under these conditions.

The usefulness of a magnifying glass lies in its ability to increase the angular size of a viewed object. *Angular magnification* is defined as follows:

Definition

M_ϕ = angular magnification (or magnifying power)

$$= \frac{\text{angular size of image (with magnifier present)}}{\text{angular size of object (without magnifier)}} \qquad (24.5)$$

Normally these sizes are compared at a distance of 25 cm from the eye, the near point of the eye.

The simplest way to increase the angular magnification of an object is to bring it closer to the eye, as shown in Fig. 24.12. However, since the human eye cannot focus at distances less than 25 cm, a magnifying glass is needed. A *magnifying glass*, or simple magnifier, is just a single converging lens which produces a large virtual image of an object placed inside the focal point of the lens. As seen from the ray diagram of Fig. 24.13, the object and the image have the same angular size as seen by the eye. The magnifying glass allows us to view the object more closely than at its normal closest distance, 25 cm, and thereby increases the angular magnification.

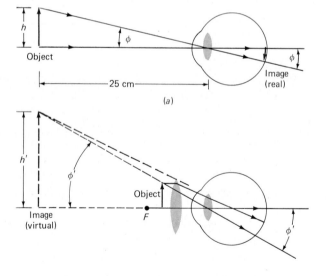

FIGURE 24.13   A magnifying glass (simple magnifier): The magnifying glass produces an enlarged, virtual image of an object. The angular magnification is $M_\phi = \phi'/\phi$, where ϕ is the angle subtended by the object at the eye lens in (a), and ϕ' is the angle subtended by the object at the eye lens in (b), in which case the magnifying glass allows the object to be placed closer to the eye. (The size of the eye is exaggerated for clarity.)

Magnifying Power of a Simple Magnifier

The angular magnification is given by Eq. (24.5) as

$$M_\phi = \frac{\phi'}{\phi} \qquad (24.6)$$

where ϕ' and ϕ are as shown in Fig. 24.13.

Let us suppose that the person using the magnifying glass relaxes the eye, so that the virtual image is seen at infinity, in which case the object is precisely at the focal point of the lens. Then $\phi' = h/f$, and so

$$M_\phi = \frac{\phi'}{\phi} = \frac{h/f}{h/25} = \frac{25}{f} \qquad \text{(where } f \text{ is in centimeters)} \tag{24.7}$$

Hence if the focal length of the magnifier is 2.5 cm and the eye is relaxed and focused on the virtual image at infinity, the magnification is 10 times.

A lens of high magnifying power must therefore be a strongly positive lens, i.e., a lens with a short focal length, since M_ϕ depends inversely on f. Aberrations in simple converging lenses limit the angular magnification M_ϕ to 2 or 3. If these aberrations are corrected by using compound lenses, magnifications up to about 20 are possible, but no higher.

Example 24.4

A pencil 6.0 mm thick is viewed by a woman with normal vision using a magnifying glass of focal length 5.0 cm. Calculate the angular magnification when the virtual image is at infinity.

SOLUTION

When the virtual image is at infinity

$$\frac{1}{p} = \frac{1}{5.0 \text{ cm}} - \frac{1}{\infty} \quad \text{and} \quad p = 5.0 \text{ cm}$$

In this case the object is exactly at the focal point of the lens.

The virtual image formed by the lens is now at infinity and is infinitely large. But its angular size is that of a 6.0-mm object at the focal distance of 5.0 cm. Hence we have, using Eq. (24.7),

$$M_\phi = \frac{(0.60 \text{ cm})/(5.0 \text{ cm})}{(0.60 \text{ cm})/(25 \text{ cm})} = \boxed{5}$$

24.4 The Compound Microscope

To increase the magnification over that possible with a simple magnifier, an additional lens can be added to a magnifying lens to form a *compound microscope*. In this case the magnifying lens is referred to as the *eyepiece* or *ocular*, and the added lens as the *objective lens*. As shown in Fig. 24.14, an object placed just outside the focal point F_1 of the objective lens forms an enlarged real image I_1 just inside the focal point F_2 of the eyepiece lens. The eyepiece then acts as a simple magnifier and produces a greatly enlarged virtual image of the real image produced by the objective lens. The final image is inverted, since the image formed by the objective lens is inverted.

The magnification of a compound microscope is simply the product of the magnifications produced by each of the two lenses. For the objective lens we have

$$M_1 = \frac{h'}{h} = \frac{q_1}{p_1} \simeq \frac{L - f_2}{f_1} \tag{24.8}$$

since the image produced by the first lens (the objective) is made to fall near the focus of the second lens (the eyepiece), and the object is presumed to be close to the focus of the first lens. For the eyepiece, which behaves like a simple magnifier, we have, from Eq. (24.7),

$$M_2 = \frac{25}{f_2}$$

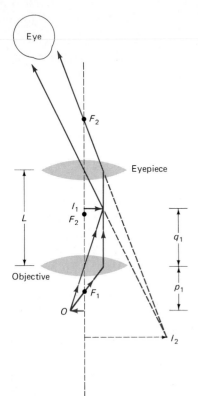

FIGURE 24.14 The compound microscope. It combines an objective lens with a simple magnifier, which in this case is called the eyepiece, or ocular.

where we have presumed that the final virtual image is at infinity. Then

$$M = M_1 M_2 \simeq \left(\frac{L - f_2}{f_1}\right)\left(\frac{25}{f_2}\right) \tag{24.9}$$

If we further assume that f_2 is small compared with L, the distance between the lenses, then $L - f_2 \simeq L$, and a reasonable approximation to the magnification of a compound microscope is

$$M = \frac{25L}{f_1 f_2} \tag{24.10}$$

where L is the distance between the two lenses. In most practical microscopes the two lenses are separated by a distance of about 18 cm for the convenience of the user. Hence $L = 18$ cm, and in this case the magnification is approximately

$$M \simeq \frac{25(18)}{f_1 f_2} \simeq \frac{450}{f_1 f_2} \tag{24.11}$$

where f_1 and f_2 are the focal lengths, in centimeters, of the two lenses.

Again we notice that the shorter the two focal lengths f_1 and f_2, the larger the magnification. Of course, larger magnifications mean thicker lenses and hence greater lens aberrations. For this reason most microscope lenses are complex, multilens structures designed to reduce aberrations as much as possible.

Equation (24.11) shows that, if the two lenses in a compound microscope have focal lengths of 1 cm each, the overall magnification is only about 450. With most light microscopes no additional detail is seen above about 400× because of limitations on the resolving power of lenses, to be discussed in Sec. 24.6.

Example 24.5

A compound microscope consists of a 10× eyepiece and a 40× objective lens at the ends of a microscope barrel 18 cm long. Find:
(a) The total magnification of the microscope.

(b) The position at which the object must be placed so that the final image is at infinity.
(c) The focal length of each lens.

SOLUTION

(a) The total magnification is

$$M = M_1 M_2 = 40 \times 10 = \boxed{400}$$

(b) For the final object to be at infinity the image produced by the objective lens must be at the focus of the eyepiece. (The diagram of Fig. 24.14 should be used here.) The image distance for the objective lens is

$$q_1 = L - f_2$$

where $L = 18$ cm, and $f_2 = \dfrac{25 \text{ cm}}{M_2} = \dfrac{25 \text{ cm}}{10} = 2.5$ cm

Hence, $q_1 = 18 - 2.5 = 15.5$ cm.

Now, from Eq. (23.10), $p_1 = q_1/M_1$, and so the object distance for the objective lens is

$$p_1 = \frac{15.5 \text{ cm}}{40} = \boxed{0.39 \text{ cm}}$$

(c) We have found in part (b) that

$$f_2 = \boxed{2.5 \text{ cm}}$$

We can find f_1 from the lens equation

$$\frac{1}{f_1} = \frac{1}{p_1} + \frac{1}{q_1} = \frac{1}{0.39 \text{ cm}} + \frac{1}{15.5 \text{ cm}}$$

or $\quad f_1 = \dfrac{0.39(15.5)\ \text{cm}}{15.9} = \boxed{0.38\ \text{cm}}$

It should be noted that Eq. (24.11) derived above is only an approximation, and hence that the values obtained for the focal lengths are also approximations.

Hence the object is just 0.01 cm outside the focus of the objective lens, which is about what we would expect.

24.5 Telescopes

Two features are particularly important in a telescope: (1) *resolving power*, or the ability to form distinct images of objects very close together; and (2) *light-gathering power*, which depends on the cross-sectional area of the objective lens on which the light from a distant object falls. Since both resolving power and light-gathering power improve with the size of the objective lens (or mirror), telescopes with very large lenses (and mirrors) are found at the great astronomical observatories of the world.

Refracting Telescopes

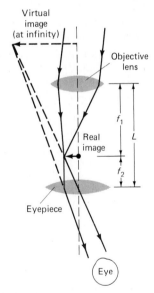

FIGURE 24.15 An astronomical telescope using two converging lenses.

In 1611 Kepler developed a refracting telescope with basically the same lens structure as a compound microscope. This is shown in Fig. 24.15. An objective lens forms a real inverted image of a distant object like a star. Since the object is so far away, the light rays entering the telescope are parallel and the image is produced at the focal plane of the objective lens. This image is much smaller than the object. An eyepiece is then used to produce an enlarged virtual image of the real image produced by the objective lens. If this virtual image is to be seen by the relaxed eye at infinity, the object being viewed through the eyepiece must be at the focal point of the eyepiece lens. Hence the length of the telescope is fixed by the fact that the focal points of the objective and the ocular lenses must coincide. The barrel length L is then the sum of the two focal lengths, as shown in Fig. 24.15.

To obtain the magnifying power of a refracting telescope, or *refractor*, of this kind, let us consider a ray which comes from one edge E of a distant star, passes through the center of the objective lens, and forms a real image E' in the focal plane of an objective lens of focal length f_1, as in Fig. 24.16. The object is of height h at a very large object distance p, and so its angular size as seen with the unaided eye is $\phi \simeq \tan\phi = h/p$. The image height h' is much smaller than h, but the angular size of the image as seen with the eyepiece is $\phi' \simeq \tan\phi' = h'/f_2$, which is larger than ϕ' if f_2 is small. Hence the magnifying power is, from Eq. (24.6),

$$M_\phi = \frac{\phi'}{\phi} = \frac{h'/f_2}{h/p}$$

But from similar triangles in Fig. 24.16, we have

$$\frac{h'}{f_1} = \frac{h}{p}$$

where h is the height of the original object at a very long object distance p, and so

$$M_\phi = \frac{h'/f_2}{h'/f_1} = \frac{f_1}{f_2} = \frac{\text{focal length of objective}}{\text{focal length of eyepiece}} \qquad (24.12)$$

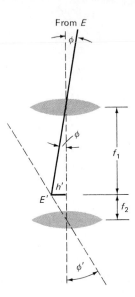

From E

ϕ

ϕ

f_1

h'

E'

f_2

ϕ'

FIGURE 24.16 Light from a star entering a refracting astronomical telescope. The angular magnification is $M_\phi = \phi'/\phi = f_1/f_2$.

The magnifying power of a refracting telescope is the ratio of the focal length of the objective lens to the focal length of the eyepiece.

Hence, for large magnifications a refractor should have an objective lens of long focal length and an eyepiece of very short focal length.

One problem with a telescope of the kind just described is that it produces an inverted image. This is particularly bothersome if the telescope is to be used for terrestrial observations, but is less troublesome in astronomical work.

Reflecting Telescopes

Most large astronomical telescopes in use today are *reflectors* rather than refractors. Astronomical telescopes need very great light-gathering power and hence the objective lenses must be very large. Since a lens can only be supported at its edges, large lenses begin to sag under their own weight and thus distort the images they produce. For this reason large concave mirrors, which can be supported over their entire back surface, are now used almost exclusively in large research telescopes. A mirror also has only one surface instead of two that must be ground, and is free from the chromatic aberration that exists in all lenses.* By using parabolic mirrors spherical aberration may also be eliminated.

Since the focus of a reflecting telescope is inside the main tube of the telescope, a variety of schemes have been developed to divert the image to a place outside the tube, where it can be observed and recorded. Two possible configurations to do this, the Newtonian focus and the Cassegrainian focus, are shown in Fig. 24.17.

The two largest telescopes in the world are the 200-in Hale reflector at Mount Palomar in California (see Fig. 24.18) and a 6-m (234-in) reflector in Russia. The largest refractor in the world, the 40-in refractor at the Yerkes Observatory in Williams Bay, Wisconsin, is considerably smaller than these reflectors.

---

*It was to solve the problem of chromatic aberration that Newton first developed the reflecting telescope.

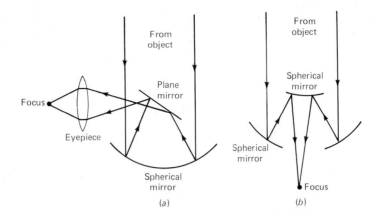

FIGURE 24.17 Two types of reflecting astronomical telescopes: (a) the Newtonian focus; (b) the Cassegrainian focus.

FIGURE 24.18 The 200-in Hale reflecting telescope at Mount Palomar, California. (*Photos courtesy of Mount Palomar Observatories.*)

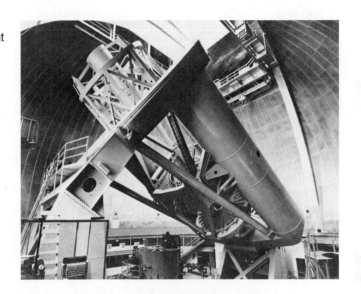

Example 24.6

The large telescope at Yerkes Observatory in Wisconsin has a 40-in-diameter objective lens of 63.5-ft focal length and an eyepiece of focal length 10 cm. What is the approximate magnifying power of this telescope?

SOLUTION

$$M_\phi = \frac{f_1}{f_2} = \frac{(63.5 \text{ ft})(12 \text{ in/ft})(2.54 \text{ cm/in})}{10 \text{ cm}} = \boxed{194\times}$$

24.6 The Resolution of Telescopes and Microscopes

The fundamental element in a telescope is the large-aperture objective lens or mirror used to focus on a very distant stellar object. Some diffraction effects occur even with the largest objective lenses or mirrors, although these effects may be quite small for large apertures.

In Sec. 22.4 we saw that, on passing through a rectangular aperture of width w, light of wavelength λ is diffracted so that the first diffraction minimum occurs at an angle θ' given by $\sin \theta' = \lambda/w$. In the case of a telescope lens we are dealing not with a rectangular aperture of fixed width w and height h but with a circular opening of diameter d. This introduces a numerical factor of 1.22 (obtained from analysis that we will not describe), so that for a circular aperture the first minimum occurs at $\sin \theta' = 1.22\lambda/d$. The larger the diameter d of the lens, the smaller θ' and the less important diffraction effects. This is one reason astronomical telescopes use such large-diameter objective lenses.

Because of diffraction, the image of a star, as formed by a telescope lens, consists of a bright central disk (known as an *Airy disk*) surrounded by a number of fainter diffraction rings, as shown in Fig. 24.19. This configuration is of crucial importance in distinguishing two distant objects such as two stars whose angular separation is small. If the stars are too close together, their diffraction patterns overlap to such an extent that it is impossible to distinguish one star from the other. We say that the two stars cannot be *resolved*. Figure 24.20 shows what happens for two stars of the same intensity when the distance between the stars is gradually reduced. In Fig. 24.20*a* the

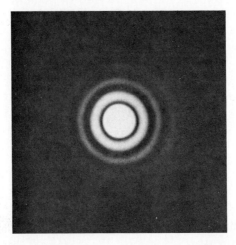

FIGURE 24.19   Image of a star, as formed by a telescope lens, showing the diffraction rings around the image. (*Photo from Cagnet, Françon, and Thrierr, Atlas of Optical Phenomena; reprinted by permission of Springer-Verlag, Heidelberg.*)

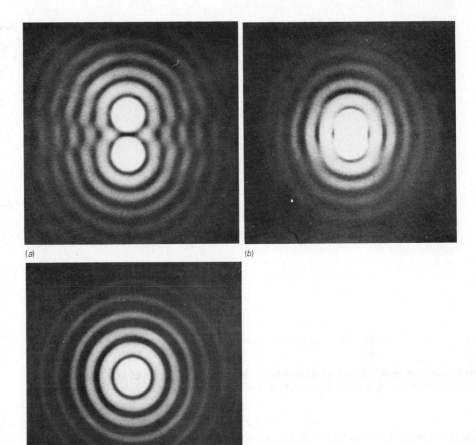

FIGURE 24.20 Variation in overlap of the images of two adjacent stars as the distance between the stars is varied. (*Photos from Cagnet, Françon, and Thrierr, Atlas of Optical Phenomena; reprinted by permission of Springer-Verlag, Heidelberg.*)

two stars can be clearly distinguished, even though their images still overlap; in (*b*) it is just barely possible to distinguish one star from the other; in (*c*), where the stars are even closer together, the two images fuse into one and the stars can no longer be resolved.

Resolving Power of a Telescope Objective

Figure 24.21 shows the light intensity on a photographic plate of the central image of the two stars. In Fig. 24.21*b* the central maximum of one star's diffraction pattern falls directly on top of the first diffraction minimum of the other star's diffraction pattern. Since the two stars are just barely resolvable under these conditions, Lord Rayleigh (1842–1919) suggested that this be taken as the criterion for resolution, a criterion now called the *Rayleigh criterion*. Since the angular separation of the two stars in this case is the same as the angular separation of the first minimum from the central image in the diffraction pattern of either star, we have for the Rayleigh criterion

$$\sin \theta_R = \frac{1.22\lambda}{d} \tag{24.13}$$

where θ_R is the minimum angular separation for resolution. Since θ_R is always small, θ_R (in radians) $\simeq \sin \theta_R$, and so

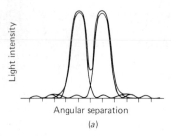

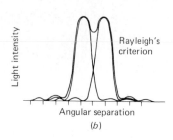

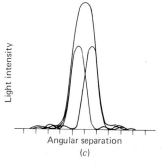

FIGURE 24.21 Rayleigh criterion for the resolution of a telescope or microscope. The angular separation in (b) is taken as the minimum angle that can be resolved. The images of the two stars in (c) overlap to form a single image.

Rayleigh Criterion for Resolution

$$\theta_R \simeq \frac{1.22\lambda}{d}$$

(24.14)

This equation sets the theoretical limit on the resolving power of a telescope. Stars with angular separations greater than θ_R in radians can be resolved, as in Fig. 24.21a; those with angular separations less than θ_R cannot be resolved, as in Fig. 24.21c. Note that the Rayleigh criterion applies to perfect lenses whose performance is limited only by diffraction and hence by the wave nature of light. Lens aberrations in imperfect lenses can drastically reduce the achievable resolution. For example, the resolving power of a high-quality camera lens is usually limited by diffraction for small lens openings but by lens aberrations for large lens openings.

Example 24.7

Experiments on the human eye indicate that its resolving power (sometimes called *visual acuity*) is about 2.0×10^{-4} rad (less than 1 minute of arc).
(a) If this resolving power is limited by the finite size of the eye opening, what would the aperture of the eye's pupil have to be to limit the resolving power of the eye to 2.0×10^{-4} rad?

(b) For larger eye apertures the ultimate limitation on the resolving power of the eye is a physiological one, the spacing of the sensitive photoreceptor cells on the retina. If the distance between the eye lens and the retina is 1.7 cm, what is the distance between the positions at which two just-resolvable light rays strike the retina?

SOLUTION

(a) Diffraction effects in the eye become important when the iris makes the pupil diameter small enough that the Rayleigh criterion leads to angles smaller than 2.0×10^{-4} rad. When this happens, we must have:

$$\theta_R = \frac{1.22\lambda}{d} = 2.0 \times 10^{-4} \text{ rad}$$

and so

$$d = \frac{1.22\lambda}{\theta_R} = \frac{1.22(500 \times 10^{-9} \text{ m})}{2.0 \times 10^{-4}} = 3.0 \times 10^{-3} \text{ m}$$

$$= \boxed{3.0 \text{ mm}}$$

It is well known that the eye's pupil is varied by the iris from about 2 to 6 mm to control the amount of light entering the eye. Hence an opening of 3.0 mm is quite reasonable.

(b) For two light rays subtending an angle of 2.0×10^{-4} rad at the eye lens, we have

$$\sin \theta \simeq \tan \theta \simeq \frac{d}{1.7 \times 10^{-2} \text{ m}}$$

where d is the separation of the two light rays on the retina, and we are assuming that $\theta \simeq \sin \theta \simeq \tan \theta$. Hence

$$d = (1.7 \times 10^{-2} \text{ m})(\theta) = (1.7 \times 10^{-2} \text{ m})(2.0 \times 10^{-4} \text{ rad})$$

$$= 3.4 \times 10^{-6} \text{ m} = \boxed{3.4 \text{ } \mu\text{m}}$$

It is known from the physiology of the eye that the average retinal photoreceptor cell has a diameter of about 1 μm.

Hence the eye can resolve light falling on two cells separated by one additional photoreceptor. This seems reasonable, since it takes more than one cell to determine the presence or absence of a boundary in an image.

For larger eye apertures spherical and chromatic aberration, and ultimately the size of the photoreceptors, limit the eye's resolving power. At smaller apertures (less than about 3 mm) diffraction is the limiting factor, as we have just seen. These two effects tend to compensate and leave the resolving power of the eye roughly constant at something less than 1 minute of arc. This corresponds to the diameter of a human hair held at arm's length!

Resolving Power of Microscopes

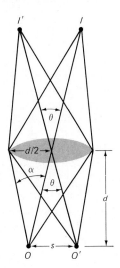

FIGURE 24.22 Resolving power of a microscope. A point O on the object produces an image at I; a point O' on the object produces an image at I'.

Unlike telescopes, whose objects of view are usually far away, microscopes are used to look at objects very close to the objective lens. We assume that the objective lens subtends an angle 2α at the object, which is located close to the focal point of the objective, as in Fig. 24.22. We wish to find the smallest distance s between two points O and O' on the object that will produce images I and I' which can just be resolved. The resolution is, of course, limited by the diffraction occurring at the lens aperture, which will lead to overlapping diffraction patterns at I and I'. The criterion for resolution is again the Rayleigh criterion $\theta_R \simeq 1.22\lambda/d$.

In this case two rays from O and O' passing through the center of the lens arrive at the image points I and I'. The angle θ subtended by these rays at the lens must be greater than or equal to θ_R for O and O' to be resolved. But

$$\theta \simeq \sin \theta \simeq \frac{s}{f}$$

Hence for resolution we need to have

$$\theta \simeq \frac{1.22\lambda}{d} \cong \frac{s}{f}$$

or $$s \simeq \frac{1.22\lambda f}{d} \tag{24.15}$$

This is the smallest object distance which will produce clearly distinct images. Notice that f/d is just the f-number of the objective lens, and hence the smaller the f-number the finer the detail that can be resolved.

If the object is approximately at the focal point of the lens, and if the angle α, which is half the angle subtended by the objective lens at the object is small, we have

$$\sin \alpha \simeq \tan \alpha = \frac{d/2}{f} \tag{24.16}$$

or $$d \simeq 2f \sin \alpha$$

and so, on combining Eqs. (24.15) and (24.17),   (24.17)

$$s = \frac{1.22\lambda f}{2f \sin \alpha} = \frac{1.22\lambda}{2 \sin \alpha} \tag{24.18}$$

This is the minimum separation that can be resolved by the microscope. It shows that the larger α (and hence the larger the aperture of the lens, or the shorter its focal length), the better the resolution.

In microscopes of high magnification the space between the object and the objective lens is often filled with oil. This increases the resolving power, since it effectively changes the wavelength of the light from λ to λ/n, and so Eq. (24.18) becomes

$$s = \frac{1.22\lambda}{2n \sin \alpha} \tag{24.19}$$

This is the resolving power of an *oil-immersion microscope*, and the product $n \sin \alpha$ is called the *numerical aperture* (NA) of the objective lens. The NA is usually specified on the side of the microscope housing, along with the magnification. The greater the NA, the greater the resolving power.

Objective lenses in air have numerical apertures of between 0.12 and 0.85, while oil-immersion lenses have numerical apertures as high as 1.3. The limit of resolution of an oil-immersion microscope using light of wavelength 500 nm is then

$$s = \frac{1.22(500 \times 10^{-9} \text{ m})}{2(1.3)} = 2.3 \times 10^{-7} \text{ m} = 230 \text{ nm}$$

Hence we see that the resolving power of a good optical microscope (230 nm) is of the same order of magnitude as the wavelength of the light used (500 nm). This leads to the important practical conclusion that *with an optical microscope it is not possible to resolve structure which is much smaller than the wavelength of the light used.*

Example 24.8

SOLUTION

The minimum distance which the lens can resolve depends on the diffraction produced by its finite aperture. This distance is given by Eq. (24.15):

$$s \approx 1.22 \frac{f}{d} \lambda$$

Here f/d is just the f-number of the lens, which is given as 16 in this case. Hence

$$s \approx 1.22(16)(500 \text{ nm}) = 10^4 \text{ nm} = 10^{-5} \text{ m} = \boxed{10^{-2} \text{ mm}}$$

The lens can therefore resolve about 0.01 mm.

Example 24.9

(a) What must be the numerical aperture of an oil-immersion microscope designed to resolve two bacteria separated by 2×10^{-4} mm, if the bacteria are illuminated by violet light at 400 nm?

(b) What is the angle subtended by the objective lens of the microscope at the object?

SOLUTION

(a) Since the numerical aperture NA $= n \sin \alpha$ and, from Eq. (24.19), the smallest distance which can be resolved is

$$s = \frac{1.22\lambda}{2\text{NA}}$$

we have

$$\text{NA} = \frac{1.22\lambda}{2s} = \frac{1.22(400 \times 10^{-9} \text{ m})}{2(2 \times 10^{-7} \text{ m})} = \boxed{1.22}$$

(b)   If we suppose that the oil has an index of refraction of 1.5, we have

$$\sin \alpha = \frac{NA}{n} = \frac{1.22}{1.5} = 0.81 \qquad \text{and} \qquad \alpha = 54°$$

The full angle subtended at the object is then

$$2\,\alpha = 2(54°) = \boxed{108°}$$

As this example indicates, often α is not a particularly small angle, and so the assumption made in the derivation of Eq. (24.19) that sin $\alpha \simeq \tan \alpha$ is only a rough approximation. A more rigorous derivation, however, leads to the same result.

24.7 Specialized Kinds of Microscopes

In biology and medicine a great variety of specialized microscopes are used for two basic purposes: (1) to decrease diffraction effects and thus increase the *resolving power* of the instrument; (2) to increase the *contrast* between the object and the background.

Increasing Resolving Power

We have already seen that the most important way to increase resolving power is to shorten the wavelength of the radiation used. An *oil-immersion microscope* accomplishes this to a limited degree by reducing the effective wavelength from λ to λ/n, where n is the index of refraction of the medium. If $n = 1.5$, this leads to a 50 percent increase in resolving power.

Another useful technique is to use only the short-wavelength components of visible light in a microscope to reduce diffraction and increase resolution; thus many microscope illuminators use blue or violet light.

An *ultraviolet (UV) microscope* reduces the wavelength used to about 100 nm and hence improves the resolving power of the microscope by a factor of about 5. The problem is that glass does not transmit UV nor can the human eye detect it (although it can be seriously burned by it). Hence quartz or fluorite lenses are needed, together with special photographic or photoelectric recording. This means added expense and reduced convenience. Hence UV microscopes are only used for special purposes. When employed, however, they have the added advantage of detecting objects transparent to visible light, since many molecules, such as proteins and nucleic acids, absorb strongly in the UV region and hence show strong contrast under UV light. In cellular biology UV microscopes have been used to observe the behavior and thus determine the function of DNA and RNA in living cells.

X-rays are so energetic that it is difficult to find ways to focus or control them. Hence *x-ray microscopes* are not in common use, although they have been used successfully in some specialized research projects.

Electron microscopes, which have greater resolving power than any of the above types, will be discussed in Sec. 27.8, after we know more about the wave properties of electrons.

Increasing Contrast

Two important ways to improve the contrast in microscopes are exemplified in the phase-contrast microscope and the polarizing microscope.

Phase-Contrast Microscope   The phase-contrast microscope, developed in 1935 by the Dutch physicist F. Zernike (1888–1966), converts *phase changes* produced by microscope specimens into *amplitude changes* in the final image. The human eye can easily detect amplitude changes since they vary the intensity of the light, but the eye does not respond directly to phase changes in the light. In the phase-contrast microscope transparent objects which differ only slightly from their surroundings in refractive index can be made visible by using light interference to vary the brightness of the background compared with that of the object being studied.

Phase-contrast microscopes are particularly useful in the study of living organisms such as bacteria, which tend to be transparent and invisible when viewed with an ordinary microscope.

Polarizing Microscope   Another somewhat related technique uses polarized light to increase the contrast between the image and its background. Even biological specimens transparent to ordinary light can be made to show great contrast with polarized light.

A polarizing microscope contains a polarizer below the microscope stage and an analyzer above it, as shown in Fig. 24.23. When the polarizer and analyzer are crossed in the absence of a specimen, no light enters the eyepiece and the field is dark. If, now, a specimen is introduced which is optically active (or *birefringent**), it will rotate the plane of polarization and will appear bright against a dark background. Fortunately many important biological substances like nucleic acids and proteins are birefringent and thus make apt specimens for study with a polarizing microscope.

Both the phase-contrast microscope and the polarizing microscope have great advantages over the older light microscope, which requires that specimens be stained to make them visible. Staining usually kills the specimen, whereas the use of phase-contrast and polarizing microscopes does not. Hence it is possible to use living cells as specimens, and to observe important biological processes like *mitosis* (cell division) as they actually occur in living cells.

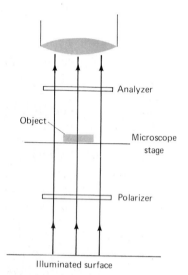

FIGURE 24.23   A polarizing microscope. The analyzer can be rotated to increase contrast between image and background.

24.8 Other Important Optical Instruments

Other optical instruments employing lenses and mirrors range from everyday devices like slide projectors to high-precision research instruments like interferometers. We will discuss only a few of these useful instruments.

The Ophthalmoscope

The *ophthalmoscope* is one of the many contributions to science and medicine made by Hermann von Helmholtz (see Fig. 12.22 and accompanying biography). It is the basic instrument for examining the inner structure of the eye. Helmholtz' original ophthalmoscope, made in 1851, consisted of a flat glass plate held at an angle of 45° to the vertical between the subject's eyes and the doctor. A light was placed over the head of the subject whose eyes were to be examined, as in Fig. 24.24. Some of this light was reflected back into the subject's eyes by the plate. The doctor could then look through the glass plate and see the illuminated interior of the eye. The lens and cornea of the eye acted as a simple magnifying glass which formed a large, virtual, upright

---

*A birefringent (or double refracting) crystal breaks up light into two beams polarized at right angles to each other. When these two beams combine on emerging from the crystal, the plane of polarization of the incident light is found to be rotated.

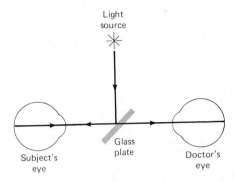

FIGURE 24.24 The original ophthalmoscope of H. von Helmholtz. The doctor can see into the patient's eye using the light reflected into the eye by the glass plate.

image of the patient's eye at a large distance behind the eye. The doctor could then examine the retina under large magnification, at no risk to the patient.

Such a device is called a *direct ophthalmoscope*. In modern versions of Helmholtz's device a totally reflecting prism with the doctor looking over its edge, or a small mirror with the doctor looking through a small hole in its center, usually replaces the glass plate.

The newest version of Helmholtz's invention is the *indirect ophthalmoscope*. This consists of a concave mirror and a battery-powered light source within a tubular handle, with the doctor sighting through a single or binocular eyepiece. The patient's eye is still used as a magnifying glass, but in addition there is an added convex glass lens which functions much as does an eyepiece in a microscope. Modern ophthalmoscopes have a rotating disk system which allows the convex lens to be changed to enable the doctor to observe the eye at varying depths and magnifications. Such eye examinations are often enhanced by administering eyedrops to dilate the pupil and thus allow more light into the eye.

Interferometers

One of the most precise and powerful research tools ever constructed is the *interferometer*, which uses the interference of two light beams to measure wavelengths of light, determine distances to very high accuracy, and measure the index of refraction of gases.

Michelson Interferometer The interferometer developed by A. A. Michelson (see Figs. 24.26 and 24.27 and accompanying biography) played a crucial role in the development of Einstein's special theory of relativity.

A sketch of Michelson's two-beam interferometer is shown in Fig. 24.25. It consists of a light source S which sends monochromatic light to a

FIGURE 24.25 The Michelson interferometer. Bright and dark fringes pass before the eye as the mirror M_2 is moved.

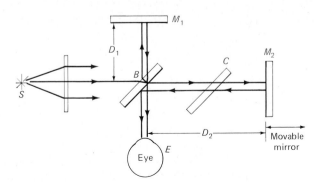

Albert Abraham Michelson (1852–1931)

FIGURE 24.26   Albert A. Michelson at his desk in the Ryerson Physical Laboratory at the University of Chicago. (*Photo courtesy of AIP Niels Bohr Library.*)

FIGURE 24.27   Photograph of A. A. Michelson (left front) with A. Einstein (front center) and R. Millikan (right front) at the California Institute of Technology, Pasadena, California, in 1931, shortly before Michelson's death. (*Photo courtesy of The California Institute of Technology Archives.*)

Although born in Europe, A. A. Michelson is usually considered the first American to receive a Nobel Prize in science, since he came to this country at the age of 4 and became a naturalized citizen. He received the 1907 Nobel Prize in physics "for his optical instruments of precision and the spectroscopic and metrologic investigations which he carried out by means of them."

Michelson was born in 1852 at what is now Strzelno in Poland. His parents were Polish Jews who immigrated to the United States in 1856. Here his father became a merchant selling supplies to the gold and silver miners of California and Nevada in the gold-rush days. Albert did very well in school, especially in mathematics and science, and, after failing to win an appointment to the Naval Academy at Annapolis, made a trip across the country to plead his case with President Grant. Owing to the kindness of the president and the authorities at Annapolis, he was finally given a special appointment to the Naval Academy.

There he did very well in science and athletics (he was a championship boxer), but less well in the humanities and seamanship. After graduating in 1873, he became a science instructor at the Naval Academy and there began his lifetime passion, the measurement of the speed of light to ever-greater precision (see Sec. 21.7). Feeling the need for better training in optics, Michelson spent the years 1880 to 1882 in Europe, studying and doing research at the great research laboratories of Berlin, Heidelberg, and Paris. He profited particularly from working with Helmholtz in Berlin and there designed and built the first of his many interferometers.

After 12 years of naval service, Michelson resigned his commission in 1881 to devote himself completely to physics. He became professor of physics at the new Case School of Applied Science in Cleveland and collaborated with E. W. Morley, a professor of chemistry at Western Reserve University, in the famous Michelson-Morley experiment (see Sec. 25.3).

In 1892 Michelson was made head of the department of physics at the University of Chicago and remained there until 1930, making it a world-famous center for research in precision optics and spectroscopy. Here he began to rule diffraction gratings which set new standards for size and resolving power. In his latter years he spent a great deal of time in California trying to improve on his already highly accurate determinations of the speed of light. He died in 1931 after several strokes suffered during preparations for a measurement of c in a mile-long evacuated metal tube.

Michelson was married twice, and had three children by each wife. He does not appear to have been an easy person to live with. He disliked social functions, and felt that university administrative tasks, and even the direction of graduate students, took too much time from his beloved research. Still, during the First World War, his patriotism drove him to abandon research and, at the age of 65, to serve his country as a reserve officer. He mellowed considerably to-

ward the end of his life, admitting in a letter to a friend that "I've also found that human nature is not so abominable as I have sometimes thought."

Michelson loved nothing more than experimental research. When asked by reporters in 1882 why he was attempting to measure the speed of light, he answered: "Because it's such great fun." Fifty years later Einstein asked Michelson a similar question, and the reply was still the same.

beam-splitting mirror B which is lightly coated with a reflecting metal so that half of the light is reflected to the stationary mirror M_1 and the other half passes through the beam splitter to the movable mirror M_2. The light striking M_1 is reflected back on itself and some of it passes through B and enters the eye at E. The light striking M_2 is also reflected back on itself, and some of it is reflected by B and also enters the eye. Hence two beams of light enter the eye at E. They differ in that one has traveled twice over the distance D_1 from B to M_1 and the other has traveled twice over the path from B to M_2, which distance we designate by D_2. (The plate C is a compensating plate introduced to ensure that the path of light through glass is the same for both beams.)

Since the light in the two beams originated at a common source S and was therefore originally in phase, the relative phase of the two beams at E will depend on the difference in the paths traveled by the two beams. Thus, if the paths differ by a whole number of wavelengths of the light used, i.e., if

$$2D_1 = 2D_2 + n\lambda \qquad (n \text{ integral}) \tag{24.20}$$

then constructive interference results. If, on the other hand, we have

$$2D_1 = 2D_2 + (m + \tfrac{1}{2})\lambda \qquad (m \text{ integral}) \tag{24.21}$$

the two beams will be 180° out of phase and will cancel. Hence the eye will see light or darkness depending on the phase difference between the two beams.

For perfect alignment of the mirrors at 90° to each other the interference pattern will consist of a series of dark and bright circles corresponding to destructive and constructive interference between the two beams, because the path difference varies with the angle of vision. If one mirror is inclined at a slight angle with respect to the other, straight fringes similar to those produced by a wedge of air between two pieces of glass are produced (see Fig. 24.28). As the mirror M_2 is moved through a distance $\lambda/2$ a dark fringe changes to a light fringe and then back again to a dark fringe. The bright and dark fringes then appear to move past the cross hairs in the observing telescope. These fringes can be counted either manually or by using photocounters, and the distance moved by the mirror can thus be measured in terms of wavelengths of light.

Thus, if 2000 bright fringes pass the cross hairs, it means that mirror M_2

FIGURE 24.28 Photograph of fringes produced with a Michelson interferometer. The straight fringes are deformed in the neighborhood of a candle flame by the change in the density of the air in the heated region. (*Photo from Cagnet, Françon, and Thrierr, Atlas of Optical Phenomena; courtesy of Springer-Verlag, Heidelberg.*)

has moved a distance of $2000(\lambda/2) = 1000\lambda$. For red light at 600 nm, this corresponds to 600 μm, which is 0.60 mm. In this way distances can be converted into wavelengths of light.

Fabry-Perot Interferometer   Another type of interferometer much used in research laboratories is the multibeam Fabry-Perot interferometer shown in Fig. 24.29.*

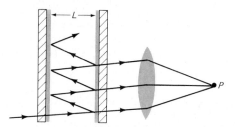

FIGURE 24.29   The Fabry-Perot interferometer.

The Fabry-Perot interferometer consists of two perfectly flat glass plates coated on their facing surfaces with a thin film of metal which reflects 99 percent and transmits only 1 percent of the incident light. A beam of light is incident on these two interferometer plates from the left, as shown. Most of the beam is trapped between the plates and is reflected back and forth a very large number of times. At each reflection by the right-hand plate, however, about 1 percent of the incident light passes through. As a result a large number of beams of light emerge at the same angle at which they entered and can be focused by a converging lens to a point P.

Any two adjacent beams emerging from the plates normal to the plates have paths differing by $2L$, where L is the distance between the reflecting surfaces of the plates. If $2L$ is equal to a whole number of wavelengths for some particular wavelength of light, then constructive interference occurs and a bright spot will appear at P. The spot will be very bright because a large number of beams will combine to form the image and they will all be perfectly in phase, since each one differs from its immediate neighbors in path length by exactly the same amount, $2L$. If, on the other hand, the path difference $2L$ is not exactly equal to a whole number of wavelengths of the light used, then the beams will cancel out to a great extent because there are so many different beams, with each one somewhat out of phase with all the others.

Hence by the interaction of many beams the two plates have become a highly selective wavelength filter. For a fixed distance between the plates, only those wavelengths which satisfy the equation $2L = n\lambda$ (n integral) will be strongly transmitted.

If the light comes from a point source, the path difference depends on the angle made by the incoming ray with the surface of the mirrors. As a result we would expect a circular interference pattern centered on a line drawn normal to the mirrors from the point source of light. This is indeed observed, as shown in Fig. 24.30. It shows great similarity to the Newton's rings pattern of Fig. 22.34.

---

*This interferometer is named after the two French physicists, Charles Fabry (1867–1945) and Alfred Pérot (1863–1925), who invented it in 1899.

FIGURE 24.30 Interference fringes produced by a Fabry-Perot interferometer using the 546-nm green line of mercury (Hg). The left side of the pattern was made with a natural Hg source, and the structure present is due to the many different isotopes found in natural Hg. The clearer and sharper fringes on the right were produced with a single-isotope $^{198}_{80}$Hg source. (*Photo courtesy of U.S. National Bureau of Standards.*)

Wavelength differences of less than 1 part in a million can be detected with a Fabry-Perot interferometer. It is therefore extremely useful for measuring the wavelengths of spectral lines and of distances in terms of spectral wavelengths.

Example 24.10

A Michelson interferometer is illuminated with red light from a He–Ne laser. An accurate screw with an attached scale measures the distance traveled by the movable mirror to the nearest 10^{-3} mm, or 10^{-6} m. It is found that moving the mirror through a distance of 0.316 mm causes a shift of 1000 fringes, as recorded by a photocell and counter. What is the wavelength of the laser light?

SOLUTION

The passage of 1000 fringes means that the total (two-way) light path has been changed by 1000 wavelengths. Hence we have:

$$1000\lambda = 2(0.316 \text{ mm})$$

or

$$\lambda = \frac{2(0.316 \times 10^{-3} \text{ m})}{1000} = 0.632 \times 10^{-6} \text{ m}$$

$$= \boxed{632 \text{ nm}}$$

Example 24.11

In the expression for a bright spot at the center of the interference pattern of a Fabry-Perot interferometer, $2L = n\lambda$, the integer n is called the order of *interference*. What is the order of interference for an interferometer with $d = 10$ cm which uses green light with $\lambda = 500$ nm?

SOLUTION

Since $2d = n\lambda$

$$n = \frac{2d}{\lambda} = \frac{2(0.10 \text{ m})}{500 \times 10^{-9} \text{ m}} = \boxed{4.0 \times 10^5}$$

Hence the path difference between two adjacent beams can be seen to correspond to a very large number of wavelengths of the light used.

Summary: Important Definitions and Equations

Cameras:

Optical instruments consisting of a lens or system of lenses used to focus light onto a photographic plate. Cameras produce real, inverted images.

f-Number of lens: The ratio of focal length to lens diameter (f/D).

Human eye: Resembles a simple camera; forms a real, inverted image of an object on the retina, which is the light-sensitive part of the eye.

Accommodation of eye: The process by which the focal length of the eye lens is automatically varied to produce a sharp image on the retina.

Defects of eye:

Nearsightedness (myopia): The inability to focus on distant objects; requires diverging lens to compensate for defect.

Farsightedness (hyperopia): The inability to focus on near objects; requires converging lens to compensate for defect.

Astigmatism: The inability to focus on horizontal and vertical patterns at the same distance at the same time; requires cylindrical lens to compensate for defect.

Simple magnifier (magnifying glass):

An optical instrument using a single converging lens to increase the angular magnification of an object. The object is placed inside the focus of the lens and produces a virtual, upright image.

Angular magnification, or magnifying power:

$$M_\phi = \frac{\text{angular size of image (with magnifier present)}}{\text{angular size of object (without magnifier)}}$$
$$= \frac{25}{f}$$

Compound microscope:

A combination of two converging lenses, an objective lens which forms an enlarged, real, inverted image, and an eyepiece or ocular, which produces a greatly enlarged virtual image of the real image produced by the objective.

Magnification: $M = 25L/f_1 f_2$, where L is the distance between the two lenses.

Telescope:

An optical instrument for collecting radiation from a distant object and producing an image which can then be magnified by an eyepiece, as in a compound microscope.

Refractor: Uses a converging lens to collect the radiation.

$$M_\phi = \frac{\text{focal length of objective}}{\text{focal length of eyepiece}}$$

Reflector: Uses a concave mirror to collect the radiation.

Resolving power: The ability of an optical instrument to form distinct images of two objects which are very close together.

Rayleigh's criterion: For a telescope the minimum angle which can be resolved is $\theta_R \approx 1.22\lambda/d$. For a microscope the minimum linear distance which can be resolved is $s \approx 1.22\lambda/(2n \sin \alpha)$, where $n \sin \alpha$ is the numerical aperture (NA) of the objective lens, n is the index of refraction of the medium, and 2α is the angle subtended by the objective lens at the object.

Specialized microscopes (used to increase either the resolving power or the contrast of the microscope):

Oil-immersion microscope: Uses oil between the object and the objective lens to decrease the effective wavelength and increase the resolving power.

Ultraviolet (UV) microscope: Uses shorter-wavelength UV light to increase the resolving power.

Phase-contrast microscope: Converts phase changes produced by microscope specimen to intensity changes to improve contrast.

Polarizing microscope: Converts changes in polarization into intensity changes to improve contrast.

Interferometer:

An optical instrument employing the interference of light beams to measure wavelengths of light and linear distances to very high accuracy.

Michelson interferometer: Uses the interference between two light beams which have traveled over two different paths usually at right angles to each other.

Fabry-Perot interferometer: Uses the multibeam interference of parallel light beams reflected back and forth many times between two parallel reflecting plates.

Questions

1 Prove from geometry that the magnification of a pinhole camera is q/p.

2 Why does a camera produce a sharper image when it is stopped down to a smaller aperture?

3 Compare the human eye with a simple camera with respect to:

(*a*) Focusing mechanism.

(*b*) Control of amount of light used to produce image.

(*c*) Type of image produced.

(*d*) Aberrations leading to distorted images.

4 (*a*) When a swimmer's eyes are open under water, distant objects seem blurred and out of focus. Why?

(b)   Why will the wearing of goggles, which keeps the water away from the swimmer's eye, solve this problem?

5   Why are bifocals worn mostly by older men and women rather than by those in their thirties?

6   Discuss the characteristics of an eye that could see electromagnetic radiation whose wavelength varied by a factor of 20, rather than 2, from one end of the visible spectrum to the other.

7   Does the use of white light in a microscope reduce the resolving power over that which could be achieved with monochromatic light? Why?

8   Prove that the distance a telescope can see into space is directly proportional to the diameter of the telescope's objective lens.

9   Why can you not use a negative lens as a simple magnifier?

10   Compare the optical system of a compound microscope with that of a reflecting telescope. How do they differ, and in what respect are they similar?

11   (a)   Atoms and molecules have dimensions of roughly 10^{-10} m. Would you expect to be able to see them with visible light? Why?

(b)   What wavelength radiation would be needed to see an atom? In what region of the spectrum would this radiation fall?

12   Describe how a Michelson interferometer might be used to measure the index of refraction of a gas.

13   Which of the following optical instruments normally produces images which can be recorded on photographic film: (a) camera, (b) compound microscope, (c) Galilean telescope, (d) projection lantern, (e) reflecting telescope with a Newtonian focus, (f) Michelson interferometer?

Problems

A 1   A converging lens with f-number equal to 2.0 and focal length 5.0 cm has a diameter of:
(a)   2.0 cm   (b)   0.40 cm   (c)   10 cm
(d)   2.5 cm   (e)   None of the above

A 2   A photograph is taken with a camera which is stopped down to f/8. Another photograph is taken with the same camera, with an f/2 lens, wide open. The ratio of the brightness of the image in the first photograph to that in the latter is:
(a)   4/1   (b)   1/4   (c)   1/16   (d)   16/1
(e)   None of the above

A 3   The magnifying power of a magnifying glass of focal length 2.0 cm is about:
(a)   2   (b)   4   (c)   50   (d)   13
(e)   None of the above

A 4   The magnification of a compound microscope containing an eyepiece of focal length 2.0 cm and an objective lens of focal length 1.5 cm is about:
(a)   300   (b)   225   (c)   8   (d)   6
(e)   None of the above

A 5   The smallest angular separation which can be resolved by a stellar telescope of lens diameter 25 cm, using a blue filter at 450 nm, is:
(a) 2.2×10^{-6} rad   (b) 2.2×10^{-6} degree
(c) 7.9×10^{-3} second of arc
(d) 2.2×10^{-6} second of arc   (e)   None of the above

A 6   Superman has the ability not merely to "see" x-rays, but also to produce them with his eyes. If he uses this power to look at the stars through a telescope, the resolving power of the telescope will be improved by a factor of:
(a)   1   (b) ∞ (c)   3   (d) 10^3 (e) 10^{-3}

A 7   A microscope uses violet light at 420 nm, and has an objective lens with f-number f/4. The smallest detail which can be resolved with such a microscope is of the order of:
(a) $\frac{1}{2}$ m   (b) $\frac{1}{2}$ cm   (c)   2 cm   (d)   2 mm
(e)   2 μm

A 8   An inventor proposes a gamma-ray microscope which uses electromagnetic radiation of wavelength 10^{-4} nm. If such a microscope could be made to work, its resolving power would be better than that of an optical microscope by about the following number of orders of magnitude:
(a) 10^6 (b)   6   (c)   2   (d) 10^2 (e)   4

A 9   In a Michelson interferometer using red light at 633 nm, a difference in the two optical paths of the following amount will lead to destructive interference of the two beams:
(a)   633 nm   (b)   2216 nm   (c)   1266 nm
(d)   1899 nm   (e)   None of the above

B 1   A pinhole camera is used to photograph a child who is 1.0 m tall. If the child is 2.5 m from the pinhole, and the pinhole is 10 cm from the photographic film, how tall will the child be in the picture taken by the pinhole camera?

B 2   How great a change in brightness is produced by a change in the f-stop used on a camera lens from f/16 to f/2?

B 3   A camera with a lens of 12-cm focal length is used to photograph a man 2.0 m tall who stands 4.0 m from the camera. How tall is the man's image?

B 4   A refracting telescope has a 1.5-D objective lens and a 20-D eyepiece. What is its magnification?

B 5   What is the magnifying power of a refracting telescope with an objective lens of 15-m focal length, and an eyepiece of focal length 5.0 cm?

B 6   A small airplane with a 20-m wing spread is flying toward an airport with two lights on each end of its wing. How close must the plane be before an air-traffic controller in the tower can distinguish these two lights as separate light sources? Assume that the human eye can resolve 1 minute of arc.

B 7   How close to a TV screen must a person sit to resolve two different color dots on the screen separated by a distance of 1.0 mm?

B 8 What is the numerical aperture of the lens in an oil-immersion microscope if the objective lens used has a diameter of 10 cm and a focal length of 7.0 cm, and the oil between the objective lens and the object has an index of refraction of 1.5?

B 9 A Michelson interferometer is being used with red He–Ne laser light at 633 nm to measure a distance of 1.00 cm in terms of wavelengths of red light. When the movable mirror is moved from one end of the 1.00-cm standard to the other, how many bright fringes pass the cross hairs of an observing telescope?

B10 A Michelson interferometer illuminated with krypton orange-red light at 606 nm is used to measure the distance between two points. If 140 dark fringes pass the telescope cross hairs as the mirror is moved from one of these points to the other, what is the distance between the two points?

C 1 A camera has a 4.0-mm lens opening. A good photograph is taken with it at a shutter speed of 1/60 s. If the camera lens is now stopped down to a 0.40-mm aperture, what would be the proper exposure time using the same object and the same film?

C 2 A camera takes a good picture with an f-stop of f/2 and a 1/50-s shutter speed. If the f-stop is changed to f/8, what must be the shutter speed to take an equally good picture?

C 3 In a simple box camera the distance from the lens to the film is 12.0 cm. If the lens has a focal length of 10.0 cm, for what position of the object will the image be sharpest?

C 4 A 70-year-old woman can a read a book only when it is held at full arm's length, in her case at 50 cm from her eyes. If she uses reading glasses to enable her to read more comfortably with the book at the near point of the eye, what kind of lenses does she need in her glasses and how strong must they be?

C 5 A man is not able to see objects clearly unless they are closer than about 1.40 m from his eyes.

(*a*) Is he nearsighted or farsighted?

(*b*) What type of correcting lenses does he need and how strong should they be?

C 6 (*a*) If in the human eye the distance from lens to retina is 1.7 cm, what is the power of the eye for distant objects?

(*b*) What is the power of the eye for objects placed at the near point of the eye, 25 cm from the eye?

C 7 A child in the first grade is seen to hold a comic book 10 cm from her eyes in order to see the book clearly.

(*a*) Is she nearsighted or farsighted?

(*b*) What power lens will she need in her glasses to enable her to read normally?

C 8 A boy uses a magnifying glass to set a piece of paper on fire and finds that putting the paper 5.0 cm from the center of the lens provides the best focus for the sun's rays. If this lens were then used to read some small newsprint, what would be the magnification if the virtual image is produced at infinity?

C 9 A small, round bug of diameter 3.0 mm is being observed under a magnifying glass which is a simple con-verging lens of focal length 4.0 cm. Calculate the angular magnification of the magnifier when the virtual image is at infinity.

C10 The length of a microscope tube is 12 cm. The focal length of the objective lens is 0.60 cm and the focal length of the eyepiece is 3.0 cm. What is the magnifying power of the microscope?

C11 A microscope has an objective lens of 1.0-cm focal length and an eyepiece of 3.0-cm focal length. An object is in sharp focus when it is 1.1 cm from the objective.

(*a*) What is the best separation of the two lenses?

(*b*) What is the magnification produced by the micro-scope?

C12 You are given a 100-cm-long mailing tube and two lenses of focal lengths 50 cm and 12 cm which will just fit inside the tube.

(*a*) Describe how you could construct a telescope from these components.

(*b*) What would be the magnification of such a tele-scope?

C13 The 40-in refractor at Yerkes Observatory has an objective lens with a focal length of about 19 m (63.5 ft). If this telescope is being used to look at the moon, which is 3.8×10^8 m from the earth, how large will a distance of 1.0 km on the moon's surface appear in the image created by the objective lens?

C14 Two stars are 20 light-years away from earth. They can be just barely resolved by the 40-in refractor at Yerkes Observatory. If the telescope is using a filter which passes only 450-nm light, how far apart are the two stars?

C15 An oil-immersion microscope uses oil with $n = 1.5$. The objective lens can accept light scattered by the object up to 50° on either side of the vertical.

(*a*) What is the numerical aperture of the objective lens?

(*b*) What is the minimum linear distance this micro-scope can resolve for 450-nm light?

C16 An ophthalmologist uses a simple ophthalmo-scope to examine the retina of a patient's eye. If the ophthalmologist uses the lens of the patient's eye, which has a focal length of 1.7 cm, as a simple magnifier, what is the angular magnification under which the doctor is able to examine the patient's eyes, if the virtual image is at infinity?

C17 An oculist is using a simple ophthalmoscope to examine the retina of a patient's eye. If the power of the lens in the patient's eye is 60 D, how large will the spacings between two adjacent photoreceptors on the retina, which are actually 2.0 μm apart, appear when the oculist focuses on the vertical image produced by the lens at infinity?

C18 How far must the movable mirror M_2 in a Michelson interferometer be moved if 750 bright fringes produced by 535-nm light are observed to move past the cross hairs of an observing telescope?

C19 A gauge block (a metal block of accurately known length used by machinists) is measured with a Michelson interferometer, first using red light at 633 nm, and then green light at 535 nm. With the red light 6352 bright fringes are counted as the mirror M_2 is moved from one end of the gauge block to the other.

(*a*) What is the length of the gauge block?

(*b*) How many fringes would be counted when the green light at 535 nm is used in the interferometer?

D 1 A woman can see objects clearly only if they lie between 50 and 100 cm from her eyes. Calculate the power of the lenses (in diopters) needed in bifocal glasses which will enable the woman to see distant objects clearly through the top half of her glasses and read a newspaper at a distance of 25 cm through the bottom half of her glasses.

D 2 A man 40 years of age is prescribed +2.0-D lenses for reading at a normal distance of 25 cm. Fifteen years later he finds that he must hold the book or newspaper 40 cm from his eyes to see the type clearly with the same +2.0-D lenses. What should be the power of the new reading lenses prescribed by the oculist?

D 3 A compound microscope has a 5× eyepiece and a 50× objective lens at the end of a microscope barrel 18 cm long. Find:

(*a*) The total magnification of the microscope.

(*b*) The position at which the object must be placed so that the final image is at 25 cm, the near point of the eye.

(*c*) The focal length of each lens.

D 4 An optical setup for laser research includes a glass lens with $n = 1.55$ and $f = 50$ cm. The apparatus must be cooled by immersing it in a water bath. If the apparatus is to continue to function properly, what must be the focal length of a lens made of the same glass and placed in contact with the $f = 50$-cm lens?

D 5 The barrel of an optical instrument consists of an objective lens of focal length 25 cm and an eyepiece of focal length 1.5 cm separated by a distance of 27 cm. The device is to be used to view objects through the eyepiece with the eye relaxed, i.e., with parallel rays entering the eye.

(*a*) Is this device intended to function as a microscope or a telescope?

(*b*) For what approximate object distance is the device intended?

(*c*) What is its magnifying power?

D 6 The lenses of an astronomical telescope are 102 cm apart when adjusted for viewing a distant object with the relaxed eye. The angular magnification of the telescope is 50. Compute the focal length of each lens.

D 7 A terrestrial telescope consists of three converging lenses in a row, as shown in Fig. 24.31. The focal lengths of the three lenses are $f_1 = 50$ cm, $f_2 = 5.0$ cm, $f_3 = 5.0$ cm.

(*a*) Find the overall magnification of this telescope.

(*b*) Describe the nature of the final image produced by this telescope.

D 8 A small evacuated glass cell of length 3.80 cm is inserted in the path of one light beam in a Michelson interferometer. A gas is slowly allowed to leak into the cell until the gas pressure is equal to atmospheric pressure, and the bright fringes passing the cross hairs of the observing telescope are counted. It is found that 28 fringes pass the cross hairs when sodium light of wavelength 589 nm is used. What is the index of refraction of the gas?

D 9 A problem occurs when a Michelson interferom-

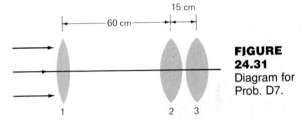

FIGURE 24.31 Diagram for Prob. D7.

eter is used in the ordinary way with yellow sodium light. For one position of the movable mirror the fringes are very sharp and clear, but as the mirror is moved the fringes become more and more blurred and finally disappear altogether. If the mirror is moved even farther, the fringes become sharp again. The explanation for this is that the yellow line is actually a doublet with wavelengths 589.0 and 589.6 nm.

(*a*) Show how this fact explains the periodic variation in the nature of the observed fringes.

(*b*) How far would the mirror M_2 in Fig. 24.27 have to be moved for the fringes to change from sharp to nonexistent and then back to sharp again?

Additional Readings

Cagnet, M., M. Françon, and J. C. Thrierr: *Atlas of Optical Phenomena*, Springer-Verlag, Heidelberg, 1962. Source for the photographs of Figs. 24.19 and 24.20.

Horridge, G. Adrian: "The Compound Eye of Insects," *Scientific American*, vol. 237, no. 1, July 1977, pp. 108–120. An interesting account of how insects see.

Jaffe, Bernard: *Michelson and the Speed of Light*, Doubleday Anchor, Garden City, N.Y., 1960. A brief, popular book on Michelson's experimental work. It is most successful in conveying the spirit which motivated Michelson, and therefore the spirit of modern science.

Livingston, Dorothy Michelson: *The Master of Light: A Biography of Albert Michelson*, University of Chicago Press, Chicago, 1973. A personal, nonscientific account by Michelson's daughter, which contains some unusual insights into Michelson's character.

Michelson, Albert A.: *Light Waves and Their Uses*, University of Chicago Press, Chicago, 1903. Michelson's fascinating 1899 Lowell Lectures at Harvard.

Price, William H.: "The Photographic Lens," *Scientific American*, vol. 235, no. 2, August 1976, pp. 72–83. A discussion of lens aberrations and the use of modern computer techniques to improve photographic lenses.

Shankland, Robert S.: "Michelson and His Interferometer," *Physics Today*, vol. 27, no. 4, April 1974, pp. 36–43. A brief account of how Michelson applied his interferometer to important problems in astronomy, atomic spectra, and metrology.

Wald, George: "Eye and Camera," *Scientific American*, vol. 183, no. 2, August 1950, pp. 32–41. A comparison of the basic physics and chemistry of the eye and the camera by the eminent Harvard biologist and 1967 Nobel Laureate.

It is no doubt that Michelson's experiment was of considerable influence upon my work insofar as it strengthened my conviction concerning the validity of the principle of the special theory of relativity. On the other side I was pretty much convinced of the validity of the principle before I did know this experiment and its result. In any case, Michelson's experiment removed practically any doubt about the validity of the principle in optics, and showed that a profound change of the basic concepts of physics was inevitable.

Albert Einstein (1879–1955)

Chapter 25

The Theory of Relativity

In 1894, A. A. Michelson delivered an address at the dedication ceremonies for the new Ryerson Physical Laboratory at the University of Chicago. In his address Michelson echoed with favor the view of his old friend Lord Kelvin (see Fig. 20.13 and accompanying biography) that all great discoveries in physics had probably already been made, and that the future progress of physics would be confined to measuring known quantities such as the velocity of light to greater and greater accuracy. Never have two great scientists been so completely wrong in their scientific judgment! The next 40 years saw the greatest revolution ever in the history of physics: Roentgen discovered x-rays and Becquerel radioactivity; atomic and nuclear physics developed and flourished; and the theories of relativity and quantum mechanics radically changed our view not only of physics but of philosophy as well. The rest of this book will be devoted to these startling developments, to what is often called *modern physics*, as opposed to the classical physics we have been studying thus far.

In this chapter we start with Albert Einstein's revolutionary theory of relativity. Most of our treatment will be confined to special relativity, i.e., to the theory that applies to one reference system moving with constant velocity with respect to another. In the last section we will say a few words about the more complicated theory of general relativity, which applies to accelerated reference systems.

25.1 Galilean Relativity

Long before Einstein's day physicists working in the field of mechanics were accustomed to dealing with the constant-velocity motion of one reference frame relative to another, something that has come to be called *Galilean relativity*.

Consider a physical event such as the collision of two particles. The event must occur at some point in space and at some instant of time. We therefore specify the event by the three space coordinates x, y, and z, measured with respect to the origin of the coordinate system, and by the time t at which the event occurs.

Frame of reference: A system in which the position of a particle is specified by three numbers (distances) measured with respect to the origin of the coordinate system, and by a single number representing the time elapsed from some chosen initial time.

Inertial frame of reference: A frame of reference in which Newton's first law of motion, i.e., the law of inertia, is valid.

An inertial reference frame is an *unaccelerated* frame. Hence any object in this frame not acted on by a net force will either remain at rest or continue to move with constant velocity. A rocket ship drifting freely in outer space with its engines cut off is an ideal inertial system. Newton assumed that the fixed stars made a good inertial frame. Once we have found an inertial frame, any other frame moving with constant speed in a straight line with respect to the first frame is also an inertial frame (Fig. 25.1).

Reference frames *accelerating* with respect to a chosen inertial frame are not inertial frames, because in such frames Newton's first law is no longer valid: the accelerating frame produces pseudoforces of the kind produced in a falling elevator, as discussed in Sec. 5.8. In practice we often neglect the small effects of the rotation and orbital motion of the earth and take the earth's surface as an inertial frame. Any reference frame moving with constant velocity with respect to the earth's surface is then also an inertial frame, but an accelerating car or a rotating merry-go-round is not, as illustrated in Fig. 25.2.

Consider two inertial frames of reference S_1 and S_2 consisting of rectangular coordinate systems specified by the axes X_1, Y_1, Z_1, and X_2, Y_2, Z_2. The second frame S_2 is moving in the positive X direction with a speed v with respect to the first frame, as in Fig. 25.3. At time $t = 0$ the two origins coincide and the clocks in the two frames are synchronized. We assume that the clocks remain synchronized throughout the motion, because the same measure of time prevails in the two reference frames. (This may seem

FIGURE 25.1 Inertial frames of reference. If we have an inertial reference frame S_1, then any other frame S_2 moving with constant speed v in a straight line with respect to the first reference frame is also an inertial frame.

FIGURE 25.2 A frame of reference which rotates with respect to an inertial frame of reference. The rotating frame is a noninertial frame.

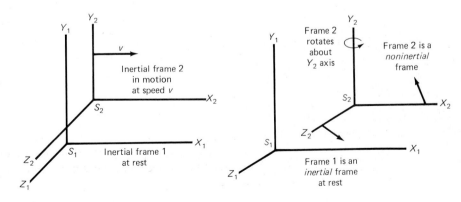

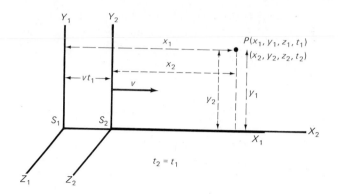

FIGURE 25.3 A Galilean transformation. Inertial frame S_2 moves with constant speed v along the X axis with respect to inertial frame S_1, which is at rest. From the diagram $x_2 = x_1 - vt_1$.

obvious, but in Einstein's theory it will prove to be an invalid assumption!)

Suppose that at time t_1 some event occurs at point P in space, say, a firecracker is exploded. The space and time coordinates of this event in frame S_1 are then (x_1, y_1, z_1, t_1) and in frame S_2 they are (x_2, y_2, z_2, t_2). From Fig. 25.3 it can be seen that the coordinates of this event in the two frames are related by the following equations, called *transformation equations*:

Equations for a Galilean Transformation

$$x_2 = x_1 - vt_1 \quad (25.1) \qquad\qquad z_2 = z_1 \quad (25.3)$$

$$y_2 = y_1 \quad (25.2) \qquad\qquad t_2 = t_1 \quad (25.4)$$

The only coordinate changed is the x coordinate, since the motion is along the X axis. Also, since the distance between the origins of the two systems is equal to vt_1, the x coordinates of the particle in the two frames differ by vt_1, and so $x_1 = x_2 + vt_1$, or $x_2 = x_1 - vt_1$.

Equations (25.1) to (25.4) are the equations for a nonrelativistic transformation between any two inertial frames of reference, usually referred to as a *Galilean transformation*. They express the common-sense assumption that space intervals and time intervals are absolute and completely independent of the motion of the observer. For example, when a firecracker goes off, an observer moving with the reference frame S_2 sees it occur at the same position in space and at the same time as an observer in the stationary frame S_1.

If we want to transform from frame S_2 back to frame S_1, the inverse transformation is obtained by interchanging 1 and 2 and reversing the sign of v, since if S_2 is assumed at rest, then S_1 is moving with a speed v along the *negative X* axis with respect to S_2. Hence the inverse transformation equations are the same as Eqs. (25.1) to (25.4) above except that Eq. (25.1) is replaced by

$$x_1 = x_2 + vt_2 \tag{25.5}$$

These transformations indicate clearly that even in classical Newtonian physics all mechanical motion is relative, i.e., that positions and velocities vary depending on the reference frame with respect to which they are measured. The speed of a person walking down the aisle of a moving train is quite different when measured with respect to the ground than it is when measured with respect to the floor of the train, as shown in Fig. 25.4. Since $KE = \frac{1}{2}mv^2$, the same is true of the person's kinetic energy.

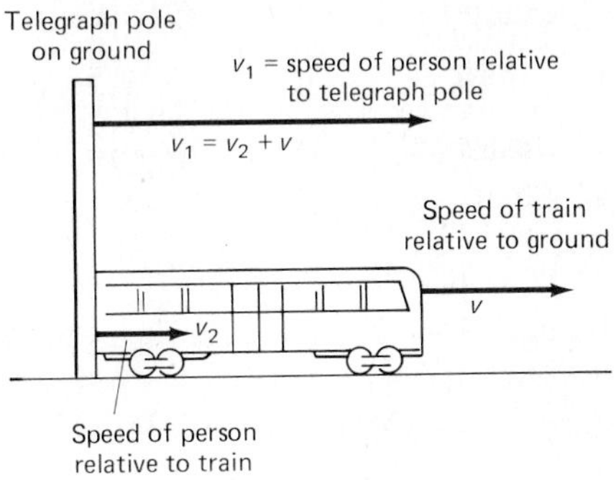

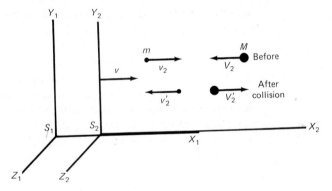

FIGURE 25.4 The speed of a person on a train depends on the frame of reference in which it is measured. The person's speed relative to the train is v_2; the person's speed relative to a telephone pole is $v_1 = v_2 + v$.

Invariance of Classical Mechanics under Galilean Transformations

It can be shown that all the laws of *mechanics* (e.g., Newton's three laws of motion) remain valid under a Galilean transformation. They also remain *invariant*; i.e., the mathematical form of the equations expressing laws like the conservation of momentum and the conservation of energy is the *same* for any two inertial frames of reference. We will prove this only for one simple case, the conservation of linear momentum.

Consider a head-on collision between two particles, as seen in two inertial frames S_1 and S_2, where S_2 is moving with a speed v along the positive X axis with respect to S_1 (see Fig. 25.5). We here use small letters for the mass

FIGURE 25.5 The conservation of linear momentum: The collision of two particles as seen in two inertial frames S_2 and S_1, where S_2 is moving with a constant speed v along the positive X axis with respect to S_1.

and velocity of the particle of mass m, and capital letters for the mass and velocity of the particle of mass M. As seen by an observer in S_2, the law of conservation of momentum takes the form

$$m\mathbf{v}_2 + M\mathbf{V}_2 = m\mathbf{v}_2' + M\mathbf{V}_2' \qquad (25.6)$$

where the velocities are vector quantities, and the unprimed velocities are those before the collision and the primed velocities those after the collision.

How does this collision appear to an observer in S_1? To find out, we note that to the observer in S_1 the velocities are increased by the velocity $\mathbf{v}$ of S_2 with respect to S_1. Hence

$$\mathbf{v}_1 = \mathbf{v}_2 + \mathbf{v} \qquad \text{or} \qquad \mathbf{v}_2 = \mathbf{v}_1 - \mathbf{v} \qquad (25.7)$$

$$\mathbf{V}_1 = \mathbf{V}_2 + \mathbf{v} \qquad \text{or} \qquad \mathbf{V}_2 = \mathbf{V}_1 - \mathbf{v} \qquad (25.8)$$

Similarly, $\qquad \mathbf{v}_2' = \mathbf{v}_1' - \mathbf{v} \qquad \text{and} \qquad \mathbf{V}_2' = \mathbf{V}_1' - \mathbf{v} \qquad (25.9)$

On substituting in Eq. (25.6), we have:

$$m(\mathbf{v}_1 - \mathbf{v}) + M(\mathbf{V}_1 - \mathbf{v}) = m(\mathbf{v}_1' - \mathbf{v}) + M(\mathbf{V}_1' - \mathbf{v}) \tag{25.10}$$

where $\mathbf{v}$, the relative velocity of the reference frames, is unchanged by the collision. Then, on canceling $-m\mathbf{v}$ and $-M\mathbf{v}$ on both sides, we have

$$m\mathbf{v}_1 + M\mathbf{V}_1 = m\mathbf{v}_1' + M\mathbf{V}_1' \tag{25.11}$$

This equation expresses the conservation of linear momentum, which therefore remains valid in reference frame S_1. Note that the total momentum measured in the two frames is *different*, but momentum is still conserved in both frames. The mathematical form of Eq. (25.11) is also identical with that of Eq. (25.6), with the subscripts merely changed from 2 to 1. Hence we have shown that the law of conservation of linear momentum is also *invariant* under a Galilean transformation.

The same kind of proof may be given for any of the other basic laws of mechanics, like Newton's second law of motion or the principle of conservation of energy. We can therefore conclude:

Principle of Galilean Relativity

> All the laws of classical mechanics are *invariant* under a Galilean transformation. Hence all inertial systems are equivalent in classical mechanics, and it is impossible by means of any experiment in mechanics to distinguish one inertial system from any other inertial system.

For this reason it is impossible by means of mechanical experiments to detect uniform rectilinear motion from inside the moving system. Consider the motion of the balls on a billiard table below deck on an ocean liner moving with constant speed in a straight line across the ocean. The balls behave exactly as they would on a billiard table in your campus student union. The motion of the ship in no way affects their motion.

The principle of Galilean relativity implies that the three basic quantities of mechanics—mass, length, and time—are all independent of the motion of the observer. We will now show that this is no longer true when we move from mechanics into the realm of electromagnetism.

Example 25.1

A man is on a train which moves along a straight track with a constant speed of 40 m/s with respect to the ground. At time $t = 0$ the man begins to walk from the rear of the car at the instant the rear of the car passes a telegraph pole on the track outside. He then walks up the aisle toward the front of the car at a speed of 2 m/s. (It may be helpful to look at Fig. 25.4 to visualize what is happening.)

(a) What is his position 10 s later with respect to the rear of the car?
(b) What is his position 10 s later with respect to the telegraph pole?
(c) What is his speed 10 s later with respect to the ground?

SOLUTION

It is convenient here to take reference frame S_2 as moving with the train and reference frame S_1 as being fixed with respect to the ground. Then

(a) $x_2 = v_2 t_2 = (2 \text{ m/s})(10 \text{ s}) = \boxed{20 \text{ m}}$

Hence the man is 20 m from the rear of the car after 10 s.

(b) Here we want to transform from the moving frame S_2 back to the rest frame S_1. The inverse Galilean transformation yields

$$x_1 = x_2 + vt_2$$

$$= 20 \text{ m} + (40 \text{ m/s})(10 \text{ s}) = \boxed{420 \text{ m}}$$

Hence at the end of 10 s the man is 420 m from the telegraph pole on the ground.

(c) The speeds are related in this case by the fact that the speed of the man with respect to the ground is the sum of his speed with respect to the train and the speed of the train with respect to the ground, since all three are in the same direction. Hence

$$v_1 = v_2 + v \qquad [\text{see Eq. (25.7)}]$$

and so

$$v_1 = 2 \text{ m/s} + 40 \text{ m/s} = \boxed{42 \text{ m/s}}$$

The man's speed with respect to the ground is therefore 42 m/s in the direction in which he is walking.

25.2 The Aether and the Speed of Light

In 1900 light was believed to consist of electromagnetic waves in a medium called the *aether*, just as sound waves are longitudinal waves in a medium called air. But we know from our consideration of the Doppler effect in sound (Sec. 12.8) that if a car moves at a speed v through the air toward a sound source which emits sound traveling with a speed u relative to the air, then the speed of the sound relative to the car is $u + v$, as in Fig. 25.6. If, now, light consists of waves in an aether, then the velocity of light waves, which presumably is fixed with respect to the aether, should change with the motion of an observer through the aether, just as in the case of sound. In classical physics there seemed to be little doubt about this. Light required some medium for its propagation ("you cannot have waves unless there is something waving"), and if the speed of light was fixed with respect to that medium by the properties of the medium, then the speed of light would be different for observers moving at different speeds with respect to that medium.

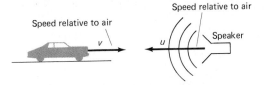

FIGURE 25.6 Speed of sound in air, as measured in a moving frame. The speed of the sound relative to the car is $v + u$.

According to this reasoning, there is one preferred frame of reference in which the speed of light is exactly 2.998×10^8 m/s in a vacuum; this is the frame of reference at rest with respect to the aether. Observers in all other frames of reference, including all inertial frames, would measure a different value for the speed of light depending on how they move with respect to the aether. Therefore the speed of light would certainly not be invariant under a Galilean transformation.

Now Maxwell's equations, as we have seen, led to the prediction that the speed of electromagnetic waves in a vacuum was related to two fundamental quantities of electricity and magnetism by the expression $c = 1/\sqrt{\epsilon_0 \mu_0}$. If the measured speed of light differs for different inertial frames, then Maxwell's equations, which are so intimately linked to the speed of light c, would also change form for two inertial frames in relative motion. Hence we appear to

Albert Einstein (1879–1955)

FIGURE 25.7 Albert Einstein. This photograph of Einstein, an avid sailor, was taken in 1936. (*Courtesy of AIP Niels Bohr Library.*)

Albert Einstein, perhaps the best-known and most-revered scientist of the twentieth century, began slowly on his spectacular scientific career. While in high school in Munich, Germany, to which his parents had moved from Ulm, his birthplace, he did so badly in Latin and Greek that he was told by a teacher: "Einstein, you will never amount to anything."

Because of his adeptness at mathematics he was admitted in 1895 to the Polytechnic Institute in Zürich, Switzerland. There he cut most of his classes to devote himself to studying theoretical physics, which had become much more fascinating to him than pure mathematics. He succeeded in graduating only with the aid of excellent lecture notes taken by a friend. Once out of school he had trouble finding a teaching position, for he did not have a strong academic record and had the added disadvantages of being Jewish and of not being a Swiss citizen.

After a few brief jobs as a tutor he was appointed an examiner in the Swiss Patent Office in Berne in 1902. His duties there were not onerous, and he had time for his beloved physics, to which he devoted most of his free time. In 1905 he published five research papers in the best German physics periodical, the *Annalen der Physik*. Three of these contributions were worthy of the Nobel Prize. One was the law of the photoelectric effect (see Sec. 26.3), which was mentioned explicitly in Einstein's 1921 Nobel Prize citation. The second was the theory of Brownian motion, the motion of tiny pollen grains when submerged in a liquid, and the third was his precedent-shattering special theory of relativity.

These contributions caused such a stir that in 1909 Einstein received his first academic appointment, a professorship at the University of Zürich, from which he had received his own Ph.D. degree in 1905. In 1911 he moved to the German university in Prague, Czechoslovakia, and in 1914 was made a professor at the University of Berlin. While in Berlin he published in 1915 his first paper on the general theory of relativity (see Sec. 25.9).

Einstein remained in Berlin until 1933. When conditions under Hitler became too difficult, he accepted a lifetime appointment as the first senior fellow at the newly formed Institute for Advanced Study in Princeton, New Jersey, where he remained until his death in 1955. The 40 years from 1915 to 1955 were devoted in great part to solitary work in developing a unified field theory which would tie together the gravitational and electromagnetic fields—an endeavor which never quite succeeded, much to Einstein's disappointment.

Einstein's personal life was not a particularly happy one despite his great fame and reasonable fortune. He was married in 1901 to a mathematician, Mileva Marec. After having two children they were divorced, and his wife took the two boys back to Switzerland. (According to one biographer, this was the only time anyone ever saw Einstein cry.) In 1917 Einstein married his cousin Elsa, who took good care of him until her death in 1936.

Einstein once wrote: "The only thing that gives me pleasure apart from my work, my violin and my sailboat, is the appreciation of my fellow workers." He did not need much contact with other people and found the accolades and invitations he constantly received wearisome and distracting.

Einstein became a U.S. citizen in 1940, and was the physicist chosen by his colleagues to alert President Roosevelt to the need for developing the nuclear bomb during World War II. Given his deep humanitarian instincts and his longing for peace and a community of nations, he lived to regret the development of the bomb and to worry about what it would mean for the future of humanity.

have a situation in which the laws of mechanics are invariant under a Galilean transformation, but the laws of electromagnetism are not.

Albert Einstein (see Fig. 25.7 and accompanying biography) found it hard to accept this conclusion because it seemed to destroy the symmetry he expected to find in the laws of physics. Before we consider his work, however, we must discuss further this mysterious aether and the attempts of Michelson and others to detect its presence.

The Luminiferous Aether

The word *aether* (or *ether*, as it is sometimes spelled) was first used by Aristotle to describe the substance of the heavens and celestial bodies. Nineteenth-century physicists postulated it as the medium for carrying light waves, and also gravitational, electric, and magnetic forces through space. In case more than one aether actually existed, the name *luminiferous aether*, or light-carrying aether, was used to specify that particular aether which transmitted *light*.

This aether had odd, and seemingly contradictory, properties. It had to be a very low density gas, since the heavenly bodies moved through it without being slowed down. Once Maxwell and Hertz had shown that electromagnetic waves were transverse waves, however, it was necessary to postulate that the aether acted like a very elastic solid, since only such a solid could transmit transverse waves at so high a speed (elastic objects such as ball bearings are very hard). The aether therefore had to have simultaneously the properties of both a very tenuous gas and a very hard solid. Physicists were not happy to accept an aether with such strange properties, but they felt they had no other choice.

The aether provided an ideal frame of reference, since it was thought to be at rest, so that all motions could be determined with respect to it. Hence in nineteenth-century physics the aether came to be really another name for "absolute space," a concept as dear to classical physics as was the idea of absolute time.

It occurred to many physicists that the existence and properties of the aether might be clarified by measuring the speed of light in several inertial systems. If the aether theory were correct, then different values of the speed of light would be obtained for inertial systems moving through the aether with different speeds and in different directions. It was thought that the speed of the earth in its orbit around the sun, which is about $10^{-4}\,c$, could be used for such a test. Any measurable effects turn out, however, to depend on v^2/c^2, where v is the speed of one reference frame with respect to the other. For the earth's motion around the sun we then have $v^2/c^2 \simeq 10^{-8}$. Such a small effect is hard to measure, but A. A. Michelson thought he could do it by using one of his interferometers to convert velocity differences into measurable shifts in interference fringes.

25.3 The Michelson-Morley Experiment

In 1887 Michelson and his colleague E. W. Morley (1838–1923) set out to explore the mysterious aether. If an aether existed, then the measured speed of light should change with the speed of the observer through the aether. The experiment Michelson devised is analogous to timing two trips by a rower in a flowing stream, one trip across the stream and back, the other trip the same distance down the stream and back. It turns out that the times for these two trips are not the same, and that the time difference increases with the speed of flow of the stream (see Example 25.2).

Example 25.2

Consider the situation shown in Fig. 25.8. Two rowers start out at the same time from a dock at the edge of a river. One rower rows down the river from A to C and back again to A. The second rower rows across the river from A to B and back again to A. Both rowers row at identical speeds of 2.0 m/s with respect to the water. There is a constant current in the river directed downstream and equal to 1.0 m/s with respect to the riverbank. The distances AB and AC are each equal to 100 m. Which rower will arrive back at the dock A first, and how much later will the second rower arrive?

SOLUTION

For the first rower who rows along the direction of the riverbank, the speed downstream will be $u + v$, where u is the rower's speed with respect to the water, and v is the speed of the water with respect to the riverbank. Similarly, the rower's speed in rowing back from C to A will be $u - v$. Hence the total time for the round trip will be:

$$t_1 = \frac{D}{u+v} + \frac{D}{u-v} = \frac{100 \text{ m}}{(2.0+1.0) \text{ m/s}} + \frac{100 \text{ m}}{(2.0-1.0) \text{ m/s}}$$

$$= 33 \text{ s} + 100 \text{ s} = 133 \text{ s}$$

The second rower must aim upstream on going from A to B, in order that the resultant velocity be in the direction from A to B. Hence the second rower's speed, as shown in Fig. 25.9, will actually be $(u^2 - v^2)^{1/2}$ from A to B. The same is true for the return trip from B to A. Hence the total time to go from A to B and back to A again will be

$$t_2 = \frac{2D}{(u^2-v^2)^{1/2}} = \frac{2(100 \text{ m})}{[(2.0 \text{ m/s})^2 - (1.0 \text{ m/s})^2]^{1/2}}$$

$$= \frac{200 \text{ m}}{1.73 \text{ m/s}} = 115 \text{ s}$$

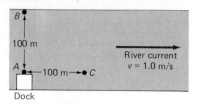

FIGURE 25.8 Diagram for Example 25.2.

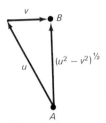

FIGURE 25.9 Composition of velocities in Example 25.2.

Hence the second rower, the one who rows across the river, arrives back at point A first, and does so $133 - 115$

$$= \boxed{18 \text{ s}} \text{ before the first rower.}$$

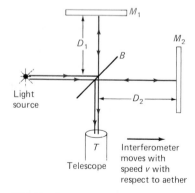

FIGURE 25.10 A Michelson interferometer used to compare the speed of light in different directions in the Michelson-Morley experiment. (For simplicity, the compensating plate shown in Fig. 24.2 is omitted.)

Michelson used his interferometer (Fig. 25.10) with the two mirrors M_1 and M_2 fixed in position, and the path lengths D_1 and D_2 the same. Then, if the interferometer were at rest in the aether, the central image of the interference pattern, as observed in the telescope T, would be bright. If, now, the interferometer moves with a speed v through the aether in the direction of the light beam, then the times for light to travel the two perpendicular paths will no longer be equal, as we will now show.

In the following derivation note carefully that:

c = the speed of light *with respect to the aether*

v = the speed of the interferometer *with respect to the aether*

For the light traveling over path D_2, the speed of light with respect to the interferometer is $c - v$ for the trip to the right, and $c + v$ for the trip to the left. Hence the time for the total trip is

$$t_2 = \frac{D_2}{c-v} + \frac{D_2}{c+v} = \frac{2cD_2}{c^2-v^2} \qquad (25.12)$$

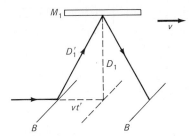

FIGURE 25.11 Light path for the perpendicular beam in the Michelson-Morley experiment. The distance traveled by this beam is actually $D_1' = \sqrt{D_1^2 + v^2 t'^2}$.

The light traveling over path D_1 does not move exactly perpendicular to D_2, since it must strike the mirror M_1 and return to the beam-splitting mirror which has in the meantime moved to the right, as shown in Fig. 25.11. Hence the light travels over the hypotenuse D_1' of a triangle, where

$$(D_1')^2 = D_1^2 + v^2 t'^2$$

and t' is the time required for the light to go from B to M_1, and vt' the distance the interferometer travels in that time, as shown in Fig. 25.11. Hence

$$(D_1')^2 = c^2 t'^2 = D_1^2 + v^2 t'^2$$

or the time for the trip from B to M_1 is, on solving for t',

$$t' = \frac{D_1}{(c^2 - v^2)^{1/2}}$$

For the round trip, the time is then

$$t_1 = 2t' = \frac{2D_1}{(c^2 - v^2)^{1/2}} \tag{25.13}$$

Since D_1 and D_2 were exactly equal in the Michelson-Morley experiment, we can use the same symbol D for both. The time difference for light to travel the two paths is then

$$\Delta t = t_2 - t_1 = \frac{2cD}{c^2 - v^2} - \frac{2D}{(c^2 - v^2)^{1/2}} \tag{25.14}$$

It can be shown* that this is approximately equal to

$$\Delta t = \frac{Dv^2}{c^3} \tag{25.15}$$

This is a very short time, even for the motion of the earth around the sun, where $v/c \simeq 10^{-4}$. Here with $D = 11$ m, a distance Michelson obtained by using multiple reflections of the beams back and forth in the interferometer, and $c = 3 \times 10^8$ m/s, we have

$$\Delta t = \frac{D}{c}\left(\frac{v}{c}\right)^2 = \frac{11\ \text{m}}{3 \times 10^8\ \text{m/s}}\ (10^{-4})^2 \simeq 4 \times 10^{-16}\ \text{s}$$

---

*The derivation involves the use of the binomial theorem (see Appendix 1D):

$$(a + b)^n = a^n + na^{n-1}b + \cdots \text{(higher-order terms)}$$

Equation (25.14) may be rewritten as

$$\Delta t = 2D[c(c^2 - v^2)^{-1} - (c^2 - v^2)^{-1/2}]$$

If we expand $(c^2 - v^2)^{-1}$ and $(c^2 - v^2)^{-1/2}$ by the binomial theorem, we obtain

$$\Delta t = 2D\left[c\left(\frac{1}{c^2} + \frac{v^2}{c^4} + \cdots\right) - \left(\frac{1}{c} + \frac{v^2}{2c^3} + \cdots\right)\right]$$

where we neglect all higher-order terms as being too small to be significant. Then we have

$$\Delta t = 2D\left(\frac{v^2}{c^3} - \frac{v^2}{2c^3}\right) = \frac{Dv^2}{c^3}$$

which is Eq. (25.15) above.

Despite the infinitesimal size of this time lag, Michelson felt confident that he could measure the fringe shift this time lag would produce in his interferometer.

Since light travels with a speed c, the path difference corresponding to the time delay Δt is

$$\Delta D = c\,\Delta t = c\left(\frac{Dv^2}{c^3}\right) = \frac{Dv^2}{c^2} \tag{25.16}$$

and the number of complete wavelengths contained in this distance is

$$n = \frac{\Delta D}{\lambda} = \frac{Dv^2}{\lambda c^2} \tag{25.17}$$

For each change in path length by λ, the fringes observed in the telescope change from light to dark and back to light again. Hence the number of fringes moving past the cross hairs in the telescope should be

$$n' = \frac{Dv^2}{\lambda c^2} \tag{25.18}$$

Michelson then rotated the interferometer through 90°. This exchanged the two light paths and doubled the expected shift. Using sodium light at 590 nm, the expected fringe shift was then:

$$2n' = 2\frac{Dv^2}{\lambda c^2} = \frac{2(11\text{ m})}{590 \times 10^{-9}\text{ m}}(10^{-8}) \simeq 0.37 \text{ fringe}$$

Hence the expected fringe shift was only a little over one-third of a fringe, but Michelson was confident that he could observe a shift as small as one-hundredth of a fringe. In the actual experiment the shift was found to be *zero* within the limits of error of the experiment. Hence Michelson and Morley were forced to conclude, very reluctantly, that

$$\Delta t \simeq 0 \simeq \frac{Dv^2}{c^3}$$

which means that $v = 0$! In other words *the interferometer did not move at all with respect to the aether*.

Hence either the aether was being carried along with the interferometer or *there was no aether*. The former conclusion was in conflict with two pieces of strong experimental evidence, that obtained by Fizeau and confirmed by Michelson for the speed of light in a moving liquid, and that of Bradley on the aberration of light from the stars. Both experiments showed clearly that the aether is not dragged along by the earth. Many have repeated the Michelson-Morley experiment in the years since 1887. In every case the results have eventually turned out to be negative.

The conclusion from the Michelson-Morley experiment and its modern equivalents is therefore clear and unambiguous: *There is no detectable aether.* (For the analogy discussed in Example 25.2, this is equivalent to saying, "There is no river.") In other words, there is no preferred frame of reference with respect to which light has a special speed c. Rather the measured speed of light in any two inertial systems is always the same, c, no matter how they are moving with respect to each other, as shown in Fig. 25.12.

FIGURE 25.12 The speed of light as measured in two different inertial frames: An observer moving with the earth around the sun and an astronaut on a spaceship traveling toward the sun will measure exactly the same speed c for the light coming from the sun.

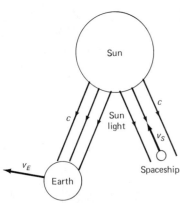

25.4 Einstein's Theory of Special Relativity

The *special theory of relativity* was published by Albert Einstein in 1905. It represented a real break with classical modes of thought in physics and signaled the end of classical nineteenth-century physics. It applies to two inertial systems, one of which is moving with constant velocity with respect to the other.

Einstein's special theory is based on two deceptively simple postulates which have most extraordinary consequences both for physics and for our perception of the universe. They are:

Postulates of Special Relativity

1 *Postulate of the constancy of the speed of light*: The speed of light in a vacuum is a constant c, independent of the inertial system in which it is measured or of the motion of the light source or observer.

2 *Principle of special relativity*: All the laws of physics, including those of mechanics and electromagnetism, have the same form in all inertial frames of reference (i.e., they are *invariant*). Hence all inertial reference frames are equivalent.

**Postulate
of the Constancy
of the Speed of Light**

The postulate of the constancy of the speed of light is completely consistent with the result of the Michelson-Morley experiment: Light travels with the same speed c both in the direction in which the interferometer is moving and at right angles to that direction. One possible conclusion from the Michelson-Morley experiment was that the aether did not exist. Einstein did not go quite that far but considered the aether a meaningless concept. Only motion relative to material bodies had physical significance; motion with respect to an aether which had never been observed made no physical sense.

The first postulate therefore seems entirely consistent with experimental facts. Whether or not Einstein arrived at this postulate on the basis of the Michelson-Morley experiment is unclear.*

**Principle of Special
Relativity**

The second postulate, the principle of special relativity, is not so firmly grounded in experiment as is Einstein's first postulate. Rather it is an extension of the principle of Galilean relativity from mechanics to electromagnetism and optics.

Einstein was led to assume the validity of this principle of special relativity on two counts:

1 If this principle were not valid, then different frames of reference would lead to different mathematical expressions for the laws of physics, some more complicated than others. The frame of reference in which the laws had their simplest expression could then be called *at rest*, and all the others *in motion*. This would then destroy the equivalence of all inertial frames of reference. This would also lead to a dependence of the form of physical laws on the

---

*Much ink has been spilled by writers trying to trace the exact historical relationship between Einstein's theory and the Michelson-Morley experiment. Even Einstein himself was not particularly consistent in discussing this relationship. Perhaps the most convincing statement is the one by Einstein quoted at the beginning of this chapter, a statement made by Einstein in a letter to Bernard Jaffe and quoted in Jaffe's book listed at the end of Chap. 24.

orientation and motion of physical systems in space. Hence empty space would have different properties in different directions—something which has never been observed.

2 The second argument in favor of Einstein's principle of special relativity is best expressed in Einstein's own words:

> There is a tremendous amount of truth in classical mechanics, since it supplied us with the actual motions of the heavenly bodies with a delicacy of detail little short of wonderful. The principle of relativity must therefore apply with great accuracy in the domain of *mechanics*. But that such a principle of such broad generality should hold with such exactness in one domain of phenomena [i.e. mechanics] and yet be invalid for another [i.e., electromagnetism] is *a priori* not very probable.

Hence Einstein postulated a principle of relativity that would be valid both in mechanics and in electromagnetism and optics, where Galilean relativity was certainly invalid. To do this he had to introduce more general transformation equations that would reduce to those for a Galilean transformation if the speed of one frame of reference with respect to the other was small compared with the speed of light.

The problem, of course, with Einstein's two postulates is that, at least at first glance, they are irreconcilable. If two inertial frames are moving toward each other with a speed v, and if the measured speed of light in one reference frame is c, then the speed of light in the second frame would seem to be $c + v$. But this contradicts the first postulate, which says that the speed must be the same, c, in both frames.

Einstein was convinced that both his postulates were correct and set out to reconcile them. He did this by introducing into physics some revolutionary new ideas, in particular, new concepts of *time* and *simultaneity*.

25.5 The Lorentz Transformation Equations

The more general transformation equations used by Einstein are called the *Lorentz transformation equations*.

It is possible to derive the Lorentz equations using simple algebra. We look for a set of equations similar to those for a Galilean transformation but which satisfy the two postulates of special relativity. We will not attempt this derivation here but will merely state the result and show that the Lorentz equations do, indeed, enable us to satisfy the two postulates of special

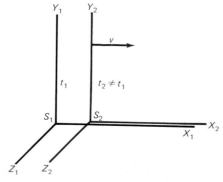

FIGURE 25.13 Two inertial frames of reference, from the viewpoint of Einstein's theory of special relativity. The major change is that the time t_2 measured in the moving frame S_2 is *different* from the time t_1 measured in the rest frame S_1.

relativity simultaneously, and do reduce to the Galilean transformation equations when v/c is small.

We first write down the equations for the Galilean and Lorentz transformations for two reference frames S_1 and S_2, where S_2 moves along the X axis with a constant speed v with respect to S_1, as in Fig. 25.13. We have:

Galilean transformation:		*Lorentz transformation:*	
$x_2 = x_1 - vt_1$	(25.1)	$x_2 = \gamma(x_1 - vt_1)$	(25.19)
$y_2 = y_1$	(25.2)	$y_2 = y_1$	(25.20)
$z_2 = z_1$	(25.3)	$z_2 = z_1$	(25.21)
$t_2 = t_1$	(25.4)	$t_2 = \gamma\left(t_1 - \dfrac{vx_1}{c^2}\right)$	(25.22)
		$\gamma = \left(1 - \dfrac{v^2}{c^2}\right)^{-1/2}$	(25.23)

Since S_2 is moving along the X axis, we would expect no change in the y and z coordinates in the relativistic case, just as there was none in the Galilean case. The main changes are the factor γ in the expression for x_2, where γ depends on v/c, the ratio of the relative speed of the coordinate systems to the speed of light. For $v/c \ll 1$, the Lorentz transformation for x_2 reduces to the Galilean transformation. The biggest change in the Lorentz transformation (compared with the Galilean transformation) is that *the times are not identical in the two frames S_1 and S_2: t_2 depends not only on t_1 but also on the position x_1* at which an event occurs, and on the ratio v/c. This time transformation is essential if the two postulates of relativity are to be satisfied simultaneously, as we will now show.

Lorentz Transformation for a Light Pulse

We want to show that Einstein's two postulates are consistent if the Lorentz transformation equations are used to transform from one reference frame to another. Consider Fig. 25.13 and imagine that at the instant $t_1 = t_2 = 0$ we produce a bright spark at the common origin of the two frames of reference. As viewed by an observer in frame 1, the resulting light pulse will travel out in all directions with the speed c. At any instant t_1, the light pulse from the spark will form a spherical wave front of radius $r_1 = ct_1$. This wave front must satisfy the equation

$$r_1^2 = x_1^2 + y_1^2 + z_1^2 = c^2t_1^2$$

or $\qquad x_1^2 + y_1^2 + z_1^2 - c^2t_1^2 = 0$ $\qquad\qquad$ (25.24)

Suppose now that an observer in S_2, who is moving with constant speed v along the positive X axis with respect to S_1, looks at this same wave front. For this observer, according to Einstein's first postulate, the measured speed of light must still be c, exactly the same as it was for an observer in S_1. Also, from the second postulate, all equations of mechanics and electromagnetism must have the same form in the two inertial reference frames. Hence, if the equation for the light ray in S_1 is Eq. (25.24) above, then the corresponding equation in frame S_2 must be:

$$x_2^2 + y_2^2 + z_2^2 - c^2t_2^2 = 0 \qquad\qquad (25.25)$$

where c is the *same* constant in both, that is, the speed of light in a vacuum.

A Galilean transformation of S_1 into S_2 cannot possibly make Eqs. (25.24) and (25.25) simultaneously valid. But the Lorentz transformation given above succeeds in doing exactly this, as we now show.

We want to prove that

$$x_2^2 + y_2^2 + z_2^2 - c^2t_2^2 = x_1^2 + y_1^2 + z_1^2 - c^2t_1^2$$

Since $y_2 = y_1$ and $z_2 = z_1$ for motion along the X axis, this reduces to

$$x_2^2 - c^2t_2^2 = x_1^2 - c^2t_1^2$$

By introducing the Lorentz transformation from frame S_2 to frame S_1, expanding out the squared terms, and canceling equal quantities, this becomes (the detailed algebra is left to Prob. D9):

$$x_1^2 - c^2t_1^2 = x_1^2 - c^2t_1^2$$

which is an identity.

Hence for this case the Lorentz transformation equations do indeed allow both postulates of special relativity to be satisfied simultaneously. This turns out to be true in general. Note the startling results here: Classical mechanics errs in implying that an observer moving through a medium toward a light source at a speed v will measure the speed of light to be $v + c$. The observer will always find the speed of light to be c, no matter how the observer or light source is moving, as illustrated in Fig. 25.12.

Equations (25.19) to (25.23) are then the desired equations which must replace the Galilean transformation equations according to Einstein's special theory of relativity. In the limit of v small compared with c, the Lorentz equations reduce to the Galilean equations, as we would expect.

A plot of the factor $\gamma = (1 - v^2/c^2)^{-1/2}$ against the ratio v/c is shown in Fig. 25.14. Note that, for values of v less than $c/2$, γ has a value very close to 1. This is because $\gamma = (1 - v^2/c^2)^{-1/2} = 1 + \frac{1}{2}v^2/c^2 + \cdots$ (by the binomial theorem). For $v = c/2$, γ is then equal to 1.125. As v increases above the value $c/2$, however, γ grows very rapidly, as can be seen in the figure.

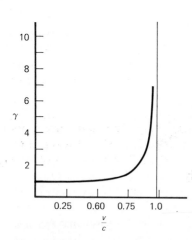

FIGURE 25.14 A plot of $\gamma = \dfrac{1}{\sqrt{1 - v^2/c^2}}$ against v/c. As $v/c \to 1$, $\gamma \to \infty$.

FIGURE 25.15 Hendrik Lorentz (1853–1928), professor of theoretical physics at the University of Leiden. He is best known for his theory of the electron and for the Lorentz transformation equations. He shared the 1902 Nobel Prize in physics with his student P. Zeeman for their work on the effect of magnetic fields on the radiation emitted by atoms. (*Photo courtesy of AIP Niels Bohr Library.*)

The French mathematician Henri Poincaré (1854–1912) first gave the name *Lorentz equations* to Eqs. (25.19) to (25.23) because the Dutch physicist Hendrik Lorentz (see Fig. 25.15) had proposed them in his work on the classical theory of electrons, which preceded Einstein's work. Since Einstein first put the derivation of the Lorentz equations on a sound basis and first saw their true significance, he deserves full credit for the special theory of relativity.

25.6 Consequences of the Lorentz Transformation Equations

The Lorentz equations lead to some very important consequences, the most surprising of which is that space and time intervals are no longer the absolutes they were in classical physics but vary with the motion of the observer measuring them.

Length Contraction

Let us consider a rod of length L_0 as measured in an inertial reference frame S_2 in which the rod is at rest, as in Fig. 25.16. Suppose, now, that reference frame S_2 moves with constant speed v in a straight line along the positive X axis with respect to another inertial frame S_1. What is the length of the rod as measured in S_1? This can be found by a simple application of the first Lorentz equation, Eq. (25.19). In the frame in which the rod is at rest, S_2, the ends of the rod are designated by x_2 and x_2', so that $L_0 = x_2' - x_2$. These coordinates are then related to the coordinates in the S_1 frame by the equations

$$x_2' = \gamma(x_1' - vt_1') \qquad \text{and} \qquad x_2 = \gamma(x_1 - vt_1)$$

To obtain the length of the rod in S_1, the observer in S_1 must measure the two ends at the same time, so that $t_1 = t_1'$, and so

FIGURE 25.16 Lorentz contraction of a moving rod in the direction of its motion: (*a*) Diagram used to derive the contraction equation $L = \sqrt{1 - v^2/c^2}\,L_0$. (*b*) Contraction of a block moving with a speed $0.87c$. Note that the contraction only occurs in the direction of the motion.

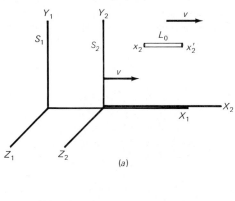

(a)

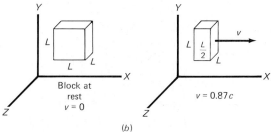

(b)

$$x_2' - x_2 = \gamma(x_1' - x_1) = L_0$$

Hence $\qquad x_1' - x_1 = \dfrac{x_2' - x_2}{\gamma} = \left(1 - \dfrac{v^2}{c^2}\right)^{1/2} L_0$

Now, $x_1' - x_1$ is the length of the rod L, as measured in the frame with respect to which the rod is moving with a speed v. Hence we have

$$L = \left(1 - \frac{v^2}{c^2}\right)^{1/2} L_0 \qquad\qquad (25.26)$$

Here L_0 is called the *proper length* of the rod, or its length in a rest frame. L is the length *as measured in a frame with respect to which the rod is in motion at a speed* v. Hence we find that all length measurements are relative and depend on the motion of the object being measured with respect to the observer.

The rod is *contracted* in the direction of motion (L is smaller than L_0), as seen by an observer with respect to whom the rod is moving, as shown in Fig. 25.16b. This is called the *Lorentz-Fitzgerald contraction* after the two physicists who had proposed before Einstein that such a contraction could explain the negative result of the Michelson-Morley experiment.

Example 25.3

An automobile 3.0 m long is moving at a speed of 25 m/s (55 mi/h). How much is the car contracted in length, as measured by an observer at rest on the road over which the car moves?

SOLUTION

The observer sees the car contracted to a length

$$L = \left(1 - \frac{v^2}{c^2}\right)^{1/2} L_0$$

$$= \left[1 - \frac{(25 \text{ m/s})^2}{(3.0 \times 10^8 \text{ m/s})^2}\right]^{1/2} (3.0 \text{ m})$$

$$= [1 - (6.9 \times 10^{-15})]^{1/2}(3.0 \text{ m})$$

For x very much less than 1, by the binomial expansion $(1 - x)^{1/2} \simeq 1 - x/2$ (see Appendix 1D). Hence

$$L = [1 - (3.5 \times 10^{-15}] (3.0 \text{ m})$$

The contraction is therefore

$$(3.5 \times 10^{-15})(3.0 \text{ m}) \simeq \boxed{10^{-14} \text{ m}}$$

This shows how small, and therefore unobservable, the Lorentz contraction is for normal speeds in everyday life.

Example 25.4

A "flying saucer" moves past a meterstick with a speed relative to the meterstick of one-half the speed of light. What length do the creatures in the flying saucer measure for the meterstick?

SOLUTION

The flying saucer creatures see the meterstick contracted to a length given by:

$$L = \left(1 - \frac{v^2}{c^2}\right)^{1/2} L_0$$

where $v/c = \frac{1}{2}$ and $v^2/c^2 = \frac{1}{4}$. Hence

$$L = \left(1 - \frac{1}{4}\right)^{1/2} (1 \text{ m}) = \left(\frac{3}{4}\right)^{1/2} (1 \text{ m}) = 0.866 \text{ m} = \boxed{86.6 \text{ cm}}$$

In this case, unlike the previous example, the effect would be observable because the relative velocity of the two inertial systems is so large.

Time Dilation

Time in moving frame is Δ

We now consider two events, the explosion of two firecrackers, at exactly the same place x_1 in a rest frame S_1, but at different times t_1 and t_1', as in Fig. 25.17. Then the time interval between t_1' and t_1 in the rest frame is

$$\Delta t_0 = t_1' - t_1$$

This time interval between two events that occur at the same place in the rest frame is called the *proper time interval*. Then, as seen in the moving frame S_2, the times for the two events are, from Eq. (25.22),

$$t_2' = \gamma \left(t_1' - \frac{vx_1'}{c^2} \right) \qquad \text{and} \qquad t_2 = \gamma \left(t_1 - \frac{vx_1}{c^2} \right)$$

But the firecrackers go off at exactly the same place in S_1, so that $x_1' = x_1$, and so

$$t_2' - t_2 = \gamma(t_1' - t_1)$$

If we designate the time interval as measured in the moving frame by $\Delta t = t_2' - t_2$, we have

$$\Delta t = t_2' - t_2 = \gamma(t_1' - t_1) = \gamma \, \Delta t_0$$

or
$$\boxed{\Delta t = \frac{\Delta t_0}{(1 - v^2/c^2)^{1/2}}} \tag{25.27}$$

Hence the time interval measured in the moving frame is expanded, or *dilated*, and therefore we speak of a *time dilation*, meaning that time seems to be slowed down. Thus, if $v = 0.98c$ and we consider a clock at rest in S_1, then, as observed in S_2, the time interval corresponding to Δt_0 in S_1 is

$$\Delta t = \frac{\Delta t_0}{(1 - 0.96)^{1/2}} = \frac{\Delta t_0}{0.2} = 5 \, \Delta t_0$$

Hence the clock at rest makes 5 ticks for every 1 tick made by an identical clock in the moving frame, since 1 s in the moving frame is 5 times as long as in the rest frame.

FIGURE 25.17 Time dilation: The time interval Δt_0 is expanded, or dilated, in the moving frame to $\Delta t = \dfrac{\Delta t_0}{\sqrt{1 - v^2/c^2}}$

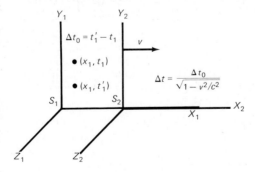

Example 25.5

Space travel from the earth to outside the sun's planetary system is difficult because the time to reach distant stars would exceed the lifetimes of the astronauts. One solution would be to have the astronauts travel so fast that the time for their trip would be greatly reduced in their reference frame because of time dilation. If a star is 100 light-years away, i.e., if it would take 100 years, as measured on earth, to reach the star (traveling at the speed of light), how long would the trip take from the astronauts' point of view if they traveled at a speed of $0.998c$?

SOLUTION

Let the time for the trip, as measured on earth, be Δt, and, as measured by the astronauts in their rest frame, be Δt_0, where, from Eq. (25.27),

$$\Delta t = \frac{\Delta t_0}{(1 - v^2/c^2)^{1/2}}$$

Then $\Delta t_0 = \left(1 - \dfrac{v^2}{c^2}\right)^{1/2} \Delta t$

$$= [1 - (0.998)^2]^{1/2} \Delta t = [1 - 0.996]^{1/2} \Delta t$$

$$= (0.004)^{1/2} \Delta t = 0.063(100 \text{ years}) = \boxed{6.3 \text{ years}}$$

Hence for the astronauts only 6.3 years would have elapsed, and they could hope to survive the trip, if the practical problem of achieving such speeds as $0.998c$ could be solved.

Another way to approach the same problem is to note that, from the astronauts' viewpoint, the earth-star distance has been contracted to 6.3 light-years by the Lorentz-Fitzgerald contraction.

Relativity of Simultaneity

A result of the time dilation predicted by the Lorentz equations is that events which are simultaneous in one inertial frame may no longer be simultaneous in a second inertial frame moving with respect to the first. Suppose we consider two firecrackers going off at exactly the same time t_1, as measured in a reference frame S_1, in which the firecrackers are at rest, but at two different positions x_1 and x_1' in that frame, as in Fig. 25.18. Then, as seen in another inertial frame S_2 moving with a speed v along the positive X axis with respect to S_1, the times of these two events are, from Eq. (25.22):

$$t_2 = \gamma\left(t_1 - \frac{vx_1}{c^2}\right) \qquad \text{and} \qquad t_2' = \gamma\left(t_1' - \frac{vx_1'}{c^2}\right)$$

In this case, t_1 and t_1', are equal, since the events are simultaneous in S_1. But then we have

$$t_2 - t_2' = \gamma\left(\frac{-vx_1}{c^2} + \frac{vx_1'}{c^2}\right)$$

and so $\qquad \Delta t_2 = t_2 - t_2' = \gamma\dfrac{v}{c^2}(x_1' - x_1)$ (25.28)

Hence, if the firecrackers go off at the same time but at different places in S_1, then the explosions are not simultaneous as measured in S_2. *Simultaneity is therefore a relative concept; events simultaneous in one inertial system may not be simultaneous in another inertial system.**

As Einstein has written, "Every reference body has its own particular time. Unless we are told the reference body to which the statement of time refers, there is no meaning in a statement of the time of an event."

---

*This led to the famous bit of doggerel which appeared in *Punch* in 1923 and was a favorite of A. A. Michelson:

There was a young lady named Bright
Whose speed was far faster than light;
She set out one day
In a relative way
And returned home the previous night.

FIGURE 25.18 The relativity of simultaneity. Two events which are simultaneous, but occur at different positions, in S_1 are not simultaneous in S_2.

FIGURE 25.19 A lightning flash striking two points A and B at the same instant as observed by Jack in a rest frame. To an observer (Jill) in a moving frame these events are no longer simultaneous.

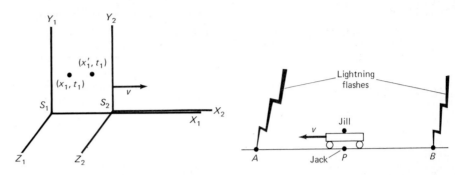

To illustrate this by an example, consider a lightning flash which at the same instant strikes two points A and B along a railroad track, as in Fig. 25.19. If Jack is at rest at point P exactly halfway between A and B, then the lightning flashes from A and B will reach his eyes at exactly the same instant. He will therefore say that the two lightning flashes are simultaneous. But if his sister Jill is on a fast train traveling from B to A at a speed v, and is at exactly the same point P when the lightning strikes, as judged by Jack, then the flashes will not appear simultaneous to Jill. Rather the light from A will reach Jill before the light from B because she is moving toward A and away from B, and the speed of light is finite. The effect will be small, since c is large, but the principle is still clear. Events simultaneous when measured in a rest frame are not simultaneous in another inertial system in motion with respect to that rest frame, if these events occur at different positions in space. This is the physical content of Eq. (25.28).

Example 25.6

Two events occur at the same time in an inertial frame S_1 and are separated by a distance of 10 km along the X axis. What is the time difference between the two events as measured by an observer in a frame S_2 moving with constant speed $0.50c$ along the positive X axis?

SOLUTION

The two events are no longer simultaneous, as seen in S_2, because of time dilation. The time delay is given by Eq. (25.28) as

$$\Delta t_2 = \gamma \frac{v}{c^2}(x_1' - x_1)$$

where here

$$\gamma = \left(1 - \frac{v^2}{c^2}\right)^{-1/2} = \left(1 - \frac{1}{4}\right)^{-1/2} = 1.15$$

Then

$$\Delta t_2 = 1.15\left(\frac{0.50c}{c^2}\right)(10 \times 10^3 \text{ m}) = \frac{0.50(1.15)(10^4 \text{ m})}{3.0 \times 10^8 \text{ m/s}}$$

$$= 1.9 \times 10^{-5} \text{ s} = \boxed{19 \ \mu s}$$

Hence events which appear simultaneous in S_1 differ by a time of 19 μs as measured by an observer in S_2.

25.7 Mass and Energy

We have seen that the Lorentz transformation equations predict that a length L_0 measured in a rest frame becomes $L = (1 - v^2/c^2)^{1/2}L_0$ when measured in another inertial frame moving with a velocity v with respect to the rest frame. Now, v cannot be greater than the speed of light c, or L will become the square root of a negative number. We can therefore conclude:

> The speed of light in a vacuum, c, is the limiting speed in the universe.

Suppose, now, that we apply a *constant* force f for a long period of time to a mass m. According to Newton's second law, we have

$$m = \frac{f}{a} \tag{25.29}$$

Since f is constant, a cannot also be constant from the viewpoint of relativity, for if it were, the speed v would increase constantly until it exceeded c—which, as we have just seen, is impossible. Hence according to relativity, the acceleration a must decrease as the speed v increases in such a way that the speed never exceeds c. If the acceleration a decreases as v increases, with f constant, then from Eq. (25.29) the mass m must increase as v increases. It turns out, as we might expect, that the desired equation for m as a function of v includes the ubiquitous factor γ. The expression for m, as found by Einstein, was

$$m = \gamma m_0 = \frac{m_0}{\sqrt{1 - v^2/c^2}} \tag{25.30}$$

Here, when $v = 0$, $m = m_0$; m_0, the mass when $v = 0$, is called the *rest mass*. It is the mass of a particle in an inertial frame in which the particle is at rest. The predicted dependence of the mass on speed v is shown in Fig. 25.20. Some

FIGURE 25.20 Values of m/m_0 as a function of the ratio v/c, as calculated by Eq. (25.30). When $v = 0.87c$, the mass has doubled from m_0 to $m = 2m_0$.

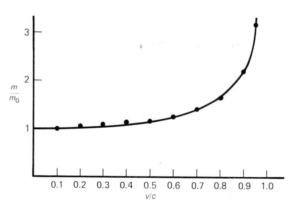

TABLE 25.1 Variation of Mass m of a Particle with its Speed v (Theoretical Values)

$\dfrac{v}{c}$	$\dfrac{m}{m_0}$
0.10	1.005
0.20	1.021
0.30	1.048
0.40	1.091
0.50	1.155
0.60	1.250
0.70	1.400
0.80	1.667
0.90	2.294
0.95	3.203
0.98	5.025
0.99	7.089
0.995	10.013
0.999	22.366
0.9995	31.627
0.9999	70.712

values for m/m_0 as a function of the ratio v/c, as calculated from Eq. (25.30), are given in Table 25.1.

The theory of relativity shows us that our intuitive concept of mass, i.e., the quantity of matter or number of atoms in an object, is inadequate. For particles moving at speeds near that of light the amount of material substance does not change, but their relativistic mass m is very different from their rest mass m_0. The operational definition of mass as inertia by Eq. (25.29), $m = f/a$, remains valid in relativistic physics and is the only truly valid definition of mass.

Einstein reasoned further that, since work done on a system normally increases the energy of the system, but in this case it rather increases the mass m by increasing v in Eq. (25.30), then in some way *mass and energy must be equivalent.* To show this we can take the expression for the mass m in Eq. (25.30) and expand the denominator by the binomial expansion (see Appendix 1D and footnote in Sec. 25.3). We obtain:

$$m = \frac{m_0}{\sqrt{1 - v^2/c^2}} = m_0 \left(1 - \frac{v^2}{c^2}\right)^{-1/2}$$

or
$$m = m_0 \left[1 - \frac{1}{2}\left(\frac{-v^2}{c^2}\right) + \cdots\right] = m_0 \left[1 + \frac{v^2}{2c^2} + \cdots\right]$$

or, neglecting higher-order terms,

$$m - m_0 = \frac{m_0 v^2}{2c^2}$$

and so
$$mc^2 - m_0 c^2 = \tfrac{1}{2}m_0 v^2 = \text{KE}$$

or
$$mc^2 = m_0 c^2 + \text{KE} \tag{25.31}$$

Dimensionally all the terms in this equation are *energies*. The term $m_0 c^2$, which contains the rest mass, must then be the energy possessed by the particle when it is at rest, a quantity we call the *rest energy* of a particle of mass m_0. This makes the term on the left the *total energy* of the particle, the sum of its *rest energy* and its kinetic energy. The total energy of a relativistic particle is therefore

$$\boxed{E = mc^2} \tag{25.32}$$

and its kinetic energy is

$$\text{KE} = mc^2 - m_0 c^2 \tag{25.33}$$

Hence, when a body is at rest, it has a minimum, or rest, energy

$$E_0 = m_0 c^2 \tag{25.34}$$

This is an enormous amount of energy, since $c^2 = 9 \times 10^{16} \text{ m}^2/\text{s}^2$, and so 1 kg of mass is the equivalent of 9×10^{16} J of energy. Hence even slight changes in the mass of particles can lead to large energy changes in aggregates of these particles, since $E_0 = m_0 c^2$. It is these mass changes which produce the tremendous energies in the sun and stars, and which fuel nuclear reactors and nuclear bombs.

Because of the possibility of converting mass into energy, the principles of conservation of mass and conservation of energy can be combined into one fundamental principle, that of the conservation of mass-energy, as indicated schematically in Fig. 25.21. As Einstein expressed it:

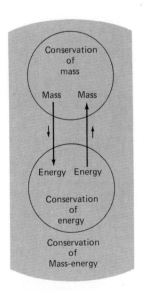

FIGURE 25.21 The principle of the conservation of mass-energy. Mass and energy need not be conserved individually, but the total amount of mass-energy in any closed system is constant.

Conservation of Mass-Energy

> Prerelativity physics contains two conservation laws of fundamental importance—namely, the law of conservation of energy and the law of conservation of mass; these two appear there as completely independent of each other. Through relativity theory they melt together into *one* principle.

Hence the famous Einstein equation $E = mc^2$ has both immense practical importance and great theoretical significance for our understanding of the fundamental laws of physics.

Example 25.7

(a) What is the kinetic energy of an electron of rest mass 9.11×10^{-31} kg which has a relativistic mass of 2.0×10^{-30} kg when in motion? Give your answer in both joules and electronvolts.

(b) What is the speed of such an electron?

SOLUTION

(a) The kinetic energy of a relativistic particle is the difference between its total energy and its rest energy. According to Eq. (25.31) this is

$$KE = mc^2 - m_0c^2$$

$$= (2.0 \times 10^{-30} \text{ kg} - 9.11 \times 10^{-31} \text{ kg})(3.0 \times 10^8 \text{ m/s})^2$$

$$= (1.1 \times 10^{-30} \text{ kg})(9.0 \times 10^{16} \text{ m}^2/\text{s}^2) = \boxed{9.9 \times 10^{-14} \text{ J}}$$

In electronvolts this is:

$$KE = (9.9 \times 10^{-14} \text{ J})\left(6.24 \times 10^{18} \, \frac{\text{eV}}{\text{J}}\right)$$

$$= \boxed{6.2 \times 10^5 \text{ eV}} = 0.62 \text{ MeV}$$

which is the energy an electron would have after being accelerated by a voltage of 0.62 million volts. Note that this is slightly more than the rest energy of the electron, which is 0.511 MeV, and larger than the energies involved in molecular bonding by about five orders of magnitude.

(b) The speed of the particle can be found from the equation:

$$m = \frac{m_0}{(1 - v^2/c^2)^{1/2}} \quad \text{and so}$$

$$1 - \frac{v^2}{c^2} = \left(\frac{m_0}{m}\right)^2 = \left(\frac{9.11 \times 10^{-31}}{2.0 \times 10^{-30}}\right)^2 = 2.1 \times 10^{-1}$$

Hence

$$\frac{v^2}{c^2} = 1 - 2.1 \times 10^{-1} = 7.9 \times 10^{-1}$$

and

$$\frac{v}{c} = 8.9 \times 10^{-1} = 0.89$$

The speed is therefore

$$v = 0.89c = \boxed{2.7 \times 10^8 \text{ m/s}}$$

This is consistent with the data in Table 25.1.

Note: In relativity theory the kinetic energy is *not* given by $\frac{1}{2}mv^2$, with m the relativistic mass, a mistake which is often made. Rather Eq. (25.33) must always be used to obtain the kinetic energy, and Eq. (25.30) to obtain the particle's speed.

Example 25.8

(a) How much energy could be obtained if the rest mass of one atom of hydrogen could be completely converted into energy?

(b) How much energy could be obtained if the rest mass of 1 mol of atomic hydrogen could be completely converted into energy?

SOLUTION

(A) Since the rest mass of a hydrogen atom is almost identical with the mass of a proton, which is 1.67×10^{-27} kg, we have

$$E_0 = m_0c^2 = (1.67 \times 10^{-27} \text{ kg})(3.0 \times 10^8 \text{ m/s})^2$$

$$= \boxed{1.5 \times 10^{-10} \text{ J}}$$

This is a very small energy.

(b) A mole of atomic hydrogen contains Avogadro's number of hydrogen atoms, or 6.02×10^{23} atoms. Hence the total energy produced is:

$$E_0 = N_A m_0 c^2 = (6.02 \times 10^{23})(1.5 \times 10^{-10} \text{ J})$$

$$= \boxed{9.0 \times 10^{13} \text{ J}}$$

This is a tremendous amount of energy, enough to light a 100-W light bulb for 9.0×10^{11} s, or 2.9×10^4 years. Hence 1 g of atomic hydrogen could keep the 100-W bulb glowing for about 30,000 years, if all its mass could be converted into energy. This indicates why the mass-energy conversion predicted by special relativity is one possible solution to the world's energy problems.

25.8 Experimental Tests of Special Relativity

In physics an elegant theory, such as that of special relativity, is a thing of beauty, but it does not endure unless it agrees with experiment. Strong experimental evidence exists for all the predictions of special relativity—the Lorentz contraction, time dilation, the change of mass with velocity, the conversion of mass into energy.

Length Contraction

There is no direct evidence for the length contraction predicted by special relativity because the only way to measure a length directly is to use a meterstick or a measuring tape in the reference frame of the object being measured. But if the object is in motion, its reference frame is also in motion and therefore the meterstick will be shortened with respect to its length when at rest. Hence we would be attempting to measure a length change with a meterstick that shortens in exactly the same way as the object being measured, as shown in Fig. 25.22. Direct verification of the Lorentz contraction is therefore impossible. Indirect confirmation comes, of course, from the negative result of the Michelson-Morley experiment. Indirect confirmation of the Lorentz contraction also comes from experiments on the lifetimes of mu mesons.

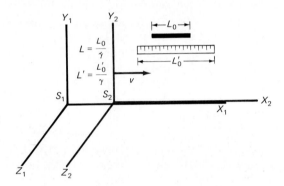

FIGURE 25.22 Using a meterstick to measure the Lorentz contraction of a moving rod. The rod and the meterstick are moving at the same speed and hence contract in the same way.

Time Dilation

A striking example of time dilation comes from the observed lifetimes of μ (mu) *mesons* (or *muons*) produced as secondary radiations from cosmic rays in the earth's upper atmosphere. Muons at rest have lifetimes of only about 2 μs on the average before they decay into other particles. Since the muons are created several thousand meters up in the atmosphere, we would expect few of them to reach earth. A muon moving with a speed of $0.995c$, for example, would only travel about 600 m in 2 μs. Few muons would therefore survive and reach the ground if they started from a height of 2000 m. The muons are, however, moving with respect to an observer on the ground, and to that observer their lifetime is given by Eq. (25.27),

$$\Delta T = \frac{\Delta T_0}{(1 - v^2/c^2)^{1/2}}$$

where ΔT_0 is their lifetime when at rest. In this case we have

$$\Delta T = \frac{2 \times 10^{-6} \text{ s}}{[1 - (0.995)^2]^{1/2}} = \frac{2 \times 10^{-6} \text{ s}}{(1 - 0.99)^{1/2}} = 20 \ \mu\text{s}$$

Hence to an observer on earth the lifetime of the muons is increased by a factor of about 10, and we would expect them to survive through a distance 10×600 m, or about 6000 m. Thus the great majority of the muons would

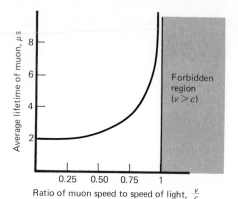

FIGURE 25.23 Dependence of average lifetimes of muons on their speed.

survive until they reached the ground. By measuring the number that reach the ground, the effective lifetime of the muons can be determined. The predicted variation of muon lifetime with speed is shown in Fig. 25.23.

Many experiments have been done on the lifetime of muons, and they all confirm the relativistic time dilation. In 1963 Frisch and Smith did an experiment in which they measured the number of muons incident at the top of Mount Washington, New Hampshire, at an altitude of 1911 m. Then they measured the number of muons incident on the ground at Cambridge, Massachusetts, where the muon flux from cosmic rays differs little from that at the foot of Mount Washington. They had predicted a time dilation of $1/\gamma = 8.4 \pm 2$; that is, the lifetimes were expected to be increased by a factor of about 8.4. The value obtained from their somewhat rough data was 8.8 ± 0.8, which agrees quite well within the limits of error of the experiment.

Another striking confirmation of time dilation has come from the work of two physicists, J. C. Hafele and R. E. Keating, who in 1971 carried four cesium atomic clocks around the world on commercial jet planes (see Fig. 2.3) and measured the time differences between their moving clocks at the end of the trip and a reference atomic clock which remained fixed at the U.S. Naval Observatory. They found excellent agreement with the predictions of relativity, as Table 25.2 shows. These results provide unambiguous proof of the reality of time dilation.

TABLE 25.2 Measured and Predicted Values for Time Differences between Stationary and Moving Clocks*

	Time differences, ns ($= 10^{-9}$ s)	
	Eastward trip around earth	Westward trip around earth
Measured values (mean for four clocks)	-59 ± 10	273 ± 7
Prediction of relativity theory (including a gravitational effect)	-40 ± 23	275 ± 21

*From J. C. Hafele and Richard E. Keating, "Around-the-World Atomic Clocks," *Science*, vol. 177, July 14, 1972, pp. 166–170.

Dependence of Mass on Particle Speed

Figure 25.24 shows the data obtained by three different experimentalists on the change in the observed mass of high-speed electrons as their speed changed. Because of their light masses electrons are easily accelerated to

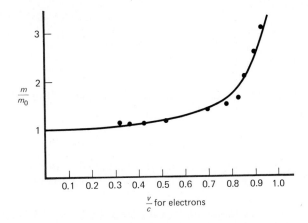

FIGURE 25.24 Experimental data on the dependence of the observed mass of high-speed electrons on their speed. Compare this experimental curve with the theoretical predictions shown in Fig. 25.20.

speeds close to that of light. These electrons are deflected in a system of electric and magnetic fields, as in the experiments of J. J. Thomson discussed in Sec. 18.4. The electron's deflection in a magnetic field determines its momentum, and its deflection in an electric field determines its kinetic energy. In this way both the speed v and the ratio e/m can be determined for electrons of a variety of speeds.

It is clear that the data agree very well with the predictions of theory, as shown in Fig. 25.20. This is all the more impressive since the source of the electrons and the deflection techniques used differed considerably in the three experiments.

Mass-Energy Conversion

The conversion of mass into energy in conformity with Einstein's theory occurs every day in phenomena such as pair production, in which electron-positron pairs are produced from pure energy in the form of gamma rays, and pair annihilation (which is the reverse of pair production), and also in fission and in fusion reactions. We will present the relevant experimental evidence for these cases in Chap. 29.

25.9 General Relativity

Whereas the special theory of relativity finds its most important application at the atomic level, the general theory of relativity is most useful in the universe of planets, stars, and galaxies, i.e., in what we usually call "the universe in the large." The general theory, first proposed by Einstein in 1915, applies to reference frames where one is in accelerated motion with respect to the other, as in Fig. 25.25. It is basically a theory of gravitation and is much more complicated and highly mathematical than is special relativity. Hence all we can do here is indicate briefly the scope and status of this theory, which many physicists regard as the most elegant and perfect of all theories in physics.

There are two basic postulates of general relativity:

1 *Principle of invariance*: All the laws of physics have the same form in all frames of reference, no matter how they are moving with respect to one another.

This principle is merely an extension of the principle of special relativity to include not merely inertial frames of reference but *all* frames of reference, including *accelerated frames*.

FIGURE 25.25 Reference frames in general relativity: Frame S_2 moves with an acceleration a along the positive X axis with respect to rest frame S_1.

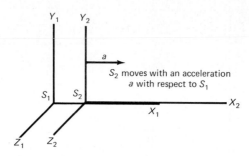

S_2 moves with an acceleration a with respect to S_1

2 *Principle of equivalence*: A uniformly accelerated reference frame is completely equivalent to a homogeneous gravitational field.

This principle is valid even in classical mechanics if the assumption is made that gravitational mass and inertial mass are identical. By *gravitational mass* we mean the mass m that occurs in Newton's law of universal gravitation, $f_g = Gmm_E/r^2$. Similarly, by *inertial mass* we mean the mass that occurs in Newton's second law, $f = ma$, as shown in Fig. 25.26. Measurements by the American physicist Robert H. Dicke (born 1916) have shown that the inertial and gravitational masses of any object are equal to better than 1 part in 10^{11}.

As long as gravitational and inertial masses are the same, no experiment in mechanics performed within an elevator or other closed compartment in space can distinguish whether the elevator is accelerating upward with an acceleration **a** or the objects in the elevator are being subjected to a gravitational force downward which produces a downward acceleration **g** = −**a**, as shown in Fig. 25.27. Einstein extended this principle of equivalence to all fields of physics, and assumed that there is no experiment of any kind that can distinguish uniformly accelerated motion from the presence of a gravitational field.

General relativity theory leads to three predictions which can be tested experimentally. The first is that a light beam should be deflected in a gravitational field. A second is that light emitted from regions in which strong gravitational fields exist should have its wavelength shifted toward the red. Thirdly, general relativity predicts an added precession of the orbit of the planet Mercury over and above that caused by the gravitational attraction of the other planets. In recent years highly precise measurements have been made of all these effects, and they all confirm the predictions of Einstein's theory of general relativity.

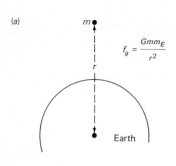

(a)

$$f_g = \frac{Gmm_E}{r^2}$$

Earth

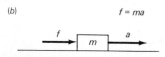

(b)

$f = ma$

FIGURE 25.26 Distinction between gravitational and inertial mass: (a) gravitational mass; (b) inertial mass. According to the general theory of relativity, the inertial and gravitational masses of any object are the same.

FIGURE 25.27 Equivalence of an accelerated frame of reference and a gravitational field: (a) The acceleration of an elevator upward with an acceleration **a** is completely equivalent to (b) a gravitational field producing an acceleration **g** downward inside the elevator, where **g** = −**a**.

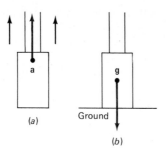

(a)

Ground

(b)

Summary: Important Definitions and Equations

Galilean relativity
Frame of reference: A system in which the position of a particle is specified by three numbers (distances) measured with respect to the origin of the coordinate system, and by a time represented by a single number with respect to a chosen initial time.
Inertial frame of reference: A frame of reference in which Newton's first law of motion is valid.

Galilean transformation equations

$$x_2 = x_1 - vt_1 \qquad z_2 = z_1$$
$$y_2 = y_1 \qquad t_2 = t_1$$

Invariance: If the equations expressing a physical law retain the same mathematical form after a coordinate transformation as they had before it, the equations are said to be *invariant* under the transformation.
Principle of Galilean relativity: All the laws of classical mechanics are invariant under a Galilean transformation. Hence all inertial systems are equivalent in classical mechanics.

Aether: A medium whose existence was postulated by classical physicists to carry light waves and gravitational, electric, and magnetic forces through space.

Michelson-Morley experiment: An experiment using a Michelson interferometer to detect the postulated influence of the aether on the speed of light.

Einstein's special theory of relativity
First postulate—Constancy of speed of light: The speed of light in a vacuum is a constant c, independent of the inertial system in which it is measured or of the motion of the light source or observer.
Second postulate—Principle of special relativity: All the laws of physics, including those of mechanics and electromagnetism, have the same form (i.e., they are invariant) in all inertial frames of reference.

Lorentz transformation equations (for S_2 moving along positive X axis with speed v with respect to S_1):

$$x_2 = \gamma(x_1 - vt_1) \qquad t_2 = \gamma\left(t_1 - \frac{vx_1}{c^2}\right)$$
$$y_2 = y_1 \qquad \gamma = \left(1 - \frac{v^2}{c^2}\right)^{-1/2}$$
$$z_2 = z_1$$

Consequences of the special theory of relativity
1. Length (Lorentz-Fitzgerald) contraction:

$$L = (1 - v^2/c^2) L_0, \text{ where } L \text{ is the proper length.}$$

2. Time dilation: $\Delta t = \dfrac{\Delta t_0}{(1 - v^2/c^2)^{1/2}}$

where Δt_0 is the proper time interval.
3. Relativity of simultaneity: Events simultaneous in one inertial system may not be simultaneous in another inertial system.
4. Limiting speed: The speed of light in a vacuum, c, is the limiting speed in the universe.
5. Variation of mass with speed:

$$m = \gamma m_0 = \frac{m_0}{(1 - v^2/c^2)^{1/2}}$$

where m_0 is the rest mass.
6. Equivalence of mass and energy: $E = mc^2$;

$$KE = mc^2 - m_0c^2$$

7. Conservation of mass-energy

Einstein's general theory of relativity
Applies to reference frames one of which is in accelerated motion with respect to the other.
First postulate—Principle of invariance: All the laws of physics have the same form in all frames of reference, no matter how they are moving with respect to one another.
Second postulate—Principle of equivalence: A uniformly accelerated reference frame is completely equivalent to a homogeneous gravitational field.

Questions

1 Can an experiment be performed in classical mechanics which will reveal whether a frame of reference is *accelerated* with respect to an inertial frame?

2 If you are shooting pool on a pool table below deck on an ocean liner and the ship has to slow down because of ocean fog, will this influence the motion of the balls on the pool table? Explain your answer.

3 In the phenomenon of time dilation, what exactly is "dilated"? Would "time retardation" be a more accurate term?

4 Two observers, one at rest in S_1 and the other at rest in S_2, which moves with a speed v along the positive X axis with respect to S_1, each carries a meterstick oriented along the X axis. Each observer finds, upon performing a measurement, that the other's meterstick is shorter than his or her meterstick. Who is right? Explain your answer.

5 Can a particle move through a medium at a speed greater than the speed of light in that medium (e.g., glass)? Explain your answer.

6 In optics it is well known that in certain crystals

light rays travel at different speeds in different directions through the crystal. Why could not the same thing be true of light traveling through the aether?

7 Why is it easy to accelerate electrons to speeds approaching that of light, whereas it is much more difficult to do this for protons or alpha particles?

8 In e/m experiments, variations of e/m with the speed of the particle are normally taken as an indication that the mass m has changed with v. Why is it not possible that the charge e varies with v? (Consider what might happen in this case to heavy atoms, whose inner electrons travel at speeds close to that of light.)

9 Discuss changes which would occur in our everyday lives if the speed of light were 30 m/s instead of 3.0×10^8 m/s.

10 (a) If an astronaut on a spaceship is able to reach a speed of $0.6c$ away from the earth, would the mass, height, or heartbeat of the astronaut change as measured with instruments inside the spaceship?

(b) Suppose observers on earth were able to measure these properties through remote-sensing devices. Would they find any changes in the astronaut's mass, height, or heartbeat?

11 (a) Why are the effects of the special theory of relativity not normally observable in day-to-day life?

(b) Suppose we lived in a universe in which everything on earth was reduced in size by a factor of 10^3. Would the effects of special relativity be any more evident than they are in our present world?

Problems

A 1 A meterstick is at rest in an inertial frame S_2. If S_2 moves with a speed of $0.50c$ with respect to a rest frame S_1, the length of the meterstick as measured in frame S_1 is, according to classical physics:
(a) 1.15 m (b) 1.00 m (c) 0.86 m
(d) 0.75 m (e) 1.33 m

A 2 A meterstick is at rest in an inertial frame S_2. If S_2 moves with a speed of $0.50c$ with respect to a rest frame S_1, the length of the meterstick as measured in frame S_1 is, according to the special theory of relativity:
(a) 1.15 m (b) 1.00 m (c) 0.86 m
(d) 0.75 m (e) 1.33 m

A 3 Two firecrackers explode at the same place in a rest frame S_1 with a time separation of 10 s in that frame. The time between explosions, as measured in a frame moving with a speed $0.80c$ with respect to the rest frame is, according to classical physics:
(a) 6.0 s (b) 16.7 s (c) 8.0 s (d) 12.5 s
(e) 10 s

A 4 Two firecrackers explode at the same place in a rest frame S_1 with a time separation of 10 s in that frame. The time between explosions, as measured in a frame moving with a speed of $0.80c$ with respect to the rest frame is, according to the special theory of relativity:
(a) 6.0 s (b) 16.7 s (c) 8.0 s (d) 12.5 s
(e) 10 s

A 5 Two firecrackers explode at the same time, but one at a place with coordinate $x_1 = 10$ m in a rest frame S_1, and the other at a place with coordinate $x_1' = 110$ m in S_1. In a frame S_2 with a speed $0.60c$ with respect to S_1, the time difference between the two explosions, according to the special theory of relativity, is:
(a) 0.33 μs (b) 20 μs (c) 16 μs
(d) 0 (e) 0.25 μs

A 6 A particle of rest mass m_0 moves at a speed of $0.60c$ relative to an observer. Its observed mass is:
(a) m_0 (b) $1.25m_0$ (c) $0.80m_0$
(d) $1.33m_0$ (e) $0.60m_0$

A 7 A particle of rest mass m_0 moves at a speed $0.80c$ with respect to a rest frame. Its momentum, as measured in the rest frame, is:
(a) $0.80m_0c$ (b) m_0c (c) $1.25m_0c$
(d) $1.33m_0c$ (e) $1.67m_0c$

A 8 To produce a change in mass of a particle by one order of magnitude, its speed as measured in a rest frame must be:
(a) $10c$ (b) c (c) $0.995c$ (d) $0.90c$
(e) $2c$

A 9 The energy associated with the rest mass of the proton is:
(a) 4.6×10^{-2} J (b) 1.5×10^{-10} J
(c) 4.9×10^{-19} J (d) 1.7×10^{-27} J
(e) 1.5×10^{-10} eV

A10 An electron and a positron (which has the same mass as an electron) collide and annihilate each other. The energy produced in the form of gamma rays by the conversion of mass into energy is:
(a) 5.0×10^{-5} J (b) 1.6×10^{-13} J
(c) 5.4×10^{-22} J (d) 1.8×10^{-30} J
(e) 1.6×10^{-13} eV

A11 The amount of electric energy consumed in New York City each year is about 3×10^{10} kWh. If all the mass in a 1-kg object could be converted into energy, it could supply the energy needs of New York for about:
(a) 1 week (b) 1 min (c) 1 day
(d) 1 year (e) 1 month

B 1 The length of a 1.0-cm division on a meterstick is measured by an observer in a reference frame moving at a speed of $0.40c$ with respect to the meterstick. How long does the 1.0-cm division appear to the observer?

B 2 Calculate the Lorentz contraction of a tennis ball traveling at 100 mi/h (45 m/s).

B 3 A physicist measures the lifetime of a neutron in a laboratory on earth and finds that it is 6.42×10^2 s. Another physicist has developed a new rocket which can travel at a speed of 1.5×10^8 m/s. If the second physicist moves at this speed with respect to the first physicist's laboratory, what lifetime does the second physicist measure for the neutron?

B 4 A particle called a pi meson, or pion, has a lifetime at rest on the earth of only about 1.8×10^{-8} s on the average.

(a) How long would such a pion live when approaching the earth at a speed of $0.95c$, as measured by an observer on earth?

(b) How long does the pion "think" the distance to the earth is?

B 5 The star closest to our solar system is Alpha Centauri, which is 4.5 light-years away.

(a) How long would it take according to a clock on earth for a spaceship to make a round trip to Alpha Centauri if the spaceship could travel at $0.995c$?

(b) How long would the trip take according to clocks on the spaceship?

B 6 Electrons in TV tubes are traveling at speeds of about $0.30c$ when they strike the fluorescent screen and produce the TV picture. What is their mass as measured in a frame in which the TV tube is at rest?

B 7 What is the fractional increase in mass of a jet plane flying at Mach 3 (i.e., 3 times the speed of sound in air), as measured in a rest frame with respect to which the plane has this speed?

B 8 It takes 80 kcal at 0°C to melt 1 kg of ice to water, as we saw in Chap. 13. By how much does the mass of the water exceed the mass of the ice because of the energy added to melt it?

B 9 (a) How much energy would be obtained if the rest mass of 1.0 kg of coal could be converted entirely into energy?

(b) How long could this much energy provide electricity for a city which requires 10^6 kWh of electric energy per year, if the energy in the coal could be converted into electric energy with 100 percent efficiency?

B10 What is the kinetic energy of an electron whose mass is 3 times its rest mass?

C 1 An event occurs in a rest frame S_1 at $x_1 = 200$ km, $y_1 = 2.0$ km, $z_1 = 2.0$ km at time $t_1 = 5.0 \times 10^{-3}$ s. Let S_2 move relative to S_1 at a speed $0.95c$ along the common X axis. Assume that the origins coincide at $t_2 = t_1 = 0$. What are the coordinates x_2, y_2, z_2, t_2 of this event in S_2?

C 2 Check your answer to Prob. C1 by carrying out the inverse Lorentz transformation to yield your original data.

C 3 A spaceship has a length L_0 in a rest frame. When in motion with a speed c with respect to that rest frame, its measured length is $L_0/2$.

(a) What is the speed of the spaceship?

(b) What is the dilation of time scale on the spaceship?

C 4 An airplane flies at a speed of 500 m/s. How long must it fly before its clock loses 1 s because of time dilation, as measured by a ground observer?

C 5 The radius of our galaxy is about 3×10^{20} m.

(a) Can a person, in principle, travel from the center of the galaxy to its edge in a normal lifetime by traveling very rapidly? Explain how he or she can accomplish this.

(b) What must the person's speed be to make the trip in 25 years?

C 6 Certain bacteria are observed to double in number every 20 days in a laboratory on earth. Two of these bacteria are placed on a spaceship and move in an orbit around the earth at a speed of $0.98c$ for a time of 1000 days as measured on earth. How many bacteria would be on the spaceship when it returns to earth?

C 7 Show that the same result can be obtained for Prob. C6 by using the Lorentz-Fitzgerald contraction instead of time dilation to solve the problem.

C 8 An astronaut spends 20 days in orbit around the earth, traveling at an average speed of 7.8×10^3 m/s. Show that, when the astronaut splashes down, he or she is about 600 μs younger than would have been the case if there had been no space voyage.

C 9 Two events occur at the same time in an inertial frame S_1, and are separated by a distance of 5.0 km along the X axis. What is the time difference between these two events as measured in frame S_2 moving with constant velocity v along X, if an observer in S_2 measures the spatial separation in S_1 to be 3.0 km?

C10 What is the speed of an electron whose kinetic energy is equal to its rest energy?

C11 An electron is observed to move at 1.8×10^8 m/s with respect to a rest frame.

(a) What is the relativistic mass of the electron?

(b) What is its kinetic energy?

C12 (a) Determine the voltage needed to accelerate electrons to a speed of $0.8c$.

(b) Determine the voltage needed to accelerate protons to $0.8c$.

C13 An electron is accelerated from rest through a potential difference of 10^6 V and thus acquires a kinetic energy of 1.0 MeV.

(a) Find its speed and its mass when it has acquired this energy.

(b) What is the electron's total energy?

C14 How much energy is required to accelerate an electron from rest to a speed of 0.99 that of light?

C15 What is the percentage error involved in using the expression $\frac{1}{2}m_0v^2$ for the kinetic energy of an electron if its speed is (a) $0.10c$; (b) $0.80c$?

C16 At the Stanford Linear Accelerator Center (SLAC), electrons are accelerated to such high speeds that their mass becomes 10^4 times their rest mass.

(a) What is the speed of the electrons when this happens?

(b) If the SLAC accelerating tube is 3.0 km long, how long is it in the reference frame of the moving electrons?

C17 (a) If classical theory were valid, what accelerating voltage would be required to produce electrons with the speed of light?

(b) What is, according to relativistic mechanics, the actual velocity of the electrons for this potential?

(c) By what fraction has their mass increased?

C18 Two 0.10-kg masses with equal and opposite speeds of 50 m/s approach each other along a straight line, collide, and then stick together. What is the additional rest mass of the system after the collision?

C19 The sun radiates energy at the rate of 3.9×10^{26} W, and this energy is produced by the conversion of the sun's mass into energy by fusion processes according to the equation $E = mc^2$.

(a) Find the rate at which the sun must convert mass into energy to produce its radiated energy.

(b) If the sun's total mass is 2.0×10^{30} kg, how long will the sun last if it continues to lose mass at this rate?

C20 (a) How much energy is released in the explosion of a uranium bomb containing 5.0 kg of fissionable material, if we assume that 1/1000 of the rest mass of the uranium is converted into energy?

(b) How much TNT would be required to produce the same amount of energy, if TNT liberates 820 kcal of energy per mole, and each mole has a mass of 0.227 kg?

(c) What is the relative energy released in the two cases, for the same initial amounts of uranium and TNT?

D 1 Show that it is possible to synchronize two clocks at different points in space so that they tell the same time. Assume that one clock is at the origin of the coordinate system of an inertial frame, and that the second clock is a distance D away from the origin in the same inertial frame of reference.

D 2 A rocket has a length of $L_0 = 500$ m as measured at rest at Cape Kennedy, Florida. It is launched, and it moves directly away from the earth and reaches a constant speed v. A radar pulse is transmitted from earth and reflected back by electronic devices at the nose and the tail of the rocket. The reflected signal from the tail is detected on earth 300 s after emission and the signal from the nose arrives 14.5 μs later. Calculate the distance of the rocket from the earth and the speed of the rocket with respect to the earth (a) nonrelativistically, and (b) relativistically.

D 3 In inertial system S_1, which is at rest, an event occurs at $x_1 = 50$ m, $y_1 = 0$, $z_1 = 0$. Exactly 10^{-6} s later a second event occurs at $x_1' = 550$ m, $y_1' = 0$, $z_1' = 0$.

(a) Does there exist another inertial frame S_2, moving with speed less than c along the X axis, such that the two events appear simultaneous as seen from S_2?

(b) If so, what is the magnitude and direction of the velocity of S_2 with respect to S_1?

(c) What is the spatial separation of x_1' and x_1, as seen in S_2?

(d) Answer parts (a), (b), and (c) for a situation where $x_1' = 150$ m.

D 4 A cube with all sides equal to a rests at the origin of an inertial frame S_1. The mass of the cube is m_0 in this rest frame. An observer moves along the positive X axis with a speed v with respect to S_1.

(a) What is the observed volume of the cube?

(b) What is the observed mass of the cube?

(c) What is the observed density of the cube?

(d) Show that, when v is much smaller than c, the density reduces to the usual value.

D 5 (a) Prove that a particle which travels at the speed of light can only have a zero rest mass.

(b) Show that for such a particle of zero rest mass, we must have: velocity $v = c$; kinetic energy KE $= E$ (total energy); momentum $p = E/c$.

D 6 Laboratory measurement shows that the speed of a particular charged particle is $0.71c$. The particle is then directed into a magnetic field $B = 1.0$ T, in which it moves in a circle of radius 3.1 m. Find the mass of the particle and identify it.

D 7 The Lorentz equations may be used to derive similar equations for transforming the components of the velocity u_1 in a rest frame S_1 to the corresponding velocity u_2 in a frame S_2. Show that the resulting equations are:

$$u_{x2} = \frac{u_{x1} - v}{1 - vu_{x1}/c^2} \qquad u_{y2} = \frac{u_{y1}/\gamma}{1 - vu_{x1}/c^2} \qquad u_{z2} = \frac{u_{z1}/\gamma}{1 - vu_{x1}/c^2}$$

where frame S_2 is moving with a velocity v along the positive X axis with respect to S_1.

D 8 Consider a reference frame S_2 moving along the positive X axis at a speed v with respect to a rest frame S_1. Use the results of Prob. D7 to show that a light ray traveling along the X axis at a speed c in S_1 has the same speed c when measured in the moving frame S_2.

D 9 Fill in the steps outlined in Sec. 25.5 to prove that Eq. (25.25) goes over into Eq. (25.24) under a Lorentz transformation.

D10 Find the total energy, kinetic energy, and momentum of a proton moving at a speed of $0.75c$.

D11 An electron is accelerated through a voltage of 10^6 V.

(a) What is its kinetic energy?

(b) What is its relativistic mass?

(c) What is its speed?

D12 Prove that the Lorentz transformation equations are able to account for the negative results of the Michelson-Morley experiment by predicting a contraction of the interferometer in the direction of its motion.

Additional Readings

Barnett, Lincoln: *The Universe and Dr. Einstein*, Mentor, New American Library, New York, 1958. An accurate popular book on of relativity theory.

Bernstein, Jeremy: *Einstein*, Penguin, New York, 1976. Probably the best brief account of Einstein's personality and ideas.

Clark, Ronald J.: *Einstein: The Life and Times*, World Publishing, New York, 1971. A standard biography of Einstein, but a bit weak on the scientific side.

Einstein, Albert: *Relativity: The Special and General Theory*, Crown, New York, 1961. A nontechnical book by the man who best understood relativity because he created it.

French, A.P.: *Special Relativity*, Norton, New York, 1968. An introductory textbook that contains a very good discussion of the historical origins of special relativity.

Gamow, George: *Mr. Tompkins in Paperback*, Cambridge University Press, New York, 1965. This collection of Gamow's popular books on physics includes one story in which Mr. Tompkins lives in a world in which the speed of light is 10 mi/h and relativistic effects become quite important.

Geroch, Robert: *General Relativity from A to B*, University of Chicago Press, Chicago, 1978. Lectures on general relativity intended for nonscience undergraduates at the University of Chicago.

Hoffman, Banesh, and Helen Dukas: *Albert Einstein, Creator and Rebel*, Viking, New York, 1972. An account of Einstein's life and work by one of his collaborators and his former secretary.

Pais, Abraham: "How Einstein Got the Nobel Prize," *American Scientist*, vol. 70, no. 4, July–August 1982, pp. 358–365. The inside story on why Einstein received his Nobel Prize for the theory of the photoelectric effect and not for relativity theory. Pais, who was a colleague of Einstein's at Princeton, has also written a more technical book on Einstein: *'Subtle is the Lord': The Science and the Life of Albert Einstein*, Oxford University Press, New York, 1982.

Shankland, Robert S.: "The Michelson-Morley Experiment," *Scientific American*, vol. 211, no. 5, November 1964, pp. 107–114. A discussion of the history and significance of the Michelson-Morley experiment. Shankland has another interesting article: "Conversations with Albert Einstein," *American Journal of Physics*, vol. 31, 1963, pp. 47–57.

Swenson, Loyd S., Jr.: *Etherial Aether*. University of Texas Press, Austin, Texas, 1972. A detailed history of the Michelson-Morley and other aether experiments from 1880 to 1930.

A lmost half a century has elapsed since Max Planck's discovery of the quantum of action, a time sufficiently long to estimate its importance for science and, more generally, for the development of human thought. There is no doubt that it was an event of the first order, comparable with the scientific revolutions brought about by Galileo and Newton, Faraday and Maxwell.

Max Born (1882–1970)

Chapter 26

The Experimental Basis of Quantum Mechanics

Modern physics traces its existence to two sharp breaks with nineteenth-century classical physics that took place at the turn of the century. As we saw in the preceding chapter, Albert Einstein made the first break with his special theory of relativity. The second, and in some ways the more dramatic break, is associated with the name of Max Planck. The quantum theory which Planck developed not only revolutionized physics but, as pointed out by Max Born in the quotation above, produced one of the greatest intellectual upheavals in human history.

If the real world is governed by the laws of relativity and quantum theory, how then do we manage to survive in our daily lives while paying little attention to these theories? For particles on the earth, relativistic effects only become important for speeds approaching c, the speed of light. So too the quantum theory only becomes important for particles the size of atoms or smaller, as Fig. 26.1 attempts to show schematically. To describe particles of the size of golf balls or rockets moving at speeds low compared with 3×10^8 m/s, neither relativity nor quantum theory is needed. Relativity (both the special and the general theories) and quantum theory are still of crucial importance, however, insofar as they govern the atomic and molecular structure of golf balls, rockets, and all other objects in our universe. These two theories also determine the properties of the universe on a cosmic scale and are essential to understanding the universe's evolution to its present condition from the "big bang" which is believed to have been its origin.

FIGURE 26.1 A rough schematic outline of the domains of validity of various physical theories for particles on the earth. The diagram is oversimplified, but is helpful in showing the relationships between classical physics, quantum theory, and special relativity.

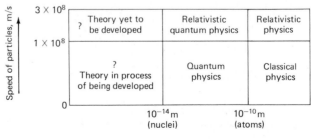

26.1 Particle Properties of Light; the Photon

In Chap. 22 we presented strong evidence that light consists of transverse electromagnetic waves traveling through space at a speed c. The phenomena of diffraction, interference, and polarization can all be explained not merely qualitatively but also quantitatively by such an assumption. But there are other experimental results—blackbody radiation, the photoelectric effect, the Compton effect—which cannot be explained on the basis of a classical wave model of light. These results also cannot be explained by the older corpuscular theories of light of the type proposed by Sir Isaac Newton. A new model for light was needed, and this was provided at the turn of the century by Planck, Einstein, and others who based their model on the concept of the *photon*.

Definition

Photon: A bundle (or "quantum") of electromagnetic energy which behaves like a particle.

As we will see in the next three sections, to explain the available experimental evidence it is necessary to postulate the existence of photons, where the energy of a photon is directly related to its frequency by the equation $E = hf$, where E is the energy of the photon (in joules), f is the frequency of the photon (in hertz or s^{-1}), and h is a constant which, to satisfy the experimental evidence, must have the value 6.63×10^{-34} J·s. Only waves or other periodic motions can have a frequency, which is, after all, a number of vibrations per second. Hence, inherent in the concept of photon are both particle properties (momentum) and wave properties (frequency). This is our first introduction to the dual wave-particle nature of reality, which will turn up again and again in this and subsequent chapters.

We will now consider the experimental evidence which led Planck and Einstein to this equation, which has turned out to be the most important equation in all of modern physics.

26.2 Blackbody Radiation

The first experimental evidence for the existence of photons came from the study of the radiation from hot objects, and particularly of what is called *blackbody radiation*. Kirchhoff's invention of the prism spectrometer in 1859 led to a great interest in analyzing the visible light emitted by hot gases and solids, and to the accumulation of a great deal of important spectroscopic data. It was the attempt by Planck (who was a student of Kirchhoff's) to understand the data on the continuous radiation from heated solids that led him to his theory of blackbody radiation and ultimately to the quantum, or photon, theory of light.

A *blackbody* is a perfect absorber which appears *black* under visible light because it absorbs all wavelengths of light falling on it. A piece of black velvet is a good approximation to a blackbody. Kirchhoff pointed out that, for all forms of radiation, an even better approximation to a blackbody is a spherical cavity in a block of metal with one tiny entrance hole in it, as shown in Fig. 26.2. Any radiation entering the hole is trapped by multiple reflections inside the cavity and cannot escape. Such a cavity is black to all visible and

FIGURE 26.2 A blackbody cavity. Any radiation entering the hole is trapped inside by successive reflections from the walls.

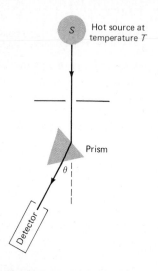

FIGURE 26.3 Experimental setup to measure the intensity of blackbody radiation as a function of temperature. The wavelength can be determined from the measured values of the angle θ.

near-visible electromagnetic radiation, since it reflects none of the radiation incident on the entrance hole, but absorbs all of it.

Just as such a blackbody cavity is a nearly perfect absorber, so when heated it is also the most efficient emitter of radiation, since its emissivity e in Stefan's radiation law [Eq. (14.3)] is equal to unity. When heated, the molecules in the walls of the cavity vibrate and emit radiation into the interior of the cavity. Experiments on the radiation from blackbodies can then be performed by putting such a cavity into a furnace and studying the amount and wavelengths of the radiation emitted from the hole in the cavity as a function of furnace temperature, as in Fig. 26.3. In the years between 1860 and 1900 physicists used spectrometers to obtain experimental curves similar to those in Fig. 26.4 for blackbody radiation. Such curves are of great theoretical importance because they have a universal character, being independent of the material out of which the emitting cavity is constructed.

Figure 26.4 shows the energy $E(\lambda)$ emitted by a blackbody in a narrow wavelength range $\Delta\lambda$ about some chosen value of the wavelength λ in the visible region of the spectrum. Note that as the temperature increases, the total amount of radiation emitted increases dramatically, as it must according to the fourth-power dependence on temperature in Stefan's law [Eq. (14.3)]. Also, we see that the peak of the curve moves to shorter and shorter wavelengths (higher frequencies) as the temperature increases, as predicted by the Wien displacement law [Eq. (14.4)], $\lambda_{max}T = k = 2.898 \times 10^{-5} \text{ m·K}$.

Although both the Stefan radiation law and the Wien displacement law were well known to physicists before 1900, no one had been able to derive both these equations satisfactorily from the basic properties of radiation. This was accomplished in 1900 by Max Planck (see Fig. 26.5 and accompanying biography).

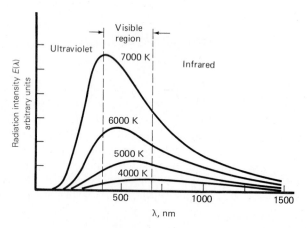

FIGURE 26.4 Blackbody radiation curves. Each graph shows the intensity of the radiation emitted by a blackbody as a function of wavelength for one particular temperature. The four curves shown here are for the four different temperatures indicated.

Planck's Quantum Theory

Planck set out to explain the form of blackbody radiation curves like those of Fig. 26.4 on the basis of first principles. There had been previous attempts to do this, but none had been very successful. One theory, due to the German physicist Wilhelm Wien (1864–1928), agreed well with experiments at short wavelengths but deviated noticeably from the experimental curves at long wavelengths. A second theory, due to the British physicists Lord Rayleigh (1842–1919) and Sir James Jeans (1877–1946), assumed that all wavelengths (and therefore all frequencies) would be radiated with equal probability.

Max Karl Ernst Ludwig Planck (1858–1947)

FIGURE 26.5 Max Planck. Planck is shown presenting a medal to Albert Einstein. *(Courtesy of AIP Niels Bohr Library, Fritz Reiche collection.)*

Max Planck, the father of the quantum theory, was born in Kiel on April 23, 1858, into a very well-known German family of lawyers, public servants, and scholars. Throughout his life he preserved a great devotion to his family, his friends, and, when possible, to his country. He studied mathematics at the University in Munich, and then spent a year at the University of Berlin working in physics under Helmholtz and Kirchhoff. He received his doctorate from Munich in 1879 for a dissertation on the second law of thermodynamics. The key to his success in explaining blackbody radiation was his deep knowledge of thermodynamics. His book on this subject, first published in 1897, still influences the way physicists approach thermodynamics today.

In 1887 Planck received an appointment to the physics faculty at the University of Kiel. Two years later his reputation had so spread that he was offered the chair of theoretical physics in Berlin. Here he developed into one of the world's greatest theoretical physicists, aided by his association with Helmholtz, and in later years with Einstein, von Laue, and Born. He remained in Berlin for the rest of his active scientific life, enjoying the stimulating contacts with colleagues in physics, mathematics, chemistry, and philosophy at the university, the culture of the great capital city, and his walks in the forests near Berlin. He was an avid walker and hiker and took yearly mountain-climbing vacations in the Alps.

On October 19, 1900, Planck presented his paper on blackbody radiation to the Berlin Physical Society. The night after the meeting his colleague H. Rubens carried out additional measurements and the next morning informed Planck that his formula fitted all Rubens' data perfectly from very short to very long wavelengths. Thus was born the quantum theory of radiation.

Planck, while modest in public, realized fully the importance of his work. His son Erwin recalled that in 1900 his father took him for a walk through the Grünewald near Berlin, in the course of which his father astounded him by saying: "Today I have made a discovery as important as that of Newton." There is no doubt that Planck was correct in this judgment, although it took time and the work of Einstein, Compton, Bohr, and others to convince the physics establishment of this fact. For his pioneering work in quantum theory Planck received the 1918 Nobel Prize in physics.

In the years before the First World War, Planck and Einstein combined to make Berlin the world's great center for theoretical physics. Not only did these two great men work together, they played chamber music together, Planck at the piano and Einstein with his violin, both finding great joy and relaxation in the music they made. Unfortunately, this somewhat idyllic existence ended with the outbreak of World War I. Three of Planck's four children by his first wife died during the war, two daughters during childbirth, and his son Karl in action in France.

The Hitler period in Germany brought increased personal tragedy to Planck. His son, Erwin, the only surviving child of his first marriage, was involved in the July 1944 plot to kill Hitler and was executed by the Nazis. Later Planck's home in Grünewald and his large personal library were destroyed in one of the allied air raids on Berlin.

After the war Planck moved to Göttingen to live with a grand-niece. He continued to write and lecture until 1947, when he died at the age of 89. His grave is marked by a simple rectangular headstone containing only his name and one additional line: $h = 6.62 \times 10^{-27}$ erg·sec. This nonzero value for Planck's constant, which distinguishes modern quantum physics from classical Newtonian physics, assures Planck a unique place in the history of science.

Since there are many more high frequencies than low frequencies within any given wavelength interval $\Delta\lambda$, they concluded that the amount of energy radiated as a function of wavelength would take the form shown by the top colored line in Fig. 26.6. Most of the energy emitted would be in the violet and ultraviolet, in total disagreement with experiment. Physicists came to talk about an "ultraviolet catastrophe," since the total energy radiated by a blackbody would be infinite, for the curve in Fig. 26.6 goes to infinity at short wavelengths in the ultraviolet.

Planck looked for one equation which would fit the experimental data at all wavelengths. After many unsuccessful attempts, he found, in 1900, the following equation, which fitted the data over the entire range from ultraviolet to infrared:

$$E(\lambda)\,\Delta\lambda = \frac{8\pi}{\lambda^4}\,\frac{hf\,\Delta\lambda}{e^{hf/kT}-1}$$

(26.1)

where $E(\lambda)\,\Delta\lambda$ is the energy per unit volume emitted by the blackbody in the wavelength interval $\Delta\lambda$ about the wavelength λ, f is the frequency, k is Boltzmann's constant, and T is the absolute temperature. This leaves the one new constant h to be explained. Planck was able to show that this constant had the dimensions of "action," where action is the product of energy and time. To reproduce the curves of Fig. 26.4, h had to have a value close to 6.6×10^{-34} J·s. Because its importance was first uncovered by Planck, the constant h, the "quantum of action," is now called *Planck's constant*.

Thus far Planck had not really explained anything. By juggling factors he knew must play a role in any explanation of blackbody radiation he had merely come up with an equation which agreed with experiment. Now he had to justify his equation logically.

To explain his result, Planck assumed that Rayleigh and Jeans were right in assuming that blackbody radiation would contain more high frequencies than low frequencies just because there are more high frequencies possible within any given $\Delta\lambda$. He further assumed, however, that for some unknown reason the probability of radiation must decrease as the frequency increased. These two assumptions would explain the experimental curves which fell to

FIGURE 26.6 Comparison of the predictions of the Rayleigh-Jeans theory and of Wien's theory with the experimental curves for blackbody radiation at 1600 K. Note how at this temperature most of the radiation is in the infrared.

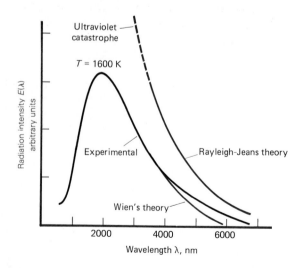

near zero at both very short and very long wavelengths, and had a maximum between these two extremes, as in Fig. 26.4.

But why should the probability of radiation fall off as the frequency increased? Planck saw that this could happen only if the energy radiated into the cavity by the walls of the blackbody was not radiated continuously, but came in discrete bundles which he called *quanta* (from the Latin *quantum*, meaning "how much"). In other words, energy was not radiated continuously from the walls like water from a hose, but rather like bullets from an automatic machine gun. Planck assumed that each quantum had an energy proportional to its frequency and hence that it was much easier for lower-frequency oscillators to radiate than it was for higher-frequency oscillators, since they were more likely at any given temperature to have enough energy to emit a whole quantum of energy. If an oscillator did not have enough energy to emit a whole quantum at the frequency corresponding to its mode of vibration, it could not emit any radiation at all. (The situation is somewhat similar to one in which you need quarters for a parking meter that accepts only quarters. The fact that you may have a pocket full of nickels and dimes does you no good, since you do not have the necessary "quantum," the quarter, which is needed.) At low temperatures there is only sufficient energy for the low-frequency oscillators to do most of the radiating (in other words, you have found a parking meter which takes nickels and dimes). As the temperature increases, however, the amount of energy available increases as T^4 (Stefan's radiation law), and high-frequency radiation becomes more likely.

On looking at his empirical equation [Eq. (26.1)], Planck saw that hf/kt in the exponential term $e^{hf/kT}$ must be dimensionless, since all exponents are pure numbers. But we know from our treatment of thermodynamics that kT has the dimensions of energy. Hence hf must also have the dimensions of energy, and so Planck for the first time wrote down the now famous equation:

$$E = hf \tag{26.2}$$

This equation states that the quantum of energy E is determined by its frequency f, and that the proportionality constant between energy and frequency is Planck's constant $h = 6.63 \times 10^{-34}$ J·s, where E is in joules and f in hertz, or s^{-1}.

Planck's blackbody equation [Eq. (26.2)], with h equal to the value just given, agreed perfectly with experiment. Planck's equation could also be shown to lead to the Stefan-Boltzmann T^4 law and to the Wien displacement law, and to reduce to the results of Wien and Rayleigh-Jeans in the proper limiting cases. These were additional proofs of its validity.

The real significance of Planck's work only dawned on the physicists of the early twentieth century, however, when Einstein carried Planck's ideas one step further in another of his famous 1905 papers. His work, and that of Compton and others, made it clear that all of physics must take the quantized nature of energy into account, i.e., that energy comes in discrete bundles of magnitude $E = hf$. This quantum approach is the dividing line separating modern physics from classical physics, and is basically due to the profound insight of one man, Max Planck.

Example 26.1

(a) What is the energy associated with a photon in the radio-frequency region, say, at a frequency of 100 kHz?
(b) What is the energy associated with a photon in the visible region of the spectrum, say, at a wavelength of 500 nm?

(c) What is the energy associated with a gamma ray of frequency 10^{20} Hz?

Express all answers in both joules and electronvolts.

SOLUTION

(a) $E = hf = (6.6 \times 10^{-34} \text{ J·s}) (100 \times 10^3 \text{ s}^{-1})$

$$= \boxed{6.6 \times 10^{-29} \text{ J}}$$

Since $1 \text{ J} = 6.24 \times 10^{18}$ eV, this is

$$E = (6.6 \times 10^{-29} \text{ J}) \left(6.24 \times 10^{18} \frac{\text{eV}}{\text{J}}\right)$$

$$= \boxed{4.1 \times 10^{-10} \text{ eV}}$$

(b) For a wavelength of 500 nm

$$f = \frac{c}{\lambda} = \frac{3.0 \times 10^8 \text{ m/s}}{500 \times 10^{-9} \text{ m}} = 6.0 \times 10^{14} \text{ Hz}$$

and

$$E = hf = (6.6 \times 10^{-34} \text{ J·s}) (6.0 \times 10^{14} \text{ s}^{-1}) = \boxed{4.0 \times 10^{-19} \text{ J}}$$

or

$$E = (4.0 \times 10^{-19} \text{ J}) \left(6.24 \times 10^{18} \frac{\text{eV}}{\text{J}}\right) = \boxed{2.5 \text{ eV}}$$

(c) $E = hf = (6.6 \times 10^{-34} \text{ J·s}) (10^{20} \text{ s}^{-1})$

$$= \boxed{6.6 \times 10^{-14} \text{ J}}$$

or

$$E = (6.6 \times 10^{-14} \text{ J}) \left(6.24 \times 10^{18} \frac{\text{eV}}{\text{J}}\right) = 4.1 \times 10^5 \text{ eV}$$

$$= \boxed{0.41 \text{ MeV}}$$

Notice how the energy of a photon differs by 15 orders of magnitude between the radio-frequency and gamma-ray regions of the spectrum. This explains why radio waves and gamma rays interact so differently with matter.

26.3 The Photoelectric Effect

The logical conclusion from Planck's work on blackbody radiation was that an oscillator in the walls of a cavity radiates electromagnetic energy into the cavity in the form of energy quanta. We might then expect that the interaction of electromagnetic radiation with matter would also be quantized rather than continuous.

The *photoelectric effect* is the name given to the release of electrons from a clean metal surface when electromagnetic radiation of the proper frequency shines on it. The photoelectric effect was discovered by Heinrich Hertz in 1887 during his attempts to produce electromagnetic waves with oscillating electric circuits. Apparatus which can be used for demonstrating the photoelectric effect is shown in Fig. 26.7. Light shines on the cathode C, which is at a negative potential with respect to the anode A. The light ejects from the cathode electrons, which are then collected by the anode. These electrons flow through the complete circuit back to the cathode and thereby constitute an electric current which can be measured with the meter M. (This is one type of circuit used in burglar-alarm systems. An intruder cutting off the light beam stops the current and triggers the alarm.)

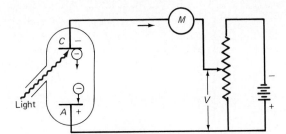

FIGURE 26.7 Apparatus used for experiments on the photoelectric effect.

The kinetic energy of the emitted electrons can be measured by reversing the battery connections between C and A. If the voltage V (the *retarding potential*) is then increased to a point where no electrons reach the anode, the initial kinetic energy of the most energetic electrons coming off the cathode is just reduced to zero by the electrons' having to climb up the potential "hill" between C and A. If this happens for an applied voltage V_0, then we have

$$eV_0 = \tfrac{1}{2}mv_{max}^2 \tag{26.3}$$

where V_0 is called the *stopping*, or *cutoff*, *potential*, and v_{max} is the speed of the most energetic electrons emitted by the cathode.

Experimental Facts about the Photoelectric Effect

When experiments on the photoelectric effect are carried out, the following results are obtained:

1 The photocurrent begins to flow almost *instantaneously*, once light is incident on the cathode, no matter how weak the light, as long as the light frequency is sufficiently high. The time delay is measured to be less than 10^{-10} s.

2 The photocurrent i is directly proportional to the intensity I of the light, as shown in Fig. 26.8 for a constant light frequency f and a constant retarding potential.

3 For any particular material, there is a cutoff frequency f_0 below which no electrons are emitted by the light, no matter how great its intensity.

4 The maximum kinetic energy of the emitted electrons depends only on the frequency of the light, not on its intensity.

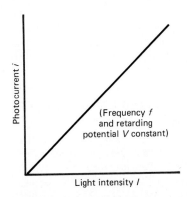

FIGURE 26.8 Dependence of photocurrent on light intensity for the photoelectric effect.

Figure 26.9 shows experimental results for three metals. The value of the voltage V_0 required to cut off the electron flow (times the constant charge e on the electron) is seen to be proportional to the frequency f. For different materials the curves remain parallel to each other, but are displaced so that the intercept on the X axis, f_0, is different for different materials. These curves may be represented by the general equation for a straight line.

$$eV_0 = hf - hf_0 \tag{26.4}$$

where h is a constant still to be determined. This equation rightly predicts that, when $f = f_0$, $eV_0 = 0$, as is clear from the figure. This frequency f_0 is called the *threshold frequency*, since frequencies below f_0 cannot eject electrons no matter how intense the light.

Turning Eq. (26.4) around, and replacing eV_0 by $\tfrac{1}{2}mv_{max}^2$ from Eq. (26.3), we have

FIGURE 26.9 Experimental results on the photoelectric effect for cesium, copper, and platinum cathodes. The value of the retarding potential needed to cut off the electron flow is plotted as a function of the frequency of the incident light. In each case the slope of the curve is $h = \dfrac{eV_0}{f - f_0}$

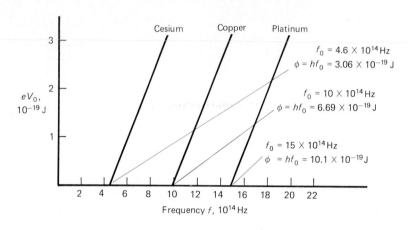

$$hf = hf_0 + \tfrac{1}{2}mv_{\max}^2 \tag{26.5}$$

Classical physics could not explain the photoelectric effect, since it predicted that, for low-intensity light, it would take a long time for an electron to accumulate enough energy to escape from the metal. Also, according to classical physics, the emission of electrons and their subsequent kinetic energies would depend only on the intensity of the light and not at all on the frequency, which is in direct conflict with Eq. (26.5).

Einstein's Quantum Explanation of the Photoelectric Effect

In 1905 Albert Einstein explained the photoelectric effect in a very simple, elegant fashion. He assumed that the incident light consisted of photons of energy $E = hf$, and that individual photons collided with individual electrons in the metal and knocked them out of the metal by giving up to the electrons their *entire* energy. Einstein then wrote a conservation-of-energy equation for this collision of a photon and an electron:

Einstein Photoelectric Equation

$$\boxed{hf = \phi + \tfrac{1}{2}mv_{\max}^2} \tag{26.6}$$

Einstein interpreted this equation in the following way. An incident photon has an energy $E = hf$ that it gives up to a *single* electron in a collision process. Some of this energy goes into overcoming the attraction of the metal for the electron, and so the electron must do an amount of work ϕ, called the *work function* of the metal, in order to escape. After escaping, an electron near the surface has a kinetic energy $\tfrac{1}{2}mv_{\max}^2$. Electrons in the interior of the metal must do additional work to escape from the metal, and come off with kinetic energies $\tfrac{1}{2}mv^2$, where v is less than the maximum speed $v_{\max}$ of the surface electrons. If the work function ϕ is set equal to hf_0, then Eq. (26.5) can be interpreted as follows for the most energetic electrons:

TABLE 26.1 Photoelectric Work Function ϕ for Clean Surfaces of Some Important Metals

Metal	Photoelectric work function ϕ	
	eV	10^{-19} J
Cesium	1.91	3.06
Rubidium	2.10	3.36
Potassium	2.26	3.62
Aluminum	2.98	4.77
Lead	4.01	6.42
Copper	4.18	6.69
Tungsten	4.58	7.33
Silver	4.73	7.57
Gold	4.82	7.71
Platinum	6.30	10.08

$$
\begin{array}{ccc}
hf & = hf_0 & + \tfrac{1}{2}mv_{\max}^2 \\
\text{Energy of} & \text{energy required to remove} & \text{kinetic energy of} \\
\text{incident photon} = & \text{electron from metal} + & \text{emerging electron}
\end{array}
$$

which is Eq. (26.5). Values of the work function, in both joules and electronvolts, for some important metals are given in Table 26.1.

It can be seen that the intensity of the light plays no role in Eq. (26.5). Intensity, however, determines the number of photoelectrons emerging, as long as the light frequency is high enough to eject electrons at all.

Table 26.2 compares the experimental facts with the predictions of both the classical and the quantum theories for the photoelectric effect. It makes clear the superiority of Einstein's theory over a classical theory in explaining the observed facts.

TABLE 26.2   Experimental Results and Theoretical Predictions for the Photoelectric Effect

Experimental facts	Classical electromagnetic theory*		Einstein's quantum theory*	
1 Photocurrent flows almost instantaneously; delay less than 10^{-10} s.	$\ominus$	For low light intensities delays of hours are to be expected.	$\oplus$	Emission of electrons is produced by photon-electron collisions which occur almost instantaneously, no matter what the light intensity.
2 Photocurrent i is directly proportional to light intensity I.	$\oplus$	The greater the light intensity, the greater the energy available to release electrons. Hence the current is greater.	$\oplus$	The greater the light intensity, the greater the number of photons. Hence the more electrons emitted (*if frequency is above threshold*) and the greater the current.
3 Cutoff potential V_0 depends only on frequency of the light and not on the intensity.	$\ominus$	Cutoff potential should depend only on the intensity of the light and not on the frequency.	$\oplus$	The energy of a photon is $E = hf$. If this is less than the work function $\phi = hf_0$, the photon has insufficient energy to eject an electron, no matter what the intensity.
4 Maximum kinetic energy of electrons depends only on the frequency, not on the intensity of the ight.	$\ominus$	Maximum kinetic energy of electrons should depend only on the intensity of the light, not on the frequency.	$\oplus$	$\frac{1}{2}mv_{max}^2 = hf - hf_0$; the intensity has no effect on $\frac{1}{2}mv_{max}^2$.

*$\oplus$ indicates agreement of theory with experiment; $\ominus$ indicates disagreement.

Millikan's Experiments

The experimental work which established Einstein's theory of the photoelectric effect was done by the American physicist Robert A. Millikan (see Fig. 16.14) during the years 1912 to 1916. The apparatus used for this work, for which Millikan received the Nobel Prize in 1923, is sometimes referred to as "a machine shop in a vacuum." Millikan continually had to slice off surface layers of atoms by remote control inside a complicated vacuum system to maintain the ultraclean metal surfaces he needed for his experiment. Millikan's results consisted of graphs similar to those of Fig. 26.9, where the slope of the curve is, from Eq. (26.4), $h = eV_0/(f - f_0)$. Hence h can be obtained from the slope of these experimental curves.

Millikan's value for h, as obtained in this way, was 6.569×10^{-34} J·s. The excellent agreement of this value for h with the value required to make Planck's equation fit the experimental data on blackbody radiation indicated the strong probability that both Planck's and Einstein's theories were correct and that light did indeed consist of photons of energy $E = hf$. A great variety of additional experimental evidence, some of which will be presented in subsequent sections, has confirmed this view and has led to a more accurate value of 6.6262×10^{-34} J·s for Planck's constant.

Example 26.2

(a) Blue light of wavelength 436 nm and intensity 1.5×10^3 W/m² falls on a piece of lithium metal in a vacuum. If the work function of lithium is 3.8×10^{-19} J, what is the maximum kinetic energy of the electrons emitted?

(b) What is the maximum speed of the emitted electrons?
(c) What is the longest wavelength which will eject electrons from the metal?

SOLUTION

The intensity of the light has no influence on the kinetic energy of the ejected electrons. The intensity of the light changes only the *number* of electrons emitted, as long as the photons have sufficient energy to eject any electrons at all from the metal.

(a) By Einstein's photoelectric equation,

$$hf = hf_0 + \tfrac{1}{2}mv_{max}^2$$

and so

$$\tfrac{1}{2}mv_{max}^2 = hf - hf_0 = \frac{hc}{\lambda} - \phi$$

$$= \frac{(6.6 \times 10^{-34} \text{ J·s})(3.0 \times 10^8 \text{ m/s})}{436 \times 10^{-9} \text{ m}} - 3.8 \times 10^{-19} \text{ J}$$

$$= 4.5 \times 10^{-19} \text{ J} - 3.8 \times 10^{-19} \text{ J} = \boxed{0.7 \times 10^{-19} \text{ J}}$$

(b) $v_{max} = \left(\dfrac{1.4 \times 10^{-19} \text{ J}}{9.1 \times 10^{-31} \text{ kg}}\right)^{1/2} = \boxed{3.9 \times 10^5 \text{ m/s}}$

(c) The longest wavelength which can eject electrons is that value of λ for which hc/λ is equal to the work function ϕ. Hence we have

$$\frac{hc}{\lambda} = \phi$$

or $\quad \lambda = \dfrac{hc}{\phi} = \dfrac{(6.6 \times 10^{-34} \text{ J·s})(3.0 \times 10^8 \text{ m/s})}{3.8 \times 10^{-19} \text{ J}}$

$$= 5.2 \times 10^{-7} \text{ m} = \boxed{522 \text{ nm}}$$

Hence only light of wavelength *less than 522 nm* (in the green) has photons with enough energy to eject electrons from lithium, irrespective of the intensity of the light.

26.4 X-Rays   omit

Other evidence for the photon nature of light (and other electromagnetic radiation) came from the continuous x-ray spectrum.

Discovery of X-Rays

In 1895 the German physicist Wilhelm Konrad Roentgen (see Fig. 26.10) was doing some experiments at the University of Würzburg on the effect of cathode rays (i.e., electron bombardment) on the luminescence of certain chemicals. He had a cathode-ray tube enclosed in black cardboard in a darkened room. When he turned on the voltage across the tube, he noticed a flash of light from a sheet of paper coated with barium platinocyanide at the other side of the room. This mystified him, since the cathode-ray beam was blocked off by the cardboard and could not possibly cause this light emission directly. A series of experiments led Roentgen to conclude that the cathode-ray tube must be producing a radiation that was invisible but very penetrating. Roentgen decided to call this radiation *x-radiation*, or *x-rays*, since x is the typical mathematical symbol for an unknown quantity.

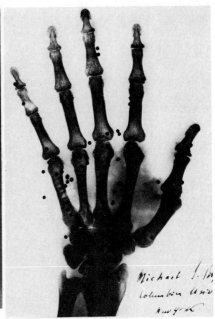

FIGURE 26.10 Wilhelm Roentgen (1845–1923), who received the first Nobel Prize in physics, in 1901, for his discovery of x-rays. He refused to patent any aspect of x-ray production so that x-rays could be used freely for the good of humanity. *(Photo courtesy of AIP Niels Bohr Library, W. F. Meggers Collection.)*

FIGURE 26.11 An x-ray photograph of a human hand. In 1896 Professor Michael I. Pupin of Columbia University first heard of Roentgen's discovery of x-rays from a physicist friend in Germany. He immediately set up equipment for x-ray experiments and in February 1896 was asked to take x-ray pictures of the hand of a New York lawyer who had been shot accidentally. This photo, which shows clearly the shotgun pellets in the hand, was one of the first x-ray photos taken in the United States. *(Photo courtesy of the Burndy Library.)*

Roentgen soon found that the x-rays could pass through paper and thin layers of metal but not through very thick layers of metal, and that they could ionize gases but were completely unaffected by electric and magnetic fields. This seemed to indicate that they were not charged particles.

On January 23, 1896, Roentgen gave his first lecture on x-rays. After his talk he asked for a volunteer from the audience, and a physiologist on the medical faculty at Würzburg stepped up. Roentgen took an x-ray photograph of his colleague's hand and passed it around the room. The audience broke into wild applause. The next day Roentgen was famous, for everyone saw the tremendous practical applications of these rays that could pass through the soft tissues of the body but cast shadows of bone or heavy foreign materials like bullets or pins embedded in human tissue.* Physics laboratories all over the world turned their attention to x-rays, with resulting gains both for pure science and for diagnostic medicine. A typical x-ray photograph is shown in Fig. 26.11.

The Continuous X-Ray Spectrum

An x-ray tube of the type developed in 1913 by the American physicist William D. Coolidge (1873–1975) is shown in Fig. 26.12. Electrons are accelerated by a high voltage from a heated filament, which acts as the cathode, to a metal target, which is the anode. The electrons plunge into the target at high speeds and produce x-rays in the collision process. The wavelength distribution of these x-rays can be determined using an x-ray spectrometer (see Fig. 26.23). The resulting plots of intensity against wavelength take the form of the continuous curves shown in Fig. 26.13. (There are sharp lines superimposed on this continuous spectrum, but we will defer a discussion of these lines until later.)

---

*Some members of the New Jersey legislature were so disturbed by the antiprivacy implications of x-rays that they tried to pass a law banning the use of x-ray opera glasses!

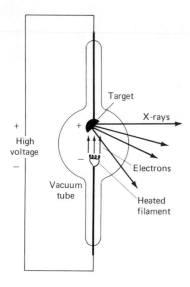

FIGURE 26.12 A Coolidge x-ray tube. Energy lost by the electrons on suddenly being brought to rest by the target is converted into x-ray photons.

Experimentally the following facts have been established about the continuous x-ray spectrum:

1 The wavelength characteristics of the *continuous spectrum* are completely independent of the target material and depend only on the voltage applied to the x-ray tube. The higher the voltage, the shorter the minimum (or cutoff) wavelength produced. The variation of cutoff wavelength with voltage can be seen in Fig. 26.13.

2 The intensity of the continuous spectrum depends both on the target material and thickness, and on the applied voltage.

3 The cutoff wavelength λ_0 is related to the applied voltage V by the empirical equation:

$$\lambda_0 = \frac{1.24 \times 10^{-6} \text{ m}}{V} \qquad (26.7)$$

where V is in volts.

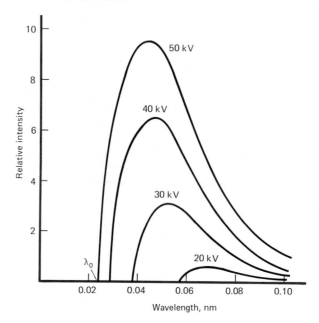

FIGURE 26.13 The continuous x-ray spectrum arising from using different accelerating voltages for the electrons producing the x-rays.

Explanation of Continuous X-Ray Spectrum

The sharp cutoff wavelength λ_0 is hard to explain on the basis of classical theory, but quantum theory clarifies the situation. The continuous x-ray spectrum is produced by what is usually called *braking radiation* (in German, *Bremsstrahlung*), which arises when the electrons plunge into the target, collide with atoms there, and rapidly decelerate. In this process much of the kinetic energy of the electrons is converted into heat, but some of it is radiated away in the form of x-ray *photons* of energy $hf = \Delta E$, where ΔE is the energy lost by a particular electron in a collision with a target atom. The amount of energy lost (ΔE) cannot exceed the total kinetic energy which the electron had before the collision, but it can have any value smaller than this,

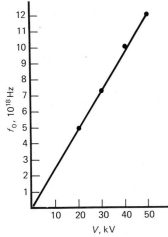

FIGURE 26.14 Graph of cutoff frequency f_0 as a function of the voltage V applied to an x-ray tube.

for multiple collisions can lead to the emission of smaller quanta in each collision. Since the largest possible value of ΔE is set by the accelerating voltage V, we have for the cutoff frequency:

$$hf_0 = \Delta E = eV \tag{26.8}$$

where V is the applied voltage to the x-ray tube. Hence

$$f_0 = \frac{e}{h}V \tag{26.9}$$

and a plot of f_0 versus V is a straight line with slope e/h, as shown in Fig. 26.14. From such a plot a value of e/h can be found and used as a check on the values of e and h already obtained from Millikan's oil-drop experiment and his work on the photoelectric effect.

If we use Eq. (26.8) in the expression for the cutoff wavelength, λ_0, we find

$$\lambda_0 = \frac{c}{f_0} = \frac{hc}{hf_0} = \frac{hc}{eV}$$

or

$$\lambda_0 = \frac{(6.63 \times 10^{-34}\ \text{J}\cdot\text{s})(3.00 \times 10^8\ \text{m/s})}{(1.60 \times 10^{-19}\ \text{C})V} = \frac{1.24 \times 10^{-6}\ \text{m}}{V} \tag{26.10}$$

where again V is in volts. This agrees perfectly with the empirical equation [Eq. (26.7)]. Hence for a voltage of 50 kV applied to the x-ray tube, the cutoff wavelength should be 2.5×10^{-11} m $= 0.025$ nm, in agreement with the experimental cutoff wavelength shown in Fig. 26.13. This wavelength is about 10^4 times shorter than that of visible light, and typical x-ray wavelengths fall in the range from 10^{-2} to 10 nm.

The sharp cutoff in the continuous x-ray spectrum at short wavelengths can therefore be well explained by a quantum or photon theory, whereas classical physics was unable to provide a convincing explanation for this phenomenon.

Example 26.3

An x-ray tube has a voltage of 100 kV applied between the cathode and the steel plate that serves as the target producing the x-rays.
(a) What is the highest-frequency photon to be expected in the x-ray beam?
(b) What is the cutoff wavelength of the continuous x-ray spectrum?

SOLUTION

(a) Since the energy of a single photon cannot be greater than the energy of an electron crashing into the target, and this electron energy is determined by the applied voltage, we must have for the highest-energy photon:

$$hf_0 = eV$$

or

$$f_0 = \frac{eV}{h} = \frac{(1.60 \times 10^{-19}\ \text{C})(100 \times 10^3\ \text{V})}{6.63 \times 10^{-34}\ \text{J}\cdot\text{s}}$$

$$= \boxed{2.41 \times 10^{19}\ \text{Hz}}$$

(b) The cutoff wavelength is the wavelength corresponding to the highest-energy photon produced in the braking radiation. Hence

$$\lambda_0 = \frac{c}{f_0} = \frac{3.00 \times 10^8\ \text{m/s}}{2.41 \times 10^{19}/\text{s}} = 1.24 \times 10^{-11}\ \text{m} = \boxed{0.0124\ \text{nm}}$$

The same result could, of course, have been obtained directly from Eq. (26.10).

$$\lambda_0 = \frac{1.24 \times 10^{-6}\ \text{m}}{V} = \frac{1.24 \times 10^{-6}\ \text{m}}{100 \times 10^3} = 1.24 \times 10^{-11}\ \text{m}$$

$$= \boxed{0.0124\ \text{nm}}$$

26.5 The Compton Effect

A fourth piece of evidence on the nature of the photon was provided by another American physicist, Arthur Holly Compton (see Fig. 26.15) in 1922. Compton was studying the scattering of x-rays by a carbon target, and directed a beam of x-rays with a discrete wavelength of 0.0710 nm at a carbon target. He found that, in addition to x-rays scattered with this original wavelength unchanged, other x-rays appeared with wavelengths of 0.0734 nm on the long-wavelength side of the original wavelength. The name *Compton effect* has been given to this increase in the wavelength of electromagnetic radiation when scattered by free electrons.

Compton could not explain this result classically and therefore set out to see if some of the newer ideas of Planck and Einstein about photons might work. He assumed that photons could be treated as particles carrying energy and momentum, and that the elastic collision between a photon and a free electron caused the photon to have its energy and hence its frequency decreased, and its wavelength therefore increased. (Notice that the binding energy of an electron in a carbon atom is so small compared with the energy of an x-ray photon that the electron can be considered *free*; note also that the photon's speed cannot change, since it always moves at the speed of light.) Compton then applied the laws of conservation of energy and momentum to the collision process* and found that the predictions of quantum theory did indeed agree with his experimental results.

---

*A major difference between the photoelectric effect and the Compton effect should be emphasized. In the photoelectric effect, in which the electron is *bound* to the metal, both energy and momentum can be conserved when the incident light is completely absorbed by the metal. In the Compton effect, where the electron is essentially *free*, on the other hand, both energy and momentum *cannot* be conserved unless the photon remains in existence after the collision.

FIGURE 26.15 Arthur Holly Compton (1892–1962). For his discovery of the effect that now bears his name he received the Nobel Prize in physics in 1927. Compton also did important work in ascertaining the nature of cosmic rays and was one of the top scientists involved in the Manhattan project to develop the atomic bomb during World War II. *(Photo courtesy of AIP Niels Bohr Library.)*

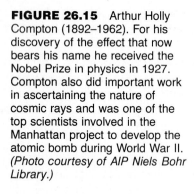

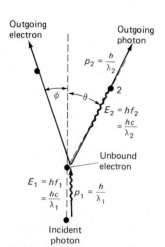

FIGURE 26.16 The Compton effect: The collision between an x-ray photon of frequency f_1 and an electron at rest. The initial frequency f_1 of the photon is decreased to f_2 after it is scattered.

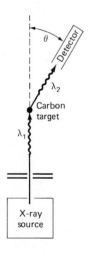

FIGURE 26.17 Sketch of Compton's apparatus for the study of the Compton effect. X-rays of wavelength λ_1 are scattered from the carbon target. The wavelength λ_2 of the scattered x-rays is measured by the detector.

To treat the photon as a particle colliding with an electron, we need formulas for the energy and momentum of a photon. The photon energy is given simply by Planck's expression, $E = hf$. The equivalent mass of the photon can be obtained by assuming that Einstein's mass-energy equation also applies here, so that $E = mc^2$, and $m = E/c^2$ is the equivalent photon mass. The momentum is then, as always, the product of mass and velocity (c for a photon), so that the photon is seen to have a momentum:

$$p = mc = \frac{Ec}{c^2} = \frac{E}{c} = \frac{hf}{c} = \frac{h}{\lambda}$$

or

$$\boxed{p = \frac{h}{\lambda}}$$

(26.11)

The momentum of a photon is equal to Planck's constant divided by the photon wavelength. (Notice how we are again mixing together particle properties like momentum and wave properties like wavelength.)

Compton then applied the laws of conservation of energy and conservation of linear momentum to the collision between a photon and an electron, as shown in Fig. 26.16, and obtained the following result:

$$\Delta\lambda = \lambda_2 - \lambda_1 = \frac{h}{m_0 c}(1 - \cos\theta)$$

(26.12)

in which λ_1 is the wavelength of the incident x-ray photon, and λ_2 is the wavelength of the scattered x-ray photon. Equation (26.12) is the equation for the Compton effect, and predicts an increase in the wavelength of the scattered photon which is a function of the angle θ at which the photon is scattered. The shift in wavelength when $\theta = 90°$ and $\cos\theta = 0$ is

$$\Delta\lambda = \frac{h}{m_0 c}(1 - 0) = \frac{6.63 \times 10^{-34} \text{ J·s}}{(9.11 \times 10^{-31} \text{ kg})(3.00 \times 10^8 \text{ m/s})}$$

$$= 2.43 \times 10^{-12} \text{ m} = 0.00243 \text{ nm}$$

The quantity $h/m_0 c = 0.00243$ nm is called the *Compton wavelength of the electron.* The Compton shift is very small compared with the wavelength of visible light (~500 nm), but for x-rays of wavelength 0.0710 nm it amounts to a few percent of the original wavelength and can be easily measured with an x-ray spectrometer of sufficient resolution. When the calculated shift of 0.00243 nm is added to $\lambda_1 = 0.0710$ nm in Compton's experiment, the resulting value is $\lambda_2 = 0.0734$ nm, in excellent agreement with experiment.

The experimental setup used by Compton is shown in Fig. 26.17. In practice both an unshifted line and a shifted line are observed. The unshifted line is due to electrons in the atoms whose binding energies are such that they are not ejected from the atom. The change in the wavelength of the shifted line increases with the scattering angle for angles from 0 to 180°, in accordance with Compton's theoretical prediction, Eq. (26.12), as shown in Fig. 26.18.

In his original paper Compton pointed out that "the wavelength and the intensity of the scattered rays are what they should be if a quantum of radiation bounced from an electron, just as one billiard ball bounces from another . . . . The obvious conclusion would be that x-rays, and so also light,

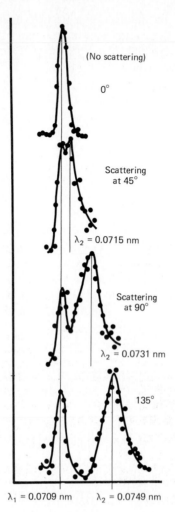

(No scattering)

0°

Scattering
at 45°

$\lambda_2 = 0.0715$ nm

Scattering
at 90°

$\lambda_2 = 0.0731$ nm

135°

$\lambda_1 = 0.0709$ nm $\lambda_2 = 0.0749$ nm

FIGURE 26.18 Experimental data on the Compton effect. These data are taken from Compton's original paper. Wavelength is plotted along the X axis and intensity along the Y axis. The top graph shows the unscattered radiation, the other graphs the radiation scattered at 45, 90, and 135°. The wavelength of the scattered x-rays increases with angle, as predicted by Compton's theory.

consist of discrete units, proceeding in definite directions, each unit possessing the energy hf and the corresponding momentum h/λ." This goes beyond the concept of the photon which emerged from the work of Planck and Einstein. The photon now is seen to have inertia and momentum and therefore many of the same properties as a billiard ball or a bullet.

Some physicists believed the Compton effect sounded the death knell for the wave theory of light, but they were underestimating the durability of that venerable theory.

Example 26.4

X-radiation of wavelength 0.112 nm is scattered from a carbon target.

(a) Calculate the wavelength of the x-rays scattered at an angle of 90° with respect to the original direction.

(b) Calculate the energy of the scattering electron after the collision.

(c) Calculate the direction of the scattering electron after the collision.

SOLUTION

(a) We presume that the scattering is due to collisions of the x-ray photons with electrons which are essentially free. Then the situation is well described by the equation for the Compton effect:

$$\Delta\lambda = \frac{h}{m_0 c}(1 - \cos\theta)$$

Here $\theta = 90°$, $\cos\theta = 1$, and so $\Delta\lambda = h/m_0 c = 0.0024$ nm.

Hence

$$\lambda_2 = \lambda_1 + \Delta\lambda = 0.112 \text{ nm} + 0.0024 \text{ nm} = \boxed{0.114 \text{ nm}}$$

(b) The energy given to an electron is equal to the energy lost by the x-ray photon, since energy must be conserved in the collision. This is

$$\Delta E = hf_1 - hf_2 = (6.6 \times 10^{-34} \text{ J·s}) \left(\frac{c}{\lambda_1} - \frac{c}{\lambda_2}\right)$$

$$= (6.6 \times 10^{-34} \text{ J·s})(3.0 \times 10^8 \text{ m/s})$$

$$\times \left(\frac{1}{0.112 \text{ nm}} - \frac{1}{0.114 \text{ nm}}\right)$$

$$= (1.98 \times 10^{-25} \text{ J·m})(0.16 \times 10^9 \text{ m}^{-1}) = \boxed{3.2 \times 10^{-17} \text{ J}}$$

Note that 3.2×10^{-17} J is equal to about 200 eV, which is roughly 200 times the binding energy of the valence electrons in carbon. Hence the assumption that the electrons are essentially free is justified.

(c) The x-ray photon has an initial momentum h/λ_1 along the positive X axis; its final momentum is h/λ_2 along the positive Y axis, since it is scattered through 90°, as shown in Fig. 26.19. To conserve momentum the electron must have a momentum after the collision with an X component $mv \cos \phi$ numerically equal to h/λ_1 along the positive X axis, and a Y component $mv \sin \phi$ numerically equal to h/λ_2 along the negative Y axis (since the initial Y component of momentum for the system is zero), as shown in the figure. Then, from the figure,

$$\tan \phi = \frac{mv \sin \phi}{mv \cos \phi} = \frac{h/\lambda_2}{h/\lambda_1} = \frac{\lambda_1}{\lambda_2} = \frac{0.112 \text{ nm}}{0.114 \text{ nm}} = 0.982$$

and so $\phi = \boxed{44.5°}$

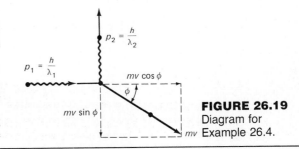

FIGURE 26.19
Diagram for
Example 26.4.

26.6 The Nature of Light

In 1920 physicists believed that the two aspects of reality represented by waves and particles were mutually exclusive and that no particle could exhibit wave properties and no wave could exhibit particle properties. The experimental results on blackbody radiation, the photoelectric effect, the continuous x-ray spectrum, and the Compton effect therefore tended to reduce physicists to a state of confusion, if not shock, because electromagnetic radiation now seemed to have both wave and particle properties.

We have seen that light undergoes diffraction and interference, which can easily be explained on the wave model of light. The phenomenon of polarization adds additional evidence that light is a wave, in fact, a particular kind of a wave—a transverse wave. The work of Planck, Einstein, and Compton seemed to show, however, that light consisted of beams of photons whose interaction with matter could be correctly predicted only by a particle model. How can we reconcile these two seemingly contradictory models of radiation?

Our approach to this very difficult question will be to postpone any attempt at an answer until we introduce a further complication with respect to particles such as electrons, protons, or hydrogen atoms. It is as if we are trying to piece together a gigantic jigsaw puzzle broken down into two separate sections. Some pieces from one section of the puzzle are mixed up with the other section, however, so that neither part can be solved by itself. If, however, the two sections are combined once again, the puzzle might be solvable. The nature of radiation and matter may be similar to our divided jigsaw puzzle. Perhaps by concentrating on radiation alone we are leaving out some key pieces of the puzzle. Perhaps we should pay more attention to the *nature of the particles* which make up matter.

26.7 Wave Properties of Matter: De Broglie Waves

As already mentioned, the concept of symmetry has played an important role in the development of physics. In 1923 to 1924 L. de Broglie (see Fig. 26.20) was writing a Ph.D. dissertation for the Sorbonne in Paris and was looking for a parallelism between particles and waves. If, as Fig. 26.21 tries to indicate, radiation has both wave and particle properties, should not matter also have wave properties? If a photon of wavelength λ has a momentum $p = h/\lambda$, why should not a particle of momentum $p = mv$ have an associated wavelength $\lambda = h/p$? This was de Broglie's line of reasoning, and it was so revolutionary that there was some skepticism on the part of his examiners. It was only after one of the examiners, Paul Langevin, sent de Broglie's thesis to Einstein and found him enthusiastic about it, that this skepticism disappeared.

The name *de Broglie wave* is given to the wave associated with every particle having a momentum p, where the wavelength is $\lambda = h/p$. It should be pointed out that there was no known experimental evidence for such an idea at the time de Broglie developed it.

FIGURE 26.20 Louis Prince de Broglie (born 1892), French theoretical physicist, who received the 1929 Nobel Prize in physics for his prediction of the wave properties of material particles. *(Photo courtesy of AIP Niels Bohr Library.)*

FIGURE 26.21 The wave and particle properties of electromagnetic radiation and of matter. The question mark means that we might be led by symmetry considerations to suspect that matter also has wave properties.

	Wave properties	Particle properties
Electromagnetic radiation (radio waves, light waves, x-rays, gamma rays)	Wavelength $\lambda f = c$ Diffraction Interference Polarization	$E = hf$ $p = \dfrac{h}{\lambda}$ Blackbody radiation Photoelectric effect Compton effect
Matter (electrons, protons, atoms, molecules)	?	$KE = \frac{1}{2}mv^2$ $p = mv$ Conservation of energy Conservation of momentum

The wavelengths which de Broglie's theory predicts for macroscopic objects are much too small to have observable effects. Thus for a 100-g ball moving at a speed of 25 m/s, the *de Broglie wavelength* is

De Broglie Wavelength

$$\lambda = \frac{h}{p} = \frac{h}{mv}$$

(26.13)

or $\quad \lambda = \dfrac{6.6 \times 10^{-34} \text{ J·s}}{(0.10 \text{ kg})(25 \text{ m/s})} = 2.6 \times 10^{-34} \text{ m}$

To produce diffraction of such de Broglie waves, diffraction gratings with separations between rulings of about 10^{-34} m would be needed. These are impossible to obtain, since the required separations are smaller than the dimensions of the atoms making up the grating by about 24 orders of magnitude!

If, however, we consider an *electron* moving with a speed of 10^6 m/s, which is 1/300 of the speed of light, then its de Broglie wavelength is:

$$\lambda = \frac{h}{mv} = \frac{6.6 \times 10^{-34} \text{ J·s}}{(9.1 \times 10^{-31} \text{ kg})(10^6 \text{ m/s})} = 7.3 \times 10^{-10} \text{ m}$$

This wavelength of 0.73 nm is of the same order of magnitude as the separation of planes of atoms in crystals, which are about 0.2 to 0.7 nm apart, and hence it seems possible to produce electron diffraction and interference with such crystals for electrons with the proper wavelength. We will see how this was actually done in Sec. 26.8.

Diffraction and Interference of X-Rays

Before discussing the diffraction and interference of de Broglie waves let us digress to consider the related phenomenon for x-rays. In 1912 the German physicist Max von Laue (1879–1960) had predicted that x-rays could be diffracted by solid crystals with a spacing between atom planes of about the same size as the x-ray wavelength, in the same way that visible light can be diffracted by gratings. This was confirmed in 1912 by the British physicists William H. Bragg (1862–1942) and his son, William L. Bragg (1890–1971).

In a crystal such as sodium chloride (NaCl) the lattice spacing between adjacent atoms can be found from the density of the crystal and the atomic masses of sodium and chlorine. It turns out to be $d = 0.282$ nm for NaCl (see Example 26.5 for a description of how this result is obtained). Now, if x-rays with wavelengths of 0.2 nm are incident on a salt crystal, each atom acts as a scattering center and radiates waves in all directions. The atoms in any one plane, however, radiate x-rays coherently in the same way a mirror reflects light. These lattice planes are called *Bragg planes*, and reflections from such planes are called *Bragg reflections*. If x-rays are reflected from two adjacent, parallel Bragg planes, constructive interference occurs for certain angles of incidence of the x-rays on the crystal planes.

Consider the Bragg planes marked 1 and 2 in Fig. 26.22. Let a beam of x-rays of wavelength λ fall on the crystal at an angle θ with respect to the Bragg planes, where here θ is the angle between the direction of the x-ray beam and the Bragg plane (*not* the normal to the plane, as in the case of a mirror). Suppose an x-ray starting at A is reflected at plane 1, and an x-ray starting at B is reflected at plane 2. Then the difference in the path lengths

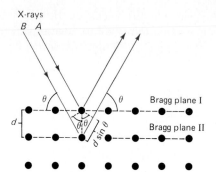

FIGURE 26.22 Reflection of x-rays by two adjacent Bragg planes in a crystal lattice.

between the two reflected x-rays is $d \sin \theta + d \sin \theta = 2d \sin \theta$, where d is the distance between the planes. Constructive interference will only occur between these two x-rays for angles for which

Bragg Equation

$$2d \sin \theta = n\lambda$$

(26.14)

where n is a small integer $(1, 2, 3, \ldots)$ and λ is the x-ray wavelength. This is the Bragg condition for constructive interference. For x-rays incident at any other angle than one which satisfies this equation, the reflected x-rays will interfere destructively and no reflected beam will be observed.

The Bragg equation thus gives us a way of determining the wavelength of x-rays if the atomic separations are known, or of determining d if the wavelength of the x-rays is known. An x-ray spectrometer similar to the one in Fig. 26.23 can be used for this purpose. The x-rays are collimated into a beam by slits in lead plates. The angle of incidence θ is gradually changed, and maximum intensities are then observed for angles θ satisfying the Bragg equation. Using x-ray diffraction photographs taken by Maurice Wilkins and Rosalind Franklin, in 1953 James D. Watson and Francis Crick worked out the famous double-helix structure of DNA.

FIGURE 26.23 An x-ray spectrometer.

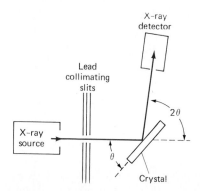

Example 26.5

The density of potassium chloride is 1.99 g/cm³ and its molecular mass 74.56 g. What is the distance between the atoms in the KCl crystal if it is a cubic structure with alternating K and Cl atoms?

SOLUTION

We first find the volume of a KCl crystal containing Avogadro's number of KCl molecules. This is, since the density $d = m/V$,

$$V = \frac{m}{d} = \frac{74.56 \text{ g}}{1.99 \text{ g/cm}^3} = 37.5 \text{ cm}^3$$

Now, this volume contains Avogadro's number of KCl molecules and hence contains 6.03×10^{23} potassium atoms and 6.03×10^{23} chlorine atoms. Hence the total number of atoms present in the crystal is $2(6.03 \times 10^{23})$. If we now consider each atom as being at the center of a cube in the crystal, the distance between atoms will be the length of a side of this cube. To find the volume of such a cube we merely divide the total volume by the total number of atoms present. We have

$$V = \frac{37.5 \text{ cm}^3}{2(6.03 \times 10^{23})} = 3.11 \times 10^{-23} \text{ cm}^3$$

Then the distance d between atoms will be the length of the side of a cube of this volume, or

$$d = (3.11 \times 10^{-23} \text{ cm}^3)^{1/3} = 3.14 \times 10^{-8} \text{ cm} = \boxed{0.314 \text{ nm}}$$

Example 26.6

A beam of x-rays is reflected from a NaCl crystal, whose lattice planes are separated by a distance $d = 0.282$ nm. The angle of incidence on the crystal planes which produces a strong reflected x-ray beam is 30°.

(a) What are the possible wavelengths of the x-rays used?
(b) How can we determine which of these possible wavelengths is the correct one?

SOLUTION

(a) The Bragg equation is applicable in this case, so that

$$n\lambda = 2d \sin \theta$$
$$= 2(0.282 \text{ nm})(0.50) = 0.282 \text{ nm}$$

The wavelength is not fully determined by this result unless the integer n, which specifies the order of interference, is known. For example, possible solutions are:

For $n = 1$, $\lambda = \boxed{0.282 \text{ nm}}$

For $n = 2$, $\lambda = \boxed{0.141 \text{ nm}}$

For $n = 3$, $\lambda = \boxed{0.094 \text{ nm}}$

and so on.
(b) To decide among the possible solutions found above, we could use the apparatus to see if any other strong reflections occur for values of θ less than 30°. If none is found, the presumption is that the beam incident at 30° corresponds to the first-order interference with $n = 1$, and that the x-ray wavelength is $\boxed{0.282 \text{ nm}}$.

26.8 Electron Diffraction and Interference

De Broglie's suggestion of wave properties for particles immediately set physicists to thinking about the possibility of electron diffraction and interference. If particles could be obtained with associated de Broglie waves of wavelength near 0.2 nm, the same kind of diffraction and interference should occur with crystals for electrons as had been observed for x-rays, as discussed in the preceding section. The wavelength of charged particles is also under the experimenter's control, for they can be accelerated to any desired speed and thus given any desired wavelength, since $\lambda = h/mv$.

Consider, for example, an electron which is accelerated by a voltage of 100 V. Its final energy is $\frac{1}{2}mv^2$, if we neglect relativistic effects, which are small in this case.* So we have

$$\frac{1}{2}mv^2 = eV$$

or $\quad v = \left(\frac{2eV}{m}\right)^{1/2} = \left[\frac{2(1.6 \times 10^{-19} \text{ C})(100 \text{ V})}{9.1 \times 10^{-31} \text{ kg}}\right]^{1/2} = 5.9 \times 10^6 \text{ m/s}$

The electron's de Broglie wavelength is therefore

$$\lambda = \frac{h}{mv} = \frac{6.6 \times 10^{-34} \text{ J·s}}{(9.1 \times 10^{-31} \text{ kg})(5.9 \times 10^6 \text{ m/s})} = 1.2 \times 10^{-10} \text{ m} = 0.12 \text{ nm}$$

which is the same order of magnitude as the lattice separation in crystals. (A general formula for the relationship between voltage and wavelength for electrons is $\lambda = 1.23$ nm/$\sqrt{V}$. See Example 26.7.)

Hence electrons can be easily produced with the proper wavelengths to make possible diffraction by crystals such as NaCl, Cu, or Ni. In this case the theory is exactly the same as for x-rays, and an equation identical with the Bragg equation results:

$$2d \sin \theta = n\lambda \tag{26.15}$$

In this equation d is the lattice spacing in the crystal, and λ is the de Broglie wavelength of the electrons.

Experimental Verification of de Broglie Waves

In 1927 an American physicist Clinton J. Davisson (1881–1952), with the assistance of his colleague L. H. Germer at what are now the Bell Telephone Laboratories, obtained data on electron beams diffracted by a single crystal of nickel in excellent agreement with de Broglie's predictions. In the same year additional confirmation of de Broglie's ideas came from the somewhat different work of George P. Thomson (1892–1975), the son of J. J. Thomson, at the University of Aberdeen in Scotland. For their verification of de Broglie's speculations Davisson and Thomson shared the 1937 Nobel Prize for

*The rest energy of an electron is 0.51 MeV. If the accelerating voltage for an electron is much less than 0.51 MV, relativistic effects are small.

FIGURE 26.24 Comparison of experimental photographs obtained by x-ray and electron diffraction: (a) The diffraction pattern obtained by scattering 0.071-nm x-rays from an aluminum foil. (b) The diffraction pattern obtained by scattering 600-eV electrons off aluminum. Since the wavelengths are different, the scales of the two have been adjusted to highlight the similarities. (Photo courtesy of Educational Development Center, Newton, Massachusetts.)

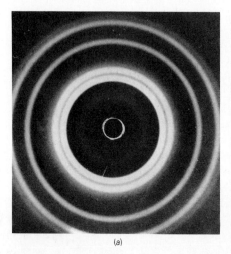

(a)

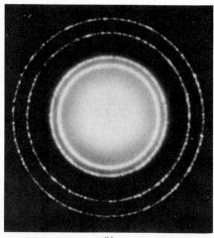

(b)

physics. The striking similarity of the patterns produced by a crystal for x-ray and electron diffraction is shown in Fig. 26.24. All this evidence points up the startling new fact that, under the proper circumstances, electrons exhibit *wave* properties.

Example 26.7

Davisson and Germer used 54-V electrons in their experiment on the electron scattering by nickel crystals. If such electrons are scattered off a nickel crystal for which the separation of the crystal planes is 0.266 nm, at what angles will maxima occur for the scattered electrons?

SOLUTION

From the de Broglie wavelength equation, the wavelength of the incident electrons is:

$$\lambda = \frac{h}{p} = \frac{h}{mv} = \frac{h}{m\sqrt{2eV/m}} = \frac{h}{\sqrt{2eVm}} = \frac{h}{\sqrt{2em}} \frac{1}{\sqrt{V}}$$

or

$$\lambda = \frac{1}{\sqrt{V}} \frac{6.63 \times 10^{-34} \text{ J·s}}{[2(1.60 \times 10^{-19} \text{ C})(9.11 \times 10^{-31} \text{ kg})]^{1/2}}$$

$$= \frac{1.23 \times 10^{-9} \text{ m}}{\sqrt{V}}$$

In this case,

$$\lambda = \frac{1.23 \times 10^{-9} \text{ m}}{\sqrt{54}} = 0.167 \text{ nm}$$

Hence the Bragg condition for constructive interference, $n\lambda = 2d \sin \theta$, yields for

$$n = 1: \quad \sin \theta_1 = \frac{\lambda}{2d} = \frac{0.167 \text{ nm}}{2(0.266 \text{ nm})} = 0.31$$

and so $\quad \theta_1 = 18°$

$$n = 2: \quad \sin \theta_2 = 2(0.31) = 0.62$$

and so $\quad \theta_2 = 38°$

$$n = 3: \quad \sin \theta_3 = 3(0.31) = 0.93$$

and so $\quad \theta_3 = 68°$

For $n = 4$ or higher, $\sin \theta > 1$, and the result is physically impossible. Hence strong electron beams are expected for angles of incidence of 18°, 38°, and 68°.

The Electron Microscope

The reality of de Broglie waves is made eminently clear in the electron microscope. Electrons accelerated through 50,000 V have wavelengths 100,000 times shorter than green light. Since the resolving power of any microscope is limited by the wavelength of the light used, electron microscopes have the great advantage of practical resolving powers better by orders of magnitude than the best optical microscopes.

Transmission Electron Microscope   The transmission electron microscope (TEM) resembles an optical microscope in its lens system except that the lenses are electrostatic and magnetic arrays which deflect and focus electrons by electromagnetic interactions in the same way that glass lenses deflect and focus light, as shown in Fig. 26.25. An electron beam from a heated cathode is accelerated to a very high speed, condensed by such a lens system on a specimen, e.g., a virus or bacterium mounted on a very thin collodion film, and then focused by an objective lens to produce an intermediate image. This image is then the object for an eyepiece lens which produces the final image on a fluorescent screen where it can be viewed by an observer, or on a photographic plate sensitive to electrons. Since the object attenuates the electron beam differently for different thicknesses and configurations, a very clear picture of the object under very high resolution can be

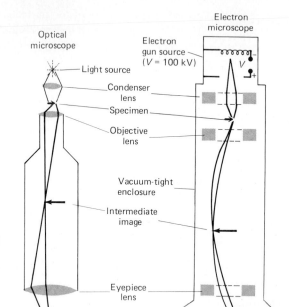

FIGURE 26.25 Comparison of a transmission electron microscope (TEM) with an optical microscope. The lenses in the TEM are magnetic lenses which use magnetic fields to focus the electrons to an image of the original object.

FIGURE 26.26 A very high resolution transmission electron microscope photograph of barium atoms: *(a)* A diagram of a complicated organic molecule which has two barium atoms ($^{137}_{56}$ Ba) at its two ends separated by a distance of only about 1.6 nm = 1.6×10^{-9} m; *(b)* electron microscope photograph in which a 5-nm distance corresponds to the length of the white line at the lower right of the photo. The two dots inside the circles are images of the two barium atoms at the end of the modecule; *(c)* an electron microscope photo under even greater magnification. The two barium atoms indicated by an arrow in *(b)* are clearly evident, and the separation between their centers is indeed about 1.6 nm Notice that the magnification in this case is about 10 million times! *(Photo courtesy of B. Jouffrey and D. Dorignac, Laboratoire d'Optique Electronique, Toulouse, France.)*

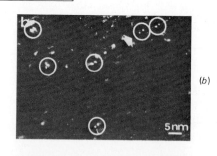

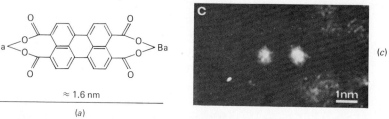

obtained, as shown in Fig. 26.26. Modern transmission electron microscopes have resolved details as small as 0.23 nm compared with the 200 nm attainable with the best optical microscopes. The limitation on the resolving power of the TEM is not the electron wavelength, which can be much shorter than 0.1 nm, but imperfections in the electrostatic and magnetic lenses which are used. Electron microscopes have the disadvantage, however, that a high vacuum is required throughout the entire instrument to reduce the scattering of electrons by air molecules.

The Scanning Electron Microscope   An extremely promising recent development is the scanning electron microscope (SEM), which is really not a true microscope at all. An SEM employs a very well focused beam of

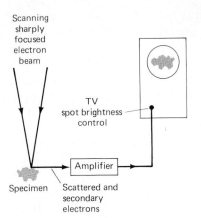

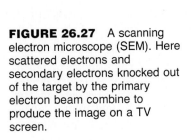

FIGURE 26.27 A scanning electron microscope (SEM). Here scattered electrons and secondary electrons knocked out of the target by the primary electron beam combine to produce the image on a TV screen.

electrons (perhaps only 0.2 nm in diameter) to scan the surface of the specimen, as in Fig. 26.27. Some of these electrons are scattered by the specimen; others eject secondary electrons from the specimen. These electrons are then collected and amplified to produce a current which controls the brightness of the spot on a conventional TV screen. The TV electron beam is scanned in synchronism with the SEM beam and is modulated in intensity by the secondary electron beam. Hence an image of the specimen is formed which varies with the shape, size, and contour of the object. The magnification of such a system can be controlled electronically by varying the sweep speed of the TV beam with respect to the SEM beam. The final image can also be transmitted immediately over conventional TV cables to distant receivers. The resolving power of an SEM is less than that of a TEM by a factor of about 10, but this is partially made up by the increased depth perception and almost three-dimensional quality of SEM pictures. It is also possible to convert brightness differences into colors to increase the contrast in SEM photographs.

Wave Properties of Other Particles

The de Broglie relation $\lambda = h/mv$ attributes a wavelength to any particle which has momentum. Hence *wave properties are intrinsic to all matter*. Research on neutron diffraction indicates that uncharged particles as well as charged particles can be diffracted. Even atoms and molecules, which have internal structure, have been diffracted. Hence the de Broglie equation is a completely general relationship that applies to all matter.

26.9 The Wave-Particle Duality

Because all energy is transported in the form of either particles or waves, the concepts of *particle* and *wave* take on great significance in physics. These two concepts provide *models* of reality which enable the physicist to understand better how the universe works. For example, a particle model that considers an ideal gas as made up of infinitesimally small particles moving along random paths through the gas provides a satisfactory description of most of the macroscopic properties of the gas—volume, pressure, temperature, etc. Similarly, a wave model for radio waves adequately describes just about all the phenomena associated with the behavior of such waves—the speed of the waves, reflection, interference, etc. But in both these cases the models are highly idealized: no unstructured particles with all their mass concentrated at one point in space actually exist; neither do radio waves of a single, precisely known frequency, as we will see in the next section. Both these highly successful models are merely that —*models*, not definitive pictures of reality; the physical world about us is too complicated to be contained within, or perfectly described by, oversimplified models of this kind.

The situation becomes even more complicated when we discuss atomic systems and radiation. For example, we have a wave model for electromagnetic radiation which well describes phenomena like interference and diffraction. But we have just discovered that we need a photon—or particle—model to describe blackbody radiation, the photoelectric effect, the continuous x-ray spectrum, and the Compton effect. Similarly, to describe atomic particles like electrons and neutrons the Newtonian particle model is inadequate, for we need de Broglie's idea of matter waves to explain electron and neutron diffraction.

Wave-Particle Duality

> An adequate description of reality at the atomic level requires the application of two different models, the wave model and the particle model, to encompass matter, radiation, and their interactions.

Bohr's Principle of Complementarity

There is an added problem in that the wave and particle models seem to be mutually incompatible. A wave must be spread out over a large region of space; a particle is confined to a very small region of space. How can we say that we need both models to describe an electron? The answer given by the Danish physicist Niels Bohr (see Fig. 27.2 and accompanying biography) in 1928 was that, if we look carefully at our interpretation of the experiments discussed in this and previous chapters, we find that we never apply the wave description and the particle description simultaneously, so that the seemingly contradictory aspects of the two models never really come into conflict. Thus, in discussing Young's double-slit experiment, electromagnetic radiation is always treated completely on a wave model, and the concept of the photon is never introduced. Similarly, in dealing with the Compton effect we always apply particle mechanics to a collision between a photon and an electron, and the wave properties of the radiation never enter the picture. In other words, we need *both* a wave and a particle model to describe the atomic universe fully, but these models never introduce logical contradictions into our analysis because they are never applied simultaneously to the same physical situation.

This led Bohr to formulate the following *principle of complementarity* for atomic particles and radiation:

Principle of Complementarity

> The wave and particle descriptions of reality are *complementary*; both models are required to describe the behavior of electromagnetic radiation and atomic particles, but the two are never applied at the same time to the same physical experiment.

In the next section we will see more clearly why the wave and particle models of reality are mutually exclusive in any given physical situation. We will find that the wave-particle duality is built into the very structure of reality and that we cannot expect to improve on the present situation no matter how many future Einsteins, Plancks, and Bohrs set their minds to the task.

Newtonian physics was based on common-sense concepts like absolute space, time, and mass. Einstein's theory of relativity showed the inadequacy of these concepts. So too our ideas of particles and waves are borrowed from our everyday experience with large-scale objects. Quantum theory has shown the inadequacy of these classical models as applied to electromagnetic radiation and to atomic particles. Hence both relativity and quantum theory represent revolutions not only in physics, but in philosophy as well.

26.10 The Heisenberg Uncertainty Principle

Further insight into the wave-particle duality inherent in the physical world can be obtained from a principle first enunciated in 1927 by the German physicist Werner Heisenberg (see Fig. 26.28) and now called the *Heisenberg uncertainty* (or *indeterminacy*) *principle*. It makes quantitative the implications of Bohr's principle of complementarity by showing to what extent the wave and particle pictures are incomplete. Heisenberg's work pointed out that any measurement of a physical system changes the system being measured to some extent and hence that a fundamental uncertainty exists in all physical measurements.

Uncertainties in Position and Momentum of an Electron

Suppose we perform a "thought experiment" using an idealized microscope to determine the position of a single electron by illuminating it with a single photon. The electron is assumed to be at rest initially, and the photon approaches the electron from below, along the positive Y axis, bounces off the electron, and enters the microscope lens, as in Fig. 26.29. As we saw in Sec. 24.6, the resolving power of a microscope is given by $1.22\lambda/(2 \sin \theta)$, where λ is the wavelength of the light and θ is the half angle subtended by the

FIGURE 26.28 Werner Heisenberg (1901–1976), the originator of the uncertainty principle. Heisenberg received the 1932 Nobel Prize in physics for the development of quantum mechanics. He also made many substantial contributions to the theory of the atomic nucleus. *(Photo courtesy of AIP Meggers Gallery of Nobel Laureates.)*

FIGURE 26.29 A thought experiment in which a single photon is used to locate an electron in an idealized light microscope: *(a)* A photon scattered by an electron and entering the lens of a microscope can have any angle from 0 to θ with respect to its original direction. *(b)* As a result, the uncertainty in the momentum in the X direction is Δp_x, Where $\Delta p_x = p \sin \theta$.

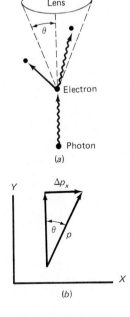

microscope lens. The uncertainty in the X position of the electron, as measured by the microscope, is then $\Delta x = 1.22\lambda/(2 \sin \theta) \simeq \lambda/(\sin \theta)$, since we are interested only in the order of magnitude of the result.

The photon has an initial momentum in the Y direction of magnitude $p = h/\lambda$, where h is Planck's constant and λ is the photon wavelength. In the collision a small amount of momentum will be transferred to the electron, and the photon will move off in a new direction with slightly decreased momentum. Depending on the direction in which the photon is scattered, it can have any X component of momentum from 0 to $p_x = p \sin \theta$, where θ is the half angle subtended by the microscope lens. Photons with X components of momentum greater than this do not enter the microscope and hence can be ignored. By the principle of conservation of momentum, the uncertainty in the X component of the electron's momentum after the collision must be equal to the uncertainty in the X component of the photon's momentum after the collision.

Hence we have for the electron:

$$\Delta p_x \simeq p \sin \theta = \frac{h}{\lambda} \sin \theta$$

Then, taking the product of these two uncertainties in the electron's position and momentum, we have

$$\Delta x \, \Delta p_x \simeq \frac{\lambda}{\sin\theta} \frac{h \sin \theta}{\lambda}$$

or $\qquad \Delta x \, \Delta p_x \simeq h$ $\qquad\qquad\qquad\qquad$ (26.16)

This is only an approximate result. Heisenberg's more careful calculation of the product of the uncertainties in position and momentum led to the more exact result:

$$\boxed{\Delta x \, \Delta p_x \geq \frac{h}{2\pi}}$$ $\qquad\qquad$ (26.17)

This result can be expressed in words as follows:

Heisenberg Uncertainty Principle

> It is impossible simultaneously to measure both the position and the corresponding momentum of a particle with complete accuracy. The product of the uncertainties in the position and the corresponding momentum is greater than, or at best equal to, $h/2\pi$.

The quantity $h/2\pi$ is sometimes written as $\hbar$, and verbalized as "h bar." According to Newtonian mechanics the position and momentum of a particle can both be measured *exactly at the same time. This is no longer true in modern quantum physics.* If we try to locate the electron more precisely by decreasing the wavelength λ, we increase the momentum of the incident photon (since $p = h/\lambda$) and thus make the electron's momentum more uncertain. The same is true for photons, as well as for all other atomic particles.

Example 26.8

(a) Find the uncertainty in the speed of a 1-g paper clip if it is desired to locate the clip to within an uncertainty in position of 10^{-4} cm.

(b) Find the uncertainty in the speed of a 1000-eV electron if it is desired to locate the electron to within an uncertainty in position of 0.10 nm, which is about the size of an atom.

(c) What fraction of the electron's initial speed is this uncertainty in its speed?

SOLUTION

(a) If the paper clip is initially at rest, the very act of measuring it will disturb it and impart to it a momentum Δp_x.

From the Heisenberg uncertainty relationship for position and momentum, we have

$$\Delta x\, \Delta p_x \geq \frac{h}{2\pi}$$

or, since $p_x = mv_x$, $\Delta x\, \Delta v_x \geq \dfrac{h}{2\pi m}$

Hence

$$\Delta v_x \geq \frac{h}{2\pi m\, \Delta x} = \frac{6.6 \times 10^{-34}\ \text{J·s}}{2(3.14)(10^{-3}\ \text{kg})(10^{-6}\ \text{m})}$$

or $\Delta v_x \geq \boxed{1.1 \times 10^{-25}\ \text{m/s}}$

This is a completely unmeasurable speed. Hence it appears that *the uncertainty principle is of little importance in dealing with large-scale objects.*

(b) For an electron we have, as in part (*a*),

$$\Delta v_x \geq \frac{h}{2\pi m\, \Delta x}$$

$$\Delta v_x = \frac{6.6 \times 10^{-34}\ \text{J·s}}{2(3.14)(9.1 \times 10^{-31}\ \text{kg})(0.10 \times 10^{-9}\ \text{m})}$$

or $\Delta v_x \geq \boxed{1.2 \times 10^{6}\ \text{m/s}}$

In this case we find a very large uncertainty in the electron's speed.

(c) To find the fraction of the actual electron speed made up by the uncertainty, note that 1000 eV = (1000 eV)(1.6 $\times$ 10^{-19} J/eV) = 1.6 $\times$ 10^{-16} J.

Hence

$$KE = \tfrac{1}{2}mv^2 = 1.6 \times 10^{-16}\ \text{J}$$

and $v = \left[\dfrac{(1.6 \times 10^{-16}\ \text{J})(2)}{9.1 \times 10^{-31}} \right]^{1/2} = 1.9 \times 10^{7}\ \text{m/s}$

The fractional uncertainty in the electron's speed is then

$$\Delta v = \frac{1.2 \times 10^{6}\ \text{m/s}}{1.9 \times 10^{7}\ \text{m/s}} = 0.063 = \boxed{6.3\%}$$

Uncertainties in Time and Energy

There is a second important version of Heisenberg's principle, which relates uncertainties in time and in energy.

Consider a classical wave. To describe it fully we must know both its wavelength and its frequency exactly, for these quantities determine its speed. If we want to determine the frequency f precisely, however, the wave must be infinitely long. To show this, suppose we choose to measure f to high accuracy by beating it against the frequency of a standard oscillator. To be sure that the measurement is highly accurate we might wait 1 min and judge that, if no beat occurs in that minute, then the two frequencies are exactly the same. But it is possible that, even if no beats were heard in the first minute, one might be heard in the second minute, or in an hour, a day, a year, or a century, which would indicate that the frequencies are not *exactly* the same. To be absolutely sure we would have to wait *forever*, in which time the wave would be spread out over an infinite distance.

Since none of us is patient enough to wait forever to measure a frequency, the question arises: What uncertainty is introduced into our knowledge of the frequency when we limit our measurement to a time Δt? Suppose that the unknown and the standard frequencies are f_1 and f_2, that the difference, or beat frequency is $\Delta f = f_2 - f_1$, and hence that Δf beats are observed per second. The time for one full beat is then $1/\Delta f$ s. Let us suppose that we count beats for a time about equal to the minimum possible time $\Delta t = 1/\Delta f$ s, which allows us to count one full beat. Then

$$\Delta t \simeq \frac{1}{\Delta f}$$

or $\Delta f\, \Delta t \simeq 1$ (26.18)

This is a purely *classical expression* for the uncertainty in a frequency Δf when measured over a time interval Δt. The shorter the time Δt, the larger the uncertainty in the frequency Δf.

We can convert this classical result into another form of the Heisenberg principle by multiplying both sides of Eq. (26.18) by Planck's constant. Then we have

$$h\, \Delta f\, \Delta t \simeq h$$

or, since $\Delta E = h\, \Delta f$

$$\Delta E\, \Delta t \simeq h \qquad (26.19)$$

Again, the more accurate calculation carried out by Heisenberg led to the result:

$$\boxed{\Delta E\, \Delta t \geq \frac{h}{2\pi}\ (= \hbar)}$$ (26.20)

Hence the uncertainties in energy and time are related in the same way as are the uncertainties in position and momentum. This second statement of the Heisenberg uncertainty principle has great importance for our knowledge of the energy levels of atoms and their spectra, as we will see in subsequent chapters.

Our conclusion, then, is that, since we can never actually determine simultaneously the position and momentum (or speed) of any "particle" with complete accuracy, we are not justified in concluding that the object under observation is actually a particle in the usual sense of that term. Similarly, since we can never determine the frequency or wavelength of a "wave" with complete accuracy, we are not justified in concluding that the object under observation is actually a wave in the usual sense of the term. Rather we are dealing with more complicated entities requiring both particle and wave models to describe them fully. This is simply another statement of Bohr's principle of complementarity.

Example 26.9

An atom remains in a particular energy state for 10^{-8} s. What is the uncertainty in our knowledge of the energy of that state?

SOLUTION

According to the Heisenberg uncertainty principle, the uncertainties in time and energy are related by the equation

$\Delta E\, \Delta t \geq \dfrac{h}{2\pi}$.   In this case $\Delta t = 10^{-8}$ s, and so

$$\Delta E \geq \frac{h}{2\pi\, \Delta t} = \frac{6.6 \times 10^{-34}\ \text{J·s}}{2(3.14)(10^{-8}\ \text{s})}$$

or $\Delta E \geq \boxed{1.1 \times 10^{-26}\ \text{J}}$

Summary: Important Definitions and Equations

Particle properties of electromagnetic radiation

Photon: A massless bundle (or quantum) of electromagnetic energy which behaves like a particle.

Energy of a photon: $E = hf$, where f is the frequency of the photon, and $h = 6.63 \times 10^{-34}$ J $\cdot$ s is Planck's constant.

Blackbody radiation

Blackbody: A perfect absorber which absorbs all the visible and near-visible electromagnetic radiation which falls on it. When heated, a blackbody becomes a perfect emitter.

Planck's equation for blackbody radiation:

$$E(\lambda)\,\Delta\lambda = \frac{8\pi}{\lambda^4}\frac{hf\,\Delta\lambda}{e^{hf/kT}-1}$$

Photoelectric effect: The release of electrons from a clean metal surface when electromagnetic radiation of the proper frequency falls on it.

Einstein's equation for the photoelectric effect:

$$hf = hf_0 + \tfrac{1}{2}mv_{\max}^2$$

where $hf_0 = \phi =$ work function of metal and $\tfrac{1}{2}mv_{\max}^2 = eV_0$, where e is the electronic charge and V_0 is the cutoff potential.

Continuous x-ray spectrum:

Cutoff frequency: $f_0 = \dfrac{e}{h}\,V$

Cutoff wavelength: $\lambda_0 = \dfrac{1.24 \times 10^{-6}\text{ m}}{V}$, where V is in volts.

Compton effect: The increase in the wavelength of electromagnetic radiation when scattered by free electrons. On a photon model the photon loses energy in the collision and hence its frequency is decreased and its wavelength increased.

Photon energy: $E = hf$

Equation for Compton effect:

$$\Delta\lambda = \lambda_2 - \lambda_1 = \frac{h}{m_0 c}(1 - \cos\theta)$$

where $h/m_0c = 0.00243$ nm is called the Compton wavelength of the electron, and $\Delta\lambda$ is the increase in the photon wavelength.

Wave properties of matter:

De Broglie waves: Every particle of momentum p has associated with it a wave of wavelength $\lambda = h/p$.

Electron diffraction: $2d \sin\theta = n\lambda$

where λ is the de Broglie wavelength of the electron.

Electron microscope: Microscopes using electrons with short de Broglie wavelengths to increase greatly the resolving power over that of optical microscopes.

The wave-particle duality

An adequate picture of reality at the atomic level requires the application of two different models, the wave model and the particle model, to describe fully matter, radiation, and their interactions.

Bohr's principle of complementarity: The wave and particle descriptions of reality are complementary; both models are required to describe the behavior of electromagnetic radiation and atomic particles, but the two are never applied at the same time to the same physical experiment.

Heisenberg uncertainty principle: It is impossible to measure simultaneously both the position and the corresponding momentum of a particle or a photon with complete accuracy.

$$\Delta x\,\Delta p_x \geq \hbar \qquad \Delta E\,\Delta t \geq \hbar$$

where $\hbar = h/2\pi$.

Questions

1 If a blackbody absorbs all the radiation incident on it, does its temperature increase? Why?

2 Try to suggest some everyday analogies with the situation where an atom only absorbs and emits energy in quantized bundles which we call photons.

3 In photographic darkrooms red, or "safety," lights are often kept on during the development process. Explain why this is possible, using the photon theory of light.

4 Explain why ultraviolet (UV) light is more dangerous to the human body (particularly to the eyes) than visible light.

5 Why could not a classical theory of light explain the results of the Compton effect?

6 Why is it impossible to observe the Compton effect using visible light?

7 Discuss changes which might occur in our daily lives if Planck's constant were four orders of magnitude larger than it is.

8 Why would gamma-ray microscopes not be particularly useful, even though they would have very high resolving powers?

9 Stanford physicist Wolfgang Panofsky is quoted as saying: "The smaller the objects, the bigger the microscope we must use to see them." Comment on the accuracy of this statement as applied to the electron microscope.

10 Why are neutrons more easily thermalized by collisions with hydrogen atoms than by collisions with nitrogen atoms?

11 If the energies associated with the valence electrons in atoms and molecules are about 1 to 2 eV, why would you expect the quantum nature of visible light to be very important when light interacts with atoms and molecules?

12 Atoms vibrating in solid crystal lattices are found to have a small, "zero-point" energy at absolute zero (0 K). Explain why this is required by the Heisenberg uncertainty principle.

Problems

A 1 The diameter of a billiard ball is larger than that of an atom by about:
(a) 12 orders of magnitude
(b) 8 orders of magnitude
(c) 4 orders of magnitude
(d) A factor of 8   (e) A factor of 12

A 2 The energy associated with a photon of red light is about:
(a) 2.8×10^{-19} eV   (b) 2.8×10^{-19} J
(c) 6.5×10^{-47} J   (d) 4.6×10^{-40} eV
(e) 4.6×10^{-40} J

A 3 In Planck's equation for blackbody radiation the exponent hf/kT is equal to 1 for visible light at a temperature of about:
(a) 250 K   (b) 25 K   (c) 10^{10} K
(d) 2500 K   (e) 25,000 K

A 4 In an experiment on the photoelectric effect it is found that the cutoff potential is 2.0 V. The speed of the maximum-energy electrons being ejected from the cathode by the incident light is:
(a) 3.2×10^{-19} J   (b) 3.2×10^{-19} m/s
(c) 8.4×10^5 m/s   (d) 7.1×10^{11} m/s
(e) 6.0×10^5 m/s

A 5 In a photoelectric effect experiment the intensity of the light is increased by a factor of 3, with the frequency held constant. The speed of the maximum-energy electrons being ejected from the cathode increases by a factor of:
(a) 9   (b) 2   (c) 3   (d) 1.73
(e) None of the above

A 6 An x-ray photon collides head-on with an electron and is scattered directly back at 180° to its original direction. The shift in the wavelength of the incident x-ray is:
(a) +0.00243 nm   (b) −0.00243 nm
(c) −0.00486 nm   (d) +0.00486 nm   (e) 0

A 7 To produce the diffraction of particles moving with a speed of 10^6 m/s with a diffraction grating ruled with 10,000 lines to the centimeter, the mass of the particles would have to be smaller than that of an electron by about:
(a) Six orders of magnitude
(b) Three orders of magnitude
(c) One order of magnitude
(d) It would be about equal to the electron mass
(e) It would have to be greater than the electron mass

A 8 The de Broglie wavelength of an electron accelerated by a voltage V is:
(a) $\dfrac{h/(2em)^{1/2}}{V^{1/2}}$ (b) $\dfrac{(2em)^{1/2}h}{V^{1/2}}$ (c) $\dfrac{(2emh)^{1/2}}{V^{1/2}}$

(d) $\dfrac{h/(2em)^{1/2}}{V}$ (e) $\dfrac{h(2em)^{1/2}}{V}$

A 9 The wavelength of electrons accelerated through 100,000 V in an electron microscope is smaller than that of blue light by about:
(a) One order of magnitude
(b) Two orders of magnitude
(c) Five orders of magnitude
(d) Three orders of magnitude
(e) Four orders of magnitude

A10 For neutrons to be diffracted by crystals, the speed of the neutrons must correspond to a temperature of roughly:
(a) 300 K   (b) 0 K   (c) 6000 K
(d) 100,000 K   (e) None of the above

B 1 Find the energy associated with the photons emitted by the antenna of WGMS, a Washington, D.C., FM radio station which broadcasts at a frequency of 103.5 MHz.

B 2 During a chest x-ray a person's body normally absorbs about 1.5×10^{-3} J of energy. About how many x-ray photons are absorbed by a person having a chest x-ray?

B 3 (a) Will red light be able to emit electrons from a clean aluminum surface? Why? (*Hint*: See Table 26.1.)
(b) What is the minimum frequency for incident radiation which will be able to eject electrons from aluminum?

B 4 (a) It is found that light of wavelength shorter than 300 nm can eject electrons from a metal surface in a vacuum. What is the work function of the metal?
(b) What is a reasonable guess as to the identity of the metal? (*Hint*: See Table 26.1.)

B 5 (a) What is the momentum of a gamma-ray photon of energy 1 MeV?
(b) What is the de Broglie wavelength of an electron moving with a speed of 2.0×10^5 m/s?

B 6 X-radiation of wavelength 0.095 nm is incident on a "free-electron gas," i.e., a collection of free electrons. Calculate the energy acquired by an electron which is struck by an x-ray photon if the photon is deflected by an angle of 60° with respect to its original direction.

B 7 When x-rays are scattered from a KCl crystal with lattice spacing $d = 0.314$ nm, it is found that the smallest angle of incidence with respect to the lattice planes at which constructive interference occurs is 25°. What is the wavelength of the incident x-rays?

B 8 A beam of 100-V electrons is incident on a KCl crystal with lattice spacing equal to 0.314 nm. What is the minimum angle of incidence with respect to the lattice plane for which a strong electron beam will be detected?

B 9 What is the uncertainty in the speed of a neutron

if it is desired to locate its position to within an uncertainty of 1.0 nm?

B10   Some unstable nuclei have lifetimes in excited states of only 10^{-14} s. What is the limit on the accuracy with which the energy of such an excited state can be determined?

C 1   A neon (Ne) lamp emits 60 W of red light at a wavelength of 680 nm. How many photons are emitted by the lamp per second?

C 2   Find the energies (in both joules and electron-volts) associated with (a) the infrared ($\lambda = 1000$ nm) and (b) the ultraviolet ($\lambda = 100$ nm) regions of the spectrum.

C 3   To cause sunburn the incident light must be able to break chemical bonds in the molecules of the human skin. To do this requires an energy of about 3.5 eV in the photons of the incident sunlight.

   (a)   To what wavelength does this energy correspond?

   (b)   In what region of the spectrum does this wavelength fall?

C 4   From each of the three graphs shown in Fig. 26.9 calculate a value for Planck's constant h. From the average of your three results obtain the best possible value for h for the data given.

C 5   What cutoff voltage must be applied to stop the fastest photoelectrons emitted by a tungsten surface under the action of ultraviolet light of wavelength 250 nm? (*Hint*: See Table 26.1.)

C 6   It is found that violet light of wavelength 350 nm is able to eject from a metal electrons which have maximum speeds of 4.0×10^5 m/s. What is the work function of the metal?

C 7   Electrons with maximum speeds of 8.4×10^5 m/s are ejected from a clean metal surface by ultraviolet radiation of wavelength 200 nm. Find:

   (a)   The work function of the metal.

   (b)   The threshold wavelength of the metal for electron emission.

   (c)   The retarding potential difference required to stop the electron flow.

C 8   For the metals given in Table 26.1 calculate enough threshold frequencies to determine from which metals electrons can be ejected by visible light.

C 9   An x-ray photon of wavelength 0.252 nm strikes a free electron and is scattered at an angle of 120°.

   (a)   What is the energy of the photon before the collision?

   (b)   What is the momentum of the photon?

   (c)   What is the wavelength of the x-ray photon after it is scattered?

   (d)   What is the energy of the scattering electron after the collision?

C10   An x-ray photon of wavelength 0.200 nm is scattered by a free electron at an angle of 90°.

   (a)   What is the speed of the photon after the collision?

   (b)   What is the wavelength of the photon after the collision?

   (c)   What is the energy of the electron after the collision?

   (d)   What is its speed, if we neglect the effects of relativity?

C11   An x-ray photon of wavelength 0.120 nm is scattered from the electrons in a carbon target.

   (a)   Calculate the wavelength of the x-rays scattered at an angle of 60° with respect to the original direction.

   (b)   Calculate the direction of the velocity of the scattering electron after the collision.

C12   (a)   How fast would an electron have to travel to have the same wavelength as a 2.0-eV photon?

   (b)   How fast would a neutron have to travel to have the same wavelength as a 2.0-eV photon?

C13   What wavelength must a photon have if its momentum is to be the same as that of an electron moving with a speed of 3.5×10^5 m/s?

C14   A beam of x-rays is reflected from the lattice planes of a KCl crystal, which are separated by a distance of 0.314 nm. The minimum angle of incidence with respect to the crystal planes which produces a strong reflected x-ray beam is 16°.

   (a)   What is the wavelength of the x-rays used?

   (b)   At what angle does the second-order reflection occur?

C15   A beam of x-rays of wavelength 0.15 nm is reflected by the lattice planes in a crystal whose structure is being investigated. The minimum angle of incidence with respect to the lattice planes which produces a strong reflected x-ray beam is 26°.

   (a)   What is the lattice separation d in the crystal?

   (b)   At what angle would you expect to find the second-order diffraction of the x-ray beam?

C16   A beam of 50-V electrons is incident on a KCl crystal, with lattice spacing between planes of 0.314 nm. For what angles of incidence with respect to the lattice planes will strong electron beams be detected?

C17   A beam of electrons is incident on a NaCl crystal with lattice spacing $d = 0.282$ nm. The minimum angle of incidence for which a strong diffracted beam is observed is 4.6°. What is the energy of the electrons in the beam?

C18   What potential difference would be required in an electron microscope to produce electrons with wavelengths of 0.010 nm? (Neglect the effects of relativity.)

C19   A metal puck of mass 0.50 kg floats without friction on a layer of air blown through holes in the bottom of an air table. A physicist measures the position of this puck to an accuracy of 10^{-4} m.

   (a)   What speed is imparted to the puck by the measurement?

   (b)   How long would it take for the puck to cross the 2.0-m table at this speed?

C20   The energy of an electron state in a hydrogen atom is 3.4 eV. If it is desired to measure this energy to an accuracy of 1 percent, what is the minimum time required for the measurement?

D 1   The wavelength of the light striking a metal surface is 250 nm less than the threshold wavelength. The measured maximum kinetic energy of the ejected photoelectrons is 2.5 eV. What is the threshold frequency for this metal?

D 2 An electron is deep inside a metal and requires 1.50 eV of energy to move to the surface of the metal before it can be ejected by an incident photon. If the metal is potassium and the incident photon has a wavelength of 400 nm:

(a) What is the kinetic energy with which the electron emerges from the metal?

(b) What is the speed of this electron?

D 3 Carry through the derivation of the equation for the Compton effect [Eq. (26.7)] from the equations describing the conservation of energy and momentum in a collision between an x-ray photon and an electron. (*Hint*: Equate the energies before and after the collision, and the X and Y components of momentum before and after the collision. The electron must be treated relativistically.)

D 4 A free proton which is at rest is struck by a photon in a Compton collision. The proton acquires a kinetic energy of 5.7 MeV in the collision. What is the minimum photon energy which can produce this result?

D 5 An inventor suggests using the momentum of photons to propel a rocket and recommends using a mercury lamp which emits 200 W of violet radiation at 356 nm. If the rocket has a mass of 100 kg and is in free space:

(a) What force can be generated in this way?

(b) What will be the acceleration of the rocket?

D 6 What is the energy and wavelength of a thermal neutron, i.e., a neutron at a temperature of about 300 K?

D 7 What is the wavelength of an electron with total energy 1.0 MeV? (Since the speed of such an electron is so high, relativistic formulas must be used.)

D 8 Use the Heisenberg uncertainty principle to estimate the kinetic energy of a neutron in a nucleus of radius 7.0×10^{-15} m.

D 9 Use the Heisenberg uncertainty principle to show that it is impossible to have electrons inside the nucleus. Assume that the nuclear radius is about 10^{-14} m.

D10 Calculate the uncertainties in position and momentum for a photon diffracted by a narrow slit of width w, and show that their product satisfies the Heisenberg uncertainty principle.

Additional Readings

Born, Max: "Max Karl Ernst Ludwig Planck," *Obituary Notices of Fellows of the Royal Society of London*, vol. 6, 1948, pp. 161–180. A brief biography of Planck by a colleague and coworker who knew him well. This is reprinted in Henry A. Boorse and Lloyd Motz (eds.), *The World of the Atom*, vol. 1, Basic Books, New York, 1966, pp. 462–484.

de Broglie, Louis: *The Revolution in Physics: A Non-Mathematical Survey of Quanta* (trans. by Ralph W. Niemeyer), Noonday Press, New York, 1953. An especially lucid account of modern physics by a physicist who contributed greatly to its development.

Everhart, Thomas E., and Thomas L. Hayes: "The Scanning Electron Microscope," *Scientific American*, vol. 226, no. 1 January 1972, pp. 54–69. Contains some remarkable photographs of biological specimens made with the scanning electron microscope.

Gamow, George: "The Principle of Uncertainty," *Scientific American*, vol. 198, no. 1, January 1958, pp. 51–57. A clear exposition of the background and significance of the Heisenberg uncertainty principle.

———: *Thirty Years That Shook Physics: The Story of Quantum Theory*, Doubleday Anchor, Garden City, N.Y., 1966. A clear and well-illustrated book on the development of the quantum theory.

Kelves, D. J.: "Robert A. Millikan," *Scientific American*, vol. 240, no. 1, January 1979, pp. 142–151. A frank and interesting account of the best-known American scientist of his day.

Medicus, H. A.: "Fifty Years of Matter Waves," *Physics Today*, vol. 27, no. 2, February 1974, pp. 38–45. A historical discussion of how de Broglie waves were first predicted by de Broglie and then found in the laboratory.

Planck, Max: *Scientific Autobiography and Other Papers* (trans. by F. Gaynor), Philosophical Library, New York, 1949. Planck's very brief and often moving autobiography is included in this volume, along with other Planck lectures on a variety of subjects.

Segré, Emilio: *From X-Rays to Quarks*, Freeman, San Francisco, 1980. A personal account of modern physicists and their discoveries, starting with Roentgen's discovery of x-rays, by a physicist who contributed substantially to this exciting period.

> *That this insecure and contradictory foundation was sufficient to enable a man of Bohr's unique instinct and tact to discover the major laws of the spectral lines and of the electron shells of the atoms together with their significance for chemistry, appeared to me like a miracle—and appears to me as a miracle even today. This is the highest form of musicality in the sphere of thought.*
>
> *Albert Einstein*

Chapter 27

The Structure of Atoms

The work of Planck and Einstein revolutionized our understanding of particles and radiation. But it indicated little about the structure of atoms and molecules, of those building blocks from which our whole universe is constructed. Ever since Kirchhoff invented the spectroscope in 1859, physicists had been accumulating data on the spectra of atoms. It had become clear that all atoms of an element emitted exactly the same series of spectral lines, and that these lines could be used for the positive identification of that element. This suggested that all atoms of any given element probably had exactly the same structure. But what kind of a structure did atoms have, a planetary structure like our solar system, with electrons in motion about a positively charged core of some sort, or another as-yet-unknown structure? And why should all hydrogen atoms, for example, have the same spectrum, and hence presumably the same structure?

The beginnings of answers to these questions were provided by Niels Bohr, who carried the work of Planck and Einstein a giant step forward and provided our first important insight into the structure of atoms.

27.1 The Spectrum of the Hydrogen Atom

Let us begin with the simplest of all atoms, hydrogen. This is the most important atom in physics because, containing only one electron and one proton, it avoids the complications inherent in heavier, more complex atoms.

Suppose we excite *atomic* hydrogen (*not* H_2, which is *molecular* hydrogen) by an electric discharge, produced by a high voltage across electrodes in an evacuated glass tube, and observe the resulting hydrogen spectrum with a prism or grating spectrometer. This *emission spectrum* of atomic hydrogen consists of a number of sharp, discrete, bright lines on a dark background, the lines being images of the spectrograph slit. Such spectra are called *line spectra*.

The line spectrum of atomic hydrogen in the visible region is shown in Fig. 27.1 and on the color plate included with this chapter. The lines are given their usual notation H_α, H_β, H_γ, . . . , with consecutive Greek letters refer-

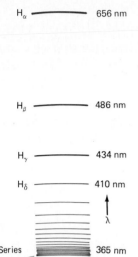

H$_\alpha$ ——— 656 nm

H$_\beta$ ——— 486 nm

H$_\gamma$ ——— 434 nm

H$_\delta$ ——— 410 nm

↑ λ

Series limit ——— 365 nm

Continuum

FIGURE 27.1 The line spectrum of the hydrogen atom in the visible region.

ring to lines at higher frequencies (or shorter wavelengths). These lines approach a *series limit* where they bunch together. Beyond the series limit is a *continuum*, or continuous light pattern, in which discrete lines can no longer be seen.

This visible spectrum of H is referred to as the *Balmer series* after the Swiss scientist J. J. Balmer (1825–1898), who in 1885 arrived at a simple empirical formula for calculating the wavelengths of its lines. The *Balmer formula* takes the form

$$\frac{1}{\lambda_n} = R\left(\frac{1}{2^2} - \frac{1}{n^2}\right) \tag{27.1}$$

where λ_n is the wavelength (in meters) of a particular line in the series, R is a constant now called the Rydberg constant after the Swedish physicist J. R. Rydberg (1854–1919), and n is an integer taking on values 3, 4, 5, . . . , for the successive lines H$_\alpha$, H$_\beta$, H$_\gamma$,. . . . For hydrogen R has to be equated to 1.0968×10^7 m^{-1} to give agreement with the observed Balmer-series lines. This Balmer formula explains nothing; it merely correlates the available experimental data in a convenient fashion. Thus, if we put $n = 4$ into Eq. (27.1) we find that the wavelength of H$_\beta$ is 486 nm, in agreement with the experimental value in Fig. 27.1. Similar agreement is found for all lines in the visible spectrum. The lines get closer together as the frequency increases. The series limit corresponds to $n = \infty$ for which $\lambda_\infty = 4/R = 365$ nm, again in agreement with experiment.

Other Spectral Series in Hydrogen

In addition to the visible Balmer series, H has a similar series in the ultraviolet, the Lyman series [named after the American physicist Theodore Lyman (1874–1954)], and a number of additional spectral series in the infrared, the most important being the Paschen series [named after the German physicist L. Paschen (1865–1947)]. The Balmer, Lyman, and Paschen series can all be represented by one equation, similar to Eq. (27.1), of the form

Rydberg Equation

$$\frac{1}{\lambda} = R\left(\frac{1}{n_f^2} - \frac{1}{n_i^2}\right) \tag{27.2}$$

where the Rydberg constant R has the same value for all the series. Here n_f and n_i are integers whose significance we will explain shortly. The three important series in H are characterized by the values of n_f and n_i given in Table 27.1. Equation (27.2) is usually called the *Rydberg equation*. Again, it organizes a great deal of data in a concise fashion, but explains nothing.

TABLE 27.1 Important Spectral Series for the Hydrogen Atom

Region	Name of series	n_f	n_i
Ultraviolet	Lyman	1	2, 3, 4, . . .
Visible	Balmer	2	3, 4, 5, . . .
Infrared	Paschen	3	4, 5, 6, . . .

Example 27.1

Calculate the wavelengths of the following lines in the spectrum of hydrogen, and indicate in what region of the spectrum they fall.

(a) The first line in the Lyman series.
(b) The second line in the Balmer series.

SOLUTION

Using the Rydberg equation and the data on the various spectral series in hydrogen from Table 27.1, we have

(a) $\dfrac{1}{\lambda} = R\left(\dfrac{1}{1^2} - \dfrac{1}{2^2}\right) = (1.0968 \times 10^7 \text{ m}^{-1})\left(\dfrac{3}{4}\right)$

or $\boxed{\lambda = 122 \text{ nm}}$

This is in the *ultraviolet*.

(b) $\dfrac{1}{\lambda} = R\left(\dfrac{1}{2^2} - \dfrac{1}{4^2}\right) = (1.0968 \times 10^7 \text{ m}^{-1})\left(\dfrac{3}{16}\right)$

or $\boxed{\lambda = 486 \text{ nm}}$

This is in the *visible*.

27.2 The Bohr-Rutherford Atom

The first physicist to succeed in explaining the observed spectrum and structure of hydrogen from first principles was Niels Bohr (see Fig. 27.3 and accompanying biography). In a daring paper published in 1913 Bohr tied together the results of classical physics with some highly original ideas to predict both the spectrum of the hydrogen atom and a precise value for the Rydberg constant R.

Among the classical ideas folded into Bohr's theory of the hydrogen atom were planetary motion, angular momentum, and centripetal force from classical mechanics, and Coulomb's law from electromagnetism. The basic model Bohr adopted for the hydrogen atom was Rutherford's nuclear model, which we must consider in more detail before we can understand how Bohr used it to explain the structure and spectrum of hydrogen.

Rutherford's Alpha-Particle Scattering Experiments

In the early years of this century alpha particles (that is, He atoms with their electrons stripped away) from radioactive materials were much used to explore the structure of atoms. Suppose, for example, that we allow alpha particles from polonium, which have unique speeds of 1.6×10^7 m/s, to fall on gold foil, which is only 5×10^{-7} m thick, as in Fig. 27.2. On the other side of

FIGURE 27.2 Alpha-particle scattering by a thin gold foil.

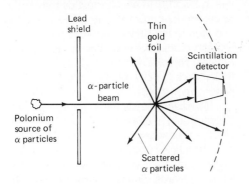

Niels Henrik David Bohr (1885–1962)

FIGURE 27.3 Niels Bohr. (*Courtesy of AIP Niels Bohr Library, Margrethe Bohr Collection.*)

FIGURE 27.4 Ernest Rutherford and Niels Bohr with their wives (photograph by Mark Oliphant, whose wife appears in front of Rutherford). Mrs. Rutherford is just visible at the left. (*Courtesy of AIP Niels Bohr Library, Margrethe Bohr collection.*)

Niels Bohr was one of the most influential physicists of the twentieth century. He was the son of a physiology professor at the University of Copenhagen. While studying physics there, he also developed into an excellent soccer player. The staunch experimentalist Lord Rutherford was once asked why he put more faith in Bohr's ideas than he did in those of other theoretical physicists. Rutherford answered: "Bohr's different. He's a football player!"

Bohr received his doctorate from Copenhagen in 1911. He resolved to go abroad for further study, since there was little physics research going on in Denmark at that time. He did research for a short time with J. J. Thomson in Cambridge, but became discouraged when Thomson could never find time to read his dissertation on the electron theory of metals. He moved to Manchester in 1912 to work with Rutherford, and there he developed his theory of atomic structure and spectra which revolutionized physics and brought Bohr the 1922 Nobel Prize in physics.

He returned to Copenhagen as a lecturer in theoretical physics in 1913, then left to assume a similar post in Manchester during the years 1914 to 1916, but returned to Copenhagen as a professor of theoretical physics in 1916. He spent the rest of his life in Denmark except for the years 1943 to 1945, when he had to flee to Sweden to avoid arrest by the Nazis. He later moved to England and then to the United States to help with the atomic bomb project at Los Alamos, New Mexico. He and his family were happily reunited in Copenhagen in 1945.

In addition to his theory of the structure and spectra of hydrogen, which he extended to explain shell structure of other atoms, Bohr made many other important contributions to physics. These include the principle of complementarity (Sec.

26.9), the correspondence principle, the liquid-drop model of the atomic nucleus, and the theory of nuclear fission.

In some ways Bohr made an even greater impact on physics by his work as director of the University Institute for Theoretical Physics, which was created for Bohr in Copenhagen in 1921 and supported by the Carlsberg Brewery. In the years between the two world wars Copenhagen became the mecca for the world's best theoretical physicists (some 600 over the years), who came to discuss and argue about physics with equally bright and dedicated colleagues from many different countries, always guided and stimulated by Bohr's penetrating understanding and deep interest in their work.

Bohr provided another great service to physics by opening his institute to Jewish physicists when they lost their academic positions in Germany in 1933. He did everything possible to help them escape and to place them in academic positions in England and the United States. This and his partly Jewish lineage did not endear him to the Nazis.

While in the United States during the war Bohr tried to convince President Roosevelt and Prime Minister Churchill that they should reveal the existence of the nuclear bomb to the world and have nuclear weapons outlawed and put under international control. Roosevelt died before anything could come of Bohr's plan. For this and other contributions to the peaceful use of atomic energy Bohr received the Atoms for Peace award in 1957.

Bohr's personal life was an exceptionally happy one, except for the death in a boating accident in 1934 of the oldest of his five sons at the age of 19. In 1912 Bohr had married

Margrethe Norlund, who was an ideal spouse and companion over 50 years until Bohr's death in 1962.

Niels Bohr was a simple, modest generous human being, much admired by all who knew him. In addition to being a great physicist he was, in the words of a colleague, "the wisest and most lovable of men."

the foil from the polonium we place a screen painted with a material which scintillates when struck by an alpha particle (this is the forerunner of the scintillation counters used in nuclear physics today). Then by moving the screen and counting the scintillations we can study the angular distribution of the alpha particles scattered by the gold foil.

What we find is that most of the alpha particles pass right through the solid foil with little or no deflection. Only 1 in 1000 is deflected through more than 10°. Occasionally, however, an alpha particle is deflected *through a very large angle*. Also, all the alphas, including the undeflected ones, are slowed down as they pass through the foil.

In 1911 the British physicist Ernest Rutherford (see Fig. 27.4, and Fig. 9.1 and accompanying biography) proposed a model of the atom to explain this behavior. His model consisted of a very small, positively charged *nucleus*, containing almost all the mass of the atom, surrounded by a number of electrons equal to the number of elementary positive charges on the nucleus, so that the atom was electrically neutral. Both the electrons and the nucleus were so small compared with the size of the atom that most of the atom was empty space. This would explain why most of the alphas passed through the foil undeflected, since it was only 2000 atoms thick. On a rare occasion, however, an alpha particle would score a direct hit on a nucleus and be deflected through a large angle, perhaps even having its direction reversed. All the alphas underwent collisions with electrons and swept them out of the atom but, since the alpha-particle mass was almost 10^4 times the electron mass, the alphas were not deflected but only slowed down by such collisions, in much the way a baseball would be on passing through a swarm of bees. On the basis of this model Rutherford calculated the expected angular distribution for the scattered alpha particles, and these predictions were completely confirmed in a series of experiments carried out by Geiger and Marsden in Rutherford's laboratory.

The Size of the Nucleus

In his nuclear model Rutherford had assumed that the force between the alpha particle and the positively charged nucleus was a repulsive Coulomb force of the form $F = k_e q_\alpha q_N / r^2$. This does not imply that the nucleus is a point charge, however. The nucleus could be a sphere of radius r_0 and the Coulomb force law would remain valid as long as the alpha particle did not penetrate the sphere. Since Geiger and Marsden's results indicated that the Coulomb force law remained valid even for head-on collisions, the alpha particle's distance of closest approach must have been greater than the radius of the nucleus r_0. Hence the distance of closest approach for a head-on collision sets an upper limit on the radius of the nucleus.

In a head-on collision an alpha particle moves in a straight line until its electric potential energy becomes exactly equal to its original kinetic energy, as in Fig. 27.5. Then it stops and retraces its path. (This is like throwing a ball

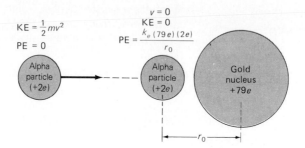

FIGURE 27.5 Distance of closest approach r_0 for a head-on collision of an alpha particle with a gold nucleus. The charge on the alpha particle is $+2$ elementary charges, and on the gold nucleus $+79$ elementary charges.

up in the air; it stops at its highest elevation when its gravitational potential energy is equal to the kinetic energy it had when it left the thrower's hand.) Suppose r_0 is the distance of closest approach to the center of the nucleus. Then the electrostatic potential energy of the alpha particle at a distance r_0 from the gold nucleus is, from Eq. (16.9),

$$PE = \frac{k_e(79e)(2e)}{r_0} = \left(9.0 \times 10^9 \ \frac{N \cdot m^2}{C^2}\right)\frac{79(2)(1.6 \times 10^{-19} \ C)^2}{r_0}$$

since the charge on the alpha is $2e$, and on the gold nucleus $79e$. Hence

$$PE = \frac{3.6 \times 10^{-26} \ J \cdot m}{r_0}$$

The original kinetic energy of the polonium alpha particles, which have masses 4 times the proton mass and speeds of 1.6×10^7 m/s, was:

$$KE = \tfrac{1}{2}(4m_p)v^2 = 2(1.67 \times 10^{-27} \ kg)(1.6 \times 10^7 \ m/s)^2$$

or $KE = 8.6 \times 10^{-13}$ J

On equating this initial kinetic energy to the potential energy when the alpha stops, we obtain

$$r_0 = \frac{3.6 \times 10^{-26} \ J \cdot m}{8.6 \times 10^{-13} \ J} = 4.2 \times 10^{-14} \ m$$

Hence the radius of a gold nucleus must be less than 4.2×10^{-14} m! Since the radii of atoms are known to be closer to 10^{-10} m, the nucleus has a radius about 10^{-4} that of an atom and therefore a volume about $(10^{-4})^3 = 10^{-12}$ that of an atom. It is clear from this that an atom is indeed mostly empty space, and that its mass is concentrated almost exclusively in a very dense, positively charged nucleus at the center of the atom.

Rutherford's model of a hydrogen atom then consists of a nucleus made up of a single proton with charge $+e$, and of a single electron with charge $-e$ outside the nucleus. Bohr added the assumption that this electron moves in an orbit about the nucleus as the earth moves in its orbit around the sun. On the basis of classical physics, however, such a model would predict that the *accelerated* electron (it is changing *direction*) would radiate energy (all accelerated charges do) and spiral into the nucleus. This would happen in less than 10^{-6} s, and so hydrogen atoms would be extremely unstable. Bohr was therefore driven to some very nonclassical ideas and revolutionary postulates to explain the spectrum of hydrogen from first principles.

27.3 Bohr's Theory of the Hydrogen Atom

While in Manchester with Rutherford in 1912 Bohr had the idea of combining Rutherford's model of the atom with the quantum ideas of Planck and Einstein to explain the spectra of atoms. About this time he chanced to find Balmer's formula for the hydrogen spectrum and decided to apply his ideas to the spectrum of this simplest of all atoms, using what is now called the *Bohr-Rutherford model* of the atom. In 1913 he published his solution in a paper in the *Philosophical Magazine*. This was the beginning of a great revolution in physics.

Postulates of the Bohr Theory

In his 1913 paper Bohr was able to predict the spectrum of hydrogen from first principles on the basis of the following postulates:

1   The hydrogen atom exists in certain *stationary states* of discrete energies. While in such states the atom is stable and emits no radiation.

2   These stationary states are those for which the electron's classical angular momentum, $L = I\omega = mr^2\omega = mvr$, is quantized in units of $\hbar = h/2\pi$:

$$L = mvr = \frac{nh}{2\pi} = n\hbar \tag{27.3}$$

where h is Planck's constant and n is an integer. (Note that the units of "action" and therefore of h, the quantum of action, are the same as the units of angular momentum.)

3   The radiation which is emitted by the atom is produced when the atom undergoes a transition from a higher-energy stationary state (E_i) to a lower-energy state (E_f). The energy difference between these two states then appears as a single photon of energy $E = hf = E_i - E_f$, from which the photon frequency is:

Bohr-Einstein Frequency Condition

$$f = \frac{E_i - E_f}{h} \tag{27.4}$$

Bohr's first postulate contradicts classical electrodynamics, according to which an accelerated electron must radiate continuously. (As Bohr indicated in his paper, this postulate "is in obvious contrast to the ordinary ideas of electrodynamics, but appears to be necessary in order to account for experimental facts.") Bohr's second postulate that the angular momentum can only take on quantized values is again a daring new idea, with no counterpart in classical physics. A planet or earth satellite can have any value of angular momentum, so long as gravitational attraction can provide the centripetal force needed to keep the planet or satellite moving in its orbit. The actual angular momentum depends on the initial velocity of the planet or satellite when it went into orbit. Bohr's postulate says that this is not the case for the hydrogen atom. No matter how the atom was formed, only certain discrete values of the angular momentum of the electron are possible.

Application of Bohr's Postulates to Hydrogen

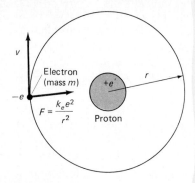

FIGURE 27.6 Bohr model of the hydrogen atom.

Bohr assumed that the centripetal force mv^2/r needed to hold the electron in a circular orbit about the nucleus must be provided by the Coulomb force, of magnitude $k_e e^2/r^2$, as shown in Fig. 27.6. Hence

$$\frac{mv^2}{r} = \frac{k_e e^2}{r^2}$$

or

$$mv^2 = \frac{k_e e^2}{r} \tag{27.5}$$

The total energy of the atom is then the sum of its kinetic and potential energies, or

$$E = \frac{mv^2}{2} - \frac{k_e e^2}{r}$$

where $-k_e e^2/r$ is the potential energy of the electron at a distance r from the proton.

From Eq. (27.5) this becomes:

$$E = \frac{k_e e^2}{2r} - \frac{k_e e^2}{r} = -\frac{k_e e^2}{2r} \tag{27.6}$$

Note that the total energy of the atom is *negative*, indicating that the atom is stable, with the electron bound to the nucleus.

This classical equation allows all possible values of r and hence all possible values of the energy E. Bohr now introduced his second postulate, according to which $L = mvr = n\hbar$. This postulate quantizes not merely the angular momentum, but also the energy E. We will specify the quantized values of the energy by E_n, since they depend on the integer n, called the *principal quantum number* of the atom. We will designate in the same way the value of the radius and the speed of the electron as r_n and v_n, since they are also quantized, as we will see.

From Bohr's second postulate we have:

$$v_n = \frac{n\hbar}{mr_n}$$

and so, from Eq. (27.5),

$$mv_n^2 = m\left(\frac{n\hbar}{mr_n}\right)^2 = \frac{k_e e^2}{r_n}$$

or

$$r_n = \frac{n^2\hbar^2}{mk_e e^2} \tag{27.7}$$

Discrete Energies in Hydrogen

To find the total energy of the system in the various energy states of H we use Eq. (27.6),

$$E_n = -\frac{k_e e^2}{2r_n}$$

or, from Eq. (27.7),

$$E_n = -\frac{k_e e^2}{2}\frac{mk_e e^2}{n^2\hbar^2} = -\frac{mk_e^2 e^4}{2n^2\hbar^2} \tag{27.8}$$

Putting in numbers and keeping five significant figures, since spectral wavelengths can be measured to very high precision, we have

$$E_n = -\frac{1}{2n^2} \frac{(9.1095 \times 10^{-31} \text{ kg})(8.9876 \times 10^9 \text{ N·m}^2/\text{C}^2)^2(1.6022 \times 10^{-19} \text{ C})^4}{(1.0546 \times 10^{-34} \text{ J·s})^2}$$

or
$$\boxed{E_n = \frac{-2.1799 \times 10^{-18} \text{ J}}{n^2} = \frac{-13.606 \text{ eV}}{n^2}}$$ (27.9)

These then are the energies of the quantized states of the hydrogen atom, as predicted by Bohr. The quantity 13.606 eV is the numerical value of the energy of the lowest ($n = 1$) state, since $E_n = -13.606$ eV when $n = 1$. We can write all the other energies in terms of the magnitude of E_1 in the form:

$$E_n = \frac{-E_1}{n^2} = \frac{-13.606 \text{ eV}}{n^2}$$ (27.10)

In Fig. 27.7 we plot the energy levels in hydrogen calculated from this equation. We also show the transitions between the various levels involved in the Lyman, Balmer, and Paschen spectral series in hydrogen. The combination of the Bohr-Einstein frequency condition with the energy values calculated from Eq. (27.10) leads to the wavelengths shown in the figure, which are all in excellent agreement with those observed spectroscopically.

FIGURE 27.7 Energy-level diagram for atomic hydrogen.

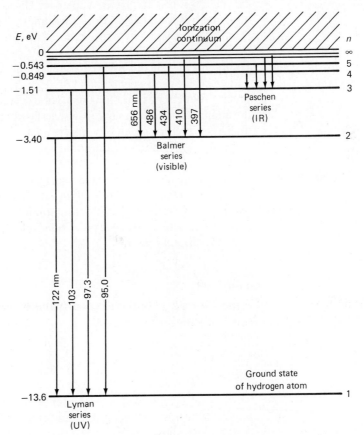

The Rydberg Constant

From the Bohr-Einstein frequency condition (Bohr's third postulate) applied to a transition from a level with $n = n_i$ to a level with $n = n_f$, we have

$$hf = E_i - E_f$$

or, from Eq. (27.10),

$$\frac{hc}{\lambda} = \frac{-E_1}{n_i^2} - \frac{-E_1}{n_f^2}$$

where E_1 is the numerical value of the energy of the atom in its lowest state, or *ground state*. Then we have

$$\frac{1}{\lambda} = \frac{E_1}{hc}\left(\frac{1}{n_f^2} - \frac{1}{n_i^2}\right)$$

This has exactly the same form as the empirical Rydberg equation [Eq. (27.2)] used to describe the spectrum of hydrogen. Hence, according to Bohr's theory, the numerical value of the Rydberg constant should be:

$$R = \frac{E_1}{hc} = \frac{2.1799 \times 10^{-18} \text{ J}}{(6.6262 \times 10^{-34} \text{ J·s})(2.9979 \times 10^8 \text{ m/s})} \qquad (27.11)$$

or $R = 1.0974 \times 10^7 \text{ m}^{-1}$

This value is in reasonable agreement with the experimental value 1.0968×10^7 m^{-1} for the hydrogen spectrum discussed in Sec. 27.1. Even better agreement is found when Bohr's assumption that the electron moves around a *stationary* nucleus is corrected to allow for the fact that both electron and proton actually move about the center of mass of the two-particle system. Hence not merely did Bohr explain well the existence of discrete energy states and the observed spectrum of hydrogen, he was also able to calculate a precise numerical value for the Rydberg constant from first principles, using no numerical constants except those flowing naturally from his theory. This ability to predict so closely the results of experiment pointed to the basic correctness of Bohr's revolutionary ideas.

Electron Radius on Bohr Model

Although the idea of quantized energy levels, or states, is the most important and lasting result of Bohr's work, it should be pointed out that the Bohr theory also predicts that discrete radii are associated with the electron's motion in these energy states. Equation (27.7) shows that these Bohr radii have the discrete values

$$r_n = \frac{n^2 \hbar^2}{m k_e e^2} \qquad (27.12)$$

Hence the radius of the electron orbits is quantized and increases as n^2. The smallest allowed radius is that for $n = 1$, the so-called *first Bohr radius*, for which

$$r_1 = \frac{\hbar^2}{m k_e e^2} = 5.28 \times 10^{-11} \text{ m} = 0.0528 \text{ nm}$$

This value agrees well as to order of magnitude with the known sizes of atoms, which are between 0.1 and 0.5 nm. The other orbits then have radii $r_n = n^2 r_1$, as shown in Fig. 27.8.

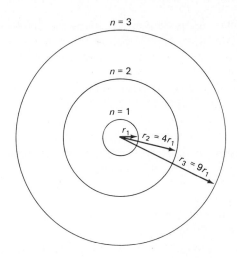

FIGURE 27.8 The first three Bohr orbits for the hydrogen atom.

Example 27.2

Calculate the values for the energy and the radius of the electron orbit, for the upper and lower energy states involved in:

(a) The first line in the Lyman series.
(b) The first line in the Balmer series.

SOLUTION

From Table 27.1 the energy levels involved in these cases are:

(a) Lyman series: $n_f = 1$ $n_i = 2$
(b) Balmer series: $n_f = 2$ $n_i = 3$

Hence we need to find the values of E_n and r_n for only three separate levels, since the $n = 2$ level is both the upper level of the first line in the Lyman series and the lower level for the first line in the Balmer series.

From the discussion in Sec. 27.3, and especially Eqs. (27.6) to (27.9), we have for these three levels:

$n = 1$: $E_1 = \boxed{-13.6 \text{ eV}}$ $r_1 = \boxed{0.0528 \text{ nm}}$

$n = 2$: $E_2 = \dfrac{-E_1}{n^2} = \dfrac{-13.6 \text{ eV}}{4} = \boxed{-3.4 \text{ eV}}$

$r^2 = n^2 r_1 = 4r_1 = 4(0.0528 \text{ nm}) = \boxed{0.211 \text{ nm}}$

$n = 3$: $E_3 = \dfrac{-E_1}{3^2} = \dfrac{-13.6 \text{ eV}}{9} = \boxed{-1.51 \text{ eV}}$

$r_3 = 3^2 r_1 = 9(0.0528 \text{ nm}) = \boxed{0.475 \text{ nm}}$

27.4 Energy Levels in Hydrogen and Other Atoms

Let us now look at the *energy-level diagram* for H, Fig. 27.7, a bit more carefully. Using the Bohr-Einstein frequency condition, we have

$$\lambda = \frac{c}{f} = \frac{hc}{hf} = \frac{hc}{E_i - E_f} = \frac{hc}{\Delta E}$$

or $$\lambda = \frac{hc}{\Delta E} = \frac{(6.63 \times 10^{-34} \text{ J·s})(3.00 \times 10^8 \text{ m/s})}{\Delta E \text{ (eV)}(1.60 \times 10^{-19} \text{ J/eV})}$$

and so $$\boxed{\lambda = \frac{1.24 \times 10^{-6} \text{ m}}{\Delta E}}$$ (27.13) where ΔE is the energy difference (*in electronvolts*) between the two states involved in the transition. This equation provides a very convenient way to obtain quickly the wavelength corresponding to a transition between any two states of known energies.

There is one point about Fig. 27.7 which can cause confusion. The only important quantities in calculating wavelengths are the energy *differences* between levels; the absolute values of these energies are of no importance. Since we are free to choose our zero of energy in any convenient way, we choose as the zero of energy for the H atom that state in which the electron has been completely removed from the atom and is at rest at a great distance from the nucleus. This occurs at the series limit ($n = \infty$) in the diagram. Hence all the energy levels below the series limit are negative, indicating that the electron is bound to the nucleus and that the atom therefore has a lower energy than do a free proton and a free electron at rest. This is the reason that all the energies in Fig. 27.7 have minus signs in front of them. Thus the energy of the lowest, or ground, state ($n = 1$) is -13.6 eV. This means that 13.6 eV of energy must be provided from outside the atom to remove the electron from the atom, or to *ionize* the atom. Hence 13.6 eV is the *ionization energy* for H. The corresponding voltage of 13.6 V is called the *ionization potential*.

Note that the continuum above $n = \infty$ corresponds to electrons having kinetic energy after they have been removed from the atom by ionization. Since kinetic energies are not quantized for free particles, any energy is possible above the series limit, and hence a continuum results. We therefore have discrete energy states below the series limit and continuous energy states above the series limit. The same is true for the energy-level diagrams of other atoms.

Photon Absorption by Atoms

The process of emission can be reversed. If UV light of wavelength 122 nm shines on H gas, atoms in the ground state E_1 in Fig. 27.7 can absorb this photon and move up to state E_2 with an overall increase in energy of 10.2 eV for the atom. For this absorption to occur, however, it is found experimentally that the incident photon must have an energy *exactly* equal to the energy difference between the states E_2 and E_1. Otherwise the photon will not be absorbed and the atom will remain in its original state.

At room temperature most gas atoms are in their ground state. Since in hydrogen a transition from this ground state to the first excited state requires a photon of wavelength 122 nm, H does not absorb at all in the visible range, but only in the ultraviolet. For this same reason most of the common gases like helium and neon are colorless since they do not absorb in the visible range. In the ultraviolet, however, they show strong absorption.

If sunlight, with a continuous distribution of wavelengths, falls on sodium (Na) vapor, the atoms will absorb out of the sunlight two very close yellow lines which satisfy the Bohr-Einstein frequency condition for the Na atom. This leads to dark lines for Na and many other elements in the sun's spectrum, as shown in the colored plate included in this chapter. The Na lines are those marked D_1 and D_2 on the top spectrum. These lines are called *Fraunhofer lines* and can be used to identify the elements in the sun's atmosphere, where most of the absorption occurs.

Excitation of Atoms

The absorption of light will lift atoms to excited states, from which they can then fall back to lower states by radiating photons of the proper frequency, as shown in Fig. 27.9. Hence the absorption of light leads to the excitation of atoms. The energy (in electronvolts) required to raise, or "excite," the atom from a lower energy state to a higher energy state is called the *excitation energy*. For example, the excitation energy for the $n = 2$ state in H is -3.4 eV $- (-13.6$ eV$) = 10.2$ eV. The voltage corresponding to this energy is called the *excitation potential*.

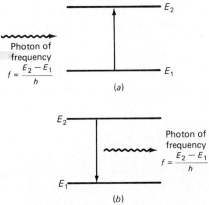

FIGURE 27.9 Excitation of atoms by photon absorption: (*a*) The atom can absorb a photon of energy $hf = E_2 - E_1$. (*b*) Once in the excited state, it can fall back to the lower state, emitting a photon of exactly the same frequency *f*.

A second way to excite atoms is by heating the gas. The inelastic collisions between atoms excite and ionize some of the gas atoms and lead to an emission spectrum. At room temperature nearly all the atoms of a gas like H are in the ground state. This is because the average kinetic energy of a gas atom at room temperature (300 K) is

$$\text{KE} = \tfrac{3}{2}kT = \tfrac{3}{2}(1.38 \times 10^{-23} \text{ J/K})(300 \text{ K})$$

$$= (6.21 \times 10^{-21} \text{ J})(6.24 \times 10^{18} \text{ eV/J}) = 0.039 \text{ eV}$$

This is only about 1/300 of the 10.2 eV required to excite a Hg atom from its ground state to its first excited state. Hence at 300 K there are very few atoms of sufficient energy to excite other atoms by inelastic collisions, and so no observable light is emitted. To excite atoms thermally requires very high temperatures. For example, $\tfrac{3}{2}kT$ is 10 eV at about 80,000 K. Some atoms (like Na) can be excited thermally at temperatures considerably lower than this, for their first excited state is much closer to their ground state. Also a small fraction of the atoms will have much larger kinetic energies than the average for a gas at temperature T, since the Boltzmann distribution (Sec. 9.9) contains atoms with energies from near zero all the way up to quite high energies.

A third way to excite atoms is to apply a high voltage across a tube containing atoms of the gas. An electric discharge then occurs through the gas, producing ions and free electrons. These free electrons are accelerated by the high voltage to such speeds that they can excite atoms by inelastic collisions. The atoms then fall back from all these excited states to the ground state and emit radiation. For this reason in electric discharges the complete spectrum of an atom is usually seen.

A variation on the electric discharge method of excitation is to bombard the gas with an electron beam of carefully controlled energy. This "electron-impact" method will be discussed in detail in the next section.

Example 27.3

What must be the temperature of a gas for the average kinetic energy of a gas atom to be 1.0 eV?

SOLUTION

The kinetic energy of gas atoms in equilibrium at temperature T is described by a Boltzmann distribution of average energy $\frac{3}{2}kT$. Here we want to have

$$\tfrac{3}{2}kT = 1.0 \text{ eV} = 1.6 \times 10^{-19} \text{ J}$$

or $$T = \frac{2}{3}\frac{1.6 \times 10^{-19} \text{ J}}{1.38 \times 10^{-23} \text{ J/K}} = \boxed{7.7 \times 10^{3} \text{ K}}$$

Hence the temperature must be close to 8000 K.

27.5 The Franck-Hertz Experiment

In 1914 two German physicists, James Franck (1882–1964) and Gustav Hertz (1887–1975), performed a crucial experiment in support of Bohr's ideas on atomic structure. They bombarded mercury (Hg) vapor with electrons of variable energy and studied how much energy was transferred from the electrons to the Hg atoms.

The apparatus used by Franck and Hertz is shown in Fig. 27.10. It consisted of a glass chamber containing mercury vapor at a gas pressure of about 1 torr (1 mmHg). A heated cathode C emitted electrons toward a grid G which was at a variable positive potential with respect to the cathode. Beyond the grid, which was a mesh structure through which electrons could pass, was a solid metal plate P maintained at about 0.5 V negative with respect to the grid. This provided the retarding potential needed to keep very low energy electrons from reaching the plate and contributing to the current measured by the meter M.

Franck and Hertz varied the voltage V_1 gradually and recorded the current through the meter as a function of this voltage. Their results are shown in Fig. 27.11. The current rose with the voltage until the voltage hit about 4.9 V. Then the current fell sharply to a low value. As the voltage V_1 between grid and cathode increased further, the current rose again, but at about 9.8 V it again fell sharply. This behavior continued at higher voltages, with the current falling at intervals of 4.9 V. The graph of current against voltage V_1 obtained by Franck and Hertz consisted of current peaks with consecutive peaks separated by an average value of 4.9 ± 0.1 V, as can be seen in the figure.

How can we explain these experimental results? One way is as follows. At voltages below 4.9 V the electrons have low kinetic energies and collide

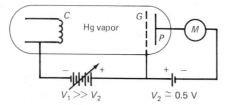

FIGURE 27.10 Apparatus for the Franck-Hertz experiment.

OPTICAL SPECTRA

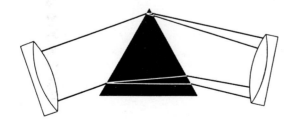

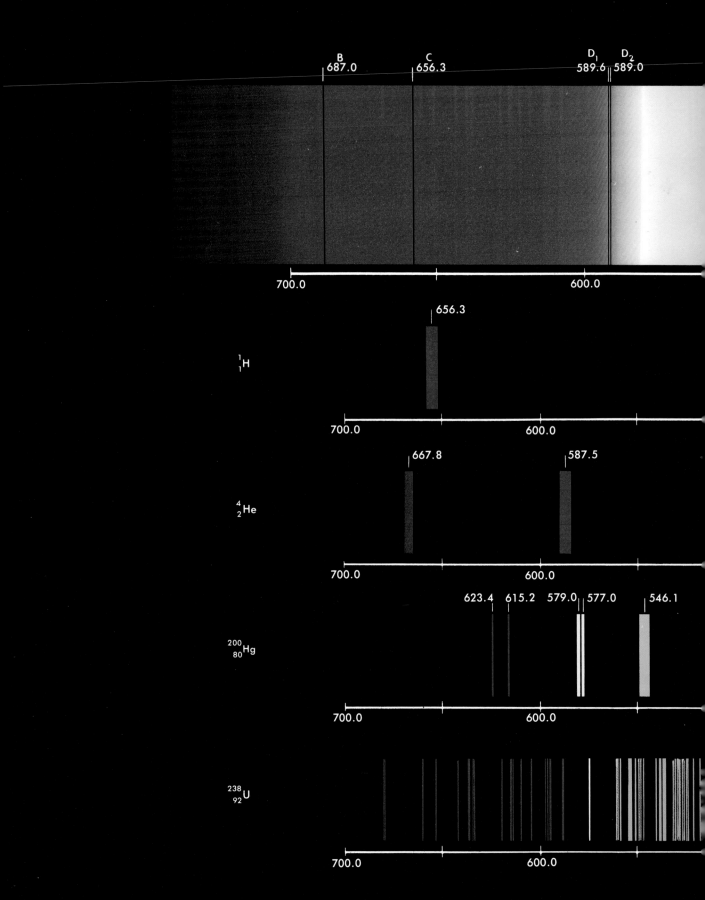

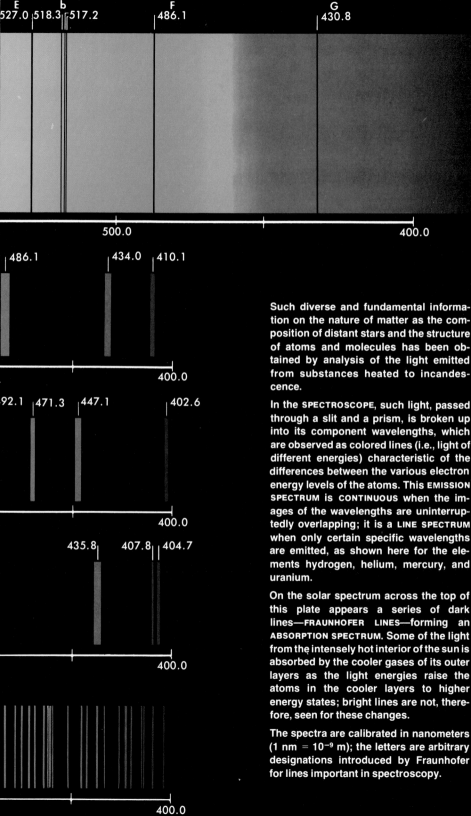

E   b 517.2   F   G
527.0 518.3   486.1   430.8

500.0   400.0

486.1   434.0 410.1

400.0

492.1 471.3 447.1   402.6

400.0

435.8   407.8 404.7

400.0

400.0

Such diverse and fundamental information on the nature of matter as the composition of distant stars and the structure of atoms and molecules has been obtained by analysis of the light emitted from substances heated to incandescence.

In the SPECTROSCOPE, such light, passed through a slit and a prism, is broken up into its component wavelengths, which are observed as colored lines (i.e., light of different energies) characteristic of the differences between the various electron energy levels of the atoms. This EMISSION SPECTRUM is CONTINUOUS when the images of the wavelengths are uninterruptedly overlapping; it is a LINE SPECTRUM when only certain specific wavelengths are emitted, as shown here for the elements hydrogen, helium, mercury, and uranium.

On the solar spectrum across the top of this plate appears a series of dark lines—FRAUNHOFER LINES—forming an ABSORPTION SPECTRUM. Some of the light from the intensely hot interior of the sun is absorbed by the cooler gases of its outer layers as the light energies raise the atoms in the cooler layers to higher energy states; bright lines are not, therefore, seen for these changes.

The spectra are calibrated in nanometers (1 nm = 10^{-9} m); the letters are arbitrary designations introduced by Fraunhofer for lines important in spectroscopy.

with the Hg atoms *elastically*, with kinetic energy conserved in the collision. At about 4.9 V something new happens: an electron loses all its kinetic energy in a single *inelastic* collision with a Hg atom. The condition of this Hg atom is therefore changed. It has been excited from its normal (or ground) energy state to an excited state which has 4.9 eV more energy than the ground state. The electrons have therefore lost 4.9 eV of energy and cannot overcome the 0.5-V retarding potential V_2 to reach the plate. The current therefore falls sharply, as observed in Fig. 27.11.

If the voltage is increased above 4.9 V, the electrons again acquire enough energy to reach the plate after losing 4.9 eV in one inelastic collision. Hence the current increases again. But at 9.8 V the electrons have enough energy to make two consecutive inelastic collisions with Hg atoms. Hence the current falls again. This falloff in current continues at intervals of 4.9 V as the voltage is increased further.

This seems like a reasonable physical explanation of what is going on in the mercury vapor, but how can we be sure that this interpretation is correct? We can look at the Hg gas with an ultraviolet grating spectrometer (see Sec. 22.6) as the voltage is increased. No spectrum at all appears until the voltage reaches 4.9 V. Then a single bright ultraviolet line at 253 nm stands out against the dark background. This should correspond to Hg atoms falling back to their ground state from the excited state to which they have been raised by a colliding electron. If we set this energy equal to hf, we can find the wavelength to be expected. It is, from Eq. (27.13),

$$\lambda = \frac{1.24 \times 10^{-6} \text{ m}}{\Delta E} = \frac{1.24 \times 10^{-6} \text{ m}}{4.9} = 253 \text{ nm}$$

which agrees perfectly with the wavelength of the observed ultraviolet line.

The Franck-Hertz experiment therefore provided direct experimental

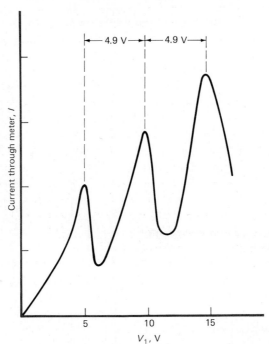

FIGURE 27.11 Results of the Franck-Hertz experiment: A plot of the current through the meter *M* in Fig. 27.10 against the applied voltage V_1.

proof of the existence of discrete energy levels in atoms, as predicted by Bohr, and of the way these energy levels lead to observed spectra. The electrons lose energy in their collisions with Hg atoms only in discrete energy chunks, or quanta, corresponding to a precise energy difference between two energy states in the atom, and these quanta are radiated away from the Hg vapor as photons of a single wavelength when the atom falls back to its ground state.

A rough energy-level diagram for Hg, as obtained from the work of Franck and Hertz, is shown in Fig. 27.12. Evidence from spectroscopy and from electron-impact studies like the Franck-Hertz experiment provides an answer to our initial question of why all atoms of a particular element are the same: All the atoms of any element have identical energy levels that determine their physical and chemical properties and distinguish that element from all other elements. Bohr provided the beginnings of an answer to the further question of why atoms have these discrete energy levels, but even his theory was not able to provide a fully satisfactory answer, as we will soon see.

FIGURE 27.12 Partial energy-level diagram for the mercury atom (Hg).

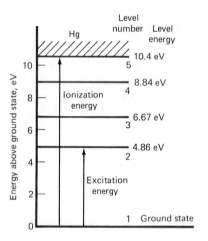

Example 27.4

In an electron-impact experiment performed on atomic hydrogen gas it is found that the atoms are excited from the ground state to excited states by electrons of energies 10.2 and 12.1 eV, respectively.
(a) What wavelengths are to be expected in the radiation from such a gas after it has been excited in this way?
(b) Are any of these wavelengths in the visible region of the spectrum? In what region of the spectrum are the other wavelengths?

SOLUTION

(a) The energy levels of the H atom, as determined from the experiment described in the problem, are shown in Fig. 27.13. There are therefore three possible transitions, one from state 3 to state 1, another from state 2 to state 1, and a third from state 3 to state 2. The corresponding wavelengths are given by Eq. (27.13) as $\lambda = (1.24 \times 10^{-6} \text{ m})/\Delta E$. Hence

$$\lambda_{31} = \frac{1.24 \times 10^{-6} \text{ m}}{E_{31}} = \frac{1.24 \times 10^{-6} \text{ m}}{12.1} = \boxed{103 \text{ nm}}$$

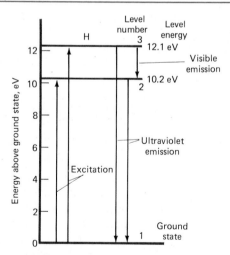

FIGURE 27.13 Diagram for Example 27.4. Two higher-energy states of the H atom are excited by electron impact. This leads to one visible (red) line and two ultraviolet lines in the emission spectrum of the H atom.

$$\lambda_{21} = \frac{1.24 \times 10^{-6} \text{ m}}{E_{21}} = \frac{1.24 \times 10^{-6} \text{ m}}{10.2} = \boxed{122 \text{ nm}}$$

$$\lambda_{32} = \frac{1.24 \times 10^{-6} \text{ m}}{E_{32}} = \frac{1.24 \times 10^{-6} \text{ m}}{12.1 - 10.2} = \boxed{653 \text{ nm}}$$

(b) Of the three expected wavelengths only λ_{32} is in the visible region of the spectrum, since 653 nm is in the red. The other two wavelengths are much shorter and are in the ultraviolet. The red line at 653 nm can be seen on the colored spectrum of hydrogen included in this chapter.

27.6 The Emission and Absorption of Radiation

Suppose we consider two energy levels E_1 and E_2 in an atom, as in Fig. 27.14. If the atom is originally in the lower state E_1, and a photon comes along of exactly the frequency $f = (E_2 - E_1)/h$, the atom can be excited from E_1 to E_2 by the *absorption* of the photon (Fig. 27.14a). Similarly, if the atom is in the excited state E_2, it can, if the transition is allowed, fall of its own accord from level 2 to level 1 and emit a photon of energy $f = (E_2 - E_1)/h$. This is called *spontaneous emission* (Fig. 27.14b).

In addition to absorption and spontaneous emission there is a third possibility. Suppose the atom is in the upper state with energy E_2. If a photon comes along of exactly the frequency $(E_2 - E_1)/h$, it can *force* the atom to fall to the lower level and emit a photon of frequency $f = (E_2 - E_1)/h$. Hence the result of this process of *forced* (or *stimulated*) *emission* is that the entering photon leaves the atom with a second photon of exactly the same frequency accompanying it, as shown in Fig. 27.14c.

FIGURE 27.14 Types of transitions possible between two energy levels in an atom: (*a*) absorption; (*b*) spontaneous emission; (*c*) forced, or stimulated, emission.

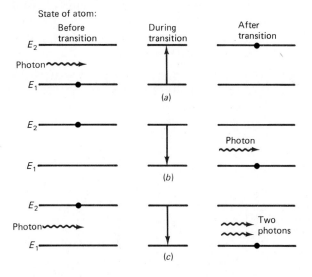

The Laser

The term *laser* is an acronym for *l*ight *a*mplification by the *s*timulated *e*mission of *r*adiation. The appropriateness of this name should be clear from the following discussion.

When a photon of the right energy is incident on an atom, it can produce either absorption or stimulated emission. Einstein showed that these two processes are equally probable, and that whether net absorption or net emission results depends on whether there are more atoms in the upper excited state or in the lower state. In general, there will be more atoms in the lower state, since the atom will tend of its own accord to fall to a lower energy state by spontaneous emission.

A laser is a device which enables stimulated emission to win out over absorption to such an extent that the atom emits very strongly at one particular frequency. To accomplish this, two criteria must be satisfied:

1 The upper state involved in the transition must be *metastable*; i.e, the atom must remain in that state for a long time before it falls of its own accord to the lower state by spontaneous emission. For laser action to occur, stimulated emission must be far greater than spontaneous emission.

2 A *population inversion* must be produced. By this we mean that there must be more atoms in the upper state than in the lower state, so that stimulated emission dominates absorption when light of the proper wavelength falls on the system. Such a population inversion must be artificially produced, since it is the reverse of what would occur naturally. A population inversion can be achieved in a variety of ways: by flooding the gas with high-intensity light and thus pumping atoms into excited states, by atomic collisions, or by other more sophisticated techniques.

A typical scheme for inverting populations using "optical pumping" is shown in Fig. 27.15. Intense radiation from a flash lamp pumps an atomic system from the ground state into an excited state of energy E_3, from which it decays to the metastable state of energy E_2. Then photons at a frequency $f = (E_2 - E_1)/h$ can cause stimulated emission from state 2 to state 1. Since the pumping radiation either is kept on continuously (continuous lasers) or is provided in short, repeated pulses (pulsed lasers), the population of the upper state is constantly replenished by the pumping process.

FIGURE 27.15 A typical scheme for producing a population inversion by optical pumping, using a three-level system. Since level E_2 is above the ground state it is normally sparsely populated at room temperature. But this level is metastable, and so atoms excited to E_3 by the pumping radiation decay and accumulate in this metastable level, leading to a population inversion between states E_2 and E_1.

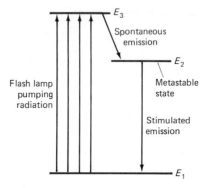

Helium-Neon Gas Laser

A common type of laboratory laser is the He–Ne gas laser shown in Fig. 27.16. In this case a mixture of gases (15% He, 85% Ne) is used to excite a metastable state of Ne by collisions with He atoms. A high voltage placed across the gas mixture produces electrical breakdown of the gas into ions and electrons. Fast-moving electrons excite the He atom to the energy level E_2 in Fig. 27.17. This level is almost identical in energy to that of the Ne atom in the state E_2'. Hence by inelastic collisions the He atoms transfer their energy to the Ne atoms, and a large population is built up in this excited metastable state of Ne. Photons at frequencies $f = (E_2' - E_1')/h$ then stimulate transitions

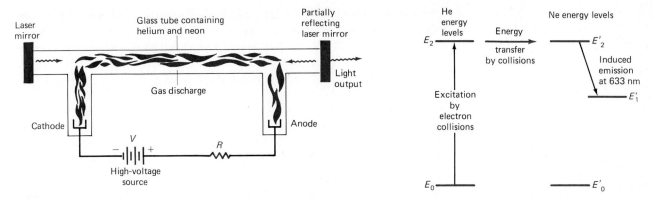

FIGURE 27.16 (left)   A schematic diagram of a He–Ne laser.

FIGURE 27.17 (right)   Energy-level diagram for He and Ne atoms, showing how energy is transferred by collisions from the E_2 level of He to the E_2' level of Ne. This leads to a population inversion between the E_2' and E_1' states of Ne.

to state E_1', with the emission of radiation at a wavelength of 633 nm. This explains why the He–Ne laser produces a very intense line in the red region of the spectrum.

In the He–Ne laser shown in Fig. 27.16 one photon of the proper frequency $f = (E_2' - E_1')/h$ moving along the axis of the tube causes forced emission of a second photon by a Ne atom. These two photons, which are at exactly the same frequency and moving in the same direction, then move along together and produce the emission of additional photons by other Ne atoms. A multiplication process results and a very large number of photons of exactly the same frequency are produced in a small fraction of a second. This effect is increased even more by the mirrors at the ends of the tube, which keep the photons moving back and forth along the axis, stimulating more emissions as they go. Some of the light emerges through the partially transparent ("leaky") mirror at one end, and a very intense beam of single-frequency, highly directional light results.

A laser beam is highly *directional* because any photons not moving exactly perpendicular to the parallel mirrors at the two ends are lost from the beam which finally emerges from one end of the tube. Laser light is also

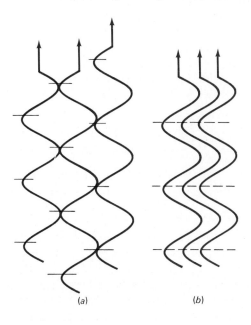

FIGURE 27.18   Difference between coherent and incoherent light: (*a*) Incoherent light; the individual photons are not in phase with one another. (*b*) Coherent light; all photons are perfectly in phase with one another.

(*a*)   (*b*)

highly *monochromatic*, since the photon causing forced emission and the one resulting from forced emission must have exactly the same frequency, or forced emission will not occur. Laser radiation is also *coherent*, which means that all the photons in the beam of light emerging from the end of the laser are perfectly *in phase*. This is illustrated in Fig. 27.18.

Uses of Lasers

Lasers have many practical uses because of their extreme monochromaticity, coherence, and directionality. They have been used to establish a perfect straight line for the laying of long pipes and tunnels, and for monitoring the distance from the earth to the moon using mirrors placed on the moon's surface by astronauts. Very powerful lasers can be used to cut and bore metals and are being developed as destructive weapons against aircraft and tanks. Lasers are the heart of new music and TV replay systems, and of the computer systems in use at supermarket checkout counters.

Medical Applications   The dream of every medical doctor is to be able to see inside a patient's body and to perform necessary surgery without having to cut open the patient. This dream is now closer to reality because of the joining of lasers to fiber-optics technology.

For diagnostic purposes an *endoscope* (meaning "seeing within") like the one in Fig. 27.19 is routinely used. The endoscope is basically a bundle of optical fibers which carries light to the inside of the body and then transfers an image of the internal organs back outside the body for viewing by the doctor, either directly or on a TV monitor. Endoscopes often use intense white-light sources, but for many purposes laser light in the visible region of the spectrum is preferable. Endoscopes can be inserted into the body through the mouth or rectum and moved along the esophagus, the gastrointestinal tract, the bronchus, or other internal passages to look for disease and blockages.

Pulsed ruby lasers have been used for many years to mend detached retinas, since the laser beam will pass through the transparent outer parts of the eye without damaging it and can then be focused on the defective area of the retina. A burst of photons generates sufficient heat to "weld" the retina to the choroid from which it has become detached.

Very high energy carbon dioxide (CO_2) lasers are now used to destroy cancerous tissue on the outside of the body or in other regions of the body accessible through natural body orifices. The 10,600-nm infrared light from a CO_2 laser is ideal for surgery because it is strongly absorbed by water, which

FIGURE 27.19   A typical endoscope for seeing inside the human body.

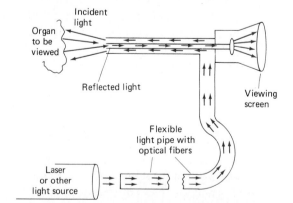

makes up 80 to 90 percent of body tissue, including tumor tissue. Sufficient energy is delivered by the laser to heat the defective tissue to a temperature high enough to destroy it. At the same time the intense heat cauterizes the wound to prevent spreading of the disease to normal tissues.

27.7 Quantum Numbers

We have seen that the energy levels of the hydrogen atom are given by Eq. (27.9),

$$E_n = \frac{-13.6 \text{ eV}}{n^2}$$

where n is called the *principal quantum number*. This quantum number n labels the discrete energy levels within the atom. All electrons that possess the same total energy E_n are said to be in the same energy *shell*. The lowest level, the ground state, corresponds to $n = 1$; the first excited state is $n = 2$; and so on. As n increases, the energy increases, since the numerical value of E_n in the above equation decreases. But the right side of the equation has a negative sign, and so the energy becomes progressively less negative as n increases. In other words, the energies of the excited states *increase with n*, as can be seen in Fig. 27.7

Nature is always more complicated than our theories. It was soon found that the Bohr theory was not adequate even for the simplest of all atoms, hydrogen. When observed under high resolution, the lines in the hydrogen spectrum show a *fine structure* in which, instead of the single lines predicted by the Bohr theory, two or more lines very close together are found. The German physicist Arnold Sommerfeld (1868–1951) suggested that this might be because the electrons can move in noncircular orbits like the ellipses in which the planets move around the sun (a possibility mentioned by Bohr in his original paper). To a first approximation all the electrons in elliptic orbits corresponding to the same value of n would have the same energy, but the orbital angular momentum of the atom would vary with the shape (the *eccentricity*) of each ellipse. When the relativistic variation of mass of the electron with its speed was taken into account (Sec. 25.8), Sommerfeld found that the different elliptical motions had slightly different energies, and that these differences corresponded well to the observed fine structure in the hydrogen spectrum.

Sommerfeld was able to show that the angular momentum for these elliptical orbits could be specified by a second quantum number l, where the angular momentum would be given by $L = \sqrt{l(l + 1)}\hbar$. For any given n, the only possible values of l, called the *orbital angular momentum quantum number*, are

$$l = 0, 1, 2, 3, \ldots, n - 1$$

Still further complications arise. If an electron has orbital angular momentum, we have a charged particle moving in a closed loop which is the equivalent of a tiny bar magnet (see Sec. 18.8). In a magnetic field this bar magnet can line up with the applied field. According to quantum theory, however, there are only a limited number of quantized directions which the orbital angular momentum vector **L** can assume with respect to the magnetic

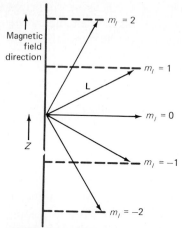

↑ Magnetic field direction

L

↕ Z

$m_l = 2$
$m_l = 1$
$m_l = 0$
$m_l = -1$
$m_l = -2$

FIGURE 27.20 Quantized orientation of the orbital angular momentum **L** in a magnetic field (for $l = 2$).

field, as shown in Fig. 27.20 for $l = 2$. These directions are specified by a third quantum number, the *magnetic quantum number* m_l, which measures the component of the orbital angular momentum in the direction of the magnetic field, which we take to be the Z direction. For this reason m_l can never be larger than l, and so the allowed values of m_l are $0, \pm 1, \pm 2, \pm 3, \ldots, \pm l$. The Z component of the angular momentum is then $L_z = m_l \hbar$.

A final complication results from an observed doubling of spectral lines which could not be explained in terms of elliptic orbits and quantized orbital angular momentum. One possible explanation rests on the assumption that the electron itself is spinning about an axis through its center as it moves about the nucleus, in the same way that the earth spins on its axis as it rotates about the sun. Such a spinning, negatively charged electron would have a tiny magnetic moment which could also line up either in the direction of an applied magnetic field or in the opposite direction to the field. These two possibilities are distinguished by a *spin quantum number* m_s which can have only one of the two values $+\frac{1}{2}$ (parallel to field) or $-\frac{1}{2}$ (antiparallel to field). It was found that this new quantum number could indeed account properly for the observed doubling of spectral lines.

In summary, then, instead of one quantum number n, we now have four quantum numbers n, l, m_l, and m_s which are needed to determine fully the energy state and the motion of the electron in the hydrogen atom. Table 27.2 summarizes the significant facts about these four quantum numbers. Each individual combination of the quantum numbers n, l, m_l, and m_s specifies a particular energy state of an electron in an atom. The number of possible electronic states increases rapidly with n. Thus for $n = 1$ there are only two different states, but for $n = 2$ there are eight different states. (Try writing out for yourself the four quantum numbers specifying each of these states.) In general the number of states can be shown to be $2n^2$ for any principal quantum number n.

TABLE 27.2 Quantum Numbers for a Single Electron in Hydrogen

Name of quantum number	Symbol	Physical significance	Allowed values
Principal quantum number	n	Determines energy shell of atom	$1, 2, 3, \ldots, \infty$
Orbital angular momentum quantum number	l	Determines angular momentum of electron in its orbit	$0, 1, 2, \ldots, n - 1$
Magnetic quantum number	m_l	Determines component of orbital angular momentum in direction of magnetic field	$0, \pm 1, \pm 2, \ldots, \pm l$
Spin quantum number	m_s	Determines component of spin angular momentum of electron in direction of magnetic field	$+\frac{1}{2}, -\frac{1}{2}$

Example 27.5

(a) How many different energy states exist in the hydrogen atom when the electron is in the energy shell specified by principal quantum number $n = 3$?

(b) Indicate how this number may be arrived at from a consideration of the possible values of the four quantum numbers n, l, m_l, and m_s.

SOLUTION

(a) As pointed out in the section above, the number of different energy states for principal quantum number n is $2n^2$. Here, with $n = 3$, we have

$2(3)^2 = \boxed{18 \text{ states}}$

(b) For $n = 3$, the possible values of l are 0, 1, and 2.

For $l = 0$: $m_l = 0$

For $l = 1$: $m_l = -1, 0, +1$

For $l = 2$: $m_l = -2, -1, 0, +1, +2$

Adding the number of possible values of m_l we obtain $1 + 3 + 5 = 9$. Now for each of these values of m_l, the spin quantum number m_s can have either the value $+\frac{1}{2}$ or $-\frac{1}{2}$. Hence the total number of possible states is $2 \times 9 = 18$, which checks with the value obtained in part (*a*).

27.8 The Pauli Exclusion Principle and the Periodic Table

Thus far we have discussed only the hydrogen atom. We now want to make use of some of the ideas obtained from our discussion of hydrogen to explain the structure of multiple-electron atoms. Such atoms contain orbiting electrons equal in number to the charge number Z of the nucleus, i.e., the number of elementary positive charges in the nucleus.

To explain, at least qualitatively, the structure of multiple-electron atoms, we will rely on two principles. The first is that, if left to itself, an atom will always arrange itself in the lowest energy state possible, just as a roller coaster will always tend to reach the lowest point on its track. The atom's overall energy state depends on the individual, or one-electron, energy states (like those for H) in which each electron finds itself. The lowest energy state for the atom is called the normal or *ground state*. This might lead us to believe that in the ground state all the electrons would crowd into the lowest possible one-electron energy states like the hydrogen ground state, with $n = 1$ and $l = 0$. That this does not happen is explained by a second principle first introduced in 1925 by the Austrian physicist Wolfgang Pauli (see Fig. 27.21) and called the Pauli exclusion principle.

FIGURE 27.21 Wolfgang Pauli (1900-1958), who received the 1945 Nobel Prize in physics for his discovery of the Pauli exclusion principle. Pauli, who became an American citizen in 1946, was a brilliant theoretical physicist, who first postulated the existence of the neutrino, and made important contributions to statistical, nuclear and elementary-particle physics. In their more whimsical moments physicists sometimes refer to "the Pauli effect," in which Pauli's mere presence in a physics laboratory was sufficient to cause valuable experimental apparatus to self-destruct. (*Photo courtesy of AIP Niels Bohr Library.*)

Pauli Exclusion Principle

> In one and the same atom no two electrons can have the same set of values for the four quantum numbers n, l, m_l, and m_s.

This means that no two electrons can be in the same one-electron state. Hence as larger and larger atoms are built up by adding electrons (and positive charges in the nucleus to keep the total charge on the atom zero), the added electrons must go into higher-energy states corresponding to larger values of n and, for a particular value of n, to larger values of l. All the electrons with the same value of n are said to be in the same *shell*. Those with the same values of n and l are said to be in the same *subshell*. When a shell or subshell contains the maximum number of electrons allowed, it is said to be *closed*. Any more electrons added after a shell or subshell is closed must therefore occupy higher-energy shells or subshells.

The Periodic Table

Let us apply these ideas to some simple atoms. In Table 27.3 we show the values of the four quantum numbers for the electrons in the first 11 atoms in the periodic table. We will discuss only the ground states of these atoms.

Hydrogen ($Z = 1$) has its only electron in the $n = 1$ shell. Hence l and m_l must both be zero; m_s can be either $+\frac{1}{2}$ or $-\frac{1}{2}$, but this has no effect on the overall energy of the atom.

TABLE 27.3 Quantum Numbers for Electrons in Some Simple Atoms[*]

Element	Number of electrons	Quantum numbers[†]				Electron configuration
		n	l	m_l	m_s	
H	1	1	0	0	$+\frac{1}{2}$	$(1s)$
He	2	1	0	0	$-\frac{1}{2}$	$(1s)^2$
		Closed shell				
Li	3	2	0	0	$+\frac{1}{2}$	$(1s)^2(2s)$
Be	4	2	0	0	$-\frac{1}{2}$	$(1s)^2(2s)^2$
		Closed subshell				
B	5	2	1	-1	$+\frac{1}{2}$	$(1s)^2(2s)^2(2p)$
C	6	2	1	-1	$-\frac{1}{2}$	$(1s)^2(2s)^2(2p)^2$
N	7	2	1	0	$+\frac{1}{2}$	$(1s)^2(2s)^2(2p)^3$
O	8	2	1	0	$-\frac{1}{2}$	$(1s)^2(2s)^2(2p)^4$
F	9	2	1	$+1$	$+\frac{1}{2}$	$(1s)^2(2s)^2(2p)^5$
Ne	10	2	1	$+1$	$-\frac{1}{2}$	$(1s)^2(2s)^2(2p)^6$
		Closed shell				
Na	11	3	0	0	$+\frac{1}{2}$	$(1s)^2(2s)^2(2p)^6(3s)$

[*]An atom on any given line contains the electron with the quantum numbers on that line and all electrons on lines above that line.

[†]Since, for any given l, all values of m_l and m_s correspond to the same energy, the values of m_l and m_s indicated are not uniquely determined for each atom. The important thing is that the values of m_l and m_s must be chosen to satisfy the Pauli exclusion principle for each atom.

Helium ($Z = 2$) must have its second electron with a spin opposite to that of the first electron to satisfy the Pauli principle. We talk about the spins being *paired*. Since this completes the $n = 1$ shell, He is a closed-shell atom and is therefore very stable, does not enter into chemical reactions, and has a very high ionization potential (24.6 V).

Lithium ($Z = 3$) has a third electron outside the He closed shell. This electron must start a new ($n = 2$) shell. Such an electron has a large energy compared with the two $n = 1$ shell electrons and is far removed from the nucleus relative to them, as should be clear from our discussion of the energy levels of H. As a result, this electron is loosely bound, with an ionization potential of only 5.4 V. Lithium is therefore a very reactive element chemically, and is monovalent, because it is this outermost electron which enters into chemical reactions.

After lithium the $n = 2$ shell fills up as shown in the table. At neon ($Z = 10$) the $n = 2$ shell has its full complement of 8 ($= 2n^2$) electrons, and the shell is closed. This makes Ne similar to He, since they both are closed-shell atoms. Ne is also very stable, with an ionization potential of 21.6 V. It is therefore inert chemically, just like He. Since He and Ne have the same closed-shell configuration of electrons, they fall in the same column of the periodic table, along with all the other inert gases like argon, krypton, and xenon.

Sodium ($Z = 11$) then must have one electron in the $n = 3$ shell outside two interior closed shells. Its electron configuration is therefore very similar to that of Li. It is reactive chemically, is monovalent, and has a low ionization potential (5.1 eV). It falls in the same column of the periodic table as Li, that of the alkali metals, and behaves chemically much like lithium, potassium, rubidium, and cesium, except for differences due to changes in the sizes of the atoms.

We could continue this process through the rest of the periodic table, but the principles on which this "building up" of the periodic table is based are the important factors here, not the details.

These principles may be summarized in the following statement:

Building-Up Principle

> Atoms are built up by adding electrons one at a time in the lowest one-electron energy states permitted by the Pauli exclusion principle.

Diagrams of the electron configurations for some of the atoms discussed above are given in Fig. 27.22. The fact that all atoms are roughly the same size is due to the exclusion principle. The increased positive charge on the nucleus reduces the radius of the inner shells, but the Pauli principle requires that some of the electrons go into outer shells of greater radius. Figure 27.23 shows the comparative sizes of He and Ne atoms with and without the Pauli exclusion principle.

The important quantum numbers in determining the energy shells and subshells of an atom are the principal quantum number n and the orbital angular momentum quantum number l. For historical reasons based on spectroscopy, values of l are specified by using the letters s, p, d, f, . . . , to signify $l = 0, 1, 2, 3, . . .$, respectively. Hence the one electron in hydrogen

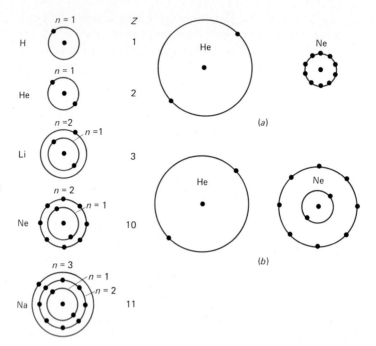

FIGURE 27.22 Schematic diagram of electron configurations for some common light atoms. (No attempt has been made to show the relative sizes of the atoms.)

FIGURE 27.23 Comparative sizes of the He and Ne atoms: (a) without the Pauli exclusion principle; (b) with the Pauli exclusion principle. Experiments clearly confirm the result in (b).

which has $n = 1$ and $l = 0$ is called a $1s$ electron. Boron, which has five electrons, then has two $1s$ electrons, two $2s$ electrons, and one $2p$ electron. Hence its electron configuration is written as $(1s)^2(2s)^2(2p)$. This symbolism is encountered all the time in dealing with atoms and molecules. The assignment of all electrons to the various energy shells and subshells by indicating the values of n and l for each electron specifies the *electron configuration* of the atom. The electron configurations for the first 11 atoms in the periodic table are given in the last column of Table 27.3. Electron configurations for other atoms can be obtained from the rules governing the allowed values of n, l, m_l, and m_s.

Example 27.6

Indicate the electron configuration for chlorine ($_{17}Cl$) and show why Cl should be similar to fluorine ($_9F$) in chemical and spectroscopic properties.

SOLUTION

From Table 27.3 we see that the electron configuration of $_{11}Na$ is $(1s)^2(2s)^2(2p)^6(3s)^1$. In the next atom, magnesium ($_{12}Mg$), the $3s$ subshell is filled, and the additional electrons in $_{17}Cl$ must therefore go into the $3p$ subshell. This can hold six electrons ($m_l = -1, 0, +1$; $m_s = \pm\frac{1}{2}$), but $_{17}Cl$ has only five electrons left for this subshell. Hence its electron configuration will be

$$(1s)^2(2s)^2(2p)^6(3s)^2(3p)^5$$

Comparing this electron configuration with that of fluorine, whose configuration is given in Table 27.3 as $(1s)^2(2s)^2(2p)^5$, we see that both have partially filled subshells containing five p electrons as the outermost electrons in the atom. Hence we would expect that $_{17}Cl$ and $_9F$ would have similar chemical and spectroscopic properties. Because they have such similar electron configurations, they appear in the same column of the periodic table.

27.9 Successes and Failures of the Bohr Theory

One of the early successes of the Bohr theory was in explaining the spectra of hydrogen-like ions, i.e., ions which have only one electron but have nuclear charges greater than 1. Examples are He^+, Li^{2+}, and Be^{3+}. The spectra of such ions should be similar to the spectrum of H except that a factor of Z^2 is introduced into the Rydberg equation because the nuclear charge is no longer e, but Ze. Good agreement was found for such one-electron ions between experimental results and those predicted by Eq. (27.2). This provided additional confirmation of the basic correctness of the Bohr theory.

The Spectrum of He⁺

For He^+ ($^4_2$He minus one electron) the wavelengths predicted by the Rydberg formula [Eq. (27.2)] are

$$\frac{1}{\lambda} = RZ^2 \left(\frac{1}{n_f^2} - \frac{1}{n_i^2} \right) = 4R\left(\frac{1}{n_f^2} - \frac{1}{n_i^2} \right)$$

Hence the same kind of spectral series should be found in He^+ as in H, but at frequencies 4 times as large, or wavelengths 4 times shorter.

This led to the understanding of a mysterious series found in 1897 by the American astronomer E. C. Pickering (1846–1919) in the spectrum of the star Zeta Puppis, and shown in Fig. 27.24. He ascribed it to hydrogen of a special form existing only in the stars, since the spectrum greatly resembled the H Balmer series but each line was slightly displaced from the corresponding Balmer line, and additional single lines appeared between every two Balmer lines.

In his 1913 paper Bohr showed that this series was actually due to He^+ and corresponded to the series with $n_f = 4$ and $n_i = 5, 6, 7, \ldots$. The Rydberg formula for He^+ with $n_f = 4$ is

$$\frac{1}{\lambda} = 4R\left(\frac{1}{4^2} - \frac{1}{n_i^2} \right) = R\left[\frac{1}{2^2} - \frac{1}{(n_i/2)^2} \right]$$

For $n_i = 6$, this is

$$\frac{1}{\lambda} = R\left(\frac{1}{2^2} - \frac{1}{3^2} \right)$$

the first term in the Balmer series for H. For $n_i = 8$, it yields the second term in the Balmer series, and so on for all even values of n_i. In between these

FIGURE 27.24 The spectrum of the ion He^+ compared with that of the H atom.

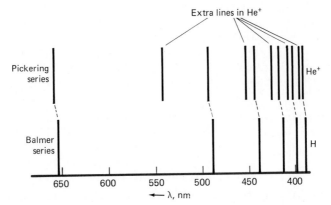

Balmer-like lines in the He$^+$ spectrum, however, are other lines with n_i odd, that is, $n_i = 5, 7, 9, \ldots$. These are the additional lines not found in the H spectrum which caused Pickering to propose that the spectrum was due to a stellar form of H.

Except for slight wavelength differences between the Balmer lines and every second line in the Pickering series, the Bohr theory gave a very convincing explanation of the mysterious Pickering series. Even those small differences were soon explained by the fact that the electron does not really move around a stationary nucleus, but about the center of mass of the two-particle system. Since the He nucleus is 4 times heavier than the H nucleus, the center of mass differs in H and He$^+$ and produces a slightly different value of the Rydberg constant R for He$^+$. When this adjusted value of R is used in Eq. (27.2), the spectral wavelengths are in complete agreement with Pickering's observations.

The Characteristic X-Ray Spectrum

In addition to the continuous x-ray spectrum it was found that sharp lines occur in the x-ray region of the spectrum in both emission and absorption. As shown in Fig. 27.25, these lines are superimposed on the continuous x-ray spectrum. These lines are similar in some ways to those produced by the outer electrons of atoms, but there are also very significant differences between x-ray line spectra and those produced by optical transitions in the visible region of the spectrum. Among these differences are the following:

X-Ray Line Spectra

1 The lines are hardly at all dependent on the chemical state of the element.
2 The line spectrum is very similar from element to element, but there is a slight shift to higher frequencies as the atomic number increases. Also, new x-ray lines appear as new energy shells are added.
3 The energy differences corresponding to the observed lines are thousands of electronvolts.

Optical Line Spectra

1 The lines are very dependent on the chemical state of the element.
2 The line spectrum differs greatly from element to element.

3 The energy differences corresponding to the observed lines are only a few electronvolts.

One possible explanation of these experimental facts is that x-ray line spectra are produced by electrons in shells deep inside the target atoms, just as the outer-shell electrons are responsible for optical line spectra. Such spectra vary with the nature of the target material (unlike continuous x-ray spectra), and are therefore called *characteristic x-ray spectra*.

Characteristic x-ray spectra can be well explained on the basis of the simple Bohr theory. Figure 27.26 shows the various electron shells corresponding to different principal quantum numbers n in a heavy atom. Shells with $n = 1, 2, 3, 4, \ldots$, are denoted as K, L, M, N, $\ldots$, shells, respectively. Suppose that, either by an electron collision process or by the absorption of an x-ray, one of the electrons in the K ($n = 1$) shell is removed from the atom. Then electrons can fall into the K shell from the L, M, N

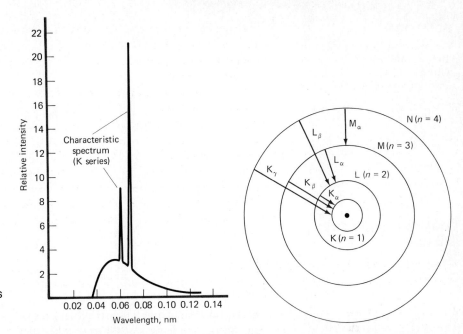

FIGURE 27.25 Characteristic lines in the x-ray spectrum of molybdenum (Mo).

FIGURE 27.26 Bohr energy levels for a heavy atom. The lines connecting these levels indicate the transitions producing x-rays.

shells, producing the lines marked K_α, K_β, K_γ in the diagram. Similarly, if an electron is missing from the L shell, electrons can fall in from the M and N shells, and the atom will radiate the L_α and L_β lines. Observed characteristic x-ray spectra agree well with this explanation. (See Fig. 27.27.)

FIGURE 27.27 X-ray lines for three elements close together in the periodic table. Notice how the spectra are identical except that they shift to shorter wavelengths (higher frequencies) as the atomic number of the atom increases. The observed wavelengths can be obtained from the Bohr energy levels for each atom, as shown in Fig. 27.26.

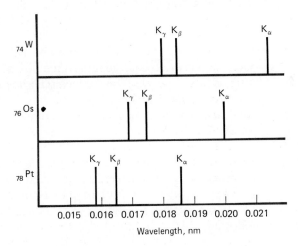

Example 27.7

Assume that the inner-shell electrons in a lead atom, for which the atomic number is $Z = 82$, have energies that are given approximately by the Rydberg formula:

$$E_n = \frac{-13.6Z^2}{n^2} \qquad eV$$

In what wavelength region would you expect to find radiation involving these inner-shell electrons?

SOLUTION

Suppose we consider the atom making a transition from a state near the series limit where $n \to \infty$ to the state with $n = 1$. This will then give us the series limit for the K series of x-ray lines. We will have

$$f = \frac{E_i - E_f}{h} = \frac{0 - [-13.6(82)^2 \text{ eV}]/1^2}{6.6 \times 10^{-34} \text{ J·s}}$$

$$= \frac{13.6(6724) \text{ eV}}{6.6 \times 10^{-34} \text{ J·s}} \left(\frac{1.6 \times 10^{-19} \text{ J}}{1 \text{ eV}} \right) = 2.2 \times 10^{19} \text{ Hz}$$

and so

$$\lambda = \frac{c}{f} = \frac{3.0 \times 10^8 \text{ m/s}}{2.2 \times 10^{19} \text{ s}^{-1}} = 1.4 \times 10^{-11} \text{ m} = \boxed{0.014 \text{ nm}}$$

This is more than four orders of magnitude shorter than visible radiation, and falls in the x-ray region of the spectrum, as is clear from Fig. 21.17.

Failures of the Bohr Theory

Although it is clear from the above that the Bohr theory was remarkably successful, it also had its share of failures:

1 The Bohr theory could not yield acceptable quantitative results for any but one-electron atoms.

2 It could not explain the structure and spectra of molecules, although Bohr devoted a long section at the end of his 1913 paper to a vain attempt to do so.

3 It said nothing about spectral intensities, which were known to differ greatly from one spectral line to another, nor could it predict the polarization of the emitted radiation.

4 Even with regard to hydrogen it failed to account for the finer details of the spectrum.

Ad Hoc Nature of Bohr Theory

A more serious objection to the pioneering theory of Bohr is that it mixed together classical and quantum physics in an odd fashion and avoided contradictions simply by introducing ad hoc postulates to exclude them.

For example, Bohr's first postulate is in direct conflict with the well-verified classical result that all accelerated charges radiate. Bohr also gave no proof of his second postulate that the angular momentum of the H atom is quantized in units of $\hbar$. He introduced it because it led to discrete energy levels, and these energy levels produced a spectrum in agreement with experiment. A theory more fundamental and more rigorous than Bohr's was needed. This was well recognized by Bohr, and he was delighted when such a theory was provided by de Broglie, Schrödinger, and Heisenberg in the years following 1923.

27.10 Wave Mechanics and Quantum Mechanics

The desired goal of a more fundamental quantum theory was achieved almost simultaneously along two very different but equivalent paths, that of wave mechanics and that of matrix, or quantum, mechanics.

Wave Mechanics

The serious study of wave mechanics began with de Broglie's insight in 1923 into the wave properties of particles (see Sec. 26.7). As we have seen, de Broglie showed that particles had both wave and particle properties interrelated by Planck's constant through the equation $\lambda = h/mv$. This led him to

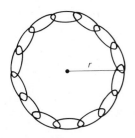

FIGURE 27.28 A circular chain consisting of links. If the chain is to be closed, there must be an integral number of links in the chain.

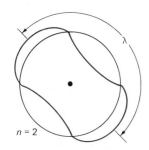

$n = 2$

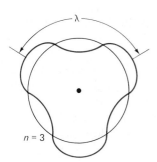

$n = 3$

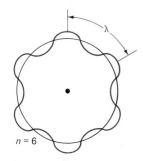

$n = 6$

FIGURE 27.29 Possible standing wave patterns for an electron in a hydrogen atom. As n increases, the wavelength decreases in size.

consider the motion of an electron in a Bohr atom in an attempt to understand why discrete energy states existed in the atom. The waves associated with discrete energy states could not be traveling waves, which are constantly changing with time, but must be *standing waves*, which constructively interfere with themselves and preserve their configuration as time goes on. De Broglie assumed, therefore, that the circumference of the circular orbit in which the electron moves in H must contain an integral number of de Broglie wavelengths, for only then is a standing wave pattern possible, that is, $2\pi r = n\lambda$. This is like forming a closed circle with a chain of equal-size links: there must be an integral number of links if the chain is to be closed, as in Fig. 27.28. So too a closed standing wave in a circle must contain an integral number of wavelengths.

Some possible standing waves of this sort are shown in Fig. 27.29. These standing waves are specified by the value of n in the equation

$$2\pi r = n\lambda = \frac{nh}{mv} \tag{27.14}$$

or

$$mvr = \frac{nh}{2\pi} = n\hbar \tag{27.15}$$

This expression for the angular momentum mvr derived by de Broglie is *exactly the same* as Eq. (27.3), which is an expression of Bohr's postulate that the angular momentum of the atom must be quantized.

Such standing waves would correspond to discrete energy states which are specified by the number of nodes in the standing wave. Hence de Broglie was able to explain both the quantized nature of the angular momentum and the discrete energy states needed to describe the hydrogen spectrum.

These ideas led de Broglie to suggest that what was needed was a new mechanics which would reduce to classical mechanics for ordinary macroscopic objects, but which would lead to observable wave properties for microscopic particles like electrons and neutrons. He saw in this an analogy with the field of optics, in which physical, or wave, optics reduces to ray optics when the wavelength of light is very small compared with the size of apertures or obstacles in the path of the light beam. As de Broglie wrote, what was needed was "a new mechanics with a wave character which would be with respect to the old mechanics what wave optics is with respect to geometrical optics."

This goal was achieved by the Austrian physicist Erwin Schrödinger (see Fig. 27.30) in a series of papers published in 1926. Schrödinger developed a single basic equation, now called the *Schrödinger wave equation*, which was similar to a classical wave equation of the kind that led Maxwell to his prediction of the speed of electromagnetic waves. The classical wave equation describes the variations in the electric field **E** as a function of time and position in space. The intensity of the wave at a particular time and place is then proportional to E^2. The Schrödinger equation, on the other hand, is the equation for something called the wave function ψ associated with an electron (or photon or other particle), and ψ^2 determines the *probability* of finding the electron at a particular point in space. Hence wave mechanics is a *probabilistic theory* which tells us, for example, where an electron in a hydrogen atom is most likely to be found, but can never tell us exactly where the electron is at any particular instant. For this reason Bohr's orbits must be abandoned as a true picture of what goes on inside an atom. Instead curves of the type shown

FIGURE 27.30 Erwin Schrödinger (1887–1961), who received the 1933 Nobel Prize in physics for his development of wave mechanics. The equation he proposed in 1926. now called the Schrödinger equation, is the single most important equation in quantum mechanics, and enables many properties of atoms and molecules to be calculated. (*Photo by Pfaundler; courtesy of AIP Niels Bohr Library.*)

in Fig. 27.31 indicate where the electron is likely to be found in the H atom in states with $n = 1, 2, 3$.

Schrödinger showed that, when applied to electrons in atoms, his equation had acceptable solutions only for certain discrete energy values, the so-called proper values, or *eigenvalues*, of the energy. Associated with each such energy was a *proper wave function*, or *eigenfunction* ψ. The solution of the Schrödinger equation led not only to the energy levels of the atom, but also to an analytical expression for the wave function ψ. This analytical wave function could then be used to calculate all other important properties of the atom, such as the intensity and polarization of the radiation the atom emits, and the change in this radiation when the atom is placed in an electric or magnetic field.

Hence the solution of Schrödinger's equation led, in principle, to all the important physical and chemical properties of atoms and molecules, without

FIGURE 27.31 Probability of finding the electron in the H atom at various distances from the nucleus in the energy states with $n = 1$, 2, and 3, according to Schrödinger's wave mechanics.

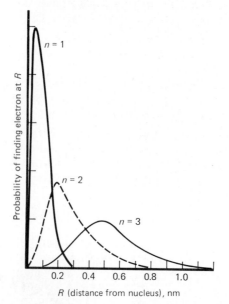

any ad hoc postulates other than the Schrödinger equation itself. In practice, of course, the Schrödinger equation can be very difficult to solve, especially for many-particle systems like heavy atoms or molecules. Successful applications of Schrödinger's equation to many different fields of physics have made clear, however, its basic correctness and importance.

Quantum Mechanics

In 1925 Werner Heisenberg (see Fig. 26.29) began to publish his first papers on matrix, or quantum, mechanics. Heisenberg's theory is much more abstract than Schrödinger's, and concentrates on the observed radiations from atoms and the knowledge they give us about energy states within the atoms. It omits completely any reference to the position or velocity of electrons in atoms, since these quantities cannot be observed and measured.

Schrödinger, Born, and others were soon able to show that the two theories of wave mechanics and quantum mechanics, so different in appearance, were in reality identical, one being merely a translation of the other into a different mathematical language. These two theories have now been blended together into what is called the *new quantum theory*, in contrast with the old quantum theory of Planck, Einstein, and Bohr. This new quantum theory, with relativistic elements introduced by P. A. M. Dirac (see Fig. 27.32), is the operative mathematical theory used by physicists in all problems involving atoms, molecules, nuclei, and aggregates of these fundamental building blocks of matter. The many great successes of physics in the past half century are due, from a theoretical point of view, in great part to this new quantum theory developed by de Broglie, Schrödinger, and Heisenberg as a continuation of the revolutionary ideas of Niels Bohr.

FIGURE 27.32   Paul A. M. Dirac (born 1902). He combined the theory of relativity with Schrödinger's wave mechanics to create relativistic wave mechanics. This led to his prediction of the existence of a positive electron (the positron), a negative proton (the antiproton), and other antiparticles. For these and other discoveries he shared with Schrödinger the 1933 Nobel Prize in physics. (*Photo courtesy of Florida State University.*)

Example 27.8

For the energy state with principal quantum number $n = 3$ in hydrogen, find:

(a) The de Broglie wavelength λ of the electron in this orbit.

(b) The value of the angular momentum for the electron in this orbit.

SOLUTION

(a) The radius of the Bohr orbit with $n = 3$ is

$$r_3 = (3)^2 r_1 = 9(0.0528) = 0.475 \text{ nm}$$

Hence, from de Broglie's condition for standing waves

$$\lambda = \frac{2\pi r_3}{n} = \frac{2\pi (0.475 \text{ nm})}{3} = \boxed{0.99 \text{ nm}}$$

There are therefore three complete wavelengths of this length in the circumference $2\pi r_3$.

(b) The angular momentum in this orbit is

$$mv_3 r_3 = (9.11 \times 10^{-31} \text{ kg})(0.73 \times 10^6 \text{ m/s})(0.475 \text{ nm})$$

$$= \boxed{3.2 \times 10^{-34} \text{ kg·m}^2/\text{s}}$$

(Check this answer by showing that it is equal to $3\hbar$.)

Summary: Important Definitions and Equations

Hydrogen spectrum

Rydberg equation: $\dfrac{1}{\lambda} = R\left(\dfrac{1}{n_f^2} - \dfrac{1}{n_i^2}\right)$

where $R = 1.0968 \times 10^7$ m^{-1}.

For Lyman series (ultraviolet): $n_f = 1$; $n_i = 2, 3, 4, \ldots$

For Balmer series (visible): $n_f = 2$; $n_i = 3, 4, 5, \ldots$

For Paschen series (infrared): $n_f = 3$; $n_i = 4, 5, 6, \ldots$

Bohr theory of hydrogen spectrum

Rutherford-Bohr atom: A model of the atom consisting of a very small, positively charged nucleus, containing almost all the mass of the atom, surrounded by a number of moving electrons equal to the number of elementary positive charges in the nucleus.

Postulates of Bohr theory:

1 Stationary states of discrete energies exist in the hydrogen atom in which the atom is stable and emits no radiation.

2 Angular momentum is quantized: $mvr = n\hbar$, where $\hbar = h/2\pi$.

3 Bohr-Einstein frequency condition: $f = (E_i - E_f)/h$.

Results of Bohr theory

Discrete energies: $E_n = \dfrac{-13.606}{n^2}$ eV

Discrete radii: $r_n = \dfrac{n^2 \hbar^2}{m k_e e^2} = n^2 r_1 = n^2(0.0528 \text{ nm})$

$$\lambda = \frac{1.24 \times 10^{-6} \text{ m}}{\Delta E}$$

where ΔE is the energy difference $E_i - E_f$, in electronvolts, between the initial and final states.

Franck-Hertz experiment: The excitation of atoms by inelastic electron collisions; a direct proof of the existence of discrete energy levels in atoms.

Excitation potential: The accelerating voltage required to give a colliding electron sufficient energy to raise atoms to an excited state. The corresponding energy is called the *excitation energy*.

Ionization potential: The accelerating voltage required to give a colliding electron sufficient energy to remove an electron completely from an atom. The corresponding energy is called the *ionization energy*.

Structure of elements heavier than hydrogen

Quantum Numbers	Allowed Values
Principal quantum number n	$1, 2, 3, \ldots$
Orbital angular momentum quantum number l	$0, 1, 2, \ldots, n-1$
Magnetic quantum number m_l	$0, \pm 1, \pm 2, \pm 3, \ldots, \pm l$
Spin quantum number m_S	$+\frac{1}{2}, -\frac{1}{2}$

Pauli exclusion principle: In one and the same atom no two electrons can have the same set of values for the four quantum numbers n, l, m_l, and m_s.

Building-up principle for periodic table of elements: Atoms are built up by adding electrons one at a time in the lowest one-electron energy states permitted by the Pauli exclusion principle.

Electron configuration of an atom: The assignment of electrons to the various energy shells and subshells by indicating the values of n and l for each electron. In so doing, values of $0, 1, 2, 3, \ldots$, for l are indicated by the letters $s, p, d, f, \ldots$.

Questions

1 Why do only gases and vapors emit line spectra, whereas solids emit a continuous spectrum?

2 Sunlight is passed through a long tube containing atomic hydrogen gas. It is found that both a strong Lyman series and a weak Balmer series appear in the absorption spectrum. What can be concluded about the temperature of the gas?

3 Why does the spectrum of hydrogen contain so many different lines and series if it is produced by only one electron?

4 Why would you expect from classical physics that an electron orbiting a nucleus in an atom would continuously lose energy by radiation?

5 Why would you expect to be able to see more lines in the Balmer series of H in the spectrum of some celestial bodies than you can in vacuum discharges in the laboratory?

6 An astronomer wants to know if there is any iron in the atmosphere of a star. Can you suggest an experiment that might determine an answer to this question?

7 Sunlight is allowed to pass through sodium vapor which has been contained in a glass flask in the laboratory. When the sunlight is looked at with a spectrometer, two dark lines characteristic of the absorption spectrum of sodium are seen in the yellow region of the spectrum.

(*a*) Where did the energy absorbed by the sodium atoms go?

(*b*) Why, then, are the absorption lines dark?

8 Prove that $2n^2$ is the number of different possible sets of four quantum numbers corresponding to a principal quantum number n.

9 Discuss the chemical behavior of the halogens (F, Cl, Br, I) on the basis of the Bohr theory and the Pauli exclusion principle.

10 Explain on the basis of de Broglie's theory of matter waves why you would expect free electrons to have a continuum of energy values instead of discrete energy values.

11 Explain, on the basis of energy considerations, why an electron in an orbit close to the nucleus must move more rapidly than an electron in an orbit farther away from the nucleus.

12 Show how the Pauli exclusion principle helps explain the solidity and impenetrability of matter.

Problems

A 1 The wavelength of the first line in the Paschen series in hydrogen in the infrared is:
(*a*)   122 nm   (*b*)   656 nm   (*c*)   959 nm
(*d*)   380 nm   (*e*)   1875 nm

A 2 An electron in a hydrogen atom moves in an orbit of radius 0.84 nm. Its speed is:
(*a*) 2.10×10^6 m/s   (*b*) 3.20×10^8 m/s
(*c*) 5.4×10^5 m/s   (*d*) 1.46×10^6 m/s
(*e*) 1.10×10^6 m/s

A 3 If we take as our zero of energy the energy of the state with the electron and proton separated and at rest, the energy of a H atom in a state characterized by the principal quantum number n equal to 10 is:
(*a*) $+2.8 \times 10^{-20}$ J   (*b*) -1.36 eV
(*c*) $+1.36$ eV   (*d*) $+0.136$ eV
(*e*) -2.18×10^{-20} J

A 4 The radius of the electron orbit in the energy state with $n = 100$ in the hydrogen atom is:
(*a*)   0.0528 nm   (*b*) 0.0428×10^{-9} m
(*c*)   528 nm   (*d*)   5.28 nm   (*e*)   5.28 m

A 5 What would the temperature of H gas have to be for the average kinetic energy of the gas atoms to correspond to the 10.2-eV energy difference between the ground state and the first excited state of hydrogen?
(*a*) 7.9×10^8 K   (*b*) 7.9×10^4 K
(*c*) 7.5×10^2 K   (*d*)   300 K
(*e*)   None of the above

A 6 In a laboratory experiment on the Franck-Hertz experiment in mercury, electrons are accelerated through a voltage V_1 (see Fig. 27.10) of 9.0 V. The average kinetic energy of the electrons when they pass through the grid G is:
(*a*)   9.0 eV   (*b*)   8.5 eV   (*c*)   0   (*d*)   4.1 eV
(*e*)   3.6 eV

A 7 In the same Franck-Hertz experiment as in Prob. A6, the average kinetic energy of the electrons when they hit the plate will be:
(*a*)   9.0 eV   (*b*)   8.5 eV   (*c*)   0   (*d*)   4.1 eV
(*e*)   3.6 eV

A 8 In a Franck-Hertz experiment on potassium (K) vapor, it is found that the current falls off rapidly at a voltage of about $V_1 = 1.6$ V. The following line would be expected to appear in the spectrum of K when that voltage is reached:
(*a*)   345 nm   (*b*)   464 nm   (*c*)   691 nm
(*d*)   764 nm   (*e*)   960 nm

A 9 In a subshell of an atom specified by $n = 4, l = 3$, the number of possible values of the magnetic quantum number m_l is:
(*a*)   1   (*b*)   3   (*c*)   6   (*d*)   7   (*e*)   4

A10 The atomic diameter of argon ($_{18}$A) is 0.382 nm and the diameter of calcium ($_{20}$Ca) is 0.393 nm. A likely value for the atomic diameter of potassium ($_{19}$K) is:
(*a*)   0.476 nm   (*b*)   0.275 nm   (*c*)   0.180 nm
(*d*)   4.76 nm   (*e*)   0.048 nm

B 1 What is the series limit for the Paschen series in the infrared spectrum of hydrogen?

B 2 What is the magnitude of the Coulomb force required to keep an electron in the hydrogen atom moving in an orbit of radius 0.84 nm?

B 3 Calculate (*a*) the energy, (*b*) the radius of the electron orbit, and (*c*) the speed of the electron in this orbit for the upper level involved in the first line in the Paschen series in hydrogen.

B 4 Ultraviolet light of wavelength 238 nm falls on mercury vapor at room temperature. Will this light be absorbed by the mercury atoms? Why?

B 5 The ionization energy of the mercury atom from its ground state is 10.4 eV. At what wavelength would you expect the lines in the spectral series involving transitions from excited states to the ground state of Hg to merge into a continuum?

B 6 A strong line in the spectrum of calcium (Ca) occurs when the atom makes a transition from an excited state 4.7 eV above the ground state to a lower excited state 1.9 eV above the ground state.

(*a*) What is the frequency of the light emitted in such a transition?

(*b*) What is the wavelength of the light?

B 7 It is found that monatomic hydrogen gas can be raised to its first excited state by absorbing 10.2-eV photons. The ionization potential of H is 13.6 V. How much energy would be required to remove the electron from the atom in its first excited state?

B 8 Why is the first ionization potential of $_4$Be (9.3 V) higher than that of its neighbors on either side, $_3$Li (5.4 V) and $_5$B (8.3 V), even though the $n = 2$ shell is not completely filled up until $_{10}$Ne is reached?

B 9 (*a*) How many electrons are there in a filled subshell with orbital angular momentum quantum number equal to 4?

(*b*) Write out the quantum numbers l, m_l, and m_s for such a subshell.

B10 What is the de Broglie wavelength of an electron in the orbit corresponding to the $n = 4$ stationary state in the H atom?

C 1 (*a*) Calculate the four longest wavelengths in the Lyman series of H and plot their positions on a linear horizontal scale.

(*b*) Indicate the series limit on the same plot. Are any of these spectral lines in the visible range?

C 2 The radius of the gold nucleus is about 7.0×10^{-15} m. What minimum energy must the alpha particle in a Rutherford scattering experiment have to come this close to the center of the nucleus and thus determine the radius of the gold nucleus?

C 3 (*a*) Calculate the value of the orbital angular momentum of the electron in the four lowest energy states of the hydrogen atom.

(*b*) Show that in each case this value is equal to $n\hbar$.

C 4 Calculate the values of (*a*) the energy, (*b*) the radius of the electron orbit, and (*c*) the speed of the electron in this orbit for the upper and lower energy states involved in the third line of the Paschen series.

C 5 (*a*) What is the radiation of greatest wavelength which will ionize hydrogen atoms in their ground state?

(*b*) What wavelength radiation will ionize a hydrogen atom in its ground state and in addition give the electron 1.2 eV of kinetic energy after it leaves the atom?

C 6 A negative mu meson, or muon, can be captured by a proton to form a so-called muonic atom, which is similar to a hydrogen atom in structure except that the muon rest mass is 207 times that of an electron.

(*a*) Calculate the radius of the first Bohr orbit for such a muonic atom.

(*b*) Calculate the energy of the ground state of such an atom.

(*c*) Find the wavelength of the first Balmer line for such an atom.

C 7 The energy separation between two states in the sodium atom is 3.36×10^{-19} J.

(*a*) Find the minimum voltage through which an electron must be accelerated to excite the Na atom from the lower to the upper state.

(*b*) Find the wavelength of the radiation which must be absorbed to excite the atom from the lower to the upper state.

(*c*) Will higher voltages than that found in (*a*) also be able to excite the Na atom from the lower to the upper state?

(*d*) Will shorter wavelengths than that found in (*b*) be able to do the same?

C 8 When ordinary table salt is put into a Bunsen-burner flame, a characteristic yellow line of sodium at about 589 nm appears. This line arises when an electron returns from the first excited state to the ground state of the sodium atom.

(*a*) What is the energy of the first excited state above the ground state in Na (in both joules and electronvolts)?

(*b*) What would the temperature of the vapor have to be to produce atoms with an average kinetic energy equal to that of this excited state?

(*c*) Since the average temperature of the molecules in the Bunsen flame is only 2100 K, how do the Na atoms obtain sufficient energy to excite the 589-nm line?

C 9 Mercury vapor is bombarded by electrons of energy 7.00 eV. What wavelengths would you expect to see in the spectrum of Hg excited in this way? (See Fig. 27.12.)

C10 How much energy is absorbed by an atom when mercury vapor is bombarded with (*a*) a 5.5-eV electron, (*b*) a 5.5-eV photon? (See Fig. 27.12.)

C11 Hydrogen gas is bombarded by an electron beam of just enough energy to excite the H_γ line in the Balmer spectrum.

(*a*) What is the least energy required in the electron beam to do this?

(*b*) What other lines would you expect to see in the visible spectrum of H under these excitation conditions?

C12 The cesium atom has excited states at 1.38 and 2.30 eV, respectively, and an ionization potential of 3.87 V. If cesium vapor is bombarded with electrons accelerated through 3.90 V, what spectral lines might you expect to see with a spectrometer?

C13 The ground state of potassium (K) has the following values of the four quantum numbers of the outermost electron: $n = 4$, $l = 0$, $m_l = 0$, $m_s = +\frac{1}{2}$.

(a) If this electron is excited to the energy state specified by the quantum numbers $n = 5$, $l = 1$, how many different electron configurations are possible in this state?

(b) What are the values of the four quantum numbers for each of these configurations?

C14 In potassium the first exception to the filling-in of the lowest possible one-electron energy levels occurs when the last electron goes into the $4s$ level while the $3d$ level is still vacant. Show that this makes the electron configuration of $_{19}$K very similar to that of $_{11}$Na and accounts for their similar physical and chemical properties.

C15 What are the wavelengths of the following spectral lines?

(a) The line in the spectrum of He$^+$ which corresponds to the same transition as the one producing the first line in the Paschen series in H.

(b) The line in the spectrum of Li^{2+} which corresponds to the same transition as that producing the first line in the Balmer series in H.

(c) The line in the spectrum of Be^{3+} that corresponds to the same transition as that producing the first line in the Lyman series of H.

(d) Indicate in what regions of the spectrum these lines occur.

C16 (a) How much energy is required to remove the electron from the ground state of He$^+$?

(b) What is the minimum frequency of an incident photon that can remove this electron from the ion?

C17 At extremely high temperatures such as exist in the stars, collisions can strip heavy atoms of almost all their electrons. Suppose that potassium ($_{19}$K) vapor has all but one of its electrons stripped away.

(a) What is the wavelength of the first line of the Lyman series for such an ion?

(b) In what region of the spectrum does this wavelength fall?

C18 (a) Calculate the first eight lines in the Pickering series for He$^+$, assuming that for He$^+$ the value of the Rydberg constant is 1.09722×10^7 m^{-1}.

(b) Compare these wavelengths with the wavelengths of the Balmer series in hydrogen, where the accurate value of the Rydberg constant is 1.09678×10^7 m^{-1}, by plotting both spectra on the same horizontal wavelength scale.

C19 A hydrogen atom exists on the average for about 10^{-8} s in an excited state before making a transition back to its ground state.

(a) About how many revolutions will the electron make in the $n = 2$ state before the atom falls back to the ground state?

(b) What is the uncertainty in the energy of the atom in this excited state?

C20 (a) What is the de Broglie wavelength of an electron in the $n = 5$ quantum state in hydrogen?

(b) How many complete de Broglie wavelengths are there in the length of the electron's orbit?

(c) What is the value of the angular momentum in this orbit?

C21 A man decides to lengthen his de Broglie wavelength by moving very slowly. If he weighs 200 lb and moves at a speed of 1 cm every 10 years:

(a) What is his de Broglie wavelength?

(b) How does this wavelength compare with the wavelength of an electron in the ground state of the hydrogen atom?

D 1 Calculate the fraction of an alpha particle's initial kinetic energy which is transferred to another particle initially at rest in a head-on elastic collision, if the other particle is (a) a gold nucleus, (b) a proton.

D 2 Show that for large values of the principal quantum number n, the frequencies of revolution for an electron in adjacent energy levels of a hydrogen atom and the frequency of light radiated in a transition between these levels all approach the same value. (This is an example of *Bohr's correspondence principle*.)

D 3 Show that for H the radius of the first Bohr orbit and the ground-state energy of the atom can be written in the following form:

$$r_1 = \frac{\hbar}{\alpha mc} = \frac{\lambda_0}{2\pi\alpha}$$

$$E_1 = \tfrac{1}{2}\alpha^2 mc^2$$

where $\lambda_0 = h/mc$ is the Compton wavelength of the electron (see Sec. 26.4) and $\alpha = k_e e^2/\hbar c$ is the fine-structure constant.

D 4 (a) Prove that when allowance is made for the motion of both electron and nucleus about the common center of mass of a one-electron atom, the result for E_n, r_n, and v_n are the same as in the simple Bohr theory but the mass of the electron m is replaced by the *reduced mass* $\mu = m/(1 + m/M)$, where M is the mass of the nucleus.

(b) Use the reduced mass to calculate accurate values of the Rydberg constant for H, He$^+$, and Li^{2+}.

(c) What is the percentage difference between the Rydberg constant for H and for He$^+$?

D 5 (a) Calculate the electric current produced by an electron moving in the first Bohr orbit.

(b) At the site of the proton in the hydrogen atom what is the approximate value of the magnetic field produced by this orbiting electron?

D 6 In an electron-impact experiment it is found that radiation with a wavelength of 589 nm is emitted by sodium vapor when bombarded by 2.11-eV electrons. Compute the value of h/e from these data.

D 7 If for some reason atoms could contain only electrons with principal quantum numbers n up to and including $n = 5$, and if all these five shells were completely filled, how many elements would there be in the periodic table?

D 8 Show that for hydrogen-like ions the wavelengths predicted by the Rydberg formula are

$$\frac{1}{\lambda} = RZ^2\left(\frac{1}{n_f^2} - \frac{1}{n_i^2}\right)$$

where Z is the number of positive elementary charges in the nucleus of the hydrogen-like ion.

D 9 An atom has a lifetime of 2.0×10^{-8} s in a particular excited state; i.e., it remains in the excited state for 2.0×10^{-8} s, on the average, before falling back to its ground state. The excited state is 1.7 eV above the ground state.

(a) What is the uncertainty in the frequency emitted by the atom in the transition back to the ground state?

(b) What is the uncertainty in the wavelength of the emitted radiation?

D10 Estimate the uncertainty in the position of the electron in the hydrogen atom in its ground ($n = 1$) state.

Additional Readings

Andrade, E. N. da C.: *Rutherford and the Nature of the Atom*, Doubleday Anchor, Garden City, N.Y., 1964. This brief, interesting life of Rutherford contains considerable material on Bohr's work in Manchester with Rutherford and on the development of Bohr's atomic theory.

———: "The Birth of the Nuclear Atom," *Scientific American*, vol. 195, no. 5, November 1956, pp. 93–104. A briefer account of the material contained in the preceding reference.

Boraiko, Allen A.: "Lasers—A Splendid Light," *National Geographic*, vol. 165, no. 3, March 1984, pp. 334–363. An up-to-date, well-illustrated article in the famous issue with the hologram on the cover.

Cline, Barbara: *The Questioners: Physicists and the Quantum Theory*, Crowell, New York, 1965. A lively account of the development of quantum physics, with considerable insight into the personalities of the physicists involved in its development.

Gamow, George: "The Exclusion Principle," *Scientific American*, vol. 201, no. 1, July 1959, pp. 74–86. A discussion of the impact of the Pauli principle on many different fields of physics.

Hänsch, Theodor W., Arthur L. Schawlow, and George W. Series: "The Structure of Atomic Hydrogen," *Scientific American*, vol. 240, no. 3, March 1979, pp. 94–110. An outstanding article by three physicists who have contributed much to our knowledge of spectra, indicating the importance and the complexity of the hydrogen spectrum.

Lasers and Light, Readings from *Scientific American*, Freeman, San Francisco, 1969. A collection of articles by some of the foremost research workers in laser optics.

Moore, Ruth: *Niels Bohr: The Man, His Science, and the World They Changed*, Knopf, New York, 1966. This is the most complete life of Bohr in English, but greater insight into his personality and scientific accomplishments can be obtained from the two following collections of essays by his friends and fellow scientists.

Physics Today, vol. 16, no. 10, October 1963, pp. 21–64. This memorial issue of a popular physics journal is devoted to Bohr and contains perceptive articles by physicists Felix Bloch, J. Rud Nielsen, Victor F. Weisskopf, and John A. Wheeler.

Rozental, S. (ed.): *Niels Bohr: His Life and Work as Seen by His Friends and Colleagues*, Interscience, New York, 1964.

To everyone who, like myself, had the good fortune to visit the physical laboratories in Cambridge and Manchester about twenty years ago and work under the inspiration of the great leaders, it was an unforgettable experience to witness almost every day the disclosure of hitherto hidden features of nature. I remember, as if it were yesterday, the enthusiasm with which the new prospects for the whole of physical and chemical science, opened by the discovery of the atomic nucleus, were discussed in the spring of 1912 among the pupils of Rutherford.

Niels Bohr (1885–1962), in his 1930 Faraday Lecture

Chapter 28
Nuclear Physics

Once physicists understood atoms and molecules reasonably well, it was inevitable that they should next tackle the atomic *nucleus*. The nucleus is four orders of magnitude smaller than an atom, however, and unraveling its secrets is correspondingly more difficult. Nevertheless, great progress has been made during this century in understanding the structure and behavior of nuclei, although many unanswered questions remain. At the same time the atomic nucleus has turned out to contain an almost infinite store of energy which is available for the use or the destruction of the world. This chapter will emphasize the basic physical principles needed to understand the nucleus and to weigh arguments concerning nuclear weapons and nuclear power.

28.1 The Structure of the Atomic Nucleus

The Bohr-Rutherford atom consists of small, light, negatively charged electrons in rapid motion about a dense, positively charged, stationary nucleus. Since the atom is electrically neutral, there must be just as many positive charges in the nucleus as there are negative extranuclear electrons. This number is the charge number, or atomic number Z.

The early theories of the nucleus, which followed from Rutherford's nuclear model of the atom, employed a proton-electron model. According to this view, the nucleus consisted of a number of protons equal to the atomic mass number A, and a number of electrons just sufficient to reduce the nuclear charge to Z. Hence each nucleus would contain A protons and $A - Z$ electrons, leaving a net positive charge of Z on the nucleus.

The proton-electron model soon ran into severe difficulties. First of all, if the electron were confined inside the nucleus, the uncertainty principle predicts that it would have a kinetic energy of about 100 MeV. Such electrons would be so energetic that they would quickly escape from the nucleus.

Also, nuclei have an associated spin angular momentum in the same way that atoms have an angular momentum produced by the orbital and spin motion of the electrons in the atom. Since the spin of each proton and each electron in the nucleus was known to be either $+\frac{1}{2}$ or $-\frac{1}{2}$, the nuclear spin would have to be the sum of these values. According to the proton-electron model a nitrogen nucleus ($^{14}_{7}N$) would have 14 protons and 7 electrons. Hence, no matter how the half-integral spins of these 21 particles were added together, the sum would have to be a half-integer like $\frac{1}{2}$, $\frac{7}{2}$, or $\frac{21}{2}$. But it was well known from the spectrum of the N_2 molecule that the spin of the N nucleus was an integer (actually 1). Hence something was clearly wrong with the proton-electron model.

Discovery of the Neutron

About 1930 evidence had accumulated that certain nuclei emitted penetrating particles which were not affected by electric and magnetic fields. These particles did not ionize atoms and molecules and hence did not appear to be photons. In 1932 the English physicist James Chadwick (see Fig. 28.1) identified such particles by allowing the nuclear radiation emitted by beryllium when bombarded by alpha particles to strike a paraffin target, as in Fig. 28.2. Protons were knocked out of the paraffin and, because of their charge, were easily detected by a Geiger counter (discussed in Sec. 28.9 below). From energy and momentum measurements and the conservation principles Chadwick was able to show that the protons in paraffin must have been hit by an uncharged particle of approximately the same mass as the proton. This particle had been knocked out of beryllium by an alpha particle. Here was the massive uncharged particle sought by physicists for a decade. Because it was electrically neutral, Chadwick named it the *neutron*.

FIGURE 28.1 James Chadwick (1891–1974). Chadwick, a student and later a colleague of Rutherford, was a great nuclear physicist in the Rutherford tradition. His discovery of the neutron in 1932 earned him the Nobel Prize in physics in 1935. (*AIP Niels Bohr Library, William G. Meggers collection.*)

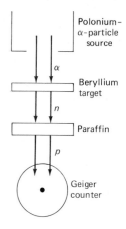

FIGURE 28.2 Chadwick's apparatus for detecting neutrons.

The Proton-Neutron Model of the Nucleus

Soon after Chadwick's discovery Heisenberg suggested a proton-neutron model of the nucleus, according to which the nucleus contains no electrons but only protons and neutrons. The charge on the nucleus is then given by the number of protons, which is equal to the atomic number (or charge number) Z, whereas the mass of the nucleus is the sum of the masses of the protons and

TABLE 28.1 Properties of Elementary Particles in Atoms (circa 1930)

Particle	Symbol	Charge, C	Mass		
			kg	u	MeV/c^2 *
Electron	$_{-1}^{0}e$	-1.60219×10^{-19}	9.11×10^{-31}	5.49×10^{-4}	0.511
Nucleons:					
Proton	$_{1}^{1}p$	$+1.60219 \times 10^{-19}$	1.67265×10^{-27}	1.00728	938.27
Neutron	$_{0}^{1}n$	0	1.67495×10^{-27}	1.00866	939.56

*See the discussion on energy units in Sec. 28.2.

neutrons. The mass number A then equals the total number of neutrons and protons in the nucleus. On this model the number of neutrons in the nucleus is $A - Z$. The protons and neutrons in the nucleus are commonly referred to collectively as *nucleons*. The total number of nucleons in a $_{7}^{14}$N nucleus would then be 14, and the overall spin would have to be integral in agreement with experiments.

The basic constituents of atoms, as known about 1930, are listed for convenience in Table 28.1. The symbols $_{-1}^{0}e$, $_{1}^{1}p$, and $_{0}^{1}n$ are used for the electron, proton, and neutron, respectively, where the superscript is the mass number of the particle and the subscript is its charge number. The masses are given in unified atomic mass units u (see Sec. 9.5) and in equivalent energy units, as well as in kilograms.

Nuclear Size

In Sec. 27.4 we showed how Rutherford was able to arrive at a maximum size for atomic nuclei by studying the distribution of the alpha particles they scattered. He found that nuclei had diameters four orders of magnitude smaller than those of atoms. Since atomic diameters are about 0.1 nm, nuclei have diameters of about 10^{-5} nm, or 10^{-14} m. Hence if an atom were expanded until it was 1 mi in diameter, the nucleus at the atom's center would be about the size of a grapefruit. Experiments on the scattering of alpha particles, neutrons, and electrons by nuclei have shown that nuclei have an approximately spherical shape and a radius that increases with the mass number A according to the approximate formula:

$$r_0 = (1.2 \times 10^{-15} \text{ m})(A)^{1/3} \tag{28.1}$$

Hence for a gold nucleus we find that

$$r_0 = (1.2 \times 10^{-15} \text{ m})(197)^{1/3} = 6.9 \times 10^{-15} \text{ m} = 0.69 \times 10^{-14} \text{ m}$$

This is consistent with the conclusions from Rutherford's scattering experiments (Sec. 27.2), which showed that the radius of a gold nucleus must be less than 4.2×10^{-14} m. The volume of a nucleus is then $V_0 = \frac{4}{3}\pi r_0^3$ or

$$V_0 = (\tfrac{4}{3})(3.14)(1.2 \times 10^{-15} \text{ m})^3(A) = (7.2 \times 10^{-45})A \quad \text{m}^3$$

and is proportional to A and hence to the number of nucleons in the nucleus.

Therefore, in the periodic table, the heavier the element, the larger the nucleus, while the nuclear density remains approximately constant. The nucleons seem to act as small marbles which are packed together as tightly as possible in the nucleus. The more marbles there are, the larger the nucleus is. Note that this is *not* what happens with atoms, where the size of the atom varies in a periodic fashion with the number of electrons in the atom.

Example 28.1

Compare the size of a uranium nucleus with that of a hydrogen nucleus.

SOLUTION

Since the radius of a nucleus is given by Eq. (28.1), we have for the ratio of the radii of the two most common isotopes of uranium and hydrogen listed in Appendix 4:

$$\frac{r_0(^{238}_{92}U)}{r_0(^1_1H)} = \frac{1.2 \times 10^{-15}\text{ m})(238)^{1/3}}{(1.2 \times 10^{-15}\text{ m})(1)^{1/3}} = (238)^{1/3} = 6.2$$

Hence the uranium nucleus has a radius about 6 times that of a hydrogen nucleus.

The volume of the uranium nucleus is 238 times larger than the volume of the hydrogen nucleus, since the volume depends directly on the mass number A.

28.2 Nuclear Forces and Binding Energies

If nuclei are composed of only protons and neutrons, what holds them together? Why does not the nucleus fly apart, since the protons must repel each other with very large forces at the small distances separating them inside a nucleus? In the nucleus there must exist some new kind of attractive force between nucleons (now called the *strong nuclear force*), which is stronger than the electrostatic repulsion between protons. (This nuclear force is clearly not a gravitational force, since the gravitational force between two protons is about 35 orders of magnitude *weaker* than the electrostatic repulsion between them.) The nature and characteristics of this nuclear force are still not completely known, and many of the experiments being carried out in laboratories all over the world are directed toward better understanding this nuclear force.

We do know from experiment that the nuclear force is the same between any two nucleons, i.e., the same for np, nn, and pp interactions. In other words, it is *charge-independent*. We also know that this very strong force only comes into play when the nucleons are very close together, say, within about 10^{-15} m of each other, and that it drops rapidly to zero for larger distances, as in Fig. 28.3. It is thus a very short-range force compared with the long-range Coulomb force which falls off as $1/r^2$.

FIGURE 28.3 The strong nuclear force between two protons: This short-range force is attractive in the range of 1 to 3 × 10^{-15} m.

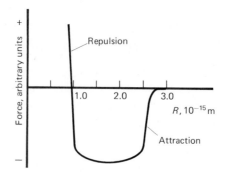

Nuclear Binding Energy

Since we cannot express the nuclear force in the form of an equation, we must rely on experiment to determine the energy of nuclei. In doing this a most useful tool is Einstein's mass-energy relationship $E = m_0c^2$. It is much easier to determine the masses of atoms than it is to measure their energies directly.

By determining the masses of the individual protons and neutrons making up a nucleus and then measuring the mass of the nucleus itself, the *mass defect* can be determined.

Definition

Mass defect Δm: The difference between the sum of the masses of the individual protons and neutrons making up a nucleus, and the mass of the nucleus itself.

The corresponding difference in energy, obtained from $E = m_0 c^2$, is called the *binding energy* of the nucleus.

Definition

Nuclear binding energy: $E = \Delta m\, c^2$, where Δm is the mass defect. This is the energy required to tear apart a nucleus into its constituent protons and neutrons.

When protons and neutrons come together to form a stable nucleus, the total energy of the nucleus is less than the sum of the energies of the isolated nucleons by the *nuclear binding energy*. For nuclei, however, this energy is of the order of some megaelectronvolts (10^6 eV) rather than the few electron-volts required for molecular binding. From the point of view of Einstein's mass-energy relationship, this means that when the nucleus was formed, the excess mass (equivalent to the binding energy) was radiated away in the form of energy. Conversely, if we want to break up a nucleus into its individual protons and neutrons, and place all these particles at rest far apart from one another, we must supply an amount of energy numerically equal to the nuclear binding energy.

It is found experimentally (mostly from precise mass-spectrometer measurements) that the binding energy varies slightly from nucleus to nucleus throughout the periodic table. To eliminate the effect of differing numbers of nucleons when comparing binding energies, we must use the *binding energy per nucleon*, i.e., the nuclear binding energy for a particular nucleus divided by the number of nucleons in that nucleus. In Fig. 28.4 we plot the nuclear binding energy per nucleon as a function of the mass number A, that is, the number of nucleons in the nucleus. Note that all the binding energies given are negative. The more negative the binding energy, the more stable the nucleus, since the binding energy represents energy which is not there, but which was radiated away when the nucleus was formed. The larger the energy required to break up the nucleus into separated protons and neutrons, the more stable the nucleus.

The curve in Fig. 28.4 has some interesting features. It has a minimum in the middle of the periodic table, in the vicinity of the iron ($^{56}_{26}$Fe) nucleus. This $^{56}_{26}$Fe nucleus has the largest binding energy per nucleon, about 8.8 MeV, of any nucleus, and is therefore unusually stable (this is the reason so much iron is found in the earth's crust). The lowest point on this binding-energy curve is the position of greatest stability, just as the lowest point on a roller coaster is the most stable position for the roller-coaster car. As we go to heavy nuclei like $^{235}_{92}$U, the binding energy changes slowly to about 7.6 MeV per nucleon, and at the low-mass end of the periodic table it changes more rapidly to values like 2.9 MeV per nucleon for tritium ($^{3}_{1}$H), indicating that both very heavy and very light nuclei are less stable than those in the middle of the periodic table. This, as we shall see, is the secret of the tremendous energies released in fission and fusion reactions.

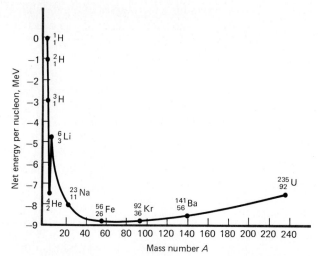

FIGURE 28.4 Curve of nuclear binding energies.

Energy Units

In dealing with nuclear reactions, more than two or three significant figures must be retained, since very small mass differences between large masses can be converted into very large amounts of energy.

The masses of the electron, proton, and neutron are given in unified atomic mass units in Table 28.1. Note that 1 u is approximately equal to the proton mass. Since masses and energies are equivalent, we can also express masses in energy units divided by c^2. Thus, using the mass of a proton, we have from Table 28.1:

$$1.00000 \text{ u} = (1.00000 \text{ u}) \left(\frac{1.67265 \times 10^{-27} \text{ kg}}{1.00728 \text{ u}} \right) = 1.66056 \times 10^{-27} \text{ kg}$$

Converting this to energy units, we have for the equivalent energy

$$E = m_0 c^2 = (1.66056 \times 10^{-27} \text{ kg})(2.99792 \times 10^8 \text{ m/s})^2$$

$$= 1.49243 \times 10^{-10} \text{ J}$$

or $\quad E = \dfrac{1.49243 \times 10^{-10} \text{ J}}{1.60219 \times 10^{-19} \text{ J/eV}} = 9.3149 \times 10^8 \text{ eV} = 931.49 \text{ MeV}$

Hence $\quad 1.00000 \text{ u} = \dfrac{931.49 \text{ MeV}}{c^2}$ (28.2)

Values for the masses of the electron, proton, and neutron in MeV/c^2 are also given in Table 28.1.

Example 28.2

Calculate the repulsive force between two protons which are separated by a distance of 10^{-15} m in a nucleus.

SOLUTION

From Coulomb's law we have

$$F_e = \frac{k_e q q'}{r^2} = \left(9.0 \times 10^9 \, \frac{\text{N} \cdot \text{m}^2}{\text{C}^2} \right) \frac{(1.6 \times 10^{-19} \text{ C})^2}{(10^{-15} \text{ m})^2}$$

$$= 2.3 \times 10^2 \text{ N} \simeq \boxed{200 \text{ N}}$$

This is approximately the force exerted by the earth's gravitational field on a 20-kg mass, and, as anyone who has ever lifted a 20-kg (~45-lb) bucket of water knows, this is quite a substantial force.

Example 28.3

Calculate the binding energy per nucleon of the $^{23}_{11}$Na (sodium) nucleus and compare the result with that indicated on the binding-energy curve of Fig. 28.4. (The mass of the $^{23}_{11}$Na atom is 22.98977 u, from the table in Appendix 4.)

The $^{23}_{11}$Na atom contains 11 protons and 12 neutrons. Hence the total mass of these nucleons is, from Table 28.1,

$$11 \times (1.00728 \text{ u}) = 11.08008 \text{ u}$$
$$12 \times (1.00866 \text{ u}) = \underline{12.10392 \text{ u}}$$

Total mass of nucleons = 23.18400 u

The mass of the ^{23}Na nucleus is the mass of the atom minus the mass of the 11 extranuclear electrons. Hence the nuclear mass is

$$22.98977 \text{ u} - 11(5.49 \times 10^{-4} \text{ u}) = 22.98373 \text{ u}$$

The mass defect Δm is therefore

$$\Delta m = 23.18400 \text{ u} - 22.98373 \text{ u} = 0.20027 \text{ u}$$

To find the binding energy we note that, from Eq. (28.2), 1.00000 u is equivalent to 931.49 MeV. Hence

$$\text{Binding energy} = (0.20027 \text{ u}) \left(\frac{931.49 \text{ MeV}}{1 \text{ u}}\right) = 186.55 \text{ MeV}$$

The binding energy per nucleon is this amount divided by the number of nucleons, or

$$\text{Binding energy per nucleon} = \frac{186.55 \text{ MeV}}{23 \text{ nucleons}}$$

$$\boxed{= 8.11 \text{ MeV per nucleon}}$$

This agrees well with the value indicated in Fig. 28.4.

28.3 Nuclear Reactions I: Natural Radioactivity

Most elements have nuclei which are stable and show no tendency to change spontaneously or "decay" into other kinds of nuclei. Some nuclei found in nature, however, especially those at the large-mass end of the periodic table, are unstable and decay into different nuclei by emitting highly energetic particles. Such nuclei are said to exhibit *natural radioactivity*, a name given to this phenomenon by the Polish-French physicist Marie Curie in 1898 (see Figs. 28.5 and 28.6).

The first evidence for natural radioactivity was found by the French physicist Antoine Henri Becquerel (1858–1908) in 1896, when he discovered that a chemical compound containing the element uranium was able to expose photographic film placed near it in a drawer, even though the film was tightly wrapped in black paper. Becquerel rightly guessed that some atoms in the uranium compound were emitting particles of sufficient energy to penetrate the wrapping paper and expose the photographic film. It was soon established that uranium atoms were the culprits. A few years later Marie Curie and her husband, Pierre Curie (1859–1906), were able to extract, using laborious chemical techniques, small quantities of the new elements polonium ($^{210}_{84}$Po) and radium ($^{226}_{88}$Ra) from tons of pitchblende and found that these elements were more highly radioactive than uranium itself. Polonium and radium are two typical naturally radioactive substances. At the present time some 30 natural elements have been found to possess radioactive isotopes, and over 50 naturally radioactive isotopes have been identified. Many additional radioactive isotopes can be produced by artificial means, as we will see.

Radioactivity is the result of the disintegration or decay of an unstable nucleus. Certain nuclei are not sufficiently stable under the combined operation of the strongly attractive nuclear force and the strongly repulsive

FIGURE 28.5 Marie Sklodowska Curie (1867–1934) with her husband Pierre Curie in their laboratory in Paris. This husband and wife team discovered the elements radium and polonium and did much of the original work on radioactivity. Madame Curie received two Nobel Prizes, the 1903 prize in physics, and the 1911 award in chemistry. (*Photo courtesy of AIP Niels Bohr Library.*)

FIGURE 28.6 Marie Curie with her daughter, Irène Joliot-Curie (1897–1956). Marie Curie so successfully trained her daughter in scientific research that Irène and Irène's husband, Frédéric Joliot-Curie (1900–1958), received the 1935 Nobel Prize in chemistry for the discovery of artificial radioactivity. (*AIP Niels Bohr Library, William G. Meggers collection.*)

Coulomb force, and spontaneously decay with the emission of nuclear particles of various kinds.

Types of Nuclear Emissions

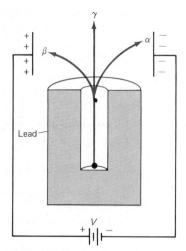

FIGURE 28.7 Particles emitted by radioactive nuclei, as separated by an electric field. Alpha and beta particles are attracted to charged plates. Gamma rays are not affected by charged plates.

What kinds of particles are emitted by naturally radioactive nuclei? It is easy to show by a simple experiment that they fall into three distinct classes. Suppose we put a piece of pitchblende (which contains uranium) at the bottom of a cylindrical cavity in a lead block, as in Fig. 28.7. At the top of the cylinder we set up two plates, with a voltage between them, so that one plate becomes positively charged and the other negatively charged. If we now move a particle detector across the top of the cylinder, we find that there are at least three different kinds of emissions from the cylinder. Some emitted particles are positively charged and attracted to the negative plate; some are negatively charged and attracted to the positive plate; and some have no charge at all, for they are completely unaffected by the charged plates. In the early days of the study of radioactivity, when the nature of these particles was as yet unclear, Lord Rutherford differentiated them by the first three letters of the Greek alphabet as alpha (α), beta (β), and gamma (γ) rays.

We now know that the alpha rays are nothing but helium nuclei stripped of their electrons, that is, $_2^4\text{He}^{2+}$. These relatively heavy particles rapidly lose their kinetic energy by collisions; they travel on the average only an inch or so in air before they combine with free electrons and become normal helium atoms.

Beta rays are nothing but ordinary electrons, $_{-1}^0e$, which differ from the electrons in the exterior of atoms only in that they are produced in the decay of radioactive nuclei. Beta rays travel only a few feet in air before they are slowed down sufficiently by collisions to combine with positive ions and form neutral atoms.

Gamma rays are high-energy photons produced by transitions between nuclear energy levels in the same way that transitions between atomic energy levels produce visible light. These bundles of electromagnetic energy have no charge and no rest mass, are highly penetrating, and are the most difficult of all radioactive emissions against which to shield.

Half-Life

We can never know how long it will take any particular nucleus to decay, just as we can never know how long it will take for a particular atom to fall spontaneously to a lower-energy state and emit light. What we can determine experimentally, however, is how long it takes, *on the average*, for a large group of radioactive nuclei of the same species to decay to some other species. Suppose, for example, that we have 100 million $^{137}_{55}$Cs nuclei in a box. If we could go away and return 30 years later, and again count the number of cesium nuclei present, we would find that we had only 50 million left. The other 50 million would have spontaneously transformed themselves into barium nuclei by the following beta-decay process*:

$$^{137}_{55}\text{Cs} \rightarrow \ ^{0}_{-1}e + \ ^{137}_{56}\text{Ba}$$

If this is the case, we say that the *half-life* of $^{137}_{55}$Cs is 30 years.

Definition

Half-life: The time required for half the nuclei in a sample of a particular nuclear species to decay radioactively.

The half-life is a measure of the stability of a particular nuclear species. Stable nuclei have infinitely long half-lives. In one half-life the amount of the radioactive substance is cut in half; in two half-lives it is reduced by a factor of 4; in three half-lives by a factor of 8, and so on, as shown in Fig. 28.8 for the decay of $^{137}_{55}$Cs. This kind of decay is plotted in Fig. 28.9, where the symbol $T_{1/2}$ denotes half-life.

The half-lives for some important radioactive nuclei are given in Table 28.2. Note the great difference in the half-lives listed, ranging from a fraction of a second to 10^{10} years. Half-lives of artificially produced radioactive nuclei can be as small as 10^{-12} s. Despite these large differences all radioactive materials decay in the same basic fashion, following the same exponential-decay law.

---

*In beta-decay processes particles called neutrinos and antineutrinos are also produced. Since they carry neither charge nor measurable mass away from the reaction, for clarity we will postpone discussion of these particles until Chap. 30.

TABLE 28.2   Half-Lives of Some Important Radioactive Isotopes

Isotope	Half-life $T_{1/2}$	Isotope	Half-life $T_{1/2}$
$^{3}_{1}$H	12.3 years	$^{90}_{38}$Sr	28 years
$^{14}_{6}$C	5730 years	$^{129}_{53}$I	1.7×10^{6} years
$^{32}_{15}$P	14.3 days	$^{131}_{53}$I	8 days
$^{40}_{19}$K	10^{9} years	$^{137}_{55}$Cs	30 years
$^{41}_{21}$Sc	0.6 s	$^{226}_{88}$Ra	1600 years
$^{60}_{27}$Co	5.24 years	$^{232}_{90}$Th	1.4×10^{10} years
$^{85}_{36}$Kr	10.8 years	$^{239}_{94}$Pu	2.4×10^{4} years
$^{88}_{36}$Kr	2.8 h		

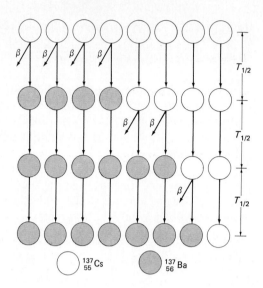

FIGURE 28.8 Reduction in amount of a radioactive sample with time. In one half-life ($T_{1/2}$) four of every eight $^{137}_{55}\text{Cs}$ nuclei on the average is converted into $^{137}_{56}\text{Ba}$ nuclei by beta decay. In the next half-life, two of the remaining four $^{137}_{55}\text{Cs}$ nuclei change into $^{137}_{56}\text{Ba}$. In the next half-life one of the remaining two Cs nuclei on the average is converted into Ba. At the end of three half-lives (i.e., about 90 years in this case), therefore, seven out of every eight Cs nuclei have been changed into Ba, on the average.

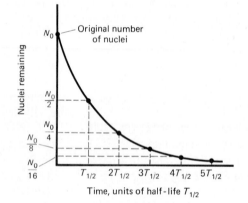

FIGURE 28.9 Exponential decay of a radioactive nucleus. In each half-life half the nuclei present at the beginning of that half-life decay.

Exponential Decay

The decay curve shown in Fig. 28.9 is a typical exponential-decay curve,* with the property that the height of the curve is reduced by a factor of 2 for each half-life along the horizontal axis. The equation of this curve is

$$N = N_0 e^{-\lambda t} \tag{28.3}$$

where N is the number of radioactive nuclei present at time t, N_0 is the number present at the beginning when $t = 0$, and λ is the so-called decay constant. The larger λ, the more rapidly the nuclei decay.

To obtain λ or t when we know N/N_0 we can transform Eq. (28.3) by taking natural logarithms (base e) of both sides of the equation and obtain

---

*See Appendix 1.E for the properties of exponential and logarithmic functions.

$$\ln N = \ln N_0 + \ln e^{-\lambda t} = \ln N_0 - \lambda t$$

Hence $\quad \lambda t = \ln N_0 - \ln N = \ln \dfrac{N_0}{N}$

Now, by the definition of the half-life $(T_{1/2})$, $N_0/N = 2$ when $t = T_{1/2}$, and so

$$T_{1/2} = \frac{\ln 2}{\lambda} = \frac{0.693}{\lambda}$$

Hence we have finally

$$\boxed{\lambda = \frac{0.693}{T_{1/2}}} \tag{28.4}$$

A reciprocal relationship therefore exists between the decay constant λ and the half-life $T_{1/2}$. If we know one, the other can be easily found. The larger λ, the more rapidly the nuclei decay and hence the shorter the half-life.

Nuclear Activity

The number of nuclei decaying in some time interval Δt is clearly proportional both to the length of the time interval and, from Eq. (28.3), to the number of nuclei remaining at that time; that is,

$$\Delta N \propto N \, \Delta t$$

The proportionality constant can be shown to be equal to the decay constant λ. Hence we have

$$\Delta N = -\lambda N \, \Delta t \tag{28.5}$$

or $\quad \dfrac{\Delta N}{\Delta t} = -\lambda N$

where the minus sign indicates that the number N decreases in time. We thus have the following definition of *nuclear activity*.

Definition

Nuclear activity A: The rate at which a particular nuclear species decays in time; $A = \Delta N/\Delta t = -\lambda N$, where λ is the decay constant and N is the number of radioactive nuclei remaining at the time at which the activity is calculated.

Since λ is a constant for any nuclear species and N decreases exponentially with time, we see that *the nuclear activity also decreases exponentially with time.*

Radiocarbon Dating

One especially useful isotope is $^{14}_{6}C$, which has a half-life of 5730 years and can be used to date fossils, old manuscripts, works of art, and other objects made from once-living materials. The isotope $^{14}_{6}C$ is continuously produced in the air by cosmic rays and is breathed in by living creatures. Since $^{14}_{6}C$ makes up a tiny but constant fraction of all the carbon in the air (about 1.5×10^{-12}), all living creatures have the same fraction of $^{14}_{6}C$ in their bodies. Once death occurs, however, no more $^{14}_{6}C$ is added to the body, and the $^{14}_{6}C$ already present decays with a half-life of 5730 years according to the equation $^{14}_{6}C \rightarrow {}^{14}_{7}N + {}^{0}_{-1}e$. This means that 5730 years after the death of an organism, the fraction of $^{14}_{6}C$ in the body will be one-half that in a living creature. Hence if the fraction of the radioactive carbon in a fossil is found to be half of that in a living creature today, the fossil is presumed to be 5730 years old. This was the technique used to date the Dead Sea scrolls, which were found wrapped in linen. From an analysis of the $^{14}_{6}C$ content of the linen, which was made from fibers of flax (a living material), it was found that the linen was 2000 years old. It could therefore be concluded that the scrolls were approximately 2000 years old.

Example 28.4

A sample of Kr gas contains 2.00×10^{20} atoms of $^{88}_{36}Kr$ at a time $t = 0$.
(a) What is the decay constant for $^{88}_{36}Kr$?
(b) At a time $t = 11.2$ h later, how many $^{88}_{36}Kr$ atoms remain?

(c) What is the initial nuclear activity of the sample?
(d) What is the nuclear activity when $t = 11.2$ h?

SOLUTION

It should be clear that not enough data are given to allow the problem to be solved. In any problem dealing with radioactive decay we need either the decay constant, the half-life, or some other data which will enable us to obtain one or the other of these two quantities. Since neither is given, the presumption is that the necessary data must be obtained from an available table. In this case, Table 28.2 gives us the half-life of $^{88}_{36}Kr$ as 2.8 h. Hence 11.2 h is just four half-lives.
(a) The decay constant is:

$$\lambda = \frac{0.693}{T_{1/2}} = \frac{0.693}{(2.8 \text{ h})(3600 \text{ s/h})} = \boxed{6.88 \times 10^{-5} \text{ s}^{-1}}$$

From Eq. (28.5), $(1/N)(\Delta N/\Delta t) = -\lambda$, and so this means that the fraction of the $^{88}_{36}Kr$ atoms which decays each second is 6.88×10^{-5}.
(b) In four half-lives (11.2 h) the number of $^{88}_{36}Kr$ atoms decreases by a factor of $2^4 = 16$. Hence the number of atoms present after 11.2 h is

$$N = \frac{N_0}{16} = \frac{2.00 \times 10^{20} \text{ atoms}}{16} = \boxed{1.25 \times 10^{19} \text{ atoms}}$$

(c) The initial nuclear activity is, from Eq. (28.5), since $N = N_0$ when $t = 0$,

$$\left(\frac{\Delta N}{\Delta t}\right)_0 = -\lambda N_0$$

$$= -(6.88 \times 10^{-5} \text{ s}^{-1})(2.00 \times 10^{20} \text{ atoms})$$

$$= \boxed{-1.38 \times 10^{16} \text{ atoms per second}}$$

(d) When $t = 11.2$ h, N is now 1.25×10^{19} atoms, and so

$$\left(\frac{\Delta N}{\Delta t}\right)_{4T_{1/2}} = -\lambda N$$

$$= -(6.88 \times 10^{-5} \text{ s}^{-1})(1.25 \times 10^{19} \text{ atoms})$$

$$= \boxed{-8.63 \times 10^{14} \text{ atoms per second}}$$

Hence in four half-lives both the number of atoms left and the nuclear activity have decreased by the same factor of 2^4, or 16.

28.4 Equations Governing Nuclear Reactions

In the radioactive decay of nuclei all the basic conservation principles of physics must be obeyed. These include the conservation of electric charge, of mass, of energy, and of linear and angular momentum.

To guarantee the conservation of electric charge and of mass (neglecting for the present the conversion of mass into energy) in any nuclear reaction, the following rules must be satisfied in the equation describing the reaction:

1 The sum of the charge numbers on the left side of the equation must be equal to the sum of the charge numbers on the right side of the equation. This guarantees the conservation of charge.

2 The sum of the mass numbers on the left side of the equation must be equal to the sum of the mass numbers on the right side of the equation. This guarantees the conservation of the number of nucleons taking part in the reaction, or the conservation of *nucleon number*.

For example, the nuclear equation describing the alpha decay of a radium nucleus is:

$$^{226}_{88}\text{Ra} \rightarrow \,^{222}_{86}\text{Rn} + \,^{4}_{2}\text{He}$$

where the radium is called the *parent nucleus*, and the radon the *daughter nucleus*. It can be seen that both the mass numbers and the charge numbers on the two sides of this equation balance out, indicating that the reaction can occur without violating the conservation principles of physics.

Conservation of Mass-Energy

Here the changes in the masses of the nuclei to be considered are very small, and the energy arising from mass-energy conversion is carried off either as kinetic energy of the interacting particles or in the form of gamma rays or other massless particles. In all cases, however, spontaneous decay of a nucleus can only occur if the particles that result from the decay have a total energy less than the energy of the decaying nucleus. In this, as in all other cases, the physical system will try to minimize its potential energy. Because of the equivalence of mass and energy, this means that the mass of the parent nucleus must be greater than the sum of the masses of the daughter nucleus and the other particles produced, since some mass will always be converted into energy. If the parent has less mass than the products, then the decay cannot occur spontaneously without violating the mass-energy conservation principle.

Example 28.5

Identify the unknown particles involved in the following nuclear reactions:

(a) The beta decay of $^{137}_{55}\text{Cs}$ (cesium).
(b) The alpha decay of $^{255}_{100}\text{Fm}$ (fermium).

SOLUTION

(a) We can write this beta-decay equation as:

$$^{137}_{55}\text{Cs} \rightarrow \,^{0}_{-1}e + \,^{A}_{Z}X$$

where the unknown nucleus $^{A}_{Z}X$ is to be determined.

From the conservation-of-charge principle we have $55 = Z - 1$, or $Z = 56$. Similarly, from the principle of conservation of nucleon number we have $137 = A + 0$, or $A = 137$. The unknown nucleus is therefore $^{137}_{56}X$. From the periodic table (Fig. 9.9) the element with atomic number 56 is seen to be barium. Hence the unknown particle is $\boxed{^{137}_{56}\text{Ba}}$.

(b) In the same way we have for the alpha decay of $^{255}_{100}\text{Fm}$,

$$^{255}_{100}\text{Fm} \rightarrow \,^{4}_{2}\text{He} + \,^{A}_{Z}X$$

Hence Z must equal 98, and A must be 251. The element with atomic number 98 is, from the periodic table, californium (Cf). Hence the unknown particle is

$$\boxed{^{251}_{98}\text{Cf}}.$$

Example 28.6

A polonium nucleus decays into lead in the following reaction:

$$^{210}_{84}\text{Po} \rightarrow \,^{4}_{2}\text{He} + \,^{206}_{82}\text{Pb}$$

The exact masses for the neutral atoms are given in Appendix 4. What is the kinetic energy of the products of this reaction?

$$KE = \tfrac{1}{2}mv^2$$

SOLUTION

The exact masses for the neutral atoms are given as: $m(^{210}_{84}\text{Po}) = 209.98288$ u; $m(^{206}_{82}\text{Pb}) = 205.97447$ u; $m(^{4}_{2}\text{He}) =$

4.002603 u. Since there are the same number (84) of extranuclear electrons in the atoms on both sides of the above equation, the electron masses cancel out (as is usually the case), and we can use the masses of the neutral atoms instead of the masses of the bare nuclei in our calculations of the mass-energy conversion.

The total mass of the product nuclei is 209.97707 u, and the mass defect is $\Delta m = 209.98288 \text{ u} - 209.97707 \text{ u} = 0.00581 \text{ u}$. This mass appears as kinetic energy of the daughter nucleus and the alpha particle, with most of it going to the alpha particle because of its smaller mass. The kinetic energy produced is:

$$KE = (0.00581 \text{ u}) \left(\frac{931.5 \text{ MeV}}{1 \text{ u}} \right) = \boxed{5.41 \text{ MeV}}$$

This then is the energy produced by the conversion of mass into energy in this particular nuclear reaction.

To balance both mass and energy in a nuclear equation like the one in this example, we should really write it as

$$^{210}_{84}\text{Po} \rightarrow {}^{4}_{2}\text{He} + {}^{206}_{82}\text{Pb} + 5.41 \text{ MeV}$$

This indicates that a small amount of mass has been converted into 5.41 MeV of energy in the reaction, and that mass-energy is therefore conserved, in addition to charge and nucleon number.

28.5 Nuclear Reactions II: Artificial Radioactivity

We have been discussing natural radioactivity, a phenomenon caused by the intrinsic instability of some nuclei which leads to their emitting alpha, beta, or gamma particles and decaying into other kinds of nuclei. It is possible, however, to take extremely stable nuclei, such as $^{14}_{7}\text{N}$, the nuclei of the atoms forming nitrogen molecules in the air, and make them unstable by bombarding them with highly energetic particles like protons, deuterons, or alpha particles. These particles can force their way inside a stable nucleus and create a new, unstable nucleus, which then decays into yet another nucleus and other smaller particles. This process we call *artificial* (or *induced*) *radioactivity*.

Suppose we fire high-energy alpha particles at a nitrogen nucleus. One reaction which occurs is found by experiment to be:

$$^{4}_{2}\text{He} + {}^{14}_{7}\text{N} \rightarrow \{{}^{18}_{9}\text{F}\} \rightarrow {}^{1}_{1}\text{H} + {}^{17}_{8}\text{O} \tag{28.6}$$

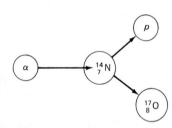

FIGURE 28.10 Alpha-particle bombardment of $^{14}_{7}\text{N}$, leading to the production of $^{17}_{8}\text{O}$.

In this reaction the alpha particle combines with the nitrogen nucleus to produce the highly unstable fluorine nucleus $^{18}_{9}\text{F}$, which then spontaneously decays into a rare isotope of oxygen and a proton, as shown in Fig. 28.10. The braces around the $^{18}_{9}\text{F}$ indicate that this "compound" nucleus cannot be observed because its lifetime is so short, but, to make sense of the experimental data, must be presumed to have existed for a very small fraction of a second. Equation (28.6) is sometimes written in a shorthand notation as $^{14}_{7}\text{N}$ (α, p) $^{17}_{8}\text{O}$.

Another possibility is to use deuterons (nuclei of heavy hydrogen, $^{2}_{1}\text{H}$) as the bombarding particles. A possible reaction in this case is the following:

$$^{2}_{1}\text{H} + {}^{238}_{92}\text{U} \rightarrow \{{}^{240}_{93}\text{Np}\} \rightarrow {}^{238}_{93}\text{Np} + 2{}^{1}_{0}n \tag{28.7}$$

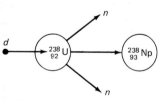

FIGURE 28.11 Deuteron bombardment of $^{238}_{92}\text{U}$, leading to the production of the transuranic element neptunium $^{238}_{93}\text{Np}$.

Here we have converted uranium, the heaviest atom to occur naturally on earth, into the "transuranic" (i.e., beyond uranium in the periodic table) element neptunium by deuteron bombardment, as shown in Fig. 28.11. Again this can be written as $^{238}_{92}\text{U}$ $(d, 2n)$ $^{238}_{93}\text{Np}$.

Many other possibilities exist, but they all require bombardment with high-energy particles which have sufficient speeds to get close enough to the bombarded nucleus to produce a new, unstable nucleus. To accomplish this goal, particle accelerators like cyclotrons and betatrons are needed. We will discuss these "atom smashers" in Chap. 30.

Neutron Bombardment of Nuclei

There is another kind of bombarding particle of great importance both in the history of nuclear physics and in the practical use of nuclear energy today. This is the neutron, which has the great advantage that it is uncharged and hence can penetrate closer to the nucleus being bombarded than can positively charged particles which are repelled electrostatically by the positively charged nucleus. On the other hand, the neutron has the compensating disadvantage that it cannot be accelerated by electric or magnetic fields, since it has no charge. Physicists desiring to use neutrons as nuclear "bullets," therefore, must rely on reactions like the one in Eq. (28.7) above, which has two energetic neutrons as end products.

In the early 1930s the Italian physicist Enrico Fermi (see Fig. 1.8 and accompanying biography) realized the importance of a careful study of the neutron bombardment of nuclei. He bombarded a series of elements from nitrogen to uranium with neutrons and studied the radioactive transformations produced. When he came to uranium, he expected to find a whole series of transuranic elements produced in reactions such as the following:

$$_{92}^{238}\text{U} + _{0}^{1}n \rightarrow _{92}^{239}\text{U} \tag{28.8}$$

The uranium nucleus would then be expected to decay spontaneously by successive beta emissions into neptunium and plutonium, by the following processes*:

$$_{92}^{239}\text{U} \rightarrow _{93}^{239}\text{Np} + _{-1}^{0}e \tag{28.9}$$

and

$$_{93}^{239}\text{Np} \rightarrow _{94}^{239}\text{Pu} + _{-1}^{0}e \tag{28.10}$$

When Fermi carried out this experiment, however, he found neither neptunium nor plutonium, but other nuclei which he could not identify. About this time he was forced to leave Italy with his wife because of the political situation under Mussolini and Hitler at the end of the 1930s. He accepted a faculty position at Columbia University in New York City and later moved to Chicago, where he directed the research that led to the first successful chain-reacting pile and to the military and commercial use of nuclear energy.

A number of European physicists, among them Irène Joliot-Curie, daughter of Marie Curie (see Fig. 28.6), continued Fermi's work. The German chemists Otto Hahn and Fritz Strassman finally identified one of the nuclei produced when uranium was bombarded by neutrons as barium ($_{56}^{141}\text{Ba}$), an element very much lighter than uranium. The presence of barium in the reaction products was so unexpected, and so impossible to explain by the then-accepted theories of nuclear transformations by alpha, beta, and gamma emissions, that Hahn and Strassman were hesitant about publishing their findings, but did so anyway in 1939. Shortly thereafter the Austrian physicist Lise Meitner (see Fig. 28.12 and accompanying biography) and her nephew, Otto Frisch, interpreted Hahn and Strassman's experimental results as proof of a new kind of nuclear reaction which they named *nuclear fission*.

---

*Again note that we are for the present omitting the neutrinos produced in such beta-decay processes.

Lise Meitner (1878–1968)

FIGURE 28.12 Lise Meitner and Otto Hahn (1879–1968) in their laboratory at Berlin-Dahlem in 1913. For over 30 years Meitner worked in Berlin with Hahn, a distinguished radiochemist and student of Rutherford. (*From Otto Hahn, A Scientific Autobiography, Scribners, New York, 1966, with permission; courtesy of AIP Niels Bohr library.*)

Lise Meitner grew up in the Vienna of Emperor Franz Joseph and horse-drawn trolley cars. She was born there in 1878 into a well-to-do Jewish family. She decided at an early age that she wanted to be a scientist like Madame Curie, and in 1901 she entered the University of Vienna. There, in an environment where serious women students were considered odd, she was treated rudely by many of her fellow students. When she received her Ph.D. in physics in 1905, she was only the second woman in history to receive a Ph.D. from the University of Vienna.

In 1907 she went to Berlin to study under Max Planck, promising her devoted parents that she would be back in Vienna in 6 months at the most. She stayed 31 years! In Berlin Meitner met Otto Hahn, a professor of chemistry, and took a research position (without salary), helping Hahn with research on the chemistry of radioactive substances. At that time women were not allowed to work in the Chemical Institute, and she had to set up her laboratory in a carpenter's workshop outside the institute.

While continuing work with Hahn at the Kaiser Wilhelm Institute for Chemistry in Berlin-Dahlem, she served as assistant to Max Planck at the Institute for Theoretical Physics at the University of Berlin from 1912 to 1915, and in 1918 was appointed head of the physics department at the Kaiser Wilhelm Institute. During World War I she enlisted as a nurse in the Austrian army, and served loyally throughout the war, as did her role model, Madame Curie, for the opposing side.

In 1922 Meitner delivered her inaugural lecture as a faculty member at the University of Berlin. The title of her talk was actually "Cosmic Physics," but it appeared in the newspapers as "Cosmetic Physics," an indication of the press's narrow-minded view of the scientific interests of women.

In the following years Hahn and Meitner did some very significant research on beta- and gamma-ray spectra. They discovered the new element protoactinium-91, and took up the work on the neutron bombardment of nuclei which Enrico Fermi had commenced in Rome. This work had to be dropped in 1938, however, when Meitner had to flee Germany after Hitler annexed Austria. She found asylum in Stockholm, Sweden, after spending a few months at Bohr's institute in Copenhagen.

At the end of 1938 Otto Hahn sent her a description of his experiments on the interaction of neutrons with uranium. He and Strassman had determined that one of the reaction products was clearly barium. Meitner was so excited about this that she showed Hahn's letter to her nephew, the physicist Otto Frisch, and suggested that they go for a long walk and sort out some of these ideas. During this walk all the pieces quickly fell into place and the idea of nuclear fission was born. Frisch then demonstrated in his laboratory the tremendous release of energy accompanying fission. A short paper by Meitner and Frisch in the British journal *Nature* in 1939 revealed the momentous idea of fission to the scientific world.

Although it was her basic insight which eventually led to the fission bomb dropped on Hiroshima, Meitner refused to work on the bomb and hoped that it would not work.

After the war, although now famous, she continued her laboratory research, interrupted only by trips to receive honorary degrees and other scientific honors. In 1946 she spent a semester as a visiting professor at Catholic University in Washington, D.C. In 1960 she retired to Cambridge, England, to be near her nephew, Otto Frisch, but continued to travel, lecture, and attend concerts, for music was one of the great joys of her life. She died in 1968 at the age of 90.

A fitting epitaph for Lise Meitner might well be her own words, spoken in a lecture at Bryn Mawr College in 1959: "Life need not be easy, provided only that it is not empty."

Example 28.7

In the nuclear reaction described by Eq. (28.6), what is the *minimum* energy an alpha particle must have to make this reaction occur? Assume that the atomic masses involved are those given in Appendix 4.

SOLUTION

In the reaction

$$^{4}_{2}\text{He} + ^{14}_{7}\text{N} \rightarrow ^{1}_{1}\text{H} + ^{17}_{8}\text{O}$$

the total mass of the reacting atoms is 4.00260 u + 14.00307 u = 18.00567 u. The total mass of the products is the mass of the hydrogen atom, 1.00783 u, plus that of the oxygen atom, which is 16.99916 u, or 18.00699 u in all.

The products therefore have a *greater* mass than the original atoms, and the reaction cannot take place unless the alpha particle brings at least enough energy to the collision to make the masses before and after the reaction the same. Otherwise mass-energy cannot be conserved in the collision.

The difference in the masses after and before is

$$18.00699 \text{ u} - 18.00567 \text{ u} = 0.00132 \text{ u}$$

Hence the minimum kinetic energy needed by the alpha particle is

$$KE = (0.00132 \text{ u}) \left(\frac{931.49 \text{ MeV}}{1 \text{ u}} \right) = 1.23 \text{ MeV}$$

Actually a kinetic energy somewhat greater than this will be needed, since the hydrogen and oxygen atoms will carry away some additional kinetic energy from the reaction.

28.6 Nuclear Fission and Nuclear Fusion

Meitner and Frisch explained the presence of $^{141}_{56}\text{Ba}$ in the decay products of uranium bombarded by neutrons by assuming that the neutrons were absorbed by uranium nuclei. Niels Bohr and the American physicist John Wheeler (born 1911) later showed that the absorbing nuclei must be not the $^{238}_{92}\text{U}$ nuclei, which make up 99.3 percent of natural uranium, but rather $^{235}_{92}\text{U}$ nuclei, which make up only 0.72 percent of natural uranium. The result was the unstable isotope $^{236}_{92}\text{U}$, which then broke up, or "fissioned," into two smaller nuclei of about equal size. One possible reaction was the following:

$$^{235}_{92}\text{U} + ^{1}_{0}n \rightarrow \{^{236}_{92}\text{U}\} \rightarrow ^{141}_{56}\text{Ba} + ^{92}_{36}\text{Kr} + 3^{1}_{0}n \qquad (28.11)$$

This would explain Hahn and Strassman's finding of barium among the product nuclei after the neutron bombardment of uranium. The fission products are highly radioactive and quickly decay into isotopes of other elements. For example, the half-life of $^{141}_{56}\text{Ba}$ is 18 min, and of $^{92}_{36}\text{Kr}$ only 2.4 s.

Physicists all over the world took up these ideas and soon showed in their laboratories that not only does the above reaction occur, but other reactions like the following also occur.

$$^{235}_{92}\text{U} + ^{1}_{0}n \rightarrow \{^{236}_{92}\text{U}\} \rightarrow ^{137}_{53}\text{I} + ^{97}_{39}\text{Y} + 2^{1}_{0}n \qquad (28.12)$$

$$^{235}_{92}\text{U} + ^{1}_{0}n \rightarrow \{^{236}_{92}\text{U}\} \rightarrow ^{143}_{57}\text{La} + ^{90}_{35}\text{Br} + 3^{1}_{0}n \qquad (28.13)$$

All these reactions conserve charge and nucleon number, and all result in two isotopes each of which has roughly half the mass of the original uranium nucleus. All three of these reactions are examples of what is now called *nuclear fission*.

Definition

Nuclear fission: The breaking up of a heavy nucleus like uranium into two nuclei of intermediate masses, together with the release of two or more neutrons and large amounts of energy.

This definition points up two other important facts about fission. First, large amounts of energy (about 200 MeV) are released in each such fission reaction, mostly in the form of the kinetic energies of the two heavy fission products and the neutrons. Second, for each neutron absorbed, at least two neutrons are produced in the fission process. These two facts immediately led Meitner, Frisch, and others to the startling conclusion that the fission of uranium might provide a tremendous source of energy which could be used for either the well-being or the destruction of humanity.

Chain Reactions

The key to releasing the tremendous store of energy in a $^{235}_{92}$U nucleus is a *chain reaction* of the kind shown in Fig. 28.13. A single neutron causes the fission of a $^{235}_{92}$U nucleus into barium and krypton. In the fission reaction shown three neutrons are produced. Even if one of these escapes from the reaction region, two are left to produce additional fission reactions. Each of these new fission processes produces enough neutrons to cause two additional fission processes, and so on. The energy thus released increases *exponentially*, since the number of fission reactions doubles at each stage of the chain. The result of this chain reaction is the release of a huge amount of energy in a very short time—the equivalent of a dynamite explosion, but much greater in size and destructive power.

The secret of a successful energy-releasing fission process is that for every neutron producing fission, at least one of the neutrons produced in that fission must be absorbed and produce fission at the next stage of the reaction.

Definition

Multiplication factor: The ratio of the number of neutrons producing fission at any stage of the reaction process to the number of neutrons producing fission in the immediately preceding stage.

When the multiplication factor becomes equal to unity, i.e., each fission produces one additional fission, the chain reaction is said to have gone *critical*. This is the case in power reactors where we want a controlled release

FIGURE 28.13 A nuclear chain reaction involving $^{235}_{92}$U.

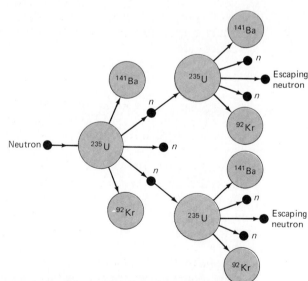

FIGURE 28.14 Critical mass required for nuclear fission: (*a*) For a small mass of ^{235}U, on the average less than one neutron emitted in a fission process is absorbed and causes another fission process. Hence the multiplication factor is less than unity, and the mass is less than critical. (*b*) For a larger mass of ^{235}U (now above the critical mass), on the average more than one neutron emitted in a fission process is absorbed and produces another fission process. In this case the multiplication factor is greater than unity.

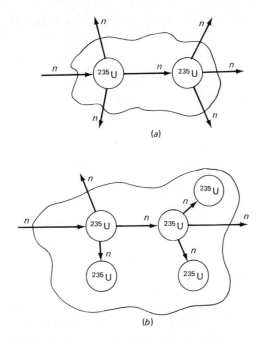

of heat. For nuclear bombs, on the other hand, the multiplication factor must be as much greater than unity as possible, to achieve the most complete energy release in the shortest possible time. For the chain reaction to achieve criticality, there is a minimum mass of fissionable material, the so-called *critical mass*, required, as shown in Fig. 28.14. This minimum or critical mass varies with the way the fissionable material is distributed in space.

That such an abstract idea could lead to a practical source of energy was shown conclusively when the first nuclear reactor (which in this case was just a *pile* of uranium and graphite blocks) went critical at Stagg Field at the University of Chicago in 1942, with Enrico Fermi directing the operation. The ultimate confirmation came when the first uranium bomb was dropped on the city of Hiroshima in Japan in 1945, destroying 90 percent of the city.

Energy Considerations in Fission Processes

The secret of the energy released in a nuclear fission process lies in the binding-energy curve for nuclei. Suppose a $^{235}_{92}$U nucleus fissions into barium and krypton nuclei, as described by Eq. (28.11). The barium and krypton nuclei are more tightly bound and stable than the uranium nucleus, as is clear from the binding-energy curve in Fig. 28.4. Hence the total energy of the barium and krypton nuclei and the neutrons produced in the reaction is less than the energy of the $^{236}_{92}$U nucleus undergoing fission, and also less than the sum of the energies of the $^{235}_{92}$U and the bombarding neutron which produced the $^{236}_{92}$U. The excess energy must, therefore, be released in the form of the kinetic energy of the particles produced in the reaction. This energy adds up to about 215 MeV per fission, as shown in Table 28.3. This corresponds to a mass of about 0.23 u, so that only about 0.1 percent of the mass of the $^{235}_{92}$U nucleus is converted into energy.

TABLE 28.3   Energy Released in the Fission of $^{235}_{92}U$ by a Neutron

Form of energy released	Amount of energy released, MeV
Kinetic energy of two fission fragments	168
Immediate gamma rays	7
Delayed gamma rays	3–12
Fission neutrons	5
Energy of decay products of fission fragments:	
Gamma rays	7
Beta rays	8
Neutrons	12
Total energy released	210–219
Average energy released	215

Although 215 MeV is only 3.4×10^{-11} J, this energy comes from a single nucleus, and so kilograms of $^{235}_{92}U$ undergoing fission can easily produce enough energy for a nuclear reactor or a nuclear bomb.

Another way to look at the same process is to note that the mass of the fission products (including the three neutrons) is less than the mass of the $^{236}_{92}U$ nucleus produced by the neutron bombardment. The mass defect (equivalent to the difference in binding energies) must therefore be converted into energy if mass-energy is to be conserved in the reaction. This is illustrated in a schematic way for the fission of $^{235}_{92}U$ in Fig. 28.15.

FIGURE 28.15 Mass-energy conversion in a fission reaction. The mass defect in the reaction is converted into the energy released in the fission process.

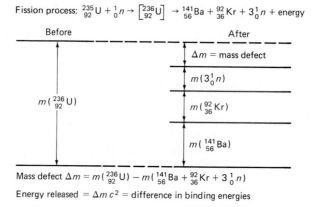

Fission process: $^{235}_{92}U + ^{1}_{0}n \rightarrow \left[^{236}_{92}U\right] \rightarrow ^{141}_{56}Ba + ^{92}_{36}Kr + 3^{1}_{0}n + \text{energy}$

Before — After

$\Delta m = \text{mass defect}$

$m(3^{1}_{0}n)$

$m(^{236}_{92}U)$

$m(^{92}_{36}Kr)$

$m(^{141}_{56}Ba)$

Mass defect $\Delta m = m(^{236}_{92}U) - m(^{141}_{56}Ba + ^{92}_{36}Kr + 3^{1}_{0}n)$

Energy released $= \Delta m c^2 = $ difference in binding energies

Nuclear Fusion Reactions

Definition

Another source of nuclear energy comes from *fusion*, which is the reverse process of fission.

Nuclear fusion: A nuclear reaction in which two light nuclei are combined, or fused together, to form a heavier nucleus, accompanied by the release of large amounts of energy.

From the binding-energy curve in Fig. 28.4 it is clear that energy can be released in such a fusion reaction, since the light elements are less tightly bound together than elements closer to the middle of the periodic table. Hence when two light nuclei like a deuterium nucleus ($^{2}_{1}H$) and a tritium nucleus ($^{3}_{1}H$) fuse to produce a helium nucleus ($^{4}_{2}He$) and a neutron, it is found experimentally that 17.6 MeV of energy is released. This energy release is to

Fusion process: $^2_1H + ^3_1H \rightarrow ^4_2He + ^1_0n$ + energy (17.6 MeV)

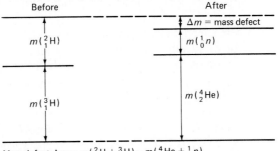

Mass defect $\Delta m = m(^2_1H + ^3_1H) - m(^4_2He + ^1_0n)$

Energy released = $\Delta m\, c^2$ = different in binding energies

FIGURE 28.16 Mass-energy conversion in a fusion reaction. The mass defect in the reaction is converted into the energy released in the fusion process.

be expected, since the helium nucleus is so much more stable than the deuterium and tritium nuclei, as is clear from the binding-energy curve.

Another way to look at this is to note that in the reaction

$$^2_1H + ^3_1H \rightarrow ^4_2He + ^1_0n \tag{28.14}$$

the combined mass of the ^2_1H and ^3_1H is greater than the mass of the ^4_2He plus the neutron. Hence to conserve mass-energy some mass must be converted into energy in the reaction. This is the source of the observed 17.6 MeV of energy released. A schematic diagram of the fusion process on this basis is shown in Fig. 28.16.

The problem with fusion processes is that they only occur at very high temperatures (10^7 to 10^8 K) such as exist in the sun or other stars. Hence controlled fusion is much more difficult to achieve than controlled fission.

Example 28.8

From the atomic masses given in Appendix 4, find the energy released in the fusion reaction described by Eq. (28.14).

SOLUTION

In the fusion reaction

$$^2_1H + ^3_1H \rightarrow ^4_2He + ^1_0n$$

the masses of the neutral atoms and of the neutron are the following:

$m(^2_1H) = 2.01410$ u $m(^4_2He) = 4.00260$ u

$m(^3_1H) = 3.01605$ u $m(^1_0n) = 1.00866$ u

The mass defect is then

$m = (2.01410 + 3.01605)$ u $- (4.00260 + 1.00866)$ u

$m = (5.03015 - 5.01126)$ u = 0.01889 u

But, since 1 u is equivalent to 931.49 MeV, this is equivalent to an energy of

$$E = (0.01889 \text{ u}) \left(\frac{931.49 \text{ MeV}}{1 \text{ u}} \right) = \boxed{17.59 \text{ MeV}}$$

This agrees well with the experimental value mentioned above. Most of this energy goes into the kinetic energy of the alpha particle and the neutron.

28.7 The Nucleus as a Practical Energy Source

Both fission and fusion have been used successfully in weapons of unbelievable destructive power. The use of these two processes as practical energy sources, however, has been marked by technical, economic, and political problems which are still far from being solved.

Fission Reactors

For a nuclear power reactor intended to produce electricity, a multiplication factor exactly equal to 1 is required, for then exactly one neutron from each fission event initiates another event, and energy is released at a steady rate. This energy is then converted into heat by a series of collisions of the fission products with the surrounding materials in the reactor, and this heat is used to produce steam and generate electricity using a conventional steam-electric turbine. The fission process is therefore used basically to boil water, so that the efficiency of a nuclear reactor is severely limited by the second law of thermodynamics.

The desired multiplication factor of unity is achieved by monitoring the kind, amount, and location of fissionable material in the reactor, by using a "moderator" to reduce the neutron velocities, and by manipulating control rods from outside the reactor.

Nuclear Fuel There are only three isotopes which readily undergo fission when bombarded by neutrons. These are $^{233}_{92}U$, $^{235}_{92}U$, and $^{239}_{94}Pu$. These are called fissionable, or *fissile*, nuclei. Most operating reactors today use uranium in which the amount of $^{235}_{92}U$ has been increased to 3 percent of the total uranium present. The fuel, in the form of small uranium dioxide cylinders, is loaded into long, thin fuel rods in the core of the reactor. The core of a reactor may contain 20,000 or more fuel rods, in all of which nuclear fission processes are occurring and heat is being generated.

Moderator The efficiency of neutrons in producing fission of $^{235}_{92}U$ nuclei depends on their speed. The more slowly the neutrons move, the longer the time they spend near a $^{235}_{92}U$ nucleus and hence the greater the chance they will be captured and produce fission. Very slow, or "thermal," neutrons are needed. In most fission reactors in the United States today the water coolant flowing in contact with the fuel rods also acts as the moderator to slow down the neutrons to thermal speeds and increase the probability of fission occurring. In Fermi's original nuclear "pile," graphite was used as the moderator*; graphite is also used in most British reactors.

Control Rods A great concern about any nuclear fission reactor is that it will "run away," i.e., get out of control. When this happens, the multiplication factor exceeds unity, an exponential buildup of heat results, and the cooling system may not be able to handle the excess heat generated. There is no possibility of a nuclear explosion, since the concentration of $^{235}_{92}U$ present (3 percent) is insufficient for such an explosion. The fuel rods could melt from

---

*Enrico Fermi and his coworkers in Rome first discovered the effectiveness of paraffin in slowing down neutrons on the morning of October 22, 1934. Fermi worked out the theory during his lunch hour, and returned to his laboratory in the afternoon with the confident prediction that water would work equally well, since, like paraffin, it contained a large number of hydrogen atoms. He and his colleagues immediately checked this out by immersing a neutron source and a silver target in the goldfish fountain in the private garden behind their laboratory. They found that the induced radioactivity of the silver was indeed increased. This was the same goldfish fountain in which the members of Fermi's group liked to sail tiny toy boats as a relaxation from their intense work in the laboratory.

the heat, however, and burn through the reactor floor, leading to what is sometimes referred to as the "China syndrome," i.e., the movement of hot fission products into the earth in the general direction of China. Functioning reactors are required to have backup cooling systems to handle such eventualities, but some of these have not worked too well in practice.

To prevent runaways, control rods are used. These are rods of cadmium or boron, which absorb neutrons strongly and can be dropped into the reactor core to reduce the multiplication factor. If at any time during operation the reactor tries to run away, the control rods are supposed to fall into place automatically and shut down the reactor.

Boiling-Water Reactor (BWR)

A schematic of a typical boiling-water reactor is shown in Fig. 28.17. The BWR uses ordinary water as the moderator and heat-exchange fluid. In the right side of this figure the steam-electric turbine, the electric generator, and the condenser are identical to those found in conventional electric power plants burning coal or oil to produce steam. The only unique feature in a nuclear power plant is the nuclear-powered boiler. The steam produced drives the turbine and is then converted back into water in the condenser, using cooling water from a river or other large body of water.

FIGURE 28.17 A boiling-water reactor. Heat produced by nuclear fission reactions generates steam to drive a steam-electric turbine.

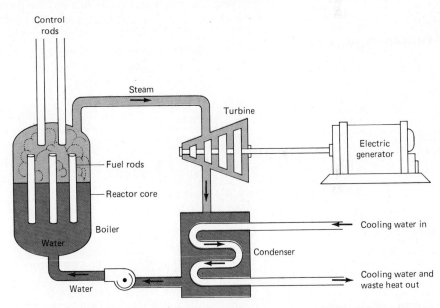

The Liquid-Metal Fast-Breeder Reactor (LMFBR)

One of the problems with conventional fission reactors is that it is unclear how long the world's supply of $^{235}_{92}U$ will last. If we could find some way to use $^{238}_{92}U$ as fuel in a nuclear reactor, we could increase the lifetime of existing uranium fuel by a factor of about 100, since $^{238}_{92}U$ makes up 99.3 percent of all natural uranium. One way to use $^{238}_{92}U$ as a nuclear fuel is to convert it into a fissionable nucleus like $^{239}_{94}Pu$. This can be done by neutron bombardment of the uranium in a reactor, which leads to the following series of reactions:

$$^{238}_{92}U + {}^{1}_{0}n \rightarrow {}^{239}_{92}U \tag{28.15}$$

$$^{239}_{92}U \rightarrow {}^{0}_{-1}e + {}^{239}_{93}Np \tag{28.16}$$

$$^{239}_{93}Np \rightarrow {}^{0}_{-1}e + {}^{239}_{94}Pu \tag{28.17}$$

In this series of reactions we have therefore produced a fissionable nucleus, plutonium, from a nonfissile isotope of uranium. Such a process is called "breeding." The plutonium can then be used as a reactor fuel, since it can be fissioned by neutrons according to the following equation:

$$\text{$_0^1$}n + \text{$_{94}^{239}$Pu} \rightarrow \{\text{$_{94}^{240}$Pu}\} \rightarrow X + Y + 2\,\text{$_0^1$}n + 200\text{ MeV} \tag{28.18}$$

Here X and Y are fission products consisting of a variety of radioactive isotopes near the middle of the periodic table. In this fission reaction the necessary energy and the required neutrons are produced to fuel a nuclear reactor similar to one fueled by $_{92}^{235}$U.

In breeder reactors the core is made up of $_{92}^{235}$U or $_{94}^{239}$Pu, surrounded by a "blanket" of $_{92}^{238}$U, as shown in Fig. 28.18. Neutrons given off by the core fission processes both sustain these processes and convert some of the $_{92}^{238}$U in the blanket into $_{94}^{239}$Pu. In principle such a breeder reactor can produce more fuel than it uses, and hence greatly extend our nuclear fuel supply.

Since fast neutrons are needed to produce the plutonium in a breeder reactor, no moderator is used, and a liquid metal like sodium is used as the coolant in place of water. The hot sodium then heats water to produce steam in a heat-exchange unit. The letters LMFBR refer to a breeder reactor (BR) using a liquid metal (LM) like sodium as the coolant and heat-transfer fluid, and fast neutrons (F) to convert $_{92}^{238}$U to $_{94}^{239}$Pu.

FIGURE 28.18 A breeder reactor. The $_{92}^{238}$U in the blanket surrounding the core is gradually converted into fissionable $_{94}^{239}$Pu by bombardment with the neutrons produced in the reactor core.

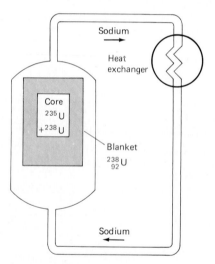

Fusion Reactors

Whereas conventional nuclear fission reactors and the LMFBR are known to work, fusion reactors are still in process of development. It is known that fusion can lead to the release of large amounts of energy, as it has in the hydrogen bomb, or H-bomb. What is still unclear is whether a controlled fusion process can produce more energy than it consumes and can work on a large enough scale to be economically feasible.

The working idea behind a fusion reactor is to cause light nuclei to fuse in processes such as the following:

$$\text{$_1^2$H} + \text{$_1^2$H} \rightarrow \text{$_2^3$He} + \text{$_0^1$}n + 3.3\text{ MeV} \tag{28.19}$$

$$\text{$_1^2$H} + \text{$_1^2$H} \rightarrow \text{$_1^3$H} + \text{$_1^1$H} + 4.0\text{ MeV} \tag{28.20}$$

$$\text{$_1^2$H} + \text{$_1^3$H} \rightarrow \text{$_2^4$He} + \text{$_0^1$}n + 17.6\text{ MeV} \tag{28.21}$$

To obtain useful amounts of energy from such processes, however, large numbers of nuclei must be brought very close together at very high temperatures and held there for a sufficient time for fusion to occur. The temperatures required are about 4×10^7 °C for the process described by Eq. (28.21) and about 2×10^8 °C for those in Eqs. (28.19) and (28.20). (Note that these are very high temperatures indeed: the highest melting point of all the elements, that of tungsten, is only 3.4×10^3 °C.)

Much research is at present underway on techniques to achieve both the high temperatures and high densities needed for controlled fusion. These techniques include confining gas particles to small regions of space using strong magnetic fields, and the compression of solid pellets containing fusionable isotopes by lasers or electron and ion beams.

Even though considerable progress has been made on various aspects of the controlled fusion problem, it is still uncertain whether practical amounts of energy can be obtained at prices which will make fusion economically viable. Even if all the technical and engineering problems related to fusion should soon be solved, it is doubtful that we will have functioning fusion reactors before the year 2000.

Example 28.9

If a nuclear-electric power plant generates 1 GW of electric power, about how many kilograms of pure $^{235}_{92}$U would be needed to produce electricity at this rate for a year, assuming that the power plant produces the electricity with an efficiency of about 33 percent?

SOLUTION

We have seen that the fission of a ^{235}U nucleus into barium and krypton releases about 215 MeV of energy. Hence 1 mol of pure $^{235}_{92}$U will produce an amount of energy equal to:

$$\mathscr{E} = (6.02 \times 10^{23})(215 \text{ MeV})$$

$$= (1.29 \times 10^{32} \text{ eV})(1.60 \times 10^{-19} \text{ J/eV}) = 2.07 \times 10^{13} \text{ J}$$

Now, the energy produced in 1 year by a 1-GW reactor operating at full capacity all year is

$$E' = \left(10^9 \, \frac{\text{J}}{\text{s}}\right) \left(\frac{3.16 \times 10^7 \text{ s}}{1 \text{ year}}\right) (1 \text{ year}) = 3.16 \times 10^{16} \text{ J}$$

Since the plant is only 33 percent efficient, the input energy to the steam-electric turbine must be

$$E = \frac{E'}{0.33} = \frac{3.16 \times 10^{16} \text{ J}}{0.33} = 9.49 \times 10^{16} \text{ J}$$

One mole of $^{235}_{92}$U contains 0.235 kg of uranium. Hence the number of kilograms of pure $^{235}_{92}$U needed is

$$N = \frac{E}{\mathscr{E}} = \frac{9.49 \times 10^{16} \text{ J}}{(2.07 \times 10^{13} \text{ J})/(0.235 \text{ kg})} = \boxed{1.08 \times 10^3 \text{ kg}}$$

Since 10^3 kg is often called a *metric ton*, a large 1-GW nuclear power requires about 1 metric ton of pure $^{235}_{92}$U a year as fuel. Since $^{235}_{92}$U is only about 0.7 percent of natural uranium, about 143 metric tons of natural uranium would be needed.

28.8 Biological Effects of Nuclear Radiations

Nuclear radiations include the alpha, beta, and gamma rays emitted by nuclei, as well as protons, neutrons, and even x-rays, which have the same ionizing properties as nuclear gamma rays even though they do not come from the nucleus.

Effects of Nuclear Radiation on Matter

Charged particles like alpha and beta rays produce *ionization* in any material through which they pass, including living matter, as shown in Fig. 28.19. Since nuclear alpha and beta particles have energies in the MeV range, and less

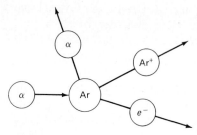

FIGURE 28.19 The ionization of an argon atom when struck by an alpha particle.

Units for Measuring Ionizing Radiation

than 20 eV is needed to remove the outermost electron from most atoms, a single alpha or beta ray can produce thousands of ions.

Gamma rays and x-rays produce ionization by photoelectric and Compton scattering processes. Neutrons, on the other hand, produce artificial radioactive isotopes, which subsequently emit high-energy beta and gamma rays. These can, in turn, produce additional ionization processes.

This ionizing property of nuclear radiation makes it harmful to living tissue. Ionization disrupts the normal chemical processes in a living cell and causes the cell to grow abnormally, as in cancer, or to die. In certain cases the radiation dose may cause *somatic* damage to an individual, that is, bodily damage observed either immediately or after a period of time in the form of cancers or leukemia. In other cases there is evidence that such radiation, especially when it is incident on the reproductive organs or on a human fetus, causes genetic mutations, i.e., changes in the genetic material passed on to human offspring, sometimes leading to abnormalities or birth defects.

There are a great variety of radiation units, but here we will confine ourselves to the most important ones. These units refer to three quite different aspects of ionizing radiation:

1 Nuclear activity. As we have seen, *nuclear activity is a measure of the number of disintegrations that take place per second in a nuclear source*, and is given by [see Eqs. (28.4) and (28.5)]:

$$\text{Nuclear activity} = \frac{\Delta N}{\Delta t} = -\lambda N = -\frac{0.693}{T_{1/2}}N \tag{28.22}$$

The SI unit for activity is the becquerel (Bq), where

1 Bq = 1 nuclear disintegration per second

One gram of radium has 3.6×10^{10} nuclei decaying per second on the average, so that the activity of a 1-g sample of radium is 3.6×10^{10} Bq.

There is an older unit for activity, the curie (Ci), which is equal to 3.7×10^{10} nuclear decays per second. Hence

1 Ci = 3.7×10^{10} Bq

2 Absorbed dose. *Absorbed dose is a measure of the energy absorbed from a beam of radiation by a given mass of absorbing material.* It is expressed in joules per kilogram. The official SI unit for absorbed dose is the *gray* (Gy), where

1 Gy = 1 J/kg

Older (and still frequently used) units for absorbed dose are the rad (*radiation absorbed dose*) and the roentgen, where

1 rad = 10^{-2} Gy

1 roentgen = 0.87 rad = 0.87×10^{-2} Gy

If, for example, a person were to stand 1 m away from a 1-Ci source of $^{60}_{27}$Co for 1 h, the dose of gamma rays received at the front surface of the body would be 1.2×10^{-2} Gy or 1.2 rad. Because gamma rays are rapidly attenuated by the body, the dose inside the body would be considerably less.

TABLE 28.4 Relative Biological Effectiveness (RBE) for Some Nuclear Radiations

Type of ionizing radiation	RBE
Gamma rays (or x-rays)	1
Electrons (beta rays)	1–2
Protons	10
Alpha particles	10
Fast neutrons	10
Thermal neutrons	3
Heavy ions	20

3 Biological effectiveness. *Biological effectiveness is a measure of the actual biological damage to be expected when the radiation is incident on the body.* The SI unit is the *sievert* (Sv), where

Absorbed dose in sieverts = absorbed dose in grays $\times$ RBE

where RBE stands for *relative biological effectiveness* and is a measure of the relative biological damage produced by equal doses of different kinds of ionizing radiation. A list of approximate RBEs for various nuclear particles is given in Table 28.4.

An older unit still frequently used is the *rem* (an acronym for *r*ad *e*quivalent *m*an), where

Absorbed dose in rems = absorbed dose in rads $\times$ RBE

Since   1 rad = 10^{-2} Gy   we have   1 rem = 10^{-2} Sv

Table 28.5 gives some data on radiation exposure from a variety of sources.

TABLE 28.5   Radiation Exposure of Individuals in the United States

Source	Average yearly dose, millirems per year (10^{-5} Sv/year)	One-time dose, millirems (10^{-5} Sv)
Natural background (cosmic rays, radioactive minerals)	100	
Medical or dental x-ray	75	100
Nuclear power plant (for average citizen)	0.005	
Nuclear power plant (for persons living very near plant)	5	
GI (gastrointestinal) series (over part of body)		2500
Plane trip, New York to California (due to high altitude)		0.01

Biological Effects of Radiation

There is no doubt that all forms of ionizing radiation are highly dangerous and should be avoided. The electrons emitted in ionization processes can damage the DNA in human cells either directly or indirectly. The immediate somatic effects of ionizing radiation are better known because human beings have been exposed to very high levels of such radiation from the nuclear bombs dropped on Hiroshima and Nagasaki at the end of World War II. Dosages of more than 7 Sv (700 rem) over the whole body are usually fatal; lesser whole-body doses above 0.1 Sv (10 rems) produce radiation sickness of a serious, but nonfatal, nature.

By *radiation sickness* we mean a collection of symptoms of excessive exposure to radiation, including nausea, vomiting, fatigue, loss of body hair, reddening of the skin, and a decrease in the white-blood-cell count.

Hard data on the delayed somatic effects (leukemia and cancers) and on genetic effects are more difficult to obtain. The consequences of exposure to radiation may be so long delayed that it is impossible to tell whether cancers and leukemia were caused by controllable ionizing radiations from nuclear bombs or reactors, by the natural background radiation to which we are all subjected, by medical or dental x-rays, or by the chemical pollutants which fill our air and which are proving in some cases to cause somatic and genetic effects not too different from those caused by nuclear radiations.

Despite these ambiguities there is general agreement that exposure to strong ionizing radiation increases the likelihood of cancers, leukemia, and genetic defects. There is less agreement on the effects of accumulated low radiation doses over long periods of time.

At the present time the natural background radiation is an overwhelmingly greater source of ionizing radiation than are well-functioning nuclear power plants (see Table 28.5). So too are medical and dental x-rays. A catastrophic accident like the one which almost occurred at Three Mile Island in Pennsylvania in 1979 could, of course, easily change this encouraging picture.

Radiation Therapy

Ionizing radiation can cause cancer. But it can also be used to cure cancer. Gamma rays from isotopes like $^{60}_{27}Co$, or x-rays produced by x-ray machines, can selectively destroy cancer cells which are fast-growing and therefore more susceptible to destruction by ionization. In so doing there is necessarily some damage to the surrounding healthy body cells, but this damage can be minimized by careful technique in administering the radiation.

Some of the more useful radioactive isotopes are listed in Table 28.3. Cobalt is particularly useful, since $^{60}_{27}Co$ decays to an excited nuclear state of $^{60}_{27}Ni$, which then immediately decays to its ground state by emitting two gamma rays in succession, each with an energy above 1 MeV, as shown in Fig. 28.20. These gamma-ray energies are much larger than those produced by most x-ray machines, and hence $^{60}_{27}Co$ is a convenient and relatively inexpensive source of highly ionizing radiation.

Sometimes radioactive isotopes can be directly introduced into cancerous tissue to destroy it. A good example is the use of radioactive iodine $^{131}_{53}I$ to treat thyroid cancer. Any iodine introduced into the body tends to be concentrated in the thyroid. Hence when radioactive iodine is injected into the blood stream, it is carried to the thyroid, where it can destroy cancerous tissue by the ionizing radiation it emits—in this case, beta and gamma rays.

Radiation therapy is most useful when the tumor or cancerous tissue is localized at one place in the body. If the cancer is spread throughout the body, radiation therapy is of little use, since in the process of killing the cancer cells the radiation will do excessive damage to healthy cells.

FIGURE 28.20 Decay of the radioactive nucleus $^{60}_{27}Co$ to the ground state of $^{60}_{28}Ni$ with the emission of two high-energy gamma rays. These gamma rays are frequently used for radiation therapy. The asterisks indicate excited nuclear states of $^{60}_{28}Ni$.

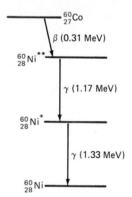

$^{60}_{27}Co$

β (0.31 MeV)

$^{60}_{28}Ni$**

γ (1.17 MeV)

$^{60}_{28}Ni$*

γ (1.33 MeV)

$^{60}_{28}Ni$

Example 28.10

A 1-mg sample of $^{60}_{27}Co$ is being used for radiation therapy. What is the activity of such a sample (*a*) in becquerels, (*b*) in curies?

SOLUTION

To solve this problem we need first to find the number of atoms N of $^{60}_{27}$Co in a 1-mg sample. Since 1 mol of ^{60}Co has a mass of 60 g and contains 6.02×10^{23} atoms, the number of atoms in 1 mg is:

$$N = \left(\frac{10^{-3}\ \text{g}}{60\ \text{g}}\right)(6.02 \times 10^{23}\ \text{atoms}) = 1.0 \times 10^{19}\ \text{atoms}$$

(a) The numerical value of the nuclear activity is, from Eq. (28.22),

$$\text{Nuclear activity} = \frac{0.693N}{T_{1/2}}$$

This is the number of nuclei decaying per second. From Table 28.3, the half-life of ^{60}Co is 5.24 years. To obtain the activity in becquerels, the time must be expressed in seconds, since 1 Bq = 1 nuclear disintegration per second. Hence

$$\text{Nuclear activity} = \frac{0.693(1.0 \times 10^{19}\ \text{atoms})}{(5.24\ \text{years})(3.16 \times 10^7\ \text{s/year})}$$

$$= \frac{0.693 \times 10^{19}\ \text{atoms disintegrating}}{1.65 \times 10^8\ \text{s}}$$

$$= \boxed{4.2 \times 10^{10}\ \text{Bq}}$$

(b) Since 1 Ci = 3.7×10^{10} Bq, we have

$$\text{Nuclear activity} = (4.2 \times 10^{10}\ \text{Bq})\left(\frac{1\ \text{Ci}}{3.7 \times 10^{10}\ \text{Bq}}\right)$$

$$= \boxed{1.1\ \text{Ci}}$$

Example 28.11

A radiologist decides to change the radiation treatment being given to a patient suffering from a cancerous tumor on the leg from x-rays to fast neutrons. If the original dosage was 10 Gy of x-rays, what must be the neutron dosage to produce the same effect on the tumor?

SOLUTION

Since the absorbed dose in Sv = absorbed dose in Gy × RBE, in the first case the effective dose delivered to the tumor was:

10 Gy × 1 = 10 Sv

To achieve the same effective dose with fast neutrons, which have an RBE of 10, we must have

$$\text{Absorbed dose in Gy} = \frac{\text{Absorbed dose in Sv}}{\text{RBE}}$$

$$= \frac{10\ \text{Sv}}{10} = 1\ \text{Gy}$$

Hence a 1-Gy absorbed dose of fast neutrons should be as effective in treating the cancer as a 10-Gy dose of x-rays.

28.9 Use of Radioactive Isotopes in Medicine and Technology

Radioactive isotopes enter into chemical reactions and form molecules in the same way as do nonradioactive stable isotopes of a given element. For this reason radioactive isotopes (or *radioisotopes*) can be used as *tracers* in chemical, biological, and industrial processes.

Definition

A tracer is a radioisotope used to follow some chemical, biological, or other process by detecting the particles emitted in radioactive decay of the isotope.

In addition to being used for therapeutic purposes, $^{131}_{53}$I can be used in the diagnosis of thyroid abnormalities. By monitoring the rate at which $^{131}_{53}$I accumulates in the thyroid, the functioning of this gland can be studied. Also by scanning the thyroid with a suitable detector, abnormalities in size or shape can be seen. Other popular tracer isotopes are $^{14}_{6}$C and $^{3}_{1}$H (tritium).

With such tracers it is possible to study how food is digested and in what parts of the body the molecules in the food concentrate. Thus the metabolism of iron in the human body has been studied by using $^{59}_{26}\text{Fe}$ as a tracer. Similar techniques can determine how the body synthesizes amino acids and other important compounds needed for human life.

Industry uses radioisotopes to study frictional wear between surfaces and moving parts by observing the transfer of radioisotopes from one surface to another. Makers of soap and detergents test the success of their products in removing dirt from clothes by adding radioactive tracers to the dirt and measuring the proportion of the tracers carried off in the soapy water.

There are countless other uses for radioactive tracers. Since most radioisotopes do not occur in nature and must be artificially produced, the use of tracers has increased greatly over the past 35 years because of the availability of nuclear reactors and high-energy particle accelerators to produce the needed radioactive isotopes.

Radiation Detectors

To monitor the passage of atoms or molecules containing tracer nuclei through the human body, and for nuclear physics research, sensitive detectors of the particles emitted by nuclei are needed. Three different kinds of detectors, described below, are available for detecting nuclear particles.

Gas-Filled Detectors *All gas-filled detectors, such as ionization chambers, proportional counters,* and *Geiger counters*, consist of a cylindrical chamber with a thin wire along the axis of the cylinder, as in Fig. 28.21. The wire is kept at a high positive electric potential compared with the outer cylinder. The gases used (often argon) and the pressure maintained vary with the desired use of the counter, but all the counters work on the same basic principles: (1) Nuclear radiation passing into the chamber through the thin window ionizes some of the gas molecules. (2) The electric field pulls electrons to the central wire and positive ions to the outer cylinder, causing a current to flow in the external circuit. (3) The resulting current through the resistor R is measured with an electric meter.

The main difference between ionization chambers, proportional counters, and Geiger counters is in the voltage applied across the electrodes, resulting in a difference in the number of ions collected, as shown in Fig. 28.22 for incident alpha particles. For proportional counters the applied voltage is so high that collisions produce an avalanche effect and a single incident particle may lead to 10^5 to 10^6 ion pairs. This makes it possible to count incident particles one by one. The size of the current pulse produced is approximately proportional to the energy of the ionizing particle, leading to the name *proportional counter*.

FIGURE 28.21 A gas-filled nuclear counter.

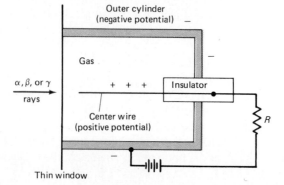

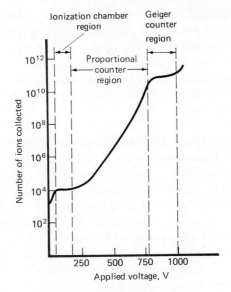

FIGURE 28.22 Different voltage regions for the gas-filled counter of Fig. 28.21. Ionization chambers use voltages below about 250 V; Geiger counters use voltages close to 1000 V. The intermediate voltage region between these values is that used for proportional counters.

In a Geiger counter [named after the German physicist Hans Geiger (1882–1945)] the applied voltage is close to 1000 V. In this case any nuclear particle can initiate an avalanche of electrons and ions, and the current pulses are independent of the energy of the incident particle. It is therefore impossible to distinguish between various types of nuclear radiations with a Geiger counter unless counter windows of different thicknesses or external absorbers are used.

Scintillation Counters   These *counters* use a phosphor or scintillator, such as ZnS or NaI, which emits light when struck by high-energy particles. This

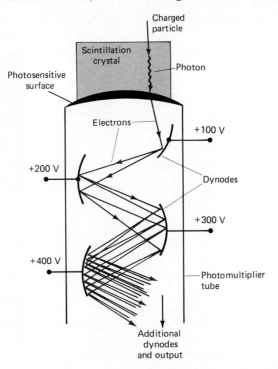

FIGURE 28.23 A scintillation crystal with attached photomultiplier tube to amplify the output signal. At each stage the number of electrons produced increases greatly.

light is then incident on a photocathode and releases an electron from this cathode by the photoelectric effect, as shown in Fig. 28.23. The photomultiplier tube multiplies the single electron into a measurable current pulse at its output. Amplifications of 10^9 are possible with photomultiplier tubes with 14 stages (each stage is called a *dynode*). Each incident particle can then be counted by a high-speed electric counter attached to the output of the photomultiplier tube.

Photographic Emulsions, Cloud Chambers, Bubble Chambers, and Spark Chambers   Such devices make the actual path of individual particles visible by causing observable changes in the gas or liquid through which the particles pass. These detectors are used more for physics research than for medical or industrial applications.

A charged particle passing through a thick *photographic emulsion* ionizes the atoms along its path. The resulting chemical changes show up the path of the particle when the film is developed, as shown in Fig. 28.24. *Cloud chambers* employ supercooled gases which produce droplets of condensed liquid along the path of a charged particle. *Bubble chambers* use a superheated liquid which evaporates and produces a vapor trail of bubbles along the path of the charged particle, as in Fig. 28.25. *Spark chambers* use high voltages across sets of closely spaced parallel plates to produce sparks when a charged particle passes through, making the path of the particle visible.

In cloud chambers, bubble chambers, and spark chambers, elaborate coincidence-counting systems outside the chamber trigger cameras to photograph the chamber only when the particles of interest are passing through. In this way permanent records of the particle's path can be obtained.

The use of *solid-state detectors* for nuclear particles will be discussed in Sec. 29.7.

FIGURE 28.24   (left) Tracks of nuclear particles in a photographic emulsion. These tracks were produced by high-energy, heavy cosmic-ray particles. These particles struck the photographic emulsion, which was carried to high altitudes by a balloon in a Brookhaven National Laboratory experiment. (*Photo courtesy of Brookhaven National Laboratory, Upton, New York.*)

FIGURE 28.25   (right) A typical bubble-chamber photograph showing the tracks of charged particles. The photograph was taken in the 80-in liquid hydrogen bubble chamber at the Brookhaven National Laboratory, Upton, New York. (*Photo courtesy of Brookhaven National Laboratory.*)

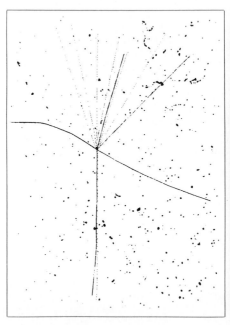
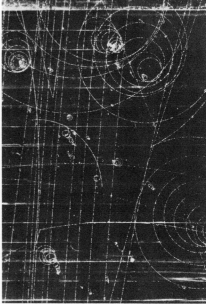

Summary: Important Definitions and Equations

Structure of the nucleus

Nucleons: The generic name given to the protons and neutrons in the nucleus.

Radius of nucleus of mass number A: $r_0 = (1.2 \times 10^{-15}$ m$)A^{1/3}$

Strong nuclear force: The short-range, charge-independent force which holds the nucleus together, despite the strong electrostatic repulsions between the protons in the nucleus.

Mass defect (Δm): The difference between the sum of the masses of the individual protons and neutrons making up a nucleus, and the mass of the nucleus itself.

Nuclear binding energy: The energy required to tear a nucleus apart into its constituent isolated protons and neutrons; $E = \Delta m \, c^2$

Unified atomic mass unit(u): A mass exactly equal to one-twelfth of the mass of the carbon ($^{12}_{6}$C) atom.

$$1.00000 \text{ u} = \frac{931.49 \text{ MeV}}{c^2}$$

Radioactivity

Natural radioactivity: The phenomenon peculiar to the nuclei of some atoms existing naturally on the earth in which they spontaneously emit alpha, beta, or gamma rays and change into other kinds of nuclei.

Alpha (α) particles = helium nuclei ($^4_2$He^{2+})

Beta (β) particles = electrons

Gamma (γ) rays = high-energy photons

Half-life($T_{1/2}$): The time required for half the nuclei in a sample of a particular nuclear species to decay radioactively.

Decay constant (λ): The constant in the exponential factor in the equation for the decay of a nuclear species: $N = N_0 e^{-\lambda t}$

$$\lambda = \frac{0.693}{T_{1/2}}$$

Nuclear activity (A): The rate at which a particular nuclear species decays in time. It is equal to $\Delta N/\Delta t = -\lambda N$, where λ is the decay constant and N is the number of radioactive nuclei remaining at the time at which the activity is calculated.

Special kinds of nuclear reactions

Nuclear fission: The breaking up of a heavy nucleus like uranium into two nuclei of intermediate masses, together with the release of two or more neutrons and large amounts of energy.

Fissile nuclei: Nuclei like $^{233}_{92}$U, $^{235}_{92}$U, and $^{239}_{94}$Pu which undergo fission when bombarded by neutrons.

Chain reaction: The rapid buildup of energy which occurs in a nuclear bomb when the multiplication factor exceeds unity and the number of neutrons producing fission at successive stages of the reaction increases exponentially.

Nuclear reactor: A large boiler in which nuclear processes generate heat which is then converted into electricity by a steam-electric turbine.

Boiling-water reactor: A nuclear reactor which uses ordinary water as the moderator and cooling fluid.

Breeder reactor: A fission reactor which both produces electric power and "breeds" additional nuclear fuel.

Liquid-metal fast-breeder reactor (LMFBR): A breeder reactor that uses a liquid metal like sodium as the coolant and heat-exchange liquid, and fast neutrons to convert $^{238}_{92}$U to $^{239}_{94}$Pu.

Nuclear fusion: A nuclear reaction in which two light nuclei are combined, or "fused" together, to form a heavier nucleus, accompanied by the release of large amounts of energy.

Nuclear radiation

Ionizing radiation: The radiation emitted in the radioactive decay of nuclei, which ionizes the atoms and molecules in living tissue and hence can damage the tissue.

Units for measuring ionizing radiation

Becquerel (Bq): A unit for measuring nuclear activity; 1 Bq = 1 nuclear disintegration per second; 1 curie (Ci) = 3.7×10^{10} Bq

Gray (Gy): A unit for measuring the energy absorbed by a given mass of material; 1 Gy = 1 J/kg; 1 rad = 10^{-2} Gy

Sievert (Sv): A unit for measuring the actual biological damage to be expected from radiation; absorbed dose in sieverts = absorbed dose in grays × RBE; 1 rem = 10^{-2} Sv

RBE (relative biological effectiveness): A measure of the relative biological damage produced by equal doses of different kinds of ionizing radiation.

Tracer: A radioisotope used to follow some chemical, biological, or other process by detecting the particles emitted in radioactive decay of the isotope.

Questions

1 (*a*) Why are the concentrations of most radioactive isotopes in nature so low?

(*b*) Suggest a process which might lead to a large concentration of a long-lived radioactive isotope in nature.

2 When a chemical reaction like $2H_2 + O_2 \rightarrow 2H_2O$ occurs, is the mass of the products exactly equal to the mass of the reactants? If not, why cannot this difference in mass be used as an energy source?

3 Discuss the difference between nuclear reactions and chemical reactions with respect to:

(*a*) The behavior of atoms of a particular element during each kind of reaction.

(*b*) The behavior of different isotopes of the same element during each kind of reaction.

(*c*) The size of the energy changes in each kind of reaction.

4 From the binding-energy curve (Fig. 28.4) would you expect fission or fusion to be the better energy-producing process?

5 (*a*) Why are the names "atom bomb" and "atomic reactor" misnomers?

(*b*) What are the more accurate names for these devices? Why?

6 What is the fundamental difference between a nuclear reactor and a nuclear bomb?

7 Why do moderators for nuclear reactors use relatively light atoms like $_1^1H$, $_1^2H$, or $_6^{12}C$ to slow down the neutrons rather than use heavy atoms like gold or lead?

8 (*a*) Would you expect some $_{94}^{239}Pu$ to be produced by a conventional boiling-water reactor? Why?

(*b*) What happens to this plutonium after it is produced?

9 (*a*) What is the real source of energy in a LMFBR?

(*b*) What is being consumed in the reaction?

(*c*) Is there any violation of conservation of energy here?

10 (*a*) What would be the advantages of using D–D fusion reactions [Eqs. (28.19) and (28.20)] in a fusion reactor rather than D–T reactions [Eq. (28.21)]?

(*b*) What problems must be solved before the D–D reactor can be made to produce useful energy?

11 Why is it that some very long-lived radioactive nuclei like $_{19}^{40}K$ ($T_{1/2} = 10^9$ years) present little danger to human life and health?

12 (*a*) Would you expect any danger to the health of miners involved in the mining of uranium ore? Why?

(*b*) How do you think this danger would compare with the dangers to which coal miners are subjected?

13 Explain in what respects gamma rays and x-rays are the same, and in what respects they differ.

Problems

Note: To solve many of these problems you will need to refer to the periodic table (Fig. 9.9) and to the table of isotopes in Appendix 4 at the back of the book.

A 1 The mass of the $_{92}^{235}U$ nucleus can be expected to be about:

(*a*) 92 u (*b*) 235 u (*c*) 235 g (*d*) 92 kg (*e*) (235 MeV)/c^2

A 2 The volume of the molybdenum nucleus $_{42}^{100}Mo$ exceeds that of the hydrogen nucleus $_1^1H$ by about:

(*a*) One order of magnitude
(*b*) Six orders of magnitude
(*c*) Two orders of magnitude
(*d*) Four orders of magnitude
(*e*) Less than one order of magnitude

A 3 When a proton and a neutron combine to form a deuteron (the nucleus of ^2H), the energy released in the process is:

(*a*) 2.41 keV (*b*) 1.72 MeV (*c*) 2.25 MeV
(*d*) 1.85 keV (*e*) None of the above.

A 4 A sample of the radioactive isotope $_{53}^{131}I$ is received by a hospital for use in locating thyroid abnormalities, but is allowed to sit on the shelf for 80 days. The fraction of the originally present $_{53}^{131}I$ remaining is about:

(*a*) 1/10 (*b*) 1/80 (*c*) 1/1000 (*d*) $1/10^6$
(*e*) $1/10^{10}$

A 5 In 20 half-lives the nuclear activity of a radioactive sample is reduced by about a factor of:

(*a*) 1/20 (*b*) 1/1000 (*c*) $1/20^2$ (*d*) $1/10^6$
(*e*) $1/e^{20}$

A 6 The nuclear activity of a sample containing 2.6×10^{18} atoms of the radioactive isotope $_{15}^{32}P$ is:

(*a*) 1.5×10^{12} Ci (*b*) 1.5×10^{12} Bq
(*c*) 1.3×10^{17} Gy (*d*) 1.3×10^{17} Bq
(*e*) 1.3×10^{17} Ci

A 7 The energy released in the reaction $_{92}^{238}U + _0^1n \rightarrow _{92}^{239}U$ would be about:

(*a*) 1.1 MeV (*b*) 2.4 keV (*c*) 4.8 MeV
(*d*) 5.16 keV (*e*) None of the above

A 8 In the nuclear equation, $_{93}^{237}Np \rightarrow _2^4He + _Z^A X$, the unknown product nucleus must be:

(*a*) $_{91}^{233}Pa$ (*b*) $_{91}^{233}U$ (*c*) $_{91}^{233}Pu$ (*d*) $_{91}^{241}Pa$
(*e*) $_{89}^{235}Ac$

A 9 To produce the same biological damage to the human body as an x-ray source producing an absorbed dose of 10^{-1} Gy, an alpha-particle source would have to produce an absorbed dose of:

(*a*) 10 Gy (*b*) 0.1 Gy (*c*) 10^{-2} Gy
(*d*) 10^{-3} Gy (*e*) 100 Gy

A10 The actual biological damage to be expected from 1.6×10^{-3} Gy of protons absorbed by a person's body can be expressed as:

(*a*) 1.6×10^{-3} Sv (*b*) 1.6×10^{-2} rem
(*c*) 1.6×10^{-2} Sv
(*d*) 1.6×10^{-4} Sv
(*e*) 1.6×10^{-3} rem

B 1 (*a*) What is the approximate radius of the $_{92}^{235}U$ nucleus?

(*b*) What is its approximate volume?

B 2 Calculate the binding energy per nucleon for the $^{56}_{26}$Fe nucleus and compare your result with that shown in the graph in Fig. 28.4.

B 3 Calculate the energy released in the beta-decay process

$$^{137}_{55}\text{Cs} \rightarrow {}^{0}_{-1}e + {}^{137}_{56}\text{Ba}$$

on the assumption that any other particles or photons produced in the reaction have zero mass.

B 4 The strontium isotope $^{90}_{38}$Sr is one of the most dangerous of the radioactive isotopes produced in nuclear explosions. How long will it take for the amount of $^{90}_{38}$Sr produced in a nuclear bomb test to decay to 10^{-6} of its original value?

B 5 From the data given in Table 28.2, what are the decay constants of (a) $^{40}_{19}$K, (b) $^{88}_{36}$Kr?

B 6 What is the nuclear activity of a sample containing 3.0×10^{15} atoms of $^{40}_{19}$K?

B 7 In the radiocarbon dating of a parchment found in an Egyptian cave, it is found that the ratio of $^{14}_{6}$C to $^{12}_{6}$C is one-fourth that found in living specimens today. About how old is the parchment?

B 8 How much energy is released in the decay of a free neutron into a proton and an electron, if the equation for the reaction is: $^{1}_{0}n \rightarrow {}^{1}_{1}p + {}^{0}_{-1}e$?

B 9 A person of mass 85 kg absorbs 5.0 J of energy from an x-ray machine. This energy is distributed over the person's entire body. What is the radiation dose absorbed by the person (a) in grays; (b) in rads?

B10 Show that for thermal neutrons the biological effectiveness of an absorbed dose of 10 rad is 0.3 Sv.

C 1 How many protons and neutrons are there in (a) $^{233}_{92}$U, (b) $^{238}_{92}$U, (c) $^{233}_{91}$Pa?

C 2 Two very important elements in the semiconductor industry are (a) silicon ($^{28}_{14}$Si), and (b) germanium ($^{72}_{32}$Ge). How many electrons, protons, and neutrons does each of these atoms contain?

C 3 (a) Find the density of the $^{206}_{82}$Pb nucleus.

(b) How much more dense is the lead nucleus than ordinary water?

C 4 By what percentage does the radius of the $^{233}_{92}$U nucleus differ from that of the $^{238}_{92}$U nucleus?

C 5 Calculate the binding energy per nucleon for (a) the $^{16}_{8}$O nucleus and (b) the $^{15}_{8}$O nucleus, and compare the results.

C 6 Calculate the atomic mass of chlorine as it is found in nature. Natural chlorine consists of 75.53% $^{35}_{17}$Cl and 24.47% $^{37}_{17}$Cl.

C 7 A sample of the frequently used radioactive isotope $^{137}_{55}$Cs is stored away in a protective lead vault and forgotten for many years. If the half-life of $^{137}_{55}$Cs is 30 years, how long will it take for the activity of the sample to decrease (a) to one-fourth of its original value, (b) to one-fifth of its original value?

C 8 The half-life of $^{13}_{7}$N is 10 min.

(a) How many nitrogen atoms decay in 1 s in a 1-mg sample of this isotope?

(b) What is the activity of this sample in becquerels (Bq)?

C 9 $^{238}_{92}$U is slightly radioactive, with a half-life of 4.5×10^{9} years. Once $^{238}_{92}$U decays radioactively, a long chain of disintegrations occurs before the original ^{238}U nucleus is transformed into a final stable nucleus. If, in this process, 8 alpha particles and 6 beta particles are emitted, what is the final stable nucleus?

C10 It has been found that the oldest rocks containing uranium on earth contain about a 50-50 mixture of $^{238}_{92}$U and $^{206}_{82}$Pb. From the data given in Prob. C9, what is the approximate age of these rocks, if we assume that all the lead was produced by the radioactive decay of uranium?

C11 What fraction of the $^{232}_{90}$Th atoms in existence at the beginning of the universe about 10 billion years ago still remain? Assume that the half-life of $^{232}_{90}$Th is 1.4×10^{10} years.

C12 An animal bone found in a cave in southern France contained 200 g of carbon and had a $^{14}_{6}$C activity of 1600 decays per minute. If the $^{14}_{6}$C activity of a living organism is 16 decays per minute for each gram of carbon, how long ago did the animal die?

C13 How many half-lives are required for any radioactive sample to drop to 1 percent of its initial activity?

C14 Since nuclear energies are in the MeV range, and gamma rays arise from transitions between two nuclear energy levels, what is roughly the wavelength of a gamma-ray photon?

C15 The two isotopes $^{137}_{55}$Cs and $^{90}_{38}$Sr are produced by fission processes followed by beta decays. Determine X, Y, X', and Y' in the following formulas for these nuclear reactions (consider only particles having rest mass):

(a) $^{235}_{92}\text{U} + {}^{1}_{0}n \rightarrow {}^{143}_{57}\text{La} + {}^{90}_{35}\text{Br} + X$

$^{90}_{35}\text{Br} \rightarrow {}^{90}_{38}\text{Sr} + Y$

(b) $^{235}_{92}\text{U} + {}^{1}_{0}n \rightarrow {}^{137}_{53}\text{I} + {}^{97}_{39}Y + X'$

$^{137}_{53}\text{I} \rightarrow {}^{137}_{55}\text{Cs} + Y'$

C16 Find the product nuclei in the following reactions:

(a) $^{12}_{6}$C (d, n) (b) $^{130}_{52}$Te $(d, 2n)$
(c) $^{59}_{27}$Co (n, α) (d) $^{11}_{5}$B (n, α)

C17 (a) Show that $^{12}_{6}$C is stable against decay into three alpha particles.

(b) Show that $^{8}_{4}$Be is unstable against decay into two alpha particles.

C18 One way to enable Geiger counters to count slow neutrons is to fill the counter with boron trifluoride gas. Then the $^{10}_{5}$B nucleus can capture the neutron and undergo the following reaction: $^{10}_{5}\text{B} + {}^{1}_{0}n \rightarrow {}^{4}_{2}\text{He} + {}^{7}_{3}\text{Li}$. The alpha particles then produce ion pairs which can be counted by the Geiger counter. Find the energy of the product nuclei in this reaction.

C19 Show that the fraction of the mass of the $^{235}_{92}$U nucleus converted into energy in a fission reaction is only about 1/1000 of the original mass.

C20 If a boiling-water reactor produces steam at 285°C, what is the maximum possible efficiency of a steam-electric turbine operating between this temperature and a water-exhaust temperature of 27°C?

C21 The simplest kind of fusion reaction is the combination of a proton and a neutron to form a heavy hydrogen, or deuterium, nucleus ($^2_1$H). This reaction is described by the equation: $^1_1p + ^1_0n \rightarrow ^2_1$H + energy. How much energy will be released in this fusion reaction?

C22 (a) In 1 mol of natural uranium, how many atoms of $^{235}_{92}$U and how many atoms of $^{238}_{92}$U are found?

(b) How many grams of $^{235}_{92}$U and $^{238}_{92}$U are found in 1 mol of natural uranium?

C23 The energy consumed in the United States each year is about 7.0×10^{19} J. How many kilograms of $^{235}_{92}$U would be needed to release an amount of energy equal to this, if all the uranium nuclei are presumed to undergo fission according to Eq. (28.11)?

C24 How many grams of $^{60}_{27}$Co are there in a sample whose activity is 2.0×10^7 Bq?

C25 A beam of fast neutrons passes through a man's body and deposits 0.052 J of energy in each kilogram of his body. Find:

(a) The absorbed dose in grays.

(b) The biological effectiveness of this dose in sieverts.

(c) Express your answers to parts (a) and (b) in rads and rems.

C26 A 1.0-μg sample of $^{131}_{53}$I is being used in the treatment of thyroid cancer. What is the activity of such a sample in (a) becquerels, (b) curies?

C27 A radiation dose of over 700 rem or 7 Sv is usually fatal to half the people exposed to it. This is equivalent to:

(a) How many grays of protons?

(b) How many grays of gamma rays?

C28 A whole-body dose of 7 Sv of gamma radiation causes death to half the people exposed to it, as noted above. Calculate the energy (in joules) delivered to a 90-kg person by such a dose.

C29 It is desired to inject enough iodine-131 into the bloodstream to produce a nuclear activity of 10^8 Bq, which can then be observed when the iodine collects in the thyroid. The $^{131}_{53}$I radioisotope sample is received directly from the reactor where it is produced, but is not used for 32 days. How many grams of iodine-131 must be injected into the bloodstream to produce the desired activity?

D 1 (a) Show that, if an electron were confined to the region inside a nucleus of diameter 10^{-14} m, the uncertainty in its kinetic energy would be greater than 100 MeV.

(b) Calculate the potential energy of such an electron in the field of a proton if the average distance between the two is about 10^{-14} m.

(c) What can you conclude about the likelihood that there are electrons inside atomic nuclei?

D 2 (a) Calculate the fraction of its rest energy made up by the binding energy of a proton in a deuterium nucleus.

(b) Repeat this calculation for the binding energy of an electron in a hydrogen atom.

(c) Compare the two results.

D 3 Find the radius of the first Bohr orbit for the innermost electrons in the $_{92}$U atom and compare its size with that of the $^{238}_{92}$U nucleus.

D 4 The silver isotope $^{108}_{47}$Ag has a half-life of 2.4 min. Initially a sample contains 2.0×10^6 nuclei of $^{108}_{47}$Ag. How many radioactive nuclei remain after 1.2 min?

D 5 Twenty percent of a certain radioactive isotope decays in 6 h.

(a) What is the decay constant of the substance?

(b) What is the half-life?

D 6 A $^{238}_{92}$U nucleus can spontaneously emit an alpha particle in the reaction

$$^{238}_{92}\text{U} \rightarrow ^4_2\text{He} + ^{234}_{90}\text{Th}$$

(a) If the uranium nucleus is originally at rest and the speed of the alpha particle is 1.5×10^7 m/s, what is the recoil velocity of the uranium nucleus?

(b) Do relativistic effects have to be taken into account here?

D 7 Use the abundance ratios and half-lives for $^{235}_{92}$U and $^{238}_{92}$U found in Appendix 4 to estimate the minimum age of the universe. Assume that when these two isotopes were formed they had the same natural abundance.

D 8 The following reaction occurs: $^1_0n + ^{10}_5\text{B} \rightarrow ^7_3\text{Li} + ^4_2\text{He}$. If the reactants are initially at rest, what is the speed of the alpha particle after the reaction?

D 9 Hydrogen atoms are much more effective than lead atoms in slowing down neutrons to thermal velocities. Prove this by calculating the final speed of a neutron which originally had a speed of 2.0×10^6 m/s, after it makes a perfectly elastic head-on collision with:

(a) A $^1_1$H nucleus which is initially at rest.

(b) A $^{206}_{82}$Pb nucleus initially at rest.

D10 Write nuclear equations similar to those in Eqs. (28.15) to (28.17) for the conversion of $^{232}_{90}$Th into $^{233}_{92}$U in a breeder reactor.

D11 The complete D–D reaction chain in a fusion reactor would be a composite of the following reactions:

$$^2_1\text{H} + ^2_1\text{H} \rightarrow ^3_2\text{He} + ^1_0n + 3.3 \text{ MeV}$$

$$^2_1\text{H} + ^2_1\text{H} \rightarrow ^3_1\text{H} + ^1_1p + 4.0 \text{ MeV}$$

$$^2_1\text{H} + ^3_1\text{H} \rightarrow ^4_2\text{He} + ^1_0n + 17.6 \text{ MeV}$$

$$^2_1\text{H} + ^3_2\text{He} \rightarrow ^4_2\text{He} + ^1_1p + 18.3 \text{ MeV}$$

where the $^3_2$He and $^3_1$H required for the last two reactions are produced in the first two reactions.

(a) Add the above equations to obtain an equation for the net reaction occurring.

(b) Obtain the average energy produced in the change of each deuterium nucleus into a different nuclear species in the reaction.

D12   In the complete D–D reaction chain in a fusion reactor, about 7.2 MeV of energy is produced from each deuterium nucleus. In the oceans the deuterium fraction in the hydrogen making up the ocean water is about 0.015 percent.

(*a*)   Calculate the energy content of 1 km^3 of ocean water.

(*b*)   Compare your answer with the energy consumed in the United States in a year, which is about 7.0×10^{19} J.

Additional Readings

Bethe, Hans A.: "The Necessity of Fission Power," *Scientific American*, vol. 234, no. 1, January 1976, pp. 21–31. A strong defense of nuclear energy by an eminent physicist and Nobel Laureate.

Cohen, Bernard L.: *The Heart of the Atom: The Structure of the Atomic Nucleus*, Doubleday Anchor, Garden City, N.Y., 1966. A brief elementary introduction to nuclear physics.

Crawford, Deborah: *Lise Meitner, Atomic Pioneer*, Crown, New York, 1969. A lively and very interesting account of a great scientist.

Curie, Eve: *Madame Curie*, Doubleday, Garden City, New York, 1937. The standard biography by Madame Curie's daughter.

Hecht, Selig: *Exploring the Atom*, Viking, New York, 1954. A simple and clear introduction to the essentials of atomic and nuclear physics.

Hohenemser, C., R. Kasperson, and R. Kates: "The Distrust of Nuclear Power," *Science*, vol. 196, 1977, pp. 25–34. A cautious and well-balanced article on the growing opposition to the use of nuclear energy.

Hughes, Donald J.: *The Neutron Story*, Doubleday Anchor, Garden City, N.Y., 1959. The history of the discovery of the neutron and its use in physics and technology.

Inglis, David R.: *Nuclear Energy—Its Physics and Its Social Challenge*, Addison-Wesley, Reading, Mass., 1973. A balanced discussion of nuclear energy, its promise and its problems.

Noz, Marilyn E., and Gerald Q. Maguire, Jr.: *Radiation Protection in the Radiologic and Health Sciences*, Lea and Febiger, Philadelphia, 1979. A clear, careful presentation of the use of radiation in the health sciences.

Nuclear Energy (Readings from *Scientific American,* with introductions by Hans A. Bethe), W. R. Freeman and Co., San Francisco, 1983. A collection of articles on the use of nuclear fission as a power source.

O'Neill, Gerald K.: "The Spark Chamber," *Scientific American*, vol. 207, no. 2, August 1962, pp. 36–43. Includes discussions of the Geiger counter, the cloud chamber, the bubble chamber, and the spark chamber.

Upton, Arthur C.: "The Biological Effects of Low-Level Ionizing Radiation," *Scientific American*, vol. 246, no. 2, February 1982, pp. 41–49. A discussion of the hazards of both background and human-produced low-level radiation.

Vandryes, Georges A.: "Superphenix: A Full-Scale Breeder Reactor," *Scientific American*, vol. 236, no. 3, March 1977, pp. 26–35. An account of France's progress in solving the country's energy problems by using breeder reactors.

Weinberg, Alvin M.: "Social Institutions and Nuclear Energy," *Science*, vol. 177, 1972, pp. 27–34. A well-balanced and frequently quoted article by the former director of the Oak Ridge National Laboratory.

Yost, Edna: *Women of Modern Science*, Dodd, Mead, New York, 1964. Includes brief biographies of Lise Meitner and of the well-known nuclear physicist C. S. Wu.

It is not in the nature of things for any one man to make a sudden violent discovery; science goes step by step, and every man depends on the work of his predecessors. . . . Scientists are not dependent on the ideas of a single man, but on the combined wisdom of thousands of men, all thinking of the same problem, and each doing his little bit to add to the great structure of knowledge which is gradually being erected.

Ernest Rutherford (1871–1937)

Chapter 29
Solid-State Physics

The properties of solid substances have always been of great interest to the intellectually curious. Our remote ancestors probably wondered why tiny bits of iron clung to a piece of lodestone. The architects and builders who planned the aqueducts and amphitheaters of ancient Rome probably debated the relative merits of stone, brick, and metal as building materials. For similar reasons physicists have always been interested in solids. A new and most successful phase in the application of physics to solids came with the invention of the transistor in 1947 and the succession of revolutionary developments in electronics it has spawned. This development was based on a deeper understanding of the properties of solids at the atomic and molecular level, using the insights of quantum mechanics. It is this more sophisticated view of solids to which we apply the name *solid-state physics*.

29.1 The Structure of Solids

There are two basic ways in which a solid substance, like a piece of iron, differs from a gas. First, the atoms or molecules in a solid are, on the average, much closer together than they are in a gas, with the separation between atoms being roughly the same size as the electronic charge cloud surrounding the nucleus of each atom. For this reason the atoms interact with one another in somewhat the same fashion they do in molecules, and the whole solid may almost be considered one gigantic molecule.

Second, there are two basic kinds of solids, *crystals* and *noncrystals*.

Definition

Crystal: A solid material with a regular geometric shape in which the particles making up the crystal arrange themselves in an ordered geometric pattern throughout.

In a crystal there is a great deal of *order* in the resulting *lattice* (the name usually given to the orderly array of points in space at which the atoms are located). Noncrystalline solids like glass and plastics, on the other hand, are amorphous and disordered.

The structure of a crystal can be explained by applying quantum mechanics to the atoms or molecules making up the crystal and calculating the arrangement of atoms or molecules which minimizes the energy of the crystal. Once this structure is determined, it is possible to predict many of the mechanical, thermal, electrical, magnetic, and optical properties of the crystal. The high degree of order present in crystalline solids simplifies the calculation of their properties and makes them better understood than noncrystalline solids.

Types of Crystalline Solids

There are four basic types of crystalline solids, two of which have direct analogues in ionic- and covalent-bonded molecules.

Ionic Crystals Ionic crystals are those in which alternate sites in the crystal lattice are occupied by ions of opposite charge. For example, in the salt (NaCl) crystal (and in all other alkali halide crystals like KCl, KI, RbCl, etc.) the Na atom loses an electron and becomes Na^+, while the Cl atom gains an electron and becomes Cl^-. These ions then arrange themselves in a cubic lattice, with the positive and negative ions occupying alternate positions in the lattice. In NaCl the positive and negative ions are separated by a lattice distance $l_0 = 0.282$ nm.

As shown in Fig. 29.1, each lattice plane (or face) of a NaCl crystal is made up of squares, with two Na^+ and two Cl^- ions at the corners of each square. An area of the plane which is two lattice distances ($2l_0 = 0.564$ nm) on a side therefore has four identical Na^+ ions at its corners, and one Na^+ ion at the center of the square. The same is true for the Cl^- ions in the lattice. For this reason this crystal structure is called a *face-centered cubic* lattice.

Calculations with regard to the NaCl crystal show that the energy of the state in which a large number of ions are linked together in a large lattice is lower than the state in which all the ions are separated by great distances. Since this energy difference is greater than the energy expended in originally creating the ions by moving electrons from the Na to the Cl atoms, the NaCl crystal is stable. Hence just as at high temperatures the ionic NaCl molecule is formed in the gaseous state, so at room temperature salt crystals will form from sodium and chlorine atoms.

All the alkali halides have similar face-centered cubic structures except for the cesium compounds. Here the Cs^+ ions are too large to fit into a

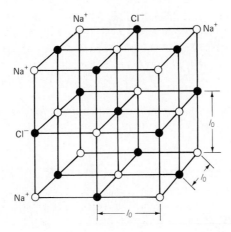

FIGURE 29.1 A face-centered cubic NaCl lattice.

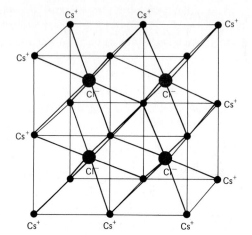

FIGURE 29.2 A body-centered cubic CsCl lattice. Notice how in this case the central atom in a cube with Cs⁺ ions at the corners is a Cl⁻ ion.

face-centered cubic structure; CsCl and the other Cs halides therefore form the *body-centered cubic* lattice shown in Fig. 29.2.

Solid ionic crystals are electrical insulators, since there are no free charges to carry electric current. If, however, salt (NaCl) is poured into water, the ions go into solution and carry an electric current. Hence ionic salts in solution are good conductors of electricity.

Covalent Crystals   The bonding of neutral atoms to form covalent crystals is similar to the bonding of two oxygen atoms to form O_2, or two nitrogen atoms to form N_2. The electrons locate themselves between the nuclei in such a way that a net attractive force results between adjacent atoms in the crystal. This lowers the total energy of the crystal and leads to a stable crystal lattice.

The symmetry of a covalent crystal depends on the valence of the atoms making up the crystal and on the size of the atoms involved. Thus atoms in column IV of the periodic table, like carbon, silicon, germanium, and tin, have valences of 4. They therefore can form covalent crystals in which each atom has four nearest neighbors. Each atom shares one of its four outer

FIGURE 29.3 The crystal structure of diamond: (*a*) The tetrahedral bonding of one atom to four neighboring atoms in diamond. (*b*) The lattice structure of diamond, with each carbon atom bonded to four neighboring carbon atoms in a tetrahedral array.

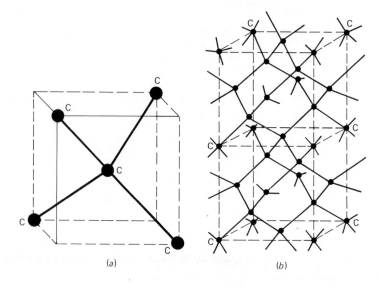

(a)   (b)

electrons with a neighbor and thus forms four strong electron-pair bonds. This leads to the tetrahedral structure for these crystals shown in Fig. 29.3 for diamond (which is made up completely of carbon atoms), in which each carbon atom is bonded to four neighboring carbon atoms. Since all four valence electrons are localized in these four covalent bonds, there are no free charges to carry electric current, and hence covalent crystals are poor conductors of electricity.

Metallic Crystals   Metallic crystals differ greatly from ionic and covalent crystals. Here the valence electrons are free to move throughout the whole crystal, while the positive ions remain fixed in position. The electrons behave, on a crude model, like a negatively charged fluid which acts as a glue to hold the positive ions in the crystal together, as shown in Fig. 29.4 for copper. This results in a lowering of energy which makes the metallic crystal stable.

Since metallic crystals contain large numbers of free electrons, which can carry both charge and energy with them, metallic crystals are good conductors of both electricity and heat. A useful model which works well in describing this fluid of free electrons is called the *free-electron gas model*. It considers the metal as made up of a "gas" of electrons free to move around in the periodic electric field provided by the stationary positive ions, as in Fig. 29.4.

FIGURE 29.4   A metallic crystal—copper, which has a face-centered cubic structure. The valence electrons of the copper atoms move freely through the crystal.

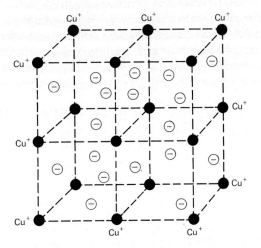

Molecular Crystals   With molecular crystals the elementary units out of which the crystal is formed are not atoms but molecules. Thus at low temperatures solid oxygen crystals can be formed. These consist of regular O_2 molecules loosely bound together by rather weak attractive forces (called *van der Waals forces*) to form a crystal. Van der Waals forces are electrical in nature, and arise when one molecule with an electric dipole moment at some instant of time induces a dipole moment in another molecule. The two molecules therefore weakly attract each other. Because the bonds in molecular crystals are so weak, the melting point of O_2 is very low (55 K). Other molecular crystals include hydrogen, nitrogen, carbon dioxide, and most organic crystals.

Example 29.1

(a) Calculate the electrostatic force between two Na^+ and Cl^- ions in a NaCl lattice for which the Na–Cl separation is 0.282 nm.
(b) Repeat this calculation when the NaCl is poured into water at a concentration such that the average separation between Na^+ and Cl^- is increased from 0.282 to 1.0 nm.
(c) How much has the attractive electrostatic force been reduced?

SOLUTION

(a) In the normal crystal lattice the electrostatic force between ions is:

$$F_e = \frac{k_e q_1 q_2}{r^2}$$

$$= \left(9.0 \times 10^9 \frac{N \cdot m^2}{C^2}\right) \frac{(1.6 \times 10^{-19} C)^2}{(0.282 \times 10^{-9} m)^2} = \boxed{2.9 \times 10^{-9} N}$$

(b) The NaCl crystal breaks up into separate Na^+ and Cl^- ions in the water. The dielectric constant for water is 80, which means that the force between the Na^+ and Cl^- charges in water is reduced by a factor of 80 below what it is in a vacuum or in air. Hence we have

$$F_e' = \frac{k_e q_1 q_2}{80 r^2}$$

$$= \left(\frac{9.0 \times 10^9}{80} \frac{N \cdot m^2}{C^2}\right) \frac{(1.6 \times 10^{-19} C)^2}{(10^{-9} m)^2} = \boxed{2.9 \times 10^{-12} N}$$

(c) The ratio in the two cases is

$$\frac{F_e'}{F_e} = \frac{2.9 \times 10^{-12} N}{2.9 \times 10^{-9} N} = \boxed{10^{-3}}$$

or the electrostatic force has been reduced by a factor of 1000 below its original value. In this case the increased distance between the ions reduces the Coulomb force by a factor of 12.5, the dielectric constant by another factor of 80.

Example 29.2

Calculate the equilibrium separation l_0 of the atoms in a potassium iodide (KI) crystal which has a density of $\rho = 3.13 \times 10^3 kg/m^2$.

SOLUTION

We assume that KI, being an alkali halide, has a cubic structure, and that each ion occupies a cubic volume of l_0^3, that is, a cube of length l_0 on a side, where l_0 is the distance between the atoms in the lattice. The mass of 1 g·mol of KI is 39.10 g + 126.9 g = 166.0 g = 0.166 kg. This mass of KI contains twice Avogadro's number of atoms, that is, N_A of K^+ and N_A of I^-, and occupies a volume of $2N_A l_0^3$, since each atom is assumed to occupy a volume equal to l_0^3. The density of the crystal is then

$$\rho = \frac{M}{V} = \frac{0.166 \text{ kg}}{2(6.02 \times 10^{23}) l_0^3}$$

But, from the statement of the problem, this density is known to be $3.13 \times 10^3 kg/m^3$. Hence

$$l_0^3 = \frac{0.166 \text{ kg}}{2(6.02 \times 10^{23})(3.13 \times 10^3 \text{ kg/m}^3)}$$

$$= 4.40 \times 10^{-29} m^3$$

and

$$l_0 = 3.53 \times 10^{-10} m = \boxed{0.353 \text{ nm}}$$

29.2 Thermal Properties of Solids

In covalent and ionic crystals the principal mechanism for heat conduction is the vibration of the atoms in the crystal lattice. If the crystal is not in thermal equilibrium, the atoms in the hotter regions vibrate more strongly than those in the cooler regions, causing a transfer of energy in the form of waves which travel through the crystal as one atom passes on vibrational energy to its neighbors. This transfer of energy from the hotter to the cooler regions of the crystal is nothing but the flow of heat.

As Table 29.1 shows, the thermal conductivities of metallic crystals exceed those of ionic and covalent crystals by one or two orders of magnitude. This is because of the contribution made to the thermal conductivity by the free electrons in the metals. Table 29.2 lists the so-called number density, or the number of free electrons per unit volume for selected metals.

TABLE 29.1 Thermal Conductivities K (at approximately 20°C)

Material	Thermal conductivity, W/(m·°C)
Metallic crystals:	
Aluminum	238
Copper	380
Silver	417
Gold	290
Tungsten	169
Nonmetallic crystals:	
NaCl	6.7
CsI	1.3
KCl	6.7
MgO	33
SiO$_2$	7.1

TABLE 29.2 Number of Free Electrons (N) per Unit Volume (V) for Selected Metals

Metal	N/V, 10^{28}/m^3
Aluminum	18.1
Silver	5.9
Gold	5.9
Copper	8.5
Iron	17.0
Sodium	2.7
Tin	14.8

In 1853 the German physicists G. H. Wiedemann and R. Franz observed that a good thermal conductor is also a good electrical conductor. This observation is incorporated into the following equation, referred to as the Wiedemann-Franz law:

$$\frac{K}{\sigma} = \frac{\pi^2}{3}\left(\frac{k}{e}\right)^2 T \tag{29.1}$$

where K is the thermal conductivity, σ the electric conductivity, k the Boltzmann constant, e the charge on the electron, and T the absolute temperature. Interestingly, this equation predicts that the ratio of K to σ should be the same for all metals at the same temperature, independent of all other properties of the metals.

Molar Heat Capacities of Solids

In Sec. 13.7 we saw that many solids have a molar heat capacity close to 6 cal/(mol·°C). This rule for a time seemed to be so universal that it was called the *law* of Dulong and Petit. We now know that this "law" is an approximation valid only at high temperatures.

To calculate the molar heat capacity of a solid classically, we consider a crystalline solid consisting of N atoms bound together in a crystal by forces of the kind discussed in the preceding section. When any atom is displaced from its equilibrium position in the crystal lattice, it is subjected to a force which, to a first approximation, is proportional to the atom's displacement. The atom therefore executes simple harmonic motion, as do all its neighbors because they are coupled to the first atom by the interaction forces in the crystal. Hence the simple harmonic motion of one atom is passed on to the next atom, and an elastic wave is propagated through the crystal in a fashion similar to a sound wave moving through air.

The total energy content (which varies with the temperature) of an ionic or covalent crystal will then consist of the vibrational energy of the nuclei, the tightly bound inner electrons, and the outer bound electrons, all of which move together as a unit. For metallic crystals there will be in addition the kinetic energy of the free electrons, but we neglect this contribution for the moment. If the temperature changes, so too does the internal energy of the solid, and the change in internal energy per unit change in temperature is what we mean by the *heat capacity* of the solid.

The so-called lattice heat capacity is then found by considering that the energy of the lattice is the sum of the energies of N simple harmonic oscillators. Since each atom in the lattice is free to move in three dimensions, we would expect it to have a kinetic energy of $\frac{3}{2}kT$, as would a gas atom at the same temperature (Sec. 9.7). As we saw in Sec. 11.2, the average kinetic energy and the average potential energy of a simple harmonic oscillator are equal, and hence the total energy should be $3kT$ per atom. Since there are N atoms, the total energy of the crystal is:

$$E = N(3kT) = 3NkT \tag{29.2}$$

The classical heat capacity, per mole, or molar heat capacity, is then

$$C_V = \frac{1}{n}\frac{\Delta E}{\Delta T} \tag{29.3}$$

where n is the number of moles. Now, from Eq. (29.2), we have

$$\Delta E = 3Nk\,\Delta T \quad \text{or} \quad \frac{\Delta E}{\Delta T} = 3Nk$$

and so
$$C_V = \frac{1}{n}(3Nk) = \frac{N}{n}(3k) = N_A(3k) = 3R \tag{29.4}$$

since on dividing the number of atoms present (N) by the number of moles present (n), we obtain the number of atoms per mole, which is Avogadro's number (N_A). Also the gas constant per mole (R) is equal to N_A times the gas constant per molecule (k). Now,

$$R = 8.31 \text{ J/(mol·°C)} = \left(8.31\,\frac{\text{J}}{\text{mol·°C}}\right)\left(\frac{1 \text{ cal}}{4.18 \text{ J}}\right) = 1.99 \text{ cal/(mol·°C)}$$

and so
$$C_V = 3R = 5.97 \text{ cal/(mol·°C)} \approx 6 \text{ cal/(mol·°C)}$$

which is the law of Dulong and Petit.

Classical physics therefore predicts that the lattice molar heat capacity of solids should be the same for all solids. This agrees with much of the room-temperature data previously given in Table 13.3, but cannot explain the low values of C_V for substances like diamond [$C_V = 1.46$ cal/(mol·°C)] and Si [$C_V = 4.73$ cal/(mol·°C)] at room temperature, or the variation in the molar heat capacity with temperature shown in Fig. 29.5.

The first successful theoretical treatment of the lattice molar heat capacity was given by Einstein in 1906. Peter Debye (1884–1966) extended Einstein's theory in 1912 and developed what is now usually referred to as the *Debye theory of specific heats*.

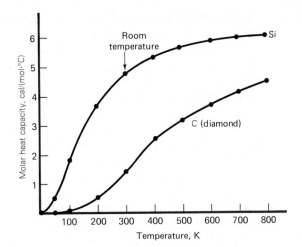

FIGURE 29.5 Variation of molar heat capacity with temperature for silicon and diamond.

Debye Theory

The Debye theory is based on the same kind of consideration used by Max Planck in explaining blackbody radiation. The atoms in a crystal can vibrate at a great variety of frequencies up to a cutoff frequency determined by the lattice spacing of the atoms and the sound velocity. The atomic oscillators have *quantized energies*, and cannot change their energies continuously, but only in quantum jumps of nhf, where n is an integer. At very low temperatures kT is very much less than hf for most of the vibrational frequencies, and the atomic oscillators cannot absorb enough energy to make a quantum jump from a lower to a higher vibrational energy level. As a result, when the temperature changes, the energy of most atoms at low temperatures does not change with it, and so $\Delta E/\Delta T$ is not equal to $3Nk$, but to some smaller value. The molar heat capacity therefore starts from zero and gradually increases with temperature since, as kT increases, more and more oscillators can absorb enough energy to increase their vibrational energies by making the necessary quantum jumps. Finally the temperature becomes so high that all the atoms are able to absorb energy, $\Delta E/\Delta T$ becomes constant, and the molar heat capacity is equal to $3R$. This kind of behavior explains the experimental curves shown in Fig. 29.5.

This result was obtained in a rigorous, quantitative fashion by Debye, with results in excellent agreement with experiment. Hence the molar heat capacity of solids provides another very important confirmation of the validity of the quantum theory. The energy quanta associated with the vibrations of a crystal lattice are called *phonons* by analogy with the photons produced by electromagnetic vibrations.

29.3 The Band Theory of Solids

When an atom is placed in close proximity to other atoms in a crystal lattice, the interactions among the atoms' electrons change drastically the energy levels in the isolated atom. If there are N atoms in the crystal, then when the atoms are far apart, each energy level can accommodate N times as many electrons as the exclusion principle allows for a single atom. When the atoms are arranged at their true interatomic distances in the crystal, however, each level is broadened by the interactions into N very closely spaced levels. Since

N is normally so large (of the order of Avogadro's number), the result is a continuous distribution of energies over a limited energy range. This range of allowed energies is called an *energy band*. Electrons in solids can only possess energies in these *allowed energy bands*. The regions between these bands, or the energy gaps, represent energies which cannot be possessed by any electron in the crystal. Such regions are called *forbidden energy bands*, since quantum-mechanical calculations show that no electron can possess such energies. This is shown schematically in Fig. 29.6. Hence whereas in the Bohr atom we have discrete, allowed energy levels, with forbidden energy regions between them, in crystals we have energy bands with forbidden energy regions (or bands) between them.

The difference between electrical conductors, insulators, and semiconductors is basically due to the arrangement of their energy bands and how these bands are filled by the available electrons. The conductivity of solid materials varies enormously from one substance to another. Thus the resistivity (the reciprocal of the conductivity) of a typical insulator such as quartz is about 10^{17} $\Omega \cdot$m, while that of most metals is of the order of 10^{-8} $\Omega \cdot$m, a ratio of 10^{25}. The range is extended even further if we include superconductors (see Sec. 29.7), which have resistivities of about 10^{-19} $\Omega \cdot$m.

FIGURE 29.6 Allowed and forbidden energy bands in crystals. The interaction of the closely spaced atoms in the lattice broadens the atomic energy levels into energy bands in the crystal.

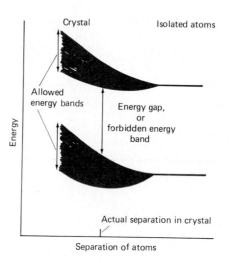

Insulators

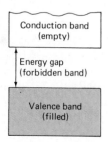

FIGURE 29.7 Energy-band structure for an insulator.

Suppose we have the situation shown in Fig. 29.7 for a particular solid crystal. The lower energy band, which is the band occupied by the outer electrons in the atoms and is called the *valence band*, is completely filled with the number of electrons permitted by the Pauli exclusion principle for an array of N atoms. Above this band is another energy band, usually called the *conduction band*, which corresponds to an excited state of the system and which is completely empty. If the energy gap between the valence band and the conduction band is large, say, 4 to 5 eV, then the crystal is an *insulator*.

If an electric field is applied to such a crystal, electrons can only be accelerated by the field if they can increase their speed and hence their energy. But if these electrons are all already in filled energy bands, there are no empty higher energy states within reach which will enable the electrons to increase their energies. They therefore cannot be accelerated, and so no current flows. A current could flow only if the electrons were given enough energy to jump the energy gap from the valence to the conduction band, and usually not enough energy is available for this to happen.

An ionic crystal such as NaCl is an example of a good insulator. The energy bands of the crystal are derived from the energy levels of the isolated Na^+ and Cl^- ions. But both these ions have a closed-shell structure, so that the valence band in NaCl is completely filled with electrons. Since there is a large energy gap between the valence band and the conduction band in the alkali halides, solid NaCl is an insulator.

Conductors

In other cases, as in some metallic crystals, the valence band is not completely filled. Hence electrons in the valence band can be raised to higher energy states in empty portions of the band by the application of an electric field. The field can therefore accelerate the electrons and produce an electric current.

The valence and conduction bands can also overlap. In this case excited states exist to which electrons can be raised by an electric field, and the material is again a conductor. Diagrams of these two kinds of band structure leading to electric conductivity are shown in Fig. 29.8.

The sodium crystal provides a good example. The sodium atom has an electron configuration $(1s)^2(2s)^2(2p)^6(3s)$. Since the 3s subshell can hold two electrons, this subshell is only half-filled. This situation carries over to the Na crystal, where the 3s energy band is also only half-filled. Hence electrons can be promoted to higher energies within this 3s valence band and thus conduct current. Also, in Na the 3p conduction band overlaps the 3s band for the equilibrium atomic separation in the Na crystal. This allows electrons from the valence band to move into the conduction band, and thus contribute to the current. Hence sodium is a good conductor. Magnesium, with a configuration $(1s)^2(2s)^2(2p)^6(3s)^2$, has its 3s band filled, but is still expected to be a conductor because its 3p conduction band overlaps the 3s valence band. Experiments confirm that magnesium is indeed a good conductor.

FIGURE 29.8 Two kinds of band structure leading to electric conductivity: (a) A crystal with a partially filled valence band. (b) A crystal in which the valence band and the conduction band overlap.

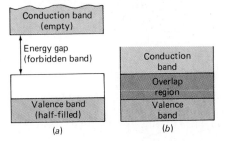

Semiconductors

A third possibility exists: There may be a gap between a filled valence band and an empty conduction band, but this gap may be very small, say, less than 1 eV, as in Fig. 29.9. In such cases we have a *semiconductor*. The energy gap is in this case small enough that an increase in temperature can excite enough electrons into the conduction band to carry a measurable current. Semiconductors therefore have electric conductivities that are large compared with those of insulators, but small compared with those of metallic conductors, as can be seen in Table 17.1

The important semiconductors silicon and germanium are found in column IV of the periodic table. Carbon, which is also in column IV, has an

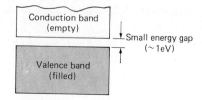

FIGURE 29.9 Band structure
of a semiconductor: The valence
band is filled, but the energy gap
between the valence band and
the conduction band is very
small.

electron configuration $(1s)^2(2s)^2(2p)^2$ and we might expect carbon to be a good
conductor because of the four unfilled $2p$ states in the atom. In the solid state,
carbon forms four covalent bonds with its four nearest neighbors, using the
two $2s$ and two $2p$ electrons. Hence the valence band is completely filled and,
since there are no electrons left over, the conduction band is completely
empty. At the small lattice spacing (0.15 nm) in the diamond crystal, the
energy gap is a large 7 eV and so diamond is a very good insulator.

For silicon, which has two $3s$ and two $3p$ electrons, the band structure is
the same as in carbon, but the binding is less tight, the lattice spacing being
0.23 nm, and the energy gap only about 1.1 eV. For germanium, with two $4s$
and two $4p$ electrons, the lattice spacing is 0.24 nm and the energy gap only
0.75 eV. Because of these small energy gaps in silicon and germanium, they
are *semiconductors*. The resistivity of Ge is about 0.5 $\Omega \cdot$m and of Si about 200
$\Omega \cdot$m, but these values vary greatly with the purity of the material. Both have
large *negative* temperature coefficients of resistivity, since the increasing
temperature creates both more electrons in the conduction band and more
holes (see below) in the valence band.

When an electron makes a transition from a normally full valence band to a
normally empty conduction band in a crystal, it leaves behind a "hole" in the
sea of electrons filling the valence band, as shown in Fig. 29.10. As a result an
electron of lower energy in the valence band can increase its energy by
moving up into this hole. This also corresponds to electrical conduction, for
the electron increases its energy by the move. Hence both the electrons and
the holes contribute to the electrical conductivity. In many ways these holes
behave in a manner similar to positively charged particles, for the deficiency
of one elementary negative charge in the valence band is the equivalent of a
positive charge there. Thus both the electrons in the conduction band and the
holes in the valence band help carry the electric current.

Holes

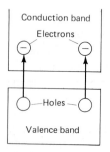

FIGURE 29.10 Production of
electron-hole pairs.

FIGURE 29.11 Parking-lot
analogy to explain motion of
holes in an electric field.

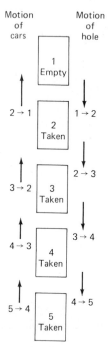

The idea of hole conduction may be clarified by the following analogy.[*] The full valence band and the empty conduction band are similar to the two levels of a parking garage, the lower level being completely filled with cars and the upper level completely empty. Hence no movement of cars can occur on either level. If one car moves from the lower to the upper level, however, cars are now free to move on both levels. Cars moving in one direction on the lower level to fill in the holes created when other cars move, are equivalent to holes moving in the opposite direction, as shown in Fig. 29.11. Similarly, the motion of electrons in one direction in the crystal is the same as the motion of holes in the opposite direction. It is for this reason that holes in the valence band of solids can always be considered as behaving in the way that positive charges would behave.

Example 29.3

(a) If we use the free-electron model and assume that the free electrons in copper behave in a similar fashion to molecules in a gas, what would be the expected rms speed of these electrons at room temperature in the absence of an electric field?

(b) Compare this rms speed with the drift velocity for a current of 1.0 A in a copper wire of diameter 0.20 cm. Assume that there is one free electron per copper atom.

SOLUTION

(a) From our discussion of gases in Chap. 9 we have for the average kinetic energy of a gas atom:

$$\overline{\tfrac{1}{2}mv^2} = \tfrac{3}{2}kT$$

or
$$v_{rms} = \sqrt{\overline{v^2}} = \sqrt{\frac{3kT}{m}}$$

For an electron with mass 9.1×10^{-31} kg at a temperature of 300 K, we therefore have:

$$v_{rms} = \left[\frac{3(1.38 \times 10^{-23}\text{ J/K})(300\text{ K})}{9.1 \times 10^{-31}\text{ kg}}\right]^{1/2} = \boxed{1.2 \times 10^5\text{ m/s}}$$

Hence we see that the average speed of these electrons is large because of their small mass, but still far from relativistic. This result is a classical one, however, and has to be modified when quantum effects are taken into account.

(b) The current is the amount of charge passing any point in the wire per second, or $I = Q/t$. But Q is the total charge in a piece of wire of cross-sectional area A and of a length numerically equal to the drift velocity v_D, since all this charge will flow past the point in 1 s. Hence

$$I = \frac{Q}{t} = nev_D A$$

where n is the number of electrons per unit volume, and e is the electronic charge.

To obtain the number density of copper atoms and hence the number of electrons per unit volume, we note that the number of atoms in 1 g of copper is:

$$\frac{6.02 \times 10^{23}\text{ atoms/mol}}{63.5\text{ g/mol}} = 9.5 \times 10^{21}\text{ atoms/g}$$

But the density of copper is 8.96 g/cm³. Hence

$$n = \left(9.5 \times 10^{21}\,\frac{\text{atoms}}{\text{g}}\right)\left(8.96\,\frac{\text{g}}{\text{cm}^3}\right) = 8.5 \times 10^{22}\text{ atoms/cm}^3$$

This is the same as the number of electrons per cubic centimeter, since each copper atom supplies one valence electron. Hence,

$$v_D = \frac{I}{neA}$$

$$= \frac{1.0\text{ A}}{(8.5 \times 10^{28}/\text{m}^{-3})(1.6 \times 10^{-19}\text{ C})(\pi)(0.10 \times 10^{-2}\text{ m})^2}$$

$$= 2.3 \times 10^{-5}\text{ m/s} = \boxed{0.023\text{ mm/s}}$$

Hence the drift velocity of the electrons due to the applied field is almost 10 orders of magnitude *smaller* than the random velocities of the free electrons in this case.

---

[*]Adapted from Hugh D. Young: *Fundamentals of Optics and Modern Physics*, McGraw-Hill, New York, 1968, p. 382.

Example 29.4

The energy gap in diamond is 7 eV and in silicon 1.2 eV. Compare these values with the value of kT (a) at room temperature, (b) at 3000 K.

(c) What conclusions can you draw from these comparisons?

SOLUTION

(a) At room temperature,

$kT = (1.38 \times 10^{-23} \text{ J/K})(300 \text{ K})$

$= (4.1 \times 10^{-21} \text{ J})\left(\dfrac{1 \text{ eV}}{1.6 \times 10^{-19} \text{ J}}\right) = \boxed{0.025 \text{ eV}}$

(b) At 3000 K,

$kT = 10(0.025 \text{ eV}) = \boxed{0.25 \text{ eV}}$

(c) It can be seen that kT is always small compared with 7 eV and hence diamond will remain a good insulator even at 3000 K. On the other hand, in Si, even at room temperature, we would expect some electrons to be thermally excited to the conduction band and hence be able to carry a current, because the electrons would have energies satisfying a Maxwell-Boltzmann distribution with a peak at 0.25 eV, and some electrons would therefore have energies above 1.2 eV. At 3000 K, where $kT \simeq \frac{1}{5}E_{\text{gap}}$, the current would be expected to grow substantially.

29.4 Semiconductors

A semiconductor, as we have seen, is a substance like silicon or germanium, which has a resistivity midway between that of a conductor and an insulator. There are two basic kinds of semiconductors, intrinsic and extrinsic. *Intrinsic semiconductors* are materials, such as silicon or germanium, in which the concentrations of negative charge carriers (electrons) and positive charge carriers (holes) are the same. *Extrinsic* (or *impurity*) *semiconductors* are materials in which the concentration of charge carriers of one sign (positive or negative) is increased by introducing impurity atoms into the crystal. Extrinsic semiconductors can be further classified as *n*-type or *p*-type.

n-Type Extrinsic Semiconductors

The conducting properties of a silicon crystal can be increased if impurity atoms are artificially forced into the crystal lattice, a process called "doping" the crystal. Suppose we dope pure silicon with phosphorus, which has an electron configuration $(1s)^2(2s)^2(2p)^6(3s)^2(3p)^3$; that is, it has one more $3p$ electron than does silicon. Only four of the five $3s$ and $3p$ electrons are needed for chemical bonding, as shown in Fig. 29.12, and one electron is therefore left over. Since such a crystal has an excess of electrons compared with the number in a pure Si crystal, a phosphorus-doped Si crystal is called an *n-type semiconductor* (where *n* means *negative*). Note that, although such a crystal has an excess of free electrons compared with pure Si, the total crystal is still electrically uncharged, for each phosphorus nucleus has one positive charge more than each of the Si nuclei, which compensates for the excess electrons.

These extra electrons occupy energy levels just slightly below the conduction band, since they are very loosely bound to their parent P atoms. These energy levels are called *donor levels* because the effect of the phosphorus impurity is to *donate* electrons to the conduction band without leaving holes in the valence band, as would happen in an intrinsic (pure) semiconductor. These donor levels are shown in Fig. 29.13. The conductivity of such a doped semiconductor can be controlled by the amount of impurity

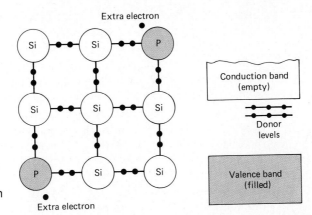

FIGURE 29.12 A silicon crystal doped with phosphorus.

FIGURE 29.13 Donor levels in an *n*-type semiconductor.

added. The addition of 1 part per million of phosphorus or arsenic can cause a significant change in the conductivity. Electrons in the donor levels can easily be raised to the conduction band by thermal or optical excitation, since the energy needed is only about 0.01 eV. These electrons then contribute to the current when an electric field is applied.

p-Type Extrinsic Semiconductors

Similarly, a boron-doped Si crystal has a deficiency of electrons compared with a normal Si crystal. The boron atom has an electron configuration $(1s)^2(2s)^2(2p)$ and hence has one less valence electron than does silicon. When a boron atom replaces a Si atom in the crystal lattice, as in Fig. 29.14, electron deficiencies, or holes, are created in the crystal. Since a deficiency of negative charge is equivalent to an excess of positive charge, such a crystal is called a *p-type semiconductor* (with *p* signifying *positive*).

The effect of doping Si with boron is to create impurity *acceptor levels* just above the filled valence band, as shown in Fig. 29.15. Electrons can then be excited thermally or optically from the filled valence band to these acceptor levels. This creates additional holes in the valence band which can be filled by other valence-band electrons moving to levels of higher energy in the band. This increase in energy of valence-band electrons when an electric field is applied leads to an electric current.

The number of extra electrons in an *n*-type semiconductor, or of holes in a *p*-type semiconductor, is usually much greater than the number of electron-hole pairs that can be created in a pure semiconductor crystal by temperature

FIGURE 29.14 A silicon crystal doped with boron.

FIGURE 29.15 Acceptor levels in a *p*-type semiconductor.

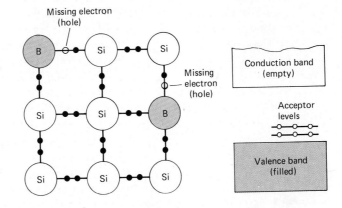

excitation of electrons from the valence band to the conduction band. Therefore in a doped semiconductor we have conduction due to both electrons and holes, with one dominating the other depending on the type impurity added. If an electric field is applied, the current will consist of both *majority carriers* (electrons in *n*-type and holes in *p*-type semiconductors) and *minority carriers* (holes in *n*-type and electrons in *p*-type semiconductors).

These ideas can be checked by a Hall-effect measurement of the type discussed in Sec. 18.3. Such a measurement verifies that the charge carriers are truly negative in *n*-type and positive in *p*-type semiconductors. The measured Hall voltage is related to the drift velocity v_D by $V_H = v_D B w$ [Eq. (18.5)], where B is the magnetic field applied perpendicular to a semiconductor ribbon of width w. Once the drift velocity is known from this equation, the number of charge carriers per unit volume (n) can be determined from the equation (see Example 29.3):

$$n = \frac{I}{v_D e A}$$

(29.5)

This provides a convenient way of evaluating the success of any doping procedure in introducing *n*- and *p*-type carriers into a crystal.

Example 29.5

A Hall-effect measurement is carried out on an impurity semiconductor consisting of doped silicon, using the setup shown in Fig. 29.16. The semiconductor ribbon is 1.0 cm wide and 0.10 cm thick; the applied magnetic field is 0.50 T applied perpendicular to the ribbon as shown; the measured current I is 3.0 mA; and the measured Hall voltage is 20 mV. In the figure the observed Hall voltage is directed from the top edge of the ribbon to the bottom edge.
(a) Is the semiconductor *p*-type or *n*-type?
(b) What is the number of charge carriers per unit volume in the semiconductor?

SOLUTION

(a) By the right-hand force rule applied to the situation shown in Fig. 29.16, the force on positive charge carriers flowing in the direction of the current I will push them to the upper edge of the ribbon. This will make the top edge positively charged and the bottom edge negatively charged, leading to a voltage from top to bottom, as observed. Hence the charge carriers are *positive charges* (i.e., holes) and the semiconductor is *p-type*. (Satisfy yourself that the result is the opposite if the conductor is *n*-type.)
(b) From Eq. (29.5) the number of charge carriers is

$$n = \frac{I}{v_D e A}$$

But, from the Hall-effect equation [Eq. (18.5)], the drift velocity is given by

$$v_D = \frac{V_H}{Bw}$$

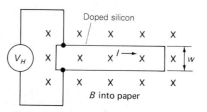

FIGURE 29.16 Diagram for Example 29.5: A Hall-effect measurement on a semiconductor. The Hall voltage is measured for a known current I and magnetic field B.

Hence the number of charge carriers is

$$n = \frac{IBw}{V_H eA}$$

$$= \frac{(3.0 \text{ mA})(0.50 \text{ T})(10^{-2} \text{ m})}{(20 \times 10^{-3} \text{ V})(1.6 \times 10^{-19} \text{ C})(10^{-2} \text{ m} \times 10^{-3} \text{ m})}$$

$$= \boxed{4.7 \times 10^{20} \text{ m}^{-3}}$$

Note that this is smaller by about eight orders of magnitude than the values for metals given in Table 29.2.

29.5 Solid-State Devices

The movement of electrons and holes in semiconductors can be manipulated for practical purposes to rectify, amplify, and detect electric signals—tasks once handled exclusively by vacuum tubes. The main difference between a vacuum tube and a *solid-state device* like a transistor is that the flow of charge occurs not in a vacuum but through solid materials. Solid-state devices include *pn* diodes, transistors, and a large variety of more complicated devices.

pn Junctions

If a region of a semiconductor doped to be *p*-type is put in close contact with a region of the same semiconductor doped to be *n*-type, we have a *pn junction*. The properties of such a junction are basic to the operation of all solid-state devices. A rough sketch of such a junction is shown in Fig. 29.17*a*. At the left side of the junction there is initially a surplus of holes to make the region *p*-type. At the right side of the junction there is a surplus of electrons, so that the region is *n*-type. Both regions, however, are still electrically neutral, as shown, because of the charges on the impurity atoms in the crystal.

When these two regions are put in intimate contact, electrons from the *n*-type semiconductor flow across the junction and fill in holes in the *p*-type region, as shown in Fig. 29.17*b*. Hence the *n*-type becomes positively charged and the *p*-type negatively charged. This creates a situation similar to that in a charged parallel-plate capacitor, and an internal electric field *E* is set up from the *n*-type to the *p*-type regions, as shown in the figure. This voltage tends to prevent any further movement of charge across the junction, and all charge flow very quickly stops.

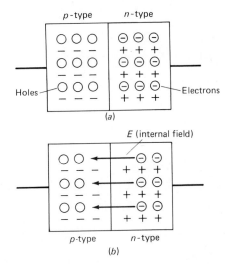

FIGURE 29.17 A *pn* junction: (*a*) Two pieces of semiconductor material—one *p*-type with an excess of holes, and the other *n*-type with an excess of electrons—are placed in contact. (*b*) After contact is made, electrons flow from the *n*-type region and fill in holes in the *p*-type region. Hence the *n*-type region becomes positively charged and the *p*-type region negatively charged, leading to an internal electric field directed from *n*-type to *p*-type.

Junction Diodes

Suppose, now, that after all charge flow has ceased, we connect a battery across the junction. If the positive terminal of the battery is connected to the *n*-side and the negative terminal to the *p*-side, so as to increase the potential difference between the two regions (a condition called *reverse bias*), as shown in Fig. 29.18*a*, the electrons in the *n*-type are attracted to the positive pole of the battery and move away from the junction. Similarly, the holes in the *p*-type are attracted to the negative pole of the battery and also move away

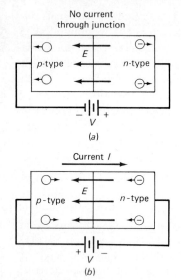

No current
through junction

p-type   *E*   *n*-type

$-$ |I| $+$
V

(*a*)

Current *I*

p-type   *E*   *n*-type

$+$ |I| $-$
V

(*b*)

FIGURE 29.18 (*a*) A *pn* junction with reverse bias: The junction acts as an insulator. (*b*) A *pn* junction with forward bias: The junction acts as a conductor.

from the junction. Hence the junction area is emptied of both free electrons and holes, and so no current can flow through the junction. Hence the *pn* junction acts as an insulator for an external voltage applied in this way.

If, however, the polarity of the battery is reversed and the positive terminal is connected to the *p*-type side of the junction (a condition called *forward bias*), as shown in Fig. 29.18*b*, the electrons are pushed by the battery in the direction of the junction, and so are the holes. Electrons therefore combine with holes at the junction, while other electrons are driven by the battery through the external circuit out of the *p*-type and into the *n*-type material, creating new holes in the *p*-type region. As long as the battery is connected, this movement of electrons in one direction and holes in the other direction continues, and we have an electric current. The dependence of the current on the magnitude and direction of the applied voltage is shown in Fig. 29.19.

A *pn* junction therefore acts like a *rectifier* which allows current to flow in one direction but not in the other. If connected to a source of alternating current, the *pn* junction will rectify the ac and convert it to pulsating dc, as shown in Fig. 29.20. Such a *pn* junction has *two* essential parts and is therefore often called a junction *diode*, or *diode rectifier*. Four diode rectifiers are often combined in the bridge-type full-wave rectifier shown in Fig. 29.21.

FIGURE 29.19 Flow of current through a *pn* junction: The junction acts as a rectifier which allows current to flow only in the direction of forward bias (note that the voltage scales differ for positive and negative voltages).

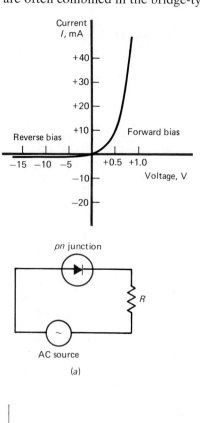

FIGURE 29.20 (*a*) Circuit diagram of a *pn* junction. The arrow in the symbol for the junction indicates the direction in which it will conduct. (*b*) The rectified current *i* flowing through the resistor *R* in the circuit of part (*a*).

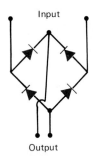

Input

Output

FIGURE 29.21 A bridge-type full-wave rectifier containing four *pn* junctions.

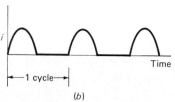

Each diode is represented by a symbol indicating the direction in which the diode will pass a conventional current.

Junction transistors consist of a crystal of one type of semiconductor (*n*-type or *p*-type) sandwiched between two crystals of the opposite type, to form transistors designated either as *pnp* or *npn* transistors. The three regions of the transistor are called the *emitter*, the *base*, and the *collector*, as shown in Fig. 29.22 for an *npn* transistor. In practice the emitter is much more heavily doped than are the base and the collector, and the base region between emitter and collector is very thin. The circuit symbols for *npn* and *pnp* transistors are given in Fig. 29.23, where the arrows indicate the direction of conventional current flow to or from the emitter.

Let us see how an *npn* transistor can be used as an amplifier. A constant voltage V_{CE} is maintained between collector and emitter by the battery of terminal voltage V_C in Fig. 29.24. Another constant voltage V_{BE} is applied between the base and the emitter by the battery V_B. Since the base is therefore maintained at a positive voltage with respect to the emitter, electrons flow from the emitter into the base. Since the base region is so thin, most of these electrons flow right across it into the collector, which is maintained at a positive voltage with respect to the emitter. Hence a large electron current flows from emitter to collector, which is equivalent to a conventional current I_C from collector to emitter. If a weak signal is applied to the base, as shown in the diagram, it results in a large change in the collector current and therefore a large change in the voltage drop across the output resistor R_L.

Junction Transistors

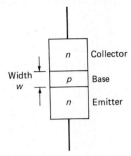

FIGURE 29.22 An *npn* transistor.

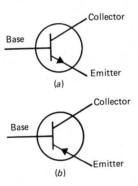

FIGURE 29.23 Circuit-diagram symbols for (*a*) an *npn* transistor; (*b*) a *pnp* transistor. The arrows indicate the direction of the conventional current.

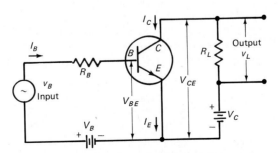

FIGURE 29.24 A transistor amplifier. A small input voltage v_B produces a large output voltage v_L.

A *pnp* transistor behaves in the same fashion except that holes move instead of electrons, and the voltages must be reversed to allow for this fact.

The junction transistor was first developed by Shockley, Bardeen, and

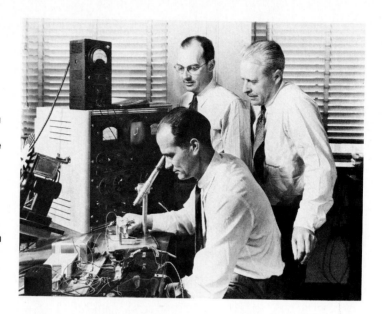

FIGURE 29.25 Left to right, William Shockley (born 1910), John Bardeen (born 1908), and Walter H. Brattain (born 1902), who did the fundamental work on semiconductors which led to the development of the transistor late in 1947 at the Bell Telephone Laboratories in Murray Hill, New Jersey. This photograph was taken in 1948. For this work they received the 1956 Nobel Prize in physics. Bardeen was the first person to win two Nobel Prizes in physics, since he also shared in the 1972 prize for his theoretical work on superconductivity. (*Courtesy of Bell Laboratories.*)

Brattain at the Bell Telephone Laboratories in New Jersey in 1948 (see Fig. 29.25). Since this solid-state device "transferred" a current across a high-resistance material (a resistor) they called it a *transfer-resistor*, abbreviated to *transistor*.

29.6 Applications of Solid-State Devices in Modern Technology

Simple transistor voltage amplifiers are used a great deal in modern communication systems. Usually a number of amplifiers are connected in series so that the output of one amplifier is fed into another amplifier, so as to multiply the voltage gain and develop large output voltages. In this way the very small voltage produced by the motion of a needle over a record can lead to the large amounts of power needed to drive a system of stereo loudspeakers. This extra energy does not, of course, come from the transistors, but rather from the dc power supplies applied to the transistor. The function of the transistor is to convert this energy input into the useful amplification of an input signal.

Because of their obvious advantages, transistors have almost completely replaced vacuum tubes for most purposes. First of all, transistors require no vacuum and no glass enclosures; they are therefore much sturdier than vacuum tubes. They require no heated filament, and hence run at lower temperatures and need no "warmup" time. They also last longer, if properly used, since there are no parts to wear out, in contrast to the filament in a vacuum tube. Most importantly, they take up much less space than do vacuum tubes.

Silicon Solar Cells

The *pn* junction (Fig. 29.17) is the basic element in a silicon solar cell. Suppose that light from the sun or some other visible radiation source is incident on such a junction. If the photons in the light have energies above 1.1 eV, which corresponds to the energy gap between the valence and conduction bands in pure silicon, they can raise electrons from the valence to the conduction band and thus create excess electrons in the conduction band and

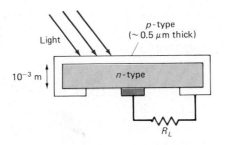

FIGURE 29.26 A photovoltaic (solar) cell. Light falling on the *p*-type region causes a current through the resistor R_L.

excess holes in the valence band. Let us suppose that these electron-hole pairs are produced by light incident on the *p*-type region of the *pn* junction, as in Fig. 29.26. In this region the extra holes produced by the light are insignificant compared with the large number of holes already present in this region (about 10^{13} cm^{-3}) due to the doping of the silicon. The extra electrons, however, are unusual in the *p*-type region, and even though some of them immediately combine with holes, others will drift to the junction boundary. Now, as we saw in Sec. 29.5, there is an internal field E at the junction directed from the *n*-type to the *p*-type region. Such a conventional electric field will therefore push electrons from the *p*-type side to the *n*-type side. Hence extra negative charge builds up in the *n*-type region and extra positive charge in the *p*-type region. The *n*-type region becomes like the negative terminal of a battery, and the *p*-type region like the positive terminal. The effect of the light on the *pn* junction is therefore to produce an electric voltage between the two sides of the junction. In practice this voltage is about 0.6 V. If an external circuit is connected between the two sides of the junction through a load resistor R_L, current continues to flow as long as light is incident on the junction.

A similar process occurs when light is incident on the *n*-type region (try to outline this process yourself).

The current that flows depends on the number of electron-hole pairs produced, and hence on the intensity of the incident light. Since any imperfection in the body of the silicon crystal provides sites at which electrons and holes can recombine without the electrons having to flow through the external circuit, very pure silicon crystals are needed in Si solar cells.

FIGURE 29.27 A weather station powered by silicon solar cells. Remote locations like this one atop Mammoth Mountain in California are ideal for such uses of solar cells. The 12 solar-cell arrays shown here produce 5 W each, or 60 W in all. (*Courtesy of NASA.*)

Solar cells seem in many ways to be the ideal solution to our growing need for energy, particularly for electric energy, since they make use of energy which would otherwise be in great part wasted, and they do so without producing thermal, chemical, or radioactive pollution. The main problems remaining with solar cells are, first, the intermittent nature of the energy provided by the sun and our failure to develop large-capacity storage mechanisms for electric energy; and, second, the inefficiency and consequent high cost of solar cells, which make them at present noncompetitive with conventional sources of electricity except in unusual situations (see Fig. 29.27), although this situation is fast improving.

Integrated Circuits

When semiconductors replaced vacuum tubes in most electronic applications, the circuits were said to be "transistorized," which was another way of saying "miniaturized," since transistors occupy so little space compared with old-fashioned vacuum tubes. A further step in miniaturization has been the development of *integrated circuits* (ICs).

The resistance of a semiconductor material depends upon its impurity concentration. The capacitance of reverse-bias diodes can also be changed by changing the voltage applied between the emitter and the base of a transistor. This means that it is possible to combine an array of transistors, resistors, and capacitors into an *integrated circuit* on a single tiny "chip" of silicon. These integrated circuits can be mass-produced to accomplish almost any desired function. Integrated circuits containing large numbers of operating components can be assembled on a square silicon chip a few millimeters on a side (see Fig. 29.28). The effort to push this miniaturization even further goes on continually.

Recently semiconductor companies have been building even more sophisticated and large-scale ICs. Such chips are very costly to develop, but once designed they can be stamped out by the millions, almost like cookies, at a very low cost per chip. This is the reason for the drastic lowering in the prices of hand calculators and digital wristwatches in recent years. ICs have improved the performance and reliability of a variety of electronic devices like computers and stereos, while at the same time lowering their prices.

FIGURE 29.28 A silicon chip containing a full 32-bit microprocessor. This dime-sized chip contains 150,000 transistors and offers processing power comparable to today's minicomputers. (*Photo courtesy of Bell Laboratories.*)

Microprocessors

In 1969 a Japanese computer company asked the American firm Intel to develop a range of medium-size ICs for use in a series of calculators of gradually increasing complexity. Intel's engineers (in particular T. Hoff and F. Faggin) had the brilliant idea that it would be simpler in the long run to design a single general-purpose IC for all the calculators, and to make this IC programmable, with the program to be unique to each kind of calculator and contained on another chip that in the final calculator would tell the IC what to do. Thus was born the *microprocessor*, an IC which can execute any sequence of coded instructions provided to it and which therefore is extremely versatile. Thus the Pac-Man game in an arcade uses exactly the same microprocessor as a Timex, Heathkit, Radio Shack, or Xerox microcomputer. The so-called personal computer is nothing but a microprocessor chip joined to an array of memory ICs, a typewriter keyboard, and a video screen. All these developments are the results of spectacular improvements in semiconductor technology, and the end of progress in this field is not yet in sight.

Example 29.6

As mentioned previously, the band gap in a pure silicon crystal is about 1.1 eV. What is the longest wavelength which is capable of moving electrons into the conduction band in such a crystal?

SOLUTION

An energy of 1.1 eV is equivalent to an energy of

$$E = (1.1 \text{ eV})\left(1.6 \times 10^{-19} \frac{\text{J}}{\text{eV}}\right) = 1.8 \times 10^{-19} \text{ J}$$

The energy of a photon is $E = hf = hc/\lambda$, and so

$$\lambda = \frac{hc}{E} = \frac{(6.6 \times 10^{-34} \text{ J·s})(3.0 \times 10^{8} \text{ m/s})}{1.8 \times 10^{-19} \text{ J}} = 1.1 \times 10^{-6} \text{ m}$$

Hence $\quad \lambda = \boxed{1.1 \times 10^3 \text{ nm}}$

This corresponds to a wavelength in the infrared region of the spectrum, since the wavelengths of visible light extend from 400 to 700 nm. An added advantage of silicon solar cells is thus that they are sensitive to red and infrared light, unlike most photographic films or photoelectric cells.

29.7 Impact of Solid-State Physics on Other Fields of Physics

As often happens in science, the deeper understanding of nature in a field such as solid-state physics leads to practical devices of great technological importance, such as transistors and integrated circuits. In turn, these devices, which were the end products of the increased understanding of nature by physicists, make significant contributions to further research in physics and other fields of pure science. We will consider only a few such spinoffs from the field of solid-state physics.

Solid-State Detectors

The development of semiconductors and ICs has led to great improvement in the accuracy and dependability of the instruments used in atomic and nuclear physics—detectors, pulse-height analyzers, microcomputers, etc. Of particular importance here is the semiconductor detector widely used in nuclear and elementary-particle physics.

The *semiconductor detector* (or *surface-barrier detector*) is simply a reverse-bias diode, or *pn* junction, in which a very thin region of *p*-type silicon is backed by a thicker region of *n*-type silicon. It is therefore very similar to the solar cell shown in Fig. 29.26. A voltage of about 100 V across a thickness of 0.1 mm produces a large electric field of 10^6 V/m in the p-type material, but no current flows because this applied voltage is opposed to the internal voltage appearing across the junction, as discussed in the preceding section. If, however, an ionizing particle produces electron-hole pairs in the p-type region, the electrons are pulled across the junction, flow through the external battery, and combine with the newly produced holes in the p-type region. We therefore have a sudden current pulse which can be used to count incident particles.

Such semiconductor detectors have the advantages of small size (about 2 cm in diameter and 1 to 2 mm thick); high efficiency, because the solid is much more dense than the gas in gas-filled counters; fast action, with response times of a few nanoseconds; and better ability to distinguish between particles of different energies than is possible with gas-filled detectors. Silicon detectors are used to detect electrons and other charged particles; germanium detectors to which lithium ions have been added [so-called Ge(Li), or "jelly," detectors] are used to detect gamma rays.

Superconductivity

This is the phenomenon in which the resistance of some metals and alloys suddenly drops to zero at temperatures within 15° or so of absolute zero, as shown in Fig. 29.29. This very complex phenomenon was explained in 1956 by the American physicists John Bardeen, Leon N. Cooper, and John R. Schrieffer (see Fig. 29.30) in a theory now called the BCS theory after its

FIGURE 29.29
Superconductivity: The electrical resistance of a mercury wire (below freezing point of mercury) falls suddenly to zero at about 4.20 K.

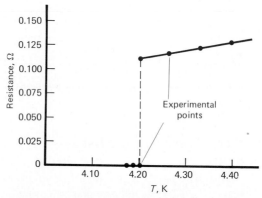

FIGURE 29.30   John Bardeen (born 1908), Leon Cooper (born 1930), and John Schrieffer (born 1931), who developed the BCS theory of superconductivity at the University of Illinois in 1956. In 1972 they received the Nobel Prize in physics for this work. It is noteworthy that 1956 was the year when Bardeen received his first Nobel Prize for his work on the transistor. (*Photo courtesy of AIP Niels Bohr Library.*)

TABLE 29.3 Superconducting Temperatures for Some Metals and Alloys

Metal	T_C, K
Al (aluminum)	1.2
In (indium)	3.4
Sn (tin)	3.7
Pb (lead)	7.2
Nb_3Sn_2	16.6
Nb_3Al	17.5
Nb_3Sn	17.9

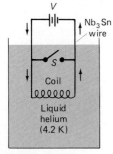

FIGURE 29.31 A superconducting coil made of Nb_3Sn wire, which has a superconducting temperature of 17.9 K. When immersed in a liquid helium bath at about 4 K, it can carry current indefinitely without any Joule heating losses.

developers. They showed how, using quantum mechanics and our knowledge of the solid state, it is possible to explain the superconducting state as a large-scale quantum state in which the motions of all the electrons in the conduction band are locked together. In this state the electrons are coupled together in pairs at sufficiently low temperatures. One electron interacts with the crystal lattice and "perturbs" it. The perturbed lattice then interacts with the second electron in the pair in such a way that there is an attraction between the two electrons which overcomes the Coulomb repulsion between them. This electron-electron interaction via the lattice produces an energy gap between the superconducting state in which the electrons act collectively, and the normal (nonsuperconducting) state in which they act individually. The energy needed to excite the system to the higher, nonsuperconducting state and destroy the superconductivity is not available at very low temperatures, and the material therefore becomes a superconductor at such temperatures.

The BCS theory has been very successful in explaining all the existing experimental facts about superconductivity, which is a very active field of research at the present time. Efforts are being directed to finding alloys which have transition temperatures higher than 17.9 K, which is the transition temperature of Nb_3Sn (niobium-3 tin) (see Table 29.3). Superconducting wire is already in use to provide the magnetic fields needed for large particle accelerators at the Brookhaven and Fermi National Laboratories (see Fig. 29.31), and there are plans for underground superconducting power lines to carry electricity from generating plants to consumers. Such lines will have negligible resistance and will thus prevent the energy losses now associated with conventional power lines at ordinary temperatures. Of course, this gain will be reduced by the costs of making and using the liquid helium needed to produce the low temperatures for the superconducting wires.

Josephson Effect

In 1962 the British physicist Brian Josephson (see Fig. 29.32) predicted this effect which was first observed in 1963. If a very thin layer (about 2 nm) of an oxide (an insulator) is sandwiched between two superconductors and a voltage V is applied across the two superconductors, a current flows (see Fig. 29.33). This current has two components: a continuous dc component which is found to persist even after the voltage source is removed, and an ac component which is present only when the voltage V is present. Remarkably, the frequency of this ac component is always exactly $f = 2eV/h$, no matter what materials are used to construct the so-called Josephson junction.

This phenomenon gives a new way of measuring the ratio e/h to high accuracy using the equation

$$\frac{e}{h} = \frac{f}{2V} \tag{29.7}$$

Since the frequency can be measured easily to 1 part in 10^{12}, the ratio of the charge on the electron to Planck's constant can be obtained to the same accuracy as that to which the voltage can be measured. This yields a value for

FIGURE 29.32 Brian Josephson (born 1940). Making use of the BCS theory of superconductivity, Josephson predicted in 1962 that both a dc current and an ac current would appear when a voltage is applied across a thin insulating layer separating two superconductors. This Josephson effect was first observed in 1963. For this work Josephson shared in the 1973 Nobel Prize in physics. (*Courtesy of AIP Meggers Gallery of Nobel Laureates.*)

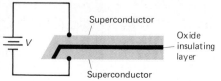

FIGURE 29.33 A Josephson junction. It consists of two superconducting layers separated by a very thin insulating layer.

e/h several orders of magnitude more precise than any other experimental method.

The reproducibility and precision of the results obtained with Josephson junctions are so great that the standard volt can now be defined in terms of a frequency, using Eq. (29.7). This eliminates the need for standard cells or other physical standards of voltage.

On a more applied level, Josephson junctions have led to the development of very sensitive detectors and of compact, fast, and efficient memory units for computers (see Fig. 29.34). SQUID (*superconducting qu*antum *interference d*evice) detectors, which are based on the Josephson effect, are now used to measure poisonous levels of iron in the liver, so that iron buildup can be treated before it does serious harm.

FIGURE 29.34 A data processing chip containing 600 Josephson junctions to be used as fast switches in computers. The shirt button indicates the tiny size of the chip. (*Courtesy of Bell Laboratories.*)

Summary: Important Definitions and Equations

Crystal: A solid material with a regular geometric shape in which the particles making up the crystal arrange themselves in an ordered geometric pattern throughout the crystal.

Lattice: The arrangement of points in space at which the particles in a crystal are located.

Types of crystalline solids:

1. Ionic crystals: Those in which alternate sites in the lattice are occupied by ions of opposite charge.

2. Covalent crystals: Those in which all lattice sites are occupied by neutral atoms.

3. Metallic crystals: Those in which positive ions occupy all lattice sites, and the valence electrons move freely through the whole crystal.

4. Molecular crystal: Those in which all lattice sites are occupied by molecules, and in which the bonding between molecules is relatively weak.

Molar heat capacity of a solid

Law of Dulong and Petit (experimental): $C_V \simeq 6$ cal/(mol·°C)

Classical theory: $C_V = 3R \simeq 6$ cal/(mol·°C), independent of temperature

Debye theory: A quantum-mechanical theory which predicts that the molar heat capacity should be equal to zero for very low temperatures, and then increase gradually until, at very high temperatures, it approaches the classical value of 6 cal/(mol·°C).

Band theory of solids

Allowed energy bands: The values of the energy which are possible for an electron in a particular crystalline solid.

Forbidden bands: The gaps between the allowed energy bands; hence values of the energy that are not possible (or are forbidden) in a particular crystal.

Valence band: The energy band in which the valence electrons of the atoms in a solid are found.

Conduction band: An energy band in which the outermost electrons of the atoms are able to move freely through the crystal.

Insulator: A crystal in which the valence band is full and the conduction band is completely empty, and where there is a large energy gap between the valence and conduction bands.

Conductor: A crystal in which the valence band is not completely filled, or in which the valence and conduction bands overlap.

Semiconductor: A crystal in which the gap between the filled valence band and the empty conduction band is small, leading to properties intermediate between those of insulators and of conductors.

Hole: The absence of an electron from a previously filled band. Such a hole is equivalent in many ways to a positive charge.

Types of semiconductors

Intrinsic: Semiconductors in which the concentrations of negative and positive charge carriers are equal.

Extrinsic: Semiconductors in which the concentration of charge carriers of one sign has been increased by introducing impurity atoms into, or doping, the crystal.

n-Type: Semiconductors with an excess of electrons which occupy donor levels just below the conduction band and which can donate electrons to the conduction band.

p-Type: Semiconductors with an excess of holes. This leads to acceptor levels just above the filled valence band, to which electrons can be excited from the valence band.

Number of charge carriers per unit volume in a semiconductor: $n = I/(v_D e A)$ where I is the current, e is the electronic charge, A is the cross-sectional area of the crystal through which the electrons flow, and v_D is the drift velocity of the electrons.

Solid-state devices

pn Junction: A region in which a *p*-type semiconductor and an *n*-type semiconductor are in close contact.

Junction diode: A *pn* junction which acts as a rectifier by allowing current to flow in only one direction through the junction.

Junction transistor: A crystal of one type of semiconductor material sandwiched in between two crystals of the opposite type to form either a *pnp* or *npn* transistor.

Silicon solar cell: A *pn* junction which produces a voltage when visible light falls on it.

Integrated circuit (IC): A circuit that incorporates numerous solid-state components like diodes and transistors into one unit, often consisting of a single chip of silicon.

Microprocessor: A single, general-purpose IC which can be programmed to perform a variety of different electronic tasks.

Semiconductor detector: A *pn* junction which produces a current pulse whenever an ionizing particle produces electron-hole pairs in the semiconductor.

Josephson effect: If a very thin layer of an insulating diode is sandwiched between two superconductors and a constant voltage V is applied across the two superconductors, the current which flows has an ac component of frequency $f = 2\ eV/h$, regardless of what material is used to construct the Josephson junction.

Questions

1 Compare the electric and thermal conductivities of copper with those of aluminum. Are the ratios of electric conductivity to thermal conductivity about the same in the two cases? Is this what you would expect? Why?

2 As the temperature increases, the lattice vibrations of a crystal increase and hence the collisions of the electrons with the positive ions in the lattice also increase. This would be expected to *increase* the resistivity, as it does in metals. Why, then, does the resistance of semiconductors like germanium and silicon *decrease* as the temperature increases?

3 If the electrons in a metallic crystal are treated as a free-electron gas, with each electron having three degrees of translational freedom, what would be the expected molar heat capacity of a metal crystal, on the basis of classical physics?

4 Why do all metals appear shiny when visible light falls on them?

5 How can an *n*-type semiconductor crystal have a surplus of free electrons if the crystal remains electrically neutral?

6 What elements are suitable to use as impurities in silicon to make it an *n*-type extrinsic semiconductor? To make it a *p*-type semiconductor?

7 An ac current is fed into the bridge-type full-wave rectifier of Fig. 29.21 at the input terminals. Draw a graph of the output current at the output terminals as a function of time.

8 Is it possible to use a junction diode as an amplifier? Why?

9 (*a*) Is the resistance of a *pn* junction the same for forward bias as for reverse bias?

(*b*) If not, in which direction is the resistance greater?

10 Why are solar cells usually painted black?

11 If the earth were covered with a huge number of solar cells to convert solar energy into electric energy, would this have any effect on the radiation balance between the earth and the sun?

12 The efficiency of silicon solar cells in converting light to electric energy is limited to about 20 percent. Can you provide some arguments to indicate what happens to the other 80 percent of the energy in the incident photons?

13 Good conductors like copper and gold are still not superconductors even at temperatures as low as 0.05 K. Such good conductors at normal temperatures have very weak interactions between the electrons and the lattice which makes them good conductors. Can you explain, on the basis of the BCS theory, why such good conductors are poor superconductors?

Problems

A 1 The electrostatic force between the two nearest Na^+ ions in a NaCl crystal is
(*a*) An attractive force of 1.5×10^{-9} N
(*b*) A repulsive force of 2.9×10^{-9} N
(*c*) A repulsive force of 0.58×10^{-9} N
(*d*) A repulsive force of 1.5×10^{-9} N
(*e*) An attractive force of 0.58×10^{-9} N

A 2 The ratio of the attractive force between two neighboring Na^+ and Cl^- ions in NaCl to the repulsive force between two neighboring Na^+ ions is:
(*a*) 2/1   (*b*) $2^{1/2}/1$ (*c*) 1/2   (*d*) 4/1   (*e*) $1/2^{1/2}$

A 3 The average classical energy of a Cu^+ ion in a copper crystal lattice at 300 K is about:
(*a*) 0.04 eV   (*b*) 0.08 eV   (*c*) 0.04 J
(*d*) 0.08 J   (*e*) 1.2×10^{-19} J

A 4 The number of charge carriers in a germanium crystal at 300 K is about 1.2×10^{19} m^{-3}. If the crystal carries a current whose density is 10^{-1} A/m^2, the drift velocity of the electrons in the crystal will be
(*a*) 5.2×10^{-2} m/s   (*b*) 5.2 m/s
(*c*) 5.2×10^{-2} cm/s   (*d*) 5.2×10^2 m/s
(*e*) None of the above

A 5 The band gap in germanium is 0.74 eV. The temperature which would provide sufficient energy to an electron to jump this gap and create an electron-hole pair is:
(*a*) 860 K   (*b*) 860°C   (*c*) 8600 K
(*d*) 300 K   (*e*) 300°C

A 6 If arsenic ($_{33}$As) atoms are forced into a crystal of pure germanium ($_{32}$Ge), the result will be:
(*a*) A *p*-type semiconductor
(*b*) An *n*-type semiconductor
(*c*) An intrinsic semiconductor   (*d*) An insulator
(*e*) A metal

A 7 At room temperature the number of charge carriers per unit volume in a good conductor exceeds the number in an intrinsic semiconductor such as silicon by about:
(*a*) 1 order of magnitude
(*b*) 5 orders of magnitude
(*c*) 10 orders of magnitude
(*d*) 20 orders of magnitude
(*e*) They are about the same

B 1 If the distance between neighboring Cl^- ions in CsCl is 0.411 nm, what is the distance between neighboring Cs^+ and Cl^- ions?

B 2 Calculate the equilibrium separation l_0 of the atoms in a rubidium chloride (RbCl) crystal which has a density of $\rho = 2.76 \times 10^3$ kg/m^3.

B 3 The energy gap between the valence and the conduction band in BaO is 4.2 eV.

(*a*) What is the minimum frequency of incident radiation which can produce an electron-hole pair in such a crystal?

(b)   In what region of the electromagnetic spectrum does this frequency fall?

(c)   What can you conclude about the electric conductivity of BaO?

B 4   The energy gap between the valence and the conduction bands in the lead sulfide (PbS) crystal is 0.37 eV. What is the maximum wavelength of radiation which can produce electron-hole pairs in such a crystal?

B 5   Find the drift velocity of the electrons in a silver wire if a current of 2.0 A flows through the wire of diameter 0.15 cm. Assume that there is one free electron per silver atom.

B 6   The donor levels in an n-type semiconductor are only about 0.01 eV below the conduction band.

(a)   To what temperature does this energy gap correspond?

(b)   What frequency photons can raise electrons from these donor levels to the conduction band?

B 7   (a)   In an experiment using a Josephson junction it is found that, when the applied dc voltage is 20.151 μV, the resulting current has an ac component at a frequency of 9.6965 GHz. From these data, what is e/h, the ratio of the charge on the electron to Planck's constant?

(b)   Compare your result with that found from the accepted values of e and h obtained in other ways.

C 1   Calculate the density of NaCl from the fact that it forms a face-centered cubic lattice with l_0 equal to 0.282 nm.

C 2   Iron has a density of 7.9×10^3 kg/m^3. Find the distance between two atoms in the iron crystal, assuming that the atoms arrange themselves in a perfect cubic array.

C 3   (a)   Compare the electrostatic force between adjacent Na$^+$ and Cl$^-$ ions in the crystal lattice of NaCl with the gravitational force between the two ions.

(b)   What can we conclude about the importance of gravitational forces in crystal formation?

C 4   Calculate the distance l_0 between the K$^+$ and Cl$^-$ ions in KCl, assuming that each ion occupies a cube of side l_0. The density of KCl is 1.98×10^3 kg/m^3.

C 5   A cube of copper ($^{63}_{29}$Cu) metal has a mass of 1.46×10^{-1} kg. The length of each edge of the cube is 2.5×10^{-2} m. If we assume that the lattice is a simple cubic lattice:

(a)   Find the number of atoms in the sample.

(b)   Find the size of a single copper atom.

C 6   For a KI crystal we found in Example 29.2 that the equilibrium separation between a K$^+$ ion and its nearest neighbor I$^-$ ion was 0.353 nm.

(a)   Assuming that each ion is attracted only by the nearest ion in the adjacent layer, calculate the force per unit area required to pull the crystal apart.

(b)   Do you expect that your answer will be too large or too small compared with the measured value? Why?

C 7   In Example 29.5 find the fraction of impurity atoms in the Si crystal.

C 8   The energy gap for a diamond crystal is 7 eV. If this amount of energy were converted into the kinetic energy of a free electron:

(a)   What would be the electron's speed?

(b)   What would be its de Broglie wavelength?

C 9   A current of 1.0 A flows in a copper wire of 2.0-mm diameter.

(a)   What is the current density?

(b)   What is the drift velocity of the electrons? (Hint: Use Table 29.2.)

C10   The drift velocity of the electrons in an aluminum wire is 5.0×10^{-1} cm/s. If the wire has a diameter of 1.5 mm, find:

(a)   The current density in the wire.

(b)   The total current.

C11   (a)   Calculate the number of free electrons per unit volume for gold ($^{197}_{79}$Au), which has a density of 19.3×10^3 kg/m^3, assuming one free electron per atom.

(b)   Compare your result with the value given in Table 29.2.

C12   The density of sodium is 0.97×10^3 kg/m^3. How many free electrons are there for each atom in a solid sodium crystal? (Hint: Use Table 29.2.)

C13   In Eq. (29.5) the quantity $1/ne = v_D/(I/A)$ is called the *Hall coefficient*. The Hall coefficient for sodium is 2.5×10^{-10} m^3/C, and the density of sodium is 0.97×10^3 kg/m^3.

(a)   Find the average number of free electrons per cubic meter in sodium.

(b)   Find the average number of free electrons per sodium atom.

C14   A gold foil 20 μm thick and 2.0 cm wide carries a current of 15 A in a Hall-effect experiment. There is a magnetic field of 1.0 T normal to the plane of the foil.

(a)   Find the Hall voltage which is produced.

(b)   What is the drift velocity of the electrons in the foil?

C15   In the full-wave bridge-type rectifier of Fig. 29.21 the ac current entering the bridge has a peak value of 10 A.

(a)   Draw a graph similar to that of Fig. 29.20b, indicating the form of the output current from the bridge.

(b)   What is the peak value of the output current?

C16   A circuit uses a transistor to amplify the output of a microphone and then feeds the output signal from the transistor to a pair of loudspeakers. The microphone has an output of 0.01 A at 0.50 V. The current gain of the transistor is 50 and the output voltage to the loudspeakers is 10 V.

(a)   Compare the power output to the speakers with the power input to the transistor amplifier.

(b)   Where does the increase in power come from?

C17   One proposed scheme for the use of solar cells is to set up a solar satellite power station (SSPS) in an orbit around the earth. This SSPS would expose 30 km^2 (that is, an area 6 by 5 km) of solar cells to an average solar power of 1.35 kW/m^2. If the solar cells are 20 percent efficient in converting the sun's radiation into electricity:

(a)   What will be the electric power output of such an SSPS?

(b)   How does this power compare with the approximately 500 GW produced by all the electric power plants in the United States?

D 1   Show that, in a body-centered cubic crystal lattice such as that shown in Fig. 29.2, a central atom of radius R will just touch each of its eight nearest neighbors which have radii r, if $R = (\sqrt{3} - 1)r$.

D 2   In a face-centered cubic lattice such as that for NaCl in Fig. 29.1, how many Na^+ ions does one Na^+ ion have as nearest neighbors, that is, Na^+ ions which are as close as possible to the first Na^+ ion and are all at equal distances from the first ion?

D 3   (*a*)   Show how well the Wiedemann-Franz law [Eq. (29.1)] is verified for copper, silver, and gold at 0°C. Use the values given elsewhere in this book, and remember that the conductivity σ is the reciprocal of the resistivity ρ.

(*b*)   Use tables from a reference book such as the *Handbook of Chemistry and Physics* to verify the Wiedemann-Franz law for the same three metals at 100°C. See if the ratio of the results for 100°C and for 0°C is indeed equal to the ratio of these two temperatures on the Kelvin scale.

D 4   When an arsenic impurity is used to dope a silicon crystal, the unbonded electron from the arsenic is attracted by the single extra positive charge on the arsenic atom. Let us assume an oversimplified model in which the unbonded electron moves in a Bohr orbit about the positively charged arsenic atom in the silicon crystal, which has a dielectric constant of 12.

(*a*)   Calculate the radius of the first Bohr orbit for this electron.

(*b*)   Compare its value with the interatomic distance of 0.23 nm for a silicon crystal.

(*c*)   What can you conclude about the motion of this electron?

D 5   Repeat Prob. D4 for a phosphorus impurity in germanium, which has a dielectric constant of 16 and an interatomic distance of 0.24 nm.

Additional Readings

Behrman, Daniel: *Solar Energy: The Awakening Science*, Little, Brown, Boston, 1976. A popular book on solar-energy research, including the production and use of solar cells, in all parts of the world.

Chalmers, Bruce: "The Photovoltaic Generation of Electricity," *Scientific American*, vol. 235, no. 4, October 1976, pp. 34–43. A clear and well-illustrated account of the physics and technology of solar cells.

Daniels, Farrington: *Direct Use of the Sun's Energy*, Yale University Press, New Haven, Conn., 1964 (paperback edition by Ballantine, 1974). A classic by the man who almost single-handedly kept solar energy research moving ahead at a time when no one was really interested.

Geballe, Theodore H., and J. K. Hulm: "Superconductors in Electric Power Technology," *Scientific American*, vol. 243, no. 5, November 1980, pp. 138–172. A discussion of the many uses to which high-field, high-current superconductors can be put in the electric-power industry.

Holden, Alan, and Phylis Singer: *Crystals and Crystal Growing*, Doubleday Anchor, Garden City, N.Y., 1966. A very informative and fascinating account of the properties and growing of crystals.

Langenberg, Donald N., Douglas J. Scalapino, and Barry N. Taylor: "The Josephson Effects," *Scientific American*, vol. 214, no. 5, May 1966, pp. 30–39. A very clear discussion of the theory of the Josephson effect and of the experiments performed to verify this theory.

Matthias, T.: "Superconductivity," *Scientific American*, vol. 197, no. 5, November 1957, pp. 92–103. An account of a field to which the author made many significant experimental contributions.

Pierce, John R.: *Quantum Electronics*, Doubleday Anchor, Garden City, N.Y., 1966. Contains a good elementary discussion of semiconductor devices.

Taylor, Barry N., Donald N. Langenberg, and William H. Parker: "The Fundamental Physical Constants," *Scientific American*, vol. 223, no. 4, October 1970, pp. 62–78. Points out how developments like the Josephson effect have improved our knowledge of fundamental atomic constants and yielded information about the overall correctness of physical theories.

Trefil, James S.: "A Future of Maglevs, SQUIDS, and Mass Drivers," *Smithsonian*, vol. 15, no. 4, July 1984, pp. 78–89. A popular, up-to-date, well-illustrated account of superconductivity.

Weber, Robert L.: *Pioneers of Science: Nobel Prize Winners in Physics*, The Institute of Physics, Bristol, England, 1980. This useful book contains brief biographies of many of the physicists whose work is described in this chapter, including Bardeen, Brattain, Shockley, Cooper, Schrieffer, and Josephson.

In addition, the September 1977 issue of *Scientific American* (vol. 237, no. 3) contains 11 excellent articles on microelectronics.

Can we ever hope to find the right way? Nay more, has this right way any existence outside our illusions? . . . I answer without hesitation that there is, in my opinion, a right way, and that we are capable of finding it. Our experience hitherto justifies us in believing that nature is the realisation of the simplest conceivable mathematical ideas. I am convinced that we can discover by means of purely mathematical constructions the concepts and the laws connecting them with each other, which furnish the key to the understanding of natural phenomena.

Albert Einstein

Chapter 30

High-Energy and Elementary-Particle Physics

Nature often turns out to be more complicated than physicists would like it to be. In the 1930s the accepted model of the atom was a marvelously simple one, consisting of only three "elementary" particles, the electron, the proton, and the neutron. We now know that both the proton and the neutron are not elementary particles at all, but are made up of other, more elementary particles, which may, in turn, be made up of additional particles not even dreamed of today. Over the last 30 years physicists have been uncovering a remarkable variety of particles which appear to make up the universe, and have been attempting to introduce order into the seeming chaos which has resulted from the discovery of so many new particles. This is the field of *elementary-particle physics*, one of the most important frontiers of current physics research.

It should be noted that the goal of physicists working in elementary-particle physics is *understanding*, not applications. It is not clear that any practical applications will ever emerge from the research being done in this field, as did the laser from quantum and electromagnetic theory, and the transistor from solid-state physics. It would, however, be presumptuous to rule out such practical applications at the present stage of our knowledge.

In what follows we will give a broad overview of this complicated and still-developing field, without spending too much time on details, many of which could change before this book is published. Since this chapter is so qualitative and so descriptive, it contains no worked examples and no problems. It is intended to give the student a realization that many questions in physics remain to be answered.

30.1 The Four Forces of Physics

In previous chapters we have discussed the four basic forces which govern all the interactions of physics. These are summarized in Table 30.1, together with their relative strengths and ranges. The two forces that most concern us here

TABLE 30.1   The Four Known Forces of Physics

Force	Present in	Relative strength	Range
Gravitational	All interactions	6×10^{-39}	$\propto \left(\dfrac{1}{r^2}\right)$
Weak nuclear	All weak interactions (e.g., beta decay)	10^{-15}	$\sim 10^{-19}$ m
Electromagnetic	All interactions involving two electrically charged particles	$\dfrac{1}{137}$ ($= \alpha$, the fine-structure constant)	$\propto \left(\dfrac{1}{r^2}\right)$
Strong nuclear	All strong interactions (e.g., nuclear binding)	1	$\sim 10^{-15}$ m

are the strong and weak nuclear forces. Both of these are very short-range forces, as is clear from experiment, but we are at present unable to write any equation describing the dependence of these forces on distance, since we do not understand them well enough to do so. For this reason elementary-particle physicists prefer to talk about "interactions" rather than the "forces" which cause these interactions. We will often follow this practice in the present chapter.

Strong interactions occur only within a class of particles called *hadrons* (to be defined later), of which the proton and the neutron are prime examples, and do not depend on the electric charge of the particles involved. *Weak interactions* occur within another class of particles called *leptons* (also to be defined later), of which the electron is the prime example. Weak interactions occur also between leptons and hadrons, and between hadrons and hadrons, and they too are independent of electric charge.

The four basic forces of physics are necessary to explain the universe around us. We need the gravitational force to hold the planets in their orbits about the sun and to keep us from flying off the earth into space. The electromagnetic force is required to hold electrons in their atoms and to bind atoms together into molecules. The strong force is needed if protons and neutrons are to coalesce to form stable nuclei. The weak force is in some ways the most mysterious of the four, but it governs the processes of radioactive beta decay of nuclei, is crucial to the energy-producing processes going on in the stars, and apparently played a very important role in the building up of the heavier elements from light nuclei at the time the universe began. Hence without any one of these forces, the universe would be a very different place than it is today.

The Strong Nuclear Force

The strong nuclear force acts between protons and neutrons in atomic nuclei to hold them together. The approximate form of the force between two protons at short distances is shown in Fig. 30.1. It can be seen that only for distances R of about 1.5×10^{-15} m is this nuclear force sufficiently strong to counterbalance the repulsive electrostatic force between the two charged protons and produce a net attraction.

To learn more of this nuclear force, we need probes capable of "seeing" details in structure of size less than 10^{-15} m. This means particles with energies greater than about 100 MeV, for only such particles have de Broglie wavelengths as small as 10^{-15} m. This is the realm of *high-energy physics*. In recent years the availability of gigantic particle accelerators to produce

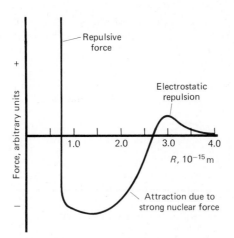

FIGURE 30.1 The force between two protons at short distances inside a nucleus. For distances greater than 3.0×10^{-15} m the electrostatic repulsion dominates the interaction. For distances between about 1.0 and 2.5 $\times$ 10^{-15} m the strong nuclear force produces a strong attraction of the two protons for each other.

energies above 100 MeV has opened up this new energy domain and led to the discovery of many new particles. For this reason physicists use the terms *high-energy physics* and *elementary-particle physics* almost interchangeably to describe this exciting field.

The Weak Nuclear Force and the Neutrino

The weak nuclear force is known to be responsible for the beta decay of radioactive nuclei. A typical beta-decay process is the one involved in radiocarbon dating:

$$^{14}_{6}C \rightarrow {}^{14}_{7}N + {}^{0}_{-1}e \tag{30.1}$$

Here both the charge and the number of nucleons are conserved, as required.

The fact that an electron is produced in this nuclear-decay process may seem to contradict a previous statement that there are no electrons in the nucleus. Electrons do not exist in nuclei any more than photons exist in atoms; both are produced just before they are observed. In beta decay a neutron in the nucleus is transformed into a proton and an electron inside the nucleus, and this electron is then immediately ejected from the nucleus as a beta ray. Free neutrons are known to decay in a similar fashion:

$$^{1}_{0}n \rightarrow {}^{1}_{1}p + {}^{0}_{-1}e \tag{30.2}$$

To determine the expected kinetic energy of the beta particle in Eq. (30.1) we will use the masses of the neutral atoms. Since there are six extranuclear electrons in the carbon atom, after beta emission the N nucleus also has only six extranuclear electrons. A seventh electron is produced in the decay process as the emitted beta particle, however, and so the mass of $^{14}_{7}N$ (minus one electron) and $_{-1}^{0}e$ together is simply the mass of one neutral $^{14}_{7}N$ atom. The mass of $^{14}_{6}C$ is found experimentally to be 14.00324 u and that of $^{14}_{7}N$ to be 14.00307 u (see Appendix 4). Hence the mass defect is

$$\Delta m = 14.00324 \text{ u} - 14.00307 \text{ u} = 0.00017 \text{ u}$$

The energy equivalent of this, using 1 u = (931.5 MeV)/c^2, is 158 keV. Most of this goes into the kinetic energy of the electron, since the daughter nucleus recoils with a very low velocity because it has a large mass compared with that of the electron. We would therefore expect to detect a very large number of electrons all with exactly the same kinetic energy, 158 keV, when a sample of $^{14}_{6}C$ decays.

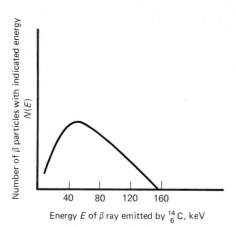

FIGURE 30.2 Observed energy spectrum of the beta particles emitted by $^{14}_{6}$C. The beta particles do not all have the same energy (158 keV, according to theoretical calculations), but a continuum of values ranging from near zero to 158 keV.

In 1914 the British physicist James Chadwick (see Fig. 28.1), who later discovered the neutron, found that beta particles from $^{14}_{6}$C did not have such well-defined energies. On the contrary their energy spectrum ranged continuously from practically zero to a maximum value close to 158 keV, as shown in Fig. 30.2. This seemed a direct violation of the principle of conservation of mass-energy in the beta-decay process, and at the time even Niels Bohr expressed doubts about the validity of the principle of conservation of energy in such processes.

Wolfgang Pauli could not accept such a violation of one of physics' most treasured laws, and in 1930 he postulated for beta decay the emission of another particle which has come to be called the *neutrino* (the name was given by Enrico Fermi and means "the little neutral one"). According to Pauli, neutrinos were particles with no electric charge and zero or very small rest mass, but they were not photons. They were emitted paired to beta particles in such a way that the sum of the energies of the neutrino and the beta particle would always be the same, for example, 158 keV for $^{14}_{6}$C. This would reinstate the principle of conservation of mass-energy.

The existence of the neutrino would also satisfy the requirements of the conservation laws for linear and angular momentum, which for beta decay could not be satisfied without the presence of this elusive particle. In 1934 Enrico Fermi worked out a detailed theory of beta decay which agreed well with experimental results. Neutrinos are designated by the Greek letter *nu* (ν). We therefore write the equation for the decay of $^{14}_{6}$C more properly as:

$$^{14}_{6}\text{C} \rightarrow {}^{14}_{7}\text{N} + {}^{0}_{-1}e + \overline{\nu}_e \tag{30.3}$$

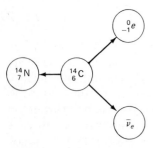

FIGURE 30.3 Beta decay of $^{14}_{6}$C. The antineutrino $\overline{\nu}_e$ carries both energy and momentum away from the reaction

where the subscript on $\overline{\nu}_e$ means that this neutrino is emitted along with an electron, and the bar indicates that it is actually an antineutrino (see Sec. 30.4). The complete reaction is then as shown in Fig. 30.3.

From the decay of tritium according to the reaction

$$^{3}_{1}\text{H} \rightarrow {}^{3}_{2}\text{He} + {}^{0}_{-1}e + \overline{\nu}_e \tag{30.4}$$

we know that the mass of the electron neutrino is less than 10^{-4} of the electron mass. Whether the neutrino mass is exactly zero is not as yet clear. Some (still disputed) experiments have indicated a neutrino mass of between 2.7×10^{-5} and 9.0×10^{-5} times the electron mass. If this turns out to be the case, it may help explain the problem of the "disappearing mass" which greatly bothers astrophysicists: There simply is not enough mass in the observable universe to account for its observed large-scale behavior. Theoretical predictions indicate that there should be about as many neutrinos as photons in the universe and 10^9 to 10^{10} times more neutrinos than protons or neutrons. Hence even a small mass for the neutrino would contribute a substantial additional mass to the universe.

Decay of the Free Neutron

Equation (30.2) for the decay of the free neutron must also be modified to include a neutrino and is written as:

$$_0^1 n \rightarrow {}_1^1 p + {}_{-1}^0 e + \bar{\nu}_e \tag{30.5}$$

The force which leads to the decay of the free neutron according to this equation is very weak. This can be seen from the fact that the average lifetime of a free neutron is about 17 min, which is extremely long on a nuclear scale, particularly when compared with lifetimes associated with strong nuclear reactions (remember that the pion's lifetime is less than 10^{-8} s). This is the reason that the force involved in beta decay is called the *weak nuclear force*.

The reason the neutrino is so hard to detect is that it interacts with matter only through this weak nuclear force. As a result a lead shield used to absorb neutrinos effectively would have to be 2×10^{14} mi or 35 light-years in thickness!

Detection of the Neutrino

As a result of its weak interaction with matter, it is very difficult to detect a neutrino, and it was in 1956, 26 years after Pauli postulated its existence, that the first neutrino was detected in the laboratory. This observation was made by the American physicists Clyde L. Cowan, Jr. (1919–1974) and Frederick Reines (born 1918), using the enormous flux of neutrinos from a nuclear reactor in which neutron-rich nuclear fission products decay by emitting beta particles and antineutrinos. Cowan and Reines worked at the Savannah River reactor in South Carolina and detected the antineutrinos by their interaction with the protons in water molecules:

$$\bar{\nu}_e + {}_1^1 p \rightarrow {}_0^1 n + {}_{+1}^0 e \tag{30.6}$$

where ${}_{+1}^0 e$ (or e^+) is a *positron*, or positive electron, and $\bar{\nu}_e$ an antineutrino produced in a beta-decay process in the reactor.

Reines and Cowan used a gigantic tank containing layers of water mixed with cadmium chloride sandwiched between layers of a clear liquid scintillator, as in Fig. 30.4. In all they used some 10 tons of liquid scintillator and 500 photomultiplier tubes. Antineutrinos first were captured by the protons in water molecules, leading to the production of neutrons and positrons, in accordance with Eq. (30.6). The positrons were slowed down by collisions and brought almost to rest in about 10^{-9} s. These positrons then combined with electrons in the water and a pair-annihilation process led to the production of two gamma rays. Since the two gamma-ray photons had to go off in opposite directions to conserve momentum, they passed through the layers of liquid scintillator on each side of the water and produced light flashes which could be seen by the photomultiplier tubes observing these layers. Then 10 μs or so later the neutron had slowed down sufficiently to be captured by a cadmium atom in the water, leading to the emission of more

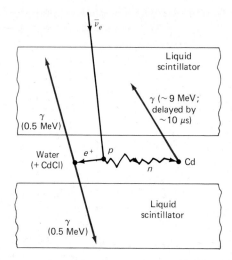

FIGURE 30.4 Schematic diagram of the apparatus used by Cowan and Reines to detect the neutrino. By observing the emitted gamma rays, they could argue back to the capture of an antineutrino by a proton in the water sandwiched between the two layers of liquid scintillator.

gamma rays, which could again be detected by the scintillators and photo-tubes.

If the antineutrinos emerging from the reactor were actually taking part in the reaction indicated in Eq. (30.6), Reines and Cowan expected to find two simultaneous scintillation pulses in adjacent layers, followed by another scintillation pulse (or pulses) 10 μs or so later. Moreover, each of these three pulses would have a characteristic identifying energy which the photocells could measure—0.5 MeV for each of the annihilation photons and about 9 MeV for the neutron-produced photon (or photons). This is exactly what they observed, with one event being observed about every 20 min, as predicted by theory. This confirmed the weakness of the neutrino's interaction with other particles.

The pioneering work of Cowan and Reines was the first convincing experimental evidence for the existence of the neutrino. Since that time, it has become possible to obtain huge numbers of neutrinos from the decay of particles produced in large accelerators and to study more easily their interactions with matter.

30.2 Particle Accelerators

The high-energy accelerators in use today are mainly of two types, linear accelerators, whose origins can be traced back to the Van de Graaff generator (Sec. 16.7), and cyclic accelerators, which are greatly improved versions of Lawrence's original cyclotron (Sec. 18.3).

Linear Accelerators

Electrons present unique problems as nuclear probes because of their small mass. If electrons are forced to move in a circular path in a cyclic particle accelerator, they soon reach such high speeds that they lose great amounts of energy by radiation, since they undergo continuous acceleration when moving in a circle. This imposes a serious limitation on the maximum energy which can be reached by electrons in a circular machine. As a result linear accelerators are the choice for accelerating electrons. The electron accelerator at the Stanford Linear Accelerator Center (SLAC) at Stanford University in Palo Alto, California (Fig. 30.5), accelerates electrons by subjecting them

FIGURE 30.5 Photo of the world's highest-energy electron accelerator at the Stanford Linear Accelerator Center (SLAC), Palo Alto, California. (*Photo courtesy of SLAC.*)

to voltage pulses as they move down a straight, 2-mi-long, evacuated tube. The SLAC accelerator can produce electrons of 20 GeV energy, and there is hope of doubling this in the near future.

Circular Accelerators: Synchrotrons

The particle accelerators which have achieved the highest energies are all of the synchrotron type. A *synchrotron* keeps particles moving in a circular evacuated tube of fixed radius, with the confining magnetic field varying in time as the particles attain higher and higher speeds. This is accomplished by an automatic feedback system which monitors the motion of the circulating bunches of particles and adjusts the accelerating field to hold the beam radius constant as the magnetic field increases.

A diagram of a synchrotron is shown in Fig. 30.6. The evacuated tube, which is in the form of a circular ring, is divided into three different kinds of

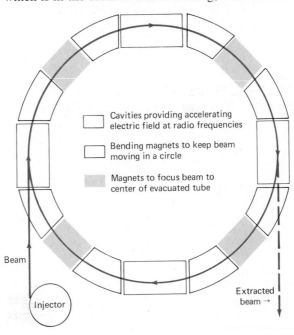

Cavities providing accelerating electric field at radio frequencies

Bending magnets to keep beam moving in a circle

Magnets to focus beam to center of evacuated tube

Beam

Injector

Extracted beam →

FIGURE 30.6 The beam tube of a synchrotron, showing the three kinds of tube sections required to accelerate and focus the protons or other particles being used as probes.

FIGURE 30.7 The proton synchrotron at Fermilab in Batavia, Illinois. The largest circle is the accelerator which has a diameter of 2.0 km. The 16-story, twin-towered central laboratory building can be seen at the bottom of the circle. (*Courtesy of Fermilab*.)

FIGURE 30.8 A part of the tunnel showing the beam tube and magnets at the CERN Super Proton Synchrotron (SPS) in Geneva, Switzerland. The tunnel is 2.2 km in diameter. (*Courtesy of CERN*.)

sections, one group of which provides the accelerating electric field, a second of which contains the magnets which bend the beam into a circular orbit, and a third of which uses different kinds of magnets to focus the particles into a narrow beam. The particles are introduced into the synchrotron by an injection system which often includes a smaller accelerator to give the beam an initial energy before it enters the synchrotron. The beam can be extracted when desired by turning on deflecting magnets to direct the beam out of the ring toward the target.

As we saw in the case of the cyclotron, the momentum of the accelerated particles is given by Eq. (18.6),

$$p = mv = Bqr$$

where B is the strength of the bending magnetic field and r is the radius of the orbit. Since in the synchrotron the particles finally move at speeds close to that of light, their energy is

$$E = pc = Bqrc \qquad (30.7)$$

Hence the maximum energy attainable is directly proportional to the radius of the fixed orbit in which the particles move. Physics laboratories all over the world are engaged in a friendly rivalry to build the world's largest synchrotron in the hopes of unraveling more of nature's mysteries.

The proton synchrotron at Fermilab in Batavia, Illinois (Fig. 30.7), has a diameter of 2.0 km and has achieved energies of 500 GeV. The Super Proton Synchrotron (SPS), built by a consortium of European nations at the CERN (Centre Européen pour la Recherche Nucléaire) laboratory in Geneva, Switzerland (Fig. 30.8), has a diameter of 2.2 km and has achieved an energy for protons of 450 GeV. These are the two largest high-energy machines of this type in the world, although plans are underway for bigger machines at CERN and in the United States and Russia.

Types of Synchrotrons

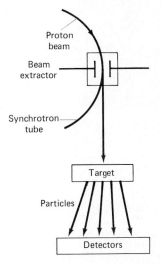

Proton beam

Beam extractor

Synchrotron tube

Target

Particles

Detectors

FIGURE 30.9 A fixed-target synchrotron. The exit beam from the synchrotron crashes into a stationary target. In this case little energy is available to be converted into the mass of new particles.

Fixed-Target Machines A *fixed-target machine* (Fig. 30.9) fires the accelerated particles, usually protons, at a fixed target, such as liquid hydrogen, and the interactions of the protons with the nucleons in the target are studied. The problem with such an approach is that most of the beam energy goes into the kinetic energy of the particles emerging from the interaction, for momentum must be conserved. Since the energy E_u useful for creating new particles is the difference between the initial and final kinetic energies, too little energy remains for creating new particles of interest to researchers.

When the beam particles and the target particles have equal masses, as when protons bombard a liquid-hydrogen target, application of the principles of conservation of momentum and energy and the ideas of special relativity lead to the result that the energy E_u useful in producing new particles is related to the total energy of the bombarding particles E and to their rest masses m_0 by the equation

$$E_u = \sqrt{2m_0c^2E} \tag{30.8}$$

where E is the energy of the beam particles. For example, for a proton of energy 500 GeV from the Fermilab accelerator, the useful energy would be only

$$E_u = \sqrt{2m_0c^2E} = \sqrt{2(931 \text{ MeV})(500 \text{ GeV})} = \sqrt{9.31 \times 10^{20} \text{ (eV)}^2} = 31 \text{ GeV}$$

and so only about 6 percent of the beam energy goes into creating new particles.

If, now, it were possible to increase the beam energy to 1000 GeV, the useful energy would be only

$$E_u = \sqrt{2(931 \text{ MeV})(1000 \text{ GeV})} = \sqrt{2} \ (31 \text{ GeV}) = 44 \text{ GeV}$$

Hence the greatly increased size and cost of the 1000-GeV accelerator would increase the useful energy by only 13 GeV. To overcome this difficulty colliding-beam machines are now used.

Colliding-Beam Machines With a *colliding-beam machine* (Fig. 30.10) two beams of high-speed particles traveling in opposite directions are collected in "storage rings" and then made to collide head on. Since the total linear momentum in the head-on collision of two particles of equal mass is zero, very little of the energy goes into the kinetic energy of the scattered particles. Hence, most of the energy is available to cause high-energy interactions which produce new particles. For example, two beams of 500-GeV protons would make available close to 1000 GeV for the creation of new particles.

FIGURE 30.10 A colliding-beam synchrotron: Diagram showing how two beams of protons or other particles traveling in opposite directions at close to the speed of light are made to collide at the intersection of two storage rings. In this case most of the energy is available for the creation of new particles.

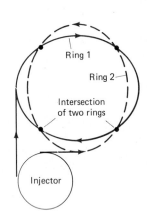

FIGURE 30.11 The UAI particle detector at the SPS accelerator at CERN. This multipurpose detector weighs over 2000 tons and is rolled into position on rails. (*Courtesy of CERN.*)

In early 1983 in the CERN interacting storage rings (ISR) machine a 270-GeV proton beam was made to collide head-on with a 270-GeV antiproton beam, making available 540 GeV for the creation of new particles. A 2000-ton particle detection (Fig. 30.11) revealed the existence of heavy particles which are especially important as tests for present theories of the elementary particles (see Sec. 30.8). Work is in progress at many other accelerator laboratories to increase the energy range of the accelerator by using colliding beams. For example, the large accelerator at Fermilab expects to reach 2000 GeV using colliding beams. Another colliding-beam accelerator, called tentatively the Desertron, has been proposed by U.S. physicists. It would be designed to reach 20,000 GeV (20 TeV).

30.3 Pions and Muons

Most high-energy collision processes of the kind discussed in the preceding section produce *pions* in addition to other particles. Pions have masses about one-seventh the mass of the proton, or approximately $275m_e$, and come in three varieties: positively charged, negatively charged, and neutral. All pions are unstable, with the charged pions having lifetimes of about 10^{-8} s and the neutral pions lifetimes of only 10^{-16} s. Pions are designated by the Greek letter pi (π).

The pion was predicted by the Japanese physicist H. Yukawa (see Fig. 30.12) in 1935. At that time Yukawa was even able to estimate its mass on the basis of the known range of the strong nuclear force, i.e., about 10^{-15} m. The pion was first seen experimentally in 1947 by the British physicist C. F. Powell (see Fig. 30.13) in photographic-emulsion tracks of cosmic-ray events, in which very high-energy particles from outer space, mainly protons, bombarded atoms in the earth's atmosphere and produced pions. Soon after Powell's discovery, for which he received the 1950 Nobel Prize, pions were also produced in high-energy particle accelerators.

FIGURE 30.12 (left) Hideki Yukawa (1907–1981), the Japanese physicist who received the 1949 Nobel Prize in physics for his 1935 prediction of an unstable particle with mass about 200 times the electron mass, which played a key role in holding the nucleus together. This photo was taken in 1950 in Yukawa's office at Columbia University in New York City. (*Photo by International News Photos; courtesy of Edward L. Bafford Photography Collection, UMBC Library.*)

FIGURE 30.13 (right) Cecil Powell (1903–1969), the British physicist who used photographic emulsions to detect the pion, the particle first predicted by Yukawa. (*Photo courtesy of AIP Meggers Gallery of Nobel Laureates.*)

Pion reactions which occur frequently when protons are bombarded by high-energy protons are the following:

$$p + p \rightarrow p + p + \pi^0 \tag{30.9}$$

$$p + p \rightarrow p + n + \pi^+ \tag{30.10}$$

Note that charge is always conserved in such interactions, as are momentum and mass-energy. The extra energy required to produce the pions, and to change a proton into a neutron, comes from the large kinetic energy of the bombarding proton.

The unstable pions have very short lives and decay into *muons* and neutrinos according to the reaction

$$\pi^+ \rightarrow \mu^+ + \nu \tag{30.11}$$

A muon is lighter than a pion, and is very similar to the electron in almost all respects except that its mass is 207 times the electron mass. Muons are sometimes called "heavy electrons" for that reason. The muon was discovered in the cosmic radiation before the pion was discovered, and was at first mistaken for the particle predicted by Yukawa. It was soon found, however, that the muon played no role in strong interactions and hence could not be the Yukawa particle.

The positively charged muon is also unstable and decays into a positron and a neutrino-antineutrino pair with an average lifetime of 2.2×10^{-6} s. The negatively charged muon decays in a similar fashion into an electron and a neutrino-antineutrino pair. Most muons found in cosmic rays at sea level are produced by the decay of pions in the upper atmosphere.

30.4 Antiparticles and Neutrinos

In 1928 P. A. M. Dirac fused special relativity and quantum theory into relativistic quantum mechanics. This theory predicted the existence of the *positron*, or antielectron, which is identical to an electron in every way except that it has a positive rather than a negative charge. The positron was first

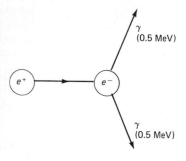

FIGURE 30.15 A pair-annihilation process. An electron and a positron collide, and all their mass is converted into energy in the form of two gamma rays, each with energy of about 0.5 MeV.

FIGURE 30.14 Carl Anderson (born 1905). In 1932 Anderson discovered the positron in a cloud-chamber photograph of cosmic rays. In 1935 he found the muon in the same way. Anderson was awarded the 1936 Nobel Prize in physics for these discoveries. (*Photo courtesy of AIP Niels Bohr Library.*)

observed in 1932 in a cloud chamber exposed to cosmic rays by the American physicist Carl Anderson (see Fig. 30.14), who a few years later also discovered the muon.

The positron is perfectly stable if left to itself, since there are no particles with lower mass than the positron except the neutrino and the photon, and these are uncharged. Hence a positron cannot decay into either of these without violating the principle of conservation of electric charge. If, however, a positron collides with an electron, they can "annihilate" each other, with their masses being entirely converted into energy in the form of two gamma rays with energies about 0.5 MeV each (Fig. 30.15). Similarly an electron-positron pair may be created out of pure energy. If gamma rays with energies above 1 MeV pass through matter, electron-positron pairs are usually produced.

Positrons are not normally observed because they soon collide with the readily available free electrons and are annihilated. If positrons could be kept away from ordinary matter, however, they would last as long as do electrons.

Other Antiparticles

Just as the electron has the positron as its antiparticle, so too all other particles like the proton, the neutron, and even the neutrino have their antiparticles, called the *antiproton*, the *antineutron*, and the *antineutrino*, respectively. These are usually designated by bars over the symbol for the particles, as $\bar{p}$, $\bar{n}$, and $\bar{\nu}$. For example, it is the antineutrino which arises from the beta decay of a neutron into a proton according to Eq. (30.5). The photon does not have an antiparticle, and is therefore said to be its own antiparticle.

In 1955 the American physicists Owen Chamberlain (born 1920) and Emilio Segrè (born in Italy in 1905) produced and detected antiprotons at the Bevatron accelerator in Berkeley, California. A short time later the antineutron was also found.

Atoms are possible which are made of nuclei containing antiprotons (which are negatively charged) and antineutrons, together with extranuclear positrons (which are positively charged). Such atoms make up what is called *antimatter*. As far as we can tell, there is no sizable amount of antimatter in the universe. If particles of matter and antimatter collide, they annihilate each other and the total mass of the two is converted into energy.

Different Kinds of Neutrinos

Neutrinos are always associated with weak interaction processes. In the beta decay of a nucleus either electrons or positrons can be emitted. Accompanying electrons are always antineutrinos, and accompanying positrons are neutrinos. There are other neutrinos which accompany the muons produced in pion decays. These muon neutrinos are different from electron neutrinos, and again are of two varieties, as shown in Table 30.2. To complicate matters even further, tau particles, which have their own neutrinos, have been discovered.

TABLE 30.2 Types of Neutrinos[*]

Kind of neutrino	Symbol	Associated with production of	Typical reaction
Electron neutrino	ν_e	Positrons	$p \rightarrow n + e^+ + \nu_e$
Electron antineutrino	$\bar{\nu}_e$	Electrons	$n \rightarrow p + e^- + \bar{\nu}_e$
Muon neutrino	ν_μ	μ^+	$\pi^+ \rightarrow \mu^+ + \nu_\mu$
Muon antineutrino	$\bar{\nu}_\mu$	μ^-	$\pi^- \rightarrow \mu^- + \bar{\nu}_\mu$

[*]For simplicity neutrinos associated with the tau particles are omitted from this table.

30.5 More New Particles

The tremendous energies available in cosmic rays and, since 1955, in particle accelerators have led to the discovery of large numbers of additional particles with exotic names like kaons, sigma particles, and eta mesons, most of which are highly unstable and have lifetimes of 10^{-8} s or less. Among these new particles are a group referred to as "strange particles" first detected in 1947 in the cosmic rays. These particles participate in strong interactions, but are strange in that they are always produced in pairs and have unexpected lifetimes.

All particles now known can be divided into three basic classes: the *photon*, which is unique since it is its own antiparticle; the *leptons*, which include the electron, the muon, and the neutrino; and the *hadrons*, which consist of a great variety of particles with masses equal to or greater than that of the proton.

Definitions

Leptons: Particles like electrons, muons, and neutrinos, with spin $\frac{1}{2}$, which do not undergo strong interactions but do undergo weak, electromagnetic, and gravitational interactions.

Hadrons: Particles which undergo strong interactions in addition to the other three. Hadrons include the mesons *and the* baryons.

Mesons: Hadrons with integral spin. Many, but not all, have masses between those of the electron and the proton.

Baryons: Hadrons with half-integral spin like the proton, neutron, and many particles heavier than the proton and neutron.

TABLE 30.3 Properties of the Hadrons

Type of particle	Name of particle	Symbol	Antiparticle	Rest mass, kg	Electric charge	Spin $\hbar$	Average lifetime
Mesons*	Pion (charged)	π^+	π^-	2.5×10^{-28}	± 1	0	2.6×10^{-8} s
	Pion (neutral)	π^0	Self	2.4×10^{-28}	0	0	0.8×10^{-16} s
Baryons†	Proton	p	$\bar{p}$	1.673×10^{-27}	± 1	$\frac{1}{2}$	More than 10^{31} years
	Neutron	n	$\bar{n}$	1.675×10^{-27}	0	$\frac{1}{2}$	1013 s

*Other mesons include the kaons and the eta meson.
†Other baryons include the lambda, sigma, xi, and omega particles.

Table 30.3 lists the more frequently observed hadrons. There are well over 100 hadrons in all, although many have very short lifetimes.

At one time nature seemed comparatively simple, with all the chemical elements made up of only three elementary particles. Now we have a bewildering array of new particles with a great variety of charges, masses, spins, and additional quantum numbers required to distinguish one from another. Where is the simplicity which physicists expect to find in nature? A ray of hope along these lines came in the early 1960s with independent proposals by the American physicists Murray Gell-Mann (see Fig. 30.16) and George Zweig that the many hadrons known to exist are actually made up of a small number of more fundamental particles which Gell-Mann called *quarks*. These quarks would, along with the photon and the leptons, be the true elementary particles.

FIGURE 30.16 Murray Gell-Mann (born 1929), an American theoretical physicist who is a world leader in elementary-particle physics at the present time. He developed a successful theory to explain the hadrons and first postulated the existence of quarks. He received the 1969 Nobel Prize in physics for his many contributions to the theory of the elementary particles. (*Photo courtesy of AIP Meggers Gallery of Nobel Laureates.*)

30.6 Quarks

Definition

A quark is a hypothetical fundamental particle of spin $\frac{1}{2}$ postulated to explain the structure of the hadrons.

Quarks are believed to have electric charges which are fractions (either one-third or two-thirds) of an elementary charge. Each quark also has its corresponding antiquark. The first quarks postulated were designated as u (up), with charge $+\frac{2}{3}$; d (down), with charge $-\frac{1}{3}$; and s (strange), with charge $-\frac{1}{3}$. The antiquarks to these three quarks were then indicated as $\bar{u}$, $\bar{d}$, and $\bar{s}$, with charges $-\frac{2}{3}$, $+\frac{1}{3}$, and $+\frac{1}{3}$, respectively.

In Gell-Mann's original scheme there were only these three distinct kinds of quark, but to explain the structure of all the hadrons more quarks than three are required, and the number is now up to at least six. Some quarks are called "charmed" and others "strange." Here *charmed* and *strange* are arbitrary terms for quantities which we cannot otherwise describe, but which are needed to remove difficulties in the theoretical formulation of the structure of the elementary particles, in the same way that the spin quantum number was needed to remove difficulties in the building up of the elements in the periodic table.

According to the quark theory, the proton and the neutron are each made up of three quarks. Thus the proton would consist of uud, and the neutron would be udd, since these combinations yield the correct charges for these two baryons.

In a similar fashion the mesons would be made up of quark-antiquark pairs. Thus the π^+ meson would be $u\bar{d}$, to yield the correct charge of $+1$ for the π^+.

Similar procedures can be used to explain the structure of hadrons heavier than the proton. The simplification introduced into physics by this quark model is equivalent to the simplification introduced into chemistry by the idea of the chemical elements, for it reduces over 100 hadrons to combinations of a much smaller number of quarks, which are presumed to be elementary particles, or at least more elementary than any of the hadrons, including the proton and the neutron.

This quark scheme, which has gradually become more complicated than the elegant one originally proposed by Gell-Mann, explains the hadrons quite well. There is also strong evidence from the scattering of electrons and neutrinos by protons and neutrons that hadrons do possess a structure which can be successfully explained on the quark model. This structure only becomes evident if the scattered particles have wavelengths small compared with the size of the proton. If, for example, the incident particle has a wavelength of about 10^{-15} m, which is about the proton size, the proton will appear to be a pointlike, featureless sphere, as in Fig. 30.17. If, however, the incident particles have wavelengths an order of magnitude smaller, that is, 10^{-16} m, then the proton might appear as shown in Fig. 30.18, where the three quarks making up the proton can be seen. The quarks themselves appear to be at most 10^{-17} m in diameter, so that the proton, with a diameter of 10^{-15} m, again seems to consist mainly of empty space, just as an atom does when we consider the size of the nucleus compared with the overall size of the atom.

Strong evidence for the presence of quarks in the proton comes from an experiment carried out on the SPS accelerator at CERN using a large bubble chamber as a detector. In the experiment 200-GeV neutrinos, which have wavelengths of slightly under 10^{-17} m, were produced from the decay of mesons obtained by allowing the main proton beam from the accelerator to hit a beryllium target. These neutrinos struck protons in the target and were scattered. The experimental results indicated an interaction between two

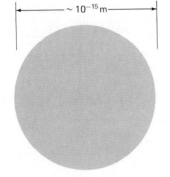

$\sim 10^{-15}$ m

FIGURE 30.17 The proton, as seen by an incident particle of wavelength about 10^{-15} m.

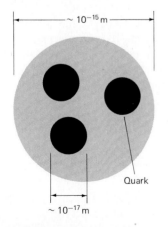

$\sim 10^{-15}$ m

Quark

$\sim 10^{-17}$ m

FIGURE 30.18 The proton, as seen by an incident particle of wavelength about 10^{-16} m.

pointlike particles, the one inside the proton being considerably smaller than the proton itself. This is presumed to be a quark. These experiments also point to the number of scattering centers inside the proton as being close to 3, in good agreement with the quark theory.

The similarity of such experiments to the one on the scattering of alpha particles by nuclei (carried out many years earlier by Rutherford and his students) is worth noting.

Additional evidence for the presence of a substructure in the proton comes from the fact that most structured systems like atoms, molecules, and nuclei can exist in excited energy states, because the particles making up the systems can organize themselves in different ways which correspond to different energies. Now scattering experiments reveal that the proton has excited states, just as does a mercury atom or a lithium nucleus, as shown in Fig. 30.19. Hence it is reasonable to assume that the proton consists of more than one particle, which can interact in different ways to produce different energy states of the system. These particles are presumed to be quarks.

FIGURE 30.19 Excited states of (a) a mercury (Hg) atom, (b) a lithium (Li) nucleus, (c) a proton.

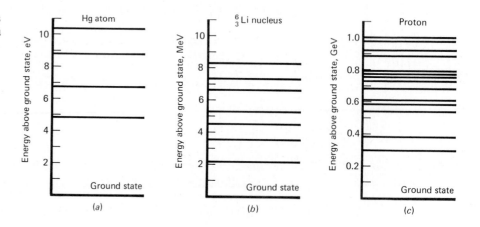

Problems with the Quark Hypothesis

There were two main difficulties with the original quark hypothesis. First, no one had ever seen a free quark; and second, the quark hypothesis seemed to conflict with the Pauli exclusion principle as applied to particular hadrons.

The existence of free quarks is still an open question. If quarks existed for even a very short time outside a nucleus, they could easily be detected because of their fractional charge. W. Fairbanks at Stanford University claims to have detected charges of one-third and two-thirds times the electron charge on niobium spheres, which would be a clear indication of the existence of free quarks, but as yet no other physicists have been able to confirm his results. Some theorists are skeptical about Fairbank's results, because they believe that current theories exclude the possibility of quarks existing freely outside of hadrons, since the force between two quarks appears to confine the quarks within the hadron.

A new version of Millikan's oil-drop experiment is being carried out at Berkeley, using the technology of ink-jet printing, in an attempt to find quarks. Small drops of a conducting liquid like mercury are ejected in a stream and fall about 10 ft in air through a region in which a strong electric field exists. The electric field deflects each drop by an amount proportional to

TABLE 30.4 Ultimate Elementary Particles (1983)

Particle doublets				Antiparticle doublets			
Particle	Symbol	Charge, units of $e =$ 1.6×10^{-19} C	Mass, units of $m_e =$ 9.1×10^{-31} kg	Particle	Symbol	Charge e	Mass m_e
Leptons:							
Electron	e^-	-1	1	Positron	e^+	$+1$	1
Electron neutrino	ν_e	0	~0 (?)	Electron antineutrino	$\bar{\nu}_e$	0	~0 (?)
Negative muon	μ^-	-1	206	Positive muon	μ^+	$+1$	206
Muon neutrino	ν_μ	0	~0 (?)	Muon antineutrino	$\bar{\nu}_\mu$	0	~0 (?)
Negative tau	τ^-	-1	~3500	Positive tau	τ^+	$+1$	~3500
Tau neutrino	ν_τ	0	~0 (?)	Tau antineutrino	$\bar{\nu}_\tau$	0	~0 (?)
Quarks:							
Up quark	u	$+\frac{2}{3}$	~6.6×10^2	Up antiquark	$\bar{u}$	$-\frac{2}{3}$	~6.6×10^2
Down quark	d	$-\frac{1}{3}$	~6.6×10^2	Down antiquark	$\bar{d}$	$+\frac{1}{3}$	~6.6×10^2
Charmed quark	c	$+\frac{2}{3}$	~3.6×10^3	Charmed antiquark	$\bar{c}$	$-\frac{2}{3}$	~3.6×10^3
Strange quark	s	$-\frac{1}{3}$	~1.1×10^3	Strange antiquark	$\bar{s}$	$+\frac{1}{3}$	~1.1×10^3
Top quark*	t	$+\frac{2}{3}$	~10^5	Top antiquark	$\bar{t}$	$-\frac{2}{3}$	~10^5
Bottom quark	b	$-\frac{1}{3}$	~10^4	Bottom antiquark	$\bar{b}$	$+\frac{1}{3}$	~10^4

*The first tentative evidence for the existence of the top quark came from a series of experiments concluded at CERN on July 4, 1983.

its charge, so that the stream of drops is fanned out into a series of lines on a sheet of paper, with each line corresponding to the same integral value of the elementary charge on a drop. If an occasional drop carries a free quark, then ink spots will occur between the lines for the integral-charge values, revealing the presence of nonintegral charges. This method can sift through thousands of drops in a few seconds, and seems to be a promising way of settling once and for all the question of whether free quarks exist.

The conflict with the Pauli exclusion principle has been removed by assigning new properties like "color" to quarks, even though we are not sure what the word *color* refers to in this context, except that, like spin, it is a useful quantum number. In other words, elementary-particle theorists have saved the theory by introducing more complications into it. There is some concern that Gell-Mann's beautifully simple and elegant quark theory is becoming excessively complicated as it tries to accommodate the ever-growing collection of new particles and new data produced by high-energy experimental physics.

Table 30.4 lists the ultimate elementary particles (exept for the photon) which are believed to constitute all of matter. Note how a nice symmetry appears to exist between the leptons and the quarks (which make up the protons and neutrons), in that in each case we have three pairs of particles, together with their associated antiparticles.

30.7 Conservation Principles for Elementary Particles

A conservation principle has the same significance for the elementary particles as for large-scale physical phenomena. It is a statement that some particular quantity of a system (charge, angular momentum, etc.) remains constant in time if the system is free from external interference. For the elementary particles there are the usual conservation principles which are absolute and must be satisfied for all reactions involving these particles. These include conservation of mass-energy, of linear momentum, of angular momentum, and of electric charge.

With the elementary particles there are other conservation principles introduced to describe what actually seems to happen in elementary-particle interactions, even though there is no fundamental understanding of why such principles should be valid. Physicists proceed on the assumption that any elementary-particle event which can happen will indeed be observed if we wait long enough and try hard enough to find it. If after all this effort the event is never seen, it is assumed that the event must be disallowed because it violates some new conservation principle, such as the principle of conservation of baryon number, which we will discuss below. In this way empirical (i.e., based on experimental observations) and somewhat tentative conservation principles are introduced as ordering principles in sorting out the elementary particles and their interactions. In many cases it turns out that, although these principles provide good working guidelines, they may not be universally valid and may only apply to one class of interactions (e.g., the strong interactions) and not to another class of interactions (e.g., the weak interactions).

We will discuss only two conservation principles of this latter, less general type—the conservation of baryon number and the conservation of parity. Other conservation principles frequently discussed in physics books and articles include the conservation of hypercharge, of strangeness, and of quark number.

Conservation of Baryon Number

The following reaction has never been observed, even though it could occur in such a way as to satisfy the four absolute conservation principles.

$$p \rightarrow e^+ + \gamma \tag{30.12}$$

To account for the absence of this seemingly possible event, physicists have posited a new conservation principle, the *conservation of baryon number*:

Conservation of Baryon Number

In any elementary-particle event, the number of baryons must be the same before and after the event.

In the event described by Eq. (30.12) the left side of the equation contains one baryon, the proton, but the right side contains no baryons. Hence this interaction is forbidden by this new conservation principle.

To allow for the presence of antiprotons, they are assigned baryon numbers of -1, whereas a proton has a baryon number of $+1$. Hence in a collision of a proton with an antiproton, the following equation might apply:

$$p + \bar{p} \rightarrow \gamma + \gamma' \tag{30.13}$$

where γ and γ' are gamma rays of energy and direction required to satisfy the absolute conservation principles. The baryon number on the left side is $1 - 1 = 0$, and on the right side it is also zero; hence baryon number is conserved and this process can occur. Indeed it does occur, and has been frequently observed.

Similarly, the decay of a neutron into a proton is permitted, since the following scheme satisfies the conservation-of-baryon-number principle:

$$n \rightarrow p + e^- + \bar{\nu}_e \qquad (30.14)$$

On the other hand, the interaction

$$p + n \rightarrow p + p + \bar{p} \qquad (30.15)$$

cannot occur, since the baryon number on the left is 2, and on the right it is only 1. The principle of conservation of baryon number is therefore violated by such an interaction, and it does not occur.

Recently physicists have been having some second thoughts about the universal validity of this principle. Theories of the elementary particles indicate that the proton may not be perfectly stable, and hence may not have an infinite lifetime, but one of only about 10^{32} years! Therefore it is possible that the proton does infrequently decay into leptons and thus violate the principle of conservation of baryon number. Eight different experiments are now being set up to detect proton decay by physicists in India, France, Italy, the United States, and the Soviet Union in mines deep below the earth's surface. From 1000 tons of water about three proton decays a year may be expected if these theories are correct. These rare events must be sorted out from a variety of background events due to cosmic rays and other sources, and deep mines are used to provide shielding against many of these sources. Experimentalists are convinced that their equipment will be sufficiently sensitive to observe proton decays if they do indeed occur.

Conservation of Parity

By *parity* we mean a fundamental property of a system which describes the observed behavior of the system when it is reflected in a mirror. The principle of conservation of parity is then stated as follows:

Conservation of Parity

For a system in which parity is conserved, the mirror image of any physical process cannot be distinguished from the process itself.

Up until 1956 physicists assumed that parity was conserved in all processes involving nuclear and elementary-particle processes. Then in 1956 the American physicists T. D. Lee and C. N. Yang (see Fig. 30.20) suggested that parity might only be conserved in strong interactions but not in weak interactions such as beta decay. In December 1956 a group of physicists at the National Bureau of Standards led by Professor C. S. Wu of Columbia University (Fig. 30.21), carried out an experiment to see if parity was indeed conserved in beta decay.

The experiment performed by Wu's group consisted of placing $^{60}_{27}\text{Co}$ nuclei in a strong magnetic field in which the magnetic moment due to the spinning nucleus was lined up in the direction of the magnetic field. Suppose the nucleus emitted a beta ray (an electron) in the direction of the magnetic field, and hence in the direction of its nuclear spin vector. As seen in a mirror, the nucleus would appear to be emitting the electron in the direction opposite to that of its spin, as shown in Fig. 30.22, since the reflection in the mirror would reverse the spin direction of the nucleus. The mirror image could only

FIGURE 30.20 C. N. Yang (left; born 1922) and T. D. Lee (born 1926). Lee came to the United States from China in 1946, a year after Yang had immigrated. In 1956 they proposed that parity was not conserved in weak interactions. When this was confirmed almost immediately in the laboratory, they shared the 1957 Nobel Prize in physics. They were the first scientists of Chinese birth to win the Nobel Prize. (*Photo by Alan Richards; courtesy of AIP Niels Bohr Library.*)

FIGURE 30.21 C. S. Wu (born in China in 1912). Shortly after the publication of Lee and Yang's article on the nonconservation of parity in weak interactions, Wu directed a group of physicists who showed experimentally that Lee and Yang were indeed right. Wu has made several other important contributions to the study of beta decay. (*Photo courtesy of AIP Niels Bohr Library.*)

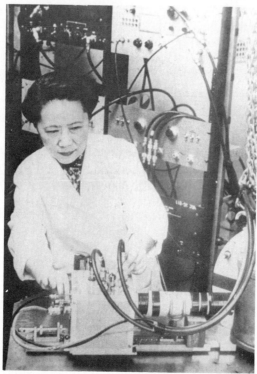

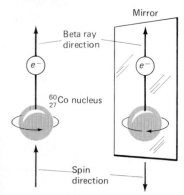

FIGURE 30.22 Conservation-of-parity experiment. $_{27}^{60}$Co nuclei were placed in a strong magnetic field at a low temperature, and the number of beta particles emitted in the spin direction and opposite to the spin direction were counted. The results indicated that more beta particles were emitted opposite to the spin direction than in the spin direction.

look the same as the original beta-decay process if the nucleus emitted as many electrons in the direction opposite to its spin as it did in the direction parallel to its spin, for the same distribution of electrons would then be observed when the spin was reversed on reflection in the mirror.

In most beta-decay experiments the emitting nuclei are oriented at random and hence it is impossible to tell if there is a preferred direction for the emerging electrons in beta decay. Wu and her colleagues overcame this problem by using a very strong magnetic field at a very low temperature (0.01

K) to align the cobalt nuclei. The experiment showed clearly that more particles were emitted opposite to the spin direction of the nucleus than in the spin direction. Hence on reflection in a mirror more electrons would be emitted in the spin direction than in the opposite direction. Thus the mirror image of the beta-decay process could be clearly distinguished from the process itself, and so parity would not be conserved.

Since beta decay is a weak interaction, the conclusion is that parity is not conserved in weak interactions. It appears that parity conservation remains perfectly valid, however, for strong interactions.

30.8 The Future of Elementary-Particle Physics

Elementary-particle physics will undoubtedly continue to progress, as do all fields of physics, by combining careful and creative experimental work with inspired model building and theoretical analysis.

Theoretical Analysis

The main problem faced by elementary-particle theory is that there are too many new particles, too many quarks, and too many unanswered questions about what goes on inside the nucleus. Considerable progress has been made in recent years in answering some of these questions by applying to high-energy interactions a theory called the *standard model*. The standard model consists of two parts, the *electroweak theory*, which applies to the weak interactions (see Table 30.5) and *quantum chromodynamics* (QCD), which applies to the strong interactions.

TABLE 30.5   Types of Interactions and Associated Quanta

Interaction	Particles involved	Quanta exchanged
Strong nuclear	Nucleons, quarks	Gluons
Electromagnetic	Leptons, nucleons	Photons
Weak nuclear	Leptons	Intermediate vector bosons ($W^{\pm}$, Z^0)
Gravitational	All particles	Gravitons

The electroweak theory has made a first giant step forward in the attempt to unify the four basic sources of physics. Steven Weinberg, Abdus Salam, and Sheldon Lee Glashow (see Figs. 30.23 to 30.25) shared the 1979 Nobel Prize in physics for developing this theory, which combines the electromagnetic and weak interactions into a single "electroweak" interaction. They were able to show that these two forces are really different aspects of the same force.

One of the basic predictions of the electroweak theory was that there should be new particles, called *intermediate vector bosons*, which act as the quanta for the weak interactions; i.e., they are the carriers of the weak nuclear force. To attempt a qualitative explanation, let us consider what holds a proton and an electron together in the hydrogen atom. We have seen that it is the electromagnetic field between the proton and the electron which binds them together. But theoretical physicists have shown that associated with every field there is a field quantum which in this case is the *photon*, the quantum of electromagnetic energy. Hence the attractive force between the proton and the electron may be considered as due to the constant interchange

FIGURE 30.23 (left) Steven Weinberg (born 1933), an American physicist who received the Nobel Prize in physics in 1979 for his theory of the unification of the electromagnetic and the weak nuclear interactions. (*Photo courtesy of AIP Niels Bohr Library.*)

FIGURE 30.24 (right) Abdus Salam (born in Pakistan in 1926). This theoretical physicist, who has done most of his work in Great Britain, worked independently on the unification of the electromagnetic and the weak nuclear interactions, and shared in the 1979 Nobel Prize in physics. Since 1964 he has also served as director of the International Center for Theoretical Physics at Trieste, Italy. (*Photo courtesy of AIP Niels Bohr Library.*)

FIGURE 30.25 Sheldon Glashow (born 1932), who shared the 1979 Nobel Prize in physics for his contributions to the electroweak theory. Glashow was a classmate of Weinberg both at the Bronx High School of Science and at Cornell University. (*Photo courtesy of AIP Niels Bohr Library, Physics Today Collection.*)

of photons between the proton and the electron, much as two ice skaters interact and exert forces on each other by throwing a softball back and forth, as in Fig. 30.26. Something similar happens in the interaction between a proton and an electron, as illustrated in a diagram called a *Feynman graph* in Fig. 30.27.

In a similar fashion the weak nuclear force, which acts in beta-decay processes and has a range of only 10^{-19} m, can be viewed as involving an *intermediate vector boson*, as shown in a Feynman diagram in Fig. 30.28. The intermediate vector bosons act as the quanta of the field corresponding to the weak nuclear force, just as the photon acts as the quantum for the electromagnetic field.

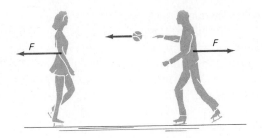

FIGURE 30.26 Two skaters exchanging a softball. In this case the resultant interaction is repulsive, as dictated by the conservation-of-momentum principle.

FIGURE 30.27 A Feynman graph of the interaction between a proton and an electron in a hydrogen atom. The exchange of photons between proton and electron produces the attractive force which holds the hydrogen atom together.

FIGURE 30.28 A Feynman graph of a weak interaction in which a neutron decays into a proton, an electron, and an antineutrino. In this case the field quantum for the weak nuclear force is the intermediate vector boson W^-.

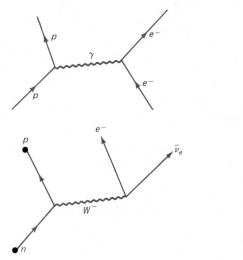

The electroweak theory predicted that there should be three intermediate vector bosons, two charged bosons designated as W^+ and W^- (W means *weak*), with masses of about 80 GeV, or 85 proton masses,* and one uncharged boson, with a mass of about 90 GeV, or 96 proton masses. The experiments at CERN using the head-on collision of a 270-GeV proton beam with a 270-GeV antiproton beam in early 1983 produced particles with almost exactly these masses. Hence the electroweak theory appears to be on very solid ground.

Quantum chromodynamics (QCD) explains the strong forces between quarks as caused by the exchange of new particles called *gluons*, where a gluon is the quantum for the strong interactions, as the photon is the quantum for the electromagnetic field. The strong force between nucleons is then explained as a spillover from the very strong forces between the quarks making up each nucleon, in somewhat the same fashion that attractive forces between atoms in molecules can be explained as spillovers from the electron-proton forces in each atom. QCD has achieved considerable success in explaining qualitatively the many new particles found by the giant accelerators.

In recent years physicists have been working on grand unified theories (GUT) to unite the electroweak force with the strong nuclear force. (Such theories lead to the prediction that the proton should have a lifetime of 10^{32}

---

*Note that this is about the mass of a krypton atom, with a nucleus $^{84}_{64}\mathrm{Kr}$.

years, as mentioned previously.) If these grand unified theories should work out, the last step remaining would be to include the gravitational force in a theory embracing all four fundamental forces of nature. Whether this can be achieved is debatable, since the gravitational force is so weak compared with the other basic forces (see Table 30.1) that it is hard to find a role for it inside the nucleus. Albert Einstein devoted 35 years of his life to developing a unified theory of the gravitational and electromagnetic fields, and died before accomplishing his goal. Of course, both more data and better theories are available now than were available in Einstein's day.

It now appears that the four basic forces of physics are related to the origin of the universe. It is thought that at the high temperatures that characterized the newborn universe, all four basic forces manifested themselves in exactly the same symmetrical form. As the universe cooled off, however, the four forces and their associated quanta became distinct, and the symmetry was "frozen out." Theoreticians refer to this as *symmetry breaking*. This connection indicates that we may need to understand better the origin of the universe before we can understand the elementary particles. Or perhaps increasing our knowledge of the elementary particles will cast additional light on how the universe began. In either case, it is clear that a trend toward unification pervades physics today.

Experimental Work

The observation in early 1983 of the intermediate vector bosons with colliding beams at the CERN accelerator by two large international groups of physicists shows the power of modern, large-scale, high-energy physics. An effective energy of 540 GeV was produced by the head-on collision of a proton and an antiproton beam. Over 1 billion collisions were automatically analyzed by counters and computers to turn up five or six events which seemed to identify the W^+ and W^- intermediate vector bosons unambiguously. Only one Z^0 was identified, but more are expected to be seen when the experiment runs for longer periods of time.

Successes of this sort indicate that we can expect an increasingly more detailed picture of the elementary particles as accelerators move to higher and higher energies, and as methods of detection, counting, and data analysis improve.

The last 50 years have shown a very rapid increase in the energies achieved by particle accelerators, and hence in their size and cost. Figure 30.29 shows this growth on a logarithmic scale. The growth rate is seen to have been much greater than linear with time, although it has slowed slightly in recent years.

Present accelerators can produce particles with sufficient energies to "see" structure at the level of about 10^{-16} m. How much farther will we be able to go? An increase in resolution by a factor of 10 will mean accelerators capable of producing particles with energies in the 5-TeV range. Such accelerators would have diameters of about 25 km, unless some entirely new principle or type of accelerator is developed. Such an accelerator, with a circumference of about 50 mi, becomes questionable from the viewpoint of both cost and land use, although the Russians have decided to build a large, fixed-target proton machine, 20 km in circumference, which they expect to produce beam energies up to 3 TeV. Even now few countries can afford such large accelerators, and it is likely that future accelerators will be limited to the United States, Russia, and the consortium of European nations operating

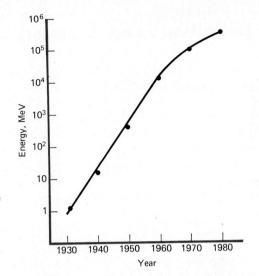

FIGURE 30.29 Maximum energy achieved by year for large particle accelerators. Note that the energy is plotted on a logarithmic scale, so that the true growth rate is far greater than linear with time.

CERN. Some day, however, the construction of larger and larger accelerators will become impossible for any single nation, and the result may well be one international center for research in elementary-particle physics, supported by all countries wishing to use the gigantic accelerator which would be the heart of such a center.

The Future

Will Einstein's faith in the simplicity of the universe, expressed at the beginning of this chapter, be justified? Only time will tell. Time is needed to decide whether elementary-particle physics is entering a period of triumphant clarification of the nature of the elementary particles or of growing confusion and disillusionment.

Physics has always progressed by asking the right questions. In elementary-particle physics these are questions such as: Why are the masses of the electron and proton what they are? Are the proton and neutron actually made up of quarks? What are the truly elementary particles? Many extremely gifted physicists—both experimentalists and theoreticians—are striving earnestly to answer these questions. Most believe, with Einstein, that the answers will be uncovered when we succeed in peeling away the layers of matter and energy cloaking nature's simplicity.

Physicists continue to believe that the study of the elementary particles is the best route to understanding our universe. Whether or not this should turn out to be true, they know from history and experience that, whatever the future of physics may be, it will certainly not be dull.

Summary: Important Definitions and Equations

The four forces of physics

Long-range forces:

    Gravitational: Present in all interactions of any kind.

    Electromagnetic: Present in all interactions involving two electrically charged particles.

Short-range forces:

    Strong nuclear: Present in all strong interactions (nuclear binding).

    Weak nuclear: Present in all weak interactions (beta decay).

Particle accelerators

Linear accelerators: Machines which accelerate charged particles (usually electrons) down a long, straight evacuated tube.

Synchrotrons (synchrocyclotrons): Machines which accelerate charged particles (usually protons) to high energies by using a magnetic field (B) to keep them moving in a circular orbit of fixed radius (r) as they acquire higher and higher speeds; $E = Bqrc$.

New particles

Hadrons:

    Pions: Particles with masses about 275 times the electron mass.

Leptons:

    Muons: Particles with masses about 207 times the electron mass.

    Positrons: The antiparticles of the electron; identical to an electron in every way except that their charge is positive.

    Electron neutrinos: Particles of very small mass and no charge which are produced in beta decay. Of the particles emitted, neutrinos accompany the positrons and antineutrinos accompany the electrons.

    Muon neutrinos: Particles of very small mass and no charge which accompany the emission of muons in pion decay. Neutrinos accompany the μ^+ and antineutrinos the μ^- produced.

Classes of elementary particles

1. Photon: A particle with zero rest mass; its own antiparticle.
2. Leptons: Particles with spin $\frac{1}{2}$, which undergo weak interactions but not strong interactions.
3. Hadrons: Particles which undergo strong interactions; mesons are hadrons with integral spin, and baryons are hadrons with half-integral spin.

Quarks: Hypothetical elementary particles of spin $\frac{1}{2}$ postulated to explain the structure of the hadrons. Quarks have electric charges of either $+\frac{2}{3}$ or $-\frac{1}{3}$.

Some conservation principles applied to elementary particles

Absolute: Conservation of mass-energy, of linear momentum, of angular momentum, and of charge.

Empirical:

    Conservation of baryon number: In any elementary-particle event, the number of baryons must be the same before and after the event.

    Conservation of parity: For a system in which parity is conserved, the mirror image of any physical process cannot be distinguished from the process itself. Parity is conserved in strong interactions but not in weak interactions.

Questions

1 If the gravitational force is by far the weakest force known in nature, why has it been known longer than the electromagnetic or the strong and weak nuclear forces, and why do we experience it more in our daily lives than any of the other four forces?

2 List those interactions (from the four listed in Table 30.1) in which (*a*) the electron participates; (*b*) the proton participates; (*c*) the neutrino participates.

3 It has been stated that, in the history of high-energy physics, less than 1 g of protons has ever been accelerated to energies above 1 GeV. Does this seem reasonable to you? Why?

4 Why are linear accelerators better than synchrotrons for accelerating electrons?

5 Why is the head-on collision of two beams of 1-GeV protons more useful to high-energy physicists than the collision of a 2-GeV proton beam with a stationary target?

6 Explain how the introduction of the idea of the neutrino in beta decay can save the laws of conservation of energy and of angular momentum.

7 What particles do the following combinations of quarks correspond to: (*a*) *uud*, (*b*) *uus*, (*c*) *du*?

8 Indicate the combinations of quarks which will produce (*a*) a neutron, (*b*) an antineutron.

9 Show that, if the photon is the quantum of the electromagnetic field, it must have zero mass.

10 Why would you expect a massless particle like the photon to be stable and, if left to itself, not to decay into other particles?

11 Comment on the following statement by the British novelist C. P. Snow: "For example, most physicists feel in their bones that the present bizarre assembly of nuclear particles, as grotesque as a stamp collection, can't possibly be, in the long run, the last word." (*The Physicists*, Little, Brown, New York, 1981, p. 182.)

(*a*) Do you think that Snow's statement is accurate?

(*b*) Do you find any convincing evidence from this chapter or from your other readings that physicists are making progress in bringing order out of the chaos of the elementary particles?

12 Do you think that relativity and quantum mechanics will remain valid as we get down to distances as small as 10^{-16}m inside the nucleus? Support your answer with evidence from the history of physics.

Additional Readings

Bernstein, Jeremy: "A Question of Parity," *The New Yorker*, vol. 38, May 12, 1961, pp. 49–104. Contains excellent profiles of Lee and Yang and an account of the development of their ideas on parity nonconservation.

Feinberg, Gerald: *What Is the World Made Of?*, Doubleday Anchor, Garden City, N.Y., 1978. An up-to-date discussion of hadrons, quarks, and other newly discovered particles.

Morrison, Philip: "The Neutrino," *Scientific American*, vol. 194, no. 1, January 1956, pp. 58–68. An account of the reasons behind physicists' confidence that the neutrino existed long before it was discovered by Reines and Cowan in 1956.

Mulvey, J. H. (ed.): *The Nature of Matter*, Oxford University Press, New York, 1981. A series of outstanding popular lectures by leading elementary-particle physicists including Nobel Laureates Salam and Gell-Mann.

Polkinghorne, J. C.: *The Particle Play*, Freeman, San Francisco, 1981. A historical overview of the search for elementary particles.

Sulak, Lawrence R.: "Waiting for the Proton to Decay," *American Scientist*, vol. 70, no. 6, 1982, pp. 616–625. An account of extremely difficult and costly experiments on the basis of which many modern theories of the elementary particles may stand or fall.

Trefil, James S.: *From Atoms to Quarks; An Introduction to the Strange World of Particle Physics*. Charles Scribner's Sons, New York, 1980. An accurate, popular book on the elementary particles.

Weber, Robert L: *Pioneers of Science*, Institute of Physics, Bristol, 1980. This book contains brief biographies of many of the physicists mentioned in this chapter.

Weinberg, Steven: *The First Three Minutes*, Basic Books, New York, 1977. A fascinating account of the connection between the first 3 min of the universe and the elementary particles which exist today.

Weisskopf, Victor F.: "The Three Spectroscopies," *Scientific American*, vol. 218, no. 5, May 1968, pp. 15–29. A stimulating discussion of the similarities that exist among the atomic, nuclear, and subnuclear domains of physics.

Yang, Chen Ning: *Elementary Particles*, Princeton University Press, Princeton, N.J., 1962. A series of lectures which includes a good discussion of parity nonconservation.

In addition, *Scientific American* contains a large number of articles by experts in the fields of high-energy and elementary-particle physics. Some of the best articles for the period 1953 to 1979 are included in *Particles and Fields* (readings from *Scientific American*, with an introduction by William J. Kaufmann III), Freeman, San Francisco, 1980. *Scientific American* articles since 1979 include Haim Harari, "The Structure of Quarks and Leptons," April 1983; Steven Weinberg, "The Decay of the Proton," June 1981; and Robert R. Wilson, "The Next Generation of Particle Accelerators," January 1980.

Appendix 1
Review of Basic Mathematics

1.A Algebra

Any *quadratic equation* may be reduced to the form $ax^2 + bx + c = 0$, with a solution

$$x = \frac{-b \pm \sqrt{b^2 - 4ac}}{2a}$$

1.B Geometry

Length of arc of a circle which subtends an angle θ (in radians): $l = r\theta$

Plane object	Area	Circumference
Rectangle with sides a and b	ab	
Triangle with base b and altitude h	$\dfrac{bh}{2}$	
Circle of radius r	πr^2	$2\pi r$
Ellipse with semiaxes a and b	πab	$2\pi\left(\dfrac{a^2 + b^2}{2}\right)^{1/2}$

Solid object	Surface area	Volume	Cross-sectional area
Sphere of radius r	$4\pi r^2$	$\dfrac{4}{3}\pi r^3$	πr^2
Cylinder of radius r and height h	$2\pi r^2 + 2\pi rh$	$\pi r^2 h$	πr^2

Useful Theorems from Geometry

1 The sum of the angles in any triangle is 180°.

2 Two triangles are *similar* if any two angles in one triangle are equal to two angles in the other triangle.

3 The corresponding sides of similar triangles are proportional.

4 The two acute angles in a right triangle add to 90° and are therefore *complementary* angles.

5 Two angles are equal if any one of the following is true:
  a Their sides are parallel and in the same direction.
  b Their sides are mutually perpendicular.
  c Each of the two angles has the same complementary angle.
  d The two are vertical angles formed by the same two straight lines.

6 *Pythagorean theorem:* In a *right triangle* with sides a, b, and c, where c is the side opposite the right angle:

$$c^2 = a^2 + b^2$$

1.C Trigonometry

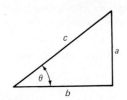

FIGURE A.1   A right triangle used to define the trigonometric functions.

In the right triangle shown in Fig. A.1, where c is the hypotenuse:

$$\text{sine } \theta = \sin \theta = \frac{\text{opposite side}}{\text{hypotenuse}} = \frac{a}{c}$$

$$\text{cosine } \theta = \cos \theta = \frac{\text{adjacent side}}{\text{hypotenuse}} = \frac{b}{c}$$

$$\text{tangent } \theta = \tan \theta = \frac{\text{opposite side}}{\text{adjacent side}} = \frac{a}{b}$$

Since, by the pythagorean theorem, $c^2 = a^2 + b^2$, we have, from the above,

$$c^2 = c^2 \sin^2 \theta + c^2 \cos^2 \theta$$

or $\sin^2 \theta + \cos^2 \theta = 1$ (for all angles)

For any angle θ:

$$\sin \theta = \cos (90° - \theta) = \sin (180° - \theta)$$

$$\cos \theta = \sin (90° - \theta) = -\cos (180° - \theta)$$

For any two angles A and B:

$$\sin (A + B) = \sin A \cos B + \cos A \sin B$$

$$\sin (A - B) = \sin A \cos B - \cos A \sin B$$

$$\cos (A + B) = \cos A \cos B - \sin A \sin B$$

$$\cos (A - B) = \cos A \cos B + \sin A \sin B$$

$$\sin 2A = 2 \sin A \cos A$$

$$\cos 2A = \cos^2 A - \sin^2 A = 2 \cos^2 A - 1 = 1 - 2 \sin^2 A$$

For any triangle with sides a, b, and c opposite the angles A, B, and C, respectively:

Law of sines: $\dfrac{a}{\sin A} = \dfrac{b}{\sin B} = \dfrac{c}{\sin C}$

Law of cosines: $c^2 = a^2 + b^2 - 2ab \cos C$

1.D Series Expansions

Binomial expansion:

$$(a + b)^n = a^n + na^{n-1}b + \frac{n(n - 1)}{2}a^{n-2}b^2 + \cdots$$

Hence $\left(\dfrac{a + b}{a}\right)^n = \left(1 + \dfrac{b}{a}\right)^n = 1 + \dfrac{nb}{a} + \dfrac{n(n - 1)}{2}\dfrac{b^2}{a^2} + \cdots$

If b/a is *small*, and we let $b/a = c$, then:

$$(1 + c)^n \simeq 1 + nc$$

$$(1 - c)^n \simeq 1 - nc$$

1.E Logarithms and Exponential Functions

By the *logarithm* or *log* of a number we mean the power to which some given base must be raised to yield that number. Ordinary or "common" logarithms use the number 10 as a base. Then, if $y = 10^x$, we have

$$\log_{10} y = \log y = x$$

since x is the power to which 10 must be raised to yield y. The subscript 10, included here to indicate clearly the base used, is usually omitted, since log is taken conventionally to mean "log to the base 10." For example, since $100 = 10^2$, we have $\log 100 = 2$, because 2 is the power to which the base 10 must be raised to yield the number 100.

There is another system of logarithms, called *natural* or *Naperian* logarithms, in which the base is not 10 but the exponential base $e = 2.718218$, which is the sum of the infinite series

$$e = 1 + \frac{1}{1} + \frac{1}{1 \cdot 2} + \frac{1}{1 \cdot 2 \cdot 3} + \frac{1}{1 \cdot 2 \cdot 3 \cdot 4} + \cdots$$

The natural logarithm is written as "ln" to distinguish it from a logarithm to the base 10. If $y = e^x$, then $\ln y = x$, for again x is the power to which e must be raised to yield y. For example, suppose $y = e^2 = (2.718)^2 = 7.39.$* Then

$$\ln y = 2 = \ln 7.39 \qquad \text{or} \qquad \ln 7.39 = 2$$

Since $\log 10 = 1$ and $\ln 10 = 2.3026$, we can write

$$\ln 10 = 2.3026 \log 10$$

This is true in general for any number N. Hence $\ln N = 2.3026 \log N$.

Rules for Use of Logarithms (to Any Base):

1 Multiplication: If $N = AB$, then $\log N = \log (AB) = \log A + \log B$

2 Division: If $N = \dfrac{A}{B}$, then $\log N = \log \dfrac{A}{B} = \log A - \log B$

3 Exponentiation: If $N = A^n$, then $\log N = n \log A$

For example:

$$\log 10^x = x \log 10 = x(1) = x$$

$$\ln 10^x = 2.3026 \log 10^x = 2.3026x$$

$$\ln e^x = x \ln e = x(1) = x$$

$$\log e^x = \frac{1}{2.3026} \ln e^x = \frac{x}{2.3026}$$

With hand and desk calculators so readily available, neither logarithms nor the engineering slide rule (which is based on logarithms) are much used any more to perform routine multiplications and divisions, or to raise numbers to powers. Logarithms are still very useful in many fields of physics, however, where important physical quantities often increase or decrease *exponentially*, and in laboratory work where the functional dependence of one physical quantity on another is often being investigated.

For example, suppose the activity of a radioactive sample decreases exponentially according to the equation $N = N_0 e^{-\lambda t}$ where N is the number of

---

*It is useful to remember that $e^3 \simeq 20$. Hence, for example, $e^{18} = (e^3)^6 \simeq (20)^6 \simeq 6.4 \times 10^7$.

counts made by a Geiger counter per second (a measure of the activity), N_0 is the initial activity of the sample, and t is the time elapsed from the time when N was equal to N_0, that is, when t was equal to zero. λ is a constant called the *decay constant* of the radioactive sample. Then, on taking natural logarithms, we have:

$$\ln N = \ln N_0 + \ln (e^{-\lambda t}) = \ln N_0 - \lambda t \qquad \text{(A.1)}$$

since $\ln e^x = x$. Equation (A.1) represents a straight line, with intercept $\ln N_0$ and slope $-\lambda$, when $\ln N$ is plotted against t. Hence both N_0 and λ can be found graphically. We will see more of this in Appendix 2 on graphs.

Appendix 2
Functional Equations and Graphs

2.A Functional Equations

In mathematics courses you often come across expressions such as $y = x^2$, which states that "y is a function of x," in this case the square of x. This means that if x is equal to 3, y is equal to 9; if x is equal to 4, y is equal to 16; and so on. For example, the formula for the area of a square of side L is $A = L^2$. Hence, if the sides of the square are 4 cm in length, then the area of the square is $A = (4 \text{ cm})^2 = 16 \text{ cm}^2$. (Note that in physics not merely the numbers but also the units, in this case "cm," have to be squared, if the answer is to make good physical sense.)

The functional equation $y = x^2$ can be graphed by plotting x along the X axis over some reasonable range and then calculating and plotting the corresponding values of y along the Y axis. The resulting graph often gives a more vivid picture of the dependence of y on x than does the simple functional equation $y = x^2$.

In some cases the functional dependence of y on x may assume a very simple form, in which y varies either directly with x or inversely with x. These are called *direct proportions*, and *inverse proportions*, respectively.

Direct Proportions

A good example of a direct proportion is the expression for the circumference of a circle, where $C = f(r)$ and r is the radius. In this case it is clear that C increases as r increases. If this dependence is linear, then we can write

$$C \propto r$$

which states that the circumference C is directly proportional to the radius r. This is not yet an equation, since it contains no equality sign; it is a *proportion*. If we graph C against r, we obtain a straight line, which can always be described by the equation

$$C = kr$$

By making use of the formula for the length of arc of a circle given in Appendix 1B, $l = r\theta$, and noting that for a complete circle $\theta = 360°$ or 2π rad, we have $C = kr = 2\pi r$, which is the usual geometric expression for the circumference of a circle.

It is true in general that any direct proportion (such as $C \propto r$) can be converted into an equation by inserting the right constant of proportionality k (as in $C = kr = 2\pi r$). The actual value of k can be obtained as the slope of the straight line relating C to r. For any two points on the straight line designated by the values C_2, r_2 and C_1, r_1, the slope is given by $k = (C_2 - C_1)/(r_2 - r_1)$, as shown in Fig. A.2. The value of k in an equation relating two physical quantities may vary in numerical value depending on the *units* used for the

FIGURE A.2 The circumference of a circle plotted against its radius. Since $C = 2\pi r$, the graph is a straight line of slope 2π.

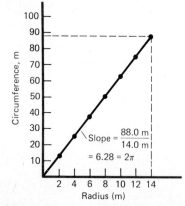

A4

two quantities in the proportion. We will see many examples of this in this course.

Inverse Proportions

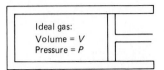

FIGURE A.3   A gas confined by a piston to a cylindrical container.

In other cases two physical quantities vary *inversely*; that is, y increases as x decreases and vice versa. For example, suppose we take a gas enclosed in a cylindrical container and use a piston to change its volume V, as in Fig. A.3. We find that the volume of the gas is inversely proportional to its pressure, if the temperature remains constant. Expressed as a proportion, we have

$$V \propto \frac{1}{P} \quad \text{(for constant } T)$$

This proportion can again be converted into an equality by inserting the proper constant of proportionality, so that

$$V = \frac{k}{P} \quad \text{or} \quad PV = k$$

In general, if $y \propto 1/x$, or $y = k/x$, then $xy = k = $ constant for an inverse proportion. If y is plotted against $1/x$, a straight line of slope k results. If y is plotted against x, the graph is a hyperbola, as in Fig. A.4. Whether a proportion is direct or inverse, this proportion can be changed to an equality by inserting the proper proportionality constant. From the data given in Fig. A.4, it can be seen that k is in this case equal to 10^5 N·m, since this is the product PV at any point on the graph.

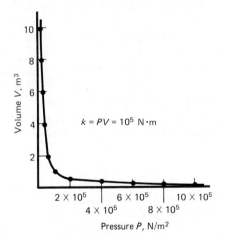

FIGURE A.4   A graph of the volume of a gas of fixed mass and at constant temperature against the pressure of the gas. Since $V \propto 1/P$, the graph is a hyperbola.

2.B Graphs

A graph is just a visual way of showing the functional dependence of one physical quantity on another, such as the speed of a car as a function of time, or the volume of a gas as a function of its pressure.

Interpolation

Figure A.6 shows how the length of a piece of steel wire varies when it is supported in a clamp and weights are attached to its free end to stretch it, as in Fig. A.5. For reasonable size weights the increased length of the wire between the clamp and the chuck is found to be directly proportional to the applied force, and we can write

$$\Delta L \propto W \quad \text{or} \quad \Delta L = kW$$

FIGURE A.5 An apparatus for measuring Young's modulus for a steel wire. By adding additional weights W at the end of the wire, the length L of the wire can be increased.

FIGURE A.6 A graph of the applied force $W = mg$ against the increase in length of the wire in Fig. A.5. It is possible to interpolate the increase in length for any applied force between values for which the increase in length has been measured.

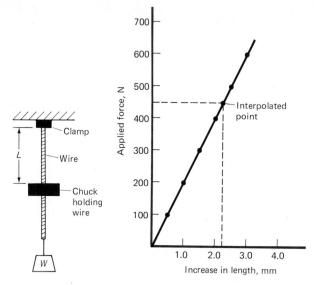

This is the equation of the straight line in the figure. The points marked on the straight line are the actual values of the applied force for which measurements of the length were made. If, now, we want to know what the length of the wire was for a weight of 450 N, we can *interpolate* between the measured points on the graph for 400 and 500 N and determine that the increased length of the wire was 2.25 mm when the applied force was 450 N.

We do this by finding the position where the line for a force of 450 N intercepts the graph, and then reading from that point down to the X axis to find the corresponding increase in length. We can be reasonably sure that the resulting length is accurate to within a small fraction of a millimeter, since we have data on both sides of the unknown point and the graph appears to be a straight line. Hence in this case interpolation appears to yield valid results. This is usually the case, unless we have a graph which is not a smooth curve but one on which the dependent variable fluctuates rapidly for small variations in the independent variable. In such cases interpolation can be a riskier proposition.

Extrapolation

Suppose, however, that we wanted the increased length of the wire for an applied force of 800 N. We could *presume* that the straight line continued beyond the last data point at 600 N and find the length of the wire from the graph by extending the straight line into this region, as in Fig. A.7. Such a process is called *extrapolation*. It is inherently risky and often leads to absurd results. For example, in the case of the wire under study, the application of a weight of 800 N would exceed the elastic limit of the wire and would begin to pull the molecules of the wire apart. As a result the graph would no longer be a straight line. Rather the curve would look something like the solid line in Fig. A.7. In this case extrapolation would lead to results in serious disagreement with experiment.

The fallacy involved in unreasonable extrapolations such as this is the presumption that the curve we are using will retain its shape even outside the region where the data points are collected. Thus the last data point on the

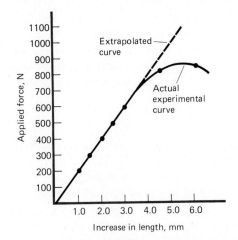

FIGURE A.7 An invalid extrapolation of the increase in length of the wire in Fig. A.5. Experiment shows that the curve ceases to be a straight line for forces above about 700 N.

curve is at an applied force of 600 N. To presume that the curve will remain a straight line even at a force of 800 N is a risky assumption, which turns out in this case to be simply wrong.

In life, just as in physics, extrapolations can lead to costly mistakes. For example, some electric utility companies are now in financial difficulties because in the 1970s they were forced to extrapolate electricity needs for the 1980s and beyond on the basis of data on electricity use in the past. Today, after investing billions of dollars in expensive nuclear power plants, they find that the extra electric power is not really needed because of conservation measures and a slowdown in the economy. In some cases we may be forced, as were the power companies, to use extrapolation, since there is simply no other approach available. If we extrapolate, however, we should be aware of the dangers involved in such a process, and of the very real possibility that it may lead to catastrophic errors.

2.C Log-Log and Semilog Scales

Two of the most powerful tools for a physicist are a piece of graph paper and a sharp pencil. By plotting experimental data in graphical form a physicist can uncover functional relationships between two physical quantities which could never be obtained merely by looking at tables of measured values for these two quantities. Graphical analysis is a way to "crack the code" beneath which nature hides its orderly behavior.

The simplest type of dependence of one physical quantity on another is the linear, or straight-line, relationship shown in Fig. A.6. If we plot measured values of the increased length of the wire against the applied force, and draw a smooth curve through the points, we can easily tell whether the functional dependence is really linear over the range of the experiment. If the functional dependence is *not* linear, however, it is very difficult to find from an ordinary graph the exact functional dependence of one measured quantity on another. Only a straight-line dependence is easy to identify unambiguously.

For this reason *the key to graphical analysis is to reduce all graphs to straight lines*. There is unfortunately no completely general method for doing this, but we will consider a few particularly useful methods.

Power Laws: Log-Log Scales

In physics, equations such as $T = 2\pi\sqrt{l/g}$, or $F = mv^2/R$, frequently occur. The first equation states that T varies directly with $\sqrt{l}$ and the second that F varies directly with v^2, all other quantities being held constant. Such equations, in which one physical quantity varies as a power of another quantity, are called *power-law equations* and can be represented in more general form as

$$y = kx^n \tag{A.2}$$

where n is the power which characterizes the functional dependence of y on x, which power we would like to know. Here n can be either positive or negative, integral or fractional, and k is a constant.

The best way to find n is to take logarithms of both sides of this equation. We then have $\log y = \log k + n \log x$. If, now, we plot $\log y$ against $\log x$, rather than y against x, we obtain a straight line of slope n, since the logarithmic equation is of the form $Y = nX + b$, where $Y = \log y$ and $X = \log x$, and $\log k = b$ is a constant. From the slope (n) of this straight line it is possible to find on what power of x the quantity y depends, since n is precisely this power. In doing this either common (base 10) logarithms, or natural (base e) logarithms may be used.

Log-Log Graph Paper

To make the technique for investigating power-law dependence even easier, graph paper is available which is so constructed that, if we plot a number on either the horizontal or vertical scales, we are actually plotting the *logarithm* of the number. In this case, since the numbers being plotted on the two axes are really logarithms, they are pure numbers without units or dimensions. Hence the slope is obtained by merely taking the ratio of the measured distances along the two axes. We have for the slope:

$$n = \frac{\log y_2 - \log y_1}{\log x_2 - \log x_1}$$

where $\log y_2 - \log y_1$ is the measured distance between y_2 and y_1 and $\log x_2 - \log x_1$ is the measured distance between x_2 and x_1, when these quantities are plotted on log-log paper. Hence the slope can be read directly from the graph paper, as in Fig. A.8, where we plot the period (T) of the planets against the

FIGURE A.8 A plot of the period of the planets against the radius of their orbit about the sun. These data are plotted on two-cycle log-log graph paper and show that $T = kR^{1.5}$.

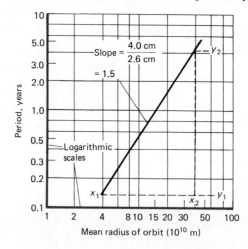

mean radius (R) of their orbits. The graph shows that $T \propto R^{3/2}$, or $T^2 \propto R^3$, which is one of Kepler's laws.

Log-log graph paper is usually described as two-cycle, three-cycle, four-cycle, etc., to describe the range of numbers which can be plotted. The logarithmic scales are so constructed that the range from 1 to 10 (one cycle) occupies the same space as from 10 to 100 (a second cycle), or from 100 to 1000 (a third cycle). Thus three-cycle log-log paper would have three regions of equal length along each axis and would allow us to plot numbers ranging from 1 to 1000. Since we want to plot our data on the largest possible scale to increase the accuracy of the graph, we should choose log-log paper with the proper number of cycles to fit the data, so that the data will fill up almost the whole sheet of graph paper rather than only a small portion of it. The graph paper used in Fig. A.8 is two-cycle log-log paper.

Exponential Relations: Semilog Scales

Many important physical quantities are related exponentially. A simple example is U.S. electric energy consumption in the period from 1955 to 1974. Each year the amount of electric energy used increased by approximately the same fractional amount. Since each year the increase was expressed as a fraction of a larger amount than the previous year, the result, when the energy consumed is plotted against time, is the rapidly increasing curve shown in Fig. A.9.

This curve can be described by the equation

$$\mathcal{W} = \mathcal{W}_0 e^{kt} \tag{A.3}$$

where $\mathcal{W}$ is the energy consumed t years after 1955, and $\mathcal{W}_0$ is the energy consumed in the year 1955. When graphed, this equation yields a rising exponential curve. The relationship can be graphed more easily by taking

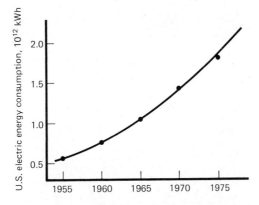

FIGURE A.9 Energy consumption in the United States for the years 1955 to 1974. There was an exponential increase in energy use during these years.

logarithms of both sides. In doing this it is convenient to use natural logarithms instead of logs to the base 10. Then, taking natural logs of both sides of Eq. (A.3), we have, since $\ln(e^{kt}) = kt$:

$$\ln \mathcal{W} = \ln \mathcal{W}_0 + kt$$

Hence, by plotting $\ln \mathcal{W}$ against t, we would expect to find a straight line of constant slope k, if the dependence of $\mathcal{W}$ on t is truly exponential. This is indeed what we find in Fig. A.10.

If we suspect an exponential dependence in a particular experiment, the

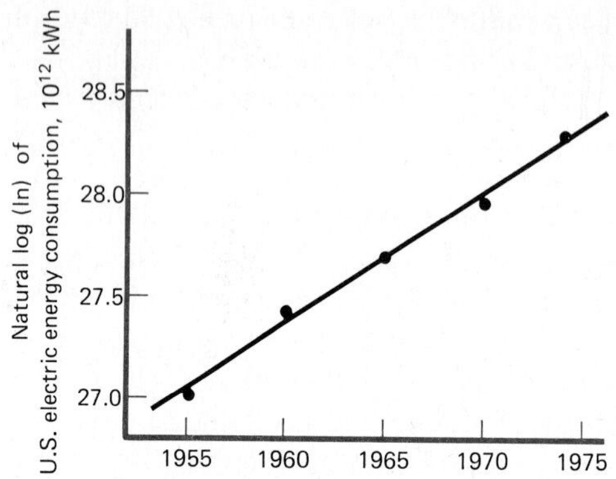

FIGURE A.10   The data of Fig. A.9, with the natural log of the energy consumed plotted along the vertical axis. The result is a straight line, indicating clearly that the dependence is exponential.

fastest approach to a solution is to plot the natural logarithm of the quantity which seems to be varying exponentially against the other quantity. If a straight line results, the dependence is truly exponential, and the exponent k is obtained from the slope of the straight line.

Semilog Graph Paper

Once again a special kind of graph paper is available to aid in the plotting of quantities which are related exponentially. Such paper is called *semilog paper*, because one scale (usually the vertical) is ruled logarithmically in the same fashion discussed above for log-log paper, and the other scale (the horizontal) is ruled in a normal linear fashion.

For example, in plotting the electric energy usage of the United States, if we plot $\mathcal{W}$ on the logarithmic scale of the semilog paper and t on the normal scale, this is completely equivalent to plotting the natural logarithm of $\mathcal{W}$ against t. A straight line results, as Fig. A.11 shows. The slope of this straight

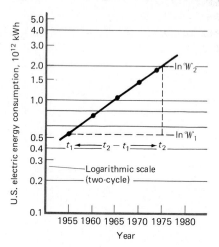

FIGURE A.11   The same data as in Fig. A.9, but plotted on semilog graph paper. Again the result is a straight line, indicating exponential dependence.

line, that is, k, can be found from the fact that, for any two values of t, say t_2 and t_1, the corresponding values of $\mathcal{W}$ are $\mathcal{W}_2$ and $\mathcal{W}_1$. The slope of the curve can then be found from the equation

$$k = \frac{\ln \mathcal{W}_2 - \ln \mathcal{W}_1}{t_2 - t_1} = \frac{\ln (\mathcal{W}_2/\mathcal{W}_1)}{t_2 - t_1}$$

Thus between the years 1955 and 1975 the amount of electricity consumed increased from 0.55×10^{12} kWh to 2.0×10^{12} kWh, as can be seen from the graph. Hence in this case we have:

$$k = \frac{\ln (2.0/0.55)}{20 \text{ years}} = \frac{1.29}{20 \text{ years}} = 0.065/\text{year}$$

The units of k must be $1/t$, or in this case 1/yr, for kt must be dimensionless since it is an exponent. Hence our equation for electrical consumption during the period 1955 to 1975 follows the equation:

$$W = W_0 e^{0.065t}$$

as is clear from Eq. (A.3).

A more complete discussion of the use of simple graphical techniques in physics may be found in Clifford Swartz, *Prelude to Physics* (Wiley, New York, 1983).

Appendix 3
Solving Physics Problems

3.A Some General Hints on Problem Solving

A frequent refrain from students in introductory physics courses is that they understand the material but cannot do the assigned physics problems. Yet if they truly understood the material, they would be able to do the problems. This is not to say that physics problems are easy, but the more complete your understanding of the material, the easier the problems will seem to you. In physics it is *understanding*, not memorization, that pays off. Hence it is a much better investment of time to study a chapter until you feel that you really understand it before tackling the assigned problems, rather than to try solving problems involving material which you have not yet mastered.

The more physics problems you do, the easier they become. At first, the plethora of units and equations introduced in a course such as this one can make problems confusing, but the more problems you do, the simpler they will seem to you. Hence you are encouraged to do problems in addition to those assigned in this course. A good source of problems, some fully worked out, some with only the answers given, is *Schaum's Outline of Theory and Problems of College Physics* (7th ed.) by Frederick J. Bueche (McGraw-Hill, New York, 1979).

The first step in tackling a set of problems is to read over the chapter in the text, your class notes, and anything else you may find helpful in understanding the material. Pay special attention to the examples worked out in the text, since the assigned problems will usually have some relationship to these sample problems. Once you feel that you really *understand* the chapter, the following steps may be useful in approaching problems. Note, however, that there are a great variety of physics problems, and that no set procedure will work equally well in all cases.

1 First read the problem very carefully and be sure that you understand the physical situation described in the statement of the problem. This is usually the most important step in arriving at a solution.

2 Form a simplified *model* of the physical situation which is capable of being treated quantitatively. Often drawing a diagram or picture of the situation will help. Textbook problems often force a simplified model of the situation by only giving you a limited number of known quantities to work with. Solving laboratory problems is more difficult, since you must decide what data are needed.

3 Make a list of the quantities which are given and those which are to be determined. In some problems additional data may be needed from tables in the text. In other cases some of the information given may not really be needed in the problem. One key to understanding physics is to learn what physical quantities are important in a particular physical situation and what quantities are irrelevant.

4 Pick out one unknown quantity and try to recall how it is related physically to the given quantities by some basic law of physics or some general equation. The choice of a relevant equation to calculate the unknown should be based on an *understanding* of the physical situation, not on a hunt through a list of equations in the hope of finding one that works. If you are unable to determine a proper approach to the solution of the problem, remember that more than one step and more than one equation may be required. Looking at some worked examples in the text may help stimulate useful ideas.

5 If it does not appear possible to calculate the first unknown from the given quantities, go on to one of the other unknowns which may be easier to handle. Perhaps this other unknown must be calculated before the first unknown can be obtained.

6 Reduce all formulas to the simplest possible analytical form before substituting numbers in them. When you have your final answer in the form of an equation, check to see that it is dimensionally correct.

7 Substitute numbers in the equations, being sure that the units are consistent (all SI) and writing the units with every number. Solve for the numerical answer, canceling units throughout to give the correct units for the answer.

8 In most numerical problems the data are given to only two or three significant figures, and answers should be to the same number of significant figures.

9 Check the problem carefully to make sure you have made no mistakes in algebra or arithmetic (we all make careless mistakes at times). Are the units in the answer correct?

10 Look back at the original physical situation to make sure the answer is physically reasonable. (If you have a car going 3×10^5 m/s, it is not reasonable!)

11 Only at this stage should you look up the answer, if given in the back of the book. If your answer agrees to within a few percent with the answer given, you can be fairly sure that your work is correct. In a text such as this, reasonably approximate answers to problems are perfectly acceptable. The method is the important thing, not great accuracy in the numerical result. Eventually you may build up enough confidence in your problem-solving ability to be convinced in certain cases that the answer given in the book is wrong. If so, discuss the matter with your instructor before handing in the assigned problems.

12 If the answer is not given in the book, try to check your work by doing the problem in a different way or working it backwards to see that it checks out.

13 After you have worked out the problem on scrap paper and are convinced that it is right, transfer the essential work to good paper in as logical and clear a fashion as possible, leaving out all false starts. Indicate the calculations carried out, but not the actual arithmetic. The problem should be so clear that a reader does not have to struggle to find out what you have done. Box the answers for the unknown, or indicate clearly in some other way what your answers are.

The best way to learn how to do physics problems is to do lots of problems. An investment of time in doing extra problems at the beginning of this course will save you much time at the end of the semester, when time is more precious. The logical thought processes and techniques learned in solving physics problems will also prove useful to you in other college courses and in your future careers. For example, the thought process followed by a medical doctor in diagnosing what is wrong with a patient does not differ too much from the thought process you will follow in successfully solving a physics problem.

3.B Powers-of-10 Notation

Most of the physics problems in this book contain data given to only two, or at most three, significant figures, and answers are therefore expected only to two or three figures. If, for example, a car travels at 20 m/s for 10 min, the distance traveled is:

$$\left(20 \ \frac{m}{s}\right) (10 \ min) \left(60 \ \frac{s}{min}\right) = 12,000 \ m$$

But it is incorrect to write the answer in this form, since it gives the answer to five figures, only two of which are really significant. Hence it is much better to express this answer in powers-of-10 notation as 1.2×10^4 m.

For problems in this book, it is expected that calculations and answers will all be expressed in powers-of-10 notation. This enables us to use the proper number of significant figures, whether the number in question is very large or exceedingly small, and to express the largeness or smallness by multiplying the number given by the correct power of 10. We shall now outline this useful tool for those students not already familiar with it.

If we raise 10 to various powers, we obtain the following results:

$10^1 = 10$	$10^{-1} = \dfrac{1}{10}$	$= 0.1$
$10^2 = 100$	$10^{-2} = \dfrac{1}{100}$	$= 0.01$
$10^3 = 1000$	$10^{-3} = \dfrac{1}{1000}$	$= 0.001$
$10^4 = 10,000$	$10^{-4} = \dfrac{1}{10,000}$	$= 0.0001$
$10^5 = 100,000$	$10^{-5} = \dfrac{1}{100,000}$	$= 0.00001$
$10^6 = 1,000,000$	$10^{-6} = \dfrac{1}{1,000,000}$	$= 0.000001$

This provides us with a shorthand method for writing both very large and very small numbers. Any number can be written as a decimal number between 1 and 10 followed by the appropriate power of 10. The power of 10 gives the number of places the decimal point must be moved to the right for positive powers and to the left for negative powers to convert to ordinary decimal form. Hence we may write large numbers as follows:

$$286,000 = 2.86 \times 100,000 = 2.86 \times 10^5$$

Note that we rewrite the number with the decimal point after the first digit, then multiply by the power of 10 corresponding to the number of digits to the right of the decimal point. Similarly, for very small numbers,

$$0.00312 = \frac{3.12}{1000} = \frac{3.12}{10^3} = 3.12 \times 10^{-3}$$

Note here that the negative power of 10 corresponds to the number of places the decimal point has to be moved to the left to convert back to ordinary decimal form.

There is an instructive relationship between powers-of-10 notation and logarithms to the base 10. In the table below we write a number of values for the variable y, first in ordinary form and then below it in powers-of-10 notation. On the third line we write $\log y$, which is equal to the exponent of y in powers-of-10 notation, since the log of y is the power to which 10 must be raised to yield y.

y	0.001	0.01	0.1	1	10	100	1000
y	10^{-3}	10^{-2}	10^{-1}	10^0	10^1	10^2	10^3
$\log y$	-3	-2	-1	0	1	2	3

This powers-of-10 notation is used throught this book and is particularly useful in multiplying and dividing large numbers. In multiplying two numbers, we multiply the initial numbers and add the powers of 10, carefully watching the signs of the powers; in dividing two numbers, we divide the initial numbers and subtract the powers of 10. For example,

$$(2.86 \times 10^5)(3.12 \times 10^{-3}) = 8.92 \times 10^{5-3} = 8.92 \times 10^2$$

and

$$\frac{2.86 \times 10^5}{3.12 \times 10^{-3}} = 0.917 \times 10^{5-(-3)} = 0.917 \times 10^8 = 9.17 \times 10^7$$

This enables us to obtain a value for 10 raised to the zero power. Since $10^1 = 10$, $10^1/10^1 = 10/10$, or $10^{(1-1)} = 1$, and so $10^0 = 1$. As a matter of fact, it can be shown in the same way that any number raised to the zero power is equal to 1.

To add or subtract two numbers in powers-of-10 notation, the power of 10 must be made the same in both cases; then the numbers are added or subtracted as usual. Thus

$$(2.76 \times 10^{-5}) + (3.46 \times 10^{-6}) = (27.6 \times 10^{-6}) + (3.46 \times 10^{-6})$$

$$= (27.6 + 3.5) \times 10^{-6} = 3.11 \times 10^{-5}$$

This corresponds to the way we normally line up the decimal points in adding or subtracting decimal numbers.

3.C Order-of-Magnitude Calculations

Often even two or three significant figures are excessive in the answer to a particular physics problem, for all we really need is the *order of magnitude* of the result. This is basically the result *to the nearest power of 10*. Thus we might say that the order of magnitude of an electron's speed is 10^5 m/s, meaning that the speed is probably somewhere between 5×10^4 and 5×10^5 m/s, but that we are not really interested in its exact value. For example, all we might want to know is whether the electron's speed is close enough to the speed of light (3 $\times$ 10^8 m/s) to make relativistic effects important. The knowledge that the speed is close to 10^5 m/s assures us that we can neglect such relativistic effects.

An order-of-magnitude calculation leads to a very approximate or, as physicists sometimes say, a "quick and dirty" answer to a physics problem. Two physicists can carry out separate order-of-magnitude calculations of the same quantity and, either because they made slightly different assumptions, or used slightly different data or approximations, come up with results in disagreement by a factor of 2 or 3, or even 5, and we would still say that their results agreed "as to order of magnitude."

Order-of-magnitude calculations are very useful in solving physics problems. At the beginning of a problem we might want to know if we are on the right track by putting in rough numbers to see if they yield a reasonable value for the desired unknown quantity. At the end of the problem, if we are suspicious that we may have made a mistake in algebra or entered the wrong number into our hand calculator, an order-of-magnitude calculation may show us at what stage of the calculation we went wrong.

In the laboratory, physicists and physics students carry out order-of-magnitude calculations all the time, to determine what range on an electric meter is needed for a particular measurement, or to ascertain the error introduced into an experiment by a drop in power-line voltage from 120 to 115 V. Here the physicist is not looking for a highly precise result, but for a "ballpark" figure to help decide the next step in the experiment. The physicist does not want to take the time to come up with an exact answer because at this stage of the research such an answer is not worth the effort involved. This is when order-of-magnitude calculations are quite useful.

For example, suppose a swimming pool owner wants to know approximately how much water the pool holds. The pool is 20.4 m long and 6.3 m wide, and its depth varies linearly between 1.1 and 4.3 m. The average depth of the water is then 2.7 m, and an order-of-magnitude calculation of the volume of water in the pool leads to:

$$V = (2 \times 10^1 \text{ m})(6 \text{ m})(3 \text{ m}) = 36 \times 10^1 \text{ m}^3 \simeq 4 \times 10^2 \text{ m}^3$$

This can be compared to the more exact value of:

$$V = (20.4 \text{ m})(6.3 \text{ m})(2.7 \text{ m}) = 3.5 \times 10^2 \text{ m}^3$$

It can be seen that the order-of-magnitude calculation leads to a result differing by about 14 percent from the exact value. This is more than enough accuracy for an order-of-magnitude calculation.

In this example the time involved in the exact calculation is not much greater than the time required for the order-of-magnitude calculation. However, with more complicated physics problems order-of-magnitude calculations often take significantly less time than exact calculations and thus are particularly useful in such cases.

Appendix 4
Masses and Abundances of Some Important Isotopes

Atomic number Z	Element	Symbol	Mass number A^*	Atomic mass,† u	Percent abundance	Half-life (if radioactive)
1	Hydrogen	H	1	1.007825	99.985	
	Deuterium	D	2	2.014102	0.015	
	Tritium	T	3*	3.016049		12.33 years
2	Helium	He	3	3.016029	0.00014	
			4	4.002603	~100	
3	Lithium	Li	6	6.015123	7.5	
			7	7.016005	92.5	
4	Beryllium	Be	7*	7.016930		53.3 days
			8*	8.005305		6.7×10^{-17} s
			9	9.012183	100	
5	Boron	B	10	10.012938	19.8	
			11	11.009305	80.2	
6	Carbon	C	11*	11.011433		20.4 min
			12	12.000000	98.89	
			13	13.003355	1.11	
			14*	14.003242		5730 years
7	Nitrogen	N	13*	13.005739		9.96 min
			14	14.003074	99.63	
			15	15.000109	0.37	
8	Oxygen	O	15*	15.003065		122 s
			16	15.994915	99.76	
			18	17.999159	0.204	
9	Fluorine	F	19	18.998403	100	
10	Neon	Ne	20	19.992439	90.51	
			22	21.991384	9.22	
11	Sodium	Na	22*	21.994435		2.602 years
			23	22.989770	100	
			24*	23.990964		15.0 h
15	Phosphorus	P	31	30.973763	100	
			32*	31.973908		14.28 days
17	Chlorine	Cl	35	34.968853	75.77	
			37	36.965903	24.23	
19	Potassium	K	39	38.963708	93.26	
			40*	39.964000		1.28×10^9 years
			41	40.961832	6.72	
26	Iron	Fe	56	55.934939	91.8	
27	Cobalt	Co	60*	59.933820		5.24 years

*An asterisk means that the isotope is radioactive.
†Masses include the electrons in the neutral atom. These atomic masses are in unified mass units u.

Atomic number Z	Element	Symbol	Mass number A*	Atomic mass,† u	Percent abundance	Half-life (if radio-active)
36	Krypton	Kr	84	83.911506	57.0	
			89*	88.917563		3.2 min
38	Strontium	Sr	86	85.909273	9.8	
			88	87.905625	82.6	
			90*	89.907746		28.8 years
53	Iodine	I	127	126.904477	100	
			131*	130.906118		8.05 days
55	Cesium	Cs	133	132.90543	100	
			137*	136.90677		30 years
56	Barium	Ba	137	136.90582	11.2	
			138	137.90524	71.7	
			144*	143.922673		11.9 s
82	Lead	Pb	206	205.97447	24.1	
84	Polonium	Po	210*	209.98288		138.4 days
			214*	213.99519		164 μs
86	Radon	Rn	222*	222.017574		3.824 days
88	Radium	Ra	226*	226.025406		1.60×10^8 years
			228*	228.031069		5.70 years
90	Thorium	Th	228*	228.02873		1.91 years
			232*	232.038054	100	1.41×10^{10} years
92	Uranium	U	233*	233.039629		1.62×10^5 years
			234*	234.040904	0.0057	2.48×10^5 years
			235*	235.043925	0.715	7.13×10^8 years
			238*	238.050786	99.27	4.51×10^9 years
			239*	239.054291		23.5 min
93	Neptunium	Np	239*	239.052932		2.35 days
94	Plutonium	Pu	239*	239.052158		2.41×10^4 years

*An asterisk means that the isotope is radioactive.
†Masses include the electrons in the neutral atom. These atomic masses are in unified mass units u.

Answers to Odd-Numbered Problems

Chapter 1
A1. (*c*)
A3. (*e*)
A5. (*e*)

A7. (*a*)
A9. (*d*)

Chapter 2
A1. (*b*)
A3. (*b*)
A5. (*e*)
A7. (*d*)
A9. (*b*)
A11. (*c*)

B1. (*a*) 91.5 m; (*b*) 4.22×10^4 m; (*c*) 5.27×10^6 m
B3. (*a*) 7.3 m/s; (*b*) 0; displacement is zero
B5. 1.9×10^{26} m
B7. (*a*) 29 m/s; (*b*) 44 m
B9. 20 m/s
B11. 99 m/s

B13. 3.6 s

B15. 38 m/s; 86 mi/h

B17. 9.8 m/s^2

C1. 0.86 m/s

C3. (a) 65 m/s; (b) 288 m

C5. (a) 0.80 m/s^2; (b) 25 s

C7. (a) 13 m/s; (b) 6.5 m/s; (c) 33 m

Chapter 3

A1. (d)

A3. (e)

A5. (e)

A7. (e)

A9. (d)

B1. 4.4×10^{20} N

B3. 0.80 m/s^2

B5. 114 kg

B7. 5.6 m/s^2

B9. 30 N

C1. 0.065 m/s^2

C3. (a) 1.3 m/s^2; 3.3 m/s

C5. (a) 21 g; (b) 4.2×10^2 m

Chapter 4

A1. (d)

A3. (d)

A5. (c)

A7. (c)

A9. (d)

B1. (a) 1.1 N at an angle of 87° above negative X axis; (b) 4.2 m/s^2 in same direction as force

B3. 1.3×10^3 m^2

B5. 4.0 N·m

B7. 300 N·m

B9. 2.5×10^3 N·m

C1. 39 m/s; 50° E of N

C3. (b) 8.8×10^3 N

Chapter 5

A1. (c)

A3. (d)

A5. (b)

A7. (e)

A9. (a)

B1. (a) 4.5 m/s^2; (b) 63°

B3. 16 m

B5. (a) 1.03 rad/s^2; (b) 0.14 m/s^2

B7. 1.1 N

B9. 450 N

C1. 3.5° above the horizontal

C3. 0.92 m

C5. (b) 1.1 m/s^2; (c) 2.2 m

C7. (a) $x = 90$ m, $y = 100$ m; (b) $t = 6.1$ s; $y = 184$ m; (c) 551 m; (d) 75 m/s

Chapter 6

A1. (a)

A3. (a)

A5. (b)

C9. (a) 7.5×10^{17} m/s^2; (b) 2.0×10^{-10} s

C11. At A: 0; B: 2.8 m/s^2; C: $+0.17$ m/s^2; D: -0.093 m/s^2

C13. 5.8 m/s

C15. 8.0 m/s^2

C17. 2.5 m/s^2

C19. 25 m

C23. 10 m/s

C7. (a) 1.2×10^3 m/s^2; (b) 1.2×10^2 m/s

C9. 6.0 m

C11. 1.2 m/s^2

C13. (a) 5.2×10^3 N; (b) 0.53

C15. 33 N

C17. (a) 1.1 s; (b) 0.56 s

C19. (a) 1.1×10^3 N; (b) 8.8×10^2 N; (c) 0

C21. (a) 98 N; (b) 113 N; (c) 83 N

D1. 3.46×10^8 m from earth

D3. (a) 1.96 m/s^2; (b) 82.4 N; (c) 58.8 N

D5. (a) 8.5 kg; (b) $T_{BC} = 68$ N; $T_{AB} = 26$ N

D7. 1.2×10^4 N

C5. (b) $T_A = 60$ N; $T_B = 92$ N

C7. 5.7×10^3 N

C9. $T_C = 1.06 \times 10^4$ N; $F = 8.25 \times 10^3$ N at 6° below $+X$ axis

C11. (a) 147 N; (b) 418 N at 69° with ground

C13. (a) 1.6×10^3 N down; (b) 2.4×10^3 N up

C15. $d_1 = 2.15$ m; $d_2 = 1.85$ m

C17. (a) 2.27×10^3 N; (b) 3.02×10^3 N at 41° to horizontal

C19. 77.6 N

D1. (a) $T_1 = 4.4 \times 10^3$ N, $T_2 = 8.8 \times 10^3$ N; (b) $\theta_1 = 63°$; $\theta_2 = 27°$

D3. $T_1 = T_2 = 592$ N; $T_3 = 391$ N; $T_4 = T_5 = 289$ N

D5. 0.48

C9. (a) 6.25 rad/s^2

C11. (a) 31.4 rad/s^2; (b) 20 rev; (c) 5.0 m

C13. (a) 5.5 m/s^2; (b) 8.3×10^3 N; (c) 29°

C15. (a) 3.0 N; (b) 8.8×10^2 N

C17. 0.75 m/s

C19. 3.1 m/s

C21. 1.6×10^3 m/s

D1. (a) 3.5 m/s^2 up the plane; (b) 7.0 m

D3. (a) No; an angle of 47° would be required; (b) 1.30×10^4 N; (c) 0.88

D5. (a) The string must provide a centripetal force; (b) 39°

D7. (a) $R = 1.7$ cm; (b) 1.6×10^2 rad/s^2

D9. (a) 0.34%; (b) 0

D11. 84 min

A7. (a)

A9. (c)

A11. (c)

B1. (a) 0; (b) 2.5×10^4 J; (c) 9.7×10^3 J
B3. 4.1×10^{-16} J
B5. 1.96×10^{10} J
B7. 5.2 J
B9. 3.3 m
B11. 5.0×10^3 hp
C1. (a) 2.5×10^3 J; (b) 1.8×10^3 J; (c) -2.5×10^3 J; (d) 0
C3. (a) 10^3 J; (b) -490 J; (c) 510 J; (d) increase of 510 J; (e) 10.1 m/s
C5. (a) 500 J; (b) 294 J; (c) 206 J
C7. 16 m/s
C9. (a) 30 J; (b) 28 J; (c) 2.0 J; (d) 19 m/s

C11. (a) 3.9 N/m; (b) 0.49 J
C13. (a) 2.5×10^5 J; (b) 1.25×10^4 N; (c) 93%
C15. (a) 11; (b) 8.3; (c) 75%
C17. 2.3×10^4 W
C19. 31 MW
C21. (a) 3.71×10^2 J; (b) 1.2×10^2 W
C23. 2.1×10^7 J or 6.0 kWh
D1. (a) 1.5×10^5 J; (b) 5.8×10^2 N
D3. 11.2 km/s
D5. $mgL(1 - \cos\theta)$
D7. $(2mgh/k)^{1/2}$
D9. (a) 44 m/s; (b) 8.3×10^5 W

Chapter 7
A1. (e)
A3. (a)
A5. (c)
A7. (d)
A9. (a)
B1. 20 kg·m/s
B3. 500 N
B5. 1.53 kg·m/s; 1.29 kg·m/s
B7. (a) 71.6° below negative X axis; (b) 1.6 kg
B9. 3.0×10^2 m/s
C1. 1.33 m/s
C3. (a) 14 kg·m/s; (b) 9.3×10^2 N
C5. 1.5 s
C7. 7.2×10^2 m/s

C9. (b) 4.7 m/s at an angle of 45° below negative X axis
C11. $v_1 = -25$ cm/s; $v_2 = +30$ cm/s
C13. 6.3 m/s at an angle of 58° above $+X$ axis
C15. 2.5×10^5 m/s
C17. (a) 0.71 m/s; (b) The earth recoils
C19. 15 N
C21. 1.8×10^5 kg
D1. (a) $m_1 = m_2$
D3. (a) 10 m/s; (b) 8.3×10^{-22} m/s; (c) 1.3×10^6 N; (d) 1.3×10^6 N; (e) 2.0 cm
D5. (a) 0.25 N; (b) 2.5×10^{-2} N·s
D7. 4.9×10^{-1} N
D9. $m = m_p$

Chapter 8
A1. (d)
A3. (b)
A5. (c)
A7. (b)
A9. (a)
B1. 7.0 rad/s²
B3. 6.2 m/s
B5. 200 hp
B7. $(3/10)^{1/2}R$
B9. 9.0×10^3 kg·m²/s
B11. 2.6 rev/s
C1. 34 rad/s
C3. (a) 5.6/s²; (b) 25 rad; (c) 4.0
C5. (a) 1.7×10^{-2} N·m; (b) 2.9×10^{-3} N·m
C7. (a) 0.063 N·m; (b) 50 s after the off switch is thrown

C9. 0.040 m
C11. (a) 32 kg·m²; (b) 1.0×10^3 N·m
C13. 6.5×10^2 rad/s
C15. 8.4 m/s
C17. 6.8 rad/s
C19. (a) 0.63 rad/s; (b) $KE_{before} = 158$ J; $KE_{after} = 79$ J; (c) Into heat
C21. 0.077 rad/s
D1. 6.9 m, with solid cylinder leading.
D3. (a) 2.31 m/s²; (b) $T_1 = 37.4$ N, $T_2 = 36.3$ N; (c) 4.62 s⁻²
D5. $\frac{2}{3}g$
D9. 4.2 rad/s
D11. (a) $\omega_1/3$; (b) No; 2/3

Chapter 9
A1. (c)
A3. (c)
A5. (c)
A7. (d)
A9. (b)
B1. (a) 7.2×10^3 kg/m³; (b) 2.2×10^{17} kg/m³
B3. 20.18 g
B5. (a) Yes; (b) 7.5×10^3 J
B7. 6.2
B9. 15%
B11. 0.112 m³

C1. 116 kg
C3. (a) 1.6×10^{-4}; (b) It agrees with the assumption that the distance between gas molecules is large
C5. (a) 4.04×10^5 N/m²; (b) 4.34×10^5 N/m²
C7. 1.4×10^4 m³
C9. 1.07 atm
C11. 47.3 lb/in²
C13. 0.45 kg
C15. (a) 8.3×10^{-21} J; (b) 6.0×10^2 m/s
C17. 1.3×10^4 m
C19. 2.4×10^3 K

C21. (*a*) % difference in the masses is greater; (*b*) H: 41 %; U: 0.21 %
D1. 17 g
D3. 2.6 atm

Chapter 10
A1. (*d*)
A3. (*e*)
A5. (*a*)
A7. (*e*)
A9. (*b*)
B3. 1.08 N/m²
B5. 3.2×10^4 N
B7. 11.1×10^3 kg/m³
B9. 0.25 mm
B11. 6.6×10^{-8} m³
C1. 36×10^{-3} m³
C3. (*a*) 2.97×10^5 N/m²; (*b*) 2.97×10^5 N/m²; (*c*) 1.99×10^5 N/m²
C5. (*a*) 1.12×10^{-5} m³; (*b*) 0.866 N

Chapter 11
A1. (*d*)
A3. (*c*)
A5. (*b*)
A7. (*d*)
A9. (*c*)
B1. 9.0×10^{-3} J
B3. $0.020 \cos 5\pi t$
B5. 0.35 Hz
B7. 7.4 N
B9. 0.028 m
C1. (*a*) 0.010 m; (*b*) 0.28 s; (*c*) 3.5 Hz; (*d*) $y = 0.010 \cos 22t$
C5. 1.6 Hz
C7. 2.1×10^{-2} s
C9. 9.90 m/s²

Chapter 12
A1. (*a*)
A3. (*a*)
A5. (*b*)
A7. (*d*)
A9. (*b*)
B1. (*a*) 340 m/s; (*b*) $X = 10^{-12} \cos (10^4 \pi t - 92x)$
B3. 106 Hz
B5. 437 Hz and 443 Hz
B7. 1.0×10^{-4} N/m²
B9. 1.7
C1. (*a*) 4.7×10^3 Hz; (*b*) $X = X_{\text{max}} \cos (9.4 \times 10^3 \pi t - 21x)$
C3. 9.7×10^{10} N/m²
C5. (*a*) Fourth harmonic; (*b*) 333 Hz
C7. 8.5 m in length
C9. (*a*) 2.2×10^3 Hz; (*b*) 7.7 cm

Chapter 13
A1. (*a*)
A3. (*b*)
A5. (*d*)

D9. (*a*) 3.7×10^3 J; (*b*) 3.0×10^2 m/s; (*c*) Zero; momentum is a vector and the atoms are all moving in different directions

C7. The hole is 19% of the volume
C9. 1.6 N
C11. (*a*) No; (*b*) 12.2 m
C13. 43 m
C15. (*a*) 7.9×10^3 N/m²; (*b*) 2.9×10^5 N
C17. (*a*) 32 m/s; (*b*) 1.4×10^5 Pa
C19. 3.9×10^5 Pa
C21. 20 Pa
C23. 9.5×10^{-4} m³
D3. 4.7×10^{-2} m³
D9. (*a*) The end wires stretch 1.2×10^{-3} m; the center wire stretches 0.2×10^{-3} m; (*b*) The end wires each support 90 N; the center wire 15 N

C11. 0.69 J
C15. (*a*) 316 m/s; (*b*) 158 Hz; (*c*) 316 Hz; 474 Hz
C17. 156 Hz
C19. (*a*) At its center; (*b*) One-fourth of the length of the string from either end; yes; (*c*) Pluck string at the center; touch string one-third of length from either end
C21. (*a*) 228 Hz; (*b*) 65th harmonic
D1. 1.28 Hz
D5. $C = n/2$
D9. (*a*) 0.030 m; (*b*) 0; (*c*) 0.030 m; (*d*) 0 (*e*) 0.030 m; (*f*) 0.015 m
D11. (*a*) $y = 0.020 \cos [2\pi(60t - 30x) + \phi]$; (*b*) $\pi/3$ rad; (*c*) $\lambda f = 2.0$ m/s; (*d*) 0.010 m

C11. (*a*) 443 Hz; (*b*) 6 beats per second
C13. 3.2×10^{-5} W/m²
C15. 3.0 dB
C17. 8.0 dB
C19. 1.5 m/s
C21. (*a*) 6.8 cm; (*b*) 6.1 cm; (*c*) 5.6×10^3 Hz; (*d*) 6.8 cm; (*e*) 5.5×10^3 Hz
C23. 0.78 m/s
C25. 479 m/s
C27. No discrete frequency; a sonic boom
D3. 47 m
D5. About 10^{13} Hz
D7. 85 Hz, 255 Hz, 425 Hz, or, in general, $(2n + 1)(85$ Hz$)$
D9. $f' = \dfrac{1 + v_0/v}{1 - v_s/v} f$

A7. (*d*)
A9. (*a*)
A11. (*c*)

B1. −460°F; 81°F; 621°F
B3. 38.9°C
B5. 2.8; 4.4
B7. 0.50°C
B9. 46 kcal; 46 kcal
B11. 5.6 kcal
B13. 3.6×10^{-4} kg/s
C1. 1.8 mm
C3. (a) 1.071×10^3 cm^3; (b) 178 kcal
C5. −145°C
C7. 4.0 min
C9. 23.9°C
C11. 0.095 kcal/(kg·°C)

C13. 1.18
C15. 0.21 kg
C17. (a) 0.080 kg; (b) 0.012 kg
C19. (a) 1.53×10^5 J; (b) 850 W
C21. 10^5 years
C23. 34 km; 21 mi
C25. (a) 1.5 kg/h; (b) 1.9%
D1. 2.4×10^4 K
D3. 125.02 cm
D5. 6.7×10^5 N
D7. 20°C
D9. $\overline{KE} = 7.7 \times 10^{-21}$ J; $E_{vap} = 6.8 \times 10^{-20}$ J

Chapter 14

A1. (a)
A3. (e)
A5. (b)
A7. (b)
A9. (d)
A11. (a)
A13. (c)
B1. (a) 1.9×10^6 J; (b) 4.2×10^3 J; (c) 50 J
B3. (a) 1.7×10^3 J; (b) Thin
B5. 5.67×10^8 J
B7. 2.9×10^7 K
B9. (a) 0; (b) −100 J
B11. (a) 100 J; (b) 0
B13. 900 J
C1. 93 W
C3. (a) 2.5×10^3 °C/m; (b) 2.5×10^{-2} °C/m; (c) Heat flows more slowly through earth to surface than through window of same area

C5. 17 s
C7. 7.6 W
C9. 1.4 W/(m^2·°C)
C11. 5.8×10^6 J
C13. 44 J
C15. 79°C
C17. 3.6 h; 4.4 h; 5.8 h
C19. (a) Work done by gas along path AB is 4.0×10^6 J; along BC is zero; along CD work done on gas is 1.6×10^6 J; along DA is zero; (b) 2.4×10^6 J
C21. (a) 575 m/s; (b) 543 m/s
C23. (a) 2.0×10^{-2} J; (b) $\Delta W/\Delta Q \approx 10^{-6}$; yes
D3. 2-mm Al, 1-mm brass
D5. (b) 1.1×10^{-4} m^2/s (Cu); 1.4×10^{-7} m^2/s (H$_2$O); 1.8×10^{-5} m^2/s (air)
D9. (a) 1.87×10^4 J; (b) 0.75×10^4 J; (c) 2.62×10^4 J
D11. (a) 50 J; (b) 10 J

Chapter 15

A1. (e)
A3. (b)
A5. (b)
A7. (d)
A9. (d)
B1. 199 K
B3. 1.2×10^3 J/K
B5. 10%
B7. (a) 50%; (b) 33%; (c) 25%
B9. (a) 12%; (b) 8.4
B11. 17
C1. (a) 1.52×10^{-3} m^3; (b) 227 K; (c) −37 J

C3. (a) 296 K; (b) Yes; the condensation of the water releases heat
C5. (a) 36.6 J/K; (b) 69.9 J/K
C7. (a) −0.020 J/K; (b) +0.028 J/K; (c) +0.008 J/K
C9. (a) 52%; (b) $Q_H = 1.92 \times 10^{12}$ J; $Q_C = 0.92 \times 10^{12}$ J
C11. 160 m
C13. (a) 2.15×10^3 J; (b) 0.15×10^3 J; (c) 14.6
C15. (a) 5.6×10^4 J; (b) 2.2×10^3 J; (c) 278; 10.9
D1. (a) 2.33×10^4 J; (b) 2.16×10^4 J; (c) 1.7×10^3 J
D3. (b) For Ne, since γ is larger for Ne
D7. (a) 67%; (b) 67%

Chapter 16

A1. (c)
A3. (a)
A5. (a)
A7. (b)
A9. (a)
B1. (a) 180 N toward the -10-μC charge; (b) 1.8×10^5 m/s^2
B3. (a) 6.8×10^6 N/C away from 30-μC charge; (b) Same toward 30-μC charge
B5. (a) 200 V; (b) Point B

B7. 1.6×10^{-13} J
B9. 0.885 pF
C1. (a) 1.49×10^3 N; (b) 79° above this line
C3. (a) 0; (b) 1.5×10^4 N normal to line joining the charges
C5. 6.4×10^7 N/C vertically downward
C7. (a) 3.4 m outside the 2.0-μC charge; (b) 0.57 m from 2.0-μC charge, on line joining charges; (c) Yes; 0.77 m from 2.0-μC charge; (d) No
C9. (a) 3.2×10^{-10} N; (b) 3.2×10^{-5} m/s^2

C11. Zero
C13. (*a*) 1.5×10^4 V/m; (*b*) 2.4×10^{-15} N; (*c*) Toward the higher-potential plate
C15. (*a*) 1.1×10^{-9} F; (*b*) 1.1×10^{-4} C; (*c*) 2.5×10^7 V/m
C17. (*a*) 8.5×10^{-11} F; (*b*) $+5.1 \times 10^{-8}$ C on one plate, -5.1×10^{-8} C on the other
C19. (*a*) 8.2×10^{-8} N; (*b*) 2.2×10^6 m/s

Chapter 17
A1. (*d*)
A3. (*a*)
A5. (*c*)
A7. (*a*)
A9. (*d*)
B1. 1.4×10^5 C
B3. (*a*) 0.60 A; (*b*) 2.2×10^3 C
B5. $0.50 \ \Omega$
B7. 5.0 V
B9. 8.6×10^8 J
C1. 6.0×10^3 J
C3. 8.5×10^7 m/s
C5. 2.1×10^2 g
C7. $0.03 \ \Omega$
C9. $468 \ \Omega$
C11. (*a*) $12 \ \Omega$; (*b*) $2.67 \ \Omega$

D1. 2.0×10^{19} N
D3. 5.9×10^{-7} C
D5. 0.28 m
D7. (*a*) 3.6×10^{-8} V·m; (*b*) The same
D9. (*a*) 1.16×10^2 V for each; (*b*) 1.16×10^{-3} C; 0.58×10^{-3} C

C13. (*a*) $10 \ \Omega$; (*b*) $I_8 = 4.0$ A, $I_6 = 1.33$ A, $I_{12} = 0.67$ A, $I_4 = 2.0$ A
C15. (*a*) $0.059 \ \mu$F; (*b*) $0.80 \ \mu$F
C17. $0.85 \ \mu$F
C19. $I_1 = 1.25$ A; $I_2 = -0.50$ A; $I_3 = 1.75$ A
C21. \$107
C23. (*a*) 23.25 W; (*b*) 23.25 W; (*c*) Yes
C25. (*a*) $R_{50} = 288 \ \Omega$; $R_{100} = 144 \ \Omega$; (*b*) $R_{150} = 96 \ \Omega$
C27. 13 h
D1. 19 min
D3. 0.40 A
D5. $16 \ \Omega$
D7. (*a*) $I_1 = 0.92$ A, $I_2 = 0.96$ A, $I_3 = -0.04$ A; (*b*) 17 V
D9. (*a*) $I_1 = 2.47$ A, $I_2 = -1.06$ A; (*b*) 1.41 A; (*c*) $V_1 = V_2 = 7.06$ V
D11. (*a*) 18 V; (*b*) *C* is at higher potential than *B*

Chapter 18
A1. (*d*)
A3. (*c*)
A5. (*e*)
A7. (*c*)
A9. (*c*)
B1. (*a*) 0.25 N; (*b*) Out of plane of paper
B3. 2.6×10^{-14} N
B5. (*a*) 5.0×10^{-6} T; (*b*) About one-tenth the magnitude of the earth's field
B7. 2.5×10^{-3} T
B9. 0.50 G
C1. 0.053 N
C3. 0.20 T
C5. (*a*) 5.7×10^{-12} N; (*b*) 5.9×10^7 m/s; The magnetic field cannot change the speed of the electron, only its direction

C7. 1.67×10^{11} C/kg
C9. 0.31 T along negative *Z* axis
C11. 2.4×10^{-5} N
C13. 0.40 A
C15. 0.55 G
C17. (*a*) Add a $2.0 \times 10^8 \ \Omega$ resistor in series with galvanometer; (*b*) $2.0 \times 10^8 \ \Omega$
D1. (*a*) 5.0×10^{-13} N; (*b*) 3.2 cm
D3. (*a*) 0.63%; (*b*) Very difficult, since process is inefficient
D5. Proton moves in a spiral path along the $+X$ axis, with radius of the circle in the *YZ* plane being 8.7 cm
D7. 92 A
D9. $\mu_0 IN/2\pi r$
D11. 2.8×10^{-3} T

Chapter 19
A1. (*b*)
A3. (*e*)
A5. (*a*)
A7. (*c*)
A9. (*a*)
A11. (*e*)
B1. 0.015 V
B3. 0.125 V
B5. (*a*) 1.0 V; (*b*) In the direction opposite that of the increasing current
B7. 2.5×10^{-5} s
B9. 10 s
B11. 70 V

C1. (*a*) 7.0×10^{-6} V; (*b*) 1.4×10^{-4} A; (*c*) 9.8×10^{-10} N; (*d*) 9.8×10^{-10} W
C3. 1.0×10^{-3} V from east to west
C5. -0.20 V from $t = 0$ to $t = 5$ s; $+0.20$ V from $t = 5$ s to $t = 10$ s
C7. (*a*) 3.33×10^{-5} T; (*b*) 3.0×10^8 m/s
C9. 1.21 A
C11. (*a*) 800; (*b*) 0.020 H
C13. (*a*) 100 s; (*b*) 6.0×10^{-6} A; (*c*) 2.0×10^{-5} C
C15. (*a*) 0.010 A; (*b*) 50 s
C17. 3000 rev/min
C19. (*a*) 900 W input; 900 W lost as heat; (*b*) 720 W input; 144 W delivered by motor, 576 W lost as heat

D1. (*a*) 8.5×10^{-3} V; (*b*) 3.6×10^{-6} J; (*c*) Into heat in the resistor

D3. 0.63 V

D5. 1.2×10^{-4} W

D9. 7.5 A

Chapter 20

A1. (*e*)
A3. (*b*)
A5. (*c*)
A7. (*b*)
A9. (*b*)
B1. (*a*) 1.20 A; (*b*) 1.70 A
B3. (*a*) 2.3×10^{-4} A; (*b*) 3.2×10^{-4} A
B5. 1.1×10^{4} Hz
B7. 1.2×10^{-4} F
B9. 4000 turns
C1. (*a*) 0; (*b*) 0.71 A; (*c*) 0.50 A
C3. (*a*) 0; (*b*) 2.3 A; (*c*) −2.3 A
C5. (*a*) 1.28 mA; (*b*) 0; (*c*) −1.28mA
C7. (*a*) 1.02×10^{-2} C; (*b*) 13.2 A; (*c*) 205 Hz
C9. (*a*) 6.8 mA; (*b*) 89.8°; (*c*) 2.82×10^{-3}; (*d*) 0

C11. 51 Ω
C13. (*a*) 0.0038; (*b*) 2.0×10^{-3} W in resistor
C15. (*a*) 2.24 MHz; (*b*) 134 m
C17. (*a*) 356 Hz; (*b*) 0.80 A; (*c*) $V_R = 120$ V, $V_L = 179$ V, $V_C = -179$ V; (*d*) 96 W
C19. (*a*) 40/1; (*b*) 50 A; (*c*) 1.25 A
C21. (*a*) 22; (*b*) 5.5
D1. (*a*) 6.0 A; (*b*) 6.3×10^{-5} A
D3. (*a*) 1.5%; (*b*) 99.4%
D5. (*a*) 6.0 Ω; (*b*) 62 mH
D7. (*a*) A resistor and an inductor, with $R = 10$ Ω, $L = 46$ mH; (*b*) 2.1×10^{-5} A
D9. (*a*) 0.11 cm/V; (*b*) 2.25 A
D11. 510 Hz

Chapter 21

A1. (*e*)
A3. (*e*)
A5. (*c*)
A7. (*b*)
A9. (*d*)
B1. 5.6×10^{4} V/(m·s)
B3. 10^{8} Hz
B5. 1.9×10^{8} m/s; less than in air
B7. (*a*) 2.9 m; (*b*) 1.5 m
B9. 230×10^{-6} s
C1. (*a*) 1.8×10^{-10} F; (*b*) 2.0×10^{-9} A (*c*) 2.0×10^{-9} A; (*d*) 2.0×10^{-15} T (*e*) 2.0×10^{-15} T
C3. (*a*) 1.5×10^{8} m/s; (*b*) 2.5×10^{6} m; (*c*) 60 Hz

C5. (*a*) 2.4 W/m²; (*b*) 2.4 W/m²
C7. (*a*) About 30 octaves; (*b*) About 10 octaves
C9. 2.3×10^{-6} H
C11. 2.2 to 3.3×10^{-12} F
C13. 1.39 to 1.72 m
C15. The husband, by about 0.12 s
D3. 1.7×10^{-2} V
D5. (*a*) $B_{\mathrm{rms}} = 2.3 \times 10^{-6}$ T; $E_{\mathrm{rms}} = 7.2 \times 10^{2}$ V/m; (*b*) 3.8×10^{26} W
D7. (*a*) $B_{\mathrm{rms}} = 2.5 \times 10^{-4}$ T; $E_{\mathrm{rms}} = 7.5 \times 10^{4}$ V/m; (*b*) These values are unrealistically high for electromagnetic waves

Chapter 22

A1. (*c*)
A3. (*a*)
A5. (*b*)
A7. (*a*)
A9. (*a*)
B1. 56 m from origin
B3. (*a*) 40°; (*b*) 25°
B5. (*a*) 7.2 mm from center; (*b*) 14 mm from center
B7. 589 nm
B9. 115 nm
B11. 25 mW/cm²
C1. At right angles to each other
C3. (*a*) 32°; (*b*) Yes
C5. 2.2 cm
C7. (*a*) The central image is 7.6 mm wide and the dark

fringes are 3.8 mm apart; (*b*) All distances in the fringe pattern are increased by a factor of 3
C9. 15
C11. 80 cm
C13. 4
C15. 210 nm
C17. (*a*) 544 nm; (*b*) 435 nm
C19. (*a*) 48°; (*b*) 51°; (*c*) 48°
D1. No angle
D3. 24.4°
D5. Central maximum for single slit includes 20 bright fringes from double slit
D7. 1.18×10^{-5}/°C
D9. (*a*) 11.2 to 22.3°; (*b*) 22.9 to 49.5°; (*c*) 35.8 to 90°

Chapter 23

A1. (*b*)
A3. (*e*)
A5. (*a*)

A7. (*c*)
A9. (*c*)
B1. 0.013 mm

B3. 2.46

B5. Virtual, upright, seven-tenths of original size, and 0.14 m behind the mirror

B7. 0.13 m from lens

B9. (*a*) 0.5 D; (*b*) 2.0 m

C1. 1.05 m

C3. 1.54

C5. (*a*) Image is real, inverted, and 12 m from mirror
(*b*) Image is real, inverted, and 30 m from mirror
(*c*) Image is virtual, upright, and 1.5 m behind the mirror
(*d*) As object moves through the focal point, image changes from real to virtual

C7. 0.56 m

C9. Virtual, upright, 0.27 m behind mirror, and 1.35 the size of the object

C11. (*a*) Virtual, upright, 0.375 m behind mirror; one-fourth original size; (*b*) Virtual, upright, 0.30 m behind mirror, 0.40 original size; (*c*) Virtual, upright, 0.167 m behind mirror, 0.67 original size; (*d*) Image gets closer to mirror and becomes larger, but is otherwise unchanged

C13. Image is 1.50 m behind mirror, virtual, upright, and enlarged 5 times

Chapter 24

A1. (*d*)

A3. (*d*)

A5. (*a*)

A7. (*e*)

A9. (*b*)

B1. 4.0 cm

B3. 6.0 cm

B5. 300

B7. 5.0 m

B9. 3.16×10^4

C1. 1.67 s

C3. 60 cm

C5. (*a*) Nearsighted; (*b*) Diverging lens of power -0.714 D

C7. (*a*) Nearsighted; (*b*) Diverging lens of power -6.0 D

Chapter 25

A1. (*b*)

A3. (*e*)

A5. (*e*)

A7. (*d*)

A9. (*b*)

A11. (*d*)

B1. 0.92 cm

B3. 7.41×10^2 s

B5. (*a*) 9.05 years; (*b*) 0.90 year

B7. 5.8×10^{-12}

B9. (*a*) 9×10^{16} J; (*b*) 2.5×10^4 years

C1. $x_2 = -3.8 \times 10^3$ km; $y_2 = 2.0$ km; $z_2 = 2.0$ km; $t_2 = 1.2 \times 10^{-3}$ s

C3. (*a*) $0.87c$; (*b*) $\Delta T = 2 \Delta T_0$

C5. Relativistically the time elapsed can be reduced by traveling close to the speed of light; (*b*) $0.9999997c$

C15. (*a*) 10; (*b*) Original direction

C17. (*a*) Real, inverted, 0.37 size of object, and 0.55 m on other side of lens; (*b*) Image at infinity, real, inverted, and very large; (*c*) Virtual, upright, twice object size, and 0.40 m from lens on object side; (*d*) Image changes from real to virtual as object passes through focal point

C19. Virtual, upright, 2.2 times object size, and 0.66 m on object size

C21. Virtual, upright image, 0.46 times object size and 0.87 m from lens on object side

C23. (*a*) 50 cm to right of second lens; (*b*) Same place

C25. (*a*) 8.0 cm to left of second lens, and two-thirds object size; (*b*) Virtual, inverted

C27. 1.50

C29. 78°

D1. 18.3 m

D3. (*a*) 21 cm; (*b*) 42 cm; -42 cm

D5. At focal point of lens

D7. (*a*) $f' \rightarrow \infty$; (*b*) $f' \rightarrow f$

D9. Rays emerging from second lens are parallel; no image is formed.

C9. 6.25

C11. (*a*) 14.0 cm; (*b*) 117

C13. 0.050 mm

C15. (*a*) 1.15; (*b*) 0.24 μm

C17. 30 μm

C19. (*a*) 2.01 mm; (*b*) 7514

D1. -1.0 D for top lens; $+2.0$ D for bottom lens

D3. (*a*) 250x; (*b*) 0.28 cm from objective; (*c*) $f_{obj} = 0.27$ cm; $f_{eye} = 5.0$ cm

D5. (*a*) Telescope; (*b*) 12.8 m; (*c*) 16.7

D7. (*a*) 10x; (*b*) An upright, virtual image at infinity

D9. (*a*) The two wavelengths 589.0 and 589.6 cause destructive interference for certain path differences; (*b*) 0.29 mm

C7. 1024

C9. 2.2×10^{-8} s

C11. (*a*) 11.4×10^{-31} kg; (*b*) 2.1×10^{-14} J

C13. (*a*) 2.82×10^8 m/s; 2.69×10^{-30} kg; (*b*) 2.42×10^{-13} J

C15. (*a*) 1%; (*b*) 67%

C17. (*a*) 2.56×10^5 V; (*b*) 2.24×10^8 m/s; (*c*) 50%

C19. (*a*) 4.33×10^9 kg/s; (*b*) 1.5×10^{13} years

D3. (*a*) Yes; (*b*) 1.8×10^8 m/s along positive X axis; (*c*) 400 m; (*d*) No such inertial frame exists; v would have to be greater than c

D11. (*a*) 1.6×10^{-13} J; (*b*) 2.7×10^{-30} kg; (*c*) $0.94c$

Chapter 26

A1. (b)
A3. (e)
A5. (e)
A7. (b)
A9. (c)
B1. 6.8×10^{-26} J
B3. (a) No, photon energy is lower than work function; (b) 7.2×10^{14} Hz
B5. (a) 5.3×10^{-22} kg·m/s; (b) 3.6 nm
B7. 0.27 nm
B9. 1.2×10^{5} m/s
C1. 2.1×10^{20} photons per second
C3. (a) 350 nm; (b) Near UV

C5. 0.37 V
C7. (a) 6.7×10^{-19} J; (b) 296 nm; (c) 4.2 V
C9. (a) 7.96×10^{-16} J; (b) 2.63×10^{-24} kg·m/s; (c) 0.256 nm; (d) 1.9×10^{-15} J
C11. (a) 0.121 nm; (b) 61°
C13. 2.1 nm
C15. (a) 0.17 nm; (b) 62°
C17. 4.4×10^{-15} J
C19. (a) 2.1×10^{-30} m/s; (b) 10^{30} s
D1. 6.0×10^{14} Hz
D5. (a) 6.7×10^{-7} N; (b) 6.7×10^{-9} m/s^2
D7. 1.4×10^{-12} m
D9. Electron energies would be above 10 MeV

Chapter 27

A1. (e)
A3. (e)
A5. (b)
A7. (b)
A9. (d)
B1. 820 nm
B3. (a) 1.06×10^{-19} J; (b) 0.845 nm; (c) 5.5×10^{5} m/s
B5. 119 nm
B7. 3.4 eV
B9. (a) 18; (b) $l = 4$; $m_l = 4, 3, 2, 1, 0, -1, -2, -3, -4$; $m_s = \pm\frac{1}{2}$
C1. (a) 122 nm, 103 nm, 97.3 nm, 95.0 nm; (b) 91.2 nm; (c) No
C3. (a) 1.05×10^{-34} J·s; 2.10×10^{-34} J·s; 3.15×10^{-34} J·s; 4.20×10^{-34} J·s
C5. (a) 91.2 nm; (b) 83.8 nm

C7. (a) 2.1 V; (b) 592 nm; (c) Yes; (d) No
C9. Wavelengths excited are 255, 185, and 685 nm; only the latter is in the visible
C11. (a) 2.86 eV; (b) 486 and 656 nm
C13. (a) 6; (b) $n, l, m_l, m_s = 5, 1, -1, \pm\frac{1}{2}$; 5, 1, 0, $\pm\frac{1}{2}$; 5, 1, +1, $\pm\frac{1}{2}$
C15. (a) 470 nm (visible); (b) 73 nm (UV); (c) 7.6 nm (far UV)
C17. (a) 0.34 nm; (b) X-ray region
C19. (a) 8.3×10^{6} rev; (b) 1.1×10^{-26} J
C21. (a) 2.4×10^{-25} m; (b) 3.3×10^{-10} m
D1. (a) 8%; (b) 64%
D5. (a) 1.1×10^{-3} A; (b) 13 T
D7. 110
D9. (a) 7.9×10^{6} Hz; (b) 1.4×10^{-5} nm

Chapter 28

A1. (b)
A3. (b)
A5. (d)
A7. (c)
A9. (c)
B1. (a) 7.4×10^{-15} m; (b) 1.7×10^{-42} m^3
B3. 0.37 MeV
B5. (a) 2.2×10^{-17} s^{-1}; (b) 6.9×10^{-5} s^{-1}
B7. 11,500 years
B9. (a) 5.9×10^{-2} Gy; (b) 5.9 rad
C1. (a) 92p, 141n; (b) 92p, 146n; (c) 91p, 142n
C3. (a) 2.3×10^{17} kg/m^3; (b) More than 10^{14} times
C5. (a) 7.98 MeV per nucleon; (b) 7.46 MeV per nucleon; $^{16}_{8}$O is more stable
C7. (a) 60 years (b) 69.7 years
C9. $^{206}_{82}$Pb (lead)
C11. 61%

C13. $6.6T_{1/2}$
C15. (a) 2n, 3e; (b) 2n, 2e
C17. (a) $m(3\alpha) > m(^{12}_{6}C)$; (b) $m(^{8}_{4}Be) > m(2\alpha)$
C19. $\Delta m/m = 0.091\%$
C21. 2.23 MeV
C23. 7.8×10^{5} kg
C25. (a) 0.052 Gy; (b) 0.52 Sv; (c) 5.2 rad, 52 rem
C27. (a) 0.7 Gy; (b) 7 Gy
C29. 3.5×10^{-6} μg
D1. (a) 6.6×10^{-11} J; (b) -2.3×10^{-14} J; (c) No likelihood
D3. $R(^{238}_{92}U) = 7.4 \times 10^{-15}$ m; $r_0 = 5.8 \times 10^{-13}$ m
D5. (a) 1.0×10^{-5} s^{-1}; (b) 6.9×10^{4} s
D7. 6×10^{9} years
D9. (a) 1.4×10^{3} m/s; (b) 1.98×10^{6} m/s
D11. (a) $6^{2}_{1}H \rightarrow 2^{4}_{2}He + 2^{1}_{1}p + 2^{1}_{0}n + 43.2$ MeV; (b) 7.2 MeV per nucleus

Chapter 29

A1. (d)
A3. (a)
A5. (c)
A7. (c)
B1. 0.356 nm
B3. (a) 1.0×10^{15} Hz; (b) UV; (c) Very low
B5. 0.12 mm/s

B7.   (a) 2.406×10^{14} C/(J·s); (b) 2.418×10^{14} C/(J·s)

C1. 2.16×10^{3} kg/m^{3}

C3.   (a) $F_e = 2.9 \times 10^{-9}$ N; $F_g = 1.9 \times 10^{-42}$ N; (c) Of no importance

C5.   (a) 1.4×10^{21} atoms; (b) 1.1×10^{-26} m^{3}

C7. 10^{-8}

C9.   (a) 3.2×10^{5} A/m^{2}; (b) 0.024 mm/s

C11.   (a) 5.7×10^{28} m^{-3}; (b) 5.9×10^{28} m^{-3}

C13.   (a) 2.5×10^{28} m^{-3}; (b) 1

C15.   (b)10 A

C17.   (a) 8.1 GW; (b) 1.6% of total

D5.   (a) 0.84 nm; (b) 3.5 times larger; (c) Free electron

Index

Page numbers in *italic* indicate illustrations or tables.

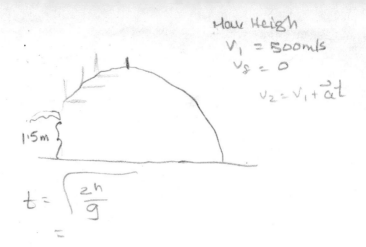

Max Heigh

$V_1 = 500 m/s$

$V_f = 0$

$V_2 = V_1 + \vec{a}t$

115 m

$t = \sqrt{\dfrac{2h}{g}}$

=

Dec 8/85.

Answers B

1 A   6   A
2 E   7   B
3 B   13   C
4 A   14   D
5 E   15   B
           19   E.

TABLE C   Useful Conversion Factors

Length
1 in = 0.0254 m; 1 m = 39.37 in
1 ft = 0.305 m; 1 m = 3.28 ft
1 mi = 1.61 km; 1 km = 0.621 mi
1 mi = 5280 ft = 1760 yd
1 Ångstrom (Å) = 10^{-10} m
1 micrometer (μm) = 10^{-6} m
1 light-year = 9.46×10^{15} m

Mass
1 u (unified mass unit) = 1.660×10^{-27} kg; 1 kg = 6.024×10^{26} u
1 metric ton = 10^3 kg
(A mass of 1 kg corresponds to a weight of 2.21 lb; a weight of 1 lb corresponds to a mass of 0.454 kg)

Time
1 h = 60 min = 3600 s
1 day = 24 h = 1440 min = 8.64×10^4 s
1 year = 365.24 days = 3.156×10^7 s

Area
1 in^2 = 6.452×10^{-4} m^2; 1 m^2 = 1550 in^2
1 ft^2 = 9.29×10^{-2} m^2; 1 m^2 = 10.76 ft^2
1 km^2 = 10^6 m^2; 1 m^2 = 10^{-6} km^2

Volume
1 in^3 = 1.64×10^{-5} m^3; 1 m^3 = 6.10×10^4 in^3
1 ft^3 = 2.83×10^{-2} m^3; 1 m^3 = 35.3 ft^3
1 km^3 = 10^9 m^3; 1 m^3 = 10^{-9} km^3
1 liter = 10^3 cm^3 = 10^{-3} m^3
1 gallon = 3.785 liters = 3.785×10^{-3} m^3

Density
1 g/cm^3 = 10^3 kg/m^3; 1 kg/m^3 = 10^{-3} g/cm^3

Angular quantities
$360° = 2\pi$ rad
1 rad = 57.3°; 1° = 1.745×10^{-2} rad
1 rev/min = 0.1047 rad/s

Velocity
1 ft/s = 0.305 m/s; 1 m/s = 3.28 ft/s
1 mi/h = 0.447 m/s; 1 m/s = 2.24 mi/h
1 mi/h = 1.61 km/h; 1 km/h = 0.621 mi/h

Force
1 lb = 4.45 N; 1 N = 0.225 lb

Pressure
1 lb/in^2 = 6.90×10^3 N/m^2 (or Pa); 1N/m^2 = 1.45×10^{-4} lb/in^2
1 lb/ft^2 = 47.9 N/m^2; 1 N/m^2 = 2.09×10^{-2} lb/ft^2
1 atm = 1.013×10^5 N/m^2; 1 N/m^2 = 9.87×10^{-6} atm
1 atm = 760 torr; 1 torr = 1.32×10^{-3} atm
1 torr = 133.3 N/m^2; 1 N/m^2 = 7.502×10^{-3} torr

Energy
1 ft·lb = 1.356 J; 1 J = 0.738 ft·lb
1 Btu = 1054 J; 1 J = 9.49×10^{-4} Btu
1 cal = 4.186 J; 1 J = 0.239 cal
1 kcal = 4186 J; 1 J = 2.39×10^{-4} kcal
1 kWh = 3.60×10^6 J; 1 J = 2.78×10^{-7} kWh
1 eV = 1.60×10^{-19} J; 1 J = 6.24×10^{18} eV
1 eV/particle = 23.1 kcal/mol

Power
1 horsepower (hp) = 746 W; 1 W = 1.34×10^{-3} hp
1 ft·lb/s = 1.356 W; 1 W = 0.738 ft·lb/s
1 hp = 550 ft·lb/s; 1 ft·lb/s = 1.82×10^{-3} hp

Mass-energy
1 u = 931.50 MeV/c^2
1 kg = 5.6096×10^{29} MeV/c^2
1 m_e = 0.5110 MeV/c^2
1 m_p = 938.28 MeV/c^2
1 m_n = 939.57 MeV/c^2

Magnetic field strength
1 T = 1 Wb/m^2 = 10^4 G

Maha 7477237